The Routledge International Handbook of Biosocial Criminology

Biosocial criminology is an interdisciplinary field that aims to explain crime and antisocial behavior by exploring both biological and environmental factors. Since the mapping of the human genome, scientists have been able to study the biosocial causes of human behavior with the greatest specificity.

After decades of almost exclusive sociological focus, criminology has undergone a paradigm shift where the field is more interdisciplinary, and this book combines perspectives from criminology and sociology with contributions from fields such as genetics, neuropsychology, and evolutionary psychology. *The Routledge International Handbook of Biosocial Criminology* is the largest and most comprehensive work of its kind, and is organized into five sections that collectively span the terrain of biosocial research on antisocial behavior.

Bringing together leading experts from around the world, this book considers the criminological, genetic, and neuropsychological foundations of offending, as well as the legal and criminal justice applications of biosocial criminological theory, and will be essential reading for students, researchers, and practitioners from across the social, behavioral, and natural sciences who are engaged in the study of antisocial behavior.

Matt DeLisi is Coordinator of Criminal Justice Studies, Professor in the Department of Sociology, and Affiliate with the Center for the Study of Violence at Iowa State University. Professor DeLisi is Fellow of the Academy of Criminal Justice Sciences, Editor-in-Chief of the *Journal of Criminal Justice*, and the author of more than 250 scholarly works.

Michael G. Vaughn is Professor in the School of Social Work at Saint Louis University. Professor Vaughn is Fellow of the Society for Social Work and Research and author of more than 250 scholarly works.

"The biosocial revolution has broken through the gates of resistance erected by traditional criminology, offering rich opportunities for transforming the criminological enterprise. For those wishing to join this revolution – or simply to understand it – DeLisi and Vaughn's *Handbook* is an indispensable resource. The breadth and depth of this volume are impressive; it is a work that should be kept close at hand and consulted regularly."

Francis T. Cullen, Distinguished Research Professor, University of Cincinnati, USA

"Biosocial criminology has expanded by leaps and bounds since its origins and the field continues to evolve. Fortunately for criminologists of all stripes, DeLisi and Vaughn have amassed a who's-who of internationally acclaimed scholars who are at the forefront of biosocial criminology. The key topics are covered and reviewed well, but more importantly the chapters are filled with nuggets for future theoretical and empirical research that is sure to serve as a model guide for students and academics alike. The editors are to be commended for spearheading a must-have volume that should be on every shelf."

Alex R. Piquero, PhD, Ashbel Smith Professor of Criminology, the University of Texas at Dallas, USA

"This comprehensive *Handbook* clearly brings out the increasing importance of biosocial research for criminology and criminal justice. Written by leading biosocial scholars, it is very wide-ranging, including not only fundamental research, but also implications for policy and practice. It should be required reading for all criminologists."

David P. Farrington, Emeritus Professor of Psychological Criminology, Cambridge University, UK

"The rise of biosocial criminology represents the most important paradigm shift criminology has ever experienced. DeLisi and Vaughn's book fully captures the scholarly diversity and intellectual excitement that accompany and propel this transition. Readers of DeLisi and Vaughn's book will not only come away better informed about biosocial criminology, they will come away inspired."

John Wright, Professor of Criminal Justice, University of Cincinnati and King Abdullhziz University, Jeddah Saudi Arabia

The Routledge International Handbook of Biosocial Criminology

Edited by Matt DeLisi and Michael G. Vaughn

Routledge
Taylor & Francis Group
LONDON AND NEW YORK

First published 2015
by Routledge
2 Park Square, Milton Park, Abingdon, Oxon OX14 4RN

and by Routledge
711 Third Avenue, New York, NY 10017

Routledge is an imprint of the Taylor & Francis Group, an informa business

British Library Cataloguing in Publication Data
A catalogue record for this book is available from the British Library

Library of Congress Cataloging-in-Publication Data
The Routledge international handbook of biosocial criminology /
edited by Matt DeLisi, Michael G Vaughn. — First Edition.
pages cm. — (Routledge international handbooks)
1. Criminology. 2. Criminal behavior—Psychological aspects.
3. Criminal behavior—Genetic aspects. 4. Human beings—Effect of
environment on. 5. Sociobiology. I. DeLisi, Matt, editor of compilation.
II. Vaughn, Michael G., editor of compilation.
HV6025.R697 2014
364—dc23
2014018248

ISBN: 978-0-415-72213-1 (hbk)
ISBN: 978-1-315-85844-9 (ebk)

Typeset in Bembo, ITC Stone Sans
by Apex CoVantage, LLC

MIX
Paper from responsible sources
FSC® C013056

Printed and bound in Great Britain by
TJ International Ltd, Padstow, Cornwall

For our family, our friends, and our biosocial buddies—Matt DeLisi and Michael G. Vaughn

Contents

Figures

Tables

Contributors

Robert Agnew is Samuel Candler Dobbs Professor of Sociology at Emory University and Past President and Fellow of the American Society of Criminology.

Bruce A. Arrigo, Ph.D. is Professor of Criminology, Law, and Society within the Department of Criminal Justice and Criminology and Professor of Public Policy at the University of North Carolina at Charlotte. He holds additional faculty appointments within the departments of Psychology and Public Health Sciences.

J.C. Barnes is an Associate Professor in the School of Criminal Justice at the University of Cincinnati. He is a biosocial criminologist whose research seeks to understand how genetic and environmental factors combine to impact criminological phenomena. Recent works have attempted to reconcile behavioral genetic findings with theoretical developments in criminology. He has published more than 70 papers and book chapters in outlets such as *Aggressive Behavior, Behavior Genetics, Criminology, Developmental Psychology, Intelligence, Journal of Marriage and Family, Justice Quarterly, Journal of Theoretical Biology*, and *PLoS One*.

Eric Beauregard is an Associate Professor in the School of Criminology at Simon Fraser University. His research interests focus on the crime-commission process, decision-making and the criminal investigation.

Kevin M. Beaver is a Professor in the College of Criminology and Criminal Justice at Florida State University and a visiting Distinguished Professor in the Center for Social and Humanities Research at King Abdulaziz University. He is a past recipient of the American Society of Criminology's Ruth Shonle Cavan Young Scholar Award and the National Institute of Justice's Graduate Research Fellowship. His research examines the biosocial contributors to antisocial behaviors.

Monic P. Behnken is Assistant Professor in the Department of Sociology and Criminal Justice Studies at Iowa State University.

Kyle Bersted received his B.A. from Creighton University in 2010, his M.A. from Southern Illinois University in 2013, and is working towards his Ph.D. He received one Masters and two Doctoral Fellowships from SIU. His dissertation will examine gene–environment interactions in the context of children's perceptions of their parents' behaviors.

Sytske Besemer is a Rubicon Fellow at the Institute of Human Development at the University of California, Berkeley. She investigates intergenerational transmission, interactions between parents and children, and other aspects around crime over the life-course. Dr. Besemer was educated at Leiden University, VU University Amsterdam and the University of Cambridge, where she was Gates Scholar.

Danielle L. Boisvert is Assistant Professor and Director of Criminal Justice and Criminology Graduate Programs at Sam Houston State University. Her research interests include life-course/developmental criminology, biosocial criminology, and behavioral genetics. Her research focuses mainly on the examination of genetic and environmental influences on a variety of delinquent and criminal behaviors throughout the life-course.

Brian B. Boutwell is an Associate Professor in the School of Social Work at Saint Louis University. His research interests span a variety of topics including human intelligence, human mating, the evolutionary origins of antisocial behavior, and the genetic underpinnings to complex behavioral traits. His current work has aimed at integrating existing criminological theory within the context of life-history theory in evolutionary biology. He has published numerous peer-reviewed articles and book chapters in outlets that include *Behavior Genetics*, *Journal of Theoretical Biology*, *Developmental Psychology*, *Intelligence*, *Criminology*, and *PLoS One*.

Katharine L. Bowen completed her Ph.D. in psychology at Cardiff University. Her doctorate examined emotion functioning in adolescent offenders. She subsequently worked as an Assistant Psychologist for Morriston Hospital's Neuropsychology speciality, and is currently part of the psychology team in the South Wales Cleft Lip and Palate Service.

Patricia A. Brennan, Ph.D. is a Professor of Psychology at Emory University. Her primary area of research is developmental psychopathology. She employs longitudinal and treatment research designs to assess biosocial risk factors for aggression, as well as the intergenerational transmission of depression and related disorders.

Charlotte Bright is Assistant Professor of Social Work at the University of Maryland.

Joshua W. Buckholtz is Assistant Professor in the Department of Psychology at Harvard University and Network Scholar for the MacArthur Foundation's Research Network on Law and Neuroscience, and serves on the faculty of the Center for Law, Brain and Behavior at Massachusetts General Hospital.

Frances R. Chen is a doctoral student in criminology at the University of Pennsylvania. She is interested in understanding antisocial behavior from a developmental and biosocial perspective. She is conducting research on heterogeneity in the development of antisocial behavior, and on how life events and their interactions with biological vulnerability affect the development of antisocial behavior.

David S. Chester is a doctoral candidate in the Department of Psychology at the University of Kentucky. His research broadly focuses on aggression, social rejection, and self-regulation, using behavioral and neuroscientific techniques (e.g. magnetic resonance imaging) to explore these topics.

Eric J. Connolly is an Assistant Professor in the Administration of Justice Program at Pennsylvania State University, Abington. He received his B.A. in criminology from West Virginia University and his M.S. in criminal justice from the University of North Alabama. His research interests include biosocial criminology, life-course/developmental criminology, and quantitative behavior genetics.

Michelle Coyne is a doctoral student in the School of Criminal Justice at the University of Cincinnati.

Matt DeLisi is Coordinator of Criminal Justice Studies and Professor of Sociology at Iowa State University. Dr. DeLisi is Fellow of the Academy of Criminal Justice Sciences.

C. Nathan DeWall is a Professor of Psychology in the Department of Psychology at the University of Kentucky. He has three main areas of expertise: interpersonal relationships, self-regulation, and aggression. His research is funded by the National Science Foundation and National Institutes of Health.

Jan Kornelis Dijkstra, Ph.D. is an Associate Professor with the Department of Sociology and the Interuniversity Center for Social Science Theory and Methodology (ICS) at the University of Groningen, the Netherlands. His research interests include delinquency, risk behaviors, peer influence, peer relations, peer status, longitudinal social network analysis, and gene–environment processes.

Lisabeth F. DiLalla completed her Ph.D. in developmental psychology at University of Virginia and her postdoctoral fellowship at the Institute for Behavioral Genetics, University of Colorado. She is a Professor at Southern Illinois University, Department of Family and Community Medicine. Her research focuses on behavioral genetics and preschool social cognitive development.

Julia Dmitrieva is Assistant Professor in the Department of Psychology at the University of Denver.

Hayley M. Dorfman is Lab Manager of the Systems Neuroscience of Psychopathology Lab at Harvard University.

Ryan M. Ellsworth is a Ph.D. candidate in the Anthropology Department of the University of Missouri. Recent research focus includes human sexuality, reproductive strategies in lowland South American cultures, and lethal violence in traditional societies.

Emma Espel is a graduate student in the Department of Psychology at the University of Denver.

Qiang "John" Fu is Associate Professor in the Division of Biostatistics Department of Community Health in the School of Public Health at Saint Louis University.

Gianni G. Geraci graduated from California State University, Long Beach with an undergraduate degree in psychology, minoring in both criminal justice and forensic science. She holds a certificate in Basic Applied Forensic Science and Crime Analysis, and is pursuing a graduate degree in psychological research at California State University, Long Beach.

Andrea L. Glenn is Assistant Professor in the Center for the Prevention of Youth Behavior Problems and the Department of Psychology at the University of Alabama. Her research focuses on understanding the biological correlates of psychopathy and using biological information in the development of interventions for youth with conduct problems.

Benjamin D. Goodlett, M.A. is a doctoral student in the clinical psychology program at Wayne State University. His research focuses on the role of parent–child interactions in developmental processes, including childhood conduct problems and adjustment to treatment for medical illnesses.

Stephanie H.M. van Goozen is a Professor in the School of Psychology at Cardiff University (UK) and Leiden University (NL). Her research is in the area of developmental psychopathology, particularly the neuropsychological and neurobiological factors that influence young children's and adolescents' antisocial behavioral development.

Erick Guerrero is an Assistant Professor at the University of Southern California, School of Social Work. Dr. Guerrero's program of research focuses on the organizational implementation and translation of culturally responsive and evidence-based treatment practices in behavioral health services.

Tracy D. Gunter is Associate Professor of Clinical Psychiatry at Indiana University School of Medicine, Adjunct Professor of Law at the Indiana University McKinney School of Law, and Adjunct Clinical Associate Professor at the University of Iowa Carver College of Medicine. Dr. Gunter is board-certified in forensic psychiatry and has an active clinical practice.

Nicholas R. Holland is an undergraduate honors student at Iowa State University.

Li Huang is an Assistant Professor in the Department of Psychology and Sociology at Tuskegee University. She received her Ph.D. from Auburn University in 2007, where her training focused mainly on adolescent development and the etiology of deviance and problem behaviors.

Peter K. Jonason is a social-personality psychologist who uses evolutionary theory to derive predictions and account for observable phenomena in personality, individual differences, mating strategies, and sexuality. Dr. Jonason works at the University of Western Sydney and has studied at the University of Connecticut and New Mexico State University.

Alice P. Jones Bartoli is a Senior Lecturer and Director of the Unit for School and Family Studies at Goldsmiths, University of London. She is also a visiting researcher at Tomsk State University. Dr. Jones Bartoli has a particular interest in the affective and empathy development of children with social, emotional, and behavioral difficulties.

Melissa Jonson-Reid is Professor of Social Work and Director of the Center for Violence and Injury Prevention at Washington University.

Atika Khurana, Ph.D. is an Assistant Professor in the College of Education at the University of Oregon, with primary research interests in the prevention of adverse forms of adolescent risk-taking. Her current work examines environmental influences on the development of working memory and executive control as predictors of adolescent risk-behavior trajectories.

Maria Koeppel recently received her Ph.D. in criminal justice from Sam Houston State University, where she worked for several years at the Crime Victims' Institute. Her research interests include gender, sexual orientation, intimate partner violence, and victimization. Maria's work has been published in a number of journals including *Justice Quarterly*, *Criminal Justice Policy Review*, *Women and Criminal Justice*, *Violence and Victims*, *European Journal of Policing Studies*, and *International Journal of Comparative and Applied Criminal Justice*.

Tina Kretschmer, Ph.D. is a postdoctoral researcher at the University Medical Centre Groningen and the Department of Sociology, University of Groningen. Her research focuses on the interplay of social and biological risk for child and adolescent behavioral psychopathology.

Eric Lacourse is Professor in the Department of Sociology at the University of Montreal. He has published widely in the areas of behavior genetics, developmental psychopathology, and lifecourse criminology.

Matthew Larson is an Assistant Professor in the Department of Criminal Justice at Wayne State University. His current research examines the effects of romantic involvement/instability on antisocial behavior, the implications of victimization for attachment, and the consequences of incarceration for relationships, families, and health.

Royce Lee is Associate Professor in the Department of Psychiatry and Behavioral Neuroscience and Associate Director of the Clinical Neuroscience and Psychopharmacology Research Unit at the University of Chicago.

Heather L. McLernon is a graduate student at California State University, Long Beach. She has co-authored four book chapters and presented research posters at various conferences. Her current research includes the relationship between psychopathology and crime, and identifying risk factors for violence within mental disorders such as schizophrenia and other schizophrenia-spectrum disorders.

Gabriel Marmolejo is a graduate student in the Department of Psychology at California State University, Long Beach.

Melissa Martineau is a senior research specialist in the area of behavioral sciences within the Royal Canadian Mounted Police. She has 12 years' experience conducting applied research and other related operational work within the area of behavioral sciences. Her research interests include the analysis of serial and violent crime, sexual offending, crime linkage, geographic profiling, and threat and risk assessment.

Brandy R. Maynard, Ph.D. is an Assistant Professor in the School of Social Work at Saint Louis University. Her research interests include the etiology, prevention, and treatment of delinquency and externalizing behavior problems; evidence-based practice; and systematic review and meta-analytic methods.

Gerben Meynen studied medicine at the VU University Amsterdam, and philosophy and theology at the University of Amsterdam. He trained as a psychiatrist, and received a Ph.D. in philosophy and in medicine. He is Endowed Professor of Forensic Psychiatry (Tilburg University) and Assistant Professor at the Faculty of Philosophy, VU University Amsterdam, and he works as a psychiatrist at GGZ in Geest, Amsterdam.

Simon C. Moore has a background in psychology, and works to translate fundamental research to inform real-world efforts to reduce harm. His interdisciplinary research includes national trials of complex interventions and fundamental research into the causes of high-risk behavior.

Joanne E. Morgan completed her Ph.D. at Cardiff University, which examined the biobehavioral factors contributing to antisocial behavior in young people. Joanne is a researcher on the Cardiff Child Development Study, a longitudinal study of children's social and emotional development.

Joseph Murray is a Wellcome Trust Research Fellow and Senior Research Associate at the Department of Psychiatry, University of Cambridge. Dr. Murray researches the development of conduct problems, crime, and violence through the life-course. Awards include the Young Scholar Award from the American Society of Criminology, Division of Corrections and Sentencing in 2008.

Joseph L. Nedelec is an Assistant Professor in the School of Criminal Justice at the University of Cincinnati. He received his Ph.D. in criminology from Florida State University. His research interests include biosocial criminology, evolutionary psychology, intelligence, quantitative behavior genetics, and cybercrime.

Jamie Newsome is Assistant Professor in the Department of Criminal Justice at the University of Texas at San Antonio. Her primary research interests center on biosocial criminology.

Isabelle Ouellet-Morin investigates the biological (genes, epigenetics, stress hormones) and cognitive mechanisms by which early victimization affects long-term trajectories of mental health problems and antisocial behavior. She holds a Ph.D. in clinical psychology (Canada) and was a postdoctoral fellow at the MRC Social, Genetic, and Developmental Psychiatry Centre (UK). Dr. Ouellet-Morin is now Assistant Professor of Criminology at the Université de Montréal (Canada).

Stéphane Paquin is a doctoral candidate in the Department of Sociology at the University of Montreal.

Leidy S. Partida is an undergraduate student completing a dual degree in psychology and criminal justice at California State University, Long Beach, and has co-authored a number of conference papers. She is interested in the biological and psychological foundations of crime.

Brian E. Perron is an Associate Professor at the University of Michigan's School of Social Work. His research focuses on services for persons with mental health and substance use disorders, with an emphasis on predictive modeling and measurement.

Jill Portnoy is a doctoral candidate in criminology at the University of Pennsylvania. Her research focuses on the biological basis of criminal behavior. She is especially interested in interactions between biological and social factors in predicting antisocial behavior.

Chad Posick is an Assistant Professor in the Department of Criminal Justice and Criminology at Georgia Southern University. He teaches in the areas of victimology and criminal behavior. His research interests include the relationships between victimization and offending, the role of emotions in behavior, and measurement issues in criminological research.

Adrian Raine is the Richard Perry University Professor of Criminology, Psychiatry, and Psychology at the University of Pennsylvania, and President of the Academy of Experimental Criminology. His interdisciplinary research focuses on the etiology and prevention of antisocial, criminal, and psychopathic behavior in children and adults.

Cesar Rebellon is Associate Professor in the Department of Sociology at the University of New Hampshire.

Rheanna J. Remmel is a graduate student in the Clinical Psychology and Law program at the University of Alabama. Her research focuses on antisocial behavior across the lifespan, the neural correlates of psychopathy, and mental health within the criminal justice system.

Alexander J. Roberts graduated from Point Loma Nazarene University with a B.A. in psychology, and received his Master's in psychological research from California State University, Long Beach. His research interests included trait displaced aggression, violent crime, and drug use. He is currently pursuing a career in law enforcement.

Joshua W. Roberts is pursuing his Ph.D. in family sciences at the University of Kentucky. He is a 2011 graduate from Eastern Kentucky University, receiving a Master's in criminal justice.

Michael Rocque is an Assistant Professor of Sociology at Bates College. His research interests include life-course criminology, race and justice, and biosocial criminology. He is working on a revised edition of *The Criminal Brain* (New York University Press, 2008) with Nicole Rafter and Chad Posick.

Danie Romer, Ph.D. is Director of the Adolescent and Health Communication Institutes of the Annenberg Public Policy Center of the University of Pennsylvania. His research focuses on the identification of preventable risks to adolescent mental and behavioral health.

Anna Rudo-Hutt is a doctoral candidate in psychology at the University of Pennsylvania. Her research focuses on the etiology of externalizing disorders and antisocial behavior in children and adolescents.

Mary Ruffolo is an Associate Professor at the University of Michigan's School of Social Work. Her current research studies focus on adapting efficacious interventions, evaluating their effectiveness, and addressing ways to disseminate interventions with at-risk populations.

Joseph Ryan is an Associate Professor at the University of Michigan's School of Social Work. His research focuses on child welfare and juvenile justice populations, with an emphasis on experiences and outcomes of families entangled with the public child welfare system and at least one other social service context.

Christopher Salas-Wright, Ph.D. is Assistant Professor in the School of Social Work at the University of Texas at Austin. His research interests include the epidemiology of antisocial behavior, high-risk Latino youth, and adolescent substance abuse and violence prevention.

Justin C. Savage is a Project Manager in the Psychology Department of Cardiff University and has carried out extensive psychological research on risk-taking propensity, decision-making, antisocial behavior, gambling behavior, neuropsychology, visual perception, virtual reality ergonomics, and human–computer interactions.

Robert A. Schug, Assistant Professor of Criminal Justice and Forensic Psychology at California State University, Long Beach, has published numerous articles in prominent psychiatric, psychological, criminal justice, and neuroscience journals. Dr. Schug has also authored and co-authored chapters for several texts in forensic psychology, criminology, and the neurosciences.

Joseph A. Schwartz is an Assistant Professor in the School of Criminology and Criminal Justice at University of Nebraska, Omaha. He obtained his Ph.D. in criminology from Florida State University. His research interests include behavior genetics, biosocial criminology, the association between intelligence and behavior, and the etiology of criminal behavior.

Stacey L. Shipley is Director of Psychology at North Texas State Hospital.

Erica L. Smearman is an M.D./Ph.D. student at Emory University pursuing her Ph.D. in the School of Public Health. She is in a fellowship for students interested in linking lab and population sciences, and is interested in the bio-psycho-social approach to the study of behavior and the application of intervention.

Jelena Todic, M.S.W. is a doctoral student in the School of Social Work at the University of Texas at Austin. Her research interests include social determinants of health and interventions to reduce health disparities.

Christopher J. Trentacosta, Ph.D. is an Associate Professor of Psychology at Wayne State University. His research interests include self-regulation, the development of psychopathology among at-risk children and families, programs to prevent early-emerging behavior problems, and the role of gene–environment interplay in early childhood development.

Chad R. Trulson is Professor and Associate Chair of the Department of Criminal Justice and the University of North Texas. Dr. Trulson is Editor-in-Chief of *Youth Violence and Juvenile Justice*.

Catherine Tuvblad, Ph.D. is a Research Associate in the Department of Psychology at the University of Southern California. She recently coedited a special issue of the *Journal of Criminal Justice* on Genetics and Antisocial Behavior.

Jamie C. Vaske is an Assistant Professor in the Department of Criminology and Criminal Justice at Western Carolina University. Her areas of interest include quantitative methods, biosocial criminology, and prevention and intervention science.

Michael G. Vaughn is Professor of Social Work in the College for Public Health and Social Justice and in the Department of Epidemiology in the School of Public Health at Saint Louis

University. Dr. Vaughn is a nationally recognized scholar in the areas of adolescent antisocial behavior, substance abuse, violence, delinquency and chronic offending, and juvenile psychopathy.

Alexander T. Vazsonyi is the John I. and Patricia J. Buster Endowed Professor of Family Studies and a Professor of Psychology at the University of Kentucky. His main research interests include adolescent development, adjustment, and well-being (deviance, violence, and health-compromising behaviors), with a focus on development in context (family, school, neighborhood, and culture).

René Veenstra, Ph.D. is Professor in the Department of Sociology, University of Groningen, the Netherlands. He has published on a variety of topics (bullying and victimization, peer relations, prosocial and antisocial behavior, social network analysis, temperament–environment interactions) in journals including *Child Development*, *Developmental Psychology*, and *Social Networks*. He is Associate Editor of the *Journal of Research on Adolescence*.

Rob S. Walker is Assistant Professor of Anthropology at the University of Missouri. He specializes in cross-cultural analyses of human culture and has been collecting comparative human databases over the past decade, particularly for hunter-gatherers and lowland South American populations.

Glenn D. Walters, Ph.D. has spent the past three years teaching classes in corrections, research methods, and drugs and crime to undergraduate students at Kutztown University. Prior to this he worked for 27 years as a clinical psychologist in the Federal Bureau of Prisons. His current research interests include criminal thinking, the latent structure of crime-related constructs, and developing a comprehensive theory of criminal behavior.

Brandon C. Welsh, Ph.D. is a Professor of Criminology at Northeastern University and a Senior Research Fellow at the Netherlands Institute for the Study of Crime and Law Enforcement in Amsterdam. His research interests include the prevention of delinquency and crime, and evidence-based policy. His latest book is *Experimental Criminology: Prospects for Advancing Science and Public Policy* (Cambridge University Press, 2013).

Guilherme W. Wendt holds a Master's degree in clinical psychology and is a Ph.D. student at Goldsmiths, University of London (sponsored by Capes Foundation, Brazil). His work focuses on bullying and psychophysiological aspects of callous-unemotional traits.

D. Anne Winiarski, M.A. is a clinical psychology graduate student at Emory University. Through the use of longitudinal research designs, she combines her backgrounds in both developmental and clinical psychology to explore physiological, emotional, and behavioral predictors of persistent aggression, antisocial behavior, and delinquency.

John Paul Wright is Professor in the School of Criminal Justice at the University of Cincinnati and visiting Distinguished Professor in the Center for Social and Humanities Research at King Abdulaziz University in Jeddah, Saudi Arabia.

Yaling Yang is Assistant Professor in the Department of Pediatrics at the University of Southern California Keck School of Medicine, and Affiliate with the Children's Hospital of Los Angeles.

Acknowledgements

We would like to thank everyone in the Routledge team for their enthusiastic support and encouragement on this project, especially Thomas Sutton, Nicola Hartley, Sarah Olney, and Heidi Lee. Thanks also to the contributors for taking time from their busy schedules to complete chapters for this volume. We appreciate your effort, and admire your scholarship.

Part I

Criminological foundations

1

A unified theory of crime and delinquency

Foundation for a biosocial criminology

Cesar J. Rebellon, J. C. Barnes, and Robert Agnew

Introduction

Biological approaches to crime were prominent in the late nineteenth century, but fell from favor in the early twentieth century. Several factors contributed to this fall. First, biological theories of that era failed to garner strong empirical support. Lombroso's (1911) theory of the born criminal, for example, suggested that the most heinous crimes tended to be committed by biologically inferior throwbacks to an earlier stage of human evolution. Over the next half-century, however, data failed to support his belief that such criminals could be systematically distinguished from the general population by various physical features (Walby and Carrier, 2010). Second, historical events in the first half of the twentieth century led many scholars and activists to question the implications of biological approaches to human behavior (Rafter, 2008). The eugenics movements of the early twentieth century, for example, suggested that some individuals should either be killed or prevented from reproducing. The Holocaust illustrated the horrific dangers of such thinking. Third, sociologists in the first half of the twentieth century had begun to observe that crime in certain geographic regions tended to remain relatively high even in the face of residential turnover. This suggested that environmental factors influenced crime independently of individual characteristics (e.g. Shaw and McKay, 1942). The combined effect of the above three factors on academic criminology was an emerging focus on the social forces that might be said to promote crime (Udry, 1995).

Since the 1930s, criminologists have therefore tended to be sociologists by training. To an extent, this tendency is understandable. Decades of macro-level research suggest that crime rates are systematically higher in geographic regions containing high levels of inequality and pockets of concentrated poverty (e.g. Shaw and McKay, 1942; Blau and Blau, 1982). Decades of micro-level research suggest that crime is higher among individuals exposed to environments lacking adequate social control, containing deviant role-models, and promoting high levels of stress or strain (e.g. Hirschi, 1969; Akers *et al.*, 1979; Elliott *et al.*, 1985; Agnew, 2001; Rebellon *et al.*, 2008). At the same time, a resurgence of biosocial research has reached something of a critical mass in recent years, making it increasingly difficult to deny that biology has important effects on crime (Cullen, 2011). Such research provides strong evidence that all human behavior, including crime, is mediated by biological factors involving a complex set of interactions among environmental stimuli, the central nervous system (brain and spinal cord); the peripheral nervous system (the somatic nervous system which controls movement and the autonomic nervous system which regulates internal functions and prepares the body for "fight or flight" as well as "rest and digest"); and the endocrine system (the organs in the body responsible for hormone secretion). Reflecting this reality, criminology is beginning to acknowledge more explicitly that both genetic and non-genetic biological forces remain very relevant to a complete understanding of interpersonal variation in crime (Walsh, 2000; Barnes *et al.*, 2013).

The present chapter attempts to provide a theoretical bridge between traditional sociological criminology and more recently emerging biosocial approaches. It begins by discussing Agnew's (2005) general theory of crime, which presents a unified framework for understanding and organizing decades of sociological research about why some individuals engage in more crime than others. In particular, it describes the manner in which five life-domains may influence delinquent/criminal behavior via their effects on the *motivations* for, and *constraints* against, crime. These life-domains are: (1) the self, which includes factors such as the personality and traits of irritability and self-control; (2) the family, which includes parenting and sibling effects; (3) the school, especially negative school experiences and limited educational achievement; (4) the peer network, particularly association with delinquent peers; and (5) the work environment, especially chronic unemployment and work in the secondary labor market.

After elaborating upon the above facets of Agnew's (2005) general theory, the present chapter expands the general theory to include numerous biosocial explanations for interpersonal similarity and variation in crime. Although Agnew's original statement of the general theory argues that biological factors influence individual traits, his treatment of biosocial influences is limited and he pays much more attention to social variables in the life-domains.

We argue that a broader consideration of biological factors can enhance the general theory in at least five ways. First, evolutionary theory helps explain why different individuals are susceptible to common sources of motivation for, and constraint against, crime. Second, it is now well known that biological factors are antecedent to many criminological variables, meaning that the effect of many biological variables is likely to be mediated by more proximal social stimuli. Third, much research now suggests that environmental factors influence the individual's non-genetic biological functioning. As such, biological factors may also be said to mediate the influence of social variables on crime. Fourth, cutting-edge research in human genetics and the etiology of complex traits has repeatedly shown that genetic and social factors interact in their influence on behavioral outcomes (Moffitt, 2005; Belsky *et al.*, 2009). We therefore argue that biological variables and the life-domains may amplify and buffer each other's influence on crime. Finally, evolutionary theory suggests biological explanations for certain group differences in crime, particularly differences by gender and age. After discussing these issues, we conclude by describing a general framework for understanding how biosocial research can inform effective, but simultaneously ethical, crime-control policy.

Agnew's general theory of crime and delinquency

Agnew's (2005) general theory attempts to explain why some people offend more than others. It conceives of crime as "*behaviors that are generally condemned* and that *carry a significant risk of sanction* by the state if detected" [italics in original] and suggests, at its core, that crime is most likely among individuals whose motivations for crime are high and whose constraints against crime are low (Agnew, 2005:12).

Motivations and constraints

With respect to motivation, Agnew's general theory suggests that individuals differ in both the enticements and pressures that they feel toward crime. Enticements include "reinforcements for crime," "exposure to successful criminal models," and "beliefs favorable to crime" (Agnew, 2005:22–26). Reinforcements refer in part to the economic rewards that often accompany criminal behavior, but also to social rewards such as admiration or esteem in the eyes of others. Successful criminal models refer to individuals from one's family, peer group, or workplace that engage in or encourage crime. Beliefs favorable to crime refer to personal attitudes that approve of, justify, or excuse crime. Beyond these enticements, Agnew suggests that some individuals experience greater levels of pressure toward crime. This pressure may include economic or social strains that promote negative emotions, for which crime may serve as a coping mechanism. Strains include the failure to achieve positively valued goals, the loss of positively valued stimuli, and the experience of noxious stimuli. Of particular importance, Agnew (2005:27) argues that the most crime-inducing strains tend to be those that "are high in magnitude, are seen as unjust, are associated with low social control, and create some pressure or incentive to engage in criminal coping" (see also Agnew, 2001).

With respect to constraints, Agnew's general theory suggests that three major types include "external control," "stake in conformity," and "internal control." External control refers to crime's negative consequences. Parents, for example, may set rules about appropriate behavior, monitor their children for rule violations, and punish rule violations. Similarly, the police and courts monitor and sanction violations of the law. Stake in conformity refers to those things that individuals jeopardize through crime. These include bonds to conventional others and investments in long-term, conventional goals. Stealing from an employer, for example, might lead not only to punishment in the criminal justice system, but also to severed ties with co-workers or the loss of one's job. Finally, internal control refers to the individual's belief that crime is wrong and to his or her exertion of restraint in the face of criminal temptation.

The five life-domains

Innumerable environmental experiences are correlated with an individual's relative involvement in crime or delinquency. Agnew organizes these variables into smaller groups based on the five life-domains in which they tend to occur, and argues that the variables in each domain tend to be correlated with one another. The life-domains include the self, the family, the school, the peer group, and the workplace.

The self refers to each individual's personality traits. Psychologists have identified a variety of such traits and criminologists have found a number of them to be correlated with crime (Wilson and Herrnstein, 1985; Caspi *et al.*, 1994). Agnew's general theory suggests that the traits of self-control and irritability are particularly important for explaining interpersonal crime differences. Broadly defined, self-control connotes an individual's tendency to behave with the long-term in mind. Individuals with low self-control tend to act in ways that are risky or impulsive and they tend not to persist in their efforts. Irritable individuals "are more likely to experience events as aversive, attribute these events to the malicious behavior of others . . . experience intense emotional reactions to these events . . . and respond to these events in an aggressive or antisocial manner" (Agnew, 2005:43–44). While all people may behave impulsively or irritably in some instances, the general theory suggests that some individuals exhibit these personality traits more frequently *compared with other individuals*. These traits affect motivation by increasing the likelihood that individuals will react to strain with strong emotions. They also lower internal control, dampen stake in conformity, and render some individuals more resistant to direct control.

The family domain includes parents and siblings during childhood, and spouses and offspring during adulthood. Agnew's general theory suggests that these factors influence motivation and constraint in several ways. First, it suggests that parental supervision and punishment exert direct control over crime. Second, it suggests that overly harsh or abusive parenting may serve as an important source of familial strain (see Agnew *et al.*, 2000) and that a minority of parents may reinforce beliefs favorable to crime (see Anderson, 2000). Third, it suggests that these motivating forces may prevail not only among parents, but also among other family members, particularly siblings. Fourth, it suggests that family members such as spouses or offspring may increase an individual's stake in conformity during adulthood (see Sampson and Laub, 1993).

The school and peer domains overlap to some degree. The school domain consists of an individual's academic performance as well as his or her interactions with school authority figures such as teachers and principals. Agnew's general theory suggests that crime tends to be higher among individuals whose school experiences involve poor teacher communication of rules, low teacher supervision, and inconsistent teacher sanctioning of misbehavior. These school experiences reflect examples of weak direct control. Beyond this, Agnew argues that variables such as poor academic performance may increase motivation by inducing strain and, simultaneously, decrease constraint insofar as they reflect low stake in conformity. In the peer domain, deviant friends may model successful criminal behavior, promote deviant beliefs, and reinforce crime with both social and non-social rewards (Akers *et al.*, 1979; Akers, 2009). Conflict within the peer group may also promote motivating strain. Deviant peers may also weaken individuals' ties to conventional others, thereby lowering stake in conformity and reducing external control.

Finally, the workplace refers to the individual's occupational successes and failures, particularly among adults. It should be noted that this domain refers to more than merely one's socioeconomic status (SES), which is traditionally measured in terms of educational attainment, income, and occupational prestige. Notwithstanding much evidence that crimes rates are higher in geographic regions containing high levels of inequality and pockets of concentrated poverty, some micro-level research fails to show that low-SES individuals are more criminal (Tittle and Meier, 1990). Agnew's general theory, however, suggests that the workplace bears a somewhat complex relationship with crime. Rather than suggesting that crime is unilaterally higher among unemployed versus employed individuals, it suggests that crime is higher among those who are chronically unemployed as well as those who work in *bad* jobs "that involve unpleasant tasks . . . little autonomy or freedom, frequent use of coercive control . . . low pay, few benefits, little prestige, limited opportunities for advancement, and high turnover" (Agnew, 2005:55). From the standpoint of the general theory, such jobs lower stake in conformity and increase strain. Likewise, they may motivate crime via exposure to criminal co-workers and may weaken direct control by failing to provide sufficient worker supervision.

The web of crime

Beyond describing the ways in which the five life-domains affect motivations and constraints, Agnew's general theory suggests that the life-domains all influence one another, but that some influences are stronger than others. The self, for example, is said to wield a relatively strong influence on an individual's experiences in all four other life-domains throughout the periods of childhood, adolescence, and adulthood. Further, each of the other life-domains is said to have at least a modest reciprocal influence on the self during each of the same three developmental periods. Similarly, children with low self-control might have a hard time focusing on schoolwork and this failure may, in turn, impede their development of high self-control. Irritable individuals may gravitate toward similar others (see Glueck and Glueck, 1950) and those others may, in turn, foster further irritability. Low self-control may hinder an individual's prospects for securing a good job which, in turn, may further impede the development of high self-control. To a greater or lesser degree, the family, school, peer, and workplace domains may also be said to bear reciprocal relations with each of the others and "[t]hese mutually reinforcing effects are what [Agnew] refers to as the 'web of crime'" (Agnew, 2005:81).

In addition to describing the ways in which each of the life-domains is related to the others, Agnew's general theory provides a number of predictions about how these domains collectively influence crime. First, the general theory predicts that the causes of interpersonal variation in crime differ by an individual's stage in the life-course. During childhood, the strongest influences are the self and the family. During adolescence, they are instead the self and the peer group. During adulthood, they are the self, the peer group, and the workplace. Second, the general theory predicts that the life-domains' effects on crime are contemporaneous, taking place "within a few months' time," rather than several years later (Agnew, 2005:124). One can imagine, for example, that an individual's crime is more likely to be influenced by his or her current level of self-control than by his or her self-control several years ago. Third, the general theory predicts that the life-domains exert non-linear effects on crime. Having one delinquent friend, for example, may not result in much crime, but having two or three may lead to a slippery slope of accelerating criminal activity. Fourth, the causes of crime interact, such that any one cause of crime has a stronger influence in the presence of other causes. Fifth, whereas the life-domains are said to exert relatively immediate influences on crime, each is said to exert a long-term influence on itself. Self-control may change over the course of many years, for example, but prior self-control will nonetheless influence future self-control over the long-term.

Finally, Agnew's general theory suggests that crime exerts influences on the life-domains and on itself. Crime may, for example, contribute to poor parenting, impede educational outcomes, promote association with other criminals who reinforce deviant beliefs, and restrict opportunities for advancement in the workplace. Aside from affecting itself indirectly through its influence on the five life-domains, crime is said to have a direct long-term effect on itself. Research clearly suggests, for example, that one of the strongest predictors of one's criminal behavior relative to other people is one's prior level of criminal behavior (Robins, 1978; Nagin and Paternoster, 2000). From the perspective of the general theory, this occurs for a number of reasons such as: the strain that crime itself can produce for the offender (e.g. strained relationships resulting from a criminal lifestyle); the reinforcement that results from the immediate pleasure of many crimes (e.g. the escape that substance use can provide from strained relationships); and the decrease in fear of external sanctions that may result from engaging in crime without being caught (e.g. a weakening of perceptual deterrence after punishment avoidance).

The general theory and biosocial criminology

Although the general theory argues that biology is one of several factors that affect crime outside the immediate life-domains, its limited attention to biological factors may preclude it from fully explaining interpersonal variation in crime. Further, most of the evidence upon which the general theory is built comes from correlational research designs that lack genetic data. Variables from the life-domains may therefore be correlated with crime in ways that reflect genetic, rather than environmental, influence. On one hand, for example, two parents whose personal genotypes contribute to their own self-control may be prone to invest much effort in setting clear rules, monitoring their children, and sanctioning rule-violations. On the other hand, the combination of their genotypes may lead their children to possess a high degree of self-control as well. In this case, parenting in one generation might be significantly correlated with crime in the next, but this correlation could be spurious owing to uncontrolled genetic factors (Wright and Beaver, 2005). On one hand, biosocial researchers tend to agree that genetics alone cannot explain human behavior (Turkheimer, 2000). On the other, compelling evidence suggests that some individuals are born with higher levels of genetically influenced characteristics that may promote crime through their influence on motivations and constraints in the life-domains. Likewise, each of the family, school, peer, and workplace domains influences crime through its effects on motivations and constraints, but motivations and constraints necessarily operate through biological and psychological mechanisms to which the general theory and mainstream criminology have paid only limited attention. Below, we draw on existing biosocial theory and research to extend Agnew's work by directly incorporating biological variables and processes into the theory.

A biosocial extension of the general theory

Biosocial theory and research suggest five ways in which Agnew's general theory can be extended to better explain interpersonal variation *and* similarity in crime. First, biosocial theory provides insight about the reasons why similar motivations and constraints influence crime among most individuals. Second, biosocial research helps illuminate the ways in which interpersonal biological differences influence motivating and constraining variables from each of the five life-domains. Third, it simultaneously suggests reason to believe that motivating and constraining variables from the life-domains influence an individual's non-genetic biology. Fourth, biosocial research suggests numerous ways in which biological variables interact with the life-domains to amplify or suppress each domain's influence on crime. Finally, evolutionary theory explains why motivations and constraints may vary by gender and age.

Why are people susceptible to similar motivations and constraints?

Our biosocial extension of the general theory begins by drawing on the field of evolutionary psychology to explain why individuals are susceptible to common sources of motivation and constraint. Evolutionary theory views human beings as biological entities that pursue self-interest which, more often than not, promotes survival and reproduction. Self-interest includes the immediate sating of biological desires such as food or sex, and the accumulation of economic resources that can assist in sating these desires long-term. Crime is often an expedient means by which to attain self-interest. Theft, for example, may provide quick access to new material resources. Violence may overcome the resistance of others who deny the individual's desires. Drug use may promote visceral gratification and alleviate physical or psychological suffering. Most fundamentally, then, our extension of the general theory argues that the primary motivation for crime is the universal human desire to pursue pleasure and avoid pain. The pleasure and pain in question may be physical or psychological. Further, psychological pleasure and pain may partly reflect the pleasure and pain that people perceive to exist among others with whom they share emotional attachments.

The above assertions are compatible with Agnew's (2005) original statement of the general theory. Agnew (2005), for example, lists reinforcement as the first of the major motivations for crime. In particular, he argues that "[p]ositive reinforcement involves the presentation of something good, like money, social approval, or the pleasurable feelings that result from drug use" and that "[n]egative reinforcement involves the removal of something bad . . ." Our extension of his theory, however, argues more explicitly than he originally did that "good" stimuli are characterized by the biological pleasure they produce, while "bad" stimuli are characterized by the biological pain they inflict. Thus we draw on evolutionary theory to argue that crime, like any form of human behavior, increases primarily as a function of the self-interest that it promotes (see also Gottfredson and Hirschi, 1990).

Evolutionary theory, however, provides a more complex view of human nature than some criminologists have tended to acknowledge. In particular, it observes that human beings are social animals whose survival often depends on collective action. People therefore tend to live and work in groups, and these groups tend to be hierarchically arranged. For this reason, we argue that people's desire to seek physical self-interest is counterbalanced by two types of pro-social constraint. First, we argue that individuals tend to possess general social concern for the wellbeing of others, particularly others to whom they are related or with whom they share long-term affective attachments (Agnew, 2001). This general social concern is evidenced by the social nature of many human emotions, including guilt, shame, and embarrassment (Tangney *et al.*, 2007). It is likewise evidenced by the fact that most human interactions do not involve forceful, fraudulent, or otherwise criminal behaviors. Second, we argue that individuals tend to seek status within collective hierarchies. In particular, individuals tend to desire respect, admiration, deference, and influence among other members of the collectives in which they participate. This concern for status is evidenced by the tendency of social institutions to be hierarchically arranged, as well as the tendency of individuals to desire and seek higher, rather than lower, rank among those to whom they compare themselves (Festinger, 1954).

The human tendency to exhibit social concern helps explain why people are generally susceptible to similar sources of constraint against crime. People's general aversion to the suffering of others, for example, helps explain why most tend to possess some amount of internal control. Similarly, concern for social standing suggests one reason why most individuals feel at least some stake in conformity. Engaging in criminal behavior, for example, might jeopardize one's standing within the family or within the workplace. To the degree that one's conventional social ties are weakened (Hirschi, 1969), concern for social standing can also explain one mechanism through which deviant peers reinforce crime (Akers, 2009). Finally, concern for social standing explains one reason why individuals are susceptible to external control, as well as one reason why they often feel compelled to exert it.

In addition to suggesting that most individuals are susceptible to common motivations and constraints, we propose that motivations and constraints necessarily operate through common biological mechanisms. A good example involves the influence that the family domain may have on the self. Building on the work of such theorists as Gottfredson and Hirschi (1990), Agnew's general theory suggests that direct control in the family domain may contribute to the trait of self-control. Self-control, however, develops much like a mental muscle via biosocial mechanisms (Baumeister *et al.*, 1994). Once developed, it continues to be regulated by the brain's absorption of glucose, particularly in the frontal cortex (Beaver *et al.*, 2007). Further, research finds that self-control can be exhausted (like a muscle) within relatively short time windows (Baumeister *et al.*, 2007) and that it may be possible to restore (or slow the reduction of) self-control levels with glucose intake (Hagger and Chatzisarantis, 2013). Contrary to the assertions of some sociological theorizing (e.g. Gottfredson and Hirschi, 1990), moreover, recent research suggests that self-control and will-power can be exercised and developed to at least some degree beyond early childhood (Muraven *et al.*, 1999). Even as individuals differ in their respective *levels* of self-control, then, common biological mechanisms may regulate most individuals' development and exertion of self-control. Similarly, common biological mechanisms may universally regulate numerous other sources of motivation or constraint, including reinforcement, peer influence, belief formation, stress, internal control, and punishment.

In sum, we draw on evolutionary theory to argue that most people are susceptible to common motivations for, and constraints against, crime. On one hand, many of the motivations for crime stem from people's biological tendency to seek pleasure and avoid pain (Pinker, 2002, 2011). On the other hand, such motivations are counterbalanced by individuals' social concern and desire for status, both of which contribute to the major constraints against crime. Further, common biological processes likely serve to regulate these motivations and constraints among most people.

How do biological differences influence the life-domains?

In his original statement of the general theory, Agnew (2005) explicitly acknowledged that biological variables may contribute to interpersonal variation in crime. His discussion of these variables, however, was somewhat limited. Of particular note, although his discussion identified a number of the major variables that biosocial research has linked to crime, it did not systematically describe the multiple ways in which these variables increase motivation and decrease constraint in each of the major life-domains. Below, we elaborate his discussion by (1) summarizing the existing research concerning whether genetics contributes to interpersonal variation in crime; (2) discussing the major biological variables through which genetic effects may operate; and (3) describing the ways in which these biological variables influence motivations and constraints across each of the major life-domains.

Two primary methods have been used to examine whether genetics might influence interpersonal variation in crime. First, researchers have compared the crime of adopted children with that of their biological parents and that of their adoptive parents. To the degree that a criminal tendency can be inherited, such children's crime should be correlated with that of their biological parents. To the degree that a criminal tendency reflects the environmental influence of child-rearing practices, such children's crime should be correlated with that of their adoptive parents. Second, researchers have compared the criminal similarity, or concordance, of monozygotic (identical) versus dizygotic (fraternal) twins. The former are said to share 100 percent of their genetic make-up, while the latter share only 50 percent of their distinguishing DNA. To the degree that criminal concordance is higher for monozygotic versus dizygotic twins, evidence suggests a genetic influence on criminal behavior. Such studies tend to suggest that a substantial proportion of interpersonal variation in crime is related to genetic factors (Cloninger *et al.*, 1982; Bouchard *et al.*, 1990; Raine, 1993; Rhee and Waldman, 2002; Moffitt, 2005; Burt, 2009a, 2009b; Ferguson, 2010). At the same time, molecular genetics makes clear that a "crime gene" does not exist. Instead, research suggests that certain *clusters* of genes may contribute to biological characteristics that increase the risk for crime via a polygenic effect (i.e. many genes; Plomin *et al.*, 2013).

The effects of gene clusters on crime, however, are not direct. Rather, they are mediated by the brain and by the various physiological mechanisms it controls (Raine, 2008). Four mechanisms have garnered particular attention in the biosocial literature. First, research suggests that low levels of arousal and impairments in the central nervous system (CNS) are associated with crime (Raine *et al.*, 1995). In particular, crime appears higher among those exhibiting low levels of beta wave activity and low levels of fast alpha wave activity, as measured via electroencephalogram (EEG) (e.g. Mednick *et al.*, 1981). Similarly, crime appears higher among those with impairments to the frontal and prefrontal cortex, which are responsible for executive functioning (see Raine *et al.*, 2005; Ogilvie *et al.*, 2011). Second, research suggests that low levels of arousal in the autonomic portion of the peripheral nervous system, as indicated via such measures as low resting heart rate or low skin conductance, are associated with a higher probability of crime (e.g. Ortiz and Raine, 2004; Raine *et al.*, 1995). Third, hormone levels have been linked to criminal propensity. Testosterone, for example, appears to be higher among violent versus non-violent criminals and among individuals who self-report higher versus lower involvement in deviant behavior (Mazur, 2009). Finally, research suggests that differential levels of neurotransmitters such as serotonin and dopamine are associated with a higher probability of crime (e.g. Brunner *et al.*, 1993; Coccaro *et al.*, 1997). Each of these mechanisms is at least partly heritable (Singh *et al.*, 1999; Wright *et al.*, 2008; Collins, 2011; Kolb and Whishaw, 2014).

We propose that the above biological variables contribute to motivations and constraints in large part through their influence on the individual traits that comprise the self. In particular, variables such as low arousal, high testosterone, and low serotonin have been linked with such traits as a "sensation-seeking" temperament, a limited ability to regulate emotion, and difficulty learning from punishment (Raine, 2002a, 2002b; Agnew, 2005; Armstrong, 2011). Because sensation-seeking may reflect a desire to increase low arousal, sensation-seeking individuals may be disproportionately prone to experience reinforcement for crime. Because they value immediate excitement so highly, sensation-seeking individuals may have lower internal control and lower long-term stakes in conformity. Those who have trouble regulating emotion may experience higher subjective strain in the face of the same objective stressors. Of particular importance, these individuals may be disproportionately prone to experience hostile-attribution bias (Crick and Dodge, 1996), whereby they interpret ambiguous social stimuli in negative ways. They may therefore be more prone than others to perceive injustice, a particularly crime-inducing strain (Agnew, 2001; Rebellon *et al.*, 2012), even where such perceptions are objectively unwarranted. Finally, whether sanctions come from informal authorities such as parents or formal authorities such as police, difficulty in learning from punishment lowers these sanctions' effectiveness in deterring crime.

Variables such as low arousal, high testosterone, and low serotonin may also exert influences on the other four life-domains. Recalling that the life-domains affect one another, however, we propose that these biological variables influence the remaining life-domains indirectly, via their influence on the self. Sensation-seeking children, for example, may be disproportionately drawn to dangerous objects or situations in the family or school domains, thus weakening the direct control they experience with average levels of supervision. In the peer group, they may be drawn to deviant associates who model crime and promote deviant beliefs. In the workplace, they may be disproportionately prone to shirk responsibilities they find boring, thus decreasing the likelihood of promotion and increasing the likelihood of termination.

Similar arguments can be made about individuals who have difficulty in regulating emotion or learning from punishment. Difficulty in regulating emotion may heighten conflict among parents and siblings. In extreme cases, it may lead to a vicious cycle whereby exasperated parents become excessively harsh, prompting further negative emotion from a child, thus further exasperating parents (see Moffitt, 1993). In the school, difficulty in regulating emotion may prompt negative reactions from teachers, and may further lead youth to interpret such reactions in excessively hostile ways. In the peer group, it may heighten social strain, particularly during the teenage years, for reasons elaborated below. In the workplace, it may likewise lead the individual to perceive innocuous supervisor instructions as hostile and their assigned responsibilities as unjustly distributed. Similarly, difficulty in learning from punishment weakens constraint against, and increases motivation for, crime. In particular, it hinders external control at the hands of authority figures in the family, school, and workplace domains. Those who have difficulty in learning from punishment may further be shunned by peers who are sensitive to the discipline of authority figures. It may therefore lead individuals to associate, often by default, with deviant peers who reinforce crime and promote deviant beliefs. Upon obtaining employment, those who have trouble learning from punishment may be more likely than co-workers to repeat mistakes on the job, further decreasing their prospects for advancement and increasing the likelihood of termination.

In sum, then, our extension of the general theory suggests that biological factors promote crime through the mediating influence of motivations and constraints across each of the major life-domains. This assertion is built on several observations derived from existing criminological theory and research. First, although a "crime gene" does not exist, adoption and twin studies provide compelling evidence that inherited gene clusters contribute to interpersonal variation in crime. Second, research suggests that these gene clusters influence crime largely via their effects on such biological variables as CNS/ANS (autonomic nervous system) arousal, hormone levels, and neurotransmitter levels. Third, extending Agnew's (2005) general theory, we propose that these variables increase motivation and decrease constraint primarily via their effects on the traits that comprise the self. Finally, we propose that while these biological variables increase motivation and decrease constraint in the family, school, peer, and workplace domains, they do so primarily via their influence on the self.

How do the life-domains influence biological differences?

We propose that a portion of each life-domain's effect on crime takes place via the mediating influence of such biological variables as prenatal and perinatal care; head injuries damaging the brain's frontal lobe; exposure to toxins such as lead; poor diet or nutrition; and the physical effects of stress. Several of the traits comprising the self, for example, are likely to affect certain of the above biological variables. Those with low self-control or with sensation-seeking temperaments may be disproportionately prone to engage in sports or dangerous physical activities. Those high in impulsivity may be more likely to engage in such behaviors without taking appropriate precautions (e.g. wearing a helmet). Such activities might increase the risk of injury to the frontal lobe which, in turn, might impair an individual's executive functioning (Raine *et al.*, 2005; Ogilvie *et al.*, 2011). Similarly, individuals with these same traits might be more prone to consume a poor diet. Over time, poor diet might in turn promote crime by impairing psychological development or further lowering of self-control (Shoenthaler, 1991).

Traits such as low self-control or sensation-seeking, then, tend to reinforce themselves over time. Agnew's original statement of the general theory is compatible with this notion, predicting explicitly that each life-domain exerts a lagged effect on itself. However, his original rationale for this prediction is based on the notions that (1) individuals form habits; (2) habits stem from repeated reinforcement; and (3) habits "close off opportunities for change" (Agnew, 2005:125). Our extension of the general theory argues that an equally important influence of prior traits on future traits is biological. In particular, traits such as low self-control may lead the individual to engage in behaviors that may further impair self-control (or at least limit opportunities for increases in self-control). A similar argument can be made about Agnew's prediction concerning the effect of crime on itself. In particular, we propose that crime not only influences itself through such motivations as reinforcement or increased strain, but also through its influence on biological sources of constraint. Drug use or violence, for example, may lead to brain injuries that damage executive functioning and lower self-control, thereby increasing the likelihood of further drug use or violence.

Domains outside the self may also influence crime, partly through their effects on non-genetic biological variables. The family domain, for example, influences prenatal and perinatal care. Research suggests that crime is higher among individuals whose mothers smoke during pregnancy or who experience delivery complications at birth (Olds, 1997, 2006; Brennan *et al.*, 1999, 2003). Poor parental supervision, sibling conflict, and association with deviant peers may increase the probability of risky behavior that may result in head injury, thus affecting executive functioning. Family homes, schools, and workplaces located in poor neighborhoods may increase one's probability of exposure to environmental toxins such as lead, which has been linked to increased criminality (Wright *et al.*, 2008). Families, schools, and peer groups may influence an individual's diet, potentially affecting psychological development and individual traits in ways that promote crime (Shoenthaler, 1991). Finally, stressful environments in the family, school, peer, or workplace domains may trigger the release of hormones such as cortisol. In cases where the environmental trigger lingers without abating quickly, these stress hormones become harmful to the body and might even lead to the elimination of synapses and neurons in the brain (Kolb and Whishaw, 2014). Successful advancement/promotion in life-domains such as the peer group may also increase a given individual's testosterone from previous levels (Mazur and Booth, 1999).

How do biology and the life-domains interact to promote crime?

In addition to mediating one another's influences on crime, biological and social variables may interact to promote crime. Numerous studies, for example, suggest the existence of *gene–environment interactions* (G × E) (e.g. Caspi *et al.*, 2002; Guo *et al.*, 2008; Simons *et al.*, 2011; Barnes and Jacobs, 2013). Some of these studies use indirect genetic measures based on the criminality of an individual's relatives. In the family domain, adoptees whose parents are criminal (i.e. an indirect indicator of genetics) and whose adoptive parents provide weak external control (i.e. an environmental variable) appear highly prone to criminality. In particular, they appear far more prone to criminality than would be expected by simply combining the separate influences that biological and adoptive parents exert individually (Cloninger *et al.*, 1982). More recently, other studies support the existence of G × E interactions using direct genetic measures. Dijkstra *et al.* (2013:1), for example, find that aggression increases popularity among males, but only among those carrying "the G-allele of a functional polymorphism in the promoter region of the $5HT_{2A}$ serotonin receptor gene (–G1438A)." In the former example, a genetic predisposition for crime may lead to criminal behavior only when external control in the family is poor. In the latter example, aggressive behavior is reinforced with social status only among those who possess certain genetic characteristics.

The most cutting-edge research concerning G × E interactions, however, suggests that they may be even more complex than the above examples imply. Specifically, *epigenetic* research (e.g. Meaney, 2010; ENCODE Project Consortium, 2011) is beginning to suggest that environmental stimuli may actually activate some genes while deactivating others (Plomin *et al.*, 2013). Thus, even if two people (e.g. twins) possess identical genotypes at a certain locus along the genome, it remains possible that their differential environmental exposures will lead to different behavioral outcomes via gene regulation. It may therefore be that variables from one life-domain contribute to the activation of certain genes, while variables from another life-domain then interact with activated genes to promote crime.

Further complicating the interpretation of G × E interactions is the fact that interactions may take two separate forms. On one hand, certain genotypes may increase "vulnerability to environmental adversity" (Simons *et al.*, 2011:883). This means that the genotype in question is associated with higher levels of a behavioral outcome such as crime when combined with adverse social environments, but that it bears *no relation* to the same outcome when combined with positive social environments. On the other hand, certain genotypes may increase "susceptibility" to the influence of social environments such that the genotype in question is associated with higher levels of crime in adverse social environments, but with *lower* levels of crime in positive/nurturing environments. Simons *et al.* (2011), for example, examined environmental interactions with genes related to dopamine and serotonin. On one hand, they found that these genes seem to be associated with higher-than-average aggression when combined with environments characterized by motivations such as high strain and peer deviance. On the other hand, however, they found that these genes seem to be associated with *lower-than-average* aggression when combined with environments characterized by constraints such as a high stake in conformity.

Whereas the above research explores G × E interactions, two further lines of research explore biosocial interactions that do not directly involve genes. First, recalling that genes exert their influence via their effects on biological variables, one line of research examines the interaction of these biological variables with social variables from the life-domains. Farrington (1997), for example, finds evidence that low resting heart rate may be associated with aggression. This association, however, appears to be contingent on variables in the family domain such as the quality of parent–child relationships. Second, recalling that many biological variables come from entirely non-genetic sources, some research examines interactions involving biological harm and social variables from the life-domains. Research suggests, for example, that birth complications are particularly prone to be associated with the individual's crime when they are combined with certain variables in the family domain, including low parent–child attachment (Raine *et al.*, 1994; Piquero and Tibbetts, 1999) and parental mental illness (Brennan *et al.*, 1993). Further, research may eventually unveil similar interactions between biological variables and social variables in the school, peer, and workplace domains. In all cases, we argue that these interactions exert their influence on crime largely via their effects on motivations and constraints.

Why do motivations and constraints vary across groups?

In his original statement of the general theory, Agnew (2005) suggested two primary reasons why groups might differ in their average criminal behavior. First, he suggested that different groups may experience different levels of motivation and constraint in the five life-domains. Adolescents, for example, might experience higher levels of stress and lower levels of external control than younger children. Second, he suggested that the same level of motivation or constraint might have a different effect on crime among members of some groups. The same level of peer delinquency, for example, might have a greater influence on males than females. Our extension of the general theory is compatible with these assertions, but draws more explicitly on evolutionary theory to elaborate a biological explanation for two of crime's most reliable correlates: gender and age.

Across time and place, males tend to be more criminal than females. This gap tends to be largest for those crimes that people tend to view as most severe. Agnew's general theory attributes part of the gender gap to the different ways in which males and females are typically socialized.

At the same time, we argue that evolutionary theory likewise explains a portion of the gender gap. On one hand, evolutionary theory suggests that females may tend to experience greater constraints against crime. In particular, females are said to bear a greater biological investment in human offspring. This occurs for a number of reasons, including their production of fewer eggs over the life-course compared with male sperm; the lengthy gestation of human offspring (females must carry children for nine months during pregnancy); and the male inability to nurse offspring. In turn, this greater female investment has been theorized to result in a lower average tendency for females to engage in crime and other risky behaviors (Trivers, 1972; Campbell, 2009). We argue that a portion of this lower tendency may reflect biologically based differences in male versus female constraint. In particular, females may tend, on average, to possess biological traits that yield higher social concern. This social concern may increase their internal control and promote a higher stake in conformity. In addition, biological relatives may tend to exert greater external control over females.

At the same time, evolutionary theory suggests that males may experience greater motivation for crime. In particular, cross-cultural evidence suggests that males and females tend to value somewhat different characteristics in prospective mates. Buss (1994), for example, finds cross-cultural evidence that males tend to place greater value in a prospective mate's physical appearance than do females, while females place greater value in a prospective mate's economic prospects and social status. As such, while the quick procurement of material wealth may sometimes serve the self-interest of both males and females, a crime such as theft may sometimes reap added benefit for males insofar as it may help attract desired mates. This may be particularly true when economic crimes are undetected, thus prompting little external control, or when such crimes victimize members of out-groups, with whom a given male is not likely to interact frequently. Similarly, even as aggression against members of in-groups can often elicit external control, aggression or other forms of risky behavior may increase social standing in certain social circumstances (Buss, 2005; Ellis *et al.*, 2012). Examples include demonstrations of violence within deviant peer groups, or the physical defense of conventional in-group members who are attacked by members of an out-group (e.g. gang conflicts). Given the higher value that females place on the status of prospective male mates, such circumstances may disproportionately reinforce male violence.

Like gender, age is reliably correlated with crime across time and place. Crime tends to increase during early adolescence, reach a peak in the late teenage years or early twenties, and decline slowly over the remainder of the life-course. Further, evidence suggests that the general shape of the age–crime relationship in modern industrial society mirrors the general relationship between age and male violence in pre-industrial societies (Pinker, 2002). Agnew's general theory (2005) acknowledges that biological forces may partly underlie this tendency. For reasons that make sense when placed in an evolutionary framework (Ellis *et al.*, 2012), the adolescent brain develops such that the "lower-order" regions affecting emotional responses to stimuli mature more quickly than "higher-order" functions such as self-regulation and impulse control (Giedd *et al.*, 1999; Steinberg, 2007). Thus adolescents may be said to suffer from a "perfect storm" of brain activity that allows them to experience intense emotional swings, but limited ability to regulate those emotional states. Further, rapid physical and emotional changes among youth at this stage of the life-course may render status hierarchies unstable, potentially increasing social competition for status among peers in general, and among prospective mates in particular.

In combination, the physical and social changes taking place during the teenage years may have a number of influences on both the motivations for, and constraints against, crime. First, hormonal changes may lower internal control by amplifying traits such as low self-control and irritability. Second, following Moffitt (1993), they may promote a desire to assert one's autonomy from conventional socializing agents such as parents or teachers. To the degree that crime promotes the feeling of autonomy, this feeling may reinforce crime while simultaneously weakening external control. Third, the hormonal changes associated with puberty may increase one's susceptibility to the influence of peers, some of whom may be older and more deviant. Insofar as such peers demonstrate their own independence through crime, these peers may provide models of criminal behavior and promote beliefs favorable to crime (see Moffitt, 1993). Fourth, navigating the new social landscape of adolescence amid unstable hierarchies may increase stress, thus heightening the motivation to cope via drug use or other forms of crime. Further, this increase in social stress takes place just as hormonal changes decrease internal control and just as the individual's desire for autonomy weakens external control.

As the teenage years give way to emerging adulthood, most individuals experience decreased motivation for crime and greater constraint against it. Individuals at this age have greater experience navigating the social stressors that accompany adulthood. Further, following Moffitt (1993), individuals at this age are more capable of asserting their autonomy in conventional ways, thus limiting the degree to which criminal behavior yields reinforcing benefits. Peers at this age likewise find themselves able to assert autonomy via legitimate channels. This decreases the degree to which they model crime, as well as the degree to which they promote criminal attitudes. At the same time, constraints against crime significantly increase. The greater autonomy that comes with adulthood is accompanied by a substantially higher stake in conformity. Many individuals at this stage of life have committed to marriage or a long-term relationship that crime could jeopardize, particularly when individuals have children of their own. Likewise, many individuals at this stage of life have accumulated educational and occupational experience that crime could also jeopardize (see also Sampson and Laub, 1993). Simultaneously, the brain reaches a state of maturity during early adulthood that was not present in the earlier phases of the life-course. At this point, the self-regulation centers in the brain are better able to handle the intense emotional feelings that are controlled by the lower-order brain regions.

Summary

Agnew's general theory of crime can be combined in a number of ways with existing biosocial theory and research to better explain interpersonal similarities and differences in crime. We have argued that (1) evolutionary theory provides insight concerning why most individuals are susceptible to common motivations and constraints; (2) genetic and non-genetic biological variables contribute to crime through their influence on motivations and constraints in the major life-domains; (3) these life-domains exert part of their influence on crime via their effects on non-genetic biological variables; (4) genetic and non-genetic biological variables may interact with social variables from the major life-domains to influence motivation and constraint; and (5) biological differences may explain a portion of crime's relationship with gender and age.

Policy implications of the extended theory

Policy-makers and the general public are often leery of biosocial criminology's implications. This reflects a combination of two primary concerns. First, if biology influences crime, some believe that intervention is not possible. Second, if biology influences crime, some believe that the only potential interventions are immoral. Further reflection, however, suggests that these concerns are misplaced (Pinker, 2002).

With respect to the possibility of intervention, it would be incorrect to suggest that an understanding of crime's biosocial underpinnings provides no useful insight about crime-control. Such assertions belie an exceedingly simplistic understanding of modern biosocial criminology. We have argued, for example, that common biological mechanisms may underlie such processes as the exertion of self-control or the social reinforcement of crime across most individuals, even as individuals differ in their levels of such variables. Understanding these biological mechanisms more fully, therefore, can help us lower motivation and increase constraint across individuals and groups. We have argued that biological differences across individuals exert much of their influence via their effects on the life-domains. Even as such differences may not be directly amenable to intervention, the social environments through which they exert their effects are. Insofar as characteristics such as sensation-seeking make parenting more difficult, for example, resources might be invested in providing social support to the parents of children exhibiting high levels of sensation-seeking. We have argued that environmental variables may wield part of their influence on crime through such mechanisms as exposure to toxins such as lead. Knowing only that neighborhood disadvantage is associated with crime, for example, crime control might hinge on ameliorating all aspects of that disadvantage. Knowing that crime in such neighborhoods may result partly from levels of lead in the air, soil, and drinking water, however, one might be able to focus on altering these ambient lead levels even in the absence of sufficient resources for altering entire neighborhoods.

With respect to the morality of intervention, two points are particularly salient. First, the research taking place today is quite different from that of theorists such as Lombroso. Like modern criminology in general, biosocial criminology does not draw a clear qualitative distinction between criminals and non-criminals. Moffitt's developmental taxonomy (1993), for example, makes clear that most individuals engage in some amount of risky or unconventional behavior during adolescence. Some individuals engage in a greater variety or frequency of risky behavior, but modern criminology does not delineate a clear-cut distinction between who is and is not a criminal. Rather, contrary to some biological perspectives of the late nineteenth century, modern criminology suggests that some individuals are *more* criminal while others are *less* criminal. This, in turn, makes modern criminology less prone to conceive of criminals as a fundamentally different type of individual, meriting inhumane treatment.

Second, even to the degree that severe or incarcerated criminals were viewed as qualitatively different from others, there is nothing about the scientific study of crime that implies how such people should be treated. The mere existence of a biological difference between male and female criminal propensities need not be interpreted as justification for male criminality. The mere existence of a genetic influence on self-control or irritability need not be interpreted as justification for higher crime among individuals with certain genetic traits. To suggest that biological influences on crime somehow legitimize such behavior reflects what Barnes (2014) and others (Pinker, 2002) have termed the "naturalistic fallacy." No finding in biosocial criminology inherently implies anything about how people should behave, or about how people should treat a given offender. This is not to say that individuals will never attempt to use biosocial findings to justify morally questionable, or outright dangerous, public policies. Biosocial criminology, however, is not unique in this regard as a form of scientific inquiry.

References

Agnew, R. (2001) 'Building on the foundation of general strain theory: specifying the types of strain most likely to lead to crime and delinquency', *Journal of Research in Crime and Delinquency*, 38: 319–361.

Agnew, R. (2005) *Why Do Criminals Offend? A General Theory of Crime and Delinquency*, Los Angeles, CA: Roxbury.

Agnew, R., Rebellon, C.J., and Thaxton, S. (2000) 'A general strain theory approach to families and crime', in Fox, G.L. and Benson, M.L. (eds), *Families and Crime*, New York: JAI Press, pp. 113–138.

Akers, R.L. (2009) *Social Learning and Social Structure: A General Theory of Crime and Deviance*, New Brunswick, NJ: Transaction.

Akers, R.L., Krohn, M.D., Lanza-Kaduce, L., and Radosevich, M. (1979) 'Social learning and deviant behavior: a specific test of a general theory', *American Sociological Review*, 44: 636–655.

Anderson, E. (2000) *Code of the Street: Decency, Violence, and the Moral Life of the Inner City*. New York: WW Norton & Co.

Armstrong, T. (2011) 'The relationship between low resting heart rate and antisocial behavior: correlation or causation', in Beaver, K. and Walsh, A. (eds), *The Ashgate Research Companion to Biosocial Theories of Crime*, Burlington, VT: Ashgate, pp. 45–68.

Barnes, J.C. (2014) 'The impact of biosocial criminology on public policy: where should we go from here?', in DeLisi, M. and Beaver, K.M. (eds), *Criminological Theory: A Life-Course Approach* (2nd edn), Farnham, UK: Jones and Bartlett.

Barnes, J.C. and Jacobs, B.A. (2013) 'Genetic risk for violent behavior and environmental exposure to disadvantage and violent crime: the case for gene–environment interaction', *Journal of Interpersonal Violence*, 28: 92–120.

Barnes, J.C., Boutwell, B.B., and Beaver, K.M. (2013) 'Contemporary biosocial criminology: a systematic review of the literature, 2000–2012', in Piquero, A.R. (ed.), *Handbook of Criminological Theory*, Hoboken, NJ: Wiley-Blackwell.

Baumeister, R.F., Heatherton, T.F., and Tice, D.M. (1994) *Losing Control*, San Diego, CA: Academic Press.

Baumeister, R.F., Vohs, K.D., and Tice, D.M. (2007) 'The strength model of self-control', *Current Directions in Psychological Science*, 16: 351–355.

Beaver, K.M., Wright, J.P., and DeLisi, M. (2007) 'Self-control as an executive function: reformulating Gottfredson and Hirschi's parental socialization thesis', *Criminal Justice and Behavior*, 34: 1345–1361.

Belsky, J., Jonassaint, C., Pluess, M., Stanton, M., Brummett, B., and Williams, R. (2009) 'Vulnerability genes or plasticity genes?', *Molecular Psychiatry*, 14: 746–754.

Blau, P. and Blau, J. (1982) 'The cost of inequality: metropolitan structure and violent crime', *American Sociological Review*, 47: 114–129.

Bouchard, T.J., Lykken, D.T., McGue, M., Segal, N., and Tellegen, A. (1990) 'Sources of human psychological differences: the Minnesota study of twins reared apart', *Science*, 250: 223–228.

Brennan, P.A., Mednick, B.R., and Mednick, S.A. (1993) 'Parental psychopathology, congenital factors, and violence', in Hodgins, S. (ed.), *Mental Disorder and Crime*, Thousand Oaks, CA: Sage, pp. 244–261.

Brennan, P.A., Grekin, E.R., and Mednick, S.A. (1999) 'Maternal smoking during pregnancy and adult male criminal outcomes', *Archives of General Psychiatry*, 56: 215–219.

Brennan, P.A., Grekin, E.R., and Mednick, S.A. (2003) 'Prenatal and perinatal influences on conduct disorder and serious delinquency', in Lahey, B.B., Moffitt, T.E., and Caspi, A. (eds), *Causes of Conduct Disorder and Juvenile Delinquency*, New York: Guilford, pp. 319–341.

Brunner, H.G., Nelen, M., Breakefield, X.O., Ropers, H.H., and van Oost, B.A. (1993) 'Abnormal behavior associated with a point mutation in the structural gene for monoamine oxidase A', *Science*, 262: 578–580.

Burt, S.A. (2009a) 'Are there meaningful etiological differences within antisocial behavior? Results from a meta-analysis', *Clinical Psychology Review*, 29: 163–178.

Burt, S.A. (2009b) 'Rethinking environmental contributions to child and adolescent psychopathology: a meta-analysis of shared environmental influences', *Psychological Bulletin*, 135: 608–637.

Buss, D. (1994) *The Evolution of Desire: Strategies of Human Mating*, New York: Basic Books.

Buss, D. (2005) *The Handbook of Evolutionary Psychology*, Hoboken, NJ: John Wiley and Sons.

Campbell, A. (2009) 'Gender and crime: an evolutionary perspective', in Walsh, A. and Beaver, K.M. (eds), *Biosocial Criminology: New Directions in Theory and Research*, New York: Routledge.

Caspi, A., Moffitt, T.E., Silva, P.A., Stouthamer-Loeber, M., Krueger, R.F., and Schmutte, P.S. (1994) 'Are some people crime-prone? Replications of the personality–crime relationship across countries, genders, races, and methods', *Criminology*, 32: 163–195.

Caspi, A., McClay, J., Moffitt, T.E., Mill, J., Martin, J., Craig, I.W., Taylor, A., and Poulton, R. (2002) 'Role of genotype in the cycle of violence in maltreated children', *Science*, 297: 851–854.

Cloninger, C.R., Sigvardsson, S., Bohman, M., and von Knorring, A.L. (1982) 'Pre-disposition to petty criminality in Swedish adoptees: II. Cross-fostering analysis of gene–environment interactions', *Archives of General Psychiatry*, 39: 1242–1247.

Coccaro, E.F., Kavoussi, R.J., Cooper, T.B., and Hauger, R.L. (1997) 'Central serotonin activity and aggression', *American Journal of Psychiatry*, 154: 1430–1435.

Collins, R.E. (2011) 'Neurotransmitters: indirect molecular invitations to aggression', in Beaver, K.M. and Walsh, A. (eds), *The Ashgate Research Companion to Biosocial Theories of Crime*, Farnham, UK: Ashgate Publishing.

Crick, N.R. and Dodge, K.A. (1996) 'Social information-processing mechanisms in reactive and proactive aggression', *Child Development*, 67: 993–1002.

Cullen, F.T. (2011) 'Beyond adolescence-limited criminology: choosing our future—The American Society of Criminology 2010 Sutherland Address', *Criminology*, 49: 287–330.

Dijkstra, J.K., Cillessen, A.H., and Borch, C. (2013) 'Popularity and adolescent friendship networks: selection and influence dynamics', *Developmental Psychology*, 49(7): 1242.

Elliott, D., Huizinga, D., and Ageton, S.S. (1985) *Explaining Delinquency and Drug Use*, Beverly Hills, CA: Sage.

Ellis, B.J., Guidice, M.D., Dishion, T J., Figueredo, A.J., Gray, P., Griskevicius, V., Hawley, P.H., Jacobs, W.J., James, J., Volk, A.A., and Wilson, D.S. (2012) 'The evolutionary basis of risky adolescent behavior: implications for science, policy, and practice', *Developmental Psychology*, 48: 598–623.

ENCODE Project Consortium (2011) 'A user's guide to the encyclopedia of DNA elements (ENCODE)', *PLoS Biology*, 9: e1001046.

Farrington, D.P. (1997) 'The relationship between low resting heart-rate and violence', in Raine, A., Brennan, P.A., Farrington, D., and Mednick, S.A. (eds), *Biological Bases of Violence*, New York: Plenum, pp. 89–105.

Ferguson, C.J. (2010) 'Genetic contributions to antisocial personality and behavior: a meta-analytic review from an evolutionary perspective', *Journal of Social Psychology*, 150: 160–180.

Festinger, L. (1954) 'A theory of social comparison processes', *Human Relations*, 7: 117–140.

Giedd, J.N., Blumenthal, J., Jeffries, N.O., Castellanos, F.X., Liu, H., Zijdenbos, A., Paus, T., Evans, A.C., and Rapoport, J.L. (1999) 'Brain development during childhood and adolescence: a longitudinal MRI study', *Nature Neuroscience*, 2: 861–863.

Glueck, S. and Glueck, E.T. (1950) *Unraveling Juvenile Delinquency*, New York: Commonwealth Fund.

Gottfredson, M.R. and Hirschi, T. (1990) *A General Theory of Crime*, Stanford, CA: Stanford University Press.

Guo, G., Roettger, M.E., and Cai, T. (2008) 'The integration of genetic propensities into social-control models of delinquency and violence among male youths', *American Sociological Review*, 73: 543–568.

Hagger, M.S. and Chatzisarantis, N.L.D. (2013) 'The sweet taste of success: the presence of glucose in the oral cavity moderates the depletion of self-control resources', *Personality and Social Psychology Bulletin*, 39: 28–42.

Hirschi, T. (1969) *Causes of Delinquency*, Berkeley, CA: University of California Press.

Kolb, B. and Whishaw, I.Q. (2014) *An Introduction to Brain and Behavior* (4th edn), New York: Worth.

Lombroso, C. (1911) *Crime: Its Causes and Remedies*, translated by H.P. Horton, Boston: Little, Brown.

Mazur, A. (2009) 'Testosterone and violence among young men', in Walsh, A. and Beaver, K.M. (eds), *Biosocial Criminology: New Directions in Theory and Research*, New York: Routledge.

Mazur, A. and Booth, A. (1999) 'The biosociology of testosterone in men', in Franks, D. and Smith, S. (eds), *Mind, Brain, and Society: Toward a Neurosociology of Emotion*, Vol. 5, Stamford, CT: JAI Press, pp. 311–338.

Meaney, M.J. (2010) 'Epigenetics and the biological definition of gene X environment interactions', *Child Development*, 81: 41–79.

Mednick, S.A., Volavka, J., Gabrielli, W.F., and Itil, T.M. (1981) 'EEG as a predictor of antisocial behavior', *Criminology*, 19: 219–229.

Moffitt, T.E. (1993) 'Adolescence-limited and life-course-persistent antisocial behavior – a developmental taxonomy', *Psychological Review*, 100: 674–701.

Moffitt, T.E. (2005) 'The new look of behavioral genetics in developmental psychopathology: gene–environment interplay in antisocial behaviors', *Psychological Bulletin*, 131: 533–554.

Muraven, M., Baumeister, R.F., and Tice, D.M. (1999) 'Longitudinal improvement of self-regulation through practice: building self-control strength through repeated exercise', *Journal of Social Psychology*, 139: 446–457.

Nagin, D. and Paternoster, R. (2000) 'Population heterogeneity and state dependence: state of the evidence and directions for future research', *Journal of Quantitative Criminology*, 16: 117–144.

Ogilvie, J.M., Stewart, A.L., Chan, R.C.K., and Shum, D.H.K. (2011) 'Neuropsychological measures of executive function and antisocial behavior: a meta-analysis', *Criminology*, 49: 1063–1107.

Olds, D.L. (1997) 'Tobacco exposure and impaired development: a review of the evidence', *Mental Retardation and Developmental Disabilities Reviews*, 3: 257–269.

Olds, D.L. (2006) 'The nurse–family partnership: an evidence-based preventive intervention', *Infant Mental Health Journal*, 27: 5–25.

Ortiz, J. and Raine, A. (2004) 'Heart rate level and antisocial behavior in children and adolescents: a meta-analysis', *Journal of the American Academy of Child and Adolescent Psychiatry*, 43: 154–162.

Pinker, S. (2002) *The Blank Slate: The Modern Denial of Human Nature*, New York: Penguin Books.

Pinker, S. (2011) *The Better Angels of Our Nature: Why Violence has Declined*, New York: Viking.

Piquero, A. and Tibbetts, S. (1999) 'The impact of pre/perinatal disturbances and disadvantaged familial environment in predicting criminal offending', *Studies on Crime and Crime Prevention*, 8: 52–70.

Plomin, R., DeFries, J.C., Knopik, V.S., and Neiderhiser, J.M. (2013) *Behavioral Genetics* (6th edn), New York: Worth.

Rafter, N. (2008) 'Criminology's darkest hour: biocriminology in Nazi Germany', *Australian and New Zealand Journal of Criminology*, 41: 287–306.

Raine, A. (1993) *The Psychopathology of Crime: Criminal Behavior as a Clinical Disorder*, San Diego, CA: Academic Press.

Raine, A. (2002a) 'The biological bases of crime', in Wilson, J.Q. and Petersilia, J. (eds), *Crime*, Oakland, CA: ICS Press, pp. 43–74.

Raine, A. (2002b) 'Biosocial studies of antisocial and violent behavior in children and adults: a review', *Journal of Abnormal Child Psychology*, 30: 311–326.

Raine, A. (2008) 'From genes to brain to antisocial behavior', *Current Directions in Psychological Science*, 17: 323–328.

Raine, A., Brennan, P.A., and Mednick, S.A. (1994) 'Birth complications combined with early maternal rejection at age 1 year predispose to violent crime at age 18 years', *Archives of General Psychiatry*, 51: 984–988.

Raine, A., Venables, P.H., and Williams, M. (1995) 'High autonomic arousal and electrodermal orienting at age 15 years as protective factors against criminal behavior at age 29 years', *American Journal of Psychiatry*, 152: 1595–1600.

Raine, A., Moffitt, T.E., Caspi, A., Loeber, R., Stouthamer-Loeber, M., and Lynam, D. (2005) 'Neurocognitive impairments in boys on the life-course-persistent antisocial path', *Journal of Abnormal Psychology*, 114: 38–49.

Rebellon, C.J., Straus, M.A., and Medeiros, R. (2008) 'Self-control in global perspective: an empirical assessment of Gottfredson and Hirschi's General Theory within and across 32 national settings', *European Journal of Criminology*, 5: 331–362.

Rebellon, C.J., Manasse, M.E., Van Gundy, K.T., and Cohn, E.S. (2012) 'Perceived injustice and delinquency: a test of general strain theory', *Journal of Criminal Justice*, 40(3): 230–237.

Rhee, S.H. and Waldman, I.D. (2002) 'Genetic and environmental influences on antisocial behavior: a meta-analysis of twin and adoption studies', *Psychological Bulletin*, 128: 490–529.

Robins, L.N. (1978) 'Sturdy childhood predictors of adult antisocial behavior: replications from longitudinal studies', *Psychological Medicine*, 8: 611–622.

Sampson, R.J. and Laub, J.H. (1993) *Crime in the Making: Pathways and Turning Points Through Life*, Cambridge, MA: Harvard University Press.

Shaw, C. and McKay, H. (1942) *Juvenile Delinquency and Urban Areas*, Chicago, IL: University of Chicago Press.

Shoenthaler, S.J. (1991) *Improve Your Child's IQ and Behavior*, London: BBC Books.

Simons, R.L., Lei, M.K., Beach, S.R.H., Brody, G.H., Philibert, R.A., and Gibbons, F.X. (2011) 'Social environmental variation, plasticity genes, and aggression: evidence for the differential susceptibility hypothesis', *American Sociological Review*, 76: 883–912.

Singh, J.P., Larson, M.G., O'Donnell, C.J., Tsuji, H., Evans, J.C., and Levy, D. (1999) 'Heritability of heart rate variability: the Framingham Heart Study', *Circulation*, 99: 2251–2254.

Steinberg, L. (2007) 'Risk taking in adolescence: new perspectives from brain and behavioral science', *Current Directions in Psychological Science*, 16: 55–59.

Tangney, J.P., Stuewig, J., and Mashek, D.J. (2007) 'Moral emotions and moral behavior', *Annual Review of Psychology*, 58: 345–372.

Tittle, C.R. and Meier, R.F. (1990) 'Specifying the SES/delinquency relationship', *Criminology*, 28: 271–299.
Trivers, R.L. (1972) 'Parental investment and sexual selection', in Campbell, B. (ed.), *Sexual Selection and the Descent of Man*, Chicago, IL: Aldine Publishing Company.
Turkheimer, E. (2000) 'Three laws of behavior genetics and what they mean', *Current Directions in Psychological Science*, 9: 160–164.
Udry, J.R. (1995) 'Sociology and biology: what biology do sociologists need to know?', *Social Forces*, 73: 1267–1278.
Walby, K. and Carrier, N. (2010) 'The rise of biocriminology: capturing observable bodily economies of "criminal man"', *Criminology and Criminal Justice*, 10: 261–285.
Walsh, A. (2000) 'Behavior genetics and anomie/strain theory', *Criminology*, 38: 1075–1108.
Wilson, J.Q. and Herrnstein, R.J. (1985) *Crime and Human Nature: The Definitive Study of the Causes of Crime*, New York: Free Press.
Wright, J.P. and Beaver, K.M. (2005) 'Do parents matter in creating self-control in their children? A genetically informed test of Gottfredson and Hirschi's theory of low self control', *Criminology*, 43: 1169–1202.
Wright, J.P., Tibbetts, S.G., and Daigle, L.E. (2008) *Criminals in the Making: Criminality across the Life Course*, Thousand Oaks, CA: Sage.

2

Biosocial criminology as a paradigm shift

Anna S. Rudo-Hutt, Jill Portnoy, Frances R. Chen, and Adrian Raine

Introduction

Early criminological research was dominated by the biological tradition as well as multifactor research (reviewed by Hirschi and Rudisill, 1976). Many scholars have named Cesare Lombroso, an Italian physician active during the late 1800s and early 1900s, as the founder of criminology (Gibson and Rafter, 2006). Lombroso is best known for his theory of the "born criminal," the idea that crime is committed by people with "atavisms," or traits found in earlier stages of human evolution (Gibson and Rafter, 2006). Using measurement of body parts (such as the skull) and observation of physical features, Lombroso created a system by which criminals could be classified into different types. The aforementioned born criminal could not help but to be criminal based on his or her biology, in contrast to the occasional criminal, who had only a slight predisposition to crime and was pushed into crime by social factors (Lombroso, 1889/2006). Thus Lombroso did incorporate biosocial principles into his work.

By applying a scientific, positivist framework to the study of criminals, Lombroso sought to make criminology a scientific, rather than philosophical, endeavor (Lombroso, 1876/2006). Despite this scientific focus, however, Lombroso was influenced by the predominant prejudices of his time, and the sexism and racism inherent in his work has been rightly criticized. In describing his studies of the skulls of 66 criminals, for instance, Lombroso stated that the features present in the skulls "recall the black American and Mongol races and, above all, prehistoric man much more than the white races" (Lombroso, 1876/2006, p. 49). By equating the features of the born criminal with nonwhite race, Lombroso's work suggested that people with those features, i.e. nonwhites, were a threat to society.

As noted above, Lombroso did allow in his theory for the possibility that some criminals were pushed into crime by social factors, such as poverty or lack of education. Most of his emphasis, however, was on biological determinants of crime, and his followers focused almost exclusively on this part of his research. One follower, Mariano Patrizi, put forward the idea that crime could be attributed to a single cause – having unbalanced and impulsive emotions – which was entirely determined by biology (Gibson, 2002). This emphasis on biological causes of crime made Lombroso's theories attractive to the burgeoning eugenics movement.

In the late 1800s, Sir Francis Galton began using the word "eugenics" to describe techniques intended to improve the genetic quality of the human race (Garver and Garver, 1991). By the early 1900s, the eugenics movement had gained supporters in intellectual circles, including among positivist criminologists (Gibson, 2002). Supporters of eugenics believed that various human afflictions, from criminality to low IQ, were purely inherited traits that could be eliminated through the use of methods that prevented carriers of these genes from procreating (Noack and Fangerau, 2007). Scholars following in Lombroso's footsteps supported these ideas, arguing that eugenics could be used to prevent "degeneration," the reversion to earlier stages of evolution which was thought to cause criminal behavior (Gibson, 2002). In fact, Benigno Di Tullio, a prominent Italian criminologist, called for the Italian government to forbid the marriage of certain groups of people, such as psychopaths, alcoholics, and people with syphilis or tuberculosis, on the grounds that these conditions could cause a descent into criminality.

Eugenics policies of forced sterilization were adopted in several countries, including the United States, where involuntary sterilization was practiced on the mentally ill, recidivistic criminals, rapists, people with low IQ, substance abusers, and other similarly afflicted inmates of state institutions (Garver and Garver, 1991). Eventually Germany began using euthanasia to eliminate unwanted genes and ethnic groups, such as children who were deformed or intellectually disabled as well as people who were Jewish (Garver and Garver, 1991; Noack and Fangerau, 2007). Upon revelation of the atrocities of the Holocaust, eugenics movements in other countries were discontinued (Joseph, 2005). Although Lombroso had argued in his work for humane treatment of criminals, due to their lack of responsibility for their biological predisposition to crime, it was clearly linked to the justifications for eugenics. As a result, the field of criminology turned sharply away from biological research and has been wary of it ever since.

The sociological shift in criminology and the re-emergence of the biosocial paradigm

Around the time criminology was moving away from biological research, the field began to move toward a more sociological paradigm. Beginning in the late 1920s, in what Laub (1983, 2004) has identified as a major turning point in the trajectory of criminological research, the work of Clifford Shaw and Henry McKay shifted the focus of criminological inquiry from the individual characteristics of offenders to the larger social processes of urbanization, immigration, and industrialization. During the 1930s, Edwin H. Sutherland and his theory of differential association, which asserted that criminal behavior is learned just like other non-criminal behaviors (Sutherland and Cressey, 1960), firmly established sociology as the dominant criminological paradigm (Laub, 2004). Importantly, Sutherland also encouraged the marginalization of criminological research that utilized individual-level or multifactor approaches (Laub and Sampson, 1991), arguably resulting in decades of research on social processes, rather than the individual characteristics of offenders. In 1969, Travis Hirschi's *Causes of Delinquency* combined multiple paradigms and was neither strictly sociological nor psychological. This departed from the predominating sociological tradition, and differentiated itself from what at the time was an emerging belief that the "criminal" is a socially constructed entity whose cause must be understood by examining society, rather than the offender (Hirschi and Rudisill, 1976). In recent years, despite the legacy of eugenics, and despite the fact that sociological research continues to have a strong presence in the field, there is increased interest in examining crime from a multidisciplinary perspective, including a biological perspective. In fact, Francis Cullen, past president of the American Society of Criminology, remarked in 2010 that as the field moves forward, criminologists "can no longer pretend that biology is not intimately implicated in human behavior and thus in criminal behavior" (Cullen, 2011, p. 311).

The 1964 publication of Hans Eysenck's book *Crime and Personality* touched off the return of biological research on crime (Rafter, 2006). In *Crime and Personality*, Eysenck, a psychologist, proposed that children who are poor conditioners (i.e. do not learn well from punishment) are more likely to develop into criminals (Rafter, 2006). Notably, Eysenck also proposed an "antisocialization process," whereby children who are poor conditioners but who grow up in criminogenic environments are less likely to become criminals than good conditioners, because they would be less likely to learn criminal behavior from the environment (Rafter, 2006). These ideas became influential in the thinking of biosocial researchers, and they received early support in a study by Raine and Venables (1981), which found that antisocial behavior was associated with poor conditionability in high-class children but with good conditionability in low-class children. Research in this vein has continued, and the echoes of Eysenck's theories can still be seen in more recent biosocial research. For instance, Blair has argued that psychopathy results from disruptions in stimulus-response and stimulus-reinforcement learning (Blair, 2004, 2008; Blair *et al.*, 2006). Although it has taken nearly five decades since the original publication of Eysenck's book, it seems that biosocial research is finally becoming accepted as a worthwhile paradigm within criminology.

Biosocial criminology in the past 10 years

The gradual (re)acceptance of biosocial research has been helped along by the neuroscience revolution. In the past two decades, there have been a number of advances in brain imaging and genetics that have allowed a greater understanding of the biological processes associated with specific illnesses, psychological disorders, and mental states. During and after the "decade of the brain" in the 1990s, not only were new techniques developed and refined (e.g. functional magnetic resonance imaging [fMRI], genome-wide association studies), but also a significant increase in interest in neuroscience led to multidisciplinary collaborations and the application of these and other biological techniques in other fields (Jones and Mendell, 1999). It is clear from the proliferation of these techniques in criminology research that the biosocial paradigm is making a comeback (Rocque *et al.*, 2012).

Nowhere is this comeback clearer than in the articles published in criminology's top journals. A survey of five top journals (*Criminology*, *Crime and Justice*, *Criminal Justice and Behavior*, *Journal of Research in Crime and Delinquency*, and the *British Journal of Criminology*) shows that the number of publications mentioning the word "biosocial" has increased steadily since the 1970s, and the 2010s are already on track to far outstrip previous decades in biosocial publications (see Figure 2.1). Biosocial research has also had a growing presence over the past decade at the annual meeting of the American Society of Criminology (ASC). In 2003, there was only one session pertaining to biological factors at the ASC annual meeting, whereas there are 24 planned for the 2013 meeting. Criminology appears to be in the midst of a shift back toward more biologically based explanations of crime.

The biosocial paradigm has led to some intriguing findings in criminological research. Because Parts II–IV of this book focus on many specific findings in biosocial research on antisocial behavior, we will touch only briefly here on some of the latest research on the interactions between psychophysiology and psychosocial factors in predicting antisocial behavior. Several studies have found that the relationship between psychophysiological measures and antisocial

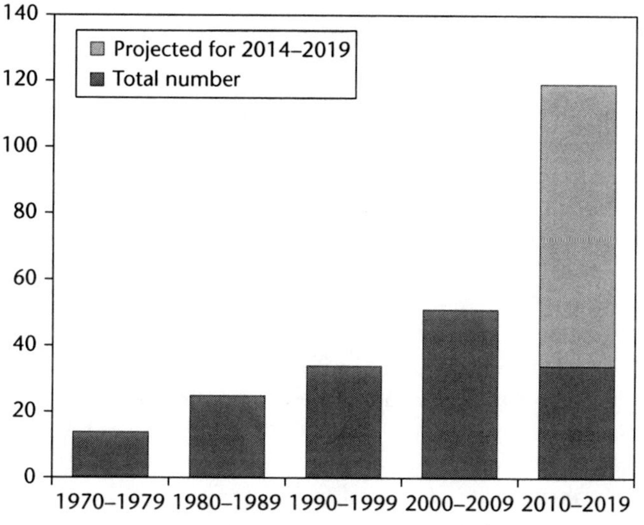

Figure 2.1 Number of biosocial publications in five top criminology journals, by decade

Note: The graph shows the total number of publications, by decade, mentioning the word "biosocial" from the following journals: *Criminology, Crime & Justice, Criminal Justice and Behavior, Journal of Research in Crime and Delinquency,* and the *British Journal of Criminology*.

behavior varies based on the home or neighborhood background of the participants (Raine, 2002). For example, Sijtsema and colleagues (2013) found that, in families with low cohesion, heightened physiological reactivity (i.e. heart rate reactivity) protected against aggressive behavior in adolescents. Likewise, Gordis and colleagues (2010) found that high respiratory sinus arrhythmia (RSA, a measure of parasympathetic regulation of heart activity) attenuated the relationship between childhood abuse and teen aggression, suggesting that high RSA is a protective factor for aggression. In another sample of children, experiencing community violence was associated with heightened proactive aggression only in those with low resting heart rate, and witnessing community violence was correlated with reactive aggression only in those with high heart rate variability (Scarpa *et al.*, 2008). Recent studies are also investigating the interaction of multiple biological systems in predicting antisocial behavior. For instance, both cortisol and alpha-amylase reactivity have been found to be negatively correlated with disruptive behavior in delinquent male teens, and the combination of the two measures accounted for more of the variance in disruptive behavior than either measure alone (de Vries-Bouw *et al.*, 2012). Additionally, in the study of child abuse and aggression mentioned above, skin conductance (SC) reactivity was found to moderate the effect of RSA on aggression, such that aggression was increased in abused girls who had RSA levels and SC reactivity that were either both high or both low (Gordis *et al.*, 2010). These studies exemplify the strength of the biosocial paradigm: it provides a way to understand why some risk factors for criminal behavior, such as abuse or having a criminal parent, do not act equally on all people. Biosocial research allows us to look beyond the environment alone, or biology alone, to predict which people, under which circumstances, will commit crime (Wright and Boisvert, 2009).

Future directions in biosocial criminology

The shift toward biosocial research in criminology has a number of possible implications for the field and for our communities more broadly, ranging from differences in the way crime is conceptualized, to the types of policy recommendations that are made in terms of treatment of criminals, to what kinds of prevention efforts we pursue. As noted by Wright and Boisvert (2009), biosocial research has the potential to add substantially to the empirical support for criminological theories already in use, by enabling us to understand the biological processes underlying the constructs proposed to be important to criminal offending (e.g. lack of self-control, reactivity to stress). These authors also note that the lens of biology allows criminologists to critically examine and reject hypotheses that are implausible based on our current understanding of human biology (Wright and Boisvert, 2009). Even more promisingly, biosocial research lends itself naturally to suggestions of how to prevent and respond to crime.

When discussing crime prevention, it is important to be aware of potential ethical concerns in singling out particular children or groups of people for treatment (Rocque *et al.*, 2012). In contrast to trying to alter specific biological factors, however, prevention efforts in use today tend to focus on improving the environment in ways that support healthy human development (Wright and Boisvert, 2009; Rocque *et al.*, 2012). These environmental interventions are much more palatable to wary criminologists – as noted by Raine (2013, p. 369), "After all, what's so heinous about investing resources in better pre- and postnatal nutrition and care for the underserved, better elementary school nutrition, reducing lead exposure, implementing education on parenting skills, and identifying children with serious behavior problems for benign interventions?" There is ample evidence to support the suggestion that interventions should focus on childhood development. Biosocial research has shown that a disruption in healthy development, whether due to *in utero* exposure to drugs, early malnutrition, obstetric complications, or a history of untreated head injuries, predisposes to antisocial behavior (Olson *et al.*, 1997; Brennan *et al.*, 1999; Fast *et al.*, 1999; Neugebauer *et al.*, 1999; Streissguth *et al.*, 1999; Hodgins *et al.*, 2001; León-Carrión *et al.*, 2003). Thus these and other factors in healthy development present prime targets for biosocial interventions.

In fact, interventions to promote healthy development have already borne fruit – one randomized control trial which incorporated pre- and postnatal nurse home visits successfully reduced delinquency at age 15 (Olds *et al.*, 1998). Despite receiving an average of only nine nurse visits during the perinatal period, these effects extended to at least age 19 for girls in the study, who continued to have fewer lifetime arrests and convictions than girls in the comparison group (Eckenrode *et al.*, 2010). In addition to improving environmental conditions, such interventions may actually work to alter the biological risk factors for crime within the treated children. For instance, a preschool enrichment program incorporating activities to promote cognitive and social development, enhance nutrition, and encourage parental involvement in school was found to increase autonomic and central nervous system arousal, both of which are protective factors for crime (Raine *et al.*, 2001). Not only do such interventions prevent crime, but also, through their focus on overall healthy development, they are likely to improve other areas of functioning for children and families participating in the interventions. For example, in another test of the intervention described by Olds and colleagues (1998), compared with children whose mothers did not receive home visits, children whose mothers received nurse home visits during the perinatal period were less likely to use substances or have internalizing disorders at age 12 (Kitzman *et al.*, 2010). Interventions that promote healthy development and prevent crime also appear to reduce the likelihood of other adverse health behaviors and psychological conditions. Therefore the use of these types of interventions seems both ethical and highly beneficial to individual children and to society as a whole.

Through biosocial research, we may be able to move beyond interventions that simply improve healthy development to those that target criminal behavior directly. In the future, it may be possible to use biomarkers to identify those who are at most risk of becoming criminals and who are, therefore, most in need of intervention (Beauchaine *et al.*, 2008). Once those at risk have been identified, we may further be able to apply interventions that work directly on the biological factors that predispose to crime. Just as molecular genetics research has led to cancer drugs that can reverse genetic mutations, we may one day be able to suppress specific genes that are involved in antisocial behavior, or increase expression of genes associated with prosocial behavior, with specially tailored medications (Raine, 2013). Interventions that operate on genes or that identify particular individuals for treatment may raise concerns about a "neo-eugenics" movement, and it is important to be mindful in the way such interventions might be implemented. For people who have already been convicted of a crime, however, genetic therapy may be greatly preferred to spending time in prison (Raine, 2013). Furthermore, although it may be preferable to remove the negative social factors that may be interacting with those genes to produce antisocial behavior, it is not always possible to do so, at least not within a reasonable time frame (Raine, 2013). Societal change to reduce the presence of criminogenic environments, such as neighborhoods plagued by crime and violence, may be the overall goal, but if it is easier and faster to alter behavior on an individual level, that is also a goal worthy of our attention.

Emphasizing biological influence should not in any way suggest abandoning the approach of exploring the influence of social factors. Instead, it provides us with a more comprehensive perspective, one that can draw on the richness of both sociological and biological research. A biological approach sheds light on individual susceptibility to social factors, and on plasticity to adapt to social factors as well. For instance, biological factors may serve as moderators of social influence: exploring protective biological factors that buffer individuals from adverse environment may help clarify mixed findings in studies focused solely on social factors. Current biosocial studies in the field in general have focused on social factors at the micro level (i.e. individual and family) and have revealed some intriguing findings (e.g. Caspi *et al.*, 2002; Murray-Close and Rellini, 2012). Future biosocial studies can integrate and build on well explored criminological theories that traditionally have an emphasis on social factors at the macro level, and inherit the legacy of the statistical sophistication that quantitative criminology cherishes. For instance, an individual's biological susceptibility may interact with neighborhood adversity to contribute to antisocial behavior or crime. The social element of this hypothesis could be adapted from research on the ecology of crime; more specifically, from social disorganization theory, which sets out to explain the relation between social structure of a neighborhood and crime through the underlying processes, referred to as social disorganization (Shaw *et al.*, 1929; Sampson *et al.*, 1997).

By integrating individual biological susceptibility, we can address how the distal environment (the neighborhood), as well as the immediate environment (the family), may exert their diverse effects on individuals with different biological wiring. This hypothesis could be tested with cross-domain interactions in a nested model (individuals nested in neighborhoods) that addresses non-independency of observations within the same neighborhood. Another example for such attempts may be to bring psychobiological stress responses (hypothalamus–pituitary–adrenal axis activity and autonomic nervous system arousal) into the general strain theory of crime framework (Agnew, 1992). Similarly, executive functioning of individuals may be incorporated with learning theory and differential association theory of crime (Sutherland, 1947) to refine our understanding of white-collar crime. Furthermore, traditional qualitative criminology research can fuel biosocial research by providing insights and hypotheses about where to next apply the tools offered by biosocial methods (DeLisi and Piquero, 2011). In sum, future biosocial research may move towards a broader scope and perfect its theoretical grounds by taking advantage of the social factors that have been well explored in traditional criminology.

Implications of the biosocial paradigm for the future of criminology

As the biosocial paradigm continues to gain ground in criminology, it is important to be mindful of why criminology turned away from biological research for so long, as well as the reservations that remain. Cole and Campbell (2013, p. 155) note that "Contemporary biocriminologists who have read their history should understand that, like Lombroso himself, if they choose to produce knowledge, they should expect to have very little control over the uses and resulting policies that rhetorically invoke that knowledge as justification." Although the eugenics movement is long dead and gone, its shadow looms large in our history. Regardless of how good our intentions, any research, especially biological research, runs the risk of being interpreted in a way that promotes injustice. For instance, when *The Bell Curve* (Herrnstein and Murray, 1994) was published in the mid-1990s, it received favorable reviews and was discussed on many prominent news programs (Naureckas, 1995). The book argues that intelligence is primarily genetically determined, and when discussing racial disparities in IQ, the authors conclude that genetic factors account for these differences (Herrnstein and Murray, 1994), arguments that eventually led to heated debates and accusations by some that the authors were practicing "scientific racism" (Naureckas, 1995). Though the American Psychological Association later established a task force to address the points raised in these debates, and though this task force reported that the cause of racial differences in IQ was unknown (Neisser *et al.*, 1996), it is unlikely that most of the nonacademic public read this report. It is very likely that misinformation or misunderstandings generated by *The Bell Curve* about the heritability of and racial differences in intelligence have not been fully eradicated.

The incorporation of social factors in biosocial research does provide a good buffer against prejudicial interpretations of research findings, and forestalls a possible resurgence of eugenics. Due to the boom in studies on gene–environment interactions, it seems that many researchers are now emphasizing the fact that genes are not as immutable as they once appeared. Ishikawa and Raine (2002), for example, take pains to explain fully that heritability research looks for a predisposition, not predestination, to crime, and that environmental factors are assumed to have a major role in determining whether genes that confer risk for criminal behavior are expressed. It is important to keep the possibility of misinterpretation in mind when preparing the results of biosocial research for publication or distribution to the public.

In addition to concerns about misinterpretation and misrepresentation of biosocial research, there are further worries about what biosocial research means in terms of free will and legal responsibility, even as there had been in Lombroso's day. If biological and social factors can explain why someone commits a crime, does that mean there is no free will? That criminals are simply at the mercy of their biology and the environments they experience? Some researchers have, in fact, argued that free will does not exist. Greene and Cohen (2004, p. 1784), for instance, describe free will as an "illusion generated by our cognitive architecture," causing some scholars to worry that neuroscientists will take the place of judges in assigning blame and meting out punishment (see e.g. Musumeci, 2013). Not all researchers, however, discredit free will – Raine (2013), for example, argues that there is a continuum of free will, and that some people have more freedom of choice in their actions than others, depending on the person's biosocial makeup. In this view, free will is more akin to self-control, reminiscent of Gottfredson and Hirschi's (1990) general theory of crime. Both views, those of Greene and Cohen (2004) and of Raine (2013), have implications for the law.

Although they acknowledge that their hard deterministic stance may lead to the conclusion that those who commit crime should not be punished, as a lack of free will implies a lack of responsibility, Greene and Cohen (2004) argue that crime should continue to be punished, just that the justification for the punishment must change. That is, our current legal system is a retributive system, in which those who are responsible for wrongdoing are punished because they deserve it. In light of their contention that free will is a fiction, Greene and Cohen (2004) advocate for a change in how punishment is conceptualized, proposing that the legal system instead take a consequentialist stance wherein punishment is predicated on the idea that it will deter future crime. In contrast, Raine (2013) argues that crime should be treated as a clinical disorder, one that society takes steps to prevent and for which it offers humane treatment when it does occur. Notably, the relevance of the philosophical notion of free will to the law has been questioned and strongly rejected by Morse (2008). In stating that free will is a "non-problem" in the law, Morse argues that the law has a "folk psychological" view which addresses responsibility in terms of "consciousness, the formation of mental states such as intention and knowledge, the capacity for rationality, and compulsion" (p. 4). Thus, in Morse's (2008) view, the discoveries of neuroscience do nothing to mitigate the responsibility of most criminals. It is likely that the issues of free will and legal responsibility will continue to be a topic of debate amongst criminology and legal scholars.

Conclusion

In sum, criminology is in the midst of a biosocial revival, one that promises to bring greater theoretical clarity, new interventions, exciting innovations, and more interdisciplinary collaboration to the field. Within the past decade, biosocial research has emerged as a model for capturing the disparate pieces of criminology's research legacy, bringing together the biological research that had been anathema for so long with the sociological research that has dominated criminology since the mid-twentieth century. Although concerns remain about the possible implications of biosocial criminology, as long as we bear in mind the hard lessons learned from Lombroso, the future of the biosocial paradigm looks bright.

References

Agnew, R. (1992) 'Foundation for a general strain theory of crime and delinquency', *Criminology*, 30(1): 47–87.

Beauchaine, T.P., Neuhaus, E., Brenner, S.L., and Gatzke-Kopp, L. (2008) 'Ten good reasons to consider biological processes in prevention and intervention research', *Development and Psychopathology*, 20(3): 745–774.

Blair, R. (2004) 'The roles of orbital frontal cortex in the modulation of antisocial behavior', *Brain and Cognition*, 55(1): 198–208.

Blair, R.J.R. (2008) 'The cognitive neuroscience of psychopathy and implications for judgments of responsibility', *Neuroethics*, 1(3): 149–157.

Blair, R.J.R., Peschardt, K.S., Budhani, S., Mitchell, D.G.V., and Pine, D.S. (2006) 'The development of psychopathy', *Journal of Child Psychology and Psychiatry*, 47(3/4): 262–276.

Brennan, P.A., Grekin, E.R., and Mednick, S.A. (1999) 'Maternal smoking during pregnancy and adult male criminal outcomes', *Archives of General Psychiatry*, 56: 215–219.

Caspi, A., McClay, J., Moffitt, T.E., Mill, J., Martin, J., Craig, I.W., *et al.* (2002) 'Role of genotype in the cycle of violence in maltreated children', *Science*, 297(5582): 851–854.

Cole, S.A. and Campbell, M.C. (2013) 'From subhumans to superhumans: criminals in the evolutionary hierarchy, or what became of Lombroso's atavistic criminals?', in P. Knepper and P.J. Ystehede (eds), *The Cesare Lombroso Handbook*, New York: Routledge, pp. 147–170.

Cullen, F.T. (2011) 'Beyond adolescent-limited criminology: choosing our future—The American Society of Criminology 2010 Sutherland Address', *Criminology*, 49: 287–330.

DeLisi, M. and Piquero, A.R. (2011) 'New frontiers in criminal careers research, 2000–2011: a state-of-the-art review', *Journal of Criminal Justice*, 39(4): 289–301.

Eckenrode, J., Campa, M., Luckey, D.W., Henderson, C.R., Cole, R., Kitzman, H., *et al.* (2010) 'Long-term effects of prenatal and infancy nurse home visitation on the life course of youths: 19-year follow-up of a randomized trial', *Archives of Pediatrics and Adolescent Medicine*, 164(1): 9–15.

Eysenck, H.J. (1964) *Crime and Personality*, London: Routledge & Kegan Paul.

Fast, D.K., Conry, J., and Loock, C.A. (1999) 'Identifying fetal alcohol syndrome among youth in the criminal justice system', *Journal of Developmental and Behavioral Pediatrics*, 20: 370–372.

Garver, K.L. and Garver, B. (1991) 'Eugenics: past, present, and the future', *American Journal of Human Genetics*, 49: 1109–1118.

Gibson, M. (2002) *Born to Crime: Cesare Lombroso and the Origins of Biological Criminology*, Westport, CT: Praeger.

Gibson, M. and Rafter, N.H. (2006) 'Editor's introduction', in M. Gibson and N.H. Rafter (eds), *Criminal Man*, Durham, NC: Duke University Press, pp. 1–36.

Gordis, E.B., Feres, N., Olezeski, C.L., Rabkin, A.N., and Trickett, P.K. (2010) 'Skin conductance reactivity and respiratory sinus arrhythmia among maltreated and comparison youth: relations with aggressive behavior', *Journal of Pediatric Psychology*, 35(5): 547–558.

Gottfredson, M.R. and Hirschi, T. (1990) *A General Theory of Crime*, Stanford, CA: Stanford University Press.

Greene, J. and Cohen, J. (2004) 'For the law, neuroscience changes nothing and everything', *Philosophical Transactions of the Royal Society of London. Series B: Biological Sciences*, 359(1451): 1775–1785.

Herrnstein, R.J. and Murray, C. (1994) *The Bell Curve: Intelligence and Class Structure in American Life*, New York: Free Press.

Hirschi, T. (1969) *Causes of Delinquency*, New Brunswick, NJ: Transaction.

Hirschi, T. and Rudisill, D. (1976) 'The great American search: causes of crime 1876–1976', *Annals of the American Academy of Political and Social Science*, 432: 14–22.

Hodgins, S., Kratzer, L., and McNeil, T.F. (2001) 'Obstetric complications, parenting, and risk of criminal behavior', *Archives of General Psychiatry*, 58(8): 746–754.

Ishikawa, S.S. and Raine, A. (2002) 'Behavioral genetics and crime', in J. Glicksohn (ed.), *The Neurobiology of Criminal Behavior*, Dordrecht, Netherlands: Kluwer, pp. 81–110.

Jones, E.G. and Mendell, L.M. (1999) 'Assessing the decade of the brain', *Science*, 284(5415): 739.

Joseph, J. (2005) 'The 1942 "euthanasia" debate in the *American Journal of Psychiatry*', *History of Psychiatry*, 16: 171–179.

Kitzman, H.J., Olds, D.L., Cole, R.E., Hanks, C.A., Anson, E.A., Arcoleo, K.J., *et al.* (2010) 'Enduring effects of prenatal and infancy home visiting by nurses on children: follow-up of a randomized trial among children at age 12 years', *Archives of Pediatrics and Adolescent Medicine*, 164(5): 412–418.

Laub, J.H. (1983) *Criminology in the Making: An Oral History*, Boston, MA: Northeastern University Press.

Laub, J.H. (2004) 'The life course of criminology in the United States: The American Society of Criminology 2003 Presidential Address', *Criminology*, 42: 1–26.

Laub, J.H. and Sampson, R.J. (1991) 'The Sutherland–Glueck debate: on the sociology of criminological knowledge', *American Journal of Sociology*, 96: 1402–1440.

León-Carrión, J. and Ramos, F.J.C. (2003) 'Blows to the head during development can predispose to violent criminal behaviour: rehabilitation of consequences of head injury is a measure for crime prevention', *Brain Injury*, 17(3): 207–216.

Lombroso, C. (1876/2006) 'Edition 1', in M. Gibson and N.H. Rafter (eds), *Criminal Man*, Durham, NC: Duke University Press, pp. 39–96.

Lombroso, C. (1889/2006) 'Edition 4', in M. Gibson and N.H. Rafter (eds), *Criminal Man*, Durham, NC: Duke University Press, pp. 222–298.

Morse, S.J. (2008) 'Determinism and the death of folk psychology: two challenges to responsibility from neuroscience', *Minnesota Journal of Law, Science, and Technology*, 9(1): 1–36.

Murray-Close, D. and Rellini, A.H. (2012) 'Cardiovascular reactivity and proactive and reactive relational aggression among women with and without a history of sexual abuse', *Biological Psychology*, 89(1): 54–62.

Musumeci, E. (2013) 'New natural born killers? The legacy of Lombroso in neuroscience and law', in P. Knepper and P.J. Ystehede (eds), *The Cesare Lombroso Handbook*, New York: Routledge, pp. 131–146.

Naureckas, J. (January/February 1995) 'Racism resurgent: how media let *The Bell Curve*'s pseudo-science define the agenda on race', *Extra!*. Retrieved December 14, 2008, from www.fair.org/index.php?page=1271

Neisser, U., Boodoo, G., Bouchard Jr., T.J., Wade, A., Brody, N., Ceci, S.J., *et al.* (1996) 'Intelligence: knowns and unknowns', *American Psychologist*, 51(2): 77–101.

Neugebauer, R., Hoek, H.W., and Susser, E. (1999) 'Prenatal exposure to wartime famine and development of antisocial personality disorder in early adulthood', *JAMA*, 282(5). 455–462.

Noack, T. and Fangerau, H. (2007) 'Eugenics, euthanasia, and aftermath', *International Journal of Mental Health*, 36: 112–124.

Olds, D., Henderson, C.R., Cole, R., Eckenrode, J., Kitzman, H., Luckey, D., *et al.* (1998) 'Long-term effects of nurse home visitation on children's criminal and antisocial behavior: 15-year follow-up of a randomized controlled trial', *JAMA*, 280(14): 1238–1244.

Olson, H.C., Streissguth, A.P., Sampson, P.D., Barr, H.M., Bookstein, F.L., and Thiede, K. (1997) 'Association of prenatal alcohol exposure with behavioral and learning problems in early adolescence', *Journal of the American Academy of Child and Adolescent Psychiatry*, 36: 1187–1194.

Rafter, N.H. (2006) 'H.J. Eysenck in Fagin's kitchen: the return to biological theory in 20th-century criminology', *History of the Human Sciences*, 19(4): 37–56.

Raine, A. (2002) 'Biosocial studies of antisocial and violent behavior in children and adults: a review', *Journal of Abnormal Child Psychology*, 30(4): 311–326.

Raine, A. (2013) *The Anatomy of Violence: The Biological Roots of Crime*, New York: Random House.

Raine, A. and Venables, P.H. (1981) 'Classical conditioning and socialization – a biosocial interaction', *Personality and Individual Differences*, 2: 273–83.

Raine, A., Venables, P.H., Dalais, C., Mellingen, K., Reynolds, C., and Mednick, S.A. (2001) 'Early educational and health enrichment at age 3–5 years is associated with increased autonomic and central nervous system arousal and orienting at age 11 years: evidence from the Mauritius Child Health Project', *Psychophysiology*, 38: 254–266.

Rocque, M., Welsh, B.C., and Raine, A. (2012) 'Biosocial criminology and modern crime prevention', *Journal of Criminal Justice*, 40(4): 306–312.

Sampson, R.J., Raudenbush, S.W., and Earls, F. (1997) 'Neighborhoods and violent crime: a multilevel study of collective efficacy', *Science*, 227: 918–924.

Scarpa, A., Tanaka, A., and Haden, S.C. (2008) 'Biosocial bases of reactive and proactive aggression: the roles of community violence exposure and heart rate', *Journal of Community Psychology*, 36(8): 969–988.

Shaw, C.R., Zorbaugh, F.M., McKay, H.D., and Cottrell, L.S. (1929) *Delinquency Areas*, Chicago, IL: University of Chicago Press.

Sijtsema, J.J., Nederhof, E., Veenstra, R., Ormel, J., Oldehinkel, A.J., and Ellis, B.J. (2013) 'Effects of family cohesion and heart rate reactivity on aggressive/rule-breaking behavior and prosocial behavior in adolescence: the Tracking Adolescents' Individual Lives Survey study', *Development and Psychopathology*, 25(3): 699–712.

Streissguth, A.P., Barr, H.M., Bookstein, F.L., Sampson, P.D., and Olson, H.C. (1999) 'The long-term neurocognitive consequences of prenatal alcohol exposure: a 14–year study', *Psychological Science*, 10: 186–190.

Sutherland, E.H. (1947) *Principles of Criminology* (4th edn), Philadelphia, PA: J.B. Lippincott.

Sutherland, E.H. and Cressey, D.R. (1960) *Principles of Criminology* (6th edn), Chicago, IL: J.B. Lippincott.

de Vries-Bouw, M., Jansen, L., Vermeiren, R., Doreleijers, T., van de Ven, P., and Popma, A. (2012) 'Concurrent attenuated reactivity of alpha-amylase and cortisol is related to disruptive behavior in male adolescents', *Hormones and Behavior*, 62(1): 77–85.

Wright, J.P. and Boisvert, D. (2009) 'What biosocial criminology offers criminology', *Criminal Justice and Behavior*, 36(11): 1228–1240.

3

An evolutionary perspective on interpersonal violence

Sex differences and personality links

Peter K. Jonason

Introduction

Interpersonal violence—violence among those who have a non-zero level of acquaintance—affects as many as one third of men and women (Koss, 1992). Interpersonal violence includes psychological aggression (e.g. verbal abuse), physical aggression (e.g. bullying), and sexual coercion (e.g. rape), and occurs in both acquaintance and intimate partner contexts (Figueredo *et al.*, 2012). There has been a considerable amount of research on interpersonal violence and thus there is no shortage of theoretical accounts that try to explain it. A sociologist might suggest interpersonal violence is a function of income inequality and other institutional factors (Ember and Ember, 2004). Geneticists might suggest interpersonal violence is a function of genes that code for aggression and testosterone (Viding, 2004). Feminists might contend interpersonal violence is part of oppressive patriarchal systems (Johnson, 1995). Social psychologists would contend that interpersonal violence is learned from models in one's life (Bandura, 1983) or in the media (Berkowitz *et al.*, 1962). And last, but not least, clinical psychologists have conceptualized interpersonal violence as one of the diagnostic criteria for personality traits such as psychopathy (Hare, 1993).

All of these are interesting perspectives, but they fail in some important ways. First, they do not answer the "why" question. That is, they are more concerned with proximal causes of interpersonal violence as opposed to ultimate ones. Second, they assume that interpersonal violence is inherently negative, but most would agree that a man defending his wife from an attacker, or a mother defending her offspring from abduction, is acceptable and even called for. Third, they fail to view interpersonal violence within the larger context of evolutionary biology; a framework that despite some objections (viz., human exceptionalism), humans fall within. In contrast, one might adopt an evolutionary paradigm to understand interpersonal violence; viewing aggression as potentially strategic and adaptive given the right circumstances (Archer, 2001; Figueredo *et al.*, 2012). Such a perspective would integrate the proximal details from above, but also provide some unique insights into the nature of interpersonal violence that are not intuitively obvious and even somewhat provocative. As a road-map, this chapter will review evolutionary psychology, discuss sex differences from this perspective, and detail how certain personality traits could represent the psychological mechanisms behind interpersonal violence. Last, these predictions will be tested in a sample of American undergraduates.

A primer on evolutionary psychology

There are numerous reviews of evolutionary psychology (Buss, 2009; Confer *et al.*, 2010) and textbooks covering the field (Gaulin and McBurney, 2001; Barrett *et al.*, 2002); the interested reader is encouraged to seek them out for a more thorough review of evolutionary psychology. What is presented here is a brief overview geared towards the topic of interpersonal violence in particular.

Evolutionary psychology starts with a simple premise. Because human beings are biological organisms that are descendants down a long chain of evolution, spanning back at least 5 million years to the last common ancestor with chimpanzees (*Pan troglydytes*) and bonobos (*Pan paniscus*), and 47 million years to the last common ancestors between anthropoids (e.g. platyrhine and strepserine monkeys) and prosimians (e.g. lemuroids and lorisoids), called "Ida", any understanding of human behavior—which is assumed to be the result of the interaction of the brain and the body—must be framed in evolutionary terms. Natural and sexual selection are thought to have shaped organisms' brains and bodies, which will then produce certain behaviors geared towards maximizing inclusive fitness. For instance, natural selection appears to have shaped the eyes of nocturnal animals (e.g. *Panthera leo*) and the personality of some people (Jonason *et al.*, 2013c) to facilitate the active exploitation of a night-time niche.

In short, the meaning of life for an evolutionary psychologist is about increasing the number of offspring one leaves on the planet. There is little concern for sentiments of "being nice" for the sake of being nice. If "being nice" increases one's offspring though helping others, then generosity and altruism will develop as evolutionarily stable strategies. However, not all individuals will adopt the same strategy insomuch as the payoff schedules may not be conducive to their liking. If exploitation (Jonason *et al.*, 2012a,b), violence (Wilson and Daly, 1985), aggression (Archer, 2001), and even rape (Thornhill and Palmer, 2004) can increase the evolutionary bottom line, those can become evolutionarily stable strategies too.

Criminality of any kind likely conforms to what is known as balancing selection. That is, the number of criminals in a population must be smaller than the number of non-criminals to minimize competition among the criminals and to have a sufficient pool of victims who are actively trying to detect and punish the criminals (Jonason and Webster, 2012). Such an equilibrium is an emergent property of the competition among conspecifics over finite resources in the environment. Remember, one of Darwin's main contributions was to remove the sentimentality from perceptions of the animal kingdom (e.g. lions lay down with lambs), highlighting the competitive nature of biological life.

There are strong objections to taking an evolutionary approach to such topics, all of which are specious, but are worth dispelling outright. One objection comes from the mistaken premise that individuals have freewill, or they act, think, and feel under conscious, volitional control. Freewill has been well dismantled elsewhere (see Harris, 2012), but the short of it is, if life is a series of random events then people have no freewill, and if one adopts a deterministic model of the world, again, there is no freewill in that our behaviors, feelings, and attitudes are determined by a stream of biological, hormonal, environmental, and physical preconditions and antecedent events that individuals (1) have no control over (e.g. genetic make-up) and (2) are barely aware of (e.g. testosterone). This idea of no freewill has major implications for our understanding of criminality in that it calls into question traditional ideas of blame and culpability. It also insults the sentiments that most people have of (1) being in charge of their life and (2) the potential for rehabilitation. Imagine a person who commits murder—who is to blame? Their personality? Their genes? Their familial environment? Unfortunate circumstances? The person has no-to-minimal control over any of these.

Another objection is that researchers, clinicians, policy-makers, and lay-people have an implicit bias towards prosocial thinking (Jonason *et al.*, 2012a,b). That is, when a behavior is seen as bad for the group, it is seen as bad for the individual. Those who have this view of undesirable behavior implicitly take a group selection approach (Ardrey, 1970), a theory with little support and few advocates today. Instead, undesirable behavior may be solely—or selfishly—optimal for the individual in the context of their life (Dawkins, 1978). That is, despite the social or even personal costs to an individual, if a behavior or tendency confers reproductive advantages, those traits, no matter how socially undesirable they are, could be heritable "alternative" strategies.

Sex differences in interpersonal violence

Evidence suggests that men are more likely than women are to commit a variety of acts of interpersonal violence. Men score higher than women on measures of aggression (Björkqvist, 1994; Archer, 2001); commit rape more than women do, a pattern that holds up across species (Thornhill and Palmer, 2004); and commit violent crimes more than women do, especially between the ages of 16 and 30 (Wilson and Daly, 1985). The tendency for men to be more violent than women is a rather uncontroversial, descriptive claim. The controversy enters when one tries to explain such a pattern from an evolutionary perspective.

To make the evolutionary case, there are at least four lines of evidence one might look to. First, biological anthropology suggests that male primates (the order to which humans belong) are more violent than female primates, and most of that violence is about food, territory, and sex (Wrangham and Peterson, 1996). Second, interpersonally violent tendencies appear to be heritable (Viding, 2004). Third, the male-aggression tendency appears to be universal and robust to method variance (Archer, 2001). And last, interpersonal violence may increase reproductive success, as seen in antisocial personality traits (Jonason *et al.*, 2009); the appeal it has to women, especially when women are ovulating and in the short-term context (Aitken *et al.*, 2013); and the apparent increased preponderance of interpersonal violence and violent crime in young men (Wilson and Daly, 1985). In short, because men pay fewer costs and gain more from being violent than women do, it could have evolved as an evolutionarily stable strategy (Buss, 2009).

Dark Triad traits and interpersonal violence

The Dark Triad—narcissism, psychopathy, and Machiavellianism (Paulhus and Williams, 2002)—is composed of three socially undesirable and arguably interpersonally aversive traits (Kowalski, 2001) that appear to be heritable (Vernon *et al.*, 2008) and that have similar correlates across at least three different nations (Jonason *et al.*, 2013a). Psychopathy is noted for its characteristic callousness and limited empathy (Ali *et al.*, 2009) and criminality (Hart, 1998), including being a sexual offender (Serin *et al.*, 1994; Kerig and Stellwagen, 2010). Machiavellianism is noted for its characteristic glib social charm and manipulativeness (Christie and Geis, 1970). Narcissism is noted for its characteristic senses of entitlement, superiority, vanity, and exploitiveness (Raskin and Terry, 1988) and is linked to partner-directed violence (Ryan *et al.*, 2008) and the acceptance of rape myths (Bushman *et al.*, 2003). Each trait is associated with different provocation to anger (Jones and Paulhus, 2010), and those high on the traits actively select volatile and self-serving environments (Jonason *et al.*, 2011; Jonason and Schmitt, 2012), use violent mate retention tactics (Jonason *et al.*, 2010a) and coercive manipulation tactics (Jonason and Webster, 2012), and tend to lack humility, to be disagreeable, dishonest, and impulsive, and to lack empathy (Lee and Ashton, 2005; Jakobwitz and Egan, 2006; Jones and Paulhus, 2011; Jonason and McCain, 2012; Jonason *et al.*, 2013b).

Research on the Dark Triad has exploded in the past five years (see Jonason *et al.*, 2012a,b). One reason is its integration within evolutionary psychology, in particular life history theory (Wilson, 1975). From this perspective, personality traits are evolutionarily stable strategies designed by natural selection and shaped by environmental contingencies (e.g. harshness and instability in one's childhood; Brumbach *et al.*, 2009) to enable reproductive fitness (Rushton, 2004; Buss, 2009). Those high on these traits might have a *fast* life history strategy geared towards maximizing immediate returns over delayed outcomes, as would be predicted by the tradeoffs expressed in life history theory seen in future discounting, short-term/exploitive mating and social strategies, and impulsivity (Jonason *et al.*, 2009, 2010b, 2011; Jonason and Tost, 2010; Jones and Paulhus, 2011; McDonald *et al.*, 2011; Jonason and Schmitt, 2012; Jonason and Webster, 2012; Jonason *et al.*, 2012a). The Dark Triad traits could be maintained in the population if some positive fitness benefit is accrued (Mealey, 1995). As Buss (2009) suggests, "what is often disparaged as a maladaptive personality marked by impulsivity and lack of self control instead can be conceptualized within life history theory as an adaptive stable strategy deployed in response to a realistic appraisal of a shorter time horizon" (p. 361; see also Wilson and Daly, 1985).

Despite these obvious potential sources of overlap between the Dark Triad traits and interpersonal violence, few attempts have been made to understand the relationship between the Dark Triad traits and interpersonal violence, especially in the general population. Beyond paradigmatic limitations, most work on the Dark Triad traits on behaviors such as aggression (Bushman *et al.*, 2003) and criminality (Hare, 1993) has not taken into account the overlap between each trait. Therefore prior links between narcissism and sexual violence could be spuriously linked by psychopathy. While most research paradigms would agree that psychopathy should be linked to interpersonal violence (Browne, 1993; Goodman *et al.*, 1993; Johnson, 1995; Kowalski, 2001; Ember and Ember, 2004), an evolutionary paradigm such as life history theory (Wilson and Daly, 1985; Jonason *et al.*, 2012b), one would predict that it is Machiavellianism that may be associated with the most forms of interpersonal violence because interpersonal violence can be one strategy used to get what one wants out of others (Figueredo *et al.*, 2012; Jonason and Webster, 2012). Therefore, when the variance among the three traits is partialled, it is Machiavellianism that should be associated with the most forms of interpersonal violence insomuch as interpersonal violence might be a strategic interpersonal tactic (Archer, 2001). In contrast, narcissism, a trait more concerned with obtaining love and admiration from others (Raskin and Terry, 1988), should not be linked to interpersonal violence when any contamination from the other traits is removed. And last, psychopathy should be best linked to "darker" (i.e. more exploitive, more physically aggressive) forms of interpersonal violence (Rauthmann, 2012).

As mentioned above, men commit more interpersonal violence than women (Archer, 2001). Men also score higher on the Dark Triad traits however they are measured (Jonason *et al.*, 2009; Jonason and Webster, 2010), and cross-culturally in the U.S.A., Singapore, and Poland (Jonason *et al.*, 2013a). Taken together, this suggests that sex differences in interpersonal violence might be mediated by the Dark Triad. Prior work suggests interpersonal violence is a downstream correlate of traits such as psychopathy (Skeem and Cooke, 2010). This builds a theoretical model, suggesting not that men are necessarily more likely to commit interpersonal violence than women, but that it is those men who score more highly on the Dark Triad traits who will. That is, the psychological dispositions underlying the Dark Triad traits facilitate interpersonal violence in men more than in women because of potential adaptive benefits linked to interpersonal violence. However, given the adaptive hypothesis offered by evolutionary psychologists for the Dark Triad, one should predict this facilitation to be not ubiquitous across types of interpersonal violence. The traits should more strongly facilitate interpersonal violence in contexts that have less risk and more reward (i.e. psychological abuse) and less strongly facilitate interpersonal violence in contexts with more risk than reward (i.e. sexual abuse and physical abuse).

Through an adaptationist perspective (Buss, 2009; Jonason *et al.*, 2012a,b), one might be able to provide a more nuanced and balanced perspective on traits such as the Dark Triad, and criminal behavior such as interpersonal violence. These predictions are put to a test in a sample of American college students. The predictions are examined in a general sense, as well as in the context of intimate relationships and, given the overlap between the Dark Triad and the Big Five personality traits (i.e. extraversion, agreeableness, emotional stability, conscientiousness, and openness; Paulhus and Williams, 2002; Jakobwitz and Egan, 2006), and overlap between these traits and interpersonal violence (Heaven, 1996), any contamination from the Big Five traits is controlled.

Method

Participants and procedure

Undergraduate psychology students (N = 302; 191 women, 111 men) aged 17–70 years (M = 21.34, SD = 5.74) from a mid-sized university in the southern U.S.A. completed a study in exchange for partial course credit. In terms of sexual orientation, 278 were heterosexual, 8 were homosexual, and 16 were bisexual. In terms of relationship status, 162 were single and 140 were involved in a serious relationship (i.e. seriously dating or married). Only those participants who completed the measures from unique IP addresses were included.

Participants were informed of the nature of the study and were asked to give consent if they wished to participate; only those who gave consent have been included. They progressed through a series of self-report measures—in the order reported below—that assessed the variables of interest. At the end of the study, participants were debriefed and thanked. Because of the potentially troubling and disturbing nature of the questions regarding interpersonal abuse, participants were provided with information about the university's counseling services.

Measures

To measure the Dark Triad traits, the Dark Triad Dirty Dozen (Jonason and Webster, 2010) was used. Participants were asked how much they agreed (1 = *not at all*; 5 = *very much*) with statements such as: "I tend to want others to admire me" (i.e. narcissism), "I tend to lack remorse" (i.e. psychopathy), and "I have used deceit or lied to get my way" (i.e. Machiavellianism). Items were averaged together to create an index of narcissism (Cronbach's α = 0.80), Machiavellianism (α = 0.81), psychopathy (α = 0.73), and a composite Dark Triad index (α = 0.88). Consistent with prior research, psychopathy was correlated with Machiavellianism ($r(302) = 0.61, P < 0.01$) and narcissism ($r(302) = 0.49, P < 0.01$) which was correlated with Machiavellianism ($r(302) = 0.61, P < 0.01$).

Experience with interpersonal conflict was measured using the interpersonal relations rating scale (Figueredo *et al.*, 2012). The scale measures psychological and physical aggression toward members of the same or the opposite sex, and whether or not they were the participants' romantic partners. It is composed of 47 parallel questions for each target's sex, asking participants the frequencies with which they performed a series of acts in the last 12 months (0 = *never*; 6 = *daily*). Annualized frequencies were estimated by converting the five Likert-scaled response options to their numerical equivalents representing the expected number of occurrences per year (i.e. *never* = 0; *only once* = 1; *6 times* = 6; *12 times* = 12; *once a week* = 52; *daily* = 365). Because count data can be skewed, and in accordance with prior work (Figueredo *et al.*, 2012), these data were log-transformed—log(x + 1). There were five subscales that were created by averaging the corresponding items for each target's sex (totaling 10 measures; $\alpha \leq 0.85$) including coercive control, psychological abuse, physical abuse, escalated (i.e. life-threatening) abuse, and sexual abuse.

Participants who reported they were in a committed relationship completed the violence assessment index (VAI) and the injury assessment index (IAI) to assess specific methods of assault, objects used in assaults, and parts of the body to which assaults occurred (Dobash *et al.*, 1995). The VAI is composed of 26 items asking how frequently a given violent act was committed by the participant (1 = *never*; 6 = *11 or more times*) and whether or not it occurred in the past month (*yes/no*). The IAI is composed of 20 items asking how frequently a given injury occurred to their romantic partner (1 = *never*; 6 = *11 or more times*) and asks whether or not the participant's partner experienced that injury in the past month (*yes/no*). By averaging items on each scale, composites were created for the VAI (α = 0.80) and the IAI (α = 0.93). An overall composite of the IAI and VAI (α = 0.86) was also created, as per prior work, by averaging the items, called the "overall violence index" (OVI; Shackelford *et al.*, 2005). Participants were also asked (*yes/no*) if they performed any of the acts or if their partners experienced any of the injuries; responses were summed for partner's experience with injuries and self-reported commission of intimate partner violence. Items intended for men only, and for those with children, were excluded to make the measures equivalent across the sample.

The ten-item personality inventory (TIPI; Gosling *et al.*, 2003) was used to measure the Big Five for statistical control. It asks two questions for each dimension. Participants were asked, for instance, how much (1 = *not at all*; 5 = *very much*) they think of themselves as "extraverted, enthusiastic" and "quiet, reserved" (reversed) to measure extraversion. Estimates of internal consistency returned low rates for extraversion (α = 0.54), agreeableness (α = 0.30), conscientiousness (α = 0.29), neuroticism (α = 0.21), and openness (α = 0.25).

Results

Table 3.1 contains sex differences and similarities, and overall descriptive statistics. Men scored higher on all measures of the Dark Triad than women did, with the largest difference in psychopathy and the smallest in narcissism. Men scored higher than women did on all but two measures of interpersonal violence, significantly so in six cases (40 percent of the total number of assessments). The sex differences ranged from zero to medium in size. For instance, men, more than women, may use sexual abuse of opposite-sex others, but the sexes were identical in their use of coercive control of opposite-sex others.

The Dark Triad traits were positively correlated with virtually every manifestation of interpersonal violence (Table 3.2). The only correlations that were not significant were between narcissism and escalation of abuse and between narcissism and physical abuse towards both same- and opposite-sex targets. Multiple regressions, which controlled for the shared variance among the Dark Triad traits, revealed that Machiavellianism was the best predictor of interpersonal

Table 3.1 Descriptive statistics and sex differences in the Dark Triad and measures of interpersonal violence

Dark Triad	*Mean (SD)*			*t*	*d*
	Overall	*Women*	*Men*		
Psychopathy[d]	1.18 (0.75)	1.67 (0.64)	2.06 (0.84)	–4.44**	–0.51
Machiavellianism[d]	1.91 (0.77)	1.81 (0.72)	2.07 (0.84)	–2.84**	–0.33
Narcissism[d]	2.27 (0.81)	2.18 (0.78)	2.43 (0.84)	–2.64**	–0.30
Dark Triad composite[d]	2.00 (0.66)	1.89 (0.60)	2.19 (0.60)	–3.90**	–0.45
Relational violence					
Number of injuries partner suffered[a]	0.80 (1.68)	0.56 (1.24)	1.10 (2.09)	–2.10*	–0.36
Number of acts of violence[b]	0.68 (1.69)	0.28 (0.58)	1.20 (2.34)	1.50	–0.25
Injury assessment index[c]	1.10 (0.29)	1.06 (0.13)	1.19 (0.50)	–2.41*	–0.42
Violence assessment index[c]	1.13 (0.22)	1.13 (0.22)	1.12 (0.24)	0.17	0.03
Overall violence index[c]	1.12 (0.19)	1.10 (0.15)	1.16 (0.27)	–1.63	–0.28
Violence towards same-sex others					
Coercive control[d]	0.17 (1.68)	0.16 (0.21)	0.20 (0.21)	–1.47	–0.17
Psychological abuse[d]	0.34 (0.23)	0.29 (0.34)	0.42 (0.42)	–2.78**	–0.33
Physical abuse[d]	0.12 (0.38)	0.08 (0.26)	0.18 (0.35)	–2.50**	–0.29
Escalation of Abuse[d]	0.08 (0.30)	0.06 (0.23)	0.11 (0.27)	–1.79	–0.21
Sexual abuse[d]	0.07 (0.25)	0.05 (0.22)	0.11 (0.33)	–2.00*	–0.22

Violence towards opposite-sex others					
Coercive control[d]	0.21 (0.27)	0.21 (0.23)	0.21 (0.30)	–0.04	0.00
Psychological abuse[d]	0.35 (0.25)	0.32 (0.34)	0.40 (0.43)	–1.59	–0.18
Physical abuse[d]	0.14 (0.37)	0.14 (0.30)	0.12 (0.35)	0.50	0.06
Escalation of abuse[d]	0.07 (0.32)	0.06 (0.23)	0.09 (0.26)	–0.86	–0.10
Sexual abuse[d]	0.10 (0.28)	0.06 (0.22)	0.17 (0.36)	–3.18**	–0.37

Notes: *d* = Cohen's *d* for effect size
* $P < 0.05$, ** $P < 0.01$
df = a, 87; b, 57; c, 140; d, 301

Table 3.2 Zero-order correlations and standardized regression coefficients using the Dark Triad to predict interpersonal violence

Same-sex others	r (β)			
	Psychopathy	*Machiavellianism*	*Narcissism*	*Dark Triad*
Coercive control[d]	0.35** (0.20**)	0.38** (0.31**)	0.21** (–0.08)	0.37**
Psychological abuse[d]	0.46** (0.27**)	0.50** (0.41**)	0.25** (–0.13)	0.47**
Physical abuse[d]	0.27** (0.21**)	0.26** (0.24**)	0.07 (–0.18*)	0.23**
Escalated abuse[d]	0.25** (0.18*)	0.26** (0.26**)	0.08 (–0.17*)	0.23**
Sexual abuse[d]	0.26** (0.17*)	0.28** (0.23**)	0.12* (–0.10)	0.26**
Opposite-sex others				
Coercive control[d]	0.36** (0.23**)	0.39** (0.24**)	0.22** (–0.04)	0.37**
Psychological abuse[d]	0.42** (0.19**)	0.49** (0.43**)	0.27** (–0.09)	0.46**
Physical abuse[d]	0.20** (0.11)	0.24** (0.24**)	0.08 (–0.13)	0.20**
Escalated abuse[d]	0.21** (0.12)	0.25** (0.27**)	0.07 (–0.16*)	0.21**
Sexual abuse[d]	0.31** (0.17*)	0.35** (0.30**)	0.17** (–0.09)	0.33**
Relational violence				
Number of injuries partner suffered[a]	0.09 (–0.08)	0.19 (0.12)	0.22* (0.20)	0.20
Number of acts of violence[b]	0.36** (0.59**)	0.09 (–0.12)	0.02 (–0.25)	0.02
Injury assessment index[c]	0.24** (0.18)	0.22* (0.14)	0.14 (0.14)	0.23**
Violence assessment index[c]	0.10 (0.06)	0.06 (–0.08)	0.13 (–0.08)	0.12
Overall violence index[c]	0.24** (0.17)	0.20* (0.06)	0.19* (0.06)	0.24**

Notes: * $P < 0.05$, ** $P < 0.01$
df = a, 87; b, 57; c, 140; d, 301

violence. Virtually all correlations with narcissism disappeared when taking into account the other two traits; making the positive, significant, zero-order correlations null or even significant and negative.

Table 3.2 also contains correlations between the Dark Triad traits and forms of relational violence as measured by the VAI, IAI, and OVI. First, Machiavellianism and psychopathy predicted injuries within relationships. Second, all three traits were correlated with overall violence. Third, when the shared variation among the Dark Triad traits was controlled, no significant associations were found. Fourth, no facets of the Dark Triad traits were correlated with the commission of violence.

A series of partial correlations were run, where all aspects of the Big Five were controlled while assessing the correlations between the Dark Triad traits and measures of interpersonal violence. None of the correlations remained significant when we assessed the IAI, VAI, and OVI; however, this may merely be a function of having too many predictors in the regression model coupled with a small number of people engaged in committed relationships. In contrast, most of the correlations among the Dark Triad traits and the interpersonal relations rating scales remained significant ($r_p = 0.23$ to 0.42, $P < 0.05$). Psychopathy, Machiavellianism, and, by extension, the Dark Triad composite were still correlated with the use of coercion, psychological abuse, and sexual abuse irrespective of the target's sex. In addition, psychopathy still correlated with the use of escalating abuse. In contrast, narcissism was correlated with the use of coercive control for both targets only after controlling for the Big Five.

Last, mediation was tested. When it came to interpersonal violence directed at same-sex others, the Dark Triad composite (Jonason *et al.*, 2009) fully mediated (i.e. the beta for participant's sex was no longer significant) the sex differences in reports of romantic partner's experience with injuries (Sobel's $z = 2.08$, $P < 0.05$) and the commission of psychological ($z = 3.47$, $P < 0.01$), physical ($z = 2.62$, $P < 0.05$), and sexual abuse ($z = 3.10$, $P < 0.01$). The Dark Triad composite also partially mediated (i.e. the beta for participant's sex decreased in magnitude but remained significant) the sex difference in the commission of sexual abuse directed towards opposite-sex others ($z = 3.31$, $P < 0.01$); the correlation between the participant's sex and this type of violence went from 0.18 ($t = 3.18$, $P < 0.01$) to 0.12 ($t = 2.03$, $P = 0.04$) after the Dark Triad ($\beta = 0.30$, $t = 5.26$, $P < 0.01$) was added to the model.

Discussion

There are good reasons to try, as a society, to discourage interpersonal violence. Interpersonal violence hurts those we care about, hurts society as a collective, and may hurt the perpetrator in terms of prison or retaliatory aggression. All of these have led well intentioned researchers to postulate a variety of viable causes, and thereby solutions, to the problems created by interpersonal violence. But the road to "hell" is paved in good intentions and, in the case of research, misguided, misinformed, or underspecified models of interpersonal violence (an arguable hell for scientists). Human beings are biological creatures subject to the same laws as the rest of the animal kingdom. That means their behavior and psychology must be framed within evolutionary biology to make any sense of the ultimate causes of phenomena of which society approves, such as generosity, and those on which society looks down, such as the Dark Triad and interpersonal violence. In this chapter, the nature of interpersonal violence has been framed using an adaptionist lens.

There is considerable evidence to lead one to expect that men should be more interpersonally violent than women (Björkqvist, 1994; Archer, 2001). While this effect was replicated, the inclusion of the Dark Triad in this study (1) provides some insights into the adaptive and psychological mechanisms that might account for interpersonal violence; and (2) suggests that men are not categorically more violent than women but, instead, it is those men with a certain disposition who are the most interpersonally violent. Based on evolutionary psychology, this might be because being interpersonally violent might be adaptive for men more than women (Wilson and Daly, 1985; Buss, 2009; Jonason *et al.*, 2009). This is especially noteworthy in that the links between interpersonal violence (1) appear to be mostly with a strategic (i.e. Machiavellianism), as opposed to impulsive (i.e. psychopathy) or vain (e.g. narcissism), disposition; and (2) are strongest where there is a fair balance of risk (e.g. retaliatory aggression; Boyd *et al.*, 2003) and (potential) reward (i.e. psychological abuse).

Much has been made out of the link between narcissism and interpersonal violence (Bushman *et al.*, 2003; Ryan *et al.*, 2008), but this study revealed that while the two are linked, those links appear to vanish upon the inclusion of the Machiavellianism and psychopathy. There are some possible reasons for this. It might be that narcissism does not have a unique correlation with interpersonal violence. Past research did not control for psychopathy and Machiavellianism. However, such a contention requires further work, given the usage of a new and concise measure of the Dark Triad and specific measures of interpersonal violence. Each of the measures may have idiosyncrasies that could make the failure to find this link a methodological artifact. Alternatively, given the manner by which the correlations with narcissism vanished and even switched direction, suppression might be present.

Limitations and conclusions

The present research has at least four limitations. First, when assessing violence in romantic relationships, results were less revealing than when using the interpersonal relations rating scale. Although the Dark Triad traits predicted injuries experienced by their partners at the zero-order level, these associations were no longer significant when shared variance among the Dark Triad was controlled. This might be the result of diminished power, with only 140 participants reporting being in romantic relationships; cutting the sample nearly in half.

Second, much of the research on psychopathy tends to be done with clinical or criminal populations (Hare, 1993; Kowalski, 2001). In contrast, this study used a student population, thus the use of a measure designed for the assessment of psychopathy in college students was warranted (Jonason and Webster, 2010). This measure has the advantage of not being contaminated by items related to violence or criminality (Skeem and Cooke, 2010) so there is little concern about predictor–criterion overlap. However, the reliance on college students may present some limitations because they might have less experience with interpersonal violence and with romantic and sexual relationships.

Third, only direct violence and its consequences were assessed. Women may use indirect violence more than men do (Björkqvist, 1994). It is possible that the sex difference in the use of indirect violence might be mediated by the Dark Triad, suggesting that the psychological system in women which facilities the adoption of this form of violence is tapped by the Dark Triad. Moreover, this study assessed only relational violence (i.e. violence in interpersonal relationships) as opposed to typological forms of violence. For instance, instrumental aggression (Bushman and Anderson, 2001), that is, aggression with a purpose as opposed to outright hostility, might be correlated with the Dark Triad. Narcissists respond with aggression to ego-threats whereas psychopaths respond to physical provocations (Jones and Paulhus, 2010), both of which may be forms of instrumental aggression; the first is instrumental in keeping one's positive self-image and the second is instrumental for protection.

Fourth, there are some notable limitations of the measures used. Both the Dirty Dozen and the TIPI are brief measures and may have limitations as a result. In the case of the TIPI, this may be a small problem given its inclusion to reduce the jangle fallacy—measuring the same thing twice—as used in prior research (Jonason *et al.*, 2011). The Dirty Dozen has been criticized for its brevity, homogeneity, and limited construct validity (Miller *et al.*, 2012), all of which might be reasonable given its extreme brevity. Nevertheless, psychometric work (Jonason and Webster, 2010; Jonason and Luévano, 2013; Webster and Jonason, 2013) and theoretical studies (Jonason *et al.*, 2011; Jonason and Webster, 2012) reveal its favorable qualities and utility. Relatedly, despite the detection of significant and predicted correlations, there were low rates of interpersonal violence, suggesting participants are experiencing or committing little interpersonal violence, but this may create range restriction. Nevertheless, future work would benefit from using alternative measures to validate and replicate the findings here (e.g. the decrease in the effect for narcissism after controlling for the other two Dark Triad traits).

In this study, different forms of interpersonal violence, in general and in the context of romantic relationships, were assessed across the sexes and in relation to the Dark Triad traits. In accordance with work on social perception (Rauthmann, 2012), psychopathy and Machiavellianism were "darker" than narcissism insomuch as they had more links to interpersonal violence. However, as opposed to the Dark Triad being seen as a dysfunction (Kowalski, 2001), an evolutionary approach would argue that such traits might be adaptive (Buss, 2009). In particular, results from Machiavellianism reveal that interpersonal violence might be one strategic means by which individuals get what they want in the world. In addition, mediation results suggest that men who are high on the Dark Triad might be afforded greater fitness via exploitive methods (Jonason *et al.*, 2012a,b). In short, the degree to which the Dark Triad traits and interpersonal violence are "bad" is a function of the perspective one takes.

Acknowledgement

The author thanks A.J. Figueredo for statistical consultations.

References

Aitken, S., Lyons, M., and Jonason, P.K. (2013) 'Dads or cads? Women's strategic decisions in the mating game', *Personality and Individual Differences*, 55: 118–122.

Ali, F., Amorim, I.S., and Chamorro-Premuzic, T. (2009) 'Empathy deficits and trait emotional intelligence in psychopathy and Machiavellianism', *Personality and Individual Differences*, 47: 758–762.

Archer, J. (2001) 'A strategic approach to aggression', *Social Development*, 10: 267–271.

Ardrey, R. (1970) *The Social Contract*, London: Collins.

Bandura, A. (1983) 'Self-efficacy determinants of anticipated fears and calamities', *Journal of Personality and Social Psychology*, 45: 464–468.

Barrett, L., Dunbar, R.I.M., and Lycett, J. (2002) *Human Evolutionary Psychology*, Princeton, NJ: Princeton University Press.

Berkowitz, L., Corwin, R., and Heironimus, M. (1962) 'Film violence and subsequent aggressive tendencies', *Public Opinion Quarterly*, 27: 217–229.

Björkqvist, K. (1994) 'Sex differences in physical, verbal, and indirect aggression: a review of recent research', *Sex Roles*, 30: 177–188.

Boyd, R., Gintis, H., Bowles, S., and Richerson, P.J. (2003) 'The evolution of altruistic punishment', *Proceedings of the National Academy of Sciences, USA*, 100: 3531–3535.

Browne, A. (1993) 'Violence against women by male partners: prevalence, outcomes, and policy implications', *American Psychologist*, 48: 1077–1087.

Brumbach, B.H., Figueredo, A.J., and Ellis, B.J. (2009) 'Effects of harsh and unpredictable environments in adolescence on development of life history strategies: a longitudinal test of an evolutionary model', *Human Nature*, 20: 25–51.

Bushman, B.J. and Anderson, C.A. (2001) 'Is it time to pull the plug on hostile versus instrumental aggression dichotomy?', *Psychological Review*, 108: 273–279.

Bushman, B.J., Bonacci, M.A., van Dijk, M., and Baumeister, R.F. (2003) 'Narcissism, sexual refusal, and aggression: testing a narcissistic reactance model of sexual coercion', *Journal of Personality and Social Psychology*, 84: 1027–1040.

Buss, D.M. (2009) 'How can evolutionary psychology explain personality and individual differences?', *Perspectives in Psychological Science*, 4: 359–366.

Christie, R. and Geis, F.L. (1970) *Studies in Machiavellianism*, New York: Academic Press.

Confer, J.C., Easton, J.A., Fleischman, D.S., Goetz, C.D., Lewis, D.M., Perilloux, C., and Buss, D.M. (2010) 'Evolutionary psychology: controversies, questions, prospects, and limitations', *American Psychologist*, 65: 110–126.

Dawkins, R. (1978) *The Selfish Gene*, London: Oxford University Press.

Dobash, R.E., Dobash, R.P., Cavanagh, K., and Lewis, R. (1995) 'Evaluating criminal justice programmes for violent men', in R.E. Dobash, R.P. Dobash, and L. Noaks (eds.), *Gender and Crime*, Cardiff, UK: University of Wales Press, pp. 358–389.

Ember, C. and Ember, M. (2004) 'War, socialization, and interpersonal violence: a cross-cultural study', *Journal of Conflict Resolution*, 38: 620–646.

Figueredo, A.J., Gladden, P.R., and Beck, C.J.A. (2012) 'Intimate partner violence and life history strategy', in A. Goetz and T. Shackelford (eds.), *The Oxford Handbook of Sexual Conflict in Humans*, New York: Oxford University Press.

Gaulin, S.J.C. and McBurney, D.H. (2001) *Psychology: An Evolutionary Approach*, Upper Saddle River, NJ: Prentice Hall.

Goodman, L.A., Koss, M.P., Fitzgerald, L.F., Russo, N.F., and Keita, G.P. (1993) 'Male violence against women: current research and future directions', *American Psychologist*, 48: 1054–1058.

Gosling, S.D., Rentfrow, P.J., and Swann, W.B. Jr. (2003) 'A very brief measure of the Big-Five personality domains', *Journal of Research in Personality*, 37: 504–528.

Hare, R.D. (1993) *Without Conscience: The Disturbing World of the Psychopaths Among Us*, New York: Pocket Books.

Harris, S. (2012) *Freewill*, New York: Free Press.

Hart, S.D. (1998) 'Psychopathy and risk for violence', in *Psychopathy: Theory, Research and Implications for Society*, Dordrecht, Netherlands: Springer, pp. 355–373.

Heaven, P.C.L. (1996) 'Personality and self-reported delinquency: analysis of the "Big Five" personality dimensions', *Personality and Individual Differences*, 20: 47–54.

Jakobwitz, S. and Egan, V. (2006) 'The dark triad and normal personality traits', *Personality and Individual Differences*, 40: 331–339.

Johnson, M.P. (1995) 'Patriarchal terrorism and common couple violence: two forms of violence', *Journal of Marriage and Family*, 57: 283–294.

Jonason, P.K. and Luévano, V.X. (2013) 'Walking the thin line between efficiency and accuracy: validity and structure of the Dirty Dozen', *Personality and Individual Differences*, 55: 76–81.

Jonason, P.K. and McCain, J. (2012) 'Using the HEXACO model to test the validity of the Dirty Dozen measure of the Dark Triad', *Personality and Individual Differences*, 53: 935–938.

Jonason, P.K. and Schmitt, D.P. (2012) 'What have you done for me lately? Friendship-selection in the shadows of Dark Triad traits', *Evolutionary Psychology*, 10: 400–421.

Jonason, P.K. and Tost, J. (2010) 'I just cannot control myself: the Dark Triad and self-control', *Personality and Individual Differences*, 49(6): 611–615.

Jonason, P.K. and Webster, G.D. (2010) 'The Dirty Dozen: a concise measure of the Dark Triad', *Psychological Assessment*, 22, 420–432.

Jonason, P.K. and Webster, G.D. (2012) 'A protean approach to social influence: Dark Triad personalities and social influence tactics', *Personality and Individual Differences*, 52: 521–526.

Jonason, P.K., Li, N.P., Webster, G.W., and Schmitt, D.P. (2009) 'The Dark Triad: facilitating short-term mating in men', *European Journal of Personality*, 23: 5–18.

Jonason, P.K., Li, N.P., and Buss, D.M. (2010a) 'The costs and benefits of the Dark Triad: implications for mate poaching and mate retention tactics', *Personality and Individual Differences*, 48: 373–378.

Jonason, P.K., Koenig, B.L., and Tost, J. (2010b) 'Living a fast life', *Human Nature*, 21(4): 428–442.

Jonason, P.K., Valentine, K.A., Li, N.P., and Harbeson, C.L. (2011) 'Mate-selection and the Dark Triad: facilitating a short-term mating strategy and creating a volatile environment', *Personality and Individual Differences*, 51: 759–763.

Jonason, P.K., Luévano, V.X., and Adams, H.M. (2012a) 'How the Dark Triad traits predict relationship choices', *Personality and Individual Differences*, 53: 180–184.

Jonason, P.K., Webster, G.W., Schmitt, D.P., Li, N.P., and Crysel, L. (2012b) 'The antihero in popular culture: a life history theory of the Dark Triad', *Review of General Psychology*, 16: 192–199.

Jonason, P.K., Li, N.P., and Czarna, A.Z. (2013a) 'Quick and dirty: some psychosocial costs associated with the Dark Triad in three countries', *Evolutionary Psychology*, 11: 172–185.

Jonason, P.K., Lyons, M., Bethell, E., and Ross, R. (2013b) 'Different routes to limited empathy in the sexes: examining the links between the Dark Triad and empathy', *Personality and Individual Differences*, 54: 572–576.

Jonason, P.K. Jones, A., and Lyons, M. (2013c) 'Creatures of the night: chronotypes and the Dark Triad traits', *Personality and Individual Differences*, 55(5): 538–541.

Jones, D.N. and Paulhus, D.L. (2010) 'Different provocations trigger aggression in narcissists and psychopaths', *Social Psychological and Personality Science*, 1: 12–18.

Jones, D.N. and Paulhus, D.L. (2011) 'The role of impulsivity in the Dark Triad of personality', *Personality and Individual Differences*, 51: 679–682.

Kerig, P.K. and Stellwagen, K.K. (2010) 'Roles of callous-unemotional traits, narcissism, and Machiavellianism in childhood aggression', *Journal of Psychopathology and Behavioral Assessment*, 32: 343–352.

Koss, M.P. (1992) 'Defending date rape', *Journal of Interpersonal Violence*, 7: 122–126.
Kowalski, R.M. (2001) *Behaving Badly: Aversive Behaviors in Interpersonal Relationships*, Washington, DC: American Psychological Association.
Lee, K. and Ashton, M.C. (2005) 'Psychopathy, Machiavellianism, and narcissism in the Five-Factor Model and the HEXACO model of personality structure', *Personality and Individual Differences*, 38: 1571–1582.
McDonald, M.M., Donnellan, M.B., and Navarrete, C.D. (2011) 'A life history approach to understanding the Dark Triad', *Personality and Individual Differences*, 52: 601–605.
Mealey, L. (1995) 'The sociobiology of sociopathy: an integrated evolutionary model', *Behavioral and Brain Sciences*, 18: 523–599.
Miller, J.D., Few, L.R., Seibert, L.A., Watts, A., Zeichner, A., and Lynam, D.R. (2012) 'An examination of the Dirty Dozen measure of psychopathy: a cautionary tale about the costs of brief measures', *Psychological Assessment*, 24: 1048–1053.
Paulhus, D.L. and Williams, K.M. (2002) 'The Dark Triad of personality: narcissism, Machiavellianism, and psychopathy', *Journal of Research in Personality*, 36: 556–563.
Raskin, R.N. and Terry, H. (1988) 'A principal components analysis of the Narcissistic Personality Inventory and further evidence of its construct validity', *Journal of Personality and Social Psychology*, 54: 890–902.
Rauthmann, J.F. (2012) 'The Dark Triad and interpersonal perception: similarities and differences in the social consequences of narcissism, Machiavellianism, and psychopathy', *Social Psychological and Personality Science*, 3: 487–496.
Rushton, J.P. (2004) 'Differential K theory: the sociobiology of individual and group differences', *Personality and Individual Differences*, 6: 441–452.
Ryan, K.M., Weikel, K., and Sprechini, G. (2008) 'Gender differences in narcissism and courtship violence in dating couples', *Sex Roles*, 58: 802–813.
Serin, R.C., Malcolm, P.B., Khanna, A., and Barbee, H.E. (1994) 'Psychopathy and deviant sexual arousal in incarcerated sexual offenders', *Journal of Interpersonal Violence*, 9: 3–11.
Shackelford, T.K., Goetz, A.T., Buss, D.M., Euler, H.A., and Hoier, S. (2005) 'When we hurt the ones we love: predicting violence against women from men's mate retention tactics', *Personal Relationships*, 12: 447–463.
Skeem, J.L., and Cooke, D.J. (2010) 'Is criminal behavior a central component of psychopathy? Conceptual directions for resolving the debate', *Psychological Assessment*, 22: 433–445.
Thornhill, R. and Palmer, C.T. (2004) 'Evolutionary life history perspective on rape', in C. Crawford and C. Salmon (eds.), *Evolutionary Psychology, Public Policy, and Personal Decisions*, Mahwah, NJ: Lawrence Erlbaum, pp. 249–274.
Vernon, P.A., Villani, V.C., Vickers, L.C., and Harris, J.A. (2008) 'A behavioral genetic investigation of the Dark Triad and the Big 5', *Personality and Individual Differences*, 44: 445–452.
Viding, E. (2004) 'On the nature and nurture of antisocial behavior and violence', *Annals of the New York Academy of Science*, 1036: 267–277.
Webster, G.D., and Jonason, P.K. (2013) 'Putting the "IRT" in "Dirty": Item Response Theory analyses of the Dark Triad Dirty Dozen—an efficient measure of narcissism, psychopathy, and Machiavellianism', *Personality and Individual Differences*, 54: 302–306.
Wilson, E.O. (1975) *Sociobiology: The New Synthesis*, Cambridge, MA: Harvard University Press.
Wilson, M. and Daly, M. (1985) 'Competitiveness, risk-taking, and violence: the young male syndrome', *Ethology and Sociobiology*, 6: 59–73.
Wrangham, R. and Peterson, D. (1996) *Demonic Males: Apes and the Origins of Human Violence*, New York: Houghton Mifflin.

4

Key findings from biosocial research and what they mean for the future of criminology

Kevin M. Beaver and Eric J. Connolly

Introduction

During the past ten years there has been an explosion of published biosocial criminological research. This body of research has focused on examining the genetic and environmental contributions to virtually every behavior, trait, and environment that is of interest to criminologists. Overall, the findings of these studies, which are summarized in other chapters in this volume, consistently show that genetic factors account for a significant amount of variation in virtually every criminogenic measure. These estimates obviously vary across studies depending on sample characteristics, the outcome under investigation, and other research design factors. However, the end result of this emerging body of empirical research is that, compared with a decade ago, there is now a better understanding of the biosocial foundations to a wide range of antisocial behaviors. In addition, the rate at which biosocial research is being produced has been quite impressive. Within a relatively short period, biosocial studies has gone from being nonexistent to commonplace in some criminology journals. By every indication, the rate at which such research is being produced is not slowing down; if anything, this area of research is continuing to grow at an exponential rate. The growth of biosocial research is due, in part, to biosocial criminologists being among some of the most productive scholars, the intense interest among new graduate students in biosocial research, and the availability of datasets that allow genetically informed analyses to be conducted.

Although there is currently an abundance of empirical information available about the biosocial influences on criminal behavior, to date there has not been much written about the substantive importance of this line of research for the future of criminological research, criminological theories, and the field of criminology in general. This chapter aims to open a dialogue on this topic by discussing the potential ways in which criminology as a discipline needs to change in order to stay current with evidence produced by biosocial criminological research. The chapter is divided into three main sections. First, a number of key biosocial findings are touched upon, followed by a more thorough discussion of what these findings mean for certain areas of criminological research. Second, the key biosocial findings are discussed as they relate to the merits of existing mainstream criminological theories. Third, what the growing number of biosocial criminologists means for the field of criminology is discussed, as well as how this growth should change criminology as a profession in order to keep pace with the burgeoning interest in biosocial criminology.

Key findings from biosocial research, and how mainstream criminological research needs to change to stay current

Given how much biosocial research has been produced, the existing findings are too voluminous to review exhaustively within this chapter and then offer recommendations on how criminological research needs to account for all of them. Instead, we focus our discussion on how criminology, as a discipline, needs to change around three key biosocial findings that are exceedingly consistent and so widely accepted that they could almost be considered axiomatic. The first key finding to emerge from the existing literature is that variation in virtually every measure studied is under some level of genetic influence (Turkheimer, 2000). For some measures, especially those directly assessing variation in antisocial behaviors and traits, a genetic effect is probably not all that surprising. Most studies, for example, have reported that genetic effects account for about 50 percent of the variance in measures of crime, aggression, and violence, as well as in traits related to crime, such as low self-control (Beaver *et al.*, 2008), impulsivity (Seroczynski *et al.*, 1999; Vierikko *et al.*, 2004), and even psychopathic personality traits (Larsson *et al.*, 2006). What is probably surprising to most criminologists, however, is that individual variability in exposure to criminogenic environments is also under significant genetic influence. For example, variation in measures of delinquent peers (Button *et al.*, 2007; Kendler *et al.*, 2007), social support (Beaver *et al.*, 2013), parental socialization (Beaver, 2011; Kendler and Baker, 2007), and even childhood maltreatment/abuse (Jaffee *et al.*, 2004) has been shown to be due, in part, to genetic influences. The genetic effect for criminogenic environments tends to be around 25 percent, though the effect appears to be significantly stronger for some environments, such as exposure to antisocial peer networks (Cleveland *et al.*, 2005).

While these findings are based on thousands of studies, a number of meta-analyses, and large reviews of the literature, they remain largely ignored by the criminological community. For example, most criminological studies fail to cite, discuss, or even mention the possibility that genetic influences have been found to impact the measures typically under investigation. Moreover, when any reference to a genetic effect is made, it is typically folded into a single sentence (e.g. previous research has detected a genetic effect) or a brief statement that is quickly sidestepped. The end result is that a large body of research suggesting that genetic influences partially explain individual differences in criminal involvement, delinquency, antisocial behavior, and environments conducive to antisocial behavior is largely ignored by the criminological community.

Perhaps all of this does not amount to much, as most mainstream criminologists focus their research efforts on estimating "pure" sociological influences on crime, which means that they leave the examination of biosocial influences to other researchers. According to this line of reasoning, the two perspectives—sociological and biosocial—can coexist without any cross-fertilization. However, from a methodological and statistical standpoint, such logic is fatally flawed. To illustrate, consider confounding, one of the most serious threats to establishing causality in social science research. As is learned in any undergraduate research methods course, confounding occurs when two variables, such as X and Y, are significantly correlated, but that the correlation is largely the result of a third, unmeasured variable, such as Z. After Z is controlled in the statistical equation, the association between X and Y falls from statistical significance.

Although it is not possible to rule out confounding in nonexperimental research designs, this does not mean that all unmeasured covariates are capable of confounding associations. Rather—and again, something that is commonly learned in undergraduate methodology courses—the only time that an unmeasured covariate could confound an association is when the unmeasured covariate is related to both *X* and *Y*. If the unmeasured covariate is related only to *X* or *Y*, then it will not confound the association and thus does not necessarily need to be included in the equation.

What this essentially means is that criminologists are always searching to identify covariates that act as potential confounders, except when it comes to genetic confounders. A quick perusal of some of the "top" criminology journals would show that, outside of a handful of studies that directly estimate genetic influences, none of them accounts for potential genetic covariance between an antisocial outcome (e.g. criminal involvement) and an environmental predictor (e.g. delinquent peers, parental socialization, neighborhood disadvantage, etc.). It is commonplace, for example, to read entire issues of criminology journals with studies analyzing measures shown to be genetically influenced, but that simply fail to account for genetic effects on any of the measures.

An example will help to clarify this. For many years there has been an extensive line of research produced examining the link between parental management techniques and variation in child levels of self-control. The results of this body of research have consistently revealed significant associations between different dimensions of parenting and levels of self-control, which most interpret as evidence of a parenting effect on self-control. What is often overlooked is that there is a lot of research showing that variation in self-control is at least 50 percent heritable and that variation in measures of parenting are also around 25 percent heritable (Kendler and Baker, 2007). If the genetic influences on parenting overlap only partially with variation in self-control—and there is a lot of evidence supporting this—then the study and the individual associations are all likely biased in some capacity.

All of this leads to the quintessential question of how these findings from studies that do not control for genetic influences should be interpreted? Without using an appropriate research design capable of controlling for genetic influences, it is difficult to say. On one hand, the findings might not be biased if genetic influences on the outcome measures are orthogonal with the genetic influences on the environmental predictor variables. Existing research suggests that this is unlikely and that there is at least some (and perhaps a substantial amount of) shared genetic influence across outcome and predictor variables such as self-control and delinquency (Boisvert *et al.*, 2012). Therefore, on the other hand, it is quite possible that the findings being produced are biased, with the associations being upwardly inflated. The degree of bias remains to be determined, but the limited available evidence investigating this issue shows that the bias appears to be quite large, with genetic influences completely accounting for some of the associations between well known criminogenic correlates and antisocial behaviors (Wright and Beaver, 2005; Koenen *et al.*, 2006; Wright *et al.*, 2008).

The issue of (genetic) confounding is not rocket science, but rather basic elementary methodological fundamentals. What does that say for criminologists who are actively publishing research that is misspecified due to genetic confounding? To us, it means that it should no longer be acceptable to use a methodological approach that does not partial out the effects of genetic factors. Instead, criminology needs to follow the lead of other disciplines and use genetically sensitive research designs as a way to eliminate genetic confounding and estimate relatively unbiased estimates of environmental influences on criminal and delinquent behaviors. Fortunately, there are several genetically informative research designs that are quite accessible to criminologists and can easily be incorporated into criminological studies. However, not using such techniques leaves criminological research open to empirical scrutiny centered on basic confounding which can easily be leveled not just by biosocial criminologists, but by even the most apathetic undergraduate criminology student.

The second key biosocial finding that criminological research needs to account for is the small effect of the shared environment. Shared environments are those environmental influences that are the same between siblings and that produce similarities between siblings. For example, environments found within the household are typically categorized as shared environments because all siblings in the same household should be exposed to the same familial environments. Other environmental influences, such as cultural and subcultural influences as well as neighborhood-level factors, can also fall under the category of shared environmental influences. Without question, almost all criminological research examines the effects of the shared environment. (As discussed below, studying the effects of another type of environment requires at least two children from the same household, something that criminologists do not do.) The problem, however, is that the shared environmental influence typically accounts for between 0 and 15 percent of the variance in antisocial outcomes (Mason and Frick, 1994; Miles and Carey, 1997; Rhee and Waldman, 2002; Ferguson, 2010). What this means is that, at most, criminologists are able to explain about 15 percent of the variance in crime and delinquency.

Shared environmental influences are frequently employed to explain variation in delinquency, crime, and other antisocial behaviors, and they are often used to explain some of the most consistent criminological findings. Take, for example, the concentration of crime within families. Recent research indicates that about 5 percent of all families produce more than 50 percent of all arrests, and that having one child who was convicted of a crime increases the odds of their siblings also being arrested for a crime by 147 percent (Beaver, 2013). There has been a lot of work done to try to determine what factors account for the similarity of siblings when it comes to crime, and most of this work focuses on shared environmental influences such as family and parenting influences. While there is some evidence supporting these types of explanation, it is clear that these findings are misspecified as they are only able to explain, at most, 15 percent of the variance. Since shared environmental influences and genetic influences are the only two factors that can account for similarity between siblings, it becomes quite feasible that virtually all detected within-family similarities in antisocial behavior are accounted for by shared genetic influences, not shared environmental influences.

Some scholars have argued that the reason the shared environmental influence typically hovers around zero is because there ceases to exist a true shared environment (Turkheimer and Waldron, 2000). The reason for this, according to such logic, is that perceptions shape the way in which ostensibly shared environments are viewed, interpreted, and ultimately acted upon. To illustrate, two siblings from the same household would experience a divorce—an environment that is supposedly shared between siblings—and thus any effect of divorce should impact them both, thus making them more similar. The problem, though, is that one sibling might be in late adolescence and the other in early adolescence, and thus their different points in maturation might produce differences in how they view the divorce and how they ultimately respond to it. Age, in this case, would just be one of many different factors that could work to transform what appears on the surface to be a shared environment into one that is really unique to each child. This same logic could easily be applied to virtually any shared environmental influence. As such, it is easy to see how and why some researchers contend that the shared environment is nonexistent.

Perhaps this is the case, perhaps it is not. From both perspectives, however, the shared environment accounts for minimal to virtually no variance in within-family criminal involvement.

If the shared environment does not exist, then criminologists need to re-evaluate how they measure environmental influences to take this into account. If, however, the shared environment does exist, then criminologists would be justified in their measurement strategies. The problem is that these measurement strategies, while perhaps reliable and accurate, will account for, at most, only 15 percent of the variance within families. Estimates that are significantly higher than this would strongly suggest that the statistical models are highly misspecified and are likely to capture only the effects of confounding influences. As a result, the only confounding influence that would be able to account for similarities within families would be shared genetic influences. Against this backdrop, we recommend that criminologists consider abandoning the futile quest for salient shared environmental contributors to crime and delinquency, and focus their attention on other factors that are shown to be much more important in explaining individual differences in antisocial behavior.

This leads to the third key finding from the current body of biosocial evidence that should result in a significant change to criminological research: the influential role of the nonshared environment. Unlike criminological research that treats all environments as the same (i.e. shared environments), biosocial research separates environments into shared and nonshared. Whereas shared environmental influences operate to make siblings similar to one another, nonshared environments are unique environmental experiences to each sibling and operate to make them different from one another. Nonshared environments are typically thought of as environmental experiences that occur outside the family, such as exposure to different peer groups and unanticipated events. Of course, nonshared environments would also include those environments that appear to be shared, but that are perceived differently or that produce divergent outcomes. Unlike the shared environment, which has been found to account for very little variance in antisocial behavior, the nonshared environment has consistently been found to account for around 40–50 percent of the variance (Mason and Frick, 1994; Miles and Carey, 1997; Rhee and Waldman, 2002; Ferguson, 2010). As a result, when trying to identify the environments that might produce individual differences in antisocial behavior, it is imperative to focus on nonshared environmental experiences.

Unfortunately, criminologists do not study the effects of the nonshared environment. In order to examine nonshared environmental influences, it is required that at least two siblings from the same household are included in the sample. The reason for this is that nonshared environments produce *differences* between siblings. A common way to examine the influence of a nonshared environment is to estimate between-sibling difference scores in environmental measures and between-sibling difference scores in the outcome measure of interest. The environmental difference scores are then examined to determine whether they are related to the difference scores in the outcome measure. If there are statistically significant differences, then this is typically viewed as evidence of a nonshared environmental influence. The most reliable method for examining the presence of a nonshared environmental influence on an outcome measure is by applying the difference-score method to samples of monozygotic (MZ) twins, since this holds all nonshared genetic influences constant.

Since criminologists typically study only one child per household, they are often unable to estimate the influence of nonshared environmental effects and thus are systematically excluding at least 40–50 percent of the variance in antisocial outcome measures. Moving forward, criminologists must analyze samples that contain two siblings per household and employ research designs capable of estimating nonshared environmental influences in a reliable and valid way. Failure to do so will produce another decade of environmental research that is misspecified and that misses the most important source of environmental variation.

Key findings from biosocial research, and how criminological theories need to change to keep pace

Just as biosocial findings should exert a significant influence on the way that criminological research is conducted, these findings should also be used to critique, modify, and perhaps even falsify criminological theories. Recall that some of the key findings to emerge from biosocial research are that genetic factors account for about 50 percent of the variance in antisocial behaviors, shared environmental influences account for about 0–15 percent of that variance, and nonshared environmental influences account for about 40–50 percent. All criminological theories, therefore, must be able to account for these findings in order to be considered correct. If they do not, then such theoretical perspectives need to be modified or discarded. While it is beyond the scope of this chapter to systematically review each criminological theory and then analyze whether or not it is able to account for these key biosocial findings, the focus instead is more general and examines the common assumptions regarding the effects of genetic influences, shared environmental influences, and nonshared environmental influences.

Although genetic influences have been found to have a substantial influence on a wide variety of antisocial behaviors, including crime and delinquency, there is not a single criminological theory that accounts for this empirical regularity. A close examination of the most influential pieces of scholarship testing mainstream criminological theories, such as social learning theory, strain theory, and social bonding theory, shows that they remain mute as to how genetic influences might matter. Moreover, perhaps the most salient criminological theory to emerge in the past 20 years—Gottfredson and Hirschi's (1990) self-control theory—does not just stand silent about genetic influences, but actually argues that the "genetic effect is near zero" (p. 60). Even theories that might be open to genetic influences, such as Agnew's general strain theory, do not directly discuss how genetic factors might explain individual variability in responses to strain, nor do they leave open the possibility that genetic factors account for about 50 percent of the variance. No matter how you cut it, there is not a single existing criminological theory that is well situated to account for about 50 percent of the variance in crime and delinquency. This is a sad state of affairs because it means that either (1) evidence suggesting that antisocial behaviors are strongly influenced by genetic factors is incorrect; or (2) existing mainstream criminological theories are not correct in their current form. Given that the heritability of antisocial behaviors is an empirical question that can be answered via quantitative methods, and given that the results of hundreds of empirical studies have revealed that the heritability of antisocial behaviors is about 50 percent, it is unlikely that the former is correct. This necessarily means that the latter is incorrect. Thus we argue that all existing criminological theories be modified to account for a heritability estimate of 50 percent or, if they are unable to be amended, that they be discarded based on the grounds of falsification. Ultimately the decision lies with the theorists and advocates of these theories to show how they can be modified, or why they should not be falsified in their current form.

Evaluating existing criminological theories in relation to environmental influences is more difficult because most criminological theories focus exclusively on environmental risk factors. Criminological theories, however, do not make the distinction between shared and nonshared environments, even though these two environments were "discovered" more than 35 years ago, and mounds of research have revealed a much stronger nonshared environmental influence versus a shared environmental influence on the antisocial behaviors. Even so, criminological theories appear to focus almost exclusively on the shared environment. Three reasons have led us to this conclusion. First, in an extensive review of the main criminological theories, we were unable to identify any single theory that discusses the differences between shared and nonshared environments, and not one of these theories even mentions the term "nonshared environment." Second, and relatedly, not one of these theories discusses the importance of examining between-sibling differences in crime and delinquency, something that is required for tests of nonshared environmental influences. Third, we are unaware of any quantitative tests of mainstream criminological theories that examine between-sibling differences or that examine the potential influences of nonshared environmental factors. When taken together, there is no evidence that criminological theories recognize the distinction between shared and nonshared environments, and there is no evidence to indicate that the nonshared environment plays a critical part in any criminological theory. This is particularly disconcerting given that, if criminological theories are focused on explaining only shared environmental influences, then at the most they are able to explain only 0–15 percent of the variance in antisocial behavior. Simply by reorienting their focus on nonshared environmental influences, it would be possible to increase the explanatory power to about 50 percent. Once again, it is on the shoulders of theorists and advocates of these theories to modify them or show empirically why a focus on the nonshared environment is unimportant and unnecessary.

In many ways, theories are considered the essence of criminology, with virtually every piece of empirical criminological research being guided and informed by at least one theoretical perspective. The problem, of course, is that if the theories are incorrect, or guide research in the wrong direction, the findings that are produced are likely to be biased or incomplete in some capacity. Theory falsification is supposed to prevent this from occurring, but in criminology very few theories have ever been falsified. However, as the biosocial criminological perspective continues to grow, there can be little doubt that any theory focusing solely on the shared environment, without taking into account genetic and nonshared environmental influences, is destined for falsification at least among the nonideologically minded.

Growing interest in biosocial criminology and some needed changes to the field

Although the amount of biosocial research that has appeared in criminological journals has grown at a rapid pace recently, the discipline in general has not kept pace with these changes. Certainly quantitative research can result in significant changes in a knowledge base, but in order for the field to embrace, or at least be open to, the biosocial perspective, more significant changes need to occur. In this section, we argue that the field of criminology must change in response to the burgeoning amount of interest in the biosocial perspective and the amount of biosocial research that is spreading across the entire discipline. Specifically, we focus on three areas that we have identified as requiring the most change: (1) the publication process; (2) training and education of students; and (3) the composition of faculty. We consider each of these below.

To begin, the publication process in criminology historically stacked the deck against any type of genetic or biological research. This is not necessarily the case today, but biosocial research faces some unique challenges that most other lines of criminological research do not. Indeed, there are still some criminology journals, including some of the "top" criminology journals, that systematically exclude any and all research that is biosocial in orientation. There are at least three reasons. First, the editors of some journals likely still hold anti-biological beliefs, and view biosocial research much more skeptically than other areas of research. While this may still occur to some extent, it is not likely a dominant force at most journals.

The second unique challenge to biosocial research is likely much more salient and much more pervasive. This challenge is that there is an absence of capable and qualified reviewers who are able to provide objective critiques of biosocial research. When a biosocial manuscript is sent for review, it is typically sent to one of four different types of reviewer. There are reviewers who despise biosocial research and will cloak their review in rhetoric and propaganda that do not have any bearing on the quality of the manuscript. These reviewers were quite commonplace a decade ago and, while they still show up frequently, they are much less typical today. Then there are reviewers who are advocates of the biosocial perspective, but who do not know much about conducting biosocial research, including the methodological and statistical approaches used in current research. The end result is that these reviewers often advocate for publication of the manuscript, but are unable to identify and correct any potential problems and errors with the manuscript. Next, there are reviewers from other disciplines, such as behavior genetics or molecular biology, who critique the paper as though it were submitted to a molecular genetics or behavior genetics journal. Oftentimes there is a rift between the standard way of carrying out and presenting analyses in criminology and those of other fields; using the standards of one discipline to critique the paper of another often produces the illusion of a sub-par biosocial paper when, in fact, the paper is bridging the gap between two very distinct disciplines. Finally, there are reviewers who are agnostic about the biosocial perspective and evaluate the manuscript based on its merits. These are likely the most helpful reviewers, but even here there is a problem: they are usually unfamiliar with the methods, statistics, content, and assumptions of biosocial research. This leaves them in a precarious situation, where they are forced to critique the paper using the same approach that they use for all other papers and may gloss over some potentially important shortcomings. This can be problematic as it likely leads to the publication of articles that produce incorrect results because of using a biosocial statistical approach incorrectly (e.g. Haynie and McHugh, 2003).

The above problems with the review process could easily be dealt with in an effective way, but that would require a small but very significant step: training undergraduate and graduate students in the biosocial perspective. Most undergraduate students are not exposed to any significant discussion of contemporary biosocial criminological research. At most, they receive a one-week discussion that centers on an outdated chapter discussing the biology–crime nexus as it relates to Lombroso or *XYY* syndrome. There are only a handful of undergraduate programs that offer even a single course devoted solely to biosocial criminology. What this necessarily boils down to is that undergraduate students are being churned out of criminology programs—even the top criminology programs—without ever learning about the findings emerging from contemporary biosocial research. As a direct result, they likely have no idea that genetic factors matter, nor do they have any idea that there exists a shared environment and a nonshared environment.

Things do not look much more promising for graduate students either, with the same type of educational training occurring. Graduate students are typically not exposed to the biosocial perspective in any sufficient detail. Once again, social/environmental explanations dominate graduate-level curricula, and these students are trained to conduct research that relates directly to these theoretical approaches. These students, therefore, are cycled through another program wherein they graduate without any research or teaching experience in biosocial criminology. The resulting outcome is that they are often skeptical of the biosocial perspective, not because of scientific evidence, but because of fear centered on not knowing the methods and statistical techniques used in biosocial studies. This is particularly problematic given that one day they may be called upon to review a biosocial manuscript submitted for publication.

These new Ph.D.s are then shuffled off to other criminology and criminal justice programs where they can serve as faculty members just to perpetuate the cycle. In order for this cycle to be interrupted, it is imperative that criminology and criminal justice departments become more diverse by including more biosocial criminologists on faculty. To put into context just how rare biosocial criminologists are, consider that across the 42 Ph.D.-granting programs in criminology and criminal justice, only four have any faculty who are actively engaged in biosocial research. This equates to less than 10 percent of all Ph.D.-granting institutions across the country. The small number of biosocial criminologists reflects, in part, that there are only a small number of criminologists worldwide who are actively engaged in this line of research. This has been changing, however, and each year there are more and more new Ph.D.s who are working from a biosocial perspective. Many of these have a strong publication record and would be considered among the top candidates for a position if it were not for what they studied. Unfortunately, criminology and criminal justice programs often are guilty of "content discrimination" where potential candidates are excluded from a position based not on qualifications or publication record, but rather on the content of their research interests. Studying the biosocial correlates to crime, in particular, is often viewed as a threat, and candidates who show an interest in biosocial research are often systematically omitted from any consideration for a position by faculty members. This type of "content discrimination" is often done behind closed doors, but a close look at the faculty interests of most criminology and criminal justice programs shows how effective this bias has become, as there are only a few biosocial criminologists currently on faculty at top-rated criminology and criminal justice programs.

Against this backdrop, criminology and criminal justice programs should be more open to hiring faculty members who are interested in biosocial criminology or, at the very least, treat their application packages just as they would any other candidate. What appears to be lost in all this is that biosocial criminologists *are* criminologists, just as macro-level criminologists are criminologists, individual-level criminologists are criminologists, and life-course/developmental criminologists are criminologists. Placing an adjective before the word "criminologist" might help to identify what theoretical camp they fall into, but in the end they are all criminologists who have the common goal of identifying the potential causes and correlates of crime. Making value judgments about the utility or value of one perspective over another certainly will not lead to a more defensible, objective, and believable knowledge base, but will likely further marginalize the field. Hiring faculty with diverse theoretical and empirical interests will go a long way towards producing well informed and well rounded criminology students, which, after all, should be the goal of any undergraduate or graduate program.

Conclusion

The overarching objective of this chapter is to discuss some of the ways in which the field of criminology needs to change in order to keep pace with evidence produced by biosocial research. In no way is the chapter meant to be exhaustive, as it does not cover other important topics such as gene–environment interactions, the neurobiology of aggression, or epigenetics. Moreover, the chapter is not meant to be definitive, but rather is designed to open a conversation on what is an important emerging topic. Many of the ideas and suggestions discussed here will be viewed as heretical and will not be agreed upon by the majority of criminologists. In short, the criticisms will be ignored and the recommendations will go unheeded. The end result is that much of what is discussed in this chapter will be repeated in future generations of criminologists and in future publications of criminological research. If that is the case, and this entire discussion is for naught, perhaps the best course of action is to follow Max Planck's insightful comments when he stated that "A scientific truth does not triumph by convincing its opponents and making them see the light, but rather because its opponents eventually die and a new generation grows up that is familiar with it." We hope that death is not needed for the biosocial perspective to be taken seriously and enjoy the same respect as other perspectives within the field of criminology.

References

Beaver, K.M. (2011) 'The effects of genetics, the environment, and low self-control on perceived maternal and paternal socialization: results from a longitudinal sample of twins', *Journal of Quantitative Criminology*, 27(1): 85–105.

Beaver, K.M. (2013) 'The familial concentration and transmission of crime', *Criminal Justice and Behavior*, 40: 139–155.

Beaver, K.M., Wright, J.P., DeLisi, M., and Vaughn, M.G. (2008) 'Genetic influences on the stability of low self-control: results from a longitudinal sample of twins', *Journal of Criminal Justice*, 36(6): 478–485.

Beaver, K.M., Boutwell, B.B., and Barnes, J.C. (2013) 'Social support or biosocial support? A genetically informative analysis of social support and its relation to self-control', *Criminal Justice and Behavior*, 41: 1–18.

Boisvert, D., Wright, J.P., Knopik, V., and Vaske, J. (2012) 'Genetic and environmental overlap between low self-control and delinquency', *Journal of Quantitative Criminology*, 28: 477–507.

Button, T.M., Corley, R.P., Rhee, S.H., Hewitt, J.K., Young, S.E., and Stallings, M.C. (2007) 'Delinquent peer affiliation and conduct problems: a twin study', *Journal of Abnormal Psychology*, 116(3): 554–564.

Cleveland, H.H., Wiebe, R.P., and Rowe, D.C. (2005) 'Sources of exposure to the smoking and drinking among adolescents: a behavioral–genetic evaluation', *Journal of Genetic Psychology*, 166: 153–169.

Ferguson, C.J. (2010) 'Genetic contributions to antisocial personality and behavior: a meta-analytic review from an evolutionary perspective', *Journal of Social Psychology*, 150: 1–21.

Gottfredson, M.R. and Hirschi, T. (1990) *A General Theory of Crime*, Stanford, CA: Stanford University Press.

Haynie, D.L. and McHugh, S. (2003) 'Sibling deviance: in the shadows of mutual and unique friendship effects?', *Criminology*, 41: 355–392.

Jaffee, S.R., Caspi, A., Moffitt, T.E., Polo-Thomas, M., Price, T.S., and Taylor, A. (2004) 'The limits of child effects: evidence for genetically mediated child effects on corporal punishment but not on physical maltreatment', *Developmental Psychology*, 40: 1047–1058.

Kendler, K.S. and Baker, J.H. (2007) 'Genetic influences on measures of the environment: a systematic review', *Psychological Medicine*, 37: 615–626.

Kendler, K.S., Jacobson, K.C., Gardner, C.O., Gillespie, N., Aggen, S.A., and Prescott, C.A. (2007) 'Creating a social world: a developmental twin study of peer-group deviance', *Archives of General Psychiatry*, 64: 958–965.

Koenen, K.C., Caspi, A., Moffitt, T.E., Rijsdijk, F., and Taylor, A. (2006) 'Genetic influences on the overlap between low IQ and antisocial behavior in young children', *Journal of Abnormal Psychology*, 155: 787–797.

Larsson, H., Andershed, H., and Lichtenstein, P. (2006) 'A genetic factor explains most of the variation in the psychopathic personality', *Journal of Abnormal Psychology*, 115: 221–230.

Mason, D.A. and Frick, P.J. (1994) 'The heritability of antisocial behavior: a meta-analysis of twin and adoption studies', *Journal of Psychopathology and Behavioral Assessment*, 16: 301–323.

Miles, D.R. and Carey, G. (1997) 'Genetic and environmental architecture of human aggression', *Journal of Personality and Social Psychology*, 72: 207–217.

Rhee, S.H. and Waldman, I.D. (2002) 'Genetic and environmental influences on antisocial behavior: a meta-analysis of twin and adoption studies', *Psychological Bulletin*, 128: 490–529.

Seroczynski, A.D., Bergeman, C.S., and Coccaro, E.F. (1999) 'Etiology of the impulsivity/aggression relationship: genes or environment?', *Psychiatry Research*, 86: 41–57.

Turkheimer, E. (2000) 'Three laws of behavior genetics and what they mean', *Current Directions in Psychological Science*, 9: 160–164.

Turkheimer, E. and Waldron, M. (2000) 'Nonshared environment: a theoretical, methodological, and quantitative review', *Psychological Bulletin*, 126: 78–108.

Vierikko, E., Pulkkinen, L., Kaprio, J., and Rose R.J. (2004) 'Genetic and environmental influences on the relationship between aggression and hyperactivity-impulsivity as rated by teachers and parents', *Twin Research and Human Genetics*, 7: 261–274.

Wright, J.P. and Beaver, K.M. (2005) 'Do parents matter in creating self-control in their children? A genetically informed test of Gottfredson and Hirschi's theory of low self-control', *Criminology*, 43: 1169–1202.

Wright, J.P., Beaver, K.M., DeLisi, M., and Vaughn, M.G. (2008) 'Evidence of negligible parenting influences on self-control, delinquent peers, and delinquency in a sample of twins', *Justice Quarterly*, 25: 544–569.

5

Sociobiology of lethal violence in small-scale societies

Ryan M. Ellsworth and Robert S. Walker

Introduction

When we take off our rose-colored glasses, we see lethal violence as a ubiquitous feature of humanity. It has been observed in virtually all times and places, from the simplest foraging societies to the largest, most complex nation-states, and all societies in between. Given its salient nature, violence has been a long-standing topic of interest to scholars in diverse disciplines. The study of violence and aggression is most useful when it is not only academic but also pragmatic. This requires that we examine evidence rather than intuition; that we be empirical rather than ideological. For a scientific understanding of the causes and consequences of lethal violence can tell us something about the human condition, as well as how, if we wish, to modify it.

Here we review lethal violence and its causes and consequences in small-scale traditional societies. In the first part of the chapter, we address interpersonal lethal violence, or homicide. The second part of the chapter is concerned with intergroup lethal violence, or warfare. Our focus on small-scale societies is not because we are convinced that these societies are superior models of ancestral social dynamics, nor representatives of some pristine human nature. Rather, such societies are the traditional domain of cultural anthropology, and the ethnographic data we present here may be of interest to criminologists otherwise unfamiliar with the literature. Throughout, our primary goal is elucidating the social contexts and associated psychology of lethal violence in small-scale societies that may prove useful in comparison with similar analyses of these phenomena in contemporary nation-states. The data presented are not comprehensive, but illustrate general patterns of lethal violence and its concomitants in traditional societies.

A survey of the ethnographic material on small-scale societies yields three main conclusions: (1) lethal interpersonal and/or intergroup violence are virtually universal characteristics of these societies; (2) males everywhere are the principal perpetrators and victims of lethal violence; and (3) the causes of lethal violence are ultimately connected to reproductive competition. These conclusions carry useful implications for understanding lethal violence in our own society.

Our theoretical approach to the topic of human lethal violence is rooted in evolutionary biology. Specifically, we are concerned with situating these two forms of violence (homicide and warfare) within a larger picture of the evolution of our species by natural selection. Theoretical assessments are developed in the course of data presentation and analysis.

Although some scholars have argued that lethal violence is a product of particular social, ecological, or political conditions (e.g. van der Dennen, 2002; Ferguson, 2011; Fry and Söderberg, 2013), this idea is called into question by evidence of lethal violence across the human societal spectrum—at all levels of social complexity and in all areas of the world (e.g. Keeley, 1996). As will be shown, despite seemingly disparate socioecological circumstances confronting small-scale foraging and agropastoral populations and those of the modern industrial world, the proximate causes of interpersonal and intergroup lethal violence are remarkably consistent, and this continuity is underpinned by a pan-human psychology; a product of an evolutionary history in which humans were, and still are, our own most potent force of nature (Alexander, 1979, 1987, 1989, 1990).

Homicide

Rates

Statistics on homicide in small-scale traditional societies are far from copious, as most ethnographic information on the topic is qualitative and descriptive, rather than quantitative. The matter is further complicated by the fact that it is not always clear whether ethnographers have lumped together deaths due to warfare *and* homicide. Despite this complication, and bearing in mind the discrepancies of authors in their categorization of violent deaths, there are lessons to be gleaned from the existing literature.

Available homicide data show considerable variability. Some societies show extremely high rates, even in the absence of warfare. For example, the Gebusi of New Guinea, between 1940 and 1962, had a homicide rate of 683 per 100,000 per annum (Knauft, 1987: 462; Nivette, 2011: 583). While the Gebusi seem extreme, comparatively high rates of homicide characterize many small-scale societies (Nivette, 2011). !Kung foragers of the Kalahari were reported to have a homicide rate of 29.3 (Lee, 1979). The Semai horticulturalists of Malaysia, renowned among some social scientists for their nonviolent culture, are estimated to have had a homicide rate of 30.3 in the period 1955–77 (Knauft, 1987: 458). Nivette's (2011) compilation of homicide rates for 100 nonstate societies ranges from 0 to 1,450 per 100,000, with an average homicide rate of 182.9. However, the distribution of these homicide rates (which include warfare deaths in some cases where present) is actually bimodal. The left part of the distribution with a mean of around five homicides per 100,000 characterizes peaceful and/or acculturated societies lacking warfare. In contrast, the right part of the distribution with a mean of nearly 500 homicides per 100,000, or two orders of magnitude larger, are those more traditional societies, and those that were sampled during ongoing warfare. Much disagreement among scholars concerns whether the right or the left side of this distribution is more representative of human evolution. In any case, homicide rates are higher overall in small-scale societies than they are among large nation-states (see Eisner, 2012).

Perpetrators and victims are overwhelmingly males. In the 22 cases of homicide recorded among the !Kung, 19 of the victims were male, as were all of the perpetrators (Lee, 1979). Statistics on homicide among African pastoralists and small-scale farmers compiled by Nivette (2011: 588) show that percentages of perpetrators who are male range from 96 percent among the Tiv, to as low as 91 percent among the Gisu, Nyoro, and Luo. The percentage of male victims in these societies are 82, 83, 83, and 75 percent, respectively. Fry and Söderberg's (2013) sample of 148 lethal aggression events in 21 foraging societies shows that in 96 percent of these events, men were the killers. What the statistics on homicide in small-scale societies reveal is that an average adult, especially a man, faced a probability of dying at the hands of another many times that of an average counterpart living in modern nation-states (UNODC, 2011: 9).

In the USA, and other nations, rates of homicide perpetration and victimization are at their highest for males between the ages of 20 and 29 (UNODC, 2011). Demographic patterns of homicide in Western societies are consistent with the young male syndrome hypothesis (Wilson and Daly, 1985). Studies showing a slightly older modal demographic for homicide and warfare perpetration in small-scale societies (Nivette, 2011: 586–588) may be partially explained by the practice of polygyny in the societies where this pattern has been found. Men tend to gain status payoffs and social capital with age, and where these things can be translated into more wives, competition extends beyond the early adult years (e.g. in the age-graded system of polygyny among the Gisu of Uganda [Nivette, 2011] and the Tiwi of northern Australia [Hart and Pilling, 1960]).

Causes of homicide

Regarding the causes of lethal interpersonal violence, many criminologists and other social science researchers have traditionally focused their studies at the level of broad social structural variables such as population density and degree of social stratification, as well as measures of abstract constructs such as social disorganization. One of the most robust and consistent findings of studies of the social structural determinants of homicide is the positive relationship between rates of homicide and indicators of social inequality (see Rosenfeld and Messner, 1991: 53). Because these researchers have largely been concerned with examining trends in modern societies, measures of social inequality have typically been operationalized in terms of income disparity or other financial variables.

In a study of the structural correlates of homicide in a sample of nonindustrial societies by Rosenfeld and Messner (1991), the authors demonstrate that, in contrast to the widely observed patterns in large-scale societies, a close relationship between measures of social inequality and disorganization and rates of homicide does *not* hold in their sample. In discussing their results, Rosenfeld and Messner declare that "the correlates of homicide observed in complex societies cannot be readily generalized to simple nonindustrial societies" (1991: 61).

While the above study has numerous limitations, and further research is needed, there may well be differences in the effects of social structural variables on rates of lethal violence in industrial and nonindustrial societies. However, examination of structural and population variables is only one level of analysis. Other levels include situational contexts precipitating acts of homicide, as well as individual motivation and justification for its execution. It is at these levels where we might observe more consistency cross-culturally, and to which we turn next.

The ethnographic evidence reveals consistency in the proximate motivations for homicide. Fry (2011: 235) has shown that for the 21 foragers included in the Standard Cross Cultural Sample, revenge is the single most commonly reported motive for homicide, followed closely by matters "over a woman." Daly and Wilson (1988: 226), in an examination of 60 societies comprising the Human Relations Area Files (HRAF) Probability Sample, found evidence for the idea of avenging a death with a retaliatory killing in 57 of those societies. Summarizing their survey of the ethnographic literature, they write "In societies from every corner of the world, we can read of vows to avenge a slain father or brother, and of rituals that sanctify those vows—of a mother raising her son to avenge a father who died in the avenger's infancy, of graveside vows, of drinking the deceased kinsman's blood as a covenant, or keeping his bloody garment as a relic" (Daly and Wilson, 1988: 225–226). Indeed, in small-scale societies retributive homicide is often conceived of as a duty, an obligation one owes to slain relatives; killing enters the realm of morality as a righteous act (Beckerman and Valentine, 2008).

Homicide resulting from disputes over women—such as wife-stealing and adultery—are reported for numerous traditional foraging societies, including the Tiwi of north Australia, Eskimos, and the !Kung (see Symons, 1979: 152; Gat, 2000a). Fry's (2011) analysis of 21 foraging societies found disputes "over a woman" to be the second most commonly reported motive for homicide. Among the Gebusi, a principal correlate of "sorcery" homicides is a lack of reciprocity in the exchange of sisters as brides between individuals in different lineages (Knauft, 1987: 465). Thus, Knauft tells us, "In this sense, sorcery homicide is ultimately about male control of marriageable women (1987: 466). As detailed by Daly and Wilson (1988), in many cases, homicides resulting from disputes over women precipitate cycles of revenge killing (Beckerman and Valentine, 2008; Chagnon, 2013).

Homicide is an outcome of conflict resulting from competition. The preponderance of revenge homicide in small-scale societies might be due, in part, to the lack of institutionalized mechanisms of dispute resolution and centralized political authority (e.g. Knauft, 1987; Rosenfeld and Messner, 1991). In small-scale societies, self-help justice for homicide consists of direct retribution—carried out against the offender by the offended parties of the victim (Daly and Wilson, 1988). Often, retributive justice (i.e. retaliatory killing) was a province of the victim's kin group; that is, justice was meted out by the personally aggrieved. An act of homicide, then, is more correctly viewed not merely as a case of interpersonal conflict with a lethal outcome, but as a matter involving the self-interest of the principals' wider network of kin. After all, it is the close kin of the slain who live to face the social consequences of the act, and displays of willingness and ability to repay such insult in kind become important in such milieus. As stated by Daly and Wilson, "it is an act of magnanimity to accept an apology for an accidental affront, but to turn the other cheek in response to deliberate aggression is mere weakness or stupidity. He who forgives a deliberate act of violence simply invites another" (1988: 237).

In many societies, revenge homicide is linked to social status and reputation. In describing blood feuding (revenge homicide) among tribal Montenegrins, Boehm writes, "the tribesmen [who] lived without any centralized governmental power up to 1840—blood feud was not merely acceptable and legitimate; for them it was a moral necessity that a man (or a clan) take vengeance, if a decent social status was to be maintained" (1984: 66). The importance of revenge and status and their relation to lethal violence are developed below.

Sociobiological perspective on demographics and causes of homicide

The demographics of homicide in both modern nation-states (Daly and Wilson, 1988) and traditional societies (Keeley, 1996) evince the same trend: men are overwhelmingly the perpetrators, and their victims are most often other men. The proximate explanation for the preponderance of males (as both victims and perpetrators) in the homicide statistics from all known cultures is multi-faceted and includes factors ranging from hormones (primarily testosterone) to social learning. Such mechanisms can be used to explain, at one level, why among most animals males engage in violent competition much more frequently and intensely than females. However, such proximate explanations do not give us an answer to the important question of *why* males possess hormonal and psychological mechanisms that lead to a greater tendency to engage in violent competitive behavior, including killing. Evolutionary biology provides the theoretical framework for understanding the ultimate causes of these sex differences via sexual selection.

The sex that invests more in the production of offspring is the limiting resource for the reproductive success of the less-investing sex (Trivers, 1972). Thus individuals of the less-investing sex are expected to compete more intensely for access to individuals of the more-investing sex. Across species, females typically invest more in offspring. Because females invest more in offspring and have a lower potential rate of reproduction (Clutton-Brock and Vincent, 1991), their reproductive capacity is a scarce resource in high demand—the object of intense competition among males for sexual access. Differential success in competition leads to higher variance in male reproductive success relative to females. When successful males can sire more offspring with more females and exclude competitors (polygyny), mating competition is a high-stakes game, favoring high-risk, high-reward tactics such as physical aggression (see Daly and Wilson, 1983, 1994; Low, 2000), of which lethal violence is sometimes a consequence. Higher risk of mortality is often the price to be paid in the gamble for high reproductive returns.

This is not to say that females are not competitive. Females do compete, sometimes violently. Female–female competitive aggression, including coalitional aggression, has been observed in several species of nonhuman primates (Hrdy, 1981). In these species, however, in contrast to males, competition is not over sexual access to multiple or specific mates, but primarily over resources crucial to rearing offspring (Daly and Wilson, 1983). Among humans, men often invest significant resources in their mates and offspring, and paternal investment often is a crucial influence on female reproductive success as well as the success of offspring (Gray and Anderson, 2010). Accordingly, human females engage in reproductive competition, including over mates. However, the tactics of women's competition differ from those of men, most notably in the frequency with which violence is used (Campbell, 1999; Benenson, 2013). Among humans, use of lethal violence as a competitive tactic is largely a male domain.

Numerous sources of evidence indicate an evolutionary history of moderate but consistent polygyny in *Homo sapiens* (see Geary, 2010). Thus reproductive competition has been a consistent selection pressure on males. Evolutionary theory supplies the framework for understanding the overrepresentation of males in lethal violent conflicts. The same theory also provides an understanding of the primary causes of lethal violence. As shown above, competition over access to women is a major cause of homicide in small-scale societies, and its logic is straightforward to understand. Females are the most direct vehicles of male reproductive success. Of course, many cases of violent aggression between human males are not so obviously interpretable as mating competition. Humans compete over access to material resources as well as intangibles such as status and respect. Nevertheless, competition for these resources should be viewed as indirect mating effort. Where property, status, or corporate interests correlate with reproductive variance, disputes over them represent reproductive competition just as much as when women themselves are directly at issue. The motivation to accumulate and defend such resources, including via the use of lethal violence, is evidence of its ancestral association with reproductive success. The pursuit and protection of social and economic resources is risky and expensive (in energy and time), and selection would not have favored the motivation and expenditure of effort to get and keep these things if they did not have reproductive payoffs. Status and increased prospects for reproductive success go hand-in-hand in small-scale societies through effects of status and resources on mating success (attracting more mates or appropriating them from competitors), and the ability to deliver superior parental investment (Chagnon and Irons, 1979; Cronk *et al.*, 2000). As elaborated below, in numerous societies there exists a close relationship between social and material resources and the use of lethal violence.

Another question arises from the fact that it is usually kin of the slain who seek retribution for a homicide. Why should it be a victim's kin who are likely to take most offense, and retaliatory action in response to the murder? There are two levels of explanation. On a proximate level, one more intuitive to human reasoning, relatives often share strong emotional bonds, take a keen interest in each others' well-being, and often are heavily dependent on one another in important social and personal matters. At an ultimate level, it is because they share genes identical by common descent. A gene's-eye view of the world (Dawkins, 1976), or inclusive fitness theory, explains why genetic relatives share a common interest in terms of gene propagation. Genes can propagate themselves through direct reproduction of the vehicles in which they reside, and through effects on copies residing in other vehicles. Selection has thus favored genes that dispose individuals to assist others who share those genes through common descent. Viewed in this light, the death of a close relative can negatively impact the success of genes residing in kin. The sentiments and bonds that tie kin together are themselves the product of kin selection. So too, presumably, are the motivations to avenge the killing of one's kin. The utility of the revenge motive may lie in deterrence of further insults to inclusive fitness (Daly and Wilson, 1988).

Warfare

Warfare differs from homicide in that, rather than describing one-on-one interpersonal killing, warfare, while including this element of individuals killing one another, occurs in a context of collective hostility and competition between coalitions. While acts of killing are common features of both homicide and war, it is facile to think of warfare as "just simply aggregated individual aggression" (van der Dennen, 1995). A concise, workable definition of warfare is *lethal aggression waged between two or more coalitions, in which the targets of such aggression are not necessarily a specific individual, but members of a coalition defined as opposed to one's own*. The advantages of this definition lie in its generality and broad applicability to particular manifestations across the societal spectrum. Synonymous terms include intergroup, or coalitionary, lethal violence or aggression (see van der Dennen, 1995: 69–94 for discussion of definitions of war).

Warfare mortality

Mortality from warfare in small-scale traditional societies is variable, but typically high compared with recent wars among nation-states (Keeley, 1996: 195–196). For example, adult mortality due to warfare was approximately 4.5, 10, and 20 percent, respectively, for Anbara, Tiwi, and Murngin foragers of Australia (Bowles, 2009: 1296, table 3). While the Anbara represent a low estimate of warfare mortality, the figures from the Tiwi and Murngin are more typical of those reported for smale-scale societies from other parts of the world. The Mae Enga, Dugum Dani, and Huli of New Guinea show comparable total mortality prevalences at 18.6, 15.5, and 13.2 percent, respectively (Keeley, 1996: 196, table 6.2). Meggitt (1977: 110) reports that 25 percent of all male fatalities are due to warfare, while those of women and children are described only as "very low" (1977: 112). Heider (1970: 128) gives a different figure for Dugum Dani war casualties: 28.5 percent for males and 2.4 percent for females, with 0.48 percent of the population killed yearly in warfare. A different population of Dani studied by Bromley (1962: 23) are reported to have lost 1 percent of the population per year through war. Among the Ache foragers of Paraguay, over the past century external warfare accounted for around 60 percent of all violent deaths (with 36 percent of all adult male deaths accounted for by external warfare) (Hill and Hurtado, 1996: 159, 163). Total warfare-related deaths among two Yanomamö groups are reported at 20.9 and

15.3 percent (Chagnon, 1974), representing 37.4 and 23.7 percent of male deaths, and 4.4 and 6.9 percent of females deaths, respectively. A survey of war-related mortality from a sample of 12 foraging groups found a median rate of 164 per 100,000 (Wrangham *et al.*, 2006).

Wherever we turn up statistical data on war-related mortality, sex differences are evident. Men are significantly more likely to die as a result of war (Keeley, 1996: 196). This sex difference is not always stark. War-related deaths among the Jivaro (Shuar) are reported at 59 percent for males and 27 percent for females (*ibid.*). The figures from the South American societies (Ache and Jivaro) include deaths resulting from significant violent conflict with invasive, nonindigenous populations, which may especially inflate the mortality rates of females compared with eras before contact. However, among the Waorani of Ecuador, death from internal warfare alone was as high as 54 percent for males and 39 percent for females (Beckerman *et al.*, 2009). Whatever the effect of violent conflict with invasive, nonindigenous invaders on female mortality rates, the sex bias is everywhere in the same direction. Adult sex ratios of war casualties in small-scale societies range from 20.8:1 among the Mae of New Guinea to 3.7:1 among Barí horticulturalists of Venezuela (Beckerman and Lizarralde, 1995: 497, table 1). In situations where wife capture is important, the sex ratio of victims is likely to be more male-biased (Walker and Bailey, 2013: 30, table 1 for lowland Amazonia).

Causes of warfare

As with homicide, an often-noted major cause of warfare is revenge (Chagnon, 1988; Keeley, 1996; van der Dennen, 2002; Beckerman and Valentine, 2008; Boehm, 2011; Walker and Bailey, 2013). Walker and Bailey (2013: 31) found that for a large sample of death events from 44 lowland South American societies, 70 percent were attributed to revenge. In an analysis of 40 raids conducted by Ecuadorian Waorani men, revenge was given as a cause in 78 percent (Erickson, 2008). Revenge is the leading cause of wars among the Auyana (Robbins, 1982), Huli (Glasse, 1968), and Dugum Dani (Heider, 1970) of New Guinea. Out of 70 battles recorded among Murngin foragers of Northeast Arnhem Land, 50 "were caused by the desire to avenge the killing of a relative, usually a clansman, by members of another clan (blood revenge)" (Warner, 1930–1931: 458). Tabulations by Keeley (1996: 199–200, tables 8.1 and 8.2) show revenge as a cause of warfare in 93.5 percent of a sample of 169 Western North American Indian societies; 75 percent of a sample of 36 nonstate (small-scale) societies had revenge reported as a cause of war.

Other principal causes of warfare include women (e.g. Keeley, 1996; Chagnon, 2013) and defense or expansion of territory and other economic resources, notably among some horticulturalists in densely populated regions of New Guinea (Meggitt, 1977: 13–14, table 1; Keeley, 1996: 200; but see Heider 1970). More generally, whatever critical resource is scarce and contestable may be a cause of intergroup conflict (see Symons, 1979; Gat, 2000a,b; Low, 2000).

Warfare resulting from conflict over tangible resources such as territory and material wealth is easily understood. Wherever there are groups of individuals who share common interests, goals, or reliance on some resource, these groups cooperate for the collective employment of force in securing and defending these contested resources against like-minded rivals. Women as the contested resource over which cooperative groups, or coalitions, compete is equally well understood in light of the above discussion of sex differences and reproductive competition. No matter how numerically abundant, to the male psychology females are always in short supply; they are defended with vigor, and great risks are taken to secure, capture, and defend a disproportionate share—either of females directly, or the resources needed to get them. Warfare, being a male enterprise, is to be viewed as a male coalitional reproductive strategy (van der Dennen,

1995). In what follows, we first focus on the motive of revenge as cause of warfare. We then turn attention to individual-level consequences of warfare in the form of social rewards.

Revenge and warfare

Revenge warfare differs from revenge homicide in an important way. While revenge homicide involves retribution through the targeted killing of the specific offending individual, revenge warfare—typically carried out by raiding parties using ambush tactics—involves retaliatory killing in which any member of the perpetrator's coalition or current community may be a suitable target; that is, any perceived ally (sometimes including women and children) of the individual offender is a legitimate substitution (Chagnon, 1988; Kelly, 2000).

The centrality of revenge as motive for warfare is in line with the "imbalance-of-power" hypothesis for coalitionary violence (Wrangham, 1999). As the name implies, this hypothesis emphasizes asymmetries, specifically in the ability to use coercion and violence to co-opt contested resources (food, territory, mates). Wrangham's (1999) original formulation of this hypothesis was intended to pertain to between-group conflict in socially cooperative species, and the size of a group's potential fighting force is the important variable underpinning this asymmetry. In addition to sheer numbers, we may add the ability of coalitions to coordinate their action in effective ways. Large differences between groups in their ability to control and defend resources make it profitable (in both proximate and ultimate senses) for numerically larger or otherwise better-equipped coalitions to attack and even eliminate competing groups (see also Otterbein, 1994: 196–197 for a similar argument derived from a different theoretical perspective). As noted by Beckerman (1991), unlike typical interspecific predator–prey dynamics, in which a period of increased predation is followed temporally by a reduction in predation as predators become satiated, the dynamics of intraspecific coalitionary killing follows a different pattern. In the latter case, the rate of killing increases with the degree to which enemies have already been killed. In other words, successful routing of a rival leads to an escalation in attacks by the aggressors.

Among humans, casualties—especially adult male casualties—resulting from intergroup aggression create imbalances of power that risk exacerbation, perhaps to the point of annihilation, if not redressed. For example, among societies such as the Yanomamö, size and integrity of male coalitions are central to military and political influence, and to defense of valued resources, as well as the securing of females as wives via exchange or coercion (Chagnon, 2013; see also Flinn and Low, 1986). Weakening of a coalition through violent deaths of its members reduces men's ability to maintain, as well as gain, these reproductively valuable resources (women, territory, political clout). A coalition that allows itself to be attacked without consequence opens the door to infliction of further injury. "There is consensus that, in situations in which coalitional violence occurs with any frequency at all, any group that does not show itself willing and able to retaliate with blood for blood finds itself at a disadvantage—its resources plundered, its wives and daughters carried off, its men bullied and dispossessed of their homes and lands" (Beckerman, 2008: 8). Revenge as a motivation for aggression is integral to maintenance of a coalition's influence, so much so that several societies have been described as locked in perpetual cycles of revenge-motivated feuding and warfare (e.g. [Dani] Heider, 1970; [Montenegro] Boehm, 1984; [Yanomamö] Chagnon, 1988; [Waorani] Beckerman *et al.*, 2009). Not only revenge raiding but pre-emptive attacks on potentially dangerous outsiders become important in this context as a mechanism of deterrence (van der Dennen, 1995). Among groups such as the Ache (Hill and Hurtado, 1996) and Andaman Islanders (Kelly, 2005), for example, strangers were shot on sight. In Erickson's (2008) study, 18 percent of 40 Waorani raids were pre-emptive attacks.

Revenge-motivated retaliation tends to increases in severity with increasing social distance between competing coalitions (Beckerman, 2008; Chernela, 2008; Valentine, 2008). Whereas internal warfare involves violent aggression waged between coalitions who nevertheless may have many and important ties with one another (through marriage, kinship, etc.), external, or interethnic, warfare can be particularly terrible as a result of the lack of kinship bonds or other ties that bind individuals and groups within a culture. Studies distinguishing between lethal intra-and intergroup violence show telling differences. For example, Walker and Bailey (2013) found that, for lowland South American societies, although internal raiding is more than twice as frequent as external warfare between ethnolinguistic groups, the latter produces higher body counts per raid. Additionally, counter-attacks tend to yield higher victim counts than the original raid. Among foragers of the Andamanese Islands, while internal relations were occasionally punctuated by periods of war, "external war is unremitting and constitutes a condition of existence that defines the boundaries of the niches exploited by two populations" (Kelly, 2000: 118).

The threat or intensity of intergroup conflict can reach levels at which intentions reach something resembling genocide. "The objective of warfare in North Alaska was to *annihilate* the members of the enemy group, men, women, and children . . . a fully successful war thus served to terminate inter-regional relations altogether" (Burch and Correll, 1972: 34, emphasis in original, cited in Gat, 2000b). Beckerman *et al.* (2009: 8137, emphasis added) described a similar pattern in Waorani warfare: "We found no tradition of standing down, even for a short period, after a rough balance of deaths was achieved. Indeed, a balance, although it might occur by chance, was never sought—the goal was to *eliminate* the other side." The Waorani illustrate that warfare is not necessarily more intense and destructive at the interethnic level, as the quotation just cited refers to war among the Waorani themselves (see also Erickson, 2008). Gat (1999) cites several ethnographic reports of complete—or near—annihilation of groups due to intense warfare among the Northwest Coast Indians, Great Plains Indians, Alaskan Eskimo, and New Guineans.

Knauft (1991), in discussing intergroup conflict in simple societies, attempts to explain its paucity relative to its occurrence in larger, more complex societies, stating that "the cost of defense and retaliation against armed aggression is typically great; it is more expeditious to move away. This tends to short-circuit patterns of feud and systematic raiding or warfare" (1991: 402; see also Kelly, 2000). While this may indeed be a strategy adopted among some groups—given there are other places to move away to, which allows resumption of a peaceful existence—it leads to marginalization and life constantly on the brink of demise. As a case in point, the Sirionó of Bolivia, upon their first encounter with Holmberg, were described as malnourished, their population small and widely scattered in isolated pockets of forest, "with a culture strikingly backward in contrast to that of their neighbors" (Holmberg, 1969: 10). The condition of the Sirionó and their culture may be due to the fact that "the entire history of the Sirionó . . . seems to reflect a strategy of retreat rather than attack" (1969: 157–158). The Sirionó were repeatedly subject to lethal raiding for the capture of their women and children by the Yanaingua from the south and by the Baure from the north.

Individual-level consequences of warfare

Warfare has consequences not only for coalitions as a whole, but also for the individuals that comprise them. A study by Bowles (2009) attempted to model the impact of warfare on the evolution of altruistic, or self-sacrificing, behavior. Bowles found support for the hypothesis that ancestral levels of intergroup violence could have played a role in the spread of altruistic traits through group selection. However, this kind of study fails to consider that, in many small-scale cultures, engaging in violent behavior, while perhaps containing the element of altruistic risk, also carries social rewards. In some cases these rewards come in the form of direct fitness benefits. For example, among the Yanomamö, men reap social and reproductive benefits through exhibiting violent behavior in both interpersonal and intergroup conflicts. *Waiteri* (fierceness) is an attribute held in high regard, and as Chagnon (1988) has demonstrated, *unokais*—men who have killed an enemy in warfare—have more wives and children than do men who are not *unokais*. Among the Ache, although men who have killed do not have more wives or higher fertility, they do have higher offspring survival rates compared with men who have not killed (Hill and Hurtado, 1996: 445). In contrast, a study of the Waorani found no reproductive advantage to men who had killed (Beckerman *et al.*, 2009).

Among the Kurelu of New Guinea, there is a term for men who have not killed: *kepu* (Matthiessen, 1962). *Kepu* men are of low social status, and their wives (if they have any) and pigs are apt to be taken from them by successful warriors. "Few kepu men have more than one wife, and many have no wives at all . . . if he has one, [she] might be raped by other men . . . the great 'kains' or war leaders have more than one wife" (Bigelow, 1969: 111). Bigelow also relates that among some Australian aboriginal cultures, men only "would be allowed to marry when they had proven their manhood through a kill in vendetta or war" (1969: 112–113).

Among the Dugum Dani of West New Guinea, men at the bottom of the social ladder are referred to as *gebu*, and as Heider (1970: 90) states, "The gebus are the men who do not fully participate in the normal male activities: they do not fight, they have never killed, they are not married . . ." In discussing leadership, and especially influential high-status men, Heider informs us that "The invariable Dani answer to the question, 'Why is (X) such a big man?', is 'He has kill [*sic*] many people in war'"(1970: 91). Heider's observations give insight into the importance of killing as a criterion of social status among the Dani. "Skill in warfare is a major factor in the prestige of a man" (1970: 92). "The most commonly voiced attribute of an important man is his success in killing; the reason given for a man's being unimportant is that he has not killed; he is *gebu*, coward" (1970: 129). Among the Siuai of Bougainville, Solomon Islands, high-status, influential men, called *mumis*, "were fierce and relentless war leaders. They laid waste to the countryside, and their club-houses were lined with the skulls of people they had slain" (Oliver, 1955: 411). According to some Siuai informants, "only fierce warriors could become mumis" (1955: 419).

What is clear from the above examples is that, in these (and many other similar) societies, men are not considered as having proven themselves *vis-à-vis* other men as either acceptable leaders or acceptable husbands unless they have demonstrated their willingness to engage and succeed in lethal combat. More generally, a recent study by Glowacki and Wrangham (2013) examining cultural rewards of participation in warfare among a sample of 20 small-scale societies (both foragers and horticulturalists) found that the most common reward bestowed on individuals for participation in war was increased prestige. Glowacki and Wrangham also found that a greater risk of dying due to participation in war was associated with a greater number of incentives (both social and material) that individuals might reap in exchange for going to war. Such rewards are argued to be powerful motivations for men to go to war. Harming the enemy can, then, be a means to the pursuit of personal prestige.

The relationship between engagement in intergroup aggression and social esteem raises the interesting question of why, in many societies (perhaps all societies in some circumstances), are individuals rewarded for displays of lethal aggression against those defined as enemies? We argue here that this phenomenon can best be understood from the view of human evolutionary history espoused herein; a history in which within-species competition was the primary selective pressure driving the evolution of those traits that make humans unique. Those individuals who take the risks to impose or redress imbalances of power to the benefit of their coalition and larger network of social alliances are assets in the deadly serious human game of cooperation for competition. As such, demonstrations of willingness and ability to vanquish those identified as collective enemies are highly valued in times of conflict and insecurity. At the same time, a reputation for ferocity and risk-taking serves a warrior in within-group status competition by deterring potential competitors from attempting too many liberties at the principal's expense (Gat, 2000b).

Analyses of the causes and consequences of warfare in small-scale societies are consistent with, and indicative of, a human evolutionary history of pervasive competition. A special and important variety of competition in humans involves the formation of male coalitions in the service of competition with other coalitions over crucial and limited resources (women, land, food, status, etc.). The importance of coalitionary competition is reflected in several physiological and psychological mechanisms that seem designed to function specifically within this context (e.g. Van Vugt, 2009; Bailey *et al.*, 2012; Flinn *et al.*, 2012). The strength and cohesion of a coalition determines the ability of its constituents to achieve the goals of procuring and defending these contested resources, which in turn translate into increased descendant-producing success. Revenge is an important strategy for the maintenance of a coalition's influence where imbalances of power drive exacerbation of competitive abilities. A corollary of the imbalance of power hypothesis is that intergroup conflict serves as a selective pressure favoring greater intragroup cooperation and larger coalitions in the service of intergroup competition (Alexander, 1989). The ubiquity of the revenge motive is testament to its significance in an evolutionary milieu of coalitionary competition. Those who were unmoved by it are not our ancestors. In addition, individuals are rewarded for demonstrating qualities associated with a coalition's success (bravery, belligerence, sacrifice, etc.). To the extent that bestowal of rewards for such exhibitions is not some recent novelty, its influence on the evolution of the human psyche must be substantial. In the next section, we sketch a scenario for the evolution of human coalitionary competition and lethal intergroup violence.

Evolution of intergroup lethal violence in humans: theoretical synthesis

If, as argued above, interpersonal homicide is an outcome of reproductive competition, then warfare must be similarly understood. Following van der Dennen (1995), we argue that warfare is best conceptualized as an aspect of male coalitional reproductive strategies. That is, war is a particularly violent form of human male reproductive competition played out in the context of intragroup cooperation in the service of intergroup competition. In this section, we lay out a scenario for the evolution of this behavior as it appears in modern humans.

Violent aggression is a feature common across primates, including all five species of great apes. The forms of violence, however, differ according to the different social structures of the great apes (Wrangham and Peterson, 1996). In chimpanzees, social structure is multi-male/multi-female troops inhabiting a specified and defended territory. Male aggression is variable in form, one manifestation of which is intergroup lethal violence; that is, attacks carried out by multiple individuals of one troop against victims of another troop (Wrangham, 1999). Leaving human cultural elaborations aside for the moment, this form of chimpanzee violence has parallels with human intergroup violence, especially as exhibited by small-scale human societies. Important parallels between chimpanzee and human intergroup aggression include: small groups of males as the principal aggressors, an apparent strategy of surprise attacks (i.e. raiding incursions with low risk to attackers), and the deliberate searching for opportunities to attack members of another social group (Manson and Wrangham, 1991: 370).

Observed similarities between humans and other apes in some manifestations of violent aggression draw attention to the question of evolutionary origins of these behaviors; that is, whether these commonalities are the result of a shared ancestry, or whether the similarities in patterns of a particular behavior are best understood as convergent solutions independently reached in response to similar adaptive challenges. Are the shared features of human and chimpanzee intergroup lethal violence the direct result of our recent common ancestry?

The parallels between chimpanzee and human intergroup violence has led some researchers to develop models for the evolution of intergroup violence that suggest this behavior in both chimpanzees and humans indeed shares a common origin. Specifically, it is argued that intergroup violence was present in the common ancestor of chimpanzees and hominins, and has been maintained in the two lineages since their divergence around 6 to 7 million years ago (Ghiglieri, 1987; Wrangham, 1987, 1999; Wrangham and Peterson, 1996; for general models of social evolution incorporating assumptions of early hominins as chimplike, see Chapais, 2008; Foley and Gamble, 2009; Grueter *et al.*, 2012).

While Wrangham has deemed such hypotheses untestable (1999: 20), the assumption of homology (or synapomorphy) is called into question upon consideration of the hominin fossil record. Fossil remains of the genus *Australopithecus* indicate a level of sexual dimorphism in body size similar to that observed in modern-day gorillas (Plavcan, 2012). A high degree of sexual dimorphism in body size is associated with one-on-one physical contest competition between males for mates, and high variance in male reproductive success. Across mammals, strong sexual dimorphism in body size is closely correlated with a mating system of harem-defense polygyny wherein single males defend sexual access to multiple females (Gordon, 2006; Plavcan, 2012). Observations of sexual dimorphism among australopithecines, combined with paleoecological evidence on likely resource distribution and home range size for Pliocene (*ca.* 5.3 to 2.5 million years ago) hominins (Foley, 1992), suggests a social structure and mating system similar to that of the gorilla. Thus, notwithstanding the question of the mating system and social structure of our most recent common ancestor shared with chimpanzees around 6 to 7 million years ago, by around 4 million years ago our hominin ancestors were likely living in social groups, the basic units of which comprised a single dominant male, small number of reproductive females, and dependent offspring. Within this milieu, coalitions would have been largely absent, and hence intergroup violence unlikely.

It is only at around 1.8 million years ago, with the appearance of *Homo erectus*, that we begin to find evidence of a significant shift in patterns of hominin sociality. Within the species *Homo erectus* there is a temporal trend of decreased sexual dimorphism in body size (see Geary, 2010). This reduction in dimorphism is consistent with a shift from single-male harem defense polygyny to multi-male/multi-female social groups. Pleistocene advancements in the sophistication of tool manufacture and use in *Homo* were accompanied by movement into a new foraging niche, exploiting energy-rich, large package, difficult to acquire food resources (e.g. roots, nuts, meat) (Kaplan *et al.*, 2000). This dietary shift is argued to have had several important consequences for the life history and demography of evolving *Homo*, including the facilitation of larger group sizes, as well as selection for cooperation in foraging pursuits—especially hunting of large game, which favors male cooperation (Kaplan *et al.*, 2000).

The change in patterns of sociality would have been attended by some degree of reduction in the skew of male reproductive success, as the incorporation of more males into a group makes it more difficult for individual males to retain reproductive monopoly. The strength of coalitional relationships among males at this stage in human evolution is debatable. We argue, however, that the shift toward multi-male social groups from a gorilla-like social structure is likely to have involved a greater degree of sociality not merely among males generally, but rather male kin, as any increase in male sociality would be favored only if it paid off genetically (through benefits to reproductive success or inclusive fitness). In addition to foraging efficiency through cooperation, genetic payoffs would have also been reaped through cooperative relationships directed increasingly at solving challenges posed by competitors. As Alexander (1989: 464) points out, "Cooperation must exist because it has aided reproductive competition, however indirectly." Once selection favors coordinated cooperative action, the stage is set for use of such alliances in intergroup competition for the means to reproductive ends (van der Dennen, 1995, 2002). It is important to note here that increases in intragroup cooperation among males can come only with a reduction in sexual competition; that is, a reduction in polygyny, with monopolization of more than one female being the province of males with particularly high social status or disproportionate control of resources (as observed ethnographically for war leaders and headmen, *unokais* and *mumis*).

Along with the reduction in sexual size dimorphism, there is also a temporal trend toward greater encephalization (brain size) within *H. erectus*. While the causes of rapid and significant encephalization from about 1.8 million years ago to modern proportions are not precisely known, it almost certainly was precipitated by selection pressures related to changes in sociality, specifically changes in the complexity of social dynamics (Humphrey, 1976; Dunbar, 1998; Geary, 2005; Bailey and Geary, 2009; Hrdy, 2009).

Trends in reduced sexual dimorphism, increased brain size, and slower life histories are accompanied by evidence of, among other things, more sophisticated tool manufacture, control and use of fire, and cooperative hunting (Flinn *et al.*, 2005). All of which support Alexander's (1990) ideas of ecological dominance involving a reduction of extrinsic forces of selection (climate, food, predators) and an increased importance of intraspecific social dynamics as the primary selective pressures driving hominin evolution. As mentioned above, increased male sociality in hominin evolution is likely to have involved cooperative relationships initially among kin, employed in the service of solving challenges posed by genetic competitors, which would have included other coalitions of males. In such cases, challenges to reproductive failure entail some measure of (genetic) success for those who have a part in causing the (genetic) failure of others. In chimpanzees, these successes come in the form of direct access to more females as a result of elimination of outgroup males (see Manson and Wrangham, 1991; Wrangham, 1999). The routes to genetic success, however, can be more circuitous than this, involving increases in status, reputations for ferocity, greater access to desirable resources, etc. In each case the ultimate rationale for pursuing the outcome is the same—genetic proliferation greater than the population average. At this level of abstraction, the parallels between chimpanzee and modern human intergroup violence achieve some resolution. But a closer examination of the differences in social structure between humans and chimpanzees show that the similarities may not be as informative as the differences.

Groups of males begin to cooperate because cooperative coalitions can easily eliminate lone individuals, and larger, more coordinated coalitions can eliminate threats from smaller, less cohesive groups. Selection thus favors male coalitions as a reproductive strategy (van der Dennen, 1995). This is the essence of Alexander and others' ideas of "cooperation to compete" (Bigelow, 1969; Alexander, 1989). Competition gives cooperation a selective advantage, while cooperation makes competition a higher-stakes endeavor. If, as mentioned earlier, imbalances of power make it profitable for strong coalitions to attack and eliminate weaker rivals, such a process, in an ecologically dominant species, would be autocatalytic, as imbalances of power created very strong selective pressures favoring ever larger, more cohesive and effective cooperative alliances able to dispatch threats from competitors.

The chasm separating the complexity of human warfare from that of chimpanzees is attributable to features of human social structure not found in chimpanzees, including, minimally, pair-bonding, high paternal certainty, bilateral kinship reckoning, and prescribed marriage involving the exchange of partners between lineages or kin groups. Such features allow vastly more complex and complicated networks of alliance, including alliances and coalitions that extend beyond the local kin group, uniting large numbers of individuals in multiple localities (Rodseth *et al.*, 1991; Chapais, 2008, 2011, 2013; Walker *et al.*, 2011). The extension of alliances beyond the local group is particularly crucial here, as it would have led to rapid escalation of the scale and scope of intergroup competition. To an unprecedented degree, recent human evolution has been characterized by a self-perpetuating runaway social arms race in which cooperation serves competition at all levels of sociality, from dyads to ethnic groups to nation-states (Alexander, 1987).

By contrast, chimpanzee alliances are insular, confined to the local groups males were born into. Another factor constraining the complexity of chimpanzee alliances and intergroup conflict is the lack of agnatic kinship structure among related males (Chapais, 2008). Kinship structure is severely limited among male chimpanzees because of promiscuous mating which results in very low paternity certainty; that is, males do not behave as if they recognize father–offspring relationships, thereby precluding identification of kin through the male line (Langergraber *et al.*, 2007).

The above listed features of human social structure underpin the complexity of human intergroup conflict and warfare, as observed, for example, among the Mae Enga of New Guinea (Meggitt, 1977). Mae Enga warfare occurs at a variety of sociopolitical levels: interphratry warfare, intraclan warfare, warfare between clans of the same phratry, and warfare between clans of different phratries (see also Warner, 1930–1931 on Murngin warfare). Reading of the complicated alliances and logistics involved with each kind of warfare gives one the impression that comparisons between humans and chimps are, in some important sense, increasingly facile.

Conclusion

In this chapter we examine the occurrence, causes, and consequences of lethal violence in small-scale societies. In the course of our examination, we attempt to shed some sociobiological light on homicide and warfare and the concomitants of these phenomena. The evolutionary scenario sketched in the preceding section attempts to place intergroup lethal violence, in particular, into the broader picture of a human evolutionary history increasingly characterized by intragroup cooperation in the service of intergroup competition.

Conclusions about human lethal violence would be questionable were they drawn from a single society, or a single point in time. However, as demonstrated in this chapter, interpersonal and intergroup lethal violence are not phenomena unique to any particular culture or any particular period of time. Lethal violence is pandemic, and its causes pervade every society, in every part of the world, and at all scales of complexity. Evolutionary theory provides us with an ultimate explanation for the demographics and causes of homicide and warfare.

Among the Murngin, a *gaingar* is the most rare, but most deadly, form of warfare. It is only employed when both factions are at their wits' end with deadly violence, "It always follows a long protracted series of killings in which each side is stimulated to an almost hysterical emotional pitch. When one group of clans finally decides to invite its enemies to a *gaingar*, the people always say that this is a spear fight to end spear fights, so that from this time on there will be peace for all the clans and tribes. This is sincerely believed at the time, since it is an effort to stop clan feuds" (Warner, 1930–1931: 473). In a footnote to this passage, Warner makes a remark that was as relevant in 1930 as it is today: "one can only smile at the seeming incongruity of a Stone Age man mouthing the meaty phrases of our twentieth century peace-makers."

Acknowledgement

We thank Martin Daly for useful discussions and feedback during the preparation of this chapter.

References

Alexander, R.D. (1979) *Darwinism and Human Affairs*, Seattle, WA: University of Washington Press.

Alexander, R.D. (1987) *The Biology of Moral Systems*, New York: Aldine de Gruyter.

Alexander, R.D. (1989) 'Evolution of the human psyche', in P. Mellars and C. Stringer (eds), *The Human Revolution*, Chicago, IL: University of Chicago Press, pp. 455–513.

Alexander, R.D. (1990) *How Did Humans Evolve? Reflections on the Uniquely Unique Species*, Museum of Zoology Special Publication No. 1, Ann Arbor, MI: University of Michigan.

Bailey, D.H. and Geary, D.C. (2009) 'Hominid brain evolution: testing climatic, ecological, and social competition models', *Human Nature*, 20: 67–79.

Bailey, D.H., Winegard, B., Oxford, J., and Geary, D.C. (2012) 'Sex differences in in-group cooperation vary dynamically with competitive conditions and outcomes', *Evolutionary Psychology*, 10: 102–119.

Beckerman, S. (1991) 'The equations of war', *Current Anthropology*, 32: 636–640.

Beckerman, S. (2008) 'Revenge: an overview', in S. Beckerman and P. Valentine (eds), *Revenge in the Cultures of Lowland South America*, Gainsville, FL: University of Florida Press, pp. 1–9.

Beckerman, S. and Lizarralde, R. (1995) 'State–tribal warfare and male-biased casualties among the Barí', *Current Anthropology*, 36: 497–500.

Beckerman, S. and Valentine, P. (2008) *Revenge in the Cultures of Lowland South America*, Gainesville, FL: University of Florida Press.

Beckerman, S., Erickson, P.I., Yost, J., *et al.* (2009) 'Life histories, blood revenge, and reproductive success among the Waorani of Ecuador', *Proceedings of the National Academy of Sciences, USA*, 106: 8134–8139.

Benenson, J.F. (2013) 'The development of human female competition: allies and adversaries', *Philosophical Transactions of the Royal Society B: Biological Sciences*, 368: 20130079.

Bigelow, R. (1969) *The Dawn Warriors: Man's Evolution toward Peace*, London: Scientific Book Club.

Boehm, C. (1984) *Blood Revenge: The Anthropology of Feuding in Montenegro and other Tribal Societies*, Lawrence, KA: University Press of Kansas.

Boehm, C. (2011) 'Retaliatory violence in human prehistory', *British Journal of Criminology*, 51: 518–534.

Bowles, S. (2009) 'Did warfare among ancestral hunter-gatherers affect the evolution of human social behaviors?', *Science*, 324: 1293–1298.

Bromley, H.M. (1962) *The Function of Fighting in Grand Valley Dani Society*, Working Papers in Dani Ethnology No. 1, Hollandia/Kota Baru: Bureau of Native Affairs.

Campbell, A. (1999) 'Staying alive: evolution, culture, and women's intrasexual aggression', *Behavioral and Brain Sciences*, 22: 203–252.

Chagnon, N.A. (1974) *Studying the Yanomamö*, New York: Holt, Rinehart, and Winston.

Chagnon, N.A. (1988) 'Life histories, blood revenge, and warfare in a tribal population', *Science*, 239: 985–992.

Chagnon, N.A. (2013) *Yanomamö*, 6th edn, Belmont, CA: Wadsworth.

Chagnon, N.A. and Irons, W. (1979) *Evolutionary Biology and Human Social Behavior*, North Scituate, NJ: Duxbury Press.

Chapais, B. (2008) *Primeval Kinship: How Pair-Bonding Gave Birth to Human Society*, Cambridge, MA: Harvard University Press.

Chapais, B. (2011) 'The evolutionary history of pair-bonding and parental collaboration', in T.K. Shackelford and C.A. Salmon (eds), *The Oxford Handbook of Evolutionary Family Psychology*, Oxford: Oxford University Press, pp. 33–50.

Chapais, B. (2013) 'Monogamy, strongly bonded social groups, and the evolution of human social structure', *Evolutionary Anthropology*, 22: 52–65.
Chernela, J. (2008) 'Guesting, feasting, and raiding: transformations of violence in the northwest Amazon', in S. Beckerman and P. Valentine (eds), *Revenge in the Cultures of Lowland South America*, Gainsville, FL: University of Florida Press, pp. 42–59.
Clutton-Brock, T.H. and Vincent, A.C.J. (1991) 'Sexual selection and the potential reproductive rates of males and females', *Nature*, 351: 58–60.
Cronk, L., Chagnon, N., and Irons, W. (2000) *Adaptation and Human Behavior*, New York: Aldine de Gruyter.
Daly, M. and Wilson, M. (1983) *Sex, Evolution, and Behavior*, 2nd edn, Belmont, CA: Wadsworth.
Daly, M. and Wilson, M. (1988) *Homicide*, New York: Aldine de Gruyter.
Daly, M. and Wilson, M. (1994) 'Evolutionary psychology of male violence', in J. Archer (ed.) *Male Violence*, London: Routledge, pp. 253–288.
Dawkins, R. (1976) *The Selfish Gene*, Oxford: Oxford University Press.
van der Dennen, J.M.G. (1995) *The Origin of War* (Vols 1 and 2), Netherlands: Origin Press.
van der Dennen, J.M.G. (2002) '(Evolutionary) theories of warfare in preindustrial (foraging) societies', *Neuroendocrinology Letters*, 23 (Suppl. 4): 55–65.
Dunbar, R.I.M. (1998) 'The social brain hypothesis', *Evolutionary Anthropology*, 6: 178–190.
Eisner, M. (2012) 'What causes large-scale variation in homicide rates?', *Aggression in Humans and Other Primates: Biology, Psychology, Sociology*, 137–161.
Erickson, P. (2008) 'Revenge, bride capture, and gender strategies for survival among the Waorani', in S. Beckerman and P. Valentine (eds), *Revenge in the Cultures of Lowland South America*, Gainsville, FL: University of Florida Press, pp. 93–116.
Ferguson, R.B. (2011) 'Born to live: challenging killer myths', in R.W. Sussman and C.R. Cloninger (eds), *Origins of Altruism and Cooperation*, New York: Springer, pp. 249–270.
Flinn, M.V. and Low, B.S. (1986) 'Resource distribution, social competition, and mating patterns in human societies', in D. Rubenstein and R. Wrangham (eds), *Ecological Aspects of Social Evolution*, Princeton, NJ: Princeton University Press, pp. 217–243.
Flinn, M.V., Geary, D.C., and Ward, C.V. (2005) 'Ecological dominance, social competition, and coalitionary arms races: why humans evolved extraordinary intelligence', *Evolution and Human Behavior*, 26: 10–46.
Flinn, M.V., Ponzi, D., and Muehlenbein, M.P. (2012) 'Hormonal mechanisms for regulation of aggression in human coalitions', *Human Nature*, 23: 68–88.
Foley, R.A. (1992) 'Evolutionary ecology of fossil hominids', in E.A. Smith and B. Winterhalder (eds), *Evolutionary Ecology and Human Behavior*, Chicago, IL: Aldine de Gruyter, pp. 131–164.
Foley, R. and Gamble, C. (2009) 'The ecology of social transitions in human evolution', *Philosophical Transactions of the Royal Society B: Biological Sciences*, 364: 3267–3279.
Fry, D.P. (2011) 'Human nature: the nomadic forager model', in R.W. Sussman and C.R. Cloninger (eds), *Origins of Altruism and Cooperation*, New York: Springer, pp. 227–248.
Fry, D.P. and Söderberg, P. (2013) 'Lethal aggression in mobile forager bands and implications for the origins of war', *Science*, 341: 270–273.
Gat, A. (1999) 'The pattern of fighting in simple, small-scale, prestate societies', *Journal of Anthropological Research*, 55: 563–583.
Gat, A. (2000a) 'The human motivational complex: evolutionary theory and the causes of hunter-gatherer fighting. Part I. Primary somatic and reproductive causes', *Anthropological Quarterly*, 73: 20–34.
Gat, A. (2000b) 'The human motivational complex: evolutionary theory and the causes of hunter-gatherer fighting. Part 2. Proximate, subordinate, and derivative causes', *Anthropological Quarterly*, 73: 74–88.
Geary, D.C. (2005) *The Origin of Mind: Evolution of Brain, Cognition, and General Intelligence*, Washington, DC: American Psychological Association.
Geary, D.C. (2010) *Male, Female: The Evolution of Human Sex Differences*, 2nd edn, Washington, DC: American Psychological Association.
Ghiglieri, M.P. (1987) 'Sociobiology of the great apes and the hominid ancestor', *Journal of Human Evolution*, 16: 319–258.
Glasse, R. (1968) *Huli of Papua: A Cognathic Descent System*, The Hague: Mouton.
Glowacki, L. and Wrangham, R.W. (2013) 'The role of rewards in motivating participation in simple warfare', *Human Nature*, 24(4): 444–460.
Gordon, A.D. (2006) 'Scaling of size and dimorphism in primates. 1. Microevolution', *International Journal of Primatology*, 27: 27–61.

Gray, P.B. and Anderson, K.G. (2010) *Fatherhood: Evolution and Human Paternal Behavior*, Cambridge, MA: Cambridge University Press.

Grueter, C.C., Chapais, B., and Zinner, D. (2012) 'Evolution of multilevel social systems in nonhuman primates and humans', *International Journal of Primatology*, 22: 1002–1037.

Hart, C.W.M. and Pilling, A.R. (1960) *The Tiwi of North Australia*, New York: Holt, Rinehart, and Winston.

Heider, K.G. (1970) *The Dugum Dani: A Papuan Culture in the Highlands of West New Guinea*, Vol. 49, New Brunswick, NJ: Transaction Publishers.

Hill, K.R. and Hurtado, A.M. (1996) *Ache Life History: The Ecology and Demography of a Foraging People*, New Brunswick, NJ: Transaction Publishers.

Holmberg, A.R. (1969) *Nomads of the Long Bow: The Sirionó of Eastern Bolivia*, New York: Natural History Press.

Hrdy, S.B. (1981) *The Woman that Never Evolved*, Cambridge, MA: Harvard University Press.

Hrdy, S.B. (2009) *Mothers and Others*, Cambridge, MA: Belknap Press.

Humphrey, N.K. (1976) 'The social function of intellect', in P.P.G. Bateson and R.A. Hinde (eds), *Growing Points in Ethology*, Cambridge: Cambridge University Press, pp. 303–317.

Kaplan, H., Hill, K., Lancaster, J., and Hurtado, A.M. (2000) 'A theory of human life history evolution: diet, intelligence, and longevity', *Evolutionary Anthropology*, 9: 156–185.

Keeley, L.H. (1996) *War Before Civilization: The Myth of the Peaceful Savage*, New York: Oxford University Press.

Kelly, R.C. (2000) *Warless Societies and the Origin of War*, Ann Arbor, MI: University of Michigan Press.

Kelly, R.C. (2005) 'The evolution of lethal intergroup violence', *Proceedings of the National Academy of Sciences, USA*, 102: 15294–15298.

Knauft, B.M. (1987) 'Reconsidering violence in simple human societies: homicide among the Gebusi of New Guinea', *Current Anthropology*, 28: 457–500.

Knauft, B.M. (1991) 'Violence and sociality in human evolution', *Current Anthropology*, 32: 391–428.

Langergraber, K.E., Mitani, J.C., and Vigilant, L. (2007) 'The limited impact of kinship on cooperation in wild chimpanzees', *Proceedings of the National Academy of Sciences, USA*, 104: 7786–7790.

Lee, R. (1979) *The !Kung San*, New York: Cambridge University Press.

Low, B.S. (2000) *Why Sex Matters*, Princeton, NJ: Princeton University Press.

Manson, J.H. and Wrangham, R.W. (1991) 'Intergroup aggression in chimpanzees and humans', *Current Anthropology*, 32: 369–377.

Matthiessen, P. (1962) *Under the Mountain Wall: A Chronicle of Two Seasons in the Stone Age*, New York: Viking.

Meggitt, M. (1977) *Blood is their Argument: Warfare among the Mae Enga Tribesmen of the New Guinea Highlands*, Palo Alto, CA: Mayfield Publishing Company.

Nivette, A.E. (2011) 'Violence in non-state societies', *British Journal of Criminology*, 51: 578–598.

Oliver, D.L. (1955) *A Solomon Island Society*, Boston, MA: Beacon Press.

Otterbein, K.F. (1994) *Feuding and Warfare*, Langhorne, PA: Gordon and Breach Science Publishers.

Plavcan, J.M. (2012) 'Sexual size dimorphism, canine dimorphism, and male–male competition in primates: Where do humans fit in?', *Human Nature*, 23: 45–67.

Robbins, S. (1982) *Auyana: Those Who Held Onto Home*, Seattle, WA: University of Washington Press.

Rodseth, L., Wrangham, R.W., Harrigan, A.M., and Smuts, B.B. (1991) 'The human community as primate society', *Current Anthropology*, 32: 221–241.

Rosenfeld, R. and Messner, S.F. (1991) 'The social sources of homicide in different types of societies', *Sociological Forum*, 6: 51–70.

Symons, D. (1979) *The Evolution of Human Sexuality*, Oxford: Oxford University Press.

Trivers, R.L. (1972) 'Parental investment and sexual selection', in B. Campbell (ed.), *Sexual Selection and the Descent of Man*, Chicago, IL: Aldine, pp. 136–179.

UNODC (2011) *Global Study on Homicide*, Vienna: United Nations Office on Drugs and Crime.

Valentine, P. (2008) 'Compelling exchanges: Curripaco revenge and warfare', in S. Beckerman and P. Valentine (eds), *Revenge in the Cultures of Lowland South America*, Gainsville, FL: University of Florida Press, pp. 22–41.

Van Vugt, M. (2009) 'Sex differences in intergroup competition, aggression, and warfare: the male warrior hypothesis', *Annals of the New York Academy of Sciences,* 1167: 124–134.

Walker, R.S. and Bailey, D.H. (2013) 'Body counts in lowland South American violence', *Evolution and Human Behavior*, 34: 29–34.

Walker, R.S., Hill, K.R., Flinn, M.V., and Ellsworth, R.M. (2011) 'Evolutionary history of hunter-gatherer marriage practices', *PloS One*, 6: e19066.

Warner, W.L. (1930–1931) 'Murngin warfare', *Oceania*, 1: 457–494.

Wilson, M. and Daly, M. (1985) 'Competitiveness, risk taking, and violence: the young male syndrome', *Ethology and Sociobiology*, 6: 59–73.

Wrangham, R.W. (1987) 'The significance of African apes for reconstructing human social evolution', in W.G. Kinzey (ed.), *The Evolution of Human Behavior: Primate Models*, Albany, NY: SUNY Press, pp. 51–71.

Wrangham, R.W. (1999) 'Evolution of coalitionary killing', *Yearbook of Physical Anthropology*, 42: 1–30.

Wrangham, R.W. and Peterson, D. (1996) *Demonic Males: Apes and the Origins of Human Violence*, Boston, MA: Houghton Mifflin.

Wrangham, R.W., Wilson, M.L., and Muller, M.N. (2006) 'Comparative rates of aggression in chimpanzees and humans', *Primates*, 47: 14–26.

Part II

Genetic foundations of antisocial behavior

6

Genetic influences on antisocial behavior over the life-course

Catherine Tuvblad

Introduction

Antisocial behavior is a broad term that encompasses a wide variety of behaviors and attitudes that violate societal norms, values and laws. Antisocial behavior includes aggression to those around them, cruelty, violence, theft, arson, and vandalism, as well as other forms of rule violations such as noncompliance, lying, intimidating, and manipulation. As such, antisocial behavior in various forms is of key significance in several psychiatric disorders, including oppositional defiant disorder and conduct disorder in children and adolescents, and antisocial personality disorder in adults (APA, 2004). This type of behavior is prevalent; self-reports have shown that between 50 and 80 percent of all youth participate in some form of antisocial behavior during their development (Rutter *et al.*, 1998). The estimated prevalence rates of oppositional defiant disorder and conduct disorder vary between 1.2 and 15.4 percent, and 1.8 and 16 percent, respectively (Loeber *et al.*, 2000). The prevalence of antisocial personality disorder is estimated to be between 1 and 3 percent in community settings, and up to 30 percent in forensic settings (APA, 2004). Moreover, antisocial behavior tends to co-occur with a range of problem behaviors and disorders, including substance abuse and dependence (Robins, 1966; White *et al.*, 2001; Miles *et al.*, 2002), psychiatric illnesses (e.g. attention deficit hyperactivity disorder, depression, anxiety) and pathological gambling (Rutter *et al.*, 1998; Kim-Cohen *et al.*, 2003; Grant and Potenza, 2007; Tuvblad *et al.*, 2009). The behavior is also associated with various types of psychosocial problems, including unstable relationships, unreliable parenting, and underachievement in education and at work (Moffitt *et al.*, 2002; Rutter *et al.*, 2006). Individuals involved in antisocial behavior not only cause physical and mental health problems for their victims, but antisocial individuals themselves are at higher risk for violence and victimization (Rutter *et al.*, 1998). In addition, antisocial behavior imposes not only a high social burden on society, but also an economic burden, including costs of the criminal justice system and compensation for victims and their families (McCollister *et al.*, 2010).

There is compelling evidence from behavioral genetic research (i.e. the study of inheritance of a specific phenotype typically using twin, sibling and/or adoption data) that heritable influences contribute to the development of antisocial behavior: approximately 50 percent of the total variance in antisocial behavior has been found to be explained by genetic or heritable influences. Most importantly, this line of research has also shown evidence of a large environmental component: environmental influences have been found to explain the remaining 50 percent of the variance in antisocial behavior (see reviews by Miles and Carey, 1997; Rhee and Waldman, 2002; Moffitt, 2005b; Waldman and Rhee, 2006; Burt, 2009). The aim of this chapter is to review the behavioral genetic literature investigating the influence of genetic and environmental factors on antisocial behavior—aggression, delinquency, rule-breaking behavior, and psychopathic personality—across the life span.

The behavioral genetic approach

Behavioral genetic research relies on the different levels of genetic relatedness between family members in order to estimate the absolute and relative contribution of heritable and environmental factors to individual differences in a phenotype of interest, in our case antisocial behavior. The behavioral genetic approach includes two major research designs: (1) studies of twins raised together, and (2) studies of adopted individuals and their biological and adoptive family members. In the past two decades, however, adoption samples have become less accessible (Moffitt, 2005a,b), and studies utilizing large twin, sibling and/or parent–child (multi-generation—based on large population-based registries) samples have instead emerged. Specifically, several large twin registries have been established, including the Minnesota Center for Twin and Family Research (https://mctfr.psych.umn.edu); The Swedish Twin Registry, which includes as many as 85,000 twin pairs (http://ki.se/en/research/the-swedish-twin-registry-1); and the Netherlands Twin Register (www.tweelingenregister.org/en).

In the classical twin design, monozygotic (MZ, identical) twins are assumed to share 100 percent of their genes, whereas dizygotic (DZ, fraternal) twins are assumed to, on average, share 50 percent of their genes. Both monozygotic and dizygotic twin pairs are assumed to share their common environment. By comparing the resemblance for antisocial behavior between groups of monozygotic and dizygotic twins, the total phenotypic variance of antisocial behavior can be divided into additive genetic factors (a^2 or heritability, h^2), shared environmental factors (c^2), and non-shared environmental factors (e^2). Shared environmental factors are experiences common to both twins in a pair and therefore affect them equally. These are influences that refer to the environmental factors that produce similarities in levels of antisocial behavior between twins living in the same home. Non-shared environmental factors, on the other hand, are experiences that affect each twin in a pair in a different way and are therefore contributing to making siblings dissimilar. These are environmental influences that produce differences in levels of antisocial behavior between twins living in the same home. Further, genetic influences or heritability can be divided into additive genetic (i.e. allelic effects add up across loci) and non-additive genetic (i.e. due to dominance and/or epistasis) factors. It is not possible to estimate dominant (non-additive) genetic effects (d^2) and shared environmental effects (c^2) simultaneously as these effects are confounded in twins raised together. Familial influences refer to the combination of genetic and shared environmental factors. Heritability is the proportion of total phenotypic variance due to genetic variation. Broad heritability involves both additive and non-additive genetic variance, whereas narrow heritability refers to additive genetic variance only. Heritability is time- and population-specific, and it is a population not an individual parameter (Neale and Cardon, 1992; Plomin *et al.*, 2001).

A crude estimation of similarity between twins can be obtained by comparing twin intraclass correlations across zygosity groups (Neale and Cardon, 1992). Twice the difference between the monozygotic and dizygotic correlations provides an estimate of the relative contribution of additive genetic influences (*A*) to the phenotypic variation in a trait (Falconer's formula: $A = 2[r_{MZ} - r_{DZ}]$). The contribution of the dominant genetic effect (*D*) is obtained by subtracting four times the dizygotic correlation from twice the monozygotic correlation ($D = 2r_{MZ} - 4r_{DZ}$). The proportion of the variance that is due to shared environmental influence (*C*) is given by subtracting the monozygotic correlation from twice the dizygotic correlation ($C = 2r_{DZ} - r_{MZ}$). Finally, the contribution of the non-shared environmental influences (*E*) can be obtained by subtracting the monozygotic correlation from unit correlation ($E = 1 - r_{MZ}$) (Plomin *et al.*, 2001; Posthuma *et al.*, 2003).

In the sibling adoption design, the phenotypic correlation between adoptive siblings is compared with the phenotypic correlation between biological siblings to estimate the influence of genetic and environmental factors on antisocial behavior. Resemblance between adoptive siblings for a phenotype is an indication of shared environmental factors, whereas the extent to which biological sibling resemblance exceeds that of adoptive siblings is evidence of genetic factors for antisocial behavior (Plomin *et al.*, 2001).

Generalizability and assumptions

Generalizability from twin studies to the general population

Several studies have examined whether twins differ in any significant way from singletons. Findings from these studies together show that twins do not differ from singletons on psychopathology (including antisocial behavior, opposition defiant disorder, and conduct disorder), personality, cognitive ability, academic performance, and brain development (van den Oord *et al.*, 1995; Simonoff *et al.*, 1997; Moilanen *et al.*, 1999; Christensen *et al.*, 2006; Ordaz *et al.*, 2010; Hjern *et al.*, 2012; Barnes and Boutwell, 2013). This suggests that findings from twin samples may be generalized to the larger population. There are, however, several assumptions related to the classical twin design that need to be considered. These assumptions include (1) the equal environment assumption, (2) random mating, and (3) lack of correlation and interaction between genetic and environmental influences.

The equal environment assumption

If the equal environment assumption is not met, then higher correlations among monozygotic twins may be due to environmental factors rather than genetic factors, and the heritability estimate could be overestimated. Several twin studies of various phenotypes have examined the equal environment assumption, and these studies generally report that the equal environment assumption holds for numerous phenotypes such as physical activity, eating behavior, psychiatric disorders (e.g. major depression, generalized anxiety disorder, phobia, alcohol and drug abuse) (Kendler *et al.*, 1993; Hettema *et al.*, 1995; Klump *et al.*, 2000; Xian *et al.*, 2000; Eriksson *et al.*, 2006), as well as attention deficit hyperactivity disorder, oppositional defiant disorder, conduct disorder, aggressive behavior (Cronk *et al.*, 2002; Jacobson *et al.*, 2002; Derks *et al.*, 2006), and antisocial behavior (Tuvblad *et al.*, 2011).

Random mating

Random mating in the parent generation can lead to increased resemblance for dizygotic but not monozygotic twin pairs.That is, non-random (assortative) mating acts to increase the resemblance between dizygotic twins, and thereby bias shared environmental estimates upward and additive genetic effects downward. A significant correlation between spouses for a particular trait or behavior is typically taken as assortative mating (Maes *et al.*, 1998). This assumption is most likely violated when it comes to antisocial behavior, as significant correlations between spouses have been found for antisocial behavior (e.g. Krueger *et al.*, 1998; Taylor *et al.*, 2000b; Maes *et al.*, 2007). For example, an early study using data from the Dunedin sample in New Zealand when the participants were 21 years old found a correlation of $r_p = 0.54$ between couple members' reports of antisocial behavior in their peers (i.e. participants were asked how many of their friends had aggression, personal, alcohol, or drug problems, or had in any way violated the law), which was equal to the correlation for couple members' reports of their own antisocial behavior as measured by a variety of offenses (e.g. theft, force, fraud, vice) (Krueger *et al.*, 1998). This finding indicates that assortative mating for antisocial behavior is considerable and that antisocial individuals seem to be clustering together in groups with similar antisocial peers. This would further suggest that assortative mating should to be taken into account when analyzing or modeling antisocial behavior data. However, the vast majority of studies report finding small or no influences from the shared environment on antisocial behavior (see meta-analyses by Miles and Carey, 1997; Rhee and Waldman, 2002; Moffitt, 2005b; Waldman and Rhee, 2006; Burt, 2009; and Table 6.1 below). This suggests that the influence of assortative mating on antisocial behavior may be minimal. Then again, to find significant shared environmental influences require a lot of power, which previous studies may have lacked. It is also possible that genetic influences on antisocial behavior have been underestimated, and could be larger than the *ca.* 50 percent that these meta-analyses show.

Gene–environment interaction (G × E) and gene–environment correlations

In the classical twin design, it is also assumed that genetic and environmental influences do not interact and are uncorrelated. Regardless, certain environmental experiences have been found to moderate the genetic and environmental influences on antisocial behavior (i.e. G × E), e.g. socio-economic status (Tuvblad *et al.*, 2006), family dysfunction (Button *et al.*, 2005), and family warmth (Rowe *et al.*, 1999). It is also possible that some genetic predispositions may be associated with certain environmental experiences, resulting in a correlation between genes and environments. Gene–environment correlations (r_{ge}) can arise in three different ways (Scarr and McCartney, 1983). (1) Passive r_{ge} occur when genes overlap between parents and their offspring. That is, a child with antisocial parents inherits genetic susceptibility for antisocial and aggressive behavior as well as experiencing an adverse home environment. (2) Evocative/reactive r_{ge} arises when a person's characteristic elicits a particular response from the environment. For example, aggressive children tend to elicit more negative affect and harsh discipline from their parents (Ge *et al.*, 1996; O'Connor *et al.*, 1998). (3) Active r_{ge} is defined as the process whereby an individual actively seeks out environmental situations that are more closely matched to the person's genotype. If the assumption of no gene–environment correlation is violated, heritability estimate for antisocial behavior could include both additive genetic effects *and* the effects of gene–environment correlation, which in turn means that the heritability estimate would be inflated.

Further, in adoption studies there are two important issues to consider: (1) generalizability to the larger population, and (2) selective placement.

Generalizability from adoption studies to the larger population

Many times, the adoptive parents tend to be in good health and from higher socio-economic backgrounds, which would indicate that results from adoption studies may not be generalized to the larger population (Rutter, 2006). Adopted children may also be at greater risk for antisocial behavior compared with non-adoptees, since birth parents giving up their children for adoption may have increased rates of externalizing behaviors, including substance use and dependence, criminal behavior, and aggression (Cloninger *et al.*, 1985; Lewis *et al.*, 2001). Several studies have shown that adopted children have higher rates of antisocial behavior compared with non-adopted children, consistent with the view that adoptees may be at higher genetic risk for antisocial behavior compared with non-adopted individuals (van den Oord *et al.*, 1994; van der Valk *et al.*, 1998; Deater-Deckard and Plomin, 1999). Adoption studies of antisocial behavior have shown that aggressive children elicit more negative environments from their parents (Ge *et al.*, 1996; O'Connor *et al.*, 1998). As mentioned above, O'Connor *et al.* (1998) found that children at genetic risk (based on the biological mother's characteristics) received more negative parenting from their adoptive parents than children not at genetic risk. Given the higher antisocial behavior scores among adoptees compared with non-adoptees, as well as the somewhat greater socio-economic status in at least some of the adoption parents, the generalizability of results from adoption studies to the wider population may be questioned.

Selective placement

Selective placement (i.e. similarities between adoptive and biological parents) for certain traits would result in overestimated correlations between adoptive siblings and thereby inflated shared environmental factors. Although selective placement may occur, specifically for physical traits such as race/ethnicity, direct selective placement is unlikely to occur for antisocial behavior. It is unlikely that children with antisocial biological parents are being placed into homes where the adoptive parents are even more antisocial. This suggests that findings from adoption studies on antisocial behavior are probably not biased as a result of selective placement.

Findings from the early adoption studies

Results from the early adoption studies generally demonstrated that the combination of a genetic predisposition (i.e. psychopathology in biological parents) with a high-risk environment (i.e. adoptive home environment) leads to greater pathology than what would be expected from either factor acting alone or both in an additive combination (Crowe, 1974; Bohman *et al.*, 1982; Cloninger *et al.*, 1982; Cadoret *et al.*, 1983; Mednick *et al.*, 1984; Cadoret *et al.*, 1995). A notable example is a study by Cloninger *et al.* (1982). The interaction of congenital (i.e. whether the biological parents were criminal) and postnatal (i.e. adverse rearing experiences in adoptive placement) background was studied in 862 Swedish men adopted at an early age by non-relatives. When both hereditary and environmental factors were present, 40 percent of the men became criminal; when only hereditary factors were present, 12.1 percent were found to be criminal; when only environmental factors were present, 6.7 percent were criminal; and when neither hereditary nor environmental factors were present, 2.7 percent were criminal. The combination of both hereditary and environmental factors (40 percent) was found to be greater than either hereditary (12.1 percent) or environmental (6.7 percent) factors, or both in an additive combination (18.8 percent), suggesting that hereditary and environmental factors interact in the development of antisocial and criminal behavior. This finding was later replicated in females (Cloninger and Gottesman, 1987).

It should be kept in mind when reviewing findings from adoption studies that in this design, the genetic or heritable risk factors are considered in a very general way. Similarly, the environmental risk factors as indexed by certain traits in the adoptive parents or characteristics of their home environment do not specify the exact nature of the child's experiences or how these result in various outcomes.

Meta-analyses examining the genetic and environmental etiology of antisocial behavior

To date, there have been only a few studies summarizing behavioral genetic studies examining genetic and environmental influences on antisocial behavior. An early study summarized findings on aggressive behavior. A total of 24 independent twin and adoption studies were included in the final analyses. These studies varied in sample size from 21 to 410 twin pairs, participants were between 4 and 68 years of age, and self-reports, parent ratings, as well as observational ratings were included. Heritable influences were found to explain about half of the total variance in aggressive behavior, and the non-shared environment explained the remaining 50 percent (Miles and Carey, 1997).

A more recent review included 19 twin and adoption studies on aggression and delinquent/rule-breaking behavior. Participants were between 1 and 18 years of age. Heritability was found to explain 65 percent, shared environment explained 5 percent, and the non-shared environment explained the remaining 30 percent of the variance in aggressive behavior. Heritability explained 48 percent of the total variance indelinquent/rule-breaking behavior, shared environment explained 18 percent, and the non-shared environment explained the remaining 34 percent. Thus genetic influences were larger for aggressive behavior than for delinquent/rule-breaking behavior, whereas shared environmental influences were larger for delinquent/rule-breaking behavior than for aggressive behavior. The pattern of higher genetic influences on aggressive behavior than for delinquent/rule-breaking behavior, and higher shared environmental factors on delinquent/rule-breaking behavior than for aggressive behavior, was consistently reported across all informants, i.e. self-reports, parent (mother and father) ratings, and teacher ratings. Aggressive and delinquent/rule-breaking behavior also showed different etiological patterns across age. Genetic influences on aggressive behavior were increasing across development, whereas shared environmental factors were decreasing (ages 1–5 years A = 55 percent, C = 19 percent, E = 26 percent; ages 6–10 years A = 63 percent, C = 14 percent, E = 23 percent; ages 11–18 years A = 63 percent, C = 3 percent, E = 34 percent, where A = genetic, C = shared environmental, and E = non-shared environmental factors). Delinquent/rule-breaking behavior, on the other hand, showed a decrease in genetic influences across development while shared environmental influences remained stable (ages 1–5 years A = 58 percent, C = 14 percent, E = 28 percent; ages

6–10 years A = 60 percent, C = 16 percent, E = 24 percent; ages 11–18 years A = 46 percent, C = 17 percent, E = 37 percent). These findings clearly suggest that there are etiological differences between aggressive and delinquent/rule-breaking behaviors; aggressive behavior is influenced primarily by genetic factors, whereas delinquent/rule-breaking behavior is influenced by both genetic and shared environmental factors. The genetic and environmental variance components did not vary markedly across sex (Burt, 2009). One limitation with this meta-analysis is the exclusion of adult samples; regardless, the majority of research examining the genetic and environmental influences on aggressive behavior and delinquency has been carried out on child and adolescent samples.

In a follow-up study, Burt (2012) examined the extent of the genetic and environmental etiological overlap between aggressive behavior and delinquent/rule-breaking behavior. Thirteen twin and sibling studies examining the co-variation between aggressive behavior and delinquent/rule-breaking behavior were collected, of which 11 (with 12 independent samples) were included in the final analyses (*n* = 12,923 twin/sibling pairs). Results showed that genetic influences on aggressive behavior were largely, but not entirely, distinct from those on delinquent/rule-breaking behavior: only 38.4 percent of the genetic influences on aggressive behavior overlapped with those on delinquent/rule-breaking behavior. Similarly, only 10.2 percent of the non-shared environmental influences on aggressive behavior overlapped with those on delinquent/rule breaking behavior. Although the genetic and environmental influences on aggressive behavior are partially distinct from those on delinquent/rule-breaking behavior, age and informant were found to moderate the extent of the overlap (Burt, 2012).

Further, Rhee and Waldman (2002) summarized the results from 51 twin and adoption studies on antisocial behavior (10 independent adoption studies and 42 independent twin studies from 51 studies in children, adolescents, and adults). The studies included examined criminal behavior, delinquency, psychopathic personality, conduct disorder, antisocial personality disorder, as well as aggressive behavior. Self-reports, report by others (i.e. parents and teachers), official records, and objective measures were compared. As antisocial behavior is considered to be a heterogeneous phenomenon, operationalization of antisocial behavior, assessment method, zygosity determination, and age and sex of the participants were examined as potential moderators. No significant differences in the magnitude of the genetic and environmental variance components between males and females were found. However, the other variables included significantly moderated the magnitude of the genetic and environmental influences on antisocial behavior. Specifically, age significantly moderated the genetic and environmental influences on antisocial behavior. A full ACE model (i.e. a model estimating A = genetic, C = shared environmental, and E = non-shared environmental factors) best described the data in children (A = 46 percent, C = 20 percent, E = 34 percent), adolescents (A = 43 percent, C = 16 percent, E = 41 percent), as well as adults (A = 41 percent, C = 9 percent, E = 50 percent). Thus, based on these findings, genetic factors seem to be fairly stable across the life span, whereas shared environmental factors seem to be slightly decreasing. Finally, results from the best-fitting overall model showed that genetic factors explained 41 percent of the variance in antisocial behavior (32 percent by additive genetic factors and 9 percent by non-additive genetic factors), 16 percent was explained by shared environmental factors, and the remaining 43 percent of variance was explained by non-shared environmental factors. A clear strength with this study is the inclusion of both twin and adoption samples which make it possible to estimate both non-additive genetic factors and shared environmental factors simultaneously.

In a follow-up study, Waldman and Rhee (2006) included only studies specifically examining genetic and environmental influences on psychopathic personality (six studies from their 2002 publication: Loehlin and Nichols, 1976; Loehlin *et al.*, 1987; Torgersen *et al.*, 1993; Gottesman and Goldsmith, 1994; Brandon and Rose 1995; DiLalla *et al.*, 1996) as well as two additional studies that were published in 2003: Blonigen *et al.*, 2003; Taylor *et al.*, 2003). Specifically, Taylor *et al.* (2003) included a sample of boys who had completed the Minnesota Temperament Inventory (MTI) (Loney *et al.*, 2002), a 19-item measure that contains two separate subscales: Antisocial and Detachment. Genetic factors explained 39 percent of the variance in the Antisocial subscale and 42 percent of the variance in the Detachment subscale, the remaining variances in these two subscales were explained by non-shared environmental factors (Taylor *et al.*, 2003). In the Blonigen *et al.* (2003) study, psychopathic personality was measured using the psychopathic personality inventory (PPI) (Lilienfeld and Andrews, 1996) in 353 male twins from the Minnesota Twin Registry. Genetic factors explained 40 percent of the variance in psychopathic personality and non-shared environmental factors explained the remaining 60 percent. Further, one shortcoming with this follow-up study by Waldman and Rhee (2006) was the exclusion of work by Livesley *et al.* (1993, 1998) showing that callousness is influenced by genetic and non-shared environmental factors. Regardless, using these nine independent samples revealed that 49 percent of the variance in psychopathic personality was due to genetic factors and the remaining 51 percent was attributable to the non-shared environmental factors (Waldman and Rhee, 2006).

Table 6.1 summarizes studies conducted after 2003 examining genetic and environmental influences on psychopathic personality in children, adolescents, and adults. The various measurement tools utilized in these studies are also summarized in Table 6.1. A wide variety of measurement tools have been used including, among others, the Antisocial Process Screening Device (APSD; Frick and Hare, 2002); Child Psychopathy Scale (CPS; Lynam, 2002); Callous-Unemotional traits (Frick *et al.*, 2003); Minnesota Multiphasic Personality Inventory (MMPI; Dahlstrom and Welsh, 1960); Multidimensional Personality Questionnaire (MPQ; Tellegen and Waller, 2008); Minnesota Temperament Inventory (MTI; Loney *et al.*, 2002); Self Report Psychopathy Scale III (Williams *et al.*, 2007); and the Youth Psychopathy Inventory (YPI; Andershed *et al.*, 2002). The studies also vary as to how psychopathic personality is being measured, i.e. self-reports, parent ratings, and teacher ratings. Most commonly self-reports are being used, with the exception of RFAB (Baker *et al.*, 2013), which in addition to self-reports also used parent ratings (see also Bezdjian *et al.*, 2011a,b and TEDS: Trouton *et al.*, 2002, in which teacher ratings were included; also Viding *et al.*, 2005, 2007; Fontaine *et al.*, 2010). Even though the prevalence of psychopathy has been found to be lower among females than males in correctional, psychiatric, and community samples (Nicholls *et al.*, 2005), the majority of the studies presented report finding no significant differences in the magnitude of the genetic and environmental variance components, but these could be constrained to be equal across males and females. So, despite the consistent sex difference in prevalence rates of psychopathy between males and females, the underlying genetic and environmental etiologies appear to be similar across the two sexes. There could, however, still be biological and social differences that could explain the higher prevalence rate in males. Certain environmental experiences could act as moderators that in turn lead to greater expression of a genetic predisposition for psychopathy in males.

According to the weighted average twin correlations summarized in Table 6.1, the heritability of psychopathic personality in males is *ca.* 64 percent [using the Falconer equation described above: $[2(r_{MZ} - r_{DZ})] \rightarrow 2(0.526 - 0.207)$], shared environmental factors account for 11 percent [$(2r_{DZ} - r_{MZ}) \rightarrow (2 \times 0.207 - 0.526)$], and the non-shared environment (e^2)

Table 6.1 Effect sizes (correlations) for psychopathic personality

Author, year	Measurement tool	Informant	Age (mean)	Project N	Male/female (N)	MZ r (N)	DZ r (N)	Subscales	A (%)	C (%)	E (%)
Beaver *et al.*, 2011	FFM	Self-report	Data collected 1994–95 in 132 middle and high schools	Add Health	M + F	0.38* (614)	0.18* (904)	Psychopathic personality	37		63
Bezdjian *et al.*, 2011a,b	CPS	Parent ratings and self-report	9–10 years (9.6)	RFAB *N* = 1219	M	0.63* (276)	0.11 (168)	Callous/disinhibited (parent)	64	–	36
					F	0.53* (278)	0.30* (194)		49	6	44
					OS		0.26* (294)				
					M	0.39* (276)	0.17 (168)	Callous/disinhibited (self-report)			
					F	0.34* (278)	0.08 (194)				
					OS		0.13 (294)				
					M	0.50* (276)	0.02 (168)	Manipulative/ deceitful (parent)	46	1	53
					F	0.58* (278)	0.35* (168)		58	4	37
					OS		0.27* (294)				
					M	0.42* (276)	0.11 (168)	Manipulative/ deceitful (self-report)			
					F	0.25* (278)	0.23* (194)				
					OS		0.04 (294)				
Blonigen *et al.*, 2005	MPQ Fearless dominance	Self-report	17 years	MTFS *N* = 1252	M	0.42 (376)	0.20 (202)	Fearless dominance	46		54
					F	0.45 (446)	0.21 (228)		45		55
					M + F	0.44 (822)	0.22 (430)		45		55
	MPQ Impulsive antisociality				M	0.51 (376)	0.17 (202)	Impulsive antisociality	51		49
					F	0.49 (446)	0.28 (228)		48		52
					M + F	0.50 (822)	0.24 (430)		49		51
Blonigen *et al.*, 2006	MPQ	Self-report	17, 24 years	MTFS *N* = 1252	M	(376)	(202)				
					F	(446)	(228)				
					M + F	(822)	(430)				

(*Continued*)

Table 6.1 Continued

Author, year	*Measurement tool*	*Informant*	*Age (mean)*	*Project N*	*Male/female (N)*	*MZ r (N)*	*DZ r (N)*	*Subscales*	*A (%)*	*C (%)*	*E (%)*
	Fearless dominance		17 years		M + F	(822)	(430)	Fearless dominance	48		52
			24 years		M + F	(822)	(430)	Fearless dominance	42		58
			24 years		M + F	(822)	(430)	Impulsive antisociality	49		51
Brook *et al.*, 2010	NZ – MPQ	Self-report	(47.8)	VETR *N* = 545	M	0.48 (269)	0.17 (276)	Fearless dominance	51		49
						0.35 (269)	0.16 (276)	Impulsive antisociality	32		68
Fontaine *et al.*, 2010	Callous-unemotional (C/U) traits	Teacher ratings	7, 9, 12 years	TEDS *N* = 9462	M (4476) F (4986) OS	0.81 0.69	0.33 0.78 0.13	C/U stable high	78 0 n.s.	1 n.s. 75	21 25
					M (4476) F (4986) OS	0.51 0.72	0.30 0.60 0.32	C/U increasing	58 26 n.s.	3 n.s. 47	39 27
					M (4476) F (4986) OS	0.62 0.80	0.32 0.53 0.36	C/U decreasing	61 54	2 n.s. 26	37 20
					M (4476) F (4986) OS	0.73 0.80	0.45 0.48 0.36	C/U stable low	68	8	24
Forsman *et al.*, 2008 (see Larsson *et al.*, 2006 for results when the twins were 16 years)	YPI	Self-report	16, 19 years	TCHAD *N* = 2196							

			19 years	$N = 1554$	M	0.46*	0.19	Grandiose/ manipulative			
					F	0.63*	0.27*				
					OS		0.23*				
					M	0.41*	0.05	Callous/unemotional			
					F	0.49*	0.24*				
					OS	–	0.12*				
					M	0.59*	0.12	Impulsive/ irresponsible			
					F	0.55*	0.07				
					OS	–	0.31*				
					M + F			Total YPI	63		37
Hicks *et al.*, 2012	MPQ	Self-report	17 years	MTFS $N = 2604$	M (1239)	0.45	0.18	Fearless dominance	46		54
					F (1365)	0.47	0.16		45		55
					M (1239)	0.52	0.22	Impulsive antisociality	52		48
					F (1365)	0.57	0.30		48	8	44
Jang *et al.*, 2000	DAPP-DQ	Self-report	18–86 years (32.5)	Vancouver and London, Canada $N = 1318$	M + F	0.50* (670)	0.25* (648)	Dissocial behavior			
Larsson *et al.*, 2006 (see also Larsson *et al.*, 2007)	YPI	Self-report	16–17 years	TCHAD $N = 2198$	M	0.48* (386)	0.30* (328)	Grandiose/ manipulative	51	3 n.s.	46
					F	0.61* (452)	0.26* (366)				
					OS		0.25* (666)				
					M	0.47* (386)	0.26* (328)	Callous/unemotional	43		57
					F	0.44* (452)	0.17* (366)				
					OS		–0.06 (666)				
					M	0.57* (386)	0.20* (328)	Impulsive/ irresponsible	56		44
					F	0.57* (452)	0.31* (366)				
					OS		0.22* (666)				
								Total YPI	63		37

(*Continued*)

Table 6.1 Continued

Author, year	*Measurement tool*	*Informant*	*Age (mean)*	*Project N*	*Male/female (N)*	*MZ r (N)*	*DZ r (N)*	*Subscales*	*A (%)*	*C (%)*	*E (%)*
Vernon *et al.*, 2008	SRP III-R12	Self-report	17–92 years (41.42)	US–Canada *N* = 278	M + F	0.54* (150)	0.44* (128)	Psychopathy	64	4 n.s.	32
Viding *et al.*, 2005	APSD + SDQ	Teacher ratings	(7.10)	TEDS *N* = 7374	M + F			Callous/unemotional scale	67	6 n.s.	27
Viding *et al.*, 2007	APSD + SDQ	Teacher ratings	7 years	TEDS *N* = 3198	M	0.72* (534)	0.32* (508)	Callous/unemotional traits	67	4 n.s.	29
					F	0.67* (612)	0.44* (562)		48	20	32
					OS		0.32* (982)				
				Weighted average	M	0.526	0.207				
					F	0.561	0.319				
					OS		0.225				
					M + F	0.472	0.266				

Notes: *$P < .05$; *MZ* = monozygotic; *DZ* = dizygotic; OS = opposite-sex dizygotic twins; r = correlation; *A* = genetic factors/heritability; *C* = shared environmental factors, *E* = non-shared environmental factors.

Assessment tools: APSD = Antisocial Process Screening Device; SDQ = Strengths and Difficulties Questionnaire; CPS = Child Psychopathy Scale; CU traits = Callous-Unemotional traits; DAPP-DQ = Dimensional Assessment of Personality Problems – Differential Questionnaire; FFM = Five Factor Model – The Big Five Personality Traits; MMPI = Minnesota Multiphasic Personality Inventory; MPQ = Multidimensional Personality Questionnaire; MTI = Minnesota Temperament Inventory; NZ MPQ = New Zealand Multidimensional Personality Inventory; SRP III-R12 = Self Report Psychopathy Scale III; YPI = Youth Psychopathy Inventory. For citation for each assessment tool see respective first-author paper in column Author, year.

Projects: Add Health = National Longitudinal Study of Adolescent Health (Udry, 2003); MTFS = Minnesota Twin and Family Study, MISTRA = Minnesota Study of Twins Reared Apart; MTR = Minnesota Twin Registry (Lykken *et al.*, 1992; Iacono and McGue, 2002; Iacono *et al.*, 2006); RFAB = University of Southern California Risk Factors for Antisocial Behavior Twin Study (Baker *et al.*, 2002, 2006, 2013); TCHAD = Twin Study of Child and Adolescent Development (Lichtenstein *et al.*, 2007); TEDS = Twins Early Development Study (Trouton *et al.*, 2002); US–Canada (Vernon *et al.*, 2008); Vancouver and London, Canada (Livesley *et al.*, 1993, 1998; Jang *et al.*, 2000); VETR = Vietnam Era Twin Registry (Eisen *et al.*, 1987; Goldberg *et al.*, 2002).

accounts for the remaining 25 percent of the variance. The heritability of psychopathic personality in females is approximately 48 percent $\{[2(r_{MZ}-r_{DZ})] \rightarrow 2(0.561 - 0.319)\}$, shared environmental factors account for 8 percent $[(2r_{DZ} - r_{MZ}) \rightarrow (2 \times 0.319 - 0.561)]$, and the non-shared environment (e^2) accounts for the remaining 44 percent of the variance. It should be kept in mind that this is a descriptive approach and that formal modeling is necessary to determine how well these estimates describe the observed data. Also, this approach does not allow for actual testing of different hypotheses: (1) to test for sex differences, i.e. to test whether the genetic and environmental variance components can be constrained to be equal in males and females; or (2) to test whether it is possible to constrain any of these effects to zero. For example, the shared environmental components for psychopathic personality are small (11 percent in males and 8 percent in females), suggesting that they may not be significant and could be set to zero.

In conclusion, based on these meta-analyses, genetic (additive and non-additive genetic effects) and non-shared environmental factors are contributing to the variance in antisocial behavior across the life span. Shared environmental factors are also of importance for antisocial behavior; however, the effect of the shared environment is generally found to be small or non-significant.

Longitudinal studies examining genetic and environmental influences on antisocial behavior

Longitudinal studies have provided information on different developmental patterns through which youth develop antisocial behavior, and various subgroups or *developmental trajectories* have received empirical support, e.g. stable high (i.e. childhood onset/life-course persistent), decreasing (i.e. childhood limited), increasing (i.e. adolescent onset/transitory), and stable low (i.e. low prevalent across development) (Broidy *et al.*, 2003; Odgers *et al.*, 2008; van Lier *et al.*, 2009; Xie *et al.*, 2011; Mata and van Dulmen, 2012). An adult onset group ('late bloomers'; DiLalla and Gottesman, 1989) has also been identified (Fontaine *et al.*, 2009). Evidence from twin studies suggests that genetic influences contribute more strongly to stable high/childhood onset than to increasing/transitory antisocial behavior (Lyons *et al.*, 1995; Taylor *et al.*, 2000a; Silberg *et al.*, 2007; Tuvblad *et al.*, 2011). For example, an early study reported that shared environmental factors were more important for antisocial behavior with an adolescent onset, whereas genetic influences were stronger for antisocial behavior occurring in adulthood (Lyons *et al.*, 1995). Early onset delinquent behavior in 11-year-old boys was found to be more strongly influenced by genetic factors than late-onset delinquent behavior (Taylor *et al.*, 2000a). Certain developmental trajectories of antisocial behavior (e.g. stable high/childhood onset) have also been found to be particularly problematic with poor long-term outcomes (APA, 2004). Further, several developmental and life-course theories aiming at explaining the development of antisocial behavior across age have been suggested (DiLalla and Gottesman, 1989; Achenbach, 1991; Frick *et al.*, 1993; Loeber and Hay, 1997), including Moffitt's developmental taxonomy of life-course persistent and adolescent-limited antisocial behavior (Moffitt, 1993) and Farrington's integrated cognitive antisocial potential theory (ICAP) (Zara, 2010).

In addition, several longitudinal twin studies have examined the genetic and environmental contribution to the *stability* in antisocial behavior. Antisocial behavior (i.e. delinquent and rule-breaking behavior (Junger-Tas *et al.*, 1994) was measured on four occasions from ages 8 to 20 years. The stability of antisocial behavior over time was explained by a common latent antisocial behavior factor. A common genetic factor accounted for 67 percent of the total variance in this latent factor, the shared environment explained 26 percent, and the remaining 7 percent was due to the non-shared environment. Significant age-specific shared environmental factors were found at ages 13–14 years, suggesting that common experiences (e.g. peers) are important for antisocial behavior at this age (Tuvblad *et al.*, 2011). Moreover, there are only a few longitudinal studies that have examined the genetic and environmental influences on psychopathic personality. One study reported that 58 percent of the stable variance in Fearless Dominance and 62 percent of the stable variance in Impulsive Antisociality (Multidimensional Personality Questionnaire (MPQ), Tellegen, n.d.) from ages 17 to 24 years was due to genetic factors and the remaining variance was explained by the non-shared environment (Blonigen *et al.*, 2006). Another study reported that the stability in psychopathic personality as measured by the Youth Psychopathy Inventory (YPI; Andershed *et al.*, 2002) from ages 16–17 to 19–20 years was explained primarily by genetic influences with no significant contribution from the shared and non-shared environment (Forsman *et al.*, 2008). Thus these results together indicate that genetic and, to a lesser extent, shared environmental influences are important in antisocial behavior that persists from childhood/adolescence through early adulthood.

Yet other twin studies have examined the genetic and environmental contribution to *change* in antisocial behavior. This line of research has generally reported that *change* or 'new' variance in antisocial behavior is largely due to non-shared environmental factors (i.e. experiences that make siblings dissimilar) (e.g. Eley *et al.*, 2003; van Beijsterveldt *et al.*, 2003; Haberstick *et al.*, 2006a). For example, a study using three waves of data from the Minnesota Twin and Family Study, utilizing biometric latent growth curve models, showed that genetic influences were largely responsible for initial level (stability) of antisocial personality disorder symptoms from late adolescence to mid-adulthood, while non-shared environmental influences were primarily responsible for change (Burt *et al.*, 2007).

Table 6.2 Summary of identified candidate genes for antisocial behavior

Dopaminergic	*Location (high-risk alleles)*	*Outcome*
MAO-A	Xp11.3 (2, 3 and 5 repeat VNTR)	Aggression (Brunner *et al.*, 1993), violence (Guo *et al.*, 2008), lower sustained attention (Manor *et al.*, 2002)
Dopamine transporter (DAT1)	5p15.3 (10 repeat VNTR)	ADHD symptoms (Waldman *et al.*, 1998; Roman *et al.*, 2001; Faraone *et al.*, 2005), violence, delinquency (Guo *et al.*, 2007), propensity to criminal career (Vaughn *et al.*, 2009)
Dopamine receptor D1 (DRD1)	5q35.1 (B2 variant of SNP)/(A-48G, G-94A, C-800T)	Aggression in patients with Alzheimers disease (Sweet *et al.*, 1998)

(*Continued*)

Table 6.2 Continued

Dopaminergic	*Location (high-risk alleles)*	*Outcome*
Dopamine receptor D2 (DRD2)	11q23 (A1 variant of the Taq SNP)/C957T SNP	Transient psychosis, alcoholism, cocaine use, opiate use (Xu *et al.*, 2004), disinihibition (Rodriguez-Jimenez *et al.*, 2006), age of first police contact and arrests, but only for youth from low-risk family environments (DeLisi *et al.*, 2008)
Catecholamine-O-methyltransferase (COMT)	22q11.21 [Val variant of resulting protein at 158th codon (SNP rs4680)]	Aggression, criminal offense convictions (Caspi *et al.*, 2008), conduct disorder (Monuteaux *et al.*, 2009)
Serotonergic		
Serotonin transporter linked promoter region (5-HTTLPR)	17q11.2 (short variant of promoter VNTR)	Aggression (Cadoret *et al.*, 2003; Haberstick *et al.*, 2006b); violence (Linnoila *et al.*, 1983; Roy *et al.*, 1988; Virkkunen *et al.*, 1989); conduct problems (Kruesi *et al.*, 1992); aggressive/delinquent behavior (Hohmann *et al.*, 2009)
Serotonin receptor (5HT1/5HT2)	5q11.2-q13/13q14-q21 (A allele of rs1406946 SNP; A161T SNP/G1438A SNP; His452Tyr SNP)	Aggression (Burt and Mikolajewski, 2008; Oades *et al.*, 2008b)/impulsivity; nonaggressive delinquency (Zouk *et al.*, 2007; Burt and Mikolajewski, 2008; Oades *et al.*, 2008b)
Tryptophan hydroxylase (TPH1/TPH2)	11p15.3-p14/12q21.1 (A218C L SNP; A779A SNP/A1473C SNP)	Violence, aggression (Manuck *et al.*, 1999; Hennig *et al.*, 2005) not non-aggressive antisocial behavior (Burt and Mikolajewski, 2008)/aggression (Popova, 2006; Oades *et al.*, 2008b)
Norepinephrine and epinephrine		
Dopamine-β-hydroxylase (DBH)	9q34 (C1021T SNP)	Impulsivity, aggression (Zhang *et al.*, 2005; Hess *et al.*, 2009)
Phenylethanolamine methyl transferase (PNMT)	17q (G148A SNP)	Cognitive and aggressive impulsivity (Oades *et al.*, 2008a)
Other candidate genes		
Estrogen (ESR1)	6q25.1 (Longer variant of TG repeat)	Aggression (Prichard *et al.*, 2007)
Androgen (AR)	Xq12 (Shorter variant of CAG repeat)	Aggression (Prichard *et al.*, 2007)

Candidate genes

Behavioral genetic research has consistently shown that heritable factors are important for the development of antisocial behavior. But with the behavioral genetic approach it is not possible to identify specific genes for antisocial behavior; this would require a different study design.

Despite large genome-wide association studies (GWAS), only a few candidate genes have been identified to be associated with antisocial behavior; the majority of these genes are summarized in Table 6.2. Many of these candidate gene findings have been replicated in both human and animal studies. A majority of these candidate genes were identified through examination of (1) the dopamine system, which is involved in mood, motivation and reward, and arousal, as well as other behaviors; (2) the serotonin system, which is involved in impulse control, affect regulation, sleep, and appetite; or (3) the epinephrine/norepinephrine system, which facilitates fight-or-flight reactions and autonomic nervous system activity (Bartels *et al.*, 2011). All three of these systems are affected by monoamine oxidase A (*MAO-A*) function (Niv and Baker, 2010; Tuvblad and Baker, 2011). The low-activity alleles of *MAO-A* have been found to interact with maladaptive childhood environment (Caspi *et al.*, 2002) and have been associated with aggression, violent delinquency, externalizing behavior, and lower inhibitory control (Brunner *et al.*, 1993; Guo *et al.*, 2008). Given that there is a sex difference in prevalence in antisocial behavior (Bartels *et al.*, 2011), genes regulating sex steroids may partly explain this sex difference. Both androgen and estrogen have also been linked to antisocial behavior in males (Prichard *et al.*, 2007). Clearly more research is warranted that can help identify genes for antisocial behavior and their relationships to the most prominent risk factors, including biological risk factors such as low resting heart rate and reduced skin conductance responses. For example, oxytocin has been implicated in the establishment of trust and affiliative behaviors (Israel *et al.*, 2008) and arginine vasopressin is related to generosity and altruistic behavior (Knafo *et al.*, 2008). Thus dysregulation of these neuropeptides may contribute to deficits in pro-social behaviors (Heinrichs and Domes, 2008) and may therefore make them worthy candidates to examine in relation to antisocial behavior.

Conclusion

Based on meta-analyses summarizing findings from twin and adoption studies on antisocial behavior across the life span, genetic, shared environmental, and non-shared environmental factors are contributing to the variance in antisocial behavior. Significant non-additive genetic effects have also been reported for antisocial behavior. Thus family similarity for antisocial behavior is primarily the result of shared genes and, to a lesser extent, common environmental factors. Longitudinal studies across development typically show that genetic influences are stable across age; there are also some indications that shared environmental factors change as a function of time. Aggressive behavior is more strongly influenced by genetic factors, whereas delinquent/rule-breaking behavior is more strongly influenced by shared environmental factors. Psychopathic personality is influenced primarily by genetic and non-shared environmental factors. However, there are to date very few longitudinal studies examining the genetic and environmental influences on psychopathy across development. Longitudinal studies are likely to have more power and would therefore better be able to detect any significant shared environmental influences on psychopathic personality. Lastly, only a handful of candidate genes have been identified for antisocial behavior, and these genes are related to the dopamine, serotonin, and epinephrine/norepinephrine systems.

Behavioral genetic research has shown that genetic *and* environmental influences are important for the development of antisocial behavior. Heritable factors may increase the propensity for antisocial behavior, but will not determine it, as there will always be some individuals with the same genetic risk factors who will not engage in antisocial behavior. At the same time, it should be kept in mind that this line of research has identified a large environmental component in antisocial behavior. By using twin and family design, it is possible to estimate not only the influence of heritable factors to a phenotype, but also the influence of environmental factors and, most importantly, the interplay between these factors in the development of antisocial behavior.

References

Achenbach, T.M. (1991) *Manual for the Child Behavior Checklist/4–18 and 1991 Profile*, Burlington, VT: University of Vermont, Department of Psychiatry.

Andershed, H., Kerr, M., Stattin, H., and Levander, S. (2002) 'Psychopathic traits in non-referred youths: a new assessment tool', in Blaauw, E., and Sheridan, L. (eds), *Psychopaths: Current International Perspectives*, The Hague: Elsevier, pp. 131–158.

APA (2004) *American Psychiatric Association DSM-IV-TR Diagnostic and Statistical Manual of Mental Disorders*, Washington, DC: American Psychiatric Association.

Baker, L.A., Barton, M., and Raine, A. (2002) 'The Southern California Twin Register at the University of Southern California', *Twin Research and Human Genetics*, 5(5): 456–459.

Baker, L.A., Barton, M., Lozano, D.I., Raine, A., and Fowler, J.H. (2006) 'The Southern California Twin Register at the University of Southern California: II', *Twin Research and Human Genetics*, 9: 933–940.

Baker, L., Tuvblad, C., Wang, P., Gomez, K., and Raine, A. (2013) 'The Southern California Twin Register at the University of Southern California: III', *Twin Research and Human Genetics*, 16(1): 336–343.

Barnes, J.C. and Boutwell, B.B. (2013) 'A demonstration of the generalizability of twin-based research on antisocial behavior', *Behavior Genetics*, 43(2): 120–131.

Bartels, M., van de Aa, N., van Beijsterveldt, C.E.M., Middeldorp, C.M., and Boomsma, D.I. (2011) 'Adolescent self-report of emotional and behavioral problems: interactions of genetic factors with sex and age', *Journal of the Canadian Academy of Child and Adolescent Psychiatry*, 20(1): 35–52.

Beaver, K., Barnes, J.C., May, J., and Schwarz, J.A. (2011) 'Psychopathic personality traits, genetic risk, and gene–environment correlations', *Criminal Justice and Behavior*, 38(9): 896–912.

van Beijsterveldt, C.E.M., Bartels, M., Hudziak, J.J., and Boomsma, D. (2003) 'Causes of stability of aggression from early childhood to adolescence: a longitudinal genetic analysis in Dutch twins', *Behavior Genetics*, 33: 591–605.

Bezdjian, S., Raine, A., Baker, L.A., and Lynam, D. (2011a) 'Psychopathic personality in children: genetic and environmental contribution', *Psychological Medicine*, 41(3): 589–600.

Bezdjian, S., Tuvblad, C., Raine, A., and Baker, L.A. (2011b) 'The genetic and environmental covariation among psychopathic personality traits, and reactive and proactive aggression in childhood', *Child Development*, 82(4): 1267–1281.

Blonigen, D.M., Carlson, S.R., Krueger, R.F., and Patrick, C.J. (2003) 'A twin study of self-reported psychopathic personality traits', *Personality and Individual Differences*, 35: 179–197.

Blonigen, D.M., Hicks, B.M., Krueger, R.F., Patrick, C.J., and Iacono, W.G. (2005) 'Psychopathic personality traits: heritability and genetic overlap with internalizing and externalizing psychopathology', *Psychological Medicine*, 35: 637–648.

Blonigen, D.M., Hicks, B.M., Krueger, R.F., Patrick, C.J., and Iacono, W.G. (2006) 'Continuity and change in psychopathic traits as measured via normal-range personality: a longitudinal-biometric study', *Journal of Abnormal Psychology*, 115(1): 85–95.

Bohman, M., Cloninger, C.R., Sigvardsson, S., and von Knorring, A.L. (1982). 'Predisposition to petty criminality in Swedish adoptees: I. Genetic and environmental heterogeneity', *Archives of General Psychiatry*, 39(11): 1233–1241.

Brandon, K. and Rose, R.J. (1995) 'A multivariate twin family study of the genetic and environmental structure of personality, beliefs, and alcohol use', *Behavior Genetics*, 25(3): 257–257.

Broidy, L.M., Tremblay, R.E., Brame, B., Fergusson, D., Horwood, J.L., Laird, R., Moffitt, T.E., Nagin, D.S., Bates, J.E., Dodge, K.A., Loeber, R., Lynam, D.R., Pettit, G.S., and Vitaro, F. (2003) 'Developmental trajectories of childhood disruptive behaviors and adolescent delinquency: a six-site, cross-national study', *Developmental Psychology*, 39(2): 222–245.

Brook, M., Panizzon, M.S., Kosson, D.S., Sullivan, E.A., Lyons, M.J., Franz, C.E., Eisen, S.A., and Kremen, W.S. (2010) 'Psychopathic personality traits in middle-aged male twins: a behavior genetic investigation', *Journal of Personality Disorders*, 24(4): 473–486.

Brunner, H.G., Nelen, M., Breakefield, X.O., Ropers, H.H., and van Oost, B.A. (1993) 'Abnormal behavior associated with a point mutation in the structural gene for monoamine oxidase A', *Science*, 262(5133): 578–580.

Burt, A.S. (2009) 'Are there meaningful etiological differences within antisocial behavior? Results of a meta-analysis', *Clinical Psychology Review*, 29: 163–178.

Burt, A.S. (2012) 'Do etiological influences on aggression overlap with those on rule breaking? A meta-analysis', *Psychological Medicine*, epub: 1–12.

Burt, S.A. and Mikolajewski, A.J. (2008) 'Preliminary evidence that specific candidate genes are associated with adolescent-onset antisocial behavior', *Aggressive Behavior*, 34: 437–445.

Burt, S.A., McGue, M., Carter, L.A., and Iacono, W.G. (2007) 'The different origins of stability and change in antisocial personality disorder symptoms', *Psychological Medicine*, 37(1): 27–38.

Button, T.M.M., Scourfield, J., Martin, N., Purcell, S., and McGuffin, P. (2005) 'Family dysfunction interacts with genes in the causation of antisocial symptoms', *Behavior Genetics*, 35: 115–120.

Cadoret, R.J., Cain, C.A., and Crowe, R.R. (1983) 'Evidence for gene–environment interaction in the development of adolescent antisocial behavior', *Behavior Genetics*, 13: 301–310.

Cadoret, R.J., Yates, W.R., Troughton, E., Woodworth, G., and Stewart, M.A. (1995) 'Genetic–environmental interaction in the genesis of aggressivity and conduct disorders', *Archives of General Psychiatry*, 52(11): 916–924.

Cadoret, R.J., Langbehn, D., Caspers, K., Troughton, E.P., Yucuis, R., Sandhu, H.K., and Philibert, R. (2003) 'Associations of the serotonin transporter promoter polymorphism with aggressivity, attention deficit, and conduct disorder in an adoptee population', *Comprehensive Psychiatry*, 44: 88–101.

Caspi, A., McClay, J., Moffitt, T.E., Mill, J., Martin, J., Craig, I.W., Taylor, A., and Poulton, R. (2002) 'Role of genotype in the cycle of violence in maltreated children', *Science*, 297(5582): 851–854.

Caspi, A., Langley, K., Milne, B., Moffitt, T.E., O'Conovan, M., Owen, M.J., Tomas, M.P., Poulton, R., Rutter, M., Taylor, A., Williams, B., and Thapar, A. (2008) 'A replicated molecular basis of subtyping antisocial behavior in children with attention-deficit/hyperactivity disorder', *Archives of General Psychiatry*, 65: 203–210.

Christensen, K., Petersen, I., Skytthe, A., Herskind, A.M., McGue, M., and Bingley, P. (2006) 'Comparison of academic performance of twins and singletons in adolescence: follow-up study', *British Medical Journal*, 6: 1–5.

Cloninger, C.R. and Gottesman, I.I. (1987) 'Genetic and environmental factors in antisocial behavior disorders', in Mednick, S.A., Moffitt, T.E., and Stack, S.A. (eds), *The Causes of Crime: New Biological Approaches*, Cambridge: Cambridge University Press.

Cloninger, C.R., Sigvardsson, S., Bohman, M., and von Knorring, A.L. (1982) 'Predisposition to petty criminality in Swedish adoptees. II. Cross-fostering analysis of gene–environment interaction', *Archives of General Psychiatry*, 39: 1242–1247.

Cloninger, C.R., Bohman, M., Sigvardsson, S., and von Knorring, A.L. (1985) 'Psychopathology in adopted-out children of alcoholics. The Stockholm Adoption Study', *Recent Developments in Alcoholism*, 3: 37–51.

Cronk, N., Slutske, W.S., Madden, P.A., Bucholz, K.K., Reich, W., and Heath, A. (2002) 'Emotional and behavioral problems among female twins: an evaluation of the equal environments assumption', *Journal of American Academy of Child and Adolescent Psychiatry*, 41: 829–837.

Crowe, R.R. (1974) 'An adoption study of antisocial personality', *Archives of General Psychiatry*, 31: 785–791.

Dahlstrom, W.G. and Welsh, G.S. (1960) *An MMPI Handbook: A Guide To Use in Clinical Practice and Research*, Oxford, UK: Minnesota University Press.

Deater-Deckard, K. and Plomin, R. (1999) 'An adoption study of the etiology of teacher and parent reports of externalizing behavior problems in middle childhood', *Child Development*, 70: 144–154.

DeLisi, M., Beaver, K.M., Wright, J.P., and Vaughn, M.G. (2008) 'The etiology of criminal onset: the enduring salience of nature and nurture', *Journal of Criminal Justice*, 36: 217–223.

Derks, E.M., Dolan, C.V., and Boomsma, D.I. (2006) 'A test of the equal environment assumption (EEA) in multivariate twin studies', *Twin Research*, 9(3): 403–411.

DiLalla, L.F. and Gottesman, I. (1989) 'Heterogeneity of causes for delinquency and criminality: lifespan perspectives', *Development and Psychopathology*, 1: 339–349.

DiLalla, D.L., Carey, G., Gottesman, I.I., and Bouchard, T.R. (1996) 'Heritability of MMPI personality indicators of psychopathology in twins reared apart', *Journal of Abnormal Psychology*, 105(4): 491–499.

Eisen, S., True, W., Goldberg, J., Henderson, W., and Robinette, C.D. (1987) 'The Vietnam Era Twin (VET) Registry: method of construction', *ACTA Geneticae Medicae et Gemellologiae (Roma)*, 36: 61–66.

Eley, T.C., Lichtenstein, P., and Moffitt, T.E. (2003) 'A longitudinal behavioral genetic analysis of the etiology of aggressive and non-aggressive antisocial behavior', *Development and Psychopathology*, 15(2): 383–402.

Eriksson, M., Rasmussen, F., and Tynelius, P. (2006) 'Genetic factors in physical activity and the equal environment assumption – the Swedish young male twin study', *Behavior Genetics*, 36(2): 238–247.

Faraone, S.V., Perlis, R.H., Doyle, A.E., Smoller, J.W., Goralnick, J.J., Holmgren, M.A., *et al.* (2005) 'Molecular genetics of attention deficit/hyperactivity disorder', *Biological Psychiatry*, 57: 1313–1323.

Fontaine, N., Carbonneau, R., Vitaro, F., Barker, E.D., and Tremblay, R.E. (2009) 'Research review: a critical review of studies on the developmental trajectories of antisocial behavior in females', *Journal of Child Psychology and Psychiatry*, 50(4): 363–385.

Fontaine, N.M., Rijsdijk, F.V., McCrory, E.J., and Viding, E. (2010) 'Etiology of different developmental trajectories of callous-unemotional traits', *Journal of the American Academy of Child Adolescent Psychiatry*, 49(7): 656–664.

Forsman, M., Lichtenstein, P., Andershed, H., and Larsson, H. (2008) 'Genetic effects explain the stability of psychopathic personality from mid- to late adolescence', *Journal of Abnormal Psychology*, 117(3): 606–617.

Frick, P.J. and Hare, R.D. (2002) *Antisocial Process Screening Device*, Toronto, Canada: Multi-Health Systems.

Frick, P.J., Lahey, B.B., Loeber, R., Tannenbaum, L., Van Horn, Y., Christ, M.A.G., Hart, E.A., and Hanson, K. (1993) 'Oppositional defiant disorder and conduct disorder: a meta-analytic review of factor analyses and cross-validation in a clinic sample', *Clinical Psychology Review*, 13(4): 319–340.

Frick, P.J., Cornell, A.H., Barry, C.T., Bodin, S.D., and Dane, H.E. (2003) 'Callous-unemotional traits and conduct problems in the prediction of conduct problem severity, aggression and self-report of delinquency', *Journal of Abnormal Child Psychology*, 31: 457–470.

Ge, X., Conger, R.D., Cadoret, R.J., Neiderhiser, J.M., Yates, W.R., Troughton, E., and Stewart, M.A. (1996) 'The developmental interface between nature and nurture: a mutual influence model of child antisocial behavior and parent behaviors', *Developmental Psychology*, 32(4): 574–589.

Goldberg, J., Curran, B., Vitek, M.E., Henderson, W.G., and Boyko, E.J. (2002) 'The Vietnam Era Twin Registry', *Twin Research and Human Genetics*, 5: 476–481.

Gottesman, I.I. and Goldsmith, H.H. (1994) 'Developmental psychopathology of antisocial behavior: inserting genes into its ontogenesis and epigenesis', in Nelson, C.A. (ed.), *The Minnesota Symposia on Child Psychology: Threats to Optimal Development: Integrating Biological, Psychological, and Social Risk Factors*, Hillsdale, NJ: Erlbaum, pp. 69–104.

Grant, J.E. and Potenza, M.N. (2007) 'Commentary: illegal behavior and pathological gambling', *Journal of the American Academy of Psychiatry and the Law*, 35(3): 302–305.

Guo, G., Roettger, M.E., and Shih, J.C. (2007) 'Contributions of the *DAT1* and *DRD2* genes to serious and violent delinquency among adolescents and young adults', *Human Genetics*, 121: 125–136.

Guo, G., Ou, X., Roettger, M., and Shih, J.C. (2008) 'The VNTR 2 repeat in *MAO-A* and delinquent behavior in adolescence and young adulthood: associations and *MAO-A* promoter activity', *European Journal of Human Genetics*, 16: 626–634.

Haberstick, B.C., Schmitz, S., Young, S.E., and Hewitt, J.K. (2006a) 'Genes and developmental stability of aggressive behavior problems at home and school in a community sample of twins aged 7–12', *Behavior Genetics*, 36: 809–819.

Haberstick, B.C., Smolen, A., and Hewitt, J.K. (2006b) 'Family-based association test of the 5HTTLPR and aggressive behavior in a general population sample of children', *Biological Psychiatry*, 59: 836–843.

Heinrichs, M. and Domes, G. (2008) 'Neuropeptides and social behaviour: effects of oxytocin and vasopressin in humans', *Progress in Brain Research*, 170: 337–350.

Hennig, J., Reuter, M., Netter, P., and Burk, C. (2005) 'Two types of aggression are differentially related to serotonergic activity and the A779C TPH polymorphism', *Behavioral Neuroscience*, 119: 16–25.

Hess, C., Reif, A., Strobel, A., Boreatti-Hummer, A., Heine, M., Lesch, K.P., and Jacob, C.P. (2009) 'A functional dopamine-b-hydroxylase gene promoter polymorphism is associated with impulsive personality styles, but not with affective disorders', *Journal of Neuronal Transmission*, 116: 121–130.

Hettema, J.M., Neale, M.C., and Kendler, K.S. (1995) 'Physical similarity and the equal-environment assumption in twin studies of psychiatric disorders', *Behavior Genetics*, 25(4): 327–335.

Hicks, B.M., Carlson, M.D., Blonigen, D.M., Patrick, C.J., Iacono, W.G., and McGue, M. (2012) 'Psychopathic personality traits and environmental contexts: differential correlates, gender differences, and genetic mediation', *Personality Disorders: Theory, Research, and Treatment*, 3(3): 209–227.

Hjern, A., Ekeus, C., Rasmussen, F., and Lindblad, F. (2012) 'Educational achievement and vocational career in twins: a Swedish national cohort study', *Acta Pædiatrica*, 10(6): 591–596.

Hohmann, S., Becker, K., Fellinger, J., Banaschewski, T., Schmidt, M.H., Esser, G., and Laucht, M. (2009) 'Evidence for epistasis between the 5-HTTLPR and the dopamine D4 receptor polymorphisms in externalizing behavior among 15-year-olds', *Journal of Neural Transmission*, 116: 1621–1629.

Iacono, W.G. and McGue, M. (2002) 'Minnesota Twin Family Study', *Twin Research and Human Genetics*, 5(5): 482–487.

Iacono, W.G., McGue, M., and Krueger, R.F. (2006) 'Minnesota Center for Twin and Family Research', *Twin Research and Human Genetics*, 9(6): 978–984.

Israel, S., Lerer, E., Shalev, I., Uzefovsky, F., Reibold, M., Bachner-Melman, R., Granot, R., Bornstein, G., Knafo, A., Yirmiya, N., and Ebstein, R.P. (2008) 'Molecular genetic studies of the arginine vasopressin 1a receptor (AVPR1a) and the oxytocin receptor (OXTR) in human behaviour: from autism to altruism with some notes in between', *Progress in Brain Research*, 170: 435–449.

Jacobson, K.C., Prescott, C.A., and Kendler, K. (2002) 'Sex differences in the genetic and environmental influences on the development of antisocial behavior', *Developmental Psychology*, 14: 395–416.

Jang, K.L., Vernon, P.A., and Livesley, W. (2000) 'Personality disorder traits, family environment, and alcohol misuse: a multivariate behavioural genetic analysis', *Addiction*, 95(6): 873–888.

Junger-Tas, J., Terlouw, G.-J., and Klein, M.W. (1994) *Delinquent Behavior Among Young People in the Western World*, Amsterdam: Kugler Publications.

Kendler, K., Neale, M.C., Kessler, R.C., Heath, A.C., and Eaves, L.J. (1993) 'A test of the equal-environment assumption in twin studies of psychiatric illness', *Behavior Genetics*, 23(1): 21–27.

Kim-Cohen, J., Caspi, A., Moffitt, T.E., Harrington, H., Milne, B.J., and Poulton, R. (2003) 'Prior juvenile diagnoses in adults with mental disorder: developmental follow-back of a prospective-longitudinal cohort', *Archives of General Psychiatry*, 60(7): 709–717.

Klump, K.L., Holly, A., Iacono, W.G., McGue, M., and Willson, L.E. (2000) 'Physical similarity and twin resemblance for eating attitudes and behaviors: a test of the equal environments assumption', *Behavior Genetics*, 30(1): 51–58.

Knafo, A., Israel, S., Darvasi, A., Bachner-Melman, R., Uzefovsky, F., Cohen, L., Feldman, E., Lerer, E., Laiba, E., Raz, Y., Nemanov, L., Gritsenko, I., Dina, C., Agam, G., Dean, B., Bornstein, G., and Ebstein, R.P. (2008) 'Individual differences in allocation of funds in the dictator game associated with length of the arginine vasopressin 1a receptor RS3 promoter region and correlation between RS3 length and hippocampal mRNA', *Genes, Brain and Behavior*, 7(3): 266–275.

Krueger, R.F., Moffitt, T.E., Caspi, A., Bleske, A., and Silva, P.A. (1998) 'Assortative mating for antisocial behavior: developmental and methodological implications', *Behavior Genetics*, 28(3): 173–186.

Kruesi, M.J., Hibbs, E.D., Zahn, T.P., Keysor, C.S., Hamburger, S.D., Bartko, J.J., and Rapoport, J.L. (1992) 'A 2-year prospective follow-up study of children and adolescents with disruptive behavior disorders. Prediction by cerebrospinal fluid 5-hydroxyindoleacetic acid, homovanillic acid, and autonomic measures?', *Archives of Genetic Psychiatry*, 49: 507–514.

Larsson, H., Andershed, H., and Lichtenstein, P. (2006) 'A genetic factor explains most of the variation in the psychopathic personality', *Journal of Abnormal Psychology*, 115(2): 221–230.

Larsson, H., Tuvblad, C., Rijsdijk, F.V., Andershed, H., Grann, M., and Lichtenstein, P. (2007) 'A common genetic factor explains the association between psychopathic personality and antisocial behavior', *Psychological Medicine*, 37(1): 15–26.

Lewis, D.O., Yeager, C.A., Gidlow, B., and Lewis, M. (2001) 'Six adoptees who murdered: neuropsychiatric vulnerabilities and characteristics of biological and adoptive parents', *Journal of the American Academy of Psychiatry and the Law*, 29(4): 390–397.

Lichtenstein, P., Tuvblad, C., Larsson, H., and Carlström, E. (2007) 'The Swedish Twin study of Child and Adolescent Development – the TCHAD-study', *Twin Research and Human Genetics*, 10(1): 67–73.

van Lier, P.A.C., Vitaro, F., Barker, E.D., Koot, H.M., and Tremblay, R. (2009) 'Developmental links between trajectories of physical violence, vandalism, theft, and alcohol–drug use from childhood to adolescence', *Journal of Abnormal Child Psychology*, 37: 481–492.

Lilienfeld, S.O. and Andrews, B.P. (1996) 'Development and preliminary validation of a measure of psychopathic personality traits in noncriminal populations', *Journal of Personality Assessment*, 66: 488–524.

Linnoila, M., Virkkunen, M., Scheinin, M., Nuutila, A., Rimon, R., and Goodwin, F.K. (1983) 'Low cerebrospinal fluid 5-hydroxyindoleacetic acid concentration differentiates impulsive from nonimpulsive violent behavior', *Life Sciences*, 33: 2609–2614.

Livesley, W., Jang, K.L., Jackson, D.N., and Vernon, P.A. (1993) 'Genetic and environmental contributions to dimensions of personality disorder', *American Journal of Psychiatry*, 150(12): 1826–1831.

Livesley, W., Jang, K.L., and Vernon, P.A. (1998) 'Phenotypic and genetic structure of traits delineating personality disorder', *Archives of General Psychiatry*, 55(10): 941–948.

Loeber, R. and Hay, D. (1997) 'Key issues in the development of aggression and violence from childhood to early adulthood', *Annual Review of Psychology*, 48: 371–410.

Loeber, R., Burke, J.D., Lahey, B.B., Winters, A., and Zera, M. (2000) 'Oppositional defiant and conduct disorder: a review of the past 10 years, part I', *Journal of the American Academy of Child and Adolescent Psychiatry*, 39(12): 1468–1484.

Loehlin, J.C. and Nichols, R.C. (1976) *Heredity, Environmental and Personality*, Austin, TX: University of Texas Press.

Loehlin, J.C., Willerman, L., and Horn, J.M. (1987) 'Personality resemblance in adoptive families: a 10 year follow-up', *Journal of Personality and Social Psychology*, 53(5): 961–969.

Loney, B.R., Taylor, J., Butler, M., and Iacono, W.G. (2002) 'The Minnesota Temperament Inventory: a psychometric study of adolescent self-reported psychopathy', Department of Psychology, University of Minnesota, unpublished manuscript.

Lykken, D.T., McGue, M., Tellegen, A., and Bouchard, T.J. Jr. (1992) 'Emergenesis – genetic traits that may not run in families', *American Psychologist*, 47(12): 1565–1577.

Lynam, D.R. (2002) 'Fledgling psychopathy: a view from personality theory', *Law and Human Behavior*, 26(2): 255–259.

Lyons, M.J., True, W.R., Eisen, S.A., Goldberg, J., Meyer, J.M., Faraone, S.V., Eaves, L.J., and Tsuang, M.T. (1995) 'Differential heritability of adult and juvenile antisocial traits', *Archives of General Psychiatry*, 52: 906–915.

McCollister, K.E., French, M.T., and Fang, H. (2010) 'The cost of crime to society: new crime-specific estimates for policy and program evaluation', *Drug and Alcohol Dependence*, 108(1/2): 98–109.

Maes, H.H.M., Neale, M.C., Kendler, K., Hewitt, J.K., Silberg, J.L., Foley, D.L., Meyer, J.M., Rutter, M., Simonoff, E., Pickles, A., and Eaves, L.J. (1998) 'Assortative mating for major psychiatric diagnosis in two population-based samples', *Psychological Medicine*, 28: 1389–1401.

Maes, H.H., Silberg, J.L., Neale, M.C., and Eaves, L.J. (2007) 'Genetic and cultural transmission of antisocial behavior: an extended twin parent model', *Twin Research and Human Genetics*, 10(1): 136–150.

Manor, I., Tyano, S., Mel, E., Eisenberg, J., Bachner-Melman, R., Kotler, M., and Ebstein, R.P. (2002) 'Family-based and association studies of monoamine oxidase A and attention deficit hyperactivity disorder (ADHD): preferential transmission of the long promoter-region repeat and its association with impaired performance on a continuous performance test (TOVA)', *Molecular Psychiatry*, 7: 626–632.

Manuck, S.B., Flory, J.D., Ferrell, R.E., Dent, K.M., Mann, J.J., and Muldoon, M.F. (1999) 'Aggression and anger-related traits associated with a polymorphism of the tryptophan hydroxylase gene', *Biological Psychiatry*, 45: 603–614.

Mata, A.D. and van Dulmen, N. (2012) 'Adult-onset antisocial behavior trajectories. Associations with adolescent family processes and emerging adulthood functioning', *Journal of Interpersonal Violence*, 27: 177–193.

Mednick, S.A., Gabrielli, W.F., and Hutchings, B. (1984) 'Genetic influence in criminal convictions: evidence from an adoption cohort', *Science*, 224: 891–894.

Miles, D.R. and Carey, G. (1997) 'Genetic and environmental architecture of human aggression', *Journal of Personality and Social Psychology*, 72(1): 207–217.

Miles, D.R., Van den Bree, M., and Pickens, R.W. (2002) 'Sex differences in shared genetic and environmental influences between conduct disorder symptoms and marijuana use in adolescents', *American Journal of Medical Genetics (Neuropsychiatric Genetics)*, 114: 159–168.

Moffitt, T.E. (1993) 'Adolescence-limited and life-course-persistent antisocial behavior: a developmental taxonomy', *Psychological Review*, 100(4): 674–701.

Moffitt, T.E. (2005a) 'Genetic and environmental influences on antisocial behaviors: evidence from behavioral-genetic research', *Advances in Genetics*, 55: 41–99.

Moffitt, T.E. (2005b) 'The new look of behavioral genetics in developmental psychopathology: gene–environment interplay in antisocial behaviors', *Psychological Bulletin*, 131(4): 533–554.

Moffitt, T.E., Caspi, A., Harrington, H., and Milne, B.J. (2002) 'Males on the life-course persistent and adolescence-limited antisocial pathways: follow-up at age 26', *Development and Psychopathology*, 14: 179–207.

Moilanen, I., Linna, S.-L., Kumpulainen, K., Tamminen, K., Piha, J., and Almqvist, F. (1999) 'Are twins' behavioural/emotional problems different from singletons'?', *European Child and Adolescent Psychiatry*, 8: 62–67.

Monuteaux, M.C., Biederman, J., Doyle, A.E., Mick, E., and Faraone, S.V. (2009) 'Genetic risk for conduct disorder symptom subtypes in an ADHD sample: specificity to aggressive symptoms', *Journal of the American Academy for Child and Adolescent Psychiatry*, 48: 757–764.

Neale, M.C. and Cardon, L.R. (1992) *Methodology for Genetic Studies of Twins and Families*, Dordrecht, Netherlands: Kluwer Academic.

Nicholls, T.L., Ogloff, J.R., Brink, J., and Spidel, A. (2005) 'Psychopathy in women: a review of its clinical usefulness for assessing risk for aggression and criminality', *Behavioral Sciences and the Law*, 23(6): 779–802.

Niv, S. and Baker, L.A. (2010) 'Genetic marker for antisocial behavior', in Thomas, C. and Pope, K. (eds), *The Origins of Antisocial Behavior: A Developmental Perspective*, New York: Oxford University Press.

O'Connor, T.G., Deater-Deckard, K., Fulker, D., Rutter, M., and Plomin, R. (1998) 'Genotype–environment correlations in late childhood and early adolescence: antisocial behavioral problems and coercive parenting', *Developmental Psychology*, 34: 970–981.

Oades, R.D., Lasky-Su, J., Christiansen, H., Faraone, S.V., Sonuga-Barke, E.J., Banaschewski, T., Chen, W., Anney, R.J., Buitelaar, J.K., Ebstein, R.P., Franke, B., Gill, M., Miranda, A., Roeyers, H., Rothenberger, A., Sergeant, J.A., Steinhausen, H.C., Taylor, E.A., Thompson, M., and Asherson, P. (2008a) 'The influence of serotonin- and other genes on impulsive behavioral aggression and cognitive impulsivity in children with attention-deficit/hyperactivity disorder (ADHD): findings from a family-based association test (FBAT) analysis', *Behavioral and Brain Functions*, epub.

Oades, R.D., Lasky-Su, J., Christiansen, H., Faraone, S.V., Sonuga-Barke, E.J.S., Banaschewski, T., Chem, W., Anney, R.J.L., Buitelaar, J.K., Ebstein, R.P., Franke, B., Gill, M., Miranda, A., Roeyers, H., Rothenberger, A., Sergeant, J.A., Steinhausen, H.C., Taylor, E.A., Thompson, M., and Asherson, P. (2008b) 'The influence of serotonin- and other genes and impulsive behavioral aggression and cognitive impulsivity in children with attention-deficit/hyperactivity disorder (ADHD): findings from a family-based association test (FBAT) analysis', *Behavioral and Brain Functions*, 48: 4–48.

Odgers, C.L., Moffitt, T.E., Broadbent, J.M., Dickson, N., Hancox, R.J., Harrington, H., Poulton, R., Sears, M.R., Thomson, W.M., and Caspi, A. (2008) 'Female and male antisocial trajectories: from childhood origins to adult outcomes', *Development and Psychopathology*, 20(2): 673–716.

van den Oord, E.J., Boomsma, D.I., and Verhulst, F.C. (1994) 'A study of problem behaviors in 10- to 15-year-old biologically related and unrelated international adoptees', *Behavior Genetics*, 24: 193–205.

van den Oord, E.J., Koot, H.M., Boomsma, D.I., Verhulst, F.C., and Orlebeke, J.F. (1995) 'A twin–singleton comparison of problem behaviour in 2–3-year-olds', *Journal of Child Psychology and Psychiatry*, 36(3): 449–458.

Ordaz, S.J., Lenroot, R.K., Wallace, G.L., Clasen, L.S., Blumenthal, J.D., Schmitt, J.E., and Giedd, J.N. (2010) 'Are there differences in brain morphometry between twins and unrelated singletons? A pediatric MRI study', *Genes, Brains and Behavior*, 9(3): 288–295.

Plomin, R., DeFries, J.C., McClearn, G.E., and McGuffin, P. (2001) *Behavioral Genetics*, New York: Worth.

Popova, N.K. (2006) 'From genes to aggressive behavior: the role of the serotonergic system', *Bioessays*, 28: 595–603.

Posthuma, D., Beem, A.L., de Geus, E.J.C., van Baal, C.M., von Hjelmborg, J.B., Iachine, I., and Boomsma, D.I. (2003) 'Theory and practice in quantitative genetics', *Twin Research and Human Genetics*, 6(5): 361–376.

Prichard, Z.M., Jorm, A.F., Mackinnon, A., and Easteal, S. (2007) 'Association analysis of 15 polymorphisms within 10 candidate genes for antisocial behavioral traits', *Psychiatric Genetics*, 17(5): 299–303.

Rhee, S.H. and Waldman, I.D. (2002) 'Genetic and environmental influences on antisocial behavior: a meta-analysis of twin and adoption studies', *Psychological Bulletin*, 128: 490–529.

Robins, L.N. (1966) *Deviant Children Grown Up*, Baltimore, MD: Williams and Wilkins.

Rodriguez-Jimenez, R., Avila, C., Ponce, G., Ibanez, M.I., Rubio, G., Jimenez-Arriero, M.A., Ampuero, I., Ramos, J.A., Hoenicka, J., and Palomo, T. (2006) 'The TaqIA polymorphism linked to the *DRD2* gene is related to lower attention and less inhibitory control in alcoholic patients', *European Psychiatry*, 21: 66–69.

Roman, T., Schmitz, M., Polanczyk, G., Eizirik, M., Rohde, L.A., and Hutz, M.H. (2001) 'Attention-deficit hyperactivity disorder: a study of association with both the dopamine transporter gene and the dopamine D4 receptor gene', *American Journal of Medical Genetics B: Neuropsychiatric Genetics*, 105: 471–478.

Rowe, D.C., Almeida, D.M., and Jacobson, K.C. (1999) 'School context and genetic influences on aggression in adolescence', *Psychological Science*, 10(3): 277–280.

Roy, A., Adinoff, B., and Linnoila, M. (1988) 'Acting out hostility in normal volunteers: negative correlation with levels of 5HIAA in cerebrospinal fluid', *Psychiatry Research*, 24: 187–194.

Rutter, M. (2006) *Genes and Behavior. Nature–Nurture Interplay Explained*, Oxford, UK: Blackwell.

Rutter, M., Giller, H., and Hagell, A. (1998) *Antisocial Behavior by Young People*, Cambridge, UK: Cambridge University Press.

Rutter, M., Kim-Cohen, J., and Maughan, B. (2006) 'Continuities and discontinuities in psychopathology between childhood and adult life', *Journal of Child Psychology and Psychiatry*, 47(3): 276–295.

Scarr, S. and McCartney, K. (1983) 'How people make their own environments: a theory of genotype–environment effects', *Child Development*, 54: 425–435.

Silberg, J.L., Rutter, M., Tracy, K., Maes, H.H., and Eaves, L. (2007) 'Etiological heterogeneity in the development of antisocial behavior: the Virginia Twin Study of Adolescent Behavioral Development and the Young Adult Follow-Up', *Psychological Medicine*, 37(8): 1193–1202.

Simonoff, E., Pickles, A., Meyer, J., Silberg, J., Maes, H., Loeber, M., Rutter, M., Hewitt, J.K., and Eaves, L. (1997) 'The Virginia Twin Study of adolescent behavioral development', *Archives of General Psychiatry*, 54: 801–808.

Sweet, R.A., Nimgaonkar, V.L., Kamboh, M.I., Lopez, O.L., Zhang, F., and Dekosky, S.T. (1998) 'Dopamine receptor genetic variation, psychosis, and aggression in Alzheimer disease', *Archives of Neurology*, 55: 1335–1340.

Taylor, J., Iacono, W.G., and McGue, M. (2000a) 'Evidence for a genetic etiology of early-onset delinquency', *Journal of Abnormal Psychology*, 109(4): 634–643.

Taylor, J., McGue, M., and Iacono, W.G. (2000b) 'Sex differences, assortative mating, and cultural transmission effects on adolescent delinquency: a twin family study', *Journal of Child Psychology and Psychiatry*, 41(4): 433–440.

Taylor, A., Loney, B.R., Bobadilla, L., Iacono, W.G., and McGue, M. (2003) 'Genetic and environmental influences on psychopathy trait dimensions in a community sample of male twins', *Journal of Abnormal Child Psychology*, 31(6): 633–645.

Tellegen, A. (n.d.) *Manual for the Multidimensional Personality Questionnaire*, Department of Psychology, University of Minnesota, unpublished manuscript.

Tellegen, A. and Waller, N.G. (2008) 'Exploring personality through test construction: development of the Multidimensional Personality Questionnaire', in Boyle, G.J., Matthews, G., and Saklofske, D.H. (eds), *The Sage Handbook of Personality Theory and Assessment: Vol. II Personality Measurement and Testing*, London: Sage, pp. 261–292.

Torgersen, S., Skre, I., Onstad, S., Edvardsen, J., and Kringlen, E. (1993) 'The psychometric genetic structure of DSM-III personality disorder criteria', *Journal of Personality Disorders*, 7: 196–213.

Trouton, A., Spinath, F.M., and Plomin, R. (2002) 'Twins early development study (TEDS): a multivariate, longitudinal genetic investigation of language, cognition and behavior problems in childhood', *Twin Research and Human Genetics*, 5: 444–448.

Tuvblad, C. and Baker, L.A. (2011) 'Human aggression across the lifespan: genetic propensities and environmental moderators', in Huber, R., Brennan, P., and Bannasch, D. (eds), *Advances in Genetics (Aggression) Vol. 75*, The Hague: Elsevier, pp. 171–214.

Tuvblad, C., Grann, M., and Lichtenstein, P. (2006) 'Heritability for adolescent antisocial behavior differs with socioeconomic status: gene–environment interaction', *Journal of Child Psychology and Psychiatry*, 47(7): 734–743.

Tuvblad, C., Zheng, M., Raine, A., and Baker, L.A. (2009) 'A common genetic factor explains the covariation among ADHD ODD and CD symptoms in 9–10 year old boys and girls', *Journal of Abnormal Child Psychology*, 37(2): 153–167.

Tuvblad, C., Narusyte, J., Grann, M., Sarnecki, J., and Lichtenstein, P. (2011) 'The genetic and environmental etiology of antisocial behavior, from early adolescence to emerging adulthood', *Behavior Genetics*, 41(5): 629–640.

Udry, J.R. (2003) *The National Longitudinal Study of Adolescent Health (Add Health), Waves I and II, 1994–1996; Wave III, 2001–2002* [machine-readable data file and documentation], Chapel Hill, NC: University of North Carolina at Chapel Hill: Carolina Population Center.

van der Valk, J.C., Verhulst, F.C., Neale, M.C., and Boomsma, D.I. (1998) 'Longitudinal genetic analysis of problem behaviors in biologically related and unrelated adoptees', *Behavior Genetics*, 28(5): 365–380.

Vaughn, M.C., DeLisi, M., Beaver, K.M., and Wright, J.P. (2009) 'DAT1 and 5HTT are associated with pathological criminal behavior in a nationally representative sample of youth', *Criminal Justice and Behavior*, 36: 1113–1324.

Vernon, P.A., Villani, V.C., Vickers, L.C., and Harris, J. (2008) 'A behavioral genetic investigation of the Dark Triad and the Big 5', *Personality and Individual Differences*, 44(2): 445–452.

Viding, E., Blair, J.R., Moffitt, T.E., and Plomin, R. (2005) 'Evidence of substantial genetic risk for psychopathy in 7-year-olds', *Journal of Child Psychology and Psychiatry*, 46(6): 592–597.

Viding, E., Frick, P.J., and Plomin, R. (2007) 'Aetiology of the relationship between callous-unemotional traits and conduct problems in childhood', *British Journal of Psychiatry*, 190: 33–38.

Virkkunen, M., De Jong, J., Bartko, J., Goodwin, F.K., and Linnoila, M. (1989) 'Relationship of psychobiological variables to recidivism in violent offenders and impulsive fire setters: a follow-up study', *Archives of General Psychiatry*, 46: 600–603.

Waldman, I.D. and Rhee, S. (2006) 'Genetic and environmental influences on psychopathy and antisocial behavior', in Patrick, C.J. (ed.), *Handbook of Psychopathy*, New York: Guilford, pp. 205–228.

Waldman, I.D., Rowe, D.C., Abramowitz, A., Kozel, S.T., Mohr, J.H., Sherman, S.L., Cleveland, H.H., Sanders, M.L., Gard, J.M.C., and Stever, C. (1998) 'Association and linkage of the dopamine transporter gene and attention-deficit hyperactivity disorder in children: heterogeneity owing to diagnostic subtype and severity', *American Journal of Human Genetics*, 63: 1767–1776.

White, H.R., Xie, M., Thompson, W., Loeber, R., and Stouthamer-Loeber, M. (2001) 'Psychopathology as a predictor of adolescent drug use trajectories', *Psychology of Addictive Behaviors*, 15(3): 210–218.

Williams, K.M., Paulhus, D.L., and Hare, R.D. (2007) 'Capturing the four-factor structure of psychopathy in college students via self-report', *Journal of Personality Assessment*, 88(2): 205–219.

Xian, H., Scherrer, J.F., Eisen, S.A., True, W.R., Heath, A.C., Goldberg, J., Lyons, M.J., and Tsuang, M.T. (2000) 'Self-reported zygosity and the equal-environments assumption for psychiatric disorders in the Vietnam Era Twin Registry', *Behavior Genetics*, 30(4): 303–310.

Xie, H., Drabick, D.A., and Chen, D. (2011) 'Developmental trajectories of aggression from late childhood through adolescence: similarities and differences across gender', *Aggressive Behavior*, 37(5): 387–404.

Xu, K., Lichtermann, D., Lipsky, R.H., Franke, P., Liu, X., Hu, Y., Cao, L., Schwab, S.G., Wildenauer, D.B., Bau, C.H., Ferro, E., Astor, W., Finch, T., Terry, J., Taubman, J., Maier, W., and Goldman, D. (2004) 'Association of specific haplotypes of D2 dopamine receptor gene with vulnerability to heroin dependence in 2 distinct populations', *Archives of General Psychiatry*, 61: 597–606.

Zara, G. (2010) 'Farrington David P: the integrated cognitive antisocial potential theory', in Cullen, F.T. and Wilcox, P. (eds), *Encyclopedia of Criminological Theory*, Thousand Oaks, CA: Sage, pp. 313–322.

Zhang, H.B., Wang, Y.F., Li, J., Wang, B., and Yang, L. (2005) 'Association between dopamine beta hydroxylase gene and attention deficit hyperactivity disorder complicated with disruptive behavior disorder', *Zhonghua Er Ke Za Zhi*, 43: 26–30.

Zouk, H., McGirr, A., Lebel, V., Benkelfat, C., Rouleau, G., and Turecki, G. (2007) 'The effect of genetic variation of the serotonin *1B* receptor gene on impulsive aggressive behavior and suicide', *American Journal of Medical Genetics Part B (Neuropsychiatric Genetics)*, 155: 996–1002.

7

Peer effects and molecular genetics in adolescent antisocial behavior

Tina Kretschmer, Jan Kornelis Dijkstra, and René Veenstra

Introduction

Adolescent antisocial behavior is detrimental to individual development and poses a huge burden to society. Historically, first nature, later nurture was held responsible for the occurrence of antisocial behavior, but contemporary research embraces genetic and social factors and focuses particularly on their interplay in trying to understand individual variation in engaging in aggression, delinquency, and substance use. What does this new orientation teach us with regard to the peer environment? Aiming to elucidate how molecular genetics and peer environment function together in increasing the risk for antisocial behavior, this chapter summarizes findings on antisocial and risk behaviors and discusses challenges for future studies.

Most past research into antecedents of antisocial behavior has attributed at least some influence to deviant peer affiliation, which makes sense given that youths spend increasingly more time interacting with peers (Larson and Richards, 1991) and are more vulnerable to peer influence than children or adults (Steinberg and Morris, 2001; Brechwald and Prinstein, 2011). Studies provide ample evidence that deviant peer affiliation increases own antisocial behavior, even after accounting for peer-related confounders such as peer rejection and family-related factors (for reviews see Monahan *et al.*, 2009; Dishion and Tipsord, 2011). In fact, one of the strongest predictors of antisocial behavior in adolescence is peer behavior (Van Lier *et al.*, 2007). Not only are adolescents influenced by their environment, they also shape their environment and actively decide whom they select as friends. Such selection effects are often driven by behavioral similarity so that antisocial adolescents will "flock together." Although studies have more often examined socialization effects, both processes are acknowledged in this chapter.

Importantly, adolescents are not uniformly vulnerable to deviant peer influence, and individual factors also affect selection of friends (Magnusson, 1985; Rutter, 1997; Caspi and Moffitt, 2006). Prior research on this person–peer environment interplay has focused on biological and psychological factors such as pubertal stage and temperament, but more recent studies reflect a growing interest in the role of specific genes. Building upon twin studies that suggested already for quite some time that heritability is important in antisocial behavior (Rhee and Waldman, 2002; Moffitt, 2005), scientific and technical advances now allow us to examine whether so-called candidate genes are linked to antisocial behavior. Candidate gene studies do not necessarily try to establish a main association between a specific gene and an outcome, but assume gene–environment correlations (rGE) and gene–environment interactions (G × E).

The term gene–environment correlation subsumes three different mechanisms in which genetic factors are associated with exposure to particular environmental conditions (Knafo and Jaffee, 2013). With regard to the peer environment, evocative rGE and active rGE are likely (Figure 7.1). In evocative rGE, individuals' genetically influenced traits evoke particular responses from the environment. For instance, a sensation-seeking child will encounter different reactions from the peer group than an anxious child. Active rGE describes that individuals' selection of environments is to some extent genetically driven. Here, the sensation-seeking child will probably select other sensation-seekers as friends, whereas the anxious child is more likely to affiliate with peers who prefer quieter spare-time activities.

Gene–environment interactions describe that genotype elevates or buffers the effect of environmental exposure on the outcome, or that a direct association between a candidate gene and an outcome is qualified by exposure to a particular environment (Figure 7.2). For instance, if peer drinking is associated with adolescents' own alcohol consumption only for carriers of a specific variant of a gene, this gene would moderate the association between environmental exposure and outcome. If, in contrast, the association between a particular genetic variant and aggressive behavior is present only if the adolescent affiliates with aggressive peers, environmental exposure has a moderating function. Theoretical assumptions about pathways should guide assignment of main predictor and moderator variables.

Since Caspi and colleagues' landmark study showed how carriers of specific variants of the monoamine-oxidase-A (*MAOA*) gene were more antisocial in adulthood after being maltreated

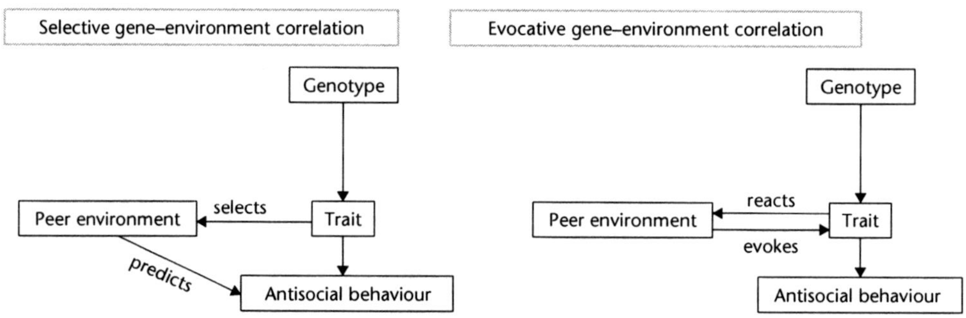

Figure 7.1 Gene–environment correlation

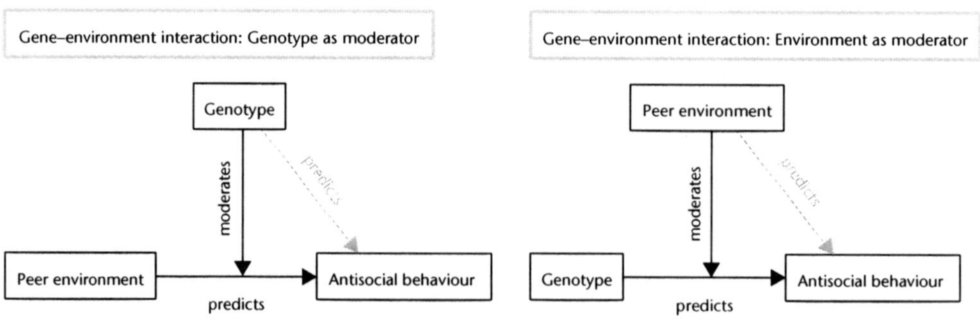

Figure 7.2 Gene–environment interaction

as children than maltreated individuals who did not carry this variant (Caspi *et al.*, 2002), studies on externalizing behavior, delinquency, aggression, and substance use have progressively employed G × E designs (Jaffee *et al.*, 2005; Bakermans-Kranenburg and van IJzendoorn, 2006; Propper *et al.*, 2007; McCrory *et al.*, 2012). Curiously, the majority of these studies remained within the family environment, that is, examined how genetic factors and negative family conditions interact in raising the risk of offspring maladjustment. While parents are crucially important for adolescent behavioral development, the significance of peers with regard to antisocial behavior is uncontested. But how can specific genes affect an adolescent's likelihood of picking up deviant behaviors from peers? Or, *vice versa*, can peers really buffer genetic effects on antisocial behavior? The steadily growing number of studies on the interplay between peer environment and molecular genetic information appears to suggest this. Following a brief introduction to molecular genetics, we focus on particular neurotransmitter systems and candidate genes relevant to them as these are implicated as biological pathways in most candidate gene studies.

Peers, molecular genetics, and antisocial behavior: where do we stand?

Human DNA consists of about 3 million pairings of adenine and thymine or guanine and cytosine that constitute the double helix. Some stretches of DNA code for protein in that they are first transcribed into messenger RNA (mRNA) that then translates into amino acids, the building blocks for protein. Traditionally, those DNA stretches that are functional in coding for proteins are called genes. The remaining DNA transcribes into messenger RNA but not further into amino acids. Current research is trying to identify the precise use of what was known as "junk DNA," suggesting that one of its functions is the regulation of protein-coding DNA.

Individual variation in some genes arises due to errors in DNA copying that result in base substitution or deletion. Although such variations in DNA are consequential (as we describe below), those of interest to social scientists are common and not eliminated by evolution, suggesting that they have some adaptive function as well. These characteristics differentiate genetic polymorphisms (as such genetic variations are called) from genetic mutations, which occur much less frequently and do not seem to serve an adaptive function. Different types of genetic polymorphism are known: Apart from quantitative differences in efficiency, as is the case in variable number tandem repeat (VNTR) polymorphisms where a DNA sequence varies in length, other genetic polymorphisms cause insertion of a different amino acid into a protein (single nucleotide polymorphisms, SNPs). Proteins are implicated in practically all processes in the human body, thus also in those chemical processes that occur in the brain. Here proteins produce, process, and recognize neurotransmitters, chemicals that direct many brain processes, including dopamine, serotonin, and acetylcholine. Neurotransmitters are important because differences in their activity in the brain appear to be directly linked to behavior and emotion regulation (Raine, 2008) and to modulate how susceptible we are to environmental conditions, including those provided by peers.

Dopamine

Dopamine plays an important role in reward-related behaviors, with brain dopamine levels increasing upon experience of reward. Many addictive substances such as cocaine and amphetamines have the same effect. Candidate genes that are implicated in the dopaminergic system are the dopamine active transporter 1 gene (*DAT1*), as well as dopamine D2 and D4 receptor genes (*DRD2* and *DRD4*). Dopamine transporter cells affect reuptake speed of dopamine, that is, the amount of time needed to clear dopamine from the synaptic cleft and essentially stop the dopamine signaling from one synapse to the next. Dopamine receptor genes are responsible for receptor density and functionality in different brain regions.

Individual variation in the *DAT1* gene occurs through a VNTR polymorphism. Most frequent variants include 9- and 10-repeat units, but between 3- and 11-repeat versions have been observed. Highest expression levels were reported for the 10-repeat variant, thus this variant works most effectively in clearing dopamine from synapses. Increased duration of dopamine activity is associated with greater reward expectancy. The 9-repeat allele has been linked to greater self-reported impulsivity and increased activity in the ventral striatum, a brain region implicated in reward processing (Forbes *et al.*, 2007), as well as to externalizing behavior in children (Young *et al.*, 2002). However, research is not conclusive and some studies report more negative outcomes for carriers of the 10-repeat allele, for example with regard to attention-deficit/hyperactivity disorder (Cornish *et al.*, 2005) and pathological criminal behavior in youth (Vaughn *et al.*, 2009). Studies have yet to establish whether the *DAT1* gene moderates associations between peer characteristics and adolescent antisocial behavior. Beaver *et al.* (2008a), however, showed that carriers of the 10-repeat allele who lived in high-risk neighborhoods were at greater risk of seeking out delinquent peers, a prime example of active gene–environment correlation.

DRD2 and *DRD4* are dopamine receptor genes that regulate receptor density and functioning in various brain regions. Variation in the *DRD2* gene results from an SNP located on the neighboring *ANKK1* gene that gives rise to the *DRD2* A1 allele. This allele has been associated with psychopathic personality traits (Wu and Barnes, 2013), alcohol dependence (Wang *et al.*, 2013), and cannabis addiction (Nacak *et al.*, 2012). *DRD4* is located on chromosome 11 and contains a VNTR polymorphism which takes the form of a long (7-repeat) or short (4-repeat) allele, with the short allele being much more common than the long variant (Oak *et al.*, 2000). Less frequent variants range from 2 to 10 repeats, and research has shown that *DRD4* is more potent in binding dopamine in the brain in the presence of the short variant. The long allele has been linked to attention-deficit/hyperactivity disorder (Faraone *et al.*, 2001) and mood disorders (López León *et al.*, 2005). Some studies have also linked the *DRD4* polymorphism to personality traits including novelty- and thrill-seeking (Ray *et al.*, 2009; Dmitrieva *et al.*, 2011), but results remain inconclusive (Schinka *et al.*, 2002; Munafò *et al.*, 2008). Finally, associations between *DRD4* and substance use have been reported (Hutchison *et al.*, 2002; Filbey *et al.*, 2008).

Important to note is a recent meta-analysis that did not support direct associations between any of the dopaminergic system polymorphisms discussed here and aggression (Vassos *et al.*, 2014). Combined with some ambiguous associations (i.e. no clarity as to which variant increases risk, inconclusive results, replication failure), research that also takes into account environmental influences on antisocial outcomes seems more fruitful than the search for direct associations between genotype and outcome. That is, genetic polymorphisms in the dopaminergic system have been found to moderate peer influence: Guo *et al.* (2008) examined whether *DRD2* and *DAT1* affect the association between peers' self-reported delinquent behavior and young adult males' violent and serious delinquency. Their analyses showed that carriers of the heterozygous variant of *DRD2* (i.e. carrying one A1 and one A2 allele rather than two A1 or two A2 alleles) were more likely to engage in serious and violent delinquency if they affiliated with delinquent friends. Using an experimental design with a drinking confederate, Larsen *et al.* (2010) showed that carriers of the long allele of *DRD4* were at increased risk for heavy alcohol use in the presence of someone who consumed large amounts of alcohol in a short period. Notably, Van der Zwaluw *et al.* (2012) did not find this interaction effect in a longitudinal study of real friends.

Dopamine system genes have also attracted interest from researchers who examined affiliation with deviant peers not as environmental measure, as in the studies discussed above, but as outcome. Beaver *et al.* (2012), for instance, used data on more than 1000 males within the Add Health sample to examine whether the *DRD2* A1 allele and *DRD4* long allele interacted with neighborhood disadvantage in the prediction of four measures of antisocial phenotype, among them presence of delinquent peers. The peer measure was assessed through adolescents' self-report referred to number of peers who smoked more than one cigarette per day, smoked marijuana more than once a month, and drank alcohol more than once a month. Beaver *et al.* (2012) showed that *DRD2* significantly predicted affiliation with delinquent peers in adolescents growing up in disadvantaged neighborhoods. They further found that *DRD2* predicted adolescent experiences of victimization (e.g. someone pulling a gun against them), but only among youth with a low number of deviant peers. That is, affiliation with deviant peers buffered the genetic risk for victimization. There is also support for the notion that dopamine polymorphisms interact with environmental characteristics in predicting desistance from delinquency. Beaver *et al.* (2008b) showed that *DRD2* and *DRD4* (as well as *MAOA*, described below) moderated the association between marriage and desistance.

Serotonin

The role of serotonin for mood, sleep, memory, and behavior is unquestioned, resulting in numerous studies on the effect of serotonergic system genes on a range of psychopathologies, mostly on the internalizing spectrum. Serotonin has also been linked to externalizing behaviors: a meta-analysis by Moore *et al.* (2002) showed that metabolites (degradation products) of brain serotonin were reduced in antisocial compared with non-antisocial individuals. A number of genes are implied in brain serotonin circuitry, the most popular one probably being the serotonin transporter polymorphism *5-HTTLPR*, which transcribes for a protein that is associated with reuptake speed of serotonin at brain synapses and constitutes an important source of variation in serotonin levels in the brain. Whereas the majority of people of Caucasian descent are carriers of the heterozygous short-long variant (*ca.* 50 percent) of this polymorphism, the homozygous short-short variant is observed less often than the homozygous long-long variant (Noskova *et al.*, 2008). Because of its reduced efficiency in terminating synaptic serotonin activity, the short allele has been treated as the risk allele in most studies (but see Glenn, 2011 for a review on psychopathology associated with the long allele). The *5-HTTLPR* short allele has been linked to personality traits such as neuroticism (Sen *et al.*, 2004) as well as mood disorders (Rosenthal *et al.*, 1998). The polymorphism is also implicated in behaviors and traits that show overlap with antisocial behavior. For instance, studies have reported associations between the short allele and substance use (Feinn *et al.*, 2005). Retz *et al.* (2004) showed that the short allele was over-represented in males displaying recurrent and overt violent behavior, whereas Vaughn *et al.* (2009) reported a significant association with pathological criminal behavior in youth. The polymorphism has further been associated with persistent, pervasive childhood aggression (Beitchman *et al.*, 2006) and antisocial traits in young adults (Lyons-Ruth *et al.*, 2007). Moreover, probably because of its importance in the amygdala, *5-HTTLPR* affects perception of emotional stimuli: Beevers *et al.* (2007) showed that the polymorphism affected the speed at which someone was able to disengage their attention from both happy and sad stimuli. This effect is interesting because it may provide a pointer as to why some adolescents react differently to cues from their peer environment. That is, peer pressure in the form of teasing and threats to ensure compliant behavior may have a stronger impact upon those adolescents who carry the more susceptible variant of *5-HTTLPR*. Surprisingly, no study has tested whether *5-HTTLPR* moderates the influence of peer characteristics on adolescent antisocial behavior, although three studies have tested whether associations between peer relationship quality (i.e. peer victimization) and later depression (Benjet *et al.*, 2009; Sugden *et al.*, 2010) and delinquency (Kretschmer *et al.*, 2014b) vary by *5-HTTLPR*. Indeed, all three studies showed that carriers of the short allele of the polymorphism were at greater risk of developing problems following victimization by peers.

Focusing on a polymorphism in a serotonin receptor gene (*5-HT2A*), Dijkstra *et al.* (2013) examined genetic moderation of the association between boys' aggression and their popularity in the peer group. The study built on two studies by Burt (2008, 2009) that showed how this polymorphism predicts which peers adolescents like and dislike, and that this association is mediated by rule-breaking behavior. Extending these findings, Dijkstra and colleagues showed that the positive association between aggressive behavior and popularity was moderated by *HT2A*.

Acetylcholine

Acetylcholine is the longest known and most common neurotransmitter, and is implicated in attention and arousal. Candidate genes that are associated with brain acetylcholine (nicotinic and muscarinic acetylcholine receptor genes) have previously been linked to personality traits (e.g. Hendershot *et al.*, 2011) and alcohol-related disorders (Jung *et al.*, 2011). Nicotinic acetylcholine receptor gene polymorphisms (e.g. *CHRNA4*, *CHRNA5*, *CHRNA7*, *CHRNB3*) have been associated with smoking (Munafò and Johnstone, 2008) and attracted some interest as moderators of peer group influence. For instance, Latendresse *et al.* (2011) examined antisocial behavior in the peer group as predictor of developmental trajectories of externalizing behavior. In this study, Latendresse and colleagues showed that adolescents who carried minor alleles of muscarinic acetylcholine receptor gene *CHRM2* and affiliated with deviant peers were at greater risk of showing stable or moderately high externalizing behavior. Treating the environmental measure as moderator, and based on a main association between *CHRNA5* and nicotine dependence, Johnson *et al.* (2010) showed that peer smoking reduced the genetic effect. Put differently, the genetic association with nicotine dependence was strongest in the absence of environmental risk.

Monoamine-oxidase-A (MAOA)

Some genetic polymorphisms affect more than one neurotransmitter system. The monoamine-oxidase-A (*MAOA*) gene codes for an enzyme that is important in degrading serotonin and dopamine. A VNTR polymorphism in this gene concerns the number of tandem repeats in the promoter region of the gene with higher expression in carriers of the 3.5- or 4-repeat variant and lower expression in carriers of the 3- (and possibly 5-) repeat variant. *MAOA* has repeatedly been linked to antisocial behavior, specifically human aggression (Buckholtz and Meyer-Lindenberg, 2008) and interacts stably with adverse environmental conditions such as maltreatment (Kim-Cohen *et al.*, 2006) in increasing the risk for antisocial outcomes. The less efficient variant of the polymorphism thereby increased the risk for aggression, and in most studies also increased vulnerability to adversity in the environment. With regard to moderation of peer effects, however, studies yielded mixed findings. Lee (2011), for instance, examined the interplay between *MAOA* and self-reported peer delinquency with regard to own antisocial behavior using a male subsample of the Add Health study, and showed that carriers of the high-activity *MAOA* variant were at greater risk of engaging in antisocial behavior if they affiliated with delinquent peers. Also using Add Health data, Beaver and Holtfreter (2009) showed that *MAOA* moderated the link between delinquent peer affiliation in adolescence and engagement in fraudulent behavior in early adulthood. Again, carriers of the high-activity variant of *MAOA* were most vulnerable to the effect of deviant peers.

Brain-derived neurotrophic factor (BDNF)

Stepping out of the neurotransmitter systems, Kretschmer *et al.* (2014a) examined the brain-derived neurotrophic factor gene (*BDNF*), which regulates the secretion of brain-derived neurotrophic factor in the brain. The *BDNF* polymorphism consists of a valine-to-methionine substitution with *BDNF* secretion being reduced in met- compared with val-alleles (Hong *et al.*, 2011). *BDNF* affects susceptibility to environmental stressors in the prediction of impulsive aggression (Wagner *et al.*, 2010) with met-allele carriers being more vulnerable to environmental risk than val-val carriers. Moreover, carriers of the met-allele show an increased risk for psychopathological disorders related to aggression (e.g. Spalletta *et al.*, 2010) and impulsivity (Oades *et al.*, 2008). In line with previous research, Kretschmer *et al.* (2014b) showed a stronger effect of peer on adolescents' own aggression in carriers of the met-met variant of this polymorphism. That is, those adolescents who had affiliated with aggressive peers in late childhood and carried two minor alleles were at greatest risk of engaging in aggressive behaviors themselves five years later.

μ-opioid receptor (OPRM1)

The *OPRM1* gene encodes for the μ-opioid receptor (MOR, receptor of endogenous opioids including heroin, morphine, and methadone). The *OPRM1* polymorphism consists of an adenine-to-guanine substitution which affects the reinforcing (thus addictive) effect of drugs in various brain regions (Ray *et al.*, 2011). This polymorphism has been linked to a variety of substance-use characteristics, including sensitivity to the effects of alcohol (Ray and Hutchison, 2004) and adolescent alcohol misuse (Miranda *et al.*, 2010), heroin addiction (Shi *et al.*, 2002), and nicotine reinforcement (Ray *et al.*, 2006). *OPRM1* has also been associated with neural sensitivity to social rejection (Way *et al.*, 2009). Moreover, the polymorphism has been found to interact with peer characteristics in predicting alcohol-use disorders in adolescence (Miranda *et al.*, 2013) in that carriers of the G-allele who affiliated with deviant peers had an almost eight times increased risk of being diagnosed with alcohol-use disorder compared with their homozygous A-allele counterparts. Another study examined whether *OPRM1* was associated with affiliation with peers with a positive attitude towards drinking (Chassin *et al.*, 2012). This study showed that male individuals homozygous for the A-allele were more likely to affiliate with alcohol-promoting peers who, in turn, increased their risk for alcohol-use disorder. A different pattern emerged for girls: G-carriers of *OPRM1* were more vulnerable to peer influence than carriers of the A-homozygous variant of the polymorphism.

Peers, molecular genetics, and antisocial behavior: where do we go from here?

Effects of peer characteristics, irrespective of whether they refer to substance use or delinquency, are not uniformly strong and some of the variation is accounted for by genetic make-up. The genetic polymorphisms that have been tested to date are theoretically meaningful in that they mostly refer to neurotransmitters implicated in impulsivity, aggression, or response to emotional stimuli, or affect how an individual processes substances such as nicotine or alcohol. However, systematic and replicated findings are still scarce. The final part of this chapter aims to provide some pointers that may help to strengthen research into the interplay of peer environment and genetic factors in antisocial behavior research.

Gene–environment correlation studies

Although we know from quantitative behavioral genetics that genetic factors are associated with delinquent peer group affiliation (Brendgen, 2012), only few studies have examined rGE using measured genotype. The studies by Burt (2009), Beaver *et al.* (2012), and Chassin *et al.* (2012) are notable exceptions, and their findings clearly suggest that adolescent peer selection is driven partly by specific genetic factors. These findings contain a word of warning: if genetic factors influence someone's preferences for affiliation, such associations need to be controlled for in G × E designs. Adolescents' peer groups are very specific environments driven by selection and choice (Veenstra and Dijkstra, 2011), more so than other environments that have played a role in gene–environment interplay studies (e.g. maltreatment, neighborhood). Given the significance of peers in adolescence, and findings from social network studies that show how peer similarity in antisocial forms of behavior is to a non-negligible extent driven by selection (individuals affiliate with those who are similar to them, Veenstra *et al.*, 2013), it is surprising that rGE studies are scarce in research on peer effects on antisocial behavior. Future studies are advised to pay more attention to the correlational mechanisms, as these may help to better understand the role of the peer environment in antisocial behavior.

Assessment of the environment

Although we have not broached the issue of validity and reliability of the peer measures, it is crucial to note that the quality of environmental assessment is a driving force in the success of a gene–environment interplay study. Particularly in the field of peer characteristics, shared-rater bias (adolescents' reports on their own and their peers' behavior), huge variability in measures used, and ambiguity as to whom "peer group" refers to affect comparability and might determine whether or not a set of findings is replicated (see Larsen *et al.*, 2010; Van der Zwaluw *et al.*, 2012). Of course, this problem is not easy to solve, and many of the studies described here are secondary analyses of existing cohort studies where measures cannot be changed in hindsight. Researchers are advised to revert to peers' own reports of their behavior as much as possible and streamline future studies with successful past ones in terms of measures and definitions (Veenstra and Steglich, 2012; Veenstra *et al.*, 2013).

Peer relationship quality as environmental measure

Importantly, adolescent development is not only affected by peer characteristics such as drinking and delinquency; the quality of relationships with peers also affects adjustment (e.g. Hanish and Guerra, 2002). Failing to establish positive relationships with peers is predictive of aggression (e.g. Boivin *et al.*, 2005) and engagement in delinquent behavior (Laird *et al.*, 2001). Peer victimization as a particularly grave form of peer negativity increases the risk for maladjustment (Arseneault *et al.*, 2010), including aggression and delinquency (Hodges *et al.*, 1999; Ostrov, 2010). Arguably, rejected or victimized adolescents are less well able to achieve social well-being through peer relationships, and may turn to other strategies. These may involve antisocial forms of behavior, especially so in adolescence, when delinquency to a certain extent contributes to status attainment and involves behaviors that may elicit confirmation from the peer group (Mayeux *et al.*, 2008). Such associations may be genetically moderated, as initial research shows (Kretschmer *et al.*, 2013). Future studies are needed to replicate and extend these findings in independent samples.

Dual risk or differential susceptibility?

This chapter focuses on peer effects in the development of adolescent antisocial behavior and how these are moderated by variations in DNA (or conversely, how associations between genetic factors and antisocial behavior differ depending on environmental conditions). The perspective taken assumes a dual-risk model in which the environment carries risk for a negative outcome, and this risk is elevated in the presence of particular genetic variants. While this model is theoretically appealing given antisocial behavior as outcome, recent accounts have begun to challenge the generalizability of the dual-risk model to person–environment interactions (Belsky and Pluess, 2009; Ellis *et al.*, 2011), for example, a differential susceptibility framework in which person factors, including but not limited to genetic polymorphisms, make individuals more susceptible to both negative and positive environmental conditions. For instance, affiliation with deviant peers may elevate the risk for antisocial behavior development in some adolescents more than in others, but these same individuals may also benefit more from prosocial peer environments than less susceptible adolescents. Although pro- and antisocial behavior do not constitute opposite ends of one continuum, taking such effects into account will further the field and contribute important information about the role of person factors in understanding peer group effects on adolescent development beyond antisocial behavior.

Re-thinking the use of DNA information

Although studying the interplay between peer environment and measured genes adds valuable information to our understanding of heterogeneity in associations between peer environment and antisocial behavior, the effects that have been published are small, and interactions explain only a tiny fraction of variance in antisocial behavior. This problem has been observed in most G × E studies as well as in social science research on main effects of genetic polymorphisms, notwithstanding quantitative genetic findings that a large proportion of population variance is genetically influenced. How can we account for this problem of "missing heritability" (Maher, 2008)? Discussions of the problem in neighboring disciplines (e.g. Plomin, 2012) suggest that we need to examine more than one polymorphism at a time. Not only are polymorphisms not necessarily inherited independently, they function additively and interact with each other. Accounting for such interdependencies should become integral in studies on peer environment and gene–environment research in general. Using polygenic scores (indexes of many polymorphisms of small effect) as well as including gene–gene interactions in our models, are steps towards this goal (Plomin, 2013). Finally, dopamine, serotonin, and acetylcholine are of course not the only human brain neurotransmitters. Norepinephrine is important in the regulation of concentration, but can also function as a hormone and as such is important in the regulation of stress and response to fight-or-flight situations. GABA is the most important inhibitory neurotransmitter in the human central nervous system. We are not suggesting that researchers should now pick single candidates from these systems instead, but the different neurotransmitter systems do not operate in isolation. Informed by molecular neuroscience, researchers are advised to account for interactions and combined effects.

Conclusion

The inclusion of molecular genetic information into studies of associations between peer environment and antisocial behavior is novel, and it may be too early to say whether genetic and environmental effects work together systematically in affecting individual risk for antisocial behavior. As with gene–environment studies in other fields, independent replications are sorely needed, as are strategies to investigate cumulative and interacting effects of several polymorphisms at a time. However, the studies reviewed in this chapter certainly show that stepping out of the family environment and examining whether individual genetic make-up qualifies the effect of fairly common environmental risks (e.g. affiliating with peers who smoke or engage in deviant behavior) is a meaningful and worthwhile strategy to understand adolescent antisocial development. The knowledge gained in such studies may not be practically applicable in the short-term, but being able to identify the adolescents most at risk for antisocial or substance-use disorders eventually can help to target prevention and intervention strategies. Whereas it is not ethical or possible to modify and adjust genetic factors, changing the environment may be particularly important for some young people. In other words, if certain environmental risks, such as peer deviance or peer drinking, are especially harmful for some adolescents (those who carry specific genetic variants), efforts should go into studying how we can prevent or intervene in these adverse environments to ensure healthy development.

References

Arseneault, L., Bowes, L., and Shakoor, S. (2010) 'Bullying victimization in youths and mental health problems: "Much ado about nothing"', *Psychological Medicine*, 40(5): 717–729.

Bakermans-Kranenburg, M.J. and van IJzendoorn, M.H. (2006) 'Gene–environment interaction of the dopamine D4 receptor (*DRD4*) and observed maternal insensitivity predicting externalizing behavior in preschoolers', *Developmental Psychobiology*, 48(5): 406–409.

Beaver, K.M. and Holtfreter, K. (2009) 'Biosocial influences on fraudulent behaviors', *Journal of Genetic Psychology*, 170: 101–114.

Beaver, K.M., Wright, J.P., and DeLisi, M. (2008a) 'Delinquent peer group formation: evidence of a gene × environment correlation', *Journal of Genetic Psychology*, 169: 227–244.

Beaver, K.M., Wright, J.P., DeLisi, M., and Vaughn, M.G. (2008b) 'Desistance from delinquency: the marriage effect revisited and extended', *Social Science Research*, 37(3): 736–752.

Beaver, K.M., Gibson, C.L., DeLisi, M., Vaughn, M.G., and Wright, J.P. (2012) 'The interaction between neighborhood disadvantage and genetic factors in the prediction of antisocial outcomes', in C.L. Gibson and M.D. Krohn (eds), *Handbook of Life-course Criminology: Emerging Trends and Directions for Future Research*, New York: Springer Science and Business Media, pp. 25–40.

Beevers, C.G., Gibb, B.E., McGeary, J.E., and Miller, I.W. (2007) 'Serotonin transporter genetic variation and biased attention for emotional word stimuli among psychiatric inpatients', *Journal of Abnormal Psychology*, 116(1): 208–212.

Beitchman, J., Baldassarra, L., Mik, H., De Luca, V., King, N., Bender, D., *et al.* (2006) 'Serotonin transporter polymorphisms and persistent, pervasive childhood aggression', *American Journal of Psychiatry*, 163(6): 1103–1105.

Belsky, J. and Pluess, M. (2009) 'Beyond diathesis stress: differential susceptibility to environmental influences', *Psychological Bulletin*, 135(6): 885–908.

Benjet, C., Thompson, R.J., and Gotlib, I.H. (2009) '*5-HTTLPR* moderates the effect of relational peer victimization on depressive symptoms in adolescent girls', *Journal of Child Psychology and Psychiatry*, 51(2): 173–179.

Boivin, M., Vitaro, F., and Poulin, F. (2005) 'Peer relationships and the development of aggressive behavior in early childhood', in R.E. Tremblay, W.W. Hartup, and J. Archer (eds), *Developmental Origins of Aggression*, New York: Guilford Press, pp. 376–397.

Brechwald, W.A. and Prinstein, M.J. (2011) 'Beyond homophily: a decade of advances in understanding peer influence processes', *Journal of Research on Adolescence*, 21(1): 166–179.

Brendgen, M. (2012) 'Genetics and peer relations: a review', *Journal of Research on Adolescence*, 22(3): 419–437.

Buckholtz, J.W. and Meyer-Lindenberg, A. (2008) '*MAOA* and the neurogenetic architecture of human aggression', *Trends in Neurosciences*, 31(3): 120–129.

Burt, S.A. (2008) 'Genes and popularity: evidence of an evocative gene–environment correlation', *Psychological Science*, 19(2): 112–113.

Burt, A. (2009) 'A mechanistic explanation of popularity: genes, rule breaking, and evocative gene–environment correlations', *Journal of Personality and Social Psychology*, 96(4): 783–794.

Caspi, A. and Moffitt, T.E. (2006) 'Gene–environment interactions in psychiatry: joining forces with neuroscience', *Nature Reviews Neuroscience*, 7(7): 583–590.

Caspi, A., McClay, J., Moffitt, T.E., Mill, J., Martin, J., Craig, I.W., *et al.* (2002) 'Role of genotype in the cycle of violence in maltreated children', *Science*, 297(5582): 851–854.

Chassin, L., Lee, M.R., Cho, Y.I.L., Wang, F.L., Agrawal, A., Sher, K.J., and Lynskey, M.T. (2012) 'Testing multiple levels of influence in the intergenerational transmission of alcohol disorders from a developmental perspective: the example of alcohol use promoting peers and mu-opioid receptor M1 variation', *Development and Psychopathology*, 24: 953–967.

Cornish, K.M., Manly, T., Savage, R., Swanson, J., Morisano, D., Butler, N., *et al.* (2005) 'Association of the dopamine transporter (*DAT1*) 10/10-repeat genotype with ADHD symptoms and response inhibition in a general population sample', *Molecular Psychiatry*, 10(7): 686–698.

Dijkstra, J.K., Lindenberg, S., Zijlstra, L., Bouma, E., and Veenstra, R. (2013) 'The secret ingredient for social success of young males: a functional polymorphism in the *5HT2A* serotonin receptor gene', *PloS One*, 8(2): e54821.

Dishion, T.J. and Tipsord, J.M. (2011) 'Peer contagion in child and adolescent social and emotional development', *Annual Review of Psychology*, 62: 189–214.

Dmitrieva, J., Chen, C., Greenberger, E., Ogunseitan, O., and Ding, Y. (2011) 'Gender-specific expression of the *DRD4* gene on adolescent delinquency, anger and thrill seeking', *Social Cognitive and Affective Neuroscience*, 6(1): 82–89.

Ellis, B.J., Boyce, W.T., Belsky, J., Bakermans-Kranenburg, M.J., and Van IJzendoorn, M.H. (2011) 'Differential susceptibility to the environment: an evolutionary–neurodevelopmental theory', *Development and Psychopathology*, 23(1): 7–28.

Faraone, S.V., Doyle, A.E., Mick, E., and Biederman, J. (2001) 'Meta-analysis of the association between the 7-repeat allele of the dopamine D4 receptor gene and attention deficit hyperactivity disorder', *American Journal of Psychiatry*, 158(7): 1052–1057.

Feinn, R., Nellissery, M., and Kranzler, H.R. (2005) 'Meta-analysis of the association of a functional serotonin transporter promoter polymorphism with alcohol dependence', *American Journal of Medical Genetics Part B: Neuropsychiatric Genetics*, 133(1): 79–84.

Filbey, F.M., Ray, L., Smolen, A., Claus, E.D., Audette, A., and Hutchison, K.E. (2008) 'Differential neural response to alcohol priming and alcohol taste cues is associated with *DRD4* VNTR and *OPRM1* genotypes', *Alcoholism*, 32(7): 1113–1123.

Forbes, E., Brown, S., Kimak, M., Ferrell, R., Manuck, S., and Hariri, A. (2007) 'Genetic variation in components of dopamine neurotransmission impacts ventral striatal reactivity associated with impulsivity', *Molecular Psychiatry*, 14(1): 60–70.

Glenn, A.L. (2011) 'The other allele: exploring the long allele of the serotonin transporter gene as a potential risk factor for psychopathy: a review of the parallels in findings', *Neuroscience and Biobehavioral Reviews*, 35(3): 612–620.

Guo, G., Roettger, M.E., and Cai, T. (2008) 'The integration of genetic propensities into social-control models of delinquency and violence among male youths', *American Sociological Review*, 73(4): 543–568.

Hanish, L.D. and Guerra, N.G. (2002) 'A longitudinal analysis of patterns of adjustment following peer victimization', *Development and Psychopathology*, 14(01): 69–89.

Hendershot, C.S., Bryan, A.D., Ewing, S.W.F., Claus, E.D., and Hutchison, K.E. (2011) 'Preliminary evidence for associations of *CHRM2* with substance use and disinhibition in adolescence', *Journal of Abnormal Child Psychology*, 39: 671–681.

Hodges, E.V., Boivin, M., Vitaro, F., and Bukowski, W.M. (1999) 'The power of friendship: protection against an escalating cycle of peer victimization', *Developmental Psychology*, 35(1): 94–106.

Hong, C., Liou, Y., and Tsai, S. (2011) 'Effects of *BDNF* polymorphisms on brain function and behavior in health and disease', *Brain Research Bulletin*, 86(5): 287–297.

Hutchison, K.E., LaChance, H., Niaura, R., Bryan, A., and Smolen, A. (2002) 'The *DRD4* VNTR polymorphism influences reactivity to smoking cues', *Journal of Abnormal Psychology*, 111(1): 134–151.

Jaffee, S.R., Caspi, A., Moffitt, T.E., Dodge, K.A., Rutter, M., Taylor, A., and Tully, L.A. (2005) 'Nature × nurture: genetic vulnerabilities interact with physical maltreatment to promote conduct problems', *Development and Psychopathology*, 17(1): 67–84.

Johnson, E.O., Chen, L., Breslau, N., Hatsukami, D., Robbins, T., Saccone, N.L., *et al.* (2010) 'Peer smoking and the nicotinic receptor genes: an examination of genetic and environmental risks for nicotine dependence', *Addiction*, 105(11): 2014–2022.

Jung, M., Park, B., Lee, B., Ro, Y., Park, R., Shin, H., *et al.* (2011) 'Association of *CHRM2* polymorphisms with severity of alcohol dependence', *Genes, Brain and Behavior*, 10(2): 253–256.

Kim-Cohen, J., Caspi, A., Taylor, A., Williams, B., Newcombe, R., Craig, I.W., and Moffitt, T.E. (2006) 'MAOA, maltreatment, and gene–environment interaction predicting children's mental health: new evidence and a meta-analysis', *Molecular Psychiatry*, 11(10): 903–913.

Knafo, A., and Jaffee, S.R. (2013) 'Gene–environment correlation in developmental psychopathology', *Development and Psychopathology*, 25(01): 1–6.

Kretschmer, T., Dijkstra, J.K., Ormel, J., Verhulst, F.C., and Veenstra, R. (2013) 'Dopamine receptor D4 gene moderates the effect of positive and negative peer experiences on later delinquency: the Tracking Adolescents' Individual Lives Survey study', *Development and Psychopathology*, 25: 1107–1117.

Kretschmer, T., Vitaro, F., and Barker, E.D. (2014a) 'Association between peer and own aggression is moderated by *BDNF* val-met polymorphism', *Journal of Research on Adolescence*, doi: 0.1111/jora.12050

Kretschmer, T., Sentse, M., Dijkstra, J.K., and Veenstra, R. (2014b) 'The interplay between peer rejection and acceptance in preadolescence and early adolescence, serotonin transporter gene, and antisocial behavior in late adolescence: the TRAILS Study', *Merrill-Palmer Quarterly*, 60(2): 193–216.

Laird, R.D., Jordan, K.Y., Dodge, K.A., Pettit, G.S., and Bates, J.E. (2001) 'Peer rejection in childhood, involvement with antisocial peers in early adolescence, and the development of externalizing behavior problems', *Development and Psychopathology*, 13(2): 337–354.

Larsen, H., van der Zwaluw, C.S., Overbeek, G., Granic, I., Franke, B., and Engels, R.C.M.E. (2010) 'A variable-number-of-tandem-repeats polymorphism in the dopamine D4 receptor gene affects social adaptation of alcohol use: investigation of a gene–environment interaction', *Psychological Science*, 21: 1064–1068.

Larson, R. and Richards, M.H. (1991) 'Daily companionship in late childhood and early adolescence: changing developmental contexts', *Child Development*, 62(2): 284–300.

Latendresse, S.J., Bates, J.E., Goodnight, J.A., Lansford, J.E., Budde, J.P., Goate, A., *et al.* (2011) 'Differential susceptibility to adolescent externalizing trajectories: examining the interplay between *CHRM2* and peer group antisocial behavior', *Child Development*, 82(6): 1797–1814.

Lee, S.S. (2011) 'Deviant peer affiliation and antisocial behavior: interaction with monoamine oxidase a (*MAOA*) genotype', *Journal of Abnormal Child Psychology*, 39(3): 321–332.

López León, S., Croes, E.A., Sayed-Tabatabaei, F.A., Claes, S., Broeckhoven, C.V., and van Duijn, C.M. (2005) 'The dopamine D4 receptor gene 48-base-pair-repeat polymorphism and mood disorders: a meta-analysis', *Biological Psychiatry*, 57(9): 999–1003.

Lyons-Ruth, K., Holmes, B.M., Sasvari-Szekely, M., Ronai, Z., Nemoda, Z., and Pauls, D. (2007) 'Serotonin transporter polymorphism and borderline/antisocial traits among low-income young adults', *Psychiatric Genetics*, 17(6): 339–354.

McCrory, E., De Brito, S.A., and Viding, E. (2012) 'The link between child abuse and psychopathology: a review of neurobiological and genetic research', *Journal of the Royal Society of Medicine*, 105(4): 151–156.

Magnusson, D. (1985) 'Implications of an interactional paradigm for research on human development', *International Journal of Behavioral Development*, 8(2): 115–137.

Maher, B. (2008) 'The case of the missing heritability', *Nature*, 456(7218): 18–21.

Mayeux, L., Sandstrom, M.J., and Cillessen, A.H. (2008) 'Is being popular a risky proposition?', *Journal of Research on Adolescence*, 18(1): 49–74.

Miranda, R., Ray, L., Justus, A., Meyerson, L.A., Knopik, V.S., McGeary, J., and Monti, P.M. (2010) 'Initial evidence of an association between *OPRM1* and adolescent alcohol misuse', *Alcoholism: Clinical and Experimental Research*, 34(1): 112–122.

Miranda, R., Reynolds, E., Ray, L., Justus, A., Knopik, V.S., McGeary, J., and Meyerson, L.A. (2013) 'Preliminary evidence for a gene–environment interaction in predicting alcohol use disorders in adolescents', *Alcoholism: Clinical and Experimental Research*, 37: 325–331.

Moffitt, T.E. (2005) 'The new look of behavioral genetics in developmental psychopathology: gene–environment interplay in antisocial behaviors', *Psychological Bulletin*, 131(4): 533–575.

Monahan, K.C., Steinberg, L., and Cauffman, E. (2009) 'Affiliation with antisocial peers, susceptibility to peer influence, and antisocial behavior during the transition to adulthood', *Developmental Psychology*, 45(6): 1520–1531.

Moore, T.M., Scarpa, A., and Raine, A. (2002) 'A meta-analysis of serotonin metabolite *5-HIAA* and antisocial behavior', *Aggressive Behavior*, 28(4): 299–316.

Munafò, M.R. and Johnstone, E.C. (2008) 'Genes and cigarette smoking', *Addiction*, 103(6): 893–904.

Munafò, M.R., Yalcin, B., Willis-Owen, S.A., and Flint, J. (2008) 'Association of the dopamine D4 receptor (*DRD4*) gene and approach-related personality traits: meta-analysis and new data', *Biological Psychiatry*, 63(2): 197–206.

Nacak, M., Isir, A.B., Balci, S.O., Pehlivan, S., Benlier, N., and Aynacioglu, S. (2012) 'Analysis of dopamine D2 receptor (*DRD2*) gene polymorphisms in cannabinoid addicts', *Journal of Forensic Sciences*, 57(6): 1621–1624.

Noskova, T., Pivac, N., Nedic, G., Kazantseva, A., Gaysina, D., Faskhutdinova, G., *et al.* (2008) 'Ethnic differences in the serotonin transporter polymorphism (*5–HTTLPR*) in several European populations', *Progress in Neuro-Psychopharmacology and Biological Psychiatry*, 32(7): 1735–1739.

Oades, R.D., Lasky-Su, J., Christiansen, H., Faraone, S.V., Sonuga-Barke, E.J., Banaschewski, T., *et al.* (2008) 'The influence of serotonin and other genes on impulsive behavioral aggression and cognitive impulsivity in children with attention-deficit/hyperactivity disorder (ADHD): findings from a family-based association test (FBAT) analysis', *Behavioral and Brain Functions*, 4(1): 48.

Oak, J.N., Oldenhof, J., and Van Tol, H.H.M. (2000) 'The dopamine D4 receptor: one decade of research', *European Journal of Pharmacology*, 405(1/3): 303–327.

Ostrov, J.M. (2010) 'Prospective associations between peer victimization and aggression', *Child Development*, 81(6): 1670–1677.

Plomin, R. (2012) 'Child development and molecular genetics: 14 years later', *Child Development*, 84: 104–120.

Plomin, R. (2013) 'Commentary: missing heritability, polygenic scores, and gene–environment correlation', *Journal of Child Psychology and Psychiatry*, 54(10): 1147–1149.

Propper, C., Willoughby, M., Halpern, C.T., Carbone, M.A., and Cox, M. (2007) 'Parenting quality, DRD4, and the prediction of externalizing and internalizing behaviors in early childhood', *Developmental Psychobiology*, 49(6): 619–632.

Raine, A. (2008) 'From genes to brain to antisocial behavior', *Current Directions in Psychological Science*, 17(5): 323–328.

Ray, L.A. and Hutchison, K.E. (2004) 'A polymorphism of the μ-opioid receptor gene (*OPRM1*) and sensitivity to the effects of alcohol in humans', *Alcoholism: Clinical and Experimental Research*, 28(12): 1789–1795.

Ray, R., Jepson, C., Patterson, F., Strasser, A., Rukstalis, M., Perkins, K., *et al.* (2006) 'Association of *OPRM1* A118G variant with the relative reinforcing value of nicotine', *Psychopharmacology*, 188(3): 355–363.

Ray, L.A., Bryan, A., MacKillop, J., McGeary, J., Hesterberg, K., and Hutchison, K.E. (2009) 'The dopamine D4 receptor (*DRD4*) gene exon III polymorphism, problematic alcohol use and novelty seeking: direct and mediated genetic effects', *Addiction Biology*, 14(2): 238–244.

Ray, R., Ruparel, K., Newberg, A., Wileyto, E.P., Loughead, J.W., Divgi, C., *et al.* (2011) 'Human mu opioid receptor (*OPRM1* A118G) polymorphism is associated with brain mu-opioid receptor binding potential in smokers', *Proceedings of the National Academy of Sciences, USA*, 108(22): 9268–9273.

Retz, W., Retz-Junginger, P., Supprian, T., Thome, J., and Rösler, M. (2004) 'Association of serotonin transporter promoter gene polymorphism with violence: relation with personality disorders, impulsivity, and childhood ADHD psychopathology', *Behavioral Sciences and the Law*, 22(3): 415–425.

Rhee, S.H. and Waldman, I.D. (2002) 'Genetic and environmental influences on antisocial behavior: a meta-analysis of twin and adoption studies', *Psychological Bulletin*, 128(3): 490–535.

Rosenthal, N.E., Mazzanti, C.M., Barnett, R.L., Hardin, T.A., Turner, E.H., Lam, G.K., *et al.* (1998) 'Role of serotonin transporter promoter repeat length polymorphism (*5-HTTLPR*) in seasonality and seasonal affective disorder', *Molecular Psychiatry*, 3(2): 175–177.

Rutter, M. (1997) 'Developmental psychopathology as an organizing research construct', in D. Magnusson (ed.), *The Lifespan Development of Individuals: Behavioral, Neurobiological, and Psychosocial Perspectives*, Cambridge: Cambridge University Press, pp. 394–413.

Schinka, J.A., Letsch, E.A., and Crawford, F.C. (2002) '*DRD4* and novelty seeking: results of meta-analyses', *American Journal of Medical Genetics*, 114(6): 643–648.

Sen, S., Burmeister, M., and Ghosh, D. (2004) 'Meta-analysis of the association between a serotonin transporter promoter polymorphism (*5-HTTLPR*) and anxiety-related personality traits', *American Journal of Medical Genetics Part B: Neuropsychiatric Genetics*, 127(1): 85–89.

Shi, J., Hui, L., Xu, Y., Wang, F., Huang, W., and Hu, G. (2002) 'Sequence variations in the mu-opioid receptor gene (*OPRM1*) associated with human addiction to heroin', *Human Mutation*, 19(4): 459–460.

Spalletta, G., Morris, D., Angelucci, F., Rubino, I., Spoletini, I., Bria, P. *et al.* (2010) 'BDNF Val66Met polymorphism is associated with aggressive behavior in schizophrenia', *European Psychiatry*, 25(6): 311–313.

Steinberg, L. and Morris, A.S. (2001) 'Adolescent development', *Journal of Cognitive Education and Psychology*, 2(1): 55–87.

Sugden, K., Arseneault, L., Harrington, H., Moffitt, T.E., Williams, B., and Caspi, A. (2010) 'Serotonin transporter gene moderates the development of emotional problems among children following bullying victimization', *Journal of the American Academy of Child and Adolescent Psychiatry*, 49(8): 830–840.

Van der Zwaluw, C.S., Larsen, H., and Engels, R.C.M.E. (2012) 'Best friends and alcohol use in adolescence: the role of the dopamine D4 receptor gene', *Addiction Biology*, 17: 1036–1045.

Van Lier, P.A., Wanner, B., Vitaro, F., Blishen, B., Carroll, W., Moore, C., *et al.* (2007) 'Onset of antisocial behavior, affiliation with deviant friends, and childhood maladjustment: a test of the childhood- and adolescent-onset models', *Development and Psychopathology*, 19(1): 167–188.

Vassos, E., Collier, D., and Fazel, S. (2014) 'Systematic meta-analyses and field synopsis of genetic association studies of violence and aggression', *Molecular Psychiatry*, doi: 10.1038/mp.2013.31

Vaughn, M.G., Beaver, K.M., and DeLisi, M. (2009) 'A general biosocial paradigm of antisocial behavior', *Youth Violence and Juvenile Justice*, 7(4): 279–298.

Veenstra, R. and Dijkstra, J.K. (2011) 'Transformations in adolescent peer networks', in B. Laursen and W.A. Collins (eds), *Relationship Pathways: From Adolescence to Young Adulthood*, Los Angeles: Sage, pp. 135–154.

Veenstra, R. and Steglich, C. (2012) 'Actor-based model for network and behavior dynamics', in B. Laursen, T.D. Little and N.A. Card (eds), *Handbook of Developmental Research Methods*, New York: Guilford Press, pp. 598–618.

Veenstra, R., Dijkstra, J.K., Steglich, C., and Van Zalk, M.H. (2013) 'Network–behavior dynamics', *Journal of Research on Adolescence*, 23(3): 399–412.

Wagner, S., Baskaya, Ö., Dahmen, N., Lieb, K., and Tadi, A. (2010) 'Modulatory role of the brain-derived neurotrophic factor Val66Met polymorphism on the effects of serious life events on impulsive aggression in borderline personality disorder', *Genes, Brain and Behavior*, 9(1): 97–102.

Wang, F., Simen, A., Arias, A., Lu, Q., and Zhang, H. (2013) 'A large-scale meta-analysis of the association between the *ANKK1/DRD2* Taq1A polymorphism and alcohol dependence', *Human Genetics*, 132(3): 347–358.

Way, B.M., Taylor, S.E., and Eisenberger, N.I. (2009) 'Variation in the μ-opioid receptor gene (OPRM1) is associated with dispositional and neural sensitivity to social rejection', *Proceedings of the National Academy of Sciences, USA*, 106(35): 15079–15084.

Wu, T. and Barnes, J.C. (2013) 'Two dopamine receptor genes (*DRD2* and *DRD4*) predict psychopathic personality traits in a sample of American adults', *Journal of Criminal Justice*, 41(3): 188–195.

Young, S.E., Smolen, A., Corley, R.P., Krauter, K.S., DeFries, J.C., Crowley, T.J., and Hewitt, J.K. (2002) 'Dopamine transporter polymorphism associated with externalizing behavior problems in children', *American Journal of Medical Genetics*, 114(2): 144–149.

8

Neurogenetics of antisocial aggression

Hayley M. Dorfman and Joshua W. Buckholtz

Introduction

Over 1 million violent crimes are committed each year in the United States alone. While the physical and emotional consequences of these crimes are severe, violent crime has substantial social costs as well. Considering the cumulative costs of medical treatment, incarceration, and lost productivity, violent crimes account for approximately 3 percent of the United States' gross domestic product (GDP) spending. The staggering costs of violent crime make understanding the etiology and pathophysiology of violence a major scientific priority. However, the causal architecture of antisocial behavior remains unclear. This is due in part to limitations in how investigators specify this construct. Considered as a behavioral phenotype, "antisociality" is multifaceted and heterogeneous, comprising several distinct phenotypes, including aggression, lack of empathy, and impulsivity. Each of these can in turn be further granulized, magnifying the already complex task of identifying specific individual-level risk factors and mechanisms. To simplify, we focus here on one specific aspect of antisociality: aggression.

Conceptualizing aggression

Other species have detailed categorizations for aggressive behavior, but the lack of a human equivalent thwarts scientific advancement. Aggression in animals is often characterized as being either "functional" or "pathological" in nature, and this distinction may also be applicable to human work. Functional aggression is contextually appropriate and constrained by species-specific rules and rituals; the ritualized violence of professional sports perhaps represents a human analog of this functional aggression. Contrastingly, pathological aggressive behaviors are atypical given the context, unstructured, and do not adhere to norms that would otherwise serve to mitigate their damaging impact on both victim and community. As noted above, human pathological aggression can be further subtyped into reactive-impulsive and proactive-instrumental variants, paralleling ethological distinctions between predatory and defensive aggression (Valzelli and Bernasconi, 1979; Valzelli *et al.*, 1981b; Krueger and South, 2009). These two forms of aggression are associated with discrete social, cognitive, and biological profiles, and likely have partially unique genetic diatheses (Valzelli *et al.*, 1981a). Since reactive-impulsive aggression is present in a multitude of psychiatric disorders, and there is overlap in symptom expression among these disorders, there may be common aggression-linked genetic and neurobiological alterations accounting for some shared genetic variance. Uncovering the neurobiological underpinnings of aggression could thus expose a core transdiagnostic risk mechanism that contributes to dysfunction across the entire spectrum of psychopathology (Buckholtz and Meyer-Lindenberg, 2012).

Heritability of aggression

The majority of violent crimes appear to be attributable to only a small number of individuals. Of note, an analysis of crime statistics in the UK reveals that approximately 10 percent of criminal offenders commit nearly 50 percent of the total crimes, implying potential heritability of antisocial behavior. There is evidence for familial clustering of criminal behavior, confirming the widely held notion that criminality has a genetic origin (Farrington *et al.*, 2001). The familial nature of antisocial behavior suggests that heritable factors may be responsible for its intergenerational transmission, and subsequent twin studies revealed that genetic factors account for a moderate proportion of the population variance in antisocial traits and behaviors. While antisociality and aggression are intertwined, they are not synonymous. However, twin and family studies have also provided evidence for a genetic structure of aggression. Aggression-specific phenotypes show evidence of heritability in mice (Miczek *et al.*, 2001) and non-human primates (Suomi, 2003), as do aggression-linked personality traits and aggressive symptoms in human adolescents and adults (Jones *et al.*, 2011). For example, aggressive antisocial behavior shows significantly higher heritability than does non-aggressive antisocial behavior (Eley *et al.*, 2003; Button *et al.*, 2004). Genetic contributions to aggression appear early, are stable throughout childhood, and genetic factors account for variance in the transition from childhood to adolescent aggression (Baker *et al.*, 2007; Beitchman *et al.*, 2004). Intriguingly, work in pre-adolescent boys suggests at least partially distinct genetic risk factors for reactive (impulsive) and proactive (instrumental) aggression (Baker *et al.*, 2008). Other data in adults suggest genetic bases for other observed aggression subtypes, such as physical vs. non-physical (Yeh *et al.*, 2010). Taken together, work to date supports the notion that genetic factors play an important role in susceptibility to aggressive behavior. However, none of the specific genetic variants involved was known until Han Brunner's landmark identification of *MAOA* in the early 1990s.

History of the "warrior gene"

MAOA was identified during a genetic analysis of an infamously violent Dutch family conducted by Han Brunner. This would become the very first candidate gene for antisocial behavior (Brunner *et al.*, 1993a,b). The characteristic behavioral phenotype, which was said to have been observed in this family for generations, included mild mental retardation; a propensity toward aggressive outbursts, often triggered by frustration, fear or anger, and typically out of proportion to the putatively inciting stimulus; and impulsively violent behavior, including rape, arson, assault, and attempted murder. Notably, affected males also exhibited marked alterations in monoamine metabolism, including a prominent reduction in urine 5-HIAA levels compared with unaffected males, the product of degradation of serotonin by the catabolic enzyme, monoamine oxidase A (*MAOA*). As the syndrome appeared to be transmitted via an X-linked mode of inheritance, Brunner's linkage analyses of this family focused on the X chromosome.

The linkage analysis isolated the *MAOA* locus on Xp11.23–11.4, and subsequent sequencing of the *MAOA* gene in probands revealed a point mutation (C936T) that leads to a premature stop codon. Consistent with this alteration in sequence, *MAOA* activity was essentially undetectable in fibroblasts cultured from affected males. Thus this X-linked mutation, present in all probands, effectively produces a functional human *MAOA* knockout. This finding was the first clear demonstration that a specific genetic variant could account for the intergenerational transmission of antisocial and criminal behavior through Mendelian inheritance. It also gave *MAOA* the inaccurate nickname by which many still know it today, the "warrior gene."

MAO-A and *MAO-B* are the two *MAO* isoforms; the genes encoding both map to adjacent sites on chromosome Xp11.23. The monoamine oxidases are localized to the outer mitochondrial membrane in the presynaptic terminal of monoamine projection neurons (*MAOA*) (Westlund *et al.*, 1993; Arai *et al.*, 2002) and in astrocytes (*MAOA* and *MAO-B*) (Levitt *et al.*, 1982; Westlund *et al.*, 1988) where they are positioned to regulate both the amount of intracellular substrate available for release and the degree of extrasynaptic monoamine inactivation. *MAOA* encodes the mitochondrial catabolic enzyme monoamine oxidase A (*MAO-A*), which catalyzes the oxidative deamination of biogenic amines. Therefore functional genetic variation in *MAOA* is likely to disturb signaling at monoaminergic synapses throughout the brain (Shih and Thompson, 1999).

Both *MAOA* and *MAOB* are comprised of 15 exons and demonstrate matching exon–intron organization, suggesting their derivation from the same ancestral gene by a duplication event (Grimsby *et al.*, 1991). However, the two isoforms show significant divergence in regional expression and activity patterns. Other differences between the two isoforms are potentially instructive for understanding the *MAOA*–aggression linkage. First, while dopamine (DA) is a good substrate for both isoforms of *MAO*, *MAOA* has higher affinity for serotonin (5-HT) and norepinephrine (NE). The relative importance of *MAOA* for regulating 5-HT and NE (i.e. vs. DA) is underscored by the finding that *MAOA* knockout mice show selective increases in brain 5HT and NE compared with wild-type, while DA is unaffected (Cases *et al.*, 1995). Moreover, the neurochemical, neuromorphological, and behavioral phenotypes produced by genetic deletion of *MAOB* are quite distinct from those of *MAOA* knockout mice. One potential explanation for the relative importance of *MAO-A* over *MAO-B* may pertain to the divergent developmental trajectories of *MAO-A* and *MAO-B* expression. *MAOA* expression precedes that of *MAO-B*, and is the more critical for regulating monoamine catabolism *ante partum*. While *MAO-A* is present at adult levels at birth, *MAO-B* appears only postnatally and subsequently exhibits a striking increase (Nicotra *et al.*, 2004).

While the C936T nonsense mutation characterized by Brunner is of course rare, being a private mutation held only by members of that specific Dutch family, many common polymorphic variants have been identified in the *MAOA* gene region. The best studied of these, by far, is a variable number of tandem repeats (VNTR) polymorphism located in the promoter region of the gene, 1.2 kilobase pairs upstream of its start site. Originally characterized by Sabol and colleagues, the *MAOA* upstream VNTR (hereafter, *MAOA* u-VNTR) is a 30 bp repeat that has been shown to impact transcription in an *in vitro* heterologous expression system and *MAOA* activity in fibroblast cultures. The presence of 3.5 or 4 repeats is associated with relatively higher *MAOA* expression and activity (therefore termed *MAOA*-H or "high" alleles), while the presence of 3 repeats results in relatively lower expression and activity (likewise termed an *MAOA*-L or "low" alleles) (Syagailo *et al.*, 2001; Mertins *et al.*, 2011). However, contradictory findings do exist, and the functional significance of the 2 and 5 repeats is still somewhat controversial (Balciuniene *et al.*, 2002; Cirulli and Goldstein, 2007).

Neurobiological mechanisms of impulsive-aggression: *MAOA*

MAOA has been implicated as an important biological regulator of impulsive-aggressive behavior. Murine genetic deletion studies generally recapitulate the findings of Brunner and colleagues, with *MAOA* knockouts showing dramatically heightened reactive aggression, poor fear learning, and increased forebrain 5-HT and NE levels (Scott *et al.*, 2008; Popova *et al.*, 2001; Godar *et al.*, 2011). Interestingly, a naturally occurring nonsense mutation (*MAOA*^{A863T}) has also been identified in mice, providing a closer parallel to the genetic anomaly described by Brunner and colleagues. Mutant allele carriers exhibit high levels of reactive aggression, including increased fighting behaviors, amplified tail rattling, and reduced attack latency (Scott *et al.*, 2008). Intriguingly, these animals also display aberrant exploratory behaviors: while they are slower to investigate objects and familiar environments, they show much quicker unconditioned escape-and-fear responses to discrete threat stimuli and threatening contexts. Significant visual, locomotor, or olfactory abnormalities do not accompany these deficits, implying a selective impairment in emotional reactivity and risk assessment.

Echoing the discoveries outlined above, molecular and behavioral pharmacology studies suggest that the relationship between *MAOA* and aggression is specifically mediated by genetically determined dysregulation in serotonin-signaling pathways. For example, murine *MAOA* knockouts show a significant increase in serotonin that is particularly dramatic early in life, with mutant pups showing a 9-fold increase compared with wild-type. Changes in DA and NE levels, while apparent, are much smaller in magnitude (Cases *et al.*, 1995). Moreover, blunting post-synaptic serotonin signaling via the 5HT2A antagonist ketanserin reduces aggressive behavior in these animals (Shih and Chen, 1999). Neuromorphological findings in *MAOA* knockouts also implicate excess serotonin signaling during development as a potential pathomechanism. *MAOA-deficient* mice show cytoarchitectonic changes in sensory cortex, which are replicated by pharmacological attenuation of *MAO-A* by clorgyline. This pharmacologically induced neuromorphological phenocopy is developmentally specific; clorgyline administration replicates the knockout phenotype only when administered during a critical developmental window. Changes in neuronal morphology (e.g. dendritic size, length, branching) have also been observed in the orbitofrontal cortex and basolateral amygdala of *MAOA* knockout mice (Bortolato *et al.*, 2011). Finally, pharmacological depletion of serotonin, but not of catecholamines, rescues the phenotype in *MAOA* deficient animals. Of note, this rescue is observed only during a specific developmental window (Cases *et al.*, 1996), which is consistent with other work showing developmentally specific effects of aberrant serotonin signaling in producing adult affective illness. Further, the observation of sensitive periods during development suggests that there may be critical windows during youth and early adolescence for mitigating the impact of genetic risk.

A body of behavioral genetic data provides qualified support for the notion that lowered *MAOA* expression (e.g. in *MAOA*-L carriers) predisposes disinhibitory psychopathology. Impulsive-aggression symptoms have been linked to reduced *MAOA* activity in platelets and brain, and numerous associations to *MAOA*-L have been reported for substance abuse, antisocial personality disorder, and other clinical diagnoses related to pathological aggression (Mukasa *et al.*, 1990; Belfrage *et al.*, 1992; Skondras *et al.*, 2004; Williams *et al.*, 2009). Of note, associations have also been seen using quantitative behavioral phenotypes indexing impulsive aggression. Gallardo-Pujol and colleagues found that *MAOA*-L allele carriers were substantially more aggressive in the point subtraction aggression paradigm than their *MAOA*-H counterparts, an effect that was magnified by provocations such as social exclusion and aversive acoustical stimuli (Gallardo-Pujol *et al.*, 2013; Kuepper *et al.*, 2013). Similarly, *MAOA*-L participants were significantly more likely than high-expressing allele carriers to administer physical punishment (hot sauce) to an opponent when they believed that 80 percent of their earnings had been taken in a power-to-take task (McDermott *et al.*, 2009). Low *MAOA* expression has also been linked to decreased cooperation in economic tasks, with *MAOA*-L males making lower contributions to community funds in public goods games. Despite the positive findings reported above, support for *MAOA*-L as a causal pathogenic variant is, on the whole, mixed. For example, some groups have reported enhanced impulsive-aggression, elevated risk for cluster B personality disorders, lower CSF 5-HIAA levels, and blunted PRL-fen in *MAOA-H* rather than *MAOA*-L allele carriers (Manuck *et al.*, 2000). Associations to the *MAOA*-H allele are particularly common in female samples, and in the presence of specific environmental factors (e.g. prenatal nicotine exposure; see below) (Sjöberg *et al.*, 2007; Wakschlag *et al.*, 2010).

Structural data

Good and colleagues (2001) conducted one of the first studies to implicate *MAOA* in the development of corticolimbic circuitry. A morphometric analysis of women with X-chromosome abnormalities (e.g. Turner syndrome) found that such women had increased gray matter volume in both amygdala and orbitofrontal cortex. Notably, these structural changes were accompanied by deficits in facial emotion recognition, particularly for fearful facial expressions (Good, 2003). Subsequent studies have shown that *MAOA*-L allele carriers show pronounced structural reductions in limbic and paralimbic regions involved in socio-affective processing. For example, Meyer-Lindenberg and colleagues (2006) reported a reduction in gray matter volume for *MAOA*-L allele carriers within the amygdala, insula, and cingulated gyrus, as well as an increase in lateral orbitofrontal volume (for males only). These findings have since been independently replicated (for cingulate and orbitofrontal cortex) (Kalin *et al.*, 2008; Roiser *et al.*, 2009; Cerasa *et al.*, 2010; Alia-Klein *et al.*, 2011).

Functional data

If *MAOA* genetic variation accounts for some of the variance in risk for impulsive-aggressive behavior, risk allele carriers should show information processing differences in brain circuits subserving emotion regulation and inhibitory control. This would be consistent with a model of risk wherein multiple genes, each of small effect size, impact susceptibility in a non-deterministic fashion by exerting a relatively deleterious influence over neural circuits that are associated with symptom-specific aspects of cognition. Supporting this notion, functional MRI studies in *MAOA*-L carriers show differences in frontoparietal and corticolimbic circuit function during tasks that index impulse control, affective arousal, and emotional memory. Early work in this area found decreased anterior cingulate and ventrolateral prefrontal cortex activation during a combined conflict-resolution and response inhibition in *MAOA*-L subjects. Building on this work, we (Meyer-Lindenberg *et al.*, 2006) used a battery of tasks to measure the impact of *MAOA* genetic variation on affective responsiveness, emotional memory, and impulse control. Across these tasks, *MAOA*-L carriers showed a pattern of enhanced subcortical (amygdala, hippocampus, insula) activation in the context of reduced cortical (ACC, orbitofrontal cortex) recruitment.

Imaging genetic studies in patient samples provide further evidence of a role for *MAOA* in impulsive-aggression and related traits. In a sample of adolescent ADHD patients, Nymberg and colleagues (2013) used fMRI to measure the impact of *MAOA* genetic variation on neural circuits for response inhibition and reward learning. They found a significant association between ADHD symptoms and frontostriatal activation during stop-signal and monetary incentive delay tasks. This activation was moderated by *MAOA* genotype: carriers of a putatively low-expressing allele at an intronic SNP (rs12843268) having reduced activation in ventral striatum and inferior frontal gyrus during the reward and response inhibition tasks, respectively. These findings suggest that alterations in motivation and inhibitory control in ADHD may be driven, in part, by *MAOA* genetic variability. Other work suggests that alterations in a frontoparietal circuit for working memory mediate the relationship between *MAOA* genetic variation and externalizing symptoms in adolescents (Ziermans *et al.*, 2012).

Connectivity data

Researchers have sought to investigate the role of *MAOA* genetic variation on brain circuits, and have found that *MAOA*-L carriers have dysregulated corticolimbic circuitry for affective arousal and inhibitory control. Specifically, *MAOA*-L expression is associated with a pattern of both amplified amygdala activation and reduced regulatory control by prefrontal cortical areas such as anterior cingulate and inferior frontal gyrus. It has been suggested that these cortical regions both exert a negative regulatory influence over amygdala and are disordered in patients with impulsive aggressive symptoms. This evidence has motivated work to investigate the impact of *MAOA* genetic variation on functional coupling within the corticostriatal network. Using functional connectivity analyses in a sample of healthy volunteers, Buckholtz and Meyer-Lindenberg (2008) showed that *MAOA*-L carriers demonstrate aberrant coupling between amygdala and ventromedial prefrontal cortex (vmPFC) during emotion tasks. Of note, connections between amygdala and vmPFC are sparse, suggesting the involvement of another region with strong anatomical connections to both; subsequent analyses identified perigenual anterior cingulate as playing this mediating role. These findings, which have been replicated by others (Dannlowski *et al.*, 2009), support the hypothesis that *MAOA* variation disrupts the functional coupling of amygdala with vmPFC to induce emotion regulation deficits.

Prior functional imaging work in patients with impulsive-aggressive symptoms, including adults diagnosed with intermittent explosive disorder, adolescents with conduct disorder, and convicted murderers, have exhibited amygdala hyperactivity and reduced prefrontal activation, particularly during emotion-processing tasks (Best *et al.*, 2002; Raine, 2013). This is especially interesting given preclinical research suggesting that amygdala and medial prefrontal cortex are wired together as a negative regulatory circuit (Cavada *et al.*, 2000). Additionally, lesion studies support the notion that disrupting this circuit produces deficits in emotion regulation and impulse control (Anderson *et al.*, 1999; Blair and Cipolotti, 2000). Together, these data indicate that genetic risk for impulsive-aggression may exert its effects by disrupting corticolimbic circuits for emotional arousal and regulation.

Genetics and environment

Reservations about the inconsistent evidence for a direct connection between *MAOA* u-VNTR and antisocial phenotypes are mitigated when we consider the contributory effect of environmental influences on impulsive-aggressive behavior. Caspi and colleagues (2002) found that early life maltreatment significantly strengthened the relationship between *MAOA* genotype and adult antisocial behavior. Specifically, *MAOA*-L men who were previously exposed to early life abuse exhibited significantly higher levels of violent behavior compared with *MAOA*-H men. These data have been consistently replicated in both men and women (Kim-Cohen *et al.*, 2006; Fergusson *et al.*, 2011).

Subsequent work in rodents provides some mechanistic insight into the nature of this gene-by-environment interaction. Stressed rats also showed a significant increase in prefrontal *MAOA* expression, along with enhanced aggression and anxiety (Márquez *et al.*, 2013). Further, pre-pubertal stress caused animals to demonstrate an increase in basal metabolic rate in the amygdala, which was accompanied by decreased activity in orbitofrontal cortex. Connectivity analyses uncovered a negative association between the medial orbital cortex and amygdala activity, which was disrupted in stressed animals, providing further evidence for corticolimbic circuit dysfunction in impulsive-aggression. On the surface, these findings would appear to contradict prior work linking aggression to lower *MAOA* expression. However, some have suggested that both extremely high and extremely low *MAOA* expression can contribute to impulsive-aggression, particularly during developmental sensitive periods. Consistent with this, clorgyline administration reduced aggression in stress-exposed animals while enhancing aggression in control animals.

Together, these data suggest that environmental factors that predispose aggression may do so, in part, by modulating *MAOA* expression during developmental critical periods. Such effects may be particularly pronounced in carriers of functional *MAOA* variants that also affect gene expression. Consistent with this, prenatal exposure to cigarette smoke, which contains compounds that affect *MAOA* expression, magnifies the impact of the *MAOA*-L allele on risk for conduct disorder symptoms. Of note, adult exposure to adverse environmental influences can also interact with *MAOA* genotype. For example, the deleterious effects of chronic cocaine use on impulsivity and emotion regulation are potentiated in *MAOA*-L carriers, and structural imaging studies suggest that corticolimbic circuits (e.g. orbitofrontal cortex and hippocampus) may be a substrate for this interaction.

Predicting violence

While there is substantial evidence to suggest that aspects of antisocial behavior are heritable, these data are not particularly useful for predictive applications. In practice, heritability metrics do not account for whether, or to what extent, genetic factors influence the expression of violent or antisocial behavior in an individual. Science is focused on understanding general phenomena and aggregating data across groups of individuals to enable inferences about the population as a whole. The relationship between *MAOA* and violence (a population-level inference obtained by averaging across many individuals) speaks only to the probative value of any given individual's *MAOA* genotype. That said, demonstrating heritability answers the fundamental threshold question for any genetic prediction effort. Given heritability, it would next be necessary to identify specific inherited genetic variants or mutations that are responsible for the intergenerational transmission of violence and antisocial behavior. As with any putative biomarker, one would need to demonstrate both high sensitivity and high specificity in order to claim validity.

Two flawed assumptions about the genetics of psychopathology influence much of the current thinking about forensic "genoprediction." First is the notion that specific genetic variants are selectively linked to specific disorders or behaviors. This belief is particularly evident in discussions of violence prediction, where *MAOA* is often considered a "violence gene," an "antisocial gene," or a "warrior gene." However, as discussed earlier, clinical associations of *MAOA* u-VNTR span multiple diagnostic categories, encompassing such "unwarriorlike" syndromes as depression and panic disorder. The current portrait of the *MAOA* u-VNTR would therefore suggest that what it predisposes is not a specific behavior (violence) or set of behaviors (antisocial conduct), but rather a broader pattern of emotional dysregulation that can, but does not necessarily, overlap with these. In fact, this would be consistent with what we know about the genetic architecture of psychopathology more generally. On the whole, genetic risk for psychiatric disorders and behavioral dispositions is pleiotropic: heritable factors confer liability to broad dimensions of symptomatically related disorders, such as schizophrenia and bipolar disorder, rather than discrete syndromes (Gejman *et al.*, 2011). This fact would seem to seriously complicate genetic prediction of antisocial behavior and violence, as no individual marker would ever come close to the phenotypic selectivity that would be desired of a genetic biomarker. This significantly constrains the value of any putative genetic marker for violence, as (in the case of *MAOA*) possessing that marker would be equally likely to signify mood or anxiety psychopathology as antisocial behavior.

The second flawed notion is that individual common genetic markers are isolated causal entities, accounting for a significant enough proportion of the variance in a target behavior to provide adequate sensitivity and specificity. However, there is increasing evidence that most aspects of psychopathology and complex traits are characterized by polygenic inheritance, meaning that many small-effect risk alleles produce a continuous distribution of genetic liability (Plomin *et al.*, 2009). Polygenicity implies that an individual's aggregate genetic profile will determine where they fall on a continuous population distribution of traits. Susceptibility to complex behaviors such as violence and aggression likely follows a similar mode of inheritance. The extremes of some of these genetically influenced trait distributions are associated with social impairment and dysfunctional behaviors such as violence. However, even assuming that we would be able to construct a reliable and valid polygenic profile for "violence" in order to establish forensically relevant thresholds for forensic prediction would be a challenging and controversial endeavor.

Buckholtz and Meyer-Lindenberg (2012) proposed that if the focus were shifted from genes themselves to the neural circuits on which they act, brain circuits may in fact turn out to be optimal descriptors of biological factors underlying psychopathology, and could therefore be better suited to understanding violence. Certainly, information of this kind would be a necessary evidence base for designing individualized therapies for violent offenders that takes causal risk profiles and neural circuit abnormalities into account, in a field where currently no accepted and proven effective therapies exist. Whether, however, such circuits would be useful for forensic prediction is again subject to similar caveats as outlined for genes above. Currently, the overlap between normal and forensic populations is too great and the variance explained too small for that to be a near-term possibility.

It is pertinent to address these issues when discussing neurogenetic mechanisms for antisocial and criminal behavior, given that conversations regarding bioprediction often assume that scientific inferences about population-level phenomena, obtained by averaging data across large numbers of individuals, can provide a valid basis for determinations about specific individuals (Treadway and Buckholtz, 2011). While the difficulty inherent in attempting to "individualize" group-level scientific data has been recognized in other domains of science, it has not received the attention it warrants in the area of forensic bioprediction. Speculation about specific potential biomarkers for violence and antisocial behavior based on neuroimaging and genetic studies in antisocial populations, especially when used for determinations of legal consequences, is very premature.

Conclusion

Considerable evidence suggests that functional genetic variation in *MAOA* disrupts corticolimbic circuitry for affective arousal and regulation and, in turn, predisposes risk for impulsive-aggression. Both preclinical and human work to date strongly suggests that these effects are due to dysregulated serotonin signaling during a critical developmental period. However, impulsive-aggression is etiologically complex, and alterations in serotonergic function make up only a single component. Given their involvement in general social behavior and aggression regulation, neuropeptides such as vasopressin and oxytocin are also likely to play a critical role in mediating impulsive-aggression. Crucially, serotonin-sensitive corticolimbic circuitry related to emotion regulation and social cognition also seems to be influenced by variability in human neuropeptide signaling (Meyer-Lindenberg *et al.*, 2009).

Genetic predisposition is not the only significant causal factor for impulsive-aggression. It is also important to consider the essential role of environment in determining how, or whether, a genetic predisposition is expressed. For example, childhood maltreatment and neglect, and other such early life stressors, are strongly associated with adolescent and adult antisocial behavior and aggression (Dodge *et al.*, 1990; Lansford *et al.*, 2006). Additionally, there is abundant evidence that genetic factors for antisocial aggression are significantly more penetrant on a background of early life stress (Hicks *et al.*, 2004), and there is compelling support for a specific interaction between several genetic variations (including *MAOA*) and childhood maltreatment in determining adult aggression.

Though the precise cellular and systems-level mechanisms underlying the interactions between genotypes and phenotypes are still imprecise, it is tempting to speculate given the available data. However, considered in isolation, effect sizes for single genetic variants such as *MAOA* u-VNTR are small. Thus, while it may be appealing to think of *MAOA* as a "violence" gene, the majority of individuals who store common risk alleles at this locus would not meet diagnostic criteria for impulsive-aggressive disorders. Alleles exert a minor influence on the development of a complex system, which may serve to magnify, unmask, or protect against the impact of other influences on this system. Neuroscience and genetics continue to shed new light on the factors and mechanisms that contribute to human variability, with a precision that was unimaginable in the recent past. However, these developments bring with them an array of ethical, legal, and social repercussions. The legal field, for example, has already begun utilizing individual brain images and genotypes as evidence in trial courts (Farisco and Petrini, 2012). While science may progress to the point where neuroscientific and genetic data can be individualized to permit valid person-level inferences, it is crucial to realize that this is not an appropriate use of the currently available scientific evidence.

References

Alia-Klein, N., Parvaz, M.A., Woicik, P.A., Konova, A.B., Maloney, T., Shumay, E., and Goldstein, R.Z. (2011) 'Gene × disease interaction on orbitofrontal gray matter in cocaine addiction', *Archives of General Psychiatry*, 68(3): 283–294.

Anderson, S.W., Bechara, A., Damasio, H., Tranel, D., and Damasio, A.R. (1999) 'Impairment of social and moral behavior related to early damage in human prefrontal cortex', *Nature Neuroscience*, 2: 1032–1037.

Arai, R., Karasawa, N., Kurokawa, K., Kanai, H., Horiike, K., and Ito, A. (2002) 'Differential subcellular location of mitochondria in rat serotonergic neurons depends on the presence and the absence of monoamine oxidase type B', *Neuroscience*, 114(4): 825–835.

Baker, L.A., Jacobson, K.C., Raine, A., Lozano, D.I., and Bezdjian, S. (2007) 'Genetic and environmental bases of childhood antisocial behavior: a multi-informant twin study', *Journal of Abnormal Psychology*, 116: 219–235.

Baker, L.A., Raine, A., Liu, J., and Jacobson, K.C. (2008) 'Differential genetic and environmental influences on reactive and proactive aggression in children', *Journal of Abnormal Child Psychology*, 36: 1265–1278.

Balciuniene, J., Emilsson, L., Oreland, L., Pettersson, U., and Jazin, E. (2002) 'Investigation of the functional effect of monoamine oxidase polymorphisms in human brain', *Human Genetics*, 110: 1–7.

Beitchman, J.H., Mik, H.M., Ehtesham, S., Douglas, L., and Kennedy, J.L. (2004) '*MAOA* and persistent, pervasive childhood aggression', *Molecular Psychiatry*, 9: 546–547.

Belfrage, H., Lidberg, L., and Oreland, L. (1992) 'Platelet monoamine oxidase activity in mentally disordered violent offenders', *Acta Psychiatrica Scandinavica*, 85: 218–221.

Best, M., Williams, J.M., and Coccaro, E.F. (2002) 'Evidence for a dysfunctional prefrontal circuit in patients with an impulsive aggressive disorder', *Proceedings of the National Academy of Sciences, USA*, 99: 8448–8453.

Blair, R.J. and Cipolotti, L. (2000) 'Impaired social response reversal. A case of "acquired sociopathy"', *Brain*, 123: 1122–1141.

Bortolato, M., Chen, K., Godar, S.C., Chen, G., Wu, W., Rebrin, I., and Shih, J.C. (2011) 'Social deficits and perseverative behaviors, but not overt aggression, in *MAO-A* hypomorphic mice', *Neuropsychopharmacology*, 36(13): 2674–2688.

Brunner, H.G., Nelen, M., Breakefield, X.O., Ropers, H.H., and van Oost, B.A. (1993a) 'Abnormal behavior associated with a point mutation in the structural gene for monoamine oxidase A', *Science*, 262: 578–580.

Brunner, H.G., Nelen, M.R., Van Zandvoort, P., Abeling, N.G., Van Gennip, A.H., Wolters, E.C., and Van Oost, B.A. (1993b) 'X-linked borderline mental retardation with prominent behavioral disturbance phenotype, genetic localization, and evidence for disturbed monoamine metabolism', *American Journal of Human Genetics*, 52(6): 1032–1039.

Buckholtz, J.W. and Meyer-Lindenberg, A. (2008) '*MAOA* and the neurogenetic architecture of human aggression', *Trends in Neurosciences*, 31(3): 120–129.

Buckholtz, J.W. and Meyer-Lindenberg, A. (2012) 'Psychopathology and the human connectome: toward a transdiagnostic model of risk for mental illness', *Neuron*, 74: 990–1004.

Button, T.M.M., Scourfield, J., Martin, N., and McGuffin, P. (2004) 'Do aggressive and non-aggressive antisocial behaviors in adolescents result from the same genetic and environmental effects?', *American Journal of Medical Genetics B, Neuropsychiatric Genetics*, 129B: 59–63.

Cases, O., Seif, I., Grimsby, J., Gaspar, P., Chen, K., Pournin, S., and De Maeyer, E. (1995) 'Aggressive behavior and altered amounts of brain serotonin and norepinephrine in mice lacking *MAOA*', *Science*, 268(5218): 1763–1766.

Cases, O., Vitalis, T., Seif, I., De Maeyer, E., Sotelo, C., and Gaspar, P. (1996) 'Lack of barrels in the somatosensory cortex of monoamine oxidase A-deficient mice: role of a serotonin excess during the critical period', *Neuron*, 16(2): 297–307.

Caspi, A., McClay, J., Moffitt, T.E., Mill, J., Martin, J., Craig, I.W., and Poulton, R. (2002) 'Role of genotype in the cycle of violence in maltreated children', *Science*, 297(5582): 851–854.

Cavada, C., Compañy, T., Tejedor, J., Cruz-Rizzolo, R.J., and Reinoso-Suárez, F. (2000) 'The anatomical connections of the macaque monkey orbitofrontal cortex. A review', *Cerebral Cortex*, 10: 220–242.

Cerasa, A., Cherubini, A., Quattrone, A., Gioia, M.C., Magariello, A., Muglia, M., and Spalletta, G. (2010) 'Morphological correlates of *MAO A* VNTR polymorphism new evidence from cortical thickness measurement', *Behavioural Brain Research*, 211(1): 118–124.

Cirulli, E.T. and Goldstein, D.B. (2007) 'In vitro assays fail to predict in vivo effects of regulatory polymorphisms', *Human Molecular Genetics*, 16: 1931–1939.

Dannlowski, U., Ohrmann, P., Konrad, C., Domschke, K., Bauer, J., Kugel, H., and Suslow, T. (2009) 'Reduced amygdala–prefrontal coupling in major depression association with *MAOA* genotype and illness severity', *International Journal of Neuropsychopharmacology*, 12(1): 11–22.

Dodge, K.A., Bates, J.E., and Pettit, G.S. (1990) 'Mechanisms in the cycle of violence', *Science*, 250(4988): 1678–1683.

Eley, T.C., Lichtenstein, P., and Moffitt, T.E. (2003) 'A longitudinal behavioral genetic analysis of the etiology of aggressive and nonaggressive antisocial behavior', *Development and Psychopathology*, 15: 383–402.

Farisco, M. and Petrini, C. (2012) 'The impact of neuroscience and genetics on the law: a recent Italian case', *Neuroethics*, 5: 317–319.

Farrington, D.P., Jolliffe, D., Loeber, R., Stouthamer-Loeber, M., and Kalb, L.M. (2001) 'The concentration of offenders in families, and family criminality in the prediction of boys' delinquency', *Journal of Adolescence*, 24: 579–596.

Fergusson, D.M., Boden, J.M., Horwood, L.J., Miller, A.L., and Kennedy, M.A. (2011) '*MAOA*, abuse exposure and antisocial behaviour 30-year longitudinal study', *British Journal of Psychiatry*, 198: 457–463.

Gallardo-Pujol, D., Andrés-Pueyo, A., and Maydeu-Olivares, A. (2013) '*MAOA* genotype, social exclusion and aggression: an experimental test of a gene–environment interaction', *Genes, Brain and Behavior*, 12(1): 140–145.

Gejman, P.V., Sanders, A.R., and Kendler, K.S. (2011) 'Genetics of schizophrenia new findings and challenges', *Annual Review of Genomics and Human Genetics*, 12: 121–144.

Godar, S.C., Bortolato, M., Frau, R., Dousti, M., Chen, K., and Shih, J.C. (2011) 'Maladaptive defensive behaviours in monoamine oxidase A-deficient mice', *International Journal of Neuropsychopharmacology*, 14(09): 1195–1207.

Good, C.D. (2003) 'Dosage-sensitive X-linked locus influences the development of amygdala and orbitofrontal cortex, and fear recognition in humans', *Brain*, 126: 2431–2446.

Good, C.D., Johnsrude, I., Ashburner, J., Henson, R.N., Friston, K.J., and Frackowiak, R.S. (2001) 'Cerebral asymmetry and the effects of sex and handedness on brain structure: a voxel-based morphometric analysis of 465 normal adult human brains', *Neuroimage*, 14(3): 685–700.

Grimsby, J., Chen, K., Wang, L.J., Lan, N.C., and Shih, J.C. (1991) 'Human monoamine oxidase A and B genes exhibit identical exon–intron organization', *Proceedings of the National Academy of Sciences, USA*, 88(9): 3637–3641.

Hicks, B.M., Krueger, R.F., Iacono, W.G., McGue, M., and Patrick, C.J. (2004) 'Family transmission and heritability of externalizing disorders: a twin-family study', *Archives of General Psychiatry*, 61: 922–928.

Jones, S.E., Miller, J.D., and Lynam, D.R. (2011) 'Personality, antisocial behavior, and aggression: a meta-analytic review', *Journal of Criminal Justice*, 39(4): 329–337.

Kalin, N.H., Shelton, S.E., Fox, A.S., Rogers, J., Oakes, T.R., and Davidson, R.J. (2008) 'The serotonin transporter genotype is associated with intermediate brain phenotypes that depend on the context of eliciting stressor', *Molecular Psychiatry*, 13(11): 1021–1027.

Kim-Cohen, J., Caspi, A., Taylor, A., Williams, B., Newcombe, R., Craig, I.W., and Moffitt, T.E. (2006) '*MAOA*, maltreatment, and gene–environment interaction predicting children's mental health: new evidence and a meta-analysis', *Molecular Psychiatry*, 11(10): 903–913.

Krueger, R.F. and South, S.C. (2009) 'Externalizing disorders cluster 5 of the proposed meta-structure for DSM-V and ICD-11', *Psychological Medicine*, 39: 2061–2070.

Kuepper, Y., Grant, P., Wielpuetz, C., and Hennig, J. (2013) '*MAOA*-uVNTR genotype predicts interindividual differences in experimental aggressiveness as a function of the degree of provocation', *Behavior and Brain Research*, 247: 73–78.

Lansford, J.E., Malone, P.S., Stevens, K.I., Dodge, K.A., Bates, J.E., and Pettit, G.S. (2006) 'Developmental trajectories of externalizing and internalizing behaviors: factors underlying resilience in physically abused children', *Development and Psychopathology*, 18(01): 35–55.

Levitt, P., Pintar, J.E., and Breakefield, X.O. (1982) 'Immunocytochemical demonstration of monoamine oxidase B in brain astrocytes and serotonergic neurons', *Proceedings of the National Academy of Sciences, USA*, 79: 6385–6389.

McDermott, R., Tingley, D., Cowden, J., Frazzetto, G., and Johnson, D.D.P. (2009) 'Monoamine oxidase A gene (*MAOA*) predicts behavioral aggression following provocation', *Proceedings of the National Academy of Sciences, USA*, 106: 2118–2123.

Manuck, S.B., Flory, J.D., Ferrell, R.E., Mann, J.J., and Muldoon, M.F. (2000) 'A regulatory polymorphism of the monoamine oxidase A gene may be associated with variability in aggression, impulsivity, and central nervous system serotonergic responsivity', *Psychiatry Research*, 95: 9–23.

Márquez, C., Poirier, G.L., Cordero, M.I., Larsen, M.H., Groner, A., Marquis, J., and Sandi, C. (2013) 'Peri-puberty stress leads to abnormal aggression, altered amygdala and orbitofrontal reactivity and increased prefrontal *MAOA* gene expression', *Translational Psychiatry*, 3(1): e216.

Mertins, V., Schote, A.B., Hoffeld, W., Griessmair, M., and Meyer, J. (2011) 'Genetic susceptibility for individual cooperation preferences the role of monoamine oxidase A gene (*MAOA*) in the voluntary provision of public goods', *PLoS One*, 6: e20959.

Meyer-Lindenberg, A., Buckholtz, J.W., Kolachana, B., Hariri, A.R., Pezawas, L., Blasi, G., and Weinberger, D.R. (2006) 'Neural mechanisms of genetic risk for impulsivity and violence in humans', *Proceedings of the National Academy of Sciences, USA*, 103: 6269–6274.

Meyer-Lindenberg, A., Kolachana, B., Gold, B., Olsh, A., Nicodemus, K.K., Mattay, V., and Weinberger, D.R. (2009) 'Genetic variants in AVPR1A linked to autism predict amygdala activation and personality traits in healthy humans', *Molecular Psychiatry*, 14(10): 968–975.

Miczek, K.A., Maxson, S.C., Fish, E.W., and Faccidomo, S. (2001) 'Aggressive behavioral phenotypes in mice', *Behavior and Brain Research*, 125: 167–181.

Mukasa, H., Nakamura, J., Yamada, S., Inoue, M., and Nakazawa, Y. (1990) 'Platelet monoamine oxidase activity and personality traits in alcoholics and methamphetamine dependents', *Drug and Alcohol Dependence*, 26: 251–254.

Nicotra, A., Pierucci, F., Parvez, H., and Senatori, O. (2004) 'Monoamine oxidase expression during development and aging', *Neurotoxicology*, 25: 155–165.

Nymberg, C., Jia, T., Lubbe, S., Ruggeri, B., Desrivieres, S., Barker, G., and Schumann, G. (2013) 'Neural mechanisms of attention-deficit/hyperactivity disorder symptoms are stratified by *MAOA* genotype', *Biological Psychiatry*, 74(8): 607–614.

Plomin, R., Haworth, C.M., and Davis, O.S. (2009) 'Common disorders are quantitative traits', *Nature Reviews Genetics*, 10(12): 872–878.

Popova, N.K., Skrinskaya, Y.A., Amstislavskaya, T.G., Vishnivetskaya, G.B., Seif, I., and De Meier, E. (2001) 'Behavioral characteristics of mice with genetic knockout of monoamine oxidase type A', *Neuroscience and Behavioral Physiology*, 31(6): 597–602.

Raine, A. (2013) *The Anatomy of Violence: The Biological Roots of Crime*, New York: Random House.

Roiser, J.P., De Martino, B., Tan, G.C., Kumaran, D., Seymour, B., Wood, N.W., and Dolan, R.J. (2009) 'A genetically mediated bias in decision making driven by failure of amygdala control', *Journal of Neuroscience*, 29(18): 5985–5991.

Scott, A.L., Bortolato, M., Chen, K., and Shih, J.C. (2008) 'Novel monoamine oxidase A knock out mice with human-like spontaneous mutation', *Neuroreport*, 19: 739–743.

Shih, J.C. and Chen, K. (1999) '*MAO-A* and -B gene knock-out mice exhibit distinctly different behavior', *Neurobiology*, 7: 235–246.

Shih, J. and Thompson, R. (1999) 'Monoamine oxidase in neuropsychiatry and behavior', *American Journal of Human Genetics*, 65: 593–598.

Sjöberg, R.L., Ducci, F., Barr, C.S., Newman, T.K., Dell'Osso, L., Virkkunen, M., and Goldman, D. (2007) 'A non-additive interaction of a functional *MAO-A* VNTR and testosterone predicts antisocial behavior', *Neuropsychopharmacology*, 33(2): 425–430.

Skondras, M., Markianos, M., Botsis, A., Bistolaki, E., and Christodoulou, G. (2004) 'Platelet monoamine oxidase activity and psychometric correlates in male violent offenders imprisoned for homicide or other violent acts', *European Archives of Psychiatry and Clinical Neuroscience*, 254: 380–386.

Suomi, S.J. (2003) 'Gene–environment interactions and the neurobiology of social conflict', *Annals of the New York Academy of Sciences*, 1008: 132–139.

Syagailo, Y.V., Stöber, G., Gräßle, M., Reimer, E., Knapp, M., Jungkunz, G., and Lesch, K.P. (2001) 'Association analysis of the functional monoamine oxidase A gene promoter polymorphism in psychiatric disorders', *American Journal of Medical Genetics*, 105(2): 168–171.

Treadway, M.T. and Buckholtz, J.W. (2011) 'On the use and misuse of genomic and neuroimaging science in forensic psychiatry: current roles and future directions', *Child and Adolescent Psychiatric Clinics of North America*, 20(3): 533–546.

Valzelli, L. and Bernasconi, S. (1979) 'Aggressiveness by isolation and brain serotonin turnover changes in different strains of mice', *Neuropsychobiology*, 5: 129–135.

Valzelli, L., Bernasconi, S., and Dalessandro, M. (1981a) 'Effect of tryptophan administration on spontaneous and P-CPA-induced muricidal aggression in laboratory rats', *Pharmacology Research Communication*, 13: 891–897.

Valzelli, L., Garattini, S., Bernasconi, S., and Sala, A. (1981b) 'Neurochemical correlates of muricidal behavior in rats', *Neuropsychobiology*, 7: 172–178.

Wakschlag, L.S., Kistner, E.O., Pine, D.S., Biesecker, G., Pickett, K.E., Skol, A.D., and Cook, E.H. (2010) 'Interaction of prenatal exposure to cigarettes and *MAOA* genotype in pathways to youth antisocial behavior', *Molecular Psychiatry*, 15(9): 928–937.

Westlund, K.N., Denney, R.M., Rose, R.M., and Abell, C.W. (1988) 'Localization of distinct monoamine oxidase A and monoamine oxidase B cell populations in human brainstem', *Neuroscience*, 25: 439–456.

Westlund, K.N., Krakower, T.J., Kwan, S.W., and Abell, C.W. (1993) 'Intracellular distribution of monoamine oxidase A in selected regions of rat and monkey brain and spinal cord', *Brain Research*, 612: 221–230.

Williams, L.M., Gatt, J.M., Kuan, S.A., Dobson-Stone, C., Palmer, D.M., Paul, R.H., and Gordon, E. (2009) 'A polymorphism of the *MAOA* gene is associated with emotional brain markers and personality traits on an antisocial index', *Neuropsychopharmacology*, 34(7): 1797–1809.

Yeh, M.T., Coccaro, E.F., and Jacobson, K.C. (2010) 'Multivariate behavior genetic analyses of aggressive behavior subtypes', *Behavior Genetics*, 40: 603–617.

Ziermans, T., Dumontheil, I., Roggeman, C., Peyrard-Janvid, M., Matsson, H., Kere, J., and Klingberg, T. (2012) 'Working memory brain activity and capacity link *MAOA* polymorphism to aggressive behavior during development', *Translational Psychiatry*, 2(2): e85.

9

Stress and antisocial behavior

The serotonin system

Jamie C. Vaske and Danielle L. Boisvert

Introduction

Stress can serve an adaptive purpose in one's life. Stress, in small to moderate doses, can motivate individuals to act or to make changes in their environment or themselves. Severe or chronic stressors, however, can lead to depression, health problems, and risky behavior. The negative consequences of stress are not only personally costly, but are also costly to society. It has been estimated that stress-related mental and physical disorders cost approximately $42 billion dollars per year, mostly due to health care costs associated with diagnosing and treating these disorders (Kalia, 2002). The prevalence of stress-related disorders and the costs associated with such disorders are expected to rise over the next decade, making stress one of the leading health concerns for practitioners and researchers.

Over the past 20 years, researchers have spent a considerable amount of time studying the psychological and behavioral consequences of stress. These consequences include (but are not limited to) anger, depression, post-traumatic stress syndrome, poor health outcomes, anxiety, poor impulse control, aggression, substance use, risky sexual behavior, and deviant or criminal behavior. In particular, Agnew (1992) put forth his general strain theory which states that stressful life circumstances and events can lead to antisocial behavior because stress leads to negative emotions (i.e. anger, depression/anxiety); thus stress → negative emotions → antisocial behavior. These stressful life circumstances could include failing to achieve a positively valued goal, removal of positive stimuli, and presentation of negative stimuli. Agnew (2001) argued that strain in the form of negative relationships with others (i.e. child maltreatment, insults from others) would have the strongest impact on antisocial behavior. While Agnew and others have focused on the impact of strain and stress on psychological functioning and behavior, researchers have begun to explore the neurobiological reasons why individuals exposed to stress would exhibit such psychological and behavioral consequences. This work suggests that stress can impact individuals down to the level of expression of particular genes. These neurobiological changes and their impact on behaviors are discussed in this chapter, with a particular focus on the serotonin system.

Research has also found that there is individual variation in one's susceptibility to the effects of stress. For instance, the diathesis stress model argues that individual characteristics, such as predisposing biological or psychological factors, moderate the effects of stress on psychological and behavioral functioning (Brown and Harris, 1978; Monroe and Simons, 1991). Thus some individuals exposed to a stressor will experience negative psychological or behavioral consequences while others do not, and the individual's characteristics (i.e. genetic or cognitive predisposition) help determine their response to the stressor. With the sequencing of the human genome, researchers have begun to explore whether genetic factors condition the effects of stress on psychological functioning and behavior. One of the first associations studied in this vein was the interaction of a serotonin transporter genetic variant (5HTTLPR) and stressful life events on depression (Caspi *et al.*, 2003). Caspi and colleagues brought to the forefront the importance of serotonin-related genes in moderating individuals' responses to stress.

Building on this prior work, this chapter has two objectives. First, it reviews the empirical evidence on the neurobiological changes that occur in response to stress. The chapter emphasizes the effects of stress on serotonin functioning, and also discusses the impact on other biological functions. The human body is a system, thus changes in one section or division have implications for the functioning of other sections. Indeed, research has shown that there are reciprocal relationships between serotonin, dopamine, and hypothalamic pituitary adrenal axis functioning in response to stress (Meaney *et al.*, 2002; Pruessner *et al.*, 2004; Champagne and Curley, 2008; Huang *et al.*, 2012). The chapter also notes the importance of the developmental timing of stressful life events, as the impact of stress on functioning varies across the life course. Next, it reviews research on the role of serotonin-related genes in increasing one's vulnerability or susceptibility to the effects of stress.

Neurobiological consequences of stress on the serotonin system

Animal and human research has shown that stress has an impact on a wide range of components within the brain including: (1) the hypothalamic–pituitary–adrenal (HPA) axis; (2) the structure of the hippocampus; (3) the structure and function of the amygdala and frontal cortex; and (4) the functioning of neurotransmitter systems including dopamine and serotonin. The HPA axis is activated when one encounters a stressful or threatening response, setting in motion a series of chemicals that prepare one for fight or flight [i.e. corticotrophin-releasing hormone (CRH), adrenocorticotropic hormone (ACTH), and arginine vasopressin (AVP)], as well as hormones that maintain equilibrium in the HPA axis (i.e. cortisol and other glucocorticoids). Numerous studies have found that the basal levels of CRH and ACTH are higher among respondents who experience stressful life events in early childhood, such as neglect, abuse, separation, or victimization (Holmes *et al.*, 2005; Carpenter *et al.*, 2007). Further, abused or neglected individuals exhibit greater CRH and ACTH responses to stress than control respondents, indicating a greater stress response among stressed individuals (Meaney, 2001; Zhang *et al.*, 2004). Respondents experiencing early stressful life events also show alterations in cortisol levels throughout development (Cicchetti and Rogosch, 2001; Tarullo and Gunnar, 2006), and lower levels of glucocorticoid receptors in the hippocampus (Champagne and Curley, 2008). One reason why the density of glucocorticoid receptors may be reduced is that abuse and neglect set in motion chemical processes called deacetylation and methylation, which prevent expression of the glucocorticoid receptor gene (Weaver *et al.*, 2004). Glucocorticoid receptors in the hippocampus are important for downregulating the HPA axis when one is in fight-or-flight mode, and fewer receptors may lead to high levels of anxiety and enhanced sensitivity to stress or perceived threats. Studies of adolescents and adults have also shown that stress can impact glucocorticoid reactivity and the density of glucocorticoid receptor levels (Lupien *et al.*, 2009), yet the impact of stress on the

HPA axis seems most pronounced during early childhood and adolescence. Thus, while stress at any point along the life course can create hyperarousal in the HPA axis, infants/young children and adolescents can be the most sensitive to stress.

Elevated levels of glucocorticoids in response to stress may lead to necrosis or the death of neurons in the hippocampus, and prevent the growth of synapses, the growth of neurons, and the development of white matter in multiple parts of the brain (Teicher *et al.*, 2002; Holmes *et al.*, 2005). This line of research supports the hypothesis that stress sets in motion a cascade of changes in multiple systems of the body, and these changes are often interrelated and have reciprocal effects. Studies have also shown that respondents who are exposed to stressful circumstances or events have lower levels of brain and glial derived neurotrophic factors (Ognibene *et al.*, 2008)—key proteins for the development and maintenance of brain cells—and a lower density of neurons in the hippocampus, amygdala, corpus callosum, frontal cortex, and cerebellar vermis (Teicher *et al.*, 2002). Similarly to the age-graded impacts of stress on the HPA axis, there is some evidence that the effects of stress on specific parts of the brain may depend upon the timing of the stressful event. Stress in very early childhood and late adulthood appears to impact the hippocampus, while stress that occurs in adolescence impacts the development of the frontal cortex (Andersen *et al.*, 2008; Lupien *et al.*, 2009).

Aside from the impacts of stress on the structure and functioning of the brain, research has shown that stress can also impact neurotransmitter activity. Stressed respondents often have higher levels of dopamine activity and lower levels of serotonin in areas such as the nucleus accumbens, striatum, amygdala, and frontal cortex (Champagne and Curley, 2008). This combination of changes may create a situation where respondents are overstimulated or aroused without proper inhibitory mechanisms to regulate the hyperarousal. In response, subjects may experience anxiety, depression, mania, and/or engage in risky or impulsive behaviors (Andersen and Teicher, 2009). Serotonin, in particular, can regulate dopamine activity, as well as HPA axis functioning and response (Champagne and Curley, 2008); thus it has far-reaching effects on multiple systems. Further, the effects of stress on serotonin functioning have been documented in investigations of rodents, non-human primates, and humans. It is due to the importance of serotonin and the consistency of findings across various types of studies that the current chapter focuses on stress and serotonin.

Before discussing the relationship between serotonin and stress, it is important to review the different components of the serotonin system, since there are various ways to assess serotonin functioning and activity. Serotonin (5HT) is a neurotransmitter that is synthesized from the oxidation of tryptophan-hydroxylase. It is released from presynaptic receptors (such as $5HT_{1D}$ and $5HT_{1A}$) where it travels across the synaptic gap to attach to postsynaptic receptors ($5HT_{1A, 1B, 1C, 1D, 1E, 1F}$, $5HT_{2A, 2B, 2C}$, $5HT_{3A, 3B}$, $5HT_4$, $5HT_{5A, 5B}$, $5HT_6$, and $5HT_7$) to transmit the inhibitory action to the adjacent neuron (Kiser *et al.*, 2012). Serotonin that is released from postsynaptic receptors and that remains in the synaptic cleft is removed by way of either transportation or degradation. A serotonin transporter (SERT) removes serotonin from the synaptic cleft and returns it to the presynaptic neuron. The density of serotonin transporters may reflect the number of serotonergic innervations in a particular area. Extracellular serotonin may also be removed from the synaptic gap by monoamine oxidase A, which degrades 5HT and produces the serotonin metabolite 5HIAA. Researchers may use cerebrospinal fluid (CSF) concentrations of 5HIAA in humans as an indirect measure of serotonin activity and turnover in the brain. Since 5HIAA is the end product of serotonin activity, 5HIAA concentrations may reflect levels of serotonin synthesis, release, receptor availability, or serotonin neuron density. It is believed that low levels of serotonin functioning and activity result from low serotonin synthesis, a lower density or binding potential of serotonin receptors or transporters, and lower 5HIAA concentrations. Many of these indices can be assessed via blood samples, lumbar punctures to collect cerebrospinal fluid, and urine samples. Researchers may also administer radioligands (i.e. [11C](+)-McN 5652) and use positron emission tomography (PET) imaging to measure transporter or receptor density and binding potential. Finally, serotonin activity can be measured indirectly through neuropharmacological challenges where researchers administer a serotonin agonist (such as fenfluramine, clomipramine, or citalopram) and record the level of change in hormones such as prolactin and cortisol.

Rodent studies

Stress during infancy, childhood, adolescence, and adulthood has been shown to impact serotonin neurotransmission in rodents. For instance, infant mice that were separated from their mother for extended periods of time exhibited lower levels of serotonin turnover (5HIAA/5HT ratio) in the nucleus accumbens, hippocampus, and frontal cortex in adolescence and adulthood than control subjects (Jones *et al.*, 1992; Matthews *et al.*, 2001; Ognibene *et al.*, 2008). Similarly, Huang and colleagues' (2012) analysis of Sprague-Dawley rats showed that mothers' exposure to chronic unpredictable stress impacted the neurotransmission of serotonin in developing fetuses. Fetuses of chronically stressed mothers had higher basal levels of serotonin (5HT) in the hippocampus and hypothalamus, but they exhibited lower levels of serotonin turnover (5HIAA/5HT ratio), fewer serotonin transporters, and lower expression of the $5HT_{1A}$ receptor in the hippocampus. These studies suggest that exposure to prenatal and postnatal stress can reduce serotonin neurotransmission, and as a result leads to a host of negative outcomes including anxiety, impulsivity, and aggression. While a number of studies show that stress decreases serotonin neurotransmission, some investigations have found that basal levels of serotonin and 5HT receptors were upregulated among rodents exposed to early life stress (Vázquez *et al.*, 2002; Ognibene *et al.*, 2008). Authors have suggested that mild stressors may lead to short increases in serotonin activity—perhaps in an attempt to inhibit or regulate fight-or-flight responses—while chronic or severe stressors lead to downregulation of serotonin activity across the life course. Further, the effects of stress on serotonin functioning may depend upon: (1) whether the rodent was housed with its peers or held in isolation, (2) the brain region of interest, (3) gender, and (4) the measure of serotonin functioning (5HT release, 5HT peak, 5HIAA, $5HT_{1A}$ receptors vs. $5HT_{1B}$ or $5HT_{2A}$) (Fone and Porkess, 2008). Thus, while there is evidence that early life stress leads to decreased serotonin functioning and maladaptive behavior (i.e. aggression) in rodents, the evidence is mixed throughout the literature and there are moderating influences that need to be accounted for when examining the association.

Studies of adolescent and adult rodents have shown that later life stress may also reduce serotonin activity levels in the brain (Kiser *et al.*, 2012). Adolescent golden hamsters that are exposed to a daily threat of attack by another hamster show higher levels of serotonin (5HT) release in adolescence and adulthood than hamsters that are not exposed to attack (Delville *et al.*, 1998). Despite the higher levels of 5HT release, the $5HT_{1A}$ receptor may be downregulated by decreasing the binding sensitivity of the receptor to 5HT neurotransmitters, resulting in lower levels of circulating serotonin and reduced regulation on the dopamine system, HPA axis, and vasopressin system (Ferris, 2000; Mahar *et al.*, 2013). This interpretation of the results is consistent with studies of rats which show that chronic and/or unpredictable exposure to stress is associated with lower concentrations of serotonin in the frontal cortex, striatum, and hippocampus (Ahmad *et al.*, 2010; Kohut *et al.*, 2012). In response, stressed rodents may be more likely to exhibit symptoms of anxiety, somatic symptoms of depression, substance use, and aggression towards other rodents, especially those that are smaller in size.

Non-human primate studies

Similarly to rodent studies, investigations of non-human primates have shown that stressful life events in early childhood can lead to decreased serotonin activity and behavioral problems (Shannon *et al.*, 2005). Maestripieri and colleagues have conducted a series of cross-fostering experiments on rhesus monkeys which reveal that maternal rejection and abuse impacts serotonin functioning early in life, and such changes may place offspring at risk for perpetuating maternal rejection and abuse on their own children later in life. In their studies, the authors had 25–43 rhesus monkeys that were raised by their biological mothers and 15 females that were removed from their mothers at birth and placed with unrelated mothers. Eight of the unrelated mothers were abusive (i.e. dragging, hitting, pushing infant to ground, pulling infant's hair) or rejecting (i.e. reduced contact with infant) while seven did not demonstrate such behaviors (Maestripieri *et al.*, 2006; Maestripieri, 2008). This type of design allowed the researchers to explore whether infants' serotonin activity levels were similar to those of their biological mother and siblings (which would indicate a role of heritability or genetics) or whether maternal abuse and rejection reduced serotonin levels across both cross-fostered and non-cross-fostered infants (i.e. was there a relationship between rejection and serotonin levels after controlling for genetics?). The authors reported that rejected or abused infants had lower levels of 5HIAA in their second year of life than those that were not rejected or abused, and the association between neglect/abuse and 5HIAA was consistent across cross-fostered and non-cross-fostered infants (Maestripieri *et al.*, 2006). Infants with lower levels of 5HIAA also exhibited more markers of anxiety (i.e. scratching) than infants with higher 5HIAA. In a later study, Maestripieri and colleagues (2007) found that cross-fostered females engaged in the same level of maternal rejection in adulthood as their adoptive mothers (relative to their biological mothers). Further, cross-fostered females who had lower levels of 5HIAA were more likely to reject and abuse their own infants than females with higher levels of 5HIAA, and females who abused their own infants had lower levels of 5HIAA (Maestripieri, 2008). This body of research suggests that maternal rejection and abuse may reduce serotonin functioning across the life course, and these changes may be associated with anxiety and violence against their own offspring later in adulthood.

Non-human primate studies have also shown that early life stress can impact the density and functioning of serotonin receptors and transporters. Spinelli *et al.* (2010) examined the association between $5HT_{1A}$ receptors and early life stress in rhesus monkeys. They randomly assigned infant monkeys to two groups: (1) monkeys were reared by their mothers for the first six months (mother-reared), or (2) infants were removed from their mothers at birth, housed in an incubator for the first 14 days, then placed in an isolated nursery from day 14 to 37, and finally placed in a cage with three peers until 24 months of age (peer-reared). Their study showed that peer-reared infants had a lower density and binding potential of $5HT_{1A}$ receptors in the hippocampus, medial cingulate cortex, anterior cingulate cortex, amygdala, dorsomedial prefrontal cortex, and raphe nuclei than mother-reared infants. The authors, however, did find that peer-rearing correlated with higher $5HT_{1A}$ density and binding potential in the dorsomedial prefrontal cortex among females, suggesting that sex may condition the effects of stress on serotonin activity. Other non-human primate studies have revealed that early life stress is associated with reduced expression of the $5HT_{1A}$ gene in the hippocampus and anterior cingulate cortex during adolescence (Law *et al.*, 2009a,b).

The density and binding potential of serotonin transporters may also be reduced among non-human primates that are isolated and/or peer-reared, relative to those who are reared by their biological mothers. A study of adolescent male rhesus monkeys found that the binding potential of serotonin transporters was 10 to 23 percent lower in the raphe nucleus, thalamus, striatum, amygdala, hippocampus, hypothalamus, and medial temporal regions for peer-reared monkeys relative to mother-reared monkeys (Ichise *et al.*, 2006). The level of 5HIAA in infancy and adolescence was also lower among peer-reared monkeys. Kinnally and colleagues (2010) extended upon Ichise *et al.*'s findings by showing that the expression of the serotonin transporter gene (i.e. mRNA) in rhesus monkeys was lower among maternally abused macaques, and that the expression of the serotonin transporter was not related to variation in the serotonin transporter gene; thus differences in the transcription of the serotonin transporter gene were more a function of early life experiences than of genetic variation, suggesting possible epigenetic effects.

Unlike studies on rodents, very few studies have investigated the impact of stress in adolescence or adulthood on serotonin activity in non-human primates. Higley and colleagues (1992) did find that CSF 5HIAA was negatively correlated with ACTH (a stress hormone from the adrenal gland) in juvenile and adolescent rhesus monkeys, indicating that higher levels of stress response were associated with lower CSF 5HIAA levels. The majority of the studies of non-human primates, however, examine the association between early childhood stress and serotonin functioning, leaving the research horizon open for investigations of later life stress and serotonin activity. Despite the lack of research in this area, it is noteworthy that the results from rodent studies on early life stress and serotonin converge with those findings from non-human primates; thus stress appears to reduce serotonergic functioning across species. The external validity and strength of this association would be bolstered by similar results in human studies. The following section addresses this concern by discussing the studies of stress and serotonin functioning in humans.

Human studies

In line with findings from rodent and non-human primate studies, a number of investigations have shown that early childhood stress is associated with reduced serotonergic functioning in adulthood. Roy's (2002) analysis of 29 adult male cocaine-dependent patients revealed that patients with higher scores on the emotional neglect subscale of the Childhood Trauma Questionnaire had lower concentrations of CSF 5HIAA than patients who reported fewer incidents of emotional neglect. Similarly, a study of suicide attempters showed that respondents who reported exposure to violence between the ages of 6 and 14 had lower levels of CSF 5HIAA, and exposure to violence as a child was related to violent behavior as an adult among respondents with CSF 5HIAA levels below the median, but there was no association among respondents with high levels of CSF 5HIAA; thus exposure to violence appeared to reduce serotonergic functioning, and low serotonergic functioning moderated the impact of exposure to childhood violence on violent behavior in adulthood (Moberg *et al.*, 2011). Childhood neglect and abuse may also impact serotonin transporter activity. Miller and colleagues' (2009) study using positron emission tomography showed that the binding potential of the serotonin transporter was significantly lower in the anterior cingulate cortex, amygdala, dorsal putamen, hippocampus, midbrain, and thalamus in major depressive disorder (MDD) patients with a history of childhood abuse, relative to MDD patients without abuse and to healthy controls. Finally, studies have found that alcohol-dependent patients and bipolar disorder patients with a history of childhood abuse have a blunted prolactin response to neuroendocrine challenge tests compared with patients who do not report childhood maltreatment (Rinne *et al.*, 2000; Berglund *et al.*, 2013). Thus findings from a variety of studies that differ in their measurement of childhood abuse, type of sample, and measure of serotonin functioning show that childhood abuse is associated with reductions in serotonin activity in adulthood.

While the association between childhood stressful life circumstances/events and serotonin functioning appears to be robust, the findings regarding later life stress and serotonin activity are less conclusive. For example, a study of champion rifle shooters showed that respondents exhibited an increase in 5HIAA when completing a shooting competition compared with their baseline levels (Helin *et al.*, 1988). Other studies have reported little to no association. Frankenhaeuser *et al.* (1986) found that students who completed a demanding examination did not experience any changes in 5HIAA concentration from baseline to follow-up. Stronks *et al.* (1998) did not find that completion of mental stress activities—as measured by solving complex mathematical problems—impacted the levels of 5HIAA and 5HT platelets in blood samples among female migraine patients and healthy controls. Further, an analysis of adults did not find an association between individuals' subjective responses to stress and 5HIAA levels from blood samples. Thus studies that examine the effects of mental stress in adulthood and the impact on serotonin indices from blood samples tend to show little to no association between stress and serotonin activity. It may be that the stress experienced in these studies is quite different from abuse and neglect, in its timing, severity, duration, and impact on other areas of life functioning.

Indeed, Manuck and colleagues (2005) found that socioeconomic status at both community and individual levels was related to an indicator of serotonin activity. In an analysis of 249 respondents from the Cholesterol and Risk Evaluation (CARE) study, the authors found that adults with lower incomes, or who resided in poorer communities, had a blunted prolactin response to administration of fenfluramine compared with respondents with higher individual or community level socioeconomic status. Further, they found that community level socioeconomic status was associated with smaller prolactin responses among respondents with one or two copies of the short allele of the 5HTTLPR serotonin transporter gene. These findings suggested that poverty—as a type of stress—can impact serotonin functioning, and it may particularly reduce serotonin activity among individuals with specific genetic variants related to the serotonin system. Thus serotonin-linked genes may increase one's sensitivity or vulnerability to stressful life events.

Serotonin gene and increased sensitivity to stress

A number of studies have found that a variant in the 5HTTLPR serotonin transporter gene may moderate the effects of stressful life events on psychopathological outcomes. The 5HTTLPR polymorphism is described as a 43 base pair insertion/deletion polymorphism, in which researchers initially assumed that there were only three genotypes (Short/Short, Short/Long, and Long/Long) (Lesch *et al.*, 1996; Wendland *et al.*, 2006). However, more recent research shows that there is an A-to-G single nucleotide polymorphism that makes the polymorphism triallelic (Hu *et al.*, 2006). There are three alleles—S, L_G, and L_A—which form six genotype groups (S/S, S/L_G, S/L_A, L_G/L_G, L_G/L_A, L_A/L_A). Most authors compare L_A/L_A with all other genotypes as the transcription levels of the S allele and L_G allele are similar (Hu *et al.*, 2006). The S and L_G alleles are associated with reduced transcription of the SERT gene and lower levels of serotonin transporters, while the L_A/L_A genotype is associated with an increased density of serotonin transporters (see Canli and Lesch, 2007; Uher and McGuffin, 2008 for reviews of 5HTTLPR). Thus, for phenotypes associated with low serotonin activity, the S and L_G alleles are considered the "risky" genotypes and are expected to increase one's sensitivity to stress.

Indeed, Caspi and colleagues' (2003) analysis of 847 Caucasian young adults from the Dunedin Multidisciplinary Health and Development Study found that adulthood stressful life events led to higher numbers of depressive symptoms among respondents with the S/S and S/L genotypes, compared with respondents with the L/L genotype. They also found that childhood maltreatment had a significant positive association with the probability of a major depressive episode among youths with the S/S or S/L genotypes, but maltreatment was not associated with depression within the L/L genotype. Since Caspi *et al.*'s (2003) study, a number of researchers have found that the short allele of 5HTTLPR interacts with childhood maltreatment or stressful life events to impact one's likelihood and severity of depression (Kendler *et al.*, 2005; Lenze *et al.*, 2005; Wilhelm *et al.*, 2006; Zalsman *et al.*, 2006; Cervilla *et al.*, 2007; Cicchetti *et al.*, 2007; Kim *et al.*, 2007; Mandelli *et al.*, 2007; Scheid *et al.*, 2007; Petersen *et al.*, 2012; Bogdan *et al.*, 2013), especially among females (Eley *et al.*, 2004; Grabe *et al.*, 2005; Sjöberg *et al.*, 2006; Taylor *et al.*, 2006; Brummett *et al.*, 2007) and among youths who have low levels of social support (Kaufman *et al.*, 2004).

While the majority of the 5HTTLPR × stressful life event literature focuses on the interaction effect on depression, other researchers have found that the short allele of 5HTTLPR interacts with stressful life events across the life course to impact substance use and criminal behavior across development (Kaufman *et al.*, 2007; Brody *et al.*, 2009; Kranzler *et al.*, 2012; Vaske *et al.*, 2012). For instance, low emotional expression by mothers was associated with higher levels of conduct problems among children and adolescents who carried one or more copies of the short allele (Sonuga-Barke *et al.*, 2009). Sulik *et al.* (2012) found that 5HTTLPR interacted with the quality of mothers' parenting to influence children's initial level of non-compliance with rules. Similarly, Reif and colleagues' (2007) analysis of 184 adult male offenders found that habitual violent behavior was highest among males with one or two copies of the short allele and who were raised in adverse environments during childhood (i.e. low social status, separation in family, exposed to violence, poor social integration). Other studies have shown that the interaction of 5HTTLPR and stressful circumstances can increase the risk of psychological endophenotypes that are related to antisocial behavior, such as self-regulation (Kochanska *et al.*, 2009), attitudes that value being 'tough' (Simons *et al.*, 2011), the internalization of morals (Kochanska *et al.*, 2011), callous unemotional traits, and narcissism (Sadeh *et al.*, 2010).

While there is a large body of research supporting the hypothesis that 5HTTLPR moderates the impact of stressful events on psychopathological outcomes, there are a number of studies that fail to detect a significant interaction effect (Gillespie *et al.*, 2005; Surtees *et al.*, 2006; Fergusson *et al.*, 2011). Most notable are meta-analyses of various studies which show null interaction effects. Risch and colleagues (2009) conducted a meta-analysis of 14 studies with a total of 14,250 respondents, and found that 5HTTLPR did not moderate the impact of stressful life events on the severity of depression. Munafó and colleagues' (2009) literature review of 15 studies showed that the findings regarding the 5HTTLPR × stressful life event interaction effect were mixed, often due to researchers employing different theoretical and statistical interaction models to probe the effect. Their meta-analysis of five of the studies showed a non-significant interaction effect between 5HTTLPR and stressful life events on depression. Other meta-analyses, however, have detected a significant interaction between 5HTTLPR and stressful life events on depression, and they have noted that methodological factors impact one's ability to replicate the interaction found in Caspi *et al.* (2003). In particular, studies have found a significant 5HTTLPR × stressful life event interaction when the stressor was child maltreatment or a specific medical condition (Karg *et al.*, 2011); the stressor was assessed via semi-structured interviews, natural disasters, or experimental stimuli (Uher and McGuffin, 2008, 2010); and in samples of adolescent females, children, and young adults (rather than adolescent boys, the middle aged, or elderly) (Uher and McGuffin, 2008). It is important to note that many of the studies in Risch *et al.* (2009) and Munafó *et al.*'s (2009) meta-analyses used self-report measures of stress, and this may be one explanation for their null results. Thus 5HTTLPR does appear to moderate the impact of stressful life events on psychopathology, but the ability to detect such an interaction may depend upon the type of sample and the mode of measuring stressful life events.

In light of the evidence that there is a significant 5HTTLPR × stressful life interaction for some groups, researchers have begun investigating the neurobiological mechanisms that may underlie the interaction effect. In particular, researchers have examined how the interaction effect may influence: (1) HPA axis functioning, (2) serotonin activity, or (3) the activity of particular brain regions. A non-human primate study of bonnet macaques showed that males with the s/l or s/s genotype raised in a stressful environment had higher concentrations of corticotrophin-releasing factor in CSF than stressed males with the l/l genotype, non-stressed males with the s/s or s/l genotype, and non-stressed males with the l/l genotype (Coplan *et al.*, 2011). Similarly, Barr *et al.* (2004) found that peer-reared females with the short allele had higher levels of adrenocorticotropic hormone in response to acute stress and lower levels of cortisol than mother-reared females or peer-reared females with the long allele. Studies of humans have produced similar results. Mueller and colleagues (2010) found that acute stress was associated with a larger cortisol response among three-day-old infants with the short or L_G allele relative to those with the L_A/L_A genotype. Way and Taylor's (2010) analysis of 182 young adults found that individuals with the short allele had the greatest rise in cortisol activity after delivering a speech and receiving negative audience feedback, relative to those with the s/l and l/l genotypes. A follow-up analysis by Mueller *et al.* (2011) found that 5HTTLPR interacted with stressful life events to predict cortisol response to acute stress (i.e. delivering a speech to an audience and solving math problems), but this association was found only among young adults (ages 19–31) and not for children or older adults. Respondents with the short allele, and who experienced three or more stressful life events during the first five years of life, had a greater cortisol response to the acute stressor than severely stressed individuals who were homozygous for the L_A allele. These findings are in line with results from the meta-analyses, which show that early life stress may have the greatest impact on functioning, and these changes may be most evident during early adulthood.

Other studies have shown that the interaction between the 5HTTLPR polymorphism and stressful life events may impact serotonin functioning. A study of rhesus macaques found that monkeys with the short allele who were reared by peers (rather than their mother) had significantly lower concentrations of CSF 5HIAA than peer-reared monkeys with the long allele or those who were mother-reared (regardless of genotype) (Bennett *et al.*, 2002). Manuck *et al.*'s (2005) study found that community level and individual levels of socioeconomic status were associated with lower prolactin response to fenfluramine among respondents with one or two copies of the short allele, but such an association was not found among those with the long allele. Contrary to Manuck *et al.*, however, a recent study of adult males in Germany failed to find that cortisol responses were higher among respondents with the short allele who had experienced negative life events (Wielpuetz *et al.*, 2014).

The combination of the short allele and stressful life events may also have implications for brain structure and functioning (Lemogne *et al.*, 2011). Respondents with the short allele and who experience stressful life events may exhibit reductions in hippocampal volumes (Frodl *et al.*, 2010) and greater activity in the occipital cortex and insula in response to fear stimuli (Klucken *et al.*, 2013). These regions are important to the regulation of the HPA axis (hippocampus), the emotional weighting of and access of memories (hippocampus), and the physiological awareness and experience of emotion (insula); thus respondents with the short allele who have a history of stressful life events may be hypervigilant in searching for and responding to threats in their environment. Other studies have found that adults with the short allele have greater activations in the amygdala, hippocampus, anterior insula, thalamus, pulvinar, caudate, precuneus, anterior cingulate cortex, and medial prefrontal cortex in response to an acute laboratory stressor than adults with the long allele. Further, there was a greater association between activation in the medial prefrontal cortex and anxiety among those with the short allele, suggesting that individuals with the short allele directed more attention to the stressor and exaggerated the nature of the threat (which in turn increased anxiety levels) more than those with the long allele (Drabant *et al.*, 2012).

Conclusion

In conclusion, research has found those with the short allele of 5HTTLPR and who have experienced stressful life events are more likely to exhibit antisocial and maladaptive behaviors than individuals who do not have both of these risk factors. One reason why these individuals may have an increased risk of psychopathology is that stressful life events set in motion a cascade of neurobiological changes that increase one's risk of maladaptive behavior, and having the short allele of 5HTTLPR exacerbates this risk; thus, in line with Agnew's (1992) hypotheses, 5HTTLPR × stressful life events → negative emotions and antisocial traits → antisocial behavior. This interpretation is consistent with many of the studies that find an effect of stressful life events on behavioral or neurobiological outcomes, and that this effect is particularly strong for those with one or more copies of the short allele, especially among individuals during childhood and adolescence. The results from the above studies suggest that individuals with the short allele and who experience stress may be at an increased risk for maladaptive behavior because such neurobiological consequences make them sensitive to anticipation of threatening experiences; thus they may be more likely to direct their attention to perceived threats, evaluate events as negative, assign more negative emotional weight to such events, and ruminate over events (Conway *et al.*, 2012). Further, the impact of stress and 5HTTLPR on other biological factors (i.e. hippocampus, serotonin activity) may reduce one's ability to inhibit responses, ability to ignore opportunities for risky behavior, and capacity to care about the negative psychosocial consequences of their behavior. Thus stress is an important correlate of antisocial cognition and behavior because it impacts our biological functioning.

References

Agnew, R. (1992) 'Foundation for a general strain theory of crime and delinquency', *Criminology*, 30: 47–89.

Agnew, R. (2001) 'Building on the foundation of general strain theory: specifying the types of strain most likely to lead to crime and delinquency', *Journal of Research in Crime and Delinquency*, 38: 319–361.

Ahmad, A., Rasheed, N., Banu, N., and Palit, G. (2010) 'Alterations in monoamine levels and oxidative systems in frontal cortex, striatum, and hippocampus of the rat brain during chronic unpredictable stress', *Stress*, 13: 355–364.

Andersen, S.L. and Teicher, M.H. (2009) 'Desperately driven and no brakes: developmental stress exposure and subsequent risk for substance abuse', *Neuroscience and Biobehavioral Reviews*, 33: 516–524.

Andersen, S.L., Tomada, A., Vincow, E.S., Valente, E., Polcari, A., and Teicher, M.H. (2008) 'Preliminary evidence for sensitive periods in the effect of childhood sexual abuse on regional brain development', *Journal of Neuropsychiatry and Clinical Neuroscience*, 20: 292–301.

Barr, C.S., Newman, T.K., Schwandt, M., Shannon, C., Dvoskin, R.L., Lindell, S.G., Taubman, J., Thompson, B., Champoux, M., Lesch, K.P., Goldman, D., Suomi, S.J., and Higley, J.D. (2004) 'Sexual dichotomy of an interaction between early adversity and the serotonin transporter gene promoter variant in rhesus macaques', *Proceedings of the National Academy of Sciences, USA*, 101: 12358–12363.

Bennett, A.J., Lesch, K.P., Heils, A., Long, J.C., Lorenz, J.G., Shoaf, S.E., Champoux, M., Suomi, S.J., Linnoila, M.V., and Higley, J.D. (2002) 'Early experience and serotonin transporter gene variation interact to influence primate CNS function', *Molecular Psychiatry*, 7: 118–122.

Berglund, K.J., Balldin, J., Berggren, U., Gerdner, A., and Fahlke, C. (2013) 'Childhood maltreatment affects serotonergic system in male alcohol-dependent individuals', *Alcoholism: Clinical and Experimental Research*, 37: 757–762.

Bogdan, R., Agrawal, A., Gaffrey, M.S., Tillman, R., and Luby, J.L. (2013) 'Serotonin transporter-linked polymorphic region (5HTTLPR) genotype and stressful life events interact to predict preschool-onset depression: a replication and developmental extension', *Journal of Child Psychology and Psychiatry*, doi: 10.1111/jcpp.12142

Brody, G.H., Beach, S.R., Philibert, R.A., Chen, Y.F., Lei, M.K., Murry, V.M., and Brown, A.C. (2009) 'Parenting moderates a genetic vulnerability factor in longitudinal increases in youths' substance use', *Journal of Consulting and Clinical Psychology*, 77: 1–11.

Brown, G.W. and Harris, T.O. (1978) *Social Origins of Depression: A Study of Psychiatric Disorder in Women*, New York: Free Press.

Brummett, B.H., Boyle, S.H., Siegler, I.C., Kuhn, C.M., Ashley-Koch, A., Jonassaint, C.R., Züchner, S., Collins, A., and Williams, R.B. (2007) 'Effects of environmental stress and gender on associations among symptoms of depression and the serotonin transporter gene linked polymorphic region (5-HTTLPR)', *Behavioral Genetics*, 38: 34–43.

Canli, T. and Lesch, K.P. (2007) 'Long story short: the serotonin transporter in emotion regulation and social cognition', *Nature Neuroscience*, 10: 1103–1109.

Carpenter, L.L., Carvalho, J.P., Tyrka, A.R., Wier, L.M., Mello, A.F., Mello, M.F., Anderson, G.M., Wilkinson, C.W., and Price, L.H. (2007) 'Decreased adrenocorticotropic hormone and cortisol responses to stress in healthy adults reporting significant childhood maltreatment', *Biological Psychiatry*, 62: 1080–1087.

Caspi, A., Sugden, K., Moffitt, T.E., Taylor, A., Craig, I.W., Harrington, H.L., McClay, J., Mill, J., Martin, J., Braithwaite, A., and Poulton, R. (2003) 'Influence of life stress on depression: moderation by a polymorphism in the 5-HTT gene', *Science*, 301: 386–389.

Cervilla, J.A., Molina, E., Rivera, M., Torres-González, F., Bellón, J.A., Moreno, B., Luna, J.D., Lorente, J.A., Mayoral, F., King, M., Nazareth, I., the PREDICT Study Core Group, and Gutiérrez, B. (2007) 'The risk for depression conferred by stressful life events is modified by variation at the serotonin transporter 5HTTLPR genotype: evidence from the Spanish PREDICT-Gene cohort', *Molecular Psychiatry*, 12: 748–755.

Champagne, F.A. and Curley, J.P. (2008) 'Epigenetic mechanisms mediating the long-term effects of maternal care on development', *Neuroscience & Biobehavioral Reviews*, 33: 593–600.

Cicchetti, D. and Rogosch, F.A. (2001) 'Diverse patterns of neuroendocrine activity in maltreated children', *Development and Psychopathology*, 13: 677–693.

Cicchetti, D., Rogosch, F.A., and Sturge-Apple, M.L. (2007) 'Interactions of child maltreatment and serotonin transporter and monoamine oxidase A polymorphisms: depressive symptomatology among adolescents from low socioeconomic status backgrounds', *Development and Psychopathology*, 19: 1161–1180.

Conway, C.C., Hammen, C., Espejo, E.P., Wray, N.R., Najman, J.M., and Brennan, P.A. (2012) 'Appraisals of stressful life events as a genetically-linked mechanism in the stress–depression relationship', *Cognitive Therapy and Research*, 36: 338–347.

Coplan, J.D., Abdallah, C.G., Kaufman, J., Gelernter, J., Smith, E.L.P., Perera, T.D., Dwork, A.J., Kaffman, A., Gorman, J.M., Rosenblum, L.A., Owens, M.S., and Nemeroff, C.B. (2011) 'Early-life stress, corticotrophin-releasing factor, and serotonin transporter gene: a pilot study', *Psychoneuroendocrinology*, 36: 289–293.

Delville, Y., Melloni, R.H., and Ferris, C.F. (1998) 'Behavioral and neurobiological consequences of social subjugation during puberty in golden hamsters', *Journal of Neuroscience*, 18: 2667–2672.

Drabant, E.M., Ramel, W., Edge, M.D., Hyde, L.W., Kuo, J.R., Goldin, P.R., Hariri, A.R., and Gross, J.J. (2012) 'Neural mechanisms underlying 5-HTTLPR-related sensitivity to acute stress', *American Journal of Psychiatry*, 169: 397–405.

Eley, T.C., Sugden, K., Corsico, A., Gregory, A.M., Sham, P., McGuffin, P., Plomin, R., and Craig, I.W. (2004) 'Gene–environment interaction analysis of serotonin system markers with adolescent depression', *Molecular Psychiatry*, 9: 908–915.

Fergusson, D.M., Horwood, L.J., Miller, A.L., and Kennedy, M.A. (2011) 'Life stress, 5-HTTLPR and mental disorder: findings from a 30-year longitudinal study', *British Journal of Psychiatry*, 198: 129–135.

Ferris, C.F. (2000) 'Adolescent stress and neural plasticity in hamsters: a vasopressin–serotonin model of inappropriate aggression behaviour', *Experimental Physiology*, 85S: 85S–90S.

Fone, K.C.F. and Porkess, M.V. (2008) 'Behavioural and neurochemical effects of post-weaning social isolation in rodents—relevance to developmental neuropsychiatric disorders', *Neuroscience and Biobehavioral Reviews*, 32: 1087–1102.

Frankenhaeuser, M., Lundberg, U., Von Wright, M.R., Von Wright, J., and Sedvall, G. (1986) 'Urinary monoamine metabolites as indices of mental stress in healthy males and females', *Pharmacology Biochemistry and Behavior*, 24: 1521–1525.

Frodl, T., Reinhold, E., Koutsouleris, N., Donohoe, G., Bondy, B., Reiser, M., Möller, H.J., and Meisenzahl, E.M. (2010) 'Childhood stress, serotonin transporter gene and brain structures in major depression', *Neuropsychopharmacology*, 35: 1383–1390.

Gillespie, N.A., Whitfield, J.B., Williams, B., Heath, A.C., and Martin, N.G. (2005) 'The relationship between stressful life events, the serotonin transporter (5-HTTLPR) genotype and major depression', *Psychological Medicine*, 35: 101–111.

Grabe, H.J., Lange, M., Wolff, B., Völzke, H., Lucht, M., Freyberger, H.J., John, U., and Cascorbi, I. (2005) 'Mental and physical distress is modulated by a polymorphism in the 5-HT transporter gene interacting with social stressors and chronic disease burden', *Molecular Psychiatry*, 10: 220–224.

Helin, P., Kuoppasalmi, K., Laakso, J., and Härkönen, M. (1988) 'Human urinary biogenic amines and some physiological responses during situation stress', *International Journal of Psychophysiology*, 6: 125–132.

Higley, J.D., Mehlman, P.T., Taub, D.M., Higley, S.B., Suomi, S.J., Linnoila, M., and Vickers, J.H. (1992) 'Cerebrospinal fluid monoamine and adrenal correlates of aggression in free-ranging rhesus monkeys', *Archives of General Psychiatry*, 49: 436–441.

Holmes, A., le Guisquet, A.M., Vogel, E., Millstein, R.A., Leman, S., and Belzung, C. (2005) 'Early life genetic, epigenetic and environmental factors shaping emotionality in rodents', *Neuroscience and Biobehavioral Reviews*, 29: 1335–1346.

Hu, X.Z., Lipsky, R.H., Zhu, G., Akhtar, L.A., Taubman, J., Greenberg, B.D., Xu, K., Arnold, P.D., Richter, M.A., Kennedy, J.L., Murphy, D.L., and Goldman, D. (2006) 'Serotonin transporter promoter gain-of-function genotypes are linked to obsessive-compulsive disorder', *American Journal of Human Genetics*, 78: 815–826.

Huang, Y., Xu, H., Li, H., Yang, H., Chen, Y., and Shi, X. (2012) 'Pre-gestational stress reduces the ratio of 5-HIAA to 5-HT and the expression of the 5-HT1A receptor and serotonin transporter in the brain of the foetal rat', *BMC Neuroscience*, 13: 22.

Ichise, M., Vines, D.C., Gura, T., Anderson, G.M., Suomi, S.J., Higley, J.D., and Innis, R.B. (2006) 'Effects of early life stress on [^{11}C] DASB positron emission tomography imaging of serotonin transporters in adolescent peer- and mother-reared rhesus monkeys', *Journal of Neuroscience*, 26: 4638–4643.

Jones, G.H., Hernandez, T.D., Kendall, D.A., Marsden, C.A., and Robbins, T.W. (1992) 'Dopaminergic and serotonergic function following isolation rearing in rats: study of behavioural responses and postmortem and in vivo neurochemistry', *Pharmacology, Biochemistry, and Behavior*, 43: 17–35.

Kalia, M. (2002) 'Assessing the economic impact of stress—the modern day hidden epidemic', *Metabolism*, 51: 49–53.

Karg, K., Burmeister, M., Shedden, K., and Sen, S. (2011) 'The serotonin transporter promoter variant (5-HTTLPR), stress, and depression meta-analysis revisited: evidence of genetic moderation', *Archives of General Psychiatry*, 68: 444–454.

Kaufman, J., Yang, B.Z., Douglas-Palumberi, H., Houshyar, S., Lipschitz, D., Krystal, J.H., and Gelernter, J. (2004) 'Social supports and serotonin transporter gene moderate depression in maltreated children', *Proceedings of the National Academy of Sciences, USA*, 101: 17316–17321.

Kaufman, J., Yang, B.Z., Douglas-Palumberi, H., Crouse-Artus, M., Lipschitz, D., Krystal, J.H., and Gelernter, J. (2007) 'Genetic and environmental predictors of early alcohol use', *Biological Psychiatry*, 61: 1228–1234.

Kendler, K.S., Kuhn, J.W., Vittum, J., Prescott, C.A., and Riley, B. (2005) 'The interaction of stressful life events and serotonin transporter polymorphism in the prediction of episodes of major depression: a replication', *Archives of General Psychiatry*, 62: 529–535.

Kim, J.M., Stewart, R., Kim, S.W., Yang, S.J., Shin, I.S., Kim, Y.H., and Yoon, J.S. (2007) 'Interactions between life stressors and susceptibility genes (5-HTTLPR and BDNF) on depression in Korean elders', *Biological Psychiatry*, 62: 423–428.

Kinnally, E.L., Tarara, E.R., Mason, W.A., Mendoza, S.P., Abel, K., Lyons, L.A., and Capitanio, J.P. (2010) 'Serotonin transporter expression is predicted by early life stress and is associated with disinhibited behavior in infant rhesus macaques', *Genes, Brain and Behavior*, 9: 45–52.

Kiser, D., Steemers, B., Branchi, I., and Homberg, J.R. (2012) 'The reciprocal interaction between serotonin and social behaviour', *Neuroscience and Biobehavioral Reviews*, 36: 786–798.

Klucken, T., Alexander, N., Schweckendiek, J., Merz, C.J., Kagerer, S., Osinsky, R., Walter, B., Vaitl, D., Hennig, J., and Stark, R. (2013) 'Individual differences in neural correlates of fear conditioning as a function of 5-HTTLPR and stressful life events', *SCAN*, 8: 318–325.

Kochanska, G., Philibert, R.A., and Barry, R.A. (2009) 'Interplay of genes and early mother–child relationship in the development of self-regulation from toddler to preschool age', *Journal of Child Psychology and Psychiatry*, 50: 1331–1338.

Kochanska, G., Kim, S., Barry, R.A., and Philibert, R.A. (2011) 'Children's genotypes interact with maternal responsive care in predicting children's competence: diathesis stress or differential susceptibility?', *Development and Psychopathology*, 23: 605–616.

Kohut, S.J., Decicco-Skinner, K.L., Johari, S., Hurwitz, Z.E., Baumann, M.H., and Riley, A.L. (2012) 'Differential modulation of cocaine's discriminative cue by repeated and variable stress exposure: relation to monoamine transporter levels', *Neuropharmacology*, 63: 330–337.

Kranzler, H.R., Scott, D., Tennen, H., Feinn, R., Williams, C., Armeli, S., Taylor, R.E., Briggs-Gowan, M.J., and Covault, J. (2012) 'The 5-HTTLPR polymorphism moderates the effect of stressful life events on drinking behavior in college students of African descent', *American Journal of Medical Genetics, B*, 159B: 484–490.

Law, A.J., Pei, Q., Feldon, J., Pryce, C.R., and Harrison, P.J. (2009a) 'Gene expression in the anterior cingulate cortex and amygdala of adolescent marmoset monkeys following parental separations in infancy', *International Journal of Neuropsychopharmacology*, 12: 761–772.

Law, A.J., Pei, Q., Walker, M., Gordon-Andrews, H., Weickert, C.S., Feldon, J., Pryce, C.R., and Harrison, P.J. (2009b) 'Early parental deprivation in the marmoset monkey produces long-term changes in hippocampal expression of genes involved in synaptic plasticity and implicated in mood disorder', *Neuropsychopharmacology*, 34: 1381–1394.

Lemogne, C., Gorwood, P., Boni, C., Pessiglione, M., Lehéricy, S., and Fossati, P. (2011) 'Cognitive appraisal and life stress moderate the effects of the 5-HTTLPR polymorphism on amygdala reactivity', *Human Brain Mapping*, 32: 1856–1867.

Lenze, E.J., Munin, M.C., Ferrell, R.E., Pollock, B.G., Skidmore, E., Lotrich, F., Rogers, J.C., Quear, T., Houck, P., and Reynolds, C.F. (2005) 'Association of the serotonin transporter gene-linked polymorphic region (5-HTTLPR) genotype with depression in elderly persons after hip fracture', *American Journal of Geriatric Psychiatry*, 13: 428–432.

Lesch, K.P., Bengel, D., Heils, A., Sabol, S.Z., Greenberg, B.D., Petri, S., Benjamin, J., Müller, C.R., Hamer, D.H., and Murphy, D.L. (1996) 'Association of anxiety-related traits with a polymorphism in the serotonin transporter gene regulatory region', *Science*, 274: 1527–1531.

Lupien, S.J., McEwan, B.S., Gunnar, M.R., and Heim, C. (2009) 'Effects of stress throughout the lifespan on the brain, behaviour and cognition', *Nature*, 10: 434–445.

Maestripieri, D. (2008) 'Neuroendocrine mechanisms underlying the intergenerational transmission of maternal behavior and infant abuse in rhesus macaques', in D.W. Pfaff, C. Kordon, and P. Chanson (eds), *Hormones and Social Behavior*, Berlin: Springer, pp. 121–130.

Maestripieri, D., McCormack, K., Lindell, S.G., Higley, J.D., and Sanchez, M.M. (2006) 'Influence of parenting style on the offspring's behaviour and CSF monoamine metabolite levels in crossfostered and noncrossfostered female rhesus macaques', *Behavioural Brain Research*, 175: 90–95.

Maestripieri, D., Lindell, S.G., and Higley, J.D. (2007) 'Intergenerational transmission of maternal behavior in rhesus macaques and its underlying mechanisms', *Developmental Psychobiology*, 49: 165–171.

Mahar, I., Bambico, F.R., Mechawar, N., and Nobrega, J.N. (2013) 'Stress, serotonin, and hippocampal neurogenesis in relation to depression and antidepressant effects', *Neuroscience & Biobehavioral Reviews*, http: // dx.doi.org/10.1016/j.neubiorev.2013.11.009

Mandelli, L., Serretti, A., Marino, E., Pirovano, A., Calati, R., and Colombo, C. (2007) 'Interaction between serotonin transporter gene, catechol-O-methyltransferase gene and stressful life events in mood disorders', *International Journal of Neuropsychopharmacology*, 10: 437–447.

Manuck, S.B., Bleil, M.E., Petersen, K.L., Flory, J.D., Mann, J.J., Ferrell, R.E., and Muldoon, M.F. (2005) 'The socio-economic status of communities predicts variation in brain serotonergic responsivity', *Psychological Medicine*, 35: 519–528.

Matthews, K., Dalley, J.W., Matthews, C., Tsai, T.H., and Robbins, T.W. (2001) 'Periodic maternal separation of neonatal rats produces region- and gender-specific effects on biogenic amine content in postmortem adult brain', *Synapse*, 40: 1–10.

Meaney, M.J. (2001) 'Maternal care, gene expression, and the transmission of individual differences in stress reactivity across generations', *Annual Review of Neuroscience*, 24: 1161–1192.

Meaney, M.J., Brake, W., and Gratton, A. (2002) 'Environmental regulation of the development of mesolimbic dopamine systems: a neurobiological mechanism for vulnerability to drug abuse?', *Psychoneuroendocrinology*, 27: 127–138.

Miller, J.M., Kinnally, E.L., Ogden, R.T., Oquendo, M.A., Mann, J.J., and Parsey, R.V. (2009) 'Reported childhood abuse is associated with low serotonin transporter binding in vivo in major depressive disorder', *Synapse*, 67: 565–573.

Moberg, T., Nordström, P., Forslund, K., Kristiansson, M., Åsberg, M., and Jokinen, J. (2011) 'CSF 5-HIAA and exposure to and expression of interpersonal violence in suicide attempters', *Journal of Affective Disorders*, 132: 173–178.

Monroe, S.M. and Simons, A.D. (1991) 'Diathesis-stress theories in the context of life stress research: implications for depressive disorders', *Psychological Bulletin*, 110: 406–425.

Mueller, A., Brocke, B., Fries, E., Lesch, K.P., and Kirschbaum, C. (2010) 'The role of the serotonin transporter polymorphism for the endocrine stress response in newborns', *Psychoneuroendocrinology*, 35: 289–296.

Mueller, A., Armbruster, D., Moser, D.A., Canli, T., Lesch, K.P., Brocke, B., and Kirschbaum, C. (2011) 'Interaction of serotonin transporter gene-linked polymorphic region and stressful life events predicts cortisol stress response', *Neuropsychopharmacology*, 36: 1332–1339.

Munafò, M.R., Durrant, C., Lewis, G., and Flint, J. (2009) 'Gene × environment interactions at the serotonin transporter locus', *Biological Psychiatry*, 65: 211–219.

Ognibene, E., Adriani, W., Caprioli, A., Ghirardi, O., Ali, S.F., Aloe, L., and Laviola, G. (2008) 'The effect of early maternal separation on brain derived neurotrophic factor and monoamine levels in adult heterozygous reeler mice', *Progress in Neuro-Psychopharmacology & Biological Psychiatry*, 32: 1269–1276.

Petersen, I.T., Bates, J.E., Goodnight, J.A., Dodge, K.A., Lansford, J.E., Pettit, G.S., Latendresse, S.J., and Dick, D.M. (2012) 'Interaction between serotonin transporter polymorphism (5-HTTLPR) and stressful life events in adolescents' trajectories of anxious/depressed symptoms', *Developmental Psychology*, 48: 1463–1475.

Pruessner, J.C., Champagne, F., Meaney, M.J., and Dagher, A. (2004) 'Dopamine release in response to psychological stress in humans and its relationship to early life maternal care: a positron emission tomography study using [^{11}C]Raclopride', *Journal of Neuroscience*, 24: 2825–2831.

Reif, A., Rösler, M., Freitag, C.M., Schneider, M., Eujen, A., Kissling, C., Wenzler, D., Jacob, C.P., Retz-Junginger, P., Thome, J., Lesch, K.P., and Retz, W. (2007) 'Nature and nurture predispose to violent behavior: serotonergic genes and adverse childhood environment', *Neuropsychopharmacology*, 32: 2375–2383.

Risch, N., Herrell, R., Lehner, T., Liang, K.Y., Eaves, L., Hoh, J., Griem, A., Kovacs, M., Ott, J., and Merikangas, K.R. (2009) 'Interaction between the serotonin transporter gene (5-HTTLPR), stressful life events, and risk of depression: a meta-analysis', *JAMA*, 301: 2462–2471.

Rinne, T., Westenberg, H.G.M., den Boer, J.A., and van den Brink, W. (2000) 'Serotonergic blunting to meta-chlorophenylpiperazine (m-CPP) highly correlates with sustained childhood abuse in impulsive and autoaggressive female borderline patients', *Biological Psychiatry*, 47: 548–556.

Roy, A. (2002) 'Self-rated childhood emotional neglect and CSF monoamine indices in abstinent cocaine-abusing adults: possible implications for suicidal behavior', *Psychiatry Research*, 112: 69–75.

Sadeh, N., Javdani, S., Jackson, J.J., Reynolds, E.K., Potenza, M.N., Gelernter, J., Lejuez, C.W., and Verona, E. (2010) 'Serotonin transporter gene associations with psychopathic traits in youth vary as a function of socioeconomic resources', *Journal of Abnormal Psychology*, 119: 604–609.

Scheid, J.M., Holzman, C.B., Jones, N., Friderici, K.H., Nummy, K.A., Symonds, L.L., Sikorskii, A., Regier, M.K., and Fisher, R. (2007) 'Depressive symptoms in mid-pregnancy, lifetime stressors and the 5-HTTLPR genotype', *Gene, Brain and Behavior*, 6: 453–464.

Shannon, C., Schwandat, M.L., Champoux, M., Shoaf, S.E., Suomi, S.J., Linnoila, M., and Higley, J.D. (2005) 'Maternal absence and stability of individual differences in CSF 5-HIAA concentrations in rhesus monkey infants', *American Journal of Psychiatry*, 162: 1658–1664.

Simons, R.L., Lei, M.K., Beach, S.R.H., Brody, G.H., Philibert, R.A., and Gibbons, F.X. (2011) 'Social environment, genes, and aggression: evidence supporting the differential susceptibility perspective', *American Sociological Review*, 76: 883–912.

Sjöberg, R.L., Nilsson, K.W., Nordquist, N., Öhrvik, J., Leppert, J., Lindström, L., and Oreland, L. (2006) 'Development of depression: sex and the interaction between environment and a promoter polymorphism of the serotonin transporter gene', *International Journal of Neuropsychopharmacology*, 9: 443–449.

Sonuga-Barke, E.J.S., Oades, R.D., Psychogiou, L., Chen, W., Franke, B., Buitelaar, J., Banaschewski, T., Ebstein, R., Gill, Anney, R., Miranda, A., Roeyers, H., Rothenberger, A., Sergeant, J., Steinhausen, H., Thompson, M., Asherson, P., and Faraone, S.V. (2009) 'Dopamine and serotonin transporter genotypes moderate sensitivity to maternal expressed emotion: the case of conduct and emotional problems in attention deficit/hyperactivity disorder', *Journal of Child Psychology and Psychiatry*, 50: 1052–1063.

Spinelli, S., Chefer, S., Carson, R.E., Jagoda, E., Lang, L., Heilig, M., Barr, C.S., Suomi, S.J., Higley, J.D., and Stein, E.A. (2010) 'Effects of early-life stress on 5-HT1A receptors in juvenile rhesus monkeys measured by PET', *Biological Psychiatry*, 67: 1146–1153.

Stronks, D.L., Tulen, J.H.M., Verheij, R., Boomsma, F., Fekkes, D., Pepplinkhuizen, L., Mantel, G.W.H., and Passchier, J. (1998) 'Serotonergic, catecholaminergic, and cardiovascular reactions to mental stress in female migraine patients: a controlled study', *Headache*, 38: 270–280.

Sulik, M.J., Eisenberg, N., Lemery-Chalfant, K., Spinrad, T.L., Silva, K.M., Eggum, N.D., Betkowski, J.A., Kupfer, A., Smith, C.L., Gaertner, B., Stover, D.A., and Verrelli, B.C. (2012) 'Interactions between serotonin transporter gene haplotypes and quality of mothers' parenting predict the development of children's noncompliance', *Developmental Psychology*, 48: 740–754.

Surtees, P.G., Wainwright, N.W.J., Willis-Owen, S.A.G., Luben, R., Day, N.E., and Flint, J. (2006) 'Social adversity, the serotonin transporter (5-HTTLPR) polymorphism and major depressive disorder', *Biological Psychiatry*, 59: 224–229.

Tarullo, A.R. and Gunnar, M.R. (2006) 'Child maltreatment and the developing HPA axis', *Hormones and Behavior*, 50: 632–639.

Taylor, S.E., Way, B.M., Welch, W.T., Hilmert, C.J., Lehman, B.J., and Eisenberger, N.I. (2006) 'Early family environment, current adversity, the serotonin transporter promoter polymorphism, and depressive symptomatology', *Biological Psychiatry*, 60: 671–676.

Teicher, M.H., Andersen, S.L., Polcari, A., Anderson, C.M., and Navalta, C.P. (2002) 'Developmental neurobiology of childhood stress and trauma', *Psychiatric Clinics of North America*, 25: 397–426.

Uher, R. and McGuffin, P. (2008) 'The moderation by the serotonin transporter gene of environmental adversity in the aetiology of mental illness: review and methodological analysis', *Molecular Psychiatry*, 13: 131–146.

Uher, R. and McGuffin, P. (2010) 'The moderation by the serotonin transporter gene of environmental adversity in the etiology of depression: 2009 update', *Molecular Psychiatry*, 15: 18–22.

Vaske, J., Newsome, J., and Wright, J.P. (2012) 'Interaction of serotonin transporter linked polymorphic region and childhood neglect on criminal behavior and substance use for males and females', *Development and Psychopathology*, 24: 181–193.

Vázquez, D.M., Eskandari, R., Zimmer, C.A., Levine, S., and López, J.F. (2002) 'Brain 5-HT receptor system in the stressed infant rat: implications for vulnerability to substance abuse', *Psychoneuroendocrinology*, 27: 245–272.

Way, B.M. and Taylor, S.E. (2010) 'The serotonin transporter promoter polymorphism (5-HTTLPR) is associated with cortisol response to psychosocial stress', *Biological Psychiatry*, 67: 487–492.

Weaver, I.C.G., Cervoni, N., Champagne, F.A., D'Alessio, A.C., Sharma, S., Seckl, J.R., Dymov, S., Szyf, M., and Meaney, M.J. (2004) 'Epigenetic programming by maternal behavior', *Nature Neuroscience*, 7: 847–854.

Wendland, J.R., Martin, B.J., Kruse, M.R., Lesch, K.P., and Murphy, D.L. (2006) 'Simultaneous genotyping of four functional loci of human SLC6A4, with a reappraisal of 5-HTTLPR and rs25531', *Molecular Psychiatry*, 11: 224–226.

Wielpuetz, C., Kuepper, Y., Grant, P., Munk, A.J.L., and Hennig, J. (2014) 'Variations in central serotonergic activity—relevance of the 5-HTTLPR, life events and their interaction', *Behavioural Brain Research*, http://dx.doi.org/10.1016/j.bbr.2013.12.023

Wilhelm, K., Mitchell, P.B., Niven, H., Finch, A., Wedgwood, L., Scimone, A., Blair, I.P., Parker, G., and Schofield, P.R. (2006) 'Life events, first depression onset and the serotonin transporter gene', *British Journal of Psychiatry*, 188: 210–215.

Zalsman, G., Huang, Y.Y., Oquendo, M.A., Burke, A.K., Hu, X.Z., Brent, D.A., Ellis, S.P., Goldman, D., and Mann, J.J. (2006) 'Gene promoter region (5-HTTLPR) polymorphism with stressful life events and severity of depression', *American Journal of Psychiatry*, 163: 1588–1593.

Zhang, K., Grady, C.J., Tsapakis, E.M., Andersen, S.L., Tarazi, F.I., and Baldessarini, R.J. (2004) 'Regulation of working memory by dopamine D_4 receptor in rats', *Neuropsychopharmacology*, 29: 1648–1655.

10

When humans prey on one another

Understanding the biosocial origins of victimization

Maria Koeppel, Brian B. Boutwell, and J.C. Barnes

Introduction

Divining a clear conception of what causes crime—while elusive and still relatively opaque—has occupied the attention and captured the imagination of scholars and lay persons alike. A cacophony of voices has weighed in over the past few centuries. Everyone, it seems, has some idea about why criminals commit crime, and explanations range from external forces that compel criminality (i.e. socio-cultural theories; Cohen, 1955; Anderson, 1999) to internal propulsions toward antisocial behavior (i.e. motivational and psychological theories; Moffitt, 1993; Patterson *et al.*, 1990). Slowly, however, scholars of crime have placed a growing recognition on the fact that criminal behavior is often a two-sided equation involving both a perpetrator *and* a victim. For the past several decades, theories of victimization have metastasized into the mainstream and found their way into the prominent journals in the field. There is a stark reality, though, that victimologists have yet to fully address. Most of their theories might be incomplete (and in some cases, incorrect).

Accompanying the proliferation of victimization research, another group of scholars has undertaken a parallel line of research aimed at illuminating the confluence of biological, environmental and social factors in the prediction of criminogenic behaviors (Beaver, 2009). A full explication of these findings is beyond the reach of the current discussion, yet one consistent result warrants mentioning because it has monumental implications for victimization research. Namely, every behavior—prosocial and antisocial—studied by scholars across the behavioral sciences is under varying degrees of genetic influence (Turkheimer, 2000; Pinker, 2002; Rhee and Waldman, 2002; Moffitt, 2005). For victimologists, then, this raises the possibility that the genetic influences that underlie every other human outcome might also spill over into the experience of victimization (Beaver *et al.*, 2009). Put more directly, one has to ask: could the outcome of victimization logically be the *only* human outcome beyond the reach of the genome?

The goal of the discussion here, then, is to delve into the possibility that both genetic and environmental forces might coalesce to influence victimization. Along the way, it will be necessary to consider the prevailing theories of victimization, a discussion that follows momentarily.

Additionally, the goal is to further broaden the current understanding of victimization via the incorporation of modern biosocial research. What should become apparent is that, while our current understanding of victimization is likely incomplete, little effort would be required to dramatically increase the potency of victimization theorizing. What follows next, however, is a brief review of mainstream victimization research.

General theories of victimization

For decades now, scholars have labored intensively in order to develop and refine theories intended to explain why some individuals (or groups) participate in crime at greater rates than other individuals. As a result, the field of criminology is flooded with a number of mainstream theories of criminal offending. Noticeably lacking for some time, however, were mainstream theories dedicated to explaining and predicting victimization. As a result, researchers have attempted to leverage dominant criminological theories such as Gottfredson and Hirschi's (1990) general theory of crime and Cohen and Felson's (1979) routine activities theory in order to explain experiences of victimization. Given that these two integrative approaches occupy much of the attention amongst scholars of victimology, we devote most of our discussion in this section to them.

Self-control and victimization

A small (but growing) arena of victimology has examined a possible relationship between low self-control and victimization. According to self-control theory, as originally developed by Gottfredson and Hirschi (1990), individuals with low self-control often act with impulse, are short-sighted, and are unable to resist temptation. These characteristics are considered to not only increase the odds of partaking in criminal activity, but—perhaps—might also elevate the risk of falling victim to criminal behavior as well. A possible relationship between self-control and victimization, then, necessarily suggests that there may be a pool of characteristics shared by both offenders and victims (in this case, the personality trait of low self-control). In particular, Schreck (1999) proposed that there are a number of qualities indicative of a lack of impulse control that may also be responsible for an increased risk of victimization.

The individual's degree of future orientation, for instance, is related to levels of impulsivity which allow individuals to seize on opportunities for deviance (Schreck, 1999). Consequently, it may also be related to an individual's involvement in activities without the forethought of making sure they or their belongings will be safe—thus increasing the potential of property victimization (Schreck, 1999). Similarly, individuals who prefer physical rather than mental activities are less likely to assess risky situations thoroughly, making themselves vulnerable to victimization in a variety of ways (Schreck, 1999). Lacking empathy and sensitivity to the feelings of others, moreover, is also a feature of low self-control, which may make it easier for offenders to commit a crime. Individuals with poor empathy might also be less invested in personal relationships, such as those with their neighbors, which decrease guardianship around the home. This may leave the individual vulnerable to property victimization, especially during long periods away from home. As a result of these possible relationships, Schreck (1999) argued that individuals demonstrating prototypical characteristics of diminished self-control were at a greater risk for victimization than other individuals.

Research that has explored the relationship between low self-control and victimization has supported Schreck's (1999) original predictions. Evidence exists, for example, suggesting that individuals with lower levels of self-control were more likely to experience victimization, including violent, personal, white-collar, and property victimization. This pattern persisted across different populations such as prison inmates, high-school students, and college students, and in both cross-sectional and longitudinal settings (Schreck *et al.*, 2002, 2006; Stewart *et al.*, 2004; Piquero *et al.*, 2005; Baron *et al.*, 2007; Kerley *et al.*, 2008, 2009; Fox *et al.*, 2009; Franklin, 2011; Franklin *et al.*, 2012). In short, traits indicative of low impulse control certainly appear correlated with the experience of victimization.

Routine activities and victimization

In addition to low self-control as a predictor of victimization, a fair amount of research has examined the routine activities of individuals as another possible etiological source of victimization. Routine activities theory suggests victimization is explained by assessing the situational characteristics of the victimization experience. Cohen and Felson (1979) first argued that offending and victimization were a result of spatial and temporal factors which would fall in line due to predictable and repetitive routine activities of members of the population. As Cohen and Felson (1979) argue, three factors must converge in time and space for a criminal act to take place (or for one to be victimized): (1) the presence of a motivated offender(s), (2) the availability of a suitable target(s), and (3) the absence of a capable guardian(s). Additionally, Cohen and Felson (1979) suggested that individuals whose routine activities lead them to spend the majority of their time outside the home would be at a higher risk for victimization than individuals whose routine activities largely take place within the home. Maintaining a public lifestyle, including working outside the home or an active nightlife away from home, would increase the chance of being seen as a suitable target by a motivated offender. Moreover, from a property crime viewpoint, a person is not able to act as a capable guardian of their own property if they are rarely at home.

Since the development of routine activities theory, scholars have spent considerable time testing the perspective's ability to predict both crime and victimization. Routine activities characteristics have been able to explain variation in property victimization, including vehicle theft and burglary (Miethe *et al.*, 1987; Kennedy and Forde, 1990; Fisher *et al.*, 1998), as well as personal victimization in the form of assault and robbery (Kennedy and Forde, 1990). Cohen and Felson's (1979) routine activities theory has also been able to explain workplace victimization (Wooldredge *et al.*, 1992; Coster *et al.*, 1999), and victimization among college students (Fisher *et al.*, 1998; Mustaine and Tewksbury, 1998; 2002; Schwartz *et al.*, 2001; Tewksbury and Mustaine, 2003; Franklin *et al.*, 2012). These studies have consistently found individuals who lead lifestyles that include an active nightlife consisting of going to bars, movies, or restaurants have a greater risk of victimization due to their increased likelihood of being viewed as a suitable target (Kennedy and Forde, 1990; Fisher *et al.*, 1998).

Biosocial explanations of victimization

The discussion of self-control and routine activities as predictors of victimization provides strong examples for how mainstream criminologists have sought to explain both criminal offending and victimization. Self-control provides an example of an individual-level criminological approach, while routine activities provides an example of an environmental approach. While these approaches to victimization have been fruitful in their efforts, often revealing empirical support, both fail to consider the notion that something other than situational factors may increase vulnerability to victimization. Biosocial explanations of victimization take into account genetic factors as possible predictors (albeit indirect) of victimization. This line of research is discussed in more detail below.

Behavioral genetics

Behavioral genetics is a field of study that seeks to explain the origin of individual differences in the population (Plomin *et al.*, 2013). More specifically, behavior genetic research typically involves partitioning the variance in some trait/behavior into that which is accounted for by genetic and environmental factors (Plomin *et al.*, 2013). Beyond simply lumping all environmental sources of variance together, however, behavior geneticists further subdivide the environment into a shared environment component (environmental factors that make siblings raised together similar to each other) and a nonshared environmental component (environmental factors that make siblings raised together different from one another) (Beaver *et al.*, 2009). Each of the individual components, then, accounts for a portion of the variance in the behavior, and when added together the three components should explain 100 percent of the variance.

Unlike practically every other social science research approach, behavioral genetic research designs are unique in that they require data drawn from kinship pairs (or other pairs of individuals with some known level of genetic relatedness). Kinship pairs are an essential element of behavior genetic designs because these methodologies exploit the known level of genetic overlap between different types of kinship pair *en route* to an estimation of the degree to which genetic factors influence the trait under investigation. A simple thought experiment reveals the logic of behavior genetic research designs. Imagine you have a data set of 100 pairs of individuals. Half of those pairs are known to be monozygotic (MZ) twins. MZ twins are unique because they share 100 percent of their DNA. Imagine that the other 50 pairs in your data set are dizygotic (DZ, fraternal) twins. DZ twins are known to share, on average, 50 percent of their distinguishing DNA. Given only this backdrop of information, you might develop a hypothesis that MZ twins should be more similar to one another on the trait of interest than DZ twins. If this were the case, you would conclude that genetic factors played a role in the etiology of the trait being studied.

As with any scientific method, the behavior genetic theory outlined here rests on a bed of assumptions. If those assumptions fail, then the estimates from this body of work may be called into question. Luckily, the most important among the assumptions made by behavior genetics has been vetted using some of the most sophisticated research designs available to modern social scientists (for an overview see Plomin *et al.*, 2013) and the overarching conclusion has been that the assumptions are met (see also Carey, 2003). Thus behavior geneticists are able to draw on information from kinship pairs to estimate the amount of variance in a range of traits that is attributable to genetic factors, shared environments, and nonshared environments. The focus of the current discussion is whether and to what extent genetic, shared environmental, and nonshared environmental factors influence variance in victimization experiences (Beaver *et al.*, 2009).

In the past several decades, a growing number of researchers have begun examining criminal and antisocial behavior through a behavioral genetics lens. Results from the continuously growing number of studies have consistently found that about 50 percent of the variance in criminal and antisocial behavior can be attributed to genetic factors, while a large portion of the remaining variance is related to nonshared environmental factors (Mason and Frick, 1994; Miles and Carey, 1997; Rhee and Waldman, 2002; Moffitt, 2005). Building on the success of this line of research, biosocial scholars in criminology have begun detailing a number of arguments as to why genetics may play a similar role in explaining the variance in victimization.

The current body of literature using behavior genetics to explain victimization is small, but growing, with results that strongly support the argument that genetic factors may play a role in victimization experiences. Beaver and colleagues (2009) found genetic factors to explain between 40 and 45 percent of the variance in victimization of adolescents. Shared environments explained none of the remaining variance, while nonshared environments accounted for 55 to 60 percent of the variance in victimization. Further results examining repeated victimization found that genetic factors accounted for 64 percent of the variance, while nonshared environments explained the remaining 36 percent. Research specific to bullying victimization of school-aged children found genetic factors explained a significant 73 percent of the variance of victimization. The remaining 27 percent was explained by nonshared environments (Ball *et al.*, 2008).

Barnes *et al.* (2011) further explored the relationship between genetics and victimization by examining the effect of gang membership on victimization. First looking at victimization assessed in the second wave of the National Longitudinal Study of Adolescent Health (Add Health), genetic factors explained 34 percent of the variance of victimization, and 66 percent of the variance was attributed to nonshared environment factors. Across later waves in the study, however, genetics effects began to wane (12 and 0 percent at waves 3 and 4, respectively), while nonshared environment emerged as the strongest predictor of victimization. Shared environment factors were not able to explain any of the variance in victimization for any of the observation points. Controlling for genetic influences on victimization, gang membership increased the risk of victimization for the latter waves of data collection, once the respondents had reached adulthood. This is an interesting result because it highlights gang membership as a possible nonshared environmental factor that may increase victimization. Identifying nonshared environment factors that influence phenotypic development is extremely important for furthering this area of research.

Research conducted by Boutwell and colleagues (2013) addressed an additional argument for a relationship between genetics and victimization, this time incorporating research stemming from Gottfredson and Hirschi's (1990) concept of self-control. As discussed above, research has supported the relationship between low self-control and victimization (Schreck *et al.*, 2002, 2006; Stewart *et al.*, 2004; Piquero *et al.*, 2005; Baron *et al.*, 2007; Kerley *et al.*, 2008, 2009; Fox *et al.*, 2009; Franklin, 2011; Franklin *et al.*, 2012). A parallel line of research has revealed that the development of self-control is significantly influenced by genetic factors (Wright and Beaver, 2005; Wright *et al.*, 2008). Results from Boutwell *et al.*'s (2013) study suggested that similar genetic factors accounted for variance in both victimization and low self-control. Interestingly, 63 percent of the covariance in self-control and victimization were due to shared genetic factors, while nonshared environments accounted for the remaining 37 percent. This research was the first to provide a glimpse into the relationship between genetics, self-control, and victimization.

Another important issue to consider—from a biosocial perspective—is the consistent finding that individuals who engage in delinquency or criminality are also more likely to be victimized (Esbensen and Huizinga, 1991; Lauritsen *et al.*, 1991; Wright and Decker, 1997). Given that genetic factors are related to individual differences in criminal behavior (Moffitt, 2005; Beaver, 2009), this raises the possibility that genetic factors related to variation in delinquency and criminality might also correspond to variation in victimization experiences. Evidence to support this line of reasoning was reported by Barnes and Beaver (2012) and by Vaske *et al.* (2012).

Indeed, both sets of authors found that the genetic influences driving variance in delinquent behavior overlapped, at least partially, with the genetic factors driving variance in victimization experiences. In other words, there is reason to suspect the genetic influences on criminality are the same genetic factors that influence victimization.

An additional body of behavioral genetics research has focused on the sensitive issue of intimate partner violence (IPV) offending and victimization. Perhaps the first study to explore the possible genetic influence on IPV was that of Hines and Saudino (2004). Using the Revised Conflict Tactic Scale, they examined both the use and receipt of psychological and physical aggression. From an offender standpoint, genetic factors explained 22 percent of the variance in the use of psychological aggression, and 16 percent of the variance in the use of physical aggression. From a victimization standpoint, genetic factors accounted for 25 percent of the variance in the receipt of psychological aggression, and 15 percent of the variance in the receipt of physical aggression. Nonshared environmental factors explained the remaining variance in all situations (Hines and Saudino, 2004). Research conducted by Barnes *et al.* (2012) explored IPV offending in three forms: hitting a partner, injuring a partner, and forcing sexual activity. The pattern of findings suggested that genetic factors accounted for 24 percent of the variance in hitting a partner, 54 percent of the variance in injuring a partner, and 51 percent of the variance in forcing sexual activity.

Despite the small number of empirical studies focusing on genetics and victimization, the results are clear. Similarly to offending patterns, genetic factors account for a significant portion of the variance of victimization. In some cases genetic factors explain up to 60 percent of the variance (Ball *et al.*, 2008; Boutwell *et al.*, 2013), although most studies found genetic factors to hover around 40 percent (Beaver *et al.*, 2009) with heritability estimates dropping to practically zero in adulthood (Barnes *et al.*, 2011). These results not only support the arguments made for the association between genetics and victimization, they also serve as additional support for the argument that behavioral genetic research methods are ripe for integration into the criminological lexicon and methodological toolkit. A brief summary of this literature might suggest that genetic factors account for variance in victimization experiences among adolescents and young adults. As focal respondents age, however, the impact of genetic influences on victimization may begin to wane as the effect of the nonshared environment becomes the most critical component. Although behavioral genetics research has begun to reveal an interesting pattern of relationships between victimization experiences and genetic factors, these designs are unable to identify which genes are implicated in the origins of victimization experiences. In the section that follows, we examine existing research concerning the molecular underpinnings of victimization experiences.

Molecular genetics

As behavior genetic research into any trait grows, a logical next step is to unpack the molecular genetic markers that underlie trait variance. In other words, if behavior genetic research has shown a trait to be heritable, then the next question asked by researchers is often "which genes affect the trait?" Enter the world of molecular genetics research. The rapidly growing field of molecular genetics research examines whether variation in certain genes is related to the development of various forms of antisocial outcomes (Beaver *et al.*, 2007). This area of inquiry is perhaps one of the fastest growing in all the sciences. Indeed, it is often claimed that we live in the "genomic age" or that we are experiencing the "post-genomic era." Both phrases highlight the importance of molecular genetics research to all of science, but especially to social science disciplines such as criminology because genetic influences have traditionally been ignored in these fields.

Before discussing molecular genetic research into victimization, it is first important to understand what a gene is and what a gene is not. To grossly oversimplify a complex biological product and process, a gene is simply a contiguous strand of DNA "letters" (i.e. the base pairs that make up the genetic alphabet: A, T, C, and G) that code for protein production. In other words, a gene is a set of instructions that can be read by the cell and can be executed within the cell toward the end goal of protein synthesis. While this is an extremely complex process that involves many loops and turns, the reality is that genes are a chemical blueprint that can be referenced by the body whenever necessary. One thing should be clear from this discussion: a gene is *not* a deterministic indicator of human complex traits such as victimization experiences. Genes do not "code for" victimization any more than they "code for" watching television. Genes do, however, code for proteins (as well as indirectly impacting the function of other genes) that may impact brain functioning in ways that make one more likely to watch television, or more likely to be victimized. The important thing to keep in mind, however, is that these genetic effects are small (sometimes infinitesimally small) and they operate in a probabilistic (not deterministic) fashion just like any other known correlate of victimization (e.g. self-control).

While empirical evidence supporting the connection between specific genes and antisocial outcomes including substance use and increased aggression is mounting (Connor *et al.*, 2002; Chen *et al.*, 2007), a small but growing line of research is seeking to examine the possible connection between genetic markers and victimization. Empirical studies assessing the relationship between molecular genetics and risk of victimization are few in number. Yet the available evidence provides compelling support for an association between certain genetic markers and probability of victimization. Beaver and colleagues (2007), for instance, examined data from the Add Health study to assess a possible relationship between different variants of the dopamine D2 receptor gene (*DRD2*) and victimization in adolescents. The findings suggested that individuals who possessed a particular variant of a gene believed to impact dopamine reception at the synapse—the A1 allele of *DRD2*—were at a significantly greater risk of adolescent victimization compared with individuals who possessed the A2 allele of the *DRD2* gene. Additional statistical analysis found the effect of the *DRD2* gene on victimization was specific to white males with a low number of delinquent peers – evidence of a gene × environment interaction (G × E). At first blush, this may seem counterintuitive as genetic effects have previously been shown to emerge when coupled with high-risk environments in a G × E fashion (Caspi *et al.*, 2002; Foley *et al.*, 2004; Rutter, 2006). Upon closer inspection, though, the finding might actually reveal a tendency for high-risk environments to blunt the influence of genetic effects on some traits (Harris, 1998; Beaver *et al.*, 2007).

The association between *DRD2* and victimization was further supported by work conducted by Vaske (2009), whose research revealed that the rate of violent victimization was higher for adult males who carried one or more copies of the A1 risk allele of *DRD2* when compared with males who did not carry the risk allele. Furthermore, Vaske *et al.* (2011) focused on variants of the *DRD2* gene and violent victimization within an offender population, again finding offenders who possessed the A1 risk allele of the *DRD2* gene were more likely to have been violently victimized than offenders who did not possess the risk allele. Both studies, in addition to Beaver *et al.*'s (2007) findings, suggest there may be a connection between carrying one or more of the A1 risk alleles of the *DRD2* gene and an increased risk of victimization. Also alluded to above, available research appears to suggest that G × E may partially underlie the true nature of the relationship between genetic markers and victimization experiences.

Conclusion

The behavior of criminals continues to capture the imagination of scholars and the general public alike. What should not be missed, though, is that the commission of a criminal act often involves the reality of a victim who bears the brunt of the behavior. It would seem, then, that fully unpacking the contributing factors to victimization would be firmly within the purview of criminological scholarship. Moreover, in light of the consistent commonality between risk factors associated with crime and risk factors associated with victimization, victimization research should pave the way for a more complete criminological discipline. Finally, biosocial scholarship is emerging as a force to be reckoned with in the field of criminology (Cullen, 2011), so it would also seem logical that the incorporation of new techniques for scholars interested in victimization will be essential. With this in mind, we can now take a step back and survey the current understanding of victimization research from a biosocial perspective.

From the perspective of behavioral genetics, virtually all human traits are—to varying degrees—heritable (Turkheimer, 2000). What this means is the reason such diversity exists for human proclivities—ranging from personality and temperament to addiction susceptibility—is the result of genetic variation that exists in the population. Genes impact every aspect of life, even if this means acting as a filter to interpret experiences as they flow in from the natural world. For victimologists, this means that heritable variation is relevant for understanding not only why some individuals are victimized in the first place, but also why some victims seem resilient to the adverse impact of a victimization experience (and others, more vulnerable).

Understanding which genes—and this could number into the hundreds—play a role in contributing to the vulnerability and susceptibility of victimization will be the task of researchers employing molecular genetics techniques (Plomin, 1990; Beaver *et al.*, 2007; Vaske *et al.*, 2011). To date, genes implicated in the functioning and regulation of dopamine have received the bulk of the attention, but emergent research will almost undoubtedly reveal other genetic factors to also be of import. As more advanced sequencing techniques become available to researchers (e.g. genome sequencing and analysis of copy number variations, to mention only two), the ever-unfolding complexity of the human genome and its relationship with human behavior will continue to penetrate social science disciplines such as criminology. Perhaps eventually findings like those discussed here will reach a critical mass where criminologists are no longer able to ignore genetic influences, but instead begin to embrace the complexity of human behavior as both a biological and environmental phenomenon. One could argue, however, that the overarching impact of biosocial victimology will be to disabuse researchers from the romantic notions surrounding not only victimization, but also crime in general. The causes of victimization likely dovetail very closely with the causes of crime, of which genetic factors play a prominent role.

References

Anderson, E. (1999) *Code of the Street: Decency, Violence, and the Moral Life of the Inner City*, New York: W.W. Norton.

Ball, H., Arseneault, L., Taylor, A., Maughan, B., Caspi, A., and Moffitt, T. (2008) 'Genetic and environmental influences on victims, bullies and bully-victims in childhood', *Journal of Child Psychology and Psychiatry*, 49: 104–112.

Barnes, J.C. and Beaver, K.M. (2012) 'Extending research on the victim–offender overlap: evidence from a genetically informative analysis', *Journal of Interpersonal Violence*, 27: 3299–3321.

Barnes, J.C., Boutwell, B.B., and Fox, K. (2011) 'The effect of gang membership on victimization: a behavioral genetic explanation', *Youth Violence and Juvenile Justice*, 10: 227–244.

Barnes, J.C., TenEyck, M., Boutwell, B.B., and Beaver, K.M. (2012) 'Indicators of domestic/intimate partner violence are structured by genetic and nonshared environmental factors', *Journal of Psychiatric Research*, 47: 371–376.

Baron, S., Forde, D., and Kay, F. (2007) 'Self-control, risky lifestyles, and situation: the role of opportunity and context in the general theory', *Journal of Criminal Justice*, 35: 119–136.

Beaver, K.M. (2009) *Biosocial Criminology: A Primer*, Dubuque, IA: Kendall/Hunt.

Beaver, K., Wright, J., DeLisi, M., Daigle, L., Swatt, M., and Gibson, C. (2007) 'Evidence of a gene × environment interaction in the creation of victimization: results from a longitudinal sample of adolescents', *International Journal of Offender Therapy and Comparative Criminology*, 51: 620–645.

Beaver, K.M., Boutwell, B.B., Barnes, J.C., and Cooper, J. (2009) 'The biosocial underpinning to adolescent victimization: results from a longitudinal sample of twins', *Youth Violence and Juvenile Justice*, 7: 223–238.

Boutwell, B., Franklin, C., Barnes, J., Tamplin, A., and Petkovsek, M. (2013) 'Unraveling the covariation of low self-control and victimization: a behavior genetic approach', *Journal of Adolescence*, 36: 657–666.

Carey, G. (2003) *Human Genetics for the Social Sciences*, Thousand Oaks, CA: Sage.

Caspi, A., McClay, J., Moffitt, T., Mill, J., Martin, J., Craig, I.W., *et al.* (2002) 'Role of the genotype in the cycle of violence in maltreated children', *Science*, 297: 851–854.

Chen, T., Blum, K., Mathews, D., Fisher, L., Schnautz, N., Braverman, E., *et al.* (2007) 'Preliminary association of both the dopamine D2 receptor (*DRD2*)[Taq1 A1 allele] and the dopamine transporter (DAT1) [480 bp allele] genes with pathological aggressive behavior, a clinical subtype of reward deficiency syndrome (RDS) in adolescents', *Gene Therapy and Molecular Biology*, 11: 93–112.

Cohen, A. (1955) *Delinquent Boys: The Culture of the Gang*, Glencoe, IL: The Free Press.

Cohen, L. and Felson, M. (1979) 'Social change and crime rate trends: a routine activity approach', *American Sociological Review*, 44: 588–608.

Connor, J., Young, R., Lawford, B., Ritchie, T., and Noble, E. (2002) 'D2 dopamine receptor (*DRD2*) polymorphism is associated with severity of alcohol dependence', *European Psychiatry*, 17: 17–23.

Coster, S., Estes, B., and Mueller, C. (1999) 'Routine activities and sexual harassment in the workplace', *Work and Occupations*, 26: 21–49.

Cullen, F. (2011) 'Beyond adolescence-limited criminology: choosing our future—The American Society of Criminology 2010 Sutherland Address', *Criminology*, 49: 287–330.

Esbensen, F.A. and Huizinga, D. (1991) 'Juvenile victimization and delinquency', *Youth and Society*, 23: 202–228.

Fisher, B., Sloan, J., Cullen, F., and Lu, C. (1998) 'Crime in the ivory tower: the level and sources of student victimization', *Criminology*, 36: 671–710.

Foley, D., Eaves, L., Wormley, B., Silberg, J., Maes, H., Kuhn, J., *et al.* (2004) 'Childhood adversity, monoamine oxidase A genotype, and risk for conduct disorder', *Archives of General Psychiatry*, 61: 738–744.

Fox, K., Gover, A., and Kaukinen, C. (2009) 'The effects of low self-control and childhood maltreatment on stalking victimization among men and women', *American Journal of Criminal Justice*, 34: 181–197.

Franklin, C. (2011) 'An investigation of the relationship between self-control and alcohol-induced sexual assault victimization', *Criminal Justice and Behavior*, 38: 263–285.

Franklin, C., Franklin, T., Nobles, M., and Kercher, G. (2012) 'Assessing the effect of routine activity theory and self-control on property, personal, and sexual assault victimization', *Criminal Justice and Behavior*, 39: 1296–1315.

Gottfredson, M. and Hirschi, T. (1990) *A General Theory of Crime*, Palo Alto, CA: Stanford University Press.

Harris, J. (1998) *The Nurture Assumption: Why Children Turn Out the Way They Do*, New York: Free Press.

Hines, D. and Saudino, K. (2004) 'Genetic and environmental influences on intimate partner aggression: a preliminary study', *Violence and Victims*, 19: 701–718.

Kennedy, L. and Forde, D. (1990) 'Routine activity and crime: an analysis of victimization in Canada', *Criminology*, 28: 137–151.

Kerley, K., Xu, X., and Sirisunyaluck, B. (2008) 'Self-control, intimate partner abuse, and intimate partner victimization: testing the general theory of crime in Thailand', *Deviant Behavior*, 29: 503–532.

Kerley, K., Hochstetler, A., and Copes, H. (2009) 'Self-control, prison victimization, and prison infractions', *Criminal Justice Review*, 34: 553–568.

Lauritsen, J., Sampson, R., and Laub, J. (1991) 'The link between offending and victimization among adolescents', *Criminology*, 29: 265–292.

Mason, D. and Frick, P. (1994) 'The heritability of antisocial behavior: a meta-analysis of twin and adoption studies', *Journal of Psychopathology and Behavioral Assessment*, 16: 301–323.

Miethe, T., Stafford, M., and Long, J. (1987) 'Social differentiation in criminal victimization: a test of routine activities and lifestyle theories', *American Sociological Review*, 52: 184–194.

Miles, D. and Carey, G. (1997) 'Genetic and environmental architecture of human aggression', *Journal of Personality and Social Psychology*, 72: 207–217.

Moffitt, T. (1993) 'Adolescence-limited and life-course-persistent antisocial behavior: a developmental taxonomy', *Psychological Review*, 100: 674–701.

Moffitt, T. (2005) 'The new look of behavioral genetics in developmental psychopathology: gene–environment interplay in antisocial behaviors', *Psychological Bulletin*, 131: 533–554.

Mustaine, E. and Tewksbury, R. (1998) 'Predicting risks of larceny theft victimization: a routine activity analysis using refined lifestyle measures', *Criminology*, 36: 829–858.

Mustaine, E. and Tewksbury, R. (2002) 'Sexual assault of college women: a feminist interpretation of a routine activities analysis', *Criminal Justice Review*, 27: 89–123.

Patterson, G., DeBaryshe, B., and Ramsey, E. (1990) 'A developmental perspective on antisocial behavior', *American Psychologist*, 44: 329–335.

Pinker, S. (2002) *The Blank Slate: The Modern Denial of Human Nature*, New York: Penguin.

Piquero, A., MacDonald, J., Dobrin, A., Daigle, L., and Cullen, F. (2005) 'Self-control, violent offending, and homicide victimization: assessing the general theory of crime', *Journal of Quantitative Criminology*, 21: 55–71.

Plomin, R. (1990) *Nature and Nurture*, Pacific Grove, CA: Brookes/Cole.

Plomin, R., DeFries, J., Knopik, V., and Neiderhiser, J. (2013) *Behavioral Genetics* (6th edn), New York: Worth.

Rhee, S. and Waldman, I. (2002) 'Genetic and environmental influences on antisocial behavior: a meta-analysis of twin and adoption studies', *Psychological Bulletin*, 128: 490–529.

Rutter, M. (2006) *Genes and Behavior: Nature–Nurture Interplay Explained*, Malden, MA: Blackwell.

Schreck, C. (1999) 'Criminal victimization and low self-control: an extension and test of a general theory of crime', *Justice Quarterly*, 16: 633–654.

Schreck, C., Wright, R., and Miller, J. (2002) 'A study of individual and situational antecedents of violent victimization', *Justice Quarterly*, 19: 159–180.

Schreck, C., Stewart, E., and Fisher, B. (2006) 'Self-control, victimization, and the influence on risky lifestyles: a longitudinal analysis using panel data', *Journal of Quantitative Criminology*, 22: 319–340.

Schwartz, M., DeKeseredy, W., Tait, D., and Alvi, S. (2001) 'Male peer support and a feminist routine activities theory: understanding sexual assault on the college campus', *Justice Quarterly*, 18: 17–21.

Stewart, E., Elifson, K., and Sterk, C. (2004) 'Integrating the general theory of crime into an explanation of violent victimization of female offenders', *Justice Quarterly*, 21: 159–181.

Tewksbury, R. and Mustaine, E. (2003) 'College students' lifestyles and self-protective behaviors: further considerations of the guardianship concept in routine activity theory', *Criminal Justice and Behavior*, 30: 302–327.

Turkheimer, E. (2000) 'Three laws of behavior genetics and what they mean', *Current Directions in Psychological Science*, 9: 160–164.

Vaske, J. (2009) 'The role of genes and abuse in the etiology of offending', PhD dissertation, University of Cincinnati.

Vaske, J., Wright, J., and Beaver, K. (2011) 'A dopamine gene (*DRD2*) distinguishes between offenders who have and have not been violently victimized', *International Journal of Offender Therapy and Comparative Criminology*, 55: 251–267.

Vaske, J., Boisvert, D., and Wright, J.P. (2012) 'Genetic and environmental contributions to the relationship between violent victimization and criminal behavior', *Journal of Interpersonal Violence*, 27: 3213–3235.

Wooldredge, J., Cullen, F., and Latessa, E. (1992) 'Victimization in the workplace: a test of routine activities theory', *Justice Quarterly*, 9: 325–335.

Wright, J. and Beaver, K. (2005) 'Do parents matter in creating self-control in their children? A genetically informed test of Gottfredson and Hirschi's theory of low self-control', *Criminology*, 43: 1169–1202.

Wright, J., Beaver, K., DeLisi, M., and Vaughn, M. (2008) 'Evidence of negligible parenting influences on self-control, delinquent peers, and delinquency in a sample of twins', *Justice Quarterly*, 25: 544–569.

Wright, R. and Decker, S. (1997) *Armed Robbers in Action: Stickups and Street Culture*, Boston, MA: Northeastern University Press.

11

Behavioral genetics and antisocial spectrum disorders

Tracy D. Gunter

Introduction

A spectrum approach to understanding mental disorder groups together those disorders with similar core features, and uses a dimensional rather than discrete categorical approach in the service of the discovery of underlying mechanisms (Krueger *et al.*, 2005; Arcos-Burgos *et al.*, 2012). Antisocial spectrum disorders represent a subset of externalizing disorders, and include constructs such as antisocial, dissocial, and severe personality disorders, as well as psychopathy. Central themes in antisocial spectrum disorders include rule-breaking, aggressive, and defiant behaviors. A number of traits have been described as risk factors for antisocial spectrum disorders: increased novelty seeking, risk-taking, boldness, nonconformity, aggression, and impulsivity; and reduced harm avoidance, guilt, empathy, conscientiousness, agreeableness, relationship stability, and sensitivity to punishment. These features may emerge in childhood, adolescence, or adulthood, and may resolve, occur episodically, or become chronic and progressive (Cadoret *et al.*, 1987, 1990, 1995; Langbehn and Cadoret, 2001; Diamantopoulou *et al.*, 2010). It is likely that biological and environmental factors do not contribute equally to all of the characteristics at all points in development (Dick *et al.*, 2006; Burt and Klump, 2012; Viding *et al.*, 2012; Hicks *et al.*, 2013). Because there remains scientific uncertainty as to whether any, or all, of these conceptualizations are biologically meaningful (Kendler and Myers, 2013), studies using categorical and dimensional definitions of antisocial spectrum disorders and traits, at various stages of development, are included in this review, as are studies examining specific behaviors and behavioral themes.

Regarding phenotypes of interest, law-breaking behavior garners much attention because of the high personal and societal cost associated with criminal behavior, adjudication, and sentencing. The rapidly expanding field of neurocriminology seeks to better understand the relationship of biological underpinnings of antisocial spectrum disorders to criminal offending, with the hope of developing interventions to decrease both the personal and societal costs related to criminal behavior (Raine, 1993, 2008, 2013; Nordstrom *et al.*, 2011). Samples drawn from forensic populations have indeed shown high frequencies of antisocial spectrum behavioral disorders (e.g. Gunter *et al.*, 2009; Black *et al.*, 2010) and provided evidence to suggest that it is the presence of the antisocial spectrum disorder that separates offenders from nonoffenders among individuals with other mental illnesses. Other study designs that have been used to investigate the biological underpinnings of antisocial spectrum disorders include samples enriched for externalizing disorders (e.g. people receiving treatment for substance-use disorders or identified as having a diagnosis of attention deficit hyperactivity disorder (ADHD), and those involved in high-risk occupations that involve fear and distress tolerance). Large population studies are frequently used to study the normal distribution of antisocial spectrum features or traits in a population.

Once the phenotype of interest and sampling strategies are selected, consideration then moves to the tools used to better understand the underlying biology of the trait or disorder of interest. Each molecular genetics strategy used to investigate the biological contributors to complex behavioral disorders has strengths and limitations. DNA sequencing is the process of determining the precise order of nucleotides (adenine, guanine, cytosine, and thymine) within a DNA molecule. Genotyping is sequencing that focuses on portions of DNA known to code for proteins. Genotyping may be performed on many different types of tissue, but not all tissues yield DNA of equal quality. Ideally, one would like to isolate DNA-containing cells from different areas of the central nervous system in order to study biobehavioral illnesses. This approach is, however, impractical in human studies. Strategies using peripheral blood, which is more readily available than brain tissue, are increasingly common in the peer-reviewed literature. These studies may provide information about biomarkers for behaviors, but the utility of these biomaterials in understanding the underlying biology of the disorders is less certain, owing to the presence of the blood–brain barrier and uncertain correlation between cell states on each side of this biological membrane. Saliva samples are the most convenient and economical sample types to obtain, but researchers have mixed views of whether the quality of the genomic DNA is sufficiently similar to that obtained from blood for researchers to truly realize an advantage (Philibert *et al.*, 2008c; Hansen *et al.*, 2011). Buccal cells tend to suffer from a number of disadvantages such as contamination, degradation, and low yield of relatively poor quality DNA (Beckett *et al.*, 2008; Hansen *et al.*, 2011).

Sequencing, however, is only part of the story. In order to proceed from gene to protein, the gene must be available for transcription, and transcribed. Epigenetic studies then examine whether the gene is available for transcription, and gene expression studies quantify how much mRNA has actually been transcribed. Technologies used to understand gene function are newer, more expensive, generate enormous amounts of data, and are more commonly subject to concerns about reliability. Since gene expression studies rely on sophisticated laboratory procedures such as cell transfection, there is also an element of uncertainty about the effects of intrinsic and extrinsic factors that might impact the cells used and testing procedures. Gene expression profiling uses a microarray to assay the RNA products of many genes simultaneously, each of which may be under complex regulatory schemes. In epigenetic studies, researchers seek to understand the mechanisms, apart from the underlying DNA sequence, that regulate gene expression. The two most common epigenetic mechanisms studied are methylation and histone modification. Studies vary in the technical platforms employed to generate the data and the analytic methods utilized to test the data so that reproducibility in this area is difficult.

There are then two basic ways to approach the interrogation of complex behavioral phenotypes in humans: genome-wide and candidate gene approaches. In genome-wide approaches, researchers generate large amounts of data from many sites in the genome, without regard to biological plausibility, and these studies are therefore prone to false positive associations (Pearson and Manolio, 2008). In candidate gene approaches, researchers select specific sites of previous signal detection, or high biological plausibility, for further study. Both false positives and negatives may be concerns unless studies are of sufficient size to have the power to detect relatively small effect sizes, and comprised of well-characterized individuals, so that an array of genetic and environmental features can be examined within relatively homogeneous groups.

Ways of describing and quantifying environmental factors (Burt *et al.*, 2008; Dunn *et al.*, 2011) are as complex as outlining the phenotype, selecting the sampling strategy, and applying an appropriate approach to the biology. Many studies rely on the retrospective recall of adult subjects, without external verification, to identify and quantify environmental risk and resilience factors during the developmental period. An ideal study design would collect biological and environmental information during the developmental period from both study subjects and family members, and then follow all of the study subjects over time. Long-term follow-up studies, however, may have high attrition rates and incomplete follow-up data (e.g. Baker *et al.*, 2013).

When approaching the analytic strategy, some studies look primarily for the effects of genes or environments on the development of, or resilience to, antisocial spectrum disorders, while others examine gene–environment interactions and correlations in the genesis and maintenance of antisocial spectrum disorders (Hicks *et al.*, 2009; Beaver *et al.*, 2010; McCrory *et al.*, 2012).

Variations in sampling strategy, sample size, phenotype definition, biotechnology used, environmental characterization, and analytic methodology make comparison across studies conducted by different research teams fraught with difficulty. Despite the challenges faced by investigators in the area of behavioral genetics, there are emerging clues concerning the possible genetic underpinnings involved in antisocial spectrum disorders. This chapter provides a summary of some of the molecular genetics work interrogating the biology of antisocial spectrum disorders, beginning with sequencing studies (genome-wide and candidate gene) and concluding with epigenetic and gene expression studies attempting to understand gene regulation and function.

Genome-wide linkage and association studies

Genome-wide linkage studies have been conducted using a variety of phenotypes related to the antisocial spectrum of disorders. Linkage studies are typically undertaken using biologically informative families to attempt to locate candidate areas for more detailed mapping. These studies likely have less power than candidate gene approaches to detect genes with relatively small effect sizes, but have been critical for the identification of regions for further fine-scale mapping (Risch and Merikangas, 1996; Chiu *et al.*, 2005).

While genome-wide linkage studies focus on alleles, genome-wide association studies (GWAS) focus on genes, individually and in functional groups. Issues in study design and interpretation of GWAS have been discussed at length in the literature (Pearson and Manolio, 2008). These genome-wide studies have provided clues about possible genomic locations that may yield findings relevant for antisocial spectrum disorders, but must be interpreted with caution, as very few findings reach genome-wide significance and even fewer findings have been replicated. The available studies are described in the text below. Table 11.1 provides an overview of studies published within 2010–13, including sample size and phenotype. The table is organized by year of publication and last name of the first author.

In examining antisocial spectrum disorders in the context of substance-use disorders, areas on chromosomes 19 (D19S714) and 2 (D2S1331) were of potential interest (Dick *et al.*, 2004; Ehlers *et al.*, 2008). There was significant evidence for linkage on chromosome 9 (9q34, near markers D9S1826 and D9S1838), and suggestive evidence of linkage on chromosomes 3 (3q24–3q25,

Table 11.1 Selected recent genome-wide scans

First author	*Year*	*Sample*	*Phenotype*
Tielbeek	2012	4816 adults from 2227 families	Antisocial behavior
Gizer	2012	1647 adults	MMPI-2 ASP
Dick	2011	3963 adults	Conduct disorder
Terracciano	2011	7860 from six samples	NEO-PI-R
Viding	2010	8374 twins (children)	Teacher-rated scales
Terracciano	2010	3972 individuals	NEO-PI-R

near D3S1279 and D3S1614) and 17 (17q12, near D17S798) (Stallings *et al.*, 2005; Ehlers *et al.*, 2008).

Antisocial spectrum disorders have also been examined in co-occurrence with ADHD. Though not significant, regions of chromosomes 3 (3p25-p24 spanning SLC6A1 or the gamma-Aminobtyric acid (GABA) transporter) and 9 (9p24 including SLC1A1 or the glutamate transporter) were noted to be areas of interest (Sonuga-Barke *et al.*, 2008), as was Synaptosomal Protein 25kDa (SNAP25) on chromosome 20. SNAP25 was also noted to be a site of interest in the study of negative emotion (Terracciano *et al.*, 2010).

Turning to personality features that might underlie, or exist on a continuum with, antisocial spectrum disorders, four single nucleotide polymorphisms (SNPs)—rs16891867, rs1861046, rs7950811, and rs11838918—reached genome-wide significance in the study of conduct disorder, but there was not replication of prior associated SNPs or genes (Dick *et al.*, 2011). Viding and colleagues studied antisocial behavior and callous unemotional traits and found no SNPs of genome-wide significance, but generated a list of 30 SNPs that they believed merited further study in psychopathy (Viding *et al.*, 2010). Looking at psychoticism, no area rose to genome-wide significance, but regions on chromosomes 1 (1p35-p34), 5 (5p13), and 13 (13q14-q21) were deemed worthy of further attention in future studies (Gillespie *et al.*, 2008). Terracciano examined the five Eysenck traits and reported a nominal association of rs2835731 (within DYRK1A) with conscientiousness, BDNF, and two cadherin genes with extroversion (Terracciano *et al.*, 2010); and CTNNA2 (rs76005653) with excitement seeking (Terracciano *et al.*, 2011).

Regarding explicit antisocial attitudes and behaviors, linkage peaks on chromosome 13 at 11 cM and chromosome 17 at 57 cM in 1647 adults, scholars noted that these sites have been associated with externalizing spectrum psychopathology in other studies (Gizer *et al.*, 2012). When using adult antisocial behavior as the phenotype of interest, no associations reached genome-wide significance, but the strongest associations were located on chromosomes 5, 14, 15, and 21. None of the leading 50 candidate SNPs or 20 candidate genes reached genome-wide significance. However, the most associated SNPs (rs12106331, rs2835702, and rs2835771) were associated with the same gene, DYRK1A, which may be involved in brain development and has been associated with learning problems in Down's syndrome (Tielbeek *et al.*, 2012).

Candidate gene studies

Candidate gene studies have been undertaken to attempt to test the associations of variability in specific pre-selected genes with complex behavioral outcomes. These candidate sites might act alone, or in concert with other genetic and environmental factors. Researchers typically select candidate sites based on findings from genome-wide studies, or select those of high perceived biological plausibility. To date, no single locus has been associated with a moderate or large effect size, and replication has been sparse (Vassos *et al.*, 2013). Candidate gene studies in this area have traditionally been small in size and considered a limited number of loci, but more recent studies examine hundreds of loci in dozens of genes in functionally related systems and attempt to consider candidate genes as parts of molecular networks or systems (Bentley *et al.*, 2013). In addition to using a molecular network or systems model to understand candidate genes (Grigorenko *et al.*, 2010), large studies using multiple levels of inquiry and consistent phenotype, with careful control for environmental factors, will be needed to make progress in the area (Vassos *et al.*, 2013).

Of particular interest to researchers examining antisocial spectrum disorders have been genes related to serotonergic and dopaminergic activity (Grigorenko *et al.*, 2010; Pavlov *et al.*, 2012; Yildrim and Derksen, 2013). Examples of genes related to metabolism include catechol-O-methyl transferase (COMT), monoamine oxidase (MAOA), and dopamine beta hydroxylase (DBH);

Table 11.2 Selected recent candidate gene: metabolic pathways

First author	*Year*	*Gene*	*Sample*	*Phenotype*
Soeiro-De-Souza	2013	COMT	82 adults	BIS non-planning
Salatino-Oliveira	2012	COMT	516 children	Disruptive behavior
Arias	2011	COMT	310 male cases*	ASPD
DeYoung	2010	COMT	174 adolescents*	CD
Langley	2010	COMT	5726 children	Antisocial in ADHD
Paloyelis	2010	COMT	36 adolescent male cases	Impulsivity
Gallardo-Pujol	2013	MAOA	57 young adult males	Aggression
Haberstick	2014	MAOA	4316 men	Antisocial behavior
Hill	2013	MAOA	209 infants	Negative emotion
Sadeh	2013	MAOA	237 men*	Psychopathy
Cicchetti	2012	MAOA	627 children	Antisocial behavior
Fergusson	2012	MAOA	399 male subjects	Antisocial behavior
Yang	2012	MAOA	588 male cases	ASPD in SUD
Ziermans	2012	MAOA	335 subjects	Problem behavior
Fergusson	2011	MAOA	398 males	Antisocial behavior
Lee	2011	MAOA	672 males	Antisocial behavior
Reti	2011	MAOA	435 adults	ASPD traits
Tikkanen	2011	MAOA	167 men*	Impulsive violence
Beach	2010	MAOA	538 adults	Antisocial behavior
Derringer	2010	MAOA	841 subjects	Antisocial behavior
Tikkanen	2010	MAOA	174 men	Impulsive recidivism
Arias	2011	TPH1	310 male cases*	ASPD
Koh	2011	TPH1	97 inpatients	Criminal behavior

Note: * Denotes a forensic sample.

Table 11.3 Selected recent candidate gene: transporters

First author	*Year*	*Sample*	*Phenotype*
Serotonin transporter (HTTLPR)			
Aslund	2013	1467 adolescents	Delinquency
Sadeh	2013	237 men*	Psychopathy
Wang	2012	273 male cases*	Antisocial alcoholism
Douglas	2011	1381 subjects	ASPD
Herman	2010	862 adults	Sociopathy
Garcia	2010	147 males*	ASPD
Li	2010	2488 adolescents	Antisocial behavior
Nederhof	2010	1032 adolescents	Low effortful control
Sadeh	2010	296 youth in two studies	Psychopathic traits
Dopamine transporter (DAT1)			
Paloyelis	2010	36 male adolescent cases	Impulsivity

Note: * Denotes a forensic sample.

those related to receptor morphology include dopamine receptor D2 (DRD2), dopamine receptor D4 (DRD4), serotonin receptor 1B (5HTR1B), serotonin receptor 2A (5HTR2A); and those related to transporter activity include the serotonin transporter (in particular a polymorphism in the promoter region of the gene referred to as the serotonin-transporter linked polymorphic regions, 5HTTLPR) and dopamine transporter (DAT). Tables 11.2 and 11.3 provide brief overviews of selected recent studies related to metabolic pathways and transporters, respectively.

Genes related to norepinephrine, glutamate, opioid, and cholinergic transmission have also been investigated, as have loci thought to mediate stress responses, such as FKBP5. Since the frequencies of antisocial spectrum disorders and psychopathy vary by gender, gender-specific sites such as androgen receptors (AR) have also garnered some attention by researchers. Novel sites such as SNAP25, mentioned in the genome-wide studies described above, have been investigated, and a few examples of novel sites are included in Table 11.4.

Metabolic pathways

MAOA is located on the X chromosome (Xp11.4-p11.3) and encodes for monoamine oxidase A, an enzyme that degrades amine neurotransmitters, such as dopamine, norepinephrine, and serotonin. It has important functions in the metabolism of neuroactive and vasoactive amines in the central nervous system and peripheral tissues. Mutation in this gene results in monoamine oxidase deficiency, or Brunner syndrome, which is characterized in part by severe impulsive behavior, caused by truncation mutations resulting in monoamine oxidase deficiency. A variable nucleotide repeat (VNTR) is found immediately upstream of the transcription start site of MAOA and has been shown to affect both gene transcription and gene methylation. This VNTR system consists of a variable number of 44 base pair (bp) repeat motifs (either 2, 3, 3.5, 4, or 5 repeats) with the shorter alleles having been associated with lower levels of transcription

Table 11.4 Selected recent candidate gene studies: other

First author	*Year*	*Site*	*Sample*	*Phenotype*
Logue	2013	ANK3	554 adults	Externalizing
Lu	2012	ANKK1	541 men	ASPD
Aluja	2011	AR	153 male cases*	Impulsive-disinhibited
Zai	2012	AVR1B	177 child cases	Aggression
Nederhof	2010	BDNF	1032 adolescents	Low effortful control
Lantendresse	2011	CHRM2	378 individuals	Externalizing behavior
Li	2012	COL25A1	2721 subjects	ASPD in SUD
Zai	2012	DRD2	144 child cases	Aggression
Schilling	2013	DRD4	192 adults	Impulsivity
Terranova	2013	GABBR2	186 men	Criminal behavior
Dick	2013	GABRA2	2128 subjects	Sensation-seeking
Strac	2012	GABRA2	355 subjects	Aggression
Moul	2013	5HTR1B	223 boys	Callous-unemotional
Moul	2013	5HTR2A	223 boys	Callous-unemotional
Arias	2011	5HTR2A	310 male cases*	ASPD
Malik	2012	OXTR	160 child cases	Childhood aggression
Dadds	2013	ROBO2	210 children	Conduct problems
Basoglou	2011	SNAP-25	129 males	Novelty seeking

Note: * Denotes a forensic sample.

and brain activity generally, but not universally (Fowler *et al.*, 2007; Pai *et al.*, 2007; Alia-Klein *et al.*, 2008). Table 11.2 presents an overview of the sample type and phenotype definition used in candidate gene studies involving MAOA reported on from 2010–13.

Caspi and colleagues (2002) published a landmark study examining why some abused children developed conduct disorder and delinquent behaviors while others did not. In that study, a low-activity variant of MAOA interacted with adverse environment to form an association with delinquency. Since, investigators have attempted to replicate this finding in phenotypes involving conduct disorder, impulsivity, negative emotionality, disruptive behavior disorder, antisocial personality disorder, and violent criminal behavior, making this perhaps the most frequently studied site in antisocial spectrum disorders. While several studies have demonstrated that the low-activity variant of MAOA in combination with environmental factors was associated with antisocial spectrum disorders (Foley *et al.*, 2004; Huang *et al.*, 2004; Contini *et al.*, 2006; Kim-Cohen *et al.*, 2006; Nilsson *et al.*, 2006; Reif *et al.*, 2007; Ducci *et al.*, 2009; Kinnally *et al.*, 2009; Prom-Wormley *et al.*, 2009; Williams *et al.*, 2009; Beach *et al.*, 2010a; Enoch *et al.*, 2010; Wakschlag *et al.*, 2010; Barnett *et al.*, 2011; Fergusson *et al.*, 2011; Barnes and Jacobs, 2013; Gallardo-Pujol *et al.*, 2013; Hill *et al.*, 2013), other studies suggest that gene effects are noncontributory to the prediction of disruptive behavior in those with a history of maltreatment or other stress during the developmental period (Haberstick *et al.*, 2005; Young *et al.*, 2006; Cicchetti *et al.*, 2012; Fergusson *et al.*, 2012). And still other investigators have noted an effect of the MAOA polymorphism on antisocial trait scores independently of consideration of adversity (Sadeh *et al.*, 2013), or only among individuals who have not suffered abuse or other environmental adversity (Lee, 2011; Reti *et al.*, 2011). While lower-activity variants of MAOA could sensitize pertinent neural circuitry to early life stress (Buckholtz and Meyer-Lindenberg, 2008), the association between the low-activity variant of MAOA and antisocial spectrum behavior could vary by the level of trauma, such that a gene effect is significant at moderate levels of trauma, but noncontributory at extreme levels of environmental trauma (Weder *et al.*, 2009).

Not inconsistent with the above, MAOA polymorphisms may also interact with other biological, psychological, and environmental factors to increase risk. For instance, MAOA genotype may interact with testosterone in predicting antisocial spectrum behavior, though no direct association between the lower-activity variant and aggressive behavior has been seen (Sjoberg *et al.*, 2008). The pathway between polymorphisms at MAOA and antisocial spectrum behaviors might be mediated through intermediary neuropsychological deficits such as impaired working memory or receptive vocabulary (Beaver *et al.*, 2008a; Ziermans *et al.*, 2012). In the context of ADHD, the low-activity variant of MAOA was associated with psychopathy trait scores in a sample of adolescent subjects (Fowler *et al.*, 2009).

In an interesting extension of gene–environment (G × E) interactions operant in the developmental period, investigators examined current stress, as opposed to stress during the developmental period, and found that the shorter MAOA allele was associated with aggression only in response to perceived financial loss within the study paradigm (McDermott *et al.*, 2009).

However, not all studies have replicated the findings discussed above (Koller *et al.*, 2003; Prichard *et al.*, 2007; Sjoberg *et al.*, 2007; Yang *et al.*, 2012; Haberstick *et al.*, 2014). A few studies suggest that the high-activity variants, or longer alleles, may be problematic alone (Tikkanen *et al.*, 2009, 2011), under the influence of unfavorable environmental factors (Sjoberg *et al.*, 2007; Vanyukov *et al.*, 2007; Tikkanen *et al.*, 2010), or in combination with a delinquent peer group (Beaver and Holtfreter, 2009; Lee, 2011), while still other studies conclude that the higher-activity variants are relatively protective in the face of environmental adversity (Widom and Brzustowicz, 2006).

Other metabolic pathways under consideration have included COMT, DBH, and tryptophan hydroxylase 1 and 2 (TPH1 and TPH2). DBH, TPH1, and TPH2 are primarily involved in synthesis of neurotransmitters, while COMT is primarily involved in degradation. Regarding COMT, several studies implicated the valine/methionone (val/met) and met/met polymorphisms in aggression and impulsivity (Rujescu *et al.*, 2003; Volavka *et al.*, 2004; Wagner *et al.*, 2010; Arias *et al.*, 2011; Singh *et al.*, 2012; Soeiro-De-Souza *et al.*, 2013), but other studies suggest that subjects homozygous for the val allele are more aggressive (Thapar *et al.*, 2005; Caspi *et al.*, 2008; Kulikova *et al.*, 2008; Monuteaux *et al.*, 2009; DeYoung *et al.*, 2010; Langley *et al.*, 2010; Salatino-Oliveira *et al.*, 2012) and psychopathic symptoms (Fowler *et al.*, 2009). In a recent study using a sample of severely affected subjects, researchers reported that the val/met polymorphism did not contribute to antisocial spectrum disorders, but did suggest that variation in other COMT regions may be important in childhood-onset aggression and impulsive violence (Vevera *et al.*, 2009; Hirata *et al.*, 2013). Sorting out trait and state impulsivity has, however, been difficult (Paloyelis *et al.*, 2010). More precise definition of the type, frequency, and severity of aggressive behavior observed may be useful in better understanding the effects of polymorphisms at COMT (Volavka, 2008); it may be that haplotypes are better determinants of final COMT activity than single SNPs (Halleland *et al.*, 2008).

Turning to other metabolic pathways, DBH polymorphisms have associated with antisocial spectrum behaviors and attitudes alone and in the context of ADHD (Barkley *et al.*, 2006; Hess *et al.*, 2009). TPH1 (Manuck *et al.*, 1999; Rujescu *et al.*, 2002; Arias *et al.*, 2011; Koh *et al.*, 2011) and TPH2 (Oades *et al.*, 2008; Perez-Rodriguez *et al.*, 2010) appear promising for phenotypes related to antisocial spectrum attitudes and behaviors alone and in the context of other disorders such as ADHD, schizophrenia, and borderline personality disorder, but study results have been mixed (Hennig *et al.*, 2005; Ke *et al.*, 2006; Mann *et al.*, 2008).

Transporters

Located on chromosome 17 (17q11.2), the serotonin transporter (5HTT or SLC6A4) encodes a transporter protein that removes serotonin from synaptic spaces into presynaptic neurons. This site is a frequent target for antidepressant medications. A variable number tandem repeat length polymorphism (VNTR) in the promoter of this gene (5HTTLPR) has been shown to affect the rate of serotonin uptake and may play a role in behavioral illness, with the short alleles in this mostly diallelic (either short or long) polymorphism being associated with lower levels of gene transcription.

The low-activity short-repeat variant of 5HTTLPR is frequently considered the risk allele for externalizing and antisocial spectrum disorders, particularly in the presence of environmental adversity or substance use (Cadoret *et al.*, 2003; Liao *et al.*, 2004; Gerra *et al.*, 2004; Retz *et al.*, 2004; Feinn *et al.*, 2005; Sakai *et al.*, 2006; Steiger *et al.*, 2008; Wu *et al.*, 2008; Aluja *et al.*, 2009; Retz and Rosler, 2009; Vaske *et al.*, 2009; Garcia *et al.*, 2010; Li and Lee, 2010; Douglas *et al.*, 2011). However, several studies have failed to replicate an association between the S-allele of 5HTTLPR and antisocial spectrum disorders (Beitchman *et al.*, 2003; Sakai *et al.*, 2007; Monuteaux *et al.*, 2009; Arias *et al.*, 2011; Wang *et al.*, 2012b), and some suggest that it is actually the long allele that is the risk allele. At the same time, other studies report a correlation between the long allele and symptoms of behavioral illness in the externalizing spectrum (Retz *et al.*, 2002; Davidge *et al.*, 2004; Glenn, 2011).

These mixed results have led some researchers to conclude that the effect of the repeat length on mental health outcomes may vary by gender, phenotype, and/or distress sensitivity. For example, it has been reported that men with long alleles and women with short alleles have lower socialization scores (Herman *et al.*, 2010). Regarding the phenotype of psychopathy, individuals homozygous for the short allele have more impulsive and irresponsible features, and those with long alleles have more affective features (Sadeh *et al.*, 2010, 2013). Other researchers suggest that the short alleles may infer a greater sensitivity to adverse behavioral outcomes after environmental adversity (Retz *et al.*, 2008). Putting together these concepts, Aslund and colleagues (2013) reported that boys with long alleles and heterozygous girls had high delinquency scores regardless of socioeconomic status (SES), but that boys with one or more short alleles and girls homozygous for short alleles demonstrated high levels of delinquency only in the context of high SES (Aslund *et al.*, 2013). Table 11.3 provides details of some of the most recent studies related to 5HTTLPR.

Another transporter site of interest has been the dopamine transporter or DAT1 (SLCA63) on chromosome 5 (5p15). The 10-repeat polymorphism is usually considered the at-risk genotype for the development of antisocial spectrum behavior (Beaver *et al.*, 2008b; Burt and Mikolajewski, 2008; Paloyelis *et al.*, 2010), although some studies suggest that the 9/10 and 10/10 polymorphisms grouped together differentiated groups of individuals with pathological aggression from healthy controls (Chen *et al.*, 2005). In other studies, the 9/10 polymorphism has been associated with higher ADHD and externalizing spectrum scores compared with the 10/10 polymorphism (Barkley *et al.*, 2006), and the 9-repeat polymorphism was found to be the risk conferring genotype for antisocial behavior in the context of heroin dependence (Gerra *et al.*, 2005).

GABA and glutamate transporters have also garnered some interest among researchers in the area of substance-use disorders, but only one involving antisocial spectrum behaviors could be located for review. In this study, variation in the glutamate transporter EAAT2 gene was associated with antisocial personality, and possibly risk-taking behaviors, in individuals with alcohol dependence (Sander *et al.*, 2000).

Dopamine receptors

Polymorphisms affecting dopamine receptors such as D2 (DRD2) have been inconsistently associated with antisocial spectrum disorders, either alone or in association with other polymorphisms such as Taq1A (Chen *et al.*, 2005; Beaver *et al.*, 2007; Eisenberg *et al.*, 2007; Wang *et al.*, 2007; Arias *et al.*, 2011; Kasiakogia-Worlley *et al.*, 2011; Zai *et al.*, 2013). The Taq1A polymorphism of ANKK1 is adjacent to the DRD2 gene and has been implicated in antisocial spectrum disorders in individuals with alcohol dependence (Ponce *et al.*, 2003, 2008, 2009; Dick *et al.*, 2007). The relevance of this finding as specific for antisocial spectrum disorders, without co-occurring substance dependence, is somewhat questionable as both DRD2 and ANKK1 are strongly implicated in substance dependence generally (Johann *et al.*, 2005; Yang *et al.*, 2008; Doehring *et al.*, 2009; Huang *et al.*, 2009), but Lu and colleagues reported that DRD2/ANKK1 and ALDH2 interacted in the antisocial group without alcohol dependence (Lu *et al.*, 2012).

The dopamine receptor, type 4 (DRD4) gene is remarkable for the presence of a variable 48 bp (or 16 amino acid) repeat motif in the third exon. Long repeats (7 or more) of this motif have been hypothesized to differentially influence the affinity of the receptor for other proteins in the membrane in the DRD4 gene, are associated with altered receptor activity, and have been correlated with conduct disorder in ADHD patients in some (Benjamin *et al.*, 1996; Faraone *et al.*, 1999; Rowe *et al.*, 2001), but not all, studies (Barkley *et al.*, 2006). Other studies have suggested that long repeats may interact with polymorphisms at other sites such as DRD2 (Beaver *et al.*, 2007; Boutwell and Beaver, 2008) and DAT (Congdon *et al.*, 2008) in impulsivity or disruptive behavior disorders. And other lines of research point to long repeats in binge drinking, smoking, and novelty seeking (Laucht *et al.*, 2005; Ray *et al.*, 2009; Vaughn *et al.*, 2009). Showing how truly mixed the candidate gene findings in this area have been, short alleles were associated with increased scores on a measure of impulsivity in a sample of healthy individuals (Schilling *et al.*, 2013) and on measures of aggression in a sample of children diagnosed with ADHD (Oades *et al.*, 2008).

Serotonin receptors

Among the serotonin receptors, 5HTR1B and 5HTR2A are probably the most frequently referenced in regards to antisocial spectrum attitudes and disorders, though the results may be best described as mixed (Lappalainen *et al.*, 1998; Preuss *et al.*, 2001; Hasegawa *et al.*, 2002; Soyka *et al.*, 2004; Giegling *et al.*, 2006; Mik *et al.*, 2007; Burt and Mikolajewski, 2008; Arias *et al.*, 2011; Moul *et al.*, 2013). 5HTR3B has been associated with antisocial alcohol dependence in an offender sample (Ducci *et al.*, 2009). See Table 11.4 for overviews of the sample type used for selected recent studies.

Androgen receptors

Since the frequencies of antisocial spectrum behaviors and psychopathy are greater in men than in women, receptors associated with sex determination are attractive sites of investigation

(Westberg and Eriksson, 2008). The androgen receptor (AR) is one such receptor, and is highly polymorphic. Although shorter CAG repeats have been associated with violent behavior, impulsive-disinhibited personality style, dominance and aggression (Jonsson *et al.*, 2001; Rajender *et al.*, 2008; Aluja *et al.*, 2011), other work indicated that highest frequency antisocial behavior was seen among men with the medium length repeats (Prichard *et al.*, 2007). When the literature is taken as a whole, It is likely that the effect of androgen receptor variation in criminal behavior is minimal or nonexistent (Loehlin *et al.*, 2003; Cheng *et al.*, 2006).

Other sites of interest

Using the phenotype of antisocial drug dependence occurring in a clinical sample, researchers conducted analyses of 1500 SNPs across 50 candidate genes and found significant associations with opioid receptor mu 1 (OPRM1) and cholinergic receptor nicotinic acid 2 (CHRNA2), but not with 5HTTLPR, MAOA, or any of the other traditional sites of interest. Variation in the muscarinic acetylcholine receptor M2 (CHRM2) gene has been associated with externalizing behavior (Dick *et al.*, 2008; Latendresse *et al.*, 2011). Also using the phenotype of externalizing behavior, Reif and colleagues reported on an association between short alleles at nitric oxide synthase (NOS), noting that the short alleles were more common in violent offenders (Reif *et al.*, 2009).

Dopamine- and cAMP-regulated phosphoprotein of relative molecular mass 32 kDa (DARPP-32) has been associated with increasing scores on a measure of anger (Reuter *et al.*, 2009). Brain-derived neurotropic factor (BDNF) regulates the expression of DARPP-32 and was identified in GWAS as a possible site of interest in antisocial spectrum disorders. However, the results of candidate gene studies provide limited support for a role for BDNF val66met in aggression occurring in the context of ADHD (Oades *et al.*, 2008) and schizophrenia (Spalletta *et al.*, 2010). At least one met allele, in combination with at least one short allele at 5HTTLPR, was hypothesized to sensitize children to stressful childhood events when examining the phenotype of low effortful control (Nederhof *et al.*, 2010).

GABA receptors have been of some interest to researchers in the area of substance-use disorders (Lind *et al.*, 2008a,b; Philibert *et al.*, 2009b; Strac *et al.*, 2012) and have been implicated in sensation-seeking and antisocial spectrum disorders (Dick *et al.*, 2009, 2013; Sakai *et al.*, 2010; Terranova *et al.*, 2013).

SNAP25 (located on 20p), mentioned in the genome-wide studies described previously, seems to be involved in the fusion of plasma membranes and synaptic vesicles, and has been examined as a possible risk factor for behavioral disorders such as ASPD and ADHD using candidate gene approaches (Faraone *et al.*, 2005; Corradini *et al.*, 2009; Arias *et al.*, 2011; Basoglu *et al.*, 2011). It is likely to exert influence in concert with other biological and environmental factors (Corradini *et al.*, 2009).

Variation at ankyrin-3 (ANK3), an anchoring protein that has been implicated in biobehavioral disorders in animal models and GWAS, was found to be associated with a measure of adult antisocial attitudes (Logue *et al.*, 2013).

Sometimes called the social neuropeptides, oxytocin and vasopressin activity has been associated with social cognition, prosocial attachment, and trust (Meyer-Lindenberg *et al.*, 2011). Deficits in oxytocin transmission have been hypothesized to play a role in the development of childhood aggression, and two OXTR SNPs may be associated with severe aggression (Malik *et al.*, 2012). Turning to the vasopressin system, AVPR1B has been associated with childhood aggression in a case-control study involving persistent and pervasive aggression (Zai *et al.*, 2012).

Looking at the HPA axis and stress pathways, FKBP5 has been associated with an increased risk of violent behavior in a forensic sample (Bevilacqua *et al.*, 2012). Antisocial personality disorder was associated with an SNP within COL25A1, a transmembrane protein previously associated with amyloid in people with Alzheimer's disease (Li *et al.*, 2012).

Not previously associated with human neuropsychological dysfunction, both genome-wide and candidate studies have provided support for an association between ROBO2 and antisocial behavior (Viding *et al.*, 2010; Dadds *et al.*, 2013). Table 11.4 provides additional information about several of these studies.

Epigenetic studies

Epigenetic modification of gene expression via methylation and histone modification is likely important in behavioral illness (Jiang *et al.*, 2008; van Ijzendoorn *et al.*, 2010). These processes are heritable, but influenced by environmental factors such as smoking (Philibert *et al.*, 2009a), alcohol and drug consumption (Philibert *et al.*, 2012; Maze *et al.*, 2013), and stress and abuse (Madlung and Comai, 2004; Beach *et al.*, 2010b; Marquez *et al.*, 2013), and may be among the mechanisms through which the environment impacts the expression of the genome (Philibert *et al.*, 2008b; Rutten and Mill, 2009).

Although sequence of the VNTR in the promoter region of MAOA has been shown to inconsistently predict brain MAOA activity, methylation status of the MAOA promoter in white blood cells can reliably predict brain MAOA activity (Shumay *et al.*, 2012). Philibert and colleagues (2008a) have demonstrated that methylation at MAOA is associated with nicotine and alcohol dependence in women. Researchers then postulated that methylation at 5HTT linked a family history of abuse to the outcome of antisocial personality disorder in women (Beach *et al.*, 2011). And, in further studies, researchers demonstrated genotype at a novel site (P2) was associated with methylation at the transcription start site (P1) for MAOA, suggesting that the overall regulation of MAOA is far more complex than previously appreciated (Philibert *et al.*, 2011).

In unpublished data, methylation at the CpG islands at the transcription start site 200 bp upstream of COMT and first exon was positively associated with increasing scores on a measure of antisocial personality, and negatively associated with scores on a measure of attention deficit hyperactivity disorder, in a nested case control analysis from the Family and Community Health Study. In looking at related externalizing spectrum disorders, DAT methylation is altered in alcohol dependence and associated with craving (Hillemacher *et al.*, 2009), and methylation at OPRM1 is associated with methadone maintenance in opiate dependent persons (Nielsen *et al.*, 2009). Given that peripheral cells appear to have a demonstrable epigenetic signature reflecting central activity (Wang *et al.*, 2012a), it is foreseeable that epigenetic profiling will become an increasingly important area of inquiry in the near future.

Gene expression

Using lymphoblast cell lines derived from peripheral blood, Philibert and colleagues (2007) found differential expression in biologically significant pathways in individuals with nicotine, alcohol, and cannabis dependence. Antisocial personality was a significant co-occurring disorder among those with alcohol and cannabis dependence, but the relatively small sample size did not permit a ferreting out of ASPD as a separate disorder. Animal studies are attempting to sort out how environmental factors such as substance ingestion impact gene expression, and how changes in expression then impact the course of substance ingestion (Spanagel, 2009).

Conclusions and future directions

Behavioral genetics researchers are increasingly using the tools of molecular genetics to extend upon discoveries from twin, family, and adoption studies concerning the heritability of antisocial spectrum disorders. Many other loci have been implicated by linkage, association, and candidate gene studies, and likely make some contribution to antisocial spectrum behavior under some circumstances, and in some populations. The bigger picture is likely in understanding complex illnesses as resulting from disturbances in molecular networks, impacting neural networks, and ultimately behavior (Thomas *et al.*, 2009). Various laboratory tests (Derijk, 2009) and imaging technologies (Demaree *et al.*, 2009) have been employed to identify intermediate phenotypes that might be attributable to specific polymorphisms, and there is no doubt that future studies in the area of behavioral genetics will use multiple levels of inquiry to better understand the long and tortuous path from the genome to behavior (Wahlsten, 2013).

References

Alia-Klein, N., Goldstein, R.Z., Kriplani, A., Logan, J., Tomasi, D., Williams, B., Telang, F., Shumay, E., Biegon, A., Craig, I.W., Henn, F., Wang, G.J., Volkow, N.D., and Fowler, J.S. (2008) 'Brain monoamine oxidase A activity predicts trait aggression', *Journal of Neuroscience*, 28: 5099–5104.

Aluja, A., Garcia, L.F., Blanch, A., De Lorenzo, D., and Fibla, J. (2009) 'Impulsive-disinhibited personality and serotonin transporter gene polymorphisms: association study in an inmate's sample', *Journal of Psychiatric Research*, 43: 906–914.

Aluja, A., Garcia, L.F., Blanch, A., and Fibla, J. (2011) 'Association of androgen receptor gene, CAG and GGN repeat length polymorphism and impulsive-disinhibited personality traits in inmates: the role of short–long haplotype', *Psychiatric Genetics*, 21: 229–239.

Arcos-Burgos, M., Velez, J., Solomon, B., and Muenke, M. (2012) 'A common genetic network underlies substance use disorders and disruptive or externalizing disorders', *Human Genetics*, 131: 917–929.

Arias, J., Acosta, C., Valencia, J.G., Montoya, G.J., Viana, J., Nieto, O.C., Florez, A.F., Medellin, B.C., Montoya, W.R., Jaramillo, C., Achury, J.G., Fuentes, C.C., Berrio, G.B., and Ruiz-Linares, A. (2011) 'Exploring epistasis in candidate genes for antisocial personality disorder', *Psychiatric Genetics*, 21: 115–124.

Aslund, C., Comasco, E., Nordquist, N., Leppert, J., Oreland, L., and Nilsson, K.W. (2013) 'Self-reported family socioeconomic status, the 5-HTTLPR genotype, and delinquent behavior in a community-based adolescent population', *Aggressive Behavior*, 39: 52–63.

Baker, L., Tuvblad, C., Wang, P., Gomez, K., Bezdijian, S., Niv, S., and Raine, A. (2013) 'The Southern California Twin Register at the University of Southern California: III', *Twin Research and Human Genetics*, 16: 336–343.

Barkley, R.A., Smith, K.M., Fischer, M., and Navia, B. (2006) 'An examination of the behavioral and neuropsychological correlates of three ADHD candidate gene polymorphisms (DRD4 7+, DBH TaqI A2, and DAT1 40 bp VNTR) in hyperactive and normal children followed to adulthood', *American Journal of Medical Genetics B, Neuropsychiatric Genetics*, 141B: 487–498.

Barnes, J.C. and Jacobs, B. (2013) 'Genetic risk for violent behavior and environmental exposure to disadvantage and violent crime: the case for gene–environment interaction', *Journal of Interpersonal Violence*, 18: 92–120.

Barnett, J.H., Xu, K., Heron, J., Goldman, D., and Jones, P.B. (2011) 'Cognitive effects of genetic variation in monoamine neurotransmitter systems: a population-based study of COMT, MAOA, and 5HTTLPR', *American Journal of Medical Genetics B, Neuropsychiatric Genetics*, 156: 158–167.

Basoglu, C., Oner, O., Ates, A., Algul, A., Bez, Y., Cetin, M., Herken, H., Erdal, M.E., and Munir, K.M. (2011) 'Synaptosomal–associated protein 25 gene polymorphisms and antisocial personality disorder: association with temperament and psychopathy', *Canadian Journal of Psychiatry*, 56: 341–347.

Beach, S.R., Brody, G.H., Gunter, T.D., Packer, H., Wernett, P., and Philibert, R.A. (2010a) 'Child maltreatment moderates the association of MAOA with symptoms of depression and antisocial personality disorder', *Journal of Family Psychology*, 24: 12–20.

Beach, S.R., Brody, G.H., Todorov, A.A., Gunter, T.D., and Philibert, R.A. (2010b) 'Methylation at SLC6A4 is linked to family history of child abuse: an examination of the Iowa Adoptee sample', *American Journal of Medical Genetics B, Neuropsychiatric Genetics*, 153B: 710–713.

Beach, S.R., Brody, G.H., Todorov, A.A., Gunter, T.D., and Philibert, R.A. (2011) 'Methylation at 5HTT mediates the impact of child sex abuse on women's antisocial behavior: an examination of the Iowa adoptee sample', *Psychosomatic Medicine*, 73: 83–87.

Beaver, K.M. and Holtfreter, K. (2009) 'Biosocial influences on fraudulent behaviors', *Journal of Genetic Psychology*, 170: 101–114.

Beaver, K.M., Wright, J.P., Delisi, M., Walsh, A., Vaughn, M.G., Boisvert, D., and Vaske, J. (2007) 'A gene × gene interaction between DRD2 and DRD4 is associated with conduct disorder and antisocial behavior in males', *Behavior and Brain Functions*, 3: 30.

Beaver, K.M., DeLisi, M., Vaughn, M.G., and Wright, J.P. (2008a) 'The intersection of genes and neuropsychological deficits in the prediction of adolescent delinquency and low self-control', *International Journal of Offender Therapy and Comparative Criminology*, 54: 22–42.

Beaver, K.M., Wright, J.P., and Walsh, A. (2008b) 'A gene-based evolutionary explanation for the association between criminal involvement and number of sex partners', *Biodemography and Social Biology*, 54: 47–55.

Beaver, K.M., Sak, A., Vaske, J., and Nilsson, J. (2010) 'Genetic risk, parent–child relations, and antisocial phenotypes in a sample of African-American males', *Psychiatry Research*, 175: 160–164.

Beckett, S.M., Laughton, S.J., Pozza, L.D., McCowage, G.B., Marshall, G., Cohn, R.J., Milne, E., and Ashton, L.J. (2008) 'Buccal swabs and treated card: methodological considerations for molecular epidemiologic studies examining pediatric populations', *American Journal of Epidemiology*, 167: 1260–1267.

Beitchman, J.H., Davidge, K.M., Kennedy, J.L., Atkinson, L., Lee, V., Shapiro, S., and Al., E. (2003) 'The serotonin transporter gene in aggressive children with and without ADHD and nonaggressive matched controls', *Annals of New York Academy of Sciences*, 1008: 248–251.

Benjamin, J., Li, L., Patterson, C., Greenberg, B.D., Murphy, D.L., and Hamer, D.H. (1996) 'Population and familial association between the D4 dopamine receptor gene and measures of novelty seeking', *Nature Genetics*, 12: 81–84.

Bentley, M., Lin, H., Fernandez, T., Lee, M., Yrigollen, C., Pakstise, A.J., Katsovich, L., Olds, D.L., Grigorenko, E.L., and Leckman, J. (2013) 'Gene variants associated with antisocial behaviour: a latent variable approach', *Journal of Child Psychology and Psychiatry*, 54, 1074–1085.

Bevilacqua, L., Carli, V., Sarchiapone, M., George, D., Goldman, D., Roy, A., and Enoch, M.A. (2012) 'Interaction between FKBP5 and childhood trauma and risk of aggressive behavior', *Archives of General Psychiatry*, 69: 62–70.

Black, D.W., Gunter, T., Loveless, P., Allen, J., and Sieleni, B. (2010) 'Antisocial personality disorder in incarcerated offenders: psychiatric comorbidity and quality of life', *Annual Review of Clinical Psychiatry*, 22: 113–120.

Boutwell, B.B. and Beaver, K.M. (2008) 'A biosocial explanation of delinquency abstention', *Criminal Behaviour and Mental Health*, 18: 59–74.

Buckholtz, J.W. and Meyer-Lindenberg, A. (2008) 'MAOA and the neurogenetic architecture of human aggression', *Trends in Neuroscience*, 31: 120–129.

Burt, S.A. and Klump, K.L. (2012) 'Etiological distinctions between aggressive and non-aggressive antisocial behavior: results from a nuclear twin family model', *Journal of Abnormal Child Psychology*, 40: 1059–1071.

Burt, S.A. and Mikolajewski, A.J. (2008) 'Preliminary evidence that specific candidate genes are associated with adolescent-onset antisocial behavior', *Aggressive Behavior*, 34: 437–445.

Burt, S.A., Barnes, A.R., McGue, M., and Iacono, W.G. (2008) 'Parental divorce and adolescent delinquency: ruling out the impact of common genes', *Developmental Psychology*, 44: 1668–1677.

Cadoret, R.J., Troughton, E., and O'Gorman, T.W. (1987) 'Genetic and environmental factors in alcohol abuse and antisocial personality', *Journal of Studies on Alcohol*, 48: 1–8.

Cadoret, R.J., Troughton, E., Bagford, J., and Woodworth, G. (1990) 'Genetic and environmental factors in adoptee antisocial personality', *European Archive of Psychiatry and Neurological Science*, 239: 231–240.

Cadoret, R.J., Yates, W.R., Troughton, E., Woodworth, G., and Stewart, M.A. (1995) 'Genetic–environmental interaction in the genesis of aggressivity and conduct disorders', *Archives of General Psychiatry*, 52: 916–924.

Cadoret, R.J., Langbehn, D., Caspers, K., Troughton, E.P., Yucuis, R., Sandhu, H.K., and Philibert, R. (2003) 'Associations of the serotonin transporter promoter polymorphism with aggressivity, attention deficit, and conduct disorder in an adoptee population', *Comprehensive Psychiatry*, 44: 88–101.

Caspi, A., McClay, J., Moffitt, T.E., Mill, J., Martin, J., Craig, I.W., Taylor, A., and Poulton, R. (2002) 'Role of genotype in the cycle of violence in maltreated children', *Science*, 297: 851–854.

Caspi, A., Langley, K., Milne, B., Moffitt, T.E., O'Donovan, M., Owen, M.J., Polo Tomas, M., Poulton, R., Rutter, M., Taylor, A., Williams, B., and Thapar, A. (2008) 'A replicated molecular genetic basis for subtyping antisocial behavior in children with attention-deficit/hyperactivity disorder', *Archives of General Psychiatry*, 65: 203–210.

Chen, T., Blum, K., Mathews, D., Fisher, L., Schnautz, N., Braverman, E.R., Schoolfield, J., Downs, B., and Comings, D. (2005) 'Are dopaminergic genes involved in a predisposition to pathological aggression? Hypothesizing the importance of "super normal controls" in psychiatricgenetic research of complex behavioral disorders', *Medical Hypotheses*, 65: 703–707.

Cheng, D., Hong, C.J., Liao, D.L., and Tsai, S.J. (2006) 'Association study of androgen receptor CAG repeat polymorphism and male violent criminal activity', *Psychoneuroendocrinology*, 31: 548–552.

Chiu, Y.F., Liu, S.Y., and Tsai, Y.Y. (2005) 'A comparison in association and linkage genome-wide scans for alcoholism susceptibility genes using single-nucleotide polymorphisms', *BMC Genetics*, 6: S89.

Cicchetti, D., Rogosch, F.A., and Thibodeau, E.L. (2012) 'The effects of child maltreatment on early signs of antisocial behavior: genetic moderation by tryptophan hydroxylase, serotonin transporter, and monoamine oxidase A genes', *Development and Psychopathology*, 24: 907–928.

Congdon, E., Lesch, K.P., and Canli, T. (2008) 'Analysis of DRD4 and DAT polymorphisms and behavioral inhibition in healthy adults: implications for impulsivity', *American Journal of Medical Genetics B, Neuropsychiatric Genetics*, 147B: 27–32.

Contini, V., Marques, F.Z., Garcia, C.E., Hutz, M.H., and Bau, C.H. (2006) 'MAOA-uVNTR polymorphism in a Brazilian sample: further support for the association with impulsive behaviors and alcohol dependence', *American Journal of Medical Genetics B, Neuropsychiatric Genetics*, 141B: 305–308.

Corradini, I., Verderio, C., Sala, M., Wilson, M.C., and Matteoli, M. (2009) 'SNAP-25 in neuropsychiatric disorders', *Annals of the New York Academy of Sciences*, 1152: 93–99.

Dadds, M., Moul, C., Avril, C., Hawes, D., and Brennan, J. (2013) 'Replication of a ROBO2 polymorphism associated with conduct problems but not psychopathic tendencies in children', *Psychiatric Genetics*, 23(6): 251–254.

Davidge, K.M., Atkinson, L., Douglas, L., Lee, V., Shapiro, S., Kennedy, J.L., and Beitchman, J.H. (2004) 'Association of the serotonin transporter and 5HT1D beta receptor genes with extreme, persistent and pervasive aggressive behaviour in children', *Psychiatric Genetics*, 14: 143–146.

Demaree, H.A., Pu, J., Jesberger, J., Feeny, N., Jeng, L., Everhart, D.E., Duerk, J., and Tkach, J. (2009) '5HTTLPR predicts left fusiform gyrus activation to positive emotional stimuli', *Magnetic Resonance Imaging*, 27: 441–448.

Derijk, R.H. (2009) 'Single nucleotide polymorphisms related to HPA axis reactivity', *Neuroimmunomodulation*, 16: 340–352.

Derringer, J., Krueger, R.F., Irons, D.E., and Iacono, W.G. (2010) 'Harsh discipline, childhood sexual assault, and MAOA genotype: an investigation of main and interactive effects on diverse clinical externalizing outcomes', *Behavior Genetics*, 40: 639–648.

DeYoung, C.G., Getchell, M., Koposov, R.A., Yrigollen, C.M., Haeffel, G.J., Klinteberg, B., Oreland, L., Ruchkin, V.V., Pakstise, A.J., and Grigorenko, E.L. (2010) 'Variation in the catechol-O-methyltransferase Val^{158}Met polymorphism associated with conduct disorder and ADHD symptoms, among adolescent male delinquents', *Psychiatric Genetics* 20: 20–24.

Diamantopoulou, S., Verhulst, F.C., and Van Der Ende, J. (2010) 'Testing developmental pathways to antisocial personality problems', *Journal of Abnormal Child Psychology*, 38: 91–103.

Dick, D.M., Li, T.K., Edenberg, H.J., Hesselbrock, V., Kramer, J., Kuperman, S., Porjesz, B., Bucholz, K., Goate, A., Nurnberger, J., and Foroud, T. (2004) 'A genome-wide screen for genes influencing conduct disorder', *Molecular Psychiatry*, 9: 81–86.

Dick, D.M., Rose, R.J., and Kaprio, J. (2006) 'The next challenge for psychiatric genetics: characterizing the risk associated with identified genes', *Annals of Clinical Psychiatry*, 18: 223–231.

Dick, D.M., Wang, J.C., Plunkett, J., Aliev, F., Hinrichs, A., Bertelsen, S., Budde, J.P., Goldstein, E.L., Kaplan, D., Edenberg, H.J., Nurnberger, J. Jr, Hesselbrock, V., Schuckit, M., Kuperman, S., Tischfield, J., Porjesz, B., Begleiter, H., Bierut, L.J., and Goate, A. (2007) 'Family-based association analyses of alcohol dependence phenotypes across DRD2 and neighboring gene ANKK1', *Alcohol Clinical Experimental Research*, 31: 1645–1653.

Dick, D.M., Aliev, F., Wang, J.C., Grucza, R.A., Schuckit, M., Kuperman, S., Kramer, J., Hinrichs, A., Bertelsen, S., Budde, J.P., Hesselbrock, V., Porjesz, B., Edenberg, H.J., Bierut, L.J., and Goate, A. (2008) 'Using dimensional models of externalizing psychopathology to aid in gene identification', *Archives of General Psychiatry*, 65: 310–318.

Dick, D.M., Latendresse, S.J., Lansford, J.E., Budde, J.P., Goate, A., Dodge, K.A., Pettit, G.S., and Bates, J.E. (2009) 'Role of GABRA2 in trajectories of externalizing behavior across development and evidence of moderation by parental monitoring', *Archives of General Psychiatry*, 66: 649–657.

Dick, D.M., Krueger, R.F., Edwards, A., Agrawal, A., Lynskey, M., Lin, P., Shuckit, M., Hesselbrock, V., Nurnberger, J.I., Jr., Almasy, L., Porjesz, B., Edenberg, H.J., Buckholz, K., Kramer, J., Kuperman, S., and Bierut, L. (2011) 'Genome-wide association study of conduct disorder symptomatology', *Molecular Psychiatry*, 16: 800–808.

Dick, D.M., Aliev, F., Lantendresse, S., Porjesz, B., Schuckit, M., Rangaswamy, M., Hesselbrock, V., Edenberg, H.J., Nurnberger, J.I. Jr, Agrawal, A., Bierut, L., Wang, J., Bucholz, K., Kuperman, S., and Kramer, J. (2013) 'How phenotype and developmental stage affect the genes we find: GABRA2 and impulsivity', *Twin Research and Human Genetics*, 16: 661–669.

Doehring, A., Hentig, N., Graff, J., Salamat, S., Schmidt, M., Geisslinger, G., Harder, S., and Lotsch, J. (2009) 'Genetic variants altering dopamine D2 receptor expression or function modulate the risk of opiate addiction and the dosage requirements of methadone substitution', *Pharmacogenetic Genomics*, 19: 407–414.

Douglas, K., Chan, G., Gelernter, J., Arias, A.J., Anton, R.F., Poling, J., Farrer, L., and Kranzler, H.R. (2011) '5-HTTLPR as a potential moderator of the effects of adverse childhood experiences on risk of antisocial personality disorder', *Psychiatric Genetics*, 21: 240–248.

Ducci, F., Enoch, M.A., Yuan, Q., Shen, P.H., White, K.V., Hodgkinson, C., Albaugh, B., Virkkunen, M., and Goldman, D. (2009) 'HTR3B is associated with alcoholism with antisocial behavior and alpha EEG power—an intermediate phenotype for alcoholism and co-morbid behaviors', *Alcohol*, 43: 73–84.

Dunn, E., Uddin, M., Subramanian, S.V., Smoller, J., Galea, S., and Koenen, K. (2011) 'Research review: gene–environment interaction research in youth depression—a systematic review with recommendations for future research', *Journal of Child Psychology and Psychiatry*, 52: 1223–1238.

Ehlers, C.L., Gilder, D.A., Slutske, W.S., Lind, P.A., and Wilhelmsen, K.C. (2008) 'Externalizing disorders in American Indians: comorbidity and a genome wide linkage analysis', *American Journal of Medical Genetics B, Neuropsychiatric Genetics*, 147B: 690–698.

Eisenberg, D., Campbell, B., Mackillop, J., Lum, J., and Wilson, D. (2007) 'Season of birth and dopamine receptor gene associations with impulsivity, sensation seeking and reproductive behaviors', *PLoS One*, 2: e1216.

Enoch, M.A., Steer, C.D., Newman, T.K., Gibson, N., and Goldman, D. (2010) 'Early life stress, MAOA, and gene–environment interactions predict behavioral disinhibition in children', *Genes, Brain, and Behavior*, 9: 65–74.

Faraone, S.V., Biederman, J., Weiffenbach, B., Keith, T., Chu, M.P., Weaver, A., Spencer, T.J., Wilens, T.E., Frazier, J., Cleves, M., and Sakai, J. (1999) 'Dopamine D4 gene 7-repeat allele and attention deficit hyperactivity disorder', *American Journal of Psychiatry*, 156: 768–770.

Faraone, S.V., Perlis, R.H., Doyle, A.E., Smoller, J.W., Goralnick, J.J., Holmgren, M.A., and Sklar, P. (2005) 'Molecular genetics of attention-deficit/hyperactivity disorder', *Biological Psychiatry*, 57: 1313–1323.

Feinn, R., Nellissery, M., and Kranzler, H.R. (2005) 'Meta-analysis of the association of a functional serotonin transporter promoter polymorphism with alcohol dependence', *American Journal of Medical Genetics B, Neuropsychiatric Genetics*, 133B: 79–84.

Fergusson, D.M., Boden, J.M., Horwood, L.J., Miller, A.L., and Kennedy, M.A. (2011) 'MAOA, abuse exposure and antisocial behaviour: 30-year longitudinal study', *British Journal of Psychiatry*, 198: 457–463.

Fergusson, D.M., Boden, J.M., Horwood, L.J., Miller, A., and Kennedy, M.A. (2012) 'Moderating role of the MAOA genotype in antisocial behaviour', *British Journal of Psychiatry*, 200: 116–123.

Foley, D.L., Eaves, L.J., Wormley, B., Silberg, J.L., Maes, H.H., Kuhn, J., and Riley, B. (2004) 'Childhood adversity, monoamine oxidase a genotype, and risk for conduct disorder', *Archives of General Psychiatry*, 61: 738–744.

Fowler, J., Alia-Klein, N., Kriplani, A., Logan, J., Williams, B., Zhu, W., Craig, I.W., Telang, F., Goldstein, R.B., Volkow, N., Vaska, P., and Wang, G. (2007) 'Evidence that brain MAOA activity does not correspond to MAOA genotype in healthy male subjects', *Biological Psychiatry*, 62: 355–358.

Fowler, T., Langley, K., Rice, F., Van Den Bree, M.B.M., Ross, K., Wilkinson, L.S., Owen, M.J., O'Donovan, M.C., and Thapar, A. (2009) 'Psychopathy trait scores in adolescents with childhood ADHD: the contribution of genotypes affecting MAOA, 5HTT and COMT activity', *Psychiatric Genetics*, 19: 312–319.

Gallardo-Pujol, D., Andres-Pueyo, A., and Maydeu-Olivares, A. (2013) 'MAOA genotype, social exclusion and aggression: an experimental test of a gene–environment interaction', *Genes, Brain, and Behavior*, 12: 140–145.

Garcia, L.F., Aluja, A., Fibla, J., Cuevas, L., and Garcia, O. (2010) 'Incremental effect for antisocial personality disorder genetic risk combining 5-HTTLPR and 5-HTTVNTR polymorphisms', *Psychiatry Research*, 177: 161–166.

Gerra, G., Garofano, L., Santoro, G., Bosari, S., Pellegrini, C., Zaimovic, A., Moi, G., Bussandri, M., Moi, A., Brambilla, F., and Donnini, C. (2004) 'Association between low-activity serotonin transporter genotype and heroin dependence: behavioral and personality correlates', *American Journal of Medical Genetics B, Neuropsychiatric Genetics*, 126B: 37–42.

Gerra, G., Garofano, L., Pellegrini, C., Bosari, S., Zaimovic, A., Moi, G., Avanzini, P., Talarico, E., Gardini, F., and Donnini, C. (2005) 'Allelic association of a dopamine transporter gene polymorphism with antisocial behaviour in heroin-dependent patients', *Addiction Biology*, 10: 275–281.

Giegling, I., Hartmann, A., Moller, H.-J., and Rujescu, D. (2006) 'Anger- and aggression-related traits are associated with polymorphisms in the 5-HT-2A gene', *Journal of Affective Disorders*, 96: 75–81.

Gillespie, N.A., Zhu, G., Evans, D.M., Medland, S.E., Wright, M.J., and Martin, N.G. (2008) 'A genome-wide scan for Eysenckian personality dimensions in adolescent twin sibships: psychoticism, extraversion, neuroticism, and lie', *Journal of Personality*, 76: 1415–1446.

Gizer, I.R., Ehlers, C.L., Vieten, C., Feiler, H.S., Gilder, D.A., and Wilhelmsen, K.C. (2012) 'Genome-wide linkage scan of antisocial behavior, depression, and impulsive substance use in the UCSF family alcoholism study', *Psychiatric Genetics*, 22: 235–244.

Glenn, A.L. (2011) 'The other allele: exploring the long allele of the serotonin transporter gene as a potential risk factor for psychopathy: a review of the parallels in findings', *Neuroscience and Biobehavioral Reviews*, 35: 612–620.

Grigorenko, E.L., DeYoung, C.G., Eastman, M., Getchell, M., Haeffel, G.J., Klinteberg, B., *et al.* (2010) 'Aggressive behavior, related conduct problems, and variation in genes affecting dopamine turnover', *Aggressive Behavior*, 36: 158–176.

Gunter, T.D., Philibert, R., and Hollenbeck, N. (2009) 'Medical and psychiatric problems among men and women in a community corrections residential setting', *Behavioral Sciences and the Law*, 27: 695–711.

Haberstick, B.C., Lessem, J.M., Hopfer, C.J., Smolen, A., Ehringer, M.A., Timberlake, D., and Hewitt, J.K. (2005) 'Monoamine oxidase A (MAOA) and antisocial behaviors in the presence of childhood and adolescent maltreatment', *American Journal of Medical Genetics B, Neuropsychiatric Genetics*, 135B: 59–64.

Haberstick, B., Lessem, J., Hewitt, J.K., Smolen, A., Hopfer, C., Halpern, C., Killeya-Jones, L., Boardman, J.D., Tabor, J., Siegler, I.C., Williams, R.B., and Harris, K.M. (2014) 'MAOA genotype, childhood maltreatment, and their interaction in the etiology of adult antisocial behaviors', *Biological Psychiatry*, 75: 25–30.

Halleland, H., Lundervold, A., Halmey, A., Haavik, J., and Johansson, S. (2008) 'Association between catechol O-methyltransferase (COMT) haplotypes and severity of hyperactivity symptoms in adults', *American Journal of Medical Genetics B, Neuropsychiatric Genetics*, 150B: 403–410.

Hansen, A.L., Waage, L., Eid, J., Johnsen, B.H., and Hart, S. (2011) 'The relationship between attachment, personality and antisocial tendencies in a prison sample: a pilot study', *Scandinavian Journal of Psychology*, 52: 268–276.

Hasegawa, Y., Higuchi, S., Matsushita, S., and Miyaoka, H. (2002) 'Association of a polymorphism of serotonin 1B receptor gene and alcohol dependence with inactive aldehyde dehydrogenase-2', *Journal of Neural Transmission*, 109: 1435–1463.

Hennig, J., Reuter, M., Netter, P., Burk, C., and Landt, O. (2005) 'Two types of aggression are differentially related to serotonergic activity and the A779C TPH polymorphism', *Behavioral Neuroscience*, 119: 16–25.

Herman, A.I., Conner, T.S., Anton, R.F., Gelernter, J., Kranzler, H.R., and Covault, J. (2010) 'Variation in the gene encoding the serotonin transporter is associated with a measure of sociopathy in alcoholics', *Addiction Biology*, 16: 124–132.

Hess, C., Reif, A., Strobel, A., Boreatti-Hummer, A., Heine, M., Lesch, K.P., and Jacob, C.P. (2009) 'A functional dopamine-beta-hydroxylase gene promoter polymorphism is associated with impulsive personality styles, but not with affective disorders', *Journal of Neural Transmission*, 116: 121–130.

Hicks, B.M., South, S., Dirago, M.A., Iacono, W.G., and McGue, M. (2009) 'Environmental adversity increases genetic risk for externalizing disorders', *Archives of General Psychiatry*, 66: 640–648.
Hicks, B.M., Carlson, M.D., Blonigen, D.M., Patrick, C.J., Iacono, W.G., and McGue, M. (2013) 'Psychopathic personality traits and environmental contexts: differential correlates, gender differences, and genetic mediation', *Personal Disorders*, 3: 209–227.
Hill, J., Quinn, J., Tibu, F., and Sharp, H. (2013) 'Evidence for interplay between genes and maternal stress in utero: monoamine oxidase A polymorphism moderates effects of life events during pregnancy on infant negative emotionality at 5 weeks', *Genes, Brain, and Behavior*, 12(4): 388–396.
Hillemacher, T., Frieling, H., Hartl, T., Wilhelm, J., Kornhuber, J., and Bleich, S. (2009) 'Promoter specific methylation of the dopamine transporter gene is altered in alcohol dependence and associated with craving', *Journal of Psychiatric Research*, 43: 388–392.
Hirata, Y., Zai, C.C., Nowrouzi, B., Beitchman, J.H., and Kennedy, J.L. (2013) 'Study of the catechol-o-methyltransferase (COMT) gene with high aggression in children', *Aggressive Behavior*, 39: 45–51.
Huang, Y.Y., Cate, S.P., Battistuzzi, C., Oquendo, M.A., Brent, D., and Mann, J.J. (2004) 'An association between a functional polymorphism in the monoamine oxidase a gene promoter, impulsive traits and early abuse experiences', *Neuropsychopharmacology*, 29: 1498–1505.
Huang, W., Payne, T.J., Ma, J.Z., Beuten, J., Dupont, R.T., Inohara, N., and Li, M.D. (2009) 'Significant association of ANKK1 and detection of a functional polymorphism with nicotine dependence in an African-American sample', *Neuropsychopharmacology*, 34: 319–330.
Jiang, Y., Langley, B., Lubin, F., Renthal, W., Wood, M., Yasui, D., Kumar, A., Nestler, E., Akbarian, S., and Beckel-Mitchener, A. (2008) 'Epigenetics in the nervous system', *Journal of Neuroscience*, 28: 11753–11759.
Johann, M., Putzhammer, A., Eichhammer, P., and Wodarz, N. (2005) 'Association of the −141C Del variant of the dopamine D2 receptor (DRD2) with positive family history and suicidality in German alcoholics', *American Journal of Medical Genetics B, Neuropsychiatric Genetics*, 132B: 46–49.
Jonsson, E.G., Von Gertten, C., Gustavsson, J.P., Yuan, Q.P., Lindblad-Toh, K., Forslund, K., Rylander, G., Mattila-Evenden, M., Asberg, M., and Schalling, M. (2001) 'Androgen receptor trinucleotide repeat polymorphism and personality traits', *Psychiatric Genetics*, 11: 19–23.
Kasiakogia-Worlley, K., McQuillan, A., Lydall, G.J., Patel, S., Kottlagi, G., Gunwardena, P., Cherian, R., Rao, H., Hillman, A., Gobikrishnan, N., Douglas, E., Qureshi, S., Jauhar, S., Ball, D., O'Kane, A., Owens, L., Dedman, A., Sharp, S., Kandaswamy, R., Guerrini, I., Thomson, A., Smith, I., Dar, K., Morgan, M., and Gurling, H. (2011) 'Lack of allelic association between markers at the DRD2 and ANKK1 gene loci with the alcohol-dependence syndrome and criminal activity', *Psychiatric Genetics*, 21: 323–324.
Ke, L., Qi, Z.Y., Ping, Y., and Ren, C.Y. (2006) 'Effect of SNP at position 40237 in exon 7 of the TPH2 gene on susceptibility to suicide', *Brain Research*, 1122: 24–26.
Kendler, K.S. and Myers, J. (2013) 'The boundaries of the internalizing and externalizing genetic spectra in men and women', *Psychological Medicine*, 20: 1–9.
Kim-Cohen, J., Caspi, A., Taylor, A., Williams, B., Newcombe, R., Craig, I.W., and Moffitt, T.E. (2006) 'MAOA, maltreatment, and gene–environment interaction predicting children's mental health: new evidence and a meta-analysis', *Molecular Psychiatry*, 11: 903–913.
Kinnally, E.L., Huang, Y.Y., Haverly, R., Burke, A.K., Galfalvy, H., Brent, D.P., Oquendo, M.A., and Mann, J.J. (2009) 'Parental care moderates the influence of MAOA-uVNTR genotype and childhood stressors on trait impulsivity and aggression in adult women', *Psychiatric Genetics*, 19: 126–133.
Koh, K.B., Choi, E., Lee, Y., Han, M., Choi, S., Kim, S., and Lee, M. (2011) 'The relation of serotonin-related gene and COMT gene polymorphisms with criminal behavior in schizophrenic disorder', *Journal of Clinical Psychiatry*, 73: 159–163.
Koller, G., Bondy, B., Preuss, U.W., Bottlender, M., and Soyka, M. (2003) 'No association between a polymorphism in the promoter region of the MAOA gene with antisocial personality traits in alcoholics', *Alcohol*, 38: 31–34.
Krueger, R.F., Markon, K.E., Patrick, C.J., and Iacono, W.G. (2005) 'Externalizing psychopathology in adulthood: a dimensional-spectrum conceptualization and its implications for DSM-V', *Journal of Abnormal Psychology*, 114: 537–550.
Kulikova, M.A., Maluchenko, N.V., Timofeeva, M.A., Shlepzova, V.A., Schegolkova, J.V., Sysoeva, O.V., Ivanitsky, A.M., and Tonevitsky, A.G. (2008) 'Effect of functional catechol-O-methyltransferase Val^{158}Met polymorphism on physical aggression', *Bulletin of Experimental Biological Medicine*, 145: 62–64.
Langbehn, D.R. and Cadoret, R.J. (2001) 'The adult antisocial syndrome with and without antecedent conduct disorder: comparisons from an adoption study', *Comprehensive Psychiatry*, 42: 272–282.

Langley, K., Heron, J., O'Donovan, M.C., Owen, M.J., and Thapar, A. (2010) 'Genotype link with extreme antisocial behavior: the contribution of cognitive pathways', *Archives of General Psychiatry*, 67: 1317–1323.

Lappalainen, J., Long, J.C., Eggert, M., Ozaki, N., Robin, R.W., Brown, G.L., Naukkarinen, H., Virkkunen, M., Linnoila, M., and Goldman, D. (1998) 'Linkage of antisocial alcoholism to the serotonin 5-HT1B receptor gene in 2 populations', *Archives of General Psychiatry*, 55: 989–994.

Latendresse, S.J., Bates, J.E., Goodnight, J.A., Lansford, J.E., Budde, J.P., Goate, A., Dodge, K.A., Pettit, G.S., and Dick, D.M. (2011) 'Differential susceptibility to adolescent externalizing trajectories: examining the interplay between CHRM2 and peer group antisocial behavior', *Child Development*, 82: 1797–1814.

Laucht, M., Becker, K., El-Faddagh, M., Hohm, E., and Schmidt, M. (2005) 'Association of the DRD4 exon III polymorphism with smoking in fifteen-year-olds: a mediating role for novelty seeking?', *Journal of the American Acaddemy of Child and Adolescent Psychiatry*, 44: 477–484.

Lee, S.S. (2011) 'Deviant peer affiliation and antisocial behavior: interaction with monoamine oxidase A (MAOA) genotype', *Journal of Abnormal Child Psychology*, 39: 321–332.

Li, D., Zhao, H., Kranzler, H.R., Oslin, D., Anton, R.F., Farrer, L.A., and Gelernter, J. (2012) 'Association of COL25A1 with comorbid antisocial personality disorder and substance dependence', *Biological Psychiatry*, 71: 733–740.

Li, J.J. and Lee, S.S. (2010) 'Latent class analysis of antisocial behavior: interaction of serotonin transporter genotype and maltreatment', *Journal of Abnormal Child Psychology*, 38: 789–801.

Liao, D.L., Hong, C.J., Shih, H.L., and Tsai, S.J. (2004) 'Possible association between serotonin transporter promoter region polymorphism and extremely violent crime in Chinese males', *Neuropsychobiology*, 50: 284–287.

Lind, P.A., Macgregor, S., Agrawal, A., Montgomery, G.W., Heath, A.C., Martin, N.G., and Whitfield, J.B. (2008a) 'The role of GABRA2 in alcohol dependence, smoking, and illicit drug use in an Australian population sample', *Alcohol Clinical Experimental Research*, 32: 1721–1731.

Lind, P.A., Macgregor, S., Montgomery, G.W., Heath, A.C., Martin, N.G., and Whitfield, J.B. (2008b) 'Effects of GABRA2 variation on physiological, psychomotor and subjective responses in the alcohol challenge twin study', *Twin Research and Human Genetics*, 11: 174–182.

Loehlin, J.C., Jonsson, E.G., Gustavsson, J.P., Schalling, M., and Stallings, M.C. (2003) 'The androgen receptor gene and psychological traits: are results consistent in Sweden and Australia?', *Twin Research*, 6: 201–208.

Logue, M.W., Solovieff, N., Leussis, M.P., Wolf, E.J., Melista, E., Baldwin, C., Koenen, K., Petryshen, T.L., and Miller, M.W. (2013) 'The ankyrin-3 gene is associated with posttraumatic stress disorder and externalizing comorbidity', *Psychoneuroendocrinology*, 38(10): 2249–2257.

Lu, R.B., Lee, J.F., Huang, S.Y., Lee, S.Y., Chang, Y.H., Kuo, P.H., Chen, S.L., Chen, S.H., Chu, C.H., Lin, W.W., Wu, P.L., and Ko, H.C. (2012) 'Interaction between ALDH2*1*1 and DRD2/ANKK1 TaqI A1A1 genes may be associated with antisocial personality disorder not co-morbid with alcoholism', *Addiction Biology*, 17: 865–874.

McCrory, E.J., De Brito, S.A., and Viding, E. (2012) 'The link between child abuse and psychopathology: a review of neurobiological and genetic research', *Journal of Research in Social Medicine*, 105: 151–156.

McDermott, R., Tingley, D., Cowden, J., Frazzetto, G., and Johnson, D.D. (2009) 'Monoamine oxidase A gene (MAOA) predicts behavioral aggression following provocation', *Proceedings of the National Academy of Sciences, USA*, 106: 2118–23.

Madlung, A. and Comai, L. (2004) 'The effect of stress on genome regulation and structure', *Annals of Botany*, 94: 481–495.

Malik, A.I., Zai, C.C., Abu, Z., Nowrouzi, B., and Beitchman, J.H. (2012) 'The role of oxytocin and oxytocin receptor gene variants in childhood-onset aggression', *Genes, Brain, and Behavior*, 11: 1–10.

Mann, J.J., Currier, D., Murphy, L., Huang, Y.Y., Galfalvy, H., Brent, D., Greenhill, L., and Oquendo, M. (2008) 'No association between a TPH2 promoter polymorphism and mood disorders or monoamine turnover', *Journal of Affective Disorders*, 106: 117–121.

Manuck, S.B., Flory, J.D., Ferrell, R.E., Dent, K.M., Mann, J.J., and Muldoon, M.F. (1999) 'Aggression and anger-related traits associated with a polymorphism of the tryptophan hydroxylase gene', *Biological Psychiatry*, 45: 603–614.

Marquez, C., Poirier, G.L., Cordero, M.I., Larsen, M.H., Groner, A., Marquis, J., Magistretti, P.J., Trono, D., and Sandi, C. (2013) 'Peripuberty stress leads to abnormal aggression, altered amygdala and orbitofrontal reactivity and increased prefrontal MAOA gene expression', *Translational Psychiatry*, 3: e216.

Maze, I., Noh, K.M., and Allis, C.D. (2013) 'Histone regulation in the CNS: basic principles of epigenetic plasticity', *Neuropsychopharmacology*, 38: 3–22.

Meyer-Lindenberg, A., Domes, G., Kirsch, P., and Heinrichs, M. (2011) 'Oxytocin and vasopressin in the human brain: social neuropeptides for translational medicine', *Nature Reviews Neuroscience*, 12: 524–538.

Mik, H.M., Ehtesham, S., Baldassarra, L., De Luca, V., Davidge, K., Bender, D., Tharmalingam, S., Kennedy, J.L., and Beitchman, J.H. (2007) 'Serotonin system genes and childhood-onset aggression', *Psychiatric Genetics*, 17: 11.

Monuteaux, M.C., Biederman, J., Doyle, A.E., Mick, E., and Faraone, S.V. (2009) 'Genetic risk for conduct disorder symptom subtypes in an ADHD sample: specificity to aggressive symptoms', *Journal of the American Academy of Child and Adolescent Psychiatry*, 48: 757–764.

Moul, C., Dobson-Stone, C., Brennan, J., Hawes, D., and Dadds, M. (2013) 'An exploration of the serotonin system in antisocial boys with high levels of callous-unemotional traits', *PLoS ONE*, 8: e56619.

Nederhof, E., Bouma, E.M., Riese, H., Laceulle, O., Ormel, J., and Oldehinkel, A.J. (2010) 'Evidence for plasticity in genotypes in a gene–environment interaction: the TRAILS study', *Genes, Brain, and Behavior*, 9: 968–973.

Nielsen, D.A., Yuferov, V., Hamon, S., Jackson, C., Ho, A., Ott, J., and Kreek, M.J. (2009) 'Increased OPRM1 DNA methylation in lymphocytes of methadone-maintained former heroin addicts', *Neuropsychopharmacology*, 34: 867–873.

Nilsson, K.W., Sjoberg, R.L., Damberg, M., Leppert, J., Ohrvik, J., Alm, P.O., Lindstrom, L., and Oreland, L. (2006) 'Role of monoamine oxidase A genotype and psychosocial factors in male adolescent criminal activity', *Biological Psychiatry*, 59: 121–127.

Nordstrom, B., Gao, Y., Glenn, A.L., Peskin, M., Rudo-Hutt, A., Schug, R., Yang, Y., and Raine, A. (2011) 'Neurocriminology', *Advances in Genetics*, 75: 255–283.

Oades, R.D., Lasky-Su, J., Christensen, H., Faraone, S.V., Sonuga-Barke, E.J., Banaschewski, T., Chen, W., Anney, R., Buitelaar, J.K., Ebstein, R.P., Franke, B., Gill, M., Miranda, A., Roeyers, H., Rothenberger, A., Sergeant, J., Steinhausen, H., Taylor, E., Thompson, M., and Asherson, P. (2008) 'The influence of serotonin- and other genes on impulsive behavioral aggression and cognitive impulsivity in children with attention-deficit/hyperactivity disorder (ADHD): findings from a family-based association test (FBAT) analysis', *Behavior and Brain Functions*, 4: 1–14.

Pai, C., Chou, S., and Huang, F. (2007) 'Assessment of the role of a functional VNTR polymorphism in MAOA gene promoter: a preliminary study', *Forensic Science Journal*, 6: 37–43.

Paloyelis, Y., Asherson, P., Mehta, M.A., Faraone, S.V., and Kuntsi, J. (2010) 'DAT1 and COMT effects on delay discounting and trait impulsivity in male adolescents with attention deficit/hyperactivity disorder and healthy controls', *Neuropsychopharmacology*, 35: 2414–2426.

Pavlov, K., Chistiakov, D., and Chekhonin, V. (2012) 'Genetic determinants of aggression and impulsivity in humans', *Journal of Applied Genetics*, 53: 61–82.

Pearson, T.A. and Manolio, T.A. (2008) 'How to interpret a genome-wide association study', *Journal of the American Medical Association*, 299: 1335–1344.

Perez-Rodriguez, M.M., Weinstein, S., New, A., Bevilacqua, L., Yuan, Q., Zhou, Z., Hodgkinson, C., Goodman, M., Koenigsberg, H., Goldman, D., and Siever, L. (2010) 'Tryptophan hydroxylase 2 haplotype association with borderline personality disorder and aggression in a sample of patients with personality disorders and healthy controls', *Journal of Psychiatric Research*, 44: 1075–1081.

Philibert, R.A., Ryu, G.Y., Yoon, J.G., Sandhu, H., Hollenbeck, N., Gunter, T., Barkhurst, A., Adams, W., and Madan, A. (2007) 'Transcriptional profiling of subjects from the Iowa adoption studies', *American Journal of Medical Genetics B, Neuropsychiatric Genetics*, 144B: 683–690.

Philibert, R.A., Gunter, T.D., Beach, S.R., Brody, G.H., and Madan, A. (2008a) 'MAOA methylation is associated with nicotine and alcohol dependence in women', *American Journal of Medical Genetics B, Neuropsychiatric Genetics*, 147B: 565–570.

Philibert, R.A., Sandhu, H., Hollenbeck, N., Gunter, T., Adams, W., and Madan, A. (2008b) 'The relationship of 5HTT (SLC6A4) methylation and genotype on mRNA expression and liability to major depression and alcohol dependence in subjects from the Iowa Adoption Studies', *American Journal of Medical Genetics B, Neuropsychiatric Genetics*, 147B: 543–549.

Philibert, R.A., Zadorozhnyaya, O., Beach, S.R., and Brody, G.H. (2008c) 'Comparison of the genotyping results using DNA obtained from blood and saliva', *Psychiatric Genetics*, 18: 275–281.

Philibert, R.A., Beach, S.R., Gunter, T.D., Brody, G.H., Madan, A., and Gerrard, M. (2009a) 'The effect of smoking on MAOA promoter methylation in DNA prepared from lymphoblasts and whole blood', *American Journal of Medical Genetics B, Neuropsychiatric Genetics*, doi: 10.1002/ajmg.b.31031

Philibert, R.A., Gunter, T.D., Beach, S.R., Brody, G.H., Hollenbeck, N., Andersen, A., and Adams, W. (2009b) 'Role of GABRA2 on risk for alcohol, nicotine, and cannabis dependence in the Iowa Adoption Studies', *Psychiatric Genetics*, 19: 91–98.

Philibert, R.A., Wernett, P., Plume, J., Packer, H., Brody, G.H., and Beach, S.R. (2011) 'Gene–environment interactions with a novel variable monoamine oxidase A transcriptional enhancer are associated with antisocial personality disorder', *Biological Psychology*, 87: 366–371.

Philibert, R.A., Plume, J., Gibbons, F.X., Brody, G., and Beach, S.R. (2012) 'The impact of recent alcohol use on genomewide DNA methylation signatures', *Frontiers in Genetics*, 3.

Ponce, G., Jimenez-Arriero, M.A., Rubio, G., Hoenicka, J., Ampuero, I., Ramos, J.A., and Palomo, T. (2003) 'The A1 allele of the DRD2 gene (TaqI A polymorphisms) is associated with antisocial personality in a sample of alcohol-dependent patients', *European Psychiatry*, 18: 356–360.

Ponce, G., Hoenicka, J., Jimenez-Arriero, M.A., Rodriguez-Jimenez, R., Aragues, M., Martin-Sune, N., Huertas, E., and Palomo, T. (2008) 'DRD2 and ANKK1 genotype in alcohol-dependent patients with psychopathic traits: association and interaction study', *British Journal of Psychiatry*, 193: 121–125.

Ponce, G., Perez-Gonzalez, R., Aragues, M., Palomo, T., Rodriguez-Jimenez, R., Jimenez-Arriero, M.A., and Hoenicka, J. (2009) 'The ANKK1 kinase gene and psychiatric disorders', *Neurotoxicology Research*, 16: 50–59.

Preuss, U.W., Koller, G., Bondy, B., Bahlmann, M., and Soyka, M. (2001) 'Impulsive traits and 5-HT2A receptor promoter polymorphism in alcohol dependents: possible association but no influence of personality disorders', *Neuropsychobiology*, 43: 186–191.

Prichard, Z.M., Jorm, A.F., Mackinnon, A., and Easteal, S. (2007) 'Association analysis of 15 polymorphisms within 10 candidate genes for antisocial behavioural traits', *Psychiatric Genetics*, 17: 299–303.

Prom-Wormley, E.C., Eaves, L.J., Foley, D.L., Gardner, C.O., Archer, K.J., Wormley, B.K., Maes, H.H., Riley, B.P., and Silberg, J.L. (2009) 'Monoamine oxidase A and childhood adversity as risk factors for conduct disorder in females', *Psychological Medicine*, 39: 579–590.

Raine, A. (1993) *The Psychopathology of Crime: Criminal Behavior as a Clinical Disorder*, SanDiego, CA: Academic Press.

Raine, A. (2008) 'From genes to brain to antisocial behavior', *Current Directions in Psychological Science*, 17: 323–328.

Raine, A. (2013) *The Anatomy of Violence: The Biological Roots of Crime*. New York: Random House.

Rajender, S., Pandu, G., Sharma, J.D., Gandhi, K.P., Singh, L., and Thangaraj, K. (2008) 'Reduced CAG repeats length in androgen receptor gene is associated with violent criminal behavior', *International Journal of Legal Medicine*, 122: 367–372.

Ray, L.A., Bryan, A., Mackillop, J., McGeary, J., Hesterberg, K., and Hutchinson, K.E. (2009) 'The dopamine D receptor (DRD4) gene exon III polymorphism, problematic alcohol use and novelty seeking: direct and mediated genetic effects', *Addiction Biology*, 14: 238–244.

Reif, A., Rosler, M., Freitag, C.M., Schneider, M., Eujen, A., Kissling, C., Wenzler, D., Jacob, C.P., Retz-Junginger, P., Thome, J., Lesch, K.P., and Retz, W. (2007) 'Nature and nurture predispose to violent behavior: serotonergic genes and adverse childhood environment', *Neuropsychopharmacology*, 32: 2375–2383.

Reif, A., Jacob, C.P., Rujescu, D., Herterich, S., Lang, S., Gutknecht, L., Baehne, C.G., Strobel, A., Freitag, C.M., Giegling, I., Romanos, M., Hartmann, A., Rosler, M., Renner, T.J., Fallgatter, A.J., Retz, W., Ehlis, A.C., and Lesch, K.P. (2009) 'Influence of functional variant of neuronal nitric oxide synthase on impulsive behaviors in humans', *Archives of General Psychiatry*, 66: 41–50.

Reti, I.M., Xu, J.Z., Yanofski, J., McKibben, J., Uhart, M., Cheng, Y.J., Zandi, P., Bienvenu, O.J., Samuels, J., Willour, V., Kasch-Semenza, L., Costa, P., Bandeen-Roche, K., Eaton, W.W., and Nestadt, G. (2011) 'Monoamine oxidase A regulates antisocial personality in whites with no history of physical abuse', *Comprehensive Psychiatry*, 52: 188–194.

Retz, W. and Rosler, M. (2009) 'The relation of ADHD and violent aggression: what can we learn from epidemiological and genetic studies?', *International Journal of Law and Psychiatry*, 32: 235–243.

Retz, W., Thome, J., Blocher, D., Baader, M., and Rosler, M. (2002) 'Association of attention deficit hyperactivity disorder-related psychopathology and personality traits with the serotonin transporter promoter region polymorphism', *Neuroscience Letters*, 319: 133–136.

Retz, W., Retz-Junginger, P., Supprian, T., Thome, J., and Rosler, M. (2004) 'Association of serotonin transporter promoter gene polymorphism with violence: relation with personality disorders, impulsivity, and childhood ADHD psychopathology', *Behavioral Sciences and the Law*, 22: 415–425.

Retz, W., Freitag, C.M., Retz-Junginger, P., Wenzler, D., Schneider, M., Kissling, C., Thome, J., and Rosler, M. (2008) 'A functional serotonin transporter promoter gene polymorphism increases ADHD symptoms in delinquents: interaction with adverse childhood environment', *Psychiatry Research*, 158: 123–131.

Reuter, M., Weber, B., Fiebach, C.J., Elger, C., and Montag, C. (2009) 'The biological basis of anger: associations with the gene coding for DARPP-32 (PPP1R1B) and with amygdala volume', *Behavior and Brain Research*, 202: 179–183.

Risch, N. and Merikangas, K. (1996) 'The future of genetic studies of complex human diseases', *Science*, 273: 1516–1517.

Rowe, D.C., Stever, C., Chase, D., Sherman, S., Abramowitz, A., and Waldman, I.D. (2001) 'Two dopamine genes related to reports of childhood retrospective inattention and conduct disorder symptoms', *Molecular Psychiatry*, 6: 429–433.

Rujescu, D., Giegling, I., Bondy, B., Gietl, A., Sill, P., and Moller, H.J. (2002) 'Association of anger-related traits with SNPs in the TPH gene', *Molecular Psychiatry*, 7: 1023–1029.

Rujescu, D., Giegling, I., Gietl, A., Hartmann, A.M., and Moller, H.J. (2003) 'A functional single nucleotide polymorphism (V158M) in the COMT gene is associated with aggressive personality traits', *Biological Psychiatry*, 54: 34–39.

Rutten, B.P. and Mill, J. (2009) 'Epigenetic mediation of environmental influences in major psychotic disorders', *Schizophrenia Bulletin*, 35: 1045–1056.

Sadeh, N., Javdani, S., Jackson, J., Reynolds, E.K., Potenza, M.N., Gelernter, J., Lejuez, C.W., and Verona, E. (2010) 'Serotonin transporter gene associations with psychopathic traits in youth vary as a function of socioeconomic resources', *Journal of Abnormal Psychology*, 119: 604–609.

Sadeh, N., Javdani, S., and Verona, E. (2013) 'Analysis of monoaminergic genes, childhood abuse, and dimensions of psychopathy', *Journal of Abnormal Psychology*, 122: 167–179.

Sakai, J.T., Young, S.E., Stallings, M.C., Timberlake, D., Smolen, A., Stetler, G.L., and Crowley, T.J. (2006) 'Case-control and within-family tests for an association between conduct disorder and 5HTTLPR', *American Journal of Medical Genetics B, Neuropsychiatric Genetics*, 141B: 825–832.

Sakai, J.T., Lessem, J.M., Haberstick, B.C., Hopfer, C.J., Smolen, A., and Ehringer, M. (2007) 'Case-control and within-family tests for associations between 5HTTLPR and conduct problems in a longitudinal adolescent sample', *Psychiatric Genetics*, 17: 207–214.

Sakai, J.T., Stallings, M.C., Crowley, T.J., Gelhorn, H.L., McQueen, M.B., and Ehringer, M.A. (2010) 'Test of association between GABRA2 (SNP rs279871) and adolescent conduct/alcohol use disorders utilizing a sample of clinic referred youth and serious substance abuse and conduct problems, controls, and available first degree relatives', *Drug and Alcohol Dependence*, 106: 199–203.

Salatino-Oliveira, A., Genro, J., Guimaraes, A.P., Chazan, R., Schmitz, M., Polanczyk, G., Roman, T., Rohde, L.A., and Hutz, M.H. (2012) 'Cathechol-O-methyltransferase Val^{158}Met polymorphism is associated with disruptive behavior disorders among children and adolescents with ADHD', *Journal of Neural Transmission*, 119: 729–733.

Sander, T., Ostapoicz, A., Samochowiec, J., Smolka, M., Winterer, G., and Schmidt, L.G. (2000) 'Genetic variation of the glutamate transporter EAAT2 gene and vulnerability to alcohol dependence', *Psychiatric Genetics*, 10: 103–107.

Schilling, C., Kuhn, S., Sander, T., and Gallinat, J. (2013) 'Association between dopamine D4 receptor genotype and trait impulsiveness', *Psychiatric Genetics*, 24(2): 82.

Shumay, E., Logan, J., Volkow, N., and Fowler, J. (2012) 'Evidence that the methylation state of the monoamine oxidase A (MAOA) gene predicts brain activity of MAOA enzyme in healthy men', *Epigenetics*, 7: 1151–1160.

Singh, J.P., Volavka, J., Czobor, P., and Van Dorn, R.A. (2012) 'A meta-analysis of the Val^{158}Met COMT polymorphism and violent behavior in schizophrenia', *PLoS One*, 7: e43423.

Sjoberg, R.L., Nilsson, K.W., Wargelius, H.L., Leppert, J., Lindstrom, L., and Oreland, L. (2007) 'Adolescent girls and criminal activity: role of MAOA-LPR genotype and psychosocial factors', *American Journal of Medical Genetics B, Neuropsychiatric Genetics*, 144B: 159–164.

Sjoberg, R.L., Ducci, F., Barr, C.S., Newman, T.K., Dell'Osso, L., Virkkunen, M., and Goldman, D. (2008) 'A non-additive interaction of a functional MAO-A VNTR and testosterone predicts antisocial behavior', *Neuropsychopharmacology*, 33: 425–430.

Soeiro-De-Souza, M.G., Stanford, M.S., Bio, D.S., Machado-Vieira, R., and Moreno, R.A. (2013) 'Association of the COMT Met(1)(5)(8) allele with trait impulsivity in healthy young adults', *Molecular Medicine Reports*, 7: 1067–1072.

Sonuga-Barke, E.J., Lasky-Su, J., Neale, B.M., Oades, R., Chen, W., Franke, B., Buitelaar, J., Banaschewski, T., Ebstein, R., Gill, M., Anney, R., Miranda, A., Mulas, F., Roeyers, H., Rothenberger, A., Sergeant, J., Steinhausen, H.C., Thompson, M., Asherson, P., and Faraone, S.V. (2008) 'Does parental expressed emotion moderate genetic effects in ADHD? An exploration using a genome wide association scan', *American Journal of Medical Genetics B, Neuropsychiatric Genetics*, 147B: 1359–1368.

Soyka, M., Preuss, U.W., Koller, G., Zill, P., and Bondy, B. (2004) 'Association of 5-HT1B receptor gene and antisocial behavior in alcoholism', *Journal of Neural Transmission*, 111: 101–109.

Spalletta, G., Morris, D.W., Angelucci, F., Rubino, I.A., Spoletini, I., Bria, P., Martinotti, G., and Siracusano (2010) 'BDNF Val66Met polymorphism is associated with aggressive behavior in schizophrenia', *European Psychiatry*, 25: 311–313.

Spanagel, R. (2009) 'Alcoholism: a systems approach from molecular physiology to addictive behavior', *Physiological Review*, 89: 649–705.

Stallings, M.C., Corley, R.P., Dennehey, B., Hewitt, J.K., Krauter, K.S., Lessem, J.M., Mikulich-Gilbertson, S.K., Rhee, S.H., Smolen, A., Young, S.E., and Crowley, T.J. (2005) 'A genome-wide search for quantitative trait Loci that influence antisocial drug dependence in adolescence', *Archives of General Psychiatry*, 62: 1042–1051.

Steiger, H., Richardson, J., Joober, R., Israel, M., Bruce, K.R., Ng Ying Kin, N.M., Howard, H., Anestin, A., Dandurand, C., and Gauvin, L. (2008) 'Dissocial behavior, the 5HTTLPR polymorphism, and maltreatment in women with bulimic syndromes', *American Journal of Medical Genetics B, Neuropsychiatric Genetics*, 147B: 128–130.

Strac, D., Nedic, G., Nikolac, M., Sviglin, K.N., Muck-Seler, D., and Pivac, N. (2012) 'The GABA receptor alpha 2 subunit gene (GABRA2) is association with alcohol-related behavior', *BMC Pharmacology and Toxicology*, 13: A8.

Terracciano, A., Sanna, S., Uda, M., Deiana, B., Usala, G., Busonero, F., Maschio, A., Scally, M., Patriciu, N., Chen, W., Distel, M., Slagboom, E., Boomsma, D.I., Villafuerte, S., Sliwerska, E., Burmeister, M., Amin, N., Janssens, J., Van Duijn, C., Schlessinger, D., Abecasis, G., and Costa, P.T. (2010) 'Genome-wide association scan for five major dimensions of personality', *Molecular Psychiatry*, 15: 647–656.

Terracciano, A., Esko, T., Sutin, A.R., De Moor, M., Meirelles, O., Zhu, G., Tanaka, T., Giegling, I., Nutile, T., Realo, A., Allik, J., Hansell, N.K., Wright, M.J., Montgomery, G., Willemsen, G., Hottenga, J., Friedl, M., Ruggiero, D., Sorice, R., Sanna, S., Cannas, A., Raikkonen, K., Widen, E., Palotie, A., Eriksson, J.G., Cucca, F., Krueger, R.F., Lahti, J., Luciano, M., Smoller, J., Van Duijn, C., Abecasis, G., Boomsma, D.I., Ciullo, M., Costa, P.T., Ferrucci, L., Martin, N.G., Metspalu, A., Rujescu, D., Schlessinger, D., and Uda, M. (2011) 'Meta-analysis of genome-wide association studies identifies common variants in CTNNA2 associated with excitement-seeking', *Translational Psychiatry*, 1: doi: e49; doi: 10.1038tp.2011.42.

Terranova, C., Tucci, M., Sartore, D., Cavarzeran, F., Di Pietra, L., Barzon, L., Palu, G., and Ferrara, S. (2013) 'GABA receptors, alcohol dependence and criminal behavior', *Journal of Forensic Sciences*, 58: 1227–1232.

Thapar, A., Langley, K., Fowler, T., Rice, F., Turic, D., Whittinger, N., Aggleton, J., Van Den Bree, M., Owen, M., and O'Donovan, M. (2005) 'Catechol O-methyltransferase gene variant and birth weight predict early-onset antisocial behavior in children with attention-deficit/hyperactivity disorder', *Archives of General Psychiatry*, 62: 1275–1278.

Thomas, P.D., Mi, H., Swan, G.E., Lerman, C., Benowitz, N., Tyndale, R.F., Bergen, A.W., and Conti, D.V. (2009) 'A systems biology network model for genetic association studies of nicotine addiction and treatment', *Pharmacogenetic Genomics*, 19: 538–551.

Tielbeek, J.J., Medland, S.E., Benyamin, B., Byrne, E.M., Heath, A.C., Madden, P.A., Martin, N.G., Wray, N.R., and Verweij, K.J. (2012) 'Unraveling the genetic etiology of adult antisocial behavior: a genome-wide association study', *PLoS One*, 7: e45086.

Tikkanen, R., Sjoberg, R.L., Ducci, F., Goldman, D., Holi, M., Tiihonen, J., and Virkkunen, M. (2009) 'Effects of MAOA-genotype, alcohol consumption, and aging on violent behavior', *Alcohol Clinical Experimental Research*, 33: 428–434.

Tikkanen, R., Ducci, F., Goldman, D., Holi, M., Lindberg, N., Tiihonen, J., and Virkkunen, M. (2010) 'MAOA alters the effects of heavy drinking and childhood physical abuse on risk for severe impulsive acts of violence among alcoholic violent offenders', *Alcoholism: Clinical and Experiential Research*, 34: 853–860.

Tikkanen, R., Auvinen-Lintunen, L., Ducci, F., Sjoberg, R., Goldman, D., Tiihonen, J., Ojansuu, I., and Virkkunen, M. (2011) 'Psychopathy, PCL-R, and MAOA genotype as predictors of violent reconvictions', *Psychiatry Research*, 185: 382–386.

Van Ijzendoorn, M.H., Caspers, K., Bakermans-Kranenburg, M.J., Beach, S.R., and Philibert, R.A. (2010) 'Methylation matters: interaction between methylation density and 5HTT genotype predicts unresolved loss or trauma', *Biological Psychiatry*, 68: 405–407.

Vanyukov, M.M., Maher, B.S., Devlin, B., Kirillova, G.P., Kirisci, L., Yu, L.M., and Ferrell, R.E. (2007) 'The MAOA promoter polymorphism, disruptive behavior disorders, and early onset substance use disorder: gene–environment interaction', *Psychiatric Genetics*, 17: 323–332.

Vaske, J., Newsome, J., Makarios, M., Wright, J.P., Boutwell, B.B., and Beaver, K.M. (2009) 'Interaction of 5HTTLPR and marijuana use on property offending', *Biodemography and Social Biology*, 55: 93–102.

Vassos, E., Collier, D., and Fazel, S. (2013) 'Systematic meta-analyses and field synopsis of genetic association studies of violence and aggression', *Molecular Psychiatry*, 19(4): 471–477.

Vaughn, M.G., Beaver, K.M., DeLisi, M., Howard, M.O., and Perron, B.E. (2009) 'Dopamine D4 receptor gene exon III polymorphism associated with binge drinking attitudinal phenotype', *Alcohol*, 43: 179–184.

Vevera, J., Stopkova, R., Bes, M., Albrecht, T., Papezova, H., Zukov, I., Raboch, J., and Stopka, P. (2009) 'COMT polymorphisms in impulsively violent offenders with antisocial personality disorder', *Neuro Endocrinology Letters*, 30: 753–756.

Viding, E., Hanscombe, K.B., Curtis, C.J.C., Davis, O.S.P., Meaburn, E.L., and Plomin, R. (2010) 'In search of genes associated with risk for psychopathic tendencies in children: a two-stage genome-wide association study of pooled DNA', *Journal of Child Psychology and Psychiatry*, 51: 780–788.

Viding, E., Fontaine, N.M., and McCrory, E.J. (2012) 'Antisocial behaviour in children with and without callous-unemotional traits', *Journal of the Royal Society of Medicine*, 105: 195–200.

Volavka, J. (2008) 'Relation between COMT and aggression: a moving target?', *Psychiatrie*, 12: 72–73.

Volavka, J., Bilder, R., and Nolan, K. (2004) 'Catecholamines and aggression: the role of COMT and MAO polymorphisms', *Annals of the New York Academy of Sciences*, 1036: 393–398.

Wagner, S., Baskaya, O., Anicker, N.J., Dahmen, N., Lieb, K., and Tadic, A. (2010) 'The catechol *o-methyltransferase* (COMT) val(158)met polymorphism modulates the association of serious life events (SLE) and impulsive aggression in female patients with borderline personality disorder (BPD)', *Acta Psychiatry Scandinavia*, 122: 110–117.

Wahlsten, D. (2013) 'A contemporary view of genes and behavior: complex systems and interactions', *Advances in Child Development and Behavior*, 44: 285–306.

Wakschlag, L.S., Kistner, E.O., Pine, D.S., Biesecker, G., Pickett, K.E., Skol, A.D., Dukic, V., Blair, R.J., Leventhal, B.L., Cox, N.J., Burns, J.L., Kasza, K.E., Wright, R.J., and Cook, E.H. Jr (2010) 'Interaction of prenatal exposure to cigarettes and MAOA genotype in pathways to youth antisocial behavior', *Molecular Psychiatry*, 15: 928–937.

Wang, D., Szyf, M., Benkelfat, C., Provencal, N., Turecki, G., Caramaschi, D., Cote, S., Vitaro, F., Tremblay, R.E., and Booij, J. (2012a) 'Peripheral SLC6A4 DNA methylation is associated with in vivo measures of human brain serotonin synthesis and childhood physical aggression', *PLoS One*, 7: e39501.

Wang, T.J., Huang, S.Y., Lin, W.W., Lo, H.Y., Wu, P.L., Wang, Y.S., Wu, Y.S., Ko, H.C., Shih, J.C., and Lu, R.B. (2007) 'Possible interaction between MAOA and DRD2 genes associated with antisocial alcoholism among Han Chinese men in Taiwan', *Progress in Neuropsychopharmacology and Biological Psychiatry*, 31: 108–114.

Wang, T.Y., Lee, S.Y., Chen, S.L., Chang, Y.H., Chen, S.H., Chu, C.H., Huang, S.Y., Tzeng, N.S., Wang, C.L., Lee, I.H., Yeh, T.L., Yang, Y.K., and Lu, R.B. (2012b) 'Interaction between serotonin transporter and serotonin receptor 1 B genes polymorphisms may be associated with antisocial alcoholism', *Behavior and Brain Functions*, 8: 18.

Weder, N., Yang, B.Z., Douglas-Palumberi, H., Massey, J., Krystal, J.H., Gelernter, J., and Kaufman, J. (2009) 'MAOA genotype, maltreatment, and aggressive behavior: the changing impact of genotype at varying levels of trauma', *Biological Psychiatry*, 65: 417–424.

Westberg, L. and Eriksson, E. (2008) 'Sex steroid-related candidate genes in psychiatric disorders', *Journal of Psychiatry and Neuroscience*, 33: 319–330.

Widom, C.S., and Brzustowicz, L.M. (2006) 'MAOA and the "cycle of violence:" childhood abuse and neglect, MAOA genotype, and risk for violent and antisocial behavior', *Biological Psychiatry*, 60: 684–689.

Williams, L.M., Gatt, J.M., Kuan, S.A., Dobson-Stone, C., Palmer, D.M., Paul, R.H., Song, L., Costa, P.T., Schofield, P.R., and Gordon, E. (2009) 'A polymorphism of the MAOA gene is associated with emotional brain markers and personality traits on an antisocial index', *Neuropsychopharmacology*, 34: 1797–1809.

Wu, C.Y., Wu, Y.S., Lee, J.F., Huang, S.Y., Yu, L., Ko, H.C., and Lu, R.B. (2008) 'The association between DRD2/ANKK1, 5-HTTLPR gene, and specific personality trait on antisocial alcoholism among Han Chinese in Taiwan', *American Journal of Medica Genetics B, Neuropsychiatric Genetics*, 147B: 447–453.

Yang, B.Z., Kranzler, H.R., Zhao, H., Gruen, J.R., Luo, X., and Gelernter, J. (2008) 'Haplotypic variants in DRD2, ANKK1, TTC12, and NCAM1 are associated with comorbid alcohol and drug dependence', *Alcohol Clinical and Experimental Research*, 32: 2117–2127.

Yang, M., Kavi, V., Wang, W., Wu, Z., and Hao, W. (2012) 'The association of 5-HTR2A-1438A/G, COMT-Val 158 Met, MAOA-LPR, DATVNTR and 5-HTTVNTR gene polymorphisms and antisocial personality disorder in male heroin-dependent Chinese subjects', *Progress in Neuropsychopharmacological and Biological Psychiatry*, 36: 282–289.

Yildrim, B. and Derksen, J.J. (2013) 'Systematic review, structural analysis, and new theoretical perspectives on the role of serotonin and associated genes in the etiology of psychopathy and sociopathy', *Neuroscience and Biobehavioral Reviews*, 37: 1254–1296.

Young, S.E., Smolen, A., Hewitt, J.K., Haberstick, B.C., Stallings, M.C., Corley, R.P., and Crowley, T.J. (2006) 'Interaction between MAO-A genotype and maltreatment in the risk for conduct disorder: failure to confirm in adolescent patients', *American Journal of Psychiatry*, 163: 1019–1025.

Zai, C.C., Muir, K.E., Nowrouzi, B., Shaikh, S., Choi, E., Berall, L., Trepanier, M., Beitchman, J.H., and Kennedy, J.L. (2012) 'Possible genetic association between vasopressin receptor 1B and child aggression', *Psychiatry Research*, 200: 784–788.

Zai, C.C., Ehtesham, S., Choi, E., Nowrouzi, B., De Luca, V., Stankovich, L., Davidge, K., Freeman, N., King, N., Kennedy, J.L., and Beitchman, J.H. (2013) 'Dopaminergic sytem genes in childhood aggression: possible role for DRD2', *World Journal of Biological Psychiatry*, 13: 65–74.

Ziermans, T., Dumontheil, I., Roggeman, C., Peyrard-Janvid, M., Matsson, H., Kere, J., and Klingberg, T. (2012) 'Working memory brain activity and capacity link MAOA polymorphism to aggressive behavior during development', *Translational Psychiatry*, 2: 1–9.

12

The warrior gene

MAOA genotype and antisocial behavior in males

Nicholas R. Holland and Matt DeLisi

Introduction

In a remarkable change of direction, the study of antisocial behavior has transformed from an almost entirely environmental approach to a biosocial paradigm where biological, psychological, and sociological constructs are studied for their multifactorial effects on crime. With the advent of social science datasets that contain biological measures—including measures genes—a dizzying array of studies have been produced linking genetic variants to endophenotypes (e.g. individual-level and personality features such as temper, impulsivity, extraversion, negative emotionality, and aggression) associated with conduct problems and even criminological phenotypes. Several scholars have commented on this dramatic paradigm shift (Moffitt, 2005; Raine, 2008, 2013; Beaver, 2009; DeLisi, 2012; Rocque *et al.*, 2012), and celebrated its role in the scientific ascendancy of criminology and criminal justice research. For instance, DeLisi (2009: 221) concluded:

> Because of this marvelous scientific accomplishment [mapping the human genome], behavioral scientists can now access the role of nature and nurture in determining complex behavioral outcomes . . . The misleading black box of 'nature or nurture' has been replaced by a transparent understanding of nature and nurture. Consequently, there has never been a more exciting time to be a criminologist.

A signal finding from this new approach relates to the complex interplay between genetic and environmental conditions to produce behavior. To date, one of the most studied susceptibility genes for antisocial behavior and cognate antisocial traits is the polymorphic monoamine oxidase A (MAOA) gene (for technical overviews see Chen *et al.*, 1991; Sabol *et al.*, 1998; Shih and Thompson, 1999; Shih *et al.*, 1999). MAOA is a catabolic enzyme or enzymatic degrader that is involved with regulating neurotransmitters including serotonin (5HT), norepinephrine (NE), dopamine (DA), and neuromodulators such as phenylethylamine (PEA) in the synapses. It is particularly involved in the regulation of serotonin and thus has generated scholarly attention for its potential role in psychiatric and behavioral traits.

Localized to Xp11.23–11.4 (Levy *et al.*, 1989), a 30-base pair variable number of tandem repeat sequence located upstream of the MAOA gene (MAOA-uVNTR) affects the transcriptional potential of the MAOA promoter region. This produces alleles with low activity and high activity. Sabol and colleagues (1998) have shown that alleles with 3.5 or 4 copies of the repeat sequence are transcribed 2 to 10 times more efficiently than those with 3 or 5 copies of the repeat (there is also a 2-repeat allele that is discussed later in this chapter). Conceptually, low-activity alleles are operationalized as genetic risk factors that reflect MAOA deficiency and, it is theorized, pose behavioral risk.

Seminal MAOA studies

A cadre of landmark studies suggests MAOA is a potentially important candidate gene for antisocial conditions.[1] Brunner *et al.* (1993a) published the seminal study of Brunner Syndrome, which identified several males in a Dutch kindred with a complete and selective deficiency of MAOA activity caused by a glutamine-to-thymine mutation that changed a glutamine (CAG) codon to a termination (TAG) codon. In other words, affected males had a knockout gene for MAOA. The result in affected males is borderline mental retardation and severe forms of aggressive behavior toward persons within their family and directed toward others in the community.

The hallmark of Brunner Syndrome was "repeated episodes of aggression, sometimes violent behavior, occurring in all eight affected males. Aggressive behavior was usually triggered by anger and was often out of proportion to the provocation. Aggressive behavior tended to cluster in periods of 1–3 days, during which the affected male would sleep very little and would experience frequent night terrors" (Brunner *et al.*, 1993b: 1035). Additionally, males with Brunner Syndrome engaged in a litany of violent acts, including rape, incest-rape, voyeurism, assault with a deadly weapon, arson, and others. It is important to note that this MAOA variant is exceedingly rare.

About a decade later, another study published in *Science* revolutionized the way that criminologists understand the role of genes in antisocial behavior. Utilizing data from the Dunedin Multidisciplinary Health and Development Study (New Zealand), Caspi and colleagues (2002) performed the first criminological gene–environment interaction study by modeling the interplay between MAOA variants and childhood maltreatment. Between the ages of 3 and 11 years, 8 percent of their sample had been severely maltreated, 28 percent experienced probable maltreatment, and 64 percent experienced no maltreatment. Dramatic interactive effects were found between low MAOA activity and maltreatment in the prediction of conduct disorder, violent convictions, violent disposition, and symptoms of antisocial personality disorder. For example, 80 percent of youth with MAOA risk alleles and severe maltreatment had conduct disorder, and 30 percent were convicted of a violent crime. The prevalence for those at genetic risk with no maltreatment was 20 and 5 percent, respectively. About 85 percent of males with both genetic and environmental risk factors displayed some form of antisocial behavior (Caspi *et al.*, 2002).[2]

Other researchers have focused on the neurological and neural substrates that are associated with MAOA activity. For instance, Alia-Klein and colleagues (2008) examined the gene products of MAOA in the brain and its association with trait aggression using positron emission tomography. This provided a direct examination of brain monoamine levels. They found significant correlations between brain MAOA activity in cortical and limbic structure and trait aggression. Moreover, brain MAOA levels explained approximately 30 percent of the variance in trait aggression. This study is significant because it shows the end result or "enzyme product" of the MAOA gene.[3]

The latter finding is significant because prominent theoretical models articulate frontal–limbic connectivity as the neurological pathways for aggression and related conditions.[4] For instance, Meyer-Lindenberg *et al.* (2006) performed an fMRI study and found that the low-expression MAOA variant was associated with increased risk for violent behavior and hyper-responsive limbic reactions during emotional arousal (a description that is sharply consistent with the behavioral repertoire of the males in Brunner's studies). Males with low-expression MAOA displayed structural and functional differences in relevant pathways.

More recently, Buckholtz and Meyer-Lindenberg (2008) developed an explanatory model for how MAOA alters the neurogenetic architecture of aggression. They suggest that MAOA modifies an individual's "socioaffective scaffold" particularly by altering serotonin and norepinephrine levels during development of the coritcolimbic circuit, which results in impairments in social decision-making and affect regulation. This compromises the ability to interpret ambiguous social interactions and perceptions of potential threat.[5]

According to Buckholtz and Meyer-Lindenberg (2008: 127), "the MAOA-L [low activity] allele labilizes this network for social decision making and affect regulation, rendering risk allele carriers more vulnerable to the influence of adverse early life experiences." In other words, their conceptual model provides a precise neurological backdrop to the gene–environment effects shown by Caspi and his collaborators. Moreover, this explanatory model squares with links of MAOA to impulsive, reactive aggression, such as domestic violence assaults or bar fight assaults. Subsequent work demonstrated that MAOA modulated the connectivity between the amygdala and ventromedial prefrontal cortex (Buckholtz *et al.*, 2008), further bolstering its role in cortical–subcortical pathways.

Endophenotypes and phenotypes

Particularly after the Caspi study, a floodgate of sorts was opened, where criminological gene hunters began to assay the ways that MAOA variants were associated with antisocial traits and behavioral phenotypes. Many of these works attempted to replicate the gene–environment interactions by particularly focusing on negative, abusive, or dysfunctional childhood factors. MAOA associations with antisocial conditions and behaviors have been produced from data selected from several large-scale studies spanning nations and continents. For instance, in their longitudinal study of nearly 400 male participants from the Christchurch (New Zealand) Health and Development Study followed through age 30, Fergusson *et al.* (2011, 2012) linked MAOA and childhood exposure to sexual and physical abuse to various forms of antisociality. These included hostility, conduct problems, property offending, and violent offending. Based on data from the Minnesota Twin Family Study, Derringer *et al.* (2010) found that persons who had experienced childhood sexual assault victimization and had low-activity alleles of MAOA displayed more antisocial behaviors and evinced more symptoms of conduct disorder.

For example, using data from the Virginia Twin Study for Adolescent Behavioral Development, Foley and colleagues (2004) largely replicated the Dunedin findings by demonstrating an MAOA–childhood adversity (defined by interparental violence, parental neglect, and inconsistent discipline) interaction in the prediction of conduct disorder. Among boys with low-activity MAOA alleles with the highest childhood adversity exposure, 100 percent had conduct disorder. Among those at genetic risk with the lowest level of environmental exposure, only about 5 percent had conduct disorder.

Drawing on data from the Environmental Risk (E-Risk) Longitudinal Twin Study, Kim-Cohen and colleagues (2006) compared the interaction between MAOA genotype, early life exposure to physical abuse, and various pediatric mental health and behavioral outcomes. On every measure, low-activity MAOA genotype among abused children was associated with significantly worse outcomes, and the effect sizes were large. Children had more mental health problems (Cohen's d = 1.55), more antisocial behaviors (d = 1.33), more attention and hyperactivity problems (d = 1.32), and more emotional problems (d = 1.15). Additionally, they meta-analyzed studies on the MAOA–maltreatment link and found significant associations between the gene–environment interaction and antisocial outcomes among males with low-activity MAOA alleles (Kim-Cohen *et al.*, 2006).

Much of this research area has shown MAOA linkages to personality and temperamental features that serve as an intermediary bridge between genotype and phenotype—a concept known as endophenotype. An assortment of studies has produced significant associations between MAOA and a multitude of antisocial traits, psychiatric conditions, and personality features that are congruent with self-regulation deficits and conduct problems. Across a variety of samples, these include significant linkages to impulsivity (Manuck *et al.*, 2000; Huang *et al.*, 2004), behavioral disinhibition (Enoch *et al.*, 2010), antisocial traits (Williams *et al.*, 2009), and anger and hostility (Beaver *et al.*, 2011).

Investigators have also found associations between MAOA variants and a variety of other antisocial conditions, including conduct disorder (Derringer *et al.*, 2010), antisocial personality disorder (Sjöberg *et al.*, 2008; Reti *et al.*, 2011), emotional dysfunction (Fowler *et al.*, 2009), low cooperativeness (Mertins *et al.*, 2011), and psychopathy (Fowler *et al.*, 2009; Tikkanen *et al.*, 2011).

In addition, several significant findings exist at the behavioral level. Indeed, MAOA variants are associated with diverse forms of crime, including total delinquency, theft, vandalism, and violence indices (Nilsson *et al.*, 2006; Widom and Brzustowicz, 2006), substance abuse disorders (Vanyukov *et al.*, 2007), alcoholism and alcohol-related problem behaviors (Contini *et al.*, 2006), and mental health problems (Nikulina *et al.*, 2012). In other words, there is considerable versatility in the range of traits and conditions that have been shown to be associated with MAOA polymorphisms.

The effects of MAOA have been shown across raters and in laboratory studies of aggression. To illustrate, Cicchetti *et al.* (2012) reported significant associations between MAOA and self-reported conduct problems, peer-reported delinquency, and teacher-rated antisocial conduct. Similarly, laboratory aggression scores assessed with the point subtraction aggression paradigm (Gallardo-Pujol *et al.*, 2013) and scores on the Aggression Questionnaire (Frazzetto *et al.*, 2007) were significantly associated with MAOA variants.

McDermott and colleagues (2009) conducted a novel experimental study to evaluate the warrior reputation of the MAOA polymorphism. In their study, subjects were able to punish those they believed have taken money from them by administering hot spicy sauce to their opponent in the course of a simple economic power-to-take game, where subjects were informed that either 20 percent or 80 percent of their money had been taken. Those with the low-activity MAOA allele reacted in a particularly aggressive manner to provocation, and displayed more intense and frequent aggressive responses. The study is noteworthy because it confirms a role of MAOA genotype to provocation and aggressive response in an experimental design. The finding also comports with findings emanating from aggression research (cf. Frazzetto *et al.*, 2007; Weder *et al.*, 2009; Gallardo-Pujol *et al.*, 2013).

Within criminology, a multitude of studies have examined the association between MAOA and a variety of antisocial phenotypes using data from the National Longitudinal Study of Adolescent Health (Add Health). Beaver *et al.* (2010a) found significant relationships between low-activity MAOA alleles and delinquent gang membership and the use of a weapon during a fight. Interestingly, these effects were direct and withstood controls for age and race. In another study, Beaver *et al.* (2010b) reported significant interactions between neuropsychological deficits and MAOA genotype. Specifically, this interaction was significantly associated with delinquency, violent delinquency, and low self-control across two waves of data collection. Although currently speculative, their findings offer potential that MAOA is associated with neuropsychological deficits (Moffitt, 1993) and low self-control (Gottfredson and Hirschi, 1990), arguably the two leading theoretical perspectives in criminology.

Recently, Stogner and Gibson (2013) reported significant interactions between low-activity alleles of MAOA and life stress in the onset of alcohol use and the onset of marijuana use. The gene–environment effect trended toward significance of cocaine onset ($P < 0.056$). Given its association with diverse forms of antisocial behavior, it is not surprising that MAOA interacting with various environmental contexts increases the likelihood of criminal justice system involvements such as arrests and convictions (Schwartz and Beaver, 2011; Beaver *et al.*, 2013). Still other studies using the Add Health reported significant linkages between MAOA and fraud (Beaver and Holtfreter, 2009) and criminal desistance (Beaver *et al.*, 2008). In other words, the Add Health studies have demonstrated MAOA effects across a wide swath of antisocial conduct spanning drug, violent, property, and white-collar offending in addition to criminal justice system involvement.[6] In sum, MAOA low-activity alleles have been linked to many endophenotypes that contribute to antisocial and delinquent behavior in addition to a range of criminal phenotypes. Almost always, it is critical to note, these effects are indirect where genetic risk factors are exacerbated by adverse or criminogenic environmental contexts.[7]

Just as MAOA has been associated with a variety of antisocial conditions, it has also been activated by several negative contexts. For instance, MAOA has been shown to interact with social exclusion or ostracism (Gallardo-Pujol *et al.*, 2013), abuse (Huang *et al.*, 2004; Nilsson *et al.*, 2006; Widom and Brzustowicz, 2006; Reif *et al.*, 2007; Åslund *et al.*, 2011; Edwards *et al.*, 2010; Fergusson *et al.*, 2011, 2012; Cicchetti *et al.*, 2012), childhood trauma (Frazzetto *et al.*, 2007; Weder *et al.*, 2009), perceived discrimination (Schwartz and Beaver, 2011), family, peer, and school networks (Guo *et al.*, 2008b), deviant peers (Beaver and Holtfreter, 2009; Lee, 2011), prenatal cigarette exposure (Wakschlag *et al.*, 2010), family stress (Enoch *et al.*, 2010), and life stress (Stogner and Gibson, 2013) in the production of antisociality. Given the salience of childhood maltreatment or abuse, several meta-analyses indicated significant associations between MAOA and antisociality when occurring in these negative environments (Kim-Cohen *et al.*, 2006; Taylor and Kim-Cohen, 2007; Byrd and Manuck, 2013), especially childhood maltreatment.

Conclusion

As is the case with any criminological effect, there is discordant evidence. Of course, the literature does not unanimously support linkages between MAOA polymorphisms, criminogenic environments, and antisocial behavior (Prichard *et al.*, 2007). There are several examples across a range of samples. In a study of schizophrenia patients, Strous and colleagues (2003) reported null effects between MAOA and aggression. In a clinical study of 50 psychiatrically referred boys (mean age 9.5 years), Beitchman *et al.* (2004) reported evidence that the high-activity 4-repeat allele was significantly associated with chronic and pervasive aggression, not low-activity variants as is commonly shown. Substantively similar findings, showing that high-activity MAOA variants and not low-activity versions were associated with antisocial behavior, were also shown in a study of adoptees (van der Vegt *et al.*, 2009).

Based on data from more than 1000 Australian men, Prichard *et al.* (2008) found no direct or interactive relationships between MAOA and antisocial behavior, but did report significant linkages between childhood adversity and antisocial conduct. There is also evidence that MAOA–antisocial behavior linkages are limited to whites, not non-whites (Widom and Brzustowicz, 2006).[8]

Non-significant findings have also been produced from large, nationally representative samples. For instance, analyses of data from the National Youth Survey Family Study (United States) failed to report significant linkages between MAOA and antisocial behavior (Huizinga *et al.*, 2006). Similarly, analyses of data from the Add Health also failed to produce significant associations between MAOA variants and conduct problems (Haberstick *et al.*, 2005, 2013). Moreover, null findings were also found based on data from male adolescents with severe behavioral/conduct problems (Young *et al.*, 2006). In other words, this research team has failed to replicate MAOA–antisocial behavior linkages across datasets.

It is unclear how MAOA is associated with conduct problems among alcoholics. In a study of the 3-repeat MAOA allele, for example, Koller and colleagues (2003) reported non-significant linkages to impulsivity and other antisocial traits among a sample of male alcoholics. Other researchers reported similarly null findings (Saito *et al.*, 2002), which potentially suggests that the MAOA–antisocial behavior linkage is not as robust in samples of alcoholics (cf. Samochowiec *et al.*, 1999; Enoch, 2006; Tikkanen *et al.*, 2010). For instance, Schmidt and colleagues (2000) found that 30 of 59 (51 percent) male alcoholics with antisocial personality disorder carried the low-activity 3-repeat allele, but only 7 of 31 (23 percent) male alcoholics with anxiety/depression symptoms were carriers—a significant difference.

In addition to these equivocal research findings, it is important to note that an understanding of the MAOA–crime relationship is complex and can potentially serve the interests of both crime control and due process models of criminal justice. For example, the MAOA gene can be used to serve the interest of both prosecution and defense, given that genetics risks could be viewed as an aggravating and mitigating circumstance (cf. Bernet *et al.*, 2007; Feresin, 2009; Forzano *et al.*, 2010). Indeed, a convicted murderer in Italy recently had his sentence reduced because of evidence of genetic risk factors, MAOA among them. Although such a decision is perhaps premature and crude, in the way that it interprets MAOA variants as a risk factor for crime to such a degree that it partially absolves criminal responsibility, it is also likely a sign of the future. There is no evidence that research on the warrior gene is slowing down, and fortunately as that research compiles, so too will an elucidation of the ways that MAOA and environmental contexts smolder into antisocial traits and behaviors.

Notes

1 In an early investigation, Ellis (1991) reviewed the evidence for linkages between MAOA genotype and eight antisocial conditions, including impulsivity, defiance to punishment, childhood hyperactivity, poor academic performance, sensation seeking and risk taking, drug and alcohol use, sexual promiscuity, and action-orientation. Based on the weight of the significant associations at the time, Ellis presciently identified MAOA as an apparent marker for antisocial behavior.

2 The prevalence of MAOA risk alleles varies considerably. In a recent study using data from a cohort of males referred for forensic assessment, 45 percent of the violent individuals in the sample possessed the low activity MAOA allele (Reif *et al.*, 2007).

3 A sizeable literature has also examined platelet MAO activity. For example, a study of males hospitalized for forensic assessment found that those with psychopathic personality had significantly lower platelet monoamine oxidase activity than control groups of male nurses and male construction workers (Lidberg *et al.*, 1985). Similarly, low MAO platelet activity was also found in a subgroup of type 2 alcoholics compared with type 1 alcoholics (Hallman *et al.*, 1996) and in drunk-driving criminal offenders

(Eensoo *et al.*, 2004). A study of former delinquents who were stratified by platelet MAO groups found that those with low platelet MAO activity were over three times more likely than those with high platelet MAO activity to continue their criminal career.

4 Other conceptual models similarly point to disruptions between limbic and frontal structures as a neurological basis for impulsive and aggressive behavior. Davidson *et al.* (2000) implicate the serotonergic system and its role in the regulation of negative emotions. Given its role in regulating serotonin, the MAOA genotype is also relevant to their theory.

5 Given this neurological framework, it is not surprising that low-activity variants of MAOA have been linked to personality dysfunction and personality disorders. For example, Jacob and colleagues (2005) examined 566 patients with personality disorders and 281 healthy controls, and found that MAOA was significantly associated with cluster b personality disorders, such as those characterized by impulsive and aggressive behavior (e.g. antisocial personality disorder). Overall, 26 percent of patients in their sample were MAOA-L carriers compared with 16.4 percent of the control group, $P = 0.005$).

6 It is common for studies to pool low-activity MAOA alleles such as the 2-repeat and 3-repeat alleles into one group and compare the effects with high-activity MAOA alleles such as the 3.5-repeat, 4-repeat, and 5-repeat alleles. However, some research has found that the 2-repeat allele is especially important given its lowest level of promoter activity. Beaver and colleagues reported associations between the 2-repeat allele and arrests, incarceration, scores on an antisocial phenotype index, and violent criminal behavior. The allele was not significantly associated with a psychopathic personality traits scale (Beaver *et al.*, 2013). An independent research team also using the Add Health data similarly reported significant associations between the 2-repeat allele, and serious and violent delinquency (Guo *et al.*, 2008a)

7 Craig's (2007) review documented several direct associations between MAOA variants and a range of phenotypes including ADHD, conduct disorder, and alcoholism.

8 The MAOA–antisociality relationship has been found using participants from Brazil (Contini *et al.*, 2006), Italy (Frazzetto *et al.*, 2007), New Zealand (Fergusson *et al.*, 2011, 2012), Spain (Gallardo-Pujol *et al.*, 2013), Germany (Reif *et al.*, 2007), Finland (Sjöberg *et al.*, 2008; Tikkanen *et al.*, 2011), Sweden (Nilsson *et al.*, 2006; Åslund *et al.*, 2011), the United Kingdom (Kim-Cohen *et al.*, 2006; Fowler *et al.*, 2009; Enoch *et al.*, 2010), and multiple samples selected from the United States.

References

Alia-Klein, N., Goldstein, R.Z., Kriplani, A., Logan, J., Tomasi, D., Williams, B., Telang, F., Shumay, E., Biegon, A., Craig, I.W., Henn, F., Wang, G.-J., Volkow, N.D., and Fowler, J.S. (2008) 'Brain monoamine oxidase A activity predicts trait aggression', *Journal of Neuroscience*, 28(19): 5099–5104.

Åslund, C., Nordquist, N., Comasco, E., Leppert, J., Oreland, L., and Nilsson, K.W. (2011) 'Maltreatment, MAOA, and delinquency: sex differences in gene–environment interaction in a large population-based cohort of adolescents', *Behavior Genetics*, 41(2): 262–272.

Beaver, K.M. (2009) *Biosocial Criminology: A Primer*, Dubuque, IA: Kendall/Hunt.

Beaver, K.M. and Holtfreter, K. (2009) 'Biosocial influences on fraudulent behaviors', *Journal of Genetic Psychology*, 170(2): 101–114.

Beaver, K.M., Wright, J.P., DeLisi, M., and Vaughn, M.G. (2008) 'Desistance from delinquency: the marriage effect revisited and extended', *Social Science Research*, 37(3): 736–752.

Beaver, K.M., DeLisi, M., Vaughn, M.G., and Barnes, J.C. (2010a) 'monoamine oxidase A genotype is associated with gang membership and weapon use', *Comprehensive Psychiatry*, 51(2): 130–134.

Beaver, K.M., DeLisi, M., Vaughn, M.G., and Wright, J.P. (2010b) 'The intersection of genes and neuropsychological deficits in the prediction of adolescent delinquency and low self-control', *International Journal of Offender Therapy and Comparative Criminology*, 54(1): 22–42.

Beaver, K.M., Nedelec, J.L., Wilde, M., Lippoff, C., and Jackson, D. (2011) 'Examining the association between MAOA genotype and incarceration, anger and hostility: the moderating influences of risk and protective factors', *Journal of Research in Personality*, 45(3): 279–284.

Beaver, K.M., Wright, J.P., Boutwell, B.B., Barnes, J.C., DeLisi, M., and Vaughn, M.G. (2013) 'Exploring the association between the 2-repeat allele of the MAOA gene promoter polymorphism and psychopathic personality traits, arrests, incarceration, and lifetime antisocial behavior', *Personality and Individual Differences*, 54: 164–168.

Beitchman, J.H., Mik, H.M., Ehtesham, S., Douglas, L., and Kennedy, J.L. (2004) 'MAOA and persistent, pervasive childhood aggression', *Molecular Psychiatry*, 9(6): 546–547.

Bernet, W., Vnencak-Jones, C.L., Farahany, N., and Montgomery, S.A. (2007) 'Bad nature, bad nurture, and testimony regarding MAOA and SLC6A4 genotyping at murder trials', *Journal of Forensic Sciences*, 52(6): 1362–1371.

Brunner, H.G., Nelen, M., Breakefield, X.O., Ropers, H.H., and van Oost, B.A. (1993a) 'Abnormal behavior associated with a point mutation in the structural gene for monoamine oxidase A', *Science*, 262: 578–580.

Brunner, H.G., Nelen, M.R., Van Zandvoort, P., Abeling, N.G., Van Gennip, A.H., Wolters, E.C., *et al.* (1993b) 'X-linked borderline mental retardation with prominent behavioral disturbance: phenotype, genetic localization, and evidence for disturbed monoamine metabolism', *American Journal of Human Genetics*, 52(6): 1032.

Buckholtz, J.W. and Meyer-Lindenberg, A. (2008) 'MAOA and the neurogenetic architecture of human aggression', *Trends in Neurosciences*, 31: 120–129.

Buckholtz, J.W., Callicott, J.H., Kolachana, B., Hariri, A.R., Goldberg, T.E., Genderson, M., Egan, M.F., Mattay, V.S., Weinberger, D.R., and Meyer-Lindenberg, A. (2008) 'Genetic variation in MAOA modulates ventromedial prefrontal circuitry mediating individual differences in human personality', *Molecular Psychiatry*, 13(3): 313–324.

Byrd, A.L. and Manuck, S.B. (2013) 'MAOA, childhood maltreatment, and antisocial behavior: meta-analysis of a gene–environment interaction', *Biological Psychiatry*, 75(1): 9–17.

Caspi, A., McCray, J., Moffitt, T.E., Mill, J., Martin, J., Craig, I.W., Taylor, A., and Poulton, R. (2002) 'Role of genotype in the cycle of violence in maltreated children', *Science*, 297(5582): 851–854.

Chen, Z.Y., Hotamisligil, G.S., Huang, J.K., Wen, L., Ezzeddine, D., Aydin-Muderrisoglu, N., *et al.* (1991) 'Structure of the human gene for monoamine oxidase type A', *Nucleic Acids Research*, 19(16): 4537–4541.

Cicchetti, D., Rogosch, F.A., and Thibodeau, E.L. (2012) 'The effects of child maltreatment on early signs of antisocial behavior: genetic moderation by tryptophan hydroxylase, serotonin transporter, and monoamine oxidase A genes', *Development and Psychopathology*, 24(3): 907–928.

Contini, V., Marques, F.Z.C., Garcia, C.E.D., Hutz, M.H., and Bau, C.H.D. (2006) 'MAOA-uVNTR polymorphism in a Brazilian sample: further support for the association with impulsive behaviors and alcohol dependence', *American Journal of Medical Genetics – Neuropsychiatric Genetics*, 141B(3): 305–308.

Craig, I.W. (2007) 'The importance of stress and genetic variation in human aggression', *BioEssays*, 29(3): 227–236.

Davidson, R.J., Putnam, K.M., and Larson, C.L. (2000) 'Dysfunction in the neural circuitry of emotion regulation – a possible prelude to violence', *Science*, 289(5479): 591–594.

DeLisi, M. (2009) 'Neuroscience and the holy grail: genetics and career criminality', in A. Walsh and K.M. Beaver (eds), *Biosocial Criminology: New Directions in Theory and Research*, New York: Routledge, pp. 209–224.

DeLisi, M. (2012) 'Genetics: *l'enfant terrible* of criminology', *Journal of Criminal Justice*, 40: 515–516.

Derringer, J., Krueger, R.F., Irons, D.E., and Iacono, W.G. (2010) 'Harsh discipline, childhood sexual assault, and MAOA genotype: an investigation of main and interactive effects on diverse clinical externalizing outcomes', *Behavior Genetics*, 40(5): 639–648.

Edwards, A.C., Dodge, K.A., Latendresse, S.J., Lansford, J.E., Bates, J.E., Pettit, G.S., *et al.* (2010) 'MAOA-uVNTR and early physical discipline interact to influence delinquent behavior', *Journal of Child Psychology and Psychiatry*, 51(6): 679–687.

Eensoo, D., Paaver, M., Pulver, A., Harro, M., and Harro, J. (2004) 'Low platelet MAO activity associated with high dysfunctional impulsivity and antisocial behavior: evidence from drunk drivers', *Psychopharmacology*, 172(3): 356–358.

Ellis, L. (1991) 'Monoamine oxidase and criminality: identifying an apparent biological marker for antisocial behavior', *Journal of Research in Crime and Delinquency*, 28: 227–251.

Enoch, M.A. (2006) 'Genetic and environmental influences on the development of alcoholism', *Annals of the New York Academy of Sciences*, 1094(1): 193–201.

Enoch, M.A., Steer, C.D., Newman, T.K., Gibson, N., and Goldman, D. (2010) 'Early life stress, MAOA, and gene–environment interactions predict behavioral disinhibition in children', *Genes, Brain and Behavior*, 9(1): 65–74.

Feresin, E. (2009) 'Lighter sentence for murderer with "bad genes"', *Nature*, 10: 1038.

Fergusson, D.M., Boden, J.M., Horwood, L.J., Miller, A.L., and Kennedy, M.A. (2011) 'MAOA, abuse exposure and antisocial behaviour: 30-year longitudinal study', *British Journal of Psychiatry*, 198(6): 457–463.

Fergusson, D.M., Boden, J.M., Horwood, L.J., Miller, A., and Kennedy, M.A. (2012) 'Moderating role of the MAOA genotype in antisocial behaviour', *British Journal of Psychiatry*, 200(2): 116–123.

Foley, D.L., Eaves, L.J., Wormley, B., Silberg, J.L., Maes, H.H., Kuhn, J., and Riley, B. (2004) 'Childhood adversity, monoamine oxidase a genotype, and risk for conduct disorder', *Archives of General Psychiatry*, 61(7): 738–746.

Forzano, F., Borry, P., Cambon-Thomsen, A., Hodgson, S.V., Tibben, A., de Vries, P., *et al.* (2010) 'Italian appeal court: a genetic predisposition to commit murder?', *European Journal of Human Genetics*, 18(5): 519–521.

Fowler, T., Langley, K., Rice, F., van den Bree, M.B., Ross, K., Wilkinson, L.S., *et al.* (2009) 'Psychopathy trait scores in adolescents with childhood ADHD: the contribution of genotypes affecting MAOA, 5HTT and COMT activity', *Psychiatric Genetics*, 19(6): 312–319.

Frazzetto, G., Di Lorenzo, G., Carola, V., Proietti, L., Sokolowska, E., Siracusano, A., Gross, C., and Troisi, A. (2007) 'Early trauma and increased risk for physical aggression during adulthood: the moderating role of MAOA genotype', *PLoS ONE*, 2(5): e486

Gallardo-Pujol, D., Andrés-Pueyo, A., and Maydeu-Olivares, A. (2013) 'MAOA genotype, social exclusion and aggression: an experimental test of a gene–environment interaction', *Genes, Brain and Behavior*, 12: 140–145.

Gottfredson, M.R. and Hirschi, T. (1990) *A General Theory of Crime*, Stanford, CA: Stanford University Press.

Guo, G., Ou, X.M., Roettger, M., and Shih, J.C. (2008a) 'The VNTR 2 repeat in MAOA and delinquent behavior in adolescence and young adulthood: associations and MAOA promoter activity', *European Journal of Human Genetics*, 16(5): 626–634.

Guo, G., Roettger, M.E., and Cai, T. (2008b) 'The integration of genetic propensities into social-control models of delinquency and violence among male youths', *American Sociological Review*, 73(4): 543–568.

Haberstick, B.C., Lessem, J.M., Hopfer, C.J., Smolen, A., Ehringer, M.A., Timberlake, D., and Hewitt, J.K. (2005) 'Monoamine oxidase A (MAOA) and antisocial behaviors in the presence of childhood and adolescent maltreatment', *American Journal of Medical Genetics – Neuropsychiatric Genetics*, 135B(1): 59–64.

Haberstick, B.C., Lessem, J.M., Hewitt, J.K., Smolen, A., Hopfer, C.J., Halpern, C.T., *et al.* (2013) 'MAOA genotype, childhood maltreatment, and their interaction in the etiology of adult antisocial behaviors', *Biological Psychiatry*, 75(1): 25–30.

Hallman, J., Knorring, L.V., and Oreland, L. (1996) 'Personality disorders according to DSM-III-R and thrombocyte monoamine oxidase activity in type 1 and type 2 alcoholics', *Journal of Studies on Alcohol and Drugs*, 57(2): 155–161.

Huang, Y.Y., Cate, S.P., Battistuzzi, C., Oquendo, M.A., Brent, D., and Mann, J.J. (2004) 'An association between a functional polymorphism in the monoamine oxidase a gene promoter, impulsive traits and early abuse experiences', *Neuropsychopharmacology*, 29: 1498–1505.

Huizinga, D., Haberstick, B.C., Smolen, A., Menard, S., Young, S.E., Corley, R.P., Stallings, M.C., Grotpeter, J., and Hewitt, J.K. (2006) 'Childhood maltreatment, subsequent antisocial behavior, and the role of monoamine oxidase A genotype', *Biological Psychiatry*, 60(7): 677–683.

Jacob, C.P., Müller, J., Schmidt, M., Hohenberger, K., Gutknecht, L., Reif, A., *et al.* (2005) 'Cluster B personality disorders are associated with allelic variation of monoamine oxidase A activity', *Neuropsychopharmacology*, 30(9): 1711–1718.

Kim-Cohen, J., Caspi, A., Taylor, A., Williams, B., Newcombe, R., Craig, I.W., and Moffitt, T.E. (2006) 'MAOA, maltreatment, and gene–environment interaction predicting children's mental health: new evidence and a meta-analysis', *Molecular Psychiatry*, 11(10): 903–913.

Koller, G., Bondy, B., Preuss, U.W., Bottlender, M., and Soyka, M. (2003) 'No association between a polymorphism in the promoter region of the MAOA gene with antisocial personality traits in alcoholics', *Alcohol and Alcoholism*, 38(1): 31–34.

Lee, S.S. (2011) 'Deviant peer affiliation and antisocial behavior: interaction with monoamine oxidase A (MAOA) genotype', *Journal of Abnormal Child Psychology*, 39(3): 321–332.

Levy, E.R., Powell, J.F., Buckle, V.J., Hsu, Y.P.P., Breakefield, X.O., and Craig, I.W. (1989) 'Localization of human monoamine oxidase-A gene to Xp11.23–11.4 by *in situ* hybridization: implications for Norrie Disease', *Genomics*, 5(2): 368–370.

Lidberg, L., Modin, I., Oreland, L., Tuck, J.R., and Gillner, A. (1985) 'Platelet monoamine oxidase activity and psychopathy', *Psychiatry Research*, 16(4): 339–343.

McDermott, R., Tingley, D., Cowden, J., Frazzetto, G., and Johnson, D.D.P. (2009) 'Monoamine oxidase A gene (MAOA) predicts behavioral aggression following provocation', *Proceedings of the National Academy of Sciences, USA*, 106: 2118–2123.

Manuck, S.B., Flory, J.D., Ferrell, R.E., Mann, J.J., and Muldoon, M.F. (2000) 'A regulatory polymorphism of the monoamine oxidase-A gene may be associated with variability in aggression, impulsivity, and central nervous system serotonergic responsivity', *Psychiatry Research*, 95(1): 9–23.

Mertins, V., Schote, A.B., Hoffeld, W., Griessmair, M., and Meyer, J. (2011) 'Genetic susceptibility for individual cooperation preferences: the role of monoamine oxidase A gene (MAOA) in the voluntary provision of public goods', *PLoS ONE*, 6(6): e20959.

Meyer-Lindenberg, A., Buckholtz, J.W., Kolachana, B., Hariri, A.R., Pezawas, L., Blasi, G., Wabnitz, A., Honea, R., Verchinski, B., Callicott, J.H., Egan, M., Mattay, V., and Weinberger, D.R. (2006) 'Neural mechanisms of genetic risk for impulsivity and violence in humans', *Proceedings of the National Academy of Sciences, USA*, 103(16): 6269–6274.

Moffitt, T.E. (1993) 'Adolescence-limited and life-course-persistent antisocial behavior: a developmental taxonomy', *Psychological Review*, 100(4): 674–701.

Moffitt, T.E. (2005) 'The new look of behavioral genetics in developmental psychopathology: gene–environment interplay in antisocial behaviors', *Psychological Bulletin*, 131(4): 533–554.

Nikulina, V., Widom, C.S., and Brzustowicz, L.M. (2012) 'Child abuse and neglect, MAOA, and mental health outcomes: a prospective examination', *Biological Psychiatry*, 71(4): 350–357.

Nilsson, K.W., Sjöberg, R.L., Damberg, M., Leppert, J., Öhrvik, J., Alm, P.O., *et al.* (2006) 'Role of monoamine oxidase A genotype and psychosocial factors in male adolescent criminal activity', *Biological Psychiatry*, 59(2): 121–127.

Prichard, Z.M., Jorm, A.F., Mackinnon, A., and Easteal, S. (2007) 'Association analysis of 15 polymorphisms within 10 candidate genes for antisocial behavioural traits', *Psychiatric Genetics*, 17(5): 299–303.

Prichard, Z., Mackinnon, A., Jorm, A.F., and Easteal, S. (2008) 'No evidence for interaction between MAOA and childhood adversity for antisocial behavior', *American Journal of Medical Genetics, Part B: Neuropsychiatric Genetics*, 147(2): 228–232.

Raine, A. (2008) 'From genes to brain to antisocial behavior', *Current Directions in Psychological Science*, 17(5): 323–328.

Raine, A. (2013) *The Anatomy of Violence: The Biological Roots of Crime*, New York: Random House.

Reif, A., Rösler, M., Freitag, C.M., Schneider, M., Eujen, A., Kissling, C., Wenzler, D., Jacob, C.P., Retz-Junginger, P., Thome, J., Lesch, K.-P., and Retz, W. (2007) 'Nature and nurture predispose to violent behavior: serotonergic genes and adverse childhood environment', *Neuropsychopharmacology*, 32(11): 2375–2383.

Reti, I.M., Xu, J.Z., Yanofski, J., McKibben, J., Uhart, M., Cheng, Y.-J., Zandi, P., Bienvenu, O.J., Samuels, J., Willour, V., Kasch-Semenza, L., Costa, P., Bandeen-Roche, K., Eaton, W.W., and Nestadt, G. (2011) 'Monoamine oxidase A regulates antisocial personality in whites with no history of physical abuse', *Comprehensive Psychiatry*, 52(2): 188–194.

Rocque, M., Welsh, B.C., and Raine, A. (2012) 'Biosocial criminology and modern crime prevention', *Journal of Criminal Justice*, 40(4): 306–312.

Sabol, S.Z., Hu, S., and Hamer, D. (1998) 'A functional polymorphism in the monoamine oxidase A gene promoter', *Human Genetics*, 103(3): 273–279.

Saito, T., Lachman, H.M., Diaz, L., Hallikainen, T., Kauhanen, J., Salonen, J.T., Ryynänen, O.-P., Karvonen, M.K., Syvälahti, E., Pohjalainen, T., Hietala, J., and Tiihonen, J. (2002) 'Analysis of monoamine oxidase A (MAOA) promoter polymorphism in Finnish male alcoholics', *Psychiatry Research*, 109(2): 113–119.

Samochowiec, J., Lesch, K.P., Rottmann, M., Smolka, M., Syagailo, Y.V., Okladnova, O., *et al.* (1999) 'Association of a regulatory polymorphism in the promoter region of the monoamine oxidase A gene with antisocial alcoholism', *Psychiatry Research*, 86(1): 67–72.

Schmidt, L.G., Sander, T., Kuhn, S., Smolka, M., Rommelspacher, H., Samochowiec, J., and Lesch, K.P. (2000) 'Different allele distribution of a regulatory MAOA gene promoter polymorphism in antisocial and anxious-depressive alcoholics', *Journal of Neural Transmission*, 107(6): 681–689.

Schwartz, J.A. and Beaver, K.M. (2011) 'Evidence of a gene × environment interaction between perceived prejudice and MAOA genotype in the prediction of criminal arrests', *Journal of Criminal Justice*, 39(5): 378–384.

Shih, J.C. and Thompson, R.F. (1999) 'Monoamine oxidase in neuropsychiatry and behavior', *American Journal of Human Genetics*, 65(3): 593–598.

Shih, J.C., Chen, K., and Ridd, M.J. (1999) 'Monoamine oxidase: from genes to behavior', *Annual Review of Neuroscience*, 22: 197–217.

Sjöberg, R.L., Ducci, F., Barr, C.S., Newman, T.K., Dell'Osso, L., Virkkunen, M., and Goldman, D. (2008) 'A non-additive interaction of a functional MAO-A VNTR and testosterone predicts antisocial behavior', *Neuropsychopharmacology*, 33(2): 425–430.

Stogner, J.M. and Gibson, C.L. (2013) 'Stressful life events and adolescent drug use: moderating influences of the MAOA gene', *Journal of Criminal Justice*, 41: 357–363.

Strous, R.D., Nolan, K.A., Lapidus, R., Diaz, L., Saito, T., and Lachman, H.M. (2003) 'Aggressive behavior in schizophrenia is associated with the low enzyme activity COMT polymorphism: a replication study', *American Journal of Medical Genetics Part B: Neuropsychiatric Genetics*, 120(1): 29–34.

Taylor, A. and Kim-Cohen, J. (2007) 'Meta-analysis of gene–environment interactions in developmental psychopathology', *Development and Psychopathology*, 19(4): 1029–1037.

Tikkanen, R., Ducci, F., Goldman, D., Holi, M., Lindberg, N., Tiihonen, J., and Virkkunen, M. (2010) 'MAOA alters the effects of heavy drinking and childhood physical abuse on risk for severe impulsive acts of violence among alcoholic violent offenders', *Alcoholism: Clinical and Experimental Research*, 34(5): 853–860.

Tikkanen, R., Auvinen-Lintunen, L., Ducci, F., Sjöberg, R.L., Goldman, D., Tiihonen, J., Ojansuu, I., and Virkkunen, M. (2011) 'Psychopathy, PCL-R, and MAOA genotype as predictors of violent reconvictions', *Psychiatry Research*, 185(3): 382–386.

Vanyukov, M.M., Maher, B.S., Devlin, B., Kirillova, G.P., Kirisci, L., Yu, L.-M., and Ferrell, R.E. (2007) 'The MAOA promoter polymorphism, disruptive behavior disorders, and early onset substance use disorder: gene–environment interaction', *Psychiatric Genetics*, 17(6): 323–332.

van der Vegt, E.J., Oostra, B.A., Arias-Vásquez, A., van der Ende, J., Verhulst, F.C., and Tiemeier, H. (2009) 'High activity of monoamine oxidase A is associated with externalizing behaviour in maltreated and nonmaltreated adoptees', *Psychiatric Genetics*, 19(4): 209–211.

Wakschlag, L.S., Kistner, E.O., Pine, D.S., Biesecker, G., Pickett, K.E., Skol, A.D., Dukic, V., Blair, R.J.R., Leventhal, B.L., Cox, N.J., Burns, J.L., Kasza, K.E., Wright, R.J., and Cook Jr, E.H. (2010) 'Interaction of prenatal exposure to cigarettes and MAOA genotype in pathways to youth antisocial behavior', *Molecular Psychiatry*, 15(9): 928–937.

Weder, N., Yang, B.Z., Douglas-Palumberi, H., Massey, J., Krystal, J.H., Gelernter, J., and Kaufman, J. (2009) 'MAOA genotype, maltreatment, and aggressive behavior: the changing impact of genotype at varying levels of trauma', *Biological Psychiatry*, 65(5): 417–424.

Widom, C.S. and Brzustowicz, L.M. (2006) 'MAOA and the "cycle of violence:" childhood abuse and neglect, MAOA genotype, and risk for violent and antisocial behavior', *Biological Psychiatry*, 60(7): 684–689.

Williams, L.M., Gatt, J.M., Kuan, S.A., Dobson-Stone, C., Palmer, D.M., Paul, R.H., *et al.* (2009) 'A polymorphism of the MAOA gene is associated with emotional brain markers and personality traits on an antisocial index', *Neuropsychopharmacology*, 34(7): 1797–1809.

Young, S., Smolen, A., Hewitt, J., Haberstick, B., Stallings, M., Corley, R., and Crowley, T. (2006) 'Interaction between MAO-A genotype and maltreatment in the risk for conduct disorder: failure to confirm in adolescent patients', *American Journal of Psychiatry*, 163(6): 1019–1025.

13

New methods in gene–environment interplay

The extended children of twin design

Qiang "John" Fu

Introduction

Criminology has traditionally studied causes of crime from an almost purely environmental perspective. Recent research has begun to elucidate genetic influences on criminal behavior (Vaughn *et al.*, 2009; Walsh and Beaver, 2009). The major shortcoming of criminological research using samples that are not genetically informed is the inability to separate genetic and family-environmental confounding effects. Behavior genetics is concerned with genetic and environmental influences on individual difference in natural populations using twin designs. The twin method is one of the major methods used in quantitative genetics to investigate genetic and environmental influences and architecture of behavioral traits (Neale and Cardon, 1982). By now, it is known that genetic effects can be found in all aspects of behavior. Since the genotype pre-exists all behavior, these genetic influences have to be considered a cause at some level. In addition, twin-family studies are uniquely able to address the complex interplay between gene and environment characterized as gene–environment interaction and correlation (Plomin *et al.*, 1977; Kendler and Eaves, 1986). We begin by introducing the logic and assumptions of the classical twin design. Then we describe the children of twins (CoT) design. Last, we expand the topic to the extended children of twins (eCoT) design.

Classical twin design

Twin studies have been used to estimate genetic and different environmental variances of behavior. If the resemblance of behavior between monozygotic (MZ) twins is greater than between dizygotic (DZ) twins, genetic influence is implicated. Any deviation from this observation suggests environmental effects on behavior. The effect size of genetic influence in a population at a particular point in time is estimated by heritability, which is the proportion of phenotypic variation that can be explained by the genetic variation. The heritability does not apply to individuals or phenotypes as a fixed feature (Rutter, 2006). The narrow-sense heritability is defined as additive genetic influences, which indicates similarity between parent and offspring. The heritability of behavior can be estimated by comparing resemblance between certain types of twin member.

For example, MZ twins share all of their genetic make-up, while DZ twins share on average half their genetic make-up. Both MZ and DZ twins share a family environment to similar degrees. The correlations in MZ and DZ twins expressed in terms of shared variance components are:

$$r_{MZ} = h^2 + c^2$$
$$r_{DZ} = (h^2/2) + c^2$$

where r_{MZ} represents MZ correlation, r_{DZ}, DZ correlation, h^2, additive genetic variance, and c^2, shared environmental variance. If the correlation observed between two siblings of MZ twin pairs for a behavior is greater than that observed between two siblings of DZ twin pairs, the heritability can be derived from twice the difference between MZ and DZ correlations [i.e. $h^2 = 2(r_{MZ} - r_{DZ})$]. Because twins grow up in a very similar environment, the shared environmental variance that is the resource to make twin siblings similar can be obtained as MZ correlation minus the heritability ($c^2 = r_{MZ} - h^2$). The nonshared environmental variance labeled e^2, which is the resource to make twin siblings different, can be calculated as subtracting MZ correlation from one ($e^2 = 1 - r_{MZ}$) given that h^2, c^2, and e^2 must sum to 1 ($h^2 + c^2 + e^2 = 1$). For example, if MZ and DZ correlations for antisocial personality and behavior were 0.67 and 0.39, respectively, the heritability estimate would be 2 × (0.67 – 0.39) = 0.56; the shared environmental variance would be 0.67 – 0.56 = 0.11; the nonshared environmental variance would be 1 – 0.67 = 0.33.

Simple comparisons between MZ and DZ twin correlations are the first step to examine whether genetic influence exists. However, this approach does not provide a statistical test for parameter estimates and 95 percent confidence intervals, and it is difficult to go beyond the heritability estimate. It is worth noting that heritability is population-specific. It can be changed in different populations and over time. Even though genetic influences are strong on some traits, environmental factors can significantly impact those traits.

The contemporary biometric modeling uses a model-fitting approach to detect genetic and environmental effects on behavior. The classical twin model is formulated as a structural equation model. For example, to examine genetic and environmental influences on antisocial personality and behavior, Figure 13.1 illustrates a path diagram for a theoretical model that partitions the total variance in antisocial personality and behavior into additive genetic (i.e. heritability), shared environmental, and nonshared environmental variances, respectively. Capital letters represent variables, and lower case letters represent parameters to be estimated. Ellipses represent latent variables, and rectangles represent observed variables. The unobserved genetic and environmental factors are typically represented by latent variables A, C, and E. Single-headed arrows represent path coefficients. Double-headed curved arrows between latent variables represent correlation coefficients. Double-headed curved arrows attached to latent variables represent that the variance of each latent variable is constrained to unity. Because the variances are fixed, path coefficients (*h*, *c*, and *e*) that represent the effects of the latent variables are estimated. This model is termed the "ACE model." In addition to additive genetic influence, the other type of genetic influence is dominant genetic effect. Because a twin model that estimates both dominant genetic variance and shared environmental variance parameters simultaneously is not identified, only one of the two parameters can be estimated in an ACE model. If dominance is not important for a trait, it may be omitted in the model. Measurement error is contained in the E variable. Several important assumptions for the ACE model should be noted. One assumption is that the environmentally caused similarity is the same for MZ and DZ twin pairs reared in the same family, termed the "equal environment assumption" (Bouchard and Propping, 1993). Empirical evidence suggests that this assumption is testable and holds up for many traits

(Bouchard and Propping, 1993; Derks *et al.*, 2006). This is why the ACE model equates the correlation for shared environmental variables in MZ and DZ twin pairs as unity, implying that MZ and DZ twins differ only in genetic variance. If MZ twin pairs were treated more similarly than DZ twin pairs, the estimate of genetic influence would be inflated. Another important assumption for the classical twin method is random mating. This means that the biological parents of twins on average do not correlate with each other on a trait of interest. This assumption is crucial because genetic correlation between DZ twin siblings that is on average a half is built on this assumption. If assortative mating had an effect on a trait, the genetic correlation in DZ twins would be greater than a half, which would result in underestimating the heritability. If the assumption of random mating is violated, the unbiased parameter estimates can be obtained by including spouses of twins in the ACE model. The final assumption is that the characteristics of twins are similar to those of singletons. Although the young twins have been found different from singletons in regard to birth weight, premature birth, and IQ, twins catch up with their counterparts in singletons in their early to middle childhood (Evans and Martin, 2000).

An ACE model can be written in expected variance–covariance matrices formulated in terms of various parameters (path coefficient h, c, and e):

$$\begin{bmatrix} \text{Variance}_{\text{MZ Twin1}} & \\ \text{Covariance}_{\text{MZ Twin1,2}} & \text{Variance}_{\text{MZ Twin2}} \end{bmatrix} = \begin{bmatrix} h^2 + c^2 + e^2 & \\ h^2 + c^2 & h^2 + c^2 + e^2 \end{bmatrix}$$

$$\begin{bmatrix} \text{Variance}_{\text{DZ Twin1}} & \\ \text{Covariance}_{\text{DZ Twin1,2}} & \text{Variance}_{\text{DZ Twin2}} \end{bmatrix} = \begin{bmatrix} h^2 + c^2 + e^2 & \\ \frac{h^2}{2} + c^2 & h^2 + c^2 + e^2 \end{bmatrix}$$

The observed data are two variance–covariance matrices that include two variances and a covariance from MZ twins, and two variances and a covariance from DZ twins. A likelihood ratio (LR) chi-square test is used to evaluate the goodness-of-fit of the ACE model to the observed data by comparing the fit between expected and observed matrices. A statistically significant LR chi-square test indicates that the ACE model does not fit the observed data well. An alternative model that omits insignificant parameters (e.g. AE or CE model) should be evaluated. In contrast, a nonsignificant test result suggests that the ACE model fits the data well. Other fit indices, such as comparative fit index (CFI), goodness-of-fit statistic (GFI), and Akaike information criterion (AIC), can also be used to evaluate model fit. Then the maximum likelihood parameter estimates (h, c, and e in Figure 13.1) are obtained to quantify the genetic (h^2), shared environmental (c^2), or nonshared environmental (e^2) variances. In modern twin analysis, the twin data can be analyzed using different statistical methods including structural equation modeling (Boker *et al.*, 2011), and mixed effects models (van den Berg *et al.*, 2006; Rabe-Hesketh *et al.*, 2008). A recent meta-analysis summarized the findings from previous twin studies on antisocial personality and behavior. The author reported that 56, 11, and 33 percent of variance in antisocial personality and behavior can be explained due to additive genetic (h^2), shared environmental (c^2), and nonshared environmental (e^2) influences, respectively (Ferguson, 2010).

The ACE model is easily extended to a bivariate genetic model for two behavioral traits by including variances of the second behavioral trait as well as cross-twin and cross-trait covariances in the expected variance–covariance matrices. This bivariate twin model used the Cholesky decomposition method to parametize genetic and environmental parameter estimates. If the cross-twin and cross-trait associations are significantly stronger in MZ than in DZ pairs, this suggests a common genetic liability that contributes to their covariation. For example, a twin study investigated the common genetic and environmental effects specific to mental wellbeing

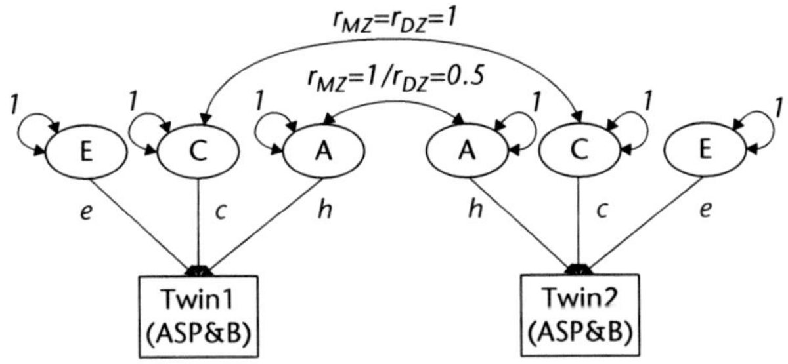

Figure 13.1 A schematic ACE model for twins' antisocial personality and behavior

(h_1, c_1, and e_1) and those shared with externalizing psychopathology (h_3, c_3, and e_3) shown in Figure 13.2 (Kendler *et al.*, 2011). A similar process of model-fitting was conducted to identify the best-fitting model (i.e. choosing the ACE or AE model). Then the maximum likelihood parameter estimates were obtained from the best-fitting model. The study did not find statistically significant shared environmental influences (c_1, c_2, and c_3) on both phenotypes. Thus the AE model that excluded C and C' latent variables was the best-fitting model shown in Figure 13.3. The study reported that the additive genetic and nonshared environmental path coefficients for externalizing psychopathology were 0.98 and 0.21, respectively. The additive genetic and nonshared environmental path coefficients specific to mental wellbeing were 0.82 and 0.47, respectively. The additive genetic and nonshared environmental path coefficients from externalizing psychopathology to mental wellbeing were −0.23, and −0.24, respectively, suggesting that externalizing psychopathology severity is negatively associated with mental wellbeing. Based on the path coefficients, additive genetic and nonshared environmental variances in externalizing psychopathology and mental wellbeing can be calculated. Additive genetic variance in externalizing psychopathology accounts for 7 percent $\{h_3^2/(h_1^2 + h_3^2) = (-0.23)^2/[0.82^2 + (-0.23)^2]\}$ of additive genetic variance in mental wellbeing. The nonshared environmental variance in externalizing psychopathology accounts for 21 percent $\{e_3^2/(e_1^2 + e_3^2) = (-0.24)^2/[0.47^2 + (-0.24)^2]\}$ of the nonshared environmental variance in mental wellbeing. The additive genetic and nonshared environmental variances in externalizing psychopathology are 96 percent ($h_2^2 = 0.98^2$) and 4 percent ($e_2^2 = 0.21^2$), respectively. The additive genetic and nonshared environmental variances specific to mental wellbeing are 67 percent $\{[h_1^2/(h_1^2 + h_3^2 + e_1^2 + e_3^2)] = 0.82^2/[0.82^2 + (-0.23)^2 + 0.47^2 + (-0.24)^2]\}$ and 22 percent $\{[e_1^2/(h_1^2 + h_3^2 + e_1^2 + e_3^2)] = 0.47^2/[0.82^2 + (-0.23)^2 + 0.47^2 + (-0.24)^2]\}$, respectively, after controlling for the genetic and nonshared environmental variances overlapping with those for externalizing psychopathology. This model can be extended to include multiple behavioral traits (Fu *et al.*, 2002, 2007).

If genetic influence on a behavior trait is conditional on an environmental factor, this implies gene and environment (G × E) interaction. A biometric model can be fitted to the observed data stratified by the environmental factor. Alternatively, an interaction term between the heritability variable and the environmental factor is formed and included in the biometric model. If the genetic effect is significantly different across different levels of the environmental factor, it indicates the interactional effect. A twin study examined whether the development of antisocial behavior varies by child maltreatment (Jaffee *et al.*, 2005). Significant G × E influence was found, implicating that part of the genetic effect on antisocial behavior operated through influences on susceptibility to the ill effects of child maltreatment. Another twin study shows that genetic

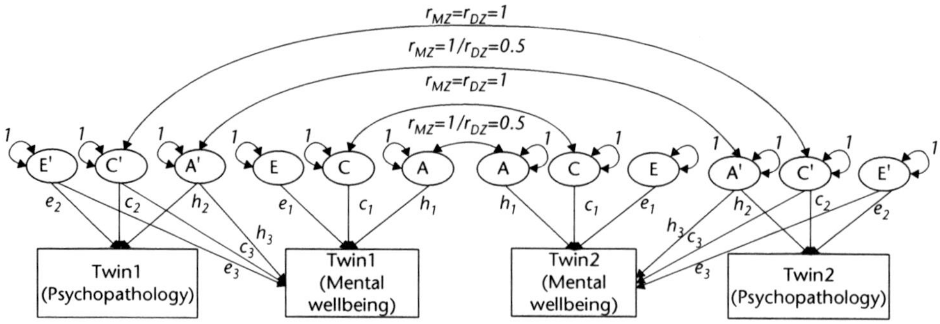

Figure 13.2 A schematic bivariate model for external psychopathology and mental wellbeing

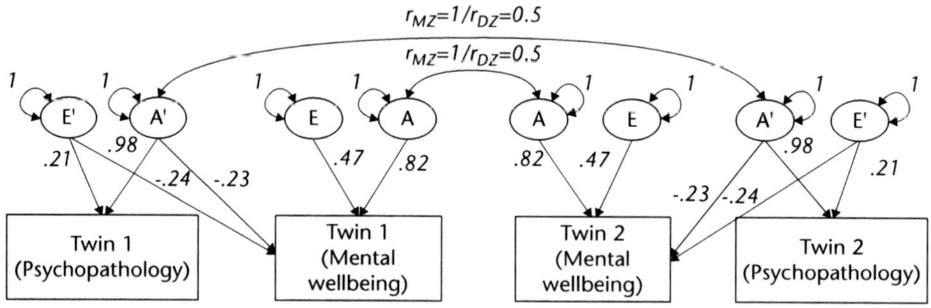

Figure 13.3 A schematic bivariate model for external psychopathology and mental wellbeing

variance is greater for children from families with more parental hostility and less parental warmth, as compared with those from families with less parental hostility and more parental warmth in the family background (Feinberg *et al.*, 2007). It has been reported that the heritability estimate of alcoholism and depression is greater for unmarried women than for married women (Heath *et al.*, 1989, 1998). These findings suggest possible interaction between genes and environmental risk factors.

The other aspect of the relationship between genes and environmental factors that has been studied is gene–environment correlation. That is, genetic propensities are correlated with the environmental factors. This correlation can further be classified into three types of gene–environment correlation: passive, active, and evocative correlations (Plomin *et al.*, 1977). Passive gene–environment correlation occurs when the genes that children inherited from their parents correlate with the family environment in which they are raised. For example, parents with antisocial behavior provide both genetic risk and adverse environments for their children. A child is raised in an abusive family when parents have antisocial personality disorder that is heritable. Active gene–environment correlation occurs when an individual actively seeks and creates an environment that correlates to his or her genetic propensity. For example, an individual with alcohol dependence that is heritable may like to find friends who are also alcohol-dependent. Evocative or reactive gene–environment correlation occurs when genetically influenced behavior elicits specific behaviors from the social and physical environment. For example, it is known that adolescent rule-breakers are popular. Adolescents with a particular form of a serotonin gene associated with rule-breaking behavior are more popular (Burt, 2008). The classical twin design cannot distinguish all three types of gene–environment correlation. Alternative approaches have been developed.

Children of twin design

The children of twin design is able to identify the genetic and environmental transmission between parental characteristics and offspring outcome. The comparison of children from twin parents differentiates genotypic and environmental factors responsible for the association between parental characteristics and child adjustment. Each twin has the same genetic relationship with his or her children as his or her identical co-twin. Therefore the genetic risk associated with twin parents can be inferred from their co-twins (Nance and Corey, 1976). The children of identical twins are cousins and share 25 percent of their genetic make-up. Each twin provides a different family environment for his or her own children. However, the children of fraternal twins share 12.5 percent of their genetic make-up. Each fraternal twin shares 50 percent of his or her genetic make-up with his or her children and 25 percent with his or her co-twin's children. When children of fraternal twins are analyzed in conjunction with children of identical twins, the genetic and environmental influences in origin on the children's outcomes can be detected (Heath *et al.*, 1985). The comparison in association of cousins from identical twin families provides a direct estimate of the nongenetic interaction between parental behavior and risk for children adjustment. The comparison in association between children and their parent's identical co-twin versus children and their parent's fraternal co-twin indicates the intergenerational transmission of genetic factors. The parents' identical co-twins as compared with fraternal co-twins represent indices of the long-term behavioral consequences of genes whose effects may initially be expressed in their nieces and nephews in childhood (Silberg *et al.*, 2010).

Figure 13.4 displays a basic model of the children of twins design (D'Onofrio *et al.*, 2003). Among path coefficients (*c, d, e, f, g, h, k*, and *p*), the *k* parameter estimate represents parental genetic influence on children, indicating passive gene–environment correlation. *p* represents the environmentally mediated parental effect on children. The additive genetic, shared environmental, and nonshared environmental effects (*g, d*, and *f*) on children are assumed to be independent of those effects (*h, c*, and *e*) on twin parents. There are alternative equivalent parameterizations of the same basic model.

D'Onofrio *et al.* (2003) examined the effect of smoking during pregnancy (SDP) on child birth weight in a sample including 279 MZ and 111 DZ families from Virginia, as well as 492 MZ and 152 DZ families from Norway, using the CoT model shown in Figure 13.4. The analysis was limited to the female twins where at least one twin has a child. To illustrate the model using the path diagram shown in Figure 13.4, Twin1 and Twin2 variables would be replaced by SDP for a mother and her co-twin, respectively. Similarly, Child1 and Child2 variables would be replaced by the birth weight measures for the first child of Twin1 and the first child of her co-twin, respectively. This study found that among all parameters specified in the model, the genetic influences on SDP ($h_{standardized} = 0.72$), nonshared environmental influences ($e_{standardized} = 0.69$), and phenotypic intergenerational effect ($p_{standardized} = -0.17$) were found to be statistically significant. The heritability estimate of mother's SDP was 52 percent ($h^2 = 0.52$). Passive gene–environment correlation was not found because path coefficient *k* was not statistically significant. Although path coefficients *g, d*, and *f* were not statistically significant, this study could not conclude that there were not additional genetic and/or environmental influences specific to the child birth

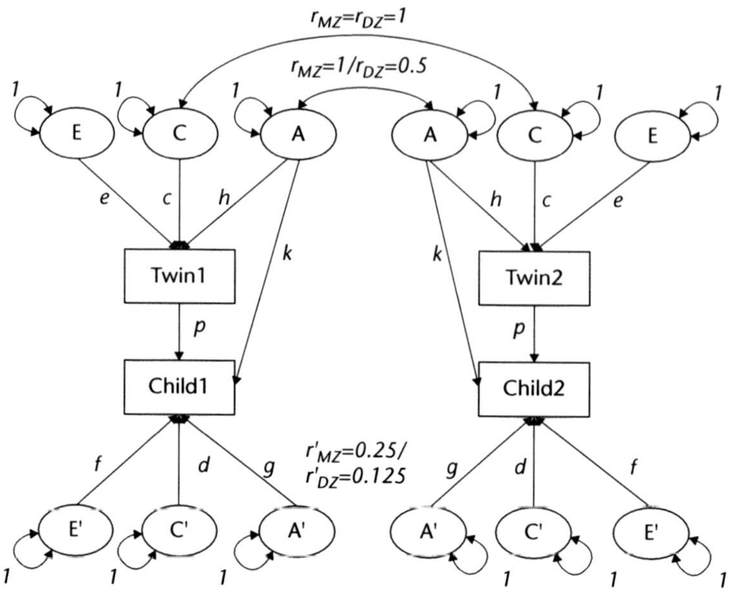

Figure 13.4 A schematic children of twin model

weight due to lack of statistical power. The study suggests that the pathway from SDP to child birth weight is solely environmental, rather than genetic. This conclusion is consistent with the report from another study showing that mothers who smoked during one pregnancy had smaller infants than the infants from the pregnancy in which they did not smoke (Naeye, 1978).

The nature of the intergenerational transmission of childhood conduct problems was investigated in an Australian twin-family sample using the CoT design (D'Onofrio *et al.*, 2007). The study has found that the intergenerational transmission of childhood conduct problems is environmentally mediated in male offspring. However, the genetic influence shared between the twin parent and the children explains the intergenerational transmission of childhood conduct problems in female offspring. The same study group reported that the relationship between parental divorce and early drug use in offspring was genetically mediated, and the relationship between parental divorce and emotional difficulties in offspring was due to a direct environmental effect using the CoT design (D'Onofrio *et al.*, 2006). In another study using the CoT design, the authors found that harsh physical punishment by parents had a direct environmental effect on behavior problems in offspring (Lynch *et al.*, 2006).

The CoT design can control for genetic confounds and unmeasured environmental factors that make twins similar. This design can study the same and different phenotypes in the parent and children. Unlike randomized controlled design, however, this design does not make a causal inference. Recently the CoT design has been extended to jointly analyze the children of twins and twin-family data, termed eCoT design.

Extended children of twin design

A schematic eCoT model is displayed in Figure 13.5. The structural model contains two parts, which are jointly estimated simultaneously in the analysis (Narusyte *et al.*, 2008). In Figure 13.5,

T_1 and T_2 on the left panel represent twin parent's phenotypes. C_1 and C_2 on the left panel represent children phenotypes in twin's families. P_1 on the right panel represents the observed phenotype of one parent of a child twin pair. T_1 and T_2 on the right panel represent phenotypes for juvenile twins. Latent variables A, C, and E represent additive genetic, shared environmental, and nonshared environmental contributions to the parent's phenotypes, respectively. A', C', and E' represent additive genetic, shared environmental, and nonshared environmental contributions to the children's phenotypes, respectively. A" represents additive genetic effects specific to the children phenotypes in the developmental stage of childhood. R and R' represent measurement error for parent and children's phenotypes, respectively. The parameter estimates are described in Table 13.1. In the twins-as-parent part of the model (left panel in Figure 13.5), parents who are MZ twins share all their genetic make-up. Therefore the correlation between parent's genetic effects (A) is equal to 1. On the other hand, DZ twins share, on average, only half of their genetic make-up, and thus the correlation is set to 0.5. Shared environment is, by definition, all nongenetic influences that make family members similar; thus the correlation between shared environmental (C) effects is set to 1.0. The correlation between the A" paths is set to 0.25 for children of MZ twins and to 0.125 for children of DZ twins because this indicates their degree of genetic relatedness. The correlation between shared environmental effects (C') for children of twins is set to 0 because they were not sharing the same rearing environment. In the twins-as-children part of the model (right panel in Figure 13.5), correlations between genetic (A) and shared environment (C) effects on parents are set to 1.0 for both twin children because the same parent is rating his or her parenting of both children. In other words, there is no variance in these factors for the parent of the twin children. Because the parent reports about each child separately, some aspects of parental behavior are distinct for each child. Therefore nonshared environment parameters (E) are left uncorrelated. For juvenile twins, the genetic correlation was set to 1 for MZ twins and to 0.5 for DZ twins. The correlation between shared environmental influences (C') is set to 1.0 for both types of juvenile twin because they are being reared in the same household.

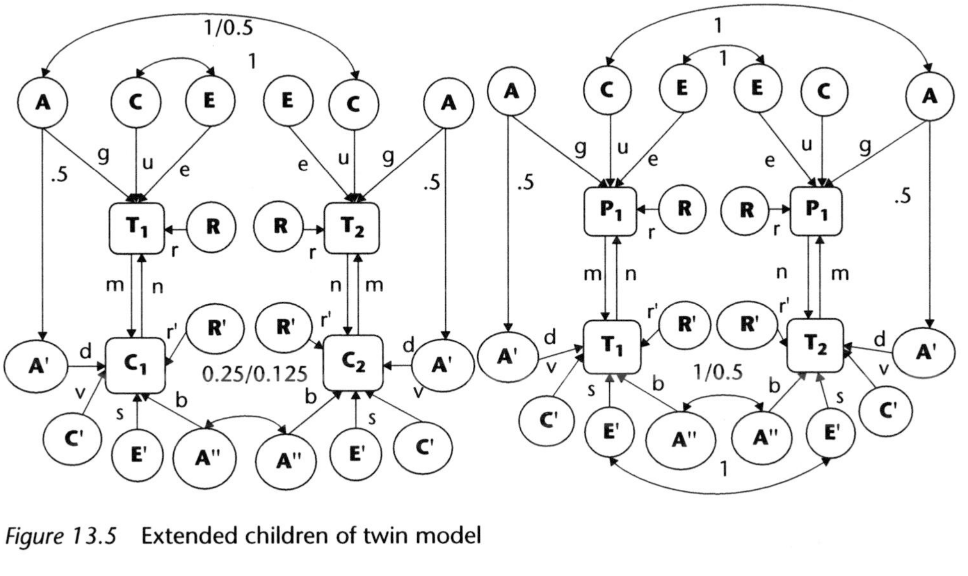

Figure 13.5 Extended children of twin model

Table 13.1 Summary of parameters of structural model for correlations between relatives

Parameter	*Description*
e	Path from nonshared environment to adult phenotype
u	Path from shared environment to adult phenotype
g	Path from persistent additive genetic effect to adult phenotype
m	A parent's environmental effect to child phenotype
n	Child environmental effect to the parent phenotype
r, r'	Measurement errors for the phenotypes
d	Path from persistent additive genetic effect to child phenotype
v	Path from shared environment to child phenotype
s	Path from nonshared environment to child phenotype
b	Path from childhood specific genetic effect to child phenotype

The power to differentiate persistent lifetime genetic effect (*d*) from developmental genetic effect (*b*) specific to childhood comes from the twins-as-children part of the model. The measured family-level variables can be included in the model (Turkheimer *et al.*, 2005). The total estimated phenotypic variance for each trait can be derived (Heath *et al.*, 1993). Path coefficients *m* and *n* reflect a reciprocal phenotypic relationship between the parental variable and the offspring variable. Passive gene–environment correlation is implicated when path coefficients *g, d*, and *m* are statistically significant, meaning that parents provide both genes and environments to their offspring. Evocative gene–environment correlation is suggested when path coefficients *n, d*, and/or *b* are significant, indicating that a child evokes a particular parent behavior through his or her inherited behavior. This model is identified given six sources of variances present in the eCoT model. The multi-groups are determined on the basis of the combination of sex by zygosity in all twins. A series of nested models based on the hypothesis are compared with the full model shown in Figure 13.5. The model identification process described previously can be used to identify the most parsimonious model. The comparison of a model with path *g* versus a model without path *g* when other parameters are held constant using an LR chi-square test can evaluate the null hypothesis that both models have a similar fit. The results can indicate whether genetic effects on the outcome in adults were significant. A heritability estimate in adults could be calculated by the proportion of g^2 in total variance of the adult variable. One can test the hypothesis that genetic variance in child outcome is significant by comparing a model with path *b* and/or path *d* to a model without path *b* and/or path *d* using an LR chi-square test. If the chi-square test were statistically significant, it would suggest significant genetic effects on the child outcome. A heritability estimate is the proportion of b^2 and/or d^2 in total variance of the child outcome. Similarly, one can test the shared environmental influences on the adult variable and the child outcome. Significant shared environmental effects on the adult variable and the child outcome are implicated in the proportion of u^2 and v^2 in total variance of the adult variable and the child outcome, respectively.

There is only one study that has adopted the eCoT approach to date, which uses a sample of Swedish twins and their families. The study investigated passive gene–environment correlation, evocative gene–environment correlation, and direct environmental effect in the relationship between parental criticism and externalizing behavior problems in adolescent offspring, controlling for other genetic and environmental confounds (Narusyte *et al.*, 2011). The study found that the heritability estimate of parental criticism was 37 and 49 percent for fathers and mothers, respectively. The nonshared environmental effect on parental criticism was 23 and 13 percent for fathers and mothers, respectively. Genetic influence on father's criticism had no effect on externalizing behavior of offspring, whereas, genetic influence on mother's criticism explained 12 percent variance in externalizing behavior of offspring. More than 50 percent of variance in externalizing behaviors of offspring was attributable to genetic influence. Mild, but significant, environmentally mediated direct relationship between father's criticism and child externalizing behavior was reported, suggesting that father's criticism increases the risk for child externalizing behavior through direct environmental effects. In contrast, strong evocative gene–environment correlation between mother's criticism and child externalizing behavior reflects mother's criticism reacting to the children's externalizing behaviors through genetic mechanism. This study not only conducted a rigorous test to unravel the differential intergenerational transmission between paternal and maternal criticism and children externalizing behavior, but also provides insights to guide the design of effective intervention programs for reducing risk for child externalizing behavior.

Conclusion

The twin-and-family designs are a powerful strategy to differentiate genetic and environmental transmission mechanisms between parents and offspring. It has fewer assumptions than the classical twin design. When the spouses of adult twins are included in the designs, the spousal interaction and assortative mating effects can be revealed. The genetically informed design can remove genetic confounds that studies using nongenetically informed human samples cannot achieve. Behavior genetics research does not diminish environmental influence; rather, the study of genetic influence advances the understanding of environmental effect. The children of twin design and its extended version enable researchers to unravel gene–environment interaction and gene–environment correlation. The twin research method has been used, beyond psychology and psychiatry, by social sciences including political science, sociology, and health services to examine gene–environment interplay (Romeis *et al.*, 2005; Guo and Tong, 2006; Hatemi *et al.*, 2010). The method can shed new light on the complex gene–environment interplay in criminal justice research. The results from the twin and family studies may have implications for intervention design and policy change in criminal justice.

References

van den Berg, S.M., Beem, L., and Boomsma, D.I. (2006) 'Fitting genetic models using MCMC algorithms with BUGS', *Twin Research and Human Genetics*, 9: 334–342.

Boker, S.M., Neale, M.C., Maes, H.H., Wilde, M.J., Spiegel, M., Brick, T.R., Spies, J., Estabrook, R., Kenny, S., Bates, T.C., Mehta, P., and Fox, J. (2011) 'OpenMx: an open source extended structural equation modeling framework', *Psychometrika*, 76: 306–317.

Bouchard, T.J. Jr and Propping, P. (1993) *Twins as a Tool of Behavioral Genetics*, Chichester, UK: John Wiley & Sons.

Burt, S.A. (2008) 'Genes and popularity: evidence of an evocative gene–environment correlation', *Psychological Science*, 19(2): 112–113.

Derks, E.M., Dolan, C.V., and Boomsma, D.I. (2006) 'A test of the equal environment assumption (EEA) in multivariate twin studies', *Twin Research and Human Genetics*, 9: 403–411.

D'Onofrio, B.M., Turkheimer, E.N., Eaves, L.J., Corey, L.A., Berg, K., Solaas, M.H., and Emery, R.E. (2003) 'The role of the children of twins design in elucidating causal relations between parent characteristics and child outcomes', *Journal of Child Psychology and Psychiatry*, 44: 1130–1144.

D'Onofrio, B.M., Turkheimer, E., Emery, R.E., Slutske, W.S., Heath, A.C., Madden, P.A., and Martin, N.G. (2006) 'A genetically informed study of the processes underlying the association between parental marital instability and offspring adjustment', *Developmental Psychology*, 42: 486–499.

D'Onofrio, B.M., Slutske, W.S., Turkheimer, E., Emery, R.E., Harden, K.P., Heath, A.C., Madden, P.A., and Martin, N.G. (2007) 'Intergenerational transmission of childhood conduct problems: a Children of Twins Study', *Archives of General Psychiatry*, 64: 820–829.

Evans, L.J. and Martin, N.G. (2000) 'The validity of twin studies', *Genescreen*, 1: 77–79.

Feinberg, M.E., Button, T.M., Neiderhiser, J.M., Reiss, D., and Hetherington, E.M. (2007) 'Parenting and adolescent antisocial behavior and depression: evidence of genotype × parenting environment interaction', *Archives of General Psychiatry*, 64: 457–65.

Ferguson, C.J. (2010) 'Genetic contributions to antisocial personality and behavior: a meta-analytic review from an evolutionary perspective', *Journal of Social Psychology*, 150: 1–21.

Fu, Q., Heath, A.C., Bucholz, K.K., Nelson, E.C., Goldberg, J., Lyons, M.J., True, W.R., Jacob, T., Tsuang, M.T., and Eisen, S.A. (2002) 'Shared genetic risk of major depression, alcohol dependence, and marijuana dependence: the contribution from antisocial personality disorder in men', *Archives of General Psychiatry*, 59: 1125–1132.

Fu, Q., Koenen, K.C., Miller, M.W., Heath, A.C., Bucholz, K.K., Lyons, M.J., Eisen, S.A., True, W., Goldberg, J., and Tsuang, M. (2007) 'Differential etiology of posttraumatic stress disorder with conduct disorder and major depression in male veterans', *Biological Psychiatry*, 62, 1088–1094.

Guo, G., and Tong, Y. (2006) 'Age at first sexual intercourse, genes, and social context: evidence from twins and the dopamine D4 receptor gene', *Demography*. 43: 747–769.

Hatemi, P.K., Hibbing, J.R., Medland, S.E., Keller, M.C., Alford, J.R., Smith, K.B., *et al.* (2010) 'Not by twins alone: using the extended family design to investigate genetic influence on political beliefs', *American Journal of Political Science*, 54(3): 798–814.

Heath, A.C., Kendler, K.S., Eaves, L.J., and Markell, D. (1985) 'The resolution of cultural and biological inheritance: informativeness of different relationships', *Behavior Genetics*, 15: 439–465.

Heath, A.C., Jardine, R., and Martin, N.G. (1989) 'Interactive effects of genotype and social environment on alcohol consumption in female twins', *Journal of Studies on Alcohol and Drugs*, 50(1): 38.

Heath, A.C., Kessler, R.C., Neale, M.C., Hewitt, J.K., Eaves, L.J., and Kendler, K.S. (1993) 'Testing hypotheses about direction of causation using cross-sectional family data', *Behavior Genetics*, 23: 29–50.

Heath, A.C., Eaves, L.J., and Martin, N.G. (1998) 'Interaction of marital status and genetic risk for symptoms of depression', *Twin Research*, 1(3): 119–122.

Jaffee, S.R., Caspi, A., Moffitt, T.E., Dodge, K.A., Rutter, M., Taylor, A., and Tully, L.A. (2005) 'Nature × nurture: genetic vulnerabilities interact with physical maltreatment to promote conduct problems', *Developmental Psychopathology*, 17: 67–84.

Kendler, K.S. and Eaves, L.J. (1986) 'Models for the joint effect of genotype and environment on liability to psychiatric illness', *American Journal of Psychiatry*, 143: 279–289.

Kendler, K.S., Myers, J.M., and Keyes, C.L. (2011) 'The relationship between the genetic and environmental influences on common externalizing psychopathology and mental wellbeing', *Twin Research and Human Genetics*, 14: 516–523.

Lynch, S.K., Turkheimer, E., D'Onofrio, B.M., Mendle, J., Emery, R.E., Slutske, W.S., and Martin, N.G. (2006) 'A genetically informed study of the association between harsh punishment and offspring behavioral problems', *Journal of Family Psychology*, 20: 190–198.

Naeye, R.L. (1978) 'Effects of maternal cigarette smoking on the fetus and placenta', *British Journal of Obstetrics and Gynecology*, 85: 732–737.

Nance, W.E. and Corey, L.A. (1976) 'Genetic models for the analysis of data from the families of identical twins', *Genetics*, 83: 811–826.

Narusyte, J., Neiderhiser, J.M., D'Onofrio, B.M., Reiss, D., Spotts, E.L., Ganiban, J., and Lichtenstein, P. (2008) 'Testing different types of genotype–environment correlation: an extended children-of-twins model', *Developmental Psychology*, 44: 1591–603.

Narusyte, J., Neiderhiser, J.M., Andershed, A.K., D'Onofrio, B.M., Reiss, D., Spotts, E., Ganiban, J., and Lichtenstein, P. (2011) 'Parental criticism and externalizing behavior problems in adolescents: the role of environment and genotype–environment correlation', *Journal of Abnormal Psychology*, 120: 365–376.

Neale, M.C., and Cardon, L.R. (1982) *Methodology for Genetic Studies of Twins and Families*, Dordrecht, the Netherlands: Kluwer Academic.

Plomin, R., DeFries, J.C., and Loehlin, J.C. (1977) 'Genotype–environment interaction and correlation in the analysis of human behavior', *Psychological Bulletin*, 84: 309–322.

Rabe-Hesketh, S., Skrondal, A., and Gjessing, H.K. (2008) 'Biometrical modeling of twin and family data using standard software for mixed models', *Biometrics*, 64: 280–288.

Romeis, J.C., Heath, A.C., Xian, H., Eisen, S.A., Scherrer, J.F., Pedersen, N.L., Fu, Q., Bucholz, K.K., Goldberg, J., Lyons, M.J., Waterman, B., Tsuang, M.T., and True, W.R. (2005) 'Heritability of SF-36 among middle-age, middle-class, male–male twins', *Medical Care*, 43: 1147–1154.

Rutter, M. (2006) *Genes and Behavior: Nature–Nurture Interplay Explained*, Oxford: Blackwell Publishing.

Silberg, J.L., Maes, H., and Eaves, L.J. (2010) 'Genetic and environmental influences on the transmission of parental depression to children's depression and conduct disturbance: an extended Children of Twins study', *Journal of Child Psychology and Psychiatry*, 51: 734–744.

Turkheimer, E., D'Onofrio, B.M., Maes, H.H., and Eaves, L.J. (2005) 'Analysis and interpretation of twin studies including measures of the shared environment', *Child Development*, 76: 1217–1233.

Vaughn, M.G., Beaver, K.M., DeLisi, M., Perron, B.E., and Schelbe, L. (2009) 'Gene–environment interplay and the importance of self-control in predicting polydrug use and substance-related problems', *Addictive Behavior*, 34: 112–116.

Walsh, A. and Beaver, K.M. (2009) *Biosocial Criminology: New Directions in Theory and Research*, New York: Taylor & Francis.

Part III

Neurological and neuropsychological foundations of antisocial behavior

14

Neuroimaging and antisocial behavior

Robert A. Schug, Gianni G. Geraci, Gabriel Marmolejo, Heather L. McLernon, Leidy S. Partida, and Alexander J. Roberts

Introduction

Biosocial approaches to criminology involve, at a very fundamental level, the examination of relationships between brain and behavior. Army medic Cesare Lombroso was the first to apply this concept to criminology, after observing unusual bone formations at the base of the skull of an Italian serial killer during a routine post-mortem examination in 1871. While Lombroso was approaching the problem of crime from an evolutionary biological perspective, his observation nonetheless became the cornerstone for an empirical understanding of the link between brain structure (in this case, the size of a specific brain region) and crime, antisociality, and violence. Fortunately, the scalpels, saws, and cutting instruments of the late-nineteenth-century autopsy surgeon have now been replaced with new tools for opening the human skull and examining the contents within—X-rays, radioactive tracer compounds, and magnetic resonance pulse signals. These exciting new tools have allowed researchers to examine the brains of living individuals (a luxury Lombroso did not have). While the technology has changed dramatically, the conceptual research question remains largely the same: Do differences in the size, shape, and functioning of different brain regions lead to—or at the very least relate to—criminal and antisocial behavior?

In recent decades, marked advances have been made by researchers in the understanding of the neurological and neuropsychological foundations of antisocial behavior. Neuroimaging—the application of techniques and technologies that generate visual representations of brain structure and functioning in living individuals—is but one of the main areas in the burgeoning discipline of neurocriminology. Maintaining an organized understanding of the current and recent studies of neuroimaging and antisocial behavior can be daunting for the consumer of this research, particularly given the growing number of studies in this area, as well as the level of technical expertise that may be necessary to interpret and synthesize results. One helpful organizational framework for understanding this material is through the classification of studies according to anatomical regions of the brain, as specific brain areas and structures have been continually implicated in various forms of antisocial behavior. Specific regions of interest (ROIs) targeted in previous research have included the whole-brain or individual hemispheres; different lobes of the cerebral cortex (e.g. frontal and temporal); subvolumes within these lobes (e.g. the prefrontal cortex within the frontal lobe—which itself may be divided into subregions such as the orbitofrontal cortex [OFC], and dorsolateral prefrontal cortex [DLPFC]); and specific structures located in subcortical regions (e.g. components of the limbic system, including the hippocampus and amygdala). Importantly, these studies have traditionally examined gray matter (consisting largely of neuronal cell bodies) and/or white matter (composed chiefly of mylenated axon tracts, and representing connectivity between various brain regions and structures).

A second helpful organizational approach is examining studies according to the imaging technique used. Neuroimaging studies of antisocial behavior may be classified as structural or functional. Structural methods examine the size (length, thickness), shape, volume, density, and other physical properties of a given brain region or component, using techniques such as computed tomography (CT) scanning, diffusion tensor imaging, and structural magnetic resonance imaging (sMRI). Functional methods measure the amount of neural activity occurring in any given brain region, utilizing techniques such as positron emission tomography (PET), single photon emission computed tomography (SPECT), and functional magnetic resonance imaging (fMRI). Finally, it is often useful to organize studies according to the type of antisocial population examined. Individuals recruited for participation in neuroimaging studies of antisocial behavior may be classified by specific crime type (e.g. murderers) or antisocial characteristics/behaviors (e.g. violence, aggression, gambling, or substance use). Participants may also be characterized by psychiatric disorders demonstrating associations with antisocial attributes, such as conduct disorder (CD), oppositional defiant disorder (ODD), substance abuse disorders, bipolar disorder, schizophrenia, antisocial personality disorder (ASPD), and psychopathy (a condition characterized by persistent antisocial, impulsive, and irresponsible behavior, and a profound lack of empathy, loyalty, and guilt).

The overarching goal of most neuroimaging studies of antisocial behavior is to make precise structural or functional measurements of ROIs in individuals characterized by crime, antisociality, or violence, and to compare these measurements with those of individuals without these characteristics (i.e. normal, healthy controls). Subsequent between-groups differences in these measurements—if identified—are considered as potential brain-based neurobiological correlates of the various forms of antisocial behaviors and traits of study. This chapter will serve as a review of current, recent, and foundational research in neuroimaging and antisocial behavior, using the aforementioned organizational frameworks (i.e. focusing on frontal and temporal lobes and limbic system, the structural and functional findings from studies of these brain areas, and—to some degree—findings from the various antisocial populations of study).

Frontal lobe and prefrontal cortex

The frontal lobe extends from the anterior (forward) tip of the brain to the central sulcus. Together, the right and left frontal lobes constitute approximately one-third of the cerebrum—by far the greatest quantity of cortical volume in humans compared with other animal species. For nearly two centuries, frontal lobe damage has been associated with problems in the control of behavior, and more recent research studies have suggested the role of the frontal lobe in the perception and appraisal of emotional stimuli, and the regulation of emotion responses (Schug *et al.*, 2009). Brain imaging studies in recent decades continue to suggest that structural and functional deficits in the frontal lobe play a crucial role in the neurobiological processes underlying antisocial and violent behavior. Earlier imaging studies of antisocial individuals often discussed more macro-level analyses of brain structure and functioning—understandable, given the limitations in spatial resolution of the imaging methods available at that time (although whole-brain volume analyses have been conducted in more recent studies, e.g. Barkataki *et al.*, 2006). Nonetheless, the frontal lobe was often implicated in associations with various forms of offending in these studies. For example, in an earlier meta-analysis of 20 structural and functional imaging studies, Mills and Raine (1994) found frontal lobe deficits to be associated with violent offending and temporal lobe deficits with sexual offending, and violent sexual offending to be associated with dysfunction in both lobes. Results from diffusion tensor imaging studies have also revealed frontal lobe white matter structural defects associated with violent behavior (e.g. Hoptman *et al.*, 2005; Li *et al.*, 2005). Additionally, several PET studies have reported significantly reduced glucose metabolism in the frontal cortex, particularly the anterior medial frontal region and left white frontal matter, in individuals characterized by violence (e.g. Goyer *et al.*, 1994; Raine *et al.*, 1997; Soderstrom *et al.*, 2000).

The surface of the frontal lobes consists of several subregions, which have distinctive patterns of connections with other areas of the cerebral cortex, along with structures deep below the surface of the brain. A number of these cortical subregions are thought to be involved in *executive functioning*, an umbrella term referring to the cognitive processes that allow for future, goal-oriented behavior (Morgan and Lilienfeld, 2000), including the ability to maintain an appropriate problem-solving set for future goal attainment (Luria, 1966). Executive functioning broadly encompasses four distinct cognitive domains: volition (capacity for intentional behavior); planning (identification and organization of the steps and elements needed to carry out an intention or achieve a goal); purposive action (translation of an intention or plan into productive, self-serving activity); and effective performance (self-monitoring and self-correction; Lezak *et al.*, 2004). Each of these domains is necessary for socially and contextually appropriate behavior and effective self-serving conduct (Lezak *et al.*, 2004), which intuitively make executive functioning and associated brain regions a promising area for investigating the etiology of antisocial behavior.

The prefrontal region of the frontal lobe (i.e. the prefrontal cortex) eventually became a primary region of study for earlier neuroimaging investigations of antisocial behavior. In terms of structural findings, Raine and colleagues (2000), using sMRI, found reductions in prefrontal gray volumes in individuals with ASPD (who were characterized by significantly more violence) compared with controls and individuals with substance dependence. In fact, individuals in the ASPD group had 11 percent less prefrontal gray matter volume compared with controls, and nearly 14 percent less prefrontal gray matter volume compared with those in the substance-dependent group. However, white matter volume in the prefrontal area was not significantly different across the three groups. More recently, Kumari and colleagues (2009) found reduced prefrontal volumes in violent schizophrenia patients compared with controls, and also that increased scores on a measure of impulsivity were associated with reduced prefrontal volumes.

In one of the classic functional imaging studies in this area, Raine *et al.* (1997) examined 41 murderers pleading not guilty by reason of insanity, and 41 age-and sex-matched controls. For this study, PET imaging was used during a continuous performance challenge task. Results indicated murderers were characterized by reduced glucose metabolism in lateral and medial prefrontal cortical regions (i.e. left and right medial superior frontal cortex, left anterior medial cortex, and right OFC). In a subsequent study, Raine and colleagues (1998) found that murderers characterized by affective (i.e. reactive and emotionally based) aggression demonstrated reduced lateral and medial prefrontal functioning when compared with controls, whereas those characterized by predatory (i.e. purposeful and goal-driven) aggression had reduced medial prefrontal but not lateral prefrontal functioning relative to the control group. Volkow and colleagues (1995) also found reduced glucose metabolism bilaterally in prefrontal regions in violent psychiatric patients with intermittent explosive disorder (IED) compared with healthy controls. Finally, a meta-analysis of 43 structural and functional neuroimaging studies (Yang and Raine, 2009) revealed significantly reduced structure and function in prefrontal regions in antisocial individuals.

The prefrontal cortex has become recognized among researchers as a composite of anatomically distinct subsystems, rather than a unitary structure (see Dinn and Harris, 2000, for a review), and differentiations have been made between the functional properties of the orbitofrontal/ventromedial and dorsolateral prefrontal sectors (Damasio, 1994; Lapierre *et al.*, 1995). These sectors may have critical yet differential associations with the emotional and behavioral components of antisocial behavior and violence. For example, evidence from multiple sources suggests that the dorsolateral prefrontal cortex (DLPFC) mediates executive functions (Dinn and Harris, 2000), and is involved with the temporal integration of behavior (Lapierre *et al.*, 1995). The DLPFC also plays a critical role in inhibitory control, which is required for the development of moral conduct and moral cognition (Kochanska *et al.*, 1997).

Structural deficits in the DLPFC have been identified in antisocial boys with substance dependence (Dalwani *et al.*, 2011); and have also been associated with emotional deficits in psychopathy (Zamboni *et al.*, 2008; Yang *et al.*, 2009a,b), and aggressive symptoms of CD (Fairchild *et al.*, 2013). Functional deficits in the DLPFC have been identified in antisocial substance-dependent boys (Crowley *et al.*, 2010); and have also been associated with behavioral inhibition and impulsivity in ASPD individuals (Völlum *et al.*, 2010), impaired decision-making ability in young offenders (Syngelaki *et al.*, 2009), and alcoholism in ASPD patients with a history of violence (Laakso *et al.*, 2002). Van Holst *et al.* (2012a,b) found that gamblers showed increased DLPFC and ACC in response to gambling-related cues relative to controls. Other researchers failed to identify functional deficits in psychopaths when performing DLPFC tasks (Roussy and Toupin, 2000; Blair *et al.*, 2006; Basoglu *et al.*, 2008). Interestingly, Glenn *et al.* (2009b), using fMRI in a community sample, found increased psychopathy scores to be related to increased activity in the right DLPFC during an emotional moral decision-making task.

The OFC (orbitofrontal cortex, also referred to as the ventromedial prefrontal cortex) plays a key role in guiding behavior by processing the reward–punishment value of stimuli, and thus is crucial to controlling violent acts (Damasio, 1994). The OFC, however, modulates sensitivity to reinforcement contingencies (see Dinn and Harris, 2000, for a review), and is involved with the inhibition of inadequately motivated actions and with the modulation of aggressive behavior and autonomic reactivity (see Lapierre *et al.*, 1995, for a review). In humans, this system contributes more to social and self-awareness than the DLPFC (Damasio, 1994; see also Lapierre *et al.*, 1995, for a review). Orbitofrontal/ventromedial prefrontal pathologies appear similar to psychopathy, whereas clinical DLPFC syndromes appear somewhat incompatible (Lapierre *et al.*, 1995).

Structural deficits in the OFC have been identified in violent males diagnosed with ASPD and type 2 alcoholism (Laakso *et al.*, 2002), violent offenders with substance abuse disorders (Schiffer *et al.*, 2011), men with ASPD (Raine *et al.*, 2011), aggressive and impulsive psychiatric inpatients (Antonucci *et al.*, 2006), and boys diagnosed with CD (Huebner *et al.*, 2008); and have been associated with gender differences (i.e. girls more so than boys) in children characterized by aggression and behavioral problems (Boes *et al.*, 2008). Additionally, increased gray matter volumes in the OFC have been associated with increases in callous-unemotional traits in girls with CD (Fairchild *et al.*, 2013). Functional deficits in the OFC have been identified in violent individuals relative to nonviolent controls (e.g. Mathews *et al.*, 2005; Joyal *et al.*, 2007; Kumari *et al.*, 2009), in boys with antisocial behavioral and substance abuse problems (Crowley *et al.*,

2010), adolescents with behavioral problems (White *et al.*, 2013), children with CD, ODD, and psychopathic traits (Finger *et al.*, 2011), and individuals with borderline personality disorder or ASPD (Völlum *et al.*, 2004); and have been associated with increased lifestyle and antisocial psychopathic traits in violent schizophrenia patients (Dolan and Fullam, 2009). Structural and functional studies have also identified abnormalities in the connectivity between the OFC and amygdala in psychopathic individuals and adolescents with CD (Marsh *et al.*, 2011; Passamonti *et al.*, 2012).

Temporal lobe

The temporal lobe of the brain is associated with several different functions: memory retention and storage, organization of sensory input, language production, visual perception, and emotional responses. Within the temporal lobes are structures of the limbic system, including the amygdala and hippocampus (discussed separately below). The temporal lobe has been a common region of interest in neuroimaging studies of antisocial behavior, which have foundations in the classic investigations of murderers using electroencephalography (EEG) that noted abnormalities in brain functioning in this region (Stafford-Clark and Taylor, 1949; Hill and Pond, 1952; Mundy-Castle, 1955). Both structural and functional neuroimaging investigations have identified deficits in temporal regions which have been related to various forms of antisocial, aggressive, and violent behavior (Dolan, 2010).

Structural deficits, including various forms of volume reductions, have been noted in the temporal lobe regions of antisocial individuals. For example, Dolan and colleagues (2002) found 20 percent reductions in volume in frontal and temporal lobes in impulsive-aggressive men with ASPD compared with controls. Similar studies examining reduced gray matter have found consistent results in individuals diagnosed with ASPD, CD, and psychopathy (Kruesi *et al.*, 2004; Pridmore *et al.*, 2005; Müller *et al.*, 2008; Gregory *et al.*, 2012). Barkataki and colleagues (2006) used structural MRI to compare volumes of numerous brain structures among 13 men with ASPD, 13 men with schizophrenia and a history of violence, 15 nonviolent men with schizophrenia, and 15 healthy nonviolent male controls. Results indicated ASPD men were characterized by reduced whole brain and temporal lobe volume, as well as increased putamen volume, compared with controls. Using a cortical thinning approach, Yang and colleagues (2009a,b) found that cortical gray matter in the right frontal and temporal cortices was significantly thinner in psychopaths compared with nonpsychopathic controls. Finally, using sMRI, Huebner *et al.* (2008) identified reduced bilateral (i.e. on both sides of the brain) temporal lobe volumes in a sample of 23 boys diagnosed with CD compared with age- and IQ-matched controls.

Several literature reviews of functional imaging studies have yielded consistent results: aggressive and violent individuals tend to be characterized by functional deficits in the temporal lobe (Bufkin and Luttrel, 2005). Regarding PET imaging findings, Volkow and colleagues (1995) identified reduced rates of glucose metabolism in medial temporal regions in eight violent psychiatric patients compared with eight normal comparison subjects; while Lencz and colleagues (1996) found an abnormal asymmetry in glucose metabolism in the medial temporal lobe (reduced activity in the left hemisphere compared with the right) in murderers relative to controls. Wong and colleagues (1997) also found that schizophrenia patients characterized by single or minor offenses demonstrated reduced activity in anterior temporal lobe structures compared with normal controls. Furthermore, Soderstrom and colleagues (2000) examined several ROIs in a structural and functional imaging study (using both sMRI and SPECT) in a sample of 21 impulsive violent offenders referred for pretrial forensic evaluations and 11 healthy controls. While results indicated no structural deficits among violent offenders, reduced regional blood flow in the right medial temporal and right angular gyri was noted in the violent offender group when compared with controls. Goethals and colleagues (2005) also found that patients with either borderline personality disorder or ASPD demonstrated reduced regional cerebral blood flow (rCBF) in both right temporal and prefrontal regions compared with controls, but that those characterized by violence also showed structural deficits. Finally, in an earlier fMRI study, Kiehl and colleagues (2001) found criminal psychopaths to be characterized by overactivation in the bilateral fronto-temporal cortex when processing emotional stimuli. In aggregate, consistent evidence from structural and functional neuroimaging studies continues to implicate the temporal lobe in various forms of criminality, violence, and antisocial behavior.

The limbic system

The limbic system consists of a horseshoe-shaped rim of cortex and adjoining structures at the center of the brain. While there is no universal agreement on the total list of structures included within this system, a general consensus in the literature has been reached on the cingulate and parahippocampal gyri, the hippocampus, the amygdala, and the septal nuclei (Nolte, 2002). In aggregate the limbic system has been shown to influence a host of different behaviors and processes, ranging from memory organization (Freund and Buzsáki, 1996) to stress regulation (Roozendaal *et al.*, 2009). The two limbic structures that have received the most attention in the neuroimaging literature on antisocial behavior are the hippocampus and the amygdala.

The hippocampus is thought to have a prominent role in learning and memory—particularly in the consolidation of explicit memories of facts and events (Nolte, 2002). With the surrounding parahippocampal gyrus, it is also believed to be involved in the regulation of aggression and contextual fear conditioning, as well as the modulation of information for impulse control and moral reasoning. As such, the hippocampus may contribute to the emotional dysregulation, impulsivity, and disinhibition observed in some antisocial groups (Raine *et al.*, 2005; Yang *et al.*, 2010). Structural imaging studies of individuals characterized by antisocial behavior have identified deficits in the hippocampal region, and psychopathy has been one key area of study. For example, Laakso and colleagues (2001), in a study of 18 violent offenders with ASPD and type 2 alcoholism, found increased levels of psychopathy to be associated with reduced volumes in the posterior half of the hippocampi—which may be related to deficits in acquisition of conditioned fear in psychopathy. Tiihonen and colleagues (2008) found offenders with psychopathy to have reduced gray matter densities in the left parahippocampal gyrus and left middle temporal gyrus compared with healthy controls. Similarly, Raine and colleagues (2005) identified an exaggerated hippocampal asymmetry (right > left) in the anterior region in unsuccessful psychopaths relative to successful (uncaught) psychopaths and controls, which may reflect a possible early disruption in normal neurodevelopment.

Regarding other types of antisocial populations, Barkataki and colleagues (2006) found decreased whole brain and hippocampal volumes and increased putamen size in violent men with schizophrenia compared with healthy controls (though nonviolent men with schizophrenia did not demonstrate these structural deficits). Yang and colleagues (2010) also identified gray matter reductions in the hippocampus and parahippocampal gyrus in murderers with schizophrenia, and in the parahippocampal gyrus in murderers without schizophrenia, compared with normal controls. Finally, Huebner *et al.* (2008), using sMRI, identified reduced left hippocampal gray matter volumes in a sample of 23 boys diagnosed with CD (17 co-morbid for attention deficit/hyperactivity disorder); and further analyses indicated that while hyperactive/impulsive symptoms were associated with structural deficits in frontospatial and temporal lobe regions, CD symptoms were associated with limbic system deficits, including the hippocampus and amygdala.

Functional imaging studies have also implicated the hippocampus in associations with various forms of antisocial behavior. For example, in their classic PET study of murderers, Raine and colleagues (1997) found reduced activity in the left medial temporal/hippocampal region and increased activity in the right medial temporal/hippocampus region in murderers relative to controls. In a subsequent study utilizing both structural MRI and SPECT, Soderstrom and colleagues (2000) identified significantly reduced regional blood flow bilaterally in the hippocampus in 21 impulsive violent offenders compared with 11 healthy controls. Similarly, Kiehl and colleagues (2001) found significantly reduced emotion-related activity in the amygdala/hippocampal formation, parahippocampal gyrus, and other limbic regions in criminal psychopaths relative to criminal nonpsychopaths and controls; and Soderstrom and colleagues (2002) found increased levels of the interpersonal features of psychopathy to be associated with reductions in rCBF in the hippocampi and caudate nuclei in a sample of 32 violent offenders. Finally, using fMRI, Lee and colleagues (2009) found male spouse batterers, in comparison with a similarly matched control group of nonbatterers, demonstrated increased neural hyperresponsivity to images representing aggressive threat in cortical and subcortical regions including the hippocampus fusiform gyrus. Taken together, results from these neuroimaging studies indicate that structural and functional deficits in the hippocampus may be associated with various forms of antisociality, though continued research is needed to clarify its exact role in the neurobiological underpinnings of antisocial behavior.

The amygdala is an almond-shaped collection of nuclei located at the forward end of the hippocampus. It receives a wide variety of sensory inputs, and is involved in emotion-related aspects of learning (Nolte, 2002). It is also considered a critical component in the neural circuitry associated with emotional information processing (Dolan and Fullam, 2009). Glenn and Raine (2014), in a recent review, discuss how deficits in amygdala structure and function may manifest in differential forms of antisocial behavior. For example, psychopathic individuals characterized by blunted emotional responding and cold, calculated aggression may show reductions in amygdala volume and activity, while those characterized by more impulsive, reactive aggressive behaviors may show increased amygdala reactivity.

Regarding structural neuroimaging findings, Wong and colleagues (1997), using sMRI, found violent schizophrenia patients demonstrated bilateral volume reductions in the amygdala compared with patient controls characterized by single or minor offenses; while Fairchild and colleagues (2011) found reduced amygdala gray matter volumes bilaterally in individuals diagnosed with either early-onset or late-onset CD compared with controls. Recent results from a subsample of 56 men from a larger longitudinal study of 503 male participants (Pardini *et al.*, 2014) indicate that men with reduced amygdala volume demonstrated increased levels of aggression and psychopathic features over the course of the lifetime at age 26. Reduced amygdala volume was also associated with aggressive and violent behavior and psychopathic traits at a 3-year follow-up. Finally, Yang and colleagues (2009a,b) found psychopathic individuals to be characterized by significant bilateral volume reductions in the amygdala compared with controls, and that surface abnormalities were localized in the basolateral, lateral, cortical, and central nuclei. Significant negative correlations between amygdala volumes and psychopathy scores were strongest for emotional and interpersonal features of psychopathy.

Earlier functional studies identified activation deficits in the amygdalya in antisocial individuals. For example, Raine and colleagues (1997), in their classic PET study, found abnormal asymmetries of activity (reduced left hemisphere relative to right hemisphere) in the amygdala and thalamus in murderers compared with controls. Later functional imaging studies employed specific methods which were driven more by hypotheses specific to the functioning of the amygdala. One such area involved examining amygdala dysfunction during tasks involving the processing of negative emotions. For example, Müller and colleagues (2003) found increased activation in the right amygdala and prefrontal regions in six psychopaths (compared with six controls) when viewing negative images (e.g. those depicting threatening, violent, or distressing scenes or objects), while Herpertz and colleagues (2008) later found increased activation in the amygdala when viewing negative images relative to neutral images in male adolescents with childhood-onset CD compared with controls. Sterzer *et al.* (2005), however, found *reduced* responsivity to emotionally negative images in the left amygdala in the more aggressive members of a group of 13 adolescents with severe CD—suggesting that aggression may be associated with deficits in emotional stimuli recognition and the cognitive control of behavior. Mixed results may reflect differences in the operationalization of antisocial behavior, as mentioned above (Glenn and Raine, 2014).

Emotional functioning in antisocial individuals has also been examined using facial expressions in functional imaging studies of the amygdala. Facial expressions may serve as powerful nonverbal communications between humans, adding an emotional framework or content to social interactions. The interpretation of emotional cues from faces expressing anger is thought to be mediated by neural circuitry involving the amygdala and prefrontal cortex. This same circuit also plays a critical role at a neurofunctional level in regulating emotion and behavior, including anger and aggression. In fact, information processing deficits (e.g. misperceiving neutral facial expressions as threatening)—perhaps associated with dysfunction in this circuitry—may serve as causal antecedents to maladaptive emotions and behaviors such as aggression (see review in Coccaro *et al.*, 2007). Individuals with IED—an illness marked by repeated unplanned outbursts of aggression which are disproportionately intense relative to situational factors and result in destruction of property and interpersonal harm—may be particularly characterized by deficits in this circuit. Coccaro and colleagues (2007) found increased amygdala reactivity and reduced OFC activation to angry faces in individuals with IED compared with healthy controls. Furthermore, individuals with a history of aggression failed to show amygdala–OFC connectivity when presented with faces expressing anger.

Passamonti and colleagues (2010) later found that individuals with either early-onset or late-onset CD were characterized by increased amygdala activation when viewing neutral but not sad faces (compared with baseline measures) relative to controls. Additionally, those with early-onset CD demonstrated reduced amygdala responses to sad faces relative to those with adolescent-onset CD, which may relate to why early-onset CD is more severe and persistent than adolescent-onset CD. In an fMRI study of 24 male schizophrenia patients with violent offense convictions, Dolan and Fullam (2009) found reduced blood oxygenation level-dependent (BOLD) responses in the amygdala in patients with high levels of psychopathy (and negative correlations between psychopathy scores and amygdala responses) during exposure to images of fearful faces. Results also indicated positive correlations between psychopathy scores and neural activation in the amygdala and inferior prefrontal regions for disgust, but also negative correlations for anger.

Amygdala dysfunction in antisocial individuals has also been examined using emotional conditioning (learning) tasks. For example, Schneider and colleagues (2000) found reduced amygdala activation during presentation of an aversive unconditioned stimulus (rotten yeast) in 12 individuals with ASPD compared with 12 healthy controls. Additionally, participants in the ASPD group were characterized by divergent neuronal network activity in cortical and subcortical (i.e. amygdala) regions. Results that showed decreased autonomic and cerebral reactions to aversive stimulation may explain deficient responsivity to punishment. Birbaumer and colleagues (2005) used an aversive differential pavlovian conditioning paradigm (slides of neutral faces presented with painful pressure) to examine fear conditioning in 10 criminal psychopaths and 10 healthy controls. Results indicated controls showed activation in the amygdala, OFC, insula, and ACC, while psychopaths showed no significant activity in this circuit.

Finally, more recent neuroimaging studies of amygdala functioning and antisocial behavior have utilized novel approaches to understand the neurobiological underpinnings of emotional experience in antisocial individuals. These novel methods include the use of moral reasoning and emotional perspective-taking tasks. For example, using fMRI in a community sample, Glenn *et al.* (2009a) found increased psychopathy scores to be related to reduced amygdala activation during an emotional moral decision-making task (a task involving [1] emotion-provoking moral personal dilemmas, such as having to choose between smothering one's crying baby to save oneself and other townspeople hiding from terrorists who may be alerted by the sound; [2] less emotional moral impersonal dilemmas, such as deciding to keep money found in a lost wallet; and [3] nonmoral dilemmas, such as choosing to take the train or the bus). Finally, in an fMRI study of medium-security prison inmates, Decety and colleagues (2013) examined the relationship between psychopathy and emotional perspective-taking. Participants were shown images of bodily injuries and asked to imagine the depicted events happening either to themselves or others. Results indicated that during the "imagine-self" perspective, individuals with high levels of psychopathy demonstrated a typical empathic neural response to pain—activations in a network consisting of the right amygdala and other cortical regions. However, during the "imagine-other" perspective, high psychopaths were characterized by an atypical pattern of neural activity; and amygdala and insula activity was inversely related to levels of the interpersonal/emotional features of psychopathy. Results may indicate a neurobiological mechanism (i.e. differential neural circuitry involving the amygdala) underlying the callousness and lack of empathy that is characteristic of many psychopathic individuals. Given the role that aberrant emotional experience may intuitively play in various forms of antisocial behavior, the amygdala remains a promising region of study in neuroimaging investigations of antisocial individuals.

Conclusion

Neuroimaging studies in recent decades have produced mounting evidence for a link between specific types of brain deficit and various forms of antisocial behavior. While the sheer volume and often highly technical nature of this area of research can be daunting for the consumer, an organizational framework based on imaging methods (structural versus functional), brain regions of interest (frontal lobe, prefrontal cortex, temporal lobe, and limbic system structures such as the hippocampus and amygdala), and operationalization of antisocial behavior (e.g. crime type, such as homicidal offenders, or criminogenic psychiatric pathologies such as psychopathy or CD) can be helpful. Findings from neuroimaging studies continue to implicate particular brain regions in crime, violence, aggression, and antisociality—specifically those associated with functions such as impulse control, planning and organization, learning, and emotional experience. Fortunately, the Lombrosian-era autopsy scalpels and saws, previously used to open the skulls of the criminal and violent for examination, have now been replaced by high-powered scanners and digital images of the brain and its components. For the neurocriminologist, this is an exciting time, as steady advances in technology (which allow for higher-resolution, more sophisticated *in vivo* investigations) coupled with an increased insight into the brain–behavior relationship will likely afford a heightened understanding of the biological underpinnings of antisocial behavior.

References

Antonucci, A. S., Gansler, D. A., Tan, S., Bhadelia, R., Patz, S., and Fulwiler, C. (2006) 'Orbitofrontal correlates of aggression and impulsivity in psychiatric patients', *Psychiatry Research: Neuroimaging*, 147: 213–220.

Barkataki, I., Kumari, V., Das, M., Taylor, P., and Sharma, T. (2006) 'Volumetric structural brain abnormalities in men with schizophrenia or antisocial personality disorder', *Behavioral Brain Research*, 169(2): 239–247.

Basoglu, C., Semiz, U., Oner, O., Gunay, H., Ebrinc, S., Cetin, M., *et al.* (2008) 'A magnetic resonance spectroscopy study of antisocial behaviour disorder, psychopathy and violent crime among military conscripts', *Acta Neuropsychiatrica*, 20(2): 72–77.

Birbaumer, N., Veir, R., Lotze, M., Erb, M., Hermann, C., Grodd, W., and Flor, H. (2005) 'Deficient fear conditioning in psychopathy: a functional magnetic resonance imaging study', *Archives of General Psychiatry*, 62(7): 799–805.

Blair, K. S., Newman, C. C., Mitchell, D. V., Richell, R. A., Leonard, A. A., Morton, J. J., and Blair, R. R. (2006) 'Differentiating among prefrontal substrates in psychopathy: neuropsychological test findings', *Neuropsychology*, 20(2): 153–165.

Boes, A.D., Tranel, D., Anderson, S.W., and Nopoulos, P. (2008) 'Right anterior cingulate: a neuroanatomical correlate of aggression and defiance in boys', *Behavioral Neuroscience*, 122(3): 677–684.

Bufkin, J. and Luttrel, V. (2005) 'Neuroimaging studies of aggressive and violent behavior current findings and implications for criminology and criminal justice', *Trauma, Violence, and Abuse*, 6(2): 176–191.

Coccaro, E. F., McCloskey, M. S., Fitzgerald, D. A., and Phan, K. L. (2007) 'Amygdala and orbitofrontal reactivity to social threat in individuals with impulsive aggression', *Biological Psychiatry*, 62: 168–178.

Crowley, T. J., Dalwani, M. S., Mikulich-Gilbertson, S. K., Du, Y. P., Lejuez, C. W., Raymond, K. M., and Banich, M. T. (2010) 'Risky decisions and their consequences: neural processing by boys with antisocial substance disorder', *PLOS ONE*, 5(9).

Dalwani, M., Sakai, J. T., Mikulich-Gilbertson, S. K., Tanabe, J., Raymond, K., McWilliams, S. K., *et al.* (2011) 'Reduced cortical gray matter volume in male adolescents with substance and conduct problems', *Drug and Alcohol Dependence*, 118(2/3): 295–305.

Damasio, A. (1994) *Descartes' Error: Emotion, Reason, and the Human Brain*, New York: GP Putnam's Sons.

Decety, J., Chen, C., Harenski, C., and Kiehl, K. A. (2013) 'An fMRI study of affective perspective taking in individuals with psychopathy: imagining another in pain does not evoke empathy', *Frontiers in Human Neuroscience*, 7: 489–501.

Dinn, W. M. and Harris, C. L. (2000) 'Neurocognitive function in antisocial personality disorder', *Psychiatry Research*, 97(2/3): 173–190.

Dolan, M. C. (2010) 'What imaging tells us about violence in anti-social men', *Criminal Behavior and Mental Health*, 20(3): 199–214.

Dolan, M. C. and Fullam, R. S. (2009) 'Psychopathy and functional magnetic resonance imaging blood oxygenation level-dependent responses to emotional faces in violent patients with schizophrenia', *Biological Psychiatry*, 66: 570–577.

Dolan, M. C., Deakin, W., Roberts, N., and Anderson, I. (2002) 'Seratonergic and cognitive impairment in impulsive aggressive personality disordered offenders: are there implications for treatment?', *Psychological Medicine*, 32: 105–117.

Fairchild, G., Passamonti, L., Hurford, G., Hagan, C., von Dem Hagen, E., van Goozen, S., *et al.* (2011) 'Brain structural abnormalities in early-onset and adolescent-onset conduct disorder', *American Journal of Psychiatry*, 168(6): 624–633.

Fairchild, G., Hagan, C. C., Walsh, N. D., Passamonti, L., Calder, A. J., and Goodyer, I. M. (2013) 'Brain structure abnormalities in adolescent girls with conduct disorder', *Journal of Child Psychology and Psychiatry*, 54(1): 86–95.

Finger, E. C., Marsh, A. A., Blair, K. S., Reid, M. E., Sims, C., Ng, P., *et al.* (2011) 'Disrupted reinforcement signaling in the orbitofrontal cortex and caudate in youths with conduct disorder or oppositional defiant disorder and a high level of psychopathic traits', *American Journal of Psychiatry*, 168(2): 152–162.

Freund, T. F. and Buzsáki, G. (1996) 'Interneurons of the hippocampus', *Hippocampus*, 6(4): 347–470.

Glenn, A. L. and Raine, A. (2014) 'Neurocriminology: implications for the punishment, prediction and prevention of criminal behavior', *Nature Reviews Neuroscience*, 15(1): 54–63.

Glenn, A. L., Raine, A., and Schug, R. A. (2009a) 'The neural correlates of moral decision-making in psychopathy', *Molecular Psychiatry*, 14: 5–6.

Glenn, A. L., Raine, A., and Schug, R. A. (2009b) 'Increased DLPFC activity during moral decision-making in psychopathy', *Molecular Psychiatry*, 14: 909–911.

Goethals, I., Audenaert, K., Jacobs, F., Van den Eynde, F., Bernagie, K., Kolindou, A., Vervaet, M., Dierckx, R., and Van Heeringen, C. (2005) 'Brain perfusion SPECT in impulsivity-related personality disorders', *Behavioural Brain Research*, 157: 187–192.

Goyer, P. F., Andreason, P. J., Semple, W. E., and Clayton, A. H. (1994) 'Positron-emission tomography and personality disorders', *Neuropsychopharmacology*, 10: 21–28.

Gregory, S., Simmons, A., Kumari, V., Howard, M., Hodgins, S., and Blackwood, N. (2012) 'The antisocial brain: psychopathy matters: a structural MRI investigation of antisocial male violent offenders', *Archives of General Psychiatry*, 69(9): 962–972.

Herpertz, S., Huebner, T., Marx, I., Vloet, T., Fink, G., Stoecker, T. *et al.* (2008) 'Emotional processing in male adolescents with childhood-onset conduct disorder', *Journal of Child Psychology and Psychiatry*, 49(7): 781–791.

Hill, D. H. and Pond, D. A. (1952) 'Reflections on one hundred capital cases submitted to electroencephalography', *Journal of Mental Sciences*, 98: 23–43.

van Holst, R. J., van der Meer, J. N., McLaren, D. G., van den Brink, W., Veltman, D. J., and Goudriaan, A. E. (2012a) 'Interactions between affective and cognitive processing systems in problematic gamblers: a functional connectivity study', *PLOS ONE*, 7(11): e49923.

van Holst, R. J., van Holstein, M., van den Brink, W., Veltman, D. J., and Goudriaan, A. E. (2012b) 'Response inhibition during cue reactivity in problem gamblers: an fMRI study', *PLOS ONE*, 7(3): e30909.

Hoptman, M. J., Volavka, J., Weiss, E. M., Czobor, P., Szeszko, P. R., Gerig, G., Chakos, M., Blocher, J., Citrome, L. L., Lindenmayer, J., Sheitman, B., Lieberman, J. A., and Bilder, R. M. (2005) 'Quantitative MRI measures of orbitofrontal cortex in patients with chronic schizophrenia or schizoaffective disorder', *Psychiatry Research: Neuroimaging*, 140: 133–145.

Huebner, T., Vloet, T., Marx, I., Konrad, K., Fink, G., Herpertz, S., and Herpertz-Dahlmann, B. (2008) 'Morphometric brain abnormalities in boys with conduct disorder', *Journal of the American Academy of Child and Adolescent Psychiatry*, 47(5): 540–547.

Joyal, C. C., Putkonen, A., Mancini-Marie, A., Hodgins, S., Kononen, M., Boulay, L., Pihlajamaki, M., Soininen, H., Stip, E., Tiihonen, J., and Aronen, H. J. (2007) 'Violent persons with schizophrenia and comorbid disorders: a functional magnetic resonance imaging study', *Schizophrenia Research*, 91: 97–102.

Kiehl, K. A., Smith, A.M., Hare, R. D., Mendrek, A., Forster, B. B., Brink, J., and Liddle, P. F. (2001) 'Limbic abnormalities in affective processing by criminal psychopaths as revealed by functional magnetic resonance imaging', *Biological Psychiatry*, 50(9): 677–684.

Kochanska, G., Murray, K., and Coy, K. C. (1997) 'Inhibitory control as a contributor to conscience in childhood: from toddler to early school age', *Child Development*, 68: 263–277.

Kruesi, M., Casanova, M., Mannheim, G. and Johnson-Bilder, A. (2004) 'Reduced temporal lobe volume in early onset conduct disorder', *Psychiatry Research: Neuroimaging*, 132: 1–11.

Kumari, V., Barkataki, I., Goswami, S., Flora, S., Das, M., and Taylor, P. (2009) 'Dysfunctional, but not functional, impulsivity is associated with a history of seriously violent behaviour and reduced orbitofrontal and hippocampal volumes in schizophrenia', *Psychiatry Research: Neuroimaging*, 173(1): 39–44.

Laakso, M. P., Vaurio, O., Koivisto, E., Savolainen, L., Eronen, M., Aronen, H. J., Hakola, P., Repo, E., Soinen, H., and Tiihonen, J. (2001) 'Psychopathy and the posterior hippocampus', *Behavioural Brain Research*, 118(2): 187–193.

Laakso, M. P., Gunning-Dixon, F., Vaurio, O., Repo-Tiihonen, E., Soininen, H., and Tiihonen, J. (2002) 'Prefrontal volumes in habitually violent subjects with antisocial personality disorder and type 2 alcoholism', *Psychiatry Research: Neuroimaging*, 114(2): 95–102.

Lapierre, D., Braun, C. M. J., and Hodgins, S. (1995) 'Ventral frontal deficits in psychopathy: neuropsychological test findings', *Neuropsychologia*, 33(2): 139–151.

Lee, T. M., Chan, S. C., and Raine, A. (2009) 'Hyperresponsivity to threat stimuli in domestic violence offenders: a functional magnetic resonance imaging study', *Journal of Clinical Psychiatry*, 70(1): 36–45.

Lencz, T., Raine, A., and Sheard, C. (1996) 'Neuroanatomical bases of electrodermal hypo-responding: a cluster analytic study', *International Journal of Psychophysiology*, 22(3): 141–153.

Lezak, M. D., Howieson, D. B., Loring, D.W., Hannay, H. J., and Fischer, J. S. (2004) *Neuropsychological Assessment*, 4th edn, New York: Oxford University Press.

Li, T., Mathews, V. P., Wang, Y., Dunn, D., and Kronenberger, W. (2005) 'Adolescents with disruptive behavior disorder investigated using an optimized MR diffusion tensor imaging protocol', *Annals of the New York Academy of Sciences*, 1064: 184–192.

Luria, A. (1966) *Higher Cortical Functions in Man*, New York: Basic Books.

Marsh, A. A., Finger, E. C., Fowler, K. A., Jurkowitz, I. T. N., Schechter, J. C., Yu, H. H., *et al.* (2011) 'Reduced amygdala–orbitofrontal connectivity during moral judgments in youths with disruptive behavior disorders and psychopathic traits', *Psychiatry Research: Neuroimaging*, 194(3): 279–286.

Mathews, V. P., Kronenberger, W. G., Wang, Y., Lurito, J. T., Lowe, M. J., and Dunn, D. W. (2005) 'Media violence exposure and frontal lobe activation measured by functional magnetic resonance imaging in aggressive and nonaggressive adolescents', *Journal of Computer Assisted Tomography*, 29: 287–292.

Mills, S. and Raine, A. (1994) 'Neuroimaging and aggression', *Journal of Offender Rehabilitation*, 21(3/4): 145–158.

Morgan, A. B. and Lilienfeld, S. O. (2000) 'A meta-analytic review of the relationship between antisocial behavior and neuropsychological measures of executive function', *Clinical Psychology Review*, 20(1): 113–136.

Müller, J. L., Sommer, M., Wagner, V., Lange, K., Taschler, H., Röder, C. H., *et al.* (2003) 'Abnormalities in emotion processing within cortical and subcortical regions in criminal psychopaths: evidence from a functional magnetic resonance imaging study using pictures with emotional content', *Biological Psychiatry*, 54(2): 152–162.

Müller, J. L., Sommer, M., Döhnel, K., Weber, T., Schmidt-Wilcke, T., and Hajak, G. (2008) 'Disturbed prefrontal and temporal brain function during emotion and cognition interaction in criminal psychopathy', *Behavioral Sciences and the Law*, 26(1): 131–150.

Mundy-Castle, A. C. (1955) 'The EEG in twenty-two cases of murder or attempted murder. Appendix on possible significance of alphoid rhythms', *Journal of the National Institute for Personnel Research*, 6: 103–120.

Nolte, J. (2002) *The Human Brain: An Introduction to its Functional Anatomy*, 5th edn, St Louis, MO: Mosby.

Pardini, D. A., Raine, A., Erickson, K., and Loeber, R. (2014) 'Lower amygdala volume in men is associated with childhood aggression, early psychopathic traits, and future violence', *Biological Psychiatry*, 75: 73–80.

Passamonti, L., Fairchild, G., Goodyer, I., Hurford, G., Hagan, C., Rowe, J., and Calder, A. (2010) 'Neural abnormalities in early-onset and adolescence-onset conduct disorder', *Archives of General Psychiatry*, 67(7): 729–738.

Passamonti, L., Fairchild, G., Fornito, A., Goodyer, I. M., Nimmo-Smith, I., Hagan, C. C., and Calder, A. J. (2012) 'Abnormal anatomical connectivity between the amygdala and orbitofrontal cortex in conduct disorder', *PLOS ONE*, 7(11): e48789.

Pridmore, S., Chambers, A., and McArthur, M. (2005) 'Neuroimaging in psychopathy', *Australian and New Zealand Journal of Psychiatry*, 39(10): 856–865.

Raine, A., Buchsbaum, M., and LaCasse, L. (1997) 'Brain abnormalities in murderers indicated by positron emission tomography', *Biological Psychiatry*, 42(6): 495–508.

Raine, A., Meloy, J. R., Bihrle, S., Stoddard, J., LaCasse, L., and Buchsbaum, M. S. (1998) 'Reduced prefrontal and increased subcortical brain functioning assessed using positron emission tomography in predatory and affective murderers', *Behavioral Sciences and the Law*, 16(3): 319–332.

Raine, A., Lencz, T., Bihrle, S., LaCasse, L., and Colletti, P. (2000) 'Reduced prefrontal gray matter volume and reduced autonomic activity in antisocial personality disorder', *Archives of General Psychiatry*, 57(2): 119–127.

Raine, A., Ishikawa, S. S., Arce, E., Lencz, T., Knuth, K. H., Bihrle, S., LaCasse, L., and Colletti, P. (2005) 'Hippocampal structural asymmetry in unsuccessful psychopaths', *Biological Psychiatry*, 55: 185–191.

Raine, A., Yang, Y., Narr, K. L., and Toga, A. W. (2011) 'Sex differences in orbitofrontal gray as a partial explanation for sex differences in antisocial personality', *Molecular Psychiatry*, 16(2): 227–236.

Roozendaal, B., McEwen, B. S., and Chattarji, S. (2009) 'Stress, memory and the amygdala', *Nature Reviews Neuroscience*, 10(6): 423–433.

Roussy, S. and Toupin, J. (2000) 'Behavioral inhibition deficits in juvenile psychopaths', *Aggressive Behavior*, 26(6): 413–424.

Schiffer, B., Müller, B. W., Scherbaum, N., Hodgins, S., Forsting, M., Wiltfang, J., *et al.* (2011) 'Disentangling structural brain alterations associated with violent behavior from those associated with substance use disorders', *Archives of General Psychiatry*, 68(10): 1039–1049.

Schneider, F., Habel, U., Kessler, C., Posse, S., Grodd, W., and Müller-Gärtner, H. W. (2000) 'Functional imaging of conditioned aversive emotional responses in antisocial personality disorder', *Neuropsychobiology*, 42(4): 192–201.

Schug, R. A., Gao, Y., Glenn, A. L., Yang, Y., and Raine, A. (2009) 'Role of the frontal lobe in violence', in *McGraw-Hill Yearbook of Science and Technology*, Hightstown, NJ: McGraw-Hill, pp. 326–328.

Soderstrom, H., Tullberg, M., Wikkelsö, C., Ekholm, S., and Forsman, A. (2000) 'Reduced regional cerebral blood flow in non-psychotic violent offenders', *Psychiatry Research: Neuroimaging*, 98(1): 29–41.

Soderstrom, H., Hultin, L., Tullberg, M., Wikkelso, C., Ekholm, S., and Forsman, A. (2002) 'Reduced frontotemporal perfusion in psychopathic personality', *Psychiatry Research Neuroimaging*, 114(2): 81–94.

Stafford Clark, D., and Taylor, F. H. (1949) 'Clinical and electro-encephalographic studies of prisoners charged with murder', *Journal of Neurology, Neurosurgery, and Psychiatry*, 12: 325–330.

Sterzer, P., Stadler, C., Krebs, A., Kleinschmidt, A., and Poustka, F. (2005) 'Abnormal neural responses to emotional visual stimuli in adolescents with conduct disorder', *Biological Psychiatry*, 57: 7–15.

Syngelaki, E. M., Moore, S. C., Savage, J. C., Fairchild, G., and Van Goozen, S. M. (2009) 'Executive functioning and risky decision making in young male offenders', *Criminal Justice and Behavior*, 36(11): 1213–1227.

Tiihonen, J., Rossi, R., Laakso, M. P., Hodgins, S., Testa, C., Perez, J., *et al.* (2008) 'Brain anatomy of persistent violent offenders: more rather than less', *Psychiatry Research: Neuroimaging*, 163(3): 201–212.

Volkow, N. D., Tancredi, L. R., Grant, C., Gillespie, H., Valentine, A., Mullani, N., *et al.* (1995) 'Brain glucose metabolism in violent psychiatric patients: a preliminary study', *Psychiatry Research: Neuroimaging*, 61(4): 243–253.

Völlum, B., Richardson, P., Stirling, J., Elliott, R., Dolan, M., Chaudhry, I., *et al.* (2004) 'Neurobiological substrates of antisocial and borderline personality disorder: preliminary results of a functional fMRI study', *Criminal Behaviour and Mental Health*, 14(1): 39–54.

Völlum, B., Richardson, P., McKie, S., Reniers, R., Elliott, R., Anderson, I. M., *et al.* (2010) 'Neuronal correlates and serotonergic modulation of behavioural inhibition and reward in healthy and antisocial individuals', *Journal of Psychiatric Research*, 44(3): 123–131.

White, S. F., Pope, K., Sinclair, S., Fowler, K. A., Brislin, S. J., Williams, W. C., *et al.* (2013) 'Disrupted expected value and prediction error signaling in youth with disruptive behavior disorders during a passive avoidance task', *American Journal of Psychiatry*, 170(3): 315–323.

Wong, M., Lumsden, J., Fenton, G., and Fenwick, P. (1997) 'Neuroimaging in mentally abnormal offenders', *Issues in Criminological and Legal Psychology*, 27: 49–58.

Yang, Y. and Raine, A. (2009) 'Prefrontal structural and functional brain imaging findings in antisocial, violent, and psychopathic individuals: a meta-analysis', *Psychiatry Research: Neuroimaging*, 174(2): 81–88.

Yang, Y. Y., Raine, A. A., Colletti, P. P., Toga, A. W., and Narr, K. L. (2009a) 'Abnormal temporal and prefrontal cortical gray matter thinning in psychopaths', *Molecular Psychiatry*, 14(6): 561–562.

Yang, Y. Y., Raine, A., Narr, K. L., Colletti, P., and Toga, A. W. (2009b) 'Localization of deformations within the amygdala in individuals with psychopathy', *Archives of General Psychiatry*, 66: 986–994.

Yang, Y. Y., Raine, A., Han, C. B., Schug, R. A., Toga, A. W., and Narr, K. L. (2010) 'Reduced hippocampal and parahippocampal volumes in murderers with schizophrenia', *Psychiatry Research: Neuroimaging*, 182: 9–13.

Zamboni, G. G., Huey, E. D., Krueger, F. F., Nichelli, P. F., and Grafman, J. J. (2008) 'Apathy and disinhibition in frontotemporal dementia: insights into their neural correlates', *Neurology*, 71(10): 736–742.

15

Neuropsychology of psychopathic traits in children

Alice P. Jones Bartoli and Guilherme W. Wendt

Introduction

This chapter begins by considering the cognitive neuropsychology associated with callous-unemotional (CU) traits, including intelligence quotient (IQ), executive functioning, and working memory. In all domains, comparisons are drawn against typically developing children and adolescents, and children with behavioral problems who do not have elevated levels of CU traits. The second part of the chapter focuses on the affective processing abilities associated with elevated CU traits.

The traditional view of an individual with psychopathy has included high levels of verbal skill; bringing to mind somebody like Hannibal Lecter: verbally sophisticated and able to manipulate others with skill and ease. Hervey Cleckley's (1976) original case studies discussed in his book *The Mask of Sanity* include descriptions like "very often indications of good sense and sound reasoning will emerge and one is likely to feel soon after meeting him that this normal and pleasant person is also one with high abilities" (p. 364). However, contrary to these popular conceptions, a number of studies suggest that there may be no differences, or an IQ deficit, in individuals who have elevated levels of CU traits.

Although early findings claiming that delinquency was associated with a lower IQ were considered controversial due to their methodology, it is now widely recognized that an inverse relationship between verbal IQ (VIQ) and antisocial behavior exists. In early childhood, a weakness in verbal learning and reasoning reliably predicts later antisocial outcomes (Henry and Moffitt, 1997). The importance of studying neurocognitive impairments is clear, and understanding such impairments may provide a key route through which genetic and psychosocial influences on antisocial behavior find expression. For example, Raine and colleagues (2005) made an investigation of neurocognitive impairments associated with antisocial behavior in a community sample of boys aged 7 to 17 years. Using subtests from the Wechsler Intelligence Scales for Children (WISC; Wechsler, 1991), their sample was divided into four groups: life-course persistent conduct disorder (CD); adolescent-limited CD; a group of individuals with childhood-limited conduct problems; and a typically developing control group. Of these four groups, the life-course persistent group was found to have a significantly lower VIQ than the other three groups, while the childhood-limited group showed a deficit in their performance IQ. Group differences were also found in a spatial memory task, where life-course persistent and adolescent-limited groups were significantly poorer than the control group. These deficits were not explained by any of the possible confounds measured, such as attention deficit hyperactivity disorder (ADHD), psychosocial adversity, child abuse, or head injury, and are consistent with earlier findings from Raine's group showing spatial impairment in another community study of young children (Raine *et al.*, 2002).

This prominence of work on cognitive deficits is also called into question by research focusing on children and adults with psychopathic traits. Cleckley hypothesized that psychopaths would have intelligence within the normal range, and noted that psychopathic individuals showed "good sense and sound reasoning" (Cleckley, 1976, p. 364), and reports that psychopaths are "often, if not usually, of superior intelligence when measured scientifically" (p. 274). More recent studies have found no evidence for verbal impairments among adult psychopaths, but significantly weaker performance on block design, a task focusing on spatial abilities (Smith *et al.*, 1992). More recently, Raine and colleagues also demonstrated spatial but not verbal impairments among adult psychopaths (Raine *et al.*, 2002). A similar pattern has been demonstrated among children with CU traits. Loney and colleagues compared children with behavioral problems with and without elevated levels of CU traits, and found that those children with behavioral difficulties without CU traits showed poorer verbal ability compared with a clinical control group, whereas those children with conduct problems with elevated CU traits showed no such deficit (Loney *et al.*, 1998). A more recent study of CU and cognitive functioning indicated a negative relationship between CU traits and both verbal and non-verbal abilities, but these did not remain significant after accounting for conduct problems and hyperactivity symptoms (Fontaine *et al.*, 2008). It seems, then, that it is important to consider the relationship between conduct problems and cognitive ability, as well as the association between cognitive ability and CU.

Neurocognitive studies relating to conduct problems, and specifically for CU individuals, are important for a number of reasons. First, understanding the relation between conduct problems, CU traits, and cognitive ability may help to understand the differing etiologies of the antisocial outcomes. More specifically, there may be scope for investigating a relation between ability and the severity of criminal or violent behavior. In the case of those individuals who show a deficit in their emotional functioning (i.e. affective empathy and capacity for remorse), but who possess good intellectual abilities, it is suggested that they may therefore be better able to plan and carry out their crimes without detection (Salekin, 2006).

Executive functioning

Executive functions (EF) refer to the higher-order cognitive processes involving initiation, planning, cognitive flexibility, and decision making that allow goal-oriented behavior appropriate to the context (Spreen and Strauss, 1998). EF can also be described as "hot" or "cool." Hot EF refers to cognitive processes that have an affective, motivational, or incentive/reward component. Tasks that could be described as hot include those involving appraisal of the affective salience of a stimulus in decision-making paradigms and reappraisal in response reversal paradigms (Rubia, 2011). Cool EF processes include those considered primarily cognitive in nature, with no substantive affective component. Examples of these include working memory, response inhibition, planning, sustained attention, and attentional set-shifting (Rubia, 2011). Hot processes are generally considered to be subsumed by ventromedial pathways connecting mesolimbic reward circuitry, including the amygdala and striatum, to the ventromedial prefrontal cortex (De Luca and

Leventer, 2008). Cool EF processes, on the other hand, refer to top-down processes associated primarily with activity in the dorsolateral prefrontal cortex and ventrolateral prefrontal cortex (De Luca and Leventer, 2008).

EF is related to intelligence in that both constructs share certain component operations (e.g. working memory). However, research has demonstrated that intelligence and EF are dissociable abilities, and early executive weaknesses have been reported to be related to childhood behavior problems independently of IQ (Nigg *et al.*, 1999). One meta-analysis of executive function and antisocial behavior found that overall effect sizes for conduct disorder and juvenile delinquency were substantial (Morgan and Lilienfeld, 2000); those with antisocial behavior were doing more poorly on tests of EF. However, these results were somewhat confounded by the definitions of antisocial behavior used in the studies included in the meta-analysis. In children, similar to the findings relating to intelligence (Speltz *et al.*, 1999), deficits in EF are more evident in those with co-morbid ADHD (Clark *et al.*, 2000).

To date, there is a scarcity of literature about EF in children with CU traits. However, some work has focused on EF in adults with antisocial personality disorder and psychopathy. Morgan and Lilienfield's meta-analysis (2000) demonstrated a robust association between EF deficits and antisocial behavior; however, due to the small number of studies carried out at that time, it was not possible to examine the association between EF and psychopathy. Since this meta-analysis, some studies have investigated the association between antisocial behavior and EF, with mixed findings. In studies focusing on adults with psychopathy, Dolan (2012) reports no greater EF impairment in adults with antisocial personality disorder (ASPD) with high levels of psychopathic traits, compared with adults with only a diagnosis of ASPD. Similarly, De Brito *et al.* (2013) report no differences in EF function between offenders with ASPD with and without psychopathy. Compared with controls, both sets of offenders showed impairments in verbal working memory and adaptive decision making. There were no differences between these two groups on either hot or cold measures of EF. However, in contrast to these findings, there is some evidence to suggest that offenders scoring high on psychopathic traits are impaired at tasks designed to test EF, compared with offending controls (Pham *et al.*, 2003). Specifically, while individuals high on psychopathic traits are deficient on tasks associated with orbitofrontal cortex (OFC) dysfunction (e.g. measures of response reversal and extinction and go–no go tasks), they do not show impairment on tasks associated with dorsolateral prefrontal cortex functioning (e.g. Wisconsin card sorting task; Roussy and Toupin, 2000; Mitchell *et al.*, 2002).

With regard to memory specifically, relatively few studies have examined memory in antisocial populations. However, what has been done suggests that physically aggressive boys have poorer working memory than non-aggressive boys, which was still significant after controlling for IQ and ADHD symptoms (Séguin *et al.*, 1999). However, other studies found no deficits in working memory for children with oppositional defiant disorder (ODD) and ODD/ADHD (van Goozen *et al.*, 2004). A recent study suggests that recall accuracy for emotional events is poorer in adolescents with high levels of CU traits (Dolan and Fullam, 2010). It is clear that there is still outstanding work to be done in better disentangling the associations between cognitive neuropsychological abilities and CU traits, separate from the influence of ADHD symptoms and conduct problems.

To date, there are few direct comparisons of the cognitive profile of children with conduct problems with and without elevated CU traits. Frick *et al.* (2003) compared groups of children with conduct problems with and without CU, and found differential profiles: poor processing of punishment information in those children with elevated CU traits; and a hostile attribution bias in those with only conduct problems. Differences have also been observed in emotional reactivity; Loney *et al.* (2003) demonstrated a slower recognition rate for negative emotional words in adolescents with conduct problems with elevated CU, compared with a faster recognition time for the same words in a group with conduct problems and low CU. Finally, Dadds *et al.* (2006) report that CU traits are uniquely related to poor recognition of fearful expressions, while conduct problems without the concomitant CU traits have been associated with hypersensitivity to hostility. In addition, a large body of research by Blair and colleagues has repeatedly demonstrated deficits in processing fear, sadness, and punishment in individuals with conduct problems and CU traits, as compared with non-CU comparison groups (Blair *et al.*, 2006). The work stemming from Blair's violence inhibition model and integrated emotion systems model (Blair, 1995; Blair *et al.*, 2005) and the work of Frick and colleagues (e.g. Frick *et al.*, 1994, 2014) has been valuable in highlighting the importance of subtyping the children with early-onset conduct problems into those with and without CU traits.

Hostile attribution bias

Children with conduct problems are characterized by the tendency to interpret ambiguous situations as being hostile, and to respond aggressively. Crick and Dodge (1994) termed this a hostile attributional bias. According to this theory, when faced with an ambiguous social situation the aggressive behavior of a child with behavioral problems is, in their view, retaliatory. According to this model, the victim is likely to consider the aggression to be unjust, and this action may lead to impaired peer relationships. The reactions of the aggressive child's peers only serve to fulfill the child's initial beliefs that those peers hold hostile intentions, which in turn acts as justification for further aggressive acts. There is some evidence that a hostile attribution bias is associated more with reactive aggression and may therefore be a marker for behavioral problems rather than CU traits (Frick *et al.*, 2003; Walters, 2007).

Emotion recognition

There is a wealth of literature reporting deficits in recognition of emotional expressions of fear in adults with psychopathy. This impairment appears to extend reliably to children with CU traits who are reported to demonstrate selective deficits in emotion recognition related to emotions associated with distress (specifically fear and sadness). This deficit has been consistently demonstrated across different age groups and across different methods of presenting emotions. This deficit is observed not only in facial expressions of emotion, but also in vocal expressions and body postures (Blair *et al.*, 2001, 2002, 2005; Kosson *et al.*, 2002; Dolan and Fullam, 2006; Muñoz, 2009). Several studies have aimed to understand the roots of these deficits better, using brain-imaging techniques. In adult psychopaths, fear recognition had been associated with hyporeactivity in amygdala, and this result has been consistently demonstrated in child and adolescent samples. Studies by Marsh *et al.* (2008) and Jones *et al.* (2009) have reported reduced amygdala activity in response to fearful facial expressions of emotion compared with age-matched comparison young people (and, in the case of Marsh *et al.*, also compared with young people with a diagnosis of ADHD). More recently, Viding *et al.* (2012) demonstrated a decreased amygdala response to fearful expressions presented pre-attentively in a CP/CU+ group compared with a sample of children with CP/CU–, regardless of levels of conduct problems (CP), ADHD or other emotional symptoms.

However, more recent work has suggested a theory that might account for the differences in amygdala reactivity and recognition performance that may be related to eye gaze. Mark Dadds' group has conducted several studies examining the patterns of eye gaze when looking at emotional faces.

Eye gaze and attention to emotional stimuli

Children with CP, CD, and also the group with CU+ are distinct in their brain function and structure, which indicates that neuropsychological factors are associated with antisocial behavior (Marsh and Blair, 2008; Sumich *et al.*, 2012). There is evidence that brain functions related to attention to emotional stimuli and in recognizing others' emotions differ in individuals with different kinds of psychopathic traits (Blair *et al.*, 2005). As mentioned earlier, youth with CU+ traits have been extensively identified by hypo-responsiveness in amygdala in response to emotional stimuli (Jones *et al.*, 2009; Viding *et al.*, 2012).

One imaging study has attempted to account for this potential attentional impairment. Sebastian *et al.* (2014) used fMRI to measure brain reactivity to fearful and neutral facial expressions of emotion in children with behavioral problems with and without elevated CU traits, and typically developing controls. On half of trials, participating children were required to look for a dot anywhere within the faces presented; on the other half, participants were directed to focus on the eye region. In the fearful eyes condition, the group of individuals with CP/CU– differ from the controls and CP/CU+ individuals. These data draw attention to different conditions in which individuals with CP present distinctions in cognitive and behavioral responses, highlighting the urgency to adopt a complex, developmental psychopathology perspective (Hyde *et al.*, 2013).

Eyes are usual a well known way to express and recognize emotions, mainly fear. Thus eyes also play an important role in the empathy process and in the development of social cognition. Particularly, deficits in eye-gaze situations are more prevalent in children and adolescents with CU+. Dadds *et al.* (2008) compared two groups of males (CU– and CU+) in a study with early adolescents. The aim of the investigation was to check the attention at different gaze conditions in males with CP. Participants were instructed to select between six different types of emotion (happiness, fear, anger, sadness, disgust, or neutral) in three conditions (free gaze, eye gaze, and mouth gaze).

Regarding the free condition, CU+ presented significant differences in recognizing fear when compared with the CU– group. When individuals were asked to focus on the eye region, there were also differences between the groups. In this condition, it was observed that CU+ demonstrated fewer first eye-gaze fixations (Dadds *et al.*, 2008). One suggestion is that youth with elevated CU traits tend to avoid looking directly in the direction of another person's eyes and, in consequence, they show deficits in accurately recognizing other people's emotions. The findings from Dadds *et al.* (2008) are congruent with a body of research that indicates a fear-blindness in the group of individuals with psychopathic traits and, more specifically, in adolescents with CU traits.

More recently, Dadds *et al.* (2011) tested the hypothesis that youth with CU+ and antisocial behaviour (AB) also tend to present deficits in maintaining eye contact with their parents, which may help to explain the development of difficulties in social and affective domains. The study used both report and observational measures in order to check the eye-gaze deficit using a naturalistic approach. The observational task was performed in two different settings. The first one took place in a free condition in which families were asked to interact for about 10–15 minutes. Following interaction, they were asked to draw a picture representing the family and also to choose a word that best describes family members. The second setting, described by the authors as "emotion talk," required participants to discuss recent episodes of happiness and sadness moments that they had experienced together.

During periods of emotion-focused talk, children with elevated levels of CU demonstrated lower levels of eye contact towards their parents compared with the group with CP only. There were no differences in the amount of eye contact or emotional interaction between the parents in these two groups. These findings help us to better understand how "fear blindness" may develop in children with CU, and may contribute towards our understanding of empathy and emotion processing in this group.

Integrated emotion systems

Related to the differences in processing emotional information in others, described above, the cognitive deficit associated with CP and CU is suggested to be related to a reduction in the salience of punishment information. Blair's integrated emotion systems (IES) model works on the assumption that individuals with elevated levels of CU traits have diminished ability to form stimulus–punishment associations, and was developed from the earlier violence inhibition mechanism model (Blair *et al.*, 1997; Blair, 2001).

In childhood, the ability to form associations between moral transgressions and the aversive outcome (e.g. the distress of others) is thought to be vital for successful socialization. Individuals with CU traits find the distress cues in others less aversive, and therefore are less likely to learn to avoid actions that bring about such a negative result. In addition, socialization by way of punishing consequences also relies on ability to form stimulus–aversive response associations. Evidence of an impairment in the understanding of "punishment" information seen in others' distress in psychopathic adults and children with CU traits can be assumed from a decreased autonomic response to fearful and sad facial expressions relative to non-CU comparisons (Blair *et al.*, 1997). In line with the brain-imaging work already discussed in this chapter, amygdala dysfunction is likely to play a role at the neural level. IES dysfunction has been demonstrated behaviorally in adults with psychopathy and CP/CU+ children using a moral/convention distinction paradigm (Smetana, 1993). In typically developing children, a distinction between moral (victim-based) and conventional (social disorder-based) transgressions can be made from 3.5 years of age and occurs across cultures (Smetana and Braeges, 1990). Studies have investigated whether individuals showing psychopathic traits can understand that, while the permissibility of conventional transgressions is rule-dependent (e.g. it is not generally acceptable for a child to talk in class), moral transgressions are not permissible regardless of the presence or absence of rules (e.g. it is not permissible to hit someone, as it will result in hurt to the other person). In contrast to neurotypical and clinical controls (children with ADHD, individuals with autism and non-psychopathic incarcerated adults), CP/CU+ children and psychopathic adults show a reduced moral/conventional distinction (Blair, 1995; Blair *et al.*, 1997).

Blair (1995) compared CP/CU+ with a group of children also recruited from schools for children with social, emotional, and behavioral difficulties (SEBD) and reported that the CP/CU+ group were found to make a significantly weaker moral/conventional distinction than children in a CP/CU– group. CP/CU+ children were also less likely to refer to a victim when asked to justify why a transgression might be wrong to do. Fisher and Blair (1998) also showed a significant correlation between moral/conventional distinction (MCD) scores and antisocial process screening device (APSD) scores in another sample of children attending SEBD school. The IES explains this impairment by suggesting that representations of moral transgressions are not efficiently paired with signs of others' distress in individuals with CU traits. It is interesting to note that, in Blair's studies, when asked to justify why a transgression might be wrong, CP/CU+ individuals tend not to suggest that moral transgressions result in victims, whereas controls report the existence of a potential victim as a reason for not committing such a transgression.

In contrast to CP/CU+ children, children with conduct problems without CU may have elevated levels of anxiety and threat-related reactive aggression and show hyper-reactivity to threat, such as angry faces (Pollak and Sinha, 2002; Blair *et al.*, 2006). This hyper-reactivity may exist as a result of hostile early life experiences, including physical maltreatment. Such children may also show a hostile attributive bias. At the cognitive level, CP/CU– children are hyper-reactive to others' displays of negative emotions; they possess an emotional memory database of maltreatment and hostility, and in combination these cognitive vulnerabilities will result in a fight-response bias. This fight-response bias is thought to be triggered in response to acute environmental stressors and to result in reactive aggression, impulsive violence, and increased propensity to make hostile attributions to ambiguous situations (Viding and Frith, 2006).

To date, there are only few direct comparisons of this affect processing profile of children with and without CU traits. Frick *et al.* (2003) compared groups of non-referred CP/CU+ and CP/CU– children, and found poor processing of punishment information in CP/CU+ and a hostile attribution bias in CP/CU–. Differences have also been observed in emotional reactivity. For example, Loney *et al.* (2003) demonstrated a slower recognition rate for negative emotional words in CP/CU+ adolescents, compared with a faster recognition time for the same words in a CP/CU– group. Finally, Dadds *et al.* (2006) report that CU traits are uniquely related to poor recognition of fearful expressions, while CP without the concomitant CU traits is associated with hypersensitivity to hostility. In addition, a large body of research by Blair and colleagues has repeatedly demonstrated deficits in processing fear, sadness, and punishment in CP/CU+ individuals, as compared with institutionalized (although not specifically CP/CU–) controls (Blair *et al.*, 2006). The work stemming from Blair's VIM and IES models (Blair, 1995; Blair *et al.*, 2005) and the work of Frick and colleagues (Frick, 1998) has been valuable in highlighting the importance of subtyping the children with early-onset CP into those with and without CU traits.

Own understanding of emotion

Work has also examined the ability of children and adolescents with CU to understand emotions within themselves. One study has reported a deficit associated with CU in children's ability to attribute feelings of fear to themselves, with a strong trend for a poverty in the self-attribution of guilt (Jones *et al.*, 2010). Similar deficits were not observed for self-attribution of embarrassment or happiness. A further study specifically investigating moral emotions has reported a deficit in processing guilt (Viding *et al.*, 2013).

One other avenue to consider when thinking about own emotions is alexithymia, which is a personality construct that describes difficulties in identifying and describing own emotional states (Sifneos, 1973). Individuals with high levels of alexithymia are characterized by difficulties relating to fantasizing or daydreaming, and a tendency toward a concrete style of thinking (Sifneos, 1973). Alexithymia has also been associated with a number of mental health and developmental disorders, including autism spectrum conditions, depression, eating disorders, and personality disorders (Hill *et al.*, 2004; Nicolo *et al.*, 2011). Studies have also demonstrated a link between high levels of alexithymia and difficulties with empathy (Bird *et al.*, 2010; Grynberg *et al.*, 2010).

Associations have been reported between alexithymia and psychopathy (Lander *et al.*, 2012), alexithymia and empathy (Guttman and Laporte, 2002), and psychopathy and empathy (Dadds *et al.*, 2009). The association between empathy and psychopathy is explored more thoroughly in the next section.

Understanding of others' emotion: empathy

Empathy has traditionally been seen as a unitary concept (Hoffman, 2000; Preston and de Waal, 2002), being defined by Hoffman (2000) as "an affective response more appropriate to another's situation than one's own" (p. 48). Defined in this way, empathy is an umbrella term encompassing phenomena such as emotional contagion, perspective-taking, and sympathy (De Vignemont and Singer, 2006). In order to be able to specify the nature of empathy deficit in any given disorder, including CP/CU+, which is the focus of this thesis, it is important to define clearly the empathic process of interest. More recent theoretical and empirical work has provided strong argument for there being different constructs underlying the conventional unitary concept of empathy (Blair, 2005; de Vignemont and Singer, 2006; Singer, 2006). De Vignemont and Singer (2006) are part of a number of researchers arguing for a narrower concept of empathy (Eisenberg and Fabes, 1990; Decety and Jackson, 2004; Blair, 2005). According to these authors, (affective) empathy exists if an individual is in an affective state and this state is isomorphic to another person's affective state. The individual's emotional state must have been elicited by the observation or imagination of that other person's affective state, and the individual should also be aware that the other person is the source of their own current emotional state. De Vignemont and Singer distinguish this from cognitive perspective-taking (or cognitive empathy), where an individual represents the mental states of others without being emotionally involved. Similarly, Blair (2005) argues for subdividing the concept of empathy. His model divides empathy into affective, cognitive, and motor subtypes. He also argues for a double dissociation between affective and cognitive types.

Although there is some disagreement about how empathy should be defined, and different terminology has been used by different researchers, there is a consensus that the act of recognizing someone's mental or emotional state is not enough in itself to be considered "empathy." However, the recent research literature supports the separation of perspective taking (analogous to theory of mind, ToM) and the concept of being able to understand others' emotions by sharing their affective states; cognitive and affective empathy, respectively. The difference between these two constructs is essentially that affective empathy allows an individual to experience what it feels like for another person to experience a certain emotion, while cognitive empathy allows an individual to understand what others' thoughts might be without necessitating a sharing of the feeling that might accompany such thoughts (Singer, 2006).

Affective empathy and CU

Singer (2006) describes affective empathy as being "the process which allows us to experience what it feels like for another person to experience a certain emotion or sensation" (p. 856). Blair argues that the recognition of emotional expressions in others, such as facial expression, vocalization, posture, or some other emotion stimulus relating to another person (e.g. being informed that a friend has suffered a death in the family), are also important for affective empathy (Blair, 2007). Affective empathy has been studied using a range of paradigms in both experimental neuropsychology and brain imaging settings.

Many of the key definitions of psychopathy include reference to affective empathy. For example, Cleckley's (1976) original work includes the description: "He has no ability to know how others feel . . . or to feel subjectively anything comparable about the situation." There is also reference to empathy impairment in the diagnostic criteria of psychopathy (PCL-R; Hare, 2003), which makes the point that the ability of psychopaths to inflict serious harm to others must be an indicator of profound disturbance in the appropriate "empathic" response to the distress of another. The APSD refers to "concern about the feelings of others" (a reverse-scored item; Frick and Hare, 2001).

Given the subjective nature of affective empathy, assessment has used several different methodologies. Tasks that have been employed to experimentally study affective empathy deficits in CU (or psychopathy) include: moral judgment/social convention tasks; emotion recognition tasks (which are argued to assess a component process of affective empathy functioning); vignette-based tasks measuring expectations and values relating to aggressive acts against another individual; and self-report questionnaires (e.g. Boldizar *et al.*, 1989; Pardini *et al.*, 2003). Moral judgment/social convention tasks work on the premise that moral transgressions impact on the rights or well-being of others, and therefore elicit a distress response, while behaviors contravening social conventions affect social order, and are less likely to cause distress to an individual (Turiel, 1977; Smetana and Braeges, 1990). A moral transgression should elicit negative affect in the observer, while social convention transgressions are less likely to do so. Experimental work on adult psychopaths has shown that they make significantly less of a distinction between moral and conventional transgressions than controls (Blair, 1995). The affective deficit associated with CU traits has been argued to be related to a reduction in the salience of punishment information (Blair *et al.*, 2006). As outlined above, Blair and colleagues (2006) proposed the integrated emotion systems model as a causal model of CU traits and antisocial behavior, where a diminished ability to form stimulus–punishment associations is associated with the development of this profile of behavior and personality. In childhood, the ability to form associations between moral transgressions and the aversive outcome (i.e. distress in others) is proposed to be fundamental to successful socialization. If individuals with elevated CU traits find distress cues in others less aversive, they may have difficulty in learning to avoid actions that elicit a negative response in another (Aniskiewicz, 1979; Blair *et al.*, 1997).

As previously discussed, studies of emotion recognition have consistently suggested that individuals with elevated CU traits and psychopathy have a specific deficit in their recognition of fearful and sad facial and vocal expressions (i.e. others' distress) (Blair *et al.*, 2001, 2002, 2005; Stevens *et al.*, 2001; Kosson *et al.*, 2002; Dolan and Fullam, 2006). Adults with psychopathy and children with CP/CU+ also show reduced autonomic response to stimuli associated with both distress and sadness in others (House and Milligan, 1976; Aniskiewicz, 1979; Blair *et al.*, 1997; Blair, 1999). It might be argued that these impairments in abilities closely related to affective empathy might represent a pathway by which individuals with elevated CU traits will fail to learn to have an appropriate empathic response. For example, Pardini *et al.* (2003) reported that CU traits in a group of incarcerated adolescents was associated with a decreased focus on the negative outcomes of hostile actions, along with lower emotional distress in response to distress in others.

Blair (1995) has also used a moral judgment/social convention task to provide evidence for intact empathy processing in autism spectrum disorder (ASD) and impairment in psychopathic individuals. Self-report measures of affective empathy are still relatively scarce. This is mainly because most empathy questionnaires cannot be divided into affective and cognitive components. Holmqvist (2008) reported a negative relationship between empathy and psychopathy as measured by the PCL-R (Hare, 1991, 2003) using the empathy inventory (Bryant, 1982).

Pardini *et al.* (2003) used the interpersonal reactivity index (Davis, 1983) in adolescents and found a negative relationship between the empathic concern subscale (part of the affective empathy component) and CU traits as measured by the APSD (Frick and Hare, 2001). Two studies carried out with children have also demonstrated an association between CU traits and a deficit in affective empathy. Our study was the first to directly compare children with CP and CU with children with CP only, and with another group with known "empathy deficits"—children with autistic spectrum conditions (ASC) (Jones *et al.*, 2010). Using the same vignette method as Pardini *et al.* (2003), we reported that children with CU traits had a selective deficit in affective empathy, but no difficulties in the associated cognitive empathy task, or theory of mind (ToM) tasks. Conversely, children with ASC performed more poorly than comparisons on ToM tasks, but showed no deficits in their affective empathy. It is worth noting that there was no need for ToM ability in the affective empathy task, which may have helped the sample with ASC to respond appropriately. This (at least partial) double dissociation was also demonstrated in a child sample, using different tasks, by Schwenck *et al.* (2012).

Neuroimaging work focusing on affective empathy in healthy adults has used various methods of eliciting an empathic response (or its components), including emotion recognition, generating an emotion, moral judgments, compassion, and pain. Studies using facial expressions of emotion have demonstrated dissociable neural systems involved in the recognition of emotion (Adolphs *et al.*, 2003; Blair, 2003). Studies have demonstrated consistently a role in emotion recognition for amygdala, OFC, insula, and anterior cingulate cortex (ACC) (Amodio and Frith, 2006; Britton *et al.*, 2006).

Some studies have also encompassed the notion of sharing others' sensations as well as emotions. Working from this premise, Singer *et al.* (2004) have used pain as a means of investigating the neural correlates of emotional empathy. This study demonstrated activation of anterior insula (AI) and ACC during observation of a loved one receiving a pain stimulus. These activations also correlated with scores on empathy measures. Similarly, Lockwood *et al.* (2013) suggest that youth with CP have less neural activity when looking at pictures of individuals in pain. Precisely, the brain regions associated with empathy (AI, inferior frontal gyrus, and ACC) were less activated in a group of boys with CP compared with controls.

Similarly to this, King *et al.* (2006) carried out an fMRI study using a first-person computer game where the participant was required to either act violently toward an aggressive assailant, or heal a passive wounded person. In both conditions, the same circuit was activated, including amygdala and ventromedial prefrontal cortex (PFC). Medial prefrontal cortex (mPFC) is also suggested to be involved in understanding social emotions such as guilt and embarrassment (Nakamura *et al.*, 1999). Völlm and colleagues (2006) have also reported activation related to affective empathy using non-verbal cartoon stimuli. Consistent with previous imaging studies, this study demonstrated increased activation in AI, and also in amygdala and mPFC.

In individuals with CP/CU+, neuroimaging studies of affective processing have tended to focus on neural correlates of facial expressions of emotion. Consistent abnormalities have been reported in amygdala and frontal regions, as well as ACC (Kiehl *et al.*, 2004; Birbaumer *et al.*, 2005; Marsh *et al.*, 2008; Yang *et al.*, 2008; De Brito *et al.*, 2009; Jones *et al.*, 2009; Han *et al.*, 2012; Sebastian *et al.*, 2012; White *et al.*, 2012, 2013). More recently, Marsh *et al.* (2013) presented photos of pain-inducing injuries during fMRI scanning, and asked participants to imagine the pain was being inflicted on them, or on someone else. The authors reported reduced brain activity in rostral ACC and amygdala in adolescents with antisocial behaviour and elevated CU traits compared with a typically developing comparison group.

Cognitive empathy

Both Blair (2005) and de Vignemont and Singer (2006) argue that perspective-taking or cognitive empathy is distinct from affective empathy. A common type of task used in the investigation of cognitive empathy abilities are tasks tapping ToM and perspective taking. Animations, cartoons, and vignettes have been used to test an individual's ability to attribute thoughts and mental states to another.

Blair *et al.* (1996) used Happé's Strange Stories (Happé, 1994) to examine ToM in adult psychopathic offenders, where correct performance involved identifying accurately the underlying intention behind a character's utterance that was not literally true. The stories used assessed the ability to understand lies, white lies, double bluff, persuasion, and misunderstanding. Compared with non-psychopathic offenders, the psychopathic adults did not show a deficit in ToM ability. In addition, Blair related the findings of this study demonstrating no functional impairment in psychopaths to the original study by Happé (1994), and compared the scores of the psychopathic group with adults having ASD. The psychopathic group was found to perform significantly better than adults with ASD, and to make greater use of mental state words in their responses. Using a different methodology, Richell *et al.* (2003) investigated cognitive empathy using the "reading the mind in the eyes" task (Baron-Cohen *et al.*, 2001) with psychopathic adults and age- and ability-matched controls, all of whom were incarcerated offenders. Psychopathic adults did not differ significantly from controls on this task. Richell *et al.* also compared their study with Baron-Cohen's original study: using the same stimuli with adults with ASD demonstrated that, while the non-psychopathic offenders' scores were commensurate with typically developing controls, the percentage of psychopathic offenders scoring above Baron-Cohen *et al.*'s original population mean did not differ from the percentage of controls from the same study (Baron-Cohen *et al.*, 2001). ToM has not been explicitly explored in children with CP/CU+. However, there is some evidence to suggest that children with behavioral difficulties are relatively unimpaired at ToM. Buitelaar and colleagues (1999) demonstrated ToM difficulties in children with ASD, but not with CD. For children with elevated CU, good theory-of-mind skills may actually facilitate antisocial behavior (e.g. lying, bullying). Sutton *et al.* (1999) have argued that some bullies (e.g. "ring-leader" bullies and relational bullies) are in fact skilled manipulators who apply their proficient social understanding to inflict suffering on others while escaping detection themselves. In addition to this work, Happé and Frith (1996) suggested that children with CD might develop a "theory of nasty minds" as a result of being raised in an aversive environment, which may prove adaptive in such situations of potential abuse or conflict. For example, hypervigilance to aggressive intentions might allow the child to escape an episode of physical or psychological abuse. The processes linking the ability to predict actions based on false belief and on-line responses to social situations (e.g. hostile attribution biases) have yet to be articulated (Crick and Dodge, 1999). However, this challenge to the traditional view of antisocial children as socially incompetent is valuable in highlighting the diverse forms and origins of behavioral difficulty (Hughes *et al.*, 2001).

A range of methods have been used to elicit mental state attributing during functional imaging of healthy adult volunteers, but very few studies to date have investigated the brain correlates of cognitive empathy in children with CU traits (or adult psychopathy). It is useful to briefly review brain-imaging studies of ToM in typically developing populations. Castelli and colleagues used animated geometric shapes to investigate neural correlates of ToM in both typically developing adults and individuals with ASD. In typically developing adults, Castelli *et al.*

(2000) showed increased activation to ToM condition (compared with animations not requiring ToM) in mPFC, temporopareital junction (TPJ), temporal poles, and extrastriate cortex. Furthermore, damage to these brain areas impairs ToM abilities (Happé *et al.*, 2001; Stuss *et al.*, 2001; Samson *et al.*, 2004). Rebecca Saxe and colleagues also examined the neural circuitry associated with ToM (Saxe, 2006; Saxe and Wexler, 2005; Saxe and Powell, 2006), and have demonstrated consistent involvement of mPFC. A recent meta-analysis of mPFC activation during cognitive empathy tasks indicates that the peak area of activation is in the dorsal region (Amodio and Frith, 2006) of the TPJ, and suggested that TPJ was selectively recruited for ToM tasks (Saxe and Wexler, 2005). However, more recent work has cast doubt on this assertion, suggesting that TPJ is also activated during other non-ToM social and non-social tasks (Mitchell, 2008). Finally, a study by Völlm *et al.* (2006) used non-verbal cartoon stimuli designed to elicit both affective and cognitive empathic responses. In the cognitive empathy condition, there was greater activation in the superior temporal gyrus, middle frontal gyrus, and lateral orbitofrontal cortex (mainly on the right side). Common activations occurred in prefrontal and temporal brain regions. These studies all concentrated on adult participants; however, recent work has demonstrated that the same areas are also activated in child and adolescent samples. Kobayashi and colleagues (2007) reported TPJ activation in both adults and children in verbal and non-verbal ToM tasks, while Moriguchi *et al.* (2007) showed activation in bilateral superior temporal sulcus (STS), TPJ and mPFC in a group of 9- to 16-year-olds. Finally, one recent study from Sebastian and colleagues (2012) has contributed to our knowledge about affective and cognitive ToM in youth with and without antisocial behavior and CU traits, as well as helping to clarify the developmental changes related to these processes. Young people with CP and CU showed differences in neural responses in these groups compared with typical individuals (Sebastian *et al.*, 2012). Especially, in an affective ToM task, reduced right amygdala activation was found in the CP group as well as in the right anterior insula. Interestingly, further investigation of the association between CP and CU and amygdala activity revealed suppressor effects between these two traits. Specifically, unique variance associated with conduct problems was positively correlated with amygdala reactivity, whereas unique variance associated with CU traits was negatively correlated with amygdala reactivity. This finding is in line with the growing body of literature reporting a decreased amygdala response to affective stimuli associated with CU. However, more recently, one study comparing children with elevated CU traits with children with ASC did not find a difference in brain activity between a CU and typically developing group, while the group with ASC were reported to have decreased activation in mPFC (O'Nions *et al.*, 2014). Although this study used the same stimuli as Sebastian *et al.* (2012), it is worth noting that only activations associated with cognitive ToM were reported (that is, understanding behaviour based on intentions). In Sebastian *et al.*'s study, the reported reduced activation associated with CU were related to affective ToM (that is, understanding behaviour based on emotions). In future, it will be important to understand better the links between brain activity and different types of emotion and empathic processing associated with CU traits.

Conclusion

It is clear from the studies outlined in this chapter that there is considerable merit in understanding the unique profile of cognitive and affective neuropsychology associated with CU traits in children. Effective treatment, which takes these strengths and weaknesses into account, needs to be differentiated from those individuals whose antisocial behavior has some different aetiology.

References

Adolphs, R., Tranel, D., and Damasio A. R. (2003) 'Dissociable neural systems for recognizing emotions', *Brain and Cognition*, 52: 61–69.

Amodio, D. and Frith, C. D. (2006) 'Meeting of minds the medial frontal cortex and social cognition', *Nature Neuroscience Reviews*, 7: 268–277.

Aniskiewicz, A. (1979) 'Autonomic components of vicarious conditioning and psychopathy', *Journal of Clinical Psychology*, 35: 60–67.

Baron-Cohen, S., Wheelwright, S., Hill, J., Raste, Y., and Plumb, I. (2001) 'The "Reading the Mind in the Eyes" Test revised version: a study with normal adults, and adults with Asperger syndrome or high-functioning autism', *Journal of Child Psychology and Psychiatry*, 42: 241–251.

Birbaumer, N., Veit, R., Lotze, M., Erb, M., Hermann, C., Grodd, W., and Flor, H. (2005) 'Deficient fear conditioning in psychopathy: a functional magnetic resonance imaging study', *Archives of General Psychiatry*, 62: 799–805.

Bird, G., Silani, G., Brindley, R., White, S., Frith, U., and Singer, T. (2010) 'Empathic brain responses in insula are modulated by levels of alexithymia but not autism', *Brain*, 133: 1515–1525.

Blair, R. (1995) 'A cognitive developmental approach to morality: investigating the psychopath', *Cognition*, 57: 1–29.

Blair, R. J. R. (1999) 'Responsiveness to distress cues in the child with psychopathic tendencies', *Personality and Individual Differences*, 27(1): 135–145.

Blair, R. J. R. (2001) 'Advances in neuropsychiatry: neurocognitive models of aggression, the antisocial personality disorders, and psychopathy', *Journal of Neurology, Neurosurgery and Psychiatry*, 71: 727–731.

Blair, R. J. R. (2003) 'Neurobiological basis of psychopathy', *British Journal of Psychiatry*, 182(1): 5–7.

Blair, R. (2005) 'Responding to the emotions of others: dissociating forms of empathy through the study of typical and psychiatric populations', *Consciousness and Cognition*, 14: 698–718.

Blair, R. (2007) 'Fine cuts of empathy and the amygdala: dissociable deficits in psychopathy and autism', *Quarterly Journal of Experimental Psychology*, 61: 157–170.

Blair, R. J. R., Sellars, C., Strickland, I., Clark, F., Williams, A., Smith, M., *et al.* (1996) 'Theory of mind in the psychopath', *Journal of Forensic Psychiatry*, 7: 15–25.

Blair, R. J., Jones, L., Clark, F., and Smith, M. (1997) 'The psychopathic individual a lack of responsiveness to distress cues?', *Psychophysiology*, 34: 192–198.

Blair, R., Colledge, E., Murray, L., and Mitchell, D. (2001) 'A selective impairment in the processing of sad and fearful expressions in children with psychopathic tendencies', *Journal of Abnormal Child Psychology*, 29: 491–498.

Blair, R., Mitchell, D. G., Richell, R. A., Kelly, S., Leonard, A., Newman, C., and Scott, S. K. (2002) 'Turning a deaf ear to fear: impaired recognition of vocal affect in psychopathic individuals', *Journal of Abnormal Psychology*, 111: 682–686.

Blair, R., Budhani, S., Colledge, E., and Scott, S. (2005) 'Deafness to fear in boys with psychopathic tendencies', *Journal of Child Psychology and Psychiatry*, 46: 327–336.

Blair, R. J. R., Peschardt, K. S., Budhani, S., Mitchell, D. G.V., and Pine, D. S. (2006) 'The development of psychopathy', *Journal of Child Psychology and Psychiatry*, 47: 262–275.

Boldizar, J. P., Perry, D. G., and Perry, L. C. (1989) 'Outcome values and aggression', *Child Development*, 60: 571–579.

Britton, J. C., Phan, K. L., Taylor, S. F., Welsh, R. C., Berridge, K. C., and Liberzon, I. (2006) 'Neural correlates of social and nonsocial emotions: an fMRI study', *Neuroimage*, 31(1): 397–409.

Bryant, B. K. (1982) 'An index of empathy for children and adolescents', *Child Development*, 53: 413–425.

Buitelaar, J., van der Wees, M., Swaab-Barneveld, H., and van der Gaag, R. (1999) 'Verbal memory and performance IQ predict theory of mind and emotion recognition ability in children with autistic spectrum disorders and in psychiatric control children', *Journal of Child Psychology and Psychiatry*, 40: 869–881.

Castelli, F., Happé, F., Frith, U., and Frith, C. (2000) 'Movement and mind: a functional imaging study of perception and interpretation of complex intentional movement patterns', *Neuroimage*, 12: 314–325.

Clark, C., Prior, M., and Kinsella, G. (2000) 'Do executive function deficits differentiate between adolescents with ADHD and oppositional defiant/conduct disorder? A neuropsychological study using the Six Elements Test and Hayling Sentence Completion Test', *Journal of Abnormal Child Psychology*, 28: 403–414.

Cleckley, H. (1976) *The Mask of Sanity*, 5th edn, St Louis, MO: Mosby.

Crick, N. R. and Dodge, K. A. (1994) 'A review and reformulation of social information-processing mechanisms in children's social adjustment', *Psychological Bulletin*, 115: 74–101.

Crick, N. R. and Dodge, K. A. (1999) '"Superiority" is in the eye of the beholder: a comment on Sutton, Smith and Swettenham', *Social Development*, 8: 128–131.

Dadds, M. R., Perry, Y., Hawes, D. J., Merz, S., Riddell, A. C., Haines, D. J., *et al.* (2006) 'Attention to the eyes and fear-recognition deficits in child psychopathy', *British Journal of Psychiatry*, 189: 280–281.

Dadds, M. R., El Masry, Y., Wimalaweera, S., and Guastella, A. J. (2008) 'Reduced eye gaze explains "fear blindness" in childhood psychopathic traits', *Journal of the American Academy of Child and Adolescent Psychiatry*, 47(4): 455–463.

Dadds, M. R., Hunter, K., Hawes, D., Frost, A. J., Vassallo, S., Bunn, P., and Merz, S. (2009) 'Learning to "talk the talk": the relationship of psychopathic traits to deficits in empathy across childhood', *Journal of Child Psychology and Psychiatry*, 50: 599–606.

Dadds, M. R., Jambrak, J., Pasalich, D., Hawes, D. J., and Brennan, J. (2011) 'Impaired attention to the eyes of attachment figures and the developmental origins of psychopathy: eye contact and origins of psychopathy', *Journal of Child Psychology and Psychiatry*, 52(3): 238–245.

Davis, M. H. (1983) 'Measuring individual differences in empathy: evidence for a multidimensional approach', *Journal of Personality and Social Psychology*, 44(1): 113.

De Brito, S. A., Mechelli, A., Wilke, M., Laurens, K. R., Jones, A. P., Barker, G. J., *et al.* (2009) 'Size matters: increased grey matter in boys with conduct problems and callous-unemotional traits', *Brain*, 132: 843–852.

De Brito, S. A., Viding, E., Kumari, V., Blackwood, N., and Hodgins, S. (2013) 'Cool and hot executive function impairments in violent offenders with antisocial personality disorder with and without psychopathy', *PLOS ONE*, 8: e65566.

De Luca, C. R. and Leventer, R. J. (2008) 'Developmental trajectories of executive functions across the lifespan', in V. Anderson, R. Jacobs, and P. Anderson (eds), *Executive Functions and the Frontal Lobes: A Lifespan Perspective*, New York: Psychology Press, pp. 3–21.

Decety, J. and Jackson, P. L. (2004) 'The functional architecture of human empathy', *Behavioral and Cognitive Neuroscience Reviews*, 3: 71–100.

Dolan, M. (2012) 'The neuropsychology of prefrontal function in antisocial personality disordered offenders with varying degrees of psychopathy', *Psychological Medicine*, 42: 1715–1725.

Dolan, M. and Fullam, R. (2006) 'Face affect recognition deficits in personality-disordered offenders: association with psychopathy', *Psychological Medicine*, 36: 1563–1569.

Dolan, M. C. and Fullam, R. S. (2010) 'Moral/conventional transgression distinction and psychopathy in conduct disordered adolescent offenders', *Personality and Individual Differences*, 49(8): 995–1000.

Eisenberg, N. and Fabes, R. (1990) 'Empathy conceptualization, measurement, and relation to prosocial behavior', *Motivation and Emotion*, 14: 131–149.

Fisher, L. and Blair, R. (1998) 'Cognitive impairment and its relationship to psychopathic tendencies in children with emotional and behavioral difficulties', *Journal of Abnormal Child Psychology*, 26: 511–519.

Fontaine, N., Barker, E. D., Salekin, R. T., and Viding, E. (2008) 'Dimensions of psychopathy and their relationships to cognitive functioning in children', *Journal of Clinical Child and Adolescent Psychology*, 37(3): 690–696.

Frick, P. J. (1998) *Conduct Disorders and Severe Antisocial Behavior*, New York: Springer

Frick, P. J. and Hare, R. D. (2001) *Antisocial Process Screening Device*, Toronto, Ontario: Multi Health Systems.

Frick, P., O'Brien, B., Wootton, J., and McBurnett, K. (1994) 'Psychopathy and conduct problems in children', *Journal of Abnormal Psychology*, 103: 700–707.

Frick, P. J., Cornell, A. H., Barry, C. T., Bodin, S., and Dane, H. E. (2003) 'Callous-unemotional traits and conduct problems in the prediction of conduct problem severity, aggression, and self-report of delinquency', *Journal of Abnormal Child Psychology*, 31: 457–470.

Frick, P., Ray, J. V., Thornton, L. C., and Kahn, R. E. (2014) 'Annual Research Review: a developmental psychopathology approach to understanding callous-unemotional traits in children and adolescents with serious conduct problems', *Journal of Child Psychology and Psychiatry*, 55(6): 532–548.

van Goozen, S. H. M., Cohen-Kettenis, P. T., Snoek, H., Matthys, W., Swaab-Barneveld, H., and Van Engeland, H. (2004) 'Executive functioning in children: a comparison of hospitalised ODD and ODD/ADHD children and normal controls', *Journal of Child Psychology and Psychiatry*, 45: 284–292.

Grynberg, D., Luminet, O., Corneille, O., Grezes, J., and Berthoz, S. (2010) 'Alexithymia in the interpersonal domain: a general deficit of empathy?', *Personality and Individual Differences*, 49: 845–850.

Guttman, H. and Laporte, L. (2002) 'Alexithymia, empathy, and psychological symptoms in a family context', *Comprehensive Psychiatry*, 43(6): 448–455.

Han, T., Alders, G. L., Greening, S. G., Neufeld, R. W., and Mitchell, D. G. (2012) 'Do fearful eyes activate empathy-related brain regions in individuals with callous traits?', *Social Cognitive and Affective Neuroscience*, 7: 958–968.

Happé, F. (1994) 'An advanced test of theory of mind understanding of story characters' thoughts and feelings by able autistic, mentally handicapped, and normal children and adults', *Journal of Autism and Developmental Disorders*, 24: 129–154.

Happé, F. and Frith, U. (1996) 'Theory of mind and social impairment in children with conduct disorder', *British Journal of Developmental Psychology*, 14: 385–398.

Happé, F., Malhi, G., and Checkley, S. (2001) 'Acquired mind-blindness following frontal lobe surgery? A single case study of impaired "theory of mind" in a patient treated with stereotactic anterior capsulotomy', *Neuropsychologia*, 39: 83–90.

Hare, R. D. (1991) *The Hare Psychopathy Checklist-Revised*, Toronto, Ontario: Multi-Health Systems.

Hare, R. D. (2003) *The Hare Psychopathy Checklist-Revised*, 2nd edn, Toronto, Ontario: Multi-Health Systems.

Henry, B. and Moffitt, T. E. (1997) 'Neuropsychological and neuroimaging studies of juvenile delinquency and adult criminal behavior', in D. Stoff, J. Breiling, and J. Maser (eds), *Handbook of Antisocial Behavior*, New York: John Wiley and Sons, pp. 280–288.

Hill, E., Berthoz, S., and Frith, U. (2004) 'Brief report cognitive processing of own emotions in individuals with autistic spectrum disorder and in their relatives', *Journal of Autism and Developmental Disorders*, 34: 229–235.

Hoffman, M. L. (2000) *Empathy and Moral Development: Implications for Caring and Justice*, New York: Cambridge University Press.

Holmqvist, R. (2008) 'Psychopathy and affect consciousness in young criminal offenders', *Journal of Interpersonal Violence*, 23: 209–224.

House, T. and Milligan, W. (1976) 'Autonomic responses to modeled distress in prison psychopaths', *Journal of Personality and Social Psychology*, 34: 556–560.

Hughes, C., Cutting, A. L., and Dunn, J. (2001) 'Acting nasty in the face of failure? Longitudinal observations of "hard-to-manage" children playing a rigged competitive game with a friend', *Journal of Abnormal Child Psychology*, 29: 403–416.

Hyde, L. W., Shaw, D. S., and Hariri, A. R. (2013) 'Understanding youth antisocial behavior using neuroscience through a developmental psychopathology lens: review, integration, and directions for research', *Developmental Review*, 33(3): 168–223.

Jones, A. P., Larsson, H., Ronald, A., Rjisdijk, F., Busfield, P., McMillan, A., and Viding, E. (2009) 'Phenotypic and aetiological relationships between psychopathic tendencies, autistic traits, and emotion attribution', *Criminal Justice and Behavior*, 36: 1198–1212.

Jones, A. P., Happé, F., Gilbert, F., Burnett, S., and Viding, E. (2010) 'Feeling, caring, knowing: different types of empathy deficit in boys with psychopathic tendencies and autism spectrum disorder', *Journal of Child Psychology and Psychiatry*, 51: 1188–1197.

Kiehl, K. A., Smith, A. M., Mendrek, A., Forster, B. B., Hare, R. D., and Liddle, P. F. (2004) 'Temporal lobe abnormalities in semantic processing by criminal psychopaths as revealed by functional magnetic resonance imaging', *Psychiatry Research: Neuroimaging*, 130(1): 27–42.

King, J. A., Blair, R. J., Mitchell, D. G. V., Dolan, R. J., and Burgess, N. (2006) 'Doing the right thing: a common neural circuit for appropriate violent or compassionate behavior', *Neuroimage*, 15: 1069–1076.

Kobayashi, C., Glover, G. H., and Temple, E. (2007) 'Cultural and linguistic effects on neural bases of "Theory of Mind" in American and Japanese children', *Brain Research*, 20: 95–107.

Kosson, D. S., Suchy, Y., Mayer, A. R., and Libby, J. (2002) 'Facial affect recognition in criminal psychopaths', *Emotion*, 2: 398–411.

Lander, G. C., Lutz-Zois, C. J., Rye, M. S., and Goodnight, J. A. (2012) 'The differential association between alexithymia and primary versus secondary psychopathy', *Personality and Individual Differences*, 52(1): 45–50.

Lockwood, P. L., Sebastian, C. L., McCrory, E. J., Hyde, Z. H., Gu, X., De Brito, S. A., and Viding, E. (2013) 'Association of callous traits with reduced neural response to others' pain in children with conduct problems', *Current Biology*, 23(10): 901–905.

Loney, B. R., Frick, P. J., Ellis, M., and McCoy, M. G. (1998) 'Intelligence, callous-unemotional traits, and antisocial behavior', *Journal of Psychopathology and Behavioral Assessment*, 20: 231–247.

Loney, B. R., Frick, P. J., Clements, C. B., Ellis, M. L., and Kerlin, K. (2003) 'Callous-unemotional traits, impulsivity, and emotional processing in adolescents with antisocial behavior problems', *Journal of Clinical Child and Adolescent Psychology*, 32: 66–80.

Marsh, A. A. and Blair, R. J. R. (2008) 'Deficits in facial affect recognition among antisocial populations: a meta-analysis', *Neuroscience & Biobehavioral Reviews*, 32(3): 454–465.

Marsh, A., Finger, E., Mitchell, D., Reid, M., Sims, C., Kosson, D., *et al.* (2008) 'Reduced amygdala response to fearful expressions in children and adolescents with callous-unemotional traits and disruptive behavior disorders', *American Journal of Psychiatry*, 165: 712–720.

Marsh, A. A., Finger, E. C., Fowler, K. A., Adalio, C. J., Jurkowitz, I. T. N., Schechter, J. C., Pine, D. S., Decety, J., and Blair, R. J. R. (2013) 'Empathic responsiveness in amygdala and anterior cingulate cortex in youths with psychopathic traits', *Journal of Child Psychology and Psychiatry*, 54: 900–910.

Mitchell, D. G., Colledge, E., Leonard, A., and Blair, R. J. (2002) 'Risky decisions and response reversal: is there evidence of orbitofrontal cortex dysfunction in psychopathic individuals?', *Neuropsychologia*, 40: 2013–2022.

Mitchell, J. (2008) 'Activity in right temporo-parietal junction is not selective for theory-of-mind', *Cerebral Cortex*, 18: 262–271.

Morgan, A. B. and Lilienfeld, S. O. (2000) 'A meta-analytic review of the relation between antisocial behavior and neuropsychological measures of executive function', *Clinical Psychology Review*, 20: 113–156.

Moriguchi, Y., Ohnishi, T., Mori, T., Matsuda, H., and Komaki, G. (2007) 'Changes of brain activity in the neural substrates for theory of mind during childhood and adolescence', *Psychiatry and Clinical Neurosciences*, 61: 355–363.

Muñoz, L. C. (2009) 'Callous-unemotional traits are related to combined deficits in recognizing afraid faces and body poses', *Journal of the American Academy of Child and Adolescent Psychiatry*, 48: 554–562.

Nakamura, K., Kawashima, R., Ito, K., Sugiura, M., Kato, T., Nakamura, A., and Kojima, S. (1999) 'Activation of the right inferior frontal cortex during assessment of facial emotion', *Journal of Neurophysiology*, 82: 1610–1614.

Nicolo, G., Semerari, A., Lysaker, P. H., Dimaggio, G., Conti, L., D'Angerio, S., and Carcione, A. (2011) 'Alexithymia in personality disorders: correlations with symptoms and interpersonal functioning', *Psychiatry Research*, 190: 37–42.

Nigg, J. T., Quamma, J. P., Greenberg, M. T., and Kusche, C. A. (1999) 'A two-year longitudinal study of neuropsychological and cognitive performance in relation to behavioral problems and competencies in elementary school children', *Journal of Abnormal Child Psychology*, 27: 51–63.

O'Nions, E., Sebastian, C. L., McCrory, E., Chantiluke, K., Happé, F., and Viding, E. (2014) 'Neural bases of Theory of Mind in children with autism spectrum disorders and children with conduct problems and callous-unemotional traits', *Developmental Science*, doi: 10.1111/desc.12167.

Pardini, D. A., Lochman, J. E., and Frick, P. J. (2003) 'Callous/unemotional traits and social-cognitive processes in adjudicated youths', *Journal of the American Academy of Child and Adolescent Psychiatry*, 42: 364–371.

Pham, T., Vanderstukken, O., Philippot, P., and Vanderlinden, M. (2003) 'Selective attention and executive functions deficits among criminal psychopaths', *Aggressive Behavior*, 29: 393–405.

Pollak, S. D. and Sinha, P. (2002) 'Effects of early experience on children's recognition of facial displays of emotion', *Developmental Psychology*, 38: 784–791.

Preston, S. and de Waal, F. (2002) 'Empathy: its ultimate and proximate bases', *Behavioral and Brain Sciences*, 25: 1–20.

Raine, A., Yaralian, P. S., Reynolds, C., Venables, P. H., and Mednick, S. A. (2002) 'Spatial but not verbal cognitive deficits at age 3 years in persistently antisocial individuals', *Development and Psychopathology*, 14: 25–44.

Raine, A., Moffit, T. E., Caspi, A., Loeber, R., Stouthamer-Loeber, M., and Lynam, D. (2005) 'Neurocognitive impairments in boys on the life-course persistent antisocial path', *Journal of Abnormal Psychology*, 114(1): 38–49.

Richell, R., Mitchell, D., Newman, C., Leonard, A., Baron-Cohen, S., and Blair, R. (2003) 'Theory of mind and psychopathy: can psychopathic individuals read the "language of the eyes"?', *Neuropsychologia*, 41: 523–526.

Roussy, S. and Toupin, J. (2000) 'Behavioral inhibition deficits in juvenile psychopaths', *Aggressive Behavior*, 26: 413–424.

Rubia, K. (2011) '"Cool" inferior frontostriatal dysfunction in attention-deficit/hyperactivity disorder versus "hot" ventromedial orbitofrontal-limbic dysfunction in conduct disorder: a review', *Biological Psychiatry*, 69: e69–e87.

Salekin, R. T. (2006) 'Psychopathy in children and adolescents: key issues in conceptualization and assessment', in C. J. Patrick (ed.), *Handbook of Psychopathy*, New York: Guilford, pp. 389–414.

Samson, D., Apperly, I., Chiavarino, C., and Humphreys, G. (2004) 'Left temporoparietal junction is necessary for representing someone else's belief', *Nature Neuroscience*, 7: 499–500.

Saxe, R. (2006) 'Why and how to study Theory of Mind with fMRI', *Brain Research*, 1079(1): 57–65.

Saxe, R. and Powell, L. (2006) 'It's the thought that counts: specific brain regions for one component of theory of mind', *Psychological Science*, 17: 692–699.

Saxe, R. and Wexler, A. (2005) 'Making sense of another mind the role of the right temporo-parietal junction', *Neuropsychologia*, 43: 1391–1399.

Schwenck, C., Mergenthaler, J., Keller, K., Zech, J., Salehi, S., Taurines, R., *et al.* (2012) 'Empathy in children with autism and conduct disorder: group-specific profiles and developmental aspects', *Journal of Child Psychology and Psychiatry*, 53: 651–659.

Sebastian, C. L., McCrory, E. J. P., Cecil, C. A. M., Lockwood, P. L., De Brito, S. A., Fontaine, N. M. G., and Viding, E. (2012) 'Neural responses to affective and cognitive theory of mind in children with conduct problems and varying levels of callous-unemotional traits', *Archives of General Psychiatry*, 69(8): 814–820.

Sebastian, C. L., McCrory, E. J., Dadds, M. R., Cecil, C. A. M., Lockwood, P. L., Hyde, Z., *et al.* (2014) 'Neural responses to fearful eyes in children with conduct problems and varying levels of callous-unemotional traits', *Psychological Medicine*, 44(1): 99–109.

Séguin, J. R., Boulerice, B., Harden, P. W., Tremblay, R. E., and Pihl, R. O. (1999) 'Executive functions and physical aggression after controlling for attention deficit hyperactivity disorder, general memory, and IQ', *Journal of Child Psychology and Psychiatry*, 40: 1197–1208.

Sifneos, P. E. (1973) 'The prevalence of alexithymic characteristics in psychosomatic patients', *Psychotherapy and Psychosomatics*, 22: 255–262.

Singer, T. (2006) 'The neuronal basis and ontogeny of empathy and mind reading review of literature and implications for future research', *Neuroscience and Biobehavioral Reviews*, 30: 855–863.

Singer, T., Seymour, B., O'Doherty, J., Kaube, H., Dolan, R. J., and Frith, C. D. (2004) 'Empathy for pain involves the affective but not sensory components of pain', *Science*, 20: 1157–1162.

Smetana, J. (1993) 'Understanding social rules', in M. Bennett (ed.), *The Child as a Psychologist: An Introduction to the Development of Social Cognition*, New York: Harvester Wheatsheaf, pp. 111–141.

Smetana, J. and Braeges, J. (1990) 'The development of toddlers' moral and conventional judgments', *Merrill-Palmer Quarterly*, 36: 329–346.

Smith, S. S., Arnett, P. A., and Newman, J. P. (1992) 'Neuropsychological differentiation of psychopathic and nonpsychopathic criminal offenders', *Personality and Individual Differences*, 13: 1233–1243.

Speltz, M. L., DeKlyen, M., and Greenberg, M. T. (1999) 'Attachment in boys with early onset conduct problems', *Development and Psychopathology*, 11: 269–285.

Spreen, O. and Strauss, E. (1998) *A Compendium of Neuropsychological Tests Administration, Norms, and Commentary*, 2nd edn, New York: Oxford University Press.

Stevens, D., Charman, T., and Blair, R. J. R. (2001) 'Recognition of emotion in facial expressions and vocal tones in children with psychopathic tendencies', *Journal of Genetic Psychology*, 162(2): 201–211.

Stuss, D., Gallup, G., and Alexander, M. (2001) 'The frontal lobes are necessary for "theory of mind"', *Brain*, 124: 279–286.

Sumich, A., Sarkar, S., Hermens, D. F., Kelesidi, K., Taylor, E., and Rubia, K. (2012) 'Electrophysiological correlates of CU traits show abnormal regressive maturation in adolescents with conduct problems', *Personality and Individual Differences*, 53(7): 862–867.

Sutton, J., Smith, P. K., and Swettenham, J. (1999) 'Social cognition and bullying: social inadequacy or skilled manipulation?', *British Journal of Developmental Psychology*, 17: 435–450.

Turiel, E. (1977) 'Distinct conceptual and developmental domains social convention and morality', in H. E. Howe and C. B. Keasey (eds), *Nebraska Symposium on Motivation*, Lincoln, NE: University of Nebraska Press.

Viding, E. and Frith, U. (2006) 'Genes for susceptibility to violence lurk in the brain', *Proceedings of the National Academy of Sciences, USA*, 103: 6085–6086.

Viding, E., Sebastian, C. L., Dadds, M. R., Lockwood, P. L., Cecil, C. A., De Brito, S. A., and McCrory, E. J. (2012) 'Amygdala response to preattentive masked fear in children with conduct problems: the role of callous-unemotional traits', *American Journal of Psychiatry*, 169(10): 1109–1116.

Viding, E., Booth, C., Rogers, J. S. C., Kuntsi, J., and Happé, F. (2013) 'Disruptive behaviors at school: information processing differentiators of callous-unemotional traits, conduct problems and ADHD', *BJEP Monograph Series II*, 9: 53–73.

de Vignemont, F. and Singer, T. (2006) 'The empathic brain: how, when and why?', *Trends in Cognitive Sciences*, 10: 435–441.

Völlm, B., Taylor, A., Richardson, P., Corcoran, R., Stirling, J., McKie, S., and Eliott, R. (2006) 'Neuronal correlates of theory of mind and empathy: a functional magnetic resonance imaging study in a nonverbal task', *Neuroimage*, 29: 90–98.

Walters, G. D. (2007) 'Measuring proactive and reactive criminal thinking with the PICTS correlations with outcome expectancies and hostile attribution biases', *Journal of Interpersonal Violence*, 22: 371–385.

Wechsler, D. (1991) *Wechsler Intelligence Scales for Children*, 3rd edn, San Antonio, TX: Harcourt Assessment.

White, S. F., Marsh, A. A., Fowler, K. A., Schechter, J. C., Adalio, C., Pope, K., *et al.* (2012) 'Reduced amygdala response in youths with disruptive behavior disorders and psychopathic traits: decreased emotional response versus increased top-down attention to nonemotional features', *American Journal of Psychiatry*, 169: 750–758.

White, S. F., Brislin, S. J., Meffert, H., Sinclair, S., and Blair, R. J. R. (2013) 'Callous-unemotional traits modulate the neural response associated with punishing another individual during social exchange: a preliminary investigation', *Journal of Personality Disorders*, 27(1): 99–112.

Yang, Y., Glenn, A., and Raine, A. (2008) 'Brain abnormalities in antisocial individuals: implications for the law', *Behavioral Sciences and the Law*, 26: 65–83.

16

The psychopharmacology of criminality and oxytocin modulation of social behavior in aggressive individuals

Royce Lee

Introduction

Psychopharmacological research focused on criminality and aggression can be characterized as having two separate goals. The first is a pragmatic one of seeking tools to apply to aggressive psychiatric patients, often treated in forensic psychiatric settings. The context of this research is distinct from the broader context of the criminal justice system, in that the individuals being treated are usually patients rather than prisoners or parolees. The second goal is primarily a scientific one, that seeks to understand criminal behavior in a biological or neuroscience framework. This goal often entails using one of the standard tools of clinical neuroscience research, the placebo-controlled medication trial. This is because it is one of the few methods with which to test causation in clinical research.

In reality, these two goals are not entirely distinct and some combination of them is usually found in individual research programs. However, one reason the distinction is useful is because it permits a more optimistic view of the possibility of progress. The overall thesis of this chapter is that the second goal of increased scientific understanding is more likely to be successful. This is a matter of good timing rather than any intrinsic virtue. As the first wave of technology-driven clinical neuroscience research has produced novel insights, it becomes more readily apparent how psychopharmacology may play role in scientific research, and possibly even point towards future approaches to the modulation of brain systems mediating criminal and aggressive behavior.

Pragmatic psychopharmacology research in criminality and aggression

There have been two excellent reviews of what could be characterized as the pragmatic psychopharmacological approach to criminality and aggression. It is best to refer readers to these rather than duplicate the work here, but given their wide availability, some discussion is justified. The first is 'Guidelines: borderline and antisocial personality disorders: summary of NICE Guidance', published by the UK National Collaborating Centre for Mental Health (Kendall *et al.*, 2009).

The section on pharmacological approaches to criminality identifies ten controlled trials of psychopharmacological medications for the treatment of antisocial personality disorder (ASPD), all focused on aggression as the target behavior. Among these ten studies are included: (1) one trial of the anticonvulsant phenytoin (Barratt *et al.*, 1997) showing reduced aggression relative to placebo in an impulsive subgroup; (2) two trials of tricyclic antidepressants in people with ASPD and substance-use disorders showing a moderate effect on alcohol dependence (nortryptilline; Powell *et al.*, 1995) and small effect for cocaine dependence (desipramine; Leal *et al.*, 1994); (3) six trials of non-phenytoin anticonvulsants versus placebo on aggression (phenytoin, Gottschalk *et al.*, 1973; valproate, Hollander *et al.*, 2003; oxcarbazepine, Mattes, 2005; levetiracetam, Mattes, 2008; topiramate, Nickel *et al.*, 2005; a combination of phenytoin, carbamazepine, and valproate, Stanford *et al.*, 2005) showing in total small and non-significant effects on aggression; (4) one trial of the selective serotonin reuptake inhibitor (SSRI) fluoxetine versus placebo in a sample of personality-disordered adults showing a medium to large effect on aggression (Coccaro and Kavoussi, 1997); (5) one trial of lithium versus placebo (Sheard, 1976) showing small, non-significant effects on aggressive violations. Although treating the anticonvulsants together in a group does not make biological sense because of their differing mechanisms of action, the small number of studies for each anticonvulsant drug, and small sample sizes, justify grouping them together for purposes of review. The document concludes that, from the perspective of efficacy, there is no strong evidence supporting the use of any pharmacological intervention to treat antisocial personality disorder, beyond treating a medical or psychiatric disorder for which they are primarily indicated. From the perspective of safety and cost, the conclusion is similarly negative, exacerbated by limited reporting of side-effects in drugs known to have narrow therapeutic windows even in highly compliant, non-psychopathic patients.

The second, more recent meta-analytic review by Khalifa and others (2010) identified an overlapping set of studies. An additional study regarding the effects of the dopamine receptor antagonist bromocriptine versus placebo on a small sample of alcohol-dependent men, a portion of whom had ASPD, was identified (Powell *et al.*, 1995). The study was largely negative though underpowered, finding no measurable effect on outcome variables of global function or alcohol abuse. The conclusion of this meta-analytic review was similar in tone to the NICE guideline recommendations, finding very little evidence for efficacy on which any treatment recommendations could be given, and even less evidence regarding side-effects to estimate risk vs. benefit tradeoffs of treatment.

The pragmatic psychopharmacological approach to criminality has several notable characteristics. First is the atheoretical rationale for the choice of medication. With the exception of the trial of the SSRI fluoxetine in personality-disordered patients (Coccaro and Kavoussi, 1997), justified by prior work demonstrating deficits in serotonergic function, the choice of medication was not based on a mature theory matching the action of the drug to a known pathophysiological mechanism. Rather, the choice of drug was ostensibly a perceived similarity between a target behavior and a known therapeutic target of the drug. Because seizures could be seen to resemble fits of rage seen in persons with intermittent explosive disorder, anticonvulsants could possibly prevent fits of rage, just as they prevent seizures. The second characteristic is a lack of concern with neural circuit-specific effects. Because these drug trials occurred before functional brain imaging technologies such as high-density EEG or functional magnetic resonance imaging were widely available, the choice of drug tested was based on neither a circuit-specific understanding of the target behavior nor a circuit-specific understanding of the effects of the pharmacological intervention. In that era, although ostensibly a biological intervention, psychopharmacology resembled behaviorism, treating the brain itself as a black box with inputs for drug and the environment, and outputs in the form of reliably measured behaviors such as counts of aggressive incidents. Finally, one may notice a lack of precision in ascertaining the phenotype in both the NICE guideline and the Cochrane review. Some of the reviewed studies are focused on ASPD, others merely include ASPD, while others are focused on personality disorder generally or borderline personality disorder specifically. While this may stem partly from the feasibility issues of conducting clinical research in forensic or high-risk settings, it may also stem from a lack of appreciation of circuit-specific abnormalities that may lead to phenotypic and behavioral differences and mistaken assumptions of biological homogeneity.

While in hindsight it may be easy to point out the methodological or theoretical weaknesses of these early efforts, their results should be seen as largely validating (and validated by) the results of recent reviews of mental health treatment approaches to criminality in individuals with mental health problems. In a meta-analytic review, Martin and colleagues (2012) found little evidence that rigorously tested mental health treatments in forensic psychiatric settings (both behavioral and pharmacological treatment) had a substantive impact on criminal behavior. Although the interpretation of a more recent meta-analysis was more optimistic, in this study there was only a small effect size on criminal recidivism, with confidence intervals encompassing zero (Morgan *et al.*, 2012). This small effect size would likely be deemed unsatisfactory by the public for outcomes such as public safety, re-arrests, and convictions. Of theoretical interest, there was a positive correlation between mental health outcome and criminal outcome. This incidental finding suggests that when treatment is effective, criminal behavior may also improve. The problem is that in real-world but suboptimal contexts, treatment is often not effective.

Similar findings emerge even when the focus narrows to a single intervention (stimulant medications) in a well defined population (attention deficit disorder), as was recently reported in a population-wide study of the effects of increased stimulant treatment availability for the treatment of attention deficit disorder in Canada (Currie *et al.*, 2013). While treatment may have been effective at reaching clinical targets, no effect was seen on rates of criminal offending.

Probabilistic outcomes are acceptable in the field of medicine, due to the recognition that attempts to fully optimize outcome with aggressive treatment results in an unacceptably high burden of side-effects. It is not clear that such interventions would be acceptable for the criminal justice system. The overall message seems to be that criminality does not necessarily improve as a "side-effect" of exposure to psychiatric medications. This points to the need for greater precision in targeting specific neural systems in future efforts to impact criminality and aggression.

Clinical neuroscience and the psychopharmacology of criminality and aggression

Criminality is a dissocial behavior that has traditionally been understood in ethical and philosophical terms. The criminal justice system is charged with its prevention and control. However, several factors have led to an increasing use of medical technology and terminology to understand criminal behavior. First is the recognition within criminology that the small proportion of the population that commits an inordinate amount of the crime is largely resistant to deterrence, the main tool that the criminal justice system wields to prevent criminal behavior. This has led to searches for more effective tools that society may use to prevent criminal behavior and protect the population. Second has been an explosive expansion in the availability of medical and psychological research technology to study persons who are at risk for repeated instances of criminal behavior. Although these tools typically have been developed and used in medical research laboratories and in clinical patient populations, they can also be applied to forensic populations within the ethical restraints required when conducting research on vulnerable individuals. Finally, and somewhat unexpectedly, there has been an increasing trend towards a merging of criminal and mentally ill populations. This is evidenced by the surprising statistic that in the United States there are currently many more persons with serious mental illness in prisons and jails than in hospitals. Regarding the latter factor, it is likely that the full impact of the shift in the mentally ill population to the prison system has not yet been fully realized in the research, clinical, or legal realm.

Recognition that a large proportion of serious crime is committed by a small portion of the population has led to research programs on individuals who exhibit relatively stable patterns of criminal behavior over the course of their lives (DeLisi and Piquero, 2011). A substantial portion of these individuals would meet criteria for antisocial personality disorder, a serious psychiatric condition found in 2–3 percent of the US population and at least 35 percent of the prison population (Black *et al.*, 2010).

Nearly two centuries ago, physicians turned their attention to individuals so prone towards criminality that doubt was raised whether these individuals truly possessed free will, the philosophical pillar of conventional perspectives on criminality. The physicians posited that psychopathic behavior was a consequence of "moral insanity" (Prichard, 1835) or a defect of the brain rendering "psychopathic inferiority." Schneider (1923) operationalized a subclass of personality disorders, as first conceptualized by Kraepelin (1905), as psychopathic. Later, Cleckley's (1964) operational definition of psychopathy as a semantic deficit preventing complex social behavior served as the conceptual frame for Hare's psychopathy checklist (1970), which remains the accepted standard in the measurement of psychopathic traits.

Despite these significant and enduring advances in our understanding of the phenomenological and psychological aspects of psychopathy, there is currently neither a medical nor psychological cure for it. Before delving into speculation about where future advances may occur, it is essential to take a step back and examine a broader aspect of criminality, impulsive aggression, which is more widely prevalent in the population and thus far has been more comprehensively studied.

Serotonin, anger, and impulsive aggression

Not all violent, interpersonal crime reflects the cold, calculated, and instrumental violence that characterizes the most severe side of psychopathic criminal behavior. In fact, a large portion, perhaps even a majority, of crime may be the result of actions inspired by impulses: a combination of contextual factors and an underlying propensity to act without adequate consideration of the consequences. Impulse control, or rather a lack of it, has been recently recognized to be broadly relevant to the field of criminology, and emerges as a superfactor (DeLisi and Vaughn, 2014).

Although impulsivity may lead to a wide variety of crimes, for our purposes we will focus on interpersonal violence because of its severity. Affective aggression, characterized by recurrent episodes of angry physical or verbal outbursts that do not have instrumental motives, is characterized in DSM-5 as intermittent explosive disorder (IED). It is not a rare disorder, with an estimated 7.3 percent lifetime prevalence in the US (Kessler *et al.*, 2006). The prevalence of IED in prisons and involved in the criminal justice system is not known. Given the substantially larger prevalence of IED compared with ASPD in the general population, it is likely that the percentage of prisoners with IED is equal to or greater than the estimated 35. 5 percent figure for ASPD (Black *et al.*, 2010).

IED is most closely associated with deficits in serotonergic signaling in the brain. Acute challenge with 5-HT receptor agonists in impulsive aggressive individuals has revealed that they exhibit sensitivity to 5-HT post-synaptic receptors, as measured by intensity of brain metabolic activity with positron emission tomography (PET) neuroimaging in the orbitofrontal, ventromedial, parietal, and cingulate cortex (Siever *et al.*, 1999; Soloff *et al.*, 2000; New *et al.*, 2002). These regions encompass neural circuits involved in executive function, working memory, and affective behavior. It is likely that the decreased dynamic range in 5-HT function detected in these pharmacological brain imaging studies reflects altered brain signaling that somehow increases the propensity for aggressive behavior. This is evidenced by the fact that administration of fluoxetine, a drug that decreases the removal of serotonin from the synaptic cleft and reverses previously identified deficits in orbitofrontal metabolic activity (New *et al.*, 2004), can actually reduce impulsive aggressive behavior in personality-disordered individuals when compared with placebo in double-blinded, randomized experiments (Salzman *et al.*, 1995; Coccaro and Kavoussi, 1997). Exactly how increasing brain 5-HT reduces aggressive behavior is not quite clear. The most likely possibility is that increased serotonin signaling reduces the intensity or likelihood of a social provocation causing anger. We demonstrated that depletion of the molecular precursor of serotonin, tryptophan, increases the subjective intensity ratings of angry face stimuli in male adults with IED but not in normal controls (Lee *et al.*, 2012). Serotonin depletion also increases the intensity of angry mood in both male and female IED subjects relative to normal controls. The results of this study suggest that decreased serotonin signaling in IED increases the likelihood of a social interaction being interpreted as antagonistic. This in turn increases the likelihood of an angry emotional reaction, increasing the motivation to act aggressively despite possible negative consequences. Angry reactions may be mediated by increased activity of the amygdala, a brain region in the medial temporal lobes involved in emotional learning and social recognition (Coccaro *et al.*, 2007).

It is not yet known if SSRI treatment reduces aggression in forensic populations, despite the fact that available data suggest that they have been frequently used in prisoners to treat depression (Baillargeon *et al.*, 2001). Research is needed to understand better the impact of SSRI treatment on pathological aggression in the prison and other forensic populations. In clinical populations, it is important to point out that, even after optimal dosing of fluoxetine, the mean levels of aggression remain high enough to warrant clinical attention (Coccaro *et al.*, 1996). Thus more research is needed to optimize pharmacological approaches to impulsive aggression.

Social neuropeptides and the psychopharmacology of aggression

Oxytocin and vasopressin are nine-amino-acid peptides expressed in the brain and periphery. They have nearly identical sequences, differing by a single amino acid, reflecting their common ancestry from the amphibian peptide vasotocin. Both hormones are released in response to environmental stress and social encounters, thus their designation as "stress" and "social" hormones. Their central and peripheral functions are dissociated from each other. In the periphery, vasopressin potentiates the effect of corticotropin-releasing factor on the pituitary, and causes the kidney to excrete sodium and conserve water. The antidiuretic effect may have evolved as a mechanism to anticipate water deprivation or hemorrhage. Oxytocin in the periphery mediates the milk "let-down" reflex, sexual function, and uterine contractility. Notably, all of these functions entail suppression of defensive aggression.

In the brain, vasopressin is synthesized in the hypothalamus, bed nucleus of the striaterminalis, and the locus coeruleus (Buijs, 1990). Vasopressin receptors are heavily expressed in periventricular regions, the forebrain, amygdala, bed nuclei of the striaterminalis, and brain stem (Brownfield and Kozlowski, 1977; Buijs, 1990; Huber *et al.*, 2005): stimulation of these receptors typically causes defensive behaviors such as aggression and/or freezing. In the rat, oxytocin is synthesized primarily in the hypothalamus, but its receptors are found in the hypothalamus, amygdala, bed nucleus of the striaterminalis, and brainstem (Uvnäs-Moberg, 2003; Huber *et al.*, 2005). Both oxytocin and vasopressin receptors utilize membrane G protein-coupled receptor kinases and protein kinase C in signal transduction (Berrada *et al.*, 2000). Oxytocin and vasopressin releasing neurons are themselves regulated by $GABA_A$ inhibitory neurons (Berrada *et al.*, 2000). Post-mortem autoradiography in humans has confirmed the presence of vasopressin and oxytocin binding sites in regions of the brain implicated in social and emotional behaviors. Vasopressin receptors are located in the amygdala, bed nuclei of the striaterminalis, thalamus, hippocampus, and medial septal nucleus. Oxytocin receptors are located in the periamygdaloid area, medial septal nucleus, lateral septum, hypothalamus, and thalamus (Loup *et al.*, 1991).

Animal studies have clearly implicated vasopressin as a neuropeptide mediator of aggressive behavior and found that resting CSF vasopressin level correlated directly with life-history measures of aggression in personality-disordered human subjects: higher levels of vasopressin predicted more lifetime incidents of aggression. Given the wealth of data regarding the effects of vasopressin on social behavior across phylogenies (Garrison *et al.*, 2012), it is likely that vasopressin affects behavioral aggression through intermediary stages of neural processing that mediate social interaction, and is less likely to be direct activation of aggressive motor behavior.

We recently reported evidence in humans supporting this assertion in a pharmacological brain imaging study (Lee *et al.*, 2014). Healthy male subjects were randomized to receive chronic treatment with either placebo or a novel small-molecule antagonist of the V1a vasopressin receptor (SRX246, Azevan Pharmaceuticals). Before randomization, all subjects completed a baseline fMRI scan during presentation of angry Ekman faces. After five days of treatment, subjects received either intranasal vasopressin or placebo, followed 45 minutes later by fMRI imaging during the presentation of angry Ekman faces. Treatment with the V1a receptor antagonist effectively blocked the effect of intranasal vasopressin on amygdala accommodation to angry faces. Neither vasopressin nor SRX246 had a direct effect on amygdala activation to angry faces; rather, vasopressin increased accommodation during the second imaging session, and SRX246 blocked this effect. The results were interpreted as evidence of a role of vasopressin V1a receptor mediation of social recognition to the highly salient stimuli of angry faces. An additional main effect of SRX246 was reduced activation of the temporoparietal junction to angry face stimuli. A theoretical implication of our results is that blockade of V1a receptors appeared to affect both of the two distinct neural systems underlying empathy that have been identified in clinical neuroscience research: (1) an emotional contagion network that includes circuits near the bottom of the brain including the insula, somatosensory, and medial frontal cortex; and (2) a mentalizing, perspective-taking, or theory of mind network that relies on coordinated activity between cortical regions near the top of the brain that include the frontal cortex, temporal lobes, and parietal lobes (Engen and Singer, 2013). It is not yet known if modulation of vasopressin V1a receptor function in aggressive humans is a potential approach to impulsive aggression, but ongoing studies of this are being done in our laboratory.

Oxytocin may also play an important role in human aggression. We previously reported that central oxytocin levels, as measured by sampling of spinal fluid, are lower in inverse proportion to lifetime history of aggression (Lee *et al.*, 2009). The sample included both normal control with no psychiatric disorder, and individuals with a personality disorder. Personality disorder itself did not determine oxytocin levels, although the subsample of subjects with repeated suicide attempts also had lower levels of oxytocin than those who had never attempted suicide. This was the first human evidence that oxytocin is related to aggressive behavior and self-aggressive behavior (suicide attempt), translating a rich preclinical literature implicating oxytocin in the suppression of aggression (Dhakar *et al.*, 2012).

A pilot study of the effects of intranasal oxytocin on aggressive behavior

If low central levels of oxytocin are associated with increased lifetime aggression, does increasing brain oxytocin have an anti-aggressive effect? When administered orally, oxytocin is rapidly degraded in the stomach before it can be absorbed. When given intravenously, oxytocin absorption into the brain is blocked by the blood–brain barrier. One way around this obstacle is intranasal administration of oxytocin, which leads to increased plasma and central levels of oxytocin in humans (Striepens *et al.*, 2013). Given the previous finding that brain vasopressin levels were positively correlated with life history of aggression (Coccaro *et al.*, 1998), we additionally wanted to test the effect of a vasopressin receptor agonist on aggressive behavior.

We hypothesized that intranasally administered oxytocin would decrease, and that intranasally administered 1-desamino-8-d-arginine vasopressin (dDAVP), a vasopressin V1b/V2 agonist (Saito *et al.*, 1997), would increase aggressive responding as measured by the point subtraction paradigm (PSAP) in a sample of adults with clinically significant impulsive aggression. An advantage of simulated laboratory aggression paradigms is that they permit the measure of aggressive behavior in a manner that is both safe and ethical. The PSAP is a computer game in which the player competes against an opponent for monetary points. The player and opponent may gain monetary points by pressing the A button. The opponent provokes the subject by aggressively subtracting monetary points from the subject. The subject is given the opportunity to ignore the provocations, aggressively subtract points with the B button, or defensively prevent point subtractions with the C button. Importantly, B button responses do not earn the subject monetary points and thus do not represent instrumental aggression. The opponent's responses are to some degree shaped by the subject's offensive and defensive responses. The research subject actually plays a computer program rather than a real opponent, but noise cues and deception are used to ensure the illusion of playing another person is maintained. The PSAP has previously been validated as a laboratory analogue of interpersonal aggressive behavior. Aggressive responding is correlated with levels of aggression in male and female adults (Dougherty *et al.*, 1999). One of the principal advantages of the PSAP over other behavioral aggression paradigms is that its rewards and punishments (monetary loss and gain) are not as susceptible to habituation as contingencies used in other paradigms, such as electric shock. Therefore it can be used to test anti- and pro-aggressive effects of drugs using repeated-measures designs.

We conducted a pilot study of the effect of intranasal oxytocin and intranasal dDAVP on laboratory aggression using the PSAP. All study procedures were approved by the Institutional Review Board (IRB) of The University of Chicago, and all subjects provided written, informed consent. Nine adults were studied: three males and six females with a mean age 37 (SD = 12). All subjects were diagnosed with intermittent explosive disorder (meeting the equivalent of DSM-5 criteria) and personality disorder by semi-structured SCID (Structured Clinical Interview for DSM Disorders) and SID-P (Structured Interview for DSM-IV Personality Disorders) interview. Exclusion criteria included current major depressive episode, substance or alcohol dependence, neurological disorder, and psychotic disorder. To ensure subjects were not inebriated or drug-abusing, all subjects provided negative urine samples for drugs of abuse and alcohol breathalyzer test at screen and before each study session. The study design was a randomized, placebo-controlled study using a within-subjects, crossover design, with subjects completing three study sessions (intranasal oxytocin, intranasal placebo, and intranasal dDAVP). Each session was separated by 7 days to minimize drug carry-over effects. After subjects were confirmed to meet inclusion/exclusion criteria, subjects were randomized to one of six drug orders, to control for order effects. Drug was administered in MAD-300 intranasal atomizers, with four puffs per nostril delivered over 10 minutes by the study physician. Our method of intranasal delivery of oxytocin has been previously validated by its demonstrated ability to increase social motivation in healthy humans (Kirkpatrick *et al.*, 2014a) as well as to increase plasma oxytocin levels (Kirkpatrick *et al.*, 2014b). Intranasal drug was administered 45 minutes before the PSAP began. Data were recorded during the PSAP and analyzed offline. The general analysis approach compared the effect of drug on PSAP parameters using repeated measures ANOVA. Main effects and interactions of the within-subjects factor of drug and between-subjects effect of gender were examined for each relevant PSAP parameter, including B button presses (as well as B to A ratio), C button presses (as well as C to A ratio), and points subtracted by the computer. Main effects and interactions were followed up with paired-samples *t-tests*. The significance level was set at $P \leq 0.05$, two-tailed. Although gender was included as a between-subjects variable, it was a non-significant factor in all analyses. Thus the reported analyses do not include gender as a factor.

Repeated-measures ANOVA revealed that there was no significant main effect of drug ($F_{2,16} = 0.273$, $P = 0.765$ on aggressive responding; mean number of B button presses). No main effects of gender or interactions of drug × gender were detected. Likewise there were no main effects of drug or gender or interactions on B to A ratio. For C button presses, there was no significant main effect of drug ($F_{2,18} = 1.308$, $P = 0.295$). No main effects were seen for gender or the interaction of drug × gender. Likewise there were no main effects of drug and gender or interaction effects on C to A ratio. Exploratory analyses revealed no main effect of drug or gender and no interaction effects on A button presses.

For the number of points subtracted by the computer opponent, repeated-measures ANOVA detected a main effect of drug at a trend level ($F_{2,16} = 2.783$, $P = 0.09$). Follow-up paired sampled *t-tests* revealed that the oxytocin condition was associated with decreased points subtracted ($t_{1,8} = -3.011$, $P = 0.02$) relative to placebo. No difference was seen between dDAVP and placebo ($t_{1,8} = 0.818$; $P > 0.1$) or between oxytocin and dDAVP ($t_{1,8} = -1.279$; $P > 0.1$).

In this study, intranasal oxytocin did not reduce aggressive responding in a sample of adults with impulsive aggression, nor was an increase in aggressive responding seen with dDAVP. There were no significant effects of drug on defensive responding. However, a significant effect of oxytocin was found whereby intranasal oxytocin was associated with decreased aggressive point subtractions by the computer relative to placebo. No effect of dDAVP was found relative to placebo. Given that aggressive responding by the computer is driven by the subject's behavior, this finding is somewhat unexpected, but interesting.

A previously reported study of the effect of intranasal oxytocin in a sample of female college undergraduate students also found no main effect of drug on aggressive responding (Campbell and Hausman, 2013). The investigators report a drug × anxiety interaction on aggressive responding, whereby oxytocin appeared to be associated with diminished aggressive responding in subjects with higher than median split state anxiety level as measured by State-Trait Anxiety

Inventory (STAI), but not in subjects with lower than median split state anxiety. Of note, the IED subjects in our study scored 44.25 (SD = 11.39) on STAI trait anxiety. Unfortunately, state anxiety was not measured in our study. Nonetheless, we failed to find an effect of intranasal oxytocin on aggressive responding. We think that the study differences are unlikely to be due to the minor differences in the technique of administration, as both approaches have been demonstrated to result in measurable plasma increases in oxytocin level (Striepens *et al.*, 2013). The critical difference between the studies is that our subjects were selected for their high levels of impulsive aggression, and were at least a decade older on average. It is possible that the effect of oxytocin is different in a population with a clinically significant aggressive disorder.

The absence of effect of dDAVP on aggressive responding could indicate that the V1b receptor does not play a role in human aggressive behavior. A previous study of the effect of intranasal arginine vasopressin (a V1a and V1b agonist) on aggressive responding during the Taylor Aggression Paradigm in a normative population found no effect, suggesting that V1a receptor stimulation likewise does not result in increased aggressive responding in laboratory conditions.

The principal limitation of our study is the small sample size that may have precluded us from detecting a meaningful effect of oxytocin and/or dDAVP. This limitation is especially salient regarding potential gender differences in the effect of oxytocin on aggressive responding. Although we failed to find a hypothesized "anti-aggressive" effect of intranasal oxytocin in adults with IED or a "pro-aggressive" effect for dDAVP, we discovered a very interesting effect of the drug on the nature of the interaction between subject and opponent. Oxytocin appeared to decrease aggression on the part of the computer, while neither increasing nor decreasing it on the part of the subject. It is most likely that the change in computer behavior is due to an increase in C button, defensive responses, and that our study lacked the statistical power to detect a significant difference in C button presses, even though the mean differences were in the right direction. Our findings of a potentially "prosocial" effect of intranasal oxytocin in a group of aggressive adults is biologically plausible and consistent with previous work examining the effect of oxytocin on social behavior. These studies have found that oxytocin increases trust during the Trust Game (Baumgartner *et al.*, 2008) as well as affective empathy (Bartz *et al.*, 2010).

There are some theoretical challenges in interpreting an effect of a drug in a human subject on the output of a computer system programmed to emulate an aggressive opponent. Qualitative data in the form of subjective assessments of the computer opponent suggested that the PSAP is a limited form of the Turing test. Without exception, across all study sessions, subjects believed that they were interacting with a human being rather than a computer program. Thus the responses of the subject were driven by neural representation of an aggressive opponent. A potentially fruitful theoretical orientation for future approaches may a cybernetic approach. Cybernetics is a subdiscipline of systems theory that is amenable to quantitative analysis of human and machine interaction. The results of this study indicate that oxytocin modulates this neural representation in a circuit-specific manner, given that oxytocin did not affect all measured behaviors, but only a subset of them. A full understanding of the psychopharmacology of human social interaction may thus require sophisticated approaches that are characteristic of systems theory.

Theoretical considerations and the future of psychopharmacological research in criminality

Because impulsivity and impulsive aggression are widely prevalent problems and significant factors in criminality, psychopharmacology research in this area is not simply an example of "looking where the light can shine." What of the more difficult pattern of the cold, calculated psychopath and so-called career criminal? This group represents a smaller piece of the pie, but remains a significant problem for the criminal justice system and public health (DeLisi and Piquero, 2011).

The recent literature on psychopharmacological approaches to this population is sparse. The older "pragmatic" literature did not usually systematically characterize the populations, making it difficult to know if the tested interventions would likely work in a person who lacked empathy or remorse. It seems likely that there are important biological differences between individuals with predominantly "hot" and "cold" aggression. We are aware of a single study parsing these two apart. It found that the arrogant–deceitful factor on the Psychopathy Checklist-Screen Version was positively correlated with prolactin response to fenfluramine challenge, a marker of post-synaptic 5-HT receptor sensitivity (Dolan and Anderson, 2003), while the impulsive–antisocial factor was negatively correlated with response to fenfluramine, replicating previous work showing that impulsive, reactive aggression is related to subsensitivity of the prolactin response to fenfluramine (Coccaro *et al.*, 1996). Based on this evidence of divergent serotonergic function, it seems highly likely that the instrumental aggression that characterizes the classic psychopath is unlikely to respond to serotonergic modulation.

What psychopharmacological approaches would work to treat psychopathy? While the answer remains out of reach, recent developments in our understanding of the neural circuit problems mediating psychopathy point to the direction where progress will be made in the near future.

The dominant paradigm in the field has been the somatic marker hypothesis (Damasio *et al.*, 1991, recently reviewed and critiqued by Dunn *et al.*, 2006), which posits that a failure to activate limbic circuits in risky situations increases the likelihood of harmful choices. Underactivation of the "somatic marker" in psychopaths has been extended into the realm of social behavior, with the finding that psychopaths also underactivate limbic brain regions such as the insula when exposed to stimuli of others in pain (Decety *et al.*, 2009, 2013; Marsh *et al.*, 2013). These new data seem to validate Cleckley's (1964) conceptualization of psychopathy as a failure of semantic networks capable of generating meaning from experience, a conceptualization that at the time was not based on any empiric data.

Is this failure due to a lesion-like deficit in neural function? A recent study suggests that this is not the case. When instructed to empathize with actors depicted in painful situations, psychopathic offenders were able to "normalize" brain activation in both of these circuits in response to viewing stimuli of others in pain (Meffert *et al.*, 2013).

This latter study raises a critical question relevant to psychopharmacological research: if psychopaths have the capacity to automatically react with empathy, why do they fail to in certain situations? There are several divergent possibilities that future research will need to address. Is the lack of activation due to an avoidant or dissociative response? Dissociation is most closely associated with exposure to a traumatic event. Empirical data thus far do find significant comorbidity of post-traumatic stress disorder (PTSD) and ASPD (Goldstein *et al.*, 2010), but studies focused on psychopathy rarely report on or investigate if post-traumatic dissociation plays a role in the blunted emotional empathy in psychopathy. Future work in this area is needed to confirm or negate this theory. Could the lack of activation be due to hostility or mistrust? One empirical question is whether or not psychopaths previously studied also met criteria for paranoid personality disorder, a disorder characterized by paranoid anxiety about the intentions of others and a pervasive lack of social engagement due to this mistrust. Paranoid personality disorder (PD) is not rare, affecting approximately 4.4 percent of the general population (Grant *et al.*, 2004), but up to 24.4 percent of the prisoner population in the UK (Ullrich *et al.*, 2008).

An interesting aspect of paranoid PD is that it may have relationships with exposure to childhood trauma. We previously found that paranoid PD was uniquely associated with elevated central levels of the stress hormone corticotropin-releasing factor (Lee *et al.*, 2006). This raises the possibility that targeting central corticotropin-releasing factor may be a future avenue of research in forensic populations. Finally, although this chapter focuses on the issue of social cognition in psychopathy, there may be other aspects of the "semantic deficit" of psychopathy that need to be considered. The kind of semantic deficit that was originally hypothesized was not so much a deficit in symbolic processing as an inability to appreciate meaning or value. A close approximation of this concept may be error processing, a well characterized, brain-based process by which positive and negative consequences of actions are automatically and rapidly evaluated. The process has been most closely observed with clinical electrophysiological EEG studies, using averaging techniques to identify the characteristic brain response to undetected and detected errors: a negative deflection of the event-related potential, maximal above the anterior cingulate and supplementary motor area, with a latency of approximately 100 ms. The error-related negativity (ERN) waveform is believed to be an alerting signal in the anterior cingulate that serves to shape future intended actions: the anterior cingulate activity is in turn triggered by a phasic decrease of dopamine signaling within the basal ganglia. A study of feedback-induced error-related negativity found reductions of this signal in females high in fearless dominance when they committed time-estimation errors (Schulreich *et al.*, 2013). On the other hand, the trait of behavioral inhibition, in many ways the polar opposite trait of psychopathy, is associated with enhanced error-related negativity (Lahat *et al.*, 2014). Punishment (to which psychopathy predicts resistance) enhances the amplitude of ERN (Riesel *et al.*, 2012). Correspondingly, the behavioral trait of punishment sensitivity predicts higher ERN amplitude to errors (Boksem *et al.*, 2008). Thus a convincing body of work implicates ERN activity in the shaping of motivated behavior and decision making. The psychopharmacology of error processing is not fully understood. Benzodiazepine sedatives and dopamine D2 receptor antagonists such as haloperidol reduce ERN amplitude to error (de Bruijn *et al.*, 2004; Riba *et al.*, 2005) while amphetamine stimulants increase it (de Bruijn *et al.*, 2004). Further work, however, is needed to understand which component process underlying generation of the ERN is affected by the specific molecular action of drugs, in order to know which approach could be used to enhance deficient error monitoring in psychopathy.

In this section we have identified at least three overlapping but functionally distinct neural circuits that are likely implicated in psychopathy: (1) a circuit that mediates emotional empathy, which includes the medial prefrontal cortex, amygdala, and insula; (2) a circuit that mediates perspective taking and theory of mind, which includes the temporoparietal junction, precuneus, and medical prefrontal cortex; and (3) a circuit that includes to basal ganglia and anterior cingulate, which mediates error and feedback processing.

Conclusion

An understanding of the interaction of these three systems in normative social interaction and abnormal dissocial interaction is currently lacking. Once achieved, this may enable predictions of how psychopharmacological interventions impact all three of these circuits in a psychopathic individual during social interaction.

We now know that vasopressin receptor modulation affects the first and second of these circuits, as does oxytocin receptor modulation. Error monitoring appears to be degraded with sedatives and enhanced with stimulants. The effect of neuropeptide modulators remains unknown.

It is conceivable that a combination of agents may be necessary to affect all three systems in order to have a powerful effect on dissocial interactions. If so, more sophisticated experimental paradigms will be needed. Our preliminary results with intranasal oxytocin suggest that cybernetic approaches may be needed. The complexity of neural systems involved will require model building based on realistic, biologically informed parameters that approximate the information processing properties of the brain. Also, in order to hold steady certain aspects of the experimental social interaction, computer simulations may be key to providing a feasible experimental approach, as well as a way of measuring affected parameters of social interaction. We think that progress on the difficult problem of psychopharmacology of psychopathy will require knowing how specific modulations of interacting neural circuits will impact specific individuals in specific social contexts.

References

Baillargeon, J., Black, S. A., Contereras, S., Grady, J., and Pulvino, J. (2001) 'Antidepressant prescribing patterns for prison inmates with depressive disorders', *Journal of Affective Disorders*, 63: 225–231.

Barratt, E. S., Stanford, M. S., Felthous, A. R., and Kent, T. A. (1997) 'The effects of phenytoin on impulsive and premeditated aggression: a controlled study', *Journal of Clinical Psychopharmacology*, 17: 341–349.

Bartz, J. A., Zaki, J., Bolger, N., Hollander, R., Ludwig, N. N., Kolevson, A., and Ochsner, K. N. (2010) 'Oxytocin selectively improves empathic accuracy', *Psychological Science*, 21: 1426–1428.

Baumgartner, T., Heinrichs, M., Vonlanthen, A., Fischbacher, U., and Fehr, E. (2008) 'Oxytocin shapes the neural circuitry of trust and trust adaptation in humans', *Neuron*, 58: 639–650.

Berrada, K., Plesnicher, C. L., Luo, X., and Thibonnier, M. (2000) 'Dynamic interaction of human vasopressin/oxytocin receptor subtypes with G protein-coupled receptor kinases and protein kinase C after agonist stimulation', *Journal of Biological Chemistry*, 275(35): 27229–27237.

Black, D. W., Gunter, T., Loveless, P., Allen, J., and Sieleni, B. (2010) 'Antisocial personality disorder in incarcerated offenders: psychiatric comorbidity and quality of life', *Annals of Clinical Psychiatry*, 22: 113–120.

Boksem, M. A., Tops, M., Kostermans, E., and De Cremer, D. (2008) 'Sensitivity to punishment and reward omission: evidence from error-related ERP components', *Biological Psychology*, 79: 185–192.

Brownfield, M. M. S. and Kozlowski, G. P. (1977) 'The hypothalamo-choroidal tract', *Cell and Tissue Research*, 178(1): 111–127.

de Bruijn, E. R., Hultstijn, W., Verkes, R. J., Ruigt, G. S., and Sabbe, B. G. (2004) 'Drug-induced stimulation and suppression of action monitoring in healthy volunteers', *Psychopharmacology*, 177: 151–160.

Buijs, R. M. (1990) 'Vasopressin and oxytocin localization and putative functions in the brain', in J. D. Pickard, F. Cohadon, and J. L. Antunes (eds), *Neuroendocrinological Aspects of Neurosurgery*, Vienna: Springer, pp. 86–89.

Campbell, A. and Hausmann, M. (2013) 'Effects of oxytocin on women's aggression depend on state anxiety', *Aggressive Behavior*, 39: 316–322.

Cleckley, H. (1964) *The Mask of Sanity*, 4th edn, St Louis, MO: Mosby.

Coccaro, E. F. and Kavoussi, R. J. (1997) 'Fluoxetine and impulsive aggressive behaviour in personality-disordered subjects', *Archives of General Psychiatry*, 54: 1081–1088.

Coccaro, E. F., Berman, M. E., Kavoussi, R. J., and Hauger, R. L. (1996) 'Relationship of prolactin response to d-fenfluramine to behavioral and questionnaire assessments of aggression in personality-disordered men', *Biological Psychiatry*, 40: 157–164.

Coccaro, E. F., Kavoussi, R. K., Hauger, R. L., and Ferris, C. F. (1998) 'Cerebrospinal fluid vasopressin correlates with aggression and serotonin function in personality disordered subjects', *Archives of General Psychiatry*, 55: 708–714.

Coccaro, E. F., McCloskey, M. S., Fitzgerald, D. A., and Phan, K. L. (2007) 'Amygdala and orbitofrontal reactivity to social threat in individuals with impulsive aggression', *Biological Psychiatry*, 62: 168–178.

Currie, J., Stabile, M., and Jones, L. E. (2013) *Do Stimulant Medications Improve Educational and Behavioral Outcomes for Children with ADHD?*, NBER No. 19105, Cambridge, MA: National Bureau of Economic Research.

Damasio, A. R., Tranel, D., and Damasio, H. (1991) 'Somatic markers and the guidance of behavior: theory and preliminary testing', in H. S. Levin, H. M. Eisenberg, and A. L. Benton (eds), *Frontal Lobe Function and Dysfunction*, New York: Oxford University Press, pp. 217–229.

Decety, J., Michalska, K. J., Akitsuki, Y., and Lahey, B. (2009) 'Atypical empathic responses in adolescents with aggressive conduct disorder: a functional MRI investigation', *Biological Psychology*, 80: 203–211.

Decety, J., Chen, C., Harenski, C., and Kiehl, K. A. (2013) 'An fMRI study of affective perspective taking in individuals with psychopathy: imagining another in pain does not evoke empathy', *Frontiers in Human Neuroscience*, 7: 489–499.

DeLisi, M. and Piquero, A. R. (2011) 'New frontiers in criminal careers research, 2000–2011: a state-of-the-art review', *Journal of Criminal Justice*, 39(4): 289–301.

DeLisi, M. and Vaughn, M. G. (2014) 'Foundation for a temperament-based theory of antisocial behavior and criminal justice system involvement', *Journal of Criminal Justice*, 42: 10–25.

Dhakar, M. B., Rich, M. E., Reno, E. L., Lee, H. J., and Caldwell, H. K. (2012) 'Heightened aggressive behavior in mice with lifelong versus postweaning knockout of the oxytocin receptor', *Hormones and Behavior*, 62: 86–92.

Dolan, M. C. and Anderson, I. M. (2003) 'The relationship between serotonergic function and the psychopathy checklist: screening version', *Journal of Psychopharmacology*, 17: 216–222.

Dougherty, D. M., Bjork, J. M., Huckabee, H. C., Moeller, F. G., and Swann, A. C. (1999) 'Laboratory measures of aggression and impulsivity in women with borderline personality disorder', *Psychiatry Research*, 85: 315–326.

Dunn, B. D., Dalgeish, T., and Lawrence, A.D. (2006) 'The somatic marker hypothesis: a critical evaluation', *Neuroscience and Biobehavioral Reviews*, 30: 239–271.

Engen, H. G. and Singer, T. (2013) 'Empathy circuits', *Current Opinion in Neurobiology*, 23: 275–282.

Garrison, J. L., Macosko, E. Z., Bernstein, S., Pokala, N., Albrecht, D. R., and Bargmann, C. I. (2012). 'Oxytocin/vasopressin-related peptides have an ancient role in reproductive behavior', *Science*, 338, 540–554.

Goldstein, R. B., Compton, W. M., and Grant, B. F. (2010) 'Antisocial behavioral syndromes and additional psychiatric comorbidity in posttraumatic stress disorder among U.S. adults: results from wave 2 of the National Epidemiologic Survey on Alcohol and Related Conditions', *Journal of the American Psychiatric Nurses Association*, 16: 145–165.

Gottschalk, L. A., Covi, L., and Uliana, R. (1973) 'Effects of diphenylhydantoin on anxiety and hostility in institutionalized prisoners', *Comprehensive Psychiatry*, 14: 503–511.

Grant, B. F., Hasin, D., Stinson, F. S., Dawson, D.A., Chou, S. P., Ruan, W. J., and Pickering, R. P. (2004) 'Prevalence, correlates, and disability of personality disorders in the United States: results from the National Epidemiologic Survey on Alcohol and Related Conditions', *Journal of Clinical Psychiatry*, 65: 948–958.

Hare, R. D. (1970) *Psychopathy: Theory and Research*, New York: Wiley and Sons.

Hollander, E., Tracy, K. A., Swann, A. C., Coccaro, E. F., McElroy, S. L., Wozniak, P., *et al.* (2003) 'Divalproex in the treatment of impulsive aggression: efficacy in cluster B personality disorders', *Neuropsychopharmacology*, 28(6): 1186–1197.

Huber, D., Veinante, P., and Stoop, R. (2005) 'Vasopressin and oxytocin excite distinct neuronal populations in the central amygdala', *Science*, 308(5719): 245–248.

Kendall, T., Pilling, S., Tyrer, P., Duggan, C., Burbeck, R., Meader, N., and Taylor, C. (2009) 'Guidelines: borderline and antisocial personality disorders: summary of NICE Guidance', *British Medical Journal*, 338: 293–295.

Kessler, R. C., Coccaro, E. F., Fava, M., Jaeger, S., Jin, R., and Walters, E. (2006). 'The prevalence and correlates of DSM-IV intermittent explosive disorder in the National Comorbidity Survey Replication', *Archives of General Psychiatry*, 63: 669–678.

Khalifa, N., Duggan, C., Stoffers, J., Huband, N., Vollm, B. A., Ferriter, M., and Lieb, K. (2010) 'Pharmacological interventions for antisocial personality disorder', *Cochrane Database Systematic Reviews*, 8: 3–21.

Kirkpatrick, M. G., Lee, R., Wardle, M. C., Jacob, S., and de Wit, H. (2014a) 'Effects of MDMA and intranasal oxytocin on social and emotional processing', *Neuropsychopharmacology*, 39(7): 1654–1663.

Kirkpatrick, M. G., Francis, S. M., Lee, R., de Wit, H., and Jacob, S. (2014b) 'Plasma oxytocin concentrations following MDMA or intranasal oxytocin in humans', *Psychoneuroendocrinology*, 46: 23–31.

Kraepelin, E. (1905) *Lectures on Clinical Psychiatry*, 2nd edn, translated by T. Johnstone, London: Balliere Tindall.

Lahat, A., Lamm, C., Chronis-Tuscano, A., Pine, D. S., Henderson, H. A., and Fox, N. A. (2014) 'Early behavioral inhibition and increased error monitoring predict later social phobia symptoms in childhood', *Journal of the American Academy of Child and Adolescent Psychiatry*, 53: 447–455.

Leal, J., Ziedonis, D., and Kosten, T. (1994) 'Antisocial personality disorder as a prognostic factor for pharmacotherapy of cocaine dependence', *Drug and Alcohol Dependence*, 35: 31–35.

Lee, R., Gollan, J., Kasckow, J., Geracioti, T., and Coccaro, E. F. (2006) 'CSF corticotropin-releasing factor in personality disorder: relationship with perceived parental care', *Neuropsychopharmacology* 31: 2289–2295.

Lee, R., Ferris, C., Van de Kar, L. D., and Coccaro, E. F. (2009) 'Cerebrospinal fluid oxytocin, life history of aggression, and personality disorder', *Psychoneuroendocrinology*, 34(10): 1567–1573.

Lee, R. J., Gill, A., Chen, B., McCloskey, M., and Coccaro, E. F. (2012) 'Modulation of central serotonin affects emotional information processing in impulsive aggressive personality disorder', *Journal of Clinical Psychopharmacology*, 32(3): 329–335.

Lee, R. J., Coccaro, E. F., Cremers, H., McCarron, R., Lu, S. F., Brownstein, M. J., and Simon, N. G. (2014) 'A novel V1a receptor antagonist blocks vasopressin-induced changes in the CNS response to emotional stimuli: an fMRI study', *Frontiers in Systems Neuroscience*, 12: doi 10.3389fnsys.2013.00100

Loup, F., Tribollet, E., Dubois-Dauphin, M., and Dreifuss, J. J. (1991) 'Localization of high-affinity binding sites for oxytocin and vasopressin in the human brain: an autoradiographic study', *Brain Research*, 555(2): 220–232.

Marsh, A. A., Finger, E. C., Fowler, K. A., Adalio, C. J., Jurkowitz, I. T., Schechter, J. C., Pine, D. S., Decety, J., and Blair, R. J. R. (2013) 'Empathic responsiveness in amygdala and anterior cingulate cortex in youths with psychopathic traits', *Journal of Child Psychology and Psychiatry*, 54: 900–910.

Martin, M. S., Dorken, S. K., Wamboldt, A. D., and Wootten, S. E. (2012) 'Stopping the revolving door: a meta-analysis of the effectiveness of interventions for criminally involved individuals with major mental disorders', *Law and Human Behavior*, 36: 1–12.

Mattes, J. A. (2005) 'Oxcarbazepine in patients with impulsive aggression: a double-blind, placebo-controlled trial', *Journal of Clinical Psychopharmacology*, 25: 575–579.

Mattes, J. A. (2008) 'Levetiracetam in patients with impulsive aggression: a double-blind, placebo-controlled trial', *Journal of Clinical Psychiatry*, 69: 311–315.

Meffert, H., Gazzola, V., den Boer, J. A., Bartels, A. A. J., and Keysers, C. (2013) 'Reduced spontaneous but relatively normal deliberate vicarious representations in psychopathy', *Brain*, 136: 2550–2562.

Morgan, R. D., Flora, D. B., Kroner, D. G., Mills, J. F., Varghese, F., and Steffan, J. S. (2012) 'Treating offenders with mental illness: a research synthesis', *Law and Human Behavior*, 36: 37–50.

New, A. S., Hazlett, E. A., Buchsbaum, M. S., *et al.* (2002) 'Blunted prefrontal cortical 18-fluorodeoxyglucose positron emission tomography response to meta-chlorophenylpiperazine in impulsive aggression', *Archives of General Psychiatry*, 59: 621–629.

New, A. S., Buchsbaum, M. S., Hazlett, E. A., *et al.* (2004) 'Fluoxetine increases relative metabolic rate in prefrontal cortex in impulsive aggression', *Psychopharmacology*, 176: 451–458.

Nickel, M. K., Nickel, C., Kaplan, P., *et al.* (2005) 'Treatment of aggression with topiramate in male borderline patients: a double-blind, placebo-controlled study', *Biological Psychiatry*, 57: 495–499.

Powell, B. J., Campbell, J. L., Landon, J. F., Liskow, B. I., Thomas, H. M., Nickel, E. J., *et al.* (1995) 'A double-blind, placebo-controlled study of nortriptyline and bromocriptine in male alcoholics subtyped by comorbid psychiatric disorders', *Alcoholism, Clinical and Experimental Research*, 19(2): 462–468.

Prichard, J. C. (1835) *Treatise on Insanity*, London: Sherwood Gilbert and Piper.

Riba, J., Rodriguez-Fornells, A., Munte, T. F., and Barbanoj, M. J. (2005) 'A neurophysiological study of the detrimental effects of alprazolam on human action monitoring', *Brain Research and Cognitive Brain Research*, 2: 554–565.

Riesel, A., Weinberg, A., Endrass, T., Kathmann, N., and Hajcak, G. (2012) 'Punishment has a lasting impact on error-related brain activity', *Psychophysiology*, 49: 239–247.

Saito, M., Tahara, A., and Sugimoto, T. (1997) '1-desamino-8-d-arginine vasopressin (DDAVP) as an agonist on V1b vasopressin receptor', *Biochemical Pharmacology*, 53: 1711–1717.

Salzman, C., Wolfson, A. N., Schatzberg, A., *et al.* (1995) 'Effect of fluoxetine on anger in symptomatic volunteers with borderline personality disorder', *Journal of Clinical Psychopharmacology*, 15: 23–31.

Schneider, K. (1923) *Die Psychopathischen Personlichkeiten*, Berlin: Springer.

Schulreich, S., Pfabigan, D. M., Derntl, B., and Sailer, U. (2013) 'Fearless dominance and reduced feedback-related negativity amplitudes in a time-estimation task – further neuroscientific evidence for dual-process models of psychopathy', *Biological Psychology*, 3: 352–363.

Sheard, M. H., Marini, J. L., and Bridges, C. I. (1976) 'The effect of lithium on impulsive aggressive behavior in man', *American Journal of Psychiatry*, 133: 1409–1413.

Siever, L. J., Buchsbaum, M. S., New, A. S., *et al.* (1999) 'd,l-Fenfluramine response in impulsive personality disorder assessed with [18F]-fluorodeoxyglucose positron emission tomography', *Neuropsychopharmacology*, 20: 413–423.

Soloff, P. H., Meltzer, C. C., Greer, P. J., *et al.* (2000) 'A fenfluramine-activated FDG-PET study of borderline personality disorder', *Biological Psychiatry*, 47: 540–547.

Stanford, M. S., Helfritz, L. E., Conklin, S. M., Villemarette-Pittman, N. R., Greve, K. W. and Adams, D. (2005) 'A comparison of anticonvulsants in the treatment of impulsive aggression', *Experimental and Clinical Psychopharmacology*, 13(1): 72–77.

Striepens, N., Kendrick, K. M., Hanking, V., Landgraf, R., Wullner, U., Maier, W. and Hurlemann, R. (2013) 'Elevated cerebrospinal fluid and blood concentrations of oxytocin following its intranasal administration in humans', *Scientific Reports*, 3: 55–63.

Ullrich, S., Deasy, D., Smith, J., Johnson, B., Clarke, M., Broughton, N., and Coid, J. (2008) 'Detecting personality disorders in the prison population of England and Wales: comparing case identification using the SCID-II screen and the SCID-II clinical interview', *Journal of Forensic Psychiatry & Psychology*, 19(3): 301–322.

Uvnäs-Moberg, K. (2003) *The Oxytocin Factor: Tapping the Hormone of Calm, Love and Healing*, Boston, MA: Perseus.

17

The neurobiology of aggression

Looking underneath the hood

C. Nathan DeWall and David S. Chester

Introduction

People become embroiled in conflict for many reasons. They might take a whiff of foul odors, perspire from high temperatures, or receive a damaging insult. All of these experiences can increase aggression. But they do not cause aggression among all people—and they fail to clarify *why* people behave aggressively. To understand why people behave aggressively, researchers may consider neurobiological processes that underlie aggression. Just as an auto mechanic will look underneath the hood to understand why a car sputters instead of drives smoothly, understanding the neurobiological mechanisms behind aggression can aid in predicting and preventing aggression.

This chapter reviews theory and research that have begun to fill this gap in the aggression literature. We divide the chapter into six sections. First, we discuss two prominent social psychological theories of aggression—the general aggression model (Anderson and Bushman, 2002; DeWall *et al.*, 2011a) and the I-cubed theory of aggression (Finkel, in press)—that incorporate neurobiology into their understanding of aggression. Second, we review research that shows how, compared with their nonaggressive counterparts, aggressive people have some structural differences in their brains. The third section takes a slightly different approach by reviewing evidence on how the brain develops over the lifespan, and how that development can influence aggression. In the fourth section, we abandon our focus on how the aggressive brain is structured, and instead concentrate on how the aggressive brain functions—how neural blood flow waxes and wanes according to environmental pressures. The fifth section describes neurobiological interventions that can reduce aggression. Finally, we offer suggestions regarding how researchers may enhance their research programs by blending their interests with insights from the neurobiological literature.

Theoretical perspectives

The aggression literature has enjoyed a litany of theoretical perspectives. To formulate and test their hypotheses, aggression researchers could choose from numerous theories, such as frustration–aggression theory (Dollard *et al.*, 1939), social learning theory (Bandura *et al.*, 1961; Bandura, 1977), excitation transfer theory (Zillmann, 1979, 1988), cognitive neoassociation theory (Berkowitz, 1989, 1990), among many others. But until recently, aggression researchers did not have readily available theories on which to base hypotheses regarding the neurobiology of aggression. Two social psychological theories, the general aggression model (GAM) and the I-cubed theory of aggression, now offer clear frameworks that extend prior theoretical perspectives and accommodate the neurobiological perspective.

The GAM is the most dominant social psychological theory of aggression (Anderson and Bushman, 2002; DeWall *et al.*, 2011a). It is a meta-theoretical framework that emphasizes the importance of aggressive knowledge structures that people develop through experience, which inform how they approach and react to aggressive situations. To understand an aggressive episode, the GAM emphasizes three critical stages: (1) person and situation inputs; (2) present internal states (i.e. cognition, arousal, affect, including brain activity); and (3) outcomes of appraisal and decision-making processes. A feedback loop can influence future cycles of aggression, which can produce a violence escalation cycle (Anderson *et al.*, 2008).

The GAM accommodates the neurobiological perspective in two ways. First, it argues for individual differences in neurobiological processes that can inform how people respond in aggressive situations. These individual differences may originate from birth, or in response to environmental situations. For example, individuals who possess genetic polymorphisms associated with lower expression of the neurotransmitter serotonin are at risk for behaving aggressively, especially if they were exposed to environmental stressors during childhood (Caspi *et al.*, 2002). Individuals may also experience a traumatic event that alters their brain activity or composition, which may have implications for their aggression. In one case study, a 56-year-old man experienced bilateral damage to his orbitofrontal cortex (OFC) and left amygdala (Blair and Cipolotti, 2000). He went on to physically attack nurses, damage property, and engage in other violent acts. The GAM accounts for these neurobiological differences in terms of how they relate to personality development.

The GAM also argues that neurobiological processes related to people's inner states can affect their aggression. Although it is possible to measure a person's thoughts, emotions, and arousal using self-report measures, more direct measures of neurobiological processes are possible. In some of our research, we use functional magnetic resonance imaging (fMRI) to assess how neural activation in specific brain regions to an environmental stressor interacts with individual differences in executive functioning to predict aggression (Chester *et al.*, 2014). Our results show that blood flow to specific brain regions in response to environmental stressors can help researchers understand the neural mechanisms underlying aggression.

The I^3 model of aggression (pronounced 'I-cubed'; Finkel, in press) applies principles of statistical moderation to understand the underlying processes involved in aggression. It argues that aggression is most likely to occur at relatively high levels of *impellance* (e.g. trait physical aggressiveness); high levels of *instigation* (e.g. provocation); and low levels of *inhibition* (e.g. poor self-control). Each component of the I^3 model has a neurobiological basis.

One method for understanding brain differences between aggressive and nonaggressive people is positron emission tomography (PET). In one study, PET scans measured glucose metabolism in the brains of 41 murderers and 41 matched control participants (Raine *et al.*, 1997). Compared with the matched controls, the murderers showed lower levels of glucose metabolism in the prefrontal cortex (PFC), which is a brain region that helps people override their aggressive urges. These findings support the I^3 model of aggression by showing that murderers, compared with non-violent controls, have a lower ability to inhibit aggressive urges.

Thus the GAM and the I^3 model of aggression provide compelling theoretical frameworks from which neurobiological hypotheses can be tested. Both models can inform neurobiological predictions based on individual differences and transient responses to environmental events. The next section delves further into neurobiological evidence regarding structural differences in the aggressive brain.

Structural differences in the aggressive brain

When confronted with examples of extreme violence, people often wonder whether the perpetrators were born that way, or if others taught them to behave violently. Both nature and nurture play a role in understanding a person's aggression. To examine these questions, researchers use a variety of methods to determine whether people with a violent history, compared with non-violent people, show differences in how their brains are structured.

One method researchers use involves measuring the size of various brain regions associated with aggression. In one recent study, researchers recruited a sample of men whose aggressive behavior had been measured since the men were in the first grade (Pardini *et al.*, 2014). Next, the researchers used MRI to measure the size of each participant's amygdala. Participants with lower amygdala volume reported engaging in more aggressive actions during childhood and adulthood. These findings mesh well with other recent findings that link lower amygdala volume to higher levels of aggression (Yang *et al.*, 2009; Matthies *et al.*, 2012).

Other research has explored the relationship between PFC size and aggression. The PFC serves many functions, including the ability to inhibit aggressive urges. Both the GAM and the I^3 theory of aggression argue that inhibitory processes play a crucial role in understanding risk and resiliency factors in aggression. For example, chronic and transient reductions in inhibition can increase aggression and other criminal behavior (Gottfredson and Hirschi, 1990; DeWall *et al.*, 2007). Consistent with these findings, people with antisocial personality disorder, compared with healthy controls, have significantly less prefrontal volume (Raine *et al.*, 2000). This finding has been replicated numerous times, including among violent prison inmates (e.g. Schiltz *et al.*, 2013).

An alternative method for understanding structural differences between aggressive and non-aggressive individuals involves assessing fractional anisotrophy (FA) using a technique called diffusion tensor imaging (DTI). FA refers to white matter structural integrity, which influences how well various brain networks can communicate with each other. In one study, people with antisocial personality disorder and psychopathy showed significantly lower FA in a white-matter tract that connects the amygdala and the orbitofrontal cortex (Craig *et al.*, 2009).

Thus people at risk for aggression not only have smaller brain volume in the amygdala and PFC, but they also possess impaired fiber tracts that undermine communication between these brain regions. These findings offer initial evidence that the neurobiological perspective can offer crucial insight into who may be at risk for aggressive and violent behavior. The next section delves further into how the aggressive brain develops.

Development and the aggressive brain

Aggression is not purely a problem among fully grown adults. Indeed, the act of harming others permeates earlier developmental stages such as toddlerhood, and peaks in adolescence and young adulthood (Archer, 2004). Why are children and adolescents so violent? Are they truly evil little creatures who are motivated to hurt each other? Or are they simply incapable of inhibiting aggressive urges? Neuroscience can help us answer this question.

The human brain takes roughly 20 years to fully mature (Giedd, 2004). A "mature" brain not only possesses fully formed subanatomical units but, more importantly, it possesses the requisite connections between various regions that enable this immensely complex organ to function. This complicated process begins around 7 weeks after fertilization, with the formation of the earliest cortical neurons, the cells that form the brain, migrating along spires of supporting cells to their proper places in the cranial vault, and then forming the innumerable connections with corresponding neurons. During the early years of life, these vast tangles of neurons and their connections become pruned in order to yield an efficiently organized neural structure. Of utmost relevance to aggressive behavior is the successful formation and pruning of the connections between the PFC and subcortical regions that may promote aggressive behavior.

The PFC is the seat of successful human self-regulation (Heatherton and Wagner, 2011). Indeed, according to the social brain hypothesis, this region evolved to its current enormity in humans because it is a crucial organ for living in groups (Dunbar, 2009). Living with others requires us to inhibit our selfish and animalistic tendencies, as well as to understand and simulate the minds of others to predict their responses to our behaviors. These functions reside in the PFC and require a great deal of neural architecture, which helps explain why the human PFC is not considered fully developed until after the age of 20 (Giedd, 2004). Indeed, human neural development is characterized by the relatively rapid development of limbic structures that generate affective and appetitive impulses that outpace a slower-developing PFC (Casey *et al.*, 2011). The self-regulation that emerges from the PFC is a crucial element to inhibiting aggressive impulses (Chester *et al.*, 2014; DeWall *et al.*, 2011b).

Adolescents already possess a relatively immature PFC (Giedd, 2004). This handicapped ability to inhibit aggressive impulses may contribute to their being one of the most violent age groups (Daly and Wilson, 1990; Archer, 2004). This notion that adolescents' diminished PFC function is, if partially, to blame for their heightened aggression is well supported. Prefrontal dysfunction theory posits that the heightened aggression and antisocial behavior observed among adolescents is due to the reduced inhibition that arises from an immature PFC's lack of control over subcortical neural regions (Raine, 2002). For instance, adolescents diagnosed with an aggressive behavior pathology showed diminished activation in the PFC when they completed the Stroop task, a classic paradigm in which participants are required to inhibit prepotent responses (Mathews *et al.*, 2005).

Because aggression comes with substantial consequences, neural development throughout childhood and adolescence prioritizes the maturation of neural connections between the PFC and subcortical regions (Barnea-Goraly *et al.*, 2005). Another area of strong neural maturation throughout childhood and adolescence is the corpus callosum. The corpus callosum refers to the thick bundle of fibers that connects the left and right hemispheres of the brain (Barnea-Goraly *et al.*, 2005). Maturation of the callosum may assist developing adolescents to inhibit aggression by avoiding the dominance of one hemisphere over the other. Frontal asymmetry, in which the left hemisphere of the PFC shows greater activity than the right hemisphere, is reliably linked with anger and aggressive behavior (Peterson *et al.*, 2008). A well developed callosum would likely allow for rapid communication between the hemispheres and avoid the PFC laterality that precedes aggression.

Early on in life, humans develop a brain that responds to the suffering of others as if it were happening to them (Singer, 2006). The empathic pain matrix consists primarily of the dorsal anterior cingulate cortex (dACC), the anterior insula, and the secondary somatosensory cortex, and is reliably recruited by seeing the misfortunes of others (Singer *et al.*, 2004). Activity in this network likely serves to prevent us from inflicting harm on others, as their pain would be vicariously experienced. However, the formation of this neural network can be delayed or prevented during development, which would release this aggression-preventing mechanism. Just so, adolescents with an aggressive disorder showed a blunted, if absent, response of the empathic pain matrix while observing the painful misfortunes of another (Decety *et al.*, 2009).

As individuals enter adolescence, they spend more time with their peers than at any other time in their lifespan (Brown, 2004). This immersion among peers is reflected in altered neurocircuity during adolescence that is not present in childhood or adulthood. Specifically, the mere presence of a peer functionally couples the medial prefrontal cortex (MPFC)—a region associated with self-focus—with the reward center, the striatum, which motivates individuals to behave in a manner that will win them approval (Somerville *et al.*, 2013). This unique life-stage of neural functioning may motivate individuals to aggress on behalf of their peers in order to win acceptance.

These findings illustrate how the brain is a phenotypic wonder, which takes approximately 20 years to fully mature. The seat of self-regulation, the PFC, takes the longest to develop and establish effective connections with other brain areas, which is the chief culprit behind the impulsive aggression of children and adolescents. However, the corpus callosum also takes a great deal of time to conjoin the left and right hemispheres of the brain, which may allow for hemispheric laterality in earlier stages of development that promotes an aggressive mindset. Empathy reduces aggression, and this region can be malformed in early development, resulting in callous children and adolescents, indifferent to the suffering aggression can cause others. Finally, the adolescent brain is geared towards satisfying peers. Therefore adolescents may behave aggressively in order to secure the rewards of inclusion. The developing brain has many potential pitfalls where normative development can be swayed towards aggressive tendencies.

Functional properties of the aggressive brain

The psychologist Steven Pinker famously said "the mind is what the brain does" (2009). Neurons, the cellular members of the brain, fire in patterns that yield every thought, feeling, and action that humans experience. Thus aggression and the psychological processes that motivate it arise in the workings of the myriad convolutions of the brain. In this section, we discuss four brain regions that function to promote aggression, one region that inhibits it, and the serial process through which a provoking event can become translated into aggression.

Aggressive impulses arise from the function of many regions, yet the dACC appears to be their central hub. Nestled beneath the PFC, the dACC sits on the mid-line of the brain atop the corpus callosum. The dACC is a likely culprit behind angry responses to provocation and subsequent aggression, given its role as the brain's alarm system (Eisenberger and Lieberman, 2004). Alarms, from those that protect cars to the smoke detectors in your house, serve to (1) detect whether the current situation does not meet a given standard (e.g. the door is opening without being unlocked), and (2) produce a signal that alerts the necessary parties to this mismatch (e.g. a blaring noise).

To understand why the dACC contributes to aggression, one must first understand the situations in which aggressive urges arise. Provocation, and the anger that follows from it, is the most robust and reliable cause of aggression (Anderson and Bushman, 2002). Seminal neuroimaging research on aggression expected that the dACC would serve just such an "alarm" role during provocation, detecting a deviation from the goal-state of agreeable social interactions and eliciting negative emotions such as anger. Confirming this expectation, an experimental induction of provocation, administered by an experimenter who angrily reprimanded each participant, was associated with greater activity in the dACC (Denson *et al.*, 2009). The response of the dACC during provocation then corresponded to greater levels of felt anger after the event, and was greater for people who tend to act aggressively on a day-to-day basis (Denson *et al.*, 2009).

The neural correlates of provocation give us great insight into this precursor of aggression, yet they fail to tell us the brain functions *while* it is being aggressive. As an initial attempt at doing so, a group of neuroscientists placed individuals in an fMRI scanner, who then competed against an opponent (Krämer *et al.*, 2007). The competition was motivated by uncomfortable noise blasts that were delivered when participants lost against their opponent. Crucially, participants were given control over how loud the noise blast was for their partner, and the times during which participants decided on the level of noise blast to administer were conceptualized as acts of aggression. During trials in which participants chose a relatively loud volume of noise to blast their opponent with, greater activity was observed in both the dACC and a portion of the brain's reward system, the striatum. Thus it appears that the brain's alarm response occurs not only during anger and provocation, but in instances of aggression as well, further reifying the notion of the dACC as a crucial hub for aggression.

The finding relating activity in the brain's pleasure center, the striatum, with aggressive behavior suggests that aggression is a rewarding behavior that may reinforce individuals for their behavior (Krämer *et al.*, 2007). Supporting this notion, neuroscientists researching human decision-making strategies demonstrated that punishing others (a form of aggression) was associated with activation in the striatum (de Quervain *et al.*, 2004; Singer *et al.* 2006). Initial data from our laboratory have shown that activation in the striatum during aggression is associated with a greater magnitude of aggressive behavior (Chester and DeWall, 2014), further suggesting that aggression is motivated not only by the brain's alarm system, but by pleasure as well.

But not all aggression occurs in immediate response to provocation. Indeed, repeatedly reflecting on an instance of provocation, a process referred to as *angry rumination*, recruits memory-related structures and the MPFC (Denson *et al.*, 2009). The MPFC is attuned to situations in which individuals reflect inwards towards themselves, particularly when they reflect on their own emotions (MacRae *et al.*, 2004). Activity in this region was associated with greater dispositional tendencies to take out one's anger on innocent parties, referred to as *displaced aggression* (Denson *et al.*, 2009). These findings suggest that people who maintain their focus on their anger long after the provocation is over take that anger with them. This maintained anger then translates to aggression towards the innocent bystanders who these people happen across.

Provocations, such as insults, are not the only social experiences that lead to greater aggression. Aggressive tendencies are also greatly magnified by instances of social exclusion (Twenge *et al.*, 2001). The dACC is reliably associated with instances of exclusion (Eisenberger *et al.*, 2003), suggesting that this region may serve as the neural mechanism behind the exclusion–aggression link. As a preliminary investigation towards this realm, dACC activation during an instance of exclusion was associated with greater dispositional levels of aggression (Eisenberger *et al.*, 2007).

In a recent study from our laboratory, participants completed a virtual ball-tossing task in the MRI scanner with two fictitious partners. For the first several minutes of the task, participants received the ball an equal number of times, but for the last segment of the task, participants' partners tossed the ball back and forth, ignoring the participant. We replicated the typical dACC activation during exclusion (Chester *et al.*, 2014). To assess whether this dACC activation motivated subsequent aggression, participants were removed from the scanner and given an opportunity to blast one of their rejecters with an uncomfortable noise at a volume and duration of their choosing. The intensity of dACC activation that participants exerted during exclusion predicted greater aggression towards the rejecters, though only among individuals who have a general inability to inhibit their impulses. Our results show support for the crucial notion of the dACC as a neural mechanism that promotes aggression across aversive social experiences. However, this effect may only hold for individuals who cannot regulate or inhibit their impulses.

The amygdala sits at the frontal terminus of the hippocampus, deep underneath the cortex, and broadly responds to emotionally and otherwise salient features of the environment, particularly to primitive threats such as snakes and angry faces (LeDoux, 2000). The notion that the amygdala played a role in aggression stems back to the early twentieth century, when patients with bilateral amygdala lesions showed enhanced docility, an inability to engage in aggression (Kluver and Bucy, 1939). More recent research shows that aggressive individuals (i.e. those diagnosed with intermittent explosive disorder) show an exaggerated amygdala response to angry faces (Coccaro *et al.*, 2007). Combined with similar findings from animal lesion studies, the amygdala appears to promote defensive-aggression in response to evolutionarily relevant threats.

Having explicated the neural regions that function to promote aggression, neuroscience's task is now to identify the neural regions that inhibit aggression. In general, successful self-regulation is seated in the prefrontal cortex (PFC; Heatherton and Wagner, 2011). The orbitofrontal cortex (OFC) lies at the bottom-most aspect of the PFC, which assists individuals with successfully regulating their behavior so as to achieve social inclusion (Beer *et al.*, 2003, 2006). Supporting this role, OFC activity has also been shown to be associated with reduced aggression (Mehta and Beer, 2009). Imagining aggressive acts tends to reduce activation in the OFC (Pietrini *et al.*, 2000). The OFC is thus the brake that inhibits aggressive behavior.

Aggression initially arises due to certain person and situation inputs (GAM; Anderson and Bushman, 2002). These situational inputs are transduced through sensory organs, processed in sensory cortices, integrated in association cortices, colored with memories in the temporal lobes (Potegal and Stemmler, 2010), and then enter the aggression matrix: the dACC, MPFC, amygdala, and striatum. If the stimulus recruits the dACC, amygdala, or striatum, individuals become motivated to aggress. If the stimulus has come and gone, the MPFC may foster rumination over the event and motivate aggression towards innocent bystanders. Finally, these aggressive signals feed forwards to the OFC, which inhibits any aggression that is deemed inappropriate through reciprocal, inhibitory connections. Person inputs, in the form of personality, genetic dispositions, and somatic variations, modulate this process at every step. Neuroscience has allowed us this dynamic view of aggressive brain function.

Neurobiological interventions to reduce aggression

The neurobiological perspective can inform interventions aimed at reducing aggression. Armed with the facts regarding neural structure, development, and functional processes, researchers can use neuroscientific techniques to alter aggressive reactions. Here, we briefly discuss two novel interventions.

The first intervention manipulated levels of the neurotransmitter serotonin (Berman *et al.*, 2009). The study recruited individuals with a history of violence and matched, non-violent control individuals. By random assignment, all study participants received either 40 mg of paroxetine (a drug that increases serotonin levels) or a placebo. Next, participants completed a laboratory measure of aggression, in which they could administer severe electric shocks to a fictitious opponent. Because lower serotonin activity is associated with aggression (e.g. Berman *et al.*, 1997), the main prediction was that increasing serotonin levels would decrease aggression, especially among individuals with a history of violence. As predicted, boosting serotonin levels reduced aggression, and this effect was strongest among individuals with a history of violence.

The second intervention manipulated activation of a brain region hypothesized to play a role in inhibiting aggressive urges (Riva *et al.*, 2014). This study used a technique called transcranial direction current stimulation (tDCS), which involves electrically stimulating target brain regions that increase the excitability of cortical neurons. Participants were exposed to a social rejection manipulation, in which some participants experienced social inclusion in a virtual ball-tossing game, whereas others experienced social rejection. During the ball-tossing game, participants had two electrodes placed on their scalp. By random assignment, half of the participants experienced stimulation to the right ventrolateral prefrontal cortex (RVLPFC), which is a brain region that aids in the down-regulation of distressing events (Eisenberger *et al.*, 2003). The other half of the participants experienced a sham stimulation. For these participants, the tDCS machine turned on briefly at the beginning and end of the ball-tossing game. After completing the ball-tossing game, participants completed the hot sauce measure of aggression, in which they could dole out hot sauce to another person who disliked hot and spicy foods (Lieberman *et al.*, 1999). As expected, social rejection increased aggression, but this relationship was found only among sham stimulation participants. Stimulating the RVLPFC eliminated the link between social rejection and aggression.

These findings offer a glimpse into how the neurobiological perspective can inform aggression interventions. Aggression interventions often struggle to achieve widespread success because aggression has multiple causes (Bushman and Huesmann, 2010). By manipulating the specific neural mechanisms involved in aggression, researchers may have a more direct means of reducing aggression.

Conclusion

This chapter shows how the neurobiological perspective plays a vital role in aggression theory, research, and prevention. Whereas neurobiological processes did not play a pivotal role in classic theories of aggression, the advent of neuroscientific methods have made it possible for researchers to use brain structure and function as a fundamental component for understanding aggression. But neuroscientific approaches to the study of aggression are still in relative infancy. Below, we offer some recommendations for how aggression researchers may utilize the neuroscientific approach, and propose some future directions.

Our first recommendation is to design studies that test competing neural mechanisms underlying a similar behavioral response. With the increasing ease of conducting neuroimaging research, there is little need for additional studies that simply paint a portrait of a known phenomenon. Neuroimaging and other neuroscientific methods provide a unique ability to tease apart competing predictions that may be difficult to examine outside of a scanning environment. For example, some of our recent research tested the hypothesis that reminders of money owed would increase aggression (DeWall *et al.*, 2014). Two hypotheses predicted the same behavioral outcome (that money reminders would increase aggression), but they predicted the behavior would result from a different neural mechanism. fMRI provided an ideal method for testing these competing hypotheses because it could compare activation in brain areas thought to underlie the two accounts.

A second recommendation is for researchers to begin conducting behavioral aggression research while measuring brain activity. By measuring behavioral aggression within a scanning environment, compared with a standard behavioral laboratory, researchers will be able to provide a more nuanced understanding of aggression. Instead of merely knowing the severity of an aggressive act, researchers will be able to assess the interconnected neural activations that occur before, during, and after participants engage in an aggressive action. For example, researchers will be able to test competing hypotheses regarding whether provocation increases aggression because it produces distress (as indicated by activation in regions such as the dACC), or because the aggressor expects to feel better by behaving aggressively (as indicated by activation in regions associated with reward, such as the nucleus accumbens).

One avenue for future research may involve combining neuroscientific methods in the study of aggression. It has become increasingly common to use electroencephalography and fMRI simultaneously, which provides researchers the ability to correlate electrical brain activity with the hemodynamic responses that occur during the electrical activity. Within the context of aggression, researchers may expose individuals to provocative situations known to induce anger and aggression. Next, researchers can correlate the hypothesized electrical activity (e.g. greater left frontal activation) with hemodynamic responses (e.g. greater activation in the dACC). Finally, researchers can examine how the EEG–fMRI correlations differ among people according to their disposition to behave aggressively (e.g. trait physical aggressiveness, psychopathy, history of violence).

Researchers may also use drugs that can alter neural responses in aggressive situations. Our preliminary research suggests that provocation-induced aggression relates most closely to activation in the brain's pleasure centers, such as the nucleus accumbens (Chester and DeWall, 2014). The implication is that people behave aggressively because they value the intrinsic reward that will accompany the aggressive act over the potential risk. Using antagonistic drugs that dampen activation in these neural regions may offer one method for reducing aggression.

This chapter only begins to scratch the surface of the exciting future of neurobiological aggression research. We hope that aggression researchers and criminologists will begin to include neurobiological approaches to their theories and research. By looking under the hood, researchers will be better able to understand why people behave aggressively, and how to prevent it.

References

Anderson, C. A. and Bushman, B. J. (2002) 'Human aggression', *Annual Review of Psychology*, 53(1): 27–51.

Anderson, C. A., Buckley, K. E., and Carnagey, N. L. (2008) 'Creating your own hostile environment: a laboratory examination of trait aggression and the violence escalation cycle', *Personality and Social Psychology Bulletin*, 34: 462–473.

Archer, J. (2004) 'Sex differences in aggression in real-world settings: a meta-analytic review', *Review of General Psychology*, 8: 291–322.

Bandura, A. (1977). *Social Learning Theory*, New York: Prentice Hall.

Bandura, A., Ross, D., and Ross, S. A. (1961) 'Transmission of aggression through imitation of aggressive models', *Journal of Abnormal and Social Psychology*, 63: 575–582.

Barnea-Goraly, N., Menon, V., Eckert, M., Tamm, L., Bammer, R., Karchemskiy, A., *et al.* (2005) 'White matter development during childhood and adolescence: a cross-sectional diffusion tensor imaging study', *Cerebral Cortex*, 15(12): 1848–1854.

Beer, J. S., Heerey, E. A., Keltner, D., Scabini, D., and Knight, R. T. (2003) 'The regulatory function of self-conscious emotion: insights from patients with orbitofrontal damage', *Journal of Personality and Social Psychology*, 85(4): 594–604.

Beer, J. S., John, O. P., Scabini, D., and Knight, R. T. (2006) 'Orbitofrontal cortex and social behavior: integrating self-monitoring and emotion–cognition interactions', *Journal of Cognitive Neuroscience*, 18(6): 871–879.

Berkowitz, L. (1989) 'Frustration–aggression hypothesis: examination and reformulation', *Psychological Bulletin*, 106: 59–73.

Berkowitz, L. (1990) 'On the formation and regulation of anger and aggression: a cognitive–neoassociationistic analysis', *American Psychologist*, 45: 494–503.

Berman, M. E., Tracy, J. I., and Coccaro, E. F. (1997) 'The serotonin hypothesis of aggression revisited', *Clinical Psychology Review*, 17: 651–665.

Berman, M. E., McCloskey, M. S., Fanning, J. R., Schumacher, J. A., and Coccaro, E. F. (2009) 'Serotonin augmentation reduces response to attack in aggressive individuals', *Psychological Science*, 20: 714–720.

Blair, R. J. and Cipolotti, L. (2000) 'Impaired social response reversal: a case of "acquired sociopathy"', *Brain*, 123: 122–141.

Brown, B. B. (2004) 'Adolescents' relationships with peers', in R. M. Lerner and L. Steinberg (eds), *Handbook of Adolescent Psychology*, 2nd edn, Hoboken, NJ: John Wiley and Sons, pp. 363–394.

Bushman, B. J. and Huesmann, L. R. (2010) 'Aggression', in S. T. Fiske, D. T. Gilbert, and G. Lindzey (eds), *Handbook of Social Psychology*, Vol 2 (5th edn), Hoboken, NJ: John Wiley and Sons, pp. 833–863.

Casey, B., Jones, R. M., and Somerville, L. H. (2011) 'Braking and accelerating of the adolescent brain', *Journal of Research on Adolescence*, 21(1): 21–33.

Caspi, A., McClay, J., Moffitt, T. E., Mill, J., Martin, J., Craig, I. W., Taylor, A., and Poulton, R. (2002) 'Role of genotype in the cycle of violence in maltreated children', *Science*, 297: 851–854.

Chester, D. S. and DeWall, C. N. (2014) 'Aggression as emotion regulation: competition between neural signatures of anger and reward', unpublished manuscript.

Chester, D. S., Eisenberger, N. I., Pond, R. S., Richman, S. B., Bushman, B. J., and DeWall, C. N. (2014) 'The interactive effect of social pain and executive functioning on aggression: an fMRI experiment', *Social Cognitive and Affective Neuroscience*, 9(5): 699–704.

Coccaro, E. F., McCloskey, M. S., Fitzgerald, D. A., and Phan, K. L. (2007) 'Amygdala and orbitofrontal reactivity to social threat in individuals with impulsive aggression', *Biological Psychiatry*, 62(2): 168–178.

Craig, M. C., Catani, M., Deeley, Q., Latham, R., Daly, E., Kanaan, R., Picchioni, M., McGuire, P. K., Fahy, T., and Murphy, D. G. M. (2009) 'Altered connections on the road to psychopathy', *Molecular Psychiatry*, 14: 946–953.

Daly, M. and Wilson, M. (1990) 'Killing the competition', *Human Nature*, 1(1): 81–107.

Decety, J., Michalska, K. J., Akitsuki, Y., and Lahey, B. B. (2009) 'Atypical empathic responses in adolescents with aggressive conduct disorder: a functional MRI investigation', *Biological Psychology*, 80(2): 203–211.

Denson, T. F., Pedersen, W. C., Ronquillo, J., and Nandy, A. S. (2009) 'The angry brain: neural correlates of anger, angry rumination, and aggressive personality', *Journal of Cognitive Neuroscience*, 21(4): 734–744.

DeWall, C. N., Baumeister, R. F., Stillman, T. F., and Gailliot, M. T. (2007) 'Violence restrained: effects of self-regulation and its depletion on aggression', *Journal of Experimental Social Psychology*, 43: 62–76.

DeWall, C. N., Anderson, C. A., and Bushman, B. J. (2011a) 'The general aggression model: theoretical extensions to violence', *Psychology of Violence*, 1: 245–258.

DeWall, C. N., Finkel, E. J., and Denson, T. F. (2011b) 'Self-control inhibits aggression', *Social and Personality Psychology Compass*, 5(7): 458–472.

DeWall, C. N., Chester, D. S., and Vohs, K. D. (2014) 'Money cues attenuate striatal and fusiform responses to anticipated social interaction: implications for aggression', unpublished manuscript.

Dollard, J., Doob, L., Miller, N., Mowrer, O., and Sears, R. (1939) *Frustration and Aggression*, New Haven, CT: Yale University Press.

Dunbar, R. I. M. (2009) 'The social brain hypothesis and its implications for social evolution', *Annals of Human Biology*, 36(5): 562–572.

Eisenberger, N. and Lieberman, M. (2004) 'Why rejection hurts: a common neural alarm system for physical and social pain', *Trends in Cognitive Sciences*, 8(7): 294–300.

Eisenberger, N. I., Lieberman, M. D., and Williams, K. D. (2003) 'Does rejection hurt? An fMRI study of social exclusion', *Science*, 302(5643): 290–292.

Eisenberger, N. I., Way, B. M., Taylor, S. E., Welch, W. T., and Lieberman, M. D. (2007) 'Understanding genetic risk for aggression: clues from the brain's response to social exclusion', *Biological Psychiatry*, 61(9): 1100–1108.

Finkel, E. J. (in press) 'The I^3 model: metatheory, theory, and evidence', In J. M. Olson and M. P. Zanna (eds), *Advances in Experimental Social Psychology*, 49, San Diego: Academic Press.

Giedd, J. N. (2004) 'Structural magnetic resonance imaging of the adolescent brain', *Annals of the New York Academy of Sciences*, 1021: 77–85.

Gottfredson, M. R. and Hirschi, T. (1990) *A General Theory of Crime*, Stanford, CA: Stanford University Press.

Heatherton, T. F. and Wagner, D. D. (2011) 'Cognitive neuroscience of self-regulation failure', *Trends in Cognitive Sciences*, 15(3): 132–139.

Kluver, H. and Bucy, P. C. (1939) 'Preliminary analysis of functions of the temporal lobes in monkeys', *Archives of Neurology and Psychiatry*, 42(6): 979–1000.

Krämer, U., Jansma, H., Tempelmann, C., and Münte, T. (2007) 'Tit-for-tat: the neural basis of reactive aggression', *NeuroImage*, 38(1): 203–211.

LeDoux, J. E. (2000) 'Emotion circuits in the brain', *Annual Review of Neuroscience*, 23(1): 155–184.

Lieberman, J. D., Solomon, S., Greenberg, J., and McGregor, H. A. (1999) 'A hot new way to measure aggression: hot sauce allocation', *Aggressive Behavior*, 25: 331–348.

Macrae, C. N., Moran, J. M., Heatherton, T. F., Banfield, J. F., and Kelley, W. M. (2004) 'Medial prefrontal activity predicts memory for self', *Cerebral Cortex*, 14(6): 647–654.

Mathews, V. P., Kronenberger, W. G., Wang, Y., Lurito, J. T., Lowe, M. J., and Dunn, D. W. (2005) 'Media violence exposure and frontal lobe activation measured by functional magnetic resonance imaging in aggressive and nonaggressive adolescents', *Journal of Computer Assisted Tomography*, 29(3): 287–292.

Matthies, S., Rüsch, N., Weber, M., Lieb, K., Philipsen, A., Tuescher, O., Ebert, D., Hennig, J., and van Elst, L. T. (2012) 'Small amygdala—high aggression? The role of the amygdala in modulating aggression in healthy subjects', *World Journal of Biological Psychiatry*, 13: 75–81.

Mehta, P. H. and Beer, J. (2009) 'Neural mechanisms of the testosterone–aggression relation: the role of orbitofrontal cortex', *Journal of Cognitive Neuroscience*, 22(10): 2357–2368.

Pardini, D. A., Raine, A., Erickson, K., and Loeber, R. (2014) 'Lower amygdala volume in men is associated with childhood aggression, early psychopathic traits, and future violence', *Biological Psychiatry*, 75: 73–80.

Peterson, C. K., Shackman, A. J., and Harmon-Jones, E. (2008) 'The role of asymmetrical frontal cortical activity in aggression', *Psychophysiology*, 45(1): 86–92.

Pietrini, P., Guazzelli, M., Basso, G., Jaffe, K., and Grafman, J. (2000) 'Neural correlates of imaginal aggressive behavior assessed by positron emission tomography in healthy subjects', *American Journal of Psychiatry*, 157(11): 1772–1781.

Pinker, S. (2009) *How the Mind Works*, New York: W. W. Norton.

Potegal, M. and Stemmler, G. (2010) 'Constructing a neurology of anger', in M. Potegal, G. Stemmler, and C. Spielberger (eds), *International Handbook of Anger*, New York: Springer, pp. 39–59.

de Quervain, D. J.-F., Fischbacher, U., Treyer, V., Schellhammer, M., Schnyder, U., Buck, A., and Fehr, E. (2004) 'The neural basis of altruistic punishment', *Science*, 305(5688): 1254–1258.

Raine, A. (2002) 'Annotation: the role of prefrontal deficits, low autonomic arousal, and early health factors in the development of antisocial and aggressive behavior in children', *Journal of Child Psychology and Psychiatry*, 43(4): 417–434.

Raine, A., Buchsbaum, M., and LaCasse, L. (1997) 'Brain abnormalities in murderers indicated by positron emission tomography', *Biological Psychiatry*, 42: 495–508.

Raine, A., Lencz, T., Bihrle, S., LaCasse, L., and Colletti, P. (2000) 'Reduced prefrontal gray matter volume and reduced autonomic activity in antisocial personality disorder', *Archives of General Psychiatry*, 57: 119–127.

Riva, P., Lauro, L. J. R., DeWall, C. N., Chester, D. S., and Bushman, B. J. (2014) 'Reducing aggressive responses to social exclusion using transcranial direct current stimulation (tDCS)', *Social Cognitive and Affective Neuroscience*, doi: 10.1093/scan/nsu053.

Schiltz, K., Witzel, J. G., Bausch-Hölterhoff, J., and Bogerts, B. (2013) 'High prevalence of brain pathology in violent prisoners: a qualitative CT and MRI study', *European Archives of Psychiatry and Clinical Neuroscience*, 263: 607–616.

Singer, T. (2006) 'The neuronal basis and ontogeny of empathy and mind reading: review of literature and implications for future research', *Neuroscience and Biobehavioral Reviews*, 30(6): 855–863.

Singer, T., Seymour, B., O'Doherty, J., Kaube, H., Dolan, R. J., and Frith, C. D. (2004) 'Empathy for pain involves the affective but not sensory components of pain', *Science*, 303(5661): 1157–1162.

Singer, T., Seymour, B., O'Doherty, J. P., Stephan, K. E., Dolan, R. J., and Frith, C. D. (2006) 'Empathic neural responses are modulated by the perceived fairness of others', *Nature*, 439(7075): 466–469.

Somerville, L. H., Jones, R. M., Ruberry, E. J., Dyke, J. P., Glover, G., and Casey, B. J. (2013) 'The medial prefrontal cortex and the emergence of self-conscious emotion in adolescence', *Psychological Science*, 24(8): 1554–1562.

Twenge, J. M., Baumeister, R. F., Tice, D. M., and Stucke, T. S. (2001) 'If you can't join them, beat them: effects of social exclusion on aggressive behavior', *Journal of Personality and Social Psychology*, 81(6): 1058–1069.

Yang, Y., Raine, A., Narr, K. L., Colletti, P., and Toga, A. W. (2009) 'Localization and deformations within the amygdala in individuals with psychopathy', *Archives of General Psychiatry*, 66: 986–994.

Zillmann, D. (1979) *Hostility and Aggression*, Hillsdale, NJ: Erlbaum.

Zillmann, D. (1988) 'Cognitive–excitation interdependencies in aggressive behavior', *Aggressive Behavior*, 14: 51–64.

18

Why focusing on nurture made and still makes sense

The biosocial development of self-control

Alexander T. Vazsonyi, Joshua W. Roberts, and Li Huang

Introduction

In this chapter we examine both theoretical and subsequent empirical work on how self-control develops in children based on Gottfredson and Hirschi's (1990) *General Theory of Crime*. Few longitudinal studies (e.g. Polakowski, 1994; Vazsonyi and Huang, 2010; for cross-sectional tests cf. Hay, 2001; Perrone *et al.*, 2004) have addressed perhaps one of the most pivotal theoretical arguments, and one of self-control theory's basic tenets, namely that early socialization processes during the first decade of a child's life effectively send a child on a life-long self-control trajectory, one highly predictive of norm violations, of deviance, and of criminal conduct. Consider the following analogy.

> An adolescent who is an incredible baseball pitcher (3 or more standard deviations above a population mean in terms of speed, accuracy and so forth) will be situated in a similar position relative to his peers at 20, 30, 40, and age 50, following a natural trajectory of improved pitching prowess during late adolescents into young adulthood, followed by an inevitable decline. The same logic applies to deviance and crime, considering low self-control. It is the inevitable decline that puzzles social and behavioral scientists, not only in crime and deviance, but a whole host of other observable behaviors (e.g. Kanazawa, 2003; "genius curve"). So, perhaps we should be most interested in developments during the first decade of life and less so in the immutable, universal, maturational growth and decline that follows.
>
> *(Vazsonyi and Huang, 2010: 255)*

Despite a large number of empirical tests focused on self-control theory, few have tackled this issue and the central question of how self-control develops, not the exclusive source of variability in deviance, but a central, probabilistic one. Most related work in criminology is focused on the stability of self-control (the stability thesis); this addresses only a very modest piece of the puzzle related to the development of self-control. Some of this (construct) stability work found evidence supporting theoretical predictions about rank order changes over time (Arneklev *et al.*,

1998; Turner and Piquero, 2002; Hay and Forrest, 2006), while others did not (Burt *et al.*, 2006; Mitchell and MacKenzie, 2006; Winfree *et al.*, 2006).

On the other hand, some researchers in the developmental sciences have focused for some time on how self-control/self-regulation develops in infants and young children (Kopp, 1982); this includes an understanding that this process was reciprocal, although most of this work was conducted recently (Kochanska *et al.*, 2001, 2007; Rothbart *et al.*, 2003; Raffaelli *et al.*, 2005; Colman *et al.*, 2006). Other developmental work has also focused on the association between self-control or self-regulation and deviance or externalizing behaviors, inspired by the work of Mischel and colleagues (1989; Eisenberg *et al.*, 2005; Zhou *et al.*, 2007; see also Brody and Ge, 2001; Crockett *et al.*, 2006). For instance, Eisenberg and colleagues (2005) also identified how socialization experiences from caregivers continue to contribute to developmental changes of self-control or self-regulation during childhood and adolescence. They recognized some time ago that different terminology and at times operationalizations of self-control and self-regulation across different disciplines or subdisciplines within social and behavioral sciences have been used, thus further contributing to how much or little scholarship in this area, including on self-control theory, has progressed (Eisenberg *et al.*, 2004).

In this chapter we examine the extent to which Gottfredson and Hirschi's self-control theory was mis-specified, as some have recently argued, in light of recent scholarship focused on biological origins of self-control, of deviance, antisocial or criminal behaviors—some based on the behavior genetic method, some focused on allele frequency or genetic markers, and some focused on gene × environment interactions. For instance, Wright and Beaver (2005) suggest Gottfredson and Hirschi (1990: 60) explicitly discount the possibility that low self-control might have a genetic component: "the magnitude of the 'genetic effect'" they say [omitted by authors: as determined by adoption studies], "is near zero." A large body of literature, however, has arrived at a very different conclusion.

The statement by Wright and Beaver is persuasive; however, Gottfredson and Hirschi (1990: 60) in fact argue the opposite, on the very same page, commenting most generally on biological positivism, on searches for physiognomy in criminals, on chromosal markers and so forth, "This result should not be surprising, *and it should not be interpreted as that biology has nothing to do with crime*" [emphasis added]. Recognizing and understanding the behavior genetic method prior to the publication of their seminal book, almost a quarter century ago, and undoubtedly due to regular collegial exchanges with David C. Rowe at the University of Arizona at the time, they also comment on the following page that:

> These correlation coefficients [path model decomposition of a genetic effect on crime shown on page 61] translate into a heritability coefficient of 0.177, which may be described as the theoretical upper limit of the correlation between father's criminality and son's criminality given our assumptions. We would not argue that this result proves the irrelevance of biology to crime. We would argue, however, that this shows the need for greater attention to conceptual matters by those interested in biology and crime.
>
> *(Gottfredson and Hirschi, 1990: 61)*

These comments appear to be mostly focused on conceptual issues, on positivism in general, which is characterized by a data-driven, reductionist approach that often lacks an idea, a conceptual framework, or even a theory. As is often the case with controversial conceptual or theoretical work—work that apparently has inspired a generation of criminologists, judging by the sheer volume of scholarship it has generated, work that challenges accepted or agreed-upon notions of how crime and deviance develop, but also work that is simply profound and parsimonious, as is the case with self-control theory—it precipitates heavy criticism.

It is clear that Gottfredson and Hirschi favored social explanations in how self-control develops, but very critically, they repeatedly acknowledged and discussed the salience of individual differences, the importance of biology, and of socialization effects. Thus the current chapter is guided by the simple question: to what extent did Gottfredson and Hirschi (1990) "have it right" a quarter century ago? Does the *General Theory* leave out, or even ignore, biology in the development of self-control and in the understanding of deviance and crime, as some have levied as a criticism against the theory? How does self-control develop, using a biosocial lens? We attempt to address these questions by providing a larger framework for the development of self-control, and also by providing a simple empirical test at the end of the chapter. This is not meant to fully address our original questions, but is simply an illustration of the importance of a reciprocal socialization process known to underlie the development of self-control and the associated likelihood of engaging in crime and deviance—an illustration (not in a genetically informed design) of reciprocity in socialization, evidence for both individual differences as well as socialization effects in the development of self-control and deviance.

On the development of self-control and the importance of biology

Few would question the impact Gottfredson and Hirschi's (1990) self-control theory has had on our understanding of crime and deviance within criminology; this is also true outside of criminology, sometimes explicitly linked to self-control theory and sometimes not, in psychology (e.g. Baumeister *et al.*, 1994; Tangney *et al.*, 2004); the developmental sciences (Eisenberg *et al.*, 2005); educational sciences (Duckworth and Seligman, 2005); or health and health-risk research (e.g. Griffin *et al.*, 2011; Miller *et al.*, 2011). Given the temporal order, it seems likely that Gottfredson and Hirschi's *General Theory* might have been an important source, something DeLisi (2013) so eloquently identifies as a "transdisciplinary" phenomenon in his recent chapter entitled "Pandora's box." The extent to which this phenomenon is truly transdisciplinary, versus simply a phenomenon appearing across a number of subdisciplines within social and behavioral sciences, is open for discussion. Gottfredson so simply, but poignantly noted:

> If theories may be judged by how much research they stimulate, control theory is doing exceptionally well. If theories may be judged by their consistency with the facts, control theory is doing exceptionally well. And if theories may be judged by the frequency with which other perspectives seek to incorporate them, control theory is, perhaps, without peer.
>
> *(Gottfredson, 2006: 96)*

Despite its success in generating a tremendous amount of scholarship, the theory has not been without heavy criticisms and has generated a tremendous amount of debate (Geis, 2000; Goode, 2008). Among others, important criticisms have been raised about the paucity of work that examines the etiology of self-control, and the extent to which self-control theory incorporates biological mechanisms or explanations:

> . . . the paucity of research examining the factors that give rise to low self-control is somewhat surprising. We have, for instance, been able to locate only a handful of empirical

studies that test them. The findings, in general, have been favorable to Gottfredson and Hirschi's position that effective child-rearing practices are predictive of self-control in children [but see Cochran *et al.*, 1998].

(Wright and Beaver, 2005: 1170)

Theoretically, Gottfredson and Hirschi identify primary socialization effects in the home during the first decade of life, by parents or caregivers, as key in the development of self-control, although they also propose secondary socialization agents, such as teachers at schools, or schools themselves. Findings on the importance of parenting processes have not been extensive, and are often based on cross-sectional work, but have provided support (Feldman and Weinberger, 1994; Polakowski, 1994; Gibbs *et al.*, 1998; Hay, 2001; Vazsonyi and Huang, 2010). This also includes work done to test the salience of "secondary" socialization contexts, such as the school (Turner *et al.*, 2005). However, recently some scholars have called into question how much parents really matter in this process (e.g. Wright and Beaver, 2005; Beaver *et al.*, 2007), with a twin criticism about self-control theory and the extent to which it is valid. In fact, Wright and Beaver conclude:

Gottfredson and Hirschi openly exclude the possibility that self-control has a genetic base. Our study, along with others from various fields, suggests that for self-control theory to be a valid theory of crime it must incorporate a more sophisticated understanding of the origins of self-control [see Pratt *et al.*, 2002, 2004].

(Wright and Beaver, 2005: 1190)

Similarly, citing the same work, Cullen *et al.* note:

Still, Wright and Beaver's research challenges the parental management thesis in suggesting that the major parental sources of self-control are likely genetic and thus cannot be traced to the style of child-rearing techniques used in a household.

(Cullen et al., 2008: 72)

There appear to be at least two misunderstandings, related to theoretical interpretation as well as to the extent to which Wright and Beaver's important work did in fact show no non-heritable effects. Together with the previous comment, Gottfredson and Hirschi in fact appear to simply deny a denial of biology:

Obviously, we do not suggest that people are born criminals, inherit a gene for criminality, or anything of the sort. In fact, we explicitly deny such notions (see Chapter 3). What we do suggest is that individual differences may have an impact on the prospects of effective socialization (or adequate control). Effective socialization is, however, always possible whatever the configuration of individual.

(Gottfredson and Hirschi, 1990: 96)

Individual differences imply biological or genetic variability which they describe as primary and as existing prior to any socialization processes or effects children might experience:

Two general sources of variation are immediately apparent in this scheme. The first is the variation among children in the degree to which they manifest such traits to begin with [. . .] Obviously, therefore, even at this threshold level the sources of low self-control are complex.

(Gottfredson and Hirschi, 1990: 96)

Importantly, and perhaps consistent with theory, Wright and Beaver show in their work, and call attention to, the fact that between 55 and 66 percent of the variability in self-control is heritable, based on hierarchical linear modeling (HLM) analysis that takes genetic relatedness into account, and based on a genetically informed estimate of heritability based on the same data. Similarly, Beaver *et al.* (2013) also found evidence which suggests that much of the observed stability in self-control over time, as well as changes in self-control, are related to genetic factors (78 to 89 percent and 74 to 92 percent, respectively). It seems, as we have argued previously, that Gottfredson and Hirschi chose to focus on positive socialization effects because these represent perhaps the only fruitful avenue for meaningful change (perhaps the essence of applied empirical work in criminology), and research has largely supported this.

In a meta-analysis of 34 rigorous studies focused on testing the malleability of self-control during the first decade of children's lives, for instance, Piquero *et al.* (2009b) concluded that (1) studies were successful in improving self-control; (2) these interventions reduced deviance; and (3) positive effects were robust across a number of potential moderators, including data sources or raters, clinical versus non-clinical approaches, and so forth. Some larger intervention effects were found for girls in comparison with boys, which the authors interpreted as being theoretically consistent because girls are less delinquent, and because they are also more receptive to intervention efforts in comparison with boys. Calling for a revised criminal justice policy, based on their findings about the importance of self-control and its malleability, they conclude:

> Our findings show that interventions aimed at improving socialization and childrearing practices in the first decade of life [e.g. Piquero *et al.*, 2009a] offer benefits for improving self-control as well as reducing delinquency/crime. It appears that investment in these efforts—in lieu of the more cost-prohibitive incarceration policies of the recent past—should be an important part of the policy response, especially because self-control is malleable and responsive to external sources of socialization.
>
> *(Piquero et al., 2009b: 28)*

Is biology really more important?

Based on the pioneering work by Wright and Beaver (2005), Beaver and colleagues similarly concluded, as a basis for their study focusing on the extent to which neuropsychological motor skill function measures co-varied with self-control, that:

> Gottfredson and Hirschi (1990) have dismissed the possibility that self-control has a biological or genetic component (Wright and Beaver, 2005) and to take these findings into consideration, the explanation of self-control set forth by Gottfredson and Hirschi needs to be revised with an explicit focus on how the biological processes of the brain affect the development of self-control.
>
> *(Beaver et al., 2007: 1346)*

The study provided evidence that measures of fine and gross motor skills were negatively associated with measures of self-control, as were socialization variables, based on a random sample of 3,000 young children. This was found for both boys and girls part of the Early Childhood Longitudinal Study, Kindergarten Class of 1998–99 (ECLS-K), although parenting measures did not reach significance for girls when prior self-control was added to model tests in the prediction of first grade self-control (only a test of developmental changes in self-control from kindergarten to first grade). There is little doubt that the frontal cortex is very important in the process of decision making (e.g. Reyna *et al.*, 2011); we know from recent brain imaging work that these decision centers continue to mature into the early 20s, and that it would be important and interesting to integrate these insights into our understanding of how to explain crime and deviance, particularly among youth, and perhaps to elaborate on self-control theory.

However, the extent to which neurological deficits *per se* are generally important for crime and deviance, something on which Moffitt and colleagues also focused in the 1990s, or to what extent brain mapping provides novel insights into why people deviate from social norms, is less clear. Mapping the brain and understanding that it continues to develop past adolescence, understanding that it is involved and which regions of it are involved in decision making, are important. But what are we are looking for exactly? This search at times is reminiscent of when biology and biological explanations were considered to be more central in our understanding of crime and deviance; yet, whether at the molecular, genetic level or physiological processes observed in the brain, having an idea of what we are looking for would seem to be a hallmark of what is necessary to building knowledge, understanding, or a coherent and testable theory of crime and deviance. Consider the following comment by expert neuroscientists who study neurological control processes:

> In sum, control functions that are typically associated with control (e.g. inhibition, task setting, and updating) are defined largely based on behavioral observations [see Miyake *et al.*, 2000]. How these functions correspond to distinct indicators of brain activity will determine whether they are truly distinct componential entities or whether they emerge from the interactions of various systems in the brain and are therefore manifest only in the minds of cognitive scientists.
>
> *(Lenartowicz et al., 2010: 690)*

One of the studies cited by Beaver *et al.* (2007) as the basis and impetus for their work was research done by Cauffman *et al.* (2005), which juxtaposed self-control measures with physiological measures related to autonomic nervous system functioning (heart rate), as well as measures of intelligence, and neuropsychological functioning (spatial span, a measure of cerebral lateralization). In a comparison of incarcerated youth (two-thirds of whom had committed a violent crime) and a sample of high school students from the general population, they found that only a measure of self-control and parental education predicted antisocial behaviors in the high school sample. On the other hand, a measure of self-control, heart rate, and the measure of brain lateralization predicted antisocial behaviors among incarcerated youth. They conclude:

> In sum, while one component of self-control, future orientation, remains significant in this full model, it does not render insignificant the effects of neuropsychological functioning (in the form of spatial span scores) and biological factors (in the form of heart rate). Thus, while certain components of self-control are important in distinguishing the two groups, it does not eliminate other factors that Gottfredson and Hirschi would expect to be eliminated. These results, then, support the contentions of those scholars attributing importance to neuropsychological and biological factors in addition to self-control [see Moffitt, 1993; Rowe, 2002].
>
> *(Cauffman et al., 2005: 155)*

There is no question that this study provided an interesting insight into potential biologically linked processes in the explanation of variability in adolescent deviance and crime. At the same time, it is simply unclear the extent to which the findings provide new and coherent information that stands, in effect, to invalidate self-control theory, as some have suggested. It is also not clear whether the theory has any predictions related to heart rate or brain lateralization, as found in the latter study; these were not found to be important for antisocial behaviors among adolescents from the general population. It is perhaps self-evident that Gottfredson and Hirschi did not discuss issues related to heart rate or brain lateralization; but does this provide the basis for a revised or different theoretical explanation?

There is no doubt that the group of incarcerated youth, where two-thirds were in jail because they killed, assaulted, or raped another human being, might suffer from neurological and other deficits. These youth seem particularly troubled, perhaps suffering from organic deficits. Yet self-control was important in understanding their antisocial behaviors as theorized. This means they suffered from self-control deficits as well as other intellectual, emotional, and perhaps neurological ones. Gottfredson and Hirschi consider low self-control a *probabilistic individual characteristic important in understanding variability in crime and deviance.* They do not seem to eliminate biology, as pointed out, nor do they address or exclude serious intellectual impairments, perhaps organic in origin or any other individual characteristics. They note on the probabilistic nature of self-control:

> In our view, lack of self-control does not require crime and can be counteracted by situational conditions or other properties of the individual. At the same time, we suggest that high self-control effectively reduces the possibility of crime—that is, those possessing it will be substantially less likely at all periods of life to engage in criminal acts.
>
> *(Gottfredson and Hirschi, 1990: 89)*

They also comment specifically on other known correlates and predictors of crime, including "low intelligence, high activity level, physical strength, and adventuresomeness" (Gottfredson and Hirschi, 1990: 96), and how these are known to interfere with or support effective socialization of self-control, with the development of self-control.

Both biology and socialization are important

Beaver *et al.* (2013) showed that mostly genetic and non-genetic (shared and non-shared) effects are important in understanding the development of self-control and its stability over time. Similarly, Beaver *et al.* (2009) illustrate that, although a molecular marker had no direct effect on self-control, the interaction between the 5HTTLPR allele and time spent with delinquent synergistically explained variability in levels of self-control. DeLisi (2011) recently wrote a poignant editorial about the importance of biology given many of these reviewed insights, calling for more genetically or biologically informed scholarship focused on the etiology of crime and deviance. He notes:

> For decades, criminology viewed genetics as startling, unconventional, and embarrassing. Genetics was *l'enfant terrible*. Today, a criminology that ignores or disputes the salience of genes toward understanding, preventing, and treating antisocial behavior risks being embarrassed.
>
> *(DeLisi, 2013: 515)*

True, *l'enfant terrible*, for many reasons; but this is also simply related to recent advances in molecular biology, in neuroimaging, or in the application of advanced quantitative techniques that are based on very narrowly defined and not readily obtained twin and adoption studies (e.g. the Add Health, which was planned over 20 years ago). But is it really so bad? Given the explanation in the present chapter about the primacy and full acknowledgement of individual differences that are present at birth, according to Gottfredson and Hirschi, this might simply be a case of misattributing a purely environmental position to their work, to their theoretical work.

Few would argue today, or even 20 years ago in the 1990s, that socialization is in fact a one-way street; rather, we have known it is a two-way street due to individual, biologically based differences; it is a dynamic and complex process, much as detailed by Sandra Scarr in her 1992 SRCD presidential address, about genotype → environment effects, about behavior genetic evidence, about "good enough parenting," or about the salience of non-shared environmental effects on child and adolescent development. These are, nevertheless, environmental effects, or as Plomin and Thomson (1987) describe them, "micro-environments within families." It is certainly true that criminologists have eschewed even speaking of biology or biologically informed explanations, and yet they were there, even in the *General Theory of Crime*, including how Gottfredson and Hirschi approach the importance or unimportance of culture, of cross-cultural, comparative criminology (chapter 8). They argue that adopting environmental, at times idiosyncratic, explanations of crime, driven by positivist thinking, has led to a fragmented field that literally did not know what it was looking for. Rather, they call for a culture-free, universalist conceptualization of crime and delinquency, one that makes a number of implicit assumptions about human nature and human behavior, more consistent with a classical conception of crime, and one that transcends local mores, practices and environments (Vazsonyi *et al.*, 2001, 2004).

In a comprehensive review of behavior genetic work on antisocial behavior, Moffitt (2005) concludes that both genetic and environmental causes are equally important for antisocial outcomes. Much of her early developmental work on the etiology of antisocial or deviant behaviors focused on neurological deficits as well as a typology of adolescence-limited versus life-course persistent offenders, related to her interpretation of the age–crime distribution of crime. Consider the following explanation:

> Gottfredson and Hirschi's (1990) general theory suggests that a single intervention strategy aimed at increasing self-control would suffice to prevent antisocial behavior at any and all ages. In contrast, Moffitt's (1993) developmental theory implies that one set of intervention strategies is needed to prevent persistent antisocial behavior that begins in childhood, while another set of strategies is needed to prevent delinquency that emerges in adolescence. Strategies for preventing childhood-onset antisocial behavior would focus on the neuro-developmental impairments that inhibit the development of verbal abilities and self-control, while strategies preventing adolescent-onset behavior would target the social influences of delinquent peers on behavior.
>
> *(Bartusch et al., 1997: 42–43)*

However, more recent work appears to have focused exclusively on the importance of self-control, with a consideration of both the 1972 birth cohort project in Dunedin, New Zealand, as well as the E-Risk same-sex twin cohort in the United Kingdom, a behavior genetic study started in 1998. In perhaps one of the most profound and far-reaching recent statements written in the social scientific literature, Moffitt, Poulton, and Caspi conclude, based on data from both the Dunedin and the E-Risk behavior genetic project, that:

> The two cohorts we are studying—born in different countries and different decades—strongly support that improving individual self-control *will prove essential for humanity's long-term health, wealth, safety, and happiness*. [emphasis added]
>
> *(Moffitt et al., 2013: 359)*

They also call to action for policies and practices that target the development of self-control among children by the age of 12:

> Improving childhood self-control is especially appealing because we find that differences in self-control between children predict their adult outcomes approximately as well as (sometimes better than) low intelligence and low social class origins, factors known to be extremely difficult to improve through intervention.
>
> *(Moffitt et al., 2013: 357)*

In the final part of this chapter, we provide a simple illustration of the extent to which there appears to be an interactive and synergistic process that characterizes how parenting, self-control, and deviance are related over time in a sample of young children followed from age 4.5 to age 10.5 years.

The current study

Vazsonyi and Huang (2010) used the National Institute of Child Health and Human Development (NICHD) longitudinal study to examine both self-control and deviance developmental trajectories from ages 4.5 to 10.5. They used simple structural equation modeling (SEM) and latent growth modeling (LGM) of developmental trajectories and found great construct stability of self-control and deviance over the 6-year period. They found evidence that children's self-control increased over the 6-year period, and at the same time their deviance trajectories decreased over the same period. However, the longitudinal interrelationships among parenting, self-control, and deviance were not examined in their study. To date, few previous studies examined longitudinal bidirectional relationships for parenting, self-control, and deviance.

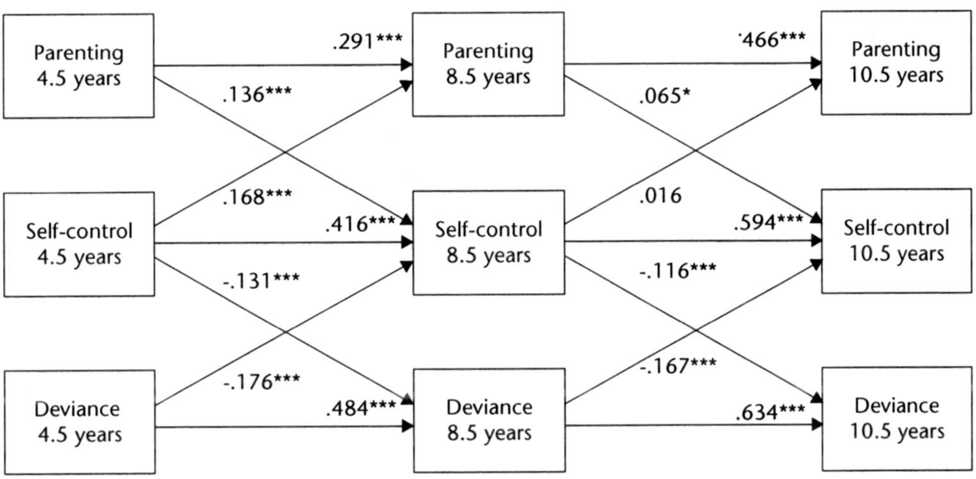

Figure 18.1 Path analysis parameter estimates

Note: $^{*}P < .05$.

We examined bidirectional relationships between parenting and self-control. By examining the bidirectional or reciprocal relationships between parenting and self-control at ages 4.5, 8.5, and 10.5, we were able to assess the degree to which differences in parenting affect differences in development of self-control at different ages (Figure 18.1). In addition to these bidirectional effects, we also tested how self-control at different ages affected children's deviant behaviors. Finally, we were also interested in sex and ethnic/race differences of bidirectional relationships among parenting, self-control, and deviance.

Methods

Participants

The data for this study were based on the NICHD Early Child Care Research Network Study of Early Child Care. Data collection commenced in 1991, when participating children were 1 month old, and has continued until middle adolescence; we focus on assessment completed at ages 4.5, 8.5, and 10.5 years. Over this period, research assistants from the 10 data-collection sites have seen each child at home, in child care, in school, and in a laboratory setting. Families lived in Little Rock, AR; Irvine, CA; Lawrence, KS; Boston, MA; Philadelphia, PA; Pittsburgh, PA; Charlottesville, VA; Morganton, NC; Seattle, WA; and Madison, WI. Of the n = 1,364 families that completed initial data collections at 1 month, complete parent and child data are available on n = 985 families, when children were in first grade. In comparing the n = 985 families with the n = 379 families where complete data were unavailable, participants' mothers were slightly older on average (28.6 versus 26.38 years), better educated (14.4 versus 13.7 years), and less likely to be of minority status (17 versus 27 percent). The final sample included n = 478 boys and n = 479 girls (51.1 percent); most children were European American n = 844 (82.3 percent), with African American n = 114 (11.1 percent), Hispanic n = 22 (1.6 percent), Native American n = 40 (4.7 percent), and Asian n = 5 (0.3 percent). Thus we coded European American as majority group; African American, Hispanic, Native American, and Asian as children part of the minority group.

Procedures

Information about parents and children was obtained from mothers using face-to-face interviews and observations of mother–child interactions. Telephone interviews were conducted numerous times, and classroom data were collected from teachers during the spring semester of first grade through sixth grade. The entire data-collection protocol was reviewed by a steering committee supervised by the NICHD, and was reviewed annually by institutional review boards of the 10 participating institutions responsible for data collection. We focused on three time points, namely when children were 4.5, 8.5, and 10.5 years old. This was largely a function of measurement issues at these time points; measurement of the key study constructs was consistent, an important prerequisite for longitudinal model tests.

Measure

Parenting

In the current study, though a number of different parenting instruments and measures are available, the Parent–Child Relationship Scale (Pianta, 1994) was selected, largely as a function of when they were employed (assessment times) and whether they were repeated subsequently, to facilitate longitudinal data analysis.

At 4.5 years old, mothers were asked to complete a 30-item questionnaire designed to assess the target child's attachment to the parent. The form was adapted from the Student–Teacher Relationship Scale (Pianta, 2001). Items were rated using a 5-point Likert scale. For children at 8.5 and 10.5 years old, parenting was measured by the 30-item Parent–Child Relationship Scale (Pianta, 1994). Initially these items were selected to measure three dimensions of warmth/security, anger/dependency, and anxiety/insecurity. Since third and fifth grade assessments only used 15-item short forms, a 15-item version was employed for the current study to maintain consistency in measurement over time. The 15 items included the parent's feelings and beliefs about his/her relationship with the child, as well as items about the child's behavior toward the parent. Based on conceptual reasons and an interest in the warmth/security dimension, as well as some data reduction analyses described subsequently, only eight items were used, each rated by mothers on a 5-point Likert scale ranging from 1 = definitely does not apply to 5 = definitely applies (e.g. "child spontaneously shares personal information with me"). The reliability estimates for the total Parent–Child Relationship Scale ranged from $\alpha = 0.81$ to $\alpha = 0.87$ in previous work (e.g. Bradley and Corwyn, 2005). The alpha level ranged from $\alpha = 0.63$ to $\alpha = 0.73$ for the current study across the three time points.

Self-control

Self-control was measured by the Social Skills Rating System (SSRS; Gresham and Elliott, 1990). Mothers completed the measure. The SSRS-Parent Form consists of three parts, namely social skills, problem behaviors, and academic competence scales. The social skills component includes three subscales: cooperation, assertion, and self-control. The current study focused only on the self-control subscale. It includes 10 items rated by mothers on a 3-point Likert-type scale (0 = never, 1 = sometimes, 2 = very often; e.g. "controls temper when arguing with other child"). The alpha ranged from $\alpha = 0.82$ to $\alpha = 0.87$ for the self-control subscale in previous research (e.g. Bradley and Corwyn, 2005). In the current study, reliabilities ranged from $\alpha = 0.79$ to $\alpha = 0.82$ over the three time points.

Deviance

The Child Behavior Checklist (CBCL, parent report; Achenbach, 1991) was used to assess children and early adolescent deviant behaviors. The measure includes 33 items that describe a variety of deviant behaviors, including "lying or cheating," "steals at home," "physically attacks people," and "gets into many fights." Mothers rated each item on a 3-point scale (0 = not true, 1 = somewhat or sometimes true; and 2 = very true or often true). Test–retest reliability estimates indicated good internal consistency in previous work ($\alpha = 0.76$ to $\alpha = 0.93$; Achenbach, 1991). In the current study, reliability estimates were adequate over the three time points ($\alpha = 0.78$ to $\alpha = 0.80$).

Results

First, descriptive statistics for the measures were computed (Table 18.1). They include means, standard deviations, skewness, kurtosis, and reliabilities for each of the measures.

Second, bivariate correlations among parenting, self-control, and deviance were computed. Listwise deletion was employed for this analysis. Results showed that all the correlations for the three time points were statistically significant and in the expected direction. Maternal parenting was positively associated with self-control and negatively associated with deviance, while self-control was negatively associated with deviance (Table 18.2).

Third, we specified and tested a longitudinal cross-lagged path model of the relationships among parenting, self-control, and deviance. Model fit was adequate: χ^2 = 186.350, df = 16, χ^2/df = 11.647, CFI = 0.951, and RMSEA = 0.088 (90 percent CI 0.077–0.1) (Figure 18.1). More importantly, results also indicated that parenting at 4.5 years predicted children's self-control at 8.5 years (β = 0.136, SE = 0.024, P < 0.001), while parenting at 8.5 years did not predict children's self-control at 10.5 years (β = 0.016, SE = 0.023, P = 0.482). Interestingly, we also found that children's self-control at 4.5 years predicted maternal parenting at 8.5 years (β = 0.168, SE = 0.033, P < 0.001) and that children's self-control at 8.5 years predicted

Table 18.1 Descriptive statistics for the main construct variables

Variable	*Number of items*	*N*	*Mean*	*SD*	*Skew*	*SE*	*Kurtosis*	*SE*	*α*
Parenting at 4.5 years	8	1,077	4.66	0.35	−1.57	0.08	3.38	0.15	0.63
Parenting at 8.5 years	8	1,028	4.25	0.32	−1.32	0.08	4.64	0.15	0.66
Parenting at 10.5 years	8	1,020	4.20	0.35	−1.39	0.07	3.41	0.15	0.73
Self-control at 4.5 years	10	1,057	1.30	0.31	0.09	0.07	−0.33	0.15	0.79
Self-control at 8.5 years	10	1,028	1.36	0.34	−0.15	0.07	−0.63	0.15	0.82
Self-control at 10.5 years	10	1,021	1.39	0.33	−0.12	0.07	−0.27	0.15	0.81
Deviance at 4.5 years	8	1,057	0.50	0.35	0.68	0.07	0.26	0.15	0.78
Deviance at 8.5 years	8	1,026	0.37	0.32	0.88	0.08	0.29	0.15	0.79
Deviance at 10.5 years	8	1,020	0.31	0.32	1.20	0.07	1.28	0.15	0.80

Table 18.2 Correlations of parenting, self-control, and deviance

Variables	1	2	3	4	5	6	7	8	9
1 Parenting (4.5 years)									
2 Parenting (8.5 years)	0.34**								
3 Parenting (10.5 years)	0.29**	0.49**							
4 Self-control (4.5 years)	0.32**	0.26**	0.21**						
5 Self-control (8.5 years)	0.30**	0.39**	0.25**	0.53**					
6 Self-control (10.5 years)	0.23**	0.29**	0.28**	0.49**	0.70**				
7 Deviance (4.5 years)	−0.19**	−0.16**	−0.11**	–	−0.38**	–			
8 Deviance (8.5 years)	−0.15**	−0.22**	−0.12**	–	−0.51**	–	0.54**		
9 Deviance (10.5 years)	−0.14**	−0.19**	−0.24**	–	−0.59**	–	0.51**	0.70**	

Note: ** Correlations are statistically significant at P < 0.01 (two-tailed).

Table 18.3 Parameter estimates for male versus female children and for ethnic/racial majority versus minority children

Variables			*Male* N = 478		*Female* N = 479		*Majority* N = 844		*Minority* N = 181	
			β	*SE*	β	*SE*	β	*SE*	β	*SE*
Parenting 4.5 years	→	Parenting 8.5 years	0.242***	0.045	0.315***	0.048	0.284***	0.035	0.258**	0.091
Parenting 8.5 years	→	Parenting 10.5 years	0.526***	0.052	0.497***	0.056	0.444***	0.041	0.467***	0.084
Self-control 4.5 years	→	Self-control 8.5 years	0.385***	0.058	0.452***	0.051	0.391***	0.041	0.412***	0.095
Self-control 8.5 years	→	Self-control 10.5 years	0.546***	0.045	0.593***	0.049	0.605***	0.035	0.623***	0.073
Deviance 4.5 years	→	Deviance 8.5 years	0.447***	0.057	0.484***	0.045	0.531***	0.033	0.443***	0.075
Deviance 8.5 years	→	Deviance 10.5 years	0.694***	0.040	0.602***	0.045	0.654***	0.034	0.700***	0.075
Parenting 4.5 years	→	Self-control 8.5 years	0.125**	0.038	0.182***	0.039	0.170***	0.030	0.168*	0.067
Parenting 8.5 years	→	Self-control 10.5 years	–0.004	0.038	0.067**	0.043	0.036	0.030	–0.086	0.056
Self-control 4.5 years	→	Parenting 8.5 years	0.117**	0.058	0.150**	0.055	0.178***	0.040	0.178**	0.111
Self-control 8.5 years	→	Parenting 10.5 years	0.031	0.049	0.024	0.056	0.090*	0.038	0.033	0.093
Self-control 4.5 years	→	Deviance 8.5 years	–0.160***	0.057	–0.151**	0.052	–0.104**	0.038	–0.146**	0.092
Self-control 8.5 years	→	Deviance 10.5 years	–0.090*	0.040	–0.140*	0.046	–0.110**	0.033	–0.103*	0.061
Deviance 4.5 years	→	Self-control 8.5 years	–0.199***	0.045	–0.149**	0.043	–0.231***	0.035	–0.092	0.054
Deviance 8.5 years	→	Self-control 10.5 years	–0.253***	0.040	–0.120**	0.046	–0.150***	0.036	–0.241***	0.061

maternal parenting at 10.5 years ($\beta = 0.065$, SE = 0.031, $P < 0.05$), thus providing evidence of a reciprocal process. Findings also showed that children's self-control at age 4.5 years negatively predicted children's deviant behaviors at 8.5 years ($\beta = -0.131$, SE = 0.032, $P < 0.001$) and that children's self-control at 8.5 years negatively predicted children's deviant behaviors at age 10.5 years ($\beta = -0.116$, SE = 0.026, $P < 0.001$). Lastly, the model also provided evidence that children's deviant behaviors at 4.5 years predicted children's self-control at 10.5 years ($\beta = -0.176$, SE = 0.026, $P < 0.001$) and that children's deviant behaviors at 8.5 years negatively predicted children's self-control at 10.5 years ($\beta = -0.167$, SE 0.027, $P < 0.001$).

Next, we also tested a series longitudinal cross-lagged path models by sex and ethnicity/race. Results showed that all the model fits were adequate. For male model, $\chi^2 = 116{,}056$, df = 16, χ^2/df = 7.254, CFI = 0.920, and RMSEA = 0.088 (90 percent CI 0.095–0.135). For female model, $\chi^2 = 71.38$, df = 16, χ^2/df = 4.461, CFI = 0.951, and RMSEA = 0.085 (90 percent CI 0.066–0.11). For majority model children the following model fit was observed: $\chi^2 = 98.52$, df = 16, χ^2/df = 6.157, CFI = 0.962, and RMSEA = 0.078 (90 percent CI 0.064–0.093). On the other hand, for minority children, the following was found: $\chi^2 = 67.798$, df = 16, χ^2/df = 4.237, CFI = 0.895, and RMSEA = 0.134 (90 percent CI 0.102–0.168).

Results included in Table 18.3 decompose the effects by sex and by ethnic/racial group. Findings provide evidence that reciprocal processes were found between parenting and self-control in both boys and girls during early and middle childhood (from 4.5 to 8.5 years old), but less so during late childhood (8.5 to 10.5 years old). Similarly, reciprocal processes were also found between self-control and deviance in both boys and girls. Finally, in comparisons by ethnic/racial group, findings indicated that reciprocal processes existed between parenting and self-control among both majority and minority group children during early and middle childhood (from 4.5 to 8.5 years old), but less so during late childhood. However, deviance at age 4.5 did not significantly predict children's self-control 4 years later at age 8.5. We found that an individual's level of self-control showed significant effects on parenting only during early childhood (between 4.5 and 8.5 years). At the same time, self-control predicted both parenting and deviance, while deviance also predicted self-control over time. In general, stability over time of each of the three constructs was greater in magnitude during late childhood, while at the same time the strength of association across them diminished.

Conclusion

In conclusion, self-control remains very important and appears to develop through an interplay between individual differences and, at least, non-heritable effects. Self-control has recently been highlighted by some as essential for the well-being and very functioning of humanity. Much of this work also supports the importance of socialization effects, above and beyond individual differences, thus reaffirming the original theoretical conceptualization of it. We are optimistic about the theory and its empirical status—the original potent concept of probabilistic self-control, tempered by situational and contextual opportunities, is important in our understanding of variability in crime and deviance. Maybe Gottfredson and Hirschi had it right, after all, when they focused on nurture (without the exclusion of nature) in the development of self-control. Others tend to agree—scholars conducting sophisticated and state-of-the-art etiological work over decades and across different continents, with and without the use of genetically informed designs (Moffitt *et al.*, 2013) as well as scholars extensively synthesizing rigorous programmatic work addressing the reduction of crime and deviance by addressing self-control (Piquero *et al.*, 2009a). Today, few would argue with the fact that socialization processes are bidirectional and that they are related to both individual (genetic) differences as much as to contextual influences. Much as DeLisi (2013) notes, work ignoring this *status quo* stands a high chance of embarrassment. However, and thankfully so, today, so far, we can do little about biological variability, but much about socialization influences and early experiences.

References

Achenbach, T. M. (1991) *Manual for the Child Behavior Checklist/4–18 and 1991 Profile*, Burlington, VT: University of Vermont, Department of Psychiatry.

Arneklev, B. J., Cochran, J. K., and Gainey, R. R. (1998) 'Testing Gottfredson and Hirschi's "low self-control" stability hypothesis: an exploratory study', *American Journal of Criminal Justice*, 23: 107–127.

Bartusch, D. R. J., Lynam, D. R., Moffitt, T. E., and Silva, P. A. (1997) 'Is age important? Testing a general versus a developmental theory of antisocial behavior', *Criminology*, 35: 13–48.

Baumeister, R. F., Heatherton, T. F., and Tice, D. M. (1994) *Losing control: how and why people fail at self-regulation*, San Diego, CA: Academic Press.

Beaver, K. M., Wright, J. P., and DeLisi, M. (2007) 'Self-control as an executive function: reformulating Gottfredson and Hirschi's parental socialization thesis', *Criminal Justice and Behavior*, 34: 1345–1361.

Beaver, K. M., Ratchford, M., and Ferguson, C. J. (2009) 'Evidence of genetic and environmental effects on the development of low self-control', *Criminal Justice and Behavior*, 36: 1158–1172.

Beaver, K. M., Connolly, E. J., Schwartz, J. A., Al-Ghamdi, M. S., and Kobeisy, A. N. (2013) 'Genetic and environmental contributions to stability and change in levels of self-control', *Journal of Criminal Justice*, 31(5): 300–308.

Bradley, R. H. and Corwyn, R. F. (2005) 'Productive activity and the prevention of behavior problems', *Developmental Psychology*, 41: 89–98.

Brody, G. H. and Ge, X. (2001) 'Linking parenting processes and self-regulation to psychological functioning and alcohol use during early adolescence', *Journal of Family Psychology*, 15: 82–94.

Burt, C. H., Simons, R. L., and Simons, L. G. (2006) 'A longitudinal test of the effects of parenting and the stability of self-control: negative evidence for the general theory of crime', *Criminology*, 44: 353–396.

Cauffman, E., Steingberg, L., and Piquero, A. R. (2005) 'Psychological, neuropsychological, and physiological correlates of serious antisocial behavior in adolescence: the role of self-control', *Criminology*, 43: 133–176.

Cochran, J. K., Wood, P. B., Sellers, C. S., Wilkerson, W., and Chamlin, M. B. (1998) 'Academic dishonesty and low self-control: an empirical test of a general theory of crime', *Deviant Behavior*, 19: 227–255.

Colman, R. A., Hardy, S. A., Albert, M., and Crocket, L. (2006) 'Early predictors of self-regulation in middle childhood', *Infant and Child Development*, 15: 421–437.

Crockett, L. J., Raffaelli, M., and Shen, Y. L. (2006) 'Linking self-regulation and risk proneness to risky sexual behavior: pathways through peer pressure and early substance use', *Journal of Research on Adolescence*, 16: 503–525.

Cullen, F. T., Unnever, J. D., Wright, J. P., and Beaver, K. M. (2008) 'Parenting and self-control', in Erich Goode (ed.), *Out of Control: Assessing the General Theory of Crime*, Stanford, CA: Stanford University Press, pp. 61–74.

DeLisi, M. (2011) 'Genetics: *l'enfant terrible* of criminology', *Journal of Criminal Justice*, 40: 515–516.

DeLisi, M. (2013) 'Pandora's box: the consequences of low self-control into adulthood', in Chris L. Gibson and Marvin D. Krohn (eds), *Handbook of Life-course Criminology*, New York: Springer-Verlag.

Duckworth, A. L. and Seligman, M. E. P. (2005) 'Self-discipline outdoes IQ in predicting academic performance of adolescents', *Psychological Science*, 16: 939–944.

Eisenberg, N., Champion, C., and Ma, Y. (2004) 'Emotion-related regulation: an emerging construct', *Merrill-Palmer Quarterly*, 50: 236–259.

Eisenberg, N., Zhou, Q., Spinrad, T. L., Valiente, C., Fabes, R. A., and Liew, J. (2005) 'Relations among positive parenting, children's effortful control, and externalizing problems: a three-wave longitudinal study', *Child Development*, 76: 1055–1071.

Feldman, S. S. and Weinberger, D. A. (1994) 'Self-restraint as a mediator of family influences on boys' delinquent behavior: a longitudinal study', *Child Development*, 65: 195–211.

Geis, G. (2000) 'On the absence of self-control as the basis for a General Theory of Crime: a critique', *Theoretical Criminology*, 4: 35–53.

Gibbs, J. J., Giever, D., and Martin, J. S. (1998) 'Parental management and self-control: an empirical test of Gottfredson and Hirschi's general theory', *Journal of Research on Crime and Delinquency*, 35: 40–70.

Goode, E. (2008) *Out of Control: Assessing the General Theory of Crime*, Stanford, CA: Stanford University Press.

Gottfredson, M. R. (2006) 'The empirical status of control theory in criminology', in F. Cullen, J. Wright, and K. Blevins (eds), *Taking Stock: The Status of Criminological Theory*, New Brunswick, NJ: Transaction Publishers.

Gottfredson, M. R. and Hirschi, T. (1990) *A General Theory of Crime*, Stanford, CA: Stanford University Press.

Gresham, F. M. and Elliot, S. N. (1990) *Social Skills Rating System Manual*, Circle Pines, MN: American Guidance Service.

Griffin, K.W., Scheier, L. M., Acevedo, B., Grenard, J. L., and Botvin, G. J. (2011) 'Long-term effects of self-control on alcohol use and sexual behavior among urban minority young women', *International Journal of Environmental Research and Public Health*, 9: 1–23.

Hay, C. (2001) 'Parenting, self-control, and delinquency: a test of self-control theory', *Criminology*, 39: 707–736.

Hay, C. and Forrest, W. (2006) 'The development of self-control: examining self-control theory's stability thesis', *Criminology*, 44: 739–774.

Kanazawa, S. (2003) 'Why productivity fades with age: the crime–genius connection', *Journal of Research in Personality*, 37: 257–272.

Kochanska, G., Coy, K. C., and Murray, K. T. (2001) 'The development of self-regulation in the first four years of life', *Child Development*, 72: 1091–1111.

Kochanska, G., Aksan, N., Penney, S. J., and Doobay, A. F. (2007) 'Early positive emotionality as a heterogeneous trait: implications for children's self-regulation', *Journal of Personality and Social Psychology*, 93: 1054–1066.

Kopp, C. B. (1982) 'Antecedents of self-regulation: a developmental perspective', *Developmental Psychology*, 18: 199–214.

Lenartowicz, A., Kalar, D. J., Congdon, E., and Poldrack, R. A. (2010) 'Towards an ontology of cognitive control', *Topics in Cognitive Science*, 2: 678–692.

Miller, H.V., Barnes, J. C., and Beaver, K. M. (2011) 'Self-control and health outcomes in a nationally representative sample', *American Journal of Health and Behavior*, 35: 15–27.

Mischel, W., Shoda, Y., and Rodriguez, M. L. (1989) 'Delay of gratification in children', *Science*, 244: 933–938.

Mitchell, O. and MacKenzie, D. L. (2006) 'The stability and resiliency of self-control in a sample of incarcerated offenders', *Crime and Delinquency*, 52: 432–449.

Miyake, A., Friedman, N. P., Emerson, M. J., Witzki, A. H., Howerter, A., and Wager, T. D. (2000) 'The unity and diversity of executive functions and their contributions to complex "frontal lobe" tasks: a latent variable analysis', *Cognitive Psychology*, 41: 49–100.

Moffitt, T. E. (1993) 'Adolescence-limited and life-course-persistent antisocial behavior: a developmental taxonomy', *Psychological Review*, 100: 674–701.

Moffitt, T. E. (2005) 'The new look of behavioral genetics in developmental psychopathology: gene–environment interplay in antisocial behaviors', *Psychological Bulletin*, 131: 533–554.

Moffitt, T. E., Poulton, R., and Caspi, A. (2013) 'Lifelong impact of early self-control: childhood self-discipline predicts adult quality of life', *American Scientist*, 101: 352–359.

Perrone D., Sullivan, C. J., Pratt, T. C., and Margaryan, S. (2004) 'Parental efficacy, self-control, and delinquency: a test of a General Theory of Crime on a nationally representative sample of youth', *International Journal of Offender Therapy in Comparative Criminology*, 48: 298–312.

Pianta, R. C. (1994) *Parent–Child Relationship Scale*, Charlottesville, VA: University of Virginia.

Pianta, R. C. (2001) *Student–Teacher Relationship Scale*, Charlottesville, VA: University of Virginia.

Piquero, A. R., Farrington, D. P., Welsh, B. C., Tremblay, R., and Jennings, W. G. (2009a) 'Effects of early family/parent training programs on antisocial behavior and delinquency', *Journal of Experimental Criminology*, 5: 83–120.

Piquero, A. R., Jennings, W. G., and Farrington, D. P. (2009b) 'On the malleability of self-control: theoretical and policy implications regarding a General Theory of Crime', *Justice Quarterly*, 27: 803–834.

Plomin, R. and Thompson, R. (1987) 'Life-span developmental behavioral genetics', in P. B. Baltes, D. L. Featherman, and R. M. Lerner (eds), *Life-Span Development and Behavior*, Hillsdale, NJ: Lawrence Erlbaum.

Polakowski, M. (1994) 'Linking self- and social control with deviance: illuminating the structure underlying a general theory of crime and its relation to deviant activity', *Journal of Quantitative Criminology*, 10: 41–78.
Pratt, T. C., Cullen, F. T. Blevins, K. R., Daigle, L., and Unnever, J. D. (2002) 'The relationship of attention deficit hyperactivity disorder to crime and delinquency: a meta-analysis', *International Journal of Police Science and Management*, 4(4): 344–360.
Pratt, T. C., Turner, M. G., and Piquero, A. R. (2004) 'Parental socialization and community context: a longitudinal analysis of the structural sources of low self-control', *Journal of Research in Crime and Delinquency*, 41(3): 219–243.
Raffaelli, M., Crockett, L. J., and Shen, Y. L. (2005) 'Developmental stability and change in self-regulation from childhood to adolescence', *Journal of Genetic Psychology Research and Theory on Human Development*, 166: 54–76.
Reyna, V. F., Chapman, S. B., Dougherty, M. R., and Confrey, J. (2011) *The Adolescent Brain: Learning, Reasoning, and Decision Making*, Washington, DC: American Psychological Association.
Rothbart, M. K., Ellis, L. K., Rueda, R., and Posner, M. I. (2003) 'Developing mechanisms of temperamental effortful control', *Journal of Personality*, 71: 1113–1144.
Rowe, D. C. (2002) *Biology and Crime*, Los Angeles, CA: Roxbury.
Scarr, S. (1992) 'Developmental theories for the 1990s: development and individual differences', *Child Development*, 63(1): 1–19.
Tangney, J. P., Baumeister, R. F., and Boone, A. L. (2004) 'High self-control predicts good adjustment, less pathology, better grades, and interpersonal success', *Journal of Personality*, 72: 271–324.
Turner, M. G. and Piquero, A. R. (2002) 'The stability of self-control', *Journal of Criminal Justice*, 30: 457–471.
Turner, M. G., Piquero A. R., and Pratt, T. C. (2005) 'The school context as a source of self-control', *Journal of Criminal Justice*, 33: 327–339.
Vazsonyi, A. T. and Huang, L. (2010) 'Where self-control comes from: on the development of self-control and its relationship to deviance over time', *Developmental Psychology*, 46: 245–257.
Vazsonyi, A. T., Pickering, L. E., Junger, M., and Hessing, D. (2001) 'An empirical test of a general theory of crime: a four-nation comparative study of self-control and the prediction of deviance', *Journal of Research in Crime and Delinquency*, 38: 91–131.
Vazsonyi, A. T., Wittekind, J. E. C., Belliston, L. M., and Van Loh, T. D. (2004) 'Extending the general theory of crime to the East: low self-control in Japanese late adolescents', *Journal of Quantitative Criminology*, 20: 189–216.
Winfree, L. T., Taylor, T. J., He, N., and Esbensen, F. A. (2006) 'Self-control and variability over time: multivariate results using a 5-year, multisite panel of youths', *Crime and Delinquency*, 52: 253–286.
Wright, J. P. and Beaver, K. M. (2005) 'Do parents matter in creating self-control in their children? A genetically informed test of Gottfredson and Hirschi's theory of low self-control', *Criminology*, 43: 1169–1202.
Zhou, Q., Hofer, C., Eisenberg, N., Reiser, M., Spinrad, T. L., and Fabes, R. A. (2007) 'The developmental trajectories of attention focusing, attentional and behavioral persistence, and externalizing problems during school-age years', *Developmental Psychology*, 43: 369–385.

Further reading

Akers, R. L. (1991) 'Self-control as a general theory of crime', *Journal of Quantitative Criminology*, 7: 201–211.
Baron, S. W. (2003) 'Self-control, social consequences, and criminal behavior: street youth and the general theory of crime', *Journal of Research in Crime and Delinquency*, 40: 403–425.
Beaver, K. M. and Wright, J. P. (2007) 'The stability of low self-control from kindergarten through first grade', *Journal of Crime and Justice*, 30: 63–86.
Beaver, K. M., Schutt, J. E., Boutwell, B. B., Ratchford, M., Roberts, K., and Barnes, J. C. (2009) 'Genetic and environmental influences on levels of self-control and delinquent peer affiliation: results from a longitudinal sample of adolescent twins', *Criminal Justice and Behavior*, 36: 41–60.
Beccaria, C. (1963) *On Crimes and Punishments*, translated by H. Paolucci, Indianapolis, IN: Bobbs-Merrill.
Bentham, J. A. (1948) *A Fragment on Government and an Introduction to the Principles of Morals and Legislation*, Oxford: Basil Blackwell.
Caspi, A., Moffitt, T. E., Silva, P. A., Stouthamer-Loeber, M., Krueger, R. F., and Schmutte, P. A. (1994) 'Are some people crime-prone? Replications of the personality–crime relationship across countries, genders, races, and methods', *Criminology*, 32: 163–196.

Cheung, N. W. and Cheung, Y. W. (2008) 'Self-control, social factors, and delinquency: a test of the general theory of crime among adolescents in Hong Kong', *Journal of Youth and Adolescence*, 37: 412–430.

DeLisi, M. (2001) 'It's all in the record: assessing self-control theory with an offender sample', *Criminal Justice Review*, 26: 1–14.

Evans, T. D., Cullen, F. T., Burton Jr, V. S., Dunaway, R. G., and Benson, M. L. (1997) 'The social consequences of self-control: testing the general theory of crime', *Criminology*, 35: 475–504.

Gibbs, J. J. and Giever, D. (1995) 'Self-control and its manifestations among university students: an empirical test of Gottfredson and Hirschi's general theory', *Justice Quarterly*, 12: 231–255.

Glueck, S. and Glueck, E. (1950) *Unraveling Juvenile Delinquency*, Cambridge, MA: Harvard University Press.

Grasmick, H. G., Tittle, C. R., Bursik Jr, R. J., and Arneklev, B. J. (1993) 'Testing the core empirical implications of Gottfredson and Hirschi's general theory of crime', *Journal of Research in Crime and Delinquency*, 30: 5–29.

Hardwick, K. H. and Brannigan, A. (2008) 'Self-control, child effects, and informal social control: a direct test of the primacy of sociogenic factors', *Canadian Journal of Criminology and Criminal Justice*, 50: 1–30.

Harris, J. R. (1995) 'Where is the child's environment? A group socialization theory of development', *Psychological Review*, 102: 458–489.

Harris, J. R. (1998) *The Nurture Assumption: Why Children Turn Out the Way They Do*, New York, NY: Free Press.

Harris, J. R. (2000) 'Socialization, personality development, and the child's environment: comment on Vandell (2000)', *Developmental Psychology*, 36: 711–723.

Hirschi, T. (2004) 'Self-control and crime', in R. F. Baumeister and K. D. Vohs (eds), *Handbook of Self-Regulation Research, Theory, and Applications*, New York: Guilford Press.

Jones, S. and Quisenberry, N. (2004) 'The general theory of crime: how general is it?', *Deviant Behavior*, 25: 401–426.

Laub, J. H. and Sampson, R. J. (1993) 'Turning points in the life course: why change matters to the study of crime', *Criminology*, 31: 301–326.

Longshore, D. (1998) 'Self-control and criminal opportunity: a prospective test of the general theory of crime', *Social Problems*, 45: 102–113.

Moffitt, T. E., Arseneault, L., Belsky, D., Dickson, N., Hancox, R. J., Harrington, H., Houts, R., Poulton, R., Roberts, B. W., Ross, S., Sears, M. R., Thomson, W. M., and Caspi, A. (2011) 'A gradient of childhood self-control predicts health, wealth, and public safety', *Proceedings of the National Academy of Sciences, USA*, 108: 2693–2698.

Ozbay, O. (2008) 'Self-control, gender, and deviance among Turkish university students', *Journal of Criminal Justice*, 36: 72–80.

Piquero, A. R., MacDonald, J., Dobrin, A., Daigle, L. E., and Cullen, F. T. (2005) 'Self-control, violent offending, and homicide victimization: assessing the general theory of crime', *Journal of Quantitative Criminology*, 21: 55–71.

Pratt, T. C. and Cullen, F. T. (2000) 'The empirical status of Gottfredson and Hirschi's general theory of crime: a meta-analysis', *Criminology*, 38: 931–964.

Ratchford, M. and Beaver, K. M. (2009) 'Neuropsychological deficits, low self-control, and delinquent involvement: toward a biosocial explanation of delinquency', *Criminal Justice and Behavior*, 36: 147–162.

Rebellon, C. J., Straus, M. A., and Medeiros, R. (2008) 'Self-control in global perspective: an empirical assessment of Gottfredson and Hirschi's general theory within and across 32 national settings', *European Journal of Criminology*, 5: 331–361.

Shreck, C. J. (1999) 'Criminal victimization and low self-control: an extension and test of a general theory of crime', *Justice Quarterly*, 16: 633–654.

Unnever, J. D., Cullen, F. T., and Pratt, T. C. (2003) 'Parental management, ADHD, and delinquent involvement: reassessing Gottfredson and Hirschi's general theory', *Justice Quarterly*, 20: 471–500.

19
Brain abnormalities in psychopathy

Yaling Yang

Introduction

Throughout history, there has been significant interest in the personality and behavior structure of psychopathy. Although the etiology of psychopathy remains unclear, mounting evidence is beginning to suggest that brain abnormalities may be responsible for several clinical features of psychopathy, including lack of fear and poor moral decision-making. With the increasing use of imaging technology in psychiatric research, studies using structural magnetic resonance imaging (sMRI), diffusion tensor imaging (DTI), and functional MRI (fMRI) have begun to locate regions in the brain that are impaired in individuals with psychopathy and related disorders. Several issues are addressed in this chapter. What are the characteristics and assessments of psychopathy? Based on the imaging findings of psychopathic individuals, what could we hypothesize regarding the neural bases of key psychopathic features? Are unique brain characteristics associated with different subgroups of psychopathy? Is there evidence supporting a neurodevelopmental contribution to the development of psychopathy? How do psychosocial factors interact with brain abnormalities in influencing the outcome of psychopathy?

The core elements of psychopathy consist of extremely low agreeableness or high interpersonal antagonism, which contribute to key characteristics of psychopathy including conning, manipulative, arrogant, callous, sensation-seeking, reward-driven, and impulsive behaviors. These individuals appear to be almost immune to embarrassment and shame as they show low levels of anxiety towards others' judgments (Cleckley, 1941). This is a psychiatric disorder that not only affects the lives of those who suffer from it, but also poses a threat to society due to the fact that psychopathic individuals are more likely to commit serious, violent crimes as well as to repeat offending after their release from incarceration (Hare and McPherson, 1984; Steadman, 2000). Although several measurements have been made available to provide detailed descriptions of traits and behavioral criteria for diagnosing psychopathy, two are most commonly used in brain-imaging studies, namely the Psychopathy Personality Inventory (PPI) (Lilienfeld and Andrews, 1996) and the Psychopathy Checklist-Revised (PCL-R) (Hare, 1991, 2003). The PPI is a self-report measurement of psychopathy, which contains 163 items assessing specific personality traits associated with psychopathy. However, the lack of measures for antisocial behaviors related to psychopathy and the problem with poor reliability as a self-report scale raise some concern over the use of the PPI. The PCL-R, the most widely used measurement of psychopathy, is a 20-item assessment checklist with good reliability and validity. Factor analysis has revealed a four-facet model for the PCL-R, including Interpersonal (e.g. superficial charming, pathological lying), Affective (e.g. shallow affect, lack of remorse), Lifestyle (e.g. impulsivity, irresponsibility), and Antisocial facets (e.g. poor behavioral control, criminal versatility) (Hare, 2003). Indeed, much of the psychopathy research today relies on the description of psychopathic features included in the PCL-R, an instrument increasingly viewed as the gold standard for assessing psychopathy.

In addition, the adult construct of psychopathy has been extended downward to children and adolescents based on a fundamental assumption that features manifested by adults with psychopathy could also be observed in children and adolescents. However, some argue that this assumption needs further validation from longitudinal data, considering that several key features of adult psychopathy, such as impulsivity and sensation seeking, can be normative during adolescence, and may be temporarily elevated before leveling off as one matures into adulthood. Nonetheless, several studies have examined psychopathic traits in children and adolescents, attempting to understand the nature, course, and etiology, and more specifically, to determine whether there is a neurodevelopmental basis to psychopathy. In that regard, Psychopathy Checklist Youth Version (PCL-YV) (Forth *et al.*, 2003), adopted from PCL-R, and the Childhood Psychopathy Scale (CPS) (Lynam, 1997) have been used most consistently to assess psychopathic characteristics among children and adolescents in several brain-imaging studies.

Neural bases of psychopathy

Research on the neural bases of psychopathy has come a long way since the report of the case of Phineas Gage, who suffered damage to his prefrontal cortex, resulting in a dramatic alteration in his personality and behavior, and who ultimately became psychopathic. However, this case had such a profound impact on the study of neuroscience that it was later credited as "the historical beginning of the study of biological basis of behavior" (Damasio, 1994). With the increasing use of brain imaging in psychopathy research, mounting data have provided a foundation for developing hypotheses regarding the neural bases of psychopathy. However, the existing hypotheses often fail to consider variation in the symptom manifestation frequently seen in psychopathic individuals. By highlighting three key features of psychopathy – emotional dysregulation (limbic and paralimbic structures), antisocial behavior (prefrontal cortex), and reward seeking (striatum) – hypotheses are provided to pinpoint the brain abnormalities underlying these core psychopathic features.

Emotional dysregulation (limbic and paralimbic structures)

The lack of strong emotion, including fear and threat, is considered one of the most distinctive features of psychopathy (Cleckley, 1941). This is evident from studies showing that individuals with psychopathy have selective impairments in recognizing distress cues (e.g. facial and vocal expression of fear and sadness), often indicated by impaired startle-reflex and diminished skin-conductance responses in psychopaths while viewing pictures depicting negative emotion (Patrick, 1994; Verona *et al.*, 2004). Thus it has been proposed that disturbances in brain regions critical for emotional regulation, including the amygdala, hippocampus, parahippocampal gyrus, and orbitofrontal cortex, may contribute to the observed diminished adverse response to distress stimuli in psychopathic individuals. Each of these brain regions is responsible for different components of emotional processing. For example, the amygdala/hippocampal complex is involved in the processing of affective stimuli, whereas the orbitofrontal cortex plays a central role in the evaluation of negative reinforcers (Kringelbach, 2005). Below, we review evidence that so far indicates structural/functional abnormalities may be present in many of these limbic and paralimbic regions in individuals with psychopathy.

Earlier studies examining the integrity of limbic and paralimbic structures have focused primarily on anatomically based tracing techniques that focus on a single region. For example, in the first study to link amygdala abnormalities to psychopathy, our group (Yang *et al.*, 2009b) showed significant volume reductions and shape deformations in bilateral amygdala in psychopaths, particularly in the basolateral and central subnuclei, compared with controls. In a subsequent study, we further showed that the reductions/deformations are more severe in psychopaths with criminal convictions compared to those without (but who reported similar level of criminal activities as those with records) (Yang *et al.*, 2010). Regarding the hippocampus, one early study (Laakso *et al.*, 2001) found strong negative correlations, up to –0.79, between the regional volumes along the anteroposterior axis of the bilateral hippocampus and psychopathy scores in a group of habitually violent offenders with antisocial personality disorder and type 2 alcoholism. However, in a recent study, findings in the opposite direction were observed (Boccardi *et al.*, 2010), showing abnormally *enlarged* hippocampus, particularly in the lateral borders in both the right and left hippocampi, in adult offenders with high psychopathy scores and substance abuse compared with controls. Furthermore, in another study by Raine *et al.* (2004), instead of volume differences, they observed exaggerated anterior hippocampal asymmetry (right larger than left) in caught violent offenders with psychopathy compared with both uncaught (successful) psychopaths and controls.

Several recent brain-imaging studies using automated segmentation methods have provided supporting evidence linking brain abnormalities in the limbic and paralimbic structures to psychopathy. For example, Ermer *et al.* (2012) found reduced regional gray matter in the bilateral parahippocampal, amygdala, hippocampus, and orbitofrontal cortex in a large sample of nearly 300 incarcerated subjects. Similarly, in a longitudinal study, Pardini *et al.* (2014) found reduced amygdala volume to be associated with higher levels of psychopathic traits in a group of men with varying histories of violence, even after controlling for earlier levels of psychopathic features. Similarly to our findings of the amygdala, Boccardi *et al.* (2011) also found reduced amygdala in the basolateral nuclei in psychopathic criminal offenders with substance abuse compared with controls. However, they also found that central and lateral nuclei were enlarged in psychopaths, compared with controls. Cope *et al.* (2012) found psychopathic traits to negatively correlate with gray matter volumes in the right hippocampus and right insula, but positively correlate with gray matter volumes in bilateral orbitofrontal cortex and right anterior cingulate cortex. Their exploratory regression analyses further indicated that gray matter volumes within right hippocampus and left orbitofrontal cortex combined explained about 22 percent of the variance in psychopathy scores. Several factors may have contributed to the differences among findings, including heterogeneous sample characteristics, various degrees of psychopathy, and co-morbid disorders. Despite the inconsistency, the majority of findings to date support a possible link between abnormalities in the limbic and paralimbic structures and psychopathy.

Given the pervasiveness of emotional problems in psychopathy, it is not surprising that fMRI studies probing dysfunction in limbic and paralimbic circuit for emotional regulation have also been increasing in recent years. Even with relatively small samples, the early fMRI studies have revealed functional impairments in psychopaths in response to affect-related information processing and regulation, with the most robust findings being decreased activation in the amygdala and orbitofrontal cortex (Kiehl *et al.*, 2001; Muller *et al.*, 2003; Birbaumer *et al.*, 2005; Finger *et al.*, 2008; Marsh and Blair, 2008). For example, an early fMRI study (Kiehl *et al.*, 2001) showed that criminal psychopaths had less affect-related activity in the amygdala/hippocampal formation and parahippocampal gyrus compared with nonpsychopathic criminal offenders and normal controls. They also showed overactivation in the fronto-temporal cortex for processing affective stimuli in psychopathic criminals, and argued that the emotional dysregulation in psychopathy may be due to insufficient input from limbic structures. This is supported by Birbaumer *et al.*, who showed a lack of differential activity in the limbic-frontal circuit including the amygdala and orbitofrontal cortex during the acquisition of fear in adults with psychopathy compared to normal controls.

More recently, a few studies have attempted to further parse out the differential effects of various affect-related processing, including selective attention and moral reasoning. In a study of relatively healthy adults (Marsh and Cardinale, 2014), increased psychopathic traits were associated with reduced activity in right amygdala during judgments of fear-evoking statements. In another study of healthy adults, during the decision-making of inflicting costly punishment on people who proposed unfair offer, psychopathic traits were associated with less amygdala activity in response to the unfairness of offer (Osumi *et al.*, 2012). Moreover, the amygdala dysfunction in psychopathic individuals was associated with reduced functional connectivity with dopaminergic-related areas, including the striatum, when punishment was available compared with when it was unavailable. Using a group of healthy adults with high and low callous-unemotional traits, Han *et al.* (2012) further demonstrated that significantly reduced amygdala and medial prefrontal cortex activity was associated with high callous-unemotional traits only when the fearful faces were occluded, not when they were isolated. This finding emphasizes that the amygdala may be activated in response to the disambiguation of stimuli rather than the representation of distress. In line with the findings of Han and colleagues, Larson *et al.* (2013) reported that psychopathy-related differences in amygdala activation appear and disappear as a function of goal-directed attention. Specifically, reduced activity in the amygdala was found in psychopathic offenders only when attention was engaged in an alternative goal-relevant task prior to presenting threat-relevant information, and that the decreased amygdala activation was mediated by increased lateral prefrontal activation.

Overall, findings suggest that disturbed function and functional connectivity of limbic and paralimbic regions may contribute to emotional dysregulation in psychopathy, and that the abnormalities can be observed across multiple classes of stimuli and various components of affect-related processing. Results are supportive of structural imaging findings in suggesting that limbic and paralimbic deficits play a significant role in the neural bases of psychopathy, particularly features related to emotional disturbances. Undoubtedly, the relationship between functional and structural brain abnormalities needs further examination to understand how they contribute jointly and/or independently to the manifestation of psychopathic traits.

Antisocial behavior (prefrontal cortex)

In addition to emotional dysregulation, a salient characteristic of psychopathy is the violation of society's rules (Hare, 1991). Similarly to individuals diagnosed with antisocial personality disorder (APD), these individuals exhibit a wide range of antisocial features, including impulsivity, irresponsibility, poor behavioral control, and criminal versatility. As indicated in neuropsychological studies, psychopaths often perform poorly on cognitive tasks that involve attention selection (e.g. Stroop task) and response inhibition (e.g. Go/No Go task) (Lapierre *et al.*, 1995; Kiehl *et al.*, 2000; Hiatt *et al.*, 2004; Munro *et al.*, 2007; Kosson *et al.*, 2013). Brain-imaging studies have identified the anterior cingulate cortex, orbitofrontal cortex, and dorsolateral prefrontal cortex to be commonly activated by these cognitive tasks. These structures are anatomically connected and share overlapping function. However, simplistically speaking, the orbitofrontal cortex calculates the reward–punishment value of the stimulus while the anterior cingulate cortex and dorsolateral prefrontal cortex make the decision and execute it (Bush *et al.*, 2000; Kringelbach and Rolls, 2004; Krain *et al.*, 2006).

Over the past few decades, evidence has been mounting indicating a robust link between prefrontal deficits, particularly in the regions mentioned above, and psychopathy and related disorders (see our previous meta-analysis for review: Yang and Raine, 2009). Structurally, several brain-imaging studies have found significant gray matter volume reduction in the prefrontal cortex in antisocial, psychopathic individuals. For example, Raine *et al.* (2000) reported an 11 percent volume reduction in the prefrontal gray matter in antisocial individuals with psychopathy. In a later study, Raine and colleagues (Raine *et al.*, 2011) further localized the frontal gray matter deficits in psychopathic individuals to the middle frontal, orbitofrontal, and ventromedial prefrontal cortex. In addition, they showed that by controlling for brain differences in the orbitofrontal and middle frontal volumes, the gender difference in antisocial behavior was reduced by 77 percent. Findings are supported by several other studies showing reduced frontal volume in association with psychopathy (Muller *et al.*, 2008; de Oliveira-Souza *et al.*, 2008; Aoki *et al.*, 2013; Bertsch *et al.*, 2013). Recently, several studies examining other brain morphological characteristics, such as cortical thickness, have also reported frontal abnormalities in adults with psychopathy. For example, one study by our group (Yang *et al.*, 2009a) showed abnormal cortical thinning in several frontal regions including the right middle frontal, right orbitofrontal, and posterior cingulate cortex in adults with psychopathy. Similarly, Ly *et al.* (2012) reported thinner cortex in the right inferior frontal, left dorsal anterior cingulate cortex, and left insula in adult criminal offenders with psychopathy compared with nonpsychopathic offenders. These findings of abnormal cortical thinning in the prefrontal cortex are in agreement with earlier studies that show volume reductions in psychopathic individuals, and were linked to dysfunction in response perseveration in psychopathy (Yang *et al.*, 2011).

Based on findings from the functional imaging studies conducted on psychopaths, the above-mentioned frontal regions are also found to be less activated in adults with psychopathy when compared with controls during various tasks, including semantic processing and social cooperation (Kiehl *et al.*, 2004; Rilling *et al.*, 2007), stress imagery task (Li *et al.*, 2006), n-back working memory task (Kumari *et al.*, 2006), motor inhibition paradigm (Barkataki *et al.*, 2008), a nonverbal Stroop task (Schiffer *et al.*, 2014), and risk taking task (Prehn *et al.*, 2013). Other more complex patterns of functional disturbances have also been observed in recent studies. For example, in a study of economic decisions, the acceptance of unfair offers was associated with neural activity in the ventromedial prefrontal cortex in adults with high psychopathy scores in contrast to activity in the dorsolateral prefrontal cortex in adults with low psychopathy scores (Vieira *et al.*, 2013). These studies presented findings that are in line with structural brain-imaging results in supporting the hypothesis that prefrontal abnormalities may be one of the most profound disturbances in psychopathy, contributing to core psychopathic features such as impulsivity, irresponsibility, risky decision-making, and response perseveration.

Most recently, studies have begun to explore network- or micro-level brain abnormalities in psychopathic individuals, particularly in the prefrontal connectivity network. Using a novel graph theory-based method to assess information flow and connectivity based on sMRI scans, our group showed significantly altered interregional connectivity patterns in adults with psychopathy compared with healthy controls. Furthermore, bilateral superior frontal cortex in the frontal network were identified as network hubs (through which the majority of the information travels between nodes) in adults with psychopathy in contrast to bilateral inferior and medial orbitofrontal cortices as network hubs in healthy controls. This is supported by findings from a DTI study by Sundram and colleagues (2012) showing reduced fractional anisotropy and increased mean diffusivity in several fiber tracts that are involved in connectivity with the frontal cortex, including the right uncinate fasciculus and inferior fronto-occipital fasciculus, to correlate with increased psychopathic traits in a group of adults with APD. These findings showed that prefrontal abnormalities *not only* result in localized dysfunction *but may also* have network-level impact due to the fact that prefrontal cortex is densely connected to many other structures critically involved in behavior control and decision-making, contributing to the complexity of psychopathy.

Reward seeking (striatum)

More recently, evidence has begun to emerge suggesting that deficits in regions densely connected with the fronto-limbic circuitry, particularly the striatum, may also contribute to traits associated with psychopathy (Glenn and Yang, 2012; Blair, 2013). Studies have linked psychopathy to increased sensitivity to reward and decreased sensitivity to punishment, suggesting potential abnormalities in the striatal regions may contribute to psychopaths being overly focused on the prospect of reward despite signals of potential later punishment (Newman and Kosson, 1986; van Honk and Schutter, 2006; Glenn *et al.*, 2009). The striatum, comprised of the caudate, putamen, and pallidum, has been linked to traits such as reward seeking, stimulus-reinforcement learning, and impulsivity (Barros-Loscertales *et al.*, 2006; Cohen *et al.*, 2009). The head of the caudate nuclei and the putamen are often referred to as the dorsal striatum, a region that is hypothesized to be associated with stimulus–response habit formation (including procedure learning, reference memory, egocentric orientation, and rule-based learning), and different behavioral response selection in the presence of changing task requirements (including motor planning and control, error correction in responding, strategy selection, and set-switching). Both the ventral striatum, consisting primarily of the nucleus accumbens, and the dorsal striatum are thought to facilitate reward-dependent behavior and are critically involved in reinforcement learning and the anticipation of reward.

To our knowledge, there have been less than a handful of brain-imaging studies conducted to date examining striatal structural integrity in psychopathic individuals. For example, Barkataki *et al.* (2006) reported increased volume in the putamen in antisocial individuals compared with healthy controls. Schiffer *et al.* (2011) found that violent offenders had larger volumes in the nucleus accumbens and the caudate head than nonoffenders, and that the volume increases were positively correlated with all core features of psychopathy. Glenn and colleagues (2009) found a 9.6 percent increase in the volume of the total striatum including the caudate, putamen, and globus pallidus in psychopathic individuals compared with a control group matched in substance use. They further revealed that increased volumes in the caudate head were associated with lifestyle features of psychopathy, whereas increased volumes in the caudate body were primarily associated with interpersonal and affective features of psychopathy. In addition, the lenticular nucleus (putamen and globus pallidus) was associated with all features except the interpersonal features. However, conflicting findings also exist, for example, one recent study by Boccardi and colleagues (2013) showed that the volume of the nucleus accumbens was 13 percent *smaller* in offenders with high psychopathy scores. They also showed that caudate and putamen local morphology *negatively* correlated with lifestyle features of psychopathy. Nevertheless, findings to date suggest that abnormal striatal morphology may play a central role in the neural underpinnings of psychopathy.

Not surprisingly, equally few studies have targeted specifically the striatal function in psychopathic individuals. However, the very few functional brain-imaging findings to date also support striatal disturbances in psychopathic individuals. For example, Völlm *et al.* (2006) found that antisocial individuals showed less differential activity in the caudate during rewarded versus nonrewarded trials compared with normal controls. In another study, children with externalizing disorders exhibited activity in the caudate during both rewarding and nonrewarding trials, in contrast to normal controls who exhibited differential activity in the caudate (i.e. activated only during rewarding, not nonrewarding trials). In a study of male prison inmates, Pujara *et al.* (2014) found neural activity in the ventral striatum during a gambling task (to assess reward and lost sensitivity) to correlate with psychopathy severity. The idea is also supported by findings of Finger *et al.* (2008), which showed failure in children with psychopathic traits to show appropriate response reversal following the change in reinforcement contingency. Specifically, children with psychopathic traits had increased activity in the caudate during the punished errors, whereas healthy children showed decreased activity. As described previously, Kiehl *et al.* (2001) found reduced activity in the ventral striatum in psychopathic offenders during an affective memory task using negatively valenced stimuli. They argued that findings may reflect the difficulty observed in psychopathic individuals that they continue to find previously rewarded responses to be rewarding, even after the feedback has changed from reinforcement to punishment. Overall, fMRI studies on striatal functioning have provided findings that are in line with sMRI studies and support the involvement of striatal abnormalities in the neural bases of psychopathy.

Subgroups of psychopathy

Although brain abnormalities have been found in several imaging studies on adults with psychopathy, drawing conclusions regarding the neuropathological patterns of this disorder has been surprisingly difficult. Many factors have been speculated to contribute to the inconsistency among the findings, such as differences in the imaging techniques, sample size, psychiatric co-morbidity, and participant recruiting methods used. The existence of different subgroups of psychopathy is another potential contributor to the differences among imaging findings. It is suggested that brain impairments may not be clearly detectible in a heterogeneous group of psychopaths, and only by dividing them into subgroups will abnormalities be more clearly revealed. Despite the lack of studies on subgroups of psychopathy, the importance of addressing this issue has been highlighted by the very few studies on unique forms of psychopathy, particularly individuals with impulsive versus predatory aggression, pathological liars, and successful versus unsuccessful psychopaths.

Impulsive versus predatory aggression

Sufficient evidence has been accumulated supporting brain abnormalities in aggressive individuals, particularly in the orbitofrontal cortex, dorsolateral prefrontal cortex, anterior cingulate cortex, and amygdala (Raine *et al.*, 1997; Coccaro *et al.*, 2007; Yang and Raine, 2009). However, a number of researchers have begun to suggest that there may be different brain bases to impulsive, affective violence versus predatory, planned violence (Raine *et al.*, 1998c; Davidson *et al.*, 2000). Specifically, frontal abnormalities may be more pronounced in individuals engaging in impulsive rather than premeditated aggression. One study found prefrontal dysfunction to be specific to affective, impulsive murderers as opposed to predatory, instrumental murderers (Raine *et al.*, 1998b). Common to both sets of impulsive and planned murderers was increased subcortical activity (midbrain, amygdala, hippocampus, and thalamus) compared with controls. While predatory, controlled murderers may have sufficient prefrontal regulation to control the excess aggressive feelings generated subcortically, this inhibitory control may be lacking in the affective, impulsive murderers. Two other studies, convergent with this frontal finding, also showed reduced glucose metabolism in the frontal cortex to be associated with frequency of impulsive aggression in patients with personality disorders (Goyer *et al.*, 1994; Raine *et al.*, 1998a) and reduced fronto-temporal cerebral blood flow in impulsive violent offenders (Soderstrom *et al.*, 2000).

In contrast to impulsive affective violent offenders, psychopaths (who are more likely than nonpsychopathic criminals to commit predatory violence) appear to show either normal or *increased* patterns of brain activation compared with nonpsychopathic offenders and controls, particularly in the frontal regions. Adults with psychopathy have been found to show *increased* bilateral blood flow in fronto-temporal regions during the processing of emotional words, *increased* activation in the dorsolateral prefrontal cortex during an aversive conditioning paradigm (Schneider *et al.*, 2000), and *overactivation* of fronto-temporal regions (Kiehl *et al.*, 2001). Similarly, predatory murderers (who may resemble psychopaths more than impulsive murderers) showed relatively normal prefrontal activation (Raine *et al.*, 1998a), again suggesting relatively *greater* brain activation to be associated with proactive aggression. This is supported by findings of one recent study on female adolescents with conduct disorders, which showed impulsive/reactive aggressive symptoms to be negatively correlated with right dorsolateral gray matter volume, whereas callous-unemotional traits were *positively* correlated with bilateral orbitofrontal volume (Fairchild *et al.*, 2013). This issue needs to be assessed more carefully in subgroups of psychopaths (i.e. predatory/planned versus impulsive psychopathic subgroups) in order to address this issue more precisely in the psychopathy domain.

Pathological lying

Several important fMRI studies have recently revealed informative clues on the brain mechanisms subserving lying in normal individuals. Spence *et al.* (2001), utilizing a computer-based interrogation procedure, found that lying was associated with both a longer response time and greater activation in bilateral ventro-lateral prefrontal cortices. Lee *et al.* (2002) found healthy adults, who feign memory impairments when tested using a force-choice format, showed greater activation in a prefrontal–parietal–subcortical circuit. A third study on two different types of lying (well rehearsed lies versus spontaneous lies) found that both types elicited more bilateral activation in the anterior prefrontal cortex, bilateral parahippocampal gyrus, right precuneus, and left cerebellum (Ganis *et al.*, 2003). Common to all of these studies is increased activation in the prefrontal cortex during lying, particularly regions involved in response conflict and inhibitory control such as the dorsolateral prefrontal cortex, ventrolateral prefrontal cortex, and anterior cingulate cortex (Marchewka *et al.*, 2012).

While these fMRI studies have shown increased bilateral activation in the prefrontal cortex when normal subjects lie, there have been no imaging studies of deceitful, manipulative, conning individuals assessing whether they have *structural* abnormalities to this same brain region. One study from our group has attempted to address this issue. Prefrontal gray and white matter volumes were assessed using sMRI data of 12 pathological liars, 16 antisocial controls and 21 normal controls (Yang *et al.*, 2005a). Subjects were defined as "pathological liars" if: (1) they fulfilled criteria for pathological lying on the Psychopathy Checklist-Revised (PCL-R), or (2) they fulfilled criteria for conning/manipulative on the PCL-R, or (3) they met the deceitfulness criterion for DSM-IV antisocial personality disorder (lifelong repeated lying, use of aliases, or conning others for personal profit or please), or (4) they admitted to malingering (i.e. telling lies to obtain sickness benefits). Liars showed a 22–26 percent increase in prefrontal white matter and a 36–42 percent reduction in prefrontal gray/white ratios compared with both antisocial controls and normal controls. Findings provide the first evidence of a structural brain deficit in liars, and indicate that excessive prefrontal white matter relative to gray may act as a predisposition to these specific features of psychopathy. Findings are supported by fMRI studies of deception among antisocial, psychopathic individuals. For example, Jiang *et al.* (2013) found neural activities in the middle frontal, superior frontal, and anterior cingulate cortex to be associated with the capacity for deception in adult offenders with APD. Overall, findings suggest that brain abnormalities, specifically in the frontal regions as mentioned above, may be specific to the lying, manipulative, and conning characteristics of psychopathy.

Successful versus unsuccessful psychopaths

Another form of psychopathy that has received increasing interest is the subgroup of "successful" psychopaths. Using newspaper advertisements, Widom (1977) identified a sample of psychopaths that was relatively successful in avoiding the negative consequences of antisocial behaviors (who he labeled "successful" psychopaths) compared with their counterparts in prisons, hospitals, and treatment programs (who he labeled "unsuccessful" psychopaths). Interestingly, there appear to be neurobiological contributions to whether or not they have the ability to avoid incarceration. One initial study indicated that unsuccessful psychopaths show reduced autonomic stress reactivity and executive function deficits compared with controls (Ishikawa *et al.*, 2001). Successful psychopaths, in contrast, showed heightened autonomic stress reactivity and performed better at the Wisconsin Card Sorting Test than unsuccessful psychopaths and controls. Because both autonomic and executive functional deficits result from structural damage to the prefrontal cortex (Damasio, 1994; Bechara *et al.*, 1997), these initial studies would lead to the prediction that unsuccessful, but not successful psychopaths would show prefrontal structural impairments. As mentioned above, one early study showed that psychopaths had an exaggerated structural anterior hippocampal asymmetry (right > left), but the feature was prominent only in unsuccessful psychopaths, not successful psychopaths, compared with controls (Raine *et al.*, 2004), suggesting that brain structural impairments may be specific to unsuccessful forms of psychopathy.

Using the same population, findings from our group revealed gray matter volume reduction in the prefrontal cortex, previously observed in a group of adults with psychopathy, to be significant only in unsuccessful psychopaths, not successful psychopaths, compared with controls (Yang *et al.*, 2005b). In a subsequent study, we further showed reduced cortical thickness in the middle frontal and orbitofrontal cortex, and reduced volume and surface shape in the amygdala, to be more prominent in unsuccessful psychopaths compared with their successful counterparts (Yang *et al.*, 2010). It was hypothesized that the hippocampal asymmetry and amygdala abnormalities may reflect an underlying neurodevelopmental abnormality, which results in emotional dysregulation, poor learning of aversive conditioning, and insensitivity to cues predicting capture in unsuccessful psychopaths. Similarly, reduced prefrontal gray matter volume and thickness is associated with poor judgment and decision-making, thus rendering unsuccessful psychopaths potentially more prone to conviction. The subgroup of successful psychopaths suggests that brain abnormalities found in one group of psychopaths may not necessarily apply to another group, hence extra caution should be taken in the interpretation and generalization of findings.

Neurodevelopmental contributions to psychopathy

Although abnormalities in the frontal-limbic-striatal structures, including the amygdala, hippocampus, parahippocampal gyrus, caudate, putamen, orbitofrontal cortex, anterior cingulate cortex, and insula, have been reported in adults with psychopathy, the possibility remains that brain impairments observed could be a consequence of a psychopathic, antisocial lifestyle rather than a cause. Nonetheless, initial evidence has begun to emerge supporting the long suspected neurodevelopmental basis to psychopathy.

One line of evidence has come from the identification of markers for fetal neural maldevelopment such as the cavum septum pellucidum. The septum pellucidum is one component of the septum and consists of a deep, midline structure made up of two translucent leaves of glia separating the lateral ventricles, forming part of the septohippocampal system. The closure of these two leaves is attributed to rapid development of midline structures such as the hippocampus, amygdala, and corpus callosum, and the lack of such limbic development may result in preservation of the cavum septum pellucidum into adulthood. In a recent study, Raine and colleagues (2010) found that a more extensively preserved cavum septum pellucidum is associated with significantly higher levels of antisocial personality and psychopathy. They further argued that, despite the use of a cross-sectional sample, this finding is consistent with prospective, longitudinal studies on babies, infants, and young children showing that prenatal substance exposure, prenatal malnutrition, and early postnatal malnutrition are associated with long-term outcomes of elevated antisocial, psychopathic traits (Raine *et al.*, 2010). These findings are in line with findings from White *et al.* (2013) showing a large cavum septum pellucidum to be associated with disruptive behavior disorders, proactive aggression, and level of psychopathic traits in children.

Another line of evidence has come largely from studies of children with traumatic brain injuries (TBIs), which showed that TBIs may result in functional impairments that increase the risk for psychopathic-related traits. For example, Hux *et al.* (1998) reported that half of the children with antisocial behavior they studied had experienced a TBI (defined as having ever received a "blow to the head"), while one-third of antisocial children with TBIs were thought (by their parents) to have suffered adverse, long-term behavioral problems including diminished attentional capacity, impaired interpersonal skills, and poor school performance. Another study conducted by Craswell *et al.* (2004) found 27.7 percent of the children with antisocial behavior to have TBIs (defined as a "significant head injury involving loss of consciousness/amnesia with ongoing cognitive or social impairment"). Several longitudinal studies using large samples have also consistently shown an elevated incidence of delinquency among children and adolescents who experienced brain trauma. Although the definition of TBIs varied between these studies, findings provided strong evidence suggesting that brain abnormalities may be present prior to the development of psychopathic traits.

Last but not least, several recent studies have begun to explore the link between brain abnormalities and psychopathic traits in children and adolescents. Despite the accumulating knowledge in the neural bases of psychopathy, studies that examine psychopathic traits in children and adolescents remain scarce. By revealing that structural abnormalities in regions overlap with those found impaired in adults with psychopathy, findings from studies of disruptive behavior disorder (including conduct disorders and oppositional defiant disorder) may provide us insight into a potential neurodevelopmental basis to psychopathy (Blair, 2006; Gao *et al.*, 2009). To date, the majority of findings linking brain abnormalities to psychopathy in developmental populations have come from studies of adolescents with conduct problems (Kruesi *et al.*, 2004; Sterzer *et al.*, 2007; Boes *et al.*, 2008; Huebner *et al.*, 2008; Dalwani *et al.*, 2011; Fairchild *et al.*, 2011; Hyatt *et al.*, 2012). For example, one early study found (statistically nonsignificant) reduced prefrontal and temporal volumes in children with early onset conduct disorder (Kruesi *et al.*, 2004). Similarly, Sterzer and colleagues (2007) showed reduced gray matter volume in bilateral temporal lobes, left amygdala, left hippocampus, orbitofrontal cortex, and ventromedial prefrontal cortex in male adolescents with early onset conduct disorder compared with healthy controls. In one recent study of typically developing children, Wallace *et al.* (2012) reported significant correlation between higher antisocial traits and thinner cortex most prominently in the bilateral anterior prefrontal cortices.

Despite the scarcity of findings from healthy developmental populations, a few studies have explored the neural correlates of psychopathic traits, specifically the callous-unemotional traits, in children and adolescents. For example, in a group of children with conduct problems, callous-unemotional traits were found to correlate negatively with reduced volume in the striatum (Fairchild *et al.*, 2013); however, the effect did not survive when controlling for conduct disorder symptoms. Using a large sample of about 200 participants, Ermer *et al.* (2013) found decreased gray matter volumes in the orbitofrontal cortex, parahippocampal gyrus, and posterior cingulate cortex to be associated with increased psychopathic traits in male adolescents. In addition, they showed that regional volume reductions in the orbitofrontal, anterior/posterior cingulate cortex, parahippocampal gyrus, and temporal pole account for 20 percent of the variance in total psychopathy scores. Although the gray matter integrity of psychopathic traits in children and adolescents remains unclear, findings to date suggest potential delays in brain maturation may contribute to increased psychopathic traits in children and adolescents, at least for the callous-unemotional traits. Overall, structural brain-imaging findings to date provide initial evidence supporting that abnormal gray matter volumes in the fronto-limbic-striatal regions may be potential neurodevelopmental biomarkers for increased psychopathic tendencies in children and adolescents.

In support of structural brain imaging findings, several recent fMRI studies have also demonstrated a possible link between psychopathic-related traits and functional disturbances in children and adolescents. One area of interest is the neural processing of empathic pains in children with conduct problems and/or elevated psychopathic traits. For example, Decety *et al.* (2009) found that children with conduct disorders showed reduced neural responses to others in pain in regions including the amygdala, striatum, insula, and anterior cingulate cortex compared with healthy controls. Similarly, Cheng *et al.* (2012) found reduced responses to others' pain in children with conduct problems and high levels of callous-unemotional traits. Lockwood *et al.* (2013) found reduced activity in the inferior frontal, insula, and anterior cingulate cortex in children with conduct problems while viewing pictures of others in pain, and that callous-unemotional traits in particular were negatively associated with the neural responses to others' pain in the insula and anterior cingulate cortex.

In another study, Marsh *et al.* (2013) found adolescents with psychopathic traits to show reduced activity in rostral anterior cingulate cortex, ventral striatum (putamen), and amygdala during viewing photographs of pain-inducing injuries (they were instructed to imagine either that the body in each photograph was their own or that it belonged to another person). They further reported that empathic pain responses within amygdala and rostral anterior cingulate cortex were negatively correlated with severity of psychopathic traits. In another study of juvenile offenders with disruptive behavior disorder, increased callous-unemotional traits were again found to be associated with lower level of activation in the anterior cingulate cortex; however, *increased* activation in the region (and the right insula) was associated with higher levels of impulsive-irresponsible and grandiose-manipulative traits.

Another area of interest is the neural activity in response to affect-related information processing in children with psychopathic traits. For example, Marsh *et al.* (2008, 2013) found that children with callous-unemotional traits showed reduced amygdala activation while processing fearful expressions (but not neural or angry expressions) compared with healthy controls and children with attention deficit hyperactivity disorder. In addition, they found that symptom severity in the callous-unemotional traits was negatively correlated with functional connectivity between amygdala and ventromedial prefrontal cortex. This is consistent with findings of Jones and colleagues (2009) which showed boys with conduct problems and elevated levels of callous-unemotional traits had reduced amygdala activity in response to fearful expressions compared with healthy controls. Similarly, Finger *et al.* (2011) reported that, relative to healthy children, children with disruptive behavior disorders and psychopathic traits showed less activity in the orbitofrontal cortex in response to both early stimulus-reinforcement exposure and rewards, and less activity in caudate in response to early stimulus-reinforcement exposure. Although there was no group difference in amygdala activity, amygdala responsiveness throughout the tasks was lower in children with psychopathic traits. More complex findings were reported by White *et al.* (2012), who showed diminished neural response in the amygdala in response to fearful expression under low versus high attentional load conditions in children with disruptive behavior disorders and psychopathic traits. Furthermore, they reported a selective inverse relationship between the neural response to fearful expression under low attentional load and the callous-unemotional features of psychopathy.

Overall, findings from these studies are suggestive that a disturbed neural circuitry involving the prefrontal cortex, amygdala, striatum, and associated regions may contribute to elevated psychopathic tendencies in children and adolescents. The overlapping in findings between children and adults presents initial evidence suggesting possible neural maldevelopment in psychopathy; however, longitudinal research is required to further test the hypothesis of psychopathy as a neurodevelopmental disorder.

Biosocial interactions

There have been very few brain-imaging studies to date that evaluate interactions between social influences and brain abnormalities in predisposing to psychopathy. For example, several studies conducted by Lewis *et al.* (1988) in young children, adolescents, and adults have found that across the lifespan, exposure to violence and abuse in the family is the strongest factor that leads to antisocial behavior and psychopathic traits in individuals with neurological impairment. Alternatively, there is also evidence that social factors can act as protective factors to prevent individuals with brain damage from becoming antisocial. Mataro *et al.* (2001) described a patient in Spain who suffered an accident to his frontal lobe in 1937, which resulted in a frontal lesion similar to that of Phineas Gage. However, unlike Gage, he did not show an increase in antisocial, psychopathic behavior. It was suggested that such a difference in outcome may be due to the fact that the Spanish man had a family who were highly protective and caring. Overall, these findings suggest that, for structurally impaired individuals, negative social factors such as child abuse may increase the risk of one engaging in criminal behavior, whereas positive social factors such as nurturing family members may dramatically reduce such risk.

Despite the increasing imaging findings in psychopathy, very few brain-imaging studies have made a direct examination of biosocial interactions in the development of psychopathy-related traits. The first one, conducted by Raine *et al.* (1998b), used positron emission tomography to address the issue of how social deficits moderate the relationship between brain function and antisocial, psychopathic behavior. In brief, they divided a sample of murderers into those with and those without psychosocial deprivation. Specifically, ratings of psychosocial deprivation in this study took into account several social risk variables, including early physical and sexual abuse, neglect, extreme poverty, foster home placement, having a criminal parent, severe family conflict, and a broken home. Compared with normal controls, murderers with psychosocial deprivation showed relatively good prefrontal functioning, whereas nondeprived murderers showed significantly reduced prefrontal functioning. Specifically, a 14.2 percent reduction in functioning in the right orbitofrontal cortex was found in murderers from good homes. These results suggest that the association between biological impairment and antisocial behavior is more prominent in those lacking social risk factors for antisocial behavior. Using fMRI, Raine *et al.* (2001) found a similar effect of biosocial interactions by comparing violent individuals with and without a child abuse history. More specifically, they found that violent offenders who had suffered severe child abuse showed reduced right hemispheric functioning, particularly in the right temporal cortex. Further analyses revealed that abused individuals who had refrained from serious violence showed relatively lower left, but higher right, activation of the temporal lobe. The results further suggest that a higher-functioning right temporal region may act as a protective factor in preventing those with social risk factors from committing criminal acts.

Although the findings are rare, these previous studies provide initial evidence suggesting that brain abnormalities, when combined with a social risk factor, may predispose an individual to psychopathy. Perhaps equally important is that the findings also raise the possibility for social protective factors to prevent a potential psychopathic outcome, even with the presence of neurobiological predispositions. Nevertheless, the findings to date are still a long way from elucidating the complicated interactive processes of the brain and environment as they relate to outcomes of psychopathy. One specific goal for future studies will be to determine more specifically which brain morphological/functional factors play an important role in this biosocial interactive process leading to psychopathy. Another goal will be to determine whether particular social risk factors interact with particular brain predispositions to lead to the manifestation of psychopathy (e.g. environmental stress may be more likely to trigger antisocial behavior if a biological predisposition of amygdala deficits is present, lowering the person's ability to handle stressful events). Future developmental studies with more sophisticated assessments on the environment and the brain will provide invaluable information for understanding the etiological equation of psychopathy.

Conclusion

By reviewing the imaging studies on psychopaths, several brain abnormalities were found to be associated with psychopathy. The emotion deficits of psychopathy are argued to be associated with impairments in limbic and paralimbic structures, particularly the amygdala, hippocampus, orbitofrontal cortex, anterior cingulate cortex, and insula. Regarding the antisocial behavior of psychopathy, it was hypothesized that prefrontal deficits, particularly in the orbitofrontal cortex, dorsolateral prefrontal cortex, and the anterior/posterior cingulate cortex could predispose individuals to poor behavioral control and decision-making, and ultimately antisocial behavior. On the other hand, structural or functional differences in the striatum and closely connected structures may contribute to reward-seeking behavior in psychopathy. Although evidence to date suggests a potential neurodevelopmental basis to psychopathic features, the issue of causality remains unclear. It is possible that brain abnormalities lead to psychopathic behavior and personality. However, it is unclear whether maintaining a psychopathic lifestyle may result in the observed brain deficits. Furthermore, empirical findings to date remain inconclusive, in part due to heterogeneity among study samples. Some initial findings suggest distinct neural underpinnings for subgroups of psychopathy (e.g. proactive or reactive aggression, successful or unsuccessful psychopaths); however, more studies are needed to understand the brain characteristics associated with different forms of psychopathy. Last but not least, findings from the very few studies examining biosocial interactions support that the association between brain abnormalities and psychopathy may be more prominent in those lacking social risk factors. Future prospective, longitudinal research on psychopathy and its subgroups, incorporating the assessment of psychosocial factors, is critical for understanding the role of brain abnormalities in the development of psychopathy.

References

Aoki, Y., Inokuchi, R., Nakao, T., and Yamasue, H. (2013) 'Neural bases of antisocial behavior: a voxel-based meta-analysis', *Social Cognition and Affective Neuroscience*, doi:10.1093/scan/nst104.

Barkataki, I., Kumari, V., Das, M., Taylor, P., and Sharma, T. (2006) 'Volumetric structural brain abnormalities in men with schizophrenia or antisocial personality disorder', *Behavior and Brain Research*, 169: 239–247.

Barkataki, I., Kumari, V., Das, M., Sumich, A., Taylor, P., and Sharma, T. (2008) 'Neural correlates of deficient response inhibition in mentally disordered violent individuals', *Behavioral Sciences and the Law*, 26: 51–64.

Barros-Loscertales, A., Meseguer, V., Sanjuan, A., Belloch, V., Parcet, M. A., Torrubia, R., and Avila, C. (2006) 'Striatum gray matter reduction in males with an overactive behavioral activation system', *European Journal of Neuroscience*, 24: 2071–2074.

Bechara, A., Damasio, H., Tranel, D., and Damasio, A. R. (1997) 'Deciding advantageously before knowing the advantageous strategy', *Science*, 275: 1293–1295.

Bertsch, K., Grothe, M., Prehn, K., Vohs, K., Berger, C., Hauenstein, K., *et al.* (2013) 'Brain volumes differ between diagnostic groups of violent criminal offenders', *European Archives of Psychiatry and Clinical Neuroscience*, 263(7): 593–606.

Birbaumer, N., Veit, R., Lotze, M., Erb, M., Hermann, C., Grodd, W., and Flor, H. (2005) 'Deficient fear conditioning in psychopathy: a functional magnetic resonance imaging study', *Archives of General Psychiatry*, 62: 799–805.

Blair, R. J. R. (2006) 'The emergence of psychopathy: implications for the neuropsychological approach to developmental disorders', *Cognition*, 101: 414–442.

Blair, R. J. R. (2013) 'The neurobiology of psychopathic traits in youths', *Nature Reviews Neuroscience*, 14: 786–799.

Boccardi, M., Ganzola, R., Rossi, R., Sabattoli, F., Laakso, M. P., Repo-Tiihonen, E., Vaurio, O., Kononen, M., Aronen, H., Thompson, P. M., Frisoni, G. B., and Tiihonen, J. (2010) 'Abnormal hippocampal shape in offenders with psychopathy', *Human Brain Mapping*, 31: 438–447.

Boccardi, M., Frisoni, G. B., Hare, R. D., Cavedo, E., Najt, P., Pievani, M., Rasser, P. E., Laakso, M. P., Aronen, H. J., Repo-Tiihonen, E., Vaurio, O., Thompson, P. M., and Tiihonen, J. (2011) 'Cortex and amygdala morphology in psychopathy', *Psychiatry Research*, 193: 85–92.

Boccardi, M., Bocchetta, M., Aronen, H., Repo-Tiihonen, E., Vaurio, O., Thompson, P., Tiihonen, J., and Frisoni, G. B. (2013) 'Atypical nucleus accumbens morphology in psychopathy: another limbic piece in the puzzle', *International Journal of Law and Psychiatry*, 36: 157–167.

Boes, A. D., Tranel, D., Andersson, S. W., and Nopoulos, P. (2008) 'Right anterior cingulate: a neuroanatomical correlate of aggression and defiance in boys', *Behavioral Neuroscience*, 122: 677–684.

Bush, G., Luu, P., and Posner, M. I. (2000) 'Cognitive and emotional influences in anterior cingulate cortex', *Trends in Cognitive Science*, 4: 215–222.

Carswell, K., Maughan, B., Davis, H., Davenport, F., and Goddard, N. (2004) 'The psychosocial needs of young offenders and adolescents from an inner city area', *Journal of Adolescence*, 27: 415–428.

Cheng, Y., Hung, A. Y., and Decety, J. (2012) 'Dissociation between affective sharing and emotion understanding in juvenile psychopaths', *Development and Psychopathology*, 24: 623–636.

Cleckley, H. (1941) *The Mask of Sanity*, St Louis, MO: Mosby.

Coccaro, E. F., McCloskey, M. S., Fitzgerald, D. A., and Phan, K. L. (2007) 'Amygdala and orbitofrontal reactivity to social threat in individuals with impulsive aggression', *Biological Psychiatry*, 62: 168–178.

Cohen, M. X., Schoene-Bake, J. C., Elger, C. E., and Weber, B. (2009) 'Connectivity-based segregation of the human striatum predicts personality characteristics', *Nature Neuroscience*, 12: 32–34.

Cope, L. M., Shane, M. S., Segall, J. M., Nyalakanti, P. K., Stevens, M. C., Pearlson, G. D., and Kiehl, K. A. (2012) 'Examining the effect of psychopathic traits on gray matter volume in a community substance abuse sample', *Psychiatry Research*, 204: 91–100.

Craswell, K., Maughan, B., Davis, H., Davenport, F., and Goddard, N. (2004) 'The psychosocial needs of young offenders and adolescents from an inner city area', *Journal of Adolescence*, 27: 415–428.

Dalwani, M., Sakai, J. T., Mikulich-Gilbertson, S. K., Tanbe, J., Raymond, K., McWilliamss, S. K., Thompson, L. L., Banich, M. T., and Crowley, T. J. (2011) 'Reduced cortical gray matter volume in male adolescents with substance and conduct problems', *Drug and Alcohol Dependence*, 118: 295–305.

Damasio, A. R. (1994) 'Descartes' error and the future of human life', *Scientific American*, 271: 144.

Davidson, R. J., Putnam, K. M., and Larson, C. L. (2000) 'Dysfunction in the neural circuitry of emotion regulation—a possible prelude to violence', *Science*, 289: 591–594.

de Oliveira-Souza, R., Hare, R. D., Bramati, I. E., Garrido, G. J., Azevedo Ignacio, F., Tovar-Moll, F., and Moll, J. (2008) 'Psychopathy as a disorder of the moral brain: fronto-temporo-limbic grey matter reductions demonstrated by voxel-based morphometry', *Neuroimage*, 40: 1202–1213.

Decety, J., Michalska, K. J., Akitsuki, Y., and Lahey, B. B. (2009) 'Atypical emphatic responses in adolescents with aggressive conduct disorder: a functional MRI investigation', *Biological Psychiatry*, 80: 203–211.

Ermer, E., Cope, L. M., Nyalakanti, P. K., Calhoun, V. D., and Kiehl, K. A. (2012) 'Aberrant paralimbic gray matter in criminal psychopathy', *Journal of Abnormal Psychology*, 121: 649–658.

Ermer, E., Cope, L. M., Nyalakanti, P. K., Calhoun, V. D., and Kiehl, K. A. (2013) 'Aberrant paralimbic gray matter in incarcerated male adolescents with psychopathic traits', *Journal of the American Academy of Child and Adolescent Psychiatry*, 52: 94–103 e3.

Fairchild, G., Passamoti, L., Hurford, G., Hagan, C. C., Von Dem Hagen, E. A., van Goozen, S. H., Goodyer, I., and Calder, A. J. (2011) 'Brain structure abnormalities in early-onset and adolescent-onset conduct disorder', *American Journal of Psychiatry*, 168: 624–633.

Fairchild, G., Hagan, C. C., Walsh, N. D., Passamonti, L., Calder, A. J., and Goodyer, I. M. (2013) 'Brain structure abnormalities in adolescent girls with conduct disorder', *Journal of Child Psychology and Psychiatry*, 54(1): 86–95.

Finger, E. C., Marsh, A. A., Mitchell, D. C., Reid, M. E., Sims, C., Budhani, S., Kosson, D., Chen, G., Towbin, K., Leibenfluft, E., Pine, D. S., and Blair, R. J. R. (2008) 'Abnormal ventromedial prefrontal cortex function in children with psychopathic traits during reversal learning', *Archives of General Psychiatry*, 65: 586–594.

Finger, E. C., Marsh, A. A., Blair, K. S., Reid, M. E., Sims, C., Ng, P., Pine, D. S., and Blair, R. J. R. (2011) 'Disrupted reinforcement signaling in the orbitofrontal cortex and caudate in youths with conduct disorder or oppositional defiant disorder and a high level of psychopathic traits', *American Journal of Psychiatry*, 168: 152–162.

Forth, A. E., Kosson, D., and Hare, R. D. (2003) *The Hare PCL: Youth Version*, Toronto, Ontario, Canada: Multi-Health Systems.

Ganis, G., Kosslyn, S. M., Stose, S., Thompson, W. L., and Yurgelun-Todd, D. A. (2003) 'Neural correlates of different types of deception: an fMRI investigation', *Cerebral Cortex*, 13: 830–836.

Gao, Y., Glenn, A. L., Schug, R. A., Yang, Y., and Raine, A. (2009) 'The neurobiology of psychopathy: a neurodevelopmental perspective', *Canadian Journal of Psychiatry*, 54: 813–823.

Glenn, A. L. and Yang, Y. (2012) 'The potential role of the striatum in antisocial behavior and psychopathy', *Biological Psychiatry*, 72: 817–822.

Glenn, A. L., Raine, A., Yaralian, P. S., and Yang, Y. (2009) 'Increased volume of the striatum in psychopathic individuals', *Biological Psychiatry*, 67: 52–58.

Gordon, H. L., Baird, A. A., and End, A. (2004) 'Functional differences among those high and low on a trait measure of psychopathy', *Biological Psychiatry*, 56: 516–521.

Goyer, P. F., Andreason, P. J., Semple, W. E., and Clayton, A. H. (1994) 'Positron-emission tomography and personality disorders', *Neuropsychopharmacology*, 20: 21–28.

Han, T., Alders, G. L., Greening, S. G., Neufeld, R. W., and Mitchell, D. G. (2012) 'Do fearful eyes activate empathy-related brain regions in individuals with callous traits?', *Social Cognition and Affective Neuroscience*, 7: 958–968.

Hare, R. D. (1991) *The Hare Psychopathy Checklist-Revised*, Toronto, Canada: Multi-Health Systems.

Hare, R. D. (2003) *Manual for the Revised Psychopathy Checklist*, Toronto, ON, Canada: Multi-Health Systems.

Hare, R. D. and McPherson, L. M. (1984) 'Violent and aggressive behavior by criminal psychopaths', *International Journal of Law and Psychiatry*, 7: 35–50.

Hiatt, K. D., Schmitt, W. A., and Newman J. P. (2004) 'Stroop tasks reveal abnormal selective attention among psychopathic offenders', *Neuropsychology*, 18: 50–59.

Huebner, T., Vloet, T. D., Marx, I., Konrad, K., Fink, G. R., Herpertz, S. C., and Herpertz-Dahlmann, B. (2008) 'Morphometric brain abnormalities in boys with conduct disorder', *Journal of the American Academy of Child and Adolescent Psychiatry*, 47: 540–547.

Hux, K., Bond, V., Skinner, S., Belau, D., and Sanger, D. (1998) 'Parental report of occurences and consequences of traumatic brain injury among delinquent and non-delinquent youth', *Brain Injury*, 12: 667–681.

Hyatt, C. J., Haney-Caron, E., and Stevens, M. C. (2012) 'Cortical thickness and folding deficits in conduct-disordered adolescents', *Biological Psychiatry*, 72: 207–214.

Ishikawa, S. S., Raine, A., Lencz, T., Bihrle, S., and Lacasse, L. (2001) 'Autonomic stress reactivity and executive functions in successful and unsuccessful criminal psychopaths from the community', *Journal of Abnormal Psychology*, 110: 423–432.

Jiang, W., Liu, H., Liao, J., Ma, X., Rong, P., Tang, Y., and Wang, W. (2013) 'A functional MRI study of deception among offenders with antisocial personality disorders', *Neuroscience*, 244: 90–98.

Jones, A. P., Laurens, K. R., Herba, C. M., Barker, G. J., and Viding, E. (2009) 'Amygdala hypoactivity to fearful faces in boys with conduct problems and callous-unemotional traits', *American Journal of Psychiatry*, 166: 95–102.

Kiehl, K. A., Smith, A. M., Hare, R. D., and Liddle, P. F. (2000) 'An event-related potential investigation of response inhibition in schizophrenia and psychopathy', *Biological Psychiatry*, 48: 210–221.

Kiehl, K. A., Smith, A. M., Hare, R. D., Mendrek, A., Forster, B. B., Brink, J., and Liddle, P. F. (2001) 'Limbic abnormalities in affective processing by criminal psychopaths as revealed by functional magnetic resonance imaging', *Biological Psychiatry*, 50: 677–684.

Kiehl, K., Smith, A. M., Mendrek, A., Forster, B. B., Hare, R. D., and Liddle, P. F. (2004) 'Temporal lobe abnormalities in semantic processing by criminal psychopaths as revealed by functional magnetic resonance imaging', *Psychiatry Research Neuroimaging*, 130: 27–42.

Kosson, D. S., Neumann, C. S., Forth, A. E., Salekin, R. T., Hare, R. D., Krischer, M. K., and Sevecke, K. (2013) 'Factor structure of the Hare Psychopathy Checklist: Youth Version (PCL: YV) in adolescent females', *Psychological Assessment*, 25: 71–83.

Krain, A. L., Wilson, A. M., Arbuckle, R., Castellanos, F. X., and Milham, M. P. (2006) 'Distinct neural mechanisms of risk and ambiguity: a meta-analysis of decision-making', *Neuroimage*, 32: 477–484.

Kringelbach, M. L. (2005) 'The human orbitofrontal cortex: linking reward to hedonic experience', *Nature Reviews Neuroscience*, 6: 691–702.

Kringelbach, M. L. and Rolls, E. T. (2004) 'The functional neuroanatomy of the human orbitofrontal cortex: evidence from neuroimaging and neuropsychology', *Progress in Neurobiology*, 72: 341–372.

Kruesi, M. J., Casanova, M. F., Mannheim, G., and Johnson-Bilder, A. (2004) 'Reduced temporal lobe volume in early onset conduct disorder', *Psychiatry Research*, 132: 1–11.

Kumari, V., Aasen, I., Taylor, P., Feytche, D. H., Das, M., Barkataki, I., Goswami, S., O'Connell, P., Howlett, M., Williams, S. C., and Sharma, T. (2006) 'Neural dysfunction and violence in schizophrenia: an fMRI investigation', *Schizophrenia Research*, 84: 144–164.

Laakso, M. P., Vaurio, O., Koivisto, E., Savolainen, L., Eronen, M., Aronen, H. J., Hakola, P., Repo, E., Soininen, H., and Tiihonen, J. (2001) 'Psychopathy and the posterior hippocampus', *Behavioral Brain Research*, 118: 187–193.

Lapierre, D., Braun, C. M., and Hodgins, S. (1995) 'Ventral frontal deficits in psychopathy: neuropsychological test findings', *Neuropsychologia*, 33: 139–151.

Larson, C. L., Baskin-Sommers, A. R., Stout, D. M., Balderston, N. L., Curtin, J. J., Schultz, D. H., Kiehl, K. A., and Newman, J. P. (2013) 'The interplay of attention and emotion: top-down attention modulates amygdala activation in psychopathy', *Cognitive and Affective Behavioral Neuroscience*, 13: 757–770.

Lee, T. M., Liu, H. L., Tan, L. H., Chan, C. C., Mahankali, S., Feng, C. M., Hou, J., Fox, P. T., and Gao, J. H. (2002) 'Lie detection by functional magnetic resonance imaging', *Human Brain Mapping*, 15: 157–164.

Lewis, D. O., Pincus, J. H., Bard, B., Richardson, E., Prichep, L. S., Feldman, M., and Yeager, C. (1988) 'Neuropsychiatric, psychoeducational, and family characteristics of 14 juveniles condemned to death in the United States', *American Journal of Psychiatry*, 145: 584–589.

Li, C. S., Kosten, T. R., and Sinha, R. (2006) 'Antisocial personality and stress-induced brain activation in cocaine-dependent patients', *Neuroreport*, 17: 243–247.

Lilienfeld, S. O. and Andrews, B. P. (1996) 'Development and preliminary validation of a self-report measure of psychopathic personality traits in noncriminal populations', *Journal of Personality Assessment*, 66: 488–524.

Lockwood, P. L., Sebastian, C. L., McCrory, E. J., Hyde, Z. H., Gu, X., De Brito, S. A., and Viding, E. (2013) 'Association of callous traits with reduced neural response to others' pain in children with conduct problems', *Current Biology*, 23: 901–905.

Ly, M., Motzkin, J. C., Philippi, C. L., Kirk, G. R., Newman, J. P., Kiehl, K. A., and Koenigs, M. (2012) 'Cortical thinning in psychopathy', *American Journal of Psychiatry*, 169: 743–749.

Lynam, D. R. (1997) 'Pursuing the psychopath: capturing the fledgling psychopath in a nomological net', *Journal of Abnormal Psychology*, 106: 425–438.

Marchewka, A., Jednorog, K., Falkiewicz, M., Szeszkowski, W., Grabowska, A., and Szatkowska, I. (2012) 'Sex, lies and fMRI—gender differences in neural basis of deception', *PLOS ONE*, 7: e43076.

Marsh, A. A. and Blair, R. J. (2008) 'Deficits in facial affect recognition among antisocial populations: a meta-analysis', *Neuroscience and Biobehavioral Reviews*, 32: 454–465.

Marsh, A. A. and Cardinale, E. M. (2014) 'When psychopathy impairs moral judgments: neural responses during judgments about causing fear', *Social Cognitive and Affective Neuroscience*, 9: 3–11.

Marsh, A. A., Finger, E. C., Mitchell, D. G., Reid, M. E., Sims, C., Kosson, D. S., Towbin, K. E., Leibenluft, E., Pine, D. S., and Blair, R. J. (2008) 'Reduced amygdala response to fearful expressions in children and adolescents with callous-unemotional traits and disruptive behavior disorders', *American Journal of Psychiatry*, 165: 712–720.

Marsh, A. A., Finger, E. C., Fowler, K. A., Adalio, C. J., Jurkowitz, I. T., Schechter, J. C., Pine, D. S., Decety, J., and Blair, R. J. (2013) 'Empathic responsiveness in amygdala and anterior cingulate cortex in youths with psychopathic traits', *Journal of Child Psychology and Psychiatry*, 54: 900–910.

Mataro, M., Jurado, M. A., Garcia-Sanchez, C., Barraquer, L., Costa-Jussa, F. R., and Junque, C. (2001) 'Long-term effects of bilateral frontal brain lesion: 60 years after injury with an iron bar', *Archives of Neurology*, 58: 1139–1142.

Muller, J. L., Sommer, M., Wagner, V., Lange, K., Taschler, H., Roder, C. H., Schuierer, G., Klein, H. E., and Hajak, G. (2003) 'Abnormalities in emotion processing within cortical and subcortical regions in criminal psychopaths: evidence from a functional magnetic imaging study using pictures with emotional content', *Psychiatry Research Neuroimaging*, 54: 152–162.

Muller, J. L., Sommer, M., Dohnel, K., Weber, T., Schmidt-Wilcke, T., and Hajak, G. (2008) 'Disturbed prefrontal and temporal brain function during emotion and cognition interaction in criminal psychopathy', *Behavioral Sciences and the Law*, 26: 131–150.

Munro, G. E., Dywan, J., Harris, G. T., McKee, S., Unsal, A., and Segalowitz, S. J. (2007) 'Response inhibition in psychopathy: the frontal N2 and P3', *Neuroscience Letters*, 418: 149–153.

Newman, J. P. and Kosson, D. S. (1986) 'Passive avoidance learning in psychopathic and nonpsychopathic offenders', *Journal of Abnormal Psychology*, 95: 252–256.

Osumi, T., Nakao, T., Kasuya, Y., Shinoda, J., Yamada, J., and Ohira, H. (2012) 'Amygdala dysfunction attenuates frustration-induced aggression in psychopathic individuals in a non-criminal population', *Journal of Affective Disorders*, 142: 331–338.

Pardini, D. A., Raine, A., Erickson, K., and Loeber, R. (2014) 'Lower amygdala volume in men is associated with childhood aggression, early psychopathic traits, and future violence', *Biological Psychiatry*, 75: 73–80.

Patrick, C. J. (1994) 'Emotion and psychopathy: startling new insights', *Psychophysiology*, 31: 319–330.

Prehn, K., Schlagenhauf, F., Schulze, L., Berger, C., Vohs, K., Fleischer, M., Hauenstein, K., Keiper, P., Domes, G., and Herpertz, S. C. (2013) 'Neural correlates of risk taking in violent criminal offenders characterized by emotional hypo- and hyper-reactivity', *Social Neuroscience*, 8: 136–147.

Pujara, M., Motzkin, J. C., Newman, J. P., Kiehl, K. A., and Koenigs, M. (2014) 'Neural correlates of reward and loss sensitivity in psychopathy', *Social Cognitive and Affective Neuroscience*, 9(6): 794–801.

Raine, A., Reynolds, C., Venables, P. H., and Mednick, S. A. (1997) 'Biosocial bases of aggressive behavior in childhood', in Raine, A., Brennan, P. A., Farrington, D. P., and Mednick, S. A. (eds), *Biosocial Bases of Violence*, New York: Plenum.

Raine, A., Meloy, J. R., Bihrle, S., Stoddard, J., Lacasse, L., and Buchsbaum, M. S. (1998a) 'Reduced prefrontal and increased subcortical brain functioning assessed using positron emission tomography in predatory and affective murderers', *Behavioral Sciences and the Law*, 16: 319–332.

Raine, A., Phil, D., Stoddard, J., Bihrle, S., and Buchsbaum, M. (1998b) 'Prefrontal glucose deficits in murderers lacking psychosocial deprivation', *Neuropsychiatry, Neuropsychology, and Behavioral Neurology*, 11: 1–7.

Raine, A., Reynolds, C., Venables, P. H., Mednick, S. A., and Farrington, D. P. (1998c) 'Fearlessness, stimulation-seeking, and large body size at age 3 years as early predispositions to childhood aggression at age 11 years', *Archives of General Psychiatry*, 55: 745–751.

Raine, A., Lencz, T., Bihrle, S., Lacasse, L., and Colletti, P. (2000) 'Reduced prefrontal gray matter volume and reduced autonomic activity in antisocial personality disorder', *Archives of General Psychiatry*, 57: 119–129.

Raine, A., Park, S., Lencz, T., Bihrle, S., Lacasse, L., Widom, C. S., Al-Dayey, L., and Singh, M. (2001) 'Reduced right hemisphere activation in severely abused violent offenders during a working memory task: an fMRI study', *Aggressive Behavior*, 27: 111–129.

Raine, A., Ishikawa, S. S., Arce, E., Lencz, T., Knuth, K. H., Bihrle, S., Lacasse, L., and Colletti, P. (2004) 'Hippocampal assymmetry in unsuccessful psychopaths', *Biological Psychiatry*, 55: 185–191.

Raine, A., Lee, L., Yang, Y., and Colletti, P. (2010) 'Neurodevelopmental marker for limbic maldevelopment in antisocial personality disorder and psychopathy', *British Journal of Psychiatry*, 197: 186–192.

Raine, A., Yang, Y., Narr, K. L., and Toga, A. W. (2011) 'Sex differences in orbitofrontal gray as a partial explanation for sex differences in antisocial personality', *Molecular Psychiatry*, 16: 227–236.

Rilling, J. K., Glenn, A. L., Jairam, M. R., Pagnoni, G., Goldsmith, D. R., Elfenbein, H. A., and Lilienfeld, S. O. (2007) 'Neural correlates of social cooperation and non-cooperation as a function of psychopathy', *Biological Psychiatry*, 61: 1260–1271.

Schiffer, B., Muller, B. W., Scherbaum, N., Hodgins, S., Forsting, M., Wiltfang, J., Gizeewski, E. R., and Leygraf, N. (2011) 'Disentangling structural brain alterations associated with violent behavior from those associated with substance use disorders', *Archives of General Psychiatry*, 68: 1039–1049.

Schiffer, B., Pawliczek, C., Forsting, M., Gizewski, E., Leygraf, N., and Hodgins, S. (2014) 'Neural mechanisms underlying cognitive control of men with lifelong antisocial behavior', *Psychiatry Research: Neuroimaging*, 222(1): 43–51.

Schneider, F., Habel, U., Kessler, C., Posse, S., Grodd, W., and Muller-Gartner, H. W. (2000) 'Functional imaging of conditioned aversive emotional responses in antisocial personality disorder', *Neuropsychobiology*, 42: 192–201.

Soderstrom, H., Tullberg, M., Wikkelso, C., Ekholm, S., and Forsman, A. (2000) 'Reduced regional cerebral blood flow in non-psychotic violent offender', *Psychiatry Research Neuroimaging*, 98: 29–41.

Spence, S. A., Farrow, T. F., Herford, A. E., Wilkinson, I. D., Zheng, Y., and Woodruff, P. W. (2001) 'Behavioural and functional anatomical correlates of deception in humans', *Neuroreport*, 12: 2849–2853.

Steadman, H. J. (2000) 'From dangerousness to risk assessment of community violence: taking stock at the turn of the century', *Journal of the American Academy of Psychiatry and Law*, 28: 265–271.

Sterzer, P., Stadler, C., Poustka, F., and Kleinschmidt, A. (2007) 'A structural neural deficit in adolescents with conduct disorder and its association with lack of empathy', *Neuroimage*, 37: 335–342.

Sundram, F., Deeley, Q., Sarkar, S., Daly, E., Latham, R., Craig, M., Raczek, M., Fahy, T., Picchioni, M., Barker, G. J., and Murphy, D. G. (2012) 'White matter microstructural abnormalities in the frontal lobe of adults with antisocial personality disorder', *Cortex*, 48: 216–229.

van Honk, J. and Schutter, D. J. (2006) 'Unmasking feigned sanity: a neurobiological model of emotion processing in primary psychopathy', *Cognitive Neuropsychiatry*, 11: 285–306.

Verona, E., Patrick, C. J., Curtin, J. J., Bradley, M. M., and Lang, P. J. (2004) 'Psychopathy and physiological response to emotionally evocative sounds', *Journal of Abnormal Psychology*, 113: 99–108.

Vieira, J. B., Almeida, P. R., Ferreira-Santos, F., Barbosa, F., Marques-Teixeria, J., and Marsh, A. (2013) 'Distinct neural activation patterns underlie economic decisions in high and low psychopathy scorers', *Social Cognition and Affective Neuroscience*, doi: 10.1093/scan/nst093.

Völlm, B., Richardson, P., McKie, S., Elliott, R., Deakin, J., and Anderson, I. M. (2006) 'Serotonergic modulation of neuronal responses to behavioural inhibition and reinforcing stimuli: an fMRI study in healthy volunteers', *European Journal of Neuroscience*, 23(2).

Wallace, G. L., Shaw, P., Lee, N. R., Clasen, L. S., Raznahan, A., Lenroot, R. K., Martin, A., and Giedd, J. N. (2012) 'Distinct cortical correlates of autistic versus antisocial traits in a longitudinal sample of typically developing youth', *Journal of Neuroscience*, 32: 4856–4860.

White, S. F., Marsh, A. A., Fowler, K. A., Schechter, J. C., Adalio, C., Pope, K., Sinclair, S., Pine, D. S., and Blair, R. J. (2012) 'Reduced amygdala response in youths with disruptive behavior disorders and psychopathic traits: decreased emotional response versus increased top-down attention to nonemotional features', *American Journal of Psychiatry*, 169: 750–758.

White, S. F., Brislin, S., Sinclair, S., Fowler, K. A., Pope, K., and Blair, R. J. (2013) 'The relationship between large cavum septum pellucidum and antisocial behavior, callous-unemotional traits and psychopathy in adolescents', *Journal of Child Psychology and Psychiatry*, 54: 575–581.

Widom, C. S. (1977) 'A methodology for studying noninstitutionalized psychopaths', *Journal of Consulting and Clinical Psychology*, 45: 674–683.

Yang, Y. and Raine, A. (2009) 'Prefrontal structural and functional brain imaging findings in antisocial, violent, and psychopathic individuals: a meta-analysis', *Psychiatry Research*, 174: 81–88.

Yang, Y., Raine, A., Lencz, T., Bihrle, S., Lacasse, L., and Colletti, P. (2005a) 'Prefrontal white matter in pathological liars', *British Journal of Psychiatry*, 187: 320–325.

Yang, Y., Raine, A., Lencz, T., Bihrle, S., Lacasse, L., and Colletti, P. (2005b) 'Volume reduction in prefrontal gray matter in unsuccessful criminal psychopaths', *Biological Psychiatry*, 57: 1103–1108.

Yang, Y., Raine, A., Colletti, P., Toga, A. W., and Narr, K. L. (2009a) 'Abnormal temporal and prefrontal cortical gray matter thinning in psychopaths', *Molecular Psychiatry*, 14: 561–2, 555.

Yang, Y., Raine, A., Narr, K. L., Colletti, P., and Toga, A. W. (2009b) 'Localization of deformations within the amygdala in individuals with psychopathy', *Archives of General Psychiatry*, 66: 986–994.

Yang, Y., Raine, A., Colletti, P., Toga, A. W., and Narr, K. L. (2010) 'Morphological alterations in the prefrontal cortex and the amygdala in unsuccessful psychopaths', *Journal of Abnormal Psychology*, 119: 546–554.

Yang, Y., Raine, A., Colletti, P., Toga, A. W., and Narr, K. L. (2011) 'Abnormal structural correlates of response perseveration in individuals with psychopathy', *Journal of Neuropsychiatry and Clinical Neuroscience*, 23: 107–110.

20

Self-reported head injuries among delinquent youth

Brian E. Perron, Michael G. Vaughn, Joseph Ryan, Christopher Salas-Wright, Mary Ruffolo, and Erick Guerrero

Introduction

Head injuries may include a number of injuries to the scalp, skull, or brain that are the result of concussions, scalp wounds, skull fractures, and other head trauma (NIH, 2013). The consequences of head injuries depend on a variety of dimensions of injury (e.g. type, location, severity, premorbid functioning, receipt and timing of medical intervention, etc.). If a head injury disrupts the normal function of the brain, it is referred to as a traumatic brain injury, TBI (CDC, 2013). Whether a head injury can be classified as a TBI is determined by the same variety of factors that influenced the consequences of the head injury (type, location, severity). This conflation of factors makes it difficult to assess accurately whether a head injury should be classified as a TBI, and also to predict any long-term outcomes.

Head injuries—specifically injuries that lead to brain impairment—are associated with a variety of behavioral and cognitive problems. For instance, impairments to the frontal lobe are linked to lessened inhibition, and can lead to a variety of risk-taking and impulsive behaviors. Individuals with damage to this portion of the brain demonstrate emotional instability, rigidity (e.g. difficulty modifying behavior), and interpersonal inappropriateness, or the inability to interpret one's effect on others or their emotional state (Hawkins and Trobst, 2000). They may act intrusively or boisterously, speak loudly with free use of profanities, and engage in elevated risk-taking (Silver and Yudofsky, 1987), all of which can lead to confrontation with others and, ultimately, offending behaviors.

Youth aged 15 to 19 years, especially delinquent males, are among the highest risk of suffering a head injury (Langlois *et al.*, 2006). As reviewed by Davies *et al.* (2012), estimates of head injuries among the general youth population range from 5 to 38 percent, whereas delinquent youth estimates are much higher, ranging from 65 to 87 percent. Through a careful meta-analysis of nine studies of juvenile offenders, Farrer *et al.* (2013) found that approximately 30 percent of those offenders had a history of a self-reported head injury. Additionally, five of those nine studies also showed that juvenile offenders were at least three times more likely to report a head injury compared with a non-offending control group.

While the epidemiologic research clearly shows that head injuries are significantly more prevalent among delinquent youth compared with their non-delinquent counterparts, the significance of head injuries is further underscored by a constellation of associated clinical and psychosocial problems.

For example, a study of incarcerated youth in Missouri (Perron and Howard, 2008) revealed that juveniles with a history of a head injury were more likely to report higher levels of mental health problems (e.g. anxiety, depression, hostility), past-year substance use, and criminal activity. Another study looked at 186 youth offenders in England (Huw Williams *et al.*, 2010) and found head injury to be associated with higher levels of violence (two more convictions on average), psychiatric distress and anxiety, and cannabis abuse. Finally, Davies *et al.* (2012) corroborated the previous studies, linking the severity of head injury symptoms with earlier onset of offending behavior and a corresponding increase in convictions and alcohol use.

This line of research provides considerable evidence that head injuries are more prevalent among the delinquent population compared with their non-delinquent counterparts, and that having a history of a head injury is associated with numerous psychosocial and clinical problems. Within a services context, a history of a head injury may be a potentially important marker for identifying delinquent adolescents who are at an elevated risk of future offending behavior. However, our confidence in this risk marker is ultimately limited by the lack of evidence demonstrating that a self-report of a head injury is a valid and reliable measurement.

While brain scans and neurocognitive assessments enable us to measure objectively various dimensions of head injuries, their use as a practical assessment among the full population, or even a subset of the population of adolescent offenders, is cost-prohibitive and impractical. Medical records are another potentially useful data source, although prior research has estimated that 75 percent of head injuries are mild and therefore never reported to a medical professional so that they would show up on a medical record (Caveness, 1979; Miller and Jones, 1985; Kraus and Nourjah, 1988); therefore using medical records to make predictive inferences is seriously problematic.

At present, the only practical alternative is a self-report, which is the most commonly used measurement strategy for studying head injury among juvenile delinquents. A common self-report item asks respondents, "In your lifetime, have you ever had a head injury that caused you to black out for more than 20 minutes" (Perron and Howard, 2008, p. 246). These questions are often followed by follow-up questions that capture different dimensions of the injury (e.g. severity, perceived consequences, receipt of medical services, etc.).

The use of a self-report as a measurement strategy of head injury is heavily criticized because of the wide range of possible biases that are common to self-reports. Moreover, it is virtually impossible to establish the reliability of a self-report through triangulation, particularly from collateral sources. For instance, some head injuries occur when nobody else is present (e.g. accidents), and therefore it is impossible to use another individual's report to corroborate what the injured party has self-reported.

The study

Although various theories of offending behavior and neurocognitive functioning suggest that a head injury increases risk of future offending behavior, the biases that can potentially affect self-reports and the absence of longitudinal research linking self-reported head injuries to offending behavior over time ultimately erode our confidence in the use of a self-reported head injury as a predictive device.

The purpose of this study is to help fill this critical gap in research by examining the empirical relationship between a self-reported head injury and offending behavior over time. We focus on characterizing the strength of this relationship to help clarify the value of the self-report as a predictive device. Within a service context, the value of any indicator as a predictive device must show a meaningful improvement over base-rate predictions—the greater the improvement, the more valuable the indicator. This methodological and theoretical approach is aligned with the work of Paul Meehl, who demonstrated in the early 1950s that statistical predictions of behavior are superior to clinical predictions (Meehl, 1996[1954])—one of the strongest and most consistent findings to date in the behavioral sciences (Kahneman, 2011).

Methods

Sample and procedures

For our study, we used data derived from the Pathways to Desistance study (Mulvey, 2012). Study participants included adolescents between the ages of 14 and 18 who were found guilty of a serious criminal offense in the juvenile or adult court systems in Phoenix, Arizona or Philadelphia, Pennsylvania, USA. Serious criminal offenses included most felonies except for minor property crimes, misdemeanor sexual assault, and weapons offenses. Given the disproportionate percentage of juvenile arrests associated with drug use, the proportion of study participants who were arrested due to drug offenses was limited to no more than 15 percent of the male sample so as to allow for a greater diversity of offenders. All female offenders who met study criteria were deemed eligible for the study regardless of whether their adjudicated offense was drug-related.

On average, participants underwent a baseline interview approximately 40 days after their initial adjudication or decertification hearing. More comprehensive descriptions of the study design and data collection procedures are available at the Pathways to Desistance website (www.pathwaysstudy.pitt.edu) and in published methodological manuscripts (Schubert *et al.*, 2004; Chung *et al.*, 2007; Mulvey and Schubert, 2012). A total of 1,354 adjudicated adolescents completed the baseline survey between November 2000 and January 2003. Ten subsequent follow-up interviews were then conducted over a 7-year (84-month) period. The average follow-up rate for each interview was 89.5 percent.

At baseline, the participants' mean age was 16.04 years old (SD = 1.14), and the majority of the participants were male (male = 86.41 percent, female = 13.59 percent). The racial/ethnic makeup of participants was relatively well distributed, as 20.24 percent were non-Hispanic white, 41.43 percent were African American, 33.53 percent were Hispanic, and 4.80 percent were of an "other" racial/ethnic background. In terms of criminal offending, slightly fewer than half of the participants reported that their most serious offense was a person crime (44 percent); the remainder reported that their most serious offense was a property offense (25 percent), drug offense (16 percent), weapons offense (10 percent), or another type of offense (4 percent).

Measurement

Primary measures

Our primary independent variable was self-reported head injury. During their baseline interview, respondents were asked whether they had ever experienced a head injury that caused unconsciousness or needed medical attention. This item was scored dichotomously (yes/no).

Our dependent measure was offending behavior, which was measured using the Self-Report of Offending (SRO; Huizinga *et al.*, 1991). The SRO contains 24 items to assess the adolescent's involvement in antisocial and illegal activities (e.g. purposely destroying or damaging property, fire setting, shoplifting, using checks or credit cards illegally, committing assault, etc.). However, two of the 24 SRO items ("ever went joyriding" and "ever broke into a car to steal") were added after a large number of subjects had completed either baseline or 6-month follow-up interviews. The introduction of these items produced a large amount of missing data at these two time points, making the calculation of a consistently meaningful score difficult across all available time points. Therefore all measurements from the SRO scores in this study utilize only the original 22 items.

From these items, four different composite measures were constructed, each of which relates to a type of offensive behavior. *Any offense* refers to whether the respondent provided an affirmative response to any of the 22 items within a specified period. *Any non-drug offense* excludes items or acts that included selling or receiving illicit drugs. *Aggressive offending* means that an affirmative response was given to items categorized as aggressive behaviors (e.g. beat someone, fought as part of a gang). *Income offending* includes only behaviors that involved illegal money or transactions (e.g. used credit card or check illegally, received stolen property). Please refer to Appendix A for a complete list of the 22 items and to see how they correspond to the composite measures (i.e. non-drug, aggressive, and income).

Control measures

In order to better isolate the relationship between our independent variable (self-reported head injury) and our dependent variable (offending behavior), we investigated other theories and prior evidence to determine which other variables might correlate with—and therefore confound the relationship of—our independent and dependent variables. We came up with three relevant categories of confounding variables: socio-demographics, clinical characteristics, and other contextual variables.

Socio-demographics are fairly straightforward: these included age, gender, and race/ethnicity (non-Hispanic white, African-American, Hispanic, and "other" race/ethnicity).

Clinical characteristics included psychiatric symptoms, lifetime history of alcohol or drug dependence, self-regulation, and impulsivity. *Psychiatric symptoms* were assessed using the Global Severity Index of the Brief Symptom Inventory (BSI) (Derogatis and Melisaratos, 1983). The BSI is a 53-item inventory in which respondents assess the degree to which they have been bothered (0 = "not at all" to 4 = "extremely") during the previous week by a list of symptoms relating to subscales such as interpersonal sensitivity, depression, and somatization. The scores of those 53 items are then summed, yielding the Global Severity Index, where higher scores represent greater levels of psychological distress. *Lifetime history of alcohol dependence and drug dependence* was assessed using the substance-use disorder modules of the Composite International Diagnostic Interview (CIDI; WHO, 1990). The CIDI is a fully structured interview schedule for establishing DSM-IV and ICD-10 substance-use disorder diagnoses. *Impulsivity*, the last of the clinical characteristics, was measured using two evaluations. The first was the impulse control subscale of the Weinberger Adjustment Inventory (WAI). This 8-item measure ($\alpha = 0.76$) is designed to assess an individual's capacity for self-control, with higher scores representing greater levels of impulsivity. The second was an adaptation of the Children's Emotion Intensity Child Report, also referred to as the Walden self-regulation scale. This was used to measure emotion regulation (Walden *et al.*, 2003). Our adaptation involved using 12 of the original 33 items, such as "I know things to do to make myself more happy," and "I can change my feelings by thinking of something else." Higher scores represent greater levels of self-regulation.

Other contextual variables included age of first offense, peer delinquency, school enrollment, violence exposure, neighborhood conditions, and lifetime gang membership. *Age of first offense* was established using the SRO (see Primary measures above), which is a 22-item questionnaire in which respondents indicated the age at which they committed their first offense for all 22 offending items. The earliest age at which they reported across all of the items was used as the age of first offense. *Peer delinquency* was measured using a subset of items from the Rochester Youth Study. The items chosen were those originally developed to assess the degree of anti-social activity among the adolescent's peers (Thornberry *et al.*, 2002)—e.g. "How many of your friends have sold drugs?" *School enrollment* reflected whether or not adolescents were enrolled in school at the time of their baseline interview (or immediately prior to entering a locked facility) using dichotomous (yes/no) scoring. *Violence exposure* was assessed using a modified version of the Exposure to Violence Inventory (ETV) (Selner-O'Hagan *et al.*, 1998). Items from the ETV document the types of violence the adolescent has experienced (e.g. "Have you ever been chased where you thought you might be seriously hurt?") and observed (e.g. "Have you ever seen someone else being raped, an attempt made to rape someone or any other type of sexual attack?"). Higher scores indicated a greater exposure to violence. *Neighborhood conditions* were assessed using an adaptation of the Neighborhood Conditions Measure, where conditions being measured were the physical disorder (e.g. "cigarettes on the street or in the gutters," and "graffiti or tags") and social disorder (e.g. "adults fighting or arguing loudly," and "people using needles or syringes to take drugs") (Elliott *et al.*, 1996). *Lifetime gang membership* was determined using a dichotomous self-report (yes/no).

Statistical analysis

Chi-square and *t-tests* were used to examine bivariate associations between head injury and other baseline control variables. Effect sizes were summarized using phi coefficients (φ) and point biserial correlations ($r_{.pb}$) (Cohen, 1988). Logistic regression analyses were used to examine associations between head injury and each offending behavior outcome measure (any offense, non-drug offense, drug offense, aggressive offense, and income offense) during the first follow-up period (6 months) for the full sample, and then separately by gender. Our first set of models examined bivariate associations. We then specified two different sets of adjusted models. The first set of adjusted models included socio-demographics. The second set of adjusted models included socio-demographics and all other variables that were significantly associated with head injury with an effect size ≥ |0.10].

We then examined the association between head injury and offending behavior over time using survival analysis. In this analysis, our primary outcome was any offense. The pattern of offending behavior over time was first examined using a Kaplan–Meier survival curve. We then used Cox proportional hazards (PH) regression to estimate the association between head injury and time to first re-offense. Similarly to logistic regression analysis, we first specified unadjusted models, and then adjusted for socio-demographics and other control variables that were associated with head injury and had an effect size ≥ |0.10].

Then, using logistic regression, we examined the association between head injury and offending behavior at 6-month follow-up for each offending outcome (any offense, any non-drug offense, aggressive offense, and income offense) to help clarify the predictive utility of head injury beyond simple base-rate estimates and then checked the stability of these models by including baseline control variables. We then looked at offending behavior over subsequent time points by examining the associations between head injury and offending behavior at each subsequent time point. All analyses were conducted using the statistical language R (R Core Team, 1997).

Results

Sample description

Of the 1,354 adolescents included in the Pathways survey, 30.1 percent (n = 411) self-report a head injury. As summarized in Table 20.1, most of the variables were statistically significant, but only a subset reached a threshold that exhibited at least a small effect size ≥ |0.10]. Gender was the only socio-demographic variable that reached this threshold. More specifically, males were significantly more likely than females to report a head injury (males = 32.1 percent versus females = 19.0 percent, $P < 0.001$). Clinical characteristics that reached a minimal effect size threshold included psychiatric symptoms ($r_{.pb} = 0.13$), (violence exposure ($r_{.pb} = 0.24$), and impulsivity ($r_{.pb} = 0.15$)). The only contextual variables that reached the threshold of |0.10| included age of first offense ($r_{.pb} = 0.10$).

Offending behavior at 6-month follow-up

Table 20.2 provides a summary of offending behavior at 6-month follow-up for the full sample and separately by gender. For the full sample, adolescents who had reported a brain injury demonstrated significantly higher rates of offending behaviors (all types) compared with those who had not. For example, 68.3 percent of adolescents with a head injury reported any type of offense compared with 54.5 percent of those without a head injury. In the unadjusted logistic regression models, the strength of the association between head injury and offending behavior was consistent across all types of offending behavior, with the exception of income offenses, which had a comparatively lower odds ratio (OR). The associations remained fairly stable when adjusting for socio-demographics (Adjusted – I models); however, the associations were all rendered non-significant when adjusting for clinical characteristics and contextual variables.

For males, the pattern of association was nearly identical to that of the full sample, which was not unexpected given that males were overrepresented in the study sample. For females, neither the unadjusted nor the adjusted models showed head injury to be significantly associated with any type of offending behavior during the 6-month follow-up period. It should be noted that two types of offending behavior could not be tested for females—drug offense and income offense—given that observed cell counts for these offenses were <5.

Offending behavior through 84-month follow-up

Survival analyses were used to examine the association between head injury and offending behavior over the full 84-month follow-up period. Because analyzing any offense showed the strongest effect at the 6-month follow-up—instead of specifying a specific type of offense—we again kept our analyses inclusive of any offense. We began our analyses by computing Kaplan–Meier survival curves separately for males and females (Figure 20.1). This analysis showed the estimated survival probability—the proportion of adolescents who make it to each subsequent follow-up period without reporting any offending behavior—at each follow-up period. For

Table 20.1 Summary of baseline characteristics of juvenile delinquents with and without self-reported head injuries

Characteristic	*Full sample N (%)*	*Head injury*		*Test statistic (effect size)*
		No N (%)	*Yes N (%)*	
Age [mean (sd)]	16.0 (1.1)	16.0 (1.1)	16.1 (1.3)	$t = 2.23$, $P = 0.03$ [$r_{.pb} = 0.06$]
Race/ethnicity				
White	274 (20.2)	172 (62.8)	102 (37.2)	$\chi^2 = 8.15$, $P = 0.04$ [$\varphi = 0.08$]
Black	561 (41.4)	406 (72.4)	155 (27.6)	
Hispanic	454 (33.5)	318 (70.0)	136 (30.0)	
Other	65 (4.85)	47 (72.3)	18 (27.7)	
Gender				
Male	1170 (86.4)	794 (67.9)	376 (32.1)	$\chi^2 = 14.0$, $P < 0.001$ [$\varphi = 0.10$]
Female	184 (13.6)	149 (81.0)	35 (19.0)	
BSI – global severity [mean (sd)]	0.56 (0.53)	0.51 (0.52)	0.67 (0.55)	$t = 4.70$, $P < 0.001$ [$r_{.pb} = 0.13$]
Lifetime history of alcohol or drug dependence				
No	1041 (80.0)	744 (71.5)	297 (28.5)	$\chi^2 = 10.6$, $P < 0.001$ [$\varphi = 0.09$]
Yes	261 (20.0)	159 (60.9)	102 (39.1)	
Age of first offense [mean (sd)]	10.4 (1.81)	10.6 (1.9)	10.0 (1.5)	$t = 5.7$, $P < 0.001$ [$r_{.pb} = -0.12$]
Peer delinquency [mean (sd)]	2.32 (0.93)	2.2 (0.90)	2.5 (0.96)	$t = 4.4$, $P < 0.001$ [$r_{.pb} = 0.06$]
Enrolled in school				
No	382 (28.2)	272 (71.2)	110 (28.8)	$\chi^2 = 0.61$, $P = 0.43$ [$\varphi = 0.02$]
Yes	972 (71.8)	671 (69.0)	301 (31.0)	
Violence exposure [mean (sd)]	1.57 (1.46)	1.35 (1.36)	2.10 (1.53)	$t = 8.59$, $P < 0.001$ [$r_{.pb} = 0.24$]

Lifetime gang membership*				
No	788 (70.2)	732 (92.9)	56 (7.1)	$\chi^2 = 1.55$, $P = 0.22$ [$\varphi = 0.04$]
Yes	334 (29.8)	303 (90.7)	31 (9.3)	
Neighborhood conditions	2.35 (0.75)	2.33 (0.75)	2.39 (0.76)	$t = -1.24$, $P = 0.22$ [$r_{.pb} = -0.03$]
Self regulation	2.75 (0.66)	2.78 (0.66)	2.71 (0.65)	$t = 1.81$, $P = 0.07$ [$r_{.pb} = 0.05$]
Impulsivity	2.96 (0.95)	2.94 (0.94)	3.26 (0.94)	$t = 5.69$, $P < 0.001$ [$r_{.pb} = -0.15$]

Notes: *$P < .05$; N for full sample is 1,354. Some variables have missing cases so some cell counts do not sum to N of the full sample. $r_{.pb}$ = point biserial correlation. φ = phi coefficient. sd = standard deviation.

Table 20.2 Effect of self-reported head injury on odds of offending behavior at 6-month follow-up

	Head injury		*Unadjusted OR (95% CI)*	*Adjusted – I* OR (95% CI)*	*Adjusted – II** OR (95% CI)*
	No N (%)	*Yes N (%)*			
Full sample					
Any offense	477 (54.5)	263 (68.3)	**1.80 (1.40–2.33)**	**1.73 (1.34–2.23)**	1.21 (0.91–1.60)
Non-drug offense	465 (53.1)	258 (67.0)	**1.79 (1.40–2.31)**	**1.73 (1.34–2.23)**	1.21 (0.92–1.60)
Drug offense	152 (17.4)	101 (26.2)	**1.69 (1.27–2.25)**	**1.56 (1.16–2.08)**	1.17 (0.86–1.61)
Aggressive offense	418 (47.7)	237 (61.6)	**1.75 (1.37–2.24)**	**1.71 (1.33–2.19)**	1.21 (0.93–1.59)
Income offense	167 (19.1)	99 (25.7)	**1.46 (1.10–1.95)**	**1.36 (1.02–1.83)**	1.05 (0.77–1.44)
Males					
Any offense	419 (56.8)	246 (69.9)	**1.77 (1.35–2.32)**	**1.76 (1.35–2.32)**	1.21 (0.89–1.63)
Non-drug offense	408 (55.3)	241 (68.5)	**1.76 (1.35–2.30)**	**1.75 (1.34–2.30)**	1.21 (0.90–1.63)
Drug offense	133 (18.0)	98 (27.8)	**1.76 (1.30–2.37)**	**1.68 (1.24–2.27)**	1.29 (0.93–1.79)
Aggressive offense	372 (50.4)	221 (62.8)	**1.66 (1.28–2.16)**	**1.68 (1.29–2.19)**	1.17 (0.88–1.56)
Income offense	143 (19.4)	97 (27.6)	**1.58 (1.17–2.13)**	**1.54 (1.12–2.06)**	1.18 (0.85–1.64)
Females					
Any offense	58 (42.0)	17 (51.5)	1.47 (0.68–3.16)	1.45 (0.66–3.21)	1.44 (0.61–3.50)
Non-drug offense	57 (41.3)	17 (51.5)	1.51 (0.70–3.26)	1.49 (0.68–3.30)	1.44 (0.61–3.49)
Drug offense	19 (13.8)	3 (9.1)	***	***	***
Aggressive offense	46 (33.3)	16 (48.5)	1.88 (0.87–4.08)	1.98 (0.88–4.45)	2.34 (0.95–5.99)
Income offense	24 (17.4)	2 (6.1)	***	***	***

Notes: Values in bold are statistically significant based on a 95% confidence interval that does not bound 1.0.

*Adjusted for socio-demographics (age, race, and gender). Gender analyses adjusted for age and race.

**Adjusted for socio-demographics (age, race, and gender), clinical characteristics (BSI–global severity, violence exposure, impulsivity), and other contextual variables (age of first offense).

***Cell count too small for analysis.

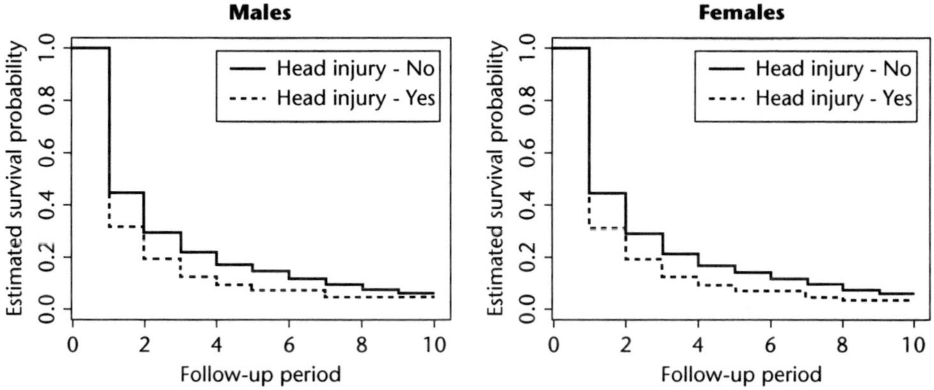

Figure 20.1 Gender comparison of estimated survival probabilities among juvenile delinquents with and without a self-reported head injury

both males and females, over half the adolescents with a head injury reported some type of offending behavior within the first follow-up period (6 months) and few adolescents with a head injury made it to subsequent follow-ups without reporting offending behaviors (Figure 20.1). This pattern of offending was nearly identical among those without a self-reported head injury, although the probability of reoffending was slightly less for the uninjured adolescents at each follow-up period.

Table 20.3 provides a summary of the hazard ratios (HRs) and 95 percent confidence intervals to characterize the association between head injury and offending behavior over time. Similarly to the logistic regression analyses (Table 20.2), head injury exhibited a statistically significant association with offending behaviors in both the unadjusted analysis (HR = 1.36, 95 percent CI = 1.20–1.54) and the models adjusted for socio-demographics (HR = 1.31, 95 percent CI = 1. 16–1. 48). The association was no longer significant in the model adjusting for socio-demographics, clinical characteristics, and other contextual variables.

Discussion

Reprise of findings

Juvenile offenders are among the highest risk groups for experiencing a head injury, but very few studies have examined the long-term consequences among this at-risk group. Most of the research to date has relied on self-reports of head injuries, which are prone to various types of bias. The objective of this study was to examine the association between self-reported head injury and offending behavior over time, which is necessary to clarify the validity of using a self-report as an indicator of future offending behaviors. Using one of the largest panel studies of youth offenders to date, we provide evidence that a self-reported head injury can substantially improve predictions of future offending behavior for males, but not females. More specifically, males with a self-reported head injury were between 1.58 and 1.77 times more likely to exhibit a range of offending behaviors at 6-month follow-up than males without a self-reported head injury. A series of survival analytic models showed that juvenile offenders with a self-reported head injury were more likely to reoffend before those without a head injury over the course of the study period.

Table 20.3 Cox proportional hazards regression showing risk of re-offense over time among juvenile delinquents with self-reported head injury

	Unadjusted HR (95% CI)	*Adjusted – I* HR (95% CI)*	*Adjusted – II** HR (95% CI)*
Head injury	**1.36 (1.20–1.54)**	**1.31 (1.16–1.48)**	1.10 (0.96–1.25)
Gender			
Male	–	Ref	Ref
Female		**0.59 (0.49–0.70)**	**0.61 (0.50–0.73)**
Age	–	**0.94 (0.90–0.99)**	**0.93 (0.88–0.98)**
Ethnicity			
White	–	Ref	Ref
Black		0.85 (0.73–1.00)	0.96 (0.82 –1.14)
Hispanic		0.87 (0.74–1.02)	0.91 (0.77–1.07)
Other		0.95 (0.71–1.28)	0.86 (0.64–1.16)
BSI–global severity	–	–	**1.25 (1.11–1.41)**
Age of first offense	–	–	**0.93 (0.90 –0.96)**
Violence exposure	–	–	**1.07 (1.02–1.12)**
Impulsivity	–	–	**1.15 (1.07–1.24)**
Wald test	24.07, df = 1, $P < 0.0001$	65.74, df = 6, $P < 0.0001$	120.5, df = 9, $P < 0.0001$
Pseudo R^2	0.018	0.053	0.096

Notes: *$P < .05$; HR = hazard ratio. Values in bold are based on a 95% confidence interval (CI) that does not bound the value 1.0.

The absence of a stable empirical relationship between self-reported head injury and offending behaviors for females deserves careful attention. One possibility is the lack of statistical power, as females comprised only 15 percent of the total sample (n = 184). In other words, this study would have only 35 percent power to detect an effect that was at least equivalent to the strongest effect size for males (OR = 1.80 at 6-month follow-up for any offense). If the expected effect size were even smaller, a significantly larger sample of females would be necessary to ensure sufficient statistical power. Another possibility is that head injuries among females may not be as severe as their male counterparts, resulting in a weaker empirical relationship between head injury and future offending behavior. (In other words, it could be possible that the weaker the head injury, the weaker the relationship between head injury and future offending behavior.) Furthermore, females who experience a head injury may have other protective factors that mitigate problems experienced among males.

Validity considerations

Criterion variables

As previously stated, we consider self-reports to be a critically important data source for studying head injuries among delinquent youth. In this study, we present evidence to suggest that a self-report is a valid and reliable device to use when attempting to predict whether a head injury might lead to future offending behavior. It is important that such validity arguments are fully grounded in a careful review of the criterion variable. A comprehensive understanding of validity requires a critical and careful review of the criterion variables. In this study, the criterion variable was offending behavior. These data were obtained by self-report and therefore are subject to the same biases that threaten the validity of our head injury measures. However, it is difficult to conceive of any single or combination source of data that would be more complete than a self-report. For example, arrest records can be used to reliably identify those who have been arrested for exhibiting offending behavior, but predicting the likelihood of future arrests based on head injury is an entirely different question from the one we addressed in this study. Collateral sources could be used to potentially triangulate any given response, but it would be impractical to locate all possible collateral sources, and failing to capture the report of any collateral source would result in distorted estimates. Furthermore, collateral sources also face the same recall biases as the primary respondent, which means that we're essentially back at square one: a method of data collection that is subject to biases.

Single versus multivariate prediction models

In this study, we found head injury to be a significant predictor of future offending behavior. This relationship remained stable when adjusting for a set of socio-demographic characteristics, but weakened when the models were adjusted for other clinical characteristics and contextual variables. Naturally, this raises the question of whether a self-reported head injury does, in fact, serve as a predictor of future offending behavior.

A few issues need to be considered. First, in the services context, a simple yet robust prediction model—where only one or two pieces of data are necessary to make a prediction—has much more practical utility than a complex model—where a variety of other factors must be evaluated and incorporated into the analysis. More specifically, it is impractical for service providers to control for a range of socio-demographics, clinical variables, and other contextual variables simply in order to determine whether a certain piece of assessment data can predict a given outcome. In fact, the validity of many fully specified regression models can be reasonably questioned purely on mathematical grounds. As Stouffer (1941) pointed out over 70 years ago, a set of ten different traits with only four values each assumes over a trillion possible individuals (see Meehl, (1996[1954]).

Second, the fully adjusted models also suggest that the observed relationship between self-reported head injury and future offending behavior may be better explained by other variables. However, it is important to note that the fully adjusted models include other clinical and contextual variables which are psychological constructs contributing significant error variance that can attenuate the observed associations. Moreover, the selection of these variables was informed by theory, but their actual inclusion in the model was based on an explicit effect size threshold (i.e. $> |0.10|$). Thus the fully adjusted models may, in fact, be misspecified (wrong variables included) or even overspecified (irrelevant variables included). These possibilities can be understood only through increased precision of the underlying theory.

Generalizability

While this study provides evidence for using a self-report as a predictive device, it is important to consider the extent to which it can generalize to other service settings. More specifically, Mulvey's Pathways to Desistance study is based on data from just two juvenile detention centers, which are not representative of the full population of delinquent youth. Thus the generalizability of such a prediction model can be reasonably questioned, although an overly strict requirement of generalizability would necessarily render any services research irrelevant to virtually all service settings.

One way to address concerns about generalizability is by collecting similar data in the service setting where the model will be used. That is, if the model will be used in a outpatient setting, then collecting similar data in outpatient settings would serve to form a reliable model for that particular setting. In fact, many administrative databases contain important attributes of the individual that can be used to create simple prediction models that may improve decision making over base-rate estimates.

Limitations and future directions

This study examines the association between head injury and future offending behavior using one of the largest panel studies of juvenile offenders ever conducted. Despite the large sample size and strong methodological design, it is important that the findings are considered in the context of the study's limitations. Most importantly, even though a self-report head injury was associated with risk of re-offense, this study has not effectively ruled out the full range of other third-variable possibilities that could explain this association (Haig, 1992). However, these alternative possibilities ultimately would not change the fact that a head injury self-report is a valid predictor of future offending behavior, because head injuries are highly correlated with any third-variable possibility. Nevertheless, the advantage of uncovering the third variable would improve the prediction model and help to further refine the theory.

Increasing the overall body of validity evidence to support any inference made from a self-reported head injury is a critical step in furthering the research that evaluates self-reports as a valid tool for collecting reliable data. While our research explores whether the self-report can be used as an accurate predictor of future behavior, other research has looked into ways of increasing the validity of self-reports themselves. The use of *think-aloud* protocols, for instance, can help clarify how respondents are interpreting self-report questions (Ericsson and Simon, 1998). In this way, researchers can determine whether subjects are interpreting the same question differently, which might ultimately indicate that the self-report item may be measuring another construct related to the construct attempting to be captured. Addressing aspects of reliability will also help build our confidence in the validity of the results, given that reliability is a necessary (but not sufficient) condition for validity. One way of doing this is within a test–retest framework to determine the consistency of responses over time. Another practical step would be correlating items in self-reports with measures of social desirability (Crowne and Marlowe, 1960), or fake-good or fake-bad measures to help uncover possible systematic biases in the data (Lees-Haley *et al.*, 1991)—that is, if a subject perceives certain items on the self-report as making them appear "good" or "bad," they may be inclined to answer inaccurately to preserve their social image.

Conclusion

In this study, we found evidence that a self-reported head injury can serve as a predictive device of future offending behavior. These predictions can help make significant improvements over simple base-rate predictions. It is important that this kind of predictive device is used solely for the purpose of helping identify delinquent youth who may be at increased risk of re-offense. It is especially important that such a device is not used for any other purpose, such as assuming any form of offending behavior in absence of evidence.

Appendix A: Self-report of offending behaviors

The following is the full list of offending behaviors that were assessed at each time point. The full list was used to construct "any offense" measure, and the superscripts show which items were used to construct: non-drug offenses (a); drug offenses (b); aggressive offenses (c); and income offenses (d).

- Purposely destroyed or damaged property that did not belong to you [a]
- Purposely set fire to a house, building, car, or vacant lot [a]
- Entered or broken into a building (home or business) to steal something [a]
- Stolen something from a store (shoplifted) [a,d]
- Bought, received, sold something you knew was stolen [a,d]
- Used checks or credit cards illegally [a,d]
- Stolen a car or motorcycle to keep or sell [a]
- Sold marijuana [b]
- Sold other illegal drugs (cocaine, crack, heroin) [b]
- Carjacked someone [a,c]
- Driven while drunk or high [b]
- Paid for sex [a]
- Forced someone to have sex with you [a,c]
- Killed someone [a,c]
- Shot someone (where bullet hit the victim) [a,c]
- Taken something from another person by force, using a weapon [a,c]
- Beaten up, threatened, or physically attacked someone as part of a gang [a,c]
- Beaten up or physically attacked somebody so badly that they probably needed a doctor [a,c]
- Been in a fight [a,c]
- Carried a gun [a,c]

References

Caveness, W. (1979) 'Incidence of craniocerebral trauma in the United States in 1976 with trend from 1970–1975', *Advances in Neurology*, 22: 1–3.

CDC (2013) 'CDC – traumatic brain injury', Centers for Disease Control and Prevention, www.cdc.gov/traumaticbraininjury

Chung, H. L., Mulvey, E. P., and Schubert, C. A. (2007) 'An empirical portrait of community reentry among serious juvenile offenders in two metropolitan cities', *Criminal Justice and Behavior*, 34(11): 1402–1426.

Cohen, J. (1988) *Statistical Power Analysis for the Behavioral Sciences*, Hillsdale, NJ: Lawrence Erlbaum Associates.

Crowne, D. P. and Marlowe, D. (1960) 'A new scale of social desirability independent of psychopathology', *Journal of Consulting Psychology*, 24(4): 34–39.

Davies, R. C., Williams, W. H., Hinder, D., Burgess, C. N. W., and Mounce, L. T. A. (2012) 'Self-reported traumatic brain injury and postconcussion symptoms in incarcerated youth', *Journal of Head Trauma Rehabilitation*, 27(3): E21–E27.

Derogatis, L. R. and Melisaratos, N. (1983) 'The brief symptom inventory: an introductory report', *Psychological Medicine*, 3: 595–605.

Elliott, D. S., Wilson, W. J., Huizinga, D., Sampson, R. J., Elliott, A., and Rankin, B. (1996) 'The effects of neighborhood disadvantage on adolescent development', *Journal of Research in Crime and Delinquency*, 33(4): 389–426.

Ericsson, K. A. and Simon, H. A. (1998) 'How to study thinking in everyday life: contrasting think-aloud protocols with descriptions and explanations of thinking', *Mind, Culture, and Activity*, 5(3): 178–186.

Farrer, T. J., Frost, R. B., and Hedges, D. W. (2013) 'Prevalence of traumatic brain injury in juvenile offenders: a meta-analysis', *Child Neuropsychology*, 19(3): 225–234.

Haig, B. D. (1992) 'From nuisance variables to explanatory theories: a reformulation of the third variable problem', *Educational Philosophy and Theory*, 24(2): 78–97.

Hawkins, K. A. and Trobst, K. K. (2000) 'Frontal lobe dysfunction and aggression: conceptual issues and research findings', *Aggression and Violent Behavior*, 5: 147–157.

Huizinga, D., Esbensen, F., and Weiher, A. W. (1991) 'Are there multiple paths to delinquency?', *Journal of Criminal Law and Criminology*, 55: 83–118.

Huw Williams, W., Cordan, G., Mewse, A. J., Tonks, J., and Burgess, C. N. (2010) 'Self-reported traumatic brain injury in male young offenders: a risk factor for re-offending, poor mental health and violence?', *Neuropsychological Rehabilitation*, 20(6): 801–812.

Kahneman, D. (2011) *Thinking, Fast and Slow*, New York: Farrar, Straus, and Giroux.

Kraus, J. and Nourjah, P. (1988) 'The epidemiology of mild, uncomplicated brain injury', *Journal of Trauma*, 28: 1637–1643.

Langlois, J., Rutland-Brown, W., and Thomas, K. (2006) *Traumatic Brain Injury in the United States: Emergency Department Visits, Hospitalizations, and Deaths*, Atlanta, GA: Centers for Disease Control and Prevention, National Center for Injury Prevention.

Lees-Haley, P. R., English, L. T., and Glenn, W. J. (1991) 'A fake bad scale on the MMPI-2 for personal injury claimants', *Psychological Reports*, 61(1): 203–210.

Meehl, P. E. (1996[1954]) 'Clinical versus statistical prediction: a theoretical analysis and a review of the evidence', Northvale: Jason Aronson.

Miller, J. and Jones, P. (1985) 'The work of a regional head injury service', *Lancet*, 1: 1141–1144.

Mulvey, E. P. (2012) *Research on Pathways to Desistance [Maricopa County, AZ and Philadelphia County, PA]: Subject Measures, 2000–2010*, Ann Arbor, MI: Inter-University Consortium for Political and Social Research (ICPSR) [distributor].

Mulvey, E. P. and Schubert, C. A. (2012) 'Some initial findings and policy implications of the pathways to desistance study', *Victims & Offenders*, 7(4): 407–427.

NIH (2013) 'Head injury – first aid', National Institutes of Health, www.nlm.nih.gov/medlineplus/ency/article/000028.htm

Perron, B. E. and Howard, M. O. (2008) 'Prevalence and correlates of traumatic brain injury among delinquent youths', *Criminal Behavior and Mental Heath*, 18(4): 243–255.

R Core Team (1997) *R Manual*, Vienna, Austria: R Foundation for Statistical Computing.

Schubert, C. A., Mulvey, E. P., Steinberg, L., Cauffman, E., Losoya, S. H., Hecker, T., Chassin, L., and Knight, G. P. (2004) 'Operational lessons from the pathways to desistance project', *Youth Violence and Juvenile Justice*, 2(3): 237–255.

Selner-O'Hagan, M. B., Kindlon, D. J., Buka, S. L., Raudenbush, S. W., and Earls, F. J. (1998) 'Assessing exposure to violence in urban youth', *Journal of Child Psychology and Psychiatry*, 39(2): 215–224.

Silver, J. M. and Yudofsky, S. C. (1987) 'Aggressive behavior in patients with neuropsychiatric disorders', *Psychiatric Annals*, 17: 367–370.

Stouffer, S. A. (1941) 'Notes on the case-study and the unique case', *Sociometry*, 4(4): 349–357.

Thornberry, T. P., Krohn, M. D., Lizotte, A. J., Smith, C. A., and Tobin, K. (2002) *Gangs and Delinquency in Developmental Perspective*, New York: Cambridge University Press.

Walden, T. A., Harris, V. S., and Catron, T. F. (2003) 'How I feel: a self-report measure of emotional arousal and regulation for children', *Psychological Assessment*, 15(3): 399–412.

WHO (1990) *Composite International Diagnostic Interview (CIDI), version 1.0*, Geneva: World Health Organization.

21

Executive functioning, reward processing, and antisocial behavior in adolescent males

Joanne E. Morgan, Katharine L. Bowen, Simon C. Moore, Justin C. Savage, and Stephanie H. M. van Goozen

Introduction

Executive function (EF) deficits are implicated in the impulsive and risk-taking behaviors exhibited by antisocial individuals (e.g. Ogilvie *et al.*, 2011), although research mostly focuses on global EF deficits rather than specific functions such as reward processing. Reward seeking is heightened during adolescence, and this contributes to a typical increase in impulsive risk-taking behaviors during this developmental stage (Steinberg, 2004, 2010). However, it is not clear to what extent reward-processing biases are a general characteristic of adolescence or particularly implicated in those who engage in antisocial behavior (ASB). The present study therefore compared EF and reward-processing biases in male adolescent offenders and matched non-offending adolescents using both a global EF measure and two measures sensitive to reward and punishment processing.

Frontal lobe and therefore EF deficits are typically observed in individuals who exhibit ASB (Gorenstein, 1982; Moffitt, 1993; Raine, 1993; Morgan and Lilienfeld, 2000). The mechanism linking EF deficits and ASB involves several regulation deficits, including reduced behavioral inhibition, an inability to anticipate the consequences of behavior, and systematic biases in estimating subjective reward and punishment values (Giancola, 1995; Ogilvie *et al.*, 2011). While such EF deficits are consistently observed in those who display ASB, this research has not considered the functional variation known to exist in the prefrontal cortex and therefore different facets of EF (Blair *et al.*, 2005, 2006). Clearly, this is a difficult task given that any measure of EF is unlikely to be entirely specific to a particular region of the prefrontal cortex or indeed the frontal lobes themselves (Morgan and Lilienfeld, 2000). Nevertheless, certain measures have more consistently been localized to specific frontal regions, for instance the Wisconsin Card Sorting Test (WCST; Heaton, 2005) is a reasonably well validated measure of EF and has long been associated with activity in the dorsolateral prefrontal cortex (DLPFC; Rezai *et al.*, 1993; Monchi *et al.*, 2001). Indeed, research using positron emission tomography shows that the dorsolateral region of the frontal cortex exhibits increased activity during performance of the WCST (Rezai *et al.*, 1993). Interestingly, this task has been used extensively in antisocial individuals with some deficits found, even though the DLPFC is not strongly implicated in ASB (Grafman *et al.*, 1996; Blair *et al.*, 2005).

DLPFC lesions often lead to lack of activity, apathy, and reduced capacity for sustained attention (Roussy and Toupin, 2000), and evidence for the association between DLPFC functioning and ASB might be explained through the presence of co-morbid conditions such as attention deficit hyperactivity disorder (ADHD; Blair *et al.*, 2005). EF deficits are more usually co-morbid with ADHD rather than conduct disorder (CD; Pennington and Ozonoff, 1996), suggesting that when investigating the role of EF function in ASB, co-morbid conditions such as ADHD should also be considered (Ogilvie *et al.*, 2011). Moreover, performance on EF measures in general, and the WCST in particular, is associated with IQ. Offenders are known to demonstrate reduced IQ compared with non-offenders (on average 8 IQ points lower in young offenders; Moffitt, 1990). Past associations between ASB and impaired WCST performance could, therefore, be confounded by IQ.

Antisocial individuals often present with risky and disinhibited behavior; a pattern that is similar to the behavior of individuals presenting with damage to the ventromedial prefrontal cortex/orbitofrontal cortex (vmPFC/OFC; Seguin, 2004; Syngelaki *et al.*, 2009. In contrast to DLPFC lesions, OFC lesions result in impulsivity, behavioral disinhibition, aggression, and antisocial tendencies (see Roussy and Toupin, 2000). Damage to this region in humans has resulted in impairments in social decision making that rely on accurately assessing the value of possible behavioral outcomes; an impairment that is referred to as a failure of high-level reward processing (Elliott *et al.*, 2004). These real-life decision-making problems have been mimicked in laboratory gambling tasks. Using the classic Iowa Gambling Task, Damasio and colleagues (e.g. Bechara *et al.*, 1994, 2000) have demonstrated that patients with OFC lesions appear to be guided by short-term gains that ultimately result in negative long-term consequences.

In addition, lesions in the OFC result in perseverative responding and an inability to achieve response reversal and extinction (Elliott, 2004). Indeed, several studies in monkeys have shown that lesions to the OFC result in perseverative responding on reversal tasks (e.g. Iversen and Mishkin, 1970; Dias *et al.*, 1996). In addition, there are findings in humans in terms of both neuropsychological and neuroimaging evidence suggesting that the OFC is involved in response reversal (Rolls *et al.*, 1994; Cools *et al.*, 2002). Imaging data have implicated Brodman's Area 47 in response reversal (Cools *et al.*, 2002; Budhani and Blair, 2005). Response reversal and extinction tasks require learning to withhold or change a behavioral response when the original response is no longer rewarded but punished. It has been argued that an inability to perform response reversal could lead to frustration, given that the individual is not easily able to adapt their behavior to achieve their goals. Frustration is a cue for aggression, and it has been argued that OFC dysfunction could lead to the heightened aggression seen in psychopathy (Blair *et al.*, 2005).

The Card Playing Task (CPT), originally used by Newman and colleagues to assess behavior in psychopaths (Newman *et al.*, 1987), is an EF task that incorporates reward and punishment contingencies and response extinction, and is therefore thought to involve the OFC. The CPT involves establishing a dominant response set for reward by initially providing participants with a high level of reward for their responses. However, as the task progresses the level of punishment increases and reward declines. To be successful participants need to take into account reinforcement changes and interrupt their dominant plan (i.e. stop playing when punishment outweighs reward). Perseveration is said to occur when the participant plays beyond the point where the rate of punishment outweighs the rate of reward.

Response perseveration on the CPT (or similar measures such as the child-modified Door-Opening Task) has been found in a number of antisocial groups, including conduct disordered children (Shapiro *et al.*, 1988); conduct disordered children with or without hyperactivity (Daugherty and Quay, 1991; Matthys *et al.*, 1998); children with emotional and behavioral difficulties (Fisher and Blair, 1998); juvenile delinquents (Fonseca and Yule, 1995); adolescents with histories of physical aggression (Seguin *et al.*, 2002); non-anxious psychopathic children (O'Brien and Frick, 1996); and adolescent substance users (Martin *et al.*, 2000). Few researchers have examined CPT performance in non-clinical community adolescents with varying degrees and types of ASB. In addition, the mechanisms underlying the behavior are unclear; perseveration could be caused by a range of factors including oversensitivity to reward, insensitivity to punishment, and/or perhaps a limitation of attention in terms of integrating the information (Newman and Wallace, 1993). The development of new measures that aim to distinguish between reward and punishment over- and under-sensitivity may be useful for understanding the processes that lead to inappropriate response perseveration such as observed on the CPT.

Finally, it is important to note that ASB is a heterogeneous concept, and specific subgroups of antisocial individuals may differ in the type and extent of EF and reward-processing deficits. For instance, LaPierre *et al.* (1995) found that psychopathic criminals, but not those without psychopathy, exhibited behavior that was characteristic of OFC/VmPFC deficits (but there was no deficit on global EF tests such as the WCST). Roussy and Toupin (2000) similarly found that juvenile psychopathic and non-psychopathic offenders did not differ on DLPFC measures, but the psychopaths committed significantly more errors on OFC measures requiring inhibition. There is also support for the notion that more serious offenders may be characterized by more pronounced executive dysfunction (Raine *et al.*, 2005).

The present study examined EF including reward sensitivity in young offenders (YOs). YO performance was compared with an age, sex, socioeconomic status, and IQ matched non-offending (normal control, NC) group (thus attempting to control for some factors that are associated with ASB and may potentially explain differences in neuropsychological functioning). Comparing performance across similarly aged groups allowed us to explore factors associated with the adolescent developmental stage in general, as well as ASB in adolescence in particular. The CPT (Newman *et al.*, 1987) was used as an extinction task that relies on OFC function. It was hypothesized that young offenders would show increased perseveration on this measure compared with the control group. The WCST (Heaton, 2005) was used to assess whether young offenders had a general EF deficit. It was hypothesized that young offenders would not show global executive problems compared with non-offending controls when IQ was controlled for (e.g. Fairchild *et al.*, 2009).

A novel gambling test, the Win–Lose Cardiff Gambling Test (Win–Lose), was developed. The Win–Lose test measures participants' relative reward and punishment preferences when they are asked to monitor windows on a computer display that show the number of points that might be won or lost on each trial of a game. It was hypothesized that young offenders would display stronger reward-seeking tendencies and therefore preferentially monitor the win window over the loss window. Of course, adolescents in general are also considered to be reward seekers and, given the novelty of this paradigm, we also compared the data of this new test with those collected in a male undergraduate sample.

Finally, to explore the effect of heterogeneity of ASB within the YO group on EF and reward processing, data on severity of offending behavior from official records, and self-report measures of psychopathic traits conduct and ADHD symptoms, were used to explore variations in behavior and/or personality in relation to neuropsychological functioning on the CPT. We predicted that impairments in reward processing would be more apparent in more extreme ASB subgroups.

Method

Participants

The YO group consisted of 40 young males aged 13–17 years old who were engaged in the criminal justice system. The young people had exhibited delinquent behavior at different levels of frequency and seriousness. They were recruited from the local youth offending service in collaboration with their case workers. Normal controls were 45 age-matched males recruited from local comprehensive schools, and were screened for the presence of ASB and involvement in the criminal justice system. The NCs were drawn from the local area in which young offenders lived in an attempt to match on social background factors. Participants were provided with vouchers in compensation for their time. The YO and NC participant details can be found in Table 21.1 below. The number of participants who completed different tasks varied (due to a variety of factors, including time limitations and motivation issues in these groups); the exact numbers are provided below each measure description.

Given the novelty of the Win–Lose test, another comparison group of male undergraduates was included. The task was initially piloted on a group of male students in the absence of normative data. This group consisted of 30 undergraduates (mean age = 20.13 years; SD = 3.12; all Caucasian) who took part for course credit. Unfortunately, information on socioeconomic status and IQ was not obtained from these individuals. All participants were tested individually, and the study was approved by Cardiff University's School of Psychology Research Ethics Committee.

Measures

Card Playing Task (Newman *et al.*, 1987)

This is a computerized version of the CPT, an extinction task that aims to assess response perseveration in the face of changing reward and punishment contingencies. In this task, the probability of a winning card decreases by 10 percent with each succeeding set of 10 cards. Participants are informed about what cards lead to win or loss, and that the aim of the game is to win as many points as possible. Participants are not informed of the decline in probability of winning. Crucially, they are told that they can stop playing the game whenever they decide to by informing the researcher. Each trial is self-paced and the task lasts approximately 10 minutes. The total number of cards played is the outcome variable of interest. This task was completed in the full sample; 40 YOs and 45 NCs.

Wisconsin Card Sorting Test (Heaton, 2005)

The computerized Wisconsin Card Sorting Test-64 Card Version (WCST-64) was used as a measure of set shifting ability. The sorting rule changed after 10 consecutive correct trials. There is no time limit on the test, and the task takes approximately 15 minutes to complete. The following measures were taken to evaluate participant performance: total number of errors made (ranging from 0 to 64), perseverative errors (errors made when continuing to sort cards according to the rule that was correct for the prior stage of the task; ranging from 0 to 62), and non-perseverative errors (range 0 to 64). This task was completed by 40 YOs and 26 NCs.

Win–Lose Cardiff Gambling Test

This is a computer-based task in which participants make gambling decisions in order to obtain as many points as possible. In this task, two windows are presented side by side on the computer screen: one of these windows displays the number of points that can be won, and the other shows the number of points that can be lost. The amounts that can be won or lost fluctuate randomly during each trial, and can go up or down (in intervals of 5 points) for both wins and losses. The maximum amount of points that can be won or lost on a single trial is 100 points and the minimum is 0 points.

Each trial lasts up to 20 seconds, during which time the participant is required to press the gamble button to decide when the trial will stop. The participant has no ability to affect whether they win or lose; however, the amount of points they win or lose is dependent on the values shown in the win and lose windows at the time the gamble button is pressed. Crucially, it is not possible for the participant to see both the win window and the lose window at the same time. In order to see a particular window, they must click on the corresponding button. There are 50 trials and the task lasts approximately 20 minutes.

The following variables were examined. Click frequency is the total frequency of clicks between the windows and provides a measure of how much time was distributed across both windows. Win_Time/Total_Time shows the proportion of time spent in the win window divided by the total time (range from 0 to 1). Values above 0.5 reflect a win window preference. The loss preference can be inferred from the remaining proportion. This task was completed by 40 YOs; 37 NCs, and 30 undergraduates.

Youth Psychopathic Traits Inventory (YPI; Andershed *et al.*, 2002)

This is a self-report measure used to assess psychopathic traits in young people aged 12 years and above. The YPI is a 50-item measure, with each item rated on a 4-point Likert-type scale that ranges from 'does not apply at all' to 'applies very well'. The scale contains three core dimensions: (1) grandiose-manipulative; (2) callous-unemotional; and (3) impulsive-irresponsible (Andershed *et al.*, 2002). The reliability and validity of the YPI have been confirmed in numerous studies (e.g. Skeem and Cauffman, 2003; Andershed *et al.*, 2007). The YPI was completed in the full sample (40 YOs; 45 NCs).

Youth Self Report (YSR; Achenbach and Rescorla, 2001)

This is a self-report questionnaire used to assess behavioral and emotional problems in young people aged 11 and 18 years. We were interested in the DSM-oriented scales of conduct disorder (CD) and attention deficit/hyperactivity disorder (AD). Participants rate how true each item is for them now or within the past 6 months on a 3-point scale ranging from 'not true' to 'very true'. Scores are converted to a standardized *t* score. Higher scores reflect higher levels of problem behavior. The reliability and validity of the YSR are well established (Achenbach and Rescorla, 2001). The YSR was completed by 39 YOs and 35 NCs.

Official crime records

Offense severity was calculated using the seriousness scale developed by the Youth Justice Board (ranging from 1 = minor offenses to 8 = serious offenses). The mean severity scores for each YO's offenses were obtained. Offense records were obtained for all 40 young offenders.

Wechsler Abbreviated Scale of Intelligence (WASI; Wechsler, 1999)

The WASI provides a brief and reliable estimate of a person's intellectual functioning. The two-subtest form was selected; this contains the vocabulary and matrix reasoning components, which tap into crystallized and fluid abilities. The WASI was completed by 40 YOs and 34 NCs.

Data analysis

Preliminary screening analyses revealed that a number of variables were positively skewed; Win–Lose click frequency, WCST total, perseverative and non-perseverative errors, YPI grandiose, YSR conduct problems, and estimated IQ were all logarithmically transformed and subsequent parametric tests were conducted. Descriptive statistics represent non-transformed data for ease of exposition. Independent samples *t-tests* were used to examine the differences between the YO and NC groups on the CPT and WCST variables. One-factor analyses of variance (ANOVA) were used to examine Win–Lose test performance across YOs, NCs, and UGs, while *post hoc* Bonferroni tests were used to explore significant group differences. Within-YO group analyses included Pearson's correlations used to examine the relationships between the CPT and various measures related to ASB.

Results

Demographic information

Data on demographic characteristics were analysed using independent samples *t-tests* or chi-square tests. Table 21.1 presents the demographic information for each group. There were no differences between the groups in ethnicity [$\chi^2(1) = 3.20$, $P = 0.074$] socioeconomic status [$\chi^2(2) = 1.02$, $P = 0.60$], age [$t(83) = 1.97$, $P = 0.052$], or WASI performance [$t(72) = -1.67$, $P = 0.099$].

Between-groups analyses

Table 21.2 shows that the YO group scored significantly higher than the NC group on YPI impulsive-irresponsibility and conduct disorder problems. There were no differences between the groups on the other psychopathy scale scores or self-reported ADHD symptom scores.

Table 21.3 shows performance on the CPT and WCST for the YO and NC groups. Independent samples *t-tests* showed that the YO group played significantly more cards than the NC group, but the groups did not differ on any of the WCST variables.

Table 21.4 displays the performance on the Win–Lose test for YOs, NCs, and UGs. One-way ANOVAs showed that there were significant differences between the groups on all variables. *Post hoc* Bonferroni *t-tests* showed that there were no significant differences between the YOs and NCs, but the YOs differed significantly from the UGs, showing less switching behavior (i.e.

Table 21.1 Participant descriptive characteristics

	Young offender		*Normal control*	
Measures	*Mean*	*SD*	*Mean*	*SD*
Age (years)	15.68	1.02	15.20	1.18
Estimated IQ (WASI)	85.68	11.02	90.26	12.82
Ethnicity (*n* and %)				
Caucasian	25	62.5	36	80.0
Non white	15	37.5	9	20.0
Socioeconomic status				
Low	24	60.0	28	62.2
Middle	12	30.0	10	22.2
High	4	10.0	7	15.6

Note: WASI = Wechsler Abbreviated Scale of Intelligence.

Table 21.2 Antisocial behavior measure scores in young offenders and normal controls

	Young offender		*Normal control;*		*t*	*P*
	Mean	*SD*	*Mean*	*SD*		
YPI grandiose manipulative	39.13	13.23	38.91	8.21	–0.41	0.685
YPI callous-unemotional	34.35	6.86	32.71	6.61	1.12	0.266
YPI impulsive-irresponsible	44.40	7.80	37.40	6.88	4.40**	<0.001
YSR conduct disorder	65.15	10.72	56.92	8.25	3.77**	<0.001
YSR attention deficit hyperactivity disorder	60.85	7.84	59.11	8.09	0.94	0.349
Offense severity	3.45	0.97	–	–	–	–

Notes: YPI = Youth Psychopathic Traits Inventory; YSR = Youth Self Report.

* $P < 0.05$; ** $P < 0.01$.

Table 21.3 Neuropsychological performance in young offender and normal control groups

	Young offender		*Normal control*		*t*	*P*
	Mean	*SD*	*Mean*	*SD*		
$CPT_{Cards\ played}$	65.67	28.96	48.20	23.65	3.06**	0.003
$WCST_{Total\ errors}$	20.28	7.51	21.12	9.82	–0.012	0.990
$WCST_{Perseverative\ errors}$	9.83	4.31	9.92	5.38	0.299	0.766
$WCST_{Non\text{-}perseverative\ errors}$	10.45	4.27	11.19	6.58	0.075	0.940

Notes: CPT = Card Playing Task; WCST = Wisconsin Card Sort Test.

* $P < 0.05$; ** $P < 0.01$.

Table 21.4 Win–Lose test performance in young offenders, normal controls and undergraduates

Parameter	*Young offender (YO)*		*Normal control (NC)*		*Undergraduate (UG)*		*F*	*Bonferroni*
	Mean	*SD*	*Mean*	*SD*	*Mean*	*SD*		
Click Freq	107.10	67.27	126.53	63.33	185.23	104.96	6.77**	UG>YO
Win_time	0.59	0.19	0.59	0.18	0.46	0.14	6.44**	UG<YO & UG<NC

Notes: Win–Lose = Win–Lose Cardiff Gambling Test; Click Freq = total frequency of clicks; Win_time = time spent in win window divided by total time spent in both windows.

** $P < 0.01$.

lower click frequency), and both YOs and NCs showed greater reward monitoring (i.e. higher Win_Time) compared with the UGs.

Within-YO groups analyses

Pearson's correlations were used to explore the relation between reward processing, CPT number of cards played and Win–Lose performance (Click frequency and Win_Time) and different approaches to ASB (offense severity, psychopathic traits, CD and ADHD problems). The results indicated a significant positive correlation between the number of cards played on the CPT and mean offense severity ($r = 0.328$, $P = 0.042$); all other correlations were non-significant (Pearson's r ranged from –0.198 to 0.058).

Discussion

This study examined EF and reward processing in offending and non-offending matched adolescent males. A global EF measure and two measures that assess reward and punishment processing were used. Results show that the YOs and matched NCs did not differ in global EF as measured by the WCST. However, the young offenders displayed an increased tendency for response perseveration for reward on the CPT compared to the non-offending controls who stopped playing sooner. However, interestingly, on the Win–Lose test the two adolescent groups did not differ in reward and punishment monitoring, with both adolescent groups spending more time monitoring the win window at the expense of monitoring the loss window compared with the group of male students.

The results on the CPT support previous research in other antisocial samples showing that antisocial youths perseverate in the face of decreasing reward and increasing punishment (e.g. Daugherty and Quay, 1991; O'Brien and Frick, 1996; Matthys *et al.*, 1998; Seguin *et al.*, 2002). Whereas the young offenders played too many cards, the control group (who were matched for IQ, age, and SES) stopped at a point where reward still outweighs punishment. This is interesting given the evidence that adolescents are reward seekers (e.g. Steinberg, 2004, 2010). Past studies have interpreted perseveration on the CPT as evidence of a reward dominant style, whereby greater sensitivity to reward than to punishment contributes to the failure to interrupt goal directed behavior to evaluate its potential negative consequences (Newman and Wallace, 1993; Goodnight *et al.*, 2006). Nevertheless, it is difficult to disentangle whether perseveration on this task is a result of increased reward sensitivity or decreased punishment sensitivity. In view of this lack of clarity, we developed the Win–Lose Cardiff Gambling Test in an attempt to further explore reward and punishment processing.

The Win–Lose test failed to show any differences between our groups of adolescent males, but interestingly, when both groups were compared with undergraduate male students, some significant findings emerged. First, the undergraduates switched significantly more between the win and lose windows compared with the young offenders, perhaps suggesting a more rational approach to playing the game whereby switching between windows and viewing the fluctuating amounts in both windows allows for a more informed decision before pressing the gamble button. The results also showed that the undergraduates spent more time on the loss compared with the win window, whereas the adolescent males showed a clear preference for viewing the win window. These findings appear to demonstrate an increased preference for the monitoring of reward information, potentially increased reward seeking/sensitivity (and a decreased preference for examining loss information) in adolescent males. Clearly, this is the first use of this measure and one must therefore be cautious in interpretation. It is important to mention that the undergraduate students differ from the YO and NC groups not only in age, but also likely in terms of other variables such as IQ, which may explain the differences. Furthermore, given that the undergraduates did not also complete the other measures (e.g. the CPT), it is difficult to draw firm conclusions about the reward sensitivity of our adolescents versus undergraduates.

Taking the Win–Lose and CPT results together, it would appear that although both adolescent groups had a preference for monitoring reward information, this reward bias resulted in deficient behavior on the CPT only for young offenders. This antisocial group appeared to have a problem extinguishing/inhibiting their reward seeking when it became maladaptive, resulting in loss/punishment. Although the control group had a preference for reward (as shown on the Win–Lose), they were able to inhibit this when necessary (as shown on the CPT). Theorists arguing for a deficit in fear in antisocial individuals would contend that this is a result of impaired processing of punishment-related information (Lykken, 1995), perhaps mediated by an insensitive behavioral inhibition system (Gray, 1987) whereby cues of punishment would be less likely to activate behavioral inhibition. This explanation does not appear to be the complete picture given the Win–Lose findings, which show that the two groups demonstrated similar levels of disinterest in punishment (lose window).

Alternatively, Newman and colleagues (Patterson and Newman, 1993) have described perseveration on the CPT as an inability to shift attention from the goal of responding to gain reward to the peripheral punishment information (the response modulation theory). The early stopping of the controls therefore could be interpreted as over-attention to peripheral cues. On the other hand, Blair (2004) devised the integrated emotion system (IES) model, which implicates both amygdala and orbitofrontal impairment and suggests that the problem lies in the detection of (or alteration of behavior in response to) contingency change. The IES model is considered to consist of a number of systems, one of which involves units that code if a contingency expectation is violated (i.e. if an expected reward does not occur), and a second system that involves preventing responses that are no longer appropriate. The theory argues for the importance of temporal difference errors, which are described as 'the difference between the expected value associated with a stimulus/action and the actual value received' (Budhani and Blair, 2005: 973). Positive temporal difference errors are said to result from unexpected rewards, while absent highly expected rewards induce negative temporal difference errors (Budhani and Blair, 2005). The theory therefore suggests that antisocial individuals are impaired in modifying their behavior as a function of negative temporal difference errors (Budhani and Blair, 2005).

Although our results do not distinguish between these theories, we are able to show that performance on the CPT does not appear to be the result of reward seeking or oversensitivity to reward alone; the two groups differed on the CPT but did not differ on the Win–Lose where both showed a reward preference. The results appear to suggest that it is the inhibition component that is the problem for antisocial youths, and this may result from punishment insensitivity, failure of response modulation, or failure in error detection.

As expected, our results show that the two groups, which were matched for age, SES, and IQ, and had similar levels of self-reported ADHD symptom, did not differ in WCST performance; this measure is related to dorsolateral functioning and is often used as a global measure of EF. The findings of impaired performance on the CPT, but intact performance on the WCST, provide further support that any EF deficit is not a global one, but rather one associated with OFC functioning. Previous associations found in the literature between WCST and ASB may have resulted from a failure to take IQ and ADHD symptoms into consideration. Finally, contrary to predictions, within the YO group, various aspects of ASB (psychopathy, CD) did not explain variations in reward processing. However, we found that offense severity was positively associated with response perseveration on the CPT, suggesting that some aspect of offending is associated with failure to inhibit responding in the face of increasing punishment. This finding suggests that the CPT in particular has associations with ASB, whereas the Win–Lose test, a test of reward monitoring, was not associated.

The present research has many strengths; our deliberate attempt at matching our young offenders and community controls is important as it allowed us to control for many factors that are associated with ASB and may potentially explain differences in neuropsychological functioning (IQ, SES, etc.) and also explore the factors associated particularly with ASB in adolescence as opposed to the adolescent period in general. The fact that we found increased response perseveration in the young offenders suggests that it is a robust finding, and an important one in the understanding of ASB.

Nevertheless, some limitations and directions for future research are worthy of mention; the measures of the various clinical disorders such as conduct disorder and psychopathy were self-report. Both the YSR and YPI have demonstrated good external validity (Achenbach and Rescorla, 2001; Poythress *et al.*, 2006; Campbell *et al.*, 2009; however, clinical diagnoses are superior to self-report and may have allowed for clearer distinctions in performance on the neuropsychological measures. In addition, time limitations meant that we were unable to administer a larger battery of neuropsychological tests; future studies should incorporate other reward-related measures such as those involving discounting procedures, and the inclusion of time parameters (e.g. immediate reward versus delayed reward) would be an interesting extension of the work. Also, the rewards used in the present study included the winning of points, and to fully explore reward mechanisms in YO it would be useful to examine performance with other types of reward such as positive social feedback. In addition, our results were not able to clearly distinguish between the different theoretical accounts of the CPT (i.e. fear dysfunction, response modulation, and error detection) and further research is required to achieve this. Finally, the non-offending group in our study were matched with the YO on a number of variables that are related to ASB. The control group could be considered an 'at risk' sample of young males, given their relatively low IQ and poor social background, yet they did not engage in ASB. It is likely that these controls had protective factors that may have prevented them from engaging in ASB. In future, it would be interesting to assess on what social factors these two groups do differ and how these interact with the neuropsychological factors measured here.

It is also important to mention the difficulties in measuring EF, in particular task impurity problems such as systematic variation attributable to non-EF processes (e.g. speed of processing). Miyake and Friedman (2012) point out that different EFs correlate with one another, thus tapping common underlying ability, but also show diversity. In effect we must bear in mind that our separate measures of EF (e.g. WCST and CPT) are likely to share similar underlying EF processes as well as non-EF related processes.

In summary, the results of this study indicate that young offenders do not differ from their non-offending counterparts in global executive functioning or reward monitoring, but have a specific deficit in reward extinction likely to be caused by a deficit in punishment monitoring or sensitivity. While both adolescent groups showed a heightened preference for reward, this became deficient on the CPT only in the antisocial youngsters. Clearly, these findings fit with other types of behavior exhibited by antisocial youngsters; seeking out the rewards of ASB (peer approval, status, possessions) and not being put off by negative consequences (official sanctions, parental disapproval). This study adds to the growing body of literature showing altered reward and punishment processing in antisocial individuals, and has implications for attempts at behavior change. One next step for future research is to explore the extent to which individual differences in reward seeking and punishment insensitivity affect the success of interventions focusing on positive and negative reinforcement.

References

Achenbach, T. M. and Rescorla, L. A. (2001) *Manual for the ASEBA School-Age Forms & Profiles*, Burlington, VT: University of Vermont, Research Center for Children, Youth, and Families.

Andershed, H., Kerr, M., Stattin, H., and Levander, S. (2002) 'Psychopathic traits in non-referred youths: a new assessment tool', in E. Blaauw and L. Sheridan (eds), *Psychopaths: Current International Perspectives*, The Hague: Elsevier, pp. 131–158.

Andershed, H., Hodgins, S., and Tengstrom, A. (2007) 'Convergent validity of the Youth Psychopathic Traits Inventory (YPI) Association with the Psychopathy Checklist Youth Version (PCL–YV)', *Assessment*, 14: 144–154.

Bechara, A., Damasio, A. R., Damasio, H., and Anderson, S. W. (1994) 'Insensitivity to future consequences following damage to the human prefrontal cortex', *Cognition*, 50: 7–15.

Bechara, A., Damasio, H., and Damasio, A. R. (2000) 'Emotion, decision making and the orbitofrontal cortex', *Cerebral Cortex*, 10: 295–307.

Blair, R. J. R. (2004) 'The roles of orbital frontal cortex in the modulation of antisocial behavior', *Brain and Cognition*, 55(1): 198–208.

Blair, J., Mitchell, D., and Blair, K. (2005) *The Psychopath: Emotion and the Brain*, Oxford: Blackwell.

Blair, K. S., Newman, C., Mitchell, D. G.V., Richell, R. A., Leonard, A., Morton, J., and Blair, R. J. R. (2006) 'Differentiating among prefrontal substrates in psychopathy: neuropsychological test findings', *Neuropsychology*, 20: 153–165.

Budhani, S. and Blair, R. J. R. (2005) 'Response reversal and children with psychopathic tendencies: success is a function of salience of contingency change', *Journal of Child Psychology and Psychiatry*, 46: 972–981.

Campbell, M. A., Doucette, N. L., and French, S. (2009) 'Validity and stability of the Youth Psychopathic Traits Inventory in a nonforensic sample of young adults', *Journal of Personality Assessment*, 91: 584–592.

Cools, R., Clark, L., Owen, A. M., and Robbins, T. W. (2002) 'Defining the neural mechanisms of probabilistic reversal learning using event-related functional magnetic resonance imaging', *Journal of Neuroscience*, 22: 4563–4567.

Daugherty, T. K. and Quay, H. C. (1991) 'Response perseveration and delayed responding in childhood behavior disorders', *Journal of Child Psychology and Psychiatry*, 32: 453–461.

Dias, E. C., Compaan, D. M., Mesulam, M. M., and Segraves, M. A. (1996) 'Selective disruption of memory-guided saccades with injection of a cholinergic antagonist in the frontal eye field of monkey', *Social Neuroscience Abstracts*, 22: 418.

Elliott, R., Newman, J. L., Longe, O. A., and William Deakin, J. F. (2004) 'Instrumental responding for rewards is associated with enhanced neuronal response in subcortical reward systems', *Neuroimage*, 21(3): 984–990.

Fairchild, G., van Goozen, S. H. M., Stollery, S. J., Aitken, M. R. F., Savage, J., Moore, S. C., *et al.* (2009) 'Decision making and executive function in male adolescents with early-onset or adolescence-onset conduct disorder and control subjects', *Biological Psychiatry*, 66: 162–168.

Fisher, L. and Blair, R. J. R. (1998) 'Cognitive impairment and its relationship to psychopathic tendencies in children with emotional and behavioral difficulties', *Journal of Abnormal Child Psychology*, 26: 511–519.

Fonseca, A. C. and Yule, W. (1995) 'Personality and antisocial behavior in children and adolescents: an inquiry into Eysenck's and Gray's theories', *Journal of Abnormal Child Psychology*, 23: 767–781.

Giancola, P. (1995) 'Evidence for dorsolateral and orbital prefrontal cortical involvement in the expression of aggressive behavior', *Aggressive Behavior*, 21: 431–450.

Goodnight, J. A., Bates, J. E., Newman, J. P., Dodge, K. A., and Pettit, G. S. (2006) 'The interactive influences of friend deviance and reward dominance on the development of externalizing behavior during middle adolescence', *Journal of Abnormal Child Psychology*, 34: 573–583.

Gorenstein, E. E. (1982) 'Frontal lobe functions in psychopathy', *Journal of Abnormal Psychology*, 91: 368–379.

Grafman, J., Schwab, K., Warden, D., Pridgen, B. S., and Brown, H. R. (1996) 'Frontal lobe injuries, violence, and aggression: a report of the Vietnam head injury', *Neurology*, 46: 1231–1238.

Gray, J. A. (1987) *The Psychology of Fear and Stress*, 2nd edn, New York: Cambridge University Press.

Heaton, R. K. (2005) *Wisconsin Card Sorting Test Manual*, Odessa, FL: Psychological Assessment Resources.

Hodgson, R., Alwyn, T., John, B., Thom, B., and Smith, A. (2002) 'The fast alcohol screening test', *Alcohol and Alcoholism*, 37: 61–66.

Iversen, S. D. and Mishkin, M. (1970) 'Perseverative interference in monkeys following selective lesions of the inferior prefrontal convexity', *Experimental Brain Research*, 11(4): 376–386.

LaPierre, D., Braun, C. M. J., and Hodgins, S. (1995) 'Ventral frontal deficits in psychopathy: neuropsychological test findings', *Neuropsychologia*, 33: 139–151.

Lykken, D. T. (1995) *The Antisocial Personalities*, Hillsdale, NJ: Lawrence Erlbaum Associates.

Martin, C. A., Rayens, M. K., Kelly, T., Hartung, C., and Leukefeld, C. (2000) 'Card perseveration task performance and post-task feeling states: relationship to drug use in adolescents', *American Journal of Drug and Alcohol Abuse*, 26: 325–333.

Matthys, W., van Goozen, S. H. M., de Vries, H., Cohen-Kettenis, P. T., and van Engeland, H. (1998) 'The dominance of behavioral activation over behavioral inhibition in conduct disordered boys with or without attention deficit hyperactivity disorder', *Journal of Child Psychology and Psychiatry*, 39: 643–651.

Miyake, A. and Friedman, N. P. (2012) 'The nature and organization of individual differences in executive functions: four general conclusions', *Current Directions in Psychological Science*, 21(1): 8–14.

Moffitt, T. E. (1990) 'The neuropsychology of juvenile delinquency: a critical review', *Crime and Justice*, 12: 99–169.

Moffitt, T. E. (1993) 'Adolescence-limited and life-course-persistent antisocial behavior: a developmental taxonomy', *Psychological Review*, 100(4): 674–701.

Monchi, O., Petrides, M., Petre, V., Worsley, K., and Dagher, A. (2001) 'Wisconsin Card Sorting revisited: distinct neural circuits participating in different stages of the task identified by event-related functional magnetic resonance imaging', *Journal of Neuroscience*, 21(19): 7733–7741.

Morgan, A. B. and Lilienfeld, S. O. (2000) 'A meta-analytic review of the relation between antisocial behavior and neuropsychological measures of executive function', *Clinical Psychology Review*, 20(1): 113–136.

Newman, J. P. and Wallace, J. F. (1993) 'Diverse pathways to deficient self-regulation: implications for disinhibitory psychopathology in children', *Clinical Psychology Review*, 13: 699–720.

Newman, J. P., Patterson, C. M., and Kosson, D. S. (1987) 'Response perseveration in psychopaths', *Journal of Abnormal Psychology*, 33: 145–148.

O'Brien, B. S. and Frick, P. J. (1996) 'Reward dominance: associations with anxiety, conduct problems, and psychopathy in children', *Journal of Abnormal Child Psychology*, 24: 223–240.

Ogilvie, J. M., Stewart, A. L., Chan, R. C. K., and Shum, D. H. K. (2011) 'Neuropsychological measures of executive function and antisocial behavior: a meta-analysis', *Criminology*, 49: 1063–1107.

Patterson, C. M. and Newman, J. P. (1993) 'Reflectivity and learning from aversive events: toward a psychological mechanism for the syndromes of reward dominance', *Psychological Review*, 100: 716–736.

Pennington, B. F. and Ozonoff, S. (1996) 'Executive functions and developmental psychopathology', *Journal of Child Psychology and Psychiatry*, 37: 51–87.

Poythress, N. G., Dembo, R., Wareham, J., and Greenbaum, P. E. (2006) 'Construct validity of the Youth Psychopathic Traits Inventory and the Antisocial Process Screening Device (APSD) with justice-involved adolescents', *Criminal Justice and Behavior*, 33: 26–55.

Raine, A. (1993) *The Psychopathology of Crime: Criminal Behavior as a Clinical Disorder*, San Diego, CA: Academic Press.

Raine, A., Moffitt, T. E., Caspi, A., Loeber, R., Stouthamer-Loeber, M., and Lynam, D. (2005) 'Neurocognitive impairments in boys on the life-course persistent antisocial path', *Journal of Abnormal Psychology*, 114(1): 38–49.

Rezai, K., Andreasen, N. C., Alliger, R., Cohen, G., Swayze, V., and O'Leary, D. S. (1993) 'The neuropsychology of the prefrontal cortex', *Archives of Neurology*, 50(6): 636–642.

Rolls, E. T., Hornak, J., Wade, D., and McGrath, J. (1994) 'Emotion-related learning in patients with social and emotional changes associated with frontal lobe damage', *Journal of Neurology, Neurosurgery & Psychiatry*, 57(12): 1518–1524.

Roussy, S. and Toupin, J. (2000) 'Behavioral inhibition deficits in juvenile psychopaths', *Aggressive Behavior*, 26: 413–424.

Seguin, J. R. (2004) 'Neurocognitive elements of antisocial behavior: relevance of an orbitofrontal cortex account', *Brain and Cognition*, 55: 185–197.

Seguin, J. R., Arseneault, L., Boulerice, B., Harden, P. W., and Tremblay, R. E. (2002) 'Response perseveration in adolescent boys with stable and unstable histories of physical aggression: the role of underlying processes', *Journal of Child Psychology and Psychiatry*, 43: 481–494.

Shapiro, S. K., Quay, H. C., Hogan, A. E., and Schwartz, K. P. (1988) 'Response perseveration and delayed responding in undersocialized aggressive conduct disorder', *Journal of Abnormal Psychology*, 97: 371–373.

Skeem, J. L. and Cauffman, E. (2003) 'Views of the downward extension: comparing the youth version of the psychopathy checklist with the youth psychopathic traits inventory', *Behavioral Sciences and the Law*, 21: 737–770.

Steinberg, L. (2004) 'Risk taking in adolescence: what changes, and why?', *Annals of the New York Academy of Sciences*, 1021: 51–58.

Steinberg, L. (2010) 'A dual systems model of adolescent risk-taking', *Developmental Psychobiology*, 52: 216–224.

Syngelaki, E. M., Moore, S. C., Savage, J. C., Fairchild, G., and van Goozen, S. H. M. (2009) 'Executive functioning and risky decision making in young male offenders', *Criminal Justice and Behavior*, 36: 1213–1227.

Wechsler, D. (1999) *Wechsler Abbreviated Scale of Intelligence (WASI)*, San Antonio, TX: Harcourt.

Part IV

Biosocial mechanisms of antisocial conditions and antisocial behavior

22

Temperament as a biosocial construct for understanding antisocial behavior

Matt DeLisi and Michael G. Vaughn

> The idea of temperament is not socially dangerous, even though the concept of inherent temperamental differences, like a speck of dirt on a fresh snow field, mars the democratic ideal by implying that some children are lightly burdened before the race of life begins, while others start with a tiny unfair advantage.
>
> (*Kagan, 1998: 18*)

Introduction

Temperament is an important concept that has rich meaning in everyday life and science. Briefly defined as the stable, largely innate tendency with which an individual experiences the environment and regulates his or her responses to it, temperament is in many respects the quintessential biosocial construct because it simultaneously embodies a person's basic nature, and the ways that environmental constraints impinge on that basic nature. As Kagan's quotation suggests, some are blessed with temperamental features that help one get along with others relatively easily, while others face difficulties because of their temperament. From a temperamental perspective, antisocial behavior is more or less likely based upon the suite of temperamental features that typify an individual's behavior. Here we review our recently advanced temperament-based theory of antisocial behavior and offer several research directions that could guide criminological research using a temperament perspective.

DeLisi and Vaughn's temperament theory

DeLisi and Vaughn (2014) suggest that two temperamental features—effortful control and negative emotionality—can explain problem behaviors from early childhood through adulthood. In addition, these temperamental features also explain the generally aversive interactions that antisocial individuals have with their parents, siblings, teachers, and (ultimately) criminal justice personnel. Thus a unique feature of DeLisi and Vaughn's theory is the use of explanatory constructs that explain not only crime or related problem behaviors, but also criminal justice system consequences that stem from these conduct problems.

Effortful control is the ability to inhibit a dominant response in favor of a subdominant, but more socially appropriate response. It is the raw material for self-regulation, and virtually all roles and responsibilities in life require effortful control. Although there is normal variance in effortful control, clinical deficits in it are consistent with childhood neurobehavioral problems. Compared with their peers with adequate effortful control, children with low effortful control tend to display higher levels of externalizing and internalizing symptoms, tend to have lower social competence, tend to have slower declines in conduct problems, and tend to have lower subjective well-being (Kiff *et al.*, 2011). A recent longitudinal study found that externalizing behaviors can be seen from infancy to middle adolescence, and striking features of youth with the most severe behavior problems were low effortful control and negative emotions (Kjeldsen *et al.*, 2014). Effortful control is required for school compliance and success, and children with severe deficits in regulating their emotional and behavioral responses rather quickly begin a career of noncompliant behavior.

Negative emotionality is the second key feature of DeLisi and Vaughn's theory, and it typifies a person who experiences the world in a generally negative way. Although there are many negative emotions, the ones most relevant to offending are anger, hostility, and irritability. The behavioral consequences of high negative emotionality are seen in the first months of life among infants with behavioral difficultness; difficulty disengaging from negative or threatening stimuli; disturbances in sleep, feeding, and self-calming; and mood dysregulation. Whereas many babies are generally happy and prone to joy, some children appear to be the opposite in terms of their overall mood. Across the journey of life, negative emotionality is relatively stable, thus persons with high levels of negative emotionality early in life are likely to continue such emotional experiences later.

Although each construct is important independently, dynamism between effortful control and negative emotionality produces the greatest likelihood of conduct problems:

> In this respect, the interaction of these specific temperamental features enhances the likelihood that an individual will produce emotional and behavioral content that is poorly received by others. As a result, these individuals are subject to social disapproval and likely to experience informal and formal social control mechanisms. Overwhelmingly, these interactions are aversive and increase the likelihood of additional antisocial behavioral responses.
> (*DeLisi and Vaughn, 2014: 14*)

According to DeLisi and Vaughn, the primary reason why effortful control and negative emotionality combine to produce conduct problems is that this interpersonal and behavioral style is consistent with what behavioral geneticists call evocative gene-by-environment correlations (evocative rGE). Evocative rGE is a concept that articulates how specific traits or features of an individual elicit or evoke particular reactions from others in the environment. In the case of a child who has very low effortful control and very high negative emotionality, others are likely to respond to that child in negative ways precisely because of their characteristics. Those who are aversive and annoying, and who cannot control themselves, are difficult to get along with. Across a variety of contexts in the family, school, and institutional environments, youth with the theorized traits are more likely to have negative exchanges and outcomes. This is particularly salient in the criminal justice context. Police officers, probation officers, social workers, parole officers, correctional officers, and others will respond punitively to those who not only cannot follow societal rules, but who also violate the law in untoward and antagonistic ways.

The latter point is the unique feature of DeLisi and Vaughn's theory, and the point that sets it apart from most criminological theories. Just as persons with low effortful control and high negative emotionality are prone to get into trouble at school and home, they are also prone to commit crime. And upon contact with criminal justice officials, they are likely to violate terms of their sentence and behave in a noncompliant way. Moreover, this noncompliance is presented in an angry, petulant, hostile emotional style. The low effortful control–high negative emotionality mixture commonly ends poorly.

In their theory, DeLisi and Vaughn (2014) advance an agenda for research in broad topical areas. Building on that approach, we offer several research directions that could test, integrate, and perhaps challenge extant criminological theory and research. In doing so, we demonstrate the compatibility of temperament, a content area with a millennia-old tradition, to criminology and criminal justice.

Temperament and criminology research

DeLisi and Vaughn's theory offers many research questions that should be tested not only to examine the value of the theory empirically, but also to provide extensions of its basic theoretical stance. These are presented in no particular order.

First, the theory offers an opportunity to examine sex differences in antisocial behavior across the life course. Although the theory focused on anger, hostility, and more externalizing-oriented emotions, less was written about internalizing-oriented emotions such as anxiety, fear, and depression. This distinction could illuminate sex differences in serious antisocial behavior. For instance, do negative emotions unfold along gender lines into internalizing problems among females and externalizing problems among males? Are the proposed effects of anger, hostility, and irritability universal or gender-dependent? Are negative emotions mediated by higher levels of effortful control generally, and among females specifically?

Second, the theory offers potential integration and challenges to self-control theory, general strain theory, and psychopathy—all of which contain elements of effortful control or negative emotionality. Does the temperament theory explain variance in antisocial behavior over and above the predictive validity of these other theories? Do temperament measures have convergent validity with measures of self-control, general strain, and psychopathy? Could extant measures of these other theories be effectively reduced to effortful control and negative emotionality? In other words, is temperament theory more parsimonious? Alternatively, do allied theories include content that could be used to extend temperament theory?

Third, and relating to the latter question, what is the role of fear in DeLisi and Vaughn's theory? Fear is an important construct with implications for psychopathy, empathy, and social behavior, but it also is found in the temperament literature. Is fear its own construct, or is it subsumed as a negative emotion?

Fourth, what are the measurement implications of temperament theory for criminology? Temperament is a topic usually covered by developmental psychologists, and samples often include very young children (who display conduct problems, but often not the serious crimes that alert the criminal justice system). Should validated temperament measures be used to study adolescents and adults? Should these measures be modified? For instance, there are attitudinal measures of the low self-control construct in Gottfredson and Hirschi's (1990) self-control theory that contain items relating to negative emotions and effortful control problems. Perhaps these could be used to examine the temperamental facets of their theory.

Fifth, do the most severe offenders also have the worst scores on effortful control and negative emotionality? Vaughn and his colleagues have consistently identified a subgroup of approximately 5 percent who are pathological offenders (Vaughn *et al.*, 2011a, 2011b, 2014): perhaps this severe group also evince the most negative emotions and lowest effortful control.

Sixth, are the effects of effortful control and negative emotionality differentially related to various forms of antisocial behavior? Conceptually, these temperamental features seem especially suited to violent crime (e.g. assault), but also should be relevant to property, white-collar, nuisance, and drug offending. Different forms of crime also could provide opportunities to extend the theory. For example, drug offending is often associated with anxiety and depression, thus these could be additional negative emotions that are relevant to temperament theory.

Seventh, longitudinal data collection efforts should span the entire life course to permit investigations of the stability of temperament across developmental swaths. This would be helpful to examine empirically whether the most clinically severe children continue their antisocial trajectory into adolescence and adulthood, and the roles of effortful control and negative emotionality in explaining delinquent development. Moreover, these types of study can pair temperament with adverse or healthy social environments over time to extend person × environment interaction research.

Eighth, can the "opposite" direction of effortful control and negative emotionality be used to study protective factors against conduct problems? Given the rich literature on high self-control, specifically high conscientiousness, it is likely that youth with high effortful control and low negative emotionality will enjoy overwhelmingly prosocial outcomes, and be at the lowest risk for antisocial outcomes.

Ninth, and similarly, are youth with high effortful control and low negative emotionality most likely to be abstainers from delinquency, as theorized by Moffitt (1993)? In what ways do temperament constructs inform the development of children who are highly prosocial, and avoid almost entirely any involvement in deviance? Answers to these questions are not only important to criminological theory, but are also essential toward prevention efforts to target temperamental deficits in antisocial youth.

Tenth, are there latent groupings of individuals based upon effortful control and negative emotionality? If so, to what degree do these temperamental features predict various life outcomes, such as educational and financial attainment, subjective well-being, health and mortality, and other outcomes?

Conclusion

In sum, temperament is a valuable construct for criminology generally. It parsimoniously links the biological to the environmental in a directly measurable way. Uniquely, DeLisi and Vaughn's temperament theory is capable of explaining not only multiple forms of antisocial behavior and crime, but also behavior in and with the criminal justice system. We urge biosocial criminologists to incorporate temperament into their future research plans.

References

DeLisi, M. and Vaughn, M. G. (2014) 'Foundation for a temperament-based theory of antisocial behavior and criminal justice system involvement', *Journal of Criminal Justice*, 42(1): 10–25.

Gottfredson, M. R. and Hirschi, T. (1990) *A General Theory of Crime*, Stanford, CA: Stanford University Press.

Kagan, J. (1998) *Galen's Prophecy: Temperament in Human Nature*, Boulder, CO: Westview.

Kiff, C. J., Lengua, L. J., and Zalewski, M. (2011) 'Nature and nurturing: parenting in the context of child temperament', *Clinical Child and Family Psychology Review*, 14(3): 251–301.

Kjeldsen, A., Janson, H., Stoolmiller, M., Torgersen, L., and Mathiesen, K. S. (2014) 'Externalising behaviour from infancy to mid-adolescence: latent profiles and early predictors', *Journal of Applied Developmental Psychology*, 35(1): 25–34.

Moffitt, T. E. (1993) 'Adolescence-limited and life-course-persistent antisocial behavior: a developmental taxonomy', *Psychological Review*, 100(4): 674–701.

Vaughn, M. G., DeLisi, M., Gunter, T., Fu, Q., Beaver, K. M., Perron, B. E., and Howard, M. O. (2011a) 'The severe 5%: a latent class analysis of the externalizing behavior spectrum in the United States', *Journal of Criminal Justice*, 39(1): 75–80.

Vaughn, M. G., Fu, Q., Wernet, S. J., DeLisi, M., Beaver, K. M., Perron, B. E., and Howard, M. O. (2011b) 'Characteristics of abstainers from substance use and antisocial behavior in the United States', *Journal of Criminal Justice*, 39(3): 212–217.

Vaughn, M. G., Salas-Wright, C. P., DeLisi, M., and Maynard, B. R. (2014) 'Violence and externalizing behavior among youth in the United States: is there a severe 5%?', *Youth Violence and Juvenile Justice*, 12(1): 3–21.

23

Why DRD4? An evolutionary-informed model of DRD4 and antisocial behavior

Julia Dmitrieva and Emma Espel

Introduction

The past two decades have seen a growing interest in genetic underpinnings of personality traits and behavior (Moffitt, 2005; Rutter, 2006). Advances in molecular biology and sequencing have brought promise of unlocking the genetic code for human behavior and psychopathology, followed by numerous reports of null findings and very small effect sizes (Maher, 2008; Makowsky *et al.*, 2011; Turkheimer, 2011). Although frustrating to molecular biologists, this lack of findings should be welcome news to social scientists—indeed, the idea that complex behavior, such as antisocial behavior, could be linked to a single gene greatly oversimplifies the complexity of human behavior. Thus the search for the missing gene (the one gene–one disorder hypothesis) has turned to the search for the missing gene × environment interaction (Beaver *et al.*, 2012). The dopamine receptor d4 gene (DRD4) has a unique evolutionary history and population distribution, making it a particularly promising gene for studying how it interacts with environment. This chapter aims to clarify the mechanisms that link the dopamine receptor gene (DRD4) and adolescent antisocial behavior.

Dopamine and the DRD4 gene

Dopamine is the key neurotransmitter that regulates brain activity, especially modulation of responses to rewards and punishments (Davidson *et al.*, 2000; Cherek and Pietras, 2003; Killcross, 2003). Novel rewards activate midbrain dopamine neurons, leading to a successful learning of cues that predict reward-occurrence (Miller and Cohen, 2001). Signals of reward initiate a phasic burst of midbrain dopamine neurons, which, in turn, elicits positive emotional states (Schultz, 1998). The function of the dopaminergic system can be illustrated with the addicting effects of cocaine and amphetamines, which enhance dopaminergic signaling and thus the feeling of reward; and of psychostimulant drugs that enhance dopamine release and are used in treatment of impulsive and attention control problems found among attention deficit hyperactivity disorder (ADHD) patients (Oak *et al.*, 2000).

There are five subtypes of dopamine receptors, with the d4 subtype being linked to novelty seeking, ADHD, and antisocial behavior. The DRD4 gene is located on chromosome 11, at 11p15.5 (Van Tol *et al.*, 1991). This gene has a number of polymorphisms, the most variable being the 48 base-pair variable number tandem repeat (48bp VNTR) in exon III. Individuals vary in the number of repeated sequences of 48 base pairs (which encode for a sequence of 16 amino acids), from having 2 to 11 48-bp repeats. The 4-repeat allele has the highest population frequency (65 percent), the 7-repeat allele is the second most frequent (19 percent), followed by the 2-repeat allele (9 percent) (Ding *et al.*, 2002). D4 receptors act as inhibitors of neuronal firing (Oak *et al.*, 2000; Rubinstein *et al.*, 2001), and the 7R allele is associated with the receptor's blunted inhibitory ability (Asghari *et al.*, 1995) and higher reward-related ventral striatum reactivity (Forbes *et al.*, 2007). At a behavioral level, the DRD4.7 allele has been linked to novelty seeking (Benjamin *et al.*, 1996; Ebstein *et al.*, 1996), impulsivity (Congdon *et al.*, 2008), and ADHD (Brookes *et al.*, 2006; Li *et al.*, 2006).

The link between the DRD4 gene and antisocial behavior

Early pharmacological and brain stimulation studies with animals showed that dopamine lowers the threshold for aggressive behavior and increases defensive rage. Similarly, early work on psychobiology of crime focused on individual variations in the dopaminergic system (Raine, 1993; Rowe, 2002). Recent studies confirm that dopamine is released not only in response to typically rewarding activities such as food, sex, and stimulant drugs, but also in response to aggression (Couppis and Kennedy, 2008). Thus a dysregulated dopaminergic system may result in individuals exhibiting greater cravings for pleasurable and exciting experiences, and seeking those experiences to alleviate the associated discomfort. Consequently, the DRD4 gene has been linked to a variety of constructs related to antisocial behavior, such as adolescent alcohol use (Guo *et al.*, 2007), pathological gambling (De Castro *et al.*, 1997), anger (Kang *et al.*, 2008), aggressive behavior (Fresan *et al.*, 2007), psychopathy (Wu and Barnes, 2013), age of first criminal arrest (DeLisi *et al.*, 2008), and lower probability of desistance from crime (Boutwell and Beaver, 2008).

The findings for the DRD4 gene, however, have been mixed. For example, Schmidt and colleagues (2002) found that the DRD4 is associated with mother-reported, but not observed, scores of aggression in children. Instead, several studies find that the effects of the DRD4 gene interact with environment. Bakermans-Kranenburg and van Ijzendoorn (2006), for example, found a sixfold increase in externalizing problems among children carrying the 7-repeat allele, but only if they were exposed to insensitive parenting. Similarly, Beaver and colleagues (2012) found that the DRD4.7 allele is associated with higher violent delinquency, but only among youth living in disadvantaged neighborhoods.

The proposed evolutionary-informed model of DRD4 and antisocial behavior

Although DRD4 is likely linked to antisocial behavior, the mechanisms that mediate this association are not entirely clear. The model that links any genetic variation to behavior should make evolutionary sense. However, it is not clear why it is evolutionarily adaptive to engage in impulsive, antisocial, and ultimately dangerous behavior. To be sure, aggressive behavior does lead to rewards (e.g. in establishing social dominance and competing for limited resources). However, there are also costly and sometimes lethal consequences to impulsive and confrontational behavior. One possibility is that DRD4 is linked to a more innocuous basic predisposition, which under some conditions is associated with impulsive and antisocial behavior, yet adaptive in other circumstances. We propose the evolutionary-informed model of the effect of the DRD4 on antisocial behavior (Figure 23.1). The DRD4.7 allele is associated with exploratory novelty seeking, fearlessness, and reward dependence, psychological traits that may lead to both adaptive and maladaptive behavior. These traits, in turn, are associated with impulse control problems and adolescent antisocial behavior in high-risk environments, but not in low-risk environments.

DRD4.7, novelty/sensation seeking, fearlessness, and reward dependence

The 7R allele has an unusual sequence architecture that suggests that it arose as a result of a rare mutation that originated relatively recently (*ca.* 40,000–50,000 years ago, right around the time of major population expansion out of Africa) and has since increased in frequency due to strong selection forces (Ding *et al.*, 2002; Wang *et al.*, 2004). The timing of this mutation and the evidence for its selection have led researchers (Chen *et al.*, 1999; Wang *et al.*, 2004) to propose that this allele is associated with behaviors that were strongly selected in those migratory environments. Indeed, Chen and colleagues (1999) demonstrated that the world distribution of the 7R alleles for 2,320 individuals from 39 indigenous populations corresponds to the historical (within the past 30,000 years) patterns of migration for these groups. Individuals whose ethnic group has moved the furthest from their probable homeland (based on archeological and historical linguistics evidence) have higher frequencies of 7R alleles ($r = 0.85$, $P < 0.001$). Furthermore, the 7R allele is also more frequent among nomadic, as compared to sedentary, populations ($r = 0.52$, $P < 0.001$). Similarly, Matthews and Butler (2011) controlled for neutral genetic processes (population drift and population admixture) in their analysis of the DRD4 gene and historical migration, and confirmed the link between the 7R frequencies in modern populations and their macro-migration histories.

At a functional level, it makes sense that blunted dopamine inhibition, which is associated with the DRD4.7 allele, would be adaptive in some contexts. Midbrain dopamine neurons are activated in response to novel rewards and are responsible for successful learning of cues that

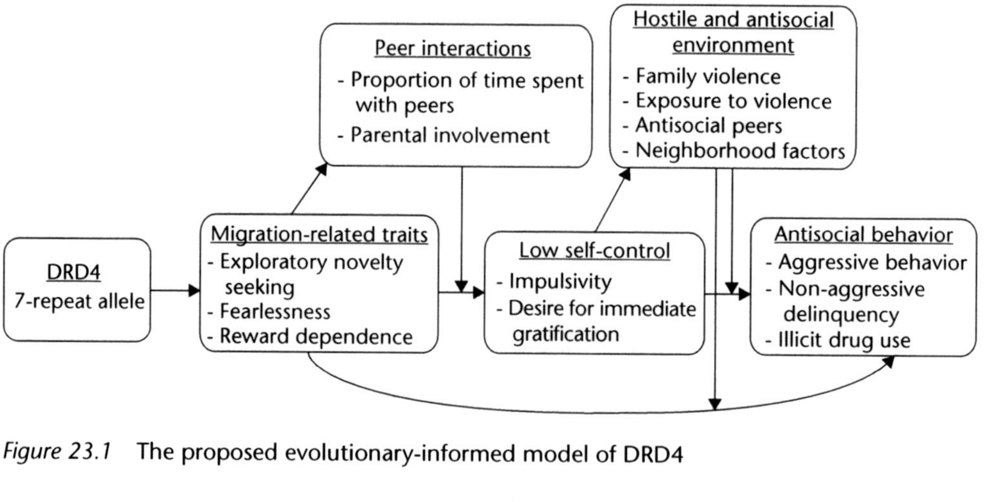

Figure 23.1 The proposed evolutionary-informed model of DRD4

predict novel reward-occurrence (Miller and Cohen, 2001). Thus blunted dopamine inhibition that is associated with the 7R alleles may result in dopamine hyperactivation that increases sensitivity to novelty and facilitates the stimulus–response link under novel conditions. In fact, DRD4 knockout mice (a line of mice that lack the DRD4 inhibitory receptors and thus serve as an extreme model of blunted DRD4 functionality that is associated with the 7R alleles) demonstrate better performance on novel complex motor tasks (Rubinstein *et al.*, 1997), and individuals with ADHD who also have a 7-repeat allele demonstrate faster reaction times (compared with individuals with ADHD but not the 7-repeat alleles) and do not exhibit reaction-time deficits on measures of attentional networks that are typically undermined by ADHD (Swanson *et al.*, 2001a,b; Langley *et al.*, 2004). This accelerated learning may be adaptive in time-critical, rapidly changing environments (e.g. migratory environments), as opposed to time-optimal stable environments (Wang *et al.*, 2004).

The findings of empirical studies attempting to link the DRD4 gene to novelty seeking and sensation seeking, however, have been mixed. Although many studies find the DRD4 novelty/sensation seeking link, a nearly equal number of studies do not find this association (Paterson *et al.*, 1999; Kluger *et al.*, 2002; Schinka *et al.*, 2002; Munafo *et al.*, 2003). Part of the problem has been in the diversity of measures and related constructs that have been interchangeably called "novelty seeking," "sensation seeking," and "impulsivity" across studies. For example, novelty seeking has been often measured with the novelty seeking subscale of the Temperament and Character Inventory (Cloninger, 1994). This subscale taps more than one personality facet and includes items that measure exploratory excitability, impulsiveness, extravagance, and disorderliness. The sensation seeking construct has been often measured with the Zuckerman Sensation Seeking Scale (Zuckerman, 1979), which taps experience seeking, boredom susceptibility, disinhibition, and thrill and adventure seeking. The evolutionary history of the DRD4.7 allele suggests its association with the exploratory excitability, experience seeking, and thrill and adventure seeking aspects of novelty/sensation seeking (representing traits that can be logically linked to exploration and migration), rather than the impulse control problems, extravagance, and disorderliness. It is also plausible that DRD4.7 would be associated with fearlessness, allowing prehistoric migrants to face the dangers of new uncertain environments. For example, Roussos *et al.* (2009) asked participants with long (7-repeat and higher) and short DRD4 alleles to participate in decision making games. DRD4-long participants not only scored higher on novelty seeking, but also were less likely to startle to affective pictures and more likely to make risky choices, whereas their problem solving and planning abilities were intact.

Reward dependence is another personality trait that has been linked to the DRD4 gene (Kühn *et al.*, 1999) and can potentially be linked to migratory environments, that is, individuals who risked to migrate in a prehistoric context would have had to be strongly motivated by the lure of potential rewards of new places in order to attempt such a challenging undertaking. Indeed, carriers of the DRD4 7-repeat allele experience greater appetitive drive, greater food cravings in response to food cues (Sobik *et al.*, 2005), smoking cravings and activation in the superior frontal gyrus and insula (areas responsible for recall of emotional states and cravings) in response to smoking cues (McClernon *et al.*, 2007), alcohol cravings and activation in the orbitofrontal cortex, striatum, and anterior cingulate gyrus in response to alcohol cues (Hutchison *et al.*, 2002; Filbey *et al.*, 2008), and heroin cravings in response to heroin cues (Shao *et al.*, 2006).

Exploratory novelty seeking, fearlessness, and reward dependence are not intrinsically negative outcomes. Youth who seek novel experiences, have high tolerance for danger, and crave rewards may raise the bar in athletic and artistic arenas, push the frontiers of human exploration and innovation, and undertake challenging and high-risk/high-pay-off projects. In fact,

Cloninger's work indicates that novelty seeking and reward dependence are associated with hedonic well-being and emotional health (Cloninger *et al.*, 1998)

Under certain circumstances, however, this propensity to seek new experiences, fearlessness, and reward dependence may be associated with impulse control problems and antisocial behavior. For example, children growing up in dangerous neighborhoods would be exposed to more antisocial experiences if they are more inclined to explore their surroundings. Similarly, the exploratory novelty seeking and fearlessness may interact with peer variables to lead to impulsive and risky behavior. For example, we know that adolescents are particularly susceptible to peer influence on risky behavior. Adolescents are more likely to crash on a mock driving test in the presence of peers, compared with adults for whom peer presence does not have a risk-facilitating effect (Gardner and Steinberg, 2005). Peer presence also elevates salience of immediate rewards among adolescents (O'Brien *et al.*, 2011). This facilitation appears to be linked to adolescents' greater activation of the brain reward circuitry (ventral striatum and orbitofrontal cortex) in the presence of peers (Chein *et al.*, 2011). Thus adolescent carriers of the DRD4.7 allele who spend a lot of time with peers could be at an especially high risk for alcohol use, as their ventral striatum and orbitofrontal cortex are activated at higher levels due to both higher peer presence and higher DRD4.7-linked hyperactivation of these areas in response to drinking cues. In another example, having antisocial peers may serve as an impetus for initiating drug use, whereas DRD4.7 carriers' higher reward dependence may serve as a factor that maintains and elevates drug seeking behavior. Thus we propose that exploratory novelty seeking and fearlessness, combined with high peer contact and low parental supervision, would be associated with more impulsive and risky behavior, which in turn would further elevate the risk for antisocial behavior.

The idea that exploratory novelty seeking and fearlessness would be directly linked to the DRD4.7 allele, whereas the link from exploratory novelty seeking to impulse control problems would be moderated by environment, is in line with previous findings. Sheese and colleagues (2007), for example, found that children carrying the DRD4.7 allele and experiencing low-quality parenting show higher levels of aggressive impulsivity. Other researchers also find the DRD4 × environment interaction in their effects on the more broadly defined (i.e. impulsive) novelty seeking (Keltikangas-Jarvinen *et al.*, 2004; Lahti *et al.*, 2005).

The role of self-control in adolescent antisocial behavior

Developmental criminology and developmental psychopathology provide a rich background for the trait × environment part of our model. Low self-control is the cornerstone of all major theories of adolescent delinquency. Gottfredson and Hirschi's (1990) general theory of crime proposes that low self-control is at the root of antisocial behavior. Criminal behavior, they theorized, is a pursuit of pleasure and avoidance of pain that an individual has failed to restrain. Criminal acts are exciting, risky, and thrilling, and provide immediate gratification. Individuals who are impulsive and seek risk and intense sensations are more likely to pursue immediate pleasures through both criminal and socially acceptable means. Social context plays a role inasmuch as ineffective and inconsistent parental socialization (e.g. poor parental monitoring and ineffective discipline) during the first half of childhood results in the failure of children to acquire effective self-control.

The signs of low self-control can be identified in individuals as young as 3 years of age, and children with difficult temperament at age 3 are more likely to commit criminal offenses during adolescence (Henry *et al.*, 1996). Studies show that low self-control is a stable characteristic (Turner and Piquero, 2002) with a strong genetic component (Beaver *et al.*, 2008). Furthermore, a number of studies demonstrate a link between childhood and adolescent low self-control and adult delinquency. However, some find only a weak link between low self-control and crime (Longshore *et al.*, 1996), and self-control does not differentiate between youths who commit serious versus minor types of offenses, whereas cognitive deficits and other physiological factors (e.g. low resting heart rate) do separate youths who show serious antisocial behavior from those who engage in minor offending (Cauffman *et al.*, 2005).

Moffitt's taxonomy of antisocial behavior (1993) proposes that there are two groups of antisocial adolescents. A life-course-persistent group is characterized by an early onset of offending (around age 8) and chronic involvement in antisocial behavior throughout adolescence into adulthood. In contrast, youths in the adolescence-limited group become involved in antisocial behavior during their adolescent years and desist by young adulthood. Family and peers play a major role in antisocial behavior of both these groups, but the onset of antisocial behavior is more heavily influenced by biological factors for the life-course-persistent group. Negative effects of early neurological risk factors, including impulsivity, are compounded by negative parental reactions to the child (e.g. due to failure to cope with a difficult child), therefore leading to an early onset of problem behaviors for the life-course-persistent group. Subsequent affiliation with deviant peers accounts for the stability and increasing severity of their antisocial behaviors. In short, Moffitt's theory assigns individual factors, such as personality and neuropsychological factors, the role of risk factors for the entry into the early onset, life-course-persistent group. Contextual factors, such as family and peer influences, compound the negative effects of their individual traits and help maintain involvement in antisocial behavior. Our assertion that hostile environment moderates the link between low impulse control and antisocial behavior is consistent with Moffitt's claim that biological factors predispose youth for early risk of offending, whereas environmental factors ensure the maintenance or desistance from crime.

Finally, Patterson's coercion training model (Patterson *et al.*, 1989) outlines how hostile parenting can contribute to an initially impulsive child turning to antisocial behavior. These researchers proposed that, for some families, there is a pattern of coercive parent–child interactions that escalates over time, undermining parental abilities to effectively discipline their child. According to this theory, children with early predispositions for antisocial behavior (i.e. children with a difficult impulsive temperament) are likely to respond to parental attempts at discipline in a hostile manner. When parents give in to their child's aggressive response, the child is conditioned (through reinforcement of negative interactions) into coercive behavior. That is, the child becomes aware of the fact that he or she can stop parental attempts at discipline through the use of hostility. With time, the hostile exchanges become increasingly aggressive and lead to harsher disciplinary attempts and more aggressive responses from the child. The effects of coercive conditioning spill over into interactions with peers, resulting in aggressive behaviors toward other children and, consequently, rejection by prosocial peers. Thus an initially antisocially prone child becomes conditioned into the use of aggression, develops deficiency in prosocial skills (due to a lack of interaction with prosocial peers), and receives further encouragement to remain antisocial through the affiliation with antisocial peers. Consistent with this theory, we propose that hostile parent–adolescent interactions would moderate the link between low self-control and antisocial behavior.

Environment as a moderator

The idea that personality traits may lead to different outcomes is not new. When it comes to antisocial behavior, Eysenck (1964) proposed that extroversion is related to biologically determined conditionability (i.e. ease of being conditioned by rewards and punishments), which is in turn associated with diverse effects. That is, "good" conditioners do well in good environments. They are easily socialized into prosocial values and norms of behavior. "Bad" conditioners, on the other hand, are not as strongly socialized into prosocial values, and therefore are likely to become engaged in antisocial behavior. The distinction is reversed in high-risk environments. Good conditioners are socialized into antisocial values of their surrounding antisocial context and become involved in antisocial behavior. Compared with them, bad conditioners are not as strongly affected by the antisocial environment and refrain from delinquency. Evidence in support of this theory demonstrates that there is an interaction in the theoretically expected direction between the effects of socioeconomic stressors and biological markers of conditionability (Raine and Venables, 1981).

Borrowing from Eysenck's model, we propose that the effects of low self-control are also influenced by social context. Impulsive individuals act on an impulse that is internal but is triggered by environmental stimuli. One needs to suppress an impulse to join friends in their delinquent behavior only if such an impulse is in fact initiated by the presence of antisocial friends. Although it is true that individuals to a considerable degree choose their friends, and the proposed model includes a trait–environment correlation, it is a lot easier to join antisocial peers when they are in abundance in one's neighborhood or school.

Similarly, there should be an interaction of the desire for immediate gratification with social context. Theories of antisocial behavior assume that crime is a shortcut toward obtaining a desirable goal. That argument is more relevant for the high-risk than the low-risk environment. Access to antisocial role models to learn from (one needs to learn how, for example, to go about stealing without being caught), availability of potential partners in crime, access to weapons, and low chance of obtaining the desired object through prosocial means (e.g. as a gift from parents) all make crime appear as a more "attractive," more effective, and less dangerous option for someone in the high-risk environment. By the same argument, someone in a low-risk environment may find it easier and more rewarding to follow the prosocial route.

It is important to note here that, although various indicators of high-risk environment are correlated, the proposed pattern of associations does not apply only to individuals who live in environments marked by high risk across all domains. Therefore adolescents living in what are considered "safe neighborhoods" may also experience pressure from antisocial peers; albeit the pervasiveness of that pressure is likely to be considerably lower than for someone living in a gang-divided neighborhood.

In summary, we propose that the DRD4.7 allele is associated with higher scores on migration-related traits (exploratory novelty seeking, fearlessness, and reward dependence), which in turn are associated with low self-control (impulsivity and desire for immediate gratification). The migration-related traits and low self-control are positively associated with offending in high-risk but not in low-risk environments. We expect that in low-risk contexts (especially fast-changing contexts, e.g. when youth relocate frequently), exploratory novelty seeking, fearlessness, and reward dependence would not only be not associated with antisocial behavior, but also be related to positive adjustment. This proposition is in line with the biological sensitivity to context (Boyce and Ellis, 2005) and differential susceptibility (Belsky, 2005) theories which suggest that some genes may be associated with stronger sensitivity to environment. As such, certain genes may contribute to individuals being more sensitive to their environment, thus being more strongly influenced by the positive and negative features of their environment.

There is ample evidence showing that adverse family conditions contribute to adolescent antisocial behavior, making this domain also key for exploration of gene × environment and trait × environment interactions. Children of parents who exhibit low warmth and acceptance, use harsh, ineffective discipline, and supervise them poorly are more likely to be involved in antisocial behavior (see Stern and Smith, 1999 for a review; Rothbaum and Weisz, 1994 for a meta-analysis of cross-sectional studies of community children; Loeber and Stouthamer-Loeber, 1986 for studies of delinquent children; and Steinberg, 2001 and Khaleque and Rohner, 2002 for reviews of cross-cultural studies). Aside from parenting, other family variables contribute to adolescent antisocial behavior. Family economic status is an overarching variable that is associated with a variety of stressors experienced by youths, including harsh physical discipline and other aspects of authoritarianism, parental emotional distress, lack of social capital, parental antisocial behavior, and child abuse and neglect (see Seccombe, 2000 for a review). Family stressful life events are also known to be related to adolescent problem behaviors (Larson and Ham, 1993; Hoffman *et al.*, 1996; McKnight *et al.*, 2002). Finally, family climate contributes to antisocial behavior, and youths who experience high levels of hostility between their parents (Gottman, 1993; Cherlin *et al.*, 1998), parental divorce (Amato and Keith, 1991), and parent–adolescent conflict (Chen *et al.*, 1998; Ary *et al.*, 1999; Dmitrieva *et al.*, 2004) exhibit higher levels of delinquency, among other psychosocial problems.

The peer domain also makes a considerable contribution to antisocial behavior. Aggressive children tend to be rejected by their peers, and this peer rejection creates a group of ostracized youths who, because of fewer opportunities to become socialized into prosocial groups, are further at risk for antisocial behavior (Moffitt, 1993). Indeed, peer rejection predicts later aggressive behavior and exacerbates the development of antisocial behavior for children with earlier conduct problems (Dodge *et al.*, 2003). There is also substantial evidence for a direct link from peer delinquency to adolescents' engagement in antisocial behavior. Having delinquent peers and peers who approve of misconduct is associated with higher involvement in problem behavior in cross-sectional (Farrell and White, 1998), longitudinal (Herrenkohl *et al.*, 2000), and cross-cultural studies (Greenberger *et al.*, 2000; Beam *et al.*, 2002).

Finally, neighborhood structure and social disorder influence antisocial behavior (Leventhal and Brooks-Gunn, 2000; Edwards, 2005). A number of studies indicate that community violence, extreme poverty, and social disorder are related to children's and adolescents' aggressive and delinquent behaviors (Gorman-Smith and Tolan, 1998; Miller *et al.*, 1999; Schwab-Stone *et al.*, 1999). Thus we propose that exposure to high-risk environments will potentiate the effects of low self-control on antisocial behavior.

Hostile family, peer, and neighborhood environments are the key candidates for the study of trait × environment and gene × environment interactions. Our model proposes two separate sets of social variables: one moderating the path to impulse control problems; the other moderating the path to antisocial behavior. We propose that the path from migration-related traits to low self-control will be predominantly moderated by high peer exposure and lack of positive parental involvement. As such, children who carry the DRD4.7 allele and have the associated high novelty seeking and reward dependence would suffer more strongly from the lack of parental socialization into self-regulating behaviors. They will also be more strongly influenced by excessive time spent with peers, due to the dysregulating influences of the peer group. Thus the moderators of the path to low self-control can be characterized as contexts that help promote or hinder successful learning of the self-control strategies.

In contrast, moderators of the paths to antisocial behavior are more directly related to antisocial influence. We propose that hostile and antisocial family, peer, and neighborhood environments will moderate the links from migration-related traits to antisocial behavior, and from low self-control to antisocial behavior. We further propose that exploratory novelty seeking, fearlessness, and reward dependence will not be associated with antisocial behavior in positive environments: this link between these traits and antisocial behavior will be present *only* among youth growing up in hostile contexts.

In summary, the bulk of the research on DRD4 supports our evolutionary-informed model of DRD4, which proposes that the DRD4.7 allele is associated with innocuous exploratory novelty seeking, fearlessness, and reward dependence, the traits that provided evolutionary advantage for the DRD4.7 carriers in the context of prehistoric migration. We assert that the DRD4.7 continues to provide advantage in certain contexts (e.g. migratory environments). With regard to antisocial behavior, our claim that exploratory novelty seeking, fearlessness, and reward dependence could lead to externalizing problems in high-risk environments is in line with major theories of adolescent offending. Specifically, we propose that DRD4.7 carriers who experience high peer contact and low parental involvement would have fewer opportunities to internalize successful self-regulation strategies, which would lead to their impulsivity and desire for immediate gratification. Adolescent DRD4.7 carriers who are exposed to violence and antisocial behavior in the family, peer, and neighborhood domains would be more strongly influenced by their environment (due to their higher novelty seeking and reward dependence) and be at greater risk for antisocial behavior. Although we provide support for various parts of the proposed model, future studies will need to test these assertions more formally and comprehensively.

References

Amato, P. R. and Keith, B. (1991) 'Parental divorce and the well-being of children a meta-analysis', *Psychological Bulletin*, 110: 26–46.

Ary, D.V., Duncan, T. E., Biglan, A., Metzler, C. W., Noell, J. W., and Smolkowski, K. (1999) 'Development of adolescent problem behavior', *Journal of Abnormal Child Psychology*, 27: 141–150.

Asghari, V., Sanyal, S., Buchwaldt, S., Paterson, A., Jovanovic, A., and Van Tol, H. H. (1995) 'Modulation of intracellular cyclic AMP levels by different human dopamine D4 receptor variants', *Journal of Neurochemistry*, 65: 1157–1165.

Bakermans-Kranenburg, M. J. and van Ijzendoorn, M. H. (2006) 'Gene–environment interaction of the dopamine D4 receptor (DRD4) and observed maternal insensitivity predicting externalizing behavior in preschoolers', *Developmental Psychobiology*, 48: 406–409.

Beam, M. R., Gil-Rivas, V., Greenberger, E., and Chen, C. (2002) 'Adolescent problem behavior and depressed mood: risk and protection within and across social contexts', *Journal of Youth and Adolescence*, 31: 343–357.

Beaver, K. M., Wright, J. P., DeLisi, M., and Vaughn, M. G. (2008) 'Genetic influences on the stability of low self-control: results from a longitudinal sample of twins', *Journal of Criminal Justice*, 36: 478–485.

Beaver, K. M., Gibson, C. L., DeLisi, M., Vaughn, M. G., and Wright, J. P. (2012) 'The interaction between neighborhood disadvantage and genetic factors in the prediction of antisocial outcomes', *Youth Violence and Juvenile Justice*, 10: 25–40.

Belsky, J. (2005) 'Differential susceptibility to rearing influence: an evolutionary hypothesis and some evidence', in B. Ellis and D. Bjorklund (eds), *Origins of the Social Mind: Evolutionary Psychology and Child Development*, New York: Guilford, pp. 139–163.

Benjamin, J., Li, L., Patterson, C., Greenberg, B. D., Murphy, D. L., and Hamer, D. H. (1996) 'Population and familial association between the D4 dopamine receptor gene and measures of novelty seeking', *Nature Genetics*, 12: 81–84.

Boutwell, B. B. and Beaver, K. M. (2008) 'A biosocial explanation of delinquency abstention', *Criminal Behaviour and Mental Health*, 18: 59–74.

Boyce, W.T. and Ellis, B. (2005) 'Biological sensitivity to context: I. An evolutionary–developmental theory of the origins and functions of stress reactivity', *Development and Psychopathology*, 17: 271–301.

Brookes, K., Xu, X., Chen, W., Zhou, K., Neale, B., Lowe, N. *et al.* (2006) 'The analysis of 51 genes in DSM-IV combined type attention deficit hyperactivity disorder association signals in DRD4, DAT1 and 16 other genes', *Molecular Psychiatry*, 11: 934–953.

Cauffman, E., Steinberg, L., and Piquero, A. (2005) 'Psychological, neuropsychological, and psychophysiological correlates of serious antisocial behavior in adolescence', *Criminology*, 43: 133–176.

Chein, J., Albert, D., O'Brien, L., Uckert, K., and Steinberg, L. (2011) 'Peers increase adolescent risk taking by enhancing activity in the brain's reward circuitry', *Developmental Science*, 14: F1–F10.

Chen, C., Greenberger, E., Lester, J., Dong, Q., and Guo, M. (1998) 'A cross-cultural study of family and peer correlates of adolescent misconduct', *Developmental Psychology*, 34: 770–781.

Chen, C., Burton, M., Greenberger, E., and Dmitrieva, J. (1999) 'Population migration and the variation of dopamine D4 receptor (DRD4) allele frequencies around the globe', *Evolution and Human Behavior*, 20: 309–324.

Cherek, D. R. and Pietras, C. J. (2003) 'Human aggression: biological correlates and environmental influences', in M. A. Ron and T. W. Robbins (eds), *Disorders of Brain and Mind 2*, New York: Cambridge University Press, pp. 375–399.

Cherlin, A. J., Chase-Lansdale, P. L., and McRae, C. (1998) 'Effects of parental divorce on mental health throughout the life course', *American Sociological Review*, 63: 239–249.

Cloninger, C. R. (1994) *The Temperament and Character Inventory (TCI): A Guide to its Development and Use*, St Louis, MO: Center for Psychobiology of Personality, Washington University.

Cloninger, C. R., Bayon, C., and Svrakic, D. M. (1998) 'Measurement of temperament and character in mood disorders a model of fundamental states as personality types', *Journal of Affective Disorders*, 51: 21–32.

Congdon, E., Lesch, K. P., and Canli, T. (2008) 'Analysis of DRD4 and DAT polymorphisms and behavioral inhibition in healthy adults: implications for impulsivity', *American Journal of Medical Genetics B, Neuropsychiatric Genetics*, 147B: 27–32.

Couppis, M. H. and Kennedy, C. H. (2008) 'The rewarding effect of aggression is reduced by nucleus accumbens dopamine receptor antagonism in mice', *Psychopharmacology*, 197: 449–456.

Davidson, R. J., Putnam, K. M., and Larson, C. L. (2000) 'Dysfunction in the neural circuitry of emotion regulation—a possible prelude to violence', *Science*, 289: 591–594.

De Castro, P. I., Torres, P., Saiz-Ruiz, J., and Fernandez-Piqueras, J. (1997) 'Genetic association study between pathological gambling and a functional DNA polymorphism at the D4 receptor', *Pharmacogenetics*, 7: 345–348.

DeLisi, M., Beaver, K., Wright, J. P., and Vaughn, M. G. (2008) 'The etiology of criminal onset: the enduring salience of nature and nurture', *Journal of Criminal Justice*, 36: 217–223.

Ding, Y.-C., Chi, H. C., Grady, D. L., Morishima, A., Kidd, J. R., Kidd, K., Flodman, P., Spence, M. A., Schuck, S., Swanson, J. M., Zhang, Y.-P., and Moyzis, R. K. (2002) 'Evidence of positive selection acting at the human dopamine receptor D4 gene locus', *Proceedings of the National Academy of Sciences, USA*, 99: 309–314.

Dmitrieva, J., Chen, C., Greenberger, E., and Gil-Rivas, V. (2004) 'Family relationships and adolescent psychosocial outcomes: converging findings from Eastern and Western cultures', *Journal of Research on Adolescence*, 14: 425–448.

Dodge, K. A., Lansford, J. E., Burks, V. S., Bates, J. E., Pettit, G. S., Fontaine, R., and Price, J. M. (2003) 'Peer rejection and social information-processing factors in the development of aggressive behavior problems in children', *Child Development*, 74: 374–393.

Ebstein, R. P., Novick, O., Umansky, R., Priel, B., Osher, Y., Blaine, D., Bennett, E. R., Nemanov, L., Katz, M., and Belmaker, R. H. (1996) 'Dopamine D4 receptor (D4DR) exon III polymorphism associated with the human personality trait of novelty seeking', *Nature Genetics*, 12: 78–80.

Edwards, B. (2005) 'Does it take a village? An investigation of neighborhood effects on Australian children's development', *Family Matters*, 72: 36–43.

Eysenck, H. J. (1964) *Crime and Personality*, Oxford: Houghton Mifflin.

Farrell, A. D. and White, K. S. (1998) 'Peer influence and drug use among urban adolescents: family structure and parent–adolescent relationship as protective factors', *Journal of Consulting and Clinical Psychology*, 66: 248–258.

Filbey, F. M., Ray, L., Smolen, A., Claus, E. D., Audette, A., and Hutchison, K. E. (2008) 'Differential neural response to alcohol priming and alcohol taste cues is associated with DRD4 VNTR and OPRM1 genotypes', *Alcoholism Clinical and Experimental Research*, 32: 1113–1123.

Forbes, E. E., Brown, S. M., Kimak, M., Ferrell, R. E., Manuck, S. B., and Hariri, A. R. (2007) 'Genetic variation in components of dopamine neurotransmission impacts ventral striatal reactivity associated with impulsivity', *Molecular Psychiatry*, 14: 60–70.

Fresan, A., Camarena, B., Apiquian, R., Aguilar, A., Urraca, N., and Nicolini, H. (2007) 'Association study of MAO-A and DRD4 genes in schizophrenic patients with aggressive behavior', *Neuropsychobiology*, 55: 171–175.

Gardner, N. and Steinberg, L. (2005) 'Peer influence on risk taking, risk preference, and risky decision making in adolescence and adulthood: an experimental study', *Developmental Psychology*, 41: 625–635.

Gottfredson, M. R. and Hirschi, T. (1990) *A General Theory of Crime*, Stanford, CA: Stanford University Press.

Gorman-Smith, D. and Tolan, P. (1998) 'The role of exposure to community violence and developmental problems among inner-city youth', *Development and Psychopathology*, 10: 101–116.

Gottman, J. M. (1993) 'The roles of conflict engagement, escalation, and avoidance in marital interaction: a longitudinal view of five types of couples', *Journal of Consulting and Clinical Psychology*, 61: 6–15.

Greenberger, E., Chen, C., Beam, M., Whang, S., and Dong, Q. (2000) 'The perceived social contexts of adolescents' misconduct: a comparative study of youths in three cultures', *Journal of Research on Adolescence*, 10: 365–388.

Guo, G., Wilhelmsen, K., and Hamilton, N. (2007) 'Gene–lifecourse interaction for alcohol consumption in adolescence and young adulthood five monoamine genes', *American Journal of Medical Genetics Part B (Neuropsychiatric Genetics)*, 144B: 417–423.

Henry, B., Caspi, A., Moffitt, T. E., and Silva, P. A. (1996) 'Temperamental and familial predictors of violent and nonviolent criminal convictions: age 3 to age 18', *Developmental Psychology*, 32: 614–623.

Herrenkohl, T. I., Maguin, E., Hill, K. G., Hawkins, J. D., Abbott, R. D., and Catalano, R. F. (2000) 'Developmental risk factors for youth violence', *Journal of Adolescent Health*, 26: 176–186.

Hoffman, M. A., Levy-Shiff, R., and Malinski, D. (1996) 'Stress and adjustment in the transition to adolescence: moderating effects of neuroticism and extroversion', *Journal of Youth and Adolescence*, 25: 161–175.

Hutchison, K. E., LaChance, H., Niaura, R., Bryan, A.D., and Smolen, A. (2002) 'The DRD4 VNTR polymorphism influences reactivity to smoking cues', *Journal of Abnormal Psychology*, 111: 134–142.

Kang, J. I., Namkoong, K., and Kim, S. J. (2008) 'Association of DRD4 and COMT polymorphisms with anger and forgiveness traits in healthy volunteers', *Neuroscience Letters*, 430: 252–257.

Keltikangas-Jarvinen, L., Raikkonen, K., Ekelund, J., and Peltonen, L. (2004) 'Nature and nurture in novelty seeking', *Molecular Psychiatry*, 9: 308–311.

Khaleque, A. and Rohner, R. P. (2002) 'Perceived parental acceptance–rejection and psychological adjustment: a meta-analysis of cross-cultural and intracultural studies', *Journal of Marriage and Family*, 64: 54–64.

Killcross, S. (2003) 'The neural substrates of anxiety', in M. A. Ron and T. W. Robbins (eds), *Disorders of Brain and Mind 2*, New York: Cambridge University Press, pp. 308–337.

Kluger, A. N., Siegfried, Z., and Ebstein, R. P. (2002) 'A meta-analysis of the association between DRD4 polymorphism and novelty seeking', *Molecular Psychiatry*, 7: 712–717.

Kühn, K., Meyer, K., Nöthen, M. M., Gänsicke, M., Papassotiropoulos, A., and Maier, W. (1999) 'Allelic variants of dopamine receptor D4 (DRD4) and serotonin receptor 5HT2c (HTR2c) and temperament factors: replication tests', *American Journal of Medical Genetics (Neuropsychiatric Genetics)*, 88: 168–172.

Lahti, J., Raikkonen, K., Ekelund, J., Peltonen, L., Raitakari, O. T., and Keltikangas-Jarvinen, L. (2005) 'Novelty seeking interaction between parental alcohol use and dopamine D4 receptor gene exon III polymorphism over 17 years', *Psychiatric Genetics*, 15: 133–139.

Langley, K., Marshall, L., van den Bree, M., Thomas, H., O'Donovan, M., and Thapar, A. (2004) 'ADHD children with and without the dopamine D4 receptor 7-repeat allele evidence of differences in performance on neuropsychological tests', *American Journal of Psychiatry*, 161: 133–138.

Larson, R. and Ham, M. (1993) 'Stress and "storm and stress" in early adolescence: the relationship of negative events with dysphoric affect', *Developmental Psychology*, 29: 130–140.

Leventhal, T. and Brooks-Gunn, J. (2000) 'The neighborhoods they live in: the effects of neighborhood residence on child and adolescent development', *Psychological Bulletin*, 126: 309–337.

Li, D., Sham, P. C., Owen, M. J., and He, L. (2006) 'Meta-analysis shows significant association between dopamine system genes and attention deficit hyperactivity disorder (ADHD)', *Human Molecular Genetics*, 15: 2276–2284.

Loeber, R. and Stouthamer-Loeber, M. (1986) 'Family factors as correlates of predictions of juvenile conduct problems and delinquency', in M. Tonry and N. Morris (eds), *Crime and Justice*, Vol. 7, Chicago, IL: University of Chicago Press, pp. 29–149.

Longshore, D., Rand, S. T., and Stein, J. A. (1996) 'Self-control in a criminal sample: an examination of construct validity', *Criminology*, 34(2): 209–228.

McClernon, F. J., Hutchison, K. E., Rose, J. E., and Kozink, R.V. (2007) 'DRD4 VNTR polymorphism is associated with transient fMRI-BOLD responses to smoking cues', *Psychopharmacology*, 194: 433–441.

McKnight, C. G., Huebner, E. S., and Suldo, S. (2002) 'Relationships among stressful life events, temperament, problem behavior, and global life satisfaction in adolescents', *Psychology in the Schools*, 39: 677–687.

Maher, B. (2008) 'Personal genomes: the case of the missing heritability', *Nature*, 456: 18–21.

Makowsky, R., Pajewski, N. M., Kilmentidis, Y. C., Vazquez, I. A., Duarter, C. W., Allison, D. B., and de los Campos, G. (2011) 'Beyond missing heritability prediction of complex traits', *PLoS Genetics*, 7: e1002051.

Matthews, L. J. and Butler, P. M. (2011) 'Novelty-seeking DRD4 polymorphisms are associated with human migration distance out-of-Africa after controlling for neutral populations gene structure', *American Journal of Physical Anthropology*, 145: 382–389.

Miller, E. K. and Cohen, J. D. (2001) 'An integrative theory of prefrontal cortex function', *Annual Review of Neuroscience*, 24: 167–202.

Miller, L. S., Wasserman, G. A., Neugebauer, R., Gorman-Smith, D., and Kamboukos, D. (1999) 'Witnessed community violence and antisocial behavior in high-risk, urban boys', *Journal of Clinical Child Psychology*, 28: 2–11.

Moffitt, T. E. (1993) 'Adolescence-limited and life-course-persistent antisocial behavior: a developmental taxonomy', *Psychological Review*, 100: 674–701.

Moffitt, T. E. (2005) 'The new look of behavioral genetics in developmental psychopathology: gene–environment interplay in antisocial behaviors', *Psychological Bulletin*, 131(4): 533–554.

Munafo, M. R., Clark, T. G., Moore, L. R., Payne, E., Walton, R., and Flint, J. (2003) 'Genetic polymorphisms and personality in healthy adults: a systematic review and meta-analysis', *Molecular Psychiatry*, 8: 471–484.

Oak, J. N., Oldenhof, J., and Van Toll, H. H. (2000) 'The dopamine D(4) receptor one decade of research', *European Journal of Pharmacology*, 405: 303–327.

O'Brien, L., Albert, D., Chein, J., and Steinberg, L. (2011) 'Adolescents prefer more immediate rewards within the presence of their peers', *Journal of Research on Adolescence*, 21: 747–753.

Paterson, A. D., Sunohara, G. A., and Kennedy, J. L. (1999) 'Dopamine D4 receptor gene: novelty or nonsense?', *Neuropsychopharmacology*, 21: 3–16.

Patterson, G. R., DeBaryshe, B. D., and Ramsey, E. (1989) 'A developmental perspective on antisocial behavior', *American Psychologist*, 44: 329–335.

Raine, A. (1993) *The Psychopathology of Crime: Criminal Behavior as a Clinical Disorder*, San Diego, CA: Academic Press.

Raine, A. and Venables, P. H. (1981) 'Classical conditioning and socialization—a biosocial interaction', *Personality and Individual Differences*, 2: 273–283.

Rothbaum, F. and Weisz, J. R. (1994) 'Parental caregiving and child externalizing behavior in nonclinical samples: a meta-analysis', *Psychological Bulletin*, 116: 55–74.

Roussos, P., Giakoumaki, S. G., and Bitsios, P. (2009) 'Cognitive and emotional processing in high novelty seeking associated with the L-DRD4 genotype', *Neuropsychologia*, 47(7): 1654–1659.

Rowe, D. C. (2002) *Biology and Crime*, Los Angeles, CA: Roxbury.

Rubinstein, M., Phillips, T. J., Bunzow, J. R., Falzone, T. L., Dziewczapolski, G., Zhang, G., Fang, Y., Larsen, J. L., McDougall, J. A., Chester, J. A., Saez, C., Pugsley, T. A., Gershanik, O., Low, M. J., and Grandy, D. K. (1997) 'Mice lacking dopamine D4 receptors are supersensitive to ethanol, cocaine, and methamphetamine', *Cell*, 90: 991–1001.

Rubinstein, M., Cepeda, C., Hurst, R. S., Flores-Hernandez, J., Ariano, M. A., Falzone, T. L., *et al.* (2001) 'Dopamine D4 receptor-deficient mice display cortical hyperexcitability', *Journal of Neuroscience*, 21(11): 3756–3763.

Rutter, M. (2006) *Genes and Behavior: Nature–Nurture Interplay Explained*, Malden, MA: Blackwell Publishing.

Schinka, J. A., Letsch, E. A., and Crawford, F. C. (2002) 'DRD4 and novelty seeking: results of meta-analyses', *American Journal of Medical Genetics*, 114: 643–648.

Schmidt, L. A., Fox, N. A., Rubin, K. H., Hu, S., and Hamer, D. H. (2002) 'Molecular genetics of shyness and aggression in preschoolers', *Personality and Individual Differences*, 33: 227–238.

Schultz, W. (1998) 'Predictive reward signal of dopamine neurons', *Journal of Neurophysiology*, 80: 1–27.

Schwab-Stone, M., Chen, C., Greenberger, E., Silver, D., Lichtman, J., and Voyce, C. (1999) 'No safe haven II: the effects of violence exposure on urban youth', *Journal of the American Academy of Child and Adolescent Psychiatry*, 38: 359–367.

Seccombe, K. (2000) 'Families in poverty in the 1990s: trends, causes, consequences, and lessons learned', *Journal of Marriage and the Family*, 62: 1094–1113.

Shao, C., Li, Y., Jiang, K., Zhang, D., Xu, Y., Lin, L., Wang, Q., Zhao, M., and Jin, L. (2006) 'Dopamine D4 receptor polymorphism modulates cue-elicited heroin craving in Chinese', *Psychopharmacology*, 186: 185–190.

Sheese, B. E., Voelker, P. M., Rothbart, M. K., and Posner, M. I. (2007) 'Parenting quality interacts with genetic variation in dopamine receptor D4 to influence temperament in early childhood', *Development and Psychopathology*, 19: 1039–1046.

Sobik, L., Hutchinson, K., and Craighead, L. (2005) 'Cue-elicited craving for food: a fresh approach to the study of binge eating', *Appetite*, 44: 253–261.

Steinberg, L. (2001) 'We know some things: parent–adolescent relationships in retrospect and prospect', *Journal of Research on Adolescence*, 11: 1–19.

Stern, S. B. and Smith, C. A. (1999) 'Reciprocal relationships between antisocial behavior and parenting: implications for delinquency intervention', *Families in Society*, 80: 169–181.

Swanson, J., Deutsch, C., Cantwell, D., Posner, M., Kennedy, J., Barr, C., Moyzis, R., Schuck, S., Flodman, P., and Spence, M. A. (2001a) 'Genes and attention-deficit hyperactivity disorder', *Clinical Neuroscience Research*, 1: 207–216.

Swanson, J. M., Kraemer, H. C., Hinshaw, S. P., Arnold, L. E., Conners, C. K., Abikoff, H. B., *et al.* (2001b) 'Clinical relevance of the primary findings of the MTA: success rates based on severity of ADHD and ODD symptoms at the end of treatment', *Journal of the American Academy of Child & Adolescent Psychiatry*, 40(2): 168–179.

Turkheimer, E. (2011) 'Still missing', *Research in Human Development*, 8: 227–241.

Turner, M. G. and Piquero, A. R. (2002) 'The stability of self-control', *Journal of Criminal Justice*, 30: 457–471.

Van Tol, H. H., Bunzow, J. R., Guan, H. C., Sunahara, R. K., Seeman, P., Niznik, H. B., and Civelli, O. (1991) 'Cloning of the gene for a human dopamine D4 receptor with high affinity for the antipsychotic clozapine', *Nature*, 18: 610–614.

Wang, E., Ding, Y.-C., Floodman, P., Kidd, J. R., Kidd, K. K., Grady, D. L., Ryder, O. A., Soence, M. A., Swanson, J. M., and Moyzis, R. K. (2004) 'The genetic architecture of selection at the human dopamine receptor D4 (DRD4) gene locus', *American Journal of Human Genetics*, 74: 931–944.

Wu, T. and Barnes J. C. (2013) 'Two dopamine receptor genes (*DRD2* and *DRD4*) predict psychopathic personality traits in a sample of American adults', *Journal of Criminal Justice*, 41: 188–195.

Zuckerman, M. (1979) *Sensation Seeking: Beyond the Optimal Level of Arousal*, Hillsdale, NJ: Lawrence Erlbaum Associates.

24

Biosocial foundations of externalizing behaviors

Lisabeth F. DiLalla and Kyle Bersted

Introduction

The etiology of externalizing problem behaviors has been a fundamental concern for researchers and professionals dealing with these problems. Many studies have examined family and peer factors that increase the likelihood of children engaging in these types of behavior, and a number of factors have been identified, such as harsh or overly permissive parenting (Patterson *et al.*, 1992; Moffitt and Caspi, 2001; Derringer *et al.*, 2010) and delinquent behaviors among peers (Farrington *et al.*, 2011). However, over the past several decades behavior genetic studies have identified the importance of biological factors as well for externalizing problem behaviors. Thus many researchers have moved to adopt an explanatory approach that includes aspects of both the environment and one's biology. As with all extremely complicated behaviors, there are a myriad possible explanatory factors, and they work in concert with each other to produce various outcomes. Additionally, there are a number of behaviors that fall under the umbrella of "externalizing" problems, and it is likely that they differ in terms of etiology. Thus a complex theoretical explanation is necessary that incorporates both biological and environmental factors. Biosocial theory is one such explanation that has been used with success in helping to pinpoint certain explanatory factors and to provide a foundation for understanding how these factors interact to produce various behaviors.

Biosocial theory

Biosocial theory asserts that both biological and social factors contribute to the development of a condition or behavior. In the realm of criminology, this is a relatively new but sorely needed perspective that unites biological (including genetic and neuropsychological), social environmental, and evolutionary factors as explanations for criminal behaviors (Beaver and Walsh, 2011). It is critical to understand that including biological factors as explanations for behavior does not in any way suggest that those behaviors are immutable or preordained. Quite the opposite, in fact. Biological influences are so complex that there are necessarily many factors that influence any given behavior, and these interact constantly with environmental influences. In fact, both biological and environmental factors influence and change each other throughout development, in addition to interacting in complicated ways. Thus we are only in the early stages of being able to understand biosocial influences on behaviors.

Biosocial theory asserts that both nature and nurture, as well as their interaction, are important factors in influencing a variety of behaviors, including externalizing behaviors. The complex interweaving of our biological make-up with the environment around us is responsible for many, if not most, of the behaviors in which we engage. Our genes code for proteins, which affect neurotransmitter and hormone activity within us, which affects our brains, which affects the ways we think. Similarly, we are exposed to various environmental influences that mold our thinking and against which we must react. Individual differences in reactions to the environment create variable behavioral responses. Thus both biology and environment contribute to behaviors, and this is especially true for complex behaviors such as externalizing behaviors (DeLisi and Vaughn, 2011).

Even variables that are considered to be biological may still have environmental influences. Although the job of the gene is to code for proteins, this does not happen continuously. Instead, genes turn on and off, depending partly on the environment within the body and partly on the external, social environment (Slavich and Cole, 2013). Some extreme environments, such as abuse, can lead to large-scale changes in gene activity throughout the body (Slavich and Cole, 2013). Similarly, environmental variables usually have biological or genetic influences behind them (Raine, 2002). For example, temperament, which has a strong biological component (Shiner *et al.*, 2012), affects children's responses to peers. Thus it is difficult to draw clear lines between biological and environmental factors, further supporting the need to consider both as predictors and influences. However, because of this complexity, it is extremely difficult to disentangle these influences, and most research to date has focused primarily on one or the other, although newer studies are beginning to attempt to tie them together. We first review some of the behavior genetic research on externalizing behaviors, demonstrating genetic influences on these behaviors, and then discuss biosocial research that examines genetic and environmental interactions and correlations. However, because the term "externalizing behavior" has so many definitions, we begin with some definitions of the various types of externalizing behavior that have been examined in the genetic and biosocial literature.

Externalizing behaviors

Children, adolescents, and adults with externalizing problems represent an extremely heterogeneous group. A number of different terms are used to describe externalizing behaviors in the literature, including antisocial behavior, aggression, disruptive behaviors, and conduct problems. These terms are further broken down into subcategories to describe more specifically the problem behaviors of interest, such as oppositionality and non-compliance. In response to the vast number of behaviors that can be classified as "externalizing," a number of different approaches to distinguishing them have been proposed (Frick *et al.*, 2013), some of which are examined in this chapter. For our purposes, *externalizing behaviors* is used as an overall descriptor to classify all behavior problems falling within the externalizing spectrum, from non-compliance and defiance to extreme forms of antisocial behavior such as violence or theft.

Oppositional defiant disorder (ODD) and conduct disorder (CD) are two common childhood and adolescent disorders that fall under the externalizing category. Despite the fact that these two disorders are often associated with one another, it is now clear that children and adolescents with ODD represent a heterogeneous group in terms of symptomology and etiology when compared with children and adolescents with CD (Kempes *et al.*, 2005). ODD is defined as a pattern of angry or irritable mood, argumentativeness or defiance, or vindictiveness (American Psychiatric Association, 2013). At least four symptoms within these emotional and behavioral categories must be present in at least one setting for a child to be diagnosed with ODD. In contrast, CD refers to a repetitive and persistent pattern of behavior in which the basic rights of others or societal rules and norms are violated, such as cruelty or violence towards people or animals, stealing, or destruction of property (American Psychiatric Association, 2013). In order for a child or adolescent to be diagnosed with CD, he or she must present with at least three symptoms. Additionally, a relatively new area of research has emerged that suggests a further distinction within CD. There appears to be a subgroup of those with CD that also present with callous and unemotional (CU) traits, meaning that they lack feelings of empathy or guilt in their relationships with others (Frick *et al.*, 2013). Research suggests that this subgroup is distinct from those without CU traits and presents clinicians with more of a challenge in treatment.

Externalizing behaviors also include other behaviors that are more common and not necessarily categorized as disorders. Regardless of whether or not the behavior warrants a diagnosis, a number of classifications and categories have been proposed to further subtype certain types of externalizing behavior. One of the major distinctions within externalizing behaviors is whether the behavior is aggressive or non-aggressive (Burt, 2012). Aggression can be defined as behavior that is deliberately aimed at hurting people or objects (Kempes *et al.*, 2005) and can be broken down into three subtypes: physical, psychological, and relational. As would be expected, physical aggression refers to any physical action taken against someone or something, such as hitting, pushing, or kicking. Psychological aggression refers to ridiculing and humiliating, whereas relational aggression includes interpersonal, social aggression such as gossiping or spreading rumors about someone. Gender appears to be an important factor when distinguishing between the three types of aggression, with boys more likely than girls to engage in physical aggression, beginning as early as 17 months old (Baillargeon *et al.*, 2007), and girls more likely to engage in relational and psychological aggression, as first proposed by Crick and Grotpeter (1995), although not all researchers have found this (Lansford *et al.*, 2012).

Another distinction within externalizing behaviors that has received attention in the literature is the difference between proactive and reactive aggression. First proposed by Dodge and Coie (1987), proactive aggression refers to goal-directed aggression that is carried out without any prior environmental instigation, whereas reactive aggression is defined by aggressive behavior that acts as a response to something in the environment, such as a threat (Kempes *et al.*, 2005). Although there does appear to be some overlap between the two subtypes, considerable evidence points to the validity of this distinction (Poulin and Boivin, 2000). Despite this ability to classify aggression, the proactive–reactive distinction has yet to be applied to other externalizing behaviors, such as oppositionality or antisocial behavior (Kempes *et al.*, 2005).

Four categories of disruptive behavior have been proposed by Frick and colleagues (1993). In this model, individual differences within externalizing behavior are based on whether the behavior is covert, overt, destructive, or non-destructive. The resulting four-quadrant approach to classifying externalizing behaviors has had a major impact on the field, especially on conceptualizations of ODD and CD. For example, externalizing behaviors that are often present in CD tend to fall under three categories: overt destructive (assault, blaming others), covert destructive (cruelty to animals, stealing, setting fires), and covert non-destructive (breaking rules, truancy, running away). The fourth quadrant, overt non-destructive (stubborn, angry, argumentative, defiant) tends to be more representative of ODD. A review by Tremblay (2010) indicates that the overt–covert and destructive–nondestructive taxonomy appears to be the most useful system to use when distinguishing between different types of externalizing behavior.

Criminality, an additional category on the externalizing spectrum, also has been extensively studied. Although heterogeneous in nature like most externalizing behaviors, criminality can be defined as behavior characterized by irresponsibility, self-indulgence, interpersonal intrusiveness, and social rule breaking (White and Walters, 1989). This definition can include anything from petty theft to murder, meaning that the continuum of criminality includes both aggressive and non-aggressive behaviors. An important distinction within criminality is the age at which the criminal acts occur. There are differences between adult crime (criminality) and juvenile crime (delinquency), some of which were outlined by Klein (1987). First, the criminal acts themselves tend to differ, as opportunities to commit certain crimes, such as fraud or embezzlement, are more common for adults than juveniles. Second, motivations behind criminal acts are usually different, with delinquency characterized more by expression or pleasure, and adult criminality more instrumental in nature. Additionally, it is believed that juveniles are more susceptible to peer pressure. Third, juvenile criminality is believed to be more malleable. Lastly, delinquency and adult crime differ in the ways in which the community respond to them. In general, adult crimes receive harsher punishments than juvenile delinquency. A meta-analysis by Leschied *et al.* (2008) indicated that a number of risk factors in childhood and adolescence have been linked with later criminality, including aggression, hyperactivity, attention problems, withdrawal, anxiety, social alienation, negative parenting, and a stressful home environment. The authors also noted that contextual variables, such as the nature of the family system and community, were more strongly linked to later criminality than were individual child factors.

The focus in the externalizing literature was shifted from the behavior itself to the onset of the behavior when the distinction was made between childhood-onset and adolescent-onset conduct problems (DiLalla and Gottesman, 1989; Moffitt, 1993). Children in the childhood-onset group become aggressive earlier and are more likely to have continued aggression, antisocial behavior, and criminality later in life, thus representing more severe cases of externalizing behavior (Moffitt, 2006). It is likely that children with these more extreme behaviors have a genetic propensity for such behaviors (DiLalla and Gottesman, 1989), whereas other adolescents with more typical delinquent behaviors limited to their adolescent years are more a product of adolescent peer pressure and development. For example, although 75 percent of children with ODD do not go on to develop CD (Hinshaw and Lee, 2003), the 25 percent who do may represent the childhood-onset group. In contrast, the adolescent-onset group is characterized by externalizing behaviors that arise during puberty, but that are more likely to disappear by adulthood. For example, although most crime and delinquency tends to emerge in late childhood and early adolescence, most individuals tend to refrain from these behaviors once they reach young adulthood (Burt, 2009). However, when externalizing behaviors begin at a young age (childhood-onset) and persist throughout adolescence, the adolescent is more likely to continue to engage in criminal activity into adulthood.

A number of studies have investigated the developmental trajectories of externalizing behaviors. Tremblay (2010) reviewed the literature and found that "overt" behaviors tend to start in infancy and decrease with age, with a peak between 2 and 4 years old. In contrast, "covert" behaviors appear to increase with age, most likely due to ongoing cognitive development. An additional review found that, although there was some variability in the developmental trajectories of violence, aggression, and delinquency across the studies reviewed, all trajectories appeared to fit into Moffitt's conceptualization. The majority of the behaviors reviewed presented with a "childhood onset" path and an "adolescent-limited" path, with the former representing more chronic and problematic effects on functioning (Jennings and Reingle, 2012).

Genetic influences on externalizing behaviors

Many studies examining genetic influences on externalizing behaviors have been conducted using twin or adoption samples, which can provide information about whether broad genetic influences are significant for behavioral outcomes. These "quasi-experimental" methods make use of the naturally occurring phenomenon of twin births, or the cultural phenomenon of adoption, to make comparisons among groups of people differing in their genetic relatedness (DiLalla *et al.*, 2012). More recently, molecular genetic studies examining specific genotypes have yielded more information about specific genetic influences on these behaviors.

Twin studies

Genetic and environmental influences on behavior can be examined by comparing correlations between monozygotic (MZ), or "identical," twins with each other, and dizygotic (DZ), or "fraternal," twins with each other. MZ twins share 100 percent of their genetic make-up, which means that they differ on a particular phenotype, or measurable characteristic, only as a function of differences in their environment. DZ twins, on the other hand, share on average approximately 50 percent of their genetic make-up, the same as any two non-twin siblings. However, they are likely to be more alike than non-twin siblings by virtue of growing up together at the same time and experiencing life events at the same age, which in this respect makes them more similar to MZ twins. The correlation between MZ twins (r_{MZ}) is a function of their shared genotype (h^2) and their shared common environmental influences (c^2). Thus $r_{MZ} = h^2 + c^2$. The correlation between DZ twins (r_{DZ}) is a function of their shared genotype ($\frac{1}{2}h^2$, because they share 50 percent of their genes) and their shared common environment (c^2). Assuming the equal environments assumption, which claims that MZ and DZ twins share common environment to the same extent as far as phenotypes of interest, then we can compare MZ and DZ correlations to arrive at an estimate of heritability (h^2). Specifically, $h^2 = \frac{1}{2}(r_{MZ} - r_{DZ})$ (see DiLalla *et al.*, 2012 for a more detailed explanation). Importantly, this equation also allows calculation of the extent to which shared and non-shared environmental influences are important for the behavior under study. The twin method has been used in many studies examining externalizing behaviors in children and adolescents.

Adoption studies

The extent to which genetic and environmental influences affect behavior can be assessed by studying adoptees and their biological and adoptive parents. Examining the similarity between adoptees and their biological parents provides an estimate of the heritability of a behavior. This is because adoptees share genes but not environment with their biological parents, assuming that prenatal environment is not an important factor (which may or may not be true depending on the behavior of interest). The similarity between adoptees and their adoptive parents gives an estimate of the influence of the family environment on the behavior, because adoptees share environment but not genes with their adoptive parents. A number of testable assumptions must be made with this method, such as assuming children are separated from their biological parents right after birth, and that the biological and adoptive parents are not correlated on variables related to the behavior of interest (such as intelligence, personality, or religiosity). Several early studies of criminality and violence utilized this methodology, especially in Denmark, where records of births and of these outcome behaviors were freely available to researchers (Mednick *et al.*, 1984).

Molecular genetics

Recent work examining specific genes has allowed researchers to examine whether behaviors of interest are more likely to occur among individuals with a "risk" allele. Linkage analyses have provided a means for determining whether a DNA marker allele is associated with a specific behavior or disorder allele (Plomin *et al.*, 2012). Specific alleles have been identified that appear to be related to certain behaviors of interest, and individuals with those alleles can be compared with those without the "risk" alleles to determine whether that association exists. However, the types of behavior that are of interest are highly complicated, and rarely does a small number of genes (or environments, for that matter) account for a significant portion of variance in a behavior. Thus it is difficult to identify significant associations between single genes and single behaviors. Nonetheless, several promising genes have been identified, which are discussed below.

Heritability of externalizing behaviors

Twin and adoption studies of externalizing behaviors routinely demonstrate genetic influences on these behaviors, regardless of which externalizing behaviors are studied. Many behavior genetic studies have focused on aggression and delinquency, or combined externalizing, in children, or on criminality and violence in adults. It is interesting to note that whereas aggression and criminality are never considered to be normative behaviors, delinquency differs in that respect because many delinquent behaviors are commonly seen in adolescence. This is why adolescent-limited or transitory delinquency appears to show little genetic influence and instead is much more influenced by peer groups and the social environment (DiLalla and Gottesman, 1989).

Research on genetic influences on childhood aggression has yielded mixed results, possibly due in part to mixed methodologies (parent reports versus direct observations). Many studies have used parent reports (e.g. Cadoret *et al.*, 1997; Miles and Carey, 1997; Deater-Deckard and Plomin, 1999; Kuntsi *et al.*, 2000) to show that childhood aggression at ages 7 to 16 years is heritable, with an average heritability estimate of 58 percent (DiLalla, 2002). More recent studies support this (Brendgen *et al.*, 2005). However, one study (Arseneault *et al.*, 2003) aggregated mother report, teacher report, examiner observations, and 5-year-old children's self-report of antisocial behavior to demonstrate a very high and significant heritability of approximately 82 percent for aggression in 5-year-olds. In an observational study of perpetrated aggression in 5-year-old twins playing with unfamiliar peers in a laboratory, we utilized observational ratings by trained coders to assess heritability of aggression (DiLalla and John, 2014). We estimated heritability at about 36 percent and showed that the genetic effects were due to dominance effects, meaning that the genes that affected these aggressive behaviors behave in a non-additive, or interactive, way. Interestingly, a meta-analysis (Bergen *et al.*, 2007) suggested that externalizing behaviors show an increase in heritability from age 10 (about 25 percent) to age 35 (about 50 percent), although the data on 5-year-olds do not fit this trend (Arseneault *et al.*, 2003; DiLalla and John, 2014).

Bullying differs from general aggression in terms of several definitional behaviors, including the power differential between bully and victim as well as the repeated nature of the aggressive acts (Greene, 2000). One study of genetic factors distinguished bullies, victims, and bully–victims (those children who both bully and are victims) in 10-year-old twins (Ball *et al.*, 2008). The authors found that both bullying and being victimized were significantly heritable (61 and 73 percent, respectively), and that there were some shared genetic effects between the two behaviors. These data were based on parent reports for victimization and parent and teacher reports for bullying. In their review of the current literature on aggression and bullying in preschool-aged children, Vlachou *et al.* (2011) cite research demonstrating that the heritability of perpetrating aggression is roughly 40 percent.

Adoption studies have demonstrated an important genetic link for both criminality and violent behaviors. For instance, Mednick and colleagues showed that sons whose biological fathers had a criminal record were significantly more likely to engage in similar behaviors themselves, and those who also were reared by adoptive fathers who had committed crimes had more than double the risk of engaging in these behaviors (Mednick *et al.*, 1984, 1987). Twin studies have yielded similarly high heritability estimates (DiLalla and Gottesman, 1989), and Cloninger and Gottesman (1987) showed that the genetic influence on violent crimes was approximately 50 percent.

Molecular genetics studies have begun to identify particular genes associated with externalizing behaviors. Two of the most frequently studied systems are genes related to the neurotransmitters dopamine and serotonin. The dopamine receptor D4 (DRD4) gene, which encodes proteins that respond to the neurotransmitter dopamine in the brain, has been associated with novelty seeking behaviors in rodents, children, and adults (Wang *et al.*, 2004; Rettew and McKee, 2005; Bailey *et al.*, 2007). The 7-repeat allele (receptors are less efficient at binding dopamine, leading to dopaminergic hypofunction) versus a 4-repeat allele appears to be related to novelty seeking (Kluger *et al.*, 2002). It also has been shown to be related to ADHD (Faraone *et al.*, 2001), especially for cases that persist into adulthood (DiLalla *et al.*, in press). The dopamine receptor D5 (DRD5) also has been shown to be related to ADHD (Wu *et al.*, 2012), as has the dopamine DAT gene (Faraone *et al.*, 2005). One study of adolescent males found that having more DAT risk alleles was related to engaging in more serious and more violent criminal activities, regardless of the degree of exposure to delinquent peers (Vaughn *et al.*, 2009). Other studies have also found a significant relation between DAT and DRD2 genes and aggressive behaviors and violence in adolescence and adulthood (Chen *et al.*, 2007). Additionally, a number of dopaminergic genes have been related to ODD and CD (Comings *et al.*, 2000).

Serotonin, specifically the 5HTTLPR polymorphism, is implicated in mood disturbances; deficits may lead to problems with mood, impulsivity, and aggression in animals and adult humans, but less is known about its relevance for children (Beitchman *et al.*, 2006). There are a number of other genes that also have been examined for their influences on externalizing behaviors. At this point in time, replication is necessary for most of these, as frequently a gene that has a significant effect in one study fails to replicate in another. However, it is clear at this point that a number of genes do indeed have significant, albeit small, effects on externalizing behaviors. What is of great interest now is to examine how these genes work within the social environment, leading us to a consideration of biosocial theory.

Biosocial influences on externalizing behaviors

In order to examine evidence of biosocial influences on externalizing behaviors, it is important first to describe ways in which biological influences and social or environmental influences work in concert. We first describe gene–environment correlations and gene–environment interactions, which are theoretically very different ways in which genes and environment co-act. We then describe two other theories, diathesis-stress theory and differential susceptibility hypothesis, that identify ways in which genes and environment non-independently contribute to externalizing behaviors.

Gene–environment correlations

Gene–environment correlations (r_{GE}) were first proposed by Plomin *et al.* (1977) and by Scarr and McCartney (1983) as a way to describe how genes and environments correlate from a developmental perspective. There are three types of r_{GE}. The first is passive r_{GE}, which begins at birth and is most prevalent when children are young. Passive r_{GE} occurs because children get both genes and environment from their parents, and therefore genes and environment are inextricably correlated. The only way to disentangle this effect is through an adoption study. For example, a highly reactive infant inherits genes from her parents that may be partially responsible for her being highly reactive, but also those same parents may provide a more chaotic environment for her because they, too, are highly reactive, and she may respond to that environment in a way that increases her reactivity. Thus both genes and environment, which are correlated, affect her temperamental behavior. The second r_{GE} is reactive or evocative correlation, which is hypothesized to be present throughout the lifespan. This occurs when something about an individual, which is genetically influenced, evokes reactions from others in the environment, thus causing the environment to become correlated with genotype. For example, if a child is aggressive, partly because of inheriting a tendency to be this way, then others in his environment might respond to him in a negative way, thus creating an environment that encourages additional aggression. Finally, the third r_{GE} is active r_{GE}, which becomes more prevalent as children get older and is hypothesized to be present later in life. Active r_{GE} occurs when an individual chooses her environment based in part on her own genotype. For example, a teenager who has inherited a propensity for behaving violently may choose to spend time with others who encourage such behavior. Thus both genes and environment would be correlated and together would increase the likelihood that she would engage in violent behaviors. This active r_{GE} has also been called "niche-picking."

Gene–environment interactions

Gene–environment interactions (g × e) occur when there is genetic sensitivity to certain environments. That is, individuals with one type of genotype respond one way in one environment, but a different way in a different environment, and other people with different genotypes might have different experiences across environments. For example, a person who is highly reactive, partly because of his genotype, might become aggressive and even violent in an environment that is chaotic and over-stimulating, but he might not show these aggressive behaviors in a calm, non-threatening environment. Another person who is temperamentally very calm might actually benefit from an over-stimulating environment, or at least might not be bothered by it. Thus genotype is hypothesized to affect how well one might respond to various environments. This also can be called "goodness of fit," meaning that one environment might "fit" one person better than another.

Diathesis-stress theory

In the diathesis-stress model, certain predispositions increase a person's risk of experiencing poor outcomes or developing a psychological disorder when they encounter a negative environment.

More generally, some people are more vulnerable than others to stress or adversity posed by a negative environment. Although the focus of diathesis or risk has usually been on genetic and biological factors, more recently other risks, such as psychological and cognitive schemas, have been shown to operate in a diathesis-stress fashion (Ingram and Luxton, 2005). The interaction between predisposition and negative environment has been supported by a number of research studies examining a variety of predispositions, environments, and outcomes. For example, vulnerabilities related to negative environments and stressors have been found in the development of depression, anxiety, personality disorders, substance abuse, eating disorders, and schizophrenia (Ingram and Price, 2001).

Another important development to come from this line of research is the concept of protective or resilience factors. These factors, such as positive parenting, intellectual functioning, cognitive schemas, personality traits, social skills, and other sources of support have the ability to reduce the risk of poor outcomes posed by negative and stressful environments and life experiences (Masten, 2001; Ingram and Luxton, 2005). The focus on risk and protective factors in the diathesis stress framework has been essential when examining why people have different reactions to negative environments. More recently, however, the investigation of gene × environment interactions has become broader to include all environments, not just those which are negative, which brings us to the next g × e theory.

Differential susceptibility

Differential susceptibility also conceptualizes how certain inherent factors and environments interact. As is the case with diathesis stress, this hypothesis suggests that people with certain alleles may be at higher risk for developing the worst outcomes when in a negative environment. However, unlike diathesis stress, the differential susceptibility hypothesis also examines more *positive* environments and makes the claim that people with these same "risk" alleles also have the potential to experience the *best* outcomes if placed in a more positive environment. In other words, "susceptibility" alleles replace the term "risk" alleles, meaning that their presence has the potential to make a person more malleable to both the positive and negative effects of the environment (Belsky and Pluess, 2009). Thus, instead of only viewing certain alleles as posing the threat of worse outcomes, differential susceptibility suggests that they operate in a "for better and for worse" fashion (Belsky *et al.*, 2007).

Although a relatively new field of research, the differential susceptibility hypothesis has been supported in a variety of studies. For example, Bakermans-Kranenburg and van IJzendoorn (2011) reviewed the literature and found a number of studies showing that children with the DRD2 and DRD4 genotype have been found to be at more risk for poor outcomes (externalizing behavior, less prosocial behavior, disorganized attachment, irritability) when experiencing a number of different negative parental environments (prenatal smoking, negative parenting, lack of maternal sensitivity). However, when other children with these genotypes have experienced more positive parental environments, they showed the most positive outcomes, more so than others in the study without their genotype. Similar patterns using a variety of outcome measures have been found with difficult temperament (e.g. Pluess and Belsky, 2010), sympathetic nervous system reactivity (e.g. El-Sheikh *et al.*, 2007), and a number of other genotypes, such as 5-HTTLPR and MAOA (e.g. Caspi *et al.*, 2003; Kim-Cohen *et al.*, 2006). Although further discussion of these results is outside the scope of this chapter, the fact that this hypothesis has been supported using such a diverse range of moderator variables (temperament, endophenotypes, genotypes), environments (parenting, childcare, stressful experiences, conflict, adversity), and outcomes (child behavior, attachment quality, symptoms of mental health disorders) suggests that the differential susceptibility hypothesis warrants continued investigation, especially when examining the biosocial foundations of externalizing behavior.

Biosocial foundations of aggression

A number of studies have begun to address how biological and environmental or social factors affect externalizing behaviors. It is becoming more clear as we study genetic influences that much of the cause of externalizing problem behaviors lies in the interaction between children's genotypes and their environment as well as their temperamental qualities (Moffitt, 2005). Biosocial theory provides a framework for examining these interactions.

A number of older adoption studies provided evidence for gene–environment interaction on criminal behaviors, showing that adopted men with criminal biological parents who also were exposed to high environmental risks (e.g. longer time in orphanage, low socioeconomic status, adoptive parent alcoholism) were at much greater risk of being criminal themselves (Cadoret and Cain, 1981; Cloninger *et al.*, 1982; Cadoret *et al.*, 1983). However, adoptive homes tend to be vetted for positive environments, making them perhaps atypical compared with non-adoptive homes, and therefore the advent of molecular genetics research has made exploration of g × e much more generalizable to a non-selected population (DiLalla and Gheyara, 2011).

One of the earliest and most oft-cited studies showing an interaction between genotype and specific environment was presented by Caspi *et al.* (2002). They showed that children with low monoamine oxidase A (MAOA) activity who had been maltreated were more likely to show antisocial and violent behaviors in adulthood, suggesting an interaction between genes and abusive environment that may lead to antisocial behaviors. Similar results were reported by Foley *et al.* (2004) and Kim-Cohen *et al.* (2006), although this was not replicated in a sample of adolescent boys (Haberstick *et al.*, 2005; Huizinga *et al.*, 2006; Young *et al.*, 2006). Using a community sample, Edwards *et al.* (2010) showed that adolescent boys with the low-active MAOA allele who had experienced physical parental discipline at age 5 were significantly more likely to show delinquent behaviors, as reported by parents, but this was not significant for aggressive behaviors. A study examining deviant peers as the risky environment also found evidence of g × e, but with high rather than low activity MAOA (Lee, 2011). Lee showed that male adolescents and young adults who interacted with deviant peers had increased levels of overt antisocial behavior if they also had high-activity MAOA genotypes. Similarly, Beaver and Holtfreter (2009) found an interaction between the high-active MAOA genotype and peer delinquency. Male adolescents with the high-risk genotype and more delinquent peers were the most likely to engage in fraudulent delinquent activities. Thus results across age groups and populations are not consistent. More probing into the effects of MAOA in difficult environments is needed.

Another genotype that has been examined in a number of biosocial studies is DRD4. An interaction between a DRD4 risk allele and parenting that is not warm or sensitive has emerged as predictive of aggressive behaviors in toddlers (Bakermans-Kranenburg and van IJzendoorn, 2006) and preschoolers (DiLalla *et al.*, 2009). In these studies, genotype risk and the negative observed parenting environment interacted to increase the likelihood of young children showing aggressive behaviors as rated by their parents (Bakermans-Kranenburg *et al.*, 2008; DiLalla *et al.*, 2009). Importantly, Bakermans-Kranenburg *et al.* (2008) found that children with the DRD4 risk allele also showed the greatest decrease in externalizing behaviors when their mother responded to an intervention increasing their sensitive behaviors, thus supporting the differential susceptibility hypothesis. A meta-analysis of studies examining dopamine-related genotypes and parenting suggests that this hypothesis may have support (Bakermans-Kranenburg and van Ijzendoorn, 2011) in that children with risk alleles may be more responsive to environmental influences, thus influencing their externalizing behaviors. More research examining the same children in both negative and positive environments is essential, however, for systematic support of the different susceptibility hypothesis (Bakermans-Kranenburg and van Ijzendoorn, 2011). An interesting study on Russian adolescents showed that for boys exposed to psychological trauma, delinquency was high regardless of whether or not they had the DRD4 risk allele (Dmitrieva *et al.*, 2010). This study did not support the differential susceptibility hypothesis, but instead demonstrated that a sufficiently negative environment had an equally negative impact on adolescent boys regardless of genotype. Interestingly, a study on adolescents found that males with a genetic risk from another dopamine gene, the DAT1, who also experienced more negative maternal environments, were more likely to interact with delinquent peers, which we know is correlated with delinquent behaviors (Beaver *et al.*, 2008). Thus a possible gene–environment correlation (DAT1 correlated with delinquent peers) may be partly responsible for increased delinquent behaviors.

Some evidence also exists for serotonin genotypes interacting with environments to affect externalizing behaviors, although there are fewer studies on this than on dopamine. Sakai *et al.* (2010) found a small 5HTTLPR genetic risk by sex effect for conduct disorders, with females who had the risk allele being more likely to be chronic offenders than non-offenders. It is unclear whether being female had a biological influence or a socialization influence, meaning that the interaction with the genetic risk could somehow be connected with the socialization of being female. One other study has shown that men with the 5HTTLPR risk allele who also experienced a stressor showed more aggressive behaviors by delivering shocks to a confederate (Verona *et al.*, 2006). Thus there does appear to be an effect of sex, but results are inconsistent regarding whether there is a greater risk for externalizing behaviors with the 5HTTLPR risk allele in being male versus female. It is possible that the effects of serotonin are quite different in stress versus non-stress conditions and in terms of affecting chronic criminal behaviors versus instructed aggression in a laboratory setting. In a review of the role that alcohol plays in increasing aggression, Heinz *et al.* (2011) noted that stressors early in life can interact with both the serotonin and dopaminergic systems to increase both alcohol consumption and impulsive aggression. Thus, in line with biosocial theory, those with a biological (genetic) risk who are exposed early in life to extreme stress are at risk for both alcoholism and aggression in adulthood.

Conclusion

Externalizing behaviors represent an extremely heterogeneous, complicated, yet important phenomenon that researchers and clinicians continue to investigate with hopes of increasing our understanding of their etiology. Given the wide scope of behaviors falling under the externalizing spectrum, and the number of factors that are associated with their onset, course, and severity, biosocial theory provides a valuable etiological explanation. Biosocial theory states that both biology and environment play a role in the etiology of any behavior, and it has been utilized by many when explaining how and why externalizing behaviors appear throughout development. Nature and nurture influence each other and interact in a variety of ways, meaning that in order to investigate a behavior's cause, both must be examined simultaneously. Twin, adoption, and molecular genetic studies have increased our understanding of how genes and environments influence externalizing behaviors, but understanding how nature and nurture *interact* over time to influence externalizing is still in its infancy, although research on g × e theories such as diathesis-stress and differential susceptibility has been promising in shedding light on some of these interactions.

As the field continues to grow, it is important to remember why we study the etiology of externalizing behaviors. Children, adolescents, and adults whose behavior falls under the externalizing spectrum represent a group of individuals whose overall functioning is undoubtedly reduced as a result of these behaviors. Moreover, their behavior often has a negative impact on their family, peers, and community as well. Understanding the etiology of these behaviors increases the likelihood of successful treatment and even prevention of these behaviors before they begin. Although knowing that a certain genotype may influence an individual's likelihood of developing an externalizing problem may not be entirely clinically practical on its own, knowledge of the endophenotypes related to the genotype, as well as the environments that may interact with the genotype, can be used to determine which individuals may benefit most from treatment and under what circumstances treatment or prevention may be most beneficial (Bakermans-Kranenburg *et al.*, 2008). Research on externalizing behaviors must continue to explore the genetic and environmental influences on these behaviors in order to increase our likelihood of finding effective treatment and prevention strategies, resulting in positive changes for both the individuals exhibiting the behaviors and the communities in which they live.

References

American Psychiatric Association (2013) *Diagnostic and Statistical Manual of Mental Disorders*, 5th edn, Arlington, VA: American Psychiatric Publishing.

Arseneault, L., Moffitt, T. E., Caspi, A., Taylor, A., Rijsdijk, F. V., Jaffee, S. R., Ablow, J. C., and Measelle, J. R. (2003) 'Strong genetic effects on cross-situational antisocial behaviour among 5-year-old children according to mothers, teachers, examiner–observers, and twins' self-reports', *Journal of Child Psychology and Psychiatry*, 44: 832–848.

Bailey, J. N., Breidenthal, S. E., Jorgensen, M. J., McCracken, J. T., and Fairbanks, L. A. (2007) 'The association of DRD4 and novelty seeking is found in a nonhuman primate model', *Psychiatric Genetics*, 17: 23–27.

Baillargeon, R. H., Zoccolillo, M., Keenan, K., Côté, S., Pérusse, D., Wu, H., and Tremblay, R. E. (2007) 'Gender differences in physical aggression: a prospective population-based survey of children before and after 2 years of age', *Developmental Psychology*, 43(1): 13–26.

Bakermans-Kranenburg, M. J. and van IJzendoorn, M. H. (2006) 'Gene–environment interaction of the dopamine D4 receptor (DRD4) and observed maternal insensitivity predicting externalizing behavior in preschoolers', *Developmental Psychobiology*, 48: 406–409.

Bakermans-Kranenburg, M. J. and van IJzendoorn, M. H. (2011) 'Differential susceptibility to rearing environment depending on dopamine-related genes: new evidence and a meta-analysis', *Development and Psychopathology*, 23(1): 39–52.

Bakermans-Kranenburg, M. J., van IJzendoorn, M. H., Pijlman, F. T. A., Mesman, J., and Juffer, F. (2008) 'Experimental evidence for differential susceptibility: dopamine D4 receptor polymorphism (DRD4 VNTR) moderates intervention effects on toddlers' externalizing behavior in a randomized control trial', *Developmental Psychology*, 44: 293–300.

Ball, H. A., Arseneault, L., Taylor, A., Maughan, B., Caspi, A., and Moffitt, T. E. (2008) 'Genetic and environmental influences on victims, bullies and bully–victims in childhood', *Journal of Child Psychology and Psychiatry*, 49(1): 104–112.

Beaver, K. M. and Holtfreter, K. (2009) 'Biosocial influences on fraudulent behaviors', *Journal of Genetic Psychology*, 170: 101–114.

Beaver, K. M. and Walsh, A. (2011) 'Biosocial criminology', in K. M. Beaver and A. Walsh (eds), *The Ashgate Research Companion to Biosocial Theories of Crime*, Farnham, UK: Ashgate.

Beaver, K. M., Wright, J. P., and DeLisi, M. (2008) 'Delinquent peer group formation: evidence of a gene × environment correlation', *Journal of Genetic Psychology*, 169(3): 227–244.

Beitchman, J. H., Baldassarra, L., Mik, H., De Luca, L., King, N., Bender, D., *et al.* (2006) 'Serotonin transporter polymorphisms and persistent, pervasive, childhood aggression', *American Journal of Psychiatry*, 163: 1103–1105.

Belsky, J. and Pluess, M. (2009) 'Beyond diathesis stress: differential susceptibility to environmental influences', *Psychological Bulletin*, 135(6): 885–908.

Belsky, J., Bakermans-Kranenburg, M. J., and van IJzendoorn, M. H. (2007) 'For better and for worse: differential susceptibility to environmental influences', *Current Directions in Psychological Science*, 16: 300–304.

Bergen, S. E., Gardner, C. O., and Kendler, K. S. (2007) 'Age-related changes in heritability of behavioral phenotypes over adolescence and young adulthood: a meta-analysis', *Twin Research and Human Genetics*, 10: 423–433.

Brendgen, M., Dionne, G., Girard, A., Boivin, M., Vitaro, F., and Perusse, D. (2005) 'Examining genetic and environmental effects on social aggression: a study of 6-year-old twins', *Child Development*, 76: 930–946.

Burt, S. A. (2009) 'Rethinking environmental contributions to child and adolescent psychopathology: a meta-analysis of shared environmental influences', *Psychological Bulletin*, 135(4): 608–637.

Burt, S. (2012) 'How do we optimally conceptualize the heterogeneity within antisocial behavior? An argument for aggressive versus non-aggressive behavioral dimensions', *Clinical Psychology Review*, 32(4): 263–279.

Cadoret, R. and Cain, C. A. (1981) 'Environmental and genetic factors in predicting antisocial behavior in adoptees', *Psychiatric Journal of the University of Ottawa*, 6: 220–225.

Cadoret, R. J., Cain, C. A., and Crowe, R. R. (1983) 'Evidence for gene–environment interaction in the development of adolescent antisocial behavior', *Behavior Genetics*, 13: 301–310.

Cadoret, R. J., Leve, L. D., and Devor, E. (1997) 'Genetics of aggressive and violent behavior', *Anger, Aggression, and Violence*, 20: 301–322.

Caspi, A., McClay, J., Moffitt, T. E., Mill, J., Martin, J., Craig, I. W., Taylor, A., and Poulton, R. (2002) 'Role of genotype in the cycle of violence in maltreated children', *Science*, 297: 851–854.

Caspi, A., Sugden, K., Moffitt, T. E., Taylor, A., Craig, I. W., Harrington, H., and Poulton, R. (2003) 'Influence of life stress on depression: moderation by a polymorphism in the 5-HTT gene', *Science*, 301: 386–389.

Chen, T. J. H., Blum, K., Mathews, D., Fisher, L., Schnautz, N., Braverman, E. R., *et al.* (2007) 'Preliminary association of both the dopamine D2 receptor (DRD2) [Taq1 A1 allele] and the dopamine transporter (DAT1) [480 bp allele] genes with pathological aggressive behavior, a clinical subtype of reward deficiency syndrome (RDS) in adolescents', *Gene Therapy and Molecular Biology*, 11: 93–112.

Cloninger, C. R. and Gottesman, I. I. (1987) 'Genetic and environmental factors in antisocial behavior disorders', in S. A. Mednick, T. E. Moffitt, and S. A. Stack (eds), *The Causes of Crime: New Biological Approaches*, New York: Cambridge University Press.

Cloninger, C. R., Sigvardsson, S., Bohman, M., and von Knorring, A. L. (1982) 'Predisposition to petty criminality in Swedish adoptees. II. Cross-fostering analysis of gene–environment interaction', *Archives of General Psychiatry*, 39: 1242–1247.

Comings, D. E., Gade-Andavolu, R., Gonzalez, N., Wu, S., Muhleman, D., Blake, H., *et al.* (2000) 'Multivariate analysis of associations of 42 genes in ADHD, ODD and conduct disorder', *Clinical Genetics*, 58: 31–40.

Crick, N. R. and Grotpeter, J. K. (1995) 'Relational aggression, gender, and social-psychological adjustment', *Child Development*, 66(3): 710–722.

Deater-Deckard, K. and Plomin, R. (1999) 'An adoption study of the etiology of teacher and parent reports of externalizing behavior problems in middle childhood', *Child Development*, 70: 144–154.

DeLisi, M. and Vaughn, M. G. (2011) 'Presaging problem behavior: the mutuality of child temperament, parenting, and family environments from gestation to age three', in K. Beaver and A. Walsh (eds), *The Ashgate Research Companion to Biosocial Theories of Crime*, Farnham, UK: Ashgate.

Derringer, J., Krueger, R. F., Irons, D. E., and Iacono, W. G. (2010) 'Harsh discipline, childhood sexual assault, and MAOA genotype: an investigation of main and interactive effects on diverse clinical externalizing outcomes', *Behavior Genetics*, 40: 639–648.

DiLalla, L. F. (2002) 'Behavior genetics of aggression in children: review and future directions', *Developmental Review*, 22: 593–622.

DiLalla, L. F. and Gheyara, S. (2011) 'The genetics of criminality and delinquency', in K. Beaver and A. Walsh (eds), *The Ashgate Research Companion to Biosocial Theories of Crime*, Farnham, UK: Ashgate.

DiLalla, L. F. and Gottesman, I. I. (1989) 'Heterogeneity of causes for delinquency and criminality: lifespan perspectives', *Development and Psychopathology*, 1: 339–349.

DiLalla, L. F. and John, S. G. (2014) 'Genetic and behavioral influences on received aggression during observed play with unfamiliar preschool-aged peers', *Merrill-Palmer Quarterly*, 60(2).

DiLalla, L. F., Elam, K. K., and Smolen, A. (2009) 'Genetic and environmental effects on preschoolers' social behaviors', *Developmental Psychobiology*, 51: 451–463.

DiLalla, L. F., Mullineaux, P.Y., and Biebl, S. J. W. (2012) 'Social-emotional development through a behavior genetics lens: infancy through preschool', in J. Benson (ed.), *Advances in Child Development and Behavior*, Vol. 42, Oxford: Elsevier, pp. 153–196.

DiLalla, L. F., Mullineaux, P. Y., and DiLalla, D. L. (in press) 'Mendelian versus molecular genetics', in R. Cautin and S. Lilienfeld (eds), *The Encyclopedia of Clinical Psychology*, Oxford: Wiley-Blackwell.

Dmitrieva, J., Chen, C., Greenberger, E., Ogunseitan, O., and Ding, Y.-C. (2010) 'Gender-specific expression of the DRD4 gene on adolescent delinquency, anger and thrill seeking', *Social Cognitive and Affective Neuroscience*, 5: 82–89.

Dodge, K. A. and Coie, J. D. (1987) 'Social-information-processing factors in reactive and proactive aggression in children's peer groups', *Journal of Personality and Social Psychology*, 53(6): 1146–1158.

Edwards, A. C., Dodge, K. A., Latendresse, S. J., Lansford, J. E., Bates, J. E., Pettit, G. S., Budde, J. P., Goate, A.M., and Dick, D. M. (2010) 'MAOA uVNTR and early physical discipline interact to influence delinquent behavior', *Journal of Child Psychology and Psychiatry*, 51: 679–687.

El-Sheikh, M., Keller, P. S., and Erath, S. A. (2007) 'Marital conflict and risk for child maladjustment over time: skin conductance level reactivity as a vulnerability factor', *Journal of Abnormal Child Psychology*, 35: 715–727.

Faraone, S.V., Doyle, A. E., Mick, E., and Biederman, J. (2001) 'Meta-analysis of the association between the 7-repeat allele of the dopamine D4 receptor gene and attention deficit hyperactivity disorder', *American Journal of Psychiatry*, 158: 1052–1057.

Faraone, S.V., Perlis, R. H., Doyle, A. E., Smoller, J. W., Goralnick, J. J., Holmgren, M. A., and Sklar, P. (2005) 'Molecular genetics of attention-deficit-hyperactivity disorder', *Biological Psychiatry*, 57: 1313–1323.

Farrington, D. P., Loeber, R., Stallings, R., and Ttofi, M. M. (2011) 'Bullying perpetration and victimization as predictors of delinquency and depression in the Pittsburgh Youth Study', *Journal of Aggression, Conflict and Peace Research*, 3: 74–81.

Foley, D. L., Eaves, L. J., Wormley, B., Silberg, J. L., Maes, H. H., Kuhn, J., *et al.* (2004) 'Childhood adversity, monoamine oxidase A genotype, and risk for conduct disorder', *Archives of General Psychiatry*, 158(7): 1052–1057.

Frick, P. J., Lahey, B. B., Loeber, R., Tannenbaum, L., Van Horn, Y., Christ, M. G., and Hanson, K. (1993) 'Oppositional defiant disorder and conduct disorder: a meta-analytic review of factor analyses and cross-validation in a clinic sample', *Clinical Psychology Review*, 13(4): 319–340.

Frick, P. J., Ray, J. V., Thornton, L. C., and Kahn, R. E. (2013) 'Can callous-unemotional traits enhance the understanding, diagnosis, and treatment of serious conduct problems in children and adolescents? A comprehensive review', *Psychological Bulletin*, 140: 1–57.

Greene, M. B. (2000) 'Bullying and harassment in schools', in R. S. Moser and C. E. Franz (eds), *Shocking Violence: Youth Perpetrators and Victims, A Multidisciplinary Perspective*, Springfield, IL: Charles C. Thomas.

Haberstick, B.C., Lessem, J. M., Hopfer, C. J., Smolen, A., Ehringer, M. A., Timberlake, D., and Hewitt, J. K. (2005) 'MAOA genotype and antisocial behaviors in the presence of childhood and adolescent maltreatment', *American Journal of Human Genetics* (*Neuropsych. Genet.*), 135B: 59–64.

Heinz, A. J., Beck, A., Meyer-Lindenberg, A., Sterzer, P., and Heinz, A. (2011) 'Cognitive and neurobiological mechanisms of alcohol-related aggression', *Nature Reviews Neuroscience*, 12: 400–413.

Hinshaw, S. P. and Lee, S. S. (2003) 'Conduct and oppositional defiant disorders', in E. Mash and R. Barkley (eds), *Child Psychopathology*, New York: Guilford Press.

Huizinga, D., Haberstick, B., Smolen, A., Menard, S., Young, S., Corley, R., Stallings, M., Grotpeter, J., and Hewitt, J. (2006) 'Childhood maltreatment, subsequent antisocial behavior, and the role of monoamine oxidase A genotype', *Biological Psychiatry*, 60: 677–683.

Ingram, R. E. and Luxton, D. D. (2005) 'Vulnerability-stress models', in B. L. Hankin and J. R. Z. Abela (eds), *Development of Psychopathology: A Vulnerability Stress Perspective*, Thousand Oaks, CA: Sage.

Ingram, R. E. and Price, J. M. (eds) (2001) *Vulnerability to Psychopathology: Risk Across the Lifespan*, New York: Guilford Press.

Jennings, W. G. and Reingle, J. M. (2012) 'On the number and shape of developmental/life-course violence, aggression, and delinquency trajectories: a state-of-the-art review', *Journal of Criminal Justice*, 40(6): 472–489.

Kempes, M., Matthys, W., de Vries, H., and van Engeland, H. (2005) 'Reactive and proactive aggression in children: a review of theory, findings and the relevance for child and adolescent psychiatry', *European Child and Adolescent Psychiatry*, 14(1): 11–19.

Kim-Cohen, J., Caspi, A., Taylor, A., Williams, B., Newcombe, R., Craig, I. W., and Moffitt, T. E. (2006) 'MAOA, maltreatment, and gene–environment interaction predicting children's mental health: new evidence and a meta-analysis', *Molecular Psychiatry*, 11: 903–913.

Klein, M. W. (1987) 'Watch out for that last variable', in S. Mednick, T. Moffitt, and S. Stack (eds), *The Cause of Crime*, New York: Cambridge University Press.

Kluger, A. N., Siegfried, Z., and Ebstein, R. P. (2002) 'A meta-analysis of the association between DRD4 polymorphism and novelty seeking', *Molecular Psychiatry*, 7: 712–717.

Kuntsi, J., Gayan, J., and Stevenson, J. (2000) 'Parents' and teachers' ratings of problem behaviours in children: genetic and contrast effect', *Twin Research*, 3: 251–258.

Lansford, J. E., Skinner, A. T., Sorbring, E., Giunta, L., Deater-Deckard, K., Dodge, K. A., and Chang, L. (2012) 'Boys' and girls' relational and physical aggression in nine countries', *Aggressive Behavior*, 38(4): 298–308.

Lee, S. S. (2011) 'Deviant peer affiliation and antisocial behavior: moderation by monoamine oxidase A (MAO-A) genotype', *Journal of Abnormal Child Psychology*, 39: 321–332.

Leschied, A., Chiodo, D., Nowicki, E., and Rodger, S. (2008) 'Childhood predictors of adult criminality: a meta-analysis drawn from the prospective longitudinal literature', *Canadian Journal of Criminology and Criminal Justice*, 50(4): 435–467.

Masten, A. S. (2001) 'Ordinary magic: resilience processes in development', *American Psychologist*, 56(3): 227–238.

Mednick, S. A., Gabrielli Jr, W. F., and Hutchings, B. (1984) 'Genetic influences in criminal behavior: evidence from an adoption cohort', *Science*, 224: 891–894.

Mednick, S. A., Gabrielli Jr, W. F., and Hutchings, B. (1987) 'Genetic factors in the etiology of criminal behavior', in S. Mednick and T. Moffitt (eds), *Biosocial Bases of Antisocial Behavior*, Cambridge: Cambridge University Press.

Miles, D. R. and Carey, G. (1997) 'Genetic and environmental architecture of human aggression', *Journal of Personality and Social Psychology*, 72: 207–217.

Moffitt, T. E. (1993) 'Adolescence-limited and life-course-persistent antisocial behavior: a developmental taxonomy', *Psychological Review*, 100: 674–701.

Moffitt, T. E. (2005) 'The new look of behavioral genetics in developmental psychopathology: gene–environment interplay in antisocial behaviors', *Psychological Bulletin*, 131: 533–554.

Moffitt, T. E. (2006) 'Life-course persistent versus adolescence-limited antisocial behavior', in D. Cicchetti and D. J. Cohen (eds), *Developmental Psychopathology Risk, Disorder, and Adaptation*, New York: Wiley.

Moffitt, T. E. and Caspi, A. (2001) 'Childhood predictors differentiate life-course persistent and adolescent-limited antisocial pathways among males and females', *Development and Psychopathology*, 13: 355–375.

Patterson, G. R., Reid, J. B., and Dishion, T. (1992) *Antisocial Boys*, Eugene, OR: Castalia.

Plomin, R., DeFries, J. G., and Loehlin, J. G. (1977) 'Genotype–environment interaction and correlation in the analysis of human behavior', *Psychological Bulletin*, 84: 309–322.

Plomin, R., DeFries, J. C., Knopik, V. S., and Neiderhiser, J. M. (2012) *Behavioral Genetics*, 6th edn, New York: Worth.

Pluess, M. and Belsky, J. (2010) 'Differential susceptibility to parenting and quality child care', *Developmental Psychology*, 46: 379–390.

Poulin F. and Boivin, M. (2000) 'Reactive and proactive aggression: evidence of a two-factor model', *Psychological Assessment*, 12: 115–122.

Raine, A. (2002) 'Biosocial studies of antisocial and violent behaviors in children and adults: a review', *Journal of Abnormal Child Psychology*, 30(4): 311–326.

Rettew, D.C. and McKee, L. (2005) 'Temperament and its role in developmental psychopathology', *Harvard Review of Psychiatry*, 13: 14–27.

Sakai, J. T., Boardman, J. D., Gelhorn, H. L., Smolen, A., Corley, R. P., Huizinga, D., Menard, S., Hewitt, J. K., and Stallings, M. C. (2010) 'Utilizing trajectory analyses to refine phenotype for genetic association: conduct problems and the serotonin transporter (5HTTLPR)', *Psychiatric Genetics*, 20(5): 199–206.

Scarr, S. and McCartney, K. (1983) 'How people make their own environments: a theory of genotype → environment effects', *Child Development*, 54: 424–435.

Shiner, R. L., Buss, K. A., McClowry, S. G., Putnam, S. P., Saudino, K. J., and Zentner, M. (2012) 'What is temperament now? Assessing progress in temperament research on the twenty-fifth anniversary of Goldsmith *et al.* (1987)', *Child Development Perspectives*, 6: 436–444.

Slavich, G. M. and Cole, S. W. (2013) 'The emerging field of human social genomics', *Clinical Psychological Science*, 1(3): 331–348.

Tremblay, R. E. (2010) 'Developmental origins of disruptive behaviour problems: the original sin hypothesis, epigenetics and their consequences for prevention', *Journal of Child Psychology and Psychiatry*, 51(4): 341–367.

Vaughn, M. G., DeLisi, M., Beaver, K. M., and Wright, J. P. (2009) 'DAT1 and 5HTT are associated with pathological criminal behavior in a nationally representative sample of youth', *Criminal Justice and Behavior*, 36: 1113–1124.

Verona, E., Joiner, T. E., Johnson, F., and Bender, T. W. (2006) 'Gender specific gene–environment interactions on laboratory-assessed aggression', *Biological Psychology*, 71(1): 33–41.

Vlachou, M., Andreou, E., Botsoglou, K., and Didaskalou, E. (2011) 'Bully/victim problems among preschool children: a review of current research evidence', *Educational Psychology Review*, 23(3): 329–358.

Wang, E., Ding, Y.-C., Flodman, J. R., Kidd, J. R., Kidd, K. K., Grady, D. L., Ryder, O. A., Spence, M. A., Swanson, J. M., and Moyzis, R. K. (2004) 'The genetic architecture of selection at the human dopamine receptor D4 (DRD4) gene locus', *American Journal of Human Genetics*, 74: 931–944.

White, T. W. and Walters, G. D. (1989) 'Lifestyle criminality and the psychology of disresponsibility', *International Journal of Offender Therapy and Comparative Criminology*, 33(3): 257–263.

Wu, J., Xiao, H., Sun, H., Zou, L., and Zhu, L.-Q. (2012) 'Role of dopamine receptors in ADHD: a systematic meta-analysis', *Molecular Neurobiology*, 45: 605–620.

Young, S. E., Smolen, A., Hewitt, J. K., Haberstick, B.C., Stallings, M. C., Corley, R. P., and Crowley, T. J. (2006) 'Interaction between MAO-A genotype and maltreatment in the risk for conduct disorder: failure to confirm in adolescent patients', *American Journal of Psychiatry*, 163: 1019–1025.

25

The biological etiology of psychopathy

Rheanna J. Remmel and Andrea L. Glenn

Introduction

Psychopathy has been widely studied within recent decades, and recently the search to determine its biological causes has intensified. Psychopathy can be described as a psychological disorder which includes exhibition of callous and antisocial behavior, lack of empathy, shallow affect, lack of guilt, and lack of remorse (Hare, 1996). It can additionally be described as a developmental disorder exhibiting lack of emotion, manipulative behavior, irresponsible interpersonal relationships, impulsivity, and possibly violence (Wu and Barnes, 2013). Its characterization as a developmental disorder implies that it is present at a relatively early age, and may involve early biological mechanisms. Although social and environmental factors have been shown to influence the development of psychopathic traits (Raine, 2002), it is important also to understand the biological factors which lead some individuals, but not others, to develop psychopathy. Before examining these factors, it is important to have an understanding of how we conceptualize and measure psychopathy.

The characteristics of psychopathy can generally be divided into four facets: interpersonal, affective, lifestyle, and antisocial behavior. Some researchers group these into two factors, so that factor 1 psychopathy refers to the interpersonal/affective traits, and factor 2 refers to the antisocial/lifestyle traits (Hare, 2003). The most frequently used instrument to measure psychopathy, especially in clinical and forensic settings, is the Psychopathy Checklist-Revised (PCL-R) (Hare, 2003). This measure consists of a 20-item semi-structured interview, as well as a comprehensive records review, resulting in a score that can range from 0 to 40. Other common measures of psychopathy include the Psychopathic Personality Inventory (PPI) (Lilienfeld and Andrews, 1996) and the Levenson Self-Report Psychopathy Scale (LSRP) (Levenson *et al.*, 1995), both of which are self-report measures and are more commonly used in community or non-incarcerated samples.

In this chapter, we focus on the biological deficits associated with psychopathy and theories as to how they may lead to the myriad of behavioral and physiological traits that make up psychopathy. We utilize a "genes to brain to antisocial behavior" framework, which posits that genetic factors result in structural or functional brain differences, which consequently result in antisocial behavior in some individuals (Raine, 2008).

Brain deficits

As brain imaging has become an increasingly common method for studying the development and manifestation of psychological disorders, recent studies have produced a wealth of information regarding how the brain may be abnormal in individuals with psychopathy. Structural and functional deficits have been found in a variety of areas of the brain that control a large number of behaviors and cognitive processes. Perhaps the most widely implicated brain regions in psychopathy are the prefrontal cortex and the limbic system.

Structural abnormalities

Various abnormalities in brain structure have been observed in psychopathy, including in gray matter volume (GMV) and gray matter concentration (GMC). GMV has been found to be reduced in the prefrontal cortex (Raine *et al.*, 2000; Yang *et al.*, 2005), which is involved in executive functioning. Ermer and colleagues (2012) additionally found that both GMV and GMC in various brain regions were significantly related to psychopathy scores. GMV refers to the amount of gray matter tissue present, whereas GMC refers to the density of the gray matter tissue. MRI scans of prison inmates showed that total PCL-R scores were negatively correlated with GMV in the right amygdala, right hippocampus, right temporal pole, left orbitofrontal cortex (OFC), and posterior cingulate cortex (PCC). PCL-R scores were also negatively correlated with GMC in the right and left amygdala, right and left hippocampus, right and left temporal pole, and left OFC. Damage to these brain areas has been associated with symptoms such as impulsivity, reactive aggression, and lack of empathy, all of which are also seen in psychopathy.

Additionally, Yang and colleagues (2011) found that psychopaths showed significantly decreased gray matter thickness in the left OFC and bilateral anterior cortex, compared with non-psychopaths. These gray matter abnormalities were significantly related to perseverative errors on the Wisconsin Card-Sorting Test (WCST), which measures the repetition of behaviors that are not, or are no longer, reinforced. This provides evidence not only for basic anatomical differences, but for the fact that these differences lead to different observable behavior in psychopaths.

White matter abnormalities have also been observed in psychopathy. Several studies have found reduced structural integrity of the uncinated fasciculus (UF) in psychopathic individuals compared with non-psychopaths (Craig *et al.*, 2009; Motzkin *et al.*, 2011). The UF is a main tract of connectivity between the OFC and the amygdala, both of which have been implicated in psychopathy (Ermer *et al.*, 2012). Some researchers hypothesize that abnormal connectivity in this network may be a contributing factor in the impulsive, antisocial, and emotionally detached behavior exhibited by psychopaths (Craig *et al.*, 2009).

Additionally, a recent study examining the morphology of brain structures in psychopaths compared with healthy controls found significant morphological differences in subregions of the striatum between groups (Boccardi *et al.*, 2013). The differences were not unidirectional and instead represented a complex pattern of local enlargement and reduction, but support past research implicating the limbic system in psychopathy (Craig *et al.*, 2009; Juárez *et al.*, 2013). Glenn *et al.* (2010) found morphological differences as well, finding that psychopaths exhibited significantly larger striatum volumes than non-psychopaths, particularly the bilateral lenticular nuclei. They also found that striatum volume was significantly related to scores on facets one (interpersonal), two (affective), and three (lifestyle) of the PCL-R. The striatum has previously been implicated in impulsivity and reward-seeking, areas that are reflected in these three facets of psychopathy. Also in the limbic system, abnormal structure of the amygdala has been observed in psychopathic prisoners compared with controls (Boccardi *et al.*, 2011).

Overall, these findings suggest a general pattern of gray matter thinning, especially in the prefrontal cortex. Both enlargement and reductions have been observed in limbic structures. As mentioned above, the prefrontal cortex is largely in control of executive functions. The limbic system is involved in emotion recognition, emotional decision-making, and conditioning, especially fear conditioning. Deficits in these areas, particularly in combination with white matter deficits that impair inter-lobe communication, and morphological anomalies of the limbic system and the striatum, may predispose individuals to developing psychopathy.

Functional abnormalities

Functional deficits in brain activity are equally as important as structural ones, as it is the function of the brain that controls our cognitions and behaviors. With the emergence of functional magnetic resonance imaging (fMRI), studying the brain activity of psychopaths has become much easier. An accumulation of evidence of how the brain performs abnormally in psychopathy is providing a clearer picture of how psychopathy may develop.

One study found abnormal functional connectivity in the limbic and paralimbic systems, especially in the frontoparietal and posterior cingulate areas, which significantly correlated with psychopathy symptoms (Juárez *et al.*, 2013). Abnormalities in the posterior cingulate, particularly, correlated significantly with PCLR-total scores and factor 2 scores, suggesting that deficits in this region may contribute directly to the development of psychopathic traits and antisocial behavior.

Further supporting the role of the limbic system in psychopathy, one study found that psychopathic criminals had significantly lower functional connectivity at rest between the right anterior ventromedial prefrontal cortex (vmPFC) and the right amygdala, compared with non-psychopathic criminals (Motzkin *et al.*, 2011). Psychopaths showed normal connectivity, however, between right and left amygdalae and between the right anterior superior temporal gyrus (aSTG) and the right amygdala, indicating that the impairment was specific to the vmPFC–right amygdala pathway. This reduced functional connectivity was also found between the right anterior vmPFC and the precuneus/PCC in the psychopathic but not the non-psychopathic group. Again, this impairment was specific and not generalized to other areas of the default mode network, of which the vmPFC and the precuneus/PCC are a part. Yang and colleagues (2012), in examining the topological characteristics of the whole brain anatomical network, found abnormal connectivity patterns in the frontal region. The wide range of connectivity deficits found in psychopathic populations suggests that there is not one single factor underlying the symptomology that psychopaths exhibit.

Further evidence indicates that abnormalities in the functioning of the amygdala may be associated with specific facets of psychopathy. Carré and colleagues (2013) found that right dorsal amygdala responsivity to fearful faces was negatively associated with the interpersonal facet of psychopathy, and reactivity to angry faces was positively associated with the lifestyle facet in men, but not in women. The lack of reactivity to viewing fearful faces in people high on psychopathy, and especially the interpersonal facet, emphasizes how psychopathic traits result in a reduced response to the distress of others (Blair, 2007; Jones *et al.*, 2009), which could partially explain the lack of empathy and remorse typically exhibited by psychopaths. Conversely, the association between heightened reactivity in the right dorsal amygdala when viewing angry faces and the lifestyle facet of psychopathy could partially explain the high level of reactive aggression seen in some psychopaths (Carré *et al.*, 2012).

Additionally, several studies have shown not decreased, but *increased* activation of the dorsolateral prefrontal cortex during tasks that require processing of emotional stimuli (Kiehl *et al.*, 2001; Gordon *et al.*, 2004; Rilling *et al.*, 2007; Glenn *et al.*, 2009). The dlPFC is involved in executive functioning and especially in emotional processing, thus increased activity in this region during emotional processing tasks may reflect that when psychopaths do not have appropriate limbic input to process an emotional stimulus, they rely more on cognitive means to process the stimulus (Glenn *et al.*, 2009).

Overall, deficits in the frontal cortex, the limbic system, and the temporal cortex have repeatedly been associated with psychopathy. These areas are broadly responsible for decision-making, complex planning, regulation of emotion, and moral judgment. Deficits or abnormalities in these areas may therefore lead individuals to exhibit psychopathic traits and antisocial behaviors.

Genetics

Because our genes guide the development of our brain, which in turn shapes our development and our behavior, the field of genetics is rich with potential to explain the etiology of psychopathic behavior. This entails not only determining whether psychopathic traits are passed down from parent to child, but also which genes in particular may be responsible for these traits.

Regarding the heritability of psychopathic traits, Beaver and colleagues (2011) conducted an adoption study and found that having a criminal biological father, but not mother, was significantly associated with scores on a psychopathic traits scale, especially for men. This suggests genetic risk based on the criminal status of the biological father. Furthermore, the same study found that respondents with a criminal biological father were significantly more likely than those without to score in the top 25 percent of the psychopathic traits scale. This effect was shown for the mixed-gender respondent group and the male respondent group, but not the female respondent group. The male group, additionally, were significantly more likely to respond in the top 10 percent on the psychopathic traits scale if they had a criminal father. Although much research has been done on the heritability of criminality and antisocial traits in general (Moffitt, 2005), this study, along with several studies using twin-based research designs (e.g. Blonigen *et al.*, 2003; Larsson *et al.*, 2007; Vernon *et al.*, 2008), provides substantial evidence for psychopathic traits, specifically, having at least some level of heritability.

Researchers have recently begun pinpointing specific genes that may predispose individuals to psychopathy and antisocial behavior. Recently, Sadeh *et al.* (2013) demonstrated positive relationships between carrying two long alleles of 5-HTTLPR (a polymorphism of the SLC6A4 gene), a gene that codes for serotonin transporter proteins, and psychopathic traits in adult males, regardless of ethnicity or income level. They found a similar relationship between carrying the low-activity variant of the MAOA (monoamine oxidase-A) gene and psychopathic traits. MAOA is an enzyme that breaks down catecholamines such as serotonin and dopamine. This evidence suggests that having higher levels of catecholamine activity in the brain may contribute to psychopathic traits. This is supported by a study by Wu and Barnes (2013), which found that the presence of "risky" alleles for two dopamine receptor genes, DRD2 and DRD4, both predicted psychopathic personality traits at a statistically significant level. Previous research has shown that dopamine levels are related to impulsivity (Retz *et al.*, 2003), implying that these genes, by influencing dopamine levels, might lead to increased reward-seeking activity and impulsive behavior.

The study of genetic influences on psychopathy is still emerging, and there have been relatively few studies that have been able to reliably identify genes associated with psychopathy.

More study in this area is needed to clarify which genes confer the most risk for developing psychopathic traits.

Psychophysiological impairments

The autonomic nervous system (ANS) serves as the connection between the central nervous system and the internal organs such as the heart, adrenal glands, lungs, and salivary glands. It is critical to the "fight-or-flight" response, which is the natural reaction to a threatening or stressful situation. Typical physiological measures of ANS activity include heart rate and skin conductance.

Skin conductance

Abnormalities in skin conductance have been well established in psychopathy and related disorders of antisocial behavior. Some of the earliest empirical research conducted with psychopaths demonstrates that they exhibit abnormal skin conductance responses to aversive stimuli (Hare, 1978). More recent research supports these findings as well. For example, Benning *et al.* (2005) found that individuals high on psychopathic personality traits exhibited significantly smaller skin conductance responses (SCRs) overall, particularly in response to aversive stimuli, than did individuals low on psychopathic traits. Fung *et al.* (2005) also demonstrated that youths with psychopathic traits were significantly less likely than youths without psychopathic traits to exhibit SCRs in anticipation of a signaled white-noise burst. They were also significantly less likely to show SCRs in response to both signaled and unsignaled white-noise bursts. These results show that individuals with psychopathic traits exhibit reduced electrodermal activity in anticipation of, and in response to, aversive stimuli.

Importantly, results from a prospective, longitudinal study of the temperamental and psychophysiological precursors to psychopathy found that adults who scored highly on a psychopathy measure at age 28 also showed significantly longer SC half-recovery times as 3-year-olds than did the adults who scored low on the psychopathy measure (Glenn *et al.*, 2007). Longer SC half-recovery times have been linked in animal studies to amygdala lesions (Pribram and McGuiness, 1975) and, as mentioned above, abnormal amygdala activity is likely to play a significant role in the development and manifestation of psychopathy. Therefore having longer SC recovery times in early childhood may be indicative of amygdala dysfunction, which could persist with age and contribute to the manifestation of psychopathic traits. These findings are important because there have been very few longitudinal studies on psychopathy, and those that have been done are not prospective in design, which means they might use proxy measures for measuring psychopathy rather than a scale designed to measure psychopathy. The results from this study show that the precursors to psychopathic traits may be present very early in childhood, as young as 3 years old.

Heart rate

Heart rate activity in psychopaths is considered to be somewhat dampened compared with non-psychopaths. One study found a negative correlation between PCL-R factor 1 scores and heart rate while viewing negative, neutral, and positive photos, in a sample of psychopathic convicted murderers (Serafim *et al.*, 2009). The authors also found that factor 1 scores explained just less than half (44.8 percent) of heart rate variance while viewing neutral and positive images. In line with these results, Casey *et al.* (2013) examined heart rate reactivity to emotional stimuli and found that psychopaths experience a lower than expected heart rate increase in response to viewing negative pictures.

In a similar type of study, Gao *et al.* (2012) found that not only did psychopaths show lower heart rate than controls before a psychosocial stressor, but psychopaths who expressed high "body sensation" (recognition of changes in their physiological activity in response to the stressor) did not exhibit heart rate increases relative to low body sensation psychopaths. This is contrary to the results in the non-psychopathic control group, wherein individuals who reported high body sensation had higher heart rates than individuals who reported low body sensation. The authors found that this discrepancy was particularly related to the interpersonal/affective facet of psychopathy.

Neuropsychological impairments

Intelligence

The role of intelligence in psychopathy is unclear. Most studies that measure intelligence in psychopaths do not do so as the primary aim of the study, and only report that the groups do not differ in IQ. There is little to no evidence to suggest that psychopaths as a group exhibit either higher or lower intelligence than non-psychopaths. However, intelligence may contribute to the development of psychopathy and its associated behaviors in more than a straightforward causal relationship. For example, Wall *et al.* (2013) found that intelligence can act as a moderating factor in psychopathy, such that more intelligent psychopaths commit less crime. These results are in line with previous studies finding that intelligence is negatively and significantly correlated with factor 2 scores, which measure criminality and antisocial behavior, in an incarcerated sample (Heinzen *et al.*, 2011).

Sellbom and Verona (2007), on the other hand, found that IQ as measured by the Shipley Institute of Living Scale correlated significantly and positively with scores on the PPI, particularly its factor 1 scale, which measures the interpersonal and affective characteristics of psychopathy. These results confirm previous findings by Salekin *et al.* (2004) of a positive relationship between intelligence and the interpersonal characteristics of psychopathy in youth.

Executive functioning

Executive functioning (EF) is a blanket term that refers to a diverse group of cognitive functions and abilities, most commonly associated with the prefrontal cortex. While some deficits in EF have been observed in psychopaths, other executive functions seem to remain intact (Ishikawa *et al.*, 2001).

In a study done with a community sample recruited from temporary employment agencies, Yang *et al.* (2011) found that level of psychopathy as measured by the PCL-R was significantly correlated with perseverative errors on the WCST. The WCST is a neuropsychological test that measures perseveration, a behavior involving the unnecessary or inappropriate repetition of a response in the absence or cessation of reward for that response. This fits with symptoms of psychopathy including impulsivity, and corresponds with the evidence presented regarding structural and functional brain abnormalities present in psychopathy.

Sellbom and Verona (2007) found that scores on a psychopathy measure (PPI-total scores) were significantly and negatively correlated with response inhibition, which is thought to be related to functioning of the prefrontal cortex, and specifically the OFC (Blair, 2003, 2004). The same study found that PPI-II (social deviance) scores correlated significantly and negatively with most of the EF tasks that were used. The opposite was true for PPI-I traits (interpersonal/affective), which correlated positively with EF. The PPI does not include any items that measure overt criminality, but the social deviance (PPI-II) factor contains the items that most closely align with criminality. This suggests that, in non-criminal, non-institutionalized samples, poor EF is related to more antisocial/criminal tendencies, whereas normal EF is associated more with the interpersonal and affective traits of psychopathy.

Hormones and neurotransmitters

Hormone imbalances

Hormones may play a promising role in explaining the biological etiology of psychopathy, but few studies to date have examined the possible relationships, and of those that have, the results have been rather inconclusive. The two hormones that have been studied most frequently in relation to psychopathy are cortisol and testosterone. Cortisol is regulated by the hypothalamus–pituitary–adrenal (HPA) axis, and is released in the presence of stressful stimuli, particularly psychosocial and uncontrollable stressors (Dickerson and Kemeny, 2004). The HPA stress response is different from the autonomic nervous system (ANS) stress response, but both systems do react to stressful stimuli. As shown above, psychopaths show abnormal ANS activity as measured by skin conductance and heart rate. Testosterone is regulated by the hypothalamus–pituitary–gonadal (HPG) axis. A series of experimental studies have demonstrated that increased testosterone levels are associated with increases in aggression (Pope *et al.*, 2000) and decreases in empathy (van Honk *et al.*, 2011), both of which have been associated with psychopathy. This, combined with the fact that testosterone is present at higher levels in males, and that psychopathy is more prevalent in males than in females, suggests that testosterone may have a role to play in psychopathy.

A few studies have demonstrated low cortisol levels in psychopaths. Holi and colleagues (2006) found low cortisol levels in psychopathic criminals with a history of violence, and Cima *et al.* (2008) also found low average cortisol levels in psychopathic offenders compared with non-psychopaths. In a study using a social stressor, O'Leary *et al.* (2007) found that having higher levels of psychopathic traits was associated with lower cortisol levels after the stressor, but only in males. In males with high psychopathic traits there was no significant difference between pre- and post-stressor cortisol levels, but in males with low psychopathic traits there was a significant difference in cortisol levels. Low cortisol levels would suggest hypoactivity of the HPA axis, which could corroborate brain evidence showing decreased responses to aversive (stressful) stimuli. Other studies, however, have failed to replicate these findings (van Honk *et al.*, 2003; Glenn *et al.*, 2011).

Few studies have been conducted examining the role of testosterone in psychopathy. Stålenheim *et al.* (1998) found that higher testosterone levels correlated significantly with factor 2 psychopathy scores in convicted criminals. Glenn *et al.* (2011) found that neither baseline testosterone nor baseline cortisol was significantly related to psychopathy scores; however, the authors found that the ratio of baseline testosterone to cortisol reactivity after stress was significantly related to psychopathy. Furthermore, they found that factor 2, but not factor 1, scores significantly predicted this testosterone to cortisol reactivity ratio.

Neurotransmitters

Neurotransmitters are chemicals that facilitate communication between nerve cells, and can transmit information about emotions, urges, and somatic sensations (Lawrence, 2005). As seen above in the section regarding genetics, the neurotransmitters serotonin and dopamine have been implicated in psychopathy. However, neurotransmitter levels are difficult to study *in vivo*, and very few studies have done so. Those that have studied neurotransmitter levels have examined levels of neurotransmitter metabolites in cerebrospinal fluid (CSF), such as HVA, a metabolite of dopamine, and 5-HIAA, a metabolite of serotonin (Freedman and Verdun-Jones, 2010).

There is evidence to suggest that serotonin levels are low in psychopathy (Dolan, 1994). However, more recent research suggests that it is important to examine both serotonin and dopamine as they relate to psychopathy. Soderstrom *et al.* (2001) found that the ratio of HVA to 5-HIAA was a significant predictor of psychopathy scores on the PCL-R, and especially factor 2 scores. The authors therefore concluded that psychopathy is related to both serotonin dysregulation and high dopamine turnover. These results were replicated in an additional forensic sample (Soderstrom *et al.*, 2003).

Additional studies on neurotransmitter levels in psychopathy should be conducted before any firm conclusions are drawn, but this evidence, in conjunction with the genetic evidence presented above, implicates genes that code for dopamine receptors (DRD2 and DRD4), serotonin transporters (5-HTTPLPR), and enzymes which metabolize monoamines (MAOA) in psychopathic traits.

Conclusion

Psychopathy is a complex disorder, which presents with a variety of symptoms at varying intensities. Based on the diverse presentation of the disorder, as well as the large amount of evidence for varying risk factors and predispositions, it is clear that there is not one single pathway for the development of psychopathy. There are, however, several factors that can be clearly linked to the development of psychopathy and the exhibition of its symptoms.

First and most basically are the genetic risk factors. They are present from the moment of conception, although their expression may be enhanced or hindered by environmental factors. These genes can influence abnormal development of the brain, as well as hinder "normal" functioning once the brain has finished developing. The brain, in turn, controls such biological mechanisms as autonomic nervous system regulation, neuropsychological functioning, and hormone levels throughout the body. Abnormalities in these areas can cause psychopaths to exhibit violent, disinhibited, and impulsive behavior, as well as a lack of empathy and emotionality.

Of course, some people may inherit genetic risk factors and still not become psychopathic. This may reflect the influence of other genes that compensate in some way, perhaps by protecting the brain against abnormal development, or by overriding the "risky" genes so that they are not expressed. Additionally, we cannot ignore the influence of the environment on the development of psychopathy. The environment may serve to exacerbate biological risk factors or to buffer against them and protect individuals from developing psychopathic traits. In individuals who do develop psychopathy, however, it seems clear that genes, brain structure, brain function, autonomic nervous system function, neuropsychological function, hormones, and neurotransmitters all play a role. Although discussed separately, these systems are all inherently linked. Future research examining the connections between these systems will be critical for improving our understanding of the developmental trajectories that lead to psychopathic traits. Not all of these factors may be present in any given individual with psychopathy, but each confers risk in a different way and, when present in some circumstances, can help to explain the etiology of psychopathy.

References

Beaver, K.M., Rowland, M.W., Schwartz, J.A., and Nedelec, J.L. (2011) 'The genetic origins of psychopathic personality traits in adult males and females: results from an adoption-based study', *Journal of Criminal Justice*, 19: 426–432.

Benning, S.D., Patrick, C.J., and Iacono, W.G. (2005) 'Psychopathy, startle blink modulation, and electrodermal reactivity in twin men', *Psychophysiology*, 42: 753–762.

Blair, R.J.R. (2003) 'Neurobiological basis of psychopathy', *British Journal of Psychiatry*, 182(1): 5–7.

Blair, R.J.R. (2004) 'The roles of orbital frontal cortex in the modulation of antisocial behavior', *Brain and Cognition*, 55(1): 198–208.

Blair, R.J.R. (2007) 'The amygdala and ventromedial prefrontal cortex in morality and psychopathy', *Trends in Cognitive Sciences*, 11: 387–391.

Blonigen, D.M., Carlson, S.R., Krueger, R.F., and Patrick, C.J. (2003) 'A twin study of self reported psychopathic personality traits', *Personality and Individual Differences*, 35: 179–197.

Boccardi, M., Frisoni, G.B., Hare, R.D., Cavedo, E., *et al.* (2011) 'Cortex and amygdala morphology in psychopathy', *Psychiatry Research: Neuroimaging*, 193: 85–92.

Boccardi, M., Bocchetta, M., Aronen, H.J., Repo-Tiihonen, E., Vaurio, O., Thompson, P.M., Tiihonen, J., and Frisoni, G.B. (2013) 'Atypical nucleus accumbens morphology in psychopathy: another limbic piece in the puzzle', *International Journal of Law and Psychiatry*, 36: 157–167.

Carré, J.M., Fisher, P.M., Manuck, S.B., and Hariri, A.R. (2012) 'Interaction between trait anxiety and trait anger predict amygdala reactivity to angry facial expressions in men but not women', *Social Cognitive and Affective Neuroscience*, 7: 213–221.

Carré, J.M., Hyde, L.W., Neumann, C.S., Viding, E., and Hariri, A.R. (2013) 'The neural signatures of distinct psychopathic traits', *Social Neuroscience*, 8: 122–135.

Casey, H., Rogers, R.D., Burns, T., and Yiend, J. (2013) 'Emotion regulation in psychopathy', *Biological Psychology*, 92: 541–548.

Cima, M., Smeets, T., and Jelicic, M. (2008) 'Self-reported trauma, cortisol levels, and aggression in psychopathic and non-psychopathic prison inmates', *Biological Psychiatry*, 78: 75–86.

Craig, M.C., Catani, M., Deeley, Q., Latham, R., Daly, E., Kanaan, R., Picchioni, M., McGuire, P.K., Fahy, T., and Murphy, D.G.M. (2009) 'Altered connections on the road to psychopathy', *Molecular Psychiatry*, 14: 946–953.

Dickerson, S.S. and Kemeny, M.E. (2004) 'Acute stressors and cortisol responses: a theoretical integration and synthesis of laboratory research', *Psychological Bulletin*, 130: 355–391.

Dolan, M. (1994) 'Psychopathy – a neurobiological perspective', *British Journal of Psychiatry*, 165: 151–159.

Ermer, E., Cope, L.M., Nyalakanti, P., Calhoun, V.D., and Kiehl, K.A. (2012) 'Aberrant paralimbic gray matter in criminal psychopathy', *Journal of Abnormal Psychology*, 121: 649–658.

Freedman, L.F. and Verdun-Jones, S.N. (2010) 'Blaming the parts instead of the person: understanding and applying neurobiological factors associated with psychopathy', *Canadian Journal of Criminology and Criminal Justice*, 52: 29–53.

Fung, M.T., Raine, A., Loeber, R., Lynam, D.R., Steinhauer, S.R., Venables, P.H., and Stouthamer-Loeber, M. (2005) 'Reduced electrodermal activity in psychopathy-prone adolescents', *Journal of Abnormal Psychology*, 114: 187–196.

Gao, Y., Raine, A., and Schug, R.A. (2012) 'Somatic aphasia: mismatch of body sensations with autonomic stress reactivity in psychopathy', *Biological Psychology*, 90: 228–233.

Glenn, A.L., Raine, A., Venables, P.H., and Mednick, S.A. (2007) 'Early temperamental and psychophysiological precursors of adult psychopathic personality', *Journal of Abnormal Psychology*, 116: 508–518.

Glenn, A.L., Raine, A., Schug, R.A., Young, L., and Hauser, M. (2009) 'Increased DLPFC activity during moral decision-making in psychopathy', *Molecular Psychiatry*, 14: 909–911.

Glenn, A.L., Raine, A., Yaralian, P.S., and Yang, Y. (2010) 'Increased volume of the striatum in psychopathic individuals', *Biological Psychiatry*, 67: 52–58.

Glenn, A.L., Raine, A., Schug, R.A., Gau, Y., and Granger, D.A. (2011) 'Increased testosterone-to-cortisol ratio in psychopathy', *Journal of Abnormal Psychology*, 120: 389–399.

Gordon, H.L., Baird, A.A., and End, A. (2004) 'Functional differences among those high and low on a trait measure of psychopathy', *Biological Psychiatry*, 56: 516–521.

Hare, R.D. (1978) 'Psychopathy and electrodermal responses to nonsignal stimulation', *Biological Psychology*, 6: 237–246.

Hare, R.D. (1996) 'Psychopathy: a clinical construct whose time has come', *Criminal Justice and Behavior*, 23: 25–54.

Hare, R.D. (2003) *Manual for the Revised Psychopathy Checklist*, 2nd edn, Toronto: Multi-Health Systems.

Heinzen, H., Köhler, D., Godt, N., Geiger, F., and Huchzermeier, C. (2011) 'Psychopathy, intelligence and conviction history', *International Journal of Law and Psychiatry*, 34: 336–340.

Holi, M., Auvinen-Lintunen, L., Lindberg, N., Tani, P., and Virkkunen, M. (2006) 'Inverse correlation between severity of psychopathic traits and serum cortisol levels in young adult violent male offenders', *Psychopathology*, 39: 102–104.

van Honk, J., Schutter, D.J.L.G., Hermans, E.J., and Putman, P. (2003) 'Low cortisol levels and the balance between punishment sensitivity and reward dependency', *NeuroReport*, 14: 1993–1996.

van Honk, J., Schutter, D.J., Bos, P.A., Kruijt, A.W., Lentjes, E.G., and Baron-Cohen, S. (2011) 'Testosterone administration impairs cognitive empathy in women depending on second-to-fourth digit ratio', *Proceedings of the National Academy of Sciences, USA*, 108(8): 3448–3452.

Ishikawa, S.S., Raine, A., Lencz, T., Bihrle, S., and Lacasse, L. (2001) 'Autonomic stress reactivity and executive functions in successful and unsuccessful criminal psychopaths from the community', *Journal of Abnormal Psychology*, 110: 423–432.

Jones, A.P., Laurens, K.R., Herba, C.M., Barker, G.J., and Viding, E. (2009) 'Amygdala hypoactivity to fearful faces in boys with conduct problems and callous-unemotional traits', *American Journal of Psychiatry*, 166: 95–102.

Juárez, M., Kiehl, K.A., and Calhoun, V.D. (2013) 'Intrinsic limbic and paralimbic networks are associated with criminal psychopathy', *Human Brain Mapping*, 34: 1921–1930.

Kiehl, K.A., Smith, A.M., Hare, R.D., Mendrek, A., Forster, B.B., and Brink, J. (2001) 'Limbic abnormalities in affective processing by criminal psychopaths as revealed by functional magnetic resonance imaging', *Biological Psychiatry*, 50: 677–684.

Larsson, H., Tuvblad, C., Rijsdijk, F., Andershed, H., Grann, M., and Lichtenstein, P. (2007) 'A common genetic factor explains the association between psychopathic personality and antisocial behavior', *Psychological Medicine*, 37: 15–26.

Lawrence, E. (ed.) (2005) *Henderson's Dictionary of Biology*, 13th edn, Harlow, UK: Prentice Hall.

Levenson, M., Kiehl, K., and Fitzpatrick, C. (1995) 'Assessing psychopathic attributes in a noninstitutionalized population', *Journal of Personality and Social Psychology*, 68: 151–158.

Lilienfeld, S.O. and Andrews, B.P. (1996) 'Development and preliminary validation of a self-report measure of psychopathic personality traits in noncriminal populations', *Journal of Personality Assessment*, 66: 488–524.

Moffitt, T.E. (2005) 'The new look of behavioral genetics in developmental psychopathology: gene–environment interplay in antisocial behaviors', *Psychological Bulletin*, 131: 533–554.

Motzkin, J.C., Newman, J.P., Kiehl, K.A., and Koenigs, M. (2011) 'Reduced prefrontal connectivity in psychopathy', *Journal of Neuroscience*, 31: 17348–17357.

O'Leary, M.M., Loney, B.R., and Eckel, L.A. (2007) 'Gender differences in the association between psychopathic personality traits and cortisol response to induced stress', *Psychoneuroendocrinology*, 32: 183–191.

Pribram, K.H. and McGuiness, D. (1975) 'Arousal, activation, and effort in the control of attention', *Psychological Review*, 82: 116–149.

Pope, H.G., Kouri, E.M., and Hudson, J.I. (2000) 'Effects of supraphysiologic doses of testosterone on mood and aggression in normal men: a randomized controlled trial', *Archives of General Psychiatry*, 57(2): 133–140.

Raine, A. (2002) 'Biosocial studies of antisocial and violent behavior in children and adults: a review', *Journal of Abnormal Child Psychology*, 30: 311–326.

Raine, A. (2008) 'From genes to brain to antisocial behavior', *Current Directions in Psychological Science*, 17: 323–328.

Raine, A., Lencz, T., Bihrle, S., LaCasse, L., and Colletti, P. (2000) 'Reduced prefrontal gray matter volume and reduced autonomic activity in antisocial personality disorder', *Archives of General Psychiatry*, 57: 119–127.

Retz, W., Rösler, M., Supprian, T., Retz-Junginger, P., and Thome, J. (2003) 'Dopamine D3 receptor gene polymorphism and violent behavior: relation to impulsiveness and ADHD-related psychopathology', *Journal of Neural Transmission*, 110: 561–572.

Rilling, J.K., Glenn, A.L., Jairam, M.R., Pagnoni, G., Goldsmith, D.R., Elfenbein, H.A., and Lilienfeld, S.O. (2007) 'Neural correlates of social cooperation and non-cooperation as a function of psychopathy', *Biological Psychiatry*, 61: 1260–1271.

Sadeh, N., Javdani, S., and Verona, E. (2013) 'Analysis of monoaminergic genes, childhood abuse, and dimensions of psychopathy', *Journal of Abnormal Psychology*, 122(1): 167–179.

Salekin, R.T., Neumann, C.S., Leistico, A.R., and Zalot, A.A. (2004) 'Psychopathy in youth and intelligence: an investigation of Cleckley's hypothesis', *Journal of Clinical Child and Adolescent Psychology*, 33: 731–742.

Sellbom, M., and Verona, E. (2007) 'Neuropsychological correlates of psychopathic traits in a non-incarcerated sample', *Journal of Research in Personality*, 41: 276–294.

Serafim, A., de Barros, D.M., Valim, A., and Gorenstein, C. (2009) 'Cardiac response and anxiety levels in psychopathic murderers', *Revista Brasileira de Psiquiatria*, 31: 214–218.

Soderstrom, H., Blennow, K., Manhem, A., and Forsman, A. (2001) 'CSF studies in violent offenders I.5-HIAA as a negative and HVA as a positive predictor of psychopathy', *Journal of Neural Transmission*, 108: 869–878.

Soderstrom, G., Blennow, K., Sjodin, A.-K., and Forsman, A. (2003) 'New evidence for an association between the CSF HVA: 5-HIAA ratio and psychopathic traits', *Journal of Neurology, Neurosurgery and Psychiatry*, 74: 918–921.

Stålenheim, E.G., Eriksson, E., von Knorring, L., and Wide, L. (1998) 'Testosterone as a biological marker in psychopathy and alcoholism', *Psychiatry Research*, 77(2): 79–88.

Vernon, P.A., Villani, V.C., Vickers, L.C., and Harris, J.A. (2008) 'A behavioral genetic investigation of the Dark Triad and the Big 5', *Personality and Individual Differences*, 44: 445–452.

Wall, T.D., Sellbom, M., and Goodwin, B.E. (2013) 'Examination of intelligence as a compensatory factor in non-criminal psychopathy in a non-incarcerated sample', *Journal of Psychopathology and Behavioral Assessment*, 35(4): 450–459.

Wu, T. and Barnes, J.C. (2013) 'Two dopamine receptor genes (DRD2 and DRD4) predict psychopathic personality traits in a sample of American adults', *Journal of Criminal Justice*, 41: 188–195.

Yang, Y., Raine, A., Lencz, T., Bihrle, S., LaCasse, L., and Colletti, P. (2005) 'Volume reduction in prefrontal gray matter in unsuccessful criminal psychopaths', *Biological Psychiatry*, 57: 1103–1108.

Yang, Y., Raine, A., Colletti, P., Toga, A.W., and Narr, K.L. (2011) 'Abnormal structural correlates of response perseveration in individuals with psychopathy', *Journal of Neuropsychiatry and Clinical Neurosciences*, 23: 107–110.

Yang, Y., Raine, A., Joshi, A.A., Joshi, S., Chang, Y.T., Schug, R.A., Wheland, D., Leahy, R., and Narr, K.L. (2012) 'Frontal information flow and connectivity in psychopathy', *British Journal of Psychiatry*, 201: 408–409.

26

Development of aggression and violence from infancy to adulthood

A biopsychosocial perspective

Stéphane Paquin, Eric Lacourse, and Isabelle Ouellet-Morin

Introduction

Although aggression and violence have greatly declined in most European societies since the end of the Middle Ages, they remain an important preoccupation in modern industrial societies. The questions surrounding the origins and evolution of criminal behavior are a centuries-old debate (Quetelet, 1835). In the mid-nineteenth century, Cesare Lombroso made his mark in the history of criminology by rejecting Locke's vision of violence. Locke's philosophical position stated that children were inherently neither good nor evil, and that violence was a normal behavior in a *state of war* (Locke, 1689). Rejecting this view, Lombroso's anthropological and physiognomical theory argued that *born criminals* had inherited traits that would make them more likely to commit violent crimes (Lombroso, 1876).

The contrast of perspectives between Locke's and Lombroso's views is a good example of the polarized debate in the scientific community about the origins of criminal behavior: do usual people commit unusual acts under certain special/rare circumstances, or are unusual acts committed by unusual people born that way? This opposition between nature and nurture as independent determinants of any behavior has been strongly criticized, and theoretical propositions that don't consider biological and environmental etiology of human behavior are mostly considered outdated and naive because they neglect the mutual interaction between these sources of influence (Raine, 2002; Rutter, 2006). The perspective on the biosocial development of aggression suggests that there is both something biological and something social about the mechanisms explaining aggressive and violent behaviors.

In the past two decades, theoretical approaches integrating rationales from genetics, biology, psychology, and sociology have been privileged to understand the ontogenesis and sociogenesis of antisocial behaviors, such as aggression and violence, that are universally less and less tolerated by citizens of modern societies (Pinker, 2011). As structuralist and cultural explanations of crime and violence were dominant in criminology, their limitations in explaining individual differences have brought scholars to investigate new ways to integrate individual and societal level explanations of criminal behavior. Some have adopted life-course or developmental perspectives that try to better describe and explain the origins and development of aggression and violence from birth to death (Moffitt, 1993; Tremblay, 2000; Granic and Patterson, 2006). Developmental criminology opened a whole new set of questions to be explained. What is the age of onset of aggressive and violent behaviors? Is the age of onset important to explain the duration, frequency, and severity of criminal careers? Are there typical trajectories that can be distinguished based on timing of onset and desistance? Are there important transitions or events affecting those trajectories? What are the ontogenic or sociogenic causal mechanisms of typical trajectories?

More than twenty years of longitudinal research across different countries have shown the importance of early developmental processes for the explanation of later interindividual differences. Studies seem to support that earlier onset of antisocial behaviors is associated with duration, frequency, and severity of criminal behaviors. Some scholars have defined this trajectory as life-course-persistent (LCP; Moffitt, 1993), which can also be often characterized by chronic physical aggression (Tremblay, 2000).

Early onset of antisocial behavior

Moffitt's (1993) developmental taxonomy had a great impact on criminology by specifying two qualitative subtypes of antisocial behavior trajectory. Moffitt's original proposition was inspired by the empirical evidence that a small proportion (around 5–7 percent) of individuals with a history of troublesome behavior in early childhood are responsible for most of the violent crimes (around 50 percent) in adulthood (Wolfgang *et al.*, 1972). She proposed two distinct types of offender. First, the LCP, high-rate offender; and second, the adolescent-limited (AL) offender who initiates minor forms of offending during adolescence and desists after a short period. Children from the LCP trajectory show a more persistent trajectory extending through adolescence and even into adulthood (Moffitt *et al.*, 2002). By adolescence, those LCPs will most likely be involved in delinquent acts at a higher rate than children following other pathways with later onset (Moffitt *et al.*, 2002).

Mechanisms to account for early onset

According to Moffitt, those two types of offending pattern need their own theoretical explanation. Moffitt proposed that the LCP trajectory is the result of the interactions of children's neuropsychological deficits (which could be inherited or acquired; Moffitt, 2006) with a criminogenic environment provided by parents living in adversity. Moffitt and Caspi (2001) showed that children following the LCP pathway had high level of neurocognitive and temperamental problems as well as inadequate parenting. Moffit's LCP offenders are explained by an interaction of neuropsychological deficits (either innate or acquired) with adverse rearing environments (Moffitt and Caspi, 2001). The AL trajectory is mostly adaptive, and the adolescents following that typical pattern are not very different from nonoffenders. Moffitt suggested that these children have had a normal childhood, but experience a 'maturity gap' during adolescence and tend to mimic the LCP, who have benefits from their antisocial activities (i.e. more money, more sexual partners, and greater popularity in some social contexts). As suggested by Moffitt, the neuropsychological deficits could be based partially on genetic factors. The largest recent meta-analysis examining the genetic and environmental etiology of constructs similar to aggression (diagnosis of antisocial personality disorder or conduct disorder, criminality measured by official records of arrest, aggression, or antisocial behavior) is that by Rhee and Waldman (2002). Their meta-analysis included 51 studies using a total of 52 independent samples involving 55,525 pairs of participants. They found that genetic factors accounted for around 50 percent of aggression variability. Shared environmental factors accounted for 15 percent, and nonshared environment accounted for around 35 percent. A genetically informative study has found support for the two types suggested by Moffit's theory (Silberg *et al.*, 2007). In a study of 10- to 17-year-old twins, they found a common genetic factor that influences conduct disorder at all ages, and a shared environmental factor that influences conduct disorder from 14 to 17 years old. Barnes *et al.* (2011) looked into the heritability of high offenders through adolescence and early adulthood. They found that heritability for individuals scoring higher than the 90th percentile was higher than what is usually found at the population level. Heritability was estimated around 70 percent while the population estimates are in the range of 50 percent. Moffit's theory also predicts an interaction between neuropsychological deficits and familial environment. Barnes (2012) tested for gene × environment interactions during adolescence and found no evidence of such an interaction.

Later work on the LCP showed that a subgroup detached from the childhood LCP in adolescence and became labeled as the recovery group (Moffitt *et al.*, 1996). With data covering more of the lifespan, a later article showed that those on the recovery trajectory were again engaging in illegal behaviors as adults (Moffitt *et al.*, 2002). This group of boys who showed intermittency had more internalizing problems than the LCP who were following a stable trajectory. They were later labeled as childhood-limited offenders (Moffitt, 2003, 2006).

Recentering on aggression and violence

Aggregating behaviors can create a lot of ambiguity when analyzing human development. Aggregating the frequency of heterogeneous behaviors such as theft, delinquency, or aggression in antisocial scales may not be well suited to identifying developmental trajectories. Nagin and Tremblay (1999) showed that chronic physical aggression during elementary school was one of the strongest factors associated with the frequency of serious and overt delinquent acts in adolescence, beyond hyperactive behaviors and opposition. When looking at heterotypic continuity, studies brought support to the heterotypic continuity principle by showing that children could manifest physical aggression in infancy and change to use violence throughout adolescence (Brame *et al.*, 2001). These results question the tendency of researchers to aggregate socially unacceptable behaviors, and that early and chronic physical aggression is a behavior that has a strong association with future delinquent and violent acts.

Behavioral manifestations of a trait may vary according to age, physiological maturity, and social contexts. An infant may have a tendency to grab and bite, while a child may punch and kick while bickering with another child. Thus some behaviors may be specific to a developmental period. Using aggregate scores may falsely suggest that a trajectory is declining while, in fact, the expression of the same underlying behavior is changing form. Very little empirical work uses appropriate measurement models that are specific to developmental periods of childhood and adolescence (Barker *et al.*, 2007; Lacourse *et al.*, 2010).

Moffit's LCP trajectory suggests it should be observable from infancy. Tremblay used a more specific measure of physical aggression, that is, onset between the ages of 2 and 4. Physical aggression is also not specific to humans; as such, an etiological definition that can establish the similarities of the behavior between animals and humans enables evolutionary theorizing. Thus Tremblay pioneered the measure of aggression in newborns without reference to intentionality by using mother-reports of three items: hits, bites, or kicks another child. Using this definition of physical aggression, Tremblay showed that aggression could be observed as early as 12 months (Tremblay *et al.*, 2004; Lacourse *et al.*, 2014). A small portion (14–17 percent) of children are already following a trajectory of frequent physical aggression in infancy and childhood (Tremblay *et al.*, 2004; Cote *et al.*, 2007).

Continuing the focus on physical aggression, Nagin and Tremblay (1999) identified clusters of trajectories of physical aggression between the ages of 6 and 15 using data from the Montreal Longitudinal and Experimental Study (MLES), a sample of 1037 French Caucasian boys from disadvantaged socio-economic neighborhoods of Montreal. They used annual teacher assessments of physical aggression made between the ages of 6 and 15 years. Researchers identified four groups of trajectories. The "low" group was a little more than 15 percent of the sampled population, and comprised children who never displayed any aggression from 6 to 15 years. The second group, "moderate desisters," were a little over 50 percent of the population. At age 6, children in this group had a modest level of physical aggression, but by 10 years old, the vast majority of them had ceased to engage in such behavior. A third group, "high desisters," represented about 30 percent of the population. At age 6, the subjects in this group showed a high level of physical aggression, but the frequency of aggressions decreased through adolescence and showed a relatively average level of physical aggression at age 15. Finally, the authors identified a small group labeled "chronic," who showed high and stable levels of physical aggression from 6 to 15 years. They also showed that physical aggression, beyond hyperactivity and opposition, had a specific predictive value on the frequency of serious and overt delinquent acts (Nagin and Tremblay, 1999). This was particularly marked in the chronic physical aggression group. This study was also replicated with remarkably similar conclusions in six international locations (from three different countries) where similar data had been collected (Broidy *et al.*, 2003).

This view of the origins of aggression led Tremblay to develop a different explanation for the LCP trajectory. He proposes that there is substantial variability in the rate at which alternatives to aggression are learned, and that the mechanism is strongly intergenerational and based on genetic and environmental factors (Tremblay *et al.*, 2005; Tremblay, 2010). The origin of the stability of interindividual differences throughout the lifespan can be explored through a genetically informed longitudinal study. Genetic and environmental influences on initial interindividual differences and on rate of change can inform on the stability of hypothesized causal effects. A recent study by Lacourse *et al.* (2014) showed that genetic factors accounted for a significant part of interindividual differences in the stability of aggression during infancy. Precisely, multiple different genetic factors accounted for interindividual differences in aggression. The results show partial support for three distinct mechanisms: genetic set-point, genetic maturation, and genetic innovation. Similar results were found during childhood for physical aggression, but also for reactive and proactive aggression (Paquin *et al.*, 2013; Lacourse *et al.*, 2014).

Tremblay's work on prevention has been highly influenced by Gerald Patterson's pioneering work at the Oregon Social Learning Center. Patterson's theoretical model focuses mainly on parenting practices and the processes that "train" the child to become an early onset offender. The causes of problem behaviors are more related to parent–child interactions than to early neurocognitive deficits of the child (Patterson *et al.*, 2000). Using a behaviorist perspective, Patterson and his team have very well described the patterns of positive and negative reinforcement that lead to an escalation in disruptive behaviors within the family, between siblings, and later with peers. Inconsistent parents tend to not reinforce positively the prosocial behaviors of their children; instead they reinforce negatively their coercive and disruptive behaviors (by withdrawing from conflictual situations). When these parenting strategies are used frequently, they teach their children to use coercive behaviors to obtain what they want. Subsequently, this social interaction style can lead to peers' rejection and to affiliation with disruptive or deviant peers as early as elementary school (Snyder *et al.*, 2005).

Usually, parenting practices are also associated with structural adversity, such as poverty, dependence on social security, or disadvantaged neighborhoods. This adversity would impact children's behavior through an increase in parents' stress levels. Several researchers have demonstrated that familial adversity contributes to almost every dimension of a suboptimal home environment, including poor parenting (Conger *et al.*, 1994). Recent studies have shown that inadequate parenting was mostly caused by children's characteristics (evocative rGE) (Marceau *et al.*, 2013), while another study found that both genetic (passive rGE) and environmental effects were simultaneously needed to account for a child's conduct disturbance. From the prevention perspective, several scholars (Lacourse *et al.*, 2002; Boisjoli *et al.*, 2007; Barker *et al.*, 2010) showed that an intervention aimed at improving children's social skills and parents' parenting skills could deflect their adolescent trajectories from a high level of aggression to the low-level trajectories. This strongly suggests that, although neuropsychological deficits are associated with high-level trajectories, other sources of influence, such as children's social skills or parents' parenting skills, do play an essential role in the developmental etiology of aggressive behavior. Other evidence shows that social relations with parents (Trentacosta *et al.*, 2011) or peers (Salvas *et al.*, 2011) can change the frequency of aggression.

Taking somewhat the same perspective as Patterson's in psychology, the sociologist Thornberry (2005) proposed an interactional theory of offending by specifying causes of onset, persistence, and desistence of offending throughout the lifespan. This theory draws ideas from the life-course perspective, namely that individuals are constantly produced, sustained, and transformed by their social context and by making use of the concept of connected lives by emphasizing intergenerational transmission of antisocial behavior. He also draws on the foundations of the developmental approach by emphasizing parameters other than the onset of trajectories. He also tries to explain the persistence of behaviors, as well as desistence from the trajectories. Individual characteristics, parenting styles, and position in the social structure are thus the important causal factors.

Empirical evidence supports the onset of violence at multiple developmental stages, as suggested by Thornberry's (2005) interactional model. One such study was done by Lacourse *et al.* (2008), who analyzed participation in serious criminal violence in a cohort of 500 young people aged 13–25 years who grew up in the city of Pittsburgh during the 1990s, where crime and violence were particularly prevalent. Four trajectories were identified: low (52 percent), decreasing moderate (28 percent), decreasing high (6 percent), and late initiation (14 percent). Moreover, usual predictors of violence during adolescence were not associated with the late-onset trajectory, suggesting that factors appearing during adolescence could explain the late-onset trajectory observed (Huesmann *et al.*, 2009).

Possible mechanisms for late onset may be informed by studies on peer affiliation. Many empirical studies show that affiliation with delinquent peers contributes to the frequency of antisocial behavior, including violence, regardless of the propensity or risk factors present in childhood. One such is a study by Lacourse *et al.* (2003) presenting developmental trajectories of delinquent group membership from age 11 to 17 years. Three trajectories were identified: a first group of nonparticipants in a delinquent peer group, comprising 74.4 percent of the sample; a second group (12.8 percent of the sample) of early participants with a high probability of being a member of a deviant peer group at 11 years but significantly reducing their participation starting at 14 years; and a last group of late-comer participants (12.8 percent of the sample) who have a null probability of being member of a deviant peer group at 11 years, but see their probability rise above the level of early participants at 15 years old, and maintain their high probability of being a member of a delinquent group. The authors showed that the frequency of violent behaviors increased when adolescents became participants in a delinquent group, even for short-term participation.

The function of the peer group is essential during childhood, but even more so in adolescence. Friends become increasingly significant in interpersonal exchanges, in self-definition, and in the empowerment process. The peer group is characterized by a high degree of social solidarity and allows adolescents to develop a sense of belonging to a social group that is different from their family. During this period, teenagers build their identity by taking some distance from their parents, and develop greater conformity with their peers. The peer group may therefore encourage the transgression of norms and facilitate the transition to crime, and this is especially marked when the relationship with parents is conflict.

Dynamic interactions between bio- and social causal factors

All theories suggest that the mechanisms driving pathways are complex, and imply factors from multiple levels of analysis. Potential causal factors of aggressive and violent behavior have been found at the genetic, neurological (neuroendocrine system, HPA axis), familial, and neighborhood levels. The study of mechanisms explaining either onset or desistence from trajectories of aggression or violence is bounded by the same limits as any other complex interactional and dynamic processes: multiple sources of confound may be present between correlates of aggressive and violent behaviors. An important source of confound is between genetic and environmental factors. It is possible that the association of aggression with a well established environmental factor (such as negative parenting or parents' aggressive behavior) could be a reflection of the association of genetic factors (which are shared by parents and children) with aggression. Another example of potential gene–environment correlation is if children's genetic propensity to aggression causes troublesome interactions with parents and peers, which in turn causes antisocial behavior (Lahey and Waldman, 2003; D'Onofrio *et al.*, 2011). To give more weight to the importance of gene–environment correlation, studies have shown that some environmental contexts are partially genetically determined. Kendler and Baker (2007) showed that genetic factors explained 28 percent of stressful life events, around 35 percent of parental warmth, and 21 percent of deviant peer affiliation. There is plenty of evidence that biological factors interact with environmental factors in either a multiplicative or moderating way (Raine, 2002). Those results suggest that the established associations of environmental factors with aggression are not necessarily purely environmental effects. To test that kind of complex mechanisms, researchers need to use genetically informative samples.

Most theories apply a diathesis-stress interpretation of the interactions between biological, psychological, and social factors associated with the onset and duration of trajectories of aggression and violence. Another interpretation of those interactions is the differential susceptibility hypothesis, which posits that some individuals are more susceptible than others to environmental influence. An implication of this perspective is that individuals most vulnerable to adverse social environments are the same people who reap the most benefit from environmental support (Belsky *et al.*, 2007; Belsky and Pluess, 2009; Simons and Burt, 2011).

Consonant with the differential susceptibility hypothesis, Simons and Burt (2011) presented data showing that social control processes deter involvement in deviant behavior by discouraging cognitive schemas and emotions that are promoted by adverse social circumstances. They suggest treating the three psychological factors proposed by strain theorists (anger and frustration from the general strain theory, hostile attribution bias, and concern with a toughness reputation) as one psychological factor labeled *hostile orientation* (Simons *et al.*, 2011). Thus social environment could affect violent behavior through mediation by cognitive factors.

Conclusion

The centuries-old debate about proponents of nature or nurture for explaining aggressive and violent behaviors is still ongoing, but with a less drastic separation of causal factors. The gold-standard for identification of causes is still the experimental design. An important contribution of contemporary twin studies lies in their ability to distinguish environmental influences with confounding factors of genetic origin (Rutter, 2006; Vitaro *et al.*, 2009; Johnson *et al.*, 2010; D'Onofrio *et al.*, 2011; Jaffee, 2011). Although these studies are still scarce, some already show that several risk factors associated with violence have a real environmental influence, such as divorce (Burt *et al.*, 2008), physical maltreatment (Jaffee *et al.*, 2004), and maternal depression (Kim-Cohen *et al.*, 2005). Conversely, it appears that the influence of parenting practices on antisocial behavior is explained mainly by heritable characteristics (Marceau *et al.*, 2013). Thus heritable characteristics of antisocial behavior would lead to negative parenting practices, not the other way around. Neighborhoods and other social circumstances such as poverty are also studied as factors influencing children's behavior through their influences on parents' parenting practices (Shaw and Shelleby, 2014).

The methodology most used to bring empirical support to the typological theories has received valid criticism. Skardhamar (2009, 2010) has shown in a simulation experiment that continuous heterogeneity, in combination with state dependence and random events, is sufficient to generate similar aggregate patterns that appear to be distinct trajectory groups when analyzed using a semi-parametric group-based model. Also, it is not at all clear if all the studies are really identifying the same high-risk group because they used a variety of outcome measures and risk factors across the lifespan. Studying developmental trajectories with genetically informed samples would allow a comparison of the importance of the numerous correlates of aggression and violence.

References

Barker, E. D., Seguin, J. R., Raskin White, H., Bates, M. E., Lacourse, E., Carbonneau, R., and Tremblay, R. E. (2007) 'Developmental trajectories of male physical violence and theft', *Archives of General Psychiatry*, 64: 592–599.

Barker, E. D., Vitaro, F., Lacourse, E., Fontaine, N. M., Carbonneau, R., and Tremblay, R. E. (2010) 'Testing the developmental distinctiveness of male proactive and reactive aggression with a nested longitudinal experimental intervention', *Aggressive Behavior*, 36(2): 127–140.

Barnes, J. C. (2012) 'Analyzing the origins of life-course-persistent offending: a consideration of environmental and genetic influences', *Criminal Justice and Behavior*, 40: 519–541.

Barnes, J. C., Beaver, K. M., and Boutwell, B. B. (2011) 'Examining the genetic underpinnings to Moffitt's developmental taxonomy: a behavioral genetic analysis', *Criminology*, 49(4): 923–954.

Belsky, J. and Pluess, M. (2009) 'Beyond diathesis stress: differential susceptibility to environmental influences', *Psychological Bulletin*, 135(6): 885–908.

Belsky, J., Bakermans-Kranenburg, M. J., and van IJzendoorn, M. H. (2007) 'For better and for worse: differential susceptibility to environmental influences', *Current Directions in Psychological Science*, 16(6): 300–304.

Boisjoli, R., Vitaro, F., Lacourse, E., Barker, E. D., and Tremblay, R. E. (2007) 'Impact and clinical significance of a preventive intervention for disruptive boys: 15-year follow-up' *British Journal of Psychiatry*, 191(5): 415–419.

Brame, B., Nagin, D. S., and Tremblay, R. E. (2001) 'Developmental trajectories of physical aggression from school entry to late adolescence', *Journal of Child Psychology and Psychiatry and Allied Disciplines*, 42(4): 503–512.

Broidy, L. M., Nagin, D. S., Tremblay, R. E., Bates, J. E., Brame, B., Dodge, K. A., *et al.* (2003) 'Developmental trajectories of childhood disruptive behaviors and adolescent delinquency: a six-site, cross-national study', *Developmental Psychology*, 39(2): 222–245.

Burt, S. A., Barnes, A. R., McGue, M., and Iacono, W. G. (2008) 'Parental divorce and adolescent delinquency: ruling out the impact of common genes', *Developmental Psychology*, 44(6): 1668–1677.

Conger, R. D., Ge, X., Elder, G. H., Lorenz, F. O., and Simons, R. L. (1994) 'Economic stress, coercive family process, and developmental problems of adolescents', *Child Development*, 65(2): 541–561.

Cote, S. M., Boivin, M., Nagin, D. S., Japel, C., Xu, Q., Zoccolillo, M., *et al.* (2007) 'The role of maternal education and nonmaternal care services in the prevention of children's physical aggression problems', *Archives of General Psychiatry*, 64(11): 1305–1312.

D'Onofrio, B., Rathouz, P. J., and Lahey, B. B. (2011) 'The importance of understanding gene–environment correlations in the development of antisocial behavior', in K. S. Kendler, S. R. Jaffee and D. Romer (eds), *The Dynamic Genome and Mental Health: The Role of Genes and Environments in Youth Development*, New York: Oxford University Press, pp. 340–364.

Granic, I. and Patterson, G. R. (2006) 'Toward a comprehensive model of antisocial development: a dynamic systems approach', *Psychological Review*, 113(1): 101–131.

Huesmann, L. R., Dubow, E. F., and Boxer, P. (2009) 'Continuity of aggression from childhood to early adulthood as a predictor of life outcomes: implications for the adolescent-limited and life-course-persistent models', *Aggressive Behavior*, 35(2): 36–149.

Jaffee, S. (2011) 'Genotype–environment correlations: definitions, methods of measurement, and implications for research on adolescent psychopathology', in K. S. Kendler, S. Jaffee, and D. Romer (eds), *The Dynamic Genome and Mental Health: The Role of Genes and Environments in Youth Development*, Oxford: Oxford University Press.

Jaffee, S. R., Caspi, A., Moffitt, T. E., and Taylor, A. (2004) 'Physical maltreatment victim to antisocial child: evidence of an environmentally mediated process', *Journal of Abnormal Psychology*, 113(1): 44–55.

Johnson, W., Turkheimer, E., Gottesman, I., and Bouchard, T. J., Jr (2010) 'Beyond heritability: twin studies in behavioral research', *Current Directions in Psychological Science*, 18(4): 217–220.

Kendler, K. S. and Baker, J. H. (2007) 'Genetic influences on measures of the environment: a systematic review', *Psychological Medicine*, 37(5): 615–626.

Kim-Cohen, J., Moffitt, T. E., Taylor, A., Pawlby, S. J., and Caspi, A. (2005) 'Maternal depression and children's antisocial behavior: nature and nurture effects', *Archives of General Psychiatry*, 62(2): 173–181.

Lacourse, E., Côté, S., Nagin, D., Vitaro, F., Brendgen, M., and Tremblay, R. E. (2002) 'A longitudinal-experimental approach to testing theories of antisocial behavior development', *Development and Psychopathology*, 14(4): 909–924.

Lacourse, E., Nagin, D., Tremblay, R. E., Vitaro, F., and Claes, M. (2003) 'Developmental trajectories of boys' delinquent group membership and facilitation of violent behaviors during adolescence', *Development and Psychopathology*, 15(1): 83–197.

Lacourse, E., Dupéré, V., and Loeber, R. (2008) 'Developmental trajectories of violence and theft', in R. Loeber, D. P. Farrington, M. Stouthamer-Loeber, and H. Raskin White (eds), *Violence and Serious Theft: Development and Prediction from Childhood to Adulthood*, New York: Taylor and Francis, pp. 231–268.

Lacourse, E., Baillargeon, R., Dupéré, V., Vitaro, F., Romano, E., and Tremblay, R. (2010) 'Two-year predictive validity of conduct disorder subtypes in early adolescence: a latent class analysis of a Canadian longitudinal sample', *Journal of Child Psychology and Psychiatry*, 51(12): 1386–1394.

Lacourse, E., Boivin, M., Brendgen, M., Petitclerc, A., Girard, A., Vitaro, F., *et al.* (2014) 'A longitudinal twin study of physical aggression during early childhood: evidence for a developmentally dynamic genome', *Psychological Medicine*, 44(12): 2617–2627.

Lahey, B. B. and Waldman, I. D. (2003) 'A developmental propensity model of the origins of conduct problems during childhood and adolescence', in B. B. Lahey, T. E. Moffitt and A. Caspi (eds), *Causes of Conduct Disorder and Juvenile Delinquency*, New York: Guilford Press, pp. 76–177.

Locke, J. (1689) *An Essay Concerning Human Understanding*, New York: Penguin Classics.

Lombroso, C. (1876) *L'uomo delinquente, (1876)*, Turin, Italy: Bocca.

Marceau, K., Horwitz, B. N., Narusyte, J., Ganiban, J. M., Spotts, E. L., Reiss, D., and Neiderhiser, J. M. (2013) 'Gene–environment correlation underlying the association between parental negativity and adolescent externalizing problems', *Child Development*, 84(6): 2031–2046.

Moffitt, T. E. (1993) 'Adolescence-limited and life-course-persistent antisocial behavior: a developmental taxonomy', *Psychological Review*, 100(4): 674–701.

Moffitt, T. E. (2003) 'Life-course-persistent and adolescence-limited antisocial behavior: a 10-year research review and a research agenda', in B. B. Lahey, T. E. Moffitt, and A. Caspi (eds), *Causes of Conduct Disorder and Juvenile Delinquency*, New York: Guilford Press, pp. 49–75.

Moffitt, T. E. (2006) 'Life-course-persistent versus adolescence-limited antisocial behavior', in D. Cicchetti and D. J. Cohen (eds), *Developmental Psychopathology, Vol 3: Risk, Disorder, and Adaptation*, 2nd edn, Hoboken, NJ: John Wiley and Sons, pp. 570–598.

Moffitt, T. E. and Caspi, A. (2001) 'Childhood predictors differentiate life-course persistent and adolescence-limited antisocial pathways among males and females', *Development and Psychopathology*, 13(2): 355–375.

Moffitt, T. E., Caspi, A., Dickson, N., Silva, P., and Stanton, W. (1996) 'Childhood-onset versus adolescent-onset antisocial conduct problems in males: natural history from ages 3 to 18 years', *Development and Psychopathology*, 8: 399–424.

Moffitt, T. E., Caspi, A., Harrington, H., and Milne, B. J. (2002) 'Males on the life-course-persistent and adolescence-limited antisocial pathways: follow-up at age 26 years', *Development and Psychopathology*, 14(1): 179–207.

Nagin, D. and Tremblay, R. E. (1999) 'Trajectories of boys' physical aggression, opposition, and hyperactivity on the path to physically violent and nonviolent juvenile delinquency', *Child Development*, 70(5): 1181–1196.

Paquin, S., Lacourse, E., Brendgen, M., Boivin, M., Dionne, G., Vitaro, F., and Tremblay, R. (2013) 'Development of physical aggression during childhood: a dynamic genome?', *Behavior Genetics*, 43: 535–545.

Patterson, G. R., Dishion, T. J., and Yoerger, K. (2000) 'Adolescent growth in new forms of problem behavior: macro- and micro-peer dynamics', *Prevention Science*, 1(1): 3–13.

Pinker, S. (2011) *The Better Angels of our Nature: Why Violence has Declined*, New York: Viking.

Quetelet, L. A. J. (1835) *Sur l'homme et le développement de ses facultés*, New York: Cambridge University Press.

Raine, A. (2002) 'Biosocial studies of antisocial and violent behavior in children and adults: a review', *Journal of Abnormal Child Psychology*, 30(4): 311–326.

Rhee, S. H. and Waldman, I. D. (2002) 'Genetic and environmental influences on antisocial behavior: a meta-analysis of twin and adoption studies', *Psychological Bulletin*, 128(3): 490–529.

Rutter, M. (2006). *Genes and Behavior: Nature/Nurture Interplay Explained*, Malden, MA: Blackwell.

Salvas, M. C., Vitaro, F., Brendgen, M., Lacourse, E., Boivin, M., and Tremblay, R. E. (2011) 'Interplay between friends' aggression and friendship quality in the development of child aggression during the early school years', *Social Development*, 20(4): 645–663.

Shaw, D. S. and Shelleby, E. C. (2014) 'Early-starting conduct problems: intersection of conduct problems and poverty', *Annual Review of Clinical Psychology*, 10: 503–528.

Silberg, J. L., Rutter, M., Tracy, K., Maes, H. H., and Eaves, L. (2007) 'Etiological heterogeneity in the development of antisocial behavior: the Virginia twin study of adolescent behavioral development and the young adult follow-up', *Psychological Medicine*, 37(8): 1193–1202.

Simons, R. L. and Burt, C. H. (2011) 'Learning to be bad: adverse social conditions, social schemas, and crime', *Criminology*, 49(2): 553–598.

Simons, R. L., Lei, M. K., Beach, S. R. H., Brody, G. H., Philibert, R. A., and Gibbons, F. X. (2011) 'Social environment, genes, and aggression: evidence supporting the differential susceptibility perspective', *American Sociological Review*, 76(6): 883–912.

Skardhamar, T. (2009) 'Reconsidering the theory on adolescent-limited and life-course persistent antisocial behaviour', *British Journal of Criminology*, 49(6): 863–878.

Skardhamar, T. (2010) 'Distinguishing facts and artifacts in group-based modeling', *Criminology*, 48(1): 295–320.

Snyder, J., Schrepferman, L., Oeser, J., Patterson, G., Stoolmiller, M., Johnson, K., and Snyder, A. (2005) 'Deviancy training and association with deviant peers in young children: occurrence and contribution to early-onset conduct problems', *Development and Psychopathology*, 17(2): 397–413.

Thornberry, T. P. (2005) 'Explaining multiple patterns of offending across the life course and across generations', *Annals of the American Academy of Political and Social Science*, 602: 156–195.

Tremblay, R. E. (2000) 'The development of aggressive behaviour during childhood: what have we learned in the past century?', *International Journal of Behavioral Development*, 24(2): 129–141.

Tremblay, R. E. (2010) 'Developmental origins of disruptive behaviour problems: the "original sin" hypothesis, epigenetics and their consequences for prevention', *Journal of Child Psychology and Psychiatry*, 51(4): 341–367.

Tremblay, R. E., Nagin, D. S., Seguin, J. R., Zoccolillo, M., Zelazo, P. D., Boivin, M., *et al.* (2004) 'Physical aggression during early childhood: trajectories and predictors', *Pediatrics*, 114(1): E43–E50.

Tremblay, R. E., Hartup, W. W., and Archer, J. (2005) *Developmental Origins of Aggression*, New York: Guilford Press.

Trentacosta, C. J., Criss, M. M., Shaw, D. S., Lacourse, E., Hyde, L. W., and Dishion, T. J. (2011) 'Antecedents and outcomes of joint trajectories of mother–son conflict and warmth during middle childhood and adolescence', *Child Development*, 82(5): 1676–1690.

Vitaro, F., Brendgen, M., and Arseneault, L. (2009) 'The discordant MZ-twin method: one step closer to the holy grail of causality', *International Journal of Behavioral Disorders*, 33(4): 376–382.

Wolfgang, M. E., Figlio, R. M., and Sellin, J. T. (1972) *Delinquency in a Birth Cohort*, Chicago, IL: University of Chicago Press.

27

Child maltreatment

An environmental pathogen?

Melissa Jonson-Reid and Charlotte Bright

Introduction

In 1962, Dr C. Henry Kempe established the modern foundation for child maltreatment research by naming the battered child syndrome, a medical condition resulting primarily from physical abuse (Kempe *et al.*, 1962). Most of the literature describing the battered child syndrome focused on physical abuse, with Kempe later calling for more attention to sexual abuse in a lecture in 1978 (Kempe, 1978). In the late 1970s we begin to see emerging discussion of emotional abuse (Garbarino, 1978). Also in the 1970s, maltreatment goes through a process of redefinition as a consequence of increasing attention to other underlying issues, such as lack of social support (Garbarino, 1977) or a mental health disorder on the part of the parent (Webster *et al.*, 1985). By 1984, researchers were increasingly calling for more emphasis on child neglect, coining the now common phrase "neglect of neglect" (Wolock and Horowitz, 1984). Since then, what we call maltreatment has expanded and contracted in response to research or public concern, ranging from a narrow focus on physical abuse to a broad focus including issues such as childhood obesity or exposure to intimate partner violence (Viner *et al.*, 2010; Trocmé *et al.*, 2013).

The theoretical connection between child maltreatment and later criminal behavior dates back to at least the early 1960s, based on the belief that victims would inevitably become abusers (Curtis, 1963). By the 1980s, however, the empirical literature raised doubts about the quality of evidence establishing this link (Garbarino and Gilliam, 1980; Widom, 1989b). Since that time, substantial research has explored the relation between maltreatment and later perpetration of criminal behavior, particularly acts of violence, establishing that maltreatment does increase the risk of offending, although the pathways may be complex (Falshaw, 2005). Is it therefore useful to equate maltreatment with a pathogen that follows a predictable course that leads to criminality?

The definition of a pathogen can be complex. Typically one thinks of a pathogen in a narrow medical frame like a particular germ leading to a particular disease. Pirofski and Casadevall (2012) argue that a given microbe can be pathogenic (cause disease) or non-pathogenic depending on contextual factors as well as the response of a given individual to that microbe. At a minimum, equating maltreatment with a pathogen would require the ability to accurately identify maltreatment regardless of whether or not it impacted a given individual. This analogy also requires that we accept criminal behavior as a type of disease with a clear causal chain leading back to the identified source. Finally, such an approach implies that by "treating" the source, one can prevent the disease. This chapter summarizes the research linking maltreatment to later criminal behavior, discusses the particular mechanisms by which researchers think maltreatment impacts the developmental course, and then reconsiders this idea of maltreatment as a pathogen in terms of its utility in informing the prevention of later delinquent and criminal behavior.

What is maltreatment?

Although local variations exist, definitions of child maltreatment in the United States typically include four elements: (1) an action or omission; (2) committed by a parent or other caregiver; (3) resulting in actual, potential, or perceived harm; (4) to a minor child (Leeb *et al.*, 2008, p. 11). Maltreatment therefore includes both abusive and neglectful behavior. In the United States, there is some consistency across states in the inclusion of emotional abuse, physical abuse, neglect, and sexual abuse as types of maltreatment (US DHHS, 2012). As mentioned in the opening paragraphs, however, the definition of maltreatment has evolved and continues to do so. For example, in 22 states, exposure to domestic violence that results in some sort of psychological or physical harm is maltreatment (Child Welfare Information Gateway, 2014). International definitions include an even broader set of conditions that interfere with a child's potential for optimum development. This variation in definitions poses the first dilemma in adopting the notion of maltreatment as pathogen.

Measurement issues

As stated in the introduction, in order to have a causal pathogenic chain one needs to be able to accurately diagnose or identify the source. The diagnosis of a given pathogen is key to selecting the appropriate treatment. Not only are there differing opinions about what should or should not be included in a definition of maltreatment, however, there is also significant variation in the measurement of maltreatment in studies of delinquent and violent behavior (see Maas *et al.*, 2008).

One longstanding issue in measurement is debate about the validity of official reports compared with self-report or scales. While studies suggest that maltreatment is under-reported (Sedlak *et al.*, 2010), and unreported cases may differ from reported cases (Walsh *et al.*, 2007; Flaherty *et al.*, 2008), studies using official reports of maltreatment do find differences when compared with similar but not reported populations (Widom, 1989b; Smith *et al.*, 2008; Jonson-Reid *et al.*, 2009). Recently there have been a few published studies comparing self-report with substantiated official reports, finding predictive utility for both forms of measurement (Shaffer *et al.*, 2008; Scott *et al.*, 2012). Smith and colleagues (2008) found that official records had stronger associations and combining the methods did not improve prediction. In contrast, Shaffer and colleagues (2008) found some additional predictive value in combining methods. Taken together, research suggests that while there are undoubtedly children who are maltreated and never noticed, their presence is not sufficient to dilute the ability to detect differences between those with reports and those without, at least in prospective studies with access to both kinds of measure over time.

Another interesting measurement issue in research specific to linking officially reported maltreatment to delinquency is the traditional reliance on substantiated cases. In essence, substantiation is a label given by child protective services that indicates that following an investigation of a report there is both harm (e.g. bruising) or serious risk of harm (e.g. a 2-year-old found wandering alone in the street) present, as well as evidence (similar to fingerprints documenting guilt in a case of a robbery) of harm present (Drake, 1996). Substantiation is a relatively rare outcome in many states, particularly those that use systems such as differential response to divert reported cases away from traditional investigations (US DHHS, 2012). Over the past decade, a significant body of research indicates that this marker used by child protection agencies is a poor predictor of outcomes (Drake *et al.*, 2003; Hussey *et al.*, 2005; Jonson-Reid *et al.*, 2009; Kohl *et al.*, 2009). Even when differences are found (e.g. Chiu *et al.*, 2011), they are not so strong as to negate the fact that unsubstantiated cases will comprise a practically significant portion of later outcomes like delinquency simply because they comprise the vast majority of reports of maltreatment. Why does this matter in terms of our understanding of maltreatment and crime? Many of the key studies related to this issue have drawn their samples based on substantiated cases. It is not clear whether those youth who were in the comparison group and became delinquent were "not maltreated," or simply unreported or unsubstantiated cases within that group.

Other researchers advocate survey or interview measures of maltreatment (Amaya-Jackson *et al.*, 2000; Straus, 2002). Scale measures of maltreatment ask for either direct self-report of prior experiences (e.g. Computer Assisted Maltreatment Inventory, CAMI; Childhood Trauma Questionnaire) or current practices (e.g. Conflict Tactics Scale), or measure what is called child abuse potential (e.g. Child Abuse Potential Inventory, CAPI) (Milner, 1986; Bernstein *et al.*, 1997; Straus, 2002; DiLillo *et al.*, 2010). While sometimes used as a proxy for maltreatment, measures assessing child abuse potential were really designed to predict possible future behavior rather than identify experiences of maltreatment, and are not discussed further here. Self-report measures include either parent report of behaviors, such as the much used Conflicts Tactics Scale, or self-report based on retrospective recall (Adverse Childhood Experiences Study, ACES—see Dube *et al.*, 2001; CAMI). Agreement between various measures is often modest at best (Widom *et al.*, 2004; Fergusson *et al.*, 2011). Studies using self-report or parent-report measures undoubtedly miss cases in which the parent or child is unwilling to report or cannot remember the incident. Further, some children may be unable to respond to self-report measures because of cognitive or communication disabilities. Despite potential weaknesses, however, both studies using official reports and self-report measures of maltreatment have detected practically significant differences in risk of later antisocial behavior (Smith, *et al.*, 2008; Fergusson *et al.*, 2011).

Clearly the issue of measurement makes it difficult to define a pathogen as cleanly as one might wish for diagnosis and treatment. On the other hand, it is interesting to note that, despite the weaknesses of the various measures employed, there is consistency in documentation of risk of later poor outcomes.

The association of maltreatment to crime

If we assume that criminal behavior represents a type of disease—an aberrant form of human development—then there must be a source of harm such as abuse or neglect. Linking some outcomes directly to maltreatment is relatively easy. For severe physical abuse, it is conceptually simple to relate subsequent physical harm (i.e. the battered child syndrome) to a specific cause (e.g. being beaten). Likewise, sexual abuse that involves physical penetration resulting in sexually transmitted disease, pregnancy, or physical injury can be traced to a given event or series of events perpetrated by an adult against a child. But this is a book about crime, not physical well-being.

There may be a substantial length of time between a given experience of maltreatment and later delinquent or criminal action, with significant opportunity for intervening events.

Despite variation in definitions and measurement of maltreatment, and the potential complication of elapsed time, the etiological literature linking maltreatment to delinquent or criminal behavior is quite substantial (McCord, 1983; Zingraff *et al.*, 1993; Maxfield and Widom, 1996; Jonson-Reid and Barth, 2000a,b; English *et al.*, 2001; Thornberry *et al.*, 2001; Caspi *et al.*, 2002; Bright and Jonson-Reid, 2008; Jonson-Reid *et al.*, 2009; Bender, 2010; Currie and Tekin, 2012). Several longitudinal studies using different measures of both maltreatment (discussed earlier) and crime (official records vs. self-report; arrest vs. adjudicated or incarcerated) have found practical and statistical differences in risk of later offending behavior between those deemed to have been abused or neglected and their comparisons (Maxfield and Widom, 1996; Maas *et al.*, 2008; Jonson-Reid *et al.*, 2009; Currie and Tekin, 2012). On the other hand, some studies using self-report of maltreatment have not found a strong link between maltreatment and later crime (Silva *et al.*, 2012). It is beyond the scope of the present chapter to provide a review of the methods and measures used in studies within this vast literature. Roughly, however, they can be divided into two groups: those that were initially designed to study the effects of maltreatment (e.g. Zingraff *et al.*, 1993; Widom, 1996; Jonson-Reid *et al.*, 2009: Silva *et al.*, 2012) and those that were studying development broadly and controlled for maltreatment (e.g. Thornberry *et al.*, 2001; Caspi *et al.*, 2002; Mersky *et al.*, 2012).

Is there a form of maltreatment that is particularly harmful?

Assuming, for a moment, that we accept maltreatment is a root cause of later criminal action, what do we do with the breadth of actions and inactions that may be included in this phenomenon? Maybe, just as a virus may have differing mutations, specific subtypes of maltreatment are the true pathogenic element. Some research has found particular forms of serious juvenile offending behavior to be more clearly associated with physical or sexual abuse (Jonson-Reid and Way, 2001; Currie and Tekin, 2012). Prospective examinations of delinquency following maltreatment have found little difference in maltreatment type and/or lower likelihood of official delinquent involvement for youth with sexual abuse histories compared with other forms (Jonson-Reid, 2004; English *et al.*, 2001; Bright and Jonson-Reid, 2008; Williams *et al.*, 2010). Indeed, there appears to be a growing consensus that neglect is at least equally likely to result in later criminal behavior (Chapple *et al.*, 2005; Yun *et al.*, 2011). There is some emerging evidence that neglect also impacts the risk of ongoing delinquent careers (Ryan *et al.*, 2013).

The cycle of violence (Widom, 1989a) or social learning theory was often used in earlier literature to explain the link between abuse and delinquency. What about neglect? The most common category of neglect is lack of supervision (US DHHS, 2012). If one assumes that neglect might impact the development of social control (Hollist *et al.*, 2009) as well as the increased likelihood for exposure to criminogenic experiences in the community (Coulton *et al.*, 1995; Nikulina *et al.*, 2011), this makes intuitive sense. Others have posited that neglect may impact later violent behavior through its effects on self-control or peer associations (Chapple *et al.*, 2005).

Developmental timing?

Argument exists regarding the relative impact of developmental timing on later criminal behavior. Currently the notion of toxic stress emphasizes early childhood as the key period for harm related to maltreatment (Shonkoff *et al.*, 2009). Researchers in this vein tend to emphasize the importance of early neural development and attachment. Others suggest that later maltreatment may be equal or more harmful in terms of externalizing behaviors (Stewart *et al.*, 2008; Thornberry *et al.*, 2010). Still other studies find that childhood and adolescent maltreatment are equally likely to lead to delinquent and criminal outcomes (Mersky *et al.*, 2012) or that cumulative or chronic maltreatment may be most predictive of untoward outcomes such as delinquency (Jonson-Reid *et al.*, 2012). Indeed for some forms of maltreatment, such as emotional abuse, chronic exposure is actually part of the definition (Drake and Jonson-Reid, in press). Similarly, studies that have measured cumulative stressors including, but not limited to, maltreatment find that more types of stress are associated with worse outcomes (e.g. Dube *et al.*, 2001). This literature is also consistent with the notion of allostatic load in biological studies, which suggests that there is a tipping point at which a person's body can no longer withstand exposure to multiple stressors and return to a healthy state. This idea has been posited in studies of health outcomes for maltreated children (Lanier *et al.*, 2010).

As is apparent in the above discussion, while there is a general consensus that abuse and neglect is a risk factor for aggressive and criminal behavior (Maxfield and Widom, 1996; Gilbert *et al.*, 2009; Mersky *et al.*, 2012), the mechanism by which maltreatment might cause this form of harm to an individual, and exact pathways from such experience to criminal actions, are still being explored. As it is, of course, unethical to randomly assign children to being maltreated, our understanding relies on theory, increasingly complex longitudinal models, and even perhaps animal models.

Caspi and Moffit (2006) suggest the need for animal models to better understand the role of genetics and environment in development because of the ability to experimentally manipulate the environment in ways not possible with humans. Veenema (2009) does a nice job of reviewing what we can learn from animal models of early stress (both neglect and harsh treatment) and the development of aggression. Johns and colleagues (2005) found an interaction between cocaine exposure, early neglecting maternal behavior, and subsequent behavior of offspring in rats. A review found that significant stress at later stages in development can also impact neurological structures (Lupien *et al.*, 2009). As multidisciplinary collaboration increases, it is likely that such reviews synthesizing findings from animal and human longitudinal studies will add to our understanding.

Overall, child abuse and neglect is a well established risk factor for delinquent and criminal behaviors. The relative magnitude of effect varies by the types of measure used and the study design. The following section explores further the context of maltreatment and what we understand about its role in the development of criminal behaviors.

A conceptual framework

Thus far, our examination of maltreatment as a pathogen has focused on the experience of abuse or neglect. The pathogenic impact of a given microbe may be dependent on the response of the individual (Pirofski and Casadevall, 2012). Similarly, the ecological model of criminal behavior would suggest that a given outcome is based on factors at multiple levels from the individual to the community (Nikulina *et al.*, 2011). This approach is similar to developmental psychopathology perspectives that examine interactions between risk and protective factors at multiple levels (Cicchetti and Toth, 2005). Researchers taking a developmental traumatology perspective also acknowledge the importance of the characteristics of the maltreatment experience at the individual and family levels (De Bellis, 2001).

Individual factors

An ecological framework for understanding the influence of maltreatment takes into account factors at the individual, family, community, and societal levels. It should be noted that different forms of the ecological model posit different kinds of interactions and address the passage of time in different ways. We focus here on the first three levels of influence (individual, family, and community) due to the difficulty of adequately addressing cultural and societal norms within the constraints of the present chapter.

All individuals are born with certain characteristics: sex, race, genetic code. Beginning with inherited characteristics, we find discussion of the interface of maltreatment with existing traits. A recent theory suggests that aspects of temperament (highly heritable) related to self-regulation may be key in the later development of antisocial behavior (DeLisi and Vaughn, 2014). When examining maltreatment and later poor outcomes, some research has suggested an interactive effect between a child's inherited characteristics and the experience of maltreatment (Caspi *et al.*, 2002). Other research has found the role of the inherited element to be specific to certain subpopulations (Widom and Brzustowicz, 2006; Åslund *et al.*, 2011) or has documented an additive rather than interactive effect between genetic risk and maltreatment and later antisocial behavior (Jonson-Reid *et al.*, 2010). Each of these studies, however, measured both maltreatment and the outcome of violent or antisocial behavior differently. Further, such studies are often unable to capture the multiple confounding and co-morbid factors typically found along with maltreatment that are discussed later in this chapter.

Other research has focused on the potential neurobiological impact of maltreatment. In other words, the experience of maltreatment alters what would have been an otherwise normative expectation of brain development. Twardosz and Lutzker (2010) provide an excellent review of this literature. Much of the work has focused on experiences in early childhood and the developing brain (see De Bellis, 2005). Recent studies of the adolescent brain, however, suggest that adolescence and puberty may bring a host of neural changes that are particularly sensitive to trauma (Johnson *et al.*, 2009; Marceau *et al.*, 2012).

There is empirical and theoretical evidence to suggest that the reaction to exposure to maltreatment may vary by gender (Bender, 2010). The relationships between maltreatment, chemical disruptions in the brain, and later psychological disorders (particularly depression and post-traumatic stress disorder) have been found to be stronger in females than in males (Shea *et al.*, 2004). Girls also display considerably higher rates of psychological symptoms and fear of re-abuse than boys (Titus *et al.*, 2003).

Gender-specific findings regarding the connection between maltreatment and offending behavior are mixed. Topitzes and colleagues (2011) found a significant relation between the two in male youth, but not female youth. This was consistent with another study of the onset of juvenile court involvement for delinquency using official reports of maltreatment (Bright and Jonson-Reid, 2008). On the other hand, maltreatment was a significant predictor of violent offending among females in two earlier studies (Widom and Maxfield, 2001; Wolfe *et al.*, 2001). A third study found an association between some types of maltreatment and violent delinquent offending, with no gender differences in the effect (Yun *et al.*, 2010). It is possible that these disparate results reflect timing issues; a longitudinal study of child behavior problems following maltreatment indicated that males had more short-term, proximal behavioral issues and females experienced problems more distally over time (Godinet *et al.*, 2014). In any case, additional research describing gender-specific outcomes following maltreatment is needed to resolve the question of whether gender differences exist in the population, or merely in some samples.

Understanding the role of race in later delinquent and criminal behavior is complicated and the literature related to race and maltreatment, and race and delinquency is vast. On the other hand, most of this literature relates to black or white children. We know comparatively little about maltreatment among other races and ethnicities, or within immigrant compared with non-immigrant populations (Zhai and Gao, 2009; Dettlaff and Johnson, 2011).

Given the above limitations, there remains debate about the role of race in relation to childhood maltreatment and later crime. In studies of low-income populations, some studies find that race is not significant once controlling for maltreatment histories (Lemmon, 2006). Other studies have found interactions between genetic inheritance and race in the study of maltreatment and behavior (Widom and Brzustowicz, 2006). Research on youth in foster care finds racial differences to be more important in models of delinquency and risk behaviors for males than females (Auslander *et al.*, 2002; Jonson-Reid, 2002).

The effect of race is often difficult to disentangle from the larger environment. Both maltreatment and crime are more common in high-poverty neighborhoods. Minority youth, particularly African American youth, are overrepresented among youth with official records of criminal behavior, and experience poverty at multiple levels. For example, Drake and Rank (2009) found that while white children from lower-income families did not commonly live in very low-resource, low-income communities, similarly poor black children were likely to live in such communities.

Each of the individual-level factors described above may well be similar to what a physician encounters in a given patient. While factors in the environment may serve to inhibit or exacerbate the disease process, the root cause can still be the microbe. A difficulty in equating maltreatment with such a pathogen model, however, is that maltreatment rarely occurs outside a context of other risks, some of which may be causes of maltreatment as well as crime.

Family dysfunction

Children who experience maltreatment are not only more likely to live in low-income families, they are also more likely to live in multi-problem families. These problems are often associated with offending behaviors as well as maltreatment (e.g. Bright *et al.*, 2014). For example, maternal mental health problems and intimate partner violence have been found to co-occur commonly with maltreatment (Kohl *et al.*, 2005, 2011; Trocmé *et al.*, 2013). A causal model that pinpoints the unique contribution of maltreatment to criminal behavior can be difficult to identify. Experiencing both maltreatment and exposure to intimate partner violence have been found to be separately related to later externalizing behaviors. Similarly, other undesirable caregiver behaviors, such as smoking during pregnancy, have been linked to heightened risk of both maltreatment and delinquency (Wu *et al.*, 2004). Perhaps maternal smoking is a unique pathogen, or perhaps it serves as a proxy for multiple risks.

Cicchetti and Toth (2005) describe maltreatment as a "pathogenic relational environment." From a disease perspective this raises an interesting question: what is the unique contribution of maltreatment in a context of broader dysfunction? Is a given adult's pre-existing mental health or substance-abuse disorder a causal factor for both poverty and maltreatment, which in turn enhances the pathogenic impact of the caregiver's problems? A recent study that examined the long-term impact of early childhood maltreatment in the context of family problems (e.g. poverty, maternal depression, maternal abuse history) found that higher-risk children without maltreatment reports had a higher level of later childhood behavior problems than lower-risk cases with maltreatment reports (MacKenzie *et al.*, 2011).

Poverty

Both criminal behavior and maltreatment are more common among low-income populations (Nikulina *et al.*, 2011; Drake and Jonson-Reid, 2014). Briefly, poverty is theorized to relate to maltreatment either through sociological or neighborhood-level mechanisms (Coulton *et al.*, 2007) that lead to a reduction of resources and increased stress on the family system. Alternatively, pre-existing individual factors may interact with family and environmental factors such as poverty (Thornberry *et al.*, 2014). Both poverty and maltreatment are linked to outcomes such as lower cognitive ability and poor school performance (Jonson-Reid *et al.*, 2010; Nikulina *et al.*, 2011; Duncan and Magnuson, 2012) that are in turn risk factors for offending (Piquero and White, 2003). Given the level of co-morbidity of poverty and maltreatment, it is difficult to disentangle the causal relationship between the two (Drake and Jonson-Reid, 2014), much less the continuing path from maltreatment to delinquent or criminal behavior.

Lower-income families are also more likely to live in disadvantaged neighborhoods, and disadvantaged neighborhoods have higher concentrations of risk factors such as exposure to violence in the community and poorly performing schools. These factors are linked to enhanced risk for criminal and violent actions for children growing up in such neighborhoods. However, even these influences may differ by individual factors such as gender (Kroneman *et al.*, 2004; Oberwittler, 2007). Residing in lower-income areas is also associated with exposure to environmental toxins such as lead, which in turn has been found to be associated with increases in later antisocial and delinquent behaviors (Miller, 2012).

Response to maltreatment

Perhaps it is not the maltreatment itself, but what happens in response, that leads to a higher or lower likelihood of criminal behavior. Relatively few studies have examined the intervening impact of services or intervention—with some evidence of a moderating role for certain groups (Widom, 1991; Jonson-Reid and Barth, 2000a; Jonson-Reid, 2002; Ryan and Testa, 2005; Lemmon, 2006; Grogan-Kaylor *et al.*, 2008; Bright and Jonson-Reid, 2010). Theoretically, services that impact recurrence of maltreatment or lead to services that can buffer the impact of maltreatment could reduce the likelihood of offending behavior (Jonson-Reid, 2004). In some studies, receipt of in-home services appeared to diminish some of the risk of later delinquent behaviors (Jonson-Reid, 2002; Lemmon, 2006; Grogan-Kaylor *et al.*, 2008). Of course, intervention following maltreatment could also have a deleterious effect if it involves exposure to new levels of risk. Early studies (Runyan and Gould, 1985; Widom, 1991), and more recent work using advanced statistical methods, suggest that once differences between youth who do not enter care and those who do are sufficiently controlled, youth who enter foster care are no more or less likely to have later problem behaviors (Lee, 2008; Berger *et al.*, 2009). On the other hand, some studies suggest that instability during care (considered a negative outcome for child welfare) may enhance risk of juvenile offending (Jonson-Reid and Barth, 2000b; Ryan and Testa, 2005). Put another way, when child welfare services can successfully address risk factors one might expect a buffering effect, but when services are ineffective or poorly delivered the opposite outcome might obtain.

Further, there is evidence that the context within which services are received may impact outcomes. One study found that services to neglected children who lived in concentrated poverty were less effective at reducing risk of recurrence than those provided to low-income children not living in a context of concentrated poverty (Jonson-Reid *et al.*, 2009). This makes intuitive sense when one considers the potential disparity in availability and quality of service systems available according to region.

Exposure and resilience

While many studies have found a link between maltreatment and criminal behavior, not all children exposed to abuse or neglect have untoward outcomes. Maxfield and Widom (1996) found that childhood experience of victimization was a significant predictor of adolescent and adult justice system involvement, as well as violence. Even so, approximately one-half of their sample had none of these problematic outcomes through adolescence, and a third continued to appear resilient in adulthood (DuMont *et al.*, 2007). Other studies have come to similar conclusions (e.g. Jonson-Reid *et al.*, 2009). McGloin and Widom (2001) determined that victims of maltreatment had approximately half the rate of resilience in adulthood as a matched control group with no reports of maltreatment. The question of why some youth are resilient, and even thrive despite adversity, is not well understood. DuMont and colleagues (2007) found several factors at the individual, family, and community levels to be important in differentiating between those with poor outcomes and those that appeared resilient. This area of research, however, is underdeveloped and lacks consistency and rigor (Cicchetti, 2013; Klika and Herrenkohl, 2013).

Regardless of the particular study or measure, it is clear that abuse or neglect does not inevitably lead to delinquent or criminal behavior. Indeed, if it did we would likely be overwhelmed with the increase in crime, given the millions of children who experience maltreatment each year. Of course, we know that individuals can be exposed to a given virus or bacterium and not develop the disease. Thus not having a universal connection does not rule out the pathogen argument.

Is crime a disease?

So far we have wrestled with defining maltreatment as a pathogen, but it is also important to consider whether criminal behavior is a disease. Although criminal behavior is not an illness in the traditional sense, it does share some characteristics with a "disease." Research has documented that, especially among children and adolescents, aggressive behavior is transmissible—a phenomenon sometimes known as "peer contagion" (Dishion and Dodge, 2005). "Deviancy training," in particular, has been found to be a mechanism through which youth displaying problem behavior influence one another's development of similar behavior, via reinforcement (Dishion and Tipsord, 2011). On the other hand, like many diseases, criminal behavior can also be disrupted. Interventions to reduce offending are most likely to be effective if they target multiple risk factors, such as economic and social disadvantage, mental health, and substance-use issues, and individual-level risk factors such as child maltreatment and other traumatic experiences (Epperson *et al.*, 2011).

Conversely, there are problems with viewing criminal behavior as a disease. Rarely is a medical condition such as cancer viewed as a choice, even when exposure to a risk factor such as smoking may have been. In contrast, delinquent or criminal acts are often viewed as behavioral choices rather than the outcome of a disease process. This view is supported by the idea that some level of rule breaking may be normative during adolescence. The choice to continue with certain behaviors or to escalate into more serious crimes may be impacted by a number of factors, including differential processing through formal systems (Ryan *et al.*, 2007), and related to other developmental needs such as belonging with peers. Further, since children in non-maltreated comparison groups also engage in delinquent behaviors (Widom and Brzustowicz, 2006; Jonson-Reid *et al.*, 2009), this means that criminal behavior at a minimum cannot be considered a disease with a discrete cause such as maltreatment.

Is maltreatment a pathogen?

There is certainly ample evidence that exposure to maltreatment increases the risk of a host of poor outcomes—one of which is criminal behavior. There is not, however, evidence that would allow us to pinpoint a specific aspect of abuse or neglect that leads to a particular disease process resulting in delinquent or criminal behavior. In many cases it remains unclear whether the maltreatment is the point of origin of risk, or is an intermediary factor. Nor is there sufficient agreement around the characteristics of maltreatment that would lend itself to a clear diagnosis that could be used across systems and apply to all age ranges, as one might expect with a biological pathogen such as the virus leading to polio.

How does a pathogen model help or hinder our response?

The answer to this question depends largely on whether one uses such a model to promote a public health approach compared with a narrow medical treatment model. For example, it is possible that viewing maltreatment as a pathogen could call attention to the damaging impact of maltreatment and leverage increased support for intervention. Few children and families reported for maltreatment receive services (Jonson-Reid, 2011). It seems intuitive that increasing both service coverage and use of evidence-based models such as Safecare (Chaffin *et al.*, 2012) may reduce recurrence and offset the risk of later offending behavior. Infusing a trauma treatment is another recommended response to maltreatment at the child level that could yield positive gains in functioning and decreases in symptoms without narrowly focusing on externalizing behaviors (Hodgdon *et al.*, 2013). On the other hand, this may make less sense for very young children. Children under the age of five years may reap benefits from quality preschool or home visitation services which, while not having a powerful impact as yet on maltreatment (Mikton and Butchart, 2009), have been found to decrease delinquency (Reynolds *et al.*, 2004; Farrington and Welsh, 2003, 2007). It is not clear that the same results would be found for children exposed to abuse or neglect, but neither is there evidence to suggest this would be harmful.

A pathogen concept could also be used to leverage primary prevention. This perspective would promote research focused on a greater understanding of epidemiology, vulnerability, and effective prevention (O'Donnell *et al.*, 2008). As yet, primary prevention of maltreatment remains a preferred but elusive goal (Mikton and Butchart, 2009; Prinz *et al.*, 2009). Much work remains to be done in identifying effective and feasible approaches that can be delivered on a large scale. There is, however, some evidence that primary or universal delinquency prevention programs have the potential to reduce offending behavior (Nieri *et al.*, 2013). At least one study of a universal school-based approach found a stronger effect among youth with a history of maltreatment (Crooks *et al.*, 2011).

Conclusion

To the extent that conceptualizing maltreatment as a pathogen leads to a narrow medical model for addressing this issue, we would argue against the utility of such a definition. While there is certainly support for the idea of abuse and neglect acting in a similar way to how a microbe causes alterations that can lead to disease, we lack the prescriptive nature of evidence that a physician would require to accurately diagnose and prescribe treatment. There remains much disagreement about the best means to define maltreatment, and much variation in the potential pathogenic impact of this experience. This makes it nearly impossible to match a treatment to a disease process. Future research, policy, and practice might be best served by viewing child maltreatment and offending behaviors as two related public health issues requiring a multi-faceted approach that places equal emphasis on primary prevention and, when that fails, on identification of vulnerable children and families matched to an array of well supported interventions.

References

Amaya-Jackson, L., Socolar, R.R., Hunter, W., Runyan, D.K., and Colindres, R. (2000) 'Directly questioning children and adolescents about maltreatment: a review of survey measures used', *Journal of Interpersonal Violence*, 15(7): 725–759.

Åslund, C., Nordquist, N., Comasco, E., Leppert, J., Oreland, L., and Nilsson, K.W. (2011) 'Maltreatment, MAOA, and delinquency: sex differences in gene–environment interaction in a large population-based cohort of adolescents', *Behavior Genetics*, 41(2): 262–272.

Auslander, W., McMillen, C., Elze, D., Thompson, R., Jonson-Reid, M., and Stiffman, A. (2002) 'Mental health problems and sexual abuse among adolescents in foster care: relationship to HIV risk behaviors and intentions', *AIDS and Behavior*, 6: 351–359.

Bender, K. (2010) 'Why do some maltreated youth become juvenile offenders? A call for further investigation and adaptation of youth services', *Children and Youth Services Review*, 32(3): 466–473.

Berger, L.M., Bruch, S.K., Johnson, E.I., James, S., and Rubin, D. (2009) 'Estimating the "impact" of out-of-home placement on child well-being: approaching the problem of selection bias', *Child Development*, 80(6): 1856–1876.

Bernstein, D.P., Ahluvalia, T., Pogge, D., and Handelsman, L. (1997) 'Validity of the Childhood Trauma Questionnaire in an adolescent psychiatric population', *Journal of the American Academy of Child and Adolescent Psychiatry*, 36: 340–348.

Bright, C. and Jonson-Reid, M. (2008) 'Onset of juvenile court involvement: exploring gender-specific influences of maltreatment and poverty', *Children and Youth Services Review*, 30: 914–927.

Bright, C.L. and Jonson-Reid, M. (2010) 'Young adult outcomes of juvenile court-involved girls', *Journal of Social Service Research*, 36(2): 94–106.

Bright, C., Kohl, P., and Jonson-Reid, M. (2014) 'Females in the juvenile justice system: who are they and how do they fare?', *Crime and Delinquency*, 60: 106–125.

Caspi, A. and Moffitt, T.E. (2006) 'Gene–environment interactions in psychiatry: joining forces with neuroscience', *Nature Reviews Neuroscience*, 7(7): 583–590.

Caspi, A., McClay, J., Moffitt, T.E., Mill, J., Martin, J., Craig, I.W., *et al.* (2002) 'Role of genotype in the cycle of violence in maltreated children', *Science*, 297(5582): 851–854.

Chaffin, M., Hecht, D., Bard, D., Silovsky, J.F., and Beasley, W.H. (2012) 'A statewide trial of the SafeCare home-based services model with parents in child protective services', *Pediatrics*, 129(3): 509–515.

Chapple, C.L., Tyler, K.A., and Bersani, B.E. (2005) 'Child neglect and adolescent violence: examining the effects of self-control and peer rejection', *Violence and Victims*, 20(1): 39–53.

Child Welfare Information Gateway (2014) State Statutes Search. https://www.childwelfare.gov/systemwide/laws_policies/state

Chiu, Y.L., Ryan, J.P., and Herz, D.C. (2011) 'Allegations of maltreatment and delinquency: does risk of juvenile arrest vary substantiation status?', *Children and Youth Services Review*, 33(6): 855–860.

Cicchetti, D. (2013) 'Annual research review: resilient functioning in maltreated children – past, present, and future perspectives', *Journal of Child Psychology and Psychiatry*, 54(4): 402–422.

Cicchetti, D. and Toth, S.L. (2005) 'Child maltreatment', *Annual Review of Clinical Psychology*, 1: 409–438.

Coulton, C.J., Korbin, J.E., Su, M., and Chow, J. (1995) 'Community level factors and child maltreatment rates', *Child Development*, 66(5): 1262–1276.

Coulton, C., Crampton, D., Irwin, M., Spilbury, J., and Korbin, J. (2007) 'How neighborhoods influence child maltreatment: a review of the literature and alternative pathways', *Child Abuse and Neglect*, 31: 1117–1142.

Crooks, C.V., Scott, K., Ellis, W., and Wolfe, D.A. (2011) 'Impact of a universal school-based violence prevention program on violent delinquency: distinctive benefits for youth with maltreatment histories', *Child Abuse and Neglect*, 35(6): 393–400.

Currie, J. and Tekin, E. (2012) 'Understanding the cycle: childhood maltreatment and future crime', *Journal of Human Resources*, 47(2): 509–549.

Curtis, G.C. (1963) 'Violence breeds violence – perhaps?', *American Journal of Psychiatry*, 120(4): 386–387.

De Bellis, M.D. (2001) 'Developmental traumatology: the psychobiological development of maltreated children and its implications for research, treatment, and policy', *Development and Psychopathology*, 13(3): 539–564.

De Bellis, M.D. (2005) 'The psychobiology of neglect', *Child Maltreatment*, 10(2): 150–172.

DeLisi, M. and Vaughn, M.G. (2014) 'Foundation for a temperament-based theory of antisocial behavior and criminal justice system involvement', *Journal of Criminal Justice*, 42(1): 10–25.

Dettlaff, A.J. and Johnson, M. (2011) 'Child maltreatment dynamics among immigrant and US born Latino children: findings from the National Survey of Child and Adolescent Well-being (NSCAW)', *Children and Youth Services Review*, 33: 936–944.

DiLillo, D., Hayes-Skelton, S.A., Fortier, M.A., Perry, A.R., Evans, S.E., Messman Moore, T.L., *et al.* (2010) 'Development and initial psychometric properties of the Computer Assisted Maltreatment Inventory (CAMI): a comprehensive self-report measure of child maltreatment history', *Child Abuse and Neglect*, 34(5): 305–317.

Dishion, T.J. and Dodge, K.A. (2005) 'Peer contagion in interventions for children and adolescents: moving towards an understanding of the ecology and dynamics of change', *Journal of Abnormal Child Psychology*, 33(3): 395–400.

Dishion, T.J. and Tipsord, J.M. (2011) 'Peer contagion in child and adolescent social and emotional development', *Annual Review of Psychology*, 62: 189–214.

Drake, B. (1996) 'Unraveling unsubstantiated', *Child Maltreatment*, 1(3): 261–271.

Drake, B. and Jonson-Reid, M. (2014) 'Poverty and child maltreatment', in *Handbook of Child Maltreatment*, Dordrecht: Springer, pp. 131–148.

Drake, B. and Jonson-Reid, M. (in press) 'Competing values and evidence: how do we evaluate mandated reporting and CPS response?', in B. Mathews and D. Bross (eds), *Mandatory Reporting Laws and the Identification of Severe Child Abuse and Neglect*, Dordrecht: Springer.

Drake, B. and Rank, M.R. (2009) 'The racial divide among American children in poverty: reassessing the importance of neighborhood', *Children and Youth Services Review*, 31(12): 1264–1271.

Drake, B., Jonson-Reid, M., Way, I., and Chung, S. (2003) 'Substantiation and recidivism', *Child Maltreatment*, 8(4): 248–260.

Dube, S.R., Anda, R.F., Felitti, V.J., Chapman, D.P., Williamson, D.F., and Giles, W.F. (2001) 'Childhood abuse, household dysfunction, and the risk of attempted suicide throughout the life span: findings from the adverse childhood experiences study', *JAMA*, 286: 3089–3096.

DuMont, K.A., Widom, C.S., and Czaja, S.J. (2007) 'Predictors of resilience in abused and neglected children grown-up: the role of individual and neighborhood characteristics', *Child Abuse and Neglect*, 31(3): 255–274.

Duncan, G.J. and Magnuson, K. (2012) 'Socioeconomic status and cognitive functioning: moving from correlation to causation', *Wiley Interdisciplinary Reviews: Cognitive Science*, 3(3): 377–386.

English, D.J., Widom, C.S., and Brandford, C. (2001) *Childhood Victimization and Delinquency, Adult Criminality, and Violent Criminal Behavior: A Replication and Extension: Final Report*, Washington, DC: National Institute of Justice.

Epperson, M., Wolff, N., Morgan, R., Fisher, W., Frueh, B.C., and Huening, J. (2011) *The Next Generation of Behavioral Health and Criminal Justice Interventions: Improving Outcomes by Improving Interventions*, New Brunswick, NJ: Center for Behavioral Health Services and Criminal Justice Research, Institute for Health, Health Care Policy, and Aging Research.

Falshaw, L. (2005) 'The link between a history of maltreatment and subsequent offending behaviour', *Probation Journal*, 52(4): 423–434.

Farrington, D.P. and Welsh, B.C. (2003) 'Family-based prevention of offending: a meta-analysis', *Australian and New Zealand Journal of Criminology*, 36: 127–151.

Farrington, D.P. and Welsh, B.C. (2007) *Saving Children from a Life of Crime: Early Risk Factors and Effective Interventions*, New York: Oxford University Press.

Fergusson, D.M., Horwood, L.J., and Boden, J.M. (2011) 'Structural equation modeling of repeated retrospective reports of childhood maltreatment', *International Journal of Methods in Psychiatric Research*, 20(2): 93–104.

Flaherty, E.G., Sege, R.D., Griffith, J., Price, L.L., Wasserman, R., Slora, E., *et al.* (2008) 'From suspicion of physical child abuse to reporting: primary care clinician decision-making', *Pediatrics*, 122(3): 611–619.

Garbarino, J. (1977) 'The human ecology of child maltreatment: a conceptual model for research', *Journal of Marriage and Family*, 39(4): 721–735.

Garbarino, J. (1978) 'The elusive "crime" of emotional abuse', *Child Abuse and Neglect*, 2(2): 89–99.

Garbarino, J. and Gilliam, G. (1980) *Understanding Abusive Families*, Lexington, MA: Lexington Books.

Gilbert, R., Widom, C.S., Browne, K., Fergusson, D., Webb, E., and Janson, S. (2009) 'Burden and consequences of child maltreatment in high-income countries', *The Lancet*, 373(9657): 68–81.

Godinet, M.T., Li, F., and Berg, T. (2014) 'Early childhood maltreatment and trajectories of behavioral problems: exploring gender and racial differences', *Child Abuse and Neglect*, 38(3): 544–556.

Grogan-Kaylor, A., Ruffolo, M.C., Ortega, R.M., and Clarke, J. (2008) 'Behaviors of youth involved in the child welfare system', *Child Abuse and Neglect*, 32(1): 35–49.

Hodgdon, H.B., Kinniburgh, K., Gabowitz, D., Blaustein, M.E., and Spinazzola, J. (2013) 'Development and implementation of trauma-informed programming in youth residential treatment centers using the ARC framework', *Journal of Family Violence*, 28(7): 679–692.

Hollist, D.R., Hughes, L.A., and Schailbe, L.M. (2009) 'Adolescent maltreatment, negative emotion, and delinquency: an assessment of general strain theory and family-based strain', *Journal of Criminal Justice*, 37(4): 379–387.

Hussey, J., Marshal, J., English, D., Knight, E., Lau, A., Dubowitz, H., *et al.* (2005) 'Defining maltreatment according to substantiation: distinction without a difference?', *Child Abuse and Neglect*, 29: 479–492.

Johns, J.M., Elliott, D.L., Hofler, V.E., Joyner, P.W., McMurray, M.S., Jarrett, T.M., *et al.* (2005) 'Cocaine treatment and prenatal environment interact to disrupt intergenerational maternal behavior in rats', *Behavioral Neuroscience*, 119(6): 1605–08.

Johnson, S., Blum, R., and Giedd, J. (2009) 'Adolescent maturity and the brain: the promise and pitfalls of neuroscience research in adolescent health policy', *Journal of Adolescent Health*, 45(3): 216–221.

Jonson-Reid, M. (2002) 'Exploring the relationship between child welfare intervention and juvenile corrections involvement', *American Journal of Orthopsychiatry*, 72: 559–576.

Jonson-Reid, M. (2004) 'Child welfare services and delinquency: the need to know more', *Child Welfare*, 83(2): 157–174.

Jonson-Reid, M. (2011) 'Disentangling system contact and services: a key pathway to evidence-based children's policy', *Children and Youth Services Review*, 33: 598–604.

Jonson-Reid, M. and Barth, R. (2000a) 'From maltreatment to juvenile incarceration: uncovering the role of child welfare services', *Child Abuse and Neglect*, 24(4): 505–520.

Jonson-Reid, M. and Barth, R. (2000b) 'From placement to prison: the path to adolescent incarceration from child welfare supervised foster or group care', *Children and Youth Services Review*, 22(7): 493–516.

Jonson-Reid, M. and Way, I. (2001) 'Adolescent sexual offenders compared to other incarcerated delinquents: childhood maltreatment, serious emotional disturbance, and offending histories', *American Journal of Orthopsychiatry*, 71(1): 1–11.

Jonson-Reid, M., Drake, B., and Kohl, P. (2009) 'Is the overrepresentation of the poor in child welfare caseloads due to bias or need?', *Children and Youth Services Review*, 31: 422–427.

Jonson-Reid, M., Presnall, N., Drake, B., Fox, L., Bierut, L., Todd, R., Kane, P.B.A., and Constantino, J. (2010) 'Joint effects of child maltreatment and inherited liability on antisocial development: an official records study', *Journal of American Academy of Child and Adolescent Psychiatry*, 49(4): 321–332.

Jonson-Reid, M., Kohl, P. L., and Drak, B. (2012) 'Child and adult outcomes of chronic child maltreatment', *Pediatrics*, 129(5): 839–845.

Kempe, C.H. (1978) 'Sexual abuse, another hidden pediatric problem: the 1977 C. Anderson Aldrich Lecture', *Pediatrics*, 62(3): 382–389.

Kempe, C.H., Silverman, F.N., Steele, B.F., Droegemueller, W., and Silver, H.K. (1962) 'The battered-child syndrome', *JAMA*, 181: 17–24.

Klika, J.B. and Herrenkohl, T.I. (2013) 'A review of developmental research on resilience in maltreated children', *Trauma, Violence, and Abuse*, 14(3): 222–234.

Kohl, P.L., Edleson, J.L., English, D.J., and Barth, R.P. (2005) 'Domestic violence and pathways into child welfare services: findings from the National Survey of Child and Adolescent Well-Being', *Children and Youth Services Review*, 27: 1167–1182.

Kohl, P., Jonson-Reid, M., and Drake, B. (2009) 'Time to leave substantiation behind', *Child Maltreatment*, 14(1): 17–26.

Kohl, P., Jonson-Reid, M., and Drake, B. (2011) 'The role of parental mental illness in the safety and stability of maltreated children', *Child Abuse and Neglect*, 35(5): 309–318.

Kroneman, L., Loeber, R., and Hipwell, A.E. (2004) 'Is neighborhood context differently related to externalizing problems and delinquency for girls compared with boys?', *Clinical Child and Family Psychology Review*, 7(2): 109–122.

Lanier, P., Jonson-Reid, M., Stahlschmidt, M.J., Drake, B., and Constantino, J. (2010) 'Child maltreatment and pediatric health outcomes: a longitudinal study of low-income children', *Journal of Pediatric Psychology*, 35(5): 511–522.

Lee, S. (2008) 'The role of foster care placement experience in later problem behavior', DPhil dissertation, Washington University in St Louis, MO.

Leeb, R.T., Paulozzi, L., Melanson, C., Simon, T., and Arieas, I. (2008) *Child Maltreatment Surveillance: Uniform Definitions for Public Health and Recommended Data Elements, version 1.0*, Atlanta, GA: Centers for Disease Control and Prevention, National Center for Injury Prevention and Control.

Lemmon, J.H. (2006) 'The effects of maltreatment recurrence and child welfare services on dimensions of delinquency', *Criminal Justice Review*, 31(1): 5–32.

Lupien, S.J., McEwen, B.S., Gunnar, M.R., and Heim, C. (2009) 'Effects of stress throughout the lifespan on the brain, behaviour and cognition', *Nature Reviews Neuroscience*, 10(6): 434–445.

Maas, C., Herrenkohl, T.I., and Sousa, C. (2008) 'Review of research on child maltreatment and violence in youth', *Trauma, Violence, and Abuse*, 9(1): 56–67.

McCord, J. (1983) 'A forty year perspective on effects of child abuse and neglect', *Child Abuse and Neglect*, 7: 265–270.

McGloin, J., and Widom, C.S. (2001) 'Resilience among abused and neglected children grown up', *Development and Psychopathology*, 13: 1021–1038.

MacKenzie, M.J., Kotch, J.B., Lee, L.C., Augsberger, A., and Hutto, N. (2011) 'A cumulative ecological-transactional risk model of child maltreatment and behavioral outcomes: reconceptualizing early maltreatment report as risk factor', *Children and Youth Services Review*, 33(11): 2392–2398.

Marceau, K., Dorn, L.D., and Susman, E.J. (2012) 'Stress and puberty-related hormone reactivity, negative emotionality, and parent–adolescent relationships', *Psychoneuroendocrinology*, 37(8): 1286–1298.

Maxfield, M.G. and Widom, C.S. (1996) 'The cycle of violence revisited 6 years later', *Archives of Pediatrics and Adolescent Medicine*, 150: 390–395.

Mersky, J.P., Topitzes, J., and Reynolds, A.J. (2012) 'Unsafe at any age linking childhood and adolescent maltreatment to delinquency and crime', *Journal of Research in Crime and Delinquency*, 49(2): 295–318.

Mikton, C. and Butchart, A. (2009) 'Child maltreatment prevention: a systematic review of reviews', *Bulletin of the World Health Organization*, 87(5): 353–361.

Miller, S.M. (2012) 'Lead poisoning: the epidemic hitting the US juvenile justice system', *International Journal of Liability and Scientific Enquiry*, 5(3): 213–220.

Milner, J.S. (1986) *The Child Abuse Potential Inventory Manual*, 2nd edn, DeKalb, IL: Psytec.

Nieri, T., Matjasko, J.L., Williams, K.R., and Guerra, N.G. (2013) 'Youth violence prevention: public health intervention and high risk populations', in B. Sanders, Y.F. Thomas, and B.G. Deeds (eds), *Crime, HIV and Health: Intersections of Criminal Justice and Public Health Concerns*, Dordrecht: Springer, pp. 225–237.

Nikulina, V., Widom, C.S., and Czaja, S. (2011) 'The role of childhood neglect and childhood poverty in predicting mental health, academic achievement and crime in adulthood', *American Journal of Community Psychology*, 48(3/4): 309–321.

Oberwittler, D. (2007) 'The effects of neighborhood poverty on adolescent problem behaviors: a multi-level analysis differentiated by gender and ethnicity', *Housing Studies*, 22(5): 781–803.

O'Donnell, M., Scott, D., and Stanley, F. (2008) 'Child abuse and neglect – is it time for a public health approach?', *Australian and New Zealand Journal of Public Health*, 32(4): 325–330.

Piquero, A.R. and White, N.A. (2003) 'On the relationship between cognitive abilities and life-course-persistent offending among a sample of African Americans: a longitudinal test of Moffitt's hypothesis', *Journal of Criminal Justice*, 31(5): 399–409.

Pirofski, L.A. and Casadevall, A. (2012) 'Q and A: what is a pathogen? A question that begs the point', *BMC Biology*, 10: 6.

Prinz, R.J., Sanders, M.R., Shapiro, C.J., Whitaker, D.J., and Lutzker, J.R. (2009) 'Population-based prevention of child maltreatment: the US Triple P system population trial', *Prevention Science*, 10(1): 1–12.

Reynolds, A.J., Ou, S.R., and Topitzes, J.W. (2004) 'Paths of effects of early childhood intervention on educational attainment and delinquency: a confirmatory analysis of the Chicago Child–Parent Centers', *Child Development*, 75(5): 1299–1328.

Runyan, D.K. and Gould, C.L. (1985) 'Foster care for child maltreatment: impact on delinquent behavior', *Pediatrics*, 75(3): 562–568.

Ryan, J.P. and Testa, M.F. (2005) 'Child maltreatment and juvenile delinquency: investigating the role of placement and placement instability', *Children and Youth Services Review*, 27(3): 227–249.

Ryan, J.P., Herz, D., Hernandez, P.M., and Marshall, J.M. (2007) 'Maltreatment and delinquency: investigating child welfare bias in juvenile justice processing', *Children and Youth Services Review*, 29(8): 1035–1050.

Ryan, J.P., Williams, A.B., and Courtney, M.E. (2013) 'Adolescent neglect, juvenile delinquency and the risk of recidivism', *Journal of Youth and Adolescence*, 42(3): 454–465.

Scott, K.M., McLaughlin, K.A., Smith, D.A., and Ellis, P.M. (2012) 'Childhood maltreatment and DSM-IV adult mental disorders: comparison of prospective and retrospective findings', *British Journal of Psychiatry*, 200(6): 469–475.

Sedlak, A., Mettenburg, J., Basena, M., Petta, I., McPherson, K., Green, A., and Lee, S. (2010) *Fourth National Incidence Study of Child Abuse and Neglect, Report to Congress*, Washington, DC: US Department of Health and Human Services.

Shaffer, A., Huston, L., and Egeland, B. (2008) 'Identification of child maltreatment using prospective and self-report methodologies: a comparison of maltreatment incidence and relation to later psychopathology', *Child Abuse and Neglect*, 32(7): 682–692.

Shea, A., Walsh, C., MacMillen, H., and Steiner, M. (2004) 'Child maltreatment and HPA axis dysregulation: relationship to major depressive disorder and post traumatic stress disorder in females', *Psychoneuroendocrinology*, 30(2): 162–178.

Shonkoff, J.P., Boyce, W.T., and McEwen, B.S. (2009) 'Neuroscience, molecular biology, and the childhood roots of health disparities: building a new framework for health promotion and disease prevention', *JAMA*, 301(21): 2252–2259.

Silva, T.C., Larm, P., Vitaro, F., Tremblay, R.E., and Hodgins, S. (2012) 'The association between maltreatment in childhood and criminal convictions to age 24: a prospective study of a community sample of males from disadvantaged neighborhoods', *European Child and Adolescent Psychiatry*, 21(7): 403–413.

Smith, C.A., Ireland, T.O., Thornberry, T.P., and Elwyn, L. (2008) 'Childhood maltreatment and antisocial behavior: comparison of self-reported and substantiated maltreatment', *American Journal of Orthopsychiatry*, 78(2): 173–186.

Stewart, A., Livingston, M., and Dennison, S. (2008) 'Transitions and turning points: examining the links between child maltreatment and juvenile offending', *Child Abuse and Neglect*, 32(1): 51–66.

Straus, M.A. (2002) 'Reviews of the CTS and examples of how the CTS has been described in journal articles', in *Handbook for the Conflict Tactics Scales (CTS) Including Revised Versions of the CTS2 and CTSPC*, Durham, NH: University of New Hampshire, Family Research Laboratory, pp. 1–4.

Thornberry, T.P., Ireland, T.O., and Smith, C.A. (2001) 'The importance of timing: the varying impact of childhood and adolescent maltreatment on multiple problem outcomes', *Development and Psychopathology*, 13(4): 957–979.

Thornberry, T.P., Henry, K.L., Ireland, T.O., and Smith, C.A. (2010) 'The causal impact of childhood-limited maltreatment and adolescent maltreatment on early adult adjustment', *Journal of Adolescent Health*, 46(4): 359–365.

Thornberry, T.P., Matsuda, M., Greenman, S.J., Augustyn, M.B., Henry, K.L., Smith, C.A., and Ireland, T.O. (2014) 'Adolescent risk factors for child maltreatment', *Child Abuse and Neglect*, 38(4): 706–722

Titus, J.C., Dennis, M.L., White, W.L., Scott, C.K., and Funk, R.R. (2003) 'Gender differences in victimization severity and outcomes among adolescents treated for substance abuse', *Child Maltreatment*, 8(1): 19–35.

Topitzes, J., Mersky, J.P., and Reynolds, A.J. (2011) 'Child maltreatment and offending behavior gender-specific effects and pathways', *Criminal Justice and Behavior*, 38(5): 492–510.

Trocmé, N., Fallon, B., Sinha, V., Van Wert, M., Kozlowski, A., and MacLaurin, B. (2013) 'Differentiating between child protection and family support in the Canadian child welfare system's response to intimate partner violence, corporal punishment, and child neglect', *International Journal of Psychology*, 48(2): 128–140.

Twardosz, S. and Lutzker, J.R. (2010) 'Child maltreatment and the developing brain: a review of neuroscience perspectives', *Aggression and Violent Behavior*, 15(1): 59–68.

US DHHS (2012) *Child Maltreatment 2011*, Washington, DC: US Department of Health and Human Services.

Veenema, A.H. (2009) 'Early life stress, the development of aggression and neuroendocrine and neurobiological correlates: what can we learn from animal models?', *Frontiers in Neuroendocrinology*, 30(4): 497–518.

Viner, R.M., Roche, E., Maguire, S.A., and Nicholls D.E. (2010) 'When does childhood obesity become a child protection issue?', *British Medical Journal*, 341(7769): 375–377.

Walsh, C.A., Jamieson, E., MacMillan, H., and Boyle, M. (2007) 'Child abuse and chronic pain in a community survey of women', *Journal of Interpersonal Violence*, 22(12): 1536–1554.

Webster, R.L., Goldstein, J., and Segall, A. (1985) 'A test of the explanatory value of alternative models of child abuse', *Journal of Comparative Family Studies*, 16: 295–317.

Widom, C.S. (1989a) 'The cycle of violence', *Science*, 244(4901): 160–166.

Widom, C.S. (1989b) 'Child abuse, neglect, and adult behavior: research design and findings on criminality, violence, and child abuse', *American Journal of Orthopsychiatry*, 59(3): 355–64.

Widom, C.S. (1991) 'The role of placement experiences in mediating the criminal consequences of early childhood victimization', *American Journal of Orthopsychiatry*, 61(2): 195–209.

Widom, C. S. (1996) 'Childhood sexual abuse and its criminal consequences', *Society*, 33(4): 47–53.

Widom, C.S. and Brzustowicz, L.M. (2006) 'MAOA and the "cycle of violence:" childhood abuse and neglect, MAOA genotype, and risk for violent and antisocial behavior', *Biological Psychiatry*, 60(7): 684–689.

Widom, C.S. and Maxfield, M.G. (2001) 'An update on the "cycle of violence"', *National Institute of Justice Research in Brief*, February, www.ncjrs.gov/pdffiles1/nij/184894.pdf

Widom, C.S., Raphael, K.G., and DuMont, K.A. (2004) 'The case for prospective longitudinal studies in child maltreatment research: commentary on Dube, Williamson, Thompson, Felitti, and Anda', *Child Abuse and Neglect*, 28(7): 715–722.

Williams, J.H., Van Dorn, R.A., Bright, C.L., Jonson-Reid, M., and Nebbitt, V.E. (2010) 'Child maltreatment and delinquency onset among African American adolescent males', *Research on Social Work Practice*, 20(3): 253–259.

Wolfe, D.A., Scott, K., Wekerle, C., and Pittman, A.L. (2001) 'Child maltreatment: risk of adjustment problems and dating violence in adolescence', *Journal of the American Academy of Child and Adolescent Psychiatry*, 40(3): 282–289.

Wolock, I. and Horowitz, B. (1984) 'Child maltreatment as a social problem: the neglect of neglect', *American Journal Orthopsychiatry*, 54: 530–543.

Wu, S.S., Ma, C.X., Carter, R.L., Ariet, M., Feaver, E.A., Resnick, M.B., and Roth, J. (2004) 'Risk factors for infant maltreatment: a population-based study', *Child Abuse and Neglect*, 28(12): 1253–1264.

Yun, I., Ball, J.D., and Lim, H. (2011) 'Disentangling the relationship between child maltreatment and violent delinquency: using a nationally representative sample', *Journal of Interpersonal Violence*, 26(1): 88–110.

Zhai, F. and Gao, Q. (2009) 'Child maltreatment among Asian Americans: characteristics and explanatory framework', *Child Maltreatment*, 14(2): 207–224.

Zingraff, M., Leiter, J., Myers, K., and Johnsen, M. (1993) 'Child maltreatment and youthful problem behavior', *Criminology*, 31: 173–202.

Further reading

Banyard, V.L., Williams, L.M., and Siegel, J.A. (2004) 'Childhood sexual abuse: a gender perspective on context and consequences', *Child Maltreatment*, 9(3): 223–238.

Brown, J., Berenson, K., and Cohen, P. (2005) 'Documented and self-reported child abuse and adult pain in a community sample', *Clinical Journal of Pain*, 21(5): 374–377.

DeLisi, M. and Vaughn, M.G. (2011) 'The importance of neuropsychological deficits relating to self-control and temperament to the prevention of serious antisocial behavior', *International Journal of Child, Youth and Family Studies*, 2(1/2): 12–35.

D'Onofrio, B.M., Van Hulle, C.A., Goodnight, J.A., Rathouz, P.J., Lahey, B.B., Achenbach, T.M., *et al.* (2012) 'Is maternal smoking during pregnancy a causal environmental risk factor for adolescent antisocial behavior? Testing etiological theories and assumptions', *Psychological Medicine*, 42(7): 1535–1545.

Dube, S.R., Anda, R.F., Whitfield, C.L., Brown, D.W., Felitti, V.J., Dong, M., and Giles, W.H. (2005) 'Long-term consequences of childhood sexual abuse by gender of victim', *American Journal of Preventive Medicine*, 28(5): 430–438.

Dubowitz, H., Villodas, M.T., Litrownik, A.J., Pitts, S.C., Hussey, J.M., Thompson, R., *et al.* (2011) 'Psychometric properties of a youth self-report measure of neglectful behavior by parents', *Child Abuse and Neglect*, 35(6): 414–424.

English, D.J., Graham, J., Newton, R.R., Lewis, T.L., Thompson, R., Kotch, J.B., and Weisbart, C. (2009) 'At-risk and maltreated children exposed to intimate partner aggression/violence: what the conflict looks like and its relationship to child outcomes', *Child Maltreatment*, 14(2): 157–171.

Grella, C.E. and Joshi, V. (2003) 'Treatment processes and outcomes among adolescents with a history of abuse who are in drug treatment', *Child Maltreatment*, 8(1): 7–18.

Ireland, T.O., Smith, C.A., and Thornberry, T.P. (2002) 'Developmental issues in the impact of child maltreatment on later delinquency and drug use', *Criminology*, 40: 359–399.

Jonson-Reid, M., Drake, B., and Zhou, P. (2013) 'Neglect subtypes, race and poverty: individual, family, and service characteristics', *Child Maltreatment*, 18: 30–41.

Moylan, C.A., Herrenkohl, T.I., Sousa, C., Tajima, E.A., Herrenkohl, R.C., and Russo, M.J. (2010) 'The effects of child abuse and exposure to domestic violence on adolescent internalizing and externalizing behavior problems', *Journal of Family Violence*, 25(1): 53–63.

Spano, R., Rivera, C., and Bolland, J. (2006) 'The impact of timing of exposure to violence on violent behavior in a high poverty sample of inner city African American youth', *Journal of Youth and Adolescence*, 35(5): 681–692.

Trickett, P.K. and Gordis, E.B. (2004) 'Aggression and antisocial behavior in sexually abused females', in M. Putallaz and K.L. Bierman (eds), *Aggression, Antisocial Behavior, and Violence among Girls: A Developmental Perspective*, New York: Guilford Press, pp. 162–185.

Twomey, J., LaGasse, L., Derauf, C., Newman, E., Shah, R., Smith, L., *et al.* (2013) 'Prenatal methamphetamine exposure, home environment, and primary caregiver risk factors predict child behavioral problems at 5 years', *American Journal of Orthopsychiatry*, 83(1): 64–72.

Webster-Stratton, C. and Taylor, T. (2001) 'Nipping early risk factors in the bud: preventing substance abuse, delinquency, and violence in adolescence through interventions targeted at young children (0–8 years)', *Prevention Science*, 2(3): 165–192.

Widom, C.S., Czaja, S.J., Bentley, T., and Johnson, M.S. (2012) 'A prospective investigation of physical health outcomes in abused and neglected children: new findings from a 30-year follow-up', *American Journal of Public Health*, 102(6): 1135–1144.

28

Early dispositional risks for conduct problems and gene–environment interplay

Christopher J. Trentacosta and Benjamin D. Goodlett

Introduction

Crime is tremendously costly to victims and to society, and involvement in serious crime is frequently rooted in individual traits and behaviors that emerge and become entrenched in the early years of life. For example, although physical aggression is relatively normative during the toddler years (Tremblay *et al.*, 1999), a nontrivial percentage of children (in the range of roughly 3 to 10 percent across several samples) exhibit chronic physical aggression across childhood and into adolescence (Nagin and Tremblay, 1999; Broidy *et al.*, 2003). This subset of children, typically referred to as "early-onset" (Patterson *et al.*, 1998) or "life-course persistent" (Moffitt, 1993), are at elevated risk for serious conduct problems and criminal misconduct in adolescence and adulthood (Moffitt *et al.*, 2002). Aspects of this developmental taxonomy have been called into question because "adolescence-onset" antisocial behavior is more similar to "life-course persistent" antisocial behavior than the taxonomy originally suggested (Fairchild *et al.*, 2013). Nonetheless, it is clear that young children exhibiting early-emerging conduct problems are especially likely to engage in criminal behaviors later in adolescence and adulthood. Therefore it is crucial to investigate the early-life traits and behavioral patterns that predispose young children to serious conduct problems and increase the likelihood that they will follow a trajectory toward criminal involvement.

Theories of criminal behavior have not always emphasized individual differences in traits and behavioral patterns. For example, Liazos documented several historical arguments in sociology against the study of individual differences, and argued that criminologists have confused the moral argument "at least as good as anyone else" with the empirical argument "the same as anyone else" (Liazos, 1972). However, during the same era, empirical evidence demonstrated differences in traits such as extraversion between people who had and had not been incarcerated (Sanocki, 1969; Eysenck and Eysenck, 1971). In more recent decades, theoretical perspectives have increasingly recognized the role of individual differences in behavioral traits as precursors to delinquency and crime. For example, the general theory of crime focuses on the association between self-control and crime (Gottfredson and Hirschi, 1990). However, the general theory of crime is limited by its exclusive focus on social determinants of self-control, whereas subsequent genetically informed studies have demonstrated the importance of both biological and social determinants in understanding the development of self-control (Wright and Beaver, 2005). Other nongenetically informed studies of precursors to youth delinquency and criminal behavior have focused on behavioral traits. For example, a single item measuring "daring" in adolescence was a robust predictor of future criminal offending (Farrington and West, 1993). In addition, the Pittsburgh Youth Study supported teacher-rated impulsiveness, self-reported under-control, motor restlessness (based on observations), and psychomotor impulsiveness (on the Trail Making Test) as predictors of delinquency at ages 10 and 13 (White *et al.*, 1994).

Theories of temperament and personality, and recent developmental psychology research, inform much of the current understanding of which dispositional traits are particularly likely to predict serious conduct problems and later criminal involvement. Temperament and personality theories focus on stable individual differences, early-appearing behavioral styles, and the regulation of emotion and attention (Rothbart *et al.*, 1995; McCrae and Costa, 1997; Wachs and Bates, 2001). One of the earliest empirical studies of temperament, the New York Longitudinal Study, found that children classified as "difficult" at ages 3–4 had poor psychiatric adjustment in early adulthood (Chess and Thomas, 1984). Within the subdiscipline of developmental psychopathology, the past decade has witnessed increased attention to the roles of temperament and personality traits in behavior problems, including conduct problems. For example, Nigg (2006) discussed several pathways from temperament traits to various forms of psychopathology, including serious conduct problems. In particular, he highlighted low fear responses, excessive approach toward novelty, and low self-regulatory effortful control as temperament pathways that may lead to serious conduct problems.

Likewise, in their overview of temperament and serious conduct problems, Frick and Morris (2004) noted poor emotional self-regulation as one pathway to the development of serious conduct problems. They also emphasized the failure to develop the emotional aspects of conscience and associated "callous-unemotional traits" as a separate pathway to conduct problems. The role of callous-unemotional traits in serious conduct problems has received increased emphasis throughout the past decade (e.g. Frick and Viding, 2009), and preparation for the fifth edition of the *Diagnostic and Statistical Manual of Mental Disorders* (DSM-5) brought further attention to the role of these traits in the most serious forms of conduct disorder (Pardini *et al.*, 2010). As a result, DSM-5 users are now able to specify whether a diagnosis of conduct disorder includes "limited prosocial emotions" if the individual presents with callousness or other characteristics (Regier *et al.*, 2013).

Lahey and Waldman's (2003) developmental propensity theory offers an especially parsimonious perspective on temperament-based precursors to serious conduct problems. The developmental propensity theory emphasizes similar traits to the criminological research of Eysenck, Farrington, and others, as well as Nigg's (2006) and Frick and Morris's (2004) perspectives. More specifically, daring, low prosociality, and negative emotionality are the three temperament "dispositions" highlighted by the developmental propensity theory. The Child and Adolescent Dispositions Scale (CADS) was created to assess each of the three dispositions included in the propensity model, and the scale was validated by Lahey, Waldman, and their colleagues (Lahey *et al.*, 2008, 2010). Studies assessing these dispositions among school-age children and adolescents with the CADS have uncovered unique associations between each dispositional dimension and symptoms of conduct disorder (Trentacosta *et al.*, 2009; Waldman *et al.*, 2011). Behavior genetic findings indicate that each disposition in the developmental propensity model is moderately heritable, with nonshared environmental influences also present, as well as shared environmental influences supported for prosociality (Waldman *et al.*, 2011). The same study also uncovered common additive genetic and nonshared environmental influences on the overlap between symptoms of conduct disorder and the dispositions. Furthermore, the study was supportive of the notion, stemming from the developmental propensity model, that subsets of youth may be at risk for conduct disorder due to genetic factors that predispose the individual to a specific disposition (e.g. genetic factors that increase risk for "daring" propensities). A separate study of school-age twins found support for genetic factors that overlapped between the dispositions measured with the CADS and a broad externalizing factor, but there was not a single genetic factor that accounted for the associations among all of the constructs (Taylor *et al.*, 2013). Thus there appeared to be three sets of genetic factors that play a role in externalizing problems, and these factors broadly corresponded to the three dispositions in the developmental propensity model (low prosociality, negative emotionality, and daring).

Several subsets of temperament and personality traits are common across many or all of the theoretical perspectives described above. These subsets of traits include: (1) low prosociality and callous-unemotional traits; (2) impulsivity, disinhibition, and poor effortful control; and (3) negative emotionality. Consistent with the notion that fully elucidating the etiology of risk for criminal involvement requires understanding the dispositional precursors to serious conduct problems, the remainder of this chapter reviews the biosocial underpinnings of these subsets of dispositional traits. Although a large and growing literature exists on each of these traits, relatively little empirical research has been conducted on the traits' biosocial underpinnings and their connections to serious conduct problems *during early childhood*. This relative lack of attention to the biosocial underpinnings of these traits in early childhood is noteworthy because extensive research shows that these traits are salient markers of early childhood development (e.g. Kochanska *et al.*, 2000), and (as reviewed above) it has long been known that serious conduct problems and criminal involvement are rooted in early childhood behavioral patterns. Thus the remainder of this chapter emphasizes research conducted in early childhood and highlights areas where further research is needed during this key period in human development. This review also highlights varieties of gene–environment interplay as key biosocial processes.

Gene–environment interplay is a broad term that encompasses several more specific mechanisms, including gene × environment interaction; variations in the role of heritability depending on environmental circumstances; gene–environment correlations; and epigenetic processes where environmental inputs influence the effects of genes (Rutter *et al.*, 2006). Mechanisms of gene–environment interplay have recently received considerable attention in studies of conduct problems and closely related constructs (e.g. antisocial behavior; Moffitt, 2005). For example, a large adolescent twin study found that genetic effects were striking for individual differences in antisocial behavior among adolescents from higher socioeconomic backgrounds, and shared environmental effects were more pronounced for adolescents from lower socioeconomic backgrounds (Tuvblad *et al.*, 2006). As another example, a landmark gene × environment interaction study showed that children possessing the low MAOA risk allele were at greater risk for antisocial behavior, but only when they had experienced maltreatment as a child (Caspi *et al.*, 2002). Not all studies have replicated this finding (Young *et al.*, 2006), but meta-analyses have generally supported the low activity MAOA genotype by maltreatment interaction when predicting antisocial behavior (Kim-Cohen *et al.*, 2006; Taylor and Kim-Cohen, 2007). Relative to these and other well known examinations of gene–environment interplay in relation to conduct problems and antisocial behavior, there is considerably less gene–environment interplay research on the three subsets of dispositional traits reviewed in this chapter, with only a relatively small portion of this literature focusing on early childhood. However, in summarizing this burgeoning literature in the following sections, we highlight several recent findings supportive of the role of gene–environment interplay when predicting these dispositional traits in early childhood. The chapter concludes with future research directions, including important gaps in the research literature and newly emerging approaches to examining gene–environment interplay that could be applied to research on the emergence of these dispositional traits during early childhood.

Low prosociality and callous-unemotional traits

In Lahey and Waldman's (2003) developmental propensity model, the prosociality disposition includes sympathy, respect for rules, and guilt in response to wrongdoing. Extensive research has connected low prosociality during childhood to conduct problems and criminal behavior. For example, Hay and Pawlby (2003) found that 11-year-old children who received a disruptive behavior disorder diagnosis such as conduct disorder were less prosocial than children who did not receive a disruptive behavior disorder diagnosis. Moreover, the emergence of prosocial behavior is considered a cornerstone of early childhood development (Eisenberg and Mussen, 1989), and longitudinal research has documented associations between markers of a prosocial personality disposition during early childhood and subsequent prosocial behavior during adolescence and adulthood (Eisenberg *et al.*, 1999).

Similar to Lahey and Waldman's (2003) conceptualization of low prosociality, Frick's characterization of callous-unemotional traits encompasses a set of features including lack of guilt and empathy and callousness towards others' needs (Frick and White, 2008). Callous-unemotional traits are closely linked, both conceptually and empirically, to psychopathy, a condition that is typically used to describe adolescents or adults with abnormal emotionality (e.g. shallow affect) and deceitful interpersonal characteristics (Hare *et al.*, 1991; Barry *et al.*, 2000). Like research showing that psychopathy helps distinguish the most persistent forms of antisocial behavior among adults (Skilling *et al.*, 2002), callous-unemotional traits help to distinguish the most serious forms of childhood conduct problems. As an example, school-age children with callous-unemotional traits and conduct problems had the highest levels of delinquency, including parent-reported police contacts, across 4 years of longitudinal follow-up (Frick *et al.*, 2005). Even though aspects of the callous-unemotional construct such as guilt and empathy are integral components of early social-emotional development beginning in the second year of life (Zahn-Waxler *et al.*, 1992; Kochanska *et al.*, 2002), very little research on connections between callous-unemotional traits and early-emerging conduct problems has focused on the early childhood years. However, a recent longitudinal study of at-risk young children supported associations between deceitful-callous traits measured at age 3 years and growth in conduct problems from ages 2 to 4 years (Hyde *et al.*, 2013).

A growing number of genetically informed studies have examined the etiology of callous-unemotional traits in children, adolescents, and young adults. A review by Viding and McCrory (2012) found that, across age groups and measures, genetic influences account for much of the variation in callous-unemotional traits. Moreover, nonshared environment influences (factors that make twins dissimilar) appear to play a larger role in callous-unemotional traits than shared environmental influences (factors that make twins more similar to each other than their genetic relation would predict). Callous-unemotional traits were also found to be stable across several years with test–retest correlations that were largely accounted for by genetic influences (Blonigen *et al.*, 2006; Forsman *et al.*, 2008). However, genetic influences on callous-unemotional traits may vary by gender, and studies have found higher heritability for males compared with females, for callous-unemotional traits (Fontaine *et al.*, 2010; Bezdjian *et al.*, 2011). Other studies found limited evidence for gender differences in the etiology of callous-unemotional traits (Blonigen *et al.*, 2006; Larsson *et al.*, 2006).

Importantly, the youngest sample used to estimate the heritability of callous-unemotional traits included 7-year-olds from the Twins Early Development Study (TEDS; Trouton *et al.*, 2002). An analysis of the TEDS sample showed that individual differences in callous-unemotional traits as rated by teachers were under strong genetic influence (Viding *et al.*, 2005). Although we are not aware of twin studies that have specifically examined callous-unemotional traits in children younger than age 7 years, the heritability of child prosocial behavior has been examined in early childhood. Knafo and Plomin (2006a) used data from the TEDS to estimate the heritability of prosocial behavior at child ages of 18 months, 2, 3, 4, and 7 years. From age 2 to 7 years, shared environmental effects decreased and genetic effects increased (Knafo and Plomin, 2006a). Both genetic factors and shared environment were contributors to change in prosocial behavior, but only genetic factors accounted for continuity in prosocial behavior (Knafo and Plomin, 2006b). In addition, the Quebec Newborn Twin Study (QNTS) examined maternal ratings of children's "disregard for rules" at four time points from 20 to 64 months old (Petitclerc *et al.*, 2011). Disregard for rules is considered an aspect of conduct disorder that shares features with several callous-unemotional traits (e.g. little guilt), and research has shown disregard for rules to be stable during early childhood (Petitclerc *et al.*, 2009). In the analysis of the QNTS, stability in disregard for rules was primarily determined by genetic factors, whereas environmental effects were age specific (Petitclerc *et al.*, 2011).

Turning to molecular genetic studies, there is emerging evidence that genes involved in the serotonin system are associated with callous-unemotional traits and related constructs. For example, Glenn (2011) reviewed wide-ranging evidence suggesting that even though the short allele polymorphism of the serotonin transporter gene is typically considered the risk allele, the long allele may be the risk allele when considering outcomes associated with psychopathy. Consistent with this suggestion, a recent study showed that carriers of the long allele of the serotonin transporter gene had affective deficits on a well established measure of psychopathy (Sadeh *et al.*, 2013). Interestingly, the same research group found evidence of a gene × environment interaction when investigating a similar research question with a sample of youth: callous-unemotional traits were inversely associated with socioeconomic status, but only among carriers of the risk (i.e. homozygous long) polymorphism of the serotonin transporter gene (Sadeh *et al.*, 2010). Moreover, one recent study of school-age children found a direct association between the serotonin transporter gene and prosociality, and prosociality also mediated the association between the serotonin transporter genotype and symptoms of oppositional defiant disorder (Brammer and Lee, 2013). Collectively, these findings support the role of serotonin system genes, especially the serotonin transporter gene, in callous-unemotional traits and low prosociality, and they also implicate the role of gene–environment interplay. It is also important to note that polymorphisms in other genes besides serotonin genes, including the oxytocin receptor gene, have been linked to callous-unemotional traits (Dadds *et al.*, 2014b).

Given the relative dearth of research examining the callous-unemotional traits construct in early childhood, it is perhaps not surprising that we are not aware of molecular genetics studies that specifically investigate callous-unemotional traits in preschool-age children. However, recent work has examined serotonin gene polymorphisms in relation to somewhat similar behavioral outcomes in young children, and provides evidence for gene–environment interplay. For example, a recent study of socioeconomically disadvantaged 2-year-olds found that maternal unresponsiveness predicted externalizing problems at a follow-up assessment 2 years later, but only among children homozygous for the long allele of the serotonin transporter gene (Davies and Cicchetti, 2014; for similar findings, see Sulik *et al.*, 2012). In another longitudinal study of parenting, the serotonin transporter gene, and several "competence" outcomes during early childhood (Kochanska *et al.*, 2011), results were consistent across study outcomes but differed somewhat from Davies and Cicchetti's findings. More specifically, children experiencing lower maternal responsiveness across early childhood showed less competent outcomes if they possessed the short allele of the serotonin transporter gene rather than the long allele. Importantly, one of the competence outcomes, moral internalization, was similar to characterizations of children low on callous-unemotional traits. In summary, recent evidence supports parenting × child serotonin transporter genotype interactions when predicting early childhood outcomes, but it would be useful to examine these gene–environment processes in connection to callous-unemotional traits or low prosociality among young children facing elevated risk for serious conduct problems.

Impulsivity, disinhibition, and poor effortful control

The second subset of traits includes disinhibition, impulsivity, and low levels of effortful control. Although Lahey and Waldman (2003) named their construct "daring" in line with Farrington's (2005) description of the construct and its connections to sensation seeking (Zuckerman, 1971), they also emphasized the construct's connection to developmental psychopathology research on disinhibition and fearlessness (e.g. Raine *et al.*, 1998). In addition, they suggested that ensuing research would likely uncover links between daring and the five-factor personality dimension of conscientiousness (John *et al.*, 1994). Because childhood difficulties with self-regulation and the five-factor model dimension conscientiousness are linked to the broad childhood temperament dimension of effortful control (Kochanska and Knaack, 2003; Rothbart, 2007), we include daring, sensation seeking, and impulsivity along with effortful control among the same subset of traits. However, it is important to note that impulsivity tends to load on the separate broad temperament dimension of extraversion/surgency (Putnam *et al.*, 2006; Rothbart, 2007). Thus this subset of traits with links to conduct problems may reflect a more heterogeneous set of traits than prosociality or negative emotionality. The underlying commonality within this subset of traits in relation to serious conduct problems appears to be the individual's underlying difficulties with self-regulation.

As with prosociality and callous-unemotional traits, the emergence of effortful control is a key component of early childhood development. Across several tasks that required suppression of a dominant response in order to execute a subdominant response, the core process of effortful control, trait-like qualities were observed for effortful control by age 2.5 years (Kochanska and Knaack, 2003). Moreover, performance on effortful control tasks in the same study was stable across early childhood, and the stability of effortful control was similar to the stability of intelligence by age 4 years. Compared with the relative lack of research on connections between callous-unemotional traits and conduct problems in early childhood, there has been more attention to associations between low effortful control and conduct problems during the preschool years, especially in relatively low-risk community samples that assess normative variation in conduct problems. For example, behavior observations and parent ratings of children's effortful control predicted externalizing behavior problems among 3-year-olds (Olson *et al.*, 2005).

Uninhibited temperament or fearlessness is another key marker of self-regulatory difficulties that can be assessed reliably during early childhood and predicts key developmental outcomes, including conduct problems. Uninhibited versus inhibited temperament can be observed and distinguished during the first 2 years of life (Kagan *et al.*, 1989), and toddlers who were classified as very uninhibited had more conduct problems during early adolescence than highly inhibited toddlers (Schwartz *et al.*, 1996). Furthermore, boys with chronic conduct problems across early childhood exhibited more fearlessness during a task designed to assess behavioral inhibition at age 2 years relative to boys who desisted from early conduct problems (Shaw *et al.*, 2003).

Turning to behavior genetic findings on this subset of traits, a recent meta-analysis focused on genetic and environmental contributors to impulsivity by examining 41 adoption, family, and twin studies that included 58 separate samples (Bezdjian *et al.*, 2011b). This meta-analysis illustrates some of the difficulties with classifying the subset of self-regulatory traits because impulsivity, behavioral inhibition, and inhibitory control are treated as a single construct for the purposes of the meta-analysis. Although performance on inhibitory control and inhibition tasks often correlates with impulsivity (e.g. Logan *et al.*, 1997), modern temperament models do not always group inhibitory control and impulsivity as belonging to the same temperament dimension (Rothbart, 2007). Keeping that caveat in mind, it is still noteworthy that the meta-analysis supported genetic and nonshared environmental influences across the diverse studies of impulsivity (Bezdjian *et al.*, 2011b). Moreover, age of study participants was a moderator of effects, such that the largest genetic effects were observed among studies with child participants. Although this meta-analysis included several studies focusing on infants and very young children, almost of all of the studies of younger children or infants focused on behavioral inhibition or inhibitory control, rather than what may be perceived as more traditional conceptualizations of impulsivity. For example, one recent study of 2-year-old twins that was included in the meta-analysis showed that genetic factors were a substantial contributor to variance in both parent-reported and observed inhibitory control (Gagne and Saudino, 2010). A separate study of toddler-aged twins that was not included in the meta-analysis also supported the role of genetic factors in inhibitory control, but shared family environment also contributed variance in observed inhibitory control (Gagne and Goldsmith, 2011).

Among this subset of traits, the largest body of molecular genetic studies and investigations of gene–environment interplay exist for dopamine system genes and children's sensation seeking or effortful control. For example, in a study of toddlers, children possessing the 7-repeat allele of the dopamine receptor D4 gene who experienced lower quality parenting had higher levels of sensation seeking, but parenting quality was not associated with sensation seeking for children without the 7-repeat allele (Sheese *et al.*, 2007). Although the 7-repeat allele was not associated with effortful control during the toddler assessment, a follow-up assessment of the same sample during the preschool period (3- to 4-year-old children) showed an interaction between this allele and parenting quality when predicting effortful control (Sheese *et al.*, 2012). Similarly, in a sample of 3-year-olds the 7-repeat allele genotype of the dopamine receptor D4 gene moderated the association between parenting and effortful control such that children with the 7-repeat allele and parents who engaged in negative parenting practices had lower effortful control (Smith *et al.*, 2012). In a sample drawn from the NICHD Study of Early Child Care, children with the 7-repeat allele were sensitive to hours spent in nonmaternal care during early childhood, such that children spending fewer hours in child care with the 7-repeat allele performed better on self-regulation measures than children without the 7-repeat allele (Berry *et al.*, 2014). Finally, in a gene × environment interaction study of the serotonin transporter gene and young children's self-regulatory effortful control, children who possessed the short allele and were insecurely attached demonstrated poorer effortful control (Kochanska *et al.*, 2009). In summary, there are a number of recent studies of gene–environment interplay indicating that the role of specific genes, especially the dopamine receptor D4 gene, on self-regulatory functioning during early childhood is contingent on environment circumstances.

Negative emotionality

The third subset of dispositional traits, negative emotionality, is described by Lahey and Waldman (2003) as "experiencing negative emotions frequently, intensely, and with little provocation." This conceptualization fits with Rothbart's (2007) broad temperament dimension of negative affectivity that includes scales tapping frustration, fear, sadness, discomfort, and low soothability. In longitudinal studies that begin in early childhood, negative emotionality is frequently examined as a potential precursor to later conduct problems and other associated behavior problems. For instance, maternal reports of toddler negative emotionality at age 24 months predicted maternal reports of youth externalizing problems in middle childhood and adolescence (Olson *et al.*, 2000). On the other hand, Lahey and Waldman acknowledge a "complex" association between negative emotionality and conduct problems, and note that several studies have failed to support this association. As one example, unlike fearlessness assessed at age 24 months, negative emotionality assessed at age 18 months did not distinguish males following a developmental trajectory of chronic conduct problems across childhood from other less problematic trajectories (Shaw *et al.*, 2005). Furthermore, longitudinal studies of school-age children conducted by Eisenberg and colleagues suggest that anger and frustration are more consistently linked to externalizing problems than other aspects of negative emotionality such as sadness (Eisenberg *et al.*, 2001, 2009). Eisenberg and colleagues' research also suggests that negative emotionality may be best conceptualized as a moderator of associations between other aspects of temperament and externalizing problems, at least among school-age children. More specifically, difficulties with attention regulation were longitudinally associated with externalizing problems, but only among children high in negative emotionality (Eisenberg *et al.*, 2000).

Negative emotionality has been shown to be genetically influenced (Goldsmith *et al.*, 1997; Lemery and Goldsmith, 2002). In a study using data from the Georgia Twin Registry, which includes children and adolescents, covariation between negative emotionality and childhood externalizing disorders was explained by additive genetic and nonshared environmental influences (Singh and Waldman, 2010). For young children (4 to 8 years old), anger/frustration during an observed challenging task was moderately influenced by genetic factors and nonshared environment (Deater-Deckard *et al.*, 2007).

Not all of the negative affectivity literature is consistent. Using a sample in middle childhood, Mullineaux *et al.* (2009) found evidence for genetic influences on mothers' and fathers' ratings of aspects of negative affectivity, but fathers' ratings on a latent scale combining the aspects of negative affectivity were not found to be influenced by genetic factors. Additionally, several studies have found evidence for gene × environment effects when predicting negative emotionality, but the studies did not uncover main effects for genetic factors (O'Connor *et al.*, 2003; Lipscomb *et al.*, 2012). Interpreting heritability studies of negative emotionality is made more difficult by evidence that the method of measurement may be related to heritability estimates. A meta-analysis of the heritability of negative emotionality found higher heritability estimates when using parent report, rather than observational ratings of child temperament (Goldsmith *et al.*, 1997).

A recent overview of molecular genetic studies of temperament showed mixed findings for direct associations between negative emotionality and candidate genes, including the dopamine receptor D4 gene and the serotonin transporter gene (Saudino and Wang, 2012). Interestingly, one recent study of school-age children found a direct association between the serotonin transporter gene and negative emotionality, and this study also demonstrated that negative emotionality mediated the association between the serotonin transporter genotype and symptoms of oppositional defiant disorder (Brammer and Lee, 2013). Even though main effect studies of candidate genes have been mixed, studies of gene–environment interplay and negative emotionality have generated some interesting insights about the development of negative emotionality in young children. In a unique gene × temperament interaction study of 3-year-old children, the serotonin transporter promoter polymorphism was not directly associated with negative emotionality (Hayden *et al.*, 2010a). However, children who were both low in positive emotionality and had a short allele polymorphism of the serotonin transporter gene had higher levels of negative emotionality than children who were high in positive emotionality or did not possess the short allele. In the same sample, the brain-derived neurotrophic factor (BDNF) gene and parental depressive disorders were associated with negative emotionality (Hayden *et al.*, 2010b). Children who possessed the BDNF methionine allele had higher levels of negative emotionality if their parent reported relationship problems or the parent had a history of depression, but negative emotionality was especially low when the child had the same allele but no relationship problems or past depression were reported by the parent. Collectively, these findings illustrate the complex interplay between candidate genes, environmental factors, negative emotionality, other temperament traits, and relevant behavioral outcomes.

Conclusion

As emphasized at the beginning of the chapter, it is crucial to understand precursors to early-emerging conduct problems because serious conduct problems that emerge early in life and persist throughout childhood are especially likely to lead to delinquency and criminal involvement. Individual differences in temperament-based dispositional traits are one important source of variation in conduct problems that are the focus of this review. The three subsets of dispositional traits reviewed in this chapter broadly correspond to the three dispositions, prosociality, daring, and negative emotionality, that are emphasized in Lahey and Waldman's (2003) developmental propensity model of risk for conduct problems. Each subset of traits appears to be uniquely linked to serious conduct problems, and associations between the dispositions and conduct problems are primarily due to genetic factors that overlap with conduct problems but are relatively unique between each subset of traits (Waldman *et al.*, 2011; Taylor *et al.*, 2013).

We have emphasized that these subsets of traits are especially important to investigate during early childhood, because related temperament factors or individual difference variables such as guilt, empathy, and effortful control show rapid development and emergent stability during the early childhood developmental period. Our review of the current literature indicates that there is comparatively little research on some of these traits and their etiology during the early childhood period, especially in relation to serious conduct problems. For example, extensive research has examined callous-unemotional traits in older children and adolescents as well as psychopathy in adults, but relatively little research has focused on characterizing callous-unemotional traits and their etiology during the years before school entry.

Another area that has seen increased attention in recent years but would benefit from further systematic investigation is the role of gene–environment interplay when predicting these dispositional traits. The existing gene–environment interplay research has tended to be heavily weighted toward gene × environment interaction studies where a candidate gene polymorphism is associated with the dispositional trait, but only in a specific environmental context (e.g. low parenting quality; Sheese *et al.*, 2007). More gene × environment interaction research is needed, particularly for dispositional traits that have received comparatively little attention in research with young children at risk for conduct problems (e.g. callous-unemotional traits).

Other mechanisms of gene–environment interplay besides candidate gene studies of gene × environment interaction deserve more attention in relation to these dispositional traits in early childhood. The mechanisms that have received little attention include gene–environment correlation and differences in heritability based on environmental circumstances, although these mechanisms have received considerably more attention when examining behavior problems rather than dispositional traits. For example, twin and adoption studies of gene–environment correlation have provided rather consistent evidence that children's genetic factors, including measured genetic risk of birth parents' antisocial behavior in the adoption studies, play an important role in the parenting they receive (see Moffitt, 2005 for a review). It would be interesting to extend these findings to studies of dispositional traits; for example, adoption studies could measure birth parents' callous-unemotional traits as a genetic risk factor that might predict the parenting provided by adoptive parents toward their young adopted child (see Leve *et al.*, 2007 for a description of the type of adoption design with young children that could address this research question).

A growing number of twin studies have examined whether heritability of behaviors varies based on environmental circumstances, including the study of antisocial behavior by Tuvblad and colleagues (2006) summarized earlier in this chapter. In another recent study, genetic factors accounted for approximately half of the variation in mental ability among 2-year-old children raised in higher socioeconomic-status homes, but genetic factors had a negligible impact among children living in lower socioeconomic-status homes (Tucker-Drob *et al.*, 2011). Applying this analytic approach in studies of young twins that assess dispositional traits could uncover evidence for an interplay between genetic factors and socioeconomic status (or other environmental factors) when predicting dispositional traits. Such evidence could have important implications for which prevention and intervention approaches might be preferred in different contexts.

As noted earlier, epigenetic mechanisms are another key aspect of gene–environment interplay. Epigenetics is a term used to describe changes that bring about alterations in gene expression but do not change the DNA sequence. DNA methylation is one key epigenetic process that involves the addition of a methyl molecule to specific sites throughout the genome. The addition of the methyl molecule to sites in a gene promoter region can alter gene expression, thus limiting the level of protein encoded by the gene. Evidence of the impact of DNA methylation for child and adolescent developmental outcomes is just beginning to emerge. For example, a recent study of DNA methylation in the oxytocin receptor gene among males with a DSM disruptive behavior disorder diagnosis showed links between higher callous-unemotional traits and greater methylation within this gene (Dadds *et al.*, 2014a). Assessing DNA methylation and other epigenetic processes in future studies is likely to further increase our understanding of environmental contributors to variation in callous-emotional traits and other dispositional traits.

Lastly, it is important to consider the implications of the existing findings and future research for prevention and intervention. For example, young boys with callous-unemotional traits were less responsive to a parent-training intervention to treat conduct problems than boys without these traits (Hawes and Dadds, 2005). Understanding the biosocial mechanisms involved in the development of callous-unemotional traits and other dispositional traits linked to conduct problems could help applied researchers refine interventions to take into account young children's temperament profile and associated biological and contextual risk factors that might increase risk for serious conduct problems.

References

Barry, C.T., Frick, P.J., DeShazo, T.M., McCoy, M.G., Ellis, M.L., and Loney, B.R. (2000) 'The importance of callous-unemotional traits for extending the concept of psychopathy to children', *Journal of Abnormal Psychology*, 109: 335–340.

Berry, D., McCartney, K., Petrill, S., Deater-Deckard, K., and Blair, C. (2014) 'Gene–environment interaction between DRD4 7-repeat VNTR and early child-care experiences predicts self-regulation abilities in prekindergarten', *Developmental Psychobiology*, 56(3): 373–391.

Bezdjian, S., Baker, L.A., and Tuvblad, C. (2011a) 'Genetic and environmental influences on impulsivity: a meta-analysis of twin, family and adoptions studies', *Clinical Psychology Review*, 31: 1209–1223.

Bezdjian, S., Raine, A., Baker, L., and Lynam, D. (2011b) 'Psychopathic personality in children: genetic and environmental contributions', *Psychological Medicine*, 41: 589–601.

Blonigen, D.M., Hicks, B.M., Krueger, R.F., Patrick, C.J., and Iacono, W.G. (2006) 'Continuity and change in psychopathic traits as measured via normal-range personality: a longitudinal-biometric study', *Journal of Abnormal Psychology*, 115, 85–95.

Brammer, W.A, and Lee, S.S. (2013) 'Prosociality and negative emotionality mediate the association of serotonin transporter genotype with childhood ADHD and ODD', *Journal of Clinical Child and Adolescent Psychology*, 42(6): 809–819.

Broidy, L.M., Tremblay, R.E., Brame, B., Fergusson, D., Horwood, J.L., Laird, R., *et al.* (2003) 'Developmental trajectories of childhood disruptive behaviors and adolescent delinquency: a six-site, cross-national study', *Developmental Psychology*, 39: 222–245.

Caspi, A., McClay, J., Moffitt, T.E., Mill, J., Martin, J., Craig, I.W., *et al.* (2002) 'Role of genotype in the cycle of violence in maltreated children', *Science*, 297(5582): 851–854.

Chess, S. and Thomas, A.M. (1984) *Origins and Evolution of Behavior Disorders: From Infancy to Early Adult Life*, New York: Brunner/Hazel.

Dadds, M.R., Moul, C., Cauchi, A., Dobson-Stone, C., Hawes, D.J., Brennan, J., and Ebstein, R.E. (2014a) 'Methylation of the oxytocin receptor gene and oxytocin blood levels in the development of psychopathy', *Development and Psychopathology*, 26(1): 33–40.

Dadds, M.R., Moul, C., Cauchi, A., Dobson-Stone, C., Hawes, D.J., Brennan, J., *et al.* (2014b) 'Polymorphisms in the oxytocin receptor gene are associated with the development of psychopathy', *Development and Psychopathology*, 26(1): 21–31.

Davies, P.T. and Cicchetti, D. (2014) 'How and why does the 5-HTTLPR gene moderate associations between maternal unresponsiveness and children's disruptive problems?', *Child Development*, 85(2): 484–500.

Deater-Deckard, K., Petrill, S.A., and Thompson, L.A. (2007) 'Anger/frustration, task persistence, and conduct problems in childhood: a behavioral genetic analysis', *Journal of Child Psychology and Psychiatry*, 48: 80–87.

Eisenberg, N. and Mussen, P.H. (1989) *The Roots of Prosocial Behavior in Children*, Cambridge: Cambridge University Press.

Eisenberg, N., Guthrie, I.K., Murphy, B.C., Shepard, S.A., Cumberland, A., and Carlo, G. (1999) 'Consistency and development of prosocial dispositions: a longitudinal study', *Child Development*, 70: 1360–1372.

Eisenberg, N., Guthrie, I.K., Fabes, R.A., Shepard, S., Losoya, S., Murphy, B.C., *et al.* (2000) 'Prediction of elementary school children's externalizing problem behaviors from attentional and behavioral regulation and negative emotionality', *Child Development*, 71: 1367–1382.

Eisenberg, N., Cumberland, A., Spinrad, T.L., Fabes, R.A., Shepard, S.A., Reiser, M., *et al.* (2001) 'The relations of regulation and emotionality to children's externalizing and internalizing problem behavior', *Child Development*, 72: 1112–1134.

Eisenberg, N., Valiente, C., Spinrad, T.L., Cumberland, A., Liew, J., Reiser, M., *et al.* (2009) 'Longitudinal relations of children's effortful control, impulsivity, and negative emotionality to their externalizing, internalizing, and co-occurring behavior problems', *Developmental Psychology*, 45: 988–999.

Eysenck, S.B. and Eysenck, H.J. (1971) 'Crime and personality: item analysis of questionnaire responses', *British Journal of Criminology*, 11: 49–63.

Fairchild, G., Van Goozen, S.H.M., Calder, A.J., and Goodyer, I.M. (2013) 'Research review: evaluating and reformulating the developmental taxonomic theory of anitsocial behaviour', *Journal of Child Psychology and Psychiatry*, 54(9): 924–940.

Farrington, D. P. (2005) 'Childhood origins of antisocial behavior', *Clinical Psychology & Psychotherapy*, 12(3): 177–190.

Farrington, D.P. and West, D.J. (1993) 'Criminal, penal and life histories of chronic offenders: risk and protective factors and early identification', *Criminal Behaviour and Mental Health*, 3: 492–523.

Fontaine, N.M.G., Rijsdijk, F.V., McCrory, E.J.P., and Viding, E. (2010) 'Etiology of different developmental trajectories of callous-unemotional traits', *Journal of the American Academy of Child and Adolescent Psychiatry*, 49: 656–664.

Forsman, M., Lichtenstein, P., Andershed, H., and Larsson, H. (2008) 'Genetic effects explain the stability of psychopathic personality from mid- to late adolescence', *Journal of Abnormal Psychology*, 117: 606–617.

Frick, P.J. and Morris, A.S. (2004) 'Temperament and developmental pathways to conduct problems', *Journal of Clinical Child and Adolescent Psychology*, 33: 54–68.

Frick, P.J. and Viding, E. (2009) 'Antisocial behavior from a developmental psychopathology perspective', *Development and Psychopathology*, 21: 1111–1131.

Frick, P.J. and White, S.F. (2008) 'Research review: the importance of callous-unemotional traits for developmental models of aggressive and antisocial behavior', *Journal of Child Psychology and Psychiatry*, 49: 359–375.

Frick, P.J., Strickle, T.R., Dandreaux, D.M., Farrell, J.M., and Kimonis, E.R. (2005) 'Callous-unemotional traits in predicting the severity and stability of conduct problems and delinquency', *Journal of Abnormal Child Psychology*, 33: 471–487.

Gagne, J.R. and Goldsmith, H.H. (2011) 'A longitudinal analysis of anger and inhibitory control in twins from 12 to 36 months of age', *Developmental Science*, 14: 112–124.

Gagne, J.R. and Saudino, K.J. (2010) 'Wait for it! A twin study of inhibitory control in early childhood', *Behavior Genetics*, 40: 327–337.

Glenn, A.L. (2011) 'The other allele: exploring the long allele of the serotonin transporter gene as a potential risk factor for psychopathy: a review of the parallels in findings', *Neuroscience and Biobehavioral Reviews*, 35: 612–620.

Goldsmith, H.H., Buss, K.A., and Lemery, K.S. (1997) 'Toddler and childhood temperament: expanded content, stronger genetic evidence, new evidence for the importance of environment', *Developmental Psychology*, 33(6): 891–905.

Gottfredson, M. R. and Hirschi, T. (1990) *A General Theory of Crime*, Stanford, CA: Stanford University Press.

Hare, R.D., Heart, S.D., and Harpur, T.J. (1991) 'Psychopathy and the DSM-IV criteria for antisocial personality disorder', *Journal of Abnormal Psychology*, 100: 391–398.

Hawes, D.J. and Dadds, M.R. (2005) 'The treatment of conduct problems in children with callous-unemotional traits', *Journal of Consulting and Clinical Psychology*, 73: 737–741.

Hay, D.F. and Pawlby, S. (2003) 'Prosocial development in relation to children's and mothers' psychological problems', *Child Development*, 74: 1314–1327.

Hayden, E.P., Klein, D.N., Dougherty, L.R., Olino, T.M., Dyson, M.W., Durbin, C.E., *et al.* (2010a) 'The role of brain-derived neurotrophic factor genotype, parental depression, and relationship discord in predicting early-emerging negative emotionality', *Psychological Science*, 21: 1678–1685.

Hayden, E.P., Klein, D.N., Olino, T.M., Dougherty, L.R., Durbin, C.E., Sheikh, H.L., and Singh, S.M. (2010b) 'The serotonin transporter promoter polymorphism and childhood positive and negative emotionality', *Emotion*, 10: 696–702.

Hyde, L.W., Shaw, D.S., Gardner, F., Cheong, J., Dishion, T.J., and Wilson, M. (2013) 'Dimensions of callousness in early childhood: links to problem behavior and family intervention effectiveness', *Development and Psychopathology*, 25: 347–363.

John, O.P., Caspi, A., Robins, R.W., Moffitt, T.E., and Stouthamer-Loeber, M. (1994) 'The Little 5 – exploring the nomological network of the 5-factor model of personality in adolescent boys', *Child Development*, 65: 160–178.

Kagan, J., Reznick, J.S., and Gibbons, J. (1989) 'Inhibited and uninhibited types of children', *Child Development*, 60: 838–845.

Kim-Cohen, J., Taylor, A., Williams, B., Newcombe, R., Craig, I.W., and Moffitt, T.E. (2006) 'MAOA, maltreatment, and gene–environment interaction prediction children's mental health: new evidence and a meta-analysis', *Molecular Psychiatry*, 11: 903–913.

Knafo, A. and Plomin, R. (2006a) 'Parental discipline and affection and children's prosocial behavior: genetic and environmental links', *Journal of Personality and Social Psychology*, 90: 147–164.

Knafo, A. and Plomin, R. (2006b) 'Prosocial behavior from early to middle childhood: genetic and environmental influences on stability and change', *Developmental Psychology*, 42: 771–786.

Kochanska, G. and Knaack, A. (2003) 'Effortful control as a personality characteristic of young children: antecedents, correlates, and consequences', *Journal of Personality*, 71: 1087–1112.

Kochanska, G., Murray, K.T., and Harlan, E.T. (2000) 'Effortful control in early childhood: continuity and change, antecedents, and implications for social development', *Developmental Psychology*, 36: 220–232.

Kochanska, G., Gross, J.N., Lin, M.H., and Nichols, K.E. (2002) 'Guilt in young children: development, determinants, and relations with a broader system of standards', *Child Development*, 73: 461–482.

Kochanska, G., Philibert, R.A., and Barry, R.A. (2009) 'Interplay of genes and early mother child relationship in the development of self-regulation from toddler to preschool age', *Journal of Child Psychology and Psychiatry*, 50: 1331–1338.

Kochanska, G., Kim, S., Barry, R.A., and Philibert, R.A. (2011) 'Children's genotypes interact with maternal responsive care in predicting children's competence: diathesis-stress or differential susceptibility?', *Development and Psychopathology*, 23: 605–616.

Lahey, B.B. and Waldman, I.D. (2003) 'A developmental propensity model of the origins of conduct problems during childhood and adolescence', in B.B. Lahey, T.E. Moffitt, and A. Caspi (eds), *The Causes of Conduct Disorder and Serious Juvenile Delinquency*, New York: Guilford Press, pp. 76–117.

Lahey, B.B., Applegate, D., Chronis, A.M., Jones, H.A., Williams, S.H., Loney, J., and Waldman, I.D. (2008) 'Psychometric characteristics of a measure of emotional dispositions developed to test a developmental propensity model of conduct disorder', *Journal of Child Clinical and Adolescent Psychology*, 37: 794–807.

Lahey, B.B., Rathouz, P.J., Applegate, B., Tackett, J.L., and Waldman, I.D. (2010) 'Psychometrics of a self-report version of the Child and Adolescent Disposition Scale', *Journal of Clinical Child and Adolescent Psychology*, 39: 351–361.

Lahey, B.B., Rathouz, P.J., Lee, S.S., Chronis–Tuscano, A., Pelham, W.E., Waldman, I.D., and Cook, E.H. (2011) 'Interactions between early parenting and a polymorphism of the child's dopamine transporter gene in predicting future child conduct disorder symptoms', *Journal of Abnormal Psychology*, 120: 33–45.

Larsson, H., Andershed, H., and Lichtenstein, P. (2006) 'A genetic factor explains most of the variation in the psychopathic personality', *Journal of Abnormal Psychology*, 115: 221–230.

Lemery, K.S., and Goldsmith, H.H. (2002) 'Genetic and environmental influences on preschool sibling cooperation and conflict: associations with difficult temperament and parenting style', *Marriage and Family Review*, 33(1): 77–99.

Leve, L.D., Neiderhiser, J.M., Ge, X., Scaramella, L.V., Conger, R.D., Reid, J.B., *et al.* (2007) 'The early growth and development study: a prospective adoption design', *Twin Research and Human Genetics*, 10: 84–95.

Liazos, A. (1972) 'The poverty of the sociology of deviance: nuts, sluts, and preverts', *Social Problems*, 20: 103–120.

Lipscomb, S.T., Leve, L.D., Shaw, D.S., Neiderhiser, J.M., Scaramella, L.V., Ge, X., *et al.* (2012) 'Negative emotionality and externalizing problems in toddlerhood: overreactive parenting as a moderator of genetic influences', *Development and Psychopathology*, 24: 167–179.

Logan, G.D., Schachar, R.J., and Tannock, R. (1997) 'Impulsivity and inhibitory control', *Psychological Science*, 8: 60–64.

McCrae, R.R. and Costa, P.T.J. (1997) 'Personality trait structure as a human universal', *American Psychologist*, 52: 509–516.

Moffitt, T.E. (1993) 'Adolescence-limited and life-course persistent antisocial behavior: a developmental taxonomy', *Psychological Review*, 100: 674–701.

Moffitt, T.E. (2005) 'The new look of behavioral genetics in developmental psychopathology: gene–environment interplay in antisocial behaviors', *Psychological Bulletin*, 131: 533–554.

Moffitt, T.E., Caspi, A., Harrington, H., and Milne, B.J. (2002) 'Males on the life-course-persistent and adolescent-limited antisocial pathways: follow-up at age 26 years', *Development and Psychopathology*, 14: 179–207.

Mullineaux, P.Y., Deater-Deckard, K., Petrill, S.A., Thompson, L.A., and DeThorne, L.S. (2009) 'Temperament in middle childhood: a behavioral genetic analysis of fathers' and mothers' reports', *Journal of Research in Personality*, 43: 737–746.

Nagin, D. and Tremblay, R.E. (1999) 'Trajectories of boys' physical aggression, opposition, and hyperactivity on the path to physically violent and non violent juvenile delinquency', *Chid Development*, 70: 1181–1196.

Nigg, J.T. (2006) 'Temperament and developmental psychopathology', *Journal of Child Psychology and Psychiatry*, 47: 395–422.

O'Connor, T.G., Caspi, A., DeFries, J.C., and Plomin, R. (2003) 'Genotype–environment interaction in children's adjustment to parental separation', *Journal of Child Psychology and Psychiatry*, 44: 849–856.

Olson, S.L., Bates, J.E., Sandy, J.M., and Lanthier, R. (2000) 'Early developmental precursors of externalizing behavior in middle childhood and adolescence', *Journal of Abnormal Child Psychology*, 28: 119–133.

Olson, S.L., Sameroff, A.J., Kerr, D.C.R., Lopez, N.L., and Wellman, H.M. (2005) 'Developmental foundations of externalizing problems in young children: the role of effortful control', *Development and Psychopathology*, 17: 25–45.

Pardini, D.A., Frick, P.J., and Moffitt, T.E. (2010) 'Building an evidence base for DSM-5 conceptualizations of oppositional defiant disorder and conduct disorder: introduction to the special section', *Journal of Abnormal Psychology*, 119: 683–688.

Patterson, G.R., Forgatch, M.S., Yoerger, K.L., and Stoolmiller, M. (1998) 'Variables that initiate and maintain an early-onset trajectory for juvenile offending', *Development and Psychopathology*, 10: 531–547.

Petitclerc, A., Boivin, M., Dionne, G., Zoccolillo, M., and Tremblay, R.E. (2009) 'Disregard for rules: the early development and predictors of a specific dimension of disruptive behavior disorders', *Journal of Child Psychology and Psychiatry*, 50: 1477–1484.

Petitclerc, A., Boivin, M., Dionne, G., Pérusse, D., and Tremblay, R.E. (2011) 'Genetic and environmental etiology of disregard for rules', *Behavior Genetics*, 41: 192–200.

Putnam, S.P., Gartstein, M.A., and Rothbart, M.K. (2006) 'Measurement of fine-grained aspects of toddler temperament: the Early Childhood Behavior Questionnaire', *Infant Behavior and Development*, 24: 386–401.

Raine, A., Reynolds, C., Venables, P.H., Mednick, S.A., and Farrington, D.P. (1998) 'Fearlessness, stimulation-seeking, and large body size at age 3 years as early predispositions to childhood aggression at age 11 years', *Archives of General Psychiatry*, 55: 745–751.

Regier, D.A., Kuhl, E.A., and Kupfer, D.J. (2013) 'The DSM-5: classification and criteria changes', *World Psychiatry*, 12: 92–98.

Rothbart, M.K. (2007) 'Temperament, development, and personality', *Current Directions in Psychological Science*, 16: 207–212.

Rothbart, M., Posner, M., and Hershey, K. (1995) 'Temperament, attention, and developmental psychopathology', in I.D.C.D. Cohen and I.D. Walman (eds), *Handbook of Temperament*, New York: Wiley, pp. 315–326.

Rutter, M., Moffitt, T.E., and Caspi, A. (2006) 'Gene–environment interplay and psychopathology: multiple varieties but real effects', *Journal of Child Psychology and Psychiatry*, 47: 226–261.

Sadeh, N., Javdani, S., Jackson, J.J., Reynolds, E.K., Potenza, M.N., Gelernter, J., *et al.* (2010) 'Serotonin transporter gene associations with psychopathic traits in youth vary as a function of socioeconomic resources', *Journal of Abnormal Psychology*, 119: 604–609.

Sadeh, N., Javdani, S., and Verona, E. (2013) 'Analysis of monoaminergic genes, childhood abuse, and dimensions of psychopathy', *Journal of Abnormal Psychology*, 122: 167–179.

Sanocki, W. (1969) 'The use of Eysenck's inventory for testing young prisoners', *Przeglad penitenciarny, (Warszawa)*, 7: 53–68.

Saudino, K.J., and Wang, M. (2012) 'Quantitative and molecular genetic studies of temperament', in M. Zetner and R.L. Shiner (eds), *Handbook of Temperament*, New York: Guilford, pp. 315–346.

Schwartz, C.E., Snidman, N., and Kagan, J. (1996) 'Early childhood temperament as a determinant of externalizing behavior in adolescence', *Development and Psychopathology*, 8: 527–537.

Shaw, D.S., Gilliom, M., Ingoldsby, E.M., and Nagin, D.N. (2003) 'Trajectories leading to school-age conduct problems', *Developmental Psychology*, 39: 189–200.

Shaw, D.S., Lacourse, E., and Nagin, D.S. (2005) 'Developmental trajectories of conduct problems and hyperactivity from ages 2 to 10', *Journal of Child Psychology and Psychiatry*, 46(9): 931–942.

Sheese, B.E., Voelker, P.M., Rothbart, M., and Posner, M.I. (2007) 'Parenting quality interacts with genetic variation in dopamine receptor D4 to influence temperament in early childhood', *Development and Psychopathology*, 19: 1039–1046.

Sheese, B.E., Rothbart, M., Voelker, P.M., and Posner, M.I. (2012) 'The dopamine receptor D4 gene 7-repeat allele interacts with parenting quality to predict effortful control in four-year-old children', *Child Development Research*, 1: 1–6.

Singh, A.L. and Waldman, I.D. (2010) 'The etiology of associations between negative emotionality and childhood externalizing disorders', *Journal of Abnormal Psychology*, 119: 376–388.

Skilling, T.A., Harris, G.T., Rice, M.E., and Quinsey, V.L. (2002) 'Identifying persistently antisocial offenders using the Hare Psychopathy Checklist and DSM antisocial personality disorder criteria', *Psychological Assessment*, 14: 27–38.

Smith, H.J., Sheikh, H.I., Dyson, M.W., Olino, T.M., Laptook, R.S., Durbin, C.E., *et al.* (2012) 'Parenting and child DRD4 genotype interact to predict children's early emerging effortful control', *Child Development*, 83: 1932–1944.

Sulik, M.J., Eisenberg, N., Lemery-Chalfant, K., Spinard, T. L., Silva, K. M., Eggum, N.D., and Verrelli, B. C. (2012) 'Interactions between serotonin transporter gene haplotypes and quality of mothers' parenting predict the development of children's noncompliance', *Developmental Psychology*, 48(3): 740.

Taylor, A. and Kim-Cohen, J. (2007) 'Meta-analysis of gene–environment interactions in developmental psychopathology', *Development and Psychopathology*, 19: 1029–1037.

Taylor, J., Allan, N., Mikolajewski, A.J., and Hart, S.A. (2013) 'Common genetic and nonshared environmental factors contribute to the association between socioemotional dispositions and the externalizing factor in children', *Journal of Child Psychology and Psychiatry*, 54: 67–76.

Tremblay, R.E., Japel, C., Perusse, D., McDuff, P., Boivin, M., Zoccolillo, M., and Montplaisir, J. (1999) 'The search for the age of "onset" of physical aggression: Rousseau and Bandura revisited', *Criminal Behaviour and Mental Health*, 9: 8–23.

Trentacosta, C.J., Hyde, L.W., Shaw, D.S., and Cheong, J. (2009) 'Adolescent dispositions for antisocial behavior in context: the roles of neighborhood dangerousness and parental knowledge', *Journal of Abnormal Psychology*, 118: 564–575.

Trouton, A., Spinath, F.M., and Plomin, R. (2002) 'Twins Early Development Study (TEDS): a multivariate, longitudinal genetic investigation of language, cognition and behaviour problems in childhood', *Twin Research*, 5: 444–448.

Tucker-Drob, E.M., Rhemtulla, M., Harden, K.P., Turkheimer, E., and Fask, D. (2011) 'Emergence of a gene × socioeconomic status interaction on infant mental ability between 10 months and 2 years', *Psychological Science*, 22: 125–133.

Tuvblad, C., Grann, M., and Lichtenstein, P. (2006) 'Heritability for adolescent antisocial behavior differs with socioeconomic status: gene–environment interaction', *Journal of Child Psychology and Psychiatry*, 47: 734–743.

Viding, E., and McCrory, E.J. (2012) 'Genetic and neurocognitive contributions to the development of psychopathy', *Development and Psychopathology*, 24: 969–990.

Viding, E., Blair, R.J.R., Moffitt, T.E., and Plomin, R. (2005) 'Evidence for substantial genetic risk for psychopathy in 7-year-olds', *Journal of Child Psychology and Psychiatry*, 46(6): 592–597.

Wachs, T.D. and Bates, J.E. (2001) 'Temperament', in I.G.B.A. Fogel (ed.), *Blackwell Handbook of Infant Development*, Oxford: Blackwell.

Waldman, I.D., Tackett, J.L., Van Hulle, C.A., Applegate, B., Pardini, D., Frick, P.J., and Lahey, B.B. (2011) 'Child and adolescent conduct disorder substantially shares genetic influences with three socioemotional dispositions', *Journal of Abnormal Psychology*, 120: 57–70.

White, J.L., Moffitt, T.E., Caspi, A., Bartusch, D.J., Needles, D.J., and Stouthamer-Loeber, M. (1994) 'Measuring impulsivity and examining its relationship to delinquency', *Journal of Abnormal Psychology*, 103: 192–208.

Wright, J.P. and Beaver, K.M. (2005) 'Do parents matter in creating self-control in their children? A genetically informed test of Gottfredson and Hirschi's theory of low self-control', *Criminology*, 43: 1169–1202.

Young, S.E., Smolen, A., Hewitt, J.K., Haberstick, B.C., Stallings, M.C., Corley, R.P., and Crowley, T.J. (2006) 'Interaction between MAO-A genotype and maltreatment in the risk for conduct disorder: failure to confirm in adolescent patients', *American Journal of Psychiatry*, 163: 1019–1025.

Zahn-Waxler, C., Radke-Yarrow, M., Wagner, E., and Chapman, M. (1992) 'Development of concern for others', *Developmental Psychology*, 28: 126–141.

Zuckerman, M. (1971) 'Dimensions of sensation seeking', *Journal of Consulting and Clinical Psychology*, 36: 45–55.

29

Blackout

The traumatic brain injury association with career criminality withstands powerful confounds

Monic P. Behnken, Matt DeLisi, Chad R. Trulson, and Michael G. Vaughn

Introduction

The association between a traumatic brain injury and behavior is well known to psychology and neurology students in the case of Phineas Gage. On September 13, 1848, Gage was injured in a railroad accident when a tamping iron blasted through his face and exited his head. Despite losing consciousness and suffering heavy bleeding, Gage miraculously not only survived the blast, but appeared to recover quickly. However, the accident caused a dramatic personality transformation of Gage from a hard-working, responsible, intelligent, prudent, and socially well adjusted person to an irreverent, impulsive, capricious, rowdy, irresponsible person whose life devolved into that of a drifter. Subsequent research of Gage's skull revealed that his injuries were consistent with persons with similar injuries who display similar impairments in rational decision making and emotional processing (Damasio *et al.*, 1994; also see Van Horn *et al.*, 2012)—a neuropsychological profile of crime.

Criminology is in the midst of a paradigm shift where biosocial factors are increasingly being implicated in the causes and correlates of crime. To date, the preponderance of this research has focused on the behavioral/molecular genetics and brain imaging correlates of antisocial behavior (Barnes and Boutwell, 2012; Rocque *et al.*, 2012; Tielbeek *et al.*, 2012; Jennings *et al.*, 2013; Portnoy *et al.*, 2013). Less research has focused on traumatic brain injury—defined as a head injury involving loss of consciousness and ongoing cognitive or social impairment—and its association with serious, violent, and chronic antisocial behavior (see Elliott, 1992). Even less research has empirically examined the robustness of an association between traumatic brain injury and serious criminality while controlling for important theoretically informed empirical confounds. The current study aims to fill this void in the literature.

Researchers from a variety of disciplines have provided evidence of a significant association between traumatic brain injury, serious antisocial behavior, and associated morbidity (Brower and Price, 2001; Kolakowsky-Hayner and Kreutzer, 2001; Timonen *et al.*, 2002; Shiroma *et al.*, 2010b; Farrer and Hedges, 2011; Perkes *et al.*, 2011). For example, meta-analytic studies indicate that traumatic brain injury is several times more prevalent in antisocial or juvenile justice system-involved samples than in the general population (Farrer and Hedges, 2011; Shiroma *et al.*, 2010a; Farrer *et al.*, 2012, 2013). Frequently, the proximal causes of traumatic brain injury relate to a fall or accident stemming from drug and alcohol intoxication, an aggressive or violent interpersonal conflict, or both (Kreutzer *et al.*, 1995; Taylor *et al.*, 2003). Upon hospitalization for a traumatic brain injury, most individuals are significantly intoxicated (Taylor *et al.*, 2003). In this way, a traumatic brain injury can be seen as an essential part of a broader behavioral repertoire characterized by risk-taking and general proneness for externalizing behaviors (Brower and Price, 2001; Mawson, 2005; Felde *et al.*, 2006; Vaughn *et al.*, 2011, 2014b) that is consistent with multiple theoretical models (Gottfredson and Hirschi, 1990; Horvath and Zuckerman, 1993; Arnett, 1996).

In recent years, investigators have begun to evaluate the traumatic brain injury–crime linkage in light of important conceptual models of antisocial behavior. Perron and Howard (2008) examined the prevalence and correlates of traumatic brain injury among confined delinquents. They reported multiple significant differences, with traumatic brain injury youth having greater prevalence of mental illness and lifetime heroin, cocaine, marijuana, and ecstasy use. In addition, traumatic brain injury youth were more likely to have been the victims of assault, victims of robbery, or to have been attacked. Those with traumatic brain injury displayed more varied and severe psychopathology including higher mean scores on depression, anxiety, psychoticism, paranoia, blame externalization, impulsiveness, fearlessness, and coldheartedness. In other words, traumatic brain injury youth had more psychiatric problems and were more psychopathic than their peers who had not incurred a traumatic brain injury. Most importantly, traumatic brain injury youth displayed higher scores on self-report delinquency measures.

Recently, Vaughn and colleagues (2014a) examined traumatic brain injury among juvenile offenders using data from youth involved in the Pathways to Desistance Study. They found that traumatic brain injury youth had worse drug and alcohol problems, more pronounced neuropsychiatric deficits, greater involvement in delinquency, and greater exposure to violence. They also performed logistic regression models with traumatic brain injury as the dependent variable, and found that male gender, psychiatric symptoms, impulsivity, and victimization were associated with traumatic brain injury status. Given the associations between traumatic brain injury, neuropsychological deficits, and self-regulation problems, their findings lent support for developmental taxonomy (Moffitt, 1993) and self-control theory (Gottfredson and Hirschi, 1990), two of the most prominent theoretical models in criminal justice.

Current aim

In short, emerging research indicates that traumatic brain injury is importantly associated with serious delinquency and antisocial behavior, and is predicted by prominent criminological models. The current study aims to further this literature by examining empirically the linkages between traumatic brain injury and career criminality among a sample of serious delinquents while also controlling for important neurological, neurobehavioral, and demographic confounds.

Method

Participants and procedures

Data are derived from structured face-to-face interviews of 723 residents in the Missouri Division of Youth Services (DYS), which is the legal guardian of youth who are adjudicated and committed to its care by the state's 45 juvenile courts. The sample represented over 97 percent of the total DYS population at the time of data collection. The sample was primarily male (87 percent), white (55.3 percent) or African American (32.9 percent), and approximately 15 years of age. These data have been used to study a variety of topics relating to delinquent youth (Vaughn *et al.*, 2007, 2008) including traumatic brain injury (Perron and Howard, 2008).

The recruitment protocol ensured that youths who had completed the interview at one facility did not complete an interview at another facility. This study was approved by DYS, the Washington University Human Studies Committee Institutional Review Board, and the federal Office of Human Research Protection, and was granted a Certificate of Confidentiality by the National Institute on Drug Abuse (NIDA). Youth received US$10 for their participation.

Interviews were conducted by a team of 15 interviewers. Seven core interviewers were graduate students and completed 530 (73.3 percent) interviews. All interviewers completed an intensive one-day training session. The training comprised an overview of the study, the consent procedures and the questionnaire, and what to do if youth reported any suicidal thoughts or unusual distress during the interview. Practice interviews were conducted with the research assistants until interviewers were comfortable and familiar with the procedures. An interview editor was on site at each facility as youth were interviewed, to minimize interviewer omissions and errors. Interviews were conducted in large rooms that provided private areas where confidential interviews could be conducted. Personal interviews were completed with all subjects over a 3-month period, and each interview lasted about 45 minutes in duration.

Measures

Traumatic brain injury

Youth were asked "In your lifetime, have you ever had a head injury that caused you to blackout for more than 20 minutes?" More than 18 percent (n = 132) of youth answered yes (coded as 1) and 81.7 percent (n = 588) answered no (coded as 0).

Neurological medical condition

Controlling for a neurological medical condition is important given evidence for their potential associations with antisocial behavior (Fazel *et al.*, 2009, 2011; Hanlon *et al.*, 2010). To assess this variable, youth were asked "Have you ever suffered from any of the following: brain infection, brain tumor, or brain disease such as epilepsy or meningitis?" Just over 2 percent of youth (n = 15) reported a neurological medical condition (no coded as 0; yes coded as 1). In the words of the youth, these included epilepsy, brain seizures, brain infection following a car accident, lead poisoning, meningitis, fluid on the spinal cord, spinal meningitis, viral meningitis, and brain seizure disorder.

Neurobehavioral conditions

Three important neurobehavioral conditions (low self-control, psychopathy, and ADHD) were included given their robust relation to antisocial behavior. There is tremendous evidence that low self-control is associated with antisocial behavior (de Ridder *et al.*, 2012; Boisvert *et al.*, 2013; Yancey *et al.*, 2013) and low self-control is also implicated in career criminality (DeLisi and Vaughn, 2008; DeLisi and Piquero, 2011). We constructed a 15-item self-control scale based on items that reflect forms of impulse control, risk-taking, difficulty in controlling one's temper, and susceptibility to boredom. All items were standardized into *z-scores* and subjected to exploratory factor analysis. We then used maximum likelihood factor analysis considering one to four factors. Results from this procedure indicated reduction in chi-square fit indices as we proceeded from one to four factors indicating unidimensionality. Internal consistency reliability for the self-control scale showed good reliability (α = 0.83).

Modified Psychopathic Personality Inventory-Short Version (mPPI-SV)

A modified 56-item version of the Psychopathic Personality Inventory Short-Form derived from the Psychopathic Personality Inventory (PPI; Lilienfeld and Andrews, 1996) was used. The PPI is a pure measure of psychopathic personality because it does not contain measures of antisocial behavior that would be tautological to crime-related outcome variables. The PPI and its variants have proven to be a leading self-report measure of psychopathic personality (for reviews see Lilienfeld and Fowler, 2006; Vaughn and Howard, 2005a,b) and is a commonly used measure for use with juvenile justice-involved adolescents (Vaughn and Howard, 2005a,b). In developing the PPI, Lilienfeld and Andrews (1996) created a measure that included eight subscales that theoretically embody the psychopathic personality. *Machiavellian Egocentricity* captures the narcissistic, self-interested, and ruthless social functioning aspect of psychopathy. *Social Potency* assesses an individual's perceived ability to manipulate and influence others. *Coldheartedness* assesses the callousness, unemotionality, and lack of sentimentality of the disorder. *Carefree Nonplanfulness* relates to indifference to planning one's actions and general irresponsibility. *Fearlessness* captures an absence of anxiety and harm avoidance and a willingness to engage in risky behaviors. *Blame Externalization* assesses an external locus of control where an individual blames others and rationalizes their behavior. *Impulsive Nonconformity* reflects a lack of concern for social rules. *Stress Immunity* encompasses an absence of reaction to stimuli that are ordinarily anxiety-inducing. Total score on the mPPI-SF (Cronbach's α = 0.76) is a summary measure of psychopathic personality traits (M = 136.42, SD = 14.04, range = 95–201).

Attention deficit hyperactivity disorder

Youth self-reported an ADHD diagnosis (n = 269 or 37.2 percent indicated yes; n = 454 or 62.8 percent indicated no). Prior research using juvenile justice-involved youth has shown that ADHD is an important correlate of serious antisocial behavior (Young *et al.*, 2009; Mordre *et al.*, 2011; Schilling *et al.*, 2011; Gordon *et al.*, 2012; DeLisi *et al.*, 2013).

Demographics

Three basic demographic variables were included as controls in the multivariate models based on their association with career criminality (DeLisi and Piquero, 2011). Age was continuously coded in years (M = 15.49, SD = 1.23, range = 11–20). In terms of sex, the sample was 87 percent male and 13 percent female. In terms of race, a dichotomous term for African American status (0 = no, 1 = yes) was included given sharp race differences in serious and violent delinquency and various forms of antisocial behavior (Elliott, 1994; Tapia, 2010; Tillyer and Engel, 2012; Beaver *et al.*, 2013). The sample was 55 percent white and 33 percent African American.

Dependent variable—Career Criminal Index

The Career Criminal Index (CCI; Vaughn and DeLisi, 2008) (M = 0.02, SD = 3.32, range = –6.63–11.48) is a unidimensional measure that encompasses the diverse forms of antisocial behavior, substance abuse problems, and criminal justice system involvement of serious, violent, and chronic offenders. It includes scale items from self-report measures that assessed various, yet overlapping, forms of delinquency and substance use, juvenile justice system involvement, antisocial onset, and personal victimizations occurring in the 12 months prior to the youth's current confinement. The CCI shows good internal consistency with the current data (α = 0.82). Items forming the CCI were subjected to exploratory factor analysis. Inspection of the scree plot, extracted sums of squares loadings (i.e. eigenvalues and percentage of cumulative variance) supported the presence of a single dominant factor (i.e. career criminality).

Analytical strategy and regression diagnostics

Difference-of-means *t-tests* and χ^2 tests, and ordinary least-squares (OLS) regression models, were executed. Post-estimation diagnostics were performed on all regression models including the full model containing all covariates. There were no problems with multicollinearity with a mean variance inflation factor (VIF) = 1.11 (VIF ranged from 1.01 to 1.24 with tolerances [1/VIF] ranging from 0.81 to 0.99). Specification error was also not detected (Ramsey RESET test F (3, 698) = 0.80, P < 0.49). There was modest evidence of heteroskedasticity (Cook–Weisberg text χ^2 = 16.57, P < 0.001). To redress this, the regression models were performed with robust HC3 standard errors that correct for potential heteroskedasticity bias (Long and Ervin, 2000).

Results

Traumatic brain injury and antisocial conditions comparisons

Difference-of-means *t-tests* examined group differences between traumatic brain injury positive and negative youth and scores on career criminality, self-control, and psychopathic personality. Chi-square measure of association crosstab was used to assess group differences in ADHD prevalence. As shown in Table 29.1, youth who reported a traumatic brain injury were significantly more antisocial and displayed greater psychopathology including higher scores on the CCI (t = –7.03, P < 0.0000), low self-control index (t = –4.67, P < 0.0000), PPI (t = –3.62, P < 0.0002), and greater prevalence of ADHD (χ^2 = 8.77, Cramér's V = 0.11, P < 0.003).

OLS regression model for CCI score

Table 29.2 reveals the four OLS regression models for the CCI. In model 1, traumatic brain injury is the only independent variable in the model and its effect is significant (β = 0.25, t = 6.73, P < 0.0001). In model 2, the neurological medical condition covariate is added. Its effect is not significant, and the inclusion of the new measure does not change the effect of traumatic brain injury, although the model F declines by about half. In model 3, three neurobehavioral conditions—low self-control, ADHD, and psychopathy—are stepped in. Although traumatic brain injury remains significant, its effect is attenuated. Although ADHD is not significantly associated with career criminality, the effects of low self-control (β = 0.32, t = 8.16, P < 0.0001) and psychopathy (β = 0.26, t = 7.56, P < 0.0001) are relatively strong. In model

Table 29.1 Traumatic brain injury and antisocial conditions

Variable	*Traumatic brain injury*		*t*	χ^2	*Cramér's V*	*P*
	Yes n = 132	*No* n = 588				
Career Criminal Index score	1.80	–0.38	–7.03			<0.0000
Low self-control	2.97	0.66	1.67			<0.0000
Psychopathic Personality Inventory total	140.39	135.53	–3.62			<0.0002
ADHD	0.48	0.35		8.77	0.11	<0.003

Table 29.2 OLS regression model for Career Criminality Index score

Variable	*Model 1*		*Model 2*		*Model 3*		*Model 4*	
	β	t	β	t	β	t	β	t
Traumatic brain injury	0.25***	6.73	0.25***	6.73	0.17***	5.00	0.16***	4.66
Neurological condition								
Neurological medical condition			0.02	0.36	0.02	0.45	0.02	0.52
Neurobehavioral condition								
Low self-control					0.32***	8.16	0.33***	8.42
ADHD					–0.05	–1.54	–0.03	–0.82
PPI total score					0.26***	7.56	0.27***	7.81
Demographics								
Age							0.02	0.69
Sex							–0.05	–1.36
African American							0.10**	3.11
F	45.24***		22.66***		48.84***		31.62***	
R^2	0.065		0.065		0.284		0.295	

Notes: *** $P < 0.0001$, ** $P < 0.01$.

4, demographic characteristics are introduced. Age and sex are not significantly associated with career criminality. However, African American youth ($\beta = 0.10$, $t = 3.11$, $P < 0.01$) are more likely to have higher scores on the CCI. Across all models, traumatic brain injury was significant at the $P < 0.0001$ significance level and uniquely accounted for 6.5 percent of the total variance in the CCI index.

Discussion

Although Phineas Gage's traumatic brain injury was extraordinary, thousands of persons each year incur a traumatic brain injury, and these injuries are associated with antisocial conduct. According to the current results, delinquents who reported a traumatic brain injury were significantly more antisocial than youth without such an injury. Specifically, those with a traumatic brain injury had lower self-control, were more psychopathic, and displayed higher scores on a measure of career criminality. In multivariate models, traumatic brain injury maintained a robust association with career criminality scores despite controls for a neurological medical condition, self-control, psychopathy, ADHD, age, race, and sex. These findings are consistent with recent research (Farrer *et al.*, 2012, 2013; Perron and Howard, 2008; Vaughn *et al.*, 2014a), but also extend the traumatic brain injury–antisocial behavior literature to career criminality.

The current study also suggests that the negative effects of a traumatic brain injury are not easily attenuated in the sense that their statistical effects withstood considerable confounds. A traumatic brain injury can be viewed as a proxy for self-regulation deficits or executive dysfunction. Through the inclusion of self-control and psychopathy, conditions that embody these deficits, these results indicate that traumatic brain injury is uniquely associated with crime. A possible explanation for the association between traumatic brain injury and career criminality relates to interpersonal violence and the resultant declines in cognition and intelligence. For instance, Schwartz and Beaver (2013) analyzed data from the National Longitudinal Study of Adolescent Health and found that fighting-related head injuries were causally related to declines in intelligence and cognitive functioning, even while controlling for age, race, sex, and changes in socioeconomic status. Serious, violent, and chronic offenders tend to have extensive violence histories as both perpetrator and victim, and the head injuries that are incurred over time likely contribute to deteriorating cognitive functioning and continued involvement in antisocial behavior.

We are unable to disentangle the causal status of traumatic brain injury in the present investigation. As such, it is unresolved whether trait self-control (low) and higher scores on psychopathy traits increase the probability of exposure behaviors and situations that can lead to suffering a traumatic brain injury. Although it does seem probable that these aforementioned traits increase the likelihood of a blow to the head, our findings suggest an additive and/or independent effect on a youth's antisocial career. It could also be conceptualized that a traumatic brain injury deepens or exacerbates existing temperament facets of poor self-regulation, antagonism, and negative emotionality (DeLisi and Vaughn, 2014). Thus traumatic brain injury possesses incremental validity and clinically meaningful risk over and above these disorders. Given the relative simplicity and time efficiency involved in querying adolescent offenders about suffering a traumatic brain injury, practitioners in the juvenile justice system may benefit from routinely incorporating this question in their assessment protocols.

There are limitations to the current study that could inform future research, particularly criminological data-collection efforts. First, the study relies on self-report data, and the typical over- and under-reporting threats apply to this type of data collection effort. An additional limitation is the use of a cross-sectional design. This approach prevented an investigation of the timing of the traumatic brain injury and whether youth who reported an injury were already more antisocial before the event, also known as premorbid antisociality (Miller, 1995; Kolakowsky-Hayner and Kreutzer, 2001; Felde *et al.*, 2006). There is considerable evidence for continuity in antisocial behavior, especially among the most severe offenders (DeLisi and Piquero, 2011; Vaughn *et al.*, 2011; Jennings and Reingle, 2012), thus longitudinal designs would permit assessments of head injury across developmental periods to assess whether offending trajectories intensify after a brain injury. This is both theoretically and practically important. Recent juvenile justice reforms that proscribe capital punishment and life imprisonment without parole for juveniles (*Roper v. Simmons*, 2005; *Graham v. Florida*, 2010; *Miller v. Alabama*, 2012) are based in part on the idea that neurological immaturity among adolescents mitigates their criminal responsibility *vis-à-vis* the harshest correctional sanctions. It would be interesting to see how traumatic brain injuries relate to antisocial conduct across adolescence and into adulthood.

The current single-item measure of traumatic brain injury lacks important information that would provide greater context to the proximal cause of the injury. For instance, a rule of thumb in the forensic sciences literature is the hat brim-line rule, which suggests that injuries located above the hat brim-line of the head are suggestive of a blow from a weapon, whereas injuries at the level of the hat brim-line are suggestive of a fall (Kremer and Sauvageau, 2009). When collecting data on traumatic brain injury, criminologists should gather data about the location of the wound to help ascertain whether the injury was incurred from a fall (such as one caused by intoxication or risk-taking behavior) or produced by a weapon. Expanded measures of traumatic brain injury and the circumstances in which they occurred will help to specify the contextual reasons for the traumatic brain injury–crime relationship. And finally, the current self-report of ADHD would ideally have been supplemented by additional raters (see Young *et al.*, 2010).

Notwithstanding these limitations, the results clearly show that traumatic brain injury possesses incremental validity over and above other neurobehavioral confounds of delinquency, which has the potential to be of tremendous value to those studying this population. The finding that delinquents with a traumatic brain injury were higher on measures of career criminality, low self-control, and ADHD has important implications for conceptualizing, treating, and adjudicating these youths. Moreover, these findings suggest that researchers and those engaged in forensic practice would benefit from understanding the connection between traumatic brain injury and career criminality, as those with this condition appear to have worse and more extensive offending histories and risk factors for offending than those without this condition. As criminologists move toward a deeper understanding of biosocial correlates of crime, these results show that understanding the impact of traumatic brain injury and its association to serious, violent, and chronic offending should receive considerable attention.

References

Arnett, J.J. (1996) 'Sensation seeking, aggressiveness, and adolescent reckless behavior', *Personality and Individual Differences*, 20(6): 693–702.

Barnes, J.C. and Boutwell, B.B. (2012) 'On the relationship of past to future involvement in crime and delinquency: a behavior genetic analysis', *Journal of Criminal Justice*, 40(1): 94–102.

Beaver, K.M., DeLisi, M., Wright, J.P., Boutwell, B.B., Barnes, J.C., and Vaughn, M.G. (2013) 'No evidence of racial discrimination in criminal justice processing: results from the National Longitudinal Study of Adolescent Health', *Personality and Individual Differences*, 55: 29–34.

Boisvert, D., Boutwell, B.B., Barnes, J.C., and Vaske, J. (2013) 'Genetic and environmental influences underlying the relationship between low self-control and substance use', *Journal of Criminal Justice*, 41(4): 262–272.

Brower, M.C. and Price, B.H. (2001) 'Neuropsychiatry of frontal lobe dysfunction in violent and criminal behaviour: a critical review', *Journal of Neurology, Neurosurgery & Psychiatry*, 71(6): 720–726.

Damasio, H., Grabowski, T., Frank, R., Galaburda, A.M., and Damasio, A.R. (1994) 'The return of Phineas Gage: clues about the brain from the skull of a famous patient', *Science*, 264: 1102–1105.

DeLisi, M. and Piquero, A.R. (2011) 'New frontiers in criminal careers research, 2000–2011: a state-of-the-art review', *Journal of Criminal Justice*, 39(4): 289–301.

DeLisi, M. and Vaughn, M.G. (2008) 'The Gottfredson–Hirschi critiques revisited: reconciling self-control theory, criminal careers, and career criminals', *International Journal of Offender Therapy and Comparative Criminology*, 52(5): 520–537.

DeLisi, M. and Vaughn, M.G. (2014) 'Foundation for a temperament-based theory of antisocial behavior and criminal justice system involvement', *Journal of Criminal Justice*, 42: 10–25.

DeLisi, M., Neppl, T.K., Lohman, B.J., Vaughn, M.G., and Shook, J.J. (2013) 'Early starters: which type of criminal onset matters most for delinquent careers?', *Journal of Criminal Justice*, 41(1): 12–17.

Elliott, F.A. (1992) 'Violence: the neurologic contribution: an overview', *Archives of Neurology*, 49(6): 595–617.
Elliott, D.S. (1994) 'Serious violent offenders: onset, developmental course and termination. The American Society of Criminology 1993 Presidential Address', *Criminology*, 32: 1–21.
Farrer, T.J. and Hedges, D.W. (2011) 'Prevalence of traumatic brain injury in incarcerated groups compared to the general population: a meta-analysis', *Progress in Neuro-Psychopharmacology and Biological Psychiatry*, 35(2): 390–394.
Farrer, T.J., Frost, R.B., and Hedges, D.W. (2012) 'Prevalence of traumatic brain injury in intimate partner violence offenders compared to the general population: a meta-analysis', *Trauma, Violence, & Abuse*, 13(2): 77–82.
Farrer, T.J., Frost, R.B., and Hedges, D.W. (2013) 'Prevalence of traumatic brain injury in juvenile offenders: a meta-analysis', *Child Neuropsychology*, 19(3): 225–234.
Fazel, S., Philipson, J., Gardiner, L., Merritt, R., and Grann, M. (2009) 'Neurological disorders and violence: a systematic review and meta-analysis with a focus on epilepsy and traumatic brain injury', *Journal of Neurology*, 256(10): 1591–1602.
Fazel, S., Lichtenstein, P., Grann, M., and Långström, N. (2011) 'Risk of violent crime in individuals with epilepsy and traumatic brain injury: a 35-year Swedish population study', *PLoS Medicine*, 8(12): e1001150.
Felde, A.B., Westermeyer, J., and Thuras, P. (2006) 'Co-morbid traumatic brain injury and substance use disorder: childhood predictors and adult correlates', *Brain Injury*, 20(1): 41–49.
Gordon, J.A., Diehl, R.L., and Anderson, L. (2012) 'Does ADHD matter? Examining attention deficit and hyperactivity disorder on the likelihood of recidivism among detained youth', *Journal of Offender Rehabilitation*, 51(8): 497–518.
Gottfredson, M.R. and Hirschi, T. (1990) *A General Theory of Crime*, Stanford, CA: Stanford University Press.
Graham v. Florida, 560 U.S. __ (2010).
Hanlon, R.E., Rubin, L.H., Jensen, M., and Daoust, S. (2010) 'Neuropsychological features of indigent murder defendants and death row inmates in relation to homicidal aspects of their crimes', *Archives of Clinical Neuropsychology*, 25(1): 1–13.
Horvath, P. and Zuckerman, M. (1993) 'Sensation seeking, risk appraisal, and risky behavior', *Personality and Individual Differences*, 14(1): 41–52.
Jennings, W.G. and Reingle, J.M. (2012) 'On the number and shape of developmental/life-course violence, aggression, and delinquency trajectories: a state-of-the-art review', *Journal of Criminal Justice*, 40(6): 472–489.
Jennings, W.G., Piquero, A.R., and Farrington, D.P. (2013) 'Does resting heart rate at age 18 distinguish general and violent offending up to age 50? Findings from the Cambridge Study in Delinquent Development', *Journal of Criminal Justice*, 41(4): 213–219.
Kolakowsky-Hayner, S.A. and Kreutzer, J.S. (2001) 'Pre-injury crime, substance abuse, and neurobehavioural functioning after traumatic brain injury', *Brain Injury*, 15(1): 53–63.
Kremer, C. and Sauvageau, A. (2009) 'Discrimination of falls and blows in blunt head trauma: assessment of predictability through combined criteria', *Journal of Forensic Sciences*, 54(4): 923–926.
Kreutzer, J.S., Marwitz, J.H., and Witol, A.D. (1995) 'Interrelationships between crime, substance abuse, and aggressive behaviours among persons with traumatic brain injury', *Brain Injury*, 9(8): 757–768.
Lilienfeld, S.O. and Andrews, B.P. (1996) 'Development and preliminary validation of a self-report measure of psychopathic personality traits in noncriminal population', *Journal of Personality Assessment*, 66(3): 488–524.
Lilienfeld, S.O. and Fowler, K.A. (2006) 'The self-report assessment of psychopathy: problems, pitfalls, and promises', in C.J. Patrick (ed.), *Handbook of Psychopathy*, New York: Guilford Press, pp. 107–132.
Long, J.S. and Ervin, L.H. (2000) 'Using heteroscedasticity consistent standard errors in the linear regression model', *American Statistician*, 54(3): 217–224.
Mawson, A.R. (2005) 'Intentional injury and the behavioral syndrome', *Aggression and Violent Behavior*, 10(3): 375–405.
Miller, L. (1995) 'Traumatic brain injury and aggression', *Journal of Offender Rehabilitation*, 21(3–4): 91–104.
Miller v. Alabama, 567 U.S. __ (2012).
Moffitt, T.E. (1993) 'Adolescence-limited and life-course-persistent antisocial behavior: a developmental taxonomy', *Psychological Review*, 100(4): 674–701.
Mordre, M., Groholt, B., Kjelsberg, E., Sandstad, B., and Myhre, A.M. (2011) 'The impact of ADHD and conduct disorder in childhood on adult delinquency: a 30 year follow-up study using official crime records', *BMC Psychiatry*, 11(1): 57–70.

Perkes, I., Schofield, P.W., Butler, T. and Hollis, S.J. (2011) 'Traumatic brain injury rates and sequelae: a comparison of prisoners with a matched community sample in Australia', *Brain Injury*, 25(2): 131–141.

Perron, B.E. and Howard, M.O. (2008) 'Prevalence and correlates of traumatic brain injury among delinquent youths', *Criminal Behaviour and Mental Health*, 18(4): 243–255.

Portnoy, J., Chen, F.R., and Raine, A. (2013) 'Biological protective factors for antisocial and criminal behavior', *Journal of Criminal Justice*, 41(5): 292–299.

de Ridder, D.T., Lensvelt-Mulders, G., Finkenauer, C., Stok, F.M., and Baumeister, R.F. (2012) 'Taking stock of self-control: a meta-analysis of how trait self-control relates to a wide range of behaviors', *Personality and Social Psychology Review*, 16(1): 76–99.

Rocque, M., Welsh, B.C., and Raine, A. (2012) 'Biosocial criminology and modern crime prevention', *Journal of Criminal Justice*, 40(4): 306–312.

Roper v. Simmons, 543 U.S. 551 (2005).

Schilling, C.M., Walsh, A., and Yun, I. (2011) 'ADHD and criminality: a primer on the genetic, neurobiological, evolutionary, and treatment literature for criminologists', *Journal of Criminal Justice*, 39(1): 3–11.

Schwartz, J.A. and Beaver, K.M. (2013) 'Serious fighting-related injuries produce a significant reduction in intelligence', *Journal of Adolescent Health*, 53(4): 520–525.

Shiroma, E.J., Ferguson, P.L., and Pickelsimer, E.E. (2010a) 'Prevalence of traumatic brain injury in an offender population: a meta-analysis', *Journal of Correctional Health Care*, 16(2): 147–159.

Shiroma, E.J., Pickelsimer, E.E., Ferguson, P.L., Gebregziabher, M., Lattimore, P.K., Nicholas, J.S., *et al.* (2010b) 'Association of medically attended traumatic brain injury and in-prison behavioral infractions: a statewide longitudinal study', *Journal of Correctional Health Care*, 16(4): 273–286.

Tapia, M. (2010) 'Untangling race and class effects on juvenile arrests', *Journal of Criminal Justice*, 38: 255–265.

Taylor, L.A., Kreutzer, J.S., Demm, S.R., and Meade, M.A. (2003) 'Traumatic brain injury and substance abuse: a review and analysis of the literature', *Neuropsychological Rehabilitation*, 13(1–2): 165–188.

Tielbeek, J.J., Medland, S.E., Benyamin, B., Byrne, E.M., Heath, A.C., Madden, P.A., *et al.* (2012) 'Unraveling the genetic etiology of adult antisocial behavior: a genome-wide association study', *PLoS ONE*, 7(10): e45086.

Tillyer, R. and Engel, R.S. (2012) 'Racial differences in speeding patterns: exploring the differential offending hypothesis', *Journal of Criminal Justice*, 40(4): 285–295.

Timonen, M., Miettunen, J., Hakko, H., Zitting, P., Veijola, J., von Wendt, L., and Räsänen, P. (2002) 'The association of preceding traumatic brain injury with mental disorders, alcoholism and criminality: the Northern Finland 1966 Birth Cohort Study', *Psychiatry Research*, 113(3): 217–226.

Van Horn, J.D., Irimia, A., Torgerson, C.M., Chambers, M.C., Kikinis, R., and Toga, A.W. (2012) 'Mapping connectivity damage in the case of Phineas Gage', *PLoS ONE*, 7(5): e37454.

Vaughn, M.G. and DeLisi, M. (2008) 'Were Wolfgang's chronic offenders psychopaths? On the convergent validity between psychopathy and career criminality', *Journal of Criminal Justice*, 36(1): 33–42.

Vaughn, M.G. and Howard, M.O. (2005a) 'The construct of psychopathy and its role in contributing to the study of serious, violent, and chronic youth offending', *Youth Violence and Juvenile Justice*, 3: 235–252.

Vaughn, M.G. and Howard, M.O. (2005b) 'Self-report measures of juvenile psychopathic personality traits: a comparative review', *Journal of Emotional and Behavioral Disorders*, 13(3): 152–162.

Vaughn, M.G., DeLisi, M., Beaver, K.M., Wright, J.P., and Howard, M.O. (2007) 'Toward a psychopathology of self-control theory: the importance of narcissistic traits', *Behavioral Sciences & the Law*, 25(6): 803–821.

Vaughn, M.G., Howard, M.O., and DeLisi, M. (2008) 'Psychopathic personality traits and delinquent careers: an empirical examination', *International Journal of Law and Psychiatry*, 31(5): 407–416.

Vaughn, M.G., DeLisi, M., Gunter, T., Fu, Q., Beaver, K.M., Perron, B.E., and Howard, M.O. (2011) 'The severe 5%: a latent class analysis of the externalizing behavior spectrum in the United States', *Journal of Criminal Justice*, 39(1): 75–80.

Vaughn, M.G., Salas-Wright, C.P., DeLisi, M., and Perron, B.E. (2014a) 'Correlates of traumatic brain injury among juvenile offenders: a multi-site study', *Criminal Behaviour and Mental Health*, 24: 188–203.

Vaughn, M.G., DeLisi, M., Salas-Wright, C., and Maynard, B.R. (2014b) 'Examining violence and externalizing behavior among youth in the United States: is there a severe 5%?', *Youth Violence and Juvenile Justice*, 12: 3–21.

Yancey, J.R., Venables, N.C., Hicks, B.M., and Patrick, C.J. (2013) 'Evidence for a heritable brain basis to deviance-promoting deficits in self-control', *Journal of Criminal Justice*, 41(5): 309–317.

Young, S., Chesney, S., Sperlinger, D., Misch, P., and Collins, P. (2009) 'A qualitative study exploring the life-course experiences of young offenders with symptoms and signs of ADHD who were detained in a residential care setting', *Criminal Behaviour and Mental Health*, 19(1): 54–63.

Young, S., Gudjonsson, G., Misch, P., Collins, P., Carter, P., Redfern, J., and Goodwin, E. (2010) 'Prevalence of ADHD symptoms among youth in a secure facility: the consistency and accuracy of self- and informant-report ratings', *Journal of Forensic Psychiatry & Psychology*, 21(2): 238–246.

30

Social stress and the gene–environment influences on crime and delinquency

D. Anne Winiarski, Erica L. Smearman, and Patricia A. Brennan

Introduction

Individuals who engage in criminal and delinquent behavior incur substantial financial and social burdens to the public. Adolescents who engage in delinquent behaviors have a higher risk of adverse consequences at both individual and societal levels, including strained interpersonal relationships with one's family and community (Zigler *et al.*, 1992), poor academic performance (Hinshaw, 1992), and an increased risk of co-morbid mental health problems (Barrett and Katsiyannis, 2009). From a financial standpoint, each child diagnosed with conduct problems costs society about $70,000 over a 7-year period (Foster and Jones, 2005). Reducing the number of individuals in prison could save state governments about $7.6 billion and the federal government about $7.2 billion annually (Schmitt *et al.*, 2010). Before researchers, clinicians, and policy-makers can begin designing effective interventions to target those at risk for engaging in delinquent and criminal behaviors, we must first come to a consensus about the kinds of factors that interact to shape their development.

Countless theories have been proposed to help explain the causes of crime and offending, ranging from perspectives that focus solely on environment ("nurture") to those that focus solely on genetic inheritance ("nature"). Other researchers have taken a more "interactionist" approach to understanding crime, and suggest that genes interact with the environment to bring about certain phenotypes, such as delinquency, persistent aggression, and related externalizing behaviors. In this chapter we use the terms delinquency and externalizing behavior interchangeably. Delinquency, defiance, impulsivity, and aggression are all core feature of externalizing behavior problems (Achenbach and Edelbrock, 1978), and children who display these behaviors can be clinically diagnosed with conduct disorder (CD) or oppositional defiant disorder (ODD).

Evaluating the expansive gene by environment (G × E) literature related to delinquency is beyond the scope of the present chapter, thus the primary focus is to summarize studies that examine how an individual's genetic predisposition interacts with social strains to bring about delinquency and criminal behavior. We begin by operationalizing social stress and reviewing the importance of strain on the development of delinquency and crime. We then explore how the field of genetics can supplement what is already known about the relationship between strain and crime, with particular emphasis on the monoamine oxidase A gene (MAOA), the serotonin transporter gene (5-HTTLPR), and the oxytocin receptor gene (OXTR). The chapter concludes with a brief discussion on the limitations of G × E research, as well as a comment on a developing area of inquiry that we believe may provide fruitful data on the interaction between genetic and environmental factors.

Understanding interpersonal stress in the context of sociological theories

Sociological theories exploring environmental contexts provide researchers with useful insight into the important role the environment can play in exacerbating genetic predispositions to delinquency. In this chapter, the sociological theory guiding our understanding of how features of the environment can interact with genetic factors to influence crime and delinquency is general strain theory (GST). The theoretical premise of GST asserts that individuals are pressured into crime as a result of experiencing adverse strains or stressors. These adverse experiences result in negative feelings, which in turn increase an individual's risk of engaging in crime to "cope" with the strain (Agnew, 2007, 2013). Agnew outlines five important characteristics that contribute to a higher likelihood of criminal coping: (1) poor coping skills and resources; (2) low levels of conventional social support; (3) low social control; (4) associations with criminal others and beliefs favorable to crime; and (5) exposure to situations where the benefits of criminal coping appear to be high, whereas the costs appear to be low. Although many studies supporting GST have been conducted, Agnew asserts that an understanding of how biological factors interact with strains would be an important addition to the empirical literature.

An individual with strained interpersonal relationships may not have access to appropriate coping resources, thereby increasing the likelihood of what Agnew refers to as "criminal coping." Although environmental factors can certainly trigger disadvantageous developmental trajectories, not all individuals exposed to social strains are likely to become criminals. The authors of this chapter argue that one possible underexplored area in GST is the potential role of genetic factors in predicting risk for criminal and delinquent outcomes. In this chapter, we demonstrate how the G × E literature can be used to inform an already well established theory of crime, and how examining biological factors related to crime may help criminologists and clinicians understand why some individuals appear so resilient in the face of adverse social environments, whereas others engage to maladaptive coping behaviors including delinquency.

GST views "strains" as factors that pressure an individual to cope with perceived stress and injustice through criminal involvement. These factors can come from a variety of sources, including economic hardship, illness, or adverse interpersonal relationships. The authors believe that disruptions in interpersonal domains of functioning play a critical role in the development and persistence of aggression, and will therefore explore how a more complete understanding of social strain can inform the G × E literature.

Before delving into a review of the research on the G × E interactions relevant to GST, it is first important to operationalize the term social stress (the variant of strain that we will be focusing on). For the purposes of this chapter, social stress is defined as stress of an interpersonal nature. In other words, social stressors are those that involve strained relationships with other people, such as conflict, disagreements, or loss. Therefore childhood abuse and/or maltreatment, maternal stress/mental health concerns, and social isolation would constitute examples of interpersonal stress. While environmental strains, such as low socioeconomic status and living in a high-crime neighborhood, may certainly be related to maladaptive developmental outcomes, these types of strain are not explored in the present chapter.

It is important to take developmental changes into account when studying social stress, and to remember that different strains become more salient across different developmental periods in life. For example, a 34-year-old may find himself less concerned with "fitting in" with the "popular crowd" than he did when he was a 16-year-old high-school student. In adolescence, particular emphasis is placed on peer groups. While this increased desire for peer acceptance forms the crux of many theories of crime, it also helps to explain why the frequency of group criminal activities is highest in adolescence (Steinberg and Morris, 2001) and should help inform G × E studies exploring aggression and delinquency during the teenage years.

Why look at genetics?

Previous studies have linked countless environmental factors, such as harsh parenting, problematic neighborhoods, and low socioeconomic status, to the development and maintenance of aggression across the lifespan (Rutter, 1981; Chang *et al.*, 2003; Barry *et al.*, 2012; Burnette *et al.*, 2012). Relatedly, others have demonstrated that some children exposed to adverse situations are quite resilient and are thus able to avoid the deleterious consequences of their negative environments (Luthar *et al.*, 2000). This body of research spawned an interest in understanding how specific risk or protective factors (Rutter, 1981) in the environment may further predispose to, or buffer against, the development of, among other pathologies, aggressive behaviors. For example, researchers examined how variables such as warm and responsive parenting buffer children in high-crime neighborhoods from engaging in delinquent behaviors. Across many of these studies, environmental factors were explored as mediators or moderators of other environmental factors, and biological factors were overlooked.

Though often overlooked, the field of genetics certainly deserves further attention, as it offers an interesting piece of this complicated puzzle by examining how one's genetic inheritance may predispose an individual to be more susceptible to harmful environmental effects. Of course, an individual's biological make-up does not necessarily determine how he or she will respond to a given situation. Humans are complex and behavioral responses are probabilistic, influenced by a number of different factors including genetic make-up as well as experiential learning over the life span. When exploring G × E interactions on behavioral outcomes, it is important to keep in mind that there is not a one-to-one correspondence between genetic risk and outcome, but that biological factors such as genetic risk may help account for individual differences in behavior arising in similar environmental contexts. In other words, genetic risk factors may help to explain why one person may be more prone to responding aggressively in a stressful situation while another person may be more prone to coping a different way.

With this in mind, there are a variety of genes that have been explored to examine whether differences in the gene's DNA sequence may be associated with variations in human behavior, including aggression and delinquency. Many of the genes explored code for proteins that are involved in the process of neuronal communication and have been implicated in psychological and behavioral processes. This chapter focuses on three genes, MAOA, 5-HTTLPR, and OXTR, and how these genes interact with social stressors to influence one's risk of engaging in crime and delinquency.

Monoamine oxidase A

Monoamine oxidase A (MAOA) is an enzyme encoded by the MAOA gene. This enzyme breaks down monoamine neurotransmitters (those that have an amine group in their structure), which includes serotonin (5-HT), dopamine (DA), and norepinephrine (NE). In general, for one neuron to communicate with the next neuron in the series, it releases a neurotransmitter into the synaptic cleft, the space between the two "pre" and "post" synaptic neurons. The synaptic cleft is a small space, and the neurotransmitter floats across this space and binds to receptors on the post-synaptic, or second, neuron. To help clear the neurotransmitters from the synaptic cleft so that this series of steps can occur at later points and communicate new information, the neurotransmitters are either degraded by enzymes or taken back up into the pre-synaptic neuron to be re-used. Several monoamine neurotransmitters (5-HT, DA, and NE) are degraded by MAOA enzymes. Therefore, when a high level of MAOA enzymes are present in the synaptic cleft (which we will refer to as "high MAOA expression"), the neurotransmitters are broken down and cleared quickly, giving them less time to impact the subsequent neuron. In contrast, when fewer MAOA enzymes are present ("low expression"), the neurotransmitters are cleared more slowly, and therefore have more time to impact the subsequent neuron. Because 5-HT, DA, and NE are critical neurotransmitters implicated in a variety of psychological processes, understanding factors that contribute to MAOA functioning, and how this functioning relates to behavioral outcomes, is essential.

Studies have explored the potential role of MAOA expression in aggressive behaviors. In the animal literature, increased aggression was identified in rats for whom the MAOA gene was knocked out (Cases *et al.*, 1995). When this gene is knocked out, the MAOA enzyme is not made. Therefore MAOA expression is completely missing or very low, and the 5-HT, DA, and NE neurotransmitters will not be cleared through this mechanism of MAOA enzymatic degradation. Interestingly, this association between lack of MAOA expression and increased aggression can be reversed (Shih and Thompson, 1999) such that aggression declines when MAOA expression is restored.

Avshalom Caspi and colleagues (2002) took MAOA research into the realm of criminology by exploring the potential interaction between MAOA expression levels and adverse social environments in the prediction of antisocial behavior. A variation in the promoter region of the MAOA gene in humans affects the level of MAOA expression. Promoter regions help regulate how much of a gene is made, and one version of the genotype results in high MAOA expression while the other results in low expression. Caspi *et al.* sought to explore whether this genotype interacted with childhood maltreatment, a source of social stress, to bring about antisocial behaviors in children. The study was motivated by the question of why certain individuals who are exposed to maltreatment in childhood develop antisocial behavior patterns later in life, whereas others do not. The authors reasoned that there must be more to the story than an adverse social environment in childhood, since not all individuals exposed to these environments develop antisocial tendencies.

The researchers found that the MAOA genotype conferring low MAOA expression (most similar to the "knock-out" condition in the rats) interacted with childhood maltreatment to predict antisocial behavior among their all-male sample. Even more intriguing was that the combination of the low-expression MAOA genotype plus exposure to maltreatment was relatively rare among this sample (12 percent of males), but that individuals with this profile accounted for about 44 percent of the cohort's violent convictions. This interaction effect has been replicated in subsequent cohorts. Among a cohort of male children, those with the low-expression MAOA genotype and past exposure to physical abuse evidenced the highest levels of antisocial behaviors (Kim-Cohen *et al.*, 2006). In an Iowa Adoptee cohort, those with the low-expression MAOA genotype and childhood maltreatment similarly evidenced higher antisocial behaviors, while those with the high-expression MAOA genotype and maltreatment instead evidenced higher depressive symptoms (Beach *et al.*, 2010).

Similar associations between MAOA expression and violent behaviors, an extreme form of aggression, have been identified, though this finding may differ across races. Higher expression of MAOA was found to "buffer" Caucasian children from the effects of abuse on the development of violence later in life. However, similar findings were not observed among the non-Caucasian participants (African American, Hispanic, Pacific Islander, and "other"; Widom and Brzustowicz, 2006). Therefore this study raises the possibility that the MAOA–maltreatment interaction may be further complicated by racial differences and/or by specification of the environmental risk context (i.e. types of environmental contexts varied across different ethnic groups in this study).

Despite these intriguing findings and their potential clinical implications, the findings linking the MAOA genotype and childhood maltreatment to antisocial outcomes or aggressive behaviors remain inconclusive. Although some research groups have replicated this finding, as discussed above, others have failed to replicate it (e.g. Huizinga *et al.*, 2006). Huizinga and colleagues found that maltreatment alone was a risk factor for adolescent and adult antisocial and violent behavior, but that the interaction between MAOA and maltreatment did not significantly predict antisocial outcomes. These discrepant findings may be partially explained by differences in how studies operationally define and measure maltreatment and antisocial outcomes. This limitation, along with several others, is discussed in detail below.

Nevertheless, the importance of these findings in the context of GST should not be discounted. As Agnew points out in many of his writings on GST, not every individual exposed to strain engages in criminal coping (e.g. Agnew, 2011). Notably, one of the individual-level factors that Agnew discusses as important in predicting a delinquent response to strain is the presence of impulsive personality traits. Variations in the MAOA genotype have been associated with impulsivity and disinhibition at the level of both brain function and outward behavior (Meyer-Lindenberg *et al.*, 2006; Passamonti *et al.*, 2006). Therefore the G × E findings on MAOA and social stress are highly compatible with strain theory and its predictions. A more in-depth understanding of how one's genetic inheritance of personality traits increases susceptibility to the influence of environmental strains could be a novel avenue for future research in GST.

Serotonin transporter gene

The serotonin transporter gene has been extensively studied in G × E interactions. This gene codes for the serotonin transporter, and variations in the promoter region of this gene lead to differences in the amount of transporter made (higher versus lower expression). Serotonin is released by neurons into the synaptic cleft, and thus communicates with subsequent neurons. Once released into the synaptic cleft, serotonin is cleared both by the MAOA enzyme, as well as by the serotonin transporter. The serotonin transporter is located on the pre-synaptic (serotonin-releasing) neuron, and functions to bring released serotonin back into the pre-synaptic neuron to be re-used.

The promoter region of the serotonin transporter gene (5-HTTLPR) helps regulate how much serotonin transporter is made. Individuals with the long or l-allele of 5-HTTLPR produce more serotonin transporters, while those with the short or s-allele produce less. Presumably, when more transporters are made and are present on the pre-synaptic neuron, such as in individuals with the l-allele, serotonin is cleared more quickly from the synaptic cleft, giving serotonin less time to act on the subsequent neuron (this resembles a similar situation to the "high MAOA expression" condition discussed above). When fewer transporters are present, such as in those with the s-allele, serotonin is cleared more slowly and thus has more time to impact the subsequent neuron (most similar to the "low MAOA expression" condition).

The serotonin system has been implicated in a variety of psychological behaviors and disorders, including externalizing and antisocial behaviors. Given the known role of social stressors on the development of these behaviors, 5-HTTLPR genotype research has explored whether the 5-HTTLPR l- and s-alleles may interact with social stressors to more closely predict behavioral outcomes. A recent study by Cicchetti *et al.* (2012) supports a 5-HTTLPR-maltreatment interaction on behavioral outcomes and highlights the importance of developmental timing of maltreatment. When comparing maltreated with non-maltreated children, the researchers concluded that early-onset maltreatment interacted with the 5-HTTLPR alleles to bring about more severe antisocial behavioral outcomes. More specifically, individuals with a history of abuse and the ll genotype had a lower risk for antisocial behavior than those who were abused but had an ss genotype. While intriguing, the researchers underscored the need to recruit large samples in order to test whether gender may moderate these effects, and also discussed the need to follow samples longitudinally in order to observe the developmental trajectories of these behaviors and more closely evaluate the influence of maltreatment at different time periods.

Our research team has similarly explored the interaction between social stressors and the 5-HTTLPR alleles, specifically on the prediction of aggression (Conway *et al.*, 2012). Participants were drawn from a larger longitudinal birth cohort study of child health and development that was conducted between 1981 and 1984 in Australia. Participants were again interviewed at age 15 and 20, along with their mothers, to assess for levels of acute and chronic stress. Participants selected for the follow-up at age 15 were selected from homes representing a range of maternal depression. This sample of youth was at high risk for chronic stress as a result of their mother's mental illness, as women with depression often generate stress in their own and their family members' lives. Findings from this G × E study indicated that 5-HTTLPR s-allele carriers demonstrated larger increases in aggressive behavior following exposure to chronic stress relative to their l-allele homozygous counterparts. Sex did not moderate these relationships.

A recent study suggesting similar interaction effects has noted sex differences, and demonstrates differential effects of 5-HTTLPR genotype based on both sex and race (Douglas *et al.*, 2011). In this study, 203 participants (out of a sample of 602 European Americans and 779 African Americans) were diagnosed with antisocial personality disorder using the DSM-IV TR diagnostic criteria. Adverse childhood events (ACEs) were conceptualized as (1) exposure to violence in the home; (2) physical abuse; and (3) sexual abuse. Participants were given an ACEs score of 0–3. Statistical analyses confirmed a significant sex by ACEs by 5-HTTLPR interaction effect. Specifically, the addition of each additional ACE increased the odds of antisocial personality disorder only among African American men, regardless of the genotype, whereas each additional ACE increased the risk for antisocial personality disorder among women only if they were homozygous for the s-allele. While male sex and ACE history predicted antisocial personality disorder among Caucasian participants, this same three-way interaction was not observed among this group. While intriguing, the authors highlight the need to replicate these findings and explore them in larger samples.

A recent biosocial criminological theory postulates that negative emotionality and effortful control play an important role in explaining life-long conduct problems (DeLisi and Vaughn, 2014). Both of these constructs comprise temperament, which is widely recognized as a heritable trait. Similarly, we argue that certain heritable genotypes may predispose an individual to react in maladaptive ways in response to environmental stressors. In the case of the 5-HTTLPR gene, s-allele carriers have been described as having higher levels of neuroticism or negative affectivity (Caspi *et al.*, 2010). This trait is highly relevant to Agnew's GST, which postulates that delinquency results from increases in anger and depression (i.e. negative affectivity) that occur in response to strain. As 5-HTTPLR s-allele carriers are already predisposed to higher levels of affectivity under all environmental conditions, they may be closer to the "tipping point" where negative affect turns into aggression and criminal behavior in response to environmental strain.

Oxytocin

When thinking about potential moderators of the association between social stress and delinquency, the oxytonergic system is another potentially important biological factor to consider. Oxytocin is a neurohormone involved in human social interaction. While oxytocin was initially cited in the literature for its prosocial effects, with higher levels of oxytocin being associated with increased trust (Kosfeld *et al.*, 2005), empathy (Bartz *et al.*, 2010), and affiliative behaviors (Lim and Young, 2006), recent findings suggest a more complex picture. For example, intranasal oxytocin administration has been associated with higher levels of non-cooperative behaviors among individuals who do not belong to the same social group (De Dreu, 2012), decreased levels of trust and cooperation among individuals sensitive to rejection (Bartz *et al.*, 2011a), and increased feelings of envy during a gambling game (Shamay-Tsoory *et al.*, 2009). Thus researchers proposed the social sensitivity hypothesis to help better explain these divergent findings concerning the effect of oxytocin on social behavior (Bartz *et al.*, 2011b).

According to the social sensitivity hypothesis, oxytocin functions to increase the perception and salience of social cues (Bartz *et al.*, 2011b), suggesting that behavioral outcomes associated with oxytocin would be dependent on the social context. Bartz and colleagues posit that higher levels of oxytocin may be associated with more positive, pro-social outcomes in positive social environments as well as more negative behavioral outcomes in negative social environments.

Oxytocin has also been noted for its unique influence on behavioral responses towards members considered to be within one's own social group (in-group) compared with those who are a member of a different social group (out-group). Among in-group members, intranasal oxytocin administration is associated with more prosocial and trusting behaviors (for a review, see De Dreu, 2012). However, this is not what is found for reactions towards out-group members. In the context of threat, oxytocin levels are associated with increased aggression towards out-group members, presumably in an effort to help protect the in-group (De Dreu, 2012). Drawing from these findings and the social sensitivity hypothesis, oxytocin may be associated with increased aggression and negative behavioral outcomes in the presence of social stress and adversity, especially when this stress occurs among out-group members.

While much of the existing research on oxytocin has focused on hormone levels, the oxytocin receptor is also critical for detecting the presence of oxytocin and allowing it to have its effects. The oxytocin receptor is encoded for by the oxytocin receptor gene (OXTR). One of the more commonly studied polymorphisms of OXTR is rs53576, which has both G and A alleles and has been evaluated in the context of social stress. This polymorphism is located in the third intron in the OXTR gene, which does not code for the protein structure but may contain regulatory areas for expression of OXTR or other genes, though the function of this polymorphism is not yet known (Gimpl and Fahrenholz, 2001).

While initial studies of rs53576 reported an overall positive influence of the G allele on social behaviors (Kogan *et al.*, 2011; Krueger *et al.*, 2012), more recent reports incorporating social stressors suggest that those carrying a G allele may demonstrate increased risk of maladaptive behaviors in the presence of social adversity. Bradley and colleagues (2011) found that individuals with a G allele were at increased risk for emotion dysregulation following early life abuse. Sturge-Apple and colleagues (2012) showed that mothers with the G allele were more influenced by the quality of their inter-partner relationship in their display of sensitivity towards their children—among those with the G allele, those with a positive inter-partner relationship exhibited the highest level of maternal sensitivity, while those with a high-conflict relationship exhibited the lowest level of sensitivity.

On the basis of these studies, our research team explored whether social stress, measured through a chronic stress interview, interacted with rs53576 in an Australian cohort to predict conduct and antisocial behavior problems at ages 15 and 20, respectively (Smearman *et al.*, 2014). We hypothesized that individuals with the G allele would evidence greater levels of these behavioral problems in the presence of social stress, compared to those with the A allele. Three informants were used to assess the behavioral outcomes: the youth, the mother, and the clinician. Data from all three were combined into a latent construct using the data analysis program Structural Equation Modeling, which better accounts for measurement error and allows for multiple measures assessing a similar construct to be combined into a single latent construct (Kline, 2011).

In our analyses, developmental differences were observed for rs53576's hypothesized association with conduct and antisocial behaviors. At age 15, rs53576 did not interact with social stress, and instead was directly related to the outcome. Those with the G allele evidenced greater levels of conduct disorder behaviors. Given the early nature of this research, replication is needed before conclusions can be confidently drawn. With this in mind, it is possible that these findings demonstrate sensitivity to peer influences, where individuals more sensitive to social cues may be more likely to engage in conduct-related behaviors that generally rise in adolescence in response to peer pressure and reward.

At age 20, there was an interaction between rs53576 and social stress in the prediction of antisocial behavior problems. Only those with the G allele exhibited higher levels of antisocial behaviors in response to chronic social stress, aligning with our hypotheses and the social sensitivity hypothesis. Overall, conduct and antisocial behaviors were relatively low among those with an A allele in our sample.

A few studies have assessed direct effects between OXTR polymorphisms and aggressive behaviors. Possible associations between certain OXTR polymorphisms and extreme, persistent, and pervasive childhood-onset aggression (Malik *et al.*, 2012), and callous-unemotional traits in aggressive children (Beitchman *et al.*, 2012), have been suggested. Given the potential role of the environment and social stress, it may be interesting to explore whether G × E interactions exist in these samples, and if they can provide enhanced explanatory power and insight into these behaviors.

Strain and interpersonal stress are driving forces in the prediction of crime and delinquency according to GST. Our findings concerning the oxytocin gene suggest that certain individuals (i.e. those with the G allele) may see these social strains as more salient, and therefore be more likely to engage in criminal coping in response to them. Another tenet of GST highlights the fact that strains may contribute to a reduction in an individual's perception of social control. Individuals with the G allele who are highly socially sensitive may, in turn, be more likely to perceive rejection from conventional social support networks, and thus might be at higher risk to engage in criminal coping for this reason as well.

Limitations of gene × environment research

Although a fascinating area, G × E studies do come with methodological and statistical limitations. For example, G × E studies are cross-sectional in nature. Cross-sectional designs are often problematic because they cannot fully take into account gene–environment correlation, whereas longitudinal, prospective studies can (Johnston *et al.*, 2013). Another concern in the field is that of theoretical specification. Research groups measure "maltreatment" and "delinquency" in different ways, which is one factor that may account for some of the discrepant findings in the field. Researchers must also be careful to take into account passive and evocative genotype–environment correlations (Kim-Cohen *et al.*, 2006). Passive G × E correlations refer to the associations between a child's genetic inheritance and the environment in which that child is raised. Reactive, or evocative, G × E correlations refer to the influence of an individual's heritable traits on his or her environment. It is also especially important to demonstrate that the risk factor and outcome are linked via a true environmental mechanism, and not unduly influenced by a third variable that could itself reflect additional genetic or alternative influences not captured by the experimental design.

Some of the limitations that have been discussed in the field have prompted journal editors to set stricter restrictions on what can be published. For example, the journal *Behavior Genetics* recently established strict publishing criteria, which included a requirement to conduct replication analyses prior to publishing a result (Hewitt, 2012). In a replication analysis, researchers test the same predictor and outcome variables using identical statistical procedures, but do so in an independent sample. The editors of the *Journal of Abnormal Child Psychology* followed suit in 2013 by publishing similar criteria requiring authors to demonstrate "sound attempts" to directly replicate previously published findings (Johnston *et al.*, 2013). The editors suggest that researchers use reliable and valid measures, assess environments and phenotypes using separate informants, and choose environments and genetic variants with known functional consequences, in addition to emphasizing the importance of longitudinal designs over cross-sectional designs.

Psychological researchers have also called for more attention to construct validation in G × E studies, rather than strict attention to statistical replication (Caspi *et al.*, 2010). Such an approach focuses on replication of G × E findings across various genes, outcomes, and environments that are relevant to the theoretical model being tested. In the case of GST, genes that influence traits such as impulsivity and negative affectivity would be expected to lead to aggression in environmental contexts that lead to high levels of strain. In addition, genes that increase attention to social cues (including environmental strain) would be expected to increase the risk for delinquency most substantially during developmental periods in which strain is less easy to control or escape, such as adolescence. G × E studies that demonstrate MAOA, 5HTTLPR, and OXT receptor gene effects on antisocial outcomes in environments of high social stress are compatible with GST, and converge in such a way that they lend credence to the value of this approach in the study and prevention of delinquency.

Future directions

An area of inquiry that is just beginning to develop is one that aims to elucidate how gene by environment studies (which generally conceptualize genes as influencing vulnerability status for a good or bad outcome) may be integrated with recent work on epigenetics and the biological embedding of environmental exposures (Belsky and Pluess, 2009; Miller *et al.*, 2011). The term "epigenetics" means "on top of" the genome, and refers to DNA modifications that do not change the DNA sequence but instead regulate which genes are made (or expressed). The epigenetic process can therefore influence gene expression without changing the DNA sequence (Tamashiro and Moran, 2010). The regulation of gene expression by epigenetic modifications is a dynamic process that has the potential to adapt to environmental stress or changes. One of the most commonly studied epigenetic processes is the addition of methyl groups to the promoter region of a gene. These methyl groups function by blocking the binding of transcription factors, which could otherwise lead to gene expression (Tamashiro and Moran, 2010). Therefore the addition of methyl groups on the promoter region of genes reduces gene expression.

Recent work has shown that exposure to certain early life experiences can alter methylation of a number of genes, such as the serotonin transporter gene, 5-HTTLPR (Beach *et al.*, 2010, 2011). Recent studies have reported an increase in methylation of the 5-HTTLPR gene among individuals who have experienced early life abuse (Beach *et al.*, 2010, 2011). Increased methylation is thought to result in reduced expression, meaning less of the serotonin transporter would be made. As discussed above, the s-allele of 5-HTTLPR, compared to the l-allele, is itself thought to result in reduced expression of the serotonin transporter (Lesch *et al.*, 1996). Therefore we now need to understand the interplay between possible risk conferred by the s-allele, and possible risk incurred through epigenetic modifications and increased methylation of the 5-HTTLPR gene following exposure to adversity. For example, will those with the s-allele *and* the greatest degree of methylation have the lowest expression of the serotonin transporter, and will this result in the highest risk for adverse outcomes following stress? This interplay needs to be studied further to understand how these processes may be related, whether they can be integrated into a theory with even greater explanatory power, and whether these processes interact with social stressors to influence and better predict aggressive and antisocial behavioral outcomes.

GST posits that individuals are "pressured" into crime when they experience a stressor. Agnew emphasizes how individuals are more likely to engage in crime when other means of coping are unavailable (e.g. Agnew, 2007, 2011). As G × E researchers continue to fine-tune their methodology, criminologists should explore whether biological factors may indeed predispose some individuals to respond to strains more adversely than others. These adverse responders, in turn, may be more likely to resort to criminal coping. Marrying the fundamental principles of GST with the G × E literature could provide a more comprehensive understanding of how environmental factors influence genetics and phenotypic expression. Clinically, this could have implications for the development of more sensitive diagnostic criteria for delinquency, and could help better identify those individuals who are most at risk for criminal coping in the context of adverse environments.

Conclusion

The burgeoning area of genetics research can be a fruitful endeavor that helps elucidate many questions about the underpinnings of behavior. As researchers explore the interactions between different genes and environments, the field will have a clearer picture of how genes interact with environmental factors to bring about different phenotypes. Furthermore, research from this burgeoning area of inquiry can be used to inform and further shape already well established sociological theories of crime such as general strain theory.

References

Achenbach, T.M. and Edelbrock, C.S. (1978) 'The classification of child psychopathology: a review and analysis of empirical efforts', *Psychological Bulletin*, 85(6): 1275–1301.

Agnew, R. (2007) *Pressured into Crime: an Overview of General Strain Theory*, New York: Oxford.

Agnew, R. (2011) 'Pressured into crime: General Strain Theory', in F.T. Cullen and R. Agnew (eds), *Criminological Theory: Past to Present*, Oxford: Oxford University Press, pp. 189–197.

Agnew, R. (2013) 'When criminal coping is likely: an extension of General Strain Theory', *Deviant Behavior*, 34(8): 653–670.

Barrett, D.E. and Katsiyannis, A. (2009) 'Predictors of offense severity, adjudication, incarceration, and repeat referrals for juvenile offenders: a multicohort replication study', *Remedial and Special Education*, 31(4): 261–275.

Barry, T.D., Lochman, J.E., Fite, P.J., Wells, K.C., and Colder, C.R. (2012) 'The influence of neighborhood characteristics and parenting practices on academic problems and aggression outcomes among moderately to highly aggressive children', *Journal of Community Psychology*, 40(3): 372–379.

Bartz, J.A., Zaki, J., Bolger, N., Hollander, E., Ludwig, N.N., Kolevzon, A., and Ochsner, K.N. (2010) 'Oxytocin selectively improves empathic accuracy', *Psychological Science*, 21(10): 1426–1428.

Bartz, J., Simeon, D., Hamilton, H., Kim, S., Crystal, S., Braun, A., Vicens, V., and Hollander, E. (2011a) 'Oxytocin can hinder trust and cooperation in borderline personality disorder', *Social Cognitive and Affective Neuroscience*, 6(5): 556–563.

Bartz, J.A., Zaki, J., Bolger, N., and Ochsner, K.N. (2011b) 'Social effects of oxytocin in humans: context and person matter', *Trends in Cognitive Sciences*, 15(7): 301–309.

Beach, S., Brody, G.H., and Todorov, A.A. (2010) 'Methylation at SLC6A4 is linked to family history of child abuse: an examination of the Iowa Adoptee sample', *American Journal of Medical Genetics Part B: Neuropsychiatric Genetics*, 153(2): 710–713.

Beach, S.R.H., Brody, G.H., Todorov, A.A, Gunter, T.D., and Philibert, R.A. (2011) 'Methylation at 5HTT mediates the impact of child sex abuse on women's antisocial behavior: an examination of the Iowa Adoptee sample', *Psychosomatic Medicine*, 73(1): 83–87.

Beitchman, J.H., Zai, C.C., Muir, K., Berall, L., Nowrouzi, B., Choi, E., and Kennedy, J.L. (2012) 'Childhood aggression, callous-unemotional traits and oxytocin genes', *European Child and Adolescent Psychiatry*, 21(3): 125–132.

Belsky, J. and Pluess, M. (2009) 'Beyond diathesis stress: differential susceptibility to environmental influences', *Psychological Bulletin*, 135(6): 885–908.

Bradley, B., Westen, D., Mercer, K.B., Binder, E.B., Jovanovic, T., Crain, D., Wingo, A., and Heim, C. (2011) 'Association between childhood maltreatment and adult emotional dysregulation in a low-income, urban, African American sample: moderation by oxytocin receptor gene', *Development and Psychopathology*, 23(2): 439–452.

Burnette, M.L., Oshri, A., Lax, R., Richards, D., and Ragbeer, S.N. (2012) 'Pathways from harsh parenting to adolescent antisocial behavior: a multidomain test of gender moderation', *Development and Psychopathology*, 24(3): 857–870.

Cases, O., Seif, I., Grimsby, J., Gaspar, P., Chen, K., Pournin, S. Müller, U., Aguet, M., Babinet, C., Chen Shih, J., and De Maeyer, E. (1995) 'Aggressive behavior and altered amounts of brain serotonin and norepinephrine in mice lacking MAOA', *Science*, 268(5218): 1763–1766.

Caspi, A., McClay, J., Moffitt, T.E., Mill, J., Martin, J., Craig, I.W., Taylor, A., and Poultin, R. (2002) 'Role of genotype in the cycle of violence in maltreated children', *Science*, 297(5582): 851–854.

Caspi, A., Hariri, A.R., Holmes, A., Uher, R., and Moffitt, T.E. (2010) 'Genetic sensitivity to the environment: the case of the serotonin transporter gene and its implications for studying complex diseases and traits', *American Journal of Psychiatry*, 167(5): 509–527.

Chang, L., Schwartz, D., Dodge, K.A., and McBride-Chang, C. (2003) 'Harsh parenting in relation to child emotion regulation and aggression', *Journal of Family Psychology*, 17(4): 598–606.

Cicchetti, D., Rogosch, F.A., and Thibodeau, E.L. (2012) 'The effects of child maltreatment on early signs of antisocial behavior: genetic moderation by tryptophan hydroxylase, serotonin transporter, and monoamine oxidase A genes', *Development and Psychopathology*, 24(3): 907–928.

Conway, C.C., Keenan-Miller, D., Hammen, C., Lind, P.A., Najman, J.M., and Brennan, P.A. (2012) 'Coaction of stress and serotonin transporter genotype in predicting aggression at the transition to adulthood', *Journal of Clinical Child and Adolescent Psychology*, 41(1): 53–63.

De Dreu, C.K.W. (2012) 'Oxytocin modulates cooperation within and competition between groups: an integrative review and research agenda', *Hormones and Behavior*, 61(3): 419–428.

DeLisi, M. and Vaughn, M.G. (2014) 'Foundation for a temperament-based theory of antisocial behavior and criminal justice system involvement', *Journal of Criminal Justice*, 42(1): 10–25.

Douglas, K., Chan, G., Gelernter, J., Arias, A.J., Anton, R.F., Poling, J., Farrer, L., and Kranzler, H.R. (2011) '5-HTTLPR as a potential moderator of the effects of adverse childhood experiences on risk of antisocial personality disorder', *Psychiatric Genetics*, 21(5): 240–248.

Foster, E.M. and Jones, D.E. (2005) 'The high costs of aggression: public expenditures resulting from conduct disorder', *American Journal of Public Health*, 95(10): 1767–1772.

Gimpl, G. and Fahrenholz, F. (2001) 'The oxytocin receptor system: structure, function, and regulation', *Physiological Reviews*, 81(2): 629–683.

Hewitt, J.K. (2012) 'Editorial policy on candidate gene association and candidate gene-by-environment interaction studies of complex traits', *Behavior Genetics*, 42(1): 1–2.

Hinshaw, S.P. (1992) 'Externalizing behavior problems and academic underachievement in childhood and adolescence: causal relationships and underlying mechanisms', *Psychological Bulletin*, 111(1): 127–155.

Huizinga, D., Haberstick, B.C., Smolen, A., Menard, S., Young, S.E., Corley, R.P., Stallings, M.C., Grotpeter, J., and Hewitt, J.K. (2006) 'Childhood maltreatment, subsequent antisocial behavior, and the role of monoamine oxidase A genotype', *Biological Psychiatry*, 60(7): 677–683.

Johnston, C., Lahey, B.B., and Matthys, W. (2013) 'Editorial policy for candidate gene studies', *Journal of Abnormal Child Psychology*, 41(4): 511–514.

Kim-Cohen, J., Caspi, A., Taylor, A., Williams, B., Newcombe, R., Craig, I.W., and Moffitt, T.E. (2006) 'MAOA, maltreatment, and gene–environment interaction predicting children's mental health: new evidence and a meta-analysis', *Molecular Psychiatry*, 11(10): 903–913.

Kline, R.B. (2011) *Principles and Practice of Structural Equation Modeling*, 3rd edn, New York: Guilford Press.

Kogan, A., Saslow, L.R., and Impett, E.A. (2011) 'Thin-slicing study of the oxytocin receptor (OXTR) gene and the evaluation and expression of the prosocial disposition', *Proceedings of the National Academy of Sciences, USA*, 108(48): 19189–19192.

Kosfeld, M., Heinrichs, M., Zak, P.J., Fischbacher, U., and Fehr, E. (2005) 'Oxytocin increases trust in humans', *Nature*, 435(7042): 673–676.

Krueger, F., Parasuraman, R., Iyengar, V., Thornburg, M., Weel, J., Lin, M., Clarke, E., McCabe, K., and Lipsky, R.H. (2012) 'Oxytocin receptor genetic variation promotes human trust behavior', *Frontiers in Human Neuroscience*, 6(4): 1–9.

Lesch, K.P. *et al.* (1996) 'Association of anxiety-related traits with a polymorphism in the serotonin transporter gene regulatory region', *Science*, 274(5292): 1527–1531.

Lim, M.M. and Young, L.J. (2006) 'Neuropeptidergic regulation of affiliative behavior and social bonding in animals', *Hormones and Behavior*, 50(4): 506–517.

Luthar, S., Cicchetti, D., and Becker, B. (2000) 'The construct of resilience: a critical evaluation and guidelines for future work', *Child Development*, 71(3): 543–562.

Malik, A.I., Zai, C.C., Abu, Z., Nowrouzi, B., and Beitchman, J.H. (2012) 'The role of oxytocin and oxytocin receptor gene variants in childhood-onset aggression', *Genes, Brain and Behavior*, 11(5): 545–551.

Meyer-Lindenberg, A., Buckholtz, J.W., Kolachana, B.R., Hariri, A., Pezawas, L., Blasi, G., Wabnitz, A., Honea, R., Verchinski, B., Callicott, J.H., Egan, M., Mattay, V., and Weinberger, D.R. (2006) 'Neural mechanisms of genetic risk for impulsivity and violence in humans', *Proceedings of the National Academy of Sciences, USA*, 103(16): 6269–6274.

Miller, G.E., Chen, E., and Parker, K.J. (2011) 'Psychological stress in childhood and susceptibility to the chronic diseases of aging: moving toward a model of behavioral and biological mechanisms', *Psychological Bulletin*, 137(6): 959–997.

Passamonti, L., Fera, F., Magariello, A., Cerasa, A., Gioia, M.C., Muglia, M., Nicoletti, G., Gallo, O., Provinciali, L., and Quattrone, A. (2006) 'Monoamine oxidase-A genetic variations influence brain activity associated with inhibitory control: new insight into the neural correlates of impulsivity', *Biological Psychiatry*, 59(4): 334–340.

Rutter, M. (1981) 'The city and the child', *American Journal of Orthopsychiatry*, 51(4): 610–625.

Schmitt, J., Warner, K., and Gupta, S. (2010) *The High Budgetary Cost of Incarceration*, Washington, DC: Center for Economic and Policy Research.

Shamay-Tsoory, S.G., Fisher, M., Dvash, J., Harari, H., Perach-Bloom, N., and Levkovitz, Y. (2009) 'Intranasal administration of oxytocin increases envy and schadenfreude (gloating)', *Biological Psychiatry*, 66(9): 864–870.

Shih, J.C. and Thompson, R.F. (1999) 'Monoamine oxidase in neuropsychiatry and behavior', *American Journal of Human Genetics*, 65: 593–598.

Smearman, E.L., Winiarski, D.A., Brennan, P.A., Najman, J., and Johnson, K.C. (2014) 'Social stress and the oxytocin receptor gene interact to predict antisocial behavior in an at-risk cohort', doi: 10.1017/S0954579414000649.

Steinberg, L. and Morris, A.S. (2001) 'Adolescent development', *Annual Review of Psychology*, 52: 83–110.

Sturge-Apple, M.L., Cicchetti, D., Davies, P.T., and Suor, J.H. (2012) 'Differential susceptibility in spillover between interparental conflict and maternal parenting practices: evidence for OXTR and 5-HTT genes', *Journal of Family Psychology*, 26(3): 431–442.

Tamashiro, K.L.K. and Moran, T.H. (2010) 'Perinatal environment and its influences on metabolic programming of offspring', *Physiology and Behavior*, 100(5): 560–566.

Widom, C.S. and Brzustowicz, L.M. (2006) 'MAOA and the "cycle of violence:" childhood abuse and neglect, MAOA genotype, and risk for violent and antisocial behavior', *Biological Psychiatry*, 60(7): 684–689.

Zigler, E., Taussig, C., and Black, K. (1992) 'Early childhood intervention: a promising preventative for juvenile delinquency', *American Psychologist*, 47(8): 997–1006.

31

Understanding families from a biosocial perspective

John Paul Wright, Jamie Newsome, and Michelle Coyne

Introduction

Thousands of studies link the family of origin to a myriad of pathologies, including low self-control, impulsivity, mental health disorders, low intelligence, drug abuse, and criminality—to name just a few (Farrington, 1978, 2010; Sayre-McCord, 2007). Indeed, no other institution has been as thoroughly researched, as thoroughly investigated, or as thoroughly critiqued as the family. Yet a focus on the volume of scholarly research into family functioning misses the even larger volume of attention paid to families in the mass media. Magazine articles, newspaper articles, and self-help books about the family and parenting abound in popular culture. The fact that so much is written on families stands as a testament to the core Western beliefs that family is the lynchpin of society and that what happens within families has dramatic consequences on the lives of children.

Core cultural beliefs about the importance of family, and about the influence of parenting in particular, are also echoed by politicians from both parties (Wright and Cullen, 2012). Liberals, for example, are apt to see families and parental functioning as linked to their position in the economic hierarchy. This is why liberals typically support policies that provide economic resources to families at the lower end of the socioeconomic distribution (Clinton, 1996). Conservatives, too, have a strong belief in the role of families in child development. Instead of providing goods and services to some families, as liberals desire, conservatives emphasize the role of parents in inculcating moral and religious values in youth, and they emphasize the need for functional, two-parent households in the socialization of children (Brooks, 2007). Again, however, focusing on how liberals and conservatives conceptualize the role of families, and especially the role of parenting in child outcomes, obscures the important point that liberals and conservatives view the family as an important institution—if not the most important institution—in American life.

Despite widespread certainty that the family environment is causally responsible for a host of youth problems, a growing number of scholars are now calling attention to the limits of family and parental influence on a range of child outcomes (Rowe, 1994; Harris, 1995; Pinker, 2003; Wright and Beaver, 2013). With thousands of studies providing statistical documentation of the deleterious role of "bad" or "ineffective" parenting in a range of child outcomes, it would seem preposterous, if not anti-scientific, to question the results of this large and growing body of evidence (Unnever *et al.*, 2006). Even so, the criticisms and charges made by these scholars present a fundamental challenge to the ways scholars have investigated family and parenting effects on youth, and they are beginning to be taken seriously as others investigate their claims empirically. In general, these scholars point out that most of the studies that reveal a link between parenting behaviors and child outcomes are based on a series of assumptions—assumptions that cannot be true, or that cannot be entirely true. These assumptions, moreover, have exerted a subtle but powerful influence on how scholars study the link between parenting behaviors and child outcomes, and they have affected the interpretation of the results produced by these studies. While these scholars do not deny that family dysfunction and that certain types of parenting behaviors can cause children to misbehave, they do call for a more nuanced understanding of the family–crime link specifically, and the family–child outcome link globally.

Here we examine the assumptions that form the methodological and logical crux of the majority of studies on families and crime. We enumerate and explain these assumptions, and show how they have been used to support a way of studying families that is largely biased in favor of finding a link between parenting behaviors and child outcomes. We then examine research on the family–crime link that, we believe, reveals important insights into the nature of offending. This body of research points to a biosocial understanding of the causal effects parenting has, and does not have, on child development. Finally, we call for a new strategy to study the family–crime link so that this research and knowledge may advance.

Assumptions underpinning family research

Any empirical research rests on a series of assumptions. Sometimes these assumptions are made explicit, and the biases resulting from the violation of assumptions are known. This is particularly true, for example, in the field of mathematical statistics, where assumptions about statistical tests are known and where research has been conducted on the violation of statistical assumptions. The same, however, cannot be said of studies into the family–crime link. For at least the past 70 years, scholars have conducted family-based studies in much the same way, and they have drawn conclusions from these studies without adequately considering if their methodology shapes their results (Rowe, 1994; Harris, 1995). Unfortunately, the assumptions underpinning these studies have not been made explicit, nor have we known the impact of violating these assumptions. Nonetheless, these assumptions have allowed scholars to create a body of research that, while important at one level, may also present a biased picture of the family–crime link.

The assumption of parental hierarchy

The assumption of "parental hierarchy" assumes that parental socialization is a unidirectional, top-down effort that produces various outcomes in offspring. This top-down approach does not recognize or prioritize the individual differences of each child within the home that can affect the quality and type of parenting received. While some studies do recognize the likelihood of "child effects"—that a child's behavior or personality traits draw out, generate, or otherwise incite parental reactions—many still do not (Lytton, 1990; Kerr and Stattin, 2000). The studies that have investigated child effects indicate that children are not passive recipients of external influences, but rather are active contributors to their rearing environment. In a study using data from the Cambridge Study in Delinquent Development, for example, Beaver and Wright (2007) found that a global measure of family risk had little direct effect on the misbehavior of the adolescent over time. However, they also found that an adolescent's involvement in a criminal lifestyle brought with it very negative consequences for family functioning. Other studies, too, have found potent "child effects" and thus serve as a note of caution against the widespread belief that parenting reflects a simple unidirectional process (Wright and Beaver, 2005; Beaver *et al.*, 2010b).

The assumption of measure purity

Second, family socialization studies assume that measures of the environment or measures assessing parental socialization efforts are uncontaminated by genetic influences. We call this the assumption of "measure purity" because scholars assume that their measures of parental behavior, parental socialization, and even parental economic standing truly reflect environmental sources of variation that are uncontaminated by other potential sources of variation—especially genetic sources of variation. The problem with this assumption, however, is that it is demonstrably false. Analyzing data on twins, studies by Plomin and his colleagues (1994) as well as Kendler and Baker (2007) and Beaver and his colleagues (2014) reveal that most, if not all, of the measures used by social scientists to assess the "environment" and "parental socialization" are themselves contaminated by genetic influences. Measures of parental rejection, for instance, are commonly associated with childhood conduct problems. Yet research finds that parental rejection is at least modestly influenced by the parents' genotype. Moreover, even measures of delinquent peers—the variable most often assumed to adequately measure and represent social learning theory—has been found to be contaminated by genetic influences (Wright *et al.*, 2008; Beaver *et al.*, 2009, 2010a).

The assumption of uniform effects

A core tenet of social information processing theory (Dodge and Crick, 1990) is that individuals exposed to the same environment will interpret that environment differently. They will not only cognitively process the environment or event differently, they will emotionally process and experience that environment differently. Harris (1995) notes that this poses a serious problem for parenting studies because scholars have traditionally assumed that children in the same family will respond similarly to the rearing strategies of their parents. In reality, children may respond very differently to the same experiences. For example, while a mother may closely supervise both of her children, one child may interpret the close supervision as a form of caring and parental love, while the other may interpret the same degree of supervision as oppressive, unnecessary, and unwanted. The assumption that children react to parenting styles the same ways, the assumption of "uniform effects," has allowed scholars to make rather bold and general claims about the impact of specific parenting styles. They claim, for instance, that a "warm and restrictive" parenting style produces children who are better adjusted behaviorally. At one level, this may be true, but it is true only if scholars ignore the full range of variation found in the traits and behaviors of other children within the household (Baumrind, 1980).

The assumption of boundless plasticity

Inherent in virtually all family-based studies of human behavior is the idea that traits and behaviors change over time. Clearly, examples abound of visible, measurable changes in human development. Infants, for example, gain mobility rapidly, allowing them the ability to first roll over from their back, then to pull up on furniture, to crawl across the floor, to walk, and finally to run. Even studies on aggression reveal a common developmental trend: the onset of aggressive behavior occurs exactly at the time mobility is gained, it peaks around age two, and declines every year thereafter (Tremblay *et al.*, 1999).

Unfortunately, many scholars assume that dramatic shifts in behaviors, traits, and abilities occur throughout the life-course. Using the above example, once children acquire the ability to run, very little change in mobility occurs across much of the rest of the life-span. Similarly, once the majority of youth age out of using physical aggression, very few will return to using direct aggression as a dominant interactional pattern. The same pattern has been found using measures of intelligence. Very early in life, test–retest reliabilities on IQ measures are moderate, but once the child reaches the age of 12, test–retest reliabilities increase precipitously and remain unchanged over much of the rest of their life (Mortensen *et al.*, 2003). Thus it should not be too surprising that a broad array of human behaviors and traits, such as cognitive ability and antisocial behavior, show very high levels of stability over time (Haberstick *et al.*, 2006; Beaver *et al.*, 2008; Vaske *et al.*, 2012).

Rarely considered, however, is the idea that some traits and behaviors reach a threshold where they become very difficult, if not in some cases impossible, to change. In this sense, many behaviors and traits become anchored, and the individuals most in need of change will resist external efforts aimed at their reform. The literature on correctional programs, academic programs, and a broad array of social interventions is littered with null or negative findings (Bierman *et al.*, 2013). In most cases, and for most people, intervention efforts do not work, and in some cases they exacerbate the problems they are designed to ameliorate (Dishion *et al.*, 1999).

Studies into parenting also frequently assume that children have the capacity for change beyond changes that occur through normal development. These same studies may also find a link between certain parenting behaviors and normative developmental trends—giving causal credit to the parenting behavior by not realizing that the change is produced by pre-programmed biological influences. When scholars assume that behaviors and traits can change, and when they do not recognize that much of what appears to be change is driven by normative biological development, they are committing what we call the assumption of "boundless plasticity." To be direct, no amount of intervention will change a youth's 85 IQ to 130, and for some youth, no amount of intervention will change their use of drugs, their use of alcohol, or their criminal propensity. Change, unfortunately, is not always possible.

The assumption that G = *0*

Lastly, many family studies make an assumption that has also been proven demonstrably false. Indeed, it is the fifth assumption that has led to the most debate and that has been the source of much bias in family studies. Because of this, we discuss this assumption in more detail. This assumption states that $G = 0$, or that the effects of genetic and biological influences are entirely absent in the development of criminal behavior, or are of insignificant importance. Indeed, thousands of studies conducted across cultures and across time periods have rested on this assumption. Unfortunately, as we will show, this assumption, like the rest, is entirely erroneous. The consequences of violating this assumption, however, have been far-reaching and likely have presented a rather distorted picture of the role and function of families in crime.

We offer a simple heuristic formula—a formula that serves more as a guide to our discussion about the assumption that $G = 0$ rather than as a mathematical proof. This formula follows:

$$\text{human behavior} = (G_{ijk} + E_{ijk});$$
$$\text{including } (G_{ijk} \times E_{ijk}) + (rG_{ijk}E_{ijk});$$
$$+ \text{random}_{ij}$$

This heuristic formula states that any human behavior, in this case criminal behavior, is the product of an individual's (G_i) genotype for family (j) with an evolutionary history (k) plus environmental influences for person (i) in family (j) with a developmental history (k). However, the equation is incomplete because genetic influences interact with environmental stimuli and because they correlate with environments. This fact is denoted in the second part of the equation. We note, however, that random events happen within a person's life and that sometimes these events occur within the family context. This is shown in the third part of the equation. We are thus left with a complex equation:

$$\text{human behavior} = (G_{ijk} + E_{ijk}) + (G_{ijk} \times E_{ijk}) + (rG_{ijk}E_{ijk}) + \text{random}_{ij}$$

The straightforward point of this initial mathematical description of human behavior is that behavior is the product of complex genetic, environmental, evolutionary, and developmental factors. These factors have their origins in families where genes are transmitted from biological parents to their offspring, where parents establish rearing environments for their children, and where children develop along ontogenic and sociogenic pathways These factors may interact with each other or they may correlate together. We recognize, too, that any number of random factors may emerge over time that influence the expression of genes, that alter environmental stimuli, or that broadly influence individual behavior and family functioning. Clearly, as this simple heuristic formula demonstrates:

$$\text{human behavior} \neq E$$

That human behavior cannot be reduced to (E) should now be obvious to anyone who has followed the dramatic scientific advances made over the past 25–50 years—especially advances made in the biological and genetic sciences (Robinson and Beaver, 2009). Even so, a broad range of studies continue to be published where genetic influences are assumed to be absent. This is especially relevant for studies into the effects families and parents have on the development of criminal behavior in their children—for no other reason than children are the genetic by-products of their parents. Even today, the possibility that genes and other biological processes account for variation in child traits and behaviors is hardly mentioned in most studies.

While the reasons for this are complex and have been covered in detail elsewhere (Walsh, 2013), it is important to recognize that during the past 70 years, fierce intellectual battles have been waged over the causal impact genes may have on social behavior and a range of human traits. In particular, scientists studying human intelligence (IQ), sex differences, race differences, and aggression and violence have been subject to career-ending protests and smear campaigns waged by those who advanced an environmental reductionist position (Walsh, 2013). Indeed, the process whereby legitimate scientists, such as James Watson and Arthur Jensen, are mercilessly and publicly humiliated by other scientists has been given the appropriate label of "academic mobbing" (Duffy and Sperry, 2012; for an excellent account see Chagnon, 2013).

Apologists are likely to argue that these battles originated before the advent of modern behavioral genetics, the mapping of the human genome, and the development of technology that allowed easy and fast coding of human DNA. Thus the battles were fought with good intentions, but scholars were handicapped by the limited state of knowledge of the day. While it is tempting to provide such a charitable appraisal, it is also entirely misleading. The principles of Darwinian evolution and Mendelian genetics have been known and accepted as scientific fact for close to 150 years. Broadly speaking, these two paradigms recognize that genes have been under constant evolutionary selection for hundreds of thousands of years, and that encoded in genes are the molecular instructions that provide human beings a broad range of phenotypes and personality characteristics. This is an important point because the "modern synthesis" of Darwinian evolution and molecular genetics logically implicated genetic and environmental influences on various forms of human conduct, including sexual selection and procreation, parenting, and, of course, antisocial behavior. Thus the historical and contemporary intellectual denial of genetic influences on human phenotypes is tantamount to a rejection of one of the greatest advances in science—that is, the modern syntheses between Darwinian evolution and molecular genetics (Wright and Cullen, 2012). In short, the intellectual argument that all human traits and behaviors could be reduced to environment, to socialization, to culture, to reinforcement, to labeling—or to parenting—boarders on academic fraud.

We believe it is important to point out that scientists have found their reputations destroyed, their characters and motives impugned, and their careers ended—not by self-serving politicians or by anti-science groups—but by academics intolerant of the idea that the human genome serves any purpose other than to create bone, sinew, and tissue (Gottfredson, 2012; Haidt, 2013). Understanding the rejection of science by academics and the power they used to ward off future investigators helps us understand how thousands of studies on the role of families and parenting have been conducted without any mention or recognition that the passive transfer of DNA from parents to offspring matters in any measurable way. Indeed, there is solid evidence that scientists received the message that if they dare study the role of biology in human behavior, their careers too would be imperiled. Interview and survey data show, for example, that controversy surrounding specific research subjects causes scholars to make a conscious decision to avoid topics of debate, to self-censor their work, and to alter the interpretation of their findings to make them comport with established dogma (Kempner, 2008).

The end result of these five assumptions is that a body of scientific findings have been generated, published, and presented as substantive fact when there remains reasonable suspicion that many of these findings are methodological artifacts. Evidence of the limits of family functioning on the problem behavior of offspring was found in a recent review of 26 meta-analyses on the role of the family in crime. After an exhaustive review, Petrosino *et al.* (2009) concluded that, contrary to much of what has been written about the association between family variables and crime, "family factors . . . are not the major reason why most persons become criminal (2009: 125)." Effect sizes, they noted, ranged from 0.12 to 0.19 for the relationship between family factors and juvenile delinquency, and they ranged from just 0.04 to 0.13 for the relationship between family factors and recidivism.

We believe the assumptions we have presented, especially the assumption that $G = 0$, have played an important but largely unrecognized role in creating a body of scientific evidence that is of questionable validity (Pinker, 2009). Moreover, the failure to account for genetic influences has allowed scholars and lay people alike to make broad claims about the alleged power of socialization, parenting, culture, and the environment. These claims, as we have discussed, can be made only when methodologies are utilized that ignore genetic influences, or that rest on other demonstrably false assumptions (Kendler and Baker, 2007; Vinkhuyzen *et al.*, 2010).

All is not lost, nor are all studies of families and crime bewitched by these types of problems. Indeed, there exist pockets of studies that directly test, or at least recognize, these assumptions. These studies provide important insights into the complex relationship between family functioning and offspring misbehavior. Moreover, these studies converge to show how the interconnection between biology and specific environmental events and processes affects certain children but does not affect others. Indeed, the good news found within the parenting literature is that researchers are now beginning to study and to understand how the unique traits, dispositions, and propensities of children are, and are not, influenced by parents. Thus in the next section we examine the biosocial research on families—research that points to the importance of parenting for some children, the limits of parenting for others, and the absence of influences parenting has on some youth.

The family–crime link from a biosocial perspective

A biosocial understanding of families involves accepting that children and parents share DNA, that the traits and behaviors of parents and children are partially shaped by their DNA, that socialization occurs within a context of genetically established limitations, and that siblings are often very different from each other because of genetic dissimilarity. Recognizing these phenomena among children and their parent(s) forces researchers to employ research designs that account for both genetic and environmental influences. In this way, many of the flawed assumptions that have guided prior research can be addressed, confidence in the findings produced by these studies can be increased, and a more complete understanding of development can be achieved. An examination of existing research on families reveals biosocial linkages to crime and offending. We begin by examining two of the most replicated findings in the social sciences: the concentration of offending in families and the stability of problem behaviors across generations.

Concentration of offending in relatively few families

Data from across the world shows that the majority of all crimes are committed by a small proportion of individuals, and those individuals are often concentrated in relatively few families. Families with one criminal child are statistically more likely to have other criminal children (Rowe *et al.*, 1996), and several studies have reported significant between-sibling correlations on delinquency and analogous behaviors such as aggression and substance abuse (Rowe and Gulley, 1992; Lauritsen, 1993; Garcia *et al.*, 2000). Farrington *et al.*'s (2001: 580) study of a large cohort of delinquent boys in London concluded that, overall, about 50 percent of all convictions in the sample was accounted for by just 6 percent of the families. Furthermore, four families (1 percent of families), containing 33 persons, totaled 448 convictions (18 percent of all convictions). Adding the next four most criminally prolific families, eight families (2 percent of families), containing 70 persons, totaled 692 convictions. Half of all convictions were accounted for by 23 families (6 percent), and about 10 per cent of the families accounted for nearly two-thirds (64 percent) of all convictions.

More recently, Beaver (2013) reported similar findings using data from the National Longitudinal Study of Adolescent Health. Taking advantage of the genetic information found in the dataset, Beaver examined the concentration of offending in families and the transmission of crime within families in a large sample of sibling pairs. After aggregating the number of arrests for sibling pairs and the number of parental arrests within families, results indicated that just 5 percent of all families accounted for over 50 percent of all arrests. Examining the number of arrests, probation sentences, incarcerations, and involvement in multiple arrests, Beaver also found substantial between-sibling similarities with effect sizes ranging from 2.16 (odds ratio, OR) for probation to 7.80 (OR) for predicting multiple arrests. In the final wave of data collection, respondents were asked about both paternal and maternal arrests. Findings indicated strong intergenerational transmission of criminal involvement, with effect sizes ranging from 2.44 (OR) for a father involved with probation to 4.49 (OR) for both parents having been incarcerated.

Extending family–crime research further, Beaver (2013) also used structural equation modeling to analyze the genetic and environmental mechanisms underlying the crime patterns observed in families. He found that parental criminality had a direct effect on child criminality that was not mediated by family environment. These results suggest that the concentration and transmission of crime in families was due partially to genetic factors and to assortative mating patterns.

Assortative mating is the phenomenon where individuals tend to select mates who are very much like themselves, who live similar lifestyles, and who share similar values and beliefs. This phenomenon has important implications for the concentration of crime in families as well as for the transmission of criminal behaviors across generations, because children born to criminal parents share criminogenic genes from both parents. Not only do these children have a concentrated genetic risk for criminal behavior, but they are also likely to be surrounded by environmental risk factors such as poverty, disadvantaged neighborhoods, and delinquent peers. As these children develop, they are likely to mate with individuals like themselves, thus perpetuating the cycle and transmission of criminogenic risk factors.

High levels of intergenerational stability in problem behaviors

A large body of research has revealed that there is a strong degree of intergenerational stability in criminal and problem behaviors, meaning essentially that children with criminal parents and grandparents are more likely to be arrested than children with noncriminal parents (Jaffee *et al.*, 2003; Farrington, 2010; Beaver, 2013). One of the earliest studies uncovering the transmission of crime across generations was Dugdale's (1887) study of a New York family, the Jukes. Dugdale noted that of the 29 directly related males in the Juke family, 17 had been arrested and 15 convicted. While a primitive study by modern standards, the pattern of findings has been observed in virtually all contemporary studies. In a study spanning 22 years, Huesman *et al.* (1984) observed aggression and antisocial behavior in 600 subjects, their parents, and their children during childhood and adulthood. The authors found substantial stability in aggressive and antisocial behavior within individuals across time, but also found stability in antisocial behavior across generations. In fact, the stability of problem behavior across generations within families was actually stronger than within individual stability across ages. Similar results were reported by Jaffee *et al.* (2003), who found that the correlation between a father and his son's criminal behavior was as high as the correlation (0.30) between criminal behavior measured at two time points for the same individual. Thornberry *et al.* (2003) examined the degree of continuity in antisocial and delinquent behavior among parents and their adolescent children, estimating separate models for mothers and fathers. Findings indicated that there was significant intergenerational continuity between adolescent delinquency and parental antisocial behavior, with slightly higher correlations for the father and child (0.31) than the mother and child (0.22).

More recent studies continue to support stability in problem behavior across generations. In many of his earlier studies using data from the Cambridge Study in Delinquent Development (Farrington *et al.*, 2009) and in a review of family literature, Farrington (2010) noted that boys who had criminal parents were significantly more likely to be arrested than boys with noncriminal parents. In another study, conducted on a large population study of 12.5 million individuals in Sweden, Frisell and his colleagues (2011) examined biological relatedness and crime, finding significant associations between parents and children as well as siblings involved in violent crime convictions. Specifically, there was a significant increase in risk of convictions for interpersonal violence among all familial pairs except for biologically unrelated siblings. Familial risk was highest for genetically related parents and children, and for full biological siblings. Furthermore, the risk between biological parents and children was higher than for adoptive parents and children, and biological siblings reared apart exhibited some of the highest familial risk patterns. There were also strong patterns of assortative mating. These findings provide strong evidence that genes are causally responsible for the transmission of crime across generations and for the concentration of violent crime in families.

Genetic contributions to the family–crime connection were also found by DeLisi *et al.* (2009a). Using a sample of 232 African American females from the National Longitudinal Study of Adolescent Health, the authors examined the effect of having a criminal father and interactions with a polymorphism of a dopamine receptor gene, DRD2, on antisocial behavior. Results indicated that females with criminal fathers have about a 20 percent increase in their likelihood of engaging in antisocial behavior and delinquency. Furthermore, this finding exists regardless of whether the child lived with, knew, or had even met her father. Interactions with having a criminal father and the dopamine receptor gene were even stronger predictors of antisocial behaviors, however. Several other studies have also found that evolutionary processes such as assortative mating and other genetic factors play a key role in the transmission of crime across generations (Krueger *et al.*, 1998; Frisell *et al.*, 2011; Beaver, 2013).

Genetically informed research designs have been successfully used to evaluate the concentration of crime and disrepute within families and the concentration of offending within families. Collectively, research across academic disciplines reveals that crime is heavily concentrated in some families but not others, and that crime and analogous behaviors are often transmitted across generations. Criminal families, moreover, exhibit a host of risk factors including poor parenting, a lack of social and financial resources, mental health issues, substance abuse, and criminal values and traditions. While environmental factors certainly play a role in these behaviors and attitudes, available research also finds that genetic influences are not only measurable, but may also be the dominant source of variation (Silberg *et al.*, 2012).

Evidence for genetic influences on antisocial behavior

Serious and persistent offenders display behavioral problems very early in the life-course, and findings indicate that antisocial behavior is highly time- and situation-stable (Moffitt, 1993; Sampson and Laub, 1993). The stability of antisocial behavior, moreover, appears to extend beyond the individual and into subsequent generations within families (Farrington *et al.*, 1996; Thornberry *et al.*, 2003). The emergence of maladaptive traits and behaviors in the first years of life, coupled with the concentration of criminality in families, suggests that familial factors may play a role in the development of antisocial behavior. Because of this, prior research has largely emphasized the investigation of early socialization factors at the expense of studying possible genetic and biological factors. Even so, there is now a greater recognition that in order to achieve a comprehensive understanding of the causes of antisocial behavior, studies must account for both biological and environmental factors at all stages of the life-course (Moffitt, 2005a).

The dominant methodologies employed to measure genetic and environmental influences use samples of adoptees and twins. On one hand, adoption studies typically compare the degree of similarity for a phenotype, or given trait or behavior, between adoptive relatives with that of biological relatives (Plomin, 2004: 41–47). Greater phenotypic similarity will be observed between biological relatives, such as adoptive children and their biological parents or siblings, if genetic factors are influencing the trait in question because some genetic material is shared and environmental conditions are not. In the event that environmental factors are influencing a phenotype, greater similarity will be observed between an individual and his or her adoptive family members because they experience the same environmental conditions but do not share any genetic material.

Twin studies, on the other hand, provide another means of assessing genetic and environmental contributions to behavior. The most simplified studies of twins use the classical twin design, which compares the phenotypic similarity of monozygotic (MZ; identical) twins with that of dizygotic (DZ; fraternal) twins reared in the same home (Mather and Jinks, 1982; Neale and Cardon, 1992). MZ twins share all of their genetic material, while DZ twins are genetically no more alike than singleton siblings (sharing about 50 percent of their genetic material). Since twins share the same home environment in this design, any increased similarity between MZ twins relative to DZ twins is attributed to genetic influences.

Over 100 studies of antisocial behavior have been published using genetically informative research designs, and meta-analyses have indicated that genetic, shared environment, and unique environmental factors all contribute to variation in antisocial behavior (Miles and Carey, 1997; Rhee and Waldman, 2002; Moffitt, 2005b). These meta-analytic reviews converge to show that genetic factors account for approximately 40–85 percent of the variation in aggression and delinquency, depending on the age of the subjects under study and the seriousness of the misbehavior analyzed. Collectively, however, meta-analytic reviews find that genetic influences account for about 50 percent of the variation in problem behavior on average. The variation left over is usually decomposed into common environmental influences—or those environmental factors shared by youth in the same household—and nonshared environmental influences. One of the major findings from a series of behavioral genetic studies, and echoed in the meta-analyses discussed here, is that nonshared environmental influences are often the dominant source of variance once genetic influences have been partitioned out. Nonshared influences range, on average, between 40 and 60 percent, again depending on the characteristics of the sample. Even more surprising, however, has been the finding that shared environmental influences—such as parenting influences—are rarely detected, often statistically insignificant, and are usually minor in magnitude. In one review, Moffitt (2005b) reported that shared environmental influences account for about 20 percent of the variation in problem behavior. However, the modal outcome in her review ranged from 0 to 10 percent.

Parenting, genes, and the family environment

The relationships between genetic and environmental factors and antisocial development are complex. Genetic factors of individuals have been found to shape the environment, and thereby the risks, to which they are exposed (Plomin *et al.*, 1977; Scarr and McCartney, 1983). Genetic factors, however, have also been found to interact with environmental factors to shape development. With respect to the family environment, this means that the genes of the parents have a significant impact on the environment and experiences of their children, and that the genes of the children influence the environment as well.

Various aspects of family functioning, such as family conflict, may impact the development of antisocial behavior in youths. However, as mentioned above in our discussion of the assumption of measurement purity, it is now clear that many familial contexts are at least partially heritable. For example, Kendler and Baker (2007) reviewed seven studies of the heritability of the family environment, and found that the heritability of family cohesion, conflict, organization, expressiveness, activity, and control ranged from 18 to 30 percent. These findings illustrate an important, and often overlooked, mechanism by which individuals are exposed to particular environments: they select or create it.

Gene–environment correlations affecting environment and behavior

Rather than being randomly subjected to an environmental condition, genetic factors appear to influence one's exposure to particular environmental contexts, including family environments. Associations between particular variants of a gene and environmental circumstances are known as gene–environment correlations (rGE; Plomin *et al.*, 1977; Scarr and McCartney, 1983). Three varieties of rGE have been identified. The first is the passive rGE, in which the antisocial behavior (or other phenotype) of a child and the environment provided by the parent are correlated as a result of both being influenced by the genotype of the parent. For example, two antisocial parents may reproduce and pass the genetic material responsible for their antisocial behavior on to their child. The parents may then raise the child using aggressive disciplinary techniques. Experiencing harsh discipline may be correlated with the child's antisocial behavior; though the relationship would be spurious because both would be due to the genotype of the parents (Moffitt, 2005b).

The second type of rGE is evocative, in which a particular genotype is correlated with an environmental condition because the carrier of the genetic variant tends to elicit a particular response from his or her environment (Scarr and McCartney, 1983). When an evocative rGE occurs, a child may exhibit antisocial behavior and stimulate negative parenting behaviors such as screaming or yelling at the child. The antisocial nature of the child is not due to the verbal altercations with the parents; instead, the parenting techniques and observed behaviors are both under the influence of the genes of the child.

Active rGE is the third type of gene–environment correlation. It is found when one's genotype influences the environments he or she elects to join (Plomin *et al.*, 1977). For example, an antisocial child may actively seek out (consciously or unconsciously) conflict with his or her parents. Though it is possible for active rGE to be present in the family environment, it is rarely investigated because children cannot actively seek out their own parents and, likewise, parents often do not seek out children to raise.

Clearly, parenting behaviors, like antisocial behaviors, can emerge through various processes. In their review, Kendler and Baker (2007) included 19 studies of parenting behaviors such as parental warmth, control, negativity, and protectiveness, and found that these parenting qualities were between 12 and 37 percent heritable. As noted previously, however, genes may be operating in additional ways to shape parenting behaviors, which can in turn influence the development of their children. It is possible to assess the etiology of parenting, particularly rGE effects, by observing pairs of twins as children or parents.

When studying twin pairs as children, significant genetic influences on parenting measures reflect the effects of the genes of the child on parenting, or evocative or active rGE. Additionally, shared environmental influences that would increase the similarity in parenting behaviors received between twins may actually reflect the influences of the parents' genotypes, or passive rGE (Klahr and Burt, 2013). It is further possible to investigate indirect genetic influences on parenting in studies of twins as parents. Significant genetic effects observed in parent-based designs indicate that the parenting behavior of the individual is under direct genetic influence and that passive rGE may be occurring because both twins are parenting different children in different households using similar behaviors. Unique environmental influences observed in such studies may partially reflect evocative rGE because the twin parents are rearing different children, and may be eliciting different parental responses (Klahr and Burt, 2014).

In an effort to integrate findings from existing research, Avinun and Knafo (2014) meta-analyzed 32 studies of children as twins that assessed child-driven effects on parenting. Overall, the heritability of parenting behaviors was estimated to be 23 percent, providing evidence of evocative rGE effects on parental positivity and negativity. Additionally, shared and unique environmental factors also accounted for some of the variation in parenting (43 and 34 percent, respectively). Significant shared environmental influences may be indicative of passive rGE, and that parents, to some extent, are treating their children similarly. The influence of nonshared environmental factors, on the other hand, may suggest that parents also treat their progeny somewhat differently, which further indicates that evocative (genetic or nongenetic) child effects may be occurring.

Similarly, Klahr and Burt (2013) conducted a meta-analysis of influences on parental warmth and acceptance, parental control, and parental negativity. Based on the results of 44 child-based and 9 parent-based studies, they concluded that genetic, shared environmental, and nonshared environmental influences at the child level all have a significant influence on all three aspects of parenting, suggesting rGE plays a role in the complex relationships between parents and their children. The parent-based analysis revealed that nonshared environmental influences contribute significantly to parenting practices, which may reflect evocative rGE. Moreover, the parent-based studies demonstrated the significance of genetic influences originating in the parent for parental warmth and negativity. These findings are particularly interesting given that parental warmth and parental negativity have both been associated with antisocial behavior in youths (Lynch *et al.*, 2006; Feinberg *et al.*, 2007).

Genes appear to play an important role, both directly and indirectly, in shaping the family environment. The interplay between genes and the environment, however, is not limited to rGE. Particular genetic variants may also interact with environmental factors in ways that influence antisocial development. The study of gene–environment interactions (G × E) can provide some insight into why individuals respond to environmental risks in different ways.

Genetic and environmental interactions influencing antisocial development

When factors in the environment moderate the expression of a genotype, or one's genotype affects the degree of influence of an environment factor, a G × E is observed (Plomin *et al.*, 1977). In short, it is particular combinations of genetic and environmental factors that are likely to produce specific outcomes. For genetic reasons, some individuals may be particularly sensitive to some risk relative to others. A heightened genetic risk, coupled with exposure to an environmental risk, is more likely to generate antisocial outcomes than having one risk or the other (Beaver, 2009). For example, Mednick *et al.* (1984) examined a sample of adoptees and found that children with both a biological and an adoptive parent who had been arrested were the most likely to be involved in criminal behaviors, suggesting the effects of genetic and environmental risks exceeded those of one risk factor.

Scientific advances have made it possible to identify some of the specific genes that may interact with environmental factors to shape development. These studies often emphasize the importance of parenting practices or other familial risk factors in the etiology of antisocial behavior (Thapar *et al.*, 2007). The earliest study that examined a measured gene and a measured environment was conducted by Caspi and his colleagues (2002). The results revealed that males possessing a variant of the MAOA gene coding for low MAOA activity who had also experienced maltreatment as children were significantly more likely to have antisocial behavioral problems than other males. A meta-analysis of studies published since Caspi and associates' initial study found evidence of significant G × E effects between MAOA and childhood maltreatment or family adversity (Kim-Cohen *et al.*, 2006). Additional studies of genotypes and familial risk factors, such as MAOA × childhood sexual assault (Derringer *et al.*, 2010; Fergusson *et al.*, 2011); MAOA × physical discipline (Edwards *et al.*, 2010); DRD2 (A1 allele) × criminal father (Beaver *et al.*, 2009); 5HTTLPR (two short alleles) × maltreatment, and TPH1 (genotypes GT and TT) × maltreatment (Cicchetti *et al.*, 2012) have produced significant effects on the development of antisocial behaviors. Taken together, these studies suggest that genetic differences between people account for differences in responses to risk factors.

Although previous research has largely investigated the ways in which genes interact with family-level risk factors to increase one's risk for antisocial outcomes, it should be noted that altering the environmental component of the interaction can reduce antisocial behaviors, further highlighting the interdependence of genes and the environment. In fact, some scholars have theorized that the genotypes frequently considered to place one at a greater risk for antisocial behavior may, in reality, reflect a heightened sensitivity that can operate in negative or positive ways (Belsky, 1997, 2005; Boyce and Ellis, 2005; Ellis *et al.*, 2011).

Rather than assuming that genes work only to increase one's vulnerability for antisocial outcomes, the differential susceptibility perspective maintains that particular genotypes are associated with greater reactivity, or "plasticity." Humans, like all living things, share an innate goal of distributing their genetic material in future generations, which can be accomplished only if natural selection favors traits that promote survival and reproduction (Belsky, 2005). However, future conditions are difficult to predict, making it impossible to strategically employ child-rearing practices that would increase the likelihood that one's offspring would survive and reproduce. The majority of parents will raise their children using techniques they believe to be optimal, but in some situations those methods may prove to be disadvantageous in the future. Children who are more malleable to parental influence may not survive, while those who do not conform to parental influences may thrive. By producing more than one child, each with a differing degree of sensitivity, the odds that at least one child will successfully reproduce increases—regardless of future conditions (Belsky, 2005).

The differential susceptibility framework emphasizes the importance of both genes and the family environment. Genes are hypothesized to determine the extent of one's malleability (Belsky, 2005). Individuals who possess genotypes that increase sensitivity will be more susceptible to the rearing environments in which they develop. The behavioral patterns of these individuals will then reflect the characteristics of those environments, whether they be positive and supportive, or antisocial in nature. Individuals who possess genotypes that decrease sensitivity, on the other hand, will be less sensitive to parental influences. Those who are less sensitive will be less likely to exhibit antisocial behaviors in the face of family adversity, but also less likely to largely benefit from promotive parenting strategies.

Research investigating the hypotheses subsumed within the differential susceptibility framework is still in its infancy, but the available evidence provides some support for the theory.

Bakermans-Kranenburg and van IJzendoorn (2006) investigated the interaction of DRD4 and maternal sensitivity on externalizing behaviors in a sample of children. They found that youths with the 7-repeat allele of DRD4 exhibited the highest levels of externalizing behaviors if mothers were insensitive, but the lowest levels when sensitivity among mothers was high. Children who did not possess the allele associated with plasticity were unaffected by maternal sensitivity. In another study, Li and Lee (2010) explored whether 5HTTLPR interacted with maltreatment to influence antisocial behavior, and found that girls with the short/short genotype who had been maltreated were 12 times more likely to engage in antisocial behaviors such as vandalism, breaking and entering, or selling marijuana.

Susceptible youths have also been found to benefit more from family- or parenting-based interventions compared with less susceptible children. In one study, families were randomly assigned to participate in an intervention program intended to reduce externalizing behaviors through increasing parental sensitivity and promoting positive disciplinary interactions (Bakermans-Kranenburg *et al.*, 2008). Children who were in the intervention group who also had the 7-repeat allele of DRD4 showed reductions in externalizing behaviors, and the effect was significantly stronger when mothers showed the greatest improvement in sensitive discipline strategies. In contrast, children without the plasticity allele did not have reductions in externalizing behaviors as a result of participating in the intervention.

In another experiment, families were randomly assigned to participate in an intervention program designed to improve various aspects of family functioning in order to reduce substance abuse in adolescents (Beach *et al.*, 2010). The results of this study indicated that after participating in the intervention, youths with the 7-repeat allele of DRD4 showed less growth in substance use between 11 and 14 years compared with youths without that genotype and with controls. Taken together, the findings from these studies suggest that those who are most likely to develop antisocial behaviors as a result of significant gene–environment interactions may also be those who are most receptive to treatment.

Conclusion

Academic research into the influence of parenting behaviors on children remained largely unchanged for decades. In study after study, scholars routinely collected data from one parent and one child within a family. And study after study, they rested on the same assumptions—namely that their measures of parenting behavior or environment were uncontaminated by genetic influences, that parenting represented a top-down approach to socialization, that parental influences could safely be generalized to other children within the same home, and that genetic influences were unimportant. Indeed, hundreds of millions of taxpayer dollars have been spent on research designs employing what is called the "standard social science model" (SSSM), a model built on the assumptions outlined above, and a model incapable of providing any greater insight into the influence and limits of parenting behaviors.

With the publication and wide dissemination of Judith Rich Harris' (1998) book *The Nurture Assumption*, however, scholars were challenged to overcome the obstacles inherent in the SSSM. Citing evidence from a broad range of disciplines, Harris argued that the nurture assumption, the idea that parenting behaviors largely determine child outcomes, had been overstated and that the evidence garnered by scholars was flimsy at best, and invalid at worst. Of course, her book was met largely with hostility by many in the academic community, but a minority of scholars took her criticisms seriously and began to employ more sophisticated study designs (DeLisi *et al.*, 2009b; Ellis *et al.*, 2011; Frisell *et al.*, 2011). These designs, while not perfect, attempted to control for genetic influences, for bi-directional effects (from child to parent and parent to child), and they attempted to measure the differential impact of parenting behaviors across children within the same home. The insights garnered through these studies highlight important facts about the family–crime relationships.

First, sometimes there is a wide range of variation between children within the same household and sometimes there is not. In certain families, crime and other social problems transcend generations, while in other families criminal behavior may be non-existent, or engaged in by only one child but not the others (Farrington, 1978; Farrington *et al.*, 2001; Belsky *et al.*, 2009; Beaver, 2013). Understanding this variation is important. Second, sometimes parents have little to no influence on the misbehavior of their offspring. Depending on a range of factors, some youth manifest problems behaviors early in life and continue to engage in crime and disrepute across their life-course. They do so despite sometimes heroic efforts made by parents to address, treat, and extinguish their child's problem behavior. Finally, it appears that certain parenting behaviors affect some youth more than others, and that part of the reason for this is that youth differ genetically (Bakermans-Kranenburg and Van IJzendoorn, 2006; Beach *et al.*, 2010; Beaver *et al.*, 2010b). For example, youth at the highest genetic risk for aggression and who are physically abused by their parent(s) seem to be at an increased risk for conduct problems. Physical abuse, however, does not appear to be as strongly connected to conduct problems in youth not already genetically at-risk (Caspi *et al.*, 2002; Edwards *et al.*, 2010; Cicchetti *et al.*, 2012). Again, recognizing that children within the same home vary, sometimes tremendously, and that part of this variation is due to genetic differences between youth, has allowed biosocial researchers to gain greater insight and greater specificity into the mechanisms that link parenting behaviors to child outcomes for some youth.

In a relatively short period, biosocial research has generated important insights into the family–crime link. These insights are just now beginning to penetrate criminology, and they have yet to fully gain traction. For example, major criminological theories continue to identify parenting behaviors as the primary source of variation in traits related to criminal behavior—namely low self-control (Gottfredson and Hirschi, 1990)—and most continue to deny the existence of genetic influences on delinquent and criminal behavior (Wright and Cullen, 2012). The tide is shifting only because biosocial criminologists continue to produce evidence revealing greater insights into the interlocking qualities of nature and nurture, because they continue to challenge long-held dominant disciplinary beliefs, and because they continue to expose the methodological and ideological biases in the social sciences (DeLisi *et al.*, 2009b). Without the advent of biosocial criminology, the useful and valid criticisms presented in a book written by a non-academic grandmother who wrote psychology textbooks for a living (Harris, 1998) would have been ignored and the important insights about the family–crime link would have remained hidden.

References

Avinun, R. and Knafo, A. (2014) 'Parenting as a reaction evoked by children's genotype: a meta-analysis of children-as-twin-studies', *Personality and Social Psychology Review*, 18: 87–102.

Bakermans-Kranenburg, M.J. and Van IJzendoorn, M.H. (2006) 'Gene–environment interaction of the dopamine D4 receptor (DRD4) and observed maternal insensitivity predicting externalizing behavior in preschoolers', *Developmental Psychobiology*, 48: 406–409.

Bakermans-Kranenburg, M.J., Van IJzendoorn, M.H., Pijlman, F.T., Mesman, J., and Juffer, F. (2008) 'Experimental evidence for differential susceptibility: dopamine D4 receptor polymorphism (DRD4 VNTR) moderates intervention effects on toddlers' externalizing behavior in a randomized controlled trial', *Developmental Psychology*, 44: 293–300.

Baumrind, D. (1980) 'New directions in socialization research', *American Psychologist*, 35: 639–652.

Beach, S.R., Brody, G.H., Lei, M.K., and Philibert, R.A. (2010) 'Differential susceptibility to parenting among African American youths: testing the DRD4 hypothesis', *Journal of Family Psychology*, 24: 513–521.

Beaver, K.M. (2009) *Biosocial Criminology: A Primer*, Dubuque, IA: Kendall/Hunt.

Beaver, K.M. (2013) 'The familial concentration and transmission of crime', *Criminal Justice and Behavior*, 40: 139–155.

Beaver, K.M. and Wright, J.P. (2007) 'A child effects explanation for the association between family risk and involvement in an antisocial lifestyle', *Journal of Adolescent Research*, 22: 640–664.

Beaver, K.M., Wright, J.P., DeLisi, M., and Vaughan, M.G. (2008) 'Genetic influences on the stability of low self-control: results from a longitudinal sample of twins', *Journal of Criminal Justice*, 36: 478–485.

Beaver, K.M., Schutt, J.E., Boutwell, B.B., Ratchford, M., Roberts, K., and Barnes, J.C. (2009) 'Genetic and environmental influences on levels of self-control and delinquent peer affiliation results from a longitudinal sample of adolescent twins', *Criminal Justice and Behavior*, 36: 41–60.

Beaver, K.M., DeLisi, M., Vaughn, M.G., and Wright, J.P. (2010a) 'The intersection of genes and neuropsychological deficits in the prediction of adolescent delinquency and low self-control', *International Journal of Offender Therapy and Comparative Criminology*, 54: 22–42.

Beaver, K.M., Ferguson, C.J., and Lynn-Whaley, J. (2010b) 'The association between parenting and levels of self-control: a genetically informative analysis', *Criminal Justice and Behavior*, 37: 1045–1065.

Beaver, K.M., Vaughn, M.G., DeLisi, M., Wright, J.P., Wiebe, R.P., Harrington, C., and Walsh, A. (2014) 'The heritability of common risk and protective factors to crime and delinquency', in M. DeLisi and K.M. Beaver (eds), *Criminological Theory: A Life-course Approach*, 2nd edn, Burlington, MA: Jones & Bartlett, pp. 99–114.

Belsky, J. (1997) 'Theory testing, effect-size evaluation, and differential susceptibility to rearing influence: the case of mothering and attachment', *Child Development*, 68: 598–600.

Belsky, J. (2005) 'Differential susceptibility to rearing influence', in B.J. Ellis and D.F. Bjorklund (eds), *Origins of the Social Mind: Evolutionary Psychology and Child Development*, New York: Guildford Press, pp. 139–163.

Belsky, J., Conger, R., and Capaldi, D.M. (2009) 'The intergenerational transmission of parenting: introduction to the special section', *Developmental Psychology*, 45: 1201–1204.

Bierman, K.L., Coie, J., Dodge, K., Greenberg, M., Lochman, J., McMohan, R., and Pinderhughes, E. (2013) 'School outcomes of aggressive-disruptive children: prediction from kindergarten risk factors and impact of the fast track prevention program', *Aggressive Behavior*, 39: 114–130.

Boyce, W.T. and Ellis, B.J. (2005) 'Biological sensitivity to context: I. An evolutionary-developmental theory of the origins and functions of stress reactivity', *Development and Psychopathology*, 17: 271–301.

Brooks, A.C. (2007) *Who Really Cares: The Surprising Truth About Compassionate Conservatism*, New York: Basic Books.

Caspi, A., McClay, J., Moffitt, T.E., Mill, J., Martin, J., Craig, I.W. *et al.* (2002) 'Role of genotype in the cycle of violence in maltreated children', *Science*, 297: 851–854.

Chagnon, N. (2013) *Noble Savages: My Life among Two Dangerous Tribes—The Yanomamo and the Anthropologists*, New York: Simon & Schuster.

Cicchetti, D., Rogosch, F.A., and Thibodeau, E.L. (2012) 'The effects of child maltreatment on early signs of antisocial behavior: genetic moderation by tryptophan hydroxylase, serotonin transporter, and monoamine oxidase A genes', *Development and Psychopathology*, 24: 907–915.

Clinton, B. (1996) *Between Hope and History: Meeting America's Challenges for the 21st Century*, New York: Random House.

DeLisi, M., Beaver, K.M., Vaughn, M.G., and Wright, J.P. (2009a) 'All in the family: gene × environment interaction between DRD2 and criminal father is associated with five antisocial phenotypes', *Criminal Justice and Behavior*, 36: 1187–1197.

DeLisi, M., Wright, J.P., Vaughn, M.G., and Beaver, K.M. (2009b) 'Copernican criminology', *The Criminologist*, 34: 14–16.

Derringer, J., Krueger, R.F., Dick, D.M., Saccone, S., Grucza, R.A., Agrawal, A., *et al.* (2010) 'Predicting sensation seeking from dopamine genes: a candidate-system approach', *Psychological Science*, 21: 1282–1290.

Dishion, T.J., McCord, J., and Poulin, F. (1999) 'When interventions harm: peer groups and problem behavior', *American Psychologist*, 54: 755–764.

Dodge, K.A. and Crick, N.R. (1990) 'Social information-processing bases of aggressive behavior in children', *Personality and Social Psychology Bulletin*, 16: 8–22.

Duffy, M. and Sperry, L. (2012) *Mobbing: Causes, Consequences, and Solutions*, New York: Oxford University Press.

Dugdale, R.L. (1887) *"The Jukes": A Study in Crime, Pauperism, Disease, and Heredity*, New York: Putnam and Sons.

Edwards, A.C., Dodge, K.A., Latendresse, S.J., Lansford, J.E., Bates, J.E., Pettit, G.S., *et al.* (2010) 'MAOA-uVNTR and early physical discipline interact to influence delinquent behavior', *Journal of Child Psychology and Psychiatry*, 51: 679–687.

Ellis, B.J., Boyce, W.T., Belsky, J., Bakermans-Kranenburg, M.J., and Van IJzendoorn, M.H. (2011) 'Differential susceptibility to the environment: an evolutionary neurodevelopmental theory', *Development and Psychopathology*, 23: 1–7.

Farrington, D.P. (1978) 'The family background of aggressive youths', in L.A. Hersov and M. Berger (eds), *Aggression and Anti-social Behavior in Childhood and Adolescence*, Oxford: Pergamon, pp.73–93.

Farrington, D.P. (2010) 'Family influences of delinquency', in D. Springer and A. Roberts (eds), *Juvenile Justice and Delinquency*, Sudbury, MA: Jones & Bartlett, pp. 203–222.

Farrington, D.P., Barnes, G., and Lambert, S. (1996) 'The concentration of offending in families', *Legal and Criminological Psychology*', 1: 47–63.

Farrington, D.P., Joliffe, D., Loeber, R., Stouthamer-Loeber, M., and Kalb, L.K. (2001) 'The concentration of offenders in families, and family criminality in the prediction of boys' delinquency', *Journal of Adolescence*, 24: 579–596.

Farrington, D.P., Coid, J.W., and Murray, J. (2009) 'Family factors in the intergenerational transmission of offending', *Criminal Behaviour and Mental Health*, 19: 109–124.

Feinberg, M.E., Kan, M.L., and Hetherington, E.M. (2007) 'The longitudinal influence of coparenting conflict on parental negativity and adolescent maladjustment', *Journal of Marriage and Family*, 69: 687–702.

Fergusson, D.M., Boden, J.M., Horwood, L.J., Miller, A.L., and Kennedy, M.A. (2011) 'MAOA, abuse exposure and antisocial behavior: 30-year longitudinal study', *British Journal of Psychiatry*, 198: 457–463.

Frisell, T., Lichtenstein, P., and Langstrom, N. (2011) 'Violent crime runs in families: a total population study of 12.5 million individuals', *Psychological Medicine*, 41: 97–105.

Garcia, M.M., Shaw, D.S., Winslow, E.B., and Yaggi, K.E. (2000) 'Destructive sibling conflict and development of conduct problems in young boys', *Developmental Psychology*, 36: 44–53.

Gottfredson, L.S. (2012) 'Resolute ignorance on race and Rushton', *Personality and Individual Differences*, 55: 218–223.

Gottfredson, M.R. and Hirschi, T. (1990) *A General Theory of Crime*, Stanford, CA: Stanford University Press.

Haberstick, B.C., Schmitz, S., Young, S.E., and Hewitt, J.K. (2006) 'Genes and developmental stability of aggressive behavior problems at home and school in a community sample of twins aged 7–12', *Behavioral Genetics*, 36: 809–819.

Haidt, J. (2013) *The Righteous Mind: Why Good People are Divided by Politics and Religion*, New York: Random House.

Harris, J.R. (1995) 'Where is the child's environment? A group socialization theory of development', *Psychological Review*, 102(3): 458–489.

Harris, J.R. (1998) *The Nurture Assumption: Why Children Turn Out the Way They Do*, New York: Free Press.

Huesman, L.R., Eron, L.D., Lefkowitz, M.M., and Walder, L.O. (1984) 'The stability of aggression over time and generations', *Developmental Psychology*, 20: 1120–1134.

Jaffee, S.R., Moffitt, T.E., Caspi, A., and Taylor, A. (2003) 'Life with (or without) father: the benefits of living with two biological parents depend on father's antisocial behavior', *Child Development*, 74: 109–126.

Kempner, J. (2008) 'The chilling effect: how do researchers react to controversy?', *PLoS Medicine*, 5: 222.

Kendler, K.S. and Baker, J.H. (2007) 'Genetic influences on measures of the environment: a systematic review', *Psychological Medicine*, 37: 615–626.

Kerr, M. and Stattin, H. (2000) 'What parents know, how they know it, and several forms of adolescent adjustment: further support for a reinterpretation of monitoring', *Developmental Psychology*, 36(3): 366–380.

Kim-Cohen, J., Caspi, A., Taylor, A., Williams, B., Newcomb, R., Craig, I.W., and Moffitt, T.E. (2006) 'MAOA, maltreatment, and gene–environment interaction predicting children's mental health: new evidence and a meta-analysis', *Molecular Psychiatry*, 11: 903–913.

Klahr, A.M. and Burt, S.A. (2013) 'Elucidating the etiology of individual differences in parenting: a meta-analysis of behavioral genetic research', *Psychological Bulletin*, 133: 1–44.

Klahr, A.M. and Burt, S.A. (2014) 'Practitioner review: evaluation of the known behavioral heterogeneity in conduct disorder to improve its assessment and treatment', *Journal of Child Psychology and Psychiatry*, doi: 10.1111/jcpp.12268.

Krueger, R.F., Moffitt, T.E., Caspi, A., Bleske, A., and Silva, P.A. (1998) 'Assortative mating for antisocial behavior: developmental and methodological implications', *Behavior Genetics*, 28: 173–186.

Lauritsen, J.L. (1993) 'Sibling resemblance in juvenile delinquency: findings from the National Youth Survey', *Criminology*, 31: 387–409.

Li, J.J. and Lee, S.S. (2010) 'Latent class analysis of antisocial behavior: interaction of serotonin transporter genotype and maltreatment', *Journal of Abnormal Child Psychology*, 38: 789–801.

Lynch, S.K., Turkheimer, E., Onofro, B.M., Mendle, J., Emery, R.E., Slutske, W.S., and Martin, N.G. (2006) 'A genetically informed study of the association between harsh punishment and offspring behavioral problems', *Journal of Family Psychology*, 20: 190–198.

Lytton, H. (1990) 'Child and parent effects in boys' conduct disorder: a reinterpretation', *Developmental Psychology*, 26: 683–697.

Mather, K. and Jinks, J.L. (1982) *Biometrical Genetics: The Study of Continuous Variation*, New York: Chapman & Hall.

Mednick, S.A., Gabrielli, W.F., and Hutchings, B. (1984) 'Genetic influences in criminal convictions: evidence from an adoption cohort', *Science*, 224: 891–894.

Miles, D.R. and Carey, G. (1997) 'Genetic and environmental architecture of human aggression', *Journal of Personality and Social Psychology*, 72: 207–217.

Moffitt, T.E. (1993) 'Adolescence-limited and life-course-persistent antisocial behavior: a developmental taxonomy', *Psychological Review*, 100: 674–701.

Moffitt, T.E. (2005a) 'The new look of behavioral genetics in developmental psychopathology: gene–environment interplay in antisocial behaviors', *Psychological Bulletin*, 131: 533–554.

Moffitt, T.E. (2005b) 'Genetic and environmental influences on antisocial behaviors: evidence from behavioral-genetic research', *Advances in Genetics*, 55: 42–104.

Mortensen, E.L., Andresen, J., Kruse, E., Sanders, S.A., and Reinisch, J.M. (2003) 'IQ stability: the relation between child and young adult intelligence test scores in low-birthweight samples', *Scandinavian Journal of Psychology*, 44: 395–398.

Neale, M. and Cardon, L.R. (eds) (1992) *Methodology for Genetic Studies of Twins and Families*, Vol. 67, Dordrecht, Netherlands: Springer.

Petrosino, A., Derzon, J., and Lavenberg, J. (2009) 'The role of the family in crime and delinquency: evidence from prior quantitative reviews', *Southwest Journal of Criminal Justice*, 6(2): 108–132.

Pinker, S. (2003) *The Blank Slate: The Modern Denial of Human Nature*, New York: Penguin Books.

Pinker, S. (2009) *Language Learnability and Language Development, with New Commentary by the Author*, Cambridge, MA: Harvard University Press.

Plomin, R. (2004) *Nature and Nurture: An Introduction to Human Behavioral Genetics*, Belmont, CA: Wadsworth.

Plomin, R., DeFries, J.C., and Loehlin, J.C. (1977) 'Genotype–environment interaction and correlation in the analysis of human behavior', *Psychological Bulletin*, 84: 309–322.

Plomin, R., Reiss, D., Hetherington, E.M., and Howe, G.W. (1994) 'Nature and nurture: genetic contributions to measures of the family environment', *Developmental Psychology*, 30: 32–43.

Rhee, S.H. and Waldman, I.D. (2002) 'Genetic and environmental influences on antisocial behavior: a meta-analysis of twin and adoption studies', *Psychological Bulletin*, 128: 490–529.

Robinson, M.B. and Beaver, K.M. (2009) *Why Crime? An Interdisciplinary Approach to Explaining Criminal Behavior*, Durham, NC: Carolina Academic Press.

Rowe, D.C. (1994) *The Limits of Family Influence: Genes, Experience, and Behavior*, New York: Guilford.

Rowe, D.C. and Gulley, B.L. (1992) 'Sibling effects on substance use and delinquency', *Criminology*, 30: 217–233.

Rowe, D.C., Linver, M.R., and Rodgers, J.L. (1996) 'Delinquency and I.Q.: using siblings to find sources of variation', in G.H. Brody (ed.), *Sibling Relationships: Their Causes and Consequences*, Norwood, NJ: Ablex Publishing, pp. 147–172.

Sampson, R.J. and Laub, J.H. (1993) *Crime in the Making: Pathways and Turning Points through Life*, Cambridge, MA: Harvard University Press.

Sayre-McCord, G. (ed.) (2007) *Crime and Family: Selected Essays of Joan McCord*, Philadelphia, PA: Temple University Press.

Scarr, S. and McCartney, K. (1983) 'How people make their own environments: a theory of genotype–environment effects', *Child Development*, 54: 424–435.

Silberg, J.L., Maes, H., and Eaves, L.J. (2012) 'Unraveling the effect of genes and environment in the transmission of parental antisocial behavior to children's conduct disturbance, depression and hyperactivity', *Journal of Child Psychology and Psychiatry*, 53: 668–677.

Thapar, A., Harold, G., Rice, F., Langley, K., and O'Donovan, M. (2007) 'The contribution of gene–environment interaction to psychopathology', *Development and Psychopathology*, 19: 989–1004.

Thornberry, T.P., Freeman-Gallant, A., Lizotte, A.J., Krohn, M.D., and Smith, C.A. (2003) 'Linked lives: the intergenerational transmission of antisocial behavior', *Journal of Abnormal Child Psychology*, 31: 171–184.

Tremblay, R.E., Japel, C., Perusse, D., McDuff, P., Boivin, M., Zoccolillo, M., and Montplaisir, J. (1999) 'The search for the age of "onset" of physical aggression: Rousseau and Bandura revisited', *Criminal Behaviour and Mental Health*, 9: 8–23.

Unnever, J.D., Cullen, F.T., and Agnew, R. (2006) 'Why is "bad" parenting criminogenic? Implications from rival theories', *Youth Violence and Juvenile Justice*, 4: 3–33.

Vaske, J., Ward, J.T., Boisvert, D., and Wright, J.P. (2012) 'The stability of risk-seeking from adolescence to emerging adulthood', *Journal of Criminal Justice*, 40: 313–322.

Vinkhuyzen, A.A.E., Van Der Sluis, S., De Geus, E.J.C., Boomsma, D.I., and Posthuma, D. (2010) 'Genetic influences on "environmental" factors', *Genes, Brain and Behavior*, 9: 276–287.

Walsh, A. (2013) *Criminological Theory: Assessing Philosophical Assumptions*, Burlington, VT: Ashgate.

Wright, J.P. and Beaver, K.M. (2005) 'Do parents matter in creating self-control in their children? A genetically informed test of Gottfredson and Hirschi's theory of low self-control', *Criminology*, 43(4): 1169–1202.

Wright, J.P. and Beaver, K.M. (2013) 'Do families matter?', in F.T. Cullen and P. Wilcox (eds), *Oxford Handbook of Criminological Theory*, Oxford: Oxford University Press.

Wright, J.P. and Cullen, F.T. (2012) 'The future of biosocial criminology beyond scholars' professional ideology', *Journal of Contemporary Criminal Justice*, 28(3): 237–253.

Wright, J.P., Beaver, K.M., DeLisi, M., and Vaughn, M.G. (2008) 'Evidence of negligible parenting influences on self-control, delinquent peers, and delinquency in a sample of twins', *Justice Quarterly*, 25: 544–569.

Part V

Legal and criminal justice applications

32

Intelligence as the quintessential biosocial variable

An examination of etiological factors and associations with criminal offending and criminal justice processing

Joseph L. Nedelec, Joseph A. Schwartz, and Eric J. Connolly

> Knowing a great deal is not the same as being smart; intelligence is not information alone but also judgment, the manner in which information is coordinated and used.
>
> *(Carl Sagan, 1980, p. 270)*

Introduction

Intelligence is one of the most widely studied concepts in the social sciences. While historically the province of psychology, the study of intelligence can be found in virtually all of the constituent fields of social science. Researchers have examined the associations between variance in intelligence and numerous social outcomes, in part because of its apparent ubiquitous influence on human endeavors. This influence has not entirely escaped those studying criminal behavior, but the infusion of intelligence into the work of criminologists is minimal compared with their social science brethren. The relative paucity of the study of intelligence in criminology is due primarily to the predominance of sociology within the discipline, and as a result criminology is relatively late to understanding the importance of intelligence as it relates to human behavior. Consequently, the purpose of this chapter is twofold: (1) to examine the empirical evidence regarding the association between intelligence, criminal behavior, and criminal justice system processing, and (2) to highlight the value of the study of intelligence to understanding the etiology of antisocial behavior. Throughout our discussions we illustrate that intelligence is the quintessential biosocial variable – it is highly influenced by biological factors, yet its importance to the human condition is a function of our social nature. The chapter is arranged in five main sections. The first section briefly surveys conceptualization and testing issues associated with intelligence. The second section examines the current evidence pertaining to the etiological factors underpinning variance in intelligence. The third provides a summary of the empirical literature regarding the association between intelligence and criminal offending. The fourth explores a relatively understudied component of the literature on intelligence: the association between intelligence and criminal justice processing. Finally, the chapter concludes with comments on the current literature regarding intelligence and criminology, and provides recommendations for future research.

What is intelligence?

As the above quote from well known astrophysicist Carl Sagan suggests, defining intelligence is not an easily accomplished task. Indeed, many writers have noted this difficulty when discussing the topic, and highlight how often the existential position is relied upon: intelligence is known when it is seen or, more appropriately, when it is absent (Sternberg, 2000; Walsh, 2003). Intelligence is conceptualized in the social sciences with reference to a variety of terms – cognitive abilities, intellectual capabilities, IQ, general intelligence, or *g*. However, it is conceptualized within individual studies. Researchers generally agree that this is a feature of the human condition related to ". . . the ability to reason, plan, solve problems, think abstractly, comprehend complex ideas, learn quickly and learn from experience [which reflects a] capability for comprehending our surroundings" (Gottfredson, 1997a: 13). It is clear from such a conceptualization that intelligence is not simply academic achievement or acquired knowledge (i.e. crystallized knowledge) but is much more broad (i.e. fluid knowledge; Walsh, 2003). An important feature of this conceptualization of intelligence, germane to the study of criminal behavior and criminal justice outcomes, is the ability to learn from experience or to understand the consequences of one's actions. This component of the definition will be fully illuminated in our discussions of the association between IQ, criminal behavior, and criminal justice processing.

With the recognition of the general characterization of intelligence, there is now considerable evidence that this general feature is measurable and that the tests used to measure intelligence do so very well (Gottfredson, 1997a; Pinker, 2002). A detailed discussion of the various manners by which intelligence is measured is beyond the scope of this chapter; for now, there are two main points pertinent to our discussion. First, as is often highlighted by many criminology textbooks, the history of intelligence testing is spotted with some regrettable policy recommendations. Infused within these discussions is often reference to the idea that intelligence tests are hopelessly biased in terms of culture and thus are of little use to the analysis of criminal behavior. Furthermore, the tiresome argument that recognition of individual differences such as IQ will inevitably lead to eugenic policies, supported with reference to Spencer or Goddard, are invariably provided as evidence of the futility of examining intelligence. These discussions are often based on a clear misunderstanding of intelligence, IQ testing, and individual differences (cf. Akers and Sellers, 2009). What is rarely recognized is the fact that IQ tests were originally designed to identify those individuals, within a scholarly setting, who could benefit most from extra pedagogical attention (Wilson and Herrnstein, 1985; Anderson, 2007). In other words, the initial use of intelligence testing was to serve a positive social function. Additionally, as outlined by Steven Pinker (2002) in *The Blank Slate*, intelligence testing in Britain was once heralded as the tool through which significant political change could be brought about. Left-leaning political groups encouraged the use of intelligence tests as a way to identify highly talented individuals, regardless of social class, and "direct them into the ruling elite" (Pinker, 2002: 301). Therefore IQ tests were seen as a method of subverting the caste system wherein rulers are provided their position through birthright rather than through ability or talent. Far from being a tool of an inevitably oppressive nature, intelligence testing has long been seen as a way of helping improve the lives of citizens within a democratic society.

Second, although various batteries designed to assess intelligence exist, it is now widely recognized in the contemporary literature on intelligence that these various tests are measuring similar constructs (Gottfredson, 1997a; Davies *et al.*, 2011). Strong support for the construct validity of intelligence testing comes not only from comparisons of different types of tests, but also from the various culturally sensitive assessments. As is also invariably mentioned in criminology textbooks, early intelligence tests were culturally biased in favor of white, middle-class, English-speaking Americans. There is not much argument on this point. However, intelligence testing has evolved considerably since these early days and, with the accumulation of decades of reliability and validity assessments, they have become unmatched in terms of psychometric value (Herrnstein and Murray, 1994; Wilson and Herrnstein, 1985; Gottfredson, 1997a). Nonetheless, the bias claim continues to be made against intelligence tests in areas of social science, and in criminology in particular (cf. Curran and Renzetti, 2001). Two points are worth making here: first, tests that have been altered to suit various cultures (e.g. Japanese and Chinese translations of American versions of IQ tests) have shown average scores that are higher than those for American samples (Rushton, 1989). It is hard to argue that intelligence tests are biased in favor of white, middle-class, English-speaking Americans when such results arise, especially when, for example, "mainland Chinese peasants show a mean of 101 using translated tests" (Walsh, 2002: 107; white American samples have a mean of 100). Second, given the debate sparked by Herrnstein and Murray's (1994) *The Bell Curve*, the Board of Scientific Affairs of the American Psychological Association (APA) convened a task force to assess what was "known and unknown" about intelligence and intelligence testing. For the purposes of the argument of bias, the task force unanimously concluded that modern assessments of intelligence are not systematically biased against groups of different racial/ethnic background or socioeconomic status (SES) (Neisser *et al.*, 1996). When one refers to the conceptualization of intelligence as outlined above, it is not surprising that these findings of nonbias emerge. It is difficult to argue that cultural differences between groups would lead to variance in terms of the ability to employ abstract thinking, to learn from experience, or to comprehend one's surroundings.

The etiological factors of intelligence

The previous section outlined the conceptualization of intelligence as it is being employed in this chapter. In this section, we explore some of the empirical evidence that has examined the etiological factors of intelligence. This literature is vast; consequently our discussion is relatively focused in nature. The etiological factors are demarcated into two groups: genetic factors and nongenetic factors (i.e. common environmental factors and unique environmental factors). We spend some time discussing these etiological factors as they are central to understanding the association between intelligence, offending, and criminal justice processing. One final note that is key to this section: our discussion of the etiological factors of intelligence focuses on those aspects that are claimed to cause variance between individuals, and not on what causes a high or low intelligence level in a single individual.

Intelligence and evolution

In describing the factors that underpin the variance in intelligence, authors often begin with the biological or sociocultural components of influence. However, in order to properly contextualize the etiological factors, it is appropriate to situate the human trait of intelligence within an evolutionary framework (Rushton, 2004). Doing so provides not only guidance in terms of what factors should be provided primacy in an etiological examination, but also justification for why such factors are focused upon. Contextualizing intelligence within an evolutionary framework also serves to highlight that human intelligence is just one type in an entire animal kingdom brimming with intelligence. As Pinker (2009: 182) notes, "A brain is a precision instrument that allows a creature to use information to solve the problems presented by its lifestyle . . . Each animal has evolved information-processing machinery to solve its problems, and we evolved machinery to solve ours." The problems to which Pinker is referring are problems related to survival and reproduction. When viewed in this way, it is obvious that intelligence refers to something more than simply exposure to and memorization of facts.

The literature examining intelligence and human evolution is voluminous, as scholars have long considered intelligence as the *sine qua non* characteristic of the human condition. An additional feature of this literature is the interesting array of hypotheses proposed to explain the emergence of human intelligence. The current stance is that human intelligence arose due to a confluence of events such as the move from an arboreal environment to the grasslands of the African savannah, an increase in the availability and consumption of meat (and the cooperative hunting necessary to obtain that meat), and the increasing social complexity of the group-based life of our ancestors (Ridley, 1993; Wright, 1994; Pinker, 2009). Other scholars indicate that the growth in human neural complexity and the associated increases in intelligence were also due in part to an arms race between the sexes (Miller, 2001) and the ever-increasing benefits of deception within the human social milieu (Trivers, 2011). Whatever the underlying evolutionary cause for the emergence of a complex human intelligence, it is clear that the structures of the human brain and its incredible neural complexity and processing arose due to the discerning forces of natural and sexual selection.

When one accepts that the human brain is a product of evolutionary process, one then must also accept that variance in neuronal structure and functioning over evolutionary time was linked to variance in genetic information. Given that evolutionary processes work at the level of the gene (Dawkins, 2006), this conclusion is necessary. Taking the argument further, given that there was genetic variance over evolutionary time which led to the development of the human brain and intelligence, there must therefore also be an underlying genetic influence to variance in the functioning of the brain among humans. In the words of Sandra Scarr (1978: 329), "Since there is a species genetic program for intellectual development there is also very likely to be genetic variability for intelligence; that is, individual differences that depend in part on genotypic differences." But, the argument does not end there; an evolutionary point of view is a biosocial point of view. Thus it is recognized that differential selection for genetic information underlying neural structure and functioning over evolutionary time was a result of the manifestation of the genetic information within an environmental arena. Thus both genetic and nongenetic factors were important to the evolution of the brain, and are important to the variance currently observed among individuals. To illustrate, we turn again to Scarr's (1978: 329) comments on the issue, "to become a functioning human being clearly requires both the species genetic program and a rearing environment within the range that was evolutionarily typical for the species."

To summarize, human intelligence evolved to help solve a suite of specific adaptive problems faced by our ancestors. Additionally, the evolutionary development of human intelligence indicates a necessity of genetic variance among individuals in neuronal structure and functioning. This variance led to differences in survival and reproductive success resulting from the unfolding interactions between genetic factors and nongenetic factors in ancestral environments.

Consequently, differential selection (i.e. evolution) occurred wherein certain genetic variants were selected over others. Finally, as Scarr (1978) and others point out, this evolutionary process provided a species-level intelligence as well as intelligence levels that vary from individual to individual. This individual-level variance is explained by both genetic and nongenetic factors. It is to these specific explanatory factors that we now turn.

The behavioral genetics of intelligence – genetic and nongenetic factors

As outlined above, our discussion of the etiological factors underlying variance in intelligence is demarcated into genetic factors, common environmental factors, and unique environmental factors. This demarcation emanates from an extensive line of work in the field of behavior genetics (BG), a fundamental component of many of the chapters in this volume. Primarily, behavior geneticists, and the methodologies they employ, are interested in assessing the various causes of differences between individuals. To be clear, this line of inquiry is concerned with explaining variance between individuals, and not with explaining phenotypic development within a single individual (Plomin *et al.*, 2013a). Briefly, BG employs a variety of sampling methodologies and statistical techniques to assess the relative influence of genetic and nongenetic factors on phenotypic variance (i.e. differences in a trait or phenotype) across individuals in a given population. By capitalizing on natural and social experiments, BG is able to produce estimates regarding the magnitude of variance due to three factors: heritability (h^2), common environmental factors (c^2), and unique or nonshared environmental factors (e^2). These three components represent the totality (100 percent) of explanatory factors to which the phenotypic variance among individuals can be ascribed. Each component therefore represents a proportion of the total amount of variance explained among individuals. Estimates of heritability indicate the degree to which phenotypic variance is due to genetic differences among individuals, while estimates related to c^2 indicate the proportion of phenotypic variance due to the effects of nongenetic factors that are shared among individuals and that serve to make siblings more similar. The final component, e^2, represents the remaining variance, which is nongenetic and serves to make siblings more dissimilar. Additionally, e^2 includes the effects of error.

As noted, BG employs natural and social experiments to estimate the proportion of variance due to these three components. An experiment provided by nature is twinning. For reasons that are yet unclear, a single fertilized egg will sometimes split into two or more zygotes that have the potential to develop into genetically identical individuals. This process of twinning produces monozygotic (MZ) twins. Another process of twinning, wherein two separate eggs are fertilized by two different sperm during the same act of conception, produces dizygotic (DZ) twins. DZ twins are no more similar genetically than nontwin siblings; however, it is the nongenetic environmental factors which DZ twins share that provides the logic for the use of twins to estimate the influences of phenotypic variance. If certain assumptions are met, MZ and DZ twins can be employed to compare the phenotypic similarities within twin pairs to assess the differential influence of genetic and nongenetic factors. Given that both twins within an MZ or DZ twin set are born at the same time, are reared by the same caregivers, attend the same schools, and so on, it is assumed that they share similar developmental environments. Therefore, any phenotypic similarity within a twin pair that is not due to similar developmental environments is due to shared genetic factors. Further, if the observed phenotypic similarity is greater in magnitude among MZ twins than it is among DZ twins, it can be assumed that the greater concordance is due to the greater similarity in genetic relatedness (MZ twins share 100 percent of their DNA while DZ twins share, on average, about 50 percent of their DNA). The complexity of statistical methods of analyses that employ twin-based samples has increased over the past half-century, but the underlying logic remains the same. The twin-based method is one of the most robust and methodologically sound ways in which to assess the relative influence of *both* genetic and nongenetic factors on phenotypic variance (Plomin *et al.*, 2013a). However, there exist other ways in which to estimate the proportional influence of genetic and nongenetic factors on phenotypic variance.

The societal experiment of adoption is an effective and powerful method to estimate proportional influence. Adoption studies are especially valuable in discerning the shared environmental component of phenotypic variation (Rowe, 1994). There are two aspects of adoption that allow researchers to employ this methodology: (1) an adoptee does not share a developmental environment with his/her biological parent, and (2) an adoptee does not share genetic material with his/her adoptive parent. Consequently, researchers can compare phenotypic similarity between an adoptee and his/her biological and adoptive parents to determine the relative influence of genetic and nongenetic factors. If the adoptee is more phenotypically similar to the biological parent, this resemblance can only be due to genetic factors. Conversely, if the adoptee is more phenotypically similar to the adoptive parent, the similarity can only be due to nongenetic factors (more specifically, shared environmental factors). The adoption method has been employed extensively by researchers and is widely considered (at least outside of criminology) to be a valid and robust method (Plomin *et al.*, 2013a,b).

A relatively rare, but exceptionally powerful, version of the adoption study methodology is the MZ twin reared apart method (Segal, 2010). In this hybrid natural–social experiment, researchers collect information from two individuals within an MZ twin pair who were adopted away to different homes. Referred to as MZA, these studies represent a particularly stringent manner by which to assess the relative influence of genetic and nongenetic factors. The underlying logic of MZA studies aligns with twin studies but with one important alteration: the twins do not share a developmental environment. Consequently, any phenotypic similarity can *only* be due to shared genetic factors.

The above methods and other variants have all been employed to assess the relative influence of genetic and nongenetic factors on the variance of intelligence among various individuals. With an understanding of the logic underlying these methods, we now turn to examples of such studies.

Etiological factors – genetic

When describing behavioral genetics methods, intelligence is often listed along with other phenotypes, such as height, as examples of highly heritable traits. Indeed, for as long as there have been systematic assessments of human intelligence, there have been illustrations of its heritability (Wilson and Herrnstein, 1985). Francis Galton, the founder of behavioral genetics, conducted his assessments on the heritability of intelligence and other abilities even before Gregor Mendel's famous work outlining the laws of heritability. Given that both specific forms of intelligence (e.g. reasoning) and general cognitive ability (i.e. *g*) have been among the most well studied topics in behavioral genetics, it not surprising that a tremendous amount of evidence has been amassed regarding the genetic influence on intelligence (Plomin *et al.*, 2013a). Given the sizeable nature of the extant literature examining the genetic influence on intelligence, our discussion is general in its focus.

As noted above, the APA task force on intelligence (Neisser *et al.*, 1996) conducted a vast review of the available literature at the time and drew a number of conclusions. Among those conclusions was the finding that, across a number of cultures and to some extent social status (the authors were careful to point out that the lowest SES levels were under-represented in the available literature), "a sizeable part of the variation in intelligence test scores is associated with genetic differences among individuals" (Neisser *et al.*, 1996: 85). The authors also highlight the difference in estimates of heritability across the life course, noting "In childhood h^2 and c^2 for IQ are of the order of 0.45 and 0.35; by late adolescence h^2 is around 0.75 and c^2 is quite low (zero in some studies)" (Neisser *et al.*, 1996: 85). In other words, during childhood approximately 40 percent of the variance in intelligence among individuals is due to differences in genetic material, whereas during early adulthood that proportion is around 75 percent (see Bouchard, 2013 for a recent review on this observation). There are two points to be derived from these observations. First, it is undeniable that variation in intelligence is influenced by genetic factors. Indeed, given that Neisser and colleagues' (1996) review included multiple cultures, and similar estimates have been found throughout the world (Jensen, 1998), this conclusion appears to be one of the most robust observations in the social sciences. Second, the increasing magnitude of heritability over the life course necessitates clarification. The most common explanation proposed emanates from the behavioral genetic concept of gene–environment correlation (denoted as rGE). Briefly, there are three types of rGE: (1) passive rGE, (2) active rGE, and (3) evocative rGE. Passive rGE refers to the fact that biological parents who rear their offspring pass on both genetic factors and a developmental environment, thus an offspring's phenotype will correlate with his/her environment. Active rGE refers to the idea that individuals will tend to systematically select into environments in which their genetic proclivities (e.g. personality or talent) are most satisfyingly expressed. Evocative rGE recognizes that a person's social environment will "react" to the manner in which the person conducts him or herself. In other words, a person's behavior or personality (i.e. genetically influenced phenotypic traits) will differentially evoke responses from others. While passive rGE is most applicable in terms of explaining the proportions of h^2 and c^2 observed during childhood, the process associated with active rGE best explains the change in proportion over time. More specifically, the increased estimates of heritability and decreased estimates of c^2 are reflective of a process of social maturity wherein an individual is less encumbered by the limitations of parental and scholastic rules, and is increasingly free to choose a lifestyle that is best suited to their own personalities and tastes (Walsh, 2003).

In an effort to update the findings reviewed by Neisser *et al.* (1996), Nisbett and colleagues (2012) reviewed portions of the literature on the heritability of intelligence. In their review, Nisbett *et al.* (2012: 132) note that the effects of genes on intelligence are "undeniable" yet the authors also indicate how the magnitude of heritability estimates varies by social class. An additional aspect of Nisbett *et al.* (2012) was discussion of the advances in brain imaging techniques since the Neisser *et al.* (1996) review. Nisbett and colleagues review the empirical evidence indicating an association between the prefrontal cortex (PFC) and components of intelligence (e.g. fluid intelligence). Additionally, the authors note the association between gray matter and cognitive ability. Given that the structure and functioning of the brain has been shown to be highly heritable (Brans *et al.*, 2010; Raine, 2013), the Nisbett *et al.* (2012) review generally substantiates the genetic effect on variance in intelligence outlined in numerous behavior genetic studies (although the authors provide some counterpoints to the observed magnitude of heritability estimates).

While studies have generally assessed the differential influence of genetic and nongenetic factors on overall intelligence (i.e. the normal distribution), researchers have also been interested in examining other aspects of intelligence, such as expertise, stability, and change. For example, Haworth *et al.* (2012) assessed the proportionate influence of genetic and nongenetic factors for those respondents scoring in the top 15 percent on tests assessing general cognitive ability (*g*) drawn from six twin studies in four different countries (Australia, the Netherlands, the United Kingdom, and the USA; N = 11,000 twin pairs). Using a liability-threshold model (which indicates the proportion of variance due to genetic and nongenetic factors that influences passing a particular threshold, in this case scoring in the top 15 percent of the distribution of *g*), the authors found that 50 percent of the variance was accounted for by genetic factors. The remaining variance was distributed relatively evenly between shared environmental effects (c^2 = 0.28) and nonshared environmental factors (e^2 = 0.22). Another example is provided by Plomin and colleagues (2013b) who assessed the influence of genetic and nongenetic factors on reading expertise among a large sample (N = 10,000) of 12-year-old twins from the UK. The authors found that over half of the variance in reading performance between expert and normal readers was accounted for by genetic factors. Thus, in addition to relative variance in the normal distribution of intelligence, genetic factors are also substantively influential in terms of variance between those who exhibit exceptional intelligence and those who do not.

A common finding in the literature on intelligence is the observation of relative stability across the life course (Deary *et al.*, 2013). Accordingly, researchers have assessed the proportional influence of genetic and nongenetic factors on the stability of intelligence. A recent analysis by Beaver and colleagues (2013a) illustrated that intelligence (verbal IQ) was generally stable across swaths of the life course, but they also noted some change in IQ scores among a subset of their analytical sample over time. Intriguingly, Beaver *et al.* (2013a) found that genetic factors accounted for the majority of the variance in terms of *both* stability (between 66 and 83 percent) and change (between 43 and 69 percent) in IQ scores. In other words, relative rank in terms of variance in IQ tends to remain, and that consistency is due primarily to differences in genetic factors; but when relative rank is altered (i.e. a change in IQ), that alteration is also due primarily to differences in genetic factors (see also Deary *et al.*, 2012, 2013).

In summary, the influence of genetic factors on the development of intelligence is indisputable. However, as noted by the Nisbett *et al.* (2012) review and the fact that h^2 never equals 1.0 (100 percent), there is also a substantive influence of nongenetic factors on the etiology of intelligence. It is to these nongenetic factors that we now turn.

Etiological factors – common and unique environmental factors

As noted, a substantial proportion of the variance in intelligence is due to nongenetic factors. These nongenetic factors can include environmental factors in the sense that most criminologists tend to conceptualize them (i.e. social variables), but they also include any nongenetic biological influence on intelligence (e.g. severe brain injury). In examining the influence of shared environmental factors such as social variables, employing the adoptive method outlined above (and variants thereof) provides the most direct assessment. There are two main variants of the adoptive method: (1) a comparison of phenotypic similarity between adoptive parent and adoptee, and (2) a comparison of phenotypic similarity between adoptive siblings. In both cases the individuals within a pair do not share genetic information but do share rearing or developmental environments. While estimates of c^2 vary from study to study, a powerful analysis conducted by Chipuer and colleagues (1990), which included multiple types of twin, family, and adoption data, provided robust estimates of the influence of shared environmental effects. The authors showed that shared environmental effects accounted for approximately 19 percent of the variance in IQ similarity between adoptive parents and their adopted children, and 29 percent of the variance in IQ similarity between adopted (unrelated) siblings living in the same developmental environment. Plomin *et al.* (2013a) provide similar estimates in their collective analysis of multiple adoption studies (where $c^2 = 0.19$ for adoptive parent-adoptee, and $c^2 = 0.32$ for unrelated adoptive siblings). Consequently, approximately a third of the variance in intelligence is attributable to shared experiences within a developmental environment (e.g. similar SES, being reared by the same parents, being exposed to the same diet choices, attending the same schools). As outlined above, however, this observed influence of the shared environment decreases considerably over the lifespan. Indeed, estimates of c^2 drop dramatically when adult samples are assessed. Nonetheless, the finding of a substantial effect of c^2 during childhood provides evidence for opportunity to affect intelligence levels in a positive way (see below).

One of the most promising components of a behavioral genetic methodology is the ability to isolate the effects of specific environmental factors that are uniquely experienced by individuals. The importance of this methodology is clear: for the vast majority of phenotypes that have been subjected to quantitative genetic analysis, the majority of the variance attributed to nongenetic factors falls within the nonshared environmental component (Rowe, 1994; Harris, 2009). In terms of intelligence, as individuals age, the proportion of variance attributed to nongenetic factors increasingly falls within the nonshared environmental component (Plomin *et al.*, 1994). Asbury and Plomin (2014: 17) provide an excellent description of this process, "Over the years nonshared experiences small and large will contribute to making identical twins into increasingly different people. The same set of DNA travels along two divergent paths of experience, making identical twins ever more different as a result." The collection of unique experiences of individuals is vast and therefore it is not a surprise that e^2 takes up a considerable amount of variance in assessing phenotypic variance.

While researchers have employed BG methods to assess the influence of nonshared environmental factors on social and emotional development, the examination of such effects on intelligence is less well represented (Petrill and Wilkerson, 2000). A stringent way in which to assess the influence of nonshared environmental factors is the MZ difference score approach (aka discordant MZ-twin method; Asbury *et al.*, 2003; Beaver, 2008; Vitaro *et al.*, 2009). The logic of the MZ difference score method is straightforward: given that MZ twins (reared together) share 100 percent of their genetic factors and, by definition, 100 percent of their shared factors, any difference between them must be due to unique experiences. Consequently, researchers employing this technique simply subtract the score of one twin on a particular item (e.g. victimization) from the co-twin's score on the same item. This method is considered so empirically sound that some scholars have indicated it is the closest social science researchers can get to experimentally controlled causality (Vitaro *et al.*, 2009). An example of such a method of assessing the nonshared environmental influence on intelligence was employed by Vitaro and colleagues (2012), who examined the association between a variety of social experiences in children in kindergarten and their academic achievement in the first grade. The authors found that, of the variety of social experiences among the 223 MZ twin pairs included in the study, within-pair differences in peer rejection and quality of teacher–student relationship predicted differences in academic achievement (an outcome highly associated with intelligence). Consequently, the authors provided evidence of nonshared environmental factors which can have an influence on variance in intelligence-related outcomes.

The molecular genetics of intelligence – SNP, GWA, and GCTA

Given the consistent finding of genetic influence on the variance in intelligence, there has also been an effort to identify the constituent genetic material that accounts for the observed influence. In molecular genetic studies, researchers assess the extent to which variance in the possession of specific genes (i.e. genetic polymorphisms such as single-nucleotide polymorphisms, SNP) accounts for phenotypic variance (referred to as candidate gene association studies). Findings in this area of research are mixed. The primary reason for the mixed results in the candidate gene association studies of intelligence is due to the fact that a phenotype such as intelligence, like all complex phenotypes, is likely influenced by a vast number of genes. Thus the likelihood of isolating the effect of a single gene is relatively low. Consequently, researchers tend to pool together numerous candidate genes in order to assess the additive influence of multiple genes. An interesting study that employed this method was conducted by Schwartz and Beaver (2012). In this study the authors created a genetic risk index by pooling together three different polymorphisms of the dopaminergic genes (i.e. DAT1, DRD2, and DRD4). Additionally, the authors created an aggregated family-level dopamine index that was used to compare the influence of the dopamine index on aggregated IQ levels between families. The results of the analyses were intriguing – the authors found that variation in dopaminergic risk scores (dopamine index) between siblings (i.e. within families) was not associated with variation in IQ between siblings. However, differences in the aggregated dopaminergic risk scores between families were associated with average differences in aggregated IQ levels between families. More specifically, those families who had higher aggregated scores on the dopaminergic risk index also had lower aggregated IQ scores, even after controlling for race (Schwartz and Beaver, 2012). This study illustrates that, while the pursuit of individual polymorphisms associated with intelligence may not always bear fruit, researchers should alter their methods rather than abandon the search.

In addition to pooled analyses such as the one conducted by Schwartz and Beaver (2012), a cutting-edge approach to examining the molecular genetics of intelligence has come in the form of genome-wide association (GWA) studies and genome-wide complex trait analyses (GCTA). Briefly, GWA studies examine the association between a phenotype and hundreds of thousands of polymorphisms (usually SNP) in the entire genome (Plomin *et al.*, 2013b). GCTA also employ scans of the entire genome, but take advantage of chance genetic similarity amongst SNP between unrelated individuals to then create genetically related pairs; this genetic similarity can then be employed to estimate heritability for phenotypic variance (Plomin *et al.*, 2013a). An important difference between the two methods is that, while GWA attempt to identify specific genes associated with phenotypic variance, GCTA do not. Given that GWA studies and GCTA are conducted on respondents who are unrelated, the analyses are not subject to the assumptions associated with twin- and adoption-based designs. However, the methods do require very large sample sizes (multiple thousands of respondents) in order to conduct the analyses.

To date, there are a handful of GWA and GCTA studies which have estimated the heritability of intelligence. For example, Davis *et al.* (2013) conducted a GWA study that included $N = 7{,}900$ children from the UK. An examination of over 350,000 SNP resulted in a modest association between nine different SNP and variation in intelligence. However, due to the very small effect sizes, the authors called for replication rather than providing a strong conclusion. A similar study was conducted by Davies *et al.* (2011), who employed GWA techniques on 549,692 SNP from over 3,500 unrelated adults from various regions in the UK. The results indicated that at a minimum (i.e. lower bound), 51 percent of the variation in crystallized intelligence and 40 percent of fluid intelligence was accounted for by additive genetic factors. Additionally, the authors found that one SNP, *FNBP1L* (which is influential in regulating neuronal morphology), was significantly associated with fluid intelligence in their sample. Nonetheless, these authors also called for continued replication rather than drawing firm conclusions. A GWA study which employed childhood respondents was conducted by Benyamin *et al.* (2013), wherein a similar association between *FNBP1L* and intelligence was observed. Additionally, Benyamin and colleagues reported that the aggregate effects of SNP accounted for between 22 and 46 percent of the variation in intelligence among the almost 18,000 participants in their analytical sample. GCTA studies provide further support for the heritability of intelligence. In their GCTA study, Plomin *et al.* (2013b) found that approximately 35 percent of the variance in general cognitive ability could be accounted for by genetic factors. It is important to recognize that this estimate, as with all estimates provided from GCTA studies, is a lower-limit estimate due to the method's inclusion of only additive genetic effects (whereas estimates derived from twin studies capture both additive and nonadditive genetic effects; Plomin *et al.*, 2013b). The GCTA method has also been employed to assess the genetic influence on stability and change in intelligence. Deary *et al.* (2012) employed multivariate GCTA on a unique sample of N = 1,940 unrelated individuals whose intelligence was assessed at age 11 and again in late adulthood (age 65, 70, or 79 years). The authors found two results that are congruent with past twin-based research: (1) there was a genetic correlation of 0.62 between intelligence measured in childhood and intelligence measured in old age (i.e. stability), and (2) approximately 24 percent of the variance in change in cognitive ability was due to genetic factors.

The available evidence from GWA and GCTA studies is preliminary. However, there are two general conclusions that can be derived from this emerging line of evidence. First, there is without question a genetic influence on intelligence, its stability, and its change over the lifespan. Second, the heritability estimates from family-based research such as twin and adoption studies are quite robust given their congruence with the estimates produced by studies that are not subject to the same assumptions (i.e. GWA and GCTA studies).

Summary

The etiology of intelligence is one of the most widely studied components of the human condition. Given the association between intelligence and a wide array of outcomes over the life course (Gottfredson, 1997a,b), understanding what influences intelligence is of vital importance. Overall, the research examined in this section indicates that, while intelligence is highly influenced by genetic factors, there is also an environmental effect. We return to this point at the end of the chapter. With a strong grasp of the etiology of intelligence, we now turn to an examination of the association between intelligence and criminal offending. As we progress through our discussion, keep in mind the various findings from biosocial criminologists which have illustrated that approximately 50 percent of the variance in antisocial behavior is due to genetic factors (Beaver, 2013). Part of the reason for this genetic influence could be attributed to the effect of intelligence on the many aspects of a person's life and the underlying impact of genetic factors on intelligence itself.

Intelligence and offending

Perhaps one of the most robust and well replicated correlates of criminal and antisocial behavior is overall intelligence. Studies have consistently revealed that, on average, individuals with overall lower levels of intelligence are significantly more likely to engage in criminal behavior (Wilson and Herrnstein, 1985; Kandel *et al.*, 1988; Herrnstein and Murray, 1994; Kratzer and Hodgins, 1999; Nedelec *et al.*, 2012; Beaver *et al.*, 2013b). As outlined above, IQ was not always considered an important correlate of criminal or delinquent behavior. Early sociological scholars such as Sutherland (1931) argued that there was no significant difference in IQ between delinquent and nondelinquent juveniles. Such arguments stemmed directly from the antiquated sociological view that all behavior was the direct result of purely social influences which ultimately manifest as behavioral variation. Scholars who advanced such arguments refused to acknowledge that any source of behavioral variation could potentially stem from the individual. These particular views ultimately resulted in a dearth of information regarding the potential association between intelligence and criminal behavior throughout the early development of the field of criminology. In the few instances where information regarding this association would appear in textbooks or journal articles, the information seemed to be from outdated psychology textbooks and was ideologically tainted (Hirschi and Hindelang, 1977).

In a now classic study, Hirschi and Hindelang (1977) attempted to illustrate the overall importance of acknowledging the association between intelligence and delinquency and incorporating such information into explanations of such behavior. They argued that including intelligence in theories of criminal behavior provides a more comprehensive and in-depth understanding of the etiological development of such behaviors. While a significant number of studies followed the suggestions set forth by Hirschi and Hindelang (Wolfgang *et al.*, 1972; Farrington, 1973; Moffitt *et al.*, 1981; Moffitt and Silva, 1988; Lynam *et al.*, 1993), research examining the association between IQ and criminal behavior came to the forefront of criminology following the publication of Herrnstein and Murray's (1994) controversial book, *The Bell Curve: Intelligence and Class Structure in American Life*. While the overarching purpose of the book was to explore the manner in which variation in intelligence ultimately results in individual differences within various aspects of behavior, an entire chapter is devoted to criminal behavior specifically. Herrnstein and Murray argued that intelligence was strongly associated with criminal behavior and directly implicated in coming into contact with the criminal justice system. In addition, the results of the empirical analyses performed by the authors revealed that many of the correlates criminologists had long considered potent predictors of criminal behavior, such as unemployment and poverty, were also more prevalent within populations of individuals with lower IQ scores. These findings resulted in a firestorm of controversy and outrage from a large number of social scientists (Dorfman, 1995; Goldberger and Manski, 1995; Hauser, 1995; Sternberg, 1995; Cullen *et al.*, 1997).

While *The Bell Curve* may have been met with a strong wave of criticism, some of which continues currently (Mears and Cochran, 2013), this comprehensive study of the association between intelligence and behavior sparked strong interest from both sides of the argument. A large number of studies employing highly sophisticated analytic methods began to examine more closely the potential role that intelligence played in the etiological development of delinquent and criminal behavior (Moffitt *et al.*, 1981; Moffitt and Silva, 1988; Lynam *et al.*, 1993; Nedelec *et al.*, 2012; Beaver *et al.*, 2013b). For example, a study performed by Moffitt and colleagues (1981) directly addressed significant limitations of previous studies in two ways. First, data drawn from a Danish birth cohort were analyzed using multivariate analyses to effectively partial out the potentially confounding influence of SES on the association between IQ scores (measured using the Børge Priens Prøve) and arrests (measured using official report data). Previous examinations had relied almost exclusively on bivariate associations, and critics had argued that comprehensive measures of social influence, such as SES, likely confound any association between IQ and criminal behavior. After controlling for SES, the association between IQ and offending was attenuated but remained statistically significant, providing evidence of a direct, causal relationship (Moffitt *et al.*, 1981). Second, Moffitt and colleagues offered a more comprehensive theoretical explanation for the association between IQ and offending behaviors. More specifically, they argued that individuals with lower verbal skills likely performed poorly in school, which in turn likely resulted in subsequent negative social experiences both within and outside the family environment. Such experiences are expected to culminate ultimately into feelings of frustration and failure, which may result in negative feelings toward authority and a greater susceptibility to peer pressure.

More comprehensive assessments

As this line of research continued to advance, studies began to provide more precise specifications of the association between IQ and offending behaviors. For example, a line of research examining the association between IQ and delinquent behavior during adolescence quickly developed throughout the 1980s and 1990s (Moffitt *et al.*, 1981; Moffitt and Silva, 1988; Lynam *et al.*, 1993). During this period, critics argued that the association between IQ and delinquency may be spurious due to differential detection (Feldman, 1977; Hirschi and Hindelang, 1977), wherein youth with lower IQ scores are no more likely to engage in delinquent behavior, but are significantly more likely to be detected by agents of the criminal justice system. Moffitt and Silva (1988) directly assessed this particular hypothesis by observing two groups of individuals within a birth cohort from Dunedin, New Zealand. The first group of youth had been detected by police, while the youth included in the comparison group had similar rates of offending but had not been detected by police. The two groups possessed nonsignificant differences in overall IQ, indicating that differential detection did not significantly contribute to the overrepresentation of individuals with low IQ within the criminal justice system. In addition, both of the observed groups (who were defined as the most delinquent in the full cohort) possessed mean IQ scores that were approximately 6 IQ points lower than the mean IQ score for the full cohort.

Another argument commonly cited by critics of intelligence research is that some unmeasured social variable is significantly influencing both IQ and delinquency, making any detected association spurious (Simons, 1978; Menard and Morse, 1984). In perhaps the most comprehensive assessment of the association between IQ and delinquency at the time, Lynam and colleagues (1993) analyzed a high-risk sample of youth from the Pittsburgh Youth Study (PYS). Using multivariate statistical models, these researchers isolated the association between IQ (measured using the Wechsler Intelligence Scale for Children-Revised [WISC-R]) and self-reported delinquency. Importantly, a host of potential confounding variables, including SES, interest and effort during the IQ test, impulsivity, academic achievement, and race, were all included in the multivariate models. The results revealed that the negative association between IQ and delinquency remained significant *even after controlling for this impressive list of covariates*. In addition, the results also revealed that impulsivity did not significantly mediate the association between verbal IQ and delinquency, even after taking into account differences between African-American and Caucasian respondents. School performance, however, did significantly moderate the association between verbal IQ and delinquency for African-American youth, but not for Caucasian youth.

In a more recent study, Beaver and colleagues (2013b) analyzed a nationally representative sample of American youth from the National Longitudinal Study of Adolescent Health (Add Health). This particular study directly addressed several limitations of previous studies, including the use of a nationally representative sample, the use of self-reported measures of contact with the criminal justice system, and a series of analyses examining subgroups based on race and gender. In addition, and perhaps most importantly, this study included a number of important potentially confounding covariates in the multivariate statistical models, including self-control and delinquent behavior. The inclusion of these covariates allows for the isolation of the association between IQ (measured using a modified version of the Peabody Picture Vocabulary Test-Revised [PPVT-R]) and contact with the criminal justice system (measured as ever arrested, ever incarcerated, and incarcerated after being arrested). The results revealed that lower levels of intelligence significantly increased the odds of being arrested, being incarcerated, and being incarcerated if arrested. The findings held across nearly all subgroups examined, with the only exceptions being the association between IQ and incarceration for white and black females. The results provide strong evidence indicating that IQ is not only directly tied to entry to the criminal justice system (via arrest), but also is intricately implicated in the retention of individuals within the criminal justice system (via incarceration; see below).

Variety of offending

During this time, studies also began to disentangle the potential role that IQ plays in the type of criminal behavior in which offenders engage (Walsh and Beyer, 1986; Walsh, 1987; Nevin, 2000; Frisell *et al.*, 2012). For example, Walsh (1987) investigated the association between verbal IQ and both violent and property offending within a sample of male probationers from Toledo, Ohio and Boise, Idaho. Rather than use a continuous measure of verbal IQ, Walsh recoded the measure into four categories ranging from highest to lowest IQ score. He then examined the mean violent and property offending scores broken down by each category. The results revealed that offenders who fell within the lowest IQ category were the most likely to engage in violent crime, but offenders who were most likely to engage in property crime fell in the highest IQ category. In addition, offenders with an IQ score between 80 and 90 appeared to be the most dangerous group, comprising just over 60 per cent of the overall sample of offenders. The overall pattern of findings is attributed to higher levels of impulsivity which accompany lower IQ scores. More specifically, offenders with lower IQ scores may prefer more spontaneous or opportunistic crimes, while offenders with higher IQ scores may take more time to plan offenses which result in a lower probability of detection (Raine, 2013).

A handful of studies have examined the potential association between intelligence and violent criminal behavior such as sexual assault (Cantor *et al.*, 2005), murder (Dwyer and Frierson, 2006), and assault (Kearns and O'Connor, 1988). For example, in a recent study, Frisell *et al.* (2012) examined the association between intelligence (measured with the Swedish Enlistment Battery [SEB80]) and being convicted of one or more violent offenses within a birth cohort in Sweden. The authors also adjusted for a number of important covariates, including income, being raised in a single parent household, and the urbanicity of the city in which the respondents grew up. In addition, a unique and powerful aspect of this study is that the authors compared the association between intelligence and being convicted of one or more violent offenses between brothers and half-brothers raised together and apart. This is an important aspect of the study since this particular design provides robust controls for any household factor that may influence both levels of intelligence and violent criminal behavior. The results indicated that full brothers and half-brothers raised together more closely resembled one another than sibling pairs reared apart, indicating that some of the detected associations were due to home environments shared by both siblings. While the association between intelligence and violent behavior was attenuated when examining siblings reared apart, the association remained statistically significant, indicating that family-level influences do not fully mediate the association.

Higher units of analysis

Another development within research examining the association between IQ and offending is directly related to the large number of countries from which data have been collected and analyzed. While a large number of studies have analyzed samples from the United States (Walsh, 1987; Beaver and Wright, 2011; Nedelec *et al.*, 2012), a sizeable number of studies have also analyzed samples from several other countries. For example, studies have analyzed samples from New Zealand (Moffitt and Silva, 1988), Denmark (Moffitt *et al.*, 1981), Sweden (Hodgins, 1992), and the United Kingdom (Farrington and West, 1971; Farrington, 1973). In perhaps the most comprehensive assessment of the potential association between intelligence and crime, Rushton and Templer (2009) examined international crime statistics between 1993 and 1996 from 116 countries. Making use of previously estimated national IQ scores (Lynn and Vanhanen, 2006), it was possible to estimate the association between the prevalence of violent criminal offending and each country's mean IQ score. Importantly, Rushton and Templer adjusted their statistical models for a number of important covariates previously identified in the literature. The results revealed a negative and statistically significant association between violent crime rates and IQ. The results provide strong evidence suggesting that such associations are not unique to any specific geographic region, but are pervasive in any region examined. In addition, these findings seem to indicate that the association between intelligence and offending is not likely the result of any specific set of social influences since the findings are consistent across a diverse set of countries with widely different social structures.

Along the same lines, studies have also examined the association between intelligence and offending across other higher-level units of analysis, including states (McDaniel, 2006; Bartels *et al.*, 2010; Pesta *et al.*, 2010) and neighborhoods (Beaver and Wright, 2011). In the first study to examine the potential association between county-level IQ and county-level crime rates, Beaver and Wright (2011) analyzed data from the Add Health. Even after controlling for concentrated disadvantage (Sampson *et al.*, 1997), the results revealed a strong and consistent association between county-level IQ and crime rates. Importantly, the results held across a number of property crimes (burglary, larceny, motor vehicle theft, and a composite property crime measure) and violent crimes (robbery, aggravated assault, and a composite violent crime measure).

Directly in line with these findings, previous studies have reported negative associations between children's blood lead levels and subsequent cognitive functioning (measured using IQ; for an overview see National Research Council, 1993). In addition, blood lead levels have also been found to be directly associated with violent criminal behavior later in life (Wright *et al.*, 2008). The results of a large number of studies indicate that blood lead levels play a significant role in the association between IQ and criminal behavior (for an overview see Nevin, 2000). For example, a significant number of studies have found that exposure to lead results in both behavioral problems early in life and lower cognitive ability or intelligence (Silva *et al.*, 1988; Thomson *et al.*, 1989). Recent studies have indicated that even low levels of lead exposure can result in significant increases in antisocial and aggressive behavior and lower overall levels of intelligence (Needleman and Bellinger, 1991; Bellinger, 2008).

Summary

While the findings vary from study to study, the overall pattern is clear: there is a statistical and substantive association between intelligence and criminal behavior. However, as noted by Walsh (2003: 121), "Low intelligence is no more a necessary and sufficient cause of antisocial behavior than high intelligence is a necessary and sufficient cause of prosocial behavior. Low intelligence is simply one risk factor among many for antisocial behavior." In a society wherein intelligence is an ever-increasing valuable resource, the risk associated with low intelligence is correspondingly increasing. The accumulated losses due to low intelligence across the lifespan

(Gottfredson, 1997b) represent a confluence of factors additively increasing the likelihood of criminal offending. The research examined in this section illustrates that criminologists should more readily incorporate measures of intelligence into their research in order to better understand the etiology of criminal behavior.

Intelligence and criminal justice processing

The previous section examined the extensive amount of research illustrating that individuals with low IQ are more likely to engage in a wide variety of criminal behaviors. Although research has focused on assessing the link between IQ and offending, comparatively less attention has been devoted to examining the relation between offender IQ and variation in criminal justice processing. Recent evidence, however, suggests that offenders with, on average, lower IQ are more likely to be arrested and commit infractions while incarcerated. Furthermore, contemporary court decisions on violent crime have revealed that offender IQ is often considered an important mitigating factor during judicial sentencing. This section of the chapter reviews this emerging line of research linking offender IQ to criminal justice processing outcomes.

IQ and likelihood of arrest

The association between IQ and contact with the criminal justice system has been a topic of much debate within criminology. Previous research has found that, on average, delinquent youth score about 8 IQ points (or half a standard deviation) lower than nondelinquent youth (Hirschi and Hindelang, 1977; Neisser *et al.*, 1996). While critics of the IQ–delinquency link have argued that the association may be spurious due to differential detection by the law (Feldman, 1977; Hirschi and Hindelang, 1977), more comprehensive evaluations of this argument have found nonsignificant differences between delinquent youth detected by law enforcement and delinquent youth not detected by law enforcement (Moffitt and Silva, 1988). Furthermore, current research has offered support for this viewpoint, finding evidence of an association between low verbal IQ and likelihood of arrest (Streissguth *et al.*, 1996; Loeber *et al.*, 2012; Beaver *et al.*, 2013b; Yun and Lee, 2013). The study by Beaver *et al.* (2013b) outlined in the previous section provides an example. As a reminder, the authors showed that low verbal IQ increased the odds of being arrested among males and females after controlling for a range of potential confounding variables, including age, self-control, and lifetime delinquency. Other research that has explored the link between childhood IQ and delinquent activity during adolescence has found similar results. For example, Loeber *et al.* (2012) analyzed data from the Pittsburgh Youth Study and found that IQ significantly predicted the probability of being arrested for a violent or nonviolent crime among boys. More specifically, their results revealed that boys with low IQ were significantly more likely to be arrested during the beginning years of adolescence (when delinquent behavior begins to steadily increase) compared to boys with high IQ.

Although a growing body of research has found a negative association between IQ and likelihood of arrest, much remains unknown about the mediating and/or moderating effects of certain behavioral traits. Some research has begun to explore the moderating effects of neighborhood factors on the association between verbal IQ and the probability of being arrested. For instance, Yun and Lee (2013) analyzed data from the Add Health and found that delinquent youth with lower verbal IQ scores were more likely to be arrested in adequate rather than disadvantaged neighborhood environments. Other studies examining male offenders who have been arrested have found significant associations between psychopathy, intelligence, and violent behavior (Holland *et al.*, 1981; Heilbrun, 1982; Walsh *et al.*, 2004). Using data from a large sample of male and female psychiatric patients, DeLisi and colleagues (2010) reported that patients with higher levels of psychopathy had lower verbal intelligence (measured by the vocabulary test of the Weschler Adult Intelligence Scale-Revised [WAIS-R]). Similar findings have been reported by Spironelli *et al.* (2014), who conducted a correlational study on a sample of violent female criminal offenders and showed a significant negative correlation ($r = -0.55$, $P < 0.001$) between fluid IQ and psychopathy. These findings suggest that perhaps violent offenders (or imprisoned psychopaths) lack cognitive control (e.g. low IQ) over emotional impulses, and tend to pursue immediate rewards without taking into account the potential consequences for their behavior. From a biological perspective, this would align with neuroscience research which has found significant associations between criminal psychopathy and altered structural and functional features of the prefrontal regions of the brain (Gordon *et al.*, 2004; Yang *et al.*, 2005; Kiehl, 2006). Since the prefrontal cortex is involved in decision-making, impulsive control, and long-term planning, the link between low IQ (which has been found to correlate with structural and functional differences in frontal lobes of the brain) and likelihood of arrest may be mediated or moderated by variation in impulsive and psychopathic personality traits.

IQ and variation in sentencing

In recent years, offender IQ has been at the forefront of several landmark court decisions regarding capital punishment. Perhaps the most prominent example comes from *Atkins v. Virginia* (June 20, 2002) where the Supreme Court held that the execution of any defendant with a diagnosis of mental retardation (or limited intellectual functioning) violated the Eighth Amendment's prohibition of cruel and unusual punishment. Central to this decision was diagnosing whether an individual was mentally retarded, which revolved around defining what was actually meant by *limited intellectual functioning*. The agreed-upon definition required that an individual's measured intelligence (based on a standardized psychometric instrument) be two standard deviations below the mean, which equates to a score of 70–75 or below on a standardized IQ scale. The court's rationale for using such a threshold was based on clinical evidence which suggests that individuals who score around 70 exhibit deficits in judgment, reasoning, and adaptive behavior (DSM-IV).

Recent research has also shown that low IQ predicts the odds of being incarcerated. Beaver *et al.* (2013b), for example, reported that low verbal IQ increased the odds of ever being incarcerated after controlling for several covariates. Interestingly, the results from their analysis revealed that the odds of being incarcerated were considerably higher for offenders with the lowest IQ (odds ratio = 0.439, $P < 0.05$) compared with offenders with the highest IQ (odds ratio = 0.153, $P < 0.05$). These results suggest that low IQ may function as an "extra-legal" factor in criminal justice sentencing where offenders with lower IQ may commit more serious criminal offenses that require incarceration rather than probation.

Another line of research has focused on uncovering the developmental origins of the low IQ and criminal justice sentencing link (Streissguth *et al.*, 2004; Fast and Conry, 2009). Findings suggest that individuals with fetal alcohol syndrome demonstrate deficits in executive functioning and tend to be impulsive, which may lead to later problems with the law (Schonfeld *et al.*, 2006) and may also contribute to variation in criminal sentencing. For example, Streissguth and colleagues (2004) analyzed a sample of adolescents and adults with fetal alcohol syndrome and found that around 60 percent of all individuals had been arrested, charged, or convicted during their lifetime. Specifically, among the individuals who had been in contact with the criminal justice system (approximately 87 percent of the sample), 45 percent had committed a violent act against another person. Another study found that children with fetal alcohol syndrome scored significantly lower on standardized verbal IQ assessments and demonstrated deficits in social skills throughout childhood and adolescence (Thomas *et al.*, 1998). This could imply that fetal alcohol syndrome is an important developmental risk factor for low IQ during the life course, which is often associated with police contact and variation in sentencing. Indeed, recent court cases have begun to use records of prenatal trauma to inform capital punishment decision-making (Winick, 2009).

IQ and prison behavior

Given that previous research has shown that IQ is negatively associated with violent delinquent behavior and contact with the criminal justice system, more contemporary research has assessed whether or not IQ is implicated in prison misconduct (Diamond *et al.*, 2012). Using data from a correctional sample, Diamond and colleagues (2012) evaluated the effect of IQ on the likelihood of violent prison misconduct. The results from their multi-level analyses revealed two important findings. First, individual variation in IQ was significantly associated with variation in violent inmate misconduct. Specifically, inmates with above-average IQ were less likely to be involved in a violent situation. Second, inmates housed in units with a higher aggregate IQ level were significantly less likely to engage in violent misconduct. These results thus suggest that IQ differences contribute to both individual-level and aggregate-level differences in inmate violent misconduct.

Despite recent research on IQ and prison behavior, the mechanisms that explain the association between low IQ and inmate violence remain unknown. Future research should use behavioral and neurological data to examine the underlying biological processes involved in inmate aggression. IQ can be helpful in identifying a general neurological basis to individual differences in inmate misconduct; however, more advanced neuroscience instruments are needed to pinpoint accurately areas of the brain that are involved in extreme cases of violent aggression among incarcerated criminals. The ability to shed light on the neurological underpinnings to inmate infractions may help develop more targeted cognitive behavioral therapy programs aimed at addressing specific psychiatric problems among male and female inmates (Vaske *et al.*, 2011).

Conclusion

This chapter reviews the etiological factors of intelligence as well as the association between intelligence, criminal offending, and components of criminal justice processing. Throughout the discussion there is recognition of the vast influence that intelligence has on an array of outcomes across the lifespan (including antisocial outcomes). Consequently, practical policy recommendations that seek to impact positively on levels of intelligence can help to improve overall quality of life and reduce potential antisocial behaviors. A unique component of the research on intelligence (relative to much of criminological research) is the proliferation of practical suggestions made by experts on intelligence. Far from being a collection of eugenicists hell-bent on ensuring the proliferation of high-IQ individuals throughout society via selective breeding, experts on intelligence clearly display a common characteristic: the desire to employ scientific knowledge to improve the lives of individuals and society overall. Our discussion of the etiological factors of intelligence highlights this point. While intelligence is influenced by genetic factors, there is also a substantive effect of nongenetic factors. Researchers have focused on this fact in the past (Gottfredson, 1997b) and continue to focus on it today. For example, in their book *G is for Genes*, Asbury and Plomin (2014) emphasize the importance of a behavioral genetic approach to education which can help to increase overall scholastic achievement (and thus decrease the likelihood of antisocial behavior). Their approach is to provide a greater level of individualized education wherein students are provided more options to choose areas of education that they most enjoy (after learning the basics of reading, writing, and mathematics). This approach recognizes the highly influential dynamic of gene–environment correlations. This approach is supported by BG research overall and by researchers who study the etiology of intelligence (e.g. Brant *et al.*, 2013; Tucker-Drob *et al.*, 2013). Additionally, it recognizes the ability to *use* the differential influence of nongenetic factors on variation in intelligence by altering components of the nonshared environment. The point, however, is that these policy recommendations are not produced or followed without an understanding of the biosocial nature of intelligence. Consequently, in order to alleviate the association between intelligence and deleterious life outcomes such as criminal behavior, a biosocial understanding is not only a suggestion, it is a necessity.

References

Akers, R.L. and Sellers, C.S. (2009) *Criminological Theories: Introduction, Evaluation, and Application*, 5th edn, New York: Oxford University Press.

Anderson, G.S. (2007) *Biological Influences on Criminal Behavior*, New York: CRC Press.

Asbury, K. and Plomin, R. (2014) *G is for Genes: the Impact of Genetics on Education and Achievement*, Malden, MA: Wiley Blackwell.

Asbury, K., Dunn, J.F., Pike, A., and Plomin, R. (2003) 'Nonshared environmental influences on individual differences in early behavioral development: a monozygotic twin differences study', *Child Development*, 74(3): 933–943.

Bartels, J.M., *et al.* (2010) 'Correlations between estimates of state IQ and FBI crime statistics', *Personality and Individual Differences*, 48: 579–583.

Beaver, K.M. (2008) 'Nonshared environmental influences on adolescent delinquent involvement and adult criminal behavior', *Criminology*, 46: 341–369.

Beaver, K.M. (2013) *Biosocial Criminology: A Primer*, 2nd edn, Dubuque, IA: Kendall Hunt.

Beaver, K.M. and Wright, J.P. (2011) 'The association between county-level IQ and county-level crime rates', *Intelligence*, 39: 22–26.

Beaver, K.M., *et al.* (2013a) 'The genetic and environmental architecture to the stability of IQ: results from two independent samples of kinship pairs', *Intelligence*, 41: 428–438.

Beaver, K.M., *et al.* (2013b) 'Intelligence is associated with criminal justice processing: arrest through incarceration', *Intelligence*, 41: 277–288.

Bellinger, D.C. (2008) 'Very low lead exposures and children's neurodevelopment', *Current Opinion in Pediatrics*, 20(2), 172–177.

Benyamin, B., *et al.* (2013) 'Childhood intelligence is heritable, highly polygenic and associated with *FNBP1L*', *Molecular Psychiatry*, 19: 253–258.

Bouchard, T.J. Jr (2013) 'The Wilson effect: the increase in heritability of IQ with age', *Twin Research and Human Genetics*, 16: 923–930.

Brans, R.G.H., *et al.* (2010) 'Brain plasticity and intellectual ability are influenced by shared genes', *Journal of Neuroscience*, 30: 5519–5524.

Brant, A.M., *et al.* (2013) 'The nature and nurture of high IQ: an extended sensitive period for intellectual development', *Psychological Science*, 24: 1487–1495.

Cantor, J.M., *et al.* (2005) 'Quantitative reanalysis of aggregate data on IQ in sexual offenders', *Psychological Bulletin*, 131: 555–568.

Chipuer, H.M., *et al.* (1990) 'LISREL modeling: genetic and environmental influences on IQ revisited', *Intelligence*, 14: 11–29.

Cullen, F.T., *et al.* (1997) 'Crime and the bell curve: lessons from intelligent criminology', *Crime & Delinquency*, 43: 387–411.

Curran, D.J. and Renzetti, C.M. (2001) *Theories of Crime*, Boston, MA: Allyn & Bacon.

Davies, G., *et al.* (2011) 'Genome-wide association studies establish that human intelligence is highly heritable and polygenic', *Molecular Psychology*, 16: 996–1005.

Davis, O.S.P., *et al.* (2013) 'A three-stage genome-wide association study of general cognitive ability: hunting the small effects', *Behavior Genetics*, 40: 759–767.

Dawkins, R. (2006) *The Selfish Gene*, 3rd edn, New York: Oxford University Press.

Deary, I.J., *et al.* (2012) 'Genetic contributions to stability and change in intelligence from childhood to old age', *Nature*, 482: 212–215.

Deary, I.J., *et al.* (2013) 'The stability of intelligence from age 11 to age 90 years: the Lothian Birth Cohort of 1921', *Psychological Science*, 24: 2361–2368.

DeLisi, M., *et al.* (2010) 'The Hannibal Lecter myth: psychopathy and verbal intelligence in the MacArthur Violence Risk Assessment Study', *Journal of Psychopathology and Behavioral Assessment*, 32: 169–177.

Diamond, B., Morris, R.G., and Barnes, J.C. (2012) 'Individual and group IQ predict inmate violence', *Intelligence*, 40: 115–122.

Dorfman, D.D. (1995) 'Soft science with a neoconservative agenda [review of *The Bell Curve: Intelligence and Class Structure in American Life*]', *Contemporary Psychology*, 40: 418–421.

Dwyer, R.G. and Frierson, R.L. (2006) 'The presence of low IQ and mental retardation among murder defendants referred for pretrial evaluation', *Journal of Forensic Sciences*, 51: 678–682.

Farrington, D.P. (1973) 'Self-reports of deviant behavior: predictive and stable?', *Journal of Criminal Law and Criminology*, 64: 99–110.

Farrington, D.P. and West, D.J. (1971) 'Comparison between early delinquents and young aggressives', *British Journal of Criminology*, 11: 341–358.

Fast, D.K. and Conry, J. (2009) 'Fetal alcohol spectrum disorders and the criminal justice system', *Developmental Disabilities*, 15: 250–257.

Feldman, M.P. (1977) *Criminal Behavior: A Psychological Analysis*, New York: John Wiley & Sons.

Frisell, T., Pawitan, Y., and Långström, N. (2012) 'Is the association between general cognitive ability and violent crime caused by family-level confounders?', *PLoS One*, 7: e41783.

Goldberger, A.S. and Manski, C.F. (1995) 'Review article: *The Bell Curve* by Herrnstein and Murray', *Journal of Economic Literature*, 33: 762–776.

Gordon, H.L., Baird, A.A., and End, A. (2004) 'Functional differences among those high and low on a trait measure of psychopathy', *Biological Psychiatry*, 56: 516–521.

Gottfredson, L.S. (1997a) 'Mainstream science on intelligence: an editorial with 52 signatories, history, bibliography', *Intelligence*, 24: 13–23.

Gottfredson, L.S. (1997b) 'Why *g* matters: the complexity of everyday life', *Intelligence*, 24: 79–132.

Harris, J.R. (2009) *The Nurture Assumption: Why Children Turn Out the Way They Do*, 2nd edn, New York: Free Press.

Hauser, R.M. (1995) 'Symposium: *The Bell Curve*', *Contemporary Sociology*, 24: 149–153.

Haworth, C.M.A., *et al.* (2012) 'A twin study of the genetics of high cognitive ability selected from 11,000 twin pairs in six studies from four countries', *Behavior Genetics*, 39: 359–370.

Heilbrun, A.B. Jr (1982) 'Cognitive models of criminal violence based upon intelligence and psychopathy levels', *Journal of Consulting and Clinical Psychology*, 50: 546–557.

Herrnstein, R.J. and Murray, C. (1994) *The Bell Curve: Intelligence and Class Structure in American Life*, New York: Free Press.

Hirschi, T. and Hindelang, M.J. (1977) 'Intelligence and delinquency: a revisionist review', *American Sociological Review*, 42: 571–587.

Hodgins, S. (1992) 'Mental disorder, intellectual deficiency, and crime: evidence from a birth cohort', *Archives of General Psychiatry*, 49: 476–483.

Holland, T.R., Beckett, G.E., and Levi, M. (1981) 'Intelligence, personality and criminal violence: a multivariate analysis', *Journal of Consulting and Clinical Psychology*, 49: 106–111.

Jensen, A. (1998) *The* g *Factor: The Science of Mental Ability*, Westport, CT: Praeger.

Kandel, E., Mednick, S.A., Kirkegaard-Sorensen, L., Hutchings, B., Knop, J., Rosenberg, R., and Schulsinger, F. (1988) 'IQ as a protective factor for subjects at high risk for antisocial behavior', *Journal of Consulting and Clinical Psychology*, 56(2): 224–226.

Kearns, A. and O'Connor, A. (1988) 'The mentally handicapped criminal offender: a 10-year study of two hospitals', *British Journal of Psychiatry*, 152: 848–851.

Kiehl, K. (2006) 'A cognitive neuroscience perspective on psychopathy: evidence for paralimbic system dysfunction', *Psychiatry Research*, 142 (2–3): 107–128.

Kratzer, L. and Hodgins, S. (1999) 'A typology of offenders: a test of Moffitt's theory among males and females from childhood to age 30', *Criminal Behaviour and Mental Health*, 9: 57–73.

Loeber, R., *et al.* (2012) 'Findings from the Pittsburgh Youth Study: cognitive impulsivity and intelligence as predictors of the age–crime curve', *Journal of the American Academy of Child and Adolescent Psychiatry*, 51: 1136–1149.

Lynam, D., Moffitt, T.E., and Stouthamer-Loeber, M. (1993) 'Explaining the relation between IQ and delinquency: class, race, test motivation, school failure, or self-control?', *Journal of Abnormal Psychology*, 102: 187–196.

Lynn, R. and Vanhanen, T. (2006) *IQ & Global Inequality*, Augusta, GA: Washington Summit Publishers.

McDaniel, M.A. (2006) 'Estimating state IQ: measurement challenges and preliminary correlates', *Intelligence*, 34: 607–619.

Mears, D.P. and Cochran, J.C. (2013) 'What is the effect of IQ on offending?' *Criminal Justice and Behavior*, 40: 1280–1300.

Menard, S. and Morse, B.J. (1984) 'A structuralist critique of the IQ–delinquency hypothesis: theory and evidence', *American Journal of Sociology*, 89: 1347–1378.

Miller, G. (2001) *The Mating Mind: How Sexual Choice Shaped the Evolution of Human Nature*, New York: Anchor.

Moffitt, T.E. and Silva, P.A. (1988) 'IQ and delinquency: a direct test of the differential detection hypothesis', *Journal of Abnormal Psychology*, 97: 330–333.

Moffitt, T.E., *et al.* (1981) 'Socioeconomic status, IQ, and delinquency', *Journal of Abnormal Psychology*, 90: 152–156.

National Research Council (1993) 'Measuring lead exposure in infants, children, and other sensitive populations', Committee on Measuring Lead in Critical Populations, Washington, DC: National Academy Press.

Nedelec, J.L., *et al.* (2012) 'Exploring the association between IQ and differential life outcomes: results from a longitudinal sample of monozygotic twins', *Temas em Psicologia*, 20: 31–43.

Needleman, H.L. and Bellinger, D. (1991) 'The health effects of low level exposure to lead', *Annual Review of Public Health*, 12(1): 111–140.

Neisser, U., *et al.* (1996) 'Intelligence: knowns and unknowns', *American Psychologist*, 51: 77–101.

Nevin, R. (2000) 'How lead exposure relates to temporal changes in IQ, violent crime, and unwed pregnancy', *Environmental Research*, 83: 1–22.

Nisbett *et al.* (2012) 'Intelligence: new findings and theoretical developments', *American Psychologist*, 67: 130–159.

Pesta, B.J., McDaniel, M.A., and Bertsch, S. (2010) 'Toward an index of well-being for the fifty US states', *Intelligence*, 38(1): 160–168.

Petrill, S.A. and Wilkerson, B. (2000) 'Intelligence and achievement: a behavioral genetic perspective', *Educational Psychology Review*, 12: 185–199.

Pinker, S. (2002) *The Blank Slate*, New York: Viking.

Pinker, S. (2009) *How the Mind Works*, New York: W.W. Norton & Co.

Plomin, R., *et al.* (1994) 'Variability and stability in cognitive abilities are largely genetic later in life', *Behavior Genetics*, 24: 207–215.

Plomin, R., *et al.* (2013a) *Behavioral Genetics*, 6th edn, New York: Worth.

Plomin, R., *et al.* (2013b) 'Common DNA markers can account for more than half of the genetic influence on cognitive abilities', *Psychological Science*, 24: 562–568.

Raine, A. (2013) *The Anatomy of Violence: The Biological Roots of Crime*, New York: Pantheon Books.

Ridley, M. (1993) *The Red Queen: Sex and the Evolution of Human Nature*, New York: Harper.

Rowe, D.C. (1994) *The Limits of Family Influence: Genes, Experience, and Behavior*, New York: Guilford Press.

Rushton, J.P. (1989) 'Genetic similarity, human altruism, and group selection', *Behavioral and Brain Sciences*, 12(3): 503–518.

Rushton, J.P. (2004) 'Placing intelligence into an evolutionary framework or how *g* fits into the *r*–*K* matrix of life-history traits including longevity', *Intelligence*, 32: 321–328.

Rushton, J.P. and Templer, D.I. (2009) 'National differences in intelligence, crime, income, and skin color', *Intelligence*, 37: 341–346.

Sagan, C. (1980) *Cosmos*, New York: Random House.

Sampson, R.J., Raudenbush, S.W., and Earls, F. (1997) 'Neighborhoods and violent crime: a multilevel study of collective efficacy', *Science*, 277: 918–924.

Scarr, S. (1978) 'From evolution to Larry P., or what shall we do about IQ tests?' *Intelligence*, 2: 325–342.

Schonfeld, A.M., Paley, B., Frankel, F., and O'Connor, M.J. (2006) 'Executive functioning predicts social skills following prenatal alcohol exposure', *Child Neuropsychology*, 12(6): 439–452.

Schwartz, J.A. and Beaver, K.M. (2012) 'Examining the effects of dopamine genes on verbal IQ within and between families', *Journal of Neural Transmission*, 120: 477–486.

Segal, N.L. (2010) 'Twins: the finest natural experiment', *Personality and Individual Differences*, 49: 317–323.

Silva, P.A., *et al.* (1988) 'Blood lead, intelligence, reading attainment, and behaviour in eleven year old children in Dunedin, New Zealand', *Journal of Child Psychology and Psychiatry*, 29: 43–52.

Simons, R.L. (1978) 'The meaning of the IQ–delinquency relationship', *American Sociological Review*, 43: 268–270.

Spironelli, C., Stegagno, D.S., and Angrilli, A. (2014) 'Intelligence and psychopathy: a correlational study on insane female offenders', *Psychological Medicine*, 44: 111–116.

Sternberg, R.J. (1995) 'For whom the bell curve tolls: a review of *The Bell Curve*', *Psychological Science*, 6: 257–261.

Sternberg, R.J. (2000) 'The concept of intelligence', in R. Sternberg (ed.), *Handbook of Intelligence*, Cambridge, UK: Cambridge University Press, pp. 10–15.

Streissguth, A.P., Barr, H.M., Kogan, J., and Bookstein, F.L. (1996) *Understanding the Occurrence of Secondary Disabilities in Clients with Fetal Alcohol Syndrome (FAS) and Fetal Alcohol Effects (FAE)*. Final report to the Centers for Disease Control and Prevention (CDC), 96–106.

Streissguth, A.P., Bookstein, F.L., Barr, H.M., Sampson, P.D., O'Malley, K., and Young, J.K. (2004) 'Risk factors for adverse life outcomes in fetal alcohol syndrome and fetal alcohol effects', *Journal of Developmental & Behavioral Pediatrics*, 25(4): 228–238.

Sutherland, E. (1931) 'Mental deficiency and crime', in K. Young (ed.), *Social Attitudes*, New York: Holt, Rinehart, and Watson, pp. 357–375.

Thomas, S.E., Kelly, S.J., Mattson, S.N., and Riley, E.P. (1998) 'Comparison of social abilities of children with fetal alcohol syndrome to those of children with similar IQ scores and normal controls', *Alcoholism: Clinical and Experimental Research*, 22(2): 528–533.

Thomson, G.O.B., *et al.* (1989) 'Blood-lead levels and children's behaviour—results from the Edinburgh Lead Study', *Journal of Child Psychology and Psychiatry*, 30: 515–528.

Trivers, R. (2011) *The Folly of Fools: The Logic of Deceit and Self-Deception in Human Life*, New York: Basic Books.

Tucker-Drob, E.M., *et al.* (2013) 'Genetic and environmental influences on cognition across development and context', *Current Directions in Psychological Science*, 22: 349–355.

Vaske, J., Galyean, K., and Cullen, F.T. (2011) 'Toward a biosocial theory of offender rehabilitation: why does cognitive-behavioral therapy work?' *Journal of Criminal Justice*, 39: 90–102.

Vitaro, F., Brendgen, M., and Arseneault, L. (2009) 'The discordant MZ-twin method: one step closer to the holy grail of causality', *International Journal of Behavioral Development*, 33: 376–382.

Vitaro, F., *et al.* (2012) 'Social experiences in kindergarten and academic achievement in grade 1: a monozygotic twin difference study', *Journal of Educational Psychology*, 2: 366–380.

Walsh, A. (1987) 'Cognitive functioning and delinquency: property versus violent offenses', *International Journal of Offender Therapy and Comparative Criminology*, 31: 285–289.

Walsh, A. (2002) 'Companions in crime: a biosocial perspective', *Human Nature Review*, 2(2): 169–178.

Walsh, A. (2003) 'Intelligence and antisocial behavior', in A. Walsh and L. Ellis (eds), *Biosocial Criminology: Challenging Environmentalism's Supremacy*, New York: Nova Science.

Walsh, A. and Beyer, J.A. (1986) 'Wechsler performance–verbal discrepancy and antisocial behavior', *Journal of Social Psychology*, 126: 419–420.

Walsh, Z., Swogger, M.T., and Kosson, D.S. (2004) 'Psychopathy, IQ, and violence in European American and African American county jail inmates', *Journal of Consulting and Clinical Psychology*, 72: 1165–1169.

Wilson, J.Q. and Herrnstein, R.J. (1985) *Crime & Human Nature: The Definitive Study of the Causes of Crime*, New York: Simon & Schuster.

Winick, B.J. (2009) 'The supreme court's evolving death penalty jurisprudence: severe mental health illness as the next frontier', *Boston College Review*, 50: 785–858.

Wolfgang, M., Figlio, R.M., and Sellin, T. (1972) *Delinquency in a Birth Cohort*, Chicago, IL: University of Chicago Press.

Wright, R. (1994) *The Moral Animal. Why We Are the Way We Are: The New Science of Evolutionary Psychology*, New York: Vintage Books.

Wright, J.P., *et al.* (2008) 'Association of prenatal and childhood blood lead concentrations with criminal arrests in early adulthood', *PLoS Medicine*, 5: e101.

Yang, Y., *et al.* (2005) 'Volume reduction in prefrontal gray matter in unsuccessful criminal psychopaths', *Biological Psychiatry*, 57: 1103–1108.

Yun, I. and Lee, J. (2013) 'IQ and delinquency: the differential detection hypothesis revisited', *Youth Violence and Juvenile Justice*, 11: 196–211.

33

Brain development and risk taking during adolescence

Implications for prevention of antisocial behavior

Daniel Romer and Atika Khurana

Introduction

Recent structural brain imaging (magnetic resonance imaging, MRI) studies have produced detailed pictures of the brain from infancy to adulthood, revealing striking patterns of maturation that continue well into the third decade of life (Giedd *et al.*, 1999). In particular, beginning in the second decade of life, grey matter volume, composed of neurons and their interconnections, starts to decline, while white matter, the insulating material that encases neuronal connections, starts to increase. These changes reflect the impact of early experiences, as unused neuronal connections are selectively pruned, while those that remain are fortified to allow greater speed of transmission. The result is a more efficient but focused brain that can make decisions in a speedy manner, especially those that have been encountered in the past (Spear, 2010).

Another observation from these studies has been the staggered maturation of different brain regions. In particular, the occipital and ventral regions develop earlier in comparison with the frontal and dorsal regions such as the prefrontal cortex (PFC), the seat of higher-order cognitive functioning. Although the implications of these changes for adolescent decision-making remain unclear, neuroscientists in collaboration with developmental psychologists prematurely concluded that these developmental trends explain the rise in risky behavior observed in adolescents. For example, Casey and colleagues (2008) suggest that the protracted development of the PFC leaves the adolescent without the cognitive resources necessary to control the already mature ventral motivational regions.

Behavioral research has supported the protracted development of the PFC by showing that executive functions, such as working memory and inhibitory control, continue to improve until late adolescence (Luna *et al.*, 2004; Luciana *et al.*, 2005). However, studies have failed to find consistent evidence regarding higher activity in the reward-processing regions (ventral striatal regions and orbital-frontal PFC) during adolescence than at other ages (e.g. Galvan *et al.*, 2007; Ernst, 2014). Instead of these developmental trends, large individual differences in adolescence were more often the observed pattern. In addition, many risk behaviors, such as use of drugs and antisocial behaviors, peak in late adolescence and early adulthood, when maturation in cognitive control is more nearly complete (Romer, 2010; Willoughby *et al.*,

2013). Thus the broad generalization that risk taking is greater during adolescence because cognitive-control systems are weak compared with motivational circuits appears not to be compatible with behavioral evidence.

Individual differences in brain function and structure: implications for impulsivity and self-control

Although early conclusions regarding adolescent brain development tended to over-generalize, it is nevertheless the case that the onset of many mental health disorders occurs during adolescence, including depression, bipolar disorder, and schizophrenia (Romer and Walker, 2007). Substance use and other antisocial behaviors, including engagement in criminally violent behavior, also emerge during this life stage. This is not surprising because adolescence is a transitional period during which multiple maturational changes are accompanied by increasing societal expectations that encourage greater independence. Although most adolescents experiment with potentially risky behavior, only a minority exhibit antisocial tendencies (Romer, 2010). Thus researchers have necessarily focused on personal characteristics that predispose some youth to develop disorders or to follow an antisocial trajectory.

One particularly important personal characteristic underlying antisocial behavior is the tendency toward impulsive behavior (Krueger *et al.*, 2002). In their longitudinal study of over 1000 New Zealand residents, Moffitt *et al.* (2011) found that weakness in self-control, observed as early as age 3, is a stable characteristic that predicts a wide range of antisocial behaviors in adolescence and beyond, including arrests, drug dependence, and school failure. In some cases, this lack of self-control only becomes manifest during adolescence and can be traced to effects of early experience on development. In other cases, such as attention deficit/hyperactivity disorder (ADHD), the symptoms appear early and can continue into adolescence (APA, 2013).

Executive attention, defined as the ability to focus on task-relevant goals while inhibiting dysfunctional impulses, is a core executive function necessary for self-control (Diamond, 2013). This ability is often assessed using tasks such as the word–color Stroop, in which participants have to inhibit the impulse to read the presented color-word and instead use the color of the presented word. Other tasks such as the flanker and go/no-go are also commonly used to assess executive attention (Banich, 2004).

Another important executive function necessary for self-control is working memory ability. Working memory is a limited-capacity system necessary for storing and processing information during planning and decision-making (Diamond, 2013). A variety of tasks, such as the n-back and reverse digit span, can be used to assess this ability. Because executive attention enables an individual to focus on goal-relevant information, it is not surprising that executive attention is critical for working memory performance (Engle, 2002). Adolescents with weak executive attention and working memory tend to be easily distracted and to report greater tendencies to act without thinking (Romer *et al.*, 2009) and to have less ability to delay gratification (Shamosh *et al.*, 2008). Individuals with ADHD are particularly subject to weakness in both executive functions (APA, 2013).

Evidence that individual differences in executive attention and working memory are important in predicting antisocial behavior comes from studies of executive function in pre-adolescents. In our research with a longitudinal cohort of over 350 Philadelphia adolescents aged 10–12 at baseline, we found clear evidence that weakness in working memory and executive attention underlies differences in self-control that are related to antisocial behaviors, such as fighting and rule breaking. In particular, individual differences at baseline in an impulsive tendency, acting without thinking (AWT), predicted all of the covariation between risky behavior and antisocial behaviors at that young age (Romer *et al.*, 2009). In subsequent waves of the study, baseline AWT continued to predict individual differences in antisocial behavior (Romer *et al.*, 2011; Khurana *et al.*, 2013). The importance of working memory ability is also evident in the finding that sensation seeking, which has often been linked with risky behavior in adolescents (Zuckerman, 2007), is positively correlated with working memory ability in adolescents (Romer *et al.*, 2011). This helps to explain the finding that impulsivity is a better predictor of risky behavior in early adolescence than sensation seeking (Romer *et al.*, 2011; Khurana *et al.*, 2013) and that impulsivity is a stronger predictor of adverse outcomes later in adolescence, such as substance dependence and other problem behavior (Magid *et al.*, 2007; Smith *et al.*, 2007).

Studies of brain structure and function suggest that self-control is enhanced by greater connectivity between cognitive control regions, such as the dorsolateral PFC and other regions that are involved in carrying out and motivating behavior, such as the striatum and limbic circuits (Davis *et al.*, 2013). Adolescents with weak functioning in cognitive control regions are more prone to initiating drug use (Whelan *et al.*, 2012). Indeed, juvenile delinquents with deficits in self-control appear to be particularly subject to less functional connectivity between cognitive control and motor regions, indicative of enhanced impulsivity (Shannon *et al.*, 2011). Furthermore, brain scans of violent offenders indicate that PFC volume is often smaller in such individuals compared with nonoffenders, also suggesting less ability to control limbic activity (Yang and Raine, 2009).

Other individual differences in brain function may also contribute to antisocial behavior. Persons with psychopathic tendencies, one of the most important predictors of antisocial behavior, exhibit less activation in some limbic regions in response to others' pain, suggesting a deficit in empathy (J. Blair, 2010; Decety *et al.*, 2013). Instead, they appear to show enhanced activity in regions indicating a rewarding response to others' pain. Such differential brain responses could help to explain their unusual pattern of callous and unemotional (CU) response to seeing others suffer.

In summary, although adolescence is a period of rapid development of the ability to exert self-control, large individual differences characterize the emergence of dysfunctional behavior, including tendencies to engage in violent antisocial behavior. It does not appear to be the case that all adolescents lack control over behavior, but rather that individual differences in self-control make some adolescents more prone to resorting to violence and other forms of antisocial behavior. It also appears that sensation seeking is a more controlled form of risk taking that does not necessarily lead to adverse outcomes.

Despite the lack of cognitive control characteristic of persons with impulsive tendencies, it is also important to recognize that impulsivity may be quite functional in some contexts. For instance, AWT may be a rational mode of behavior when resources are scarce and waiting to consider risks leads to suboptimal outcomes (Korte *et al.* 2005). Similarly, inability to wait for large rewards, another symptom of weak self-control, may be functional in environments in which rewards are uncertain and immediate gratification is normative (McGuire and Kable, 2013). Finally, sensation seeking is likely to be functional for adolescents in many circumstances, given the need to separate from parents and to try out new identities and skills (Spear, 2010). As we detail below, early experiences may predispose certain forms of brain development that lead to impulsive patterns of behavior in adolescence and adulthood.

Trajectories of serious antisocial behavior begin in childhood

Developmental studies of antisocial behavior indicate that early markers, such as aggressive tendencies, appear in the first years of life (e.g. Broidy *et al.*, 2003). Several longitudinal studies have confirmed the existence of at least four trajectories from early childhood to adulthood. The one most likely to involve violence is the life-course-persistent antisocial trajectory first identified by Moffitt (1993). In her famous Dunedin study, both male and female residents (about 7–8 percent) exhibited this trajectory of early and continuing aggressive and delinquent behavior (Odgers *et al.*, 2008). Another group (about 20 percent), called the adolescent onset trajectory, did not exhibit antisocial behavior until adolescence, but in male youth, their levels of antisocial behavior approached those of the life-course-persistent group. Nevertheless, serious cases of violence that resulted in convictions at age 32 were far more prevalent in the life-course-persistent group (33 vs. 10 percent). In females, serious cases of violence (3 percent) were far lower than in males, and were only evident in the life-course-persistent group. Thus the adolescent-onset group was relatively small in size and did not engage in as serious and violent offenses as the life-course-persistent group who showed antisocial tendencies during early childhood. Another trajectory group of about 20–25 percent exhibited antisocial behavior early in childhood, but displayed a declining trajectory as the group approached adolescence and became comparable with the "normative" group (55–60 percent) who exhibited very little antisocial involvement. Overall, the majority of youth transitioned through adolescence with little evidence of antisocial behavior.

Patterns of mostly benign risk-taking behavior have also been observed in the US using the very large and nationally representative Add Health dataset (Zheng and Cleveland, 2013). Although the study did not start until youth were in adolescence, 50 percent of males and 59 percent of females from the ages of 15 to 22 did not exhibit antisocial behavior. A small group of males (12 percent) reported high levels of violent behavior that nevertheless declined over the period of the study. Another group of male youth with violent tendencies (13 percent) exhibited their highest level of antisocial behavior in late adolescence before dramatically declining at ages 21–22. Female youth displayed far lower levels of violent behavior that largely declined over the period of the study.

Increasing evidence points toward a small subgroup of youth in the life-course trajectory that exhibits a particularly stable and harmful tendency toward violence characteristic of psychopathy (Frick and Viding, 2009). These youth exhibit early signs of CU responses to others' suffering or their own wrongdoing, and exhibit signs of early impulsivity (Frick *et al.*, 2003). Adolescent boys with high levels of CU traits exhibit a different pattern of amygdala response to fearful faces than youth with antisocial tendencies (Viding *et al.*, 2012). Adolescent boys with CU traits also exhibit brain responses indicative of reduced empathy to others' suffering (Lockwood *et al.*, 2013), resembling patterns exhibited by adult psychopaths, namely under-response to threat and others' suffering (J. Blair, 2010; Decety *et al.*, 2013). Antisocial youth without CU traits tend to exhibit heightened response to threat and greater empathic response to others' suffering (Frick and Viding, 2009). Thus the presence of CU traits appears to represent early symptoms of psychopathy that remain relatively stable as youth age (Lynam *et al.*, 2007; McMahon *et al.*, 2010). Similar to the life-course-persistent group in the Dunedin Study, a recent twin study in the UK found a small group (3 percent) of children ages 7, 9, and 12 years who exhibited high and stable levels of CU traits (Fontaine *et al.*, 2010).

Parenting and family predictors of antisocial trajectories

Several longitudinal studies, including those described above, have found that parental behavior during the prenatal period as well as later in development exerts a very large influence on antisocial trajectories. This was particularly evident in the Avon, UK longitudinal cohort that was studied prior to birth (Barker and Maugham, 2009). Mothers who reported greater affective disorder during the third trimester of pregnancy as well as greater anxiety postpartum were more likely to have a male child in the early-onset persistent trajectory. These mothers were also more likely to experience domestic violence and receive lower material support from their partners. They reported using harsher parenting methods and not enjoying having the child. The negative impact of these dysfunctional interactional processes on the child's temperament was evident in the very first year of life, with high levels of under-controlled behavior, suggestive of impulsive tendencies. Similar patterns were found in the Dunedin study, where children in the persistent trajectory were more likely to exhibit impulsive temperament and had parents experiencing higher levels of partner physical abuse (Odgers *et al.*, 2008). Results from the longitudinal Pittsburgh study of low-income male youth found that those with violent convictions by age 30 were more likely to have exhibited antisocial tendencies prior to age 10 and were more likely to exhibit CU traits as rated by parents and teachers at an early age (Loeber *et al.*, 2005; Loeber and Pardini, 2008). In addition, factors associated with poverty, such as maternal depression, low caregiving quality, and neighborhood risk, were significant predictors of antisocial involvement at later ages. The Add Health study also found high levels of neighborhood disadvantage and weak family support in the most violent trajectories (Zheng and Cleveland, 2013).

The twin study conducted in the UK by Fontaine *et al.* (2010) observed high levels of heritability (between 0.58 and 0.78) of CU traits in boys in all four trajectory groups; however, the remaining variance was attributable to nonshared environmental factors. For girls, heritability was high only in the decreasing and stable low groups (0.54 and 0.68), while for the early and late onset groups, shared environmental influence was the largest contributor for CU traits (0.74 and 0.47). Thus, although family influences appeared to differ for the two genders, parents had a strong influence on the emergence of CU traits.

In general, longitudinal studies indicate that although most youth pass through adolescence with low or declining levels of antisocial behavior, those exposed to harsh and neglectful parenting and impoverished backgrounds are at greater risk for antisocial behaviors. Nevertheless, longitudinal studies can be criticized regarding the conclusions they can reach on the effects of family functioning on child development, because children are genetically related to their parents. That is, effects attributable to parenting behavior may merely result from shared genetic influences. For example, parents who are neglectful or use harsh discipline may have offspring who exhibit similar tendencies. However, a recent review of this body of research, including designs that can separate the effects of genetics from experience, suggests that parenting behaviors do affect child antisocial development apart from shared genetic background (Jaffee *et al.*, 2012).

Does early experience influence brain development and self-control?

Considerable evidence suggests that early experience, specifically continued exposure to stress, influences brain development related to the ability to exert self-control (Champagne, 2010). Stress exposure in the form of child maltreatment, including parental neglect and abuse, is found to disrupt the stress response and regulatory systems in children (Cicchetti and Toth, 2005; Gunnar and Fisher, 2006) by impeding the development of frontal grey matter (Tomoda *et al.*, 2009; De Brito *et al.* 2013) and the hippocampus (Teicher *et al.*, 2003; Luby *et al.*, 2013). Frontal brain regions are critical for the ability to exert cognitive control over behavior, while the hippocampi play an important role in regulating the HPA response to stress by modulating the down-regulation of the system following a stressful event (Shonkoff *et al.*, 2009).

Animal models also verify the adverse effects of child maltreatment on brain development (Duque *et al.*, 2012). Poor parental nurturance in rats, shortly following birth, is found to have epigenetic effects on gene methylation, with particular effects on hippocampal function (Champagne, 2010). Thus the effects of maltreatment need not involve abuse, but can be transmitted, as much if not more, by parental neglect. Human studies also indicate that neglect may be a greater source of risk to brain development and later antisocial behavior than abuse *per se* (Kotch *et al.*, 2008). Indeed, lack of nurturant parenting alone has been linked to deficits in executive functioning in human children (Bernier *et al.*, 2010).

Another important marker of adverse family functioning is partner abuse, typically of the mother. As already noted, children in families with partner abuse are at increased risk of progressing on the persistent trajectory of antisocial behavior (Sternberg *et al.*, 2006; Howell, 2011). They are also found to exhibit reduced levels of inhibitory control as assessed with the stop signal task, a variant of the go/no-go paradigm (Nigg *et al.*, 2007). Exposure to family violence also predicts heightened amygdala response to threatening faces (McCrory *et al.*, 2011). Even children exposed to neighborhood violence exhibit reduced ability to exert self-control in executive function tasks, again suggesting adverse effects on brain function resulting from violence (Sharkey *et al.*, 2012).

Exposure to poverty also leads to impaired development of self-control (C. Blair, 2010; Zalewski *et al.*, 2012) and executive functions (Evans and Schamberg, 2009; Raver *et al.*, 2013). The detrimental effects of poverty on hippocampal volume (Luby *et al.*, 2013) and executive functions (Evans and Schamberg, 2009) are found to be mediated by parental stress and neglectful parenting. Other studies have found that "positive" parenting behaviors, such as use of praise to reward desirable behavior, can buffer children from the deleterious effects of contextual stressors and promote better child outcomes (Odgers *et al.*, 2012), whereas "negative" parenting, such as use of harsh and punitive discipline, is predictive of children's involvement in problem behaviors (Scaramella *et al.*, 2008). High levels of parental stress, associated with poverty, can diminish parent–child relationship quality, detract from positive parenting behaviors, and increase child vulnerability to internalizing symptoms (Rodriguez, 2011) and problem behaviors (Crnic *et al.*, 2005; Buodo *et al.*, 2013).

In sum, the evidence suggests that early contextual influences involving disruptive family process and neighborhood risk can impact the developing brain with adverse consequences for the ability to exert self-control. All of these conditions are also found to be associated with early progression on the persistent and increasing antisocial trajectories. On the positive side, human and animal studies indicate that the adverse effects of early neglect can be reversed by later exposure to more nurturant rearing conditions (Champagne, 2010). The most powerful evidence for this is found in studies of children raised in orphanages who showed significant improvements in cognitive functioning when placed in more nurturant foster care settings (Nelson *et al.* 2007). As we detail below, family-based interventions targeting parenting behaviors find that positive parenting support can help to reduce the effects of cumulative stress on child outcomes. Nevertheless, it is also important to consider the possibility that antisocial traits and behavior may be functional for youth raised in hostile environments.

Can antisocial behavior be adaptive?

Although antisocial behavior would appear to be maladaptive on its face, there is reason to believe that impulsive and antisocial behavior confers benefits in certain environments. This argument has been convincingly advanced by McEwen and colleagues (Korte *et al.*, 2005) in their model contrasting two personality types, "hawks" and "doves," that are observed across a wide range of species. Hawks are characterized by fearlessness, impulsivity, and aggressive behavior, while doves are more fearful, sensitive to environmental contingencies, and cooperative. These types flourish under different environments, and in the context of human ecology, doves would be expected to predominate, especially in developed economies. However, under adverse and harsh conditions, one might expect hawks to do quite well. It is not surprising therefore that violence and aggression are more common in socially and economically disadvantaged families and neighborhoods. Cross-country comparisons also indicate that violence is more common in countries with high levels of inequality, such as the USA (Wilkinson, 2004).

One potential mechanism underlying manifestation of either a hawk- or a dove-like personality is the epigenetic effect of environmental influences on gene expression. Epigenetic effects of parenting are mediated by methylation of DNA in the hippocampus, which silences genes that influence the HPA response to stress (Champagne, 2010). Epigenetic animal models suggest that early maternal behavior toward offspring sends a signal to the developing brain to expect either a harsh or benign environment and prepares the newborn with behavioral tendencies that can maximize success in those circumstances. Thus mothers who engage in neglectful parenting produce offspring with altered hippocampal function that is then transmitted by female offspring to their children. This epigenetic mechanism could produce intergenerational consistencies that promote hawkish or dovish offspring. It is thus possible that epigenetic mechanisms are partially responsible for the persistence of hawk-like personalities in human populations despite the apparent dysfunction of this type of behavior.

Despite the appeal of the hawk–dove dichotomy, it is likely that there is an intermediate personality type that shares some characteristics with hawks, but is not as fearless. This subtype would correspond to the antisocial personality that is as impulsive but considerably more reactive to provocation than the psychopathic hawk (Frick and Viding, 2009). This antisocial type displays greater amygdala reactivity to threat and is thought to behave aggressively primarily out of fear of others. Nevertheless, it is likely that both types of antisocial person are raised under adverse conditions, with parenting that leads to aggressive behavior in offspring.

Given the powerful evidence surrounding effects of early adversity, one would expect that interventions that attempt to reduce these influences on children would be effective in reducing antisocial behavior. It is for this reason that we turn to parenting interventions as a potential point of intervention to prevent the development of antisocial behavior.

Models of parenting and antisocial behavior in children

Developmental theories emphasize the role of parenting practices as important influences in shaping the trajectories of antisocial behaviors. One approach focuses on the effects of harsh parenting on the training of antisocial tendencies. In particular, Patterson's coercion theory (1982) posits that disruptive and coercive family processes, such as harsh and inconsistent discipline, can lead to the development of antisocial behaviors in children by training them to employ aggression rather than cooperative behaviors. Guided by the principles of social learning and operant conditioning, coercion theory proposes that aggressive and antisocial behaviors persist because they serve a functional role in coercive families.

Another theoretical framework for understanding family influences on child antisocial behavior is the self-control theory of Gottfredson and Hirschi (1990). This theory views the development of self-control as an important predictor of antisocial behavior in children. Parents are thought to play a critical role in the development of self-control through active monitoring and use of positive disciplining strategies to rectify deviant acts. In addition, strong emotional bonds between the parent and child facilitate the adoption of self-control over antisocial tendencies. Thus, in contrast to the coercion framework, which assumes that deviant behaviors are extensions of negative interactions observed in the family context, self-control theory asserts that ineffective parenting leads to antisocial outcomes by impeding the development of the child's ability to exert self-control.

Clearly both approaches provide guidance for interventions, with Patterson's theory focusing on what is learned by coercive family interactions and Gottfredson and Hirschi focusing on the mediating role of self-control. Given their utility in predicting antisocial behaviors, both theories have guided the development of several successful family-centered preventive interventions, with a common emphasis on parent management training (Kazdin, 2005). Prominent family-based interventions that have garnered the most empirical support are described below. Common characteristics include parent management training components, such as use of consistent and appropriate disciplining strategies, positive reinforcement of prosocial behaviors, effective conflict management, and adaptive problem-solving skills.

Interventions with promising results in the first 10 years of life

Given the importance of early onset of antisocial behavior, interventions that target the causes of such behavior would appear to be critical. One of the early interventions with that aim is the home visitation program designed by David Olds (Olds *et al.*, 1998). This program targets low-income first-time mothers with visits by a public health nurse prior to birth and for a period of up to 2 years after the birth of the child. The program provides guidance to the mother for more positive parenting as well as use of social services. The strategy of early and sustained intervention is consistent with research indicating that nurturant parenting in early development can reduce later antisocial behavior. Follow-ups of children prior to age 15 in the original trial revealed less child neglect and abuse, and lower conviction rates in families that were randomly assigned to receive the full 2-year program (Olds *et al.*, 1998). However, a follow-up at age 19 indicated that the reduction in criminal behavior remained significant only in female adolescents (Eckenrode *et al.*, 2010). Females were also less likely to experience pregnancy. Given the success of this intervention and its receipt of federal funding, a national research program has been established to identify critical characteristics of program success to assist future intervention development efforts (Duggan *et al.*, 2013).

Other parent training interventions have targeted high-risk families starting at age 2 and later. Parent Management Training developed by Patterson *et al.* (1982) was one of the first parenting skill-based interventions that demonstrated reductions in problem behaviors among children aged 3–12 years. As part of the intervention, parents are taught skills related to consistent disciplining, monitoring/supervision, responsive parenting, and problem solving. Later adaptations of this model include the Multidimensional Treatment Foster Care Program (Chamberlain *et al.*, 2008), which is effective in reducing recidivism rates among high-risk chronic youth offenders in foster care settings.

Another popular parenting program, the Incredible Years Program (IYP; Webster-Stratton, 1998; Webster-Stratton *et al.*, 2001) has been tested extensively with 2–7-year-old children attending Head Start programs. Parents are taught everyday parenting skills through videotape modeling. Similar video vignettes are also shown to children, with an emphasis on prosocial behaviors. This program is found to be effective in promoting effective parenting skills, reducing conduct problems, and improving social competence in children (Webster-Stratton and

Hammond, 1997). A recent community-based trial in Ireland targeting disadvantaged urban areas showed that this intervention can be successfully implemented outside of clinical settings (McGilloway *et al.*, 2012b).

Other interventions have taken a public health approach using a tiered intervention strategy in which more intensive intervention is provided to families with more serious child behavior problems. A good example of this approach is the Triple P (Positive Parenting Program) intervention, in which mass media are used to attract the widest possible range of at-risk families, and more intensive training is delivered depending on the seriousness of the parenting problem. A large randomized trial in South Carolina found that the intervention reduced the rate of child maltreatment in counties that received the program compared with a matched set of counties that were left untreated (Prinz *et al.*, 2009). A second approach, Family Check Up (FCU), uses an annual check-up model in which families self-select from a tiered program of intervention services depending on the family's needs (Dishion *et al.*, 2014). An intensive study of the program with low-income children aged 2 at baseline found that it was successful even for children exhibiting early CU traits (Hyde *et al.*, 2013). Similar large-scale follow-ups of parenting programs have been conducted in the UK and found to show substantial reductions in child problem behavior and better mental health in parents (Lindsay and Strand, 2013). This study also suggests that a population-based strategy for improving child outcomes can be successfully implemented.

In some cases, interventions that target parental involvement have been delivered in school settings. Two famous cases are the HighScope/Perry Preschool Program (Schweinhart *et al.*, 2005) and the Chicago Child-Parent Center (Reynolds *et al.*, 2001). The Perry Program included a home visitation component that involved mothers in the child's educational process and encouraged them to reinforce the program components taught at school in the home setting. Preschoolers in the treatment group had substantially lower rates of lifetime arrests and criminal involvement assessed at age 40. Similar strategies and outcomes were observed in the Chicago program, although that program provided services beyond the preschool period. Another remarkably simple intervention successful in high-poverty schools in Baltimore is the Good Behavior Game. This intervention uses a positive disciplining approach to encourage children to stay on task and avoid disruptive classroom behavior. When implemented during the early elementary years, the program was found to be effective in enhancing classroom discipline, reducing delinquent involvement during adolescence, and reducing arrests as adults (Petras *et al.*, 2008). Finally, the Fast Track Program (Slough *et al.*, 2008) uses a multi-pronged approach by targeting social-cognitive skills, school environment, academic performance, and negative peer relations, in addition to parent training, from first grade into high school. It reduced conduct disorders and antisocial behaviors at 3- and 6-year follow-ups.

Interventions during later years that may be promising

Although interventions that begin during early childhood are likely to be most effective, those that begin later, especially during adolescence, are also found to be protective. The most dramatic demonstration of the reversal of effects of early deprivation is the naturalistic experiment reported by Copeland and Costello (2010) of American Indian children whose families received income supplements ($4,000 per year) with the introduction of casino gambling on reservations. At age 9, children in families who received the supplement were less likely to have a criminal record by age 16 and were less likely to have committed a minor crime by age 21 than children in families who did not receive the supplement. Children older than age 9 at the time when the supplement was introduced did not seem to benefit. However, it was noteworthy that the supplement did not increase the amount of time spent by parents in the home; rather it improved the level of positive parental interaction with the child, consistent with the hypothesis that poverty increases negative parenting behavior.

Interventions that are directed at changing parenting practices during adolescence have found effects on antisocial behaviors. For instance, the FCU intervention is effective in reducing antisocial behaviors in youth from early–late adolescence, with effects mediated by reductions in family conflict (Van Ryzin and Dishion, 2012). Other successful large-scale interventions for early adolescents and their families, including 'Guiding Good Choices'/'Preparing for drug free years' and 'Iowa Strengthening Families Program' (Spoth *et al.*, 2000), provide skill training to parents in appropriate disciplining strategies, positive reinforcement of prosocial behaviors, and conflict management, as well as training adolescents to resist peer pressure. Intervention-related reductions have been reported on substance use and delinquency.

Conclusion

Our review of interventions finds that efforts to reduce adverse parenting behavior can reduce antisocial behavior in children as well as adolescents, with some studies showing reductions in criminal arrests in adulthood. Reviews of intervention effects conclude that parenting interventions are effective (Piquero *et al.*, 2009; McGilloway *et al.*, 2012a), and that they can reduce antisocial behavior in youth who exhibit both reactive and CU characteristics (Waller *et al.* 2013). Some of the studies find that the interventions reduce overt impulsive behavior typically classified as hyperactivity. Although there is little evidence that these interventions improve the executive functions that underlie impulsive behavior, they are certainly consistent with that hypothesis. Future research would benefit from pinpointing the mediators that are effective in improving self-control at both the neurobiological and behavioral levels. In addition, the mechanisms by which parenting behaviors interact with individual characteristics (temperament, developmental changes) to influence behavioral outcomes need to be examined in greater detail.

The findings also suggest that the emergence of antisocial behavior in adolescence is conditional on earlier experiences that predispose youth to this developmental trajectory. Thus preventing those earlier experiences should help to reduce antisocial behavior in adolescence and into adulthood. In view of the finding that most policies directed toward child maltreatment in developed countries have not had observable effects at the population level (Gilbert *et al.*, 2012), the challenge that remains is to refine and implement parenting interventions on a large scale in the years to come.

References

APA (2013) *Attention Deficit/Hyperactivity Disorder*, Arlington, VA: American Psychiatric Publishing.

Banich, M.T. (2004) *Cognitive Neuroscience and Neuropsychology*, 2nd edn, Boston, MA: Houghton Mifflin.

Barker, E.D. and Maugham, B. (2009) 'Differentiating early-onset persistent versus childhood-limited conduct problem youth', *American Journal of Psychiatry*, 166(8): 900–908.

Bernier, A., Carlson, S.M., and Whipple, N. (2010) 'From external regulation to self-regulation: early parenting precursors of young children's executive functioning', *Child Development*, 81(1): 326–339.

Blair, C. (2010) 'Stress and the development of self-regulation in context', *Child Development Perspectives*, 4(3): 181–188.

Blair, J.R. (2010) 'Neuroimaging of psychopathy and antisocial behavior: a targeted review', *Current Psychiatry Reports*, 12: 76–82.

Broidy, L.M., Nagin, D., Tremblay, R.E., Bates, J.E., Brame, B., Dodge, K.A., Fergusson, D., Horwood, J.L., Loeber, R., Laird, R., Lynam, D.R., Moffitt, T.E., Pettit, G.S., and Vitaro, F. (2003) 'Developmental trajectories of childhood disruptive behaviors and adolescent delinquency: a six-site, cross-national study', *Developmental Psychology*, 39(2): 222–245.

Buodo, G., Moscardino, U., Scrimin, S., Altoè, G., and Palomba, D. (2013) 'Parenting stress and externalizing behavior symptoms in children: the impact of emotional reactivity', *Child Psychiatry & Human Development*, 21: 1–12.

Casey, B.J., Getz, S., and Galvan, A. (2008) 'The adolescent brain', *Developmental Review*, 28(1): 62–77.

Chamberlain, P., Price, J., Leve, L.D., Laurent, H., Landsverk, J.A., and Reid, J.B. (2008) 'Prevention of behavior problems for children in foster care: outcomes and mediation effects', *Prevention Science*, 9(1): 17–27.

Champagne, F.A. (2010) 'Early adversity and developmental outcomes: interaction between genetics, epigenetics, and social experiences across the life span', *Perspectives on Psychological Science*, 5(5): 564–574.

Cicchetti, D. and Toth, S.L. (2005) 'Child maltreatment', *Annual Review of Clinical Psychology*, 1: 409–438.

Copeland, W. and Costello, E.J. (2010) 'Parents' incomes and children's outcomes: a quasi-experiment', *American Economic Journal of Applied Economics*, 2(1): 86–115.

Crnic, K.A., Gaze, C., and Hoffman, C. (2005) 'Cumulative parenting stress across the preschool period: relations to maternal parenting and child behaviour at age 5', *Infant and Child Development*, 14(2): 117–132.

Davis, F.C., Knodt, A.R., Sporns, O., Lahey, B.B., Zald, D.H., Brigidi, B.D., and Hariri, A.R. (2013) 'Impulsivity and the modular organization of resting-state neural networks', *Cerebral Cortex*, 23(6): 1444–1452.

De Brito, S.A., Viding, E., Sebastian, C.L., Kelly, P.A., Mechelli, A., Maris, H., and McCrory, E.J.P. (2013) 'Reduced orbitofrontal and temporal grey matter in a community sample of maltreated children', *Journal of Child Psychology and Psychiatry*, 54(1): 105–112.

Decety, J., Chen, C., Harenski, C., and Kiehl, K.A. (2013) 'An fMRI study of affective perspective taking in individuals with psychopathy: imagining another in pain does not evoke empathy', *Frontiers in Human Neuroscience*, 7: 1–12.

Diamond, A. (2013) 'Executive functions', *Annual Review of Psychology*, 64: 135–168.

Dishion, T.J., Brennan, L.M., Shaw, D.S., McEachern, A.D., Wilson, M.N., and Jo, B. (2014) 'Prevention of problem behavior through annual family check-ups in early childhood: intervention effects from home to early elementary school', *Journal of Abnormal Child Psychology*, 42(3): 343–354.

Duggan, A., Minkowitz, C.S., Chaffin, M., Korfmacher, J., Brooks-Gunn, J., Crowne, S., Filene, J., Gonsalves, K., Landsverk, J., and Harwood, J. (2013) 'Creating a national home visiting research network', *Pediatrics*, 132: S82–S89.

Duque, A., Coman, D., Carlyle, B.C., Bordner, K.A., George, E.D., Papademetris, X., Hyder, F., and Simen, A.A. (2012) 'Neuroanatomical change in a mouse model of early life neglect', *Brain Structure and Function*, 217: 459–472.

Eckenrode, J., Campa, M., Luckey, D.W., Henderson, C.R., Cole, R., Kitzman, H., Anson, E., Sidoro-Arcoleo, K., Powers, J., and Olds, D. (2010) 'Long-term effects of prenatal and infancy nurse home visitation on the life course of youths', *Archives of Pediatric and Adolescent Medicine*, 164(1): 9–15.

Engle, R.W. (2002) 'Working memory capacity as executive attention', *Current Directions in Psychological Science*, 11: 19–23.

Ernst, M. (2014) 'The triadic model perspective for the study of adolescent motivated behavior', *Brain and Cognition*, doi 10.1016/j.bandc.2014.01.006.

Evans, G.W. and Schamberg, M.A. (2009) 'Childhood poverty, chronic stress, and adult working memory', *Proceedings of the National Academy of Sciences, USA*, 106(16): 6545–6549.

Fontaine, N.M.G., Rijsdijk, F.V., McCrory, E.J.P., and Viding, E. (2010) 'Etiology of different developmental trajectories of callous-unemotional traits', *Journal of the American Academy of Child and Adolescent Psychiatry*, 49(7): 656–664.

Frick, P.J. and Viding, E. (2009) 'Antisocial behavior from a developmental psychopathological perspective', *Development and Psychopathology*, 21: 1111–1131.

Frick, P.J., Cornell, A.H., Bodin, S.D., Dane, H.E., Barry, C.T., and Loney, B.R. (2003) 'Callous-unemotional traits and developmental pathways to severe conduct problems', *Developmental Psychology*, 39(2): 246–260.

Galvan, A., Hare, T., Voss, H., Glover, G., and Casey, B.J. (2007) 'Risk-taking and the adolescent brain: who is at risk?', *Developmental Science*, 10(2): F8–F14.

Giedd, J.N., Blumenthal, J., Jeffries, N.O., Castellanos, F.X., Liu, H., Zijdenbos, A., Paus, T., Evans, A.C., and Rapoport, J.L. (1999) 'Brain development during childhood and adolescence: a longitudinal MRI study', *Nature Neuroscience*, 2(10): 861–3.

Gilbert, R., Fluke, J., O'Donnell, M., Gonzalez-Izquierdo, A., Brownell, M., Gulliver, P., Janson, S., and Sidebotham, P. (2012) 'Child maltreatment: variation in trends and policies in six developed countries', *Lancet*, 379: 758–772.

Gottfredson, M.R. and Hirschi, T. (1990) *A General Theory of Crime*, Stanford, CA: Stanford University Press.

Gunnar, M.R. and Fisher, P.A. (2006) 'Bringing basic research on early experience and stress neurobiology to bear on preventive interventions for neglected and maltreated children', *Development and Psychopathology*, 18(3): 651–677.

Howell, K.H. (2011) 'Resilience and psychopathology in children exposed to family violence', *Aggression and Violent Behavior*, 16: 562–569.

Hyde, L.W., Shaw, D.S., Gardner, F., Cheong, J., Dishion, T.J., and Wilson, M.N. (2013) 'Dimensions of callousness in early childhood: links to problem behavior and family intervention effectiveness', *Development and Psychopathology*, 25: 347–363.

Jaffee, S.R., Strait, L.B., and Odgers, C. (2012) 'From correlates to causes: can quasi-experimental studies and statistical innovations bring us closer to identifying the causes of antisocial behavior?', *Psychological Bulletin*, 138(2): 272–295.

Kazdin, A.E. (2005) *Parent Management Training: Treatment for Oppositional, Aggressive, and Antisocial Behavior in Children and Adolescents*, New York: Oxford University Press.

Khurana, A., Romer, D., Betancourt, L.M., Brodsky, N.L., Giannetta, J.M., and Hurt, H. (2013) 'Working memory ability predicts trajectories of early alcohol use in adolescents: the mediational role of impulsivity', *Addiction*, 108(3): 506–515.

Korte, S.M., Koolhaas, J.M., Wingfield, J.C., and McEwen, B.S. (2005) 'The Darwinian concept of stress: benefits of allostasis and costs of allostatic load and the trade-offs in health and disease', *Neuroscience and Biobehavioral Reviews*, 29: 3–38.

Kotch, J.B., Lewis, T., Hussey, J.M., English, D., Thompson, R., Litrownik, A.J., Runyan, D.K., Bangdiwala, S.I., Margolis, B., and Dubowitz, H. (2008) 'Importance of early neglect for childhood aggression', *Pediatrics*, 121(4): 725–731.

Krueger, R.F., Hicks, B.M., Patrick, C.J., Carlson, S.R., Iacono, W.G., and McGue, M. (2002) 'Etiologic connections among substance dependence, antisocial behavior, and personality: modeling the externalizing spectrum', *Journal of Abnormal Psychology*, 11(3): 411–424.

Lindsay, G. and Strand, S. (2013) 'Evaluation of the national roll-out of parenting programmes across England: the parenting early intervention programme (PEIP)', *BMC Public Health*, 13: 972–989.

Lockwood, P.L., Sebastian, C.L., McCrory, E.J.P., Hyde, Z.H., Gu, X., De Brito, S.A., and Viding, E. (2013) 'Association of callous traits with reduced neural response to others' pain in children with conduct problems', *Current Biology*, 23: 901–905.

Loeber, R. and Pardini, D. (2008) 'Neurobiology and the development of violence: common assumptions and controversies', *Philosophical Transactions of the Royal Society B*, 362: 2491–2503.

Loeber, R., Pardini, D., Homish, D.L., Wei, E.H., Crawford, A.M., Farrington, D.P., *et al.* (2005) 'The prediction of violence and homicide in young men', *Journal of Consulting and Clinical Psychology*, 73(6): 1074–1088.

Luby, J., Belden, A., Botteron, K., Marrus, N., Harms, M.P., Babb, C., Nishino, T., and Barch, D.M. (2013) 'The effects of poverty on childhood brain development: the mediating effect of caregiving and stressful life events', *JAMA Pediatrics*, 167(12): 1135–1142.

Luciana, M., Conklin, H.M., Hooper, C.J., and Yarger, R.S. (2005) 'The development of nonverbal working memory and executive control processes in adolescents', *Child Development*, 76(3): 697–712.

Luna, B., Garver, K.E., Urban, T.M., Lazar, N.A., and Sweeney, J.A. (2004) 'Maturation of cognitive processes from late childhood to adulthood', *Child Development*, 75: 1357–1372.

Lynam, D.R., Caspi, A., Moffitt, T.E., Loeber, R., and Stouthamer-Loeber, M. (2007) 'Longitudinal evidence that psychopathy scores in early adolescence predict adult psychopathy', *Journal of Abnormal Psychology*, 116(1): 155–165.

McCrory, E.J.P., De Brito, S.A., Sebastian, C.L., Mechelli, A., Bird, G., Kelly, P.A., and Viding, E. (2011) 'Heightened neural reactivity to threat in child victims of family violence', *Current Biology*, 21(23): R947–R948.

McGilloway, F.M., Bywater, T., Hutchings, J., Smith, S.M., and Donnelly, M. (2012a) 'Behavioral and cognitive-behavioral group-based parenting programmes for early-onset conduct problems in children aged 3 to 12 years (Review)', *Evidence-Based Child Health*, 2: 318–692.

McGilloway, S., Mhaille, G.N., Bywater, T., Furlong, M., Leckey, Y., Kelly, P., Comiskey, C., and Donnelly, M. (2012b) 'A parenting intervention for childhood behavioral problems: a randomized controlled trial in disadvantaged community-based settings', *Journal of Consulting and Clinical Psychology*, 80(1): 116–127.

McGuire, J.T. and Kable, J.W. (2013) 'Rational predictions can underlie apparent failures to delay gratification', *Psychological Review*, 120(2): 395–410.

McMahon, R.J., Witkiewitz, K., Kotler, J.S., and the Conduct Problems Prevention Research Group (2010) 'Predictive validity of callous-unemotional traits measured in early adolescence with respect to multiple antisocial outcomes', *Journal of Abnormal Psychology*, 119(4): 752–763.

Magid, V., MacLean, M.G., and Colder, C.G. (2007) 'Differentiating between sensation seeking and impulsivity through their mediated relations with alcohol use and problems', *Addictive Behaviors*, 32: 2046–2061.

Moffitt, T.E. (1993) 'Adolescence-limited and life-course-persistent antisocial behavior: a developmental taxonomy', *Psychological Review*, 100(4): 674–701.

Moffitt, T.E., Arseneault, L., Belsky, D., Dickson, N., Hancox, R.J., Harrington, H., Houts, R., *et al.* (2011) 'A gradient of childhood self-control predicts health, wealth, and public safety', *Proceedings of the National Academy of Sciences, USA*, 108(7): 2693–2698.

Nelson, C.A., Zeanah, C.H., Fox, N.A., Marshall, P.J., Smyke, A.T., and Guthrie, D. (2007) 'Cognitive recovery in socially deprived young children: the Bucharest Early Intervention Project', *Science*, 318: 1937–1940.

Nigg, J.T., Nikolas, M., Friderici, K., Park, L., and Zucker, R.A. (2007) 'Genotype and neuropsychological response inhibition as resilience promoters for attention-deficit/hyperactivity disorder, oppositional defiant disorder, and conduct disorder under conditions of psychological adversity', *Development and Psychopathology*, 19: 767–787.

Odgers, C., Moffitt, T.E., Broadbent, J.M., Dickson, N., Hancox, R.J., Harrington, H., Poulton, R., Sears, M.R., Thomson, W.M., and Caspi, A. (2008) 'Female and male antisocial trajectories: from childhood origins to adult outcomes', *Development and Psychopathology*, 20: 673–716.

Odgers, C.L., Caspi, A., Russell, M.A., Sampson, R.J., Arsenault, L., and Moffitt, T.E. (2012) 'Supportive parenting mediates widening neighborhood socioeconomic disparities in children's antisocial behavior from ages 5 to 12', *Development and Psychopathology*, 24(3): 705–721.

Olds, D., Henderson, C.R., and Cole, R. (1998) 'Long-term effects of home visitation on maternal life course and child abuse and neglect: fifteen-year follow-up of a randomized trial', *JAMA*, 280(14): 1238–1244.

Patterson, G.R. (1982) *Coercive Family Process*, Vol. 3, Eugene, OR: Castalia Publishing Co.

Petras, H., Kellam, S.G., Brown, H.C., Muthen, B.O., Ialongo, N.S., and Poduska, J.M. (2008) 'Developmental epidemiological courses leading to antisocial personality disorder and violent and criminal behavior: effects by young adulthood of a universal preventive intervention in first- and second-grade classrooms', *Drug and Alcohol Dependence*, 955: S45–S59.

Piquero, A.R., Farrington, D.P., Welsh, B.C., Tremblay, R.E., and Jennings, W.G. (2009) 'Effects of early family/parent training programs on antisocial behavior and delinquency', *Journal of Experimental Criminology*, 5: 83–120.

Prinz, R.J., Sanders, M.R., Shapiro, C.J., Whitaker, D.J., and Lutzker, J.R. (2009) 'Population-based prevention of child maltreatment: the U.S. Triple P System population trial', *Prevention Science*, 10: 1–12.

Raver, C.C., Blair, C., and Willoughby, M. (2013) 'Poverty as a predictor of 4-year-olds' executive function: new perspectives on models of differential susceptibility', *Developmental Psychology*, 49(2): 292–304.

Reynolds, A., Temple, J.A., Robertson, D.L., and Mann, E.A. (2001) 'Long-term effects of an early childhood intervention on educational achievement and juvenile arrest', *JAMA*, 285(18): 2339–2346.

Rodriguez, C.M. (2011) 'Association between independent reports of maternal parenting stress and children's internalizing symptomatology', *Journal of Child and Family Studies*, 20(5): 631–639.

Romer, D. (2010) 'Adolescent risk taking, impulsivity, and brain development: implications for prevention', *Developmental Psychobiology*, 52(3): 263–276.

Romer, D. and Walker, E. (2007) *Adolescent Psychopathology and the Developing Brain: Integrating Brain and Prevention Science*, New York: Oxford University Press.

Romer, D., Betancourt, L., Giannetta, J.M., Brodsky, N.L., Farah, M.J., and Hurt, H. (2009) 'Executive cognitive functions and impulsivity as correlates of risk taking and problem behavior in preadolescents', *Neuropsychologia*, 47(13): 2916–2926.

Romer, D., Betancourt, L.M., Brodsky, N.L., Giannetta, J.M., Yang, W., and Hurt, H. (2011) 'Does adolescent risk taking imply weak executive function? A prospective study of relations between working memory performance, impulsivity, and risk taking in early adolescence', *Developmental Science*, 14(5): 1119–1133.

Scaramella, L.V., Neppl, T.K., Ontai, L.L., and Conger, R.D. (2008) 'Consequences of socioeconomic disadvantage across three generations: parenting behavior and child externalizing problems', *Journal of Family Psychology*, 22(5): 725–733.

Schweinhart, L.J., Montie, J., Xiang, Z., Barnett, W.S., Belfield, C.R., and Nores, M. (2005) *Lifetime Effects: The High/Scope Perry Preschool Study through Age 40*, Ypsilanti, MI: High/Scope Press.

Shamosh, N.A., Deyoung, C.G., Green, A.E., Reis, D.L., Johnson, M.R., Conway, A.R., Engle, R.W., Braver, T.S., and Gray, J.R. (2008) 'Individual differences in delay discounting: relation to intelligence, working memory, and anterior prefrontal cortex', *Psychological Science*, 19(9): 904–11.

Shannon, B.J., Raichle, M.E., Snyder, A.Z., Fair, D.A., Mills, K.L., Zhang, D., Bache, K., Calhoun, V.D., Nigg, J.T., Nagel, B.J., Stevens, A.A., and Kiehl, K.A. (2011) 'Premotor functional connectivity predicts impulsivity in juvenile offenders', *Proceedings of the National Academy of Sciences, USA*, 108(27): 11241–11245.

Sharkey, P.T., Tirado-Strayer, N., Papachristos, A.V., and Raver, C.C. (2012) 'The effect of local violence on children's attention and impulse control', *American Journal of Public Health*, 102(12): 2287–2293.

Shonkoff, J.P., Boyce, W.T., and McEwen, B.S. (2009) 'Neuroscience, molecular biology, and the childhood roots of health disparities: building a new framework for health promotion and disease prevention', *JAMA*, 301(21): 2252–2259.

Slough, N.M., McMahon, R.J., and the Conduct Problems Prevention Research Group (2008) 'Preventing serious conduct problems in school-age youth: the fast track program', *Cognitive and Behavioral Practice*, 15: 3–17.

Smith, G.T., Fischer, S., Cyders, M.A., Annus, A.M., Spillane, N.S., and McCarthy, D.M. (2007) 'On the validity and utility of discriminating among impulsivity-like traits', *Assessment*, 14(2): 155–170.

Spear, L. (2010) *The Behavioral Neuroscience of Adolescence*, New York: W.W. Norton & Co.

Spoth R.L., Redmond C., and Shin C. (2000) 'Reducing adolescents' aggressive and hostile behaviors: randomized trial effects of a brief family intervention 4 years past baseline', *Archives of Pediatrics & Adolescent Medicine*, 154(12): 1248–1257.

Sternberg, K.J., Baradaran, L.P., Abbott, C.B., Lamb, M.E., and Guterman, E. (2006) 'Type of violence, age, and gender differences in the effects of family violence on children's behavior problems: a meta-analysis', *Developmental Review*, 26: 89–112.

Teicher, M.H., Anderson, S.L., Polcari, A., Anderson, C.M., Navalta, C.P., and Kim, D.M. (2003) 'The neurobiological consequences of early stress and childhood maltreatment', *Neuroscience and Biobehavioral Reviews*, 27: 33–44.

Tomoda, A., Suzuki, H., Rabi, K., Sheu, Y.-S., Polcari, A., and Teicher, M.H. (2009) 'Reduced prefrontal cortical gray matter volume in young adults exposed to harsh punishment', *Neuroimage*, 47: t66–t71.

Van Ryzin, M.J. and Dishion, T.J. (2012) 'The impact of a family-centered intervention on the ecology of adolescent antisocial behavior: modeling developmental sequelae and trajectories during adolescence', *Development and Psychopathology*, 24(3): 1139–1155.

Viding, E., Sebastian, C.L., Dadds, M.R., Lockwood, P.L., Cecil, C.A.M., De Brito, S.A., and McCrory, E.J.P. (2012) 'Amygdala response to preattentive masked fear in children with conduct problems: the role of callous-unemotional traits', *American Journal of Psychiatry*, 169(10): 1109–1116.

Waller, R., Garner, F., and Hyde, L.W. (2013) 'What are the associations between parenting, callous-unemotional traits, and antisocial behavior in youth? A systematic review of evidence', *Clinical Psychology Review*, 33: 593–608.

Webster-Stratton, C. (1998) 'Preventing conduct problems in Head Start children: strengthening parenting competencies', *Journal of Consulting and Clinical Psychology*, 66(5): 715–730.

Webster-Stratton, C. and Hammond, M. (1997) 'Treating children with early-onset conduct problems: a comparison of child and parent training interventions', *Journal of Consulting and Clinical Psychology*, 65(1): 93–109.

Webster-Stratton, C., Reid, M.J., and Hammond, M. (2001) 'Preventing conduct problems, promoting social competence: a parent and teacher training partnership in Head Start', *Journal of Clinical Child Psychology*, 30(3): 283–302.

Whelan, R., Conrod, P.J., Poline, J.-B., Lourdusamy, A., Banaschewski, T., Barker, G.J., Bellgrove, M.A., Buchel, C., *et al.* (2012) 'Adolescent impulsivity phenotypes characterized by distinct brain networks', *Nature Neuroscience*, 15(6): 920–925.

Wilkinson, R. (2004) 'Why is violence more common where inequality is greater?', *Annals of the New York Academy of Science*, 1036: 1–12.

Willoughby, T., Good, M., Adachi, P.J.C., Hamza, C., and Tavernier, R. (2013) 'Examining the link between adolescent brain development and risk taking from a social-developmental perspective', *Brain and Cognition*, 83: 315–323.

Yang, Y. and Raine, A. (2009) 'Prefrontal structural and functional brain imaging findings in antisocial, violent, and psychopathic individuals: a meta-analysis', *Psychiatry Research: Neuroimaging*, 174: 81–88.

Zalewski, M., Lengua, L.J., Fisher, P.A., Trancik, A., Bush, N.R., and Meltzoff, A.N. (2012) 'Poverty and single parenting: relations with preschoolers' cortisol and effortful control', *Infant and Child Development*, 21(5): 537–554.

Zheng, Y. and Cleveland, H.H. (2013) 'Identifying gender-specific developmental trajectories of nonviolent and violent delinquency from adolescence to young adulthood', *Journal of Adolescence*, 36: 371–381.

Zuckerman, M. (2007) *Sensation Seeking and Risky Behavior*, Washington, DC: APA Books.

34

Offender rehabilitation from a maturation/biosocial perspective

Michael Rocque and Brandon C. Welsh

Introduction

In recent years, the field of criminology has borne witness to the emergence of several areas of research that have pushed the study of crime and justice to exciting frontiers. Research on crime over the life-course, offender rehabilitation, and biosocial theories has burgeoned, offering new insights on why people offend and, perhaps more importantly, how crime can be reduced.

These developments are by no means unrelated. The (re)new(ed) interest in crime/delinquency prior to and after adolescence has coincided with a resurgence of research interest and belief in rehabilitation (see Andrews and Bonta, 2010; Cullen, 2013) as well as how physiological factors affect behavior (Moffitt, 1993; Perry *et al.*, 1995; Steinberg, 2008; Walsh and Beaver, 2009; Raine, 2013). Thus ideas that previously were not part of the academic conversation—such as the influence of brain development on behavior, or the ability to reform offenders—are no longer relegated to the background, but form part of the larger discussion of crime over the life-course.

Developmental/life-course criminology, biosocial work, and research on rehabilitation each have contributed to a greater understanding of crime and crime control. The current chapter brings together each of these areas by discussing a perspective on crime over the life-course from a biosocial view, focusing on the reduction of offending in emerging and full adulthood. In recent years, explanations for why offenders slow down and eventually stop committing crimes (which criminologists refer to as "desistance") have proliferated. Yet most theories fail to take biological processes into account, and each remains ostensibly mutually exclusive or competing. From our perspective, however, most of the popular accounts of desistance, as well as emerging neurological findings, can be seen as part of a larger process of development over the life-course. Drawing on this life-course and biosocial research, we develop and present a maturation theory of desistance. We then discuss the policy implications of this perspective, focusing on offender rehabilitation.

Desistance from crime: What is it?

Research in criminology and criminal justice has increasingly begun to focus on the "brute fact" that nearly all offenders eventually stop committing crimes (Hirschi and Gottfredson, 1983; Laub and Sampson, 2001). But what is desistance? In recent years, researchers have begun to seriously consider the definition of desistance in their work. Desistance is not an observable phenomenon that is amenable to empirical study (see Maruna, 1997, 2001). As Laub and Sampson point out (2001), desistance is not as easy to operationalize as other outcomes because it is not the presence of something, but rather its absence, that defines it. According to Kazemian (2007), the differing definitions of desistance, and differing types of data (e.g. relying on self-report or official records) used to measure desistance often lead to inconsistent results (for a list of different definitions used in the literature, see Kazemian, 2007: 9; see also Massoglia and Uggen, 2007).

Researchers have used differing definitions of desistance. For example, Maruna (2001) defines desistance as "the long-term abstinence from crime among individuals who had previously engaged in persistent patterns of offending" (2001: 26). In this sense, desistance is not something that can be adequately measured in the short term or by using cross-sectional data. How much time is required to "see" desistance is not a settled question (Laub and Sampson, 2001, 2003; Maruna, 2001).

Others have defined desistance as a decline in the level of offending over time (Mulvey *et al.*, 2004; Massoglia and Uggen, 2007). Paternoster and Bushway (2009) formally define desistance as the point at which offending reaches a level that is "not significantly different from zero" (2009: 1110; see also Bushway *et al.*, 2001). This suggests that desistance occurs when the likelihood of crime for offenders is indistinguishable from that of non-offenders.

Desistance as a process

The work of Bushway and colleagues (2001, 2003) in particular has helped to formally and statistically define and measure desistance using longitudinal trajectory models. They use Poisson-based semiparametric models which examine offending trajectories for groups of individuals. Desistance, then, is modeled as a reduction (to a near-zero level) in offending over time. The method was based on that of Nagin and Land (1993), which makes no assumptions about the shape of the trajectories and allows an explicit modeling of factors related to change and to group-based differences. This provides an empirical method of modeling desistance rather than devising arbitrary cut-off points and assigning those crime-free for that period as desisters.

Bushway and colleagues view desistance as a "developmental process that unfolds over time rather than a static state that is achieved" (Bushway *et al.*, 2003: 133). This method essentially removes the problem of how long one must be "crime free" to be considered a "desister." Bushway *et al.* (2003) are also clear that desistance involves decreases in "criminality" (the individual's potentiality of committing criminal acts) rather than just incidents of criminal behavior. In other words, something about the *person* changes over time, leading to a cessation of offending. Examining desistance as a process also highlights the importance of understanding what changes are occurring at the same time as desistance. Thus it is essential to study offenders as they desist, rather than solely after the fact (Maruna, 2001; Kazemian, 2007). By doing so, we can better understand desistance as a process, including the ability to examine between-individual variation in how and when individuals move away from crime. This suggests that the way in which desistance is defined also has implications for the type of analysis used.

Explanations of desistance from crime

Much has been accomplished since Laub and Sampson (2001: 5) lamented that "there is relatively little theoretical conceptualization about crime cessation, the various reasons for desistance, and the mechanisms underlying the desistance process." These scholars among others have developed—if not fully fledged theories—compelling and empirically supported explanations for why people stop offending. We briefly discuss these explanations in this section.

Sociological theories

Early research on desistance or the "age-effect" argued that desistance could not be explained with available sociological variables (Hirschi and Gottfredson, 1983). Perhaps seeing this as a challenge, two of Hirschi's own students, John H. Laub and Robert J. Sampson, developed a theory of crime over the life-course—including desistance—that suggested social bonds/ relationships were the reason people eventually stopped offending. The age-graded theory of informal social control argued that the difference between persisting and desisting offenders was the development of social bonds—quality marriages and rewarding employment. Subsequent work has demonstrated that these bonds do in fact predict desistance from crime (Laub *et al.*, 1998; Laub and Sampson, 2003; Sampson *et al.*, 2006; Bersani *et al.*, 2009).

Other sociological work has focused on the issue of civic involvement. Uggen and colleagues have argued that participation in society and feeling as if one is a fully functioning citizen is integral in the movement away from crime (Uggen *et al.*, 2004; Uggen and Inderbitzen, 2010). Thus the act of voting, paying taxes, and other forms of civic involvement may help reduce offending in adulthood (see also Farrall and Calverley, 2006).

Psychological/cognitive explanations of desistance

Next to sociological explanations, psychological theories of desistance are the most prominent in the literature. While a comprehensive review is beyond the scope of this chapter, there are three main categories of psychological desistance theory: identity, psychosocial maturation, and personality theories. Identity or cognitive explanations of desistance center around the notion that, over time (perhaps because of some integral event), people begin to think of themselves differently. Instead of "hell raisers" they start to see themselves as more grown up, as "family men" (Hill, 1971). Several prominent theories of desistance have been offered based on this premise, from Giordano and colleagues' (2002) theory of cognitive transformation, to Paternoster and Bushway's theory of the "feared self" (2009), to Maruna's theory of the self (2001). What these and other studies have in common is the notion that people desist from crime when their attitudes about such behavior and who they are as people become inconsistent with engaging in crime.

Psychosocial maturation is a term that was developed by researchers in the 1970s, and revised recently. According to Cauffman and Steinberg (2000), psychosocial maturity is comprised of three main constructs: responsibility (being able to rely on oneself), temperance (impulsivity, controlling aggressive behavior), and perspective taking (taking others' point of view into consideration, future or present orientation). While psychosocial maturity is not often examined as a "desistance theory," some work has shown that it is negatively related to crime over the life-course (Modecki, 2008; Monahan *et al.*, 2009).

Finally, research in personality constructs has indicated that what previously were thought of as unchanging traits can and do change over time—in such ways that may affect behavior. The "big five" personality traits—neuroticism, openness to new experience, agreeableness, conscientiousness, and extraversion—tend to mature over time in ways that would be consistent with reduced offending (Block, 1971; Adams, 2004; Caspi *et al.*, 2005; Blonigen *et al.*, 2006, 2008).

Blonigen (2010: 98) argues that these personality changes are the basis for desistance from crime. He says "the age–crime curve, specifically the component of desistance from late adolescence to early adulthood, derives from normative maturation in personality traits linked to antisocial behavior, and [. . .] changes in these constructs should be conceptualized within a theoretical framework that emphasizes their co-development during this critical stage in the life-course."

Neurological explanations of desistance

At the present time, few accounts of desistance in criminology have considered neurological research (Blonigen, 2010; but see Collins, 2004; DeLisi and Beaver, 2014). In large part, this may be due to a lack of data on cognitive processes over the life-course. However, with the rise of biosocial perspectives in the field, knowledge about the functioning and development of the brain has begun to influence our understanding of offending and culpability (Steinberg, 2009; Raine, 2013). The recent scholarship on the development of the brain seems to have direct implications for life-course criminology, especially desistance from crime (Restak, 2001).

In a landmark study, Giedd and colleagues (1999) found that the brain does not reach "full" maturity until the individual reaches their mid-20s. Interestingly, this maps closely the point at which desistance becomes the norm for general populations. What changes are taking place after childhood? Researchers have pointed specifically to the prefrontal cortex, which is the area responsible for impulse control and risk-taking behavior (Steinberg, 2008, 2010). Two changes may be contributing to the increase in antisocial behavior in adolescence and the decline thereafter: increases in white matter (the fatty substance coating the axon, which helps in information processing); and consequent decreases in grey matter (Gogtay *et al.*, 2004; Spear, 2007; Giedd, 2008). Interestingly, there are sex-based differences in brain maturation (Giedd, 2008) that may have implications for differential desistance patterns. The state of knowledge on changes in the brain structure through adolescence and childhood is still somewhat nascent, and research has uncovered many possibilities, making it difficult to know which makes a greater contribution to behavioral reform.

How can we make sense of all these seemingly disparate explanations for changes in behavior in emerging adulthood? While it is true that these explanations have often been seen as competing or mutually exclusive (e.g. is it the personality/identity that changes prior to acceptance of adult social institutions such as marriage, or does marriage change identity?), they may not be. It is possible to view these explanations as each contributing to an integrated understanding of development through adulthood. Each of the theories described above is related to what we might think of as adult maturation—what it means to be an adult.

In scholarly as well as public discourse, the term maturation is bandied about frequently with no clear definition. Perhaps because of this lack of clarity regarding the notion of maturation, it has yet to be featured in the desistance literature. Interestingly, however, it may have been the earliest criminological theory of desistance. The Gluecks, as early as the 1930s, argued that when offenders settle down and stop committing crimes, it is due to maturation (or what some have called maturational reform) (see Glueck and Glueck, 1937, 1940, 1945, 1970; Glueck, 1964). What they meant by maturation was not fully specified, but they were clear it was not an analogue for age (meaning people "mature" at different rates). According to them, "not age per se, but rather the acquisition of a certain degree of what we have called 'maturation' regardless of age at which this is achieved among different groups of offenders, is significantly related to changes in criminalistic behavior once embarked upon" (Glueck and Glueck, 1945: 84). Because maturation was not well defined, it was seen as tautological by some (see Wootton, 1962), and mostly forgotten as an explanation of desistance. Yet the Gluecks were clear that work remained to be done in terms of identifying the "components" of maturation (Glueck and Glueck, 1940: 270). And it was also clear that they viewed this search as an interdisciplinary task for biological, sociological, and psychological researchers; a "collaboration between imaginative, open-minded representatives of all such disciplines" (Glueck and Glueck, 1974: 175). We argue that the recent interdisciplinary scholarship on desistance has identified several "components" of maturation, which we describe below.

The components of maturation

As an exploratory first step in taking up the torch of the Gluecks' call for further work on defining the components of maturation, we identify five components or what we call "domains" of maturation. Each domain is drawn from the literature reviewed above.

Adult social role maturation

The first domain or component is based on the work by Sampson, Laub, and others, which has consistently found that exposure to prosocial adult institutions such as marriage and employment is related to desistance from crime. Including this as a domain of maturation makes clear our stance that maturation is not solely a physiological phenomenon, but is marked by social relationships as well. It is considered normative in today's society to attain meaningful romantic relationships as well as full-time employment and completion of formal education (Uggen *et al.*, 2004). Those who have not attained these statuses are considered "off-time" in adulthood (Uggen and Massoglia, 2003), and may be more vulnerable to offending. Participation in these institutions creates pressures to be less selfish and to think of others (Yamaguchi and Kandel, 1985; Adams, 2004).

Civic maturation

Civic maturation is the idea that part of reaching adulthood is the attendant feeling that it is important to "give back" and to be a fully contributing member of society. This component of maturation draws from the work of Uggen and colleagues, as well as Farrall and Calverley (2006). Like social maturation, civic maturation (marked by a change in attitude toward the state and citizenship obligations such as voting, volunteering, or paying taxes) implies the individual has come to think less of him/herself and more about the welfare of others in society. Individuals who show signs of desistance seem to want to become "productive members of society" and "good taxpayers" (Uggen *et al.*, 2004). The notion of "good citizenship" has been emphasized in other desistance scholarship (see also Maruna, 2001). As Farrall and Calverley (2006: 137) state: "These processes of socialization and resocialization are important for [a] consideration of desistance and citizenship values as they suggest that common changes (i.e. from offender to non-offender) are associated with shifts in values."

Psychosocial maturation

Psychosocial maturation draws from the work of Cauffman and Steinberg (2000) and Blonigen *et al.* (2008). Thus it includes personality changes in adulthood as well as the three facets (temperance, perspective, and responsibility) of psychosocial maturation identified by Cauffman and Steinberg (2000). Here, the idea is that changes in psychological processes lead to changes in behavior. This domain of maturation may be most recognizable from a layperson's perspective of what maturation means; for example, being more agreeable, conscientious, less impulsive, and more independent.

Cognitive transformation/identity maturation

Cognitive transformation or identity maturation draws from the work of Paternoster and Bushway, Giordano and colleagues, and Maruna. These scholars argue that changes in how a person views themself, as well as criminal behavior, are integral in promoting desistance from crime. This is a component of maturation, we argue, because part of becoming an adult is accepting adult social roles. Shirking responsibility, seeing oneself as a "troublemaker," and the like, are not consistent with adulthood. There is much evidence that changes in identity are related to changes in behavior among desisting individuals (see Maruna, 2001; Paternoster and Bushway, 2009; King, 2013), thus such changes may be part of an overall process of becoming an adult.

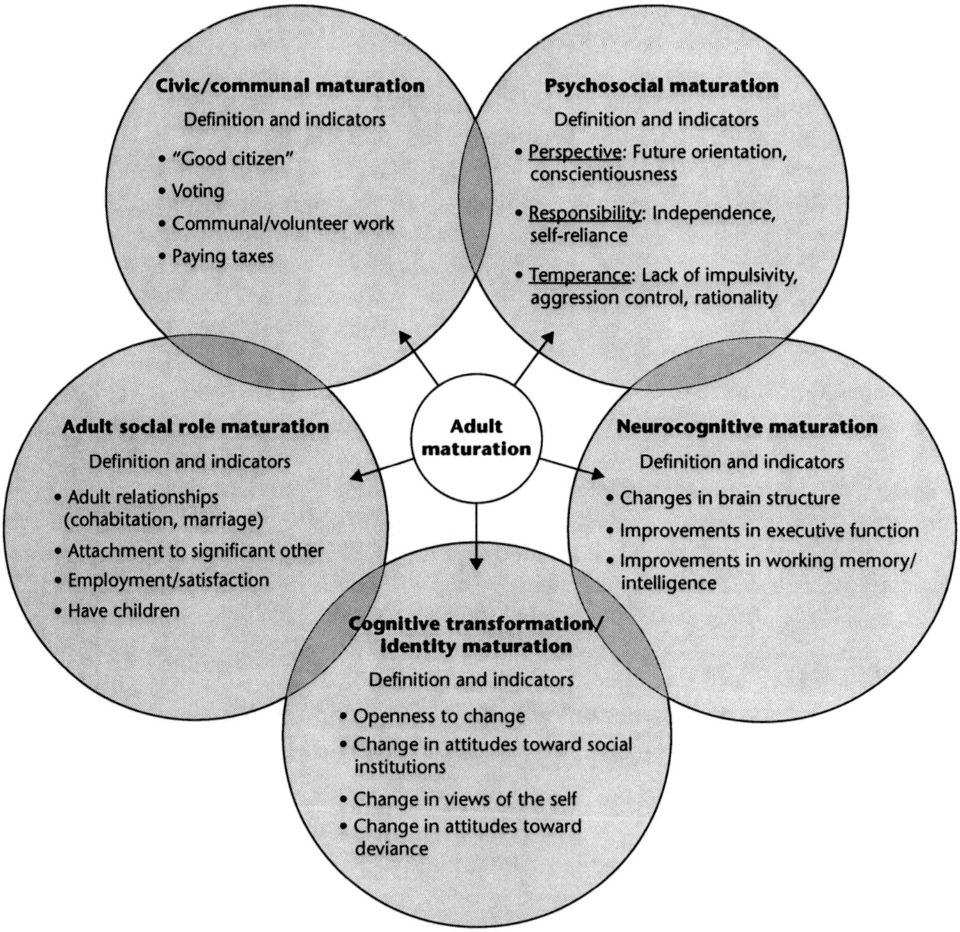

Figure 34.1 Maturation domains and relationships

Neurological/cognitive maturation

The last domain of maturation we identify is the most biologically based of the five. Much remains unknown regarding brain maturation or development, but accumulating evidence suggests that a confluence of changes through adolescence and emerging adulthood may have an influence on desistance. For example, as previously stated, Giedd *et al.* (1999) found that full brain maturation does not occur until the mid-20s, which is the end point of the "crime-prone" ages. Increasing cognitive control, improving decision-making, and reducing risk-taking—which are processes that occur in adulthood—may be related to changes in brain chemistry. Future work has much to do to uncover the exact changes that are pertinent for desistance from crime, but this perspective clearly has implications for offender rehabilitation. Figure 34.1 illustrates the domains graphically.

Viewing desistance from crime from a maturation/biosocial perspective opens up a vista of opportunity for offender rehabilitation. Using the term "maturation" by no means implies an "ontogenetic" machine-like process of development that occurs at the same rate for each individual. As life-course criminology has demonstrated, individual trajectories of crime vary by individual. Some persist longer than others, while some barely engage in offending at all (Moffitt, 1993; Laub and Sampson, 2003). Thus it may be possible to "hasten maturation" by encouraging "benign maturation [to] occur earlier" (Glueck and Glueck, 1937: 205). It is to the policy implications of this maturation perspective that we now turn.

Offender rehabilitation from a maturation perspective

Clearly maturation is a "normative" process, meaning that for some (perhaps most) individuals, attainment of full maturation will be reached without intervention (see e.g. Levinson, 1978; Gove, 1985). This may be the basis for understanding Moffitt's (1993) notion of the adolescent-limited offender, a group of individuals who engage in crime but naturally desist. However, for some, more persistent offenders, intervention and treatment can be beneficial, as a large body of research has shown (Lipsey, 2009; Cullen, 2013). In addition, it may be possible to encourage maturation earlier than it would occur naturally for certain offenders. We describe offender rehabilitation implications and current programming from a maturation perspective, illustrating how viewing desistance and/or persistence from such a perspective clarifies how and why treatment may work.

The first domain of maturation focuses on the importance of adult social roles. As we have indicated, research has long shown that meaningful romantic relationships (e.g. marriage) are related to a decrease in offending for emerging adults (Sampson and Laub, 1993; Horney *et al.*, 1995; Laub and Sampson, 2003; Bersani *et al.*, 2009). While clearly it is not ethically feasible to assign romantic partners to offenders as part of treatment, it is possible to create policies and programs that encourage the sustainment of ties that already exist. Correctional practices that facilitate the maintenance of social bonds may help reduce recidivism. For example, recent work in corrections shows that more frequent visitation by family members reduces not only in-prison problem behavior, but also recidivism (see Bales and Mears, 2008; Mears *et al.*, 2012; Duwe and Clark, 2013). Spousal visitation may be especially important here (Mears *et al.*, 2012). Changing policy so that visitation becomes more acceptable can thus have a beneficial impact on post-release behavior, and state departments of corrections are taking notice (G. Duwe, personal communication).

Employment has long been a focus for prison rehabilitation efforts, as it is one of the strongest predictors of successful reintegration (Visher *et al.*, 2005). The maturation perspective demonstrates why meaningful employment may be related to later behavior. However, it is interesting to note that the literature on the effectiveness of work programs on recidivism is rather equivocal. Some work has found that work programs do not reduce recidivism (Visher *et al.*, 2005; Moses, 2012). However, an interesting finding in the employment–recidivism literature is that employment programs may not "take root" for more youthful offenders, who have not yet reached maturation in other domains (see Uggen, 2000). Thus understanding how and why employment may be related to desistance can help guide programming. Further, the desistance literature indicates that it is *meaningful* employment that matters for behavioral reform. Simply having a job may not have much substantive impact.

For this reason, perhaps vocational training and education (the completion of which is itself hypothesized to be a marker of adult status) may be more important than mere job placement. Vocational and educational training may assist ex-offenders in finding more interesting and stable work. The literature indicates that vocational training and education can be effective in reducing recidivism (Hull *et al.*, 2000; Wilson *et al.*, 2000; Gordon and Weldon, 2003). A meta-analysis conducted by the Rand Corporation demonstrated that inmates participating in education programs had a 42 percent lower odds of recidivism than those who did not, with cost-savings of roughly $8,000 per inmate (Davis *et al.*, 2013). Further, the effect of education and vocational training while in prison worked at least partially by increasing employability of the ex-offenders—which directly links this treatment strategy to adult social role maturation. Finally, a recent study finds that higher education programs (resulting in a college degree) may also reduce recidivism among confined offenders (Kim and Clark, 2013). In sum, from a maturational perspective, it is not just employment itself or even education that is important for recidivism, but rather meaningful employment and a certain level of educational attainment that sends a signal of adult status.

With respect to civic or citizenship maturation, several implications for offender rehabilitation can be explicated. Certain states have policies which, in effect, remove citizenship from offenders once they have been convicted of a felony. Convicted felons may not be able to hold office, vote, or serve on juries (Uggen *et al.*, 2004). The biosocial maturation perspective advanced here leads to the policy implication that these barriers to citizenship or "civic reintegration" are impediments to desistance. While research is scarce on the effect of these barriers on recidivism, some work has shown that they are a source of frustration to ex-offenders wanting to "make good" (Maruna, 2001; Uggen and Manza, 2004; Uggen *et al.*, 2004; Farrall and Calverley, 2006; Manza and Uggen, 2006). Thus it is logical that a policy implication of this perspective is to remove barriers to citizenship for ex-offenders. Another implication would be the development of programs for prisoners that encourage civic engagement and volunteer work. This work would help create the tie to society that may "bind" the individual to it.

Perhaps the most convincing evidence from the effectiveness research on offender rehabilitation literature relates to the psychosocial and cognitive transformation/identity maturation domains. It is no exaggeration to state that cognitive behavioral therapy (CBT) techniques represent the most effective tools in the rehabilitation toolbox (Lipsey *et al.*, 2007; Andrews and Bonta, 2010). Cognitive behavioral therapy comes in many varieties, but each stems from the notion that in order to change individual behavior, it is necessary to change thinking patterns and attitudes (Landenberger and Lipsey, 2005; Lipsey *et al.*, 2007). One example of such a program is the reasoning and rehabilitation program (Tong and Farrington, 2008). This program seeks to adjust offenders' maladaptive thinking styles to consider the consequences of their actions for themselves and others before acting. Thus this program targets impulsivity and perspective (components of psychosocial maturation) in an attempt to change behavior. In addition, CBT programs often try to change the way offenders see criminal behavior itself (from acceptable or justified to unacceptable), which is consistent with cognitive transformations described by Giordano *et al.* (2002).

We argue that therapy programs (including CBT) should focus specifically on identity in addition to attitudes toward crime. Research has demonstrated people often act in concordance with their self-perceptions (Matsueda, 1992; Sparks and Shepherd, 1992; Reynolds and Ceranic, 2007). This suggests that a key risk factor that rehabilitation programs may be able to address is the way in which offenders view themselves. To the extent that therapy can help instill prosocial identities, they will be more effective. Thus we argue that a primary goal of treatment should be efforts to modify entrenched antisocial identities. This could be done in the same vein as a psychotherapeutic intervention developed by Dewar and Gracey (2007), which helped improve the sense of self of a patient diagnosed with herpes simplex viral encephalitis. In this case, the patient had experienced memory loss and "threatened identity" (p. 604), and thus had a desire to seek treatment to adjust her identity. This may imply that offenders must be amenable to therapy to change their identity if such techniques are to be successful. This is consistent with the desistance scholarship which argues that personal agency or choice is a major factor in whether or not an individual stops offending.

Related to this notion of identity change, Ward and Marshall (2007) discuss what they call "narrative identity" in the context of offender rehabilitation. According to these authors, if an individual does not have a "clear narrative identity," they are less likely to desist from crime (p. 289). They argue that rehabilitation programs that help provide meaningful life activities and core values can assist in the prosocial identity building necessary for desistance. Supporting this argument, Stevens (2012) suggests that one reason "therapeutic communities" help to reduce recidivism is through providing a safe and secure environment in which the offender can shed their former antisocial identity and "a 'new' and 'better' person emerges for whom long-term desistance from crime is a viable and personally desirable achievement" (Stevens, 2012: 528). Therapeutic communities utilize group therapy in which offenders reflect together on the reasons they ended up in prison, and are given support to change their character in a prosocial manner. Stevens (2012) analyzes interviews of therapeutic community residents and finds that this treatment is successful in helping offenders establish new prosocial identities.

The fifth maturation domain focuses on cognitive or neurological factors. While this may be viewed initially as the least amenable to treatment (at least ethically) (see Vaske *et al.*, 2011), recent research has shown that this view is not an accurate assessment. For example, the budding field of epigenetics and work on neural plasticity has demonstrated that the brain adapts to changes in environment, even through adulthood, though brain plasticity is much reduced by then (Walsh, 2009; Bavelier *et al.*, 2010; Vaske *et al.*, 2011). Bavelier and colleagues suggest that certain activities and treatment may "remove the brakes" on brain plasticity that develop with age (such as the aforementioned change in white matter). It is this type of treatment that Fishbein (2000) may have been referring to when suggesting that neurorehabilitation could be useful as a crime control policy. For example, Fishbein describes the use of strategies typically used to measure neuropsychological deficits as methods of improving cognitive function. There is increasing interest in the use of cognitive exercises to strengthen prefrontal performance (Peretz *et al.*, 2011). These techniques may have implications for offender rehabilitation as well.

Cognitive behavioral therapy may be a key form of neurorehabilitation, impacting neurological maturation. While CBT has often been thought of as a purely psychological form of treatment, Vaske and colleagues (2011) demonstrate that it is likely effective, at least in part, because of changes such programs elicit in brain regions, particularly the prefrontal cortex:

> [A] wide range of studies implicate regions of the frontal cortex, parietal lobe, and temporal lobe in social skills, coping skills, and decision-making skills. More specifically, regions such as the medial prefrontal cortex (MPFC), dorsolateral prefrontal cortex (DLPFC), ventromedial prefrontal cortex (VMPFC) including the orbitofrontal cortex (OFC), cingulate cortex, temporoparietal junction (TPJ), and insula seem particularly important for a number of social skills, coping skills, and decision-making skills.
>
> *(Vaske et al., 2011: 92)*

These brain areas are also implicated in impulsivity, perspective taking, and other components of psychosocial maturation, indicating that any treatment targeting that maturation domain will likely have an influence on neurological functioning. This sort of overlap between which domain treatments impact should not be surprising, given that the maturation domains themselves are likely considerably interconnected. For example, it is likely that getting married or attaining a meaningful job would affect not only one's identity, but also sense of responsibility and self-control. Thus treatments targeting one domain logically should affect others.

Ward and Maruna's (2007) recent rehabilitation perspective, which they call the "Good Lives Model" (GLM), is compatible with the biosocial perspective advanced in this chapter (see also Ward and Stewart, 2003; Ward and Marshall, 2007). The GLM was developed as an alternative to the Risk Need Responsivity (RNR) approach popularized by Andrews and Bonta (2010). While addressing risk is seen as important in the GLM, this perspective also sees value in promoting individual "goods," which are defined as goals that offenders have for their lives. Thus there is a sense of agency and strength-based orientation in the GLM that is absent from the RNR approach. These "goods" may be tangible things such as a meaningful relationship or employment (adult social maturation), or the desire to be part of the larger community (civic maturation). As Ward and Marshall (2007) argue, the focus on goods in the GLM can help the individual construct a life and narrative identity that allows them to follow a path toward desistance.

We view the GLM approach as complementary to the maturation/developmental theory advanced in this chapter. Rather than focusing solely on containing risk, the GLM seeks to encourage individual growth in terms of goals and relationships. These goals and relationships are part of what creates self-value for the ex-offender, and allow him/her to be a fully functioning adult in the modern world. In other words, the GLM, from our perspective, helps to promote adult maturation.

Conclusion

In this chapter we present elements of a biosocial maturation perspective of desistance, and demonstrate how it can inform offender rehabilitation. To date, theories of desistance have been quite narrow in scope and consequently their relevance to policy has been circumscribed. Many theories focus on either social or psychological factors that are related to desistance from crime. These perspectives, we argue, have often been viewed as mutually exclusive or competing. However, it is possible to see each—along with recent neurological research—as pieces of the larger developmental process that can be called maturation. Thus ours is an integrative, multidisciplinary perspective that allows a greater understanding of why individuals eventually stop offending. This information is particularly relevant to offender rehabilitation.

Drawing on desistance literature and neurological research, we identify five "domains" of maturation. These domains include (1) adult social role maturation; (2) civic or citizenship maturation; (3) psychosocial maturation; (4) identity/cognitive transformation maturation; and (5) neurological maturation. While we specifically describe these domains separately, this is not intended to convey the impression that they are independent; on the contrary, it is likely that each is interdependent such that changes in one domain lead to, or are associated with, changes in another. Rehabilitation programs addressing these domains may very well have the largest impact on recidivism. Existing programs, including education, CBT, and vocational training, may be effective because of their influence on adult development. Programs that specifically address these domains, such as those which attempt to change an individual's antisocial identity, should be a focus of future research.

Because our perspective is most relevant to desistance from crime and is thus an adult-oriented theory, we do not focus here on crime prevention and other programs which may be effective in preventing individuals from initiating involvement in delinquency. However, certain crime prevention programs are important in ensuring normal development and "on-time" maturation.

Farrington and Welsh (2007) discuss developmental crime-prevention programs, which seek to prevent early delinquency by reducing risk factors and promoting protective factors such as healthy familial relationships and positive educational experiences. Because crime or delinquency is itself often a barrier to normal adult development (e.g. incarceration is negatively associated with higher education, marriage, and meaningful employment), crime prevention at an earlier life-stage is likely to encourage adult maturation, thus reducing crime throughout the life-course. Indeed, early crime prevention programs such as David Olds' Nurse Family Partnership, which helps educate at-risk parents about proper health, nutrition, and infancy development, have been found to prevent their children's crime through adolescence (Eckenrode *et al.*, 2010).

Further, as we have argued elsewhere (Rocque *et al.*, 2012, 2013, 2014), developmental programs work to prevent crime at least partially because of their effect on physiological processes such as neurological development. Thus these programs are likely to promote maturation in adulthood by ensuring healthy brain as well as social development which, in turn, reduces such things as impulsivity and school/occupational failure. Other, less researched strategies may also be relevant for crime prevention and brain maturation. Several studies have indicated the possibility that nutrition may influence problem behavior (Rocque *et al.*, 2013), and nutrition has been a key component of early childhood developmental programs (Raine *et al.*, 2003; Olds *et al.*, 2007). In addition, several studies have demonstrated in randomized trials that fish oil supplements reduce aggression (Itomura *et al.*, 2005; Hamazaki *et al.*, 2008; Kirby *et al.*, 2010). Future research should examine the effect of nutritional supplements on brain/cognitive maturation to determine the viability of incorporating nutrition in offender rehabilitation strategies.

This is indeed an exciting time in criminology. New areas of investigation are emerging and informing the field, including crime over the life-course, desistance from crime, and offender rehabilitation. We have presented an integrated biosocial view of desistance, and demonstrated how this view can inform offender rehabilitation. We anticipate that as new advances take place, our knowledge of the processes underlying desistance and rehabilitation will be expanded. The concept of maturation, we believe, should play an integral role in this work.

Acknowledgement

The authors would like to express their gratitude to Cheryl Miller for her assistance with the design of Figure 34.1.

References

Adams, K. (2004) 'Developmental aspects of adult crime', in T. Thornberry (ed.), *Developmental Theories of Crime and Delinquency*, Newark, NJ: Transaction Publishers.

Andrews, D.A. and Bonta, J. (2010) *The Psychology of Criminal Conduct*, New Providence, NJ: Matthew Bender.

Bales, W.D. and Mears, D.P. (2008) 'Inmate social ties and the transition to society: does visitation reduce recidivism?', *Journal of Research in Crime and Delinquency*, 45: 287–321.

Bavelier, D., Levi, D.M., Li, R.W., Dan, Y., and Hensch, T.K. (2010) 'Removing brakes on adult brain plasticity: from molecular to behavioral interventions', *Journal of Neuroscience*, 30: 14964–14971.

Bersani, B.E., Laub, J.H., and Nieuwbeerta, P. (2009) 'Marriage and desistance from crime in the Netherlands: do gender and socio-historical context matter?', *Journal of Quantitative Criminology*, 25: 3–24.

Block, J. (1971) *Lives Through Time*, Berkeley, CA: Bancroft Books.

Blonigen, D.M. (2010) 'Explaining the relationship between age and crime: contributions from the developmental literature on personality', *Clinical Psychology Review*, 30: 89–100.

Blonigen, D.M., Hicks, B.M., Krueger, R.F., Patrick, C.J., and Iacono, W.G. (2006) 'Continuity and change in psychopathic traits as measured via normal-range personality: a longitudinal-biometric study', *Journal of Abnormal Psychology*, 115: 85–95.

Blonigen, D.M., Carlson, M.D., Hicks, B.M., Krueger, R.F., and Iacono, W.G. (2008) 'Stability and change in personality traits from late adolescence to early adulthood: a longitudinal twin study', *Journal of Personality*, 76: 229–266.

Bushway, S.D., Piquero, A.R., Broidy, L.M., Cauffman, E., and Mazerolle, P. (2001) 'An empirical framework for studying desistance as a process', *Criminology*, 39: 491–516.

Bushway, S.D., Thornberry, T.P., and Krohn, M.D. (2003) 'Desistance as a developmental process: a comparison of static and dynamic approaches', *Journal of Quantitative Criminology*, 19: 129–153.

Caspi, A., Roberts, B., and Shiner, R.L. (2005) 'Personality development: stability and change', *Annual Review of Psychology*, 56: 453–484.

Cauffman, E. and Steinberg, L. (2000) '(Im)maturity of judgment in adolescence: why adolescents may be less culpable than adults', *Behavioral Sciences and the Law*, 18: 741–760.

Collins, R.E. (2004) 'Onset and desistance in criminal careers: neurobiology and the age–crime relationship', *Journal of Offender Rehabilitation*, 39(3): 1–19.

Cullen, F.T. (2013) 'Rehabilitation: beyond nothing works', *Crime and Justice*, 42: 299–376.

Davis, L.M., Bozick, R., Steele, J.L., Saunders, J., and Miles, J.N.V. (2013) *Evaluating the Effectiveness of Correctional Education: A Meta-Analysis of Programs that Provide Education to Incarcerated Adults*, Santa Monica, CA: RAND Corporation.

DeLisi, M. and Beaver, K.M. (2014) *Criminological Theory: A Life-course Approach*, Burlington, MA: Jones & Bartlett.

Dewar, B.K. and Gracey, F. (2007) '"Am not was": cognitive-behavioural therapy for adjustment and identity change following herpes simplex encephalitis', *Neuropsychological Rehabilitation*, 17: 602–620.

Duwe, G. and Clark, V. (2013) 'Blessed be the social tie that binds: the effects of prison visitation on offender recidivism', *Criminal Justice Policy Review*, 24: 271–296.

Eckenrode, J., Campa, M., Luckey, D.W., Henderson Jr, C.R., Cole, R., Kitzman, H., *et al.* (2010) 'Long-term effects of prenatal and infancy nurse home visitation on the life course of youths: 19-year follow-up of a randomized trial', *Archives of Pediatrics and Adolescent Medicine*, 164: 9–15.

Farrall, S. and Calverley, A. (2006) *Understanding Desistance from Crime: Theoretical Directions in Rehabilitation and Resettlement*, Maidenhead: Open University Press.

Farrington, D.P. and Welsh, B. (2007) *Saving Children from a Life of Crime: Early Risk Factors and Effective Interventions*, Oxford: Oxford University Press.

Fishbein, D. (2000) 'The importance of neurobiological research to the prevention of psychopathology', *Prevention Science*, 1: 89–106.

Giedd, J.N. (2008) 'The teen brain: insights from neuroimaging', *Journal of Adolescent Health*, 42: 335–343.
Giedd, J.N., Blumenthal, J., Jeffries, N.O., Castellanos, F.X., Liu, H., Zijdenbos, A., *et al.* (1999) 'Brain development during childhood and adolescence: a longitudinal MRI study', *Nature Neuroscience*, 2: 861–862.
Giordano, P.C., Cernkovich, S.A., and Rudolph, J.L. (2002) 'Gender, crime, and desistance: toward a theory of cognitive transformation', *American Journal of Sociology*, 107: 990–1064.
Glueck, S. (1964) 'Wanted: a comparative criminology', in S. Glueck and E. Glueck (eds), *Ventures in Criminology: Selected Recent Papers*, Cambridge: Cambridge University Press, pp. 306–322.
Glueck, S. and Glueck, E.T. (1937) *Later Criminal Careers*, New York: The Commonwealth Fund.
Glueck, S. and Glueck, E.T. (1940) *Juvenile Delinquents Grown Up*, New York: The Commonwealth Fund.
Glueck, S. and Glueck, E.T. (1945) *After-conduct of Discharged Offenders*, London: Macmillan.
Glueck, S. and Glueck, E.T. (1970) *Toward a Typology of Juvenile Offenders: Implications for Therapy and Prevention*, New York: Grune and Stratton.
Glueck, S. and Glueck, E.T. (1974) *Of Delinquency and Crime: A Panorama of Years of Search and Research*, Springfield, IL: Charles C. Thomas.
Gogtay, N., Giedd, J.N., Lusk, L., Hayashi, K.M., Greenstein, D., Vaituzis, A.C., *et al.* (2004) 'Dynamic mapping of human cortical development during childhood through early adulthood', *Proceedings of the National Academy of Sciences, USA*, 101: 8174–8179.
Gordon, H.R. and Weldon, B. (2003) 'The impact of career and technical education programs on adult offenders: learning behind bars', *Journal of Correctional Education*, 54: 200–209.
Gove, W.R. (1985) 'The effect of age and gender on deviant behavior: a biopsychosocial perspective', in A.S. Rossi (ed.), *Gender and the Life Course*, Chicago, IL: Aldine, pp. 115–144.
Hamazaki, K., Syafruddin, D., Tunru, I.S., Fazwir, M., Asih, P.B.S., Sawazaki, S., and Hamazaki, T. (2008) 'The effects of docosahexaenoic acid-rich fish oil on behavior, school attendance rate and malaria infection in school children – a double-blind, randomized, placebo-controlled trial in Lampung, Indonesia', *Asia Pacific Journal of Clinical Nutrition*, 17: 258–263.
Hill, T.W. (1971) 'From hell-raiser to family man', in J. Spradley and D.W. McCurdy (eds), *Conformity and Conflict: Readings in Cultural Anthropology*, Boston, MA: Little, Brown, pp. 186–200.
Hirschi, T. and Gottfredson, M.R. (1983) 'Age and the explanation of crime', *American Journal of Sociology*, 89: 552–584.
Horney, J., Osgood, D.W., and Marshall, I.H. (1995) 'Criminal careers in the short-term: intra-individual variability in crime and its relation to local life circumstances', *American Sociological Review*, 60: 655–673.
Hull, K.A., Forrester, S., Brown, J., Jobe, D., and McCullen, C. (2000) 'Analysis of recidivism rates for participants of the academic/vocational/transition education programs offered by the Virginia Department of Correctional Education', *Journal of Correctional Education*, 51: 256–261.
Itomura, M., Hamazaki, K., Sawazaki, S., Kobayashi, M., Terasawa, K., Watanabe, S., and Hamazaki, T. (2005) 'The effect of fish oil on physical aggression in schoolchildren—a randomized, double-blind, placebo-controlled trial', *Journal of Nutritional Biochemistry*, 16: 163–171.
Kazemian, L. (2007) 'Desistance from crime: theoretical, empirical, methodological, and policy considerations', *Journal of Contemporary Criminal Justice*, 23: 5–27.
Kim, R.H. and Clark, D. (2013) 'The effect of prison-based college education programs on recidivism: propensity score matching approach', *Journal of Criminal Justice*, 41: 196–204.
King, S. (2013) 'Early desistance narratives: a qualitative analysis of probationers' transitions towards desistance', *Punishment and Society*, 15: 147–165.
Kirby, A., Woodward, A., Jackson, S., Wang, Y., and Crawford, M. (2010) 'A double-blind, placebo-controlled study investigating the effects of omega-3 supplementation in children aged 8–10 years from a mainstream school population', *Research in Developmental Disabilities*, 31: 718–730.
Landenberger, N.A. and Lipsey, M.W. (2005) 'The positive effects of cognitive-behavioral programs for offenders: a meta-analysis of factors associated with effective treatment', *Journal of Experimental Criminology*, 1: 451–476.
Laub, J.H. and Sampson, R.J. (2001) 'Understanding desistance from crime', *Crime and Justice*, 28: 1–69.
Laub, J.H. and Sampson, R.J. (2003) *Shared Beginnings, Divergent Lives: Delinquent Boys to Age 70*, Cambridge, MA: Harvard University Press.
Laub, J.H., Nagin, D.S., and Sampson, R.J. (1998) 'Trajectories of change in criminal offending: good marriages and the desistance process', *American Sociological Review*, 63: 225–238.
Levinson, D. (1978) *The Seasons of a Man's Life*, New York: Ballantine Books.

Lipsey, M.W. (2009) 'The primary factors that characterize effective interventions with juvenile offenders: a meta-analytic overview', *Victims and Offenders*, 4: 124–147.

Lipsey, M.W., Landenberger, N.A., and Wilson, S.J. (2007) 'Effects of cognitive-behavioral programs for criminal offenders', *Campbell Systematic Reviews*, 6: 1–27.

Manza, J, and Uggen, C. (2006) *Locked Out: Felon Disenfranchisement and American Democracy*, New York: Oxford University Press.

Maruna, S. (1997) 'Going straight: desistance from crime and life narratives of reform', in A. Lieblich and R. Josselson (eds), *The Narrative Study of Lives*,Vol. 5,Thousand Oaks, CA: Sage, pp. 55–93.

Maruna, S. (2001) *Making Good: How Ex-convicts Reform and Rebuild their Lives*,Washington, DC: American Psychological Association.

Massoglia, M. and Uggen, C. (2007) 'Subjective desistance and the transition to adulthood', *Journal of Contemporary Criminal Justice*, 23: 90–103.

Matsueda, R.L. (1992) 'Reflected appraisals, parental labeling, and delinquency: specifying a symbolic interactionist theory', *American Journal of Sociology*, 55: 1577–1611.

Mears, D.P., Cochran, J.C., Siennick, S.E., and Bales, W.D. (2012) 'Prison visitation and recidivism', *Justice Quarterly*, 29: 888–918.

Modecki, K.L. (2008) 'Addressing gaps in the maturity of judgment literature: age differences and delinquency', *Law and Human Behavior*, 32: 78–91.

Moffitt, T.E. (1993) 'Adolescence-limited and life-course-persistent antisocial behavior: a developmental taxonomy', *Psychological Review*, 100(4): 674–701.

Monahan, K.C., Steinberg, L., Cauffman, E., and Mulvey, E.P. (2009) 'Trajectories of antisocial behavior and psychosocial maturity from adolescence to young adulthood', *Developmental Psychology*, 45: 1654–1668.

Moses, M.C. (2012) 'Ex-offender job placement programs do not reduce recidivism', *Corrections Today*, August/September: 106–108.

Mulvey, E.P., Steinberg, L., Fagan, J., Cauffman, E., Piquero, A.R., Chassin, L., *et al.* (2004) 'Theory and research on desistance from antisocial activity among serious adolescent offenders', *Youth Violence and Juvenile Justice*, 2(3): 213–236.

Nagin, D.S. and Land, K.C. (1993) 'Age, criminal careers, and population heterogeneity: specification and estimation of a nonparametric, mixed Poisson model', *Criminology*, 31: 327–362.

Olds, D.L., Kitzman, H., Hanks, C., Cole, R., Anson, E., Sidora-Arcoleo, K., *et al.* (2007) 'Effects of nurse home visiting on maternal and child functioning: age-9 follow-up of a randomized trial', *Pediatrics*, 120: e832–e845.

Paternoster, R. and Bushway, S. (2009) 'Desistance and the "feared self": toward an identity theory of criminal desistance', *Journal of Criminal Law and Criminology*, 99: 1103–1156.

Peretz, C., Korczyn, A.D., Shatil, E., Aharonson, V., Birnboim, S., and Giladi, N. (2011) 'Computer-based, personalized cognitive training versus classical computer games: a randomized double-blind prospective trial of cognitive stimulation', *Neuroepidemiology*, 36(2): 91–99.

Perry, B.D., Pollard, R.A., Blakley, T.L., Baker, W.L., and Vigilante, D. (1995) 'Childhood trauma, the neurobiology of adaptation, and use dependent development of the brain: how states become traits', *Infant Mental Health Journal*, 16: 271–291.

Raine, A. (2013) *The Anatomy of Violence: The Biological Roots of Crime*, New York: Random House.

Raine, A., Liu, J., Venables, P., and Mednick, S.A. (2003) 'Preventing crime and schizophrenia using early environmental enrichment', in A. Raine (ed.), *Crime and Schizophrenia: Causes and Cures*, New York: Nova Sciences.

Restak, R.M. (2001) *The Secret Life of the Brain*, New York: Joseph Henry Press.

Reynolds, S.J. and Ceranic, T.L. (2007) 'The effects of moral judgment and moral identity on moral behavior: an empirical examination of the moral individual', *Journal of Applied Psychology*, 92: 1610–1624.

Rocque, M., Welsh, B.C., and Raine, A. (2012) 'Biosocial criminology and modern crime prevention', *Journal of Criminal Justice*, 40: 306–312.

Rocque, M., Raine, A. and Welsh, B.C. (2013) 'Experimental neurocriminology: etiology and treatment', in B.C. Welsh, A.A. Braga, and G.J.N. Bruinsma (eds), *Experimental Criminology: Prospects for Advancing Science and Public Policy*, New York: Cambridge University Press, pp. 43–64.

Rocque, M., Welsh, B.C., and Raine, A. (2014) 'The biosocial explanation: policy implications of biosocial criminology: crime prevention and offender rehabilitation', in K.M. Beaver, J.C. Barnes, and B. Boutwell (eds), *The Nature Versus Biosocial Debate in Criminology: On the Origins of Criminal Behavior and Criminality*, Thousand Oaks, CA: Sage, ch. 26.

Sampson, R.J. and Laub, J.H. (1993) *Crime in the Making: Pathways and Turning Points through Life*, Cambridge, MA: Harvard University Press.
Sampson, R.J., Laub, J.H., and Wimer, C. (2006) 'Does marriage reduce crime? A counterfactual approach to within individual causal effects', *Criminology*, 44: 465–508.
Sparks, P. and Shepherd, R. (1992) 'Self-identity and the theory of planned behavior: assessing the role of identification with "green consumerism"', *Social Psychology Quarterly*, 55: 388–399.
Spear, L.P. (2007) 'The developing brain and adolescent-typical behavior patterns: an evolutionary approach', in D. Romer and E.F. Walker (eds), *Adolescent Psychopathology and the Developing Brain: Integrating Brain and Prevention Science*, Oxford: Oxford University Press, pp. 9–30.
Steinberg, L. (2008) 'A social neuroscience perspective on adolescent risk-taking', *Developmental Review*, 28: 78–106.
Steinberg, L. (2009) 'Should the science of adolescent brain development inform public policy?', *American Psychologist*, 64: 67–78.
Steinberg, L. (2010) 'Commentary: a behavioral scientist looks at the science of adolescent brain development', *Brain and Cognition*, 72: 160–164.
Stevens, A. (2012) '"I am the person now I was always meant to be": identity reconstruction and narrative reframing in therapeutic community prisons', *Criminology and Criminal Justice*, 12: 527–547.
Tong, J. and Farrington, D.P. (2008) 'The effectiveness of "reasoning and rehabilitation" in reducing reoffending', *Psicothema*, 20: 20–28.
Uggen, C. (2000) 'Work as a turning point in the life course of criminals: a duration model of age, employment, and recidivism', *American Sociological Review*, 65: 529–546.
Uggen, C. and Inderbitzin, M. (2010) 'Public criminologies', *Criminology and Public Policy*, 9: 725–749.
Uggen, C. and Manza, J. (2004) 'Voting and subsequent crime and arrest: evidence from a community sample', *Columbia Human Rights Law Review*, 36: 193–215.
Uggen, C. and Massoglia, M. (2003) 'Desistance from crime and deviance as a turning point in the life course', in J.T. Mortimer and M.J. Shanahan (eds), *Handbook of Life Course*, New York: Springer, pp. 311–329.
Uggen, C., Manza, J., and Behrens, A. (2004) 'Less than the average citizen: stigma, role transition, and the civic reintegration of convicted felons', in S. Maruna and R. Immarigeon (eds), *After Crime and Punishment: Pathways to Offender Reintegration*, Cullompton, UK: Willan Publishing, pp. 258–290.
Vaske, J., Galyean, K., and Cullen, F.T. (2011) 'Toward a biosocial theory of offender rehabilitation: why does cognitive-behavioral therapy work?', *Journal of Criminal Justice*, 39: 90–102.
Visher, C.A., Winterfield, L., and Coggeshall, M.B. (2005) 'Ex-offender employment programs and recidivism: a meta-analysis', *Journal of Experimental Criminology*, 1: 295–316.
Walsh, A. (2009) 'Crazy by design: a biosocial approach to the age–crime curve', in A. Walsh and K.M. Beaver (eds), *Biosocial Criminology: New Directions in Theory and Research*, New York: Springer, pp. 154–175.
Walsh, A. and Beaver, K.M. (2009) *Biosocial Criminology*, New York: Springer.
Ward, T. and Marshall, B. (2007) 'Narrative identity and offender rehabilitation', *International Journal of Offender Therapy and Comparative Criminology*, 51: 279–297.
Ward, T. and Maruna, S. (2007) *Rehabilitation*, New York: Routledge.
Ward, T. and Stewart, C. (2003) 'Criminogenic needs and human needs: a theoretical model', *Psychology, Crime and Law*, 9: 125–143.
Wilson, D.B., Gallagher, C.A., and MacKenzie, D.L. (2000) 'A meta-analysis of corrections-based education, vocation, and work programs for adult offenders', *Journal of Research in Crime and Delinquency*, 37: 347–368.
Wootton, B. (1962) *Social Science and Social Pathology*, London: Allen & Unwin.
Yamaguchi, K. and Kandel, D.B. (1985) 'On the resolution of role incompatibility: a life event history analysis of family roles and marijuana use', *American Journal of Sociology*, 90: 1284–1325.

35

Antisocial cognition and criminal thinking

Theoretical interpretations and biosocial implications

Glenn D. Walters

Introduction

Antisocial cognition is both a risk factor and a criminogenic need. Along with antisocial history, antisocial personality processes, and antisocial peer associations, it is one of the "big four" predictors of recidivism (Andrews and Bonta, 2010). It is also a major criminogenic need and, as such, a principal target for correctional intervention (Andrews *et al.*, 2006). There is more to antisocial cognition, however, than its ability to predict recidivism and guide intervention. It is also capable of informing and shaping criminological theory—specifically, biosocial theory. The purpose of this chapter is to highlight five contributions antisocial cognition and criminal thinking make to criminological theory in general and to biosocial theories of crime in particular.

Content versus process

Theoretical interpretations

Much of the research and theory on antisocial cognition has focused principally on criminal thinking, a construct that varies as a function of both content (what an offender thinks) and process (how an offender thinks). Criminal thought content can be assessed with the three attitudes toward the law, court, and police subscales of the Criminal Sentiments Scale-Modified (CSS-M; Gendreau *et al.*, 1979; Simourd, 1997) or three of the four scales (violence, antisocial intent, and associates) from Section Two of the Measures of Criminal Attitudes and Associates (MCAA; Mills *et al.*, 2002) inventory. Criminal thought process can best be measured with the criminal thinking or higher-order scales of the Psychological Inventory of Criminal Thinking Styles (PICTS; Walters, 1995). All three measures—the CSS-M, MCAA, and PICTS—are based on respondent self-report. A thorough evaluation of criminal thinking requires that both criminal thought content and criminal thought process be measured, compared, and integrated.

Research conducted on the CSS-M has shown that this 41-item measure correlates modestly with future recidivism, although results have varied widely across groups and outcomes

(Simourd and Olver, 2002). Simourd and van de Ven (1999), for instance, determined that the CSS-M predicted recidivism in violent offenders but not in non-violent offenders, whereas Witte *et al.* (2006) ascertained that the CSS-M predicted non-sexual recidivism but not sexual recidivism in a group of sex offenders. The MCAA has also been found to correlate with subsequent offending in released prisoners. In the only prospective study to date, Mills *et al.* (2004) ascertained that the MCAA possessed incremental validity relative to an actuarial risk measure for the purpose of predicting both general and violent recidivism.

The PICTS is composed of eight thinking-style scales, two higher-order scales (proactive and reactive criminal thinking), and a superordinate scale (general criminal thinking). Brief descriptions of each scale are provided in Table 35.1. A recent item response theory (IRT) analysis of the PICTS revealed that the sentimentality (Sn) scale failed to load onto a general criminal thinking factor or either of two (proactive and reactive) subfactors (Walters *et al.*, 2011). Accordingly, only seven of the original eight thinking styles are construed to be measures of criminal thinking. More importantly, these seven scales are not as effective or consistent in predicting future recidivism as are the higher-order (P, R) and superordinate (GCT) scales. In a meta-analysis of recidivism prediction using the P, R, and GCT scales, Walters (2012b) ascertained that all three scales achieved low–medium effect sizes ($r = 0.17–0.20$) in predicting general and violent recidivism.

Table 35.1 Thinking, higher-order, and superordinate scales for the psychological inventory of criminal thinking styles

Scale	*Description*
Mollification (Mo)	Blaming external circumstances, including others, for the negative consequences of one's own actions and life decisions.
Cutoff (Co)	Rapid elimination of common deterrents to crime via drugs, poor frustration tolerance, or a simple phrase (e.g. "fuck it").
Entitlement (En)	Belief that one is above the law and deserving of special treatment, often accompanied by a misidentification of wants as needs.
Power orientation (Po)	Drive for control over the interpersonal environment coupled with a tendency to put others down in order to feel better about oneself.
Sentimentality (Sn)	Performing good deeds in an effort to erase the harm caused by one's involvement in a criminal lifestyle.
Superoptimism (So)	Unrealistic belief that one can indefinitely escape the negative consequences of one's long-term involvement in criminal activity.
Cognitive indolence (Ci)	Failure to critically evaluate one's ideas and plans, thus leading to recklessness, impulsivity, and poor decision making.
Discontinuity (Ds)	Tendency to be sidetracked by outside events, making it difficult to achieve goals and follow through on good initial intentions.
Proactive (P)	Criminal thinking that is planful, instrumental, goal-directed, and cold-blooded.
Reactive (R)	Criminal thinking that is emotional, spontaneous, impulsive, and hot-blooded.
General (GCT)	General criminal thinking encompassing both the proactive and reactive dimensions.

I know of only one study in which participants were administered both a measure of criminal thought content and a measure of criminal thought process. In that study, Morgan *et al.* (2010) administered the CSS-M and PICTS to a large sample of mentally disturbed male and female inmates housed in several Texas Department of Criminal Justice facilities. Crossing the three best content measures from the CSS-M (attitudes toward the law, attitudes toward the court, attitudes toward the police) with the two higher-order scales from the PICTS (P, R) revealed significant correlations between all five scales, although the within-measure correlations were three times higher than between-measure correlations (see Table 35.2). This suggests that content and process measures of criminal thinking are related but far from identical.

Biosocial implications

Criminologists have long been skeptical of biological interpretations of crime. We should not let this blind us, however, to the possibility that biological factors play a significant role in the etiology of antisocial behavior. There is a growing body of research, in fact, to suggest that biological factors, genetics in particular, contribute to the development of delinquency, criminality, and general antisocial behavior (Walters, 1992; Rowe and Farrington, 1997; Rhee and Waldman, 2002). In a recent study addressing this issue, Beaver (2013) discovered that 50 percent of the adult crimes committed by participants in a nationally representative survey were committed by participants from just 5 percent of surveyed families. Beaver also identified support for the intergenerational transmission of crime, an effect he attributed to genetics and assortative mating.

With respect to criminal thought content versus criminal thought process, it is speculated that the latter may be more influenced by genetic factors than the former. I base this supposition on the notion that where thought content is shaped largely by experience (nurture), thought process is a function of both heredity and experience (nature and nurture). Unfortunately, there is virtually no research on the genetic foundations of thought content and process, though Williams (2012) has identified genetically based sex differences in two cognitive processes: counting and time perception. In extending this line of research to antisocial cognition, it will be important to use measures that optimally assess criminal thought content—for example, the attitudes scales of the CSS-M, and criminal thought process—perhaps something along the lines of the PICTS P scale.

Table 35.2 Descriptive statistics and correlational matrix for the PICTS P and R scales and CSS attitudes subscales

Statistic	*M*	*SD*	*PICTS-R*	*CSS-L*	*CSS-C*	*CSS-P*
PICTS proactive (P) scale	66.14	18.42	0.69*	0.23*	0.14*	0.18*
PICTS reactive (R) scale	60.97	15.14		0.12*	0.16*	0.18*
CSS attitudes toward law	6.07	4.30			0.50*	0.60*
CSS attitudes toward court	9.44	3.75				0.60*
CSS attitudes toward police	5.92	3.49				

Notes: *M* = mean; SD = standard deviation; PICTS-R = Psychological Inventory of Criminal Thinking Styles Reactive Scale; CSS-L = Criminal Sentiments Scale-Attitudes Toward the Law subscale; CSS-C = Criminal Sentiments Scale-Attitudes Toward the Court subscale; CSS-P = Criminal Sentiments Scale-Attitudes Toward the Police subscale; *N* = 512.

*$P < 0.005$ (significant using a Bonferroni correction: 0.05 alpha level/10 individual correlations).

Latent structure

Theoretical interpretations

There are two questions that must be answered whenever one seeks information on the latent or underlying structure of constructs. These two questions can be labeled latent structure type (categorical versus continuous) and latent structure model (the number of categories or the number of dimensions). Different statistical procedures are required for each question. According to the results of a large Monte Carlo investigation, Meehl's (1995) taxometric method is optimal for answering questions about latent structure type, whereas latent class analysis and factor analysis are required for answering questions about latent structure model (McGrath and Walters, 2012). Hence, if latent structure type is found to be categorical, then latent class analysis should be used to identify the number of categories; and if latent structure type is found to be continuous, then exploratory or confirmatory factor analysis should be used to identify the number of dimensions.

Two published reports have examined the latent structure type of criminal thought process using the taxometric method. In the first of these studies, Walters (2007a) ascertained that criminal thought process, as measured by the eight PICTS thinking-style scales, showed consistent evidence of dimensional latent structure in 771 male prisoners from three different institutions (a maximum security federal penitentiary, a medium security federal correctional institution, and a minimum security federal prison camp). Dimensional latent structure also surfaced when the PICTS thinking-style scales were analyzed in mixed groups of male and female prison inmates and college students (Walters and McCoy, 2007). It should be noted that these dimensional results persisted even when the taxometric analyses were restricted to the seven PICTS thinking-style scales that have been found to load onto a general criminal thinking factor (i.e. the eight criminal thinking-style scales minus sentimentality).

Although no published studies investigating the latent structure type of the CSS-M are available, data from the previously mentioned Morgan *et al.* (2010) study were subjected to taxometric analysis for the purposes of this chapter. Participants were 517 mentally ill male and female inmates from the Morgan *et al.* sample who completed the CSS-M. Indicators for this analysis were the three CSS-M attitudes subscales (attitudes toward the law, attitudes toward the court, and attitudes toward the police). These data were analyzed in three stages using standard non-redundant taxometric procedures: mean above minus below a cut (MAMBAC), maximum covariance (MAXCOV), and latent model factor analysis (L-mode).

First, cases were divided into putative taxon and complement groups using a cut score obtained by averaging the base rate estimates achieved with the MAMBAC and MAXCOV procedures (taxon BR = 0.535). Second, pre-taxometric analyses were conducted, the results of which indicated significantly higher inter-indicator correlations in the full sample (r = 0.56) compared with the putative taxon (r = 0.17) and complement (r = 0.21) subsamples. This, along with adequate indicator validity (d = 1.73–2.08), denotes that the current data were appropriate for taxometric analysis. Third, the comparison curve fix index (CCFI) results (MAMBAC CCFI = 0.283; MAXCOV CCFI = 0.341; L-mode CCFI = 0.102; mean CCFI = 0.242) fell into the range associated with continuous or dimensional latent structure (CCFI < 0.450; Ruscio *et al.*, 2010). The actual and comparison curves are reproduced in Figure 35.1.

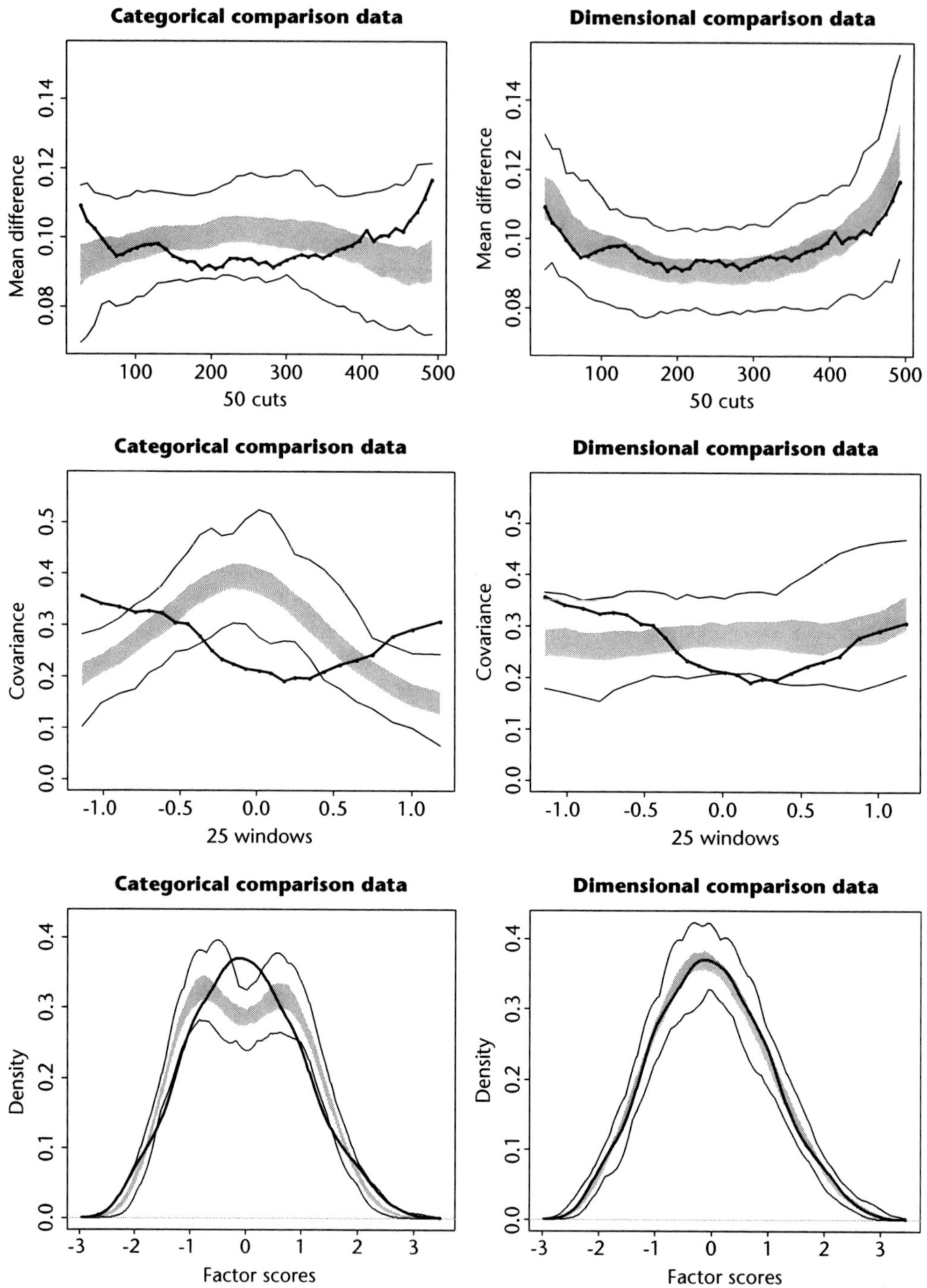

Figure 35.1 Taxometric graphs obtained when the three attitudes subscales of the Criminal Sentiments Scale-Modified (CSS-M) served as indicators and a putative taxon base rate of 0.535 was used

Notes: Results are shown for MAMBAC (mean above minus below a cut; top panel); MAXCOV (maximum covariance; middle panel); and L-mode (latent mode; bottom panel) procedures. In each graph, the dark line shows results for the empirical data, lighter lines connect the minimum and maximum values at each data point, and the gray area represents the middle 50 percent of values. For each procedure, the results for the empirical data resemble those for the dimensional (continuous) comparison data more closely than those for the categorical comparison data.

Biosocial implications

The results of taxometric research have important implications for theory, research, and practice. For instance, taxometric results can provide clues on a construct's etiology. Whereas knowing the latent structure type of a construct tells us nothing about causal content, it has the capacity to inform causal process. According to Meehl (1992), discontinuous or categorical latent structure supports one of four major causal pathways: specific etiology, a threshold effect, nonlinear interaction, or developmental bifurcation. The causal origins of a continuous or dimensional construct, by comparison, are additive. In the case of continuous latent structure, a construct is caused by multiple factors, which in isolation or interaction increase a person's risk of involvement in the pattern. Hence, if criminal thinking is a continuous construct, then individual differences in criminal thinking are quantitative rather than qualitative (a matter of degree rather than a difference in kind) and caused by an accumulation of many risk factors, each of which contributes only a small fraction to the overall liability for criminal thinking.

Risk factors for criminal thinking run the gamut from the purely biologic to the wholly environmental. Interactions between risk factors may be even more important than individual risk factors in determining liability for criminal thinking. This is particularly true of the biologically based person × situation interactions that are of cardinal significance in biosocial theory. Childhood temperament, the product of ongoing gene–environment interactions, may be a particularly salient risk factor for criminal thinking. First, research indicates that childhood temperament has a continuous rather than discontinuous latent structure, which, in turn, suggests that temperament is more appropriately organized into dimensions than into types (Walters, 2011). Second, several temperament/personality dimensions have been found to correlate with criminal thought process: high neuroticism and low agreeableness and conscientiousness, in particular (Otter and Egan, 2007; Bulten *et al.*, 2009).

Lifestyle theory—the model upon which the PICTS and the concept of criminal thought process are based—postulates the existence of two higher-order temperament dimensions: fearlessness and disinhibition. Each higher-order dimension is composed of several lower-level subdimensions. Fearlessness, for instance, is composed of the lower-level temperament dimensions of low negative emotionality (anxiety, depression, and worry, but not anger), callousness, and sociability. Disinhibition, on the other hand, is composed of the lower-level temperament dimensions of weak effortful control, impulsivity, and sensation-seeking (Walters, 2012a, 2014). Comprehending the additive effect of divergent risk factors in the etiology, development, and progression of criminal thinking and behavior requires a thorough understanding of these higher-order temperament dimensions and their lower-level subdimensions.

Proactive and reactive criminal thinking

Theoretical interpretations

Taxometrics may be optimal for determining whether the latent structure of a construct is continuous or categorical, but it is of little use in identifying the number of dimensions or categories in a construct (McGrath and Walters, 2012). Once it has been determined through taxometric analysis that a construct such as criminal thinking is dimensional, the next step is to use factor analysis to verify the number of factors. In a series of confirmatory factor analyses, Walters and colleagues identified two dimensions: a proactive criminal thinking dimension, and a reactive criminal thinking dimension (Walters, 2008, 2009b; Walters *et al.*, 2009). These two factors, while highly correlated, appear to measure different aspects of general criminal thought process. Whereas proactive criminal thinking is instrumental, calculated, goal-directed, and cold-blooded, reactive criminal thinking is hostile, spontaneous, impulsive, and hot-blooded.

Although the proactive (P) and reactive (R) scales of the PICTS correlate strongly (r = 0.69–0.75; Walters and Geyer, 2005; Walters, 2008; Walters *et al.*, 2009, 2011), they typically display divergent patterns of association with sundry precursor, concurrent, and outcome measures. For instance, whereas P correlates significantly with prior arrests for proactive aggression (robbery, burglary), R correlates significantly with prior arrests for reactive aggression (assault, domestic violence: Walters *et al.*, 2007). Likewise, P has been shown to correlate with positive outcome expectancies for crime but not with hostile attribution biases, whereas R correlates with hostile attribution biases but not positive outcome expectancies for crime (Walters, 2007b). With respect to intervention, the R scale but not the P scale dropped significantly over the course of a 6-week anger management class (Walters, 2009a). Finally, R was found to correlate significantly with program dropout in one study (Walters, 2004) and P was found to correlate with program completion in a second study (Walters and Geyer, 2005).

Walters (2007b) modeled proactive and reactive adult criminal thinking after Dodge and Coie's (1987) work on proactive and reactive childhood aggression. It is noteworthy that, like proactive and reactive criminal thinking, proactive and reactive childhood aggression correlate highly (Poulin and Boivin, 2000) and demonstrate countervailing relationships with positive outcome expectancies for aggression and hostile attribution biases (i.e. proactive childhood aggression correlates with positive outcome expectancies for aggression but not hostile attribution biases, whereas reactive childhood aggression correlates with hostile attribution biases but not positive outcome expectancies for aggression; Crick and Dodge, 1996). Developmental research is required, however, to determine whether childhood patterns of proactive and reactive aggression lead to adult patterns of proactive and reactive criminal thinking and behavior, and whether both are differentially linked to fearlessness and disinhibition as proposed by lifestyle theory (Walters, 2014; see Figure 35.2).

Biosocial implications

Haller (2013) recently reviewed animal and human research on the neuroanatomical and neurophysiological correlates of proactive or hypoarousal-related aggression and reactive or hyperarousal-related aggression. In hypoarousal-related aggression, the central amygdala–lateral hypothalamus–ventral periaqueductal gray pathway is activated in a manner similar to what occurs when an animal initiates a predatory attack (Tulogdi *et al.*, 2010). Neurophysiologically,

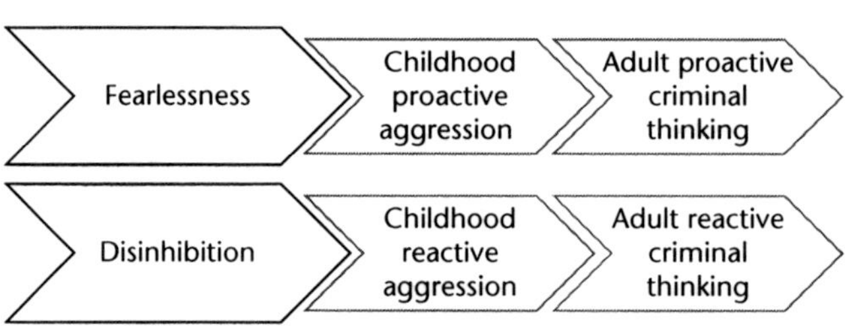

Figure 35.2 Developmental foundations of adult proactive and reactive criminal thinking proposed by lifestyle theory

vasopressinergic neurotransmission is diminished and serotonin levels are elevated in hypoarousal-related aggression. Conversely, in hyperarousal-related aggression the medial amygdala–mediobasal hypothalamus–dorsal periaqueductal gray pathway is activated, vasopressinergic neurotransmission is elevated, and serotonergic control over the hypothalamus is weak (Toth *et al.*, 2012).

From the results of his review, Haller (2013) concludes that abnormal aggression cannot be considered a unitary phenomenon. He goes on to state that the "neural backgrounds of various forms of abnormal aggression show qualitative differences, undermining the tacit assumption that the brain mechanisms of normal and abnormal aggression differ in quantitative terms only" (p. 107). This conclusion conflicts, however, with a growing body of taxometric research showing that in addition to criminal thinking, childhood aggression (Walters *et al.*, 2010), child and adolescent antisocial behavior (Walters and Ruscio, 2013), and psychopathy (Edens *et al.*, 2006) all display continuous latent structure and, as such, individual differences on these variables are quantitative rather than qualitative in nature. Reconciling the taxometric and neurological results on proactive and reactive aggression and criminality is therefore a prime area for future research.

Hierarchical organization

Theoretical interpretations

Like the steps on a ladder, objects, items, or concepts are positioned at different levels of a hierarchy to form superordinate, subordinate, and parallel relationships. A subordinate item or construct derives from a superordinate item or construct, whereas a parallel relationship exists between two items or constructs at the same level. A hierarchy was previously described in the latent structure section of this chapter on temperament. As the reader may recall, fearlessness and disinhibition sit at the top of the temperament hierarchy and low negative emotionality (anxiety and depression), callousness, and sociability fall under fearlessness, whereas weak effortful control, impulsivity, and sensation-seeking fall under disinhibition. A second hierarchy was also briefly mentioned in this chapter, this one involving the PICTS. Three levels were suggested: general criminal thinking, proactive and reactive criminal thinking, and the seven individual thinking styles.

Before discussing the PICTS hierarchy, it may be helpful to examine a general hierarchy of cognitive processes. As portrayed in Figure 35.3, belief systems are situated at the top of the hierarchy, under which can be found various schematic subnetworks. Schematic subnetworks are groups of thoughts connected by a common purpose or theme (one of which is thinking styles). The other five schematic subnetworks proposed by lifestyle theory are attributions, efficacy expectancies, outcome expectancies, goals, and values. Individual criminal thoughts or schemes sit at the bottom of the hierarchy. Schemes join to form schematic subnetworks, which then combine to form belief systems. Influence can go from higher to lower levels, a process known as deduction, or from lower to higher levels, a process known as induction.

General criminal thinking is situated at the top of the PICTS hierarchy. This is the superordinate factor in the hierarchy. At the next level of the PICTS hierarchy, just below the superordinate factor, are the two higher-order factors: proactive and reactive criminal thinking. The seven individual thinking-style scales that load onto the superordinate factor and one of the two higher-order factors fall at the bottom of the PICTS hierarchy (see Figure 35.4). As is the case with the general cognitive hierarchy, the PICTS hierarchy holds that influence can travel from

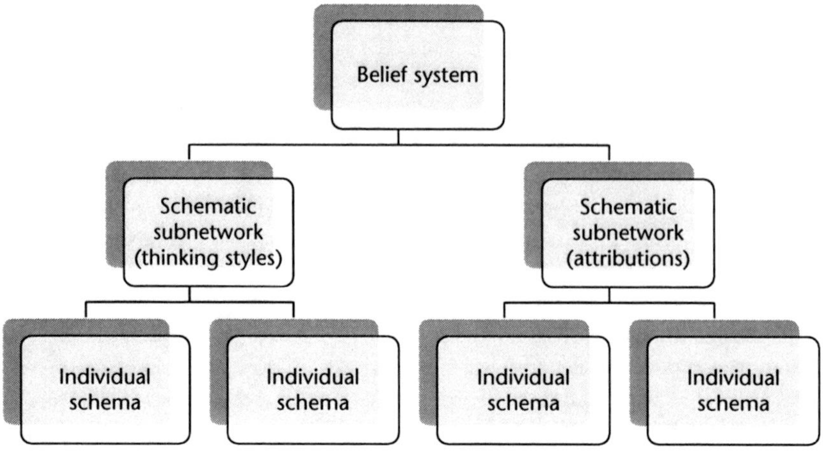

Figure 35.3 Hierarchical organization of general criminal cognition

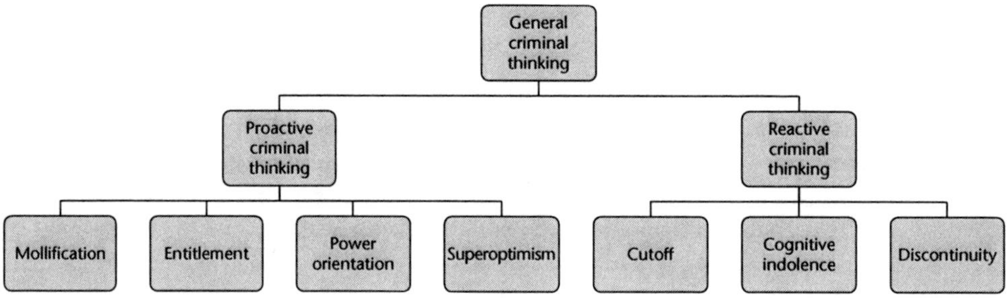

Figure 35.4 Hierarchical organization of the psychological inventory of criminal thinking styles (PICTS)

lower levels to higher levels (induction), or from higher levels to lower levels (deduction). When interpreting the PICTS clinically, one is advised to start at the superordinate level and then work down to the higher-order factors, followed by the individual thinking styles.

Biosocial implications

Hierarchical models clearly serve a heuristic function in that they make it easier to understand and conceptualize complex relationships and patterns. Beyond their heuristic value, however, is there something about hierarchies that speaks to the nature of human cognition? The results of a recent study by Farooqui *et al.* (2012) suggest that hierarchies may possess more than just heuristic value. In a series of three experiments, Farooqui *et al.* (2012) determined that activation of the frontoparietal region of the human cortex corresponded directly with the hierarchical cognitive demands of a visual target-detection task. This suggests that human cognition may be inherently hierarchical, and that greater amounts of cortical activity are required to deal with tasks at the higher levels of the hierarchy (general criminal thinking) than at the lower levels of the hierarchy (mollification or discontinuity). Of course, this is a hypothesis that requires additional study, but if found to be true it could lead to increased understanding of the mechanisms behind antisocial cognition.

Antisocial cognition as a third variable

The conceptual fragmentation and overreliance on single-variable theories that led Ericson and Carriere (1994) to criticize the fields of criminology and criminal justice is just as true today as it was 20 years ago. A solution exists, nevertheless, in the form of mediation and moderation analysis whereby third variables are integrated into existing single variable theories. The first order of business when conducting a mediation or moderation analysis is understanding the difference between mediation and moderation (Baron and Kenney, 1986). Whereas the third variable in a mediation analysis interposes itself between the independent and dependent variables, the third variable in a moderation analysis alters the strength or direction of the relationship between the independent and dependent variables. In his social cognitive theory of personality, Bandura (1986) positioned cognitive factors principally, though not entirely, in the role of mediating variables, and social factors in the role of both mediating and moderating variables. Biological factors, on the other hand, could serve as independent, mediating, or moderating variables.

One of the "brute facts" of crime is that past criminality is among the best predictors of future criminality. Although this may prove the old psychological adage that the best predictor of future behavior is past behavior, it still does not qualify as an explanation. In searching for an explanation, Walters and colleagues tested four of the six previously mentioned schematic subnetworks as possible mediators of crime continuity. In one study, criminal thinking (PICTS GCT score) and low self-efficacy to avoid future police contact partially mediated the relationship between past and future criminality (Walters, 2013a). In a second study, hedonistic values and short-term goals partially mediated the relationship between past and future criminality (Walters, 2013b). In a third study, general antisocial cognition partially mediated the relationship between juvenile delinquency and adult crime when pre-existing differences on variables central to four major criminological theories (low self-control, delinquent peers, maternal support, intelligence) were controlled (Walters and DeLisi, 2013).

The mediating effect of antisocial cognition is a special case of a more general process known as psychological inertia (Walters, 2012a). Newton's first law of motion holds that a body at rest will remain at rest and a body in motion will remain in motion unless acted upon by an outside force. Psychological inertia is the psychological expression of this law, meaning that once a process starts it is self-maintaining until acted upon by an outside force (change). The six schematic subnetworks of lifestyle theory (i.e. thinking styles, attributions, efficacy expectancies, outcome expectancies, goals, and values) contribute to psychological inertia, but so do non-cognitive factors like enabling, conditioning, and reinforcement. Hence a parent's enabling behavior or a pattern's reinforcing nature can be just as effective in maintaining a criminal lifestyle as an attitude of entitlement (thinking style), self-labeling (attribution), or lack of confidence in one's ability to work a regular job (efficacy expectancies). Behavioral change requires that these agents of psychological inertia be neutralized.

Biosocial implications

Mediation and moderation do not necessarily operate independently. There are some designs, in fact, where mediation and moderation coincide. Moderation can be mediated in a process known as mediated moderation, and mediation can be moderated in a process known as moderated mediation. Musher-Eizenman *et al.* (2004) observed a moderated mediated effect in a study exploring the relationship between environmental and emotional regulation and aggressive behavior. Social cognition was found to mediate the environmental/emotional regulation-aggressive behavior relationship, and biological sex was found to moderate this mediated relationship, such that retaliation approval beliefs served as the mediator for girls and self-evaluation mediated the relationship in boys. The moderated mediation design offers researchers the opportunity to examine biological (sex) and social/cognitive (social cognition) variables simultaneously, thus opening the door to more complex analyses and more sophisticated theories.

A mediated moderation analysis also provides researchers with the opportunity to examine interactions and mediated relationships between biological, social/cognitive, and criminal justice variables. Mediated moderation, in fact, was supported by the results of a study in which researchers examined the effect of an interaction between genotype (presence of the DRD4 l-allele and 5-HTTLPR s-allele) and adverse social environment on subsequent aggression (Simons *et al.*, 2011). In this study, the moderated effect of genotype on the adverse social environment–aggression relationship was mediated by a latent construct referred to as hostile orientation. Several conditions are required when conducting moderated mediation and mediated moderation analyses. First, proper temporal order must be established by means of a longitudinal design. Second, special statistical analyses are required to determine whether a proposed moderated mediation or mediated moderation effect exists (Preacher *et al.*, 2007). While both conditions were satisfied in the Simons *et al.* (2011) study, neither condition was satisfied in the Musher-Eizenman *et al.* (2004) cross-sectional study. This illustrates that while moderated mediation and mediated moderation analyses can be helpful in clarifying relationships between biological and social/cognitive variables, they need to be properly implemented to provide trustworthy results.

Conclusion

As the information presented in this chapter indicates, antisocial cognition and criminal thinking have a great deal to offer biosocial theories of crime. To reach its full potential, however, research on antisocial cognition will need to expand beyond criminal thought content and process to include other important schematic subnetworks: attributions, efficacy expectancies, outcome expectancies, goals, and values. In addition, more complex research designs, such as those involving moderated mediation and mediated moderation, are required to model the complex relationships that exist between biological, social, and cognitive variables in the formation of early antisocial behavior, adolescent delinquency, and adult criminality.

Acknowledgements

I would like to thank Robert Morgan for allowing the use of data he and his colleagues have collected on the Criminal Sentiments Scale-Modified (CSS-M) and Psychological Inventory of Criminal Thinking Styles (PICTS) to clarify certain points made in this chapter.

References

Andrews, D.A. and Bonta, J. (2010) 'Rehabilitating criminal justice policy and practice', *Psychology, Public Policy, and Law*, 16: 39–55.

Andrews, D.A., Bonta, J., and Wormith, J.S. (2006) 'The recent past and near future of risk and/or need assessment', *Crime & Delinquency*, 52: 7–27.

Bandura, A. (1986) *Social Foundations of Thought and Action: A Social Cognitive Theory*, Englewood Cliffs, NJ: Prentice Hall.

Baron, R.M. and Kenney, D.A. (1986) 'The moderator–mediator variable distinction in social psychological research: conceptual, strategic, and statistical considerations', *Journal of Personality and Social Psychology*, 51: 1173–1182.

Beaver, K.M. (2013) 'The familial concentration and transmission of crime', *Criminal Justice and Behavior*, 40: 35–49.

Bulten, E., Nijman, H., and van der Staak, C. (2009) 'Measuring criminal thinking styles: the construct validity and utility of the PICTS in a Dutch prison sample', *Legal and Criminological Psychology*, 14: 35–49.

Crick, N.R. and Dodge, K.A. (1996) 'Social information-processing mechanisms in reactive and proactive aggression', *Child Development*, 67: 993–1002.

Dodge, K.A. and Coie, J.D. (1987) 'Social-information processing factors in reactive and proactive aggression in children's peer groups', *Journal of Personality and Social Psychology*, 53: 1146–1158.

Edens, J.F., Marcus, D.K., Lilienfeld, S.O., and Poythress, N.G. (2006) 'Psychopathic, not psychopath: taxometric evidence for the dimensional structure of psychopathy', *Journal of Abnormal Psychology*, 115: 131–144.

Ericson, R. and Carriere, K. (1994) 'The fragmentation of criminology', in D. Nelken (ed.), *The Futures of Criminology*, London: Sage.

Farooqui, A.A., Mitchell, D., Thompson, R., and Duncan, J. (2012) 'Hierarchical organization of cognition reflected in distributed frontoparietal activity', *Journal of Neuroscience*, 32: 17373–17381.

Gendreau, P., Grant, B.A., Leipciger, M., and Collins, S. (1979) 'Norms and recidivism for the MMPI and selected experimental scales on a Canadian delinquent sample', *Canadian Journal of Behavioural Science*, 11(1): 21–31.

Haller, J. (2013) 'The neurobiology of abnormal manifestations of aggression—a review of hypothalamic mechanisms in cats, rodents, and humans', *Brain Research Bulletin*, 93: 97–109.

McGrath, R.E. and Walters, G.D. (2012) 'Taxometric analysis as a general strategy for distinguishing categorical from dimensional latent structure', *Psychological Methods*, 17: 284–293.

Meehl, P.E. (1992) 'Factors and taxa, traits and types, differences of degree and differences in kind', *Journal of Personality*, 60: 117–174.

Meehl, P.E. (1995) 'Bootstraps taxometrics: solving the classification problem in psychopathology', *American Psychologist*, 50: 266–275.

Mills, J.F., Kroner, D.G., and Forthe, A.E. (2002) 'Measures of Criminal Attitudes and Associations (MCAA): development, factor structure, reliability, and validity', *Assessment*, 9: 240–253.

Mills, J.F., Kroner, D.G., and Hemmati, T. (2004) 'The Measures of Criminal Attitudes and Associates (MCAA): the prediction of general and violent recidivism', *Criminal Justice and Behavior*, 31: 717–733.

Morgan, R.D., Fisher, W.H., Duan, N., Mandracchia, J.T., and Murray, D. (2010) 'Prevalence of criminal thinking among state prison inmates with serious mental illness', *Law and Human Behavior*, 34: 324–335.

Musher-Eizenman, D.R., Boxer, P., Danner, S., Dubow, E.F., Goldstein, S.E., and Donna, M.L. (2004) 'Social-cognitive mediators of the relation of environmental and emotion regulation factors to children's aggression', *Aggressive Behavior*, 30: 389–408.

Otter, Z. and Egan, V. (2007) 'The evolutionary role of self-deception enhancement as a protective factor against antisocial cognitions', *Personality and Individual Differences*, 43: 2258–2269.

Poulin, F. and Boivin, M. (2000) 'Reactive and proactive aggression: evidence of a two-factor model', *Psychological Assessment*, 12: 115–122.

Preacher, K.J., Rucker, D.D., and Hayes, A.F. (2007) 'Addressing moderated mediation hypotheses: theory, methods, and prescriptions', *Multivariate Behavioral Research*, 42: 185–227.

Rhee, S.H. and Waldman, I.D. (2002) 'Genetic and environmental influences on antisocial behavior: a meta-analysis of twin and adoption studies', *Psychological Bulletin*, 128: 490–529.

Rowe, D.C. and Farrington, D.P. (1997) 'The familial transmission of criminal convictions', *Criminology*, 35: 177–202.

Ruscio, J., Walters, G.D., Markus, D., and Kaczetow, W. (2010) 'Comparing the relative fit of categorical and dimensional latent variable models using consistency tests', *Psychological Assessment*, 22: 5–21.

Simons, R.L, Lei, M.K., Beach, S.R.H., Brody, G.H., Philibert, R.A., and Gibbons, F.X. (2011) 'Social environment, genes, and aggression: evidence supporting the differential susceptibility perspective', *American Sociological Review*, 76: 883–912.

Simourd, D.J. (1997) 'The Criminal Sentiments Scale–Modified and Pride in Delinquency Scale: psychometric properties and construct validity of two measures of criminal attitudes', *Criminal Justice and Behavior*, 24: 52–70.

Simourd, D.J. and Olver, M.E. (2002) 'The future of criminal attitudes research and practice', *Criminal Justice and Behavior*, 29: 427–446.

Simourd, D.J. and van de Ven, J. (1999) 'Assessment of criminal attitudes: criterion-related validity of the Criminal Sentiments Scale–Modified and Pride in Delinquency Scale', *Criminal Justice and Behavior*, 26: 90–106.

Toth, M., Tulogdi, A., Biro, L., Soros, P., Mikics, E., and Haller, J. (2012) 'The neural background of hyper-emotional aggression induced by post-weaning social isolation', *Behavioural Brain Research*, 233: 120–129.

Tulogdi, A., Toth, M., Halasz, J., Mikics, E., Fuzesi, T., and Haller, J. (2010) 'Brain mechanisms involved in predatory aggression are activated in a laboratory model of violent intra-specific aggression', *European Journal of Neuroscience*, 32: 1744–1753.

Walters, G.D. (1992) 'A meta-analysis of the gene-crime relationship', *Criminology*, 30: 595–613.

Walters, G.D. (1995) 'The Psychological Inventory of Criminal Thinking Styles: part I. Reliability and preliminary validity', *Criminal Justice and Behavior*, 22: 307–325.

Walters, G.D. (2004) 'Predictors of early termination in a prison-based program of psychoeducation', *Prison Journal*, 84: 171–183.

Walters, G.D. (2007a) 'The latent structure of the criminal lifestyle: a taxometric analysis of the Lifestyle Criminality Screening Form and Psychological Inventory of Criminal Thinking Styles', *Criminal Justice and Behavior*, 34: 1623–1637.

Walters, G.D. (2007b) 'Measuring proactive and reactive criminal thinking with the PICTS: correlations with outcome expectancies and hostile attribution biases', *Journal of Interpersonal Violence*, 22: 371–385.

Walters, G.D. (2008) 'Self-report measures of psychopathy, antisocial personality, and criminal lifestyle: testing and validating a two-dimensional model', *Criminal Justice and Behavior*, 35: 1459–1483.

Walters, G.D. (2009a) 'Anger management training in incarcerated male offenders: differential impact on proactive and reactive criminal thinking', *International Journal of Forensic Mental Health*, 8: 214–217.

Walters, G.D. (2009b) 'Latent structure of a two-dimensional model of antisocial personality disorder: construct validation and taxometric analysis', *Journal of Personality Disorders*, 23: 647–660.

Walters, G.D. (2011) 'Childhood temperament: dimensions or types?', *Personality and Individual Differences*, 50: 1168–1173.

Walters, G.D. (2012a) *Crime in a Psychological Context: From Career Criminals to Criminal Careers*, Thousand Oaks, CA: Sage.

Walters, G.D. (2012b) 'Criminal thinking and recidivism: meta-analytic evidence on the predictive and incremental validity of the Psychological Inventory of Criminal Thinking Styles (PICTS)', *Aggression and Violent Behavior*, 17: 272–278.

Walters, G.D. (2013a) 'Cognitive mediation of crime continuity: a causal mediation analysis of the past crime–future crime relationship', *Crime and Delinquency*, doi: 0011128712470987.

Walters, G.D. (2013b) 'Short-term goals and physically hedonistic values as mediators of the past crime–future crime relationship', *Legal and Criminological Psychology*, doi: 10.1111/lcrp.12014.

Walters, G.D. (2014) *Drugs, Crime, and their Relationships: Theory, Research, Practice, and Policy*, Sudbury, MA: Jones & Bartlett.

Walters, G.D. and DeLisi, M. (2013) 'Antisocial cognition and crime continuity: cognitive mediation of the past crime–future crime relationship', *Journal of Criminal Justice*, 41: 135–140.

Walters, G.D. and Geyer, M.D. (2005) 'Construct validity of the Psychological Inventory of Criminal Thinking Styles in relationship to the PAI, disciplinary adjustment, and program completion', *Journal of Personality Assessment*, 84: 252–260.

Walters, G.D. and McCoy, K. (2007) 'Taxometric analysis of the Psychological Inventory of Criminal Thinking Styles in incarcerated offenders and college students', *Criminal Justice and Behavior*, 34: 781–793.

Walters, G.D. and Ruscio, J. (2103) 'Trajectories of youthful antisocial behavior: categories or continua?', *Journal of Abnormal Child Psychology*, 41: 653–666.

Walters, G.D., Frederick, A.A., and Schlauch, C. (2007) 'Postdicting arrests for proactive and reactive aggression with the PICTS proactive and reactive scales', *Journal of Interpersonal Violence*, 22: 1415–1430.

Walters, G.D., Felix, C.M., and Reinoehl, R. (2009) 'Replicability and cross-gender invariance of a two-dimensional model of antisociality in male and female college students', *Personality and Individual Differences*, 46: 704–708.

Walters, G.D., Ronen, T., and Rosenbaum, M. (2010) 'The latent structure of childhood aggression: a taxometric analysis of self-reported and teacher-rated aggression in Israeli schoolchildren', *Psychological Assessment*, 22: 628–637.

Walters, G.D., Hagman, B.T., and Cohn, A.M. (2011) 'Towards a hierarchical model of criminal thinking: evidence from item response theory and confirmatory factor analysis', *Psychological Assessment*, 23: 925–935.

Williams, C.L. (2012) 'Sex differences in counting and timing', *Frontiers in Integrative Neuroscience*, January 5, doi: 10.3389/fnint.2011.00088

Witte, T.D., Di Placido, C., Gu, D., and Wong, S.C. (2006) 'An investigation of the validity and reliability of the Criminal Sentiments Scale in a sample of treated sex offenders', *Sexual Abuse: A Journal of Research and Treatment*, 18: 249–258.

36

Sexual homicide

A clinical and investigative analysis

Stacey L. Shipley and Bruce A. Arrigo

Introduction

Sexual homicide is considered among the most depraved of human behaviors (Palermo, 2004; Purcell and Arrigo, 2006). By undertaking an examination of it, we seek to understand the darkest side of aberrant thought and degenerate conduct. How can one person inflict such callous brutality upon another? Some argue that the very instinct of self-preservation or survival necessitates aggressive and sexual impulses (e.g. Zillmann, 1998; Schlesinger, 2003; Pennington, 2005). However, when these two powerful drives are fused without the requisite prosocial attachments and restraints, various forms of sexual sadism can surface and various types of sexual homicide can occur.

Here, we first define sexual murder as a clinical and investigative construct. This discussion addresses the construct's relationship to sadistic rape and serial homicide. Next, we review the relevant empirical and anecdotal evidence on the bio-psychosocial correlates of serial murder. We emphasize how these correlates help to explain the etiology of criminal motive and the ritualized complexities of ongoing sadistic victimization and eroticized violence. We then examine the classic FBI research on (serial) sexual homicide, mindful of those risk factors that contribute to the manifestation and maintenance of this offender type. For the subjects of the FBI study, these factors consisted of a chaotic family history, childhood sexual and/or physical abuse, past traumatic and/or humiliating events, deviant fantasies, escalating paraphilic behavior, a preoccupation with murderous thoughts, the decision to kill, and victim profiling. We conclude by offering several provisional recommendations for future clinical science and investigative practice directions regarding the under-examined phenomenon of sexual homicide.

Sexual murder: clinical and investigative dynamics

What is sexual murder? Based on the definition developed by Ressler *et al.*:

> *sexual homicide* describes murders with evidence or observations that indicate that the murder was sexual in nature. These include: victim attire or lack of attire; exposure of the sexual

> parts of the victim's body; sexual positioning of the victim's body; insertion of foreign objects into the victim's body cavities; evidence of sexual intercourse (oral, anal, vaginal); and evidence of substitute sexual activity, interest, or sadistic fantasy ... the killing of a person in the context of power, sexuality, and brutality.
>
> *(Ressler et al., 1992: xiii)*

The actual extent of sexual homicide is difficult to discern, largely because of how crimes are investigated and categorized. Sexual homicides can appear to lack a motive, to be altogether random, and can be misclassified as insufficiently driven by, or altogether devoid of, erotic impulses (Hazelwood, 2009). Meloy (2000) estimates that sexual murder occurs in fewer than 1 percent of homicides in the United States, but its sensational nature creates a disproportionate amount of public fear (also see Hickey, 2006).

Sexual homicides are almost exclusively perpetrated by men. However, sexual sadists sometimes use their consensual sex partner as a compliant accomplice to assist them in luring and killing unsuspecting female victims (Meloy, 2000). Very few scientific studies exist that examine and/or explain the etiology, onset, and maintenance of sexual murder (Purcell and Arrigo, 2006). However, based on a review of the empirical and anecdotal scholarship to date, two prominent types of sexual homicide are prevalent. One type consists of rapists who murder as a forensic countermeasure or rapists whose displaced anger amounts to retaliatory killing (Groth *et al.*, 1977; Rada, 1978; Ressler *et al.*, 1992; Pardue and Arrigo, 2008; Shipley and Arrigo, 2008; Hazelwood, 2009). The other type includes sexual sadists who kill as a necessary condition for the experience of sexual arousal and/or gratification to occur. Individuals placed within this offender classification are called lust murderers (Ressler *et al.*, 1992; Hazelwood *et al.*, 1993; Hazelwood and Michaud, 2001; Purcell and Arigo, 2006; Hazelwood, 2009). For the purpose of this section, we specifically review the lust murderer type, given its relationship to sadistic rape and serial homicide.

The sexually sadistic or lust murderer gains eroticized satisfaction from the victim's suffering, as well as from the violent aspects of the murder itself. For example, the sadistic attacker may participate in or otherwise perform postmortem sexual acts with the victim, especially if the killing involves playing out (staging) a well rehearsed and ritualized fantasy (Groth *et al.*, 1977; Rada, 1978; Ressler *et al.*, 1992). As Ressler *et al.* (1992: 6) reported, "For this murderer, aggression and sexuality become fused into a single psychological experience—sadism—in which aggression is eroticized."

The sadistic rapist experiences eroticized violence as satisfying and, as a category of offending, it is most closely correlated with sexual homicide. Sexual and aggressive urges are joined together as one, in which the would-be killer is aroused by the victim's degradation, injury, suffering, torture, and mutilation (Groth and Birnbaum, 2001; Purcell and Arrigo, 2006; Shipley and Arrigo, 2008). Psychopathy and sadism are typically observed in the sexual homicide offender. In this instance, the assailant's actions—including those behaviors that are deviant and sexual in nature—exhibit an absence of empathy, guilt, or remorse because the motivation to kill is compelled, in part, by the rush of sensation-seeking conduct (Shipley and Arrigo, 2008, 2013; Shipley and Russell, 2013), especially conduct that is of a paraphilic nature (Hickey, 2005). Paraphilias can be defined as "... recurrent, intense, sexually arousing fantasies, sexual urges, or behaviors generally involving (1) nonhuman objects, (2) the suffering of oneself or one's partner, or (3) children or other nonconsenting partners ..." (Hazelwood and Warren, 2009: 57). This offender typically uses restraints, relies on bondage, and inflicts significant bodily harm on the victim (Purcell and Arrigo, 2006). In some cases, the eroticized deviance can escalate resulting in death (Kocsis *et al.*, 2002). This attacker's victim-instructions are cold and detached. The offender is obsessively ritualistic and routinely habitual in his behavior.

For sadistic rapists whose cycle of violence results in the victim's death, mutilation is commonly featured on sexually significant areas of the woman's corpse (e.g. vagina, breasts) (Purcell and Arrigo, 2006). Bizarre acts of torture, dismemberment, necrophilia, and other forms of extreme paraphilia are consistently noted (Hazelwood and Burgess, 1987; NCWP, 2004; Hazelwood, 2009). The assault is prolonged and the weapon of choice is a knife. Drug or alcohol use during the offense is minimal so the attacker remains in control. The sadistic rapist usually possesses average to above-average intelligence, and he may work as a white-collar employee. It is not uncommon for this victimizer to be married; however, it is likely that his wife is firmly under his control, if not a compliant victim herself (NCWP, 2004). This offender is typically an outgoing, white male, who is respected by his peers. His sadistic fantasies are compulsive and covert.

It is not uncommon for the sexual homicide offender to have his acts of brutality audiotaped, videotaped, digitally reproduced, and/or documented in a journal in order to repeatedly relive the experience of violent and morbid sadistic sexuality (Hazelwood and Michaud, 2001; Hickey, 2005; Hazelwood, 2009). The offender is almost never psychotic; in other words, he rarely experiences delusions, hallucinations, or disorganized thinking. The depravity inherent in sexual homicides is extremely disturbing, and it is this aberrance that is considered outside the bounds of a developed human conscience (Zillmann, 1998; Meloy, 2000; Pennington, 2005). The biopsychosocial source for this degeneracy is generally traceable by malignant personality patterns, maladaptive coping styles, and destructive interactions with the social and interpersonal world (Hickey, 2005, 2006)

It should be pointed out that rape murderers arguably are distinct from lust murders in terms of individual motives and personality dynamics (Rada, 1978; Liebert, 1985; Hickey, 2005). According to Rada (1978):

> The rapist who murders rarely reports any sexual satisfaction from the murder and does not perform sexual acts on the dead victim. Rape murder occurs subsequent to sexual assault and does not appear to have sexual connotations itself. The lust murderer, on the other hand, frequently needs the murder to arouse his sexual interest and desires. Often he does not perform intercourse with the victim, dead or alive. He may, however, experience intense sexual pleasure and orgasm at the time of the murder – some compulsively disembowel the victim and masturbate to orgasm ... Rapists are capable of murder but typically for different reasons. In some rapists, however, there appears to be a progressive increase in aggressive fantasies toward women which over time may eventually lead to murder.
>
> *(cited in Liebert, 1985: 189–190)*

Notwithstanding the above distinction, the most brutal dimension of these rape crimes is the role of sexual sadism, in which the assailant is not only dismissive of his victim's agonizing suffering, but is also exhilarated, aroused, and propelled by it (Hickey, 2005; Shipley and Arrigo, 2008; Hazelwood, 2009). With each successive rape, sadistic deviance and sexual violence will likely increase and intensify until, eventually, it results in murder. When these actions are repetitively sustained over time, the behavior becomes serial sexual homicide.

Holmes and Holmes (1999) developed a typology for serial murderers that included the *hedonistic type*. These offenders kill as a result of sensation-seeking or they otherwise derive some sort of sexually charged pleasure from their homicides. The researchers divided this type of offender into two subcategories: the *lust killer* and the *thrill killer*.

The lust killer murders principally for sexual gratification, even if this does not entail traditional intercourse (Holmes and Holmes, 1999). However, sex or multiple sadistically erotic acts with a live victim are common (Arrigo and Purcell, 2001). Orgasm or sexually arousing behavior (i.e. masturbation) is the driving force for this offender, even after the person has killed his victim (Canter and Wentink, 2004; Hickey, 2005). Moreover, this attacker may also be sexually excited and/or satisfied from the murder itself. Ritualistic displays of sexual mutilation, facial disfigurement, cannibalism, body dismemberment, vampirism, and necrophilia are routinely featured in this type of homicidal act (Purcell and Arrigo, 2006). The body is often concealed and the murder weapon removed. Close contact homicide, specifically beating or manual strangulation, are most frequently reported (Holmes and DeBurger, 1988).

The thrill killer murders for the visceral experience, for the rush of excitement. However, once the victim is dead, the offender loses interest in the deceased subject (Canter and Wentink, 2004). As a result, the process of killing is prolonged for as long as possible through extended acts of torture (Holmes and Holmes, 2002). Characteristic behaviors for this type of killer include a reliance on restraints, the infliction of bite marks, and the placement of burns on the victim's body. The pattern of behaviors that indicates motive for this type of assailant consists of sadistic acts whose duration is attenuated as long as possible prior to death, a concealed corpse, manual or ligature strangulation, and an animated victim during multiple sexual acts.

Serial murder: on motive and bio-psychosocial correlates

The Federal Bureau of Investigation (FBI) defines serial murder as the act of committing "three or more separate [killings] with an emotional cooling-off period between homicides" (Ressler *et al.*, 1992: 139). Serial homicides can span in time from a period of hours to years. The motive is frequently fantasy driven, and the perpetrator's behavior often suggests sadistic, sexual underpinnings as evidenced by the crime scene. The attacker's debasement of the victim and the sexual overtones of the killing can be far from subtle, especially when the victim's legs are set in a spread-eagled position to deliberately shock those who discover her.

The FBI has reported that murder accounts for only 2 percent of all violent crime, and serial murder accounts for only a fraction of that percentage (Morton and Hilts, 2013). Because of their skill at remaining undetected, exact figures on the prevalence of serial killers and the incidence of their victimization are unavailable. The motivation to kill serially can be traced to an overwhelming need to satisfy internal wishes. These desires stem from violent sexual fantasies the attacker has rehearsed repeatedly since his youth (Hazelwood, 2009). For the serial sexual murderer, witnessing the pain and humiliation of the victim and ultimately controlling the person's life or death is extremely arousing, enjoyable, and exhilarating. The crimes are identified as sexual in nature, even if actual intercourse does not occur (e.g. Hazelwood, 2009).

Revitch and Schlesinger (1989) developed the concept of a motivational spectrum or continuum in classifying serial sexual murder. In their classification schema, one end of the spectrum included homicides committed as a product of external (sociogenic or environmental) factors; the other end of the continuum consisted of homicides committed as a product of internal (endogenous or psychogenic/psychological) pressures that compel the offender to act (Schlesinger, 1998). These compulsive homicides are typically ritualistic, with fantasies that likely preceded the career killing by many years. Revitch and Schlesinger (1989) maintained that it is quite typical of compulsive serial murderers to harbor underlying and unresolved sexual conflict. As Schlesinger noted:

> [Compulsive serial murderers typically exhibit] . . . a combination of hostility to[ward] women, preoccupation with maternal sexual conduct, overt or covert incestuous preoccupation, guilt over sex and rejection of sex as impure, and feelings of sexual inferiority. Some unhealthy emotional involvement with the mother has been found in many of these cases. The mother may be rejecting and punitive, or, to the contrary, seductive, at times openly so. In some cases, the child may have experienced or witnessed the mother's promiscuity. In adult serial murderers, hostility to women is the outstanding characteristic.
>
> *(Schlesinger, 1998: 213)*

The role of fantasy in serial murder is frequently recognized in the literature as a central component to the crime's ongoing commission (Prentky *et al.*, 1989; Myers *et al.*, 1993; Hazelwood, 2009). Image-conjuring is rehearsal for action, and the intensification of the visualizations compels the assailant to engage in criminal behavior. When fantasies that connect violence with sexual arousal are normalized for the purpose of masturbation and orgasm, then this coupling fuels the all-consuming compulsion to act aggressively and sadistically (i.e. to kill repetitively). Typically, this fusion of violence and sexual drive takes place at a young age, and it is most often associated with eroticized experiences of extreme humiliation, shame, and/or physical injury whose onset took place in early childhood (Hickey, 2006).

Consider the following example. The link between sex and degradation can inadvertently form from the combination of bedwetting and destructive parenting (DeHart and Mahoney, 1994). Enuresis or bedwetting is frequently identified in the past histories of serial killers (Hickey, 2006); however, it is not correlated with serial homicide *per se*. For serial murderers, enuresis is a product of psychological conditioning, sourced in the experience of intense parental shaming that occurred in childhood. A boy who wets his bed and endures severe physical beatings and/or harsh verbal assaults may feel, as a result, both hatred for and sexualized feelings toward his mother. These feelings surface given how the mother forcibly and painfully washes the child's genitals while, at the same time, verbally berating him. Additionally, he may have retaliatory fantasies, where the imagined act of violent murder is intensely arousing sexually. In this instance, a conditioned combination of sex and aggression surfaces whereby the individual learns to anticipate sexuality as a complement to violence (DeHart and Mahoney, 1994). Money (1990: 28) described this phenomenon in the context of neurobiology, stating, ". . . the brain becomes pathologically activated to transmit messages of attack simultaneously with messages of sexual arousal and mating behavior."

Hale (1994) argued that the role of humiliation and embarrassment are critical motivational features for the serial murderer. The victim elicits for the assailant the memories of someone who was responsible for demeaning and diminishing him as a child. The feelings of humiliation build, until eventually they are transformed into rage. It is at this point that the offender destroys the victim in an attempt to restore a state of emotional equilibrium and to re-establish a sense of power and control. Despite the grandiose and callous nature of most serial murderers, their demeanor and behaviors are likely a defense to guard against extreme feelings of inadequacy (Hickey, 2006). As Drukteinis (1992: 534) observed, "The aberrant sexual activity of the adult is seen as a triumph, converting an original childhood trauma into success and mastery." And as Kernberg (1982: 512) notes, [A sense of] ". . . triumph over fear and pain [emerges,] achieved by inflicting fear and pain on others . . . Some of these narcissistic personalities may pursue joyful types of cruelty." Generally, the profile of the serial homicide offender indicates that he is motivated by God-like (omnipotent) feelings, especially given his power to sustain or finish another person's life (Hickey, 2006). He may come to idealize his pathology as he revels in his feelings of prowess and superiority (Carlisle, 1993).

Lust murderers report experiencing escalating and increasingly violent, compulsive fantasies (Blanchard, 1995; Hickey, 2005; Hazelwood, 2009); rejection and abandonment fears (Norris, 1989; Blanchard, 1995; Hickey, 2005); impaired attachments, loneliness and isolation; and intense feelings of inadequacy (Hickey, 2005). Some lust murderers have a history of child abuse, and many come from chaotic home environments (Norris, 1989; Blanchard, 1995; Hazelwood, 2009). Their need for dominance is prominent (Hickey, 2005). Their violent personalities are often compartmentalized, and they can keep their aggressiveness effectively hidden from others (Liebert, 1985; Holmes and DeBurger, 1988; Blanchard, 1995; Purcell and Arrigo, 2006; Hazelwood, 2009).

Malignant narcissism has been described as pathological grandiosity resulting from rejection and humiliation and an overwhelming need for power and control (Meloy, 2000). Liebert (1985: 192) described the role of malignant narcissism in serial murder from a psychodynamic perspective. He stated that the narcissistic individual has:

> . . . incorporated too much of the bad from the maternal relationship and can split this introjected badness from his own personality and perceive it as originating from the outside . . . the individual no longer possesses the "badness" – it is the other person, the female victim who "has it". He may either project his introjected, dissociated badness on to his victim and justify his own violence or displace his violence toward his bad mother on to the victim and destroy the mother's badness.
>
> *(Liebert, 1985: 192)*

Much like the serial rapist, the serial murderer can blend in with those around him. His ability to project a facade of normalcy and respectability allows this assailant to avoid detection. Liebert described this dual life in the following manner:

> . . . the unobtrusive, superficially adjusted individual in society who appears to behave and relate normally but is incapable of true social integration, true intimacy or consistent psychological integration. It is this individual who may appear to be a "good family man" and a "good citizen" but one driven by a crescendoing evolution of antisocial, aggressive, sexual impulsivity into a repetitious pattern of lust murder.
>
> *(Liebert, 1985: 194)*

Moreover, while sadistic serial rapists and lust murderers are quite similar in several important respects (Arrigo and Purcell, 2001; Graney and Arrigo, 2002; Shipley and Arrigo, 2008), for serial killers the homicide is primary; the sexual assault is secondary (Holmes and Holmes, 2002; Shipley and Arrigo, 2008). The elaborate scenes of death erotically stimulate and satisfy this offender (Hickey, 2005). Still, for other serial murderers, the act of killing is a necessity. It is a natural and essential progression of the mutilation and torture that takes place as linked to the sexual assault.

Serial homicide offenders almost always target strangers manipulatively or through a ruse or pretense, in order to gain the confidence of and access to their victims. For example, serial killers such as Ted Bundy, Jeffrey Dahmer, John Wayne Gacy, and Gary Ridgway all primarily used deception as the preferred method by which to interact with their respective victims (Egger,

2002). This strategy was employed in order to guarantee the victim's trust and to establish a situation where the attack could take place (Holmes and Holmes 2002; Purcell and Arrigo, 2006). In cases of lust murder, the victim's face may be covered in order to establish dehumanization and/or depersonalization (Kocsis *et al.*, 2002; Hickey, 2006). The facial concealment may also be an act of undoing, especially if post-offense remorse exists. The vast majority of serial murderers engage in ritualistic behaviors that satisfy a psychological need; however, these behaviors are not essential to the crime's completion or to the avoidance of detection. Often, underlying sexual motives exist that are inextricably fused with violence. These groups likely consist of a disproportionate number of male offenders who experience sexual dysfunctions (Schlesinger, 2003; Palermo, 2004). In these cases, it is not uncommon for a foreign object to be inserted into a vaginal and/or anal cavity as emblematic of the sex act (Hazelwood, 2005).

A large number of serial murders are committed by a small number of offenders (Morton and Hilts, 2013). The sexual homicide assailant relies on a *modus operandi* (MO) or behavioral methods that are needed to conduct the offense and to avoid apprehension. These behaviors change based on criminal experience, education, situational needs, and relatively static rituals (Shipley and Arrigo, 2008; Hazelwood, 2009). These rituals constitute a repeated pattern of behavior that is unnecessary to the commission of the crime, but provides psychosexual arousal and gratification, while complementing the motive (Hazelwood, 2009).

The FBI study: classic research on sexual homicide

The rare occurrence or low frequency of sexual homicide makes predictive statistical analysis impossible (Morton and Hilts, 2013). Research on sexual homicide is also limited by the use of small, nonrandom samples and retrospective data; yet the information gleaned over the past 100 years continues to be relevant (Meloy, 2000). In 1982, with a grant from the National Institute of Justice, an FBI study was conducted based primarily on a qualitative research design (Ressler *et al.*, 1992). Specifically, the researchers used a case-review method, direct observation, and first-hand investigative interviews, to examine 36 convicted and incarcerated sexual murderers. The purpose of the study was to determine general offender characteristics. Data were collected between 1979 and 1983 by agents from the FBI's Behavioral Science Unit (BSU) and by agents trained by the BSU. Most of the respondents from this all-male sample were white, seven had been convicted of murdering a single person, and the remaining 29 were found guilty of killing multiple subjects. Based on the self-report responses, 118 fatalities (mostly female) occurred; nine women survived their attacks and were identified as attempted-murder victims. The FBI study was considered to be the first to examine sexual murderers collectively as a distinct subgroup of killers from a law enforcement perspective (Ressler *et al.*, 1992; Hickey, 2005; Purcell and Arrigo, 2006). Because self-reporting was the primary means of data collection, the following information regarding the experiences of these sexual murderers must be positioned within the context of *their perspective*.

The family histories for the subjects of the study were overwhelmingly determined to be chaotic and troubled, with substance abuse listed as a major problem. Seventy percent of the respondents had a home life involving alcohol abuse, and one-third had a drug abuse history. The presence of mental illness in the family was also noted for over half of the sample. The men in the study also identified aggression in their home life as an important contributor to their sexualized acts of killing. Twenty-six of the 36 men had a history of early psychiatric problems, as well as a history of aggressive acts. Twenty-five of the 36 sexual murderers had some type of psychiatric assessment or confinement as a child or adolescent. The development of cognitive distortions and negative attitudes and beliefs became the "ultimate justification for violent acts toward others" (Ressler *et al.*, 1992: 32). Criminal history was present in half of the perpetrators' families, and half of the reported cases of criminal activity involved sexual inappropriateness and abuse amongst family members.

As children, respondents indicated significant problems with attachment, inconsistent contact with some family members, and dysfunctional patterns of social interaction. When describing their home life, a high level of instability was noted. One-third of the men reported growing up in one location; 17 indicated they had occasional instability, and six said they had chronic instability or that they frequently moved. Additionally, 40 percent resided outside of the family residence before age 18 in places such as foster care, detention centers, state homes, and psychiatric hospitals. The researchers concluded that the presence of community stability or prosocial attachment could not mitigate the lack of adequate family stability and the absence of psychological bonding. The majority of the sample also described limited peer relationships, indicating that they were shy, aloof, or a loner.

The FBI study also explored the impact of direct childhood trauma (i.e. physical and sexual abuse) and indirect childhood trauma (i.e. witnessing disturbing interactions). Prior research suggested that children exposed to acts of rape, murder, and suicide would be susceptible to disturbing images of mutilation or scenes of violence, would experience decreased impulse control, would have difficulty assigning responsibility for violent acts, and would likely experience intrusive revenge fantasies (e.g. Pynoos and Eth, 1985). The sexual murderers in the FBI study frequently reported that their experience of parental discipline was unreasonable, intimidating, unpredictable, and abusive, consistent with direct and indirect trauma (Ressler *et al.*, 1992).

This included witnessing displays of sexualized violence or being subjected to the same. However, it should be noted that the events recounted by some sexual homicide offenders were much less severe than many individuals who, in society, never become violent, suggesting the potentially complicated role of biological and other social and psychological correlates in the commission of sexual homicide.

Many of the respondents reported psychological trauma (e.g. witnessing disturbing sex acts committed by parents or other adults), physical abuse (e.g. subjected to sexual violence/injury), and interpersonal conflict or general neglect during childhood (e.g. exposure to pornography, participation in voyeurism). In 23 cases, psychological abuse was reported and it typically included humiliation (e.g. shaming for bedwetting). Thirteen perpetrators reported a childhood history of physical abuse. Sexual problems, fantasies of eroticized dominance, and violent experiences in childhood and adolescence were frequently noted among sample participants (Ressler *et al.*, 1992: 24).

The FBI investigators queried respondents about sexual injury (i.e. surgery) or disease (i.e. sexually transmitted diseases or STDs). Six men reported contracting an STD during adolescence. Other experiences, including "self-mutilation to the genitals, [were performed] as punishment for 'bad thoughts,' including, in one case, shaving the hair from the offender's body because it made him 'feel dirty'" (Ressler *et al.*, 1992: 24–25). Furthermore, 19 of the study's participants reported experiencing sexually stressful situations in childhood that involved negative parental reaction to masturbation, verbal attacks and public embarrassment in regard to gender identity, and/or ridicule and judgment for participation in homosexual activity with peers.

With respect to sexual abuse, 43 percent of the respondents (12/28) reported victimization during childhood (ages one to 12); 32 percent ($N = 9$) reported victimization during adolescence (ages 13 to 18), and 37 percent ($N = 10$) indicated abuse during adulthood (ages 18 and over). These histories of victimization occurred both inside and outside the family, as well as in detention and related custodial environments.

Ressler *et al.* (1992) also examined the reported adult sexual attitudes of the 36 participants. The researchers found that respondents reported performance that was often highly visual and autoerotic. Twenty men reported an age when they first consented to sex and experienced orgasm; however, limited self-reports of peer-related, consensual sexual history were noted. Nearly half of the subjects indicated that as adults they had an aversion to sex, almost 70 percent felt sexually incompetent, 56 percent experienced some form of intimacy dysfunction, and 30 percent expressed concern with the size of their genitalia. According to Ressler *et al.* (1992: 25), "Many [subjects] were concerned with other sexual problems ('I don't think I'm like other people'); some men preferred sex only with dead women. The interviewers suspected that of the sixteen offenders who did not report an age for first consenting sex, most never experienced 'normal' sex."

When the respondents were asked to rank their sexual interests, they reported a strong dependence on visual sexual stimuli. Pornography was ranked highest (81 percent), followed by compulsive masturbation (79 percent), then fetishism (an attraction to inanimate objects, e.g. shoes) (72 percent), and finally voyeurism (71 percent). Although noteworthy, a far less significant number of respondents identified their primary sexual interests to include bondage, indecent exposure, bestiality, obscene phone calls, frotteurism (when a person derives sexual pleasure from rubbing against another nonconsenting person), or cross-dressing. Three offenders indicated participation with sexual asphyxia practices, beginning at age 12 and continuing through adulthood. According to Ressler *et al.* (1992: 26), "In one case the man described his early sexual fantasies (starting at age 4 to 5) and how they were intensified by his piercing his abdomen with such instruments as pins, small knife blades, and fishhooks while he was masturbating."

The FBI study additionally noted that many respondents were preoccupied with murderous thoughts, and that a pathway to this preoccupation could be specified (Ressler *et al.* 1992; see also Hickey, 2006). Rather than cultivating prosocial, peer-related interests and activities as juveniles (e.g. normal intimacy relationships), subjects retreated into a fantasy world of their own making, fraught with eroticized violence. This was how the sexual homicide offenders coped with their childhood trauma and tumultuous home life during their youth. Escape enabled them to visually exert control over, and possibly revenge on, those around them. As the need for escape became increasingly pressing, victimizing and erotic imagery toward and sadistic and deadly mastery over people became essential, although maladaptive, dimensions of survival. Ultimately, these fantasies propelled the respondents to commit acts of sexual homicide. Thus, as the investigators noted, "Murder is compensatory in the fantasy world of the murderer. Because these offenders believe they are entitled to whatever they want and that they live in an unjust world, fantasy emerges as an important escape and a place in which to express emotion and control regarding other human beings" (Ressler *et al.*, 1992: 34).

In order to assess how respondents went from fantasy to action, the FBI researchers explored antecedent behaviors (i.e. pre-crime stress factors, frame of mind, and planning). Stress factors present just before the murder(s) consisted of: conflict with females, parental discord, financial strain, marital problems, conflict with males, birth of a child, physical injury, legal problems, employment problems, and stress from a recent death. The frame of mind or primary emotional state around the time of the murder(s) influenced the sexual homicide offenders' interpretation of external events. The FBI study noted that highly negative emotional states occurred during these critical times, and they included frustration (50 percent), hostility and anger (46 percent), agitation (43 percent), and excitement (41 percent) (Ressler *et al.*, 1992). No apparent compassion for the victim's pain, fear, and/or vulnerability was noted, suggesting that the respondents possessed a severe lack of empathy and a high degree of callousness symptomatic among psychopaths and sadists (Shipley and Arrigo, 2001; Shipley and Russell, 2013).

With respect to premeditation and the decision to kill, 50 percent of the offenders in the FBI study described their murder(s) as intentional, especially in regard to who would be victimized and killed, when the sexual homicide would occur, and where the crime would take place (Ressler *et al.*, 1992). Thirty-four percent identified themselves as in a ready-to-kill mindset that made them susceptible to offending and in search of opportunities to act upon their murderous impulses. It should be pointed out that for sexual homicide assailants whose sexually violent fantasies are so rehearsed, victim selection and scenarios for abduction are well thought out, even if the offender seemingly acts impulsively (Hazelwood and Warren, 2009). When the right combination of variables present themselves (e.g. a potential victim who matches fantasy characteristics, and opportunity for gaining control over that subject), the decision to murder follows (Hickey, 2005). Finally, 16 percent of the sample viewed their murder(s) as solely unplanned. As a matter of premeditation, these offenders put considerable thought and preparation into murder; however, their plans are edited, altered, or otherwise improved to avoid detection with each killing. The FBI study also revealed that the pre-crime actions of the sexual homicide offenders sometimes included alcohol and/or drug usage, although it is not typically reported among sexual sadists as a cohort (Purcell and Arrigo, 2006).

Victim selection also helps to understand the structure and process of sexual homicide. Many perpetrators seek out a victim that meets specific criteria for their violent and eroticized fantasy (Hickey, 2005). Sometimes a victim is symbolic of someone who is significant to the offender, such as a mother figure or an ex-lover. Other murderers are triggered by certain activities or mannerisms of their victims. A victim may remind the killer of a belief, such as the belief that the world is unjust, and that the assailant has been treated unfairly. A victim's behavior may unknowingly do something to disturb an offender's feeling of dominance and control. If a victim attempts to run away or escape to save her life or to stop a sexual assault, it may infuriate the offender and escalate the violence to homicide. If the offender's fantasy is based on complete dominance, he murders the victim to preserve his fantasy (Hazelwood, 2009).

Other sexual murderers wait until a victim who meets the specifications of his fantasy becomes available. He may troll or hunt night after night for the victim who will meet his specifications and bring his fantasy into reality. For some offenders, the actual sexual homicide does not compare to the fantasy (Hazelwood, 2009). For others, the act is exhilarating, the fantasy is reinforced, and they are fueled by feelings of power. In the FBI study, one murderer described feelings of excitement after committing sexual homicide:

> I drive up to my apartment with two murdered girls in my car. The trunk is a mess, with one body stabbed to death. The other [body] is on the back seat. The landlord is [at my apartment] with two friends. I [drove] right up and they kept on talking and I thought, wow, would they freak out if I just got out and opened my trunk and back seat and just threw bodies out in front of them . . . I took the heads up to my room. I could sit there looking at the heads on an overstuffed chair, tripping on them on my bed, looking at them [when] one of them somehow becomes unsettled, comes rolling down the chair, very grisly. Tumbling down the chair, rolls across the cushion and hits the rug, bonk. The neighbor downstairs hates my guts. I'm always making noise late at night. He gets a broom and whacks on the ceiling. 'Buddy,' I say, 'I'm sorry for that, dropped my head, sorry.' That helped bring me out of the depressions. I would trip on that.
>
> *Ressler et al. (1992: 53)*

In the sample of sexual murderers in the FBI study, 56 percent of the 108 victim cases were raped prior to death (Ressler *et al.*, 1992). In their sample, postmortem sexual behavior was evident in 42 percent of the 92 cases. In one case, after the victim was dead, the offender ejaculated into a knife wound he had created; and in another, a victim was found with her genitals mutilated and her breasts amputated. The perpetrator had returned to the crime scene 14 hours after the murder to amputate the breasts. Elements of torture, mutilation, and overkill were apparent in many of these homicides. As Ressler *et al.* (1992: 55) observed, "Unusual or bizarre imagery or acts may be necessary to elicit feelings of sexual excitement in the killer. The level of complexity and bizarreness that is needed to obtain and sustain emotional arousal suggests that the ultimate expression of the murderer's perversion may be his mutilation of the victim" (also see Hazelwood, 2005).

Conclusion

The fantasy life of the sexual homicide offender is often rich and complex, and it provides the blueprint for later (serial) action. According to Hazelwood and Michaud:

> . . . there are no boundaries to what a particular individual might do to other people or to him- or herself . . . When it comes to sexual behavior, there are no limits to what a person might find erotically stimulating. Among violent sexual offenders, often the only logic to their crimes is internal . . . Although we can find patterns and common elements among them, no two offenders ever commit exactly the same sexual crime.
>
> *Hazelwood and Michaud (2001: 3)*

With these concerns in mind, directions for both clinical research and investigative practice are warranted. While the incidence of sexual homicide offending is considerably lower than all other types of murder (Morton and Hilts, 2013), the sadistically violent and morbidly victimizing dimensions of these fantasy-driven crimes leave families devastated and communities in disbelief (Egger, 2002; Hazelwood, 2005; Hickey, 2006).

Specifically in the area of future research inquiries, it is important to further develop the idiographic and nomethetic details of the sexual homicide offender. This includes better measures of, and/or more accurate predictors for, discrete behavioral and personality characteristics; thinking patterns, cognitions and distortions; fantasy dynamics and structure; etiological and motivational factors; victim selection and profiling; and crime scene methods to avoid detection and apprehension. Indeed, as Hazelwood and Michaud (2001) warned, the sexual element of a homicide might not be so readily apparent, and its subtlety could be missed by forensic psychologists, psychiatrists, law enforcement personnel, or even those who are trained experts in sexual homicide offending. At present, the clinical data on and the investigative science of this crime, its violent manifestations, and its deadly outcomes remain mostly under-developed.

Additionally, the repetition of sexual murder is thought to be driven by thinking patterns and internal processing that perpetuate and reinforce sexually violent fantasies (Hazelwood, 2005; Hickey, 2005). These visualizations represent rehearsals for action. Thus it is essential that clinicians carefully document the nature, frequency, and content of sadistic or any deviant sexual fantasies expressed by their patients and any resultant behaviors that follow from these visualizations or distortions. Moreover, the importance of familiarizing oneself with case files and knowing the right questions to ask in a stress interview cannot be understated.

Still further, when mental health professionals and law enforcement officials interact with children and/or adolescents who brutally or gruesomely victimize, then the youths should be questioned about their inner fantasy worlds and motivational cues for delinquent conduct. The FBI study provided the following example of an offender who shot and killed his grandparents as an adolescent, and then shot and killed numerous women as an adult following release from a juvenile facility. The correlation between sadistic behavior toward animals and the sometimes quick progression to act out homicidal fantasies with people is made disturbingly evident by the following statement supplied by one of the study's respondents:

> My stepbrother and I are running around the yard with sticks chasing the giant bluejay, trying to snuff him. He's running around squawking and we've got him wounded and we're trying to kill him before my stepdad finds out, because he'd skin us for snuffing his fine feathered friends. And my grandparents were the same way about small furry animals or the feathery animals, so I snuffed everything that flew, when I could get my gun trained on them.
> *(Ressler et al., 1992: 39).*

At the age of 12, this youth was living with his grandparents in a rural area and was given a gun. However, by the age of 14, the boy's parental surrogates took it away because of his indiscriminate and dangerous shooting. Shortly thereafter, the boy shot and killed his grandparents. We suggest that parents, teachers, mental health professionals, and law enforcement personnel should be mindful to not accept a "boys will be boys" approach to animal cruelty or animal mutilation. These activities can indicate a lack of empathy toward or an inability to form healthy attachments with others. In the most extreme of cases where such tendencies remain unchecked and/or unresolved, a much more elaborately destructive fantasy life can emerge with the potential to generalize to human targets.

The crime of sexual homicide is among the most devastating and deplorable acts of criminal victimization imaginable. It is also one of the most difficult to predict, treat, and prevent. This chapter examines sexual homicide from a clinical and investigative perspective. To accomplish this, we delineate how this type of killing is related to sadistic rape and serial murder. Because our focus is on the etiology, manifestation, and maintenance of sexual homicide, we present the empirical and anecdotal evidence on the bio-psychosocial correlates of serial murder. A review of these correlates helps to explain motive for the sexual homicide offender, and they help to clarify the ritualized complexities of repetitive, fantasy-based, and erotically charged killing. We also review the classic FBI research on serialized sexual homicide. The principal risk factors for the onset, development, and repetition of this crime type are discussed. We conclude by offering several tentative recommendations in the areas of future clinical science and investigative practice. Future investments in both areas would further the mental health and justice systems' understanding regarding the under-examined world of sexual homicide.

References

Arrigo, B.A. and Purcell, C.E. (2001) 'Explaining paraphilias and lust murder: an integrated model', *International Journal of Offender Therapy and Comparative Criminology*, 45: 6–31.

Blanchard, G.T. (1995) 'Sexually addicted lust murderers', *Sexual Addiction and Compulsivity*, 2(1): 62–71.

Canter, D. and Wentink, N. (2004) 'An empirical test of Holmes and Holmes's serial murder typology', *Criminal Justice and Behavior*, 20(10): 26–37.

Carlisle, A.L. (1993) 'The divided self: toward an understanding of the dark side of the serial killer', *American Journal of Criminal Justice*, 17(2): 23–36.

DeHart, D.D. and Mahoney, J.M. (1994) 'The serial murderer's motivations: an interdisciplinary review', *Omega*, 29(1): 29–45.

Drukteinis, A.M. (1992) 'Serial murder: the heart of darkness', *Annals*, 22(10): 532–538.

Egger, S.A. (2002) *The Killers Among Us: An Examination of Serial Murder and its Investigation*, 2nd edn, Upper Saddle River, NJ: Prentice-Hall.

Graney, D.J. and Arrigo, B.A. (2002) *The Power Serial Rapist: A Criminology Victimology Typology of Female Victim Selection*, Springfield, IL: Charles C. Thomas.

Groth, A.N. and Birnbaum, H.J. (2001) *Men Who Rape: The Psychology of the Offender*, New York: Perseus.

Groth, A.N., Burgess, A., and Holmstrom, L. (1977) 'Rape, power, anger and sexuality', *American Journal of Psychiatry*, 134: 1239–1243.

Hale, R. (1994) 'The role of humiliation and embarrassment in serial murder', *Psychology, A Journal of Human Behavior*, 31: 17–23.

Hazelwood, R. (2005) *Key Motives and Behaviors in Sexual Homicide*, presentation at 'Homicide: Behaviors, Motives and Psychology: A Gathering of Leading Experts Conference', San Diego, California.

Hazelwood, R. (2009) 'Analyzing the rape and profiling the offender', in R. Hazelwood and A. Burgess (eds), *Practical Aspects of Rape Investigation: A Multidisciplinary Approach*, Boca Raton, FL: Taylor & Francis, pp. 97–122.

Hazelwood, R. and Burgess, A. (1987) 'An introduction to the serial rapist: research by the FBI', *FBI Law Enforcement Bulletin*, September: 16–24.

Hazelwood, R. and Michaud, S.G. (2001) *Dark Dreams*, New York: St Martin's Press.

Hazelwood, R. and Warren, J. (2009) 'The relevance of fantasy in serial sexual crimes investigation', in R. Hazelwood and A. Burgess (eds), *Practical Aspects of Rape Investigation: A Multidisciplinary Approach*. Boca Raton, FL: Taylor & Francis, pp. 55–64.

Hazelwood, R., Warren, J., and Dietz, P.E. (1993) 'Compliant victims of the sexual sadist', *Australia Family Physician*, 22: 474–479.

Hickey, E. (2005) *Sex Crimes and Paraphilia*, Upper Saddle River, NJ: Prentice-Hall.

Hickey, E.W. (2006) *Serial Murderers and their Victims*, 4th edn, Belmont, CA: Wadsworth.

Holmes, R.M. and DeBurger, J.E. (1988) *Serial Murder*, Newbury Park, CA: Sage.

Holmes, R.M. and Holmes, S.T. (1999) *Serial Murder*, 2nd edn, Thousand Oaks, CA: Sage.

Holmes, R.M. and Holmes, S.T. (2002) *Sex Crimes: Patterns and Behavior*, 2nd edn, Thousand Oaks, CA: Sage.

Kernberg, O. (1982) 'An ego psychology and object relations approach to the narcissistic personality', in APA (ed.), *Psychiatry: Annual Review*, Washington, DC: American Psychiatric Publishing, pp. 510–523.

Kocsis, R.N., Cooksey, R.W., and Irwin, H.J. (2002) 'Psychological profiling of offender characteristics from crime behaviors in serial rape offences', *International Journal of Offender Therapy and Comparative Criminology*, 46(2): 144–169.

Liebert, J. (1985) 'Contributions to psychiatric consultation in the investigation of serial murder', *International Journal of Offender Therapy and Comparative Criminology*, 28: 187–200.

Meloy, J.R. (2000) 'The nature and dynamics of sexual homicide: an integrative review', *Aggression and Violent Behavior*, 5(1): 1–22.

Money, J. (1990) 'Forensic sexology: paraphilic serial rape (biastophilia) and lust murder', *American Journal of Psychotherapy*, 44: 26–36.

Morton, R.J. and Hilts, A.M. (2013) 'Serial murder: multi-disciplinary perspectives for investigators', Washington, DC: Behavioral Analysis Unit, National Center for the Analysis of Violent Crime, Critical Incident Response Group, Federal Bureau of Investigation, www.fbi.gov/stats-services/publications/serial-murder

Myers, W.C., Reccoppa, L., Burton, K., and McElroy, R. (1993) 'Malignant sex and aggression: an overview of serial sexual homicide', *Bulletin of the American Academy of Psychiatry and Law*, 21: 435–451.

NCWP (2004) *Suspect Typology; Profiling the Sex Offender*, Beverly Hills, CA: National Center for Women and Policing.

Norris, J. (1989) *Serial Killers*, New York: Anchor Books.

Palermo, G.B. (2004) *The Faces of Violence*, 2nd edn, Springfield, IL: Charles C. Thomas.

Pardue, A. and Arrigo, B.A. (2008) 'Power, anger, and sadistic rapists: towards a differentiated model of offender personality', *International Journal of Offender Therapy and Comparative Criminology*, 52(4): 378–400.

Pennington, B.F. (2005) *The Development of Psychopathology: Nature and Nurture*, New York: Guilford Press.

Prentky, R.A., Burgess, A.W., Rokous, F., Lee, A., Hartman, C., Ressler, R., and Douglas, J. (1989) 'The presumptive role of fantasy in serial sexual homicide', *American Journal of Psychiatry*, 146: 887–891.

Purcell, C.E. and Arrigo, B.A. (2006) *The Psychology of Lust Murder: Paraphilia, Sexual Killing and Serial Homicide*, San Diego, CA: Elsevier.

Pynoos, R.S. and Eth, S. (1985) 'Children traumatized by witnessing acts of personal violence: homicide, rape, or suicidal behavior', in S. Eth and R.S. Pynoos (eds), *Post-traumatic Stress Disorders in Children*, Washington, DC: American Psychiatric Publishing, pp. 17–44.

Rada, R. (1978) *Clinical Aspects of the Rapists*, New York: Grune & Stratton.

Ressler, R.K., Burgess, A.W., and Douglas, J.E. (1992) *Sexual Homicide: Patterns and Motives*, New York: Free Press.

Revitch, E. and Schlesinger, L.B. (1989) *Sex Murder and Sex Aggression*, Springfield, IL: Charles C. Thomas.

Schlesinger, L.B. (1998) 'Pathological narcissism and serial homicide: review and case study', *Current Psychology*, 17(2/3): 212–221.

Schlesinger, L.B. (2003) *Sexual Murder: Catathymic and Compulsive Homicides*, Boca Raton. FL: CRC Press.

Shipley, S.L. and Arrigo, B.A. (2001) 'The confusion over psychopathy (II): implications for forensic (correctional) practice', *International Journal of Offender Therapy and Comparative Criminology*, 45(4): 407–420.

Shipley, S.L. and Arrigo, B.A. (2008) 'Serial killers and serial rapists: dichotomy or continuum – an examination of commonalities and comparison of typologies', in R. Kocsis (ed.), *Serial Murder and the Psychology of Violence*, Totowa, NJ: Humana Press.

Shipley, S.L. and Arrigo, B.A. (2013) *Introduction to Forensic Psychology, 3E: Issues and Controversies in Law, Law Enforcement and Corrections*, New York: Elsevier/Academic Press.

Shipley, S.L. and Russell, J.L. (2013) 'Predatory and affective aggression: calculated and explosive pathways to violence', in J.B. Helfgott (ed.), *Criminal Psychology*, Santa Barbara, CA: ABC–CLIO.

Zillmann, D. (1998) *Connections between Aggression and Sexuality*, Washington, DC: Psychology Press.

37

Neuroethics of criminal responsibility

Mental disorders influencing behavior

Gerben Meynen

Introduction

Neuroethics has been defined in different ways. Basically, it consists of two branches of research. First, it concerns the ethics of performing neuroscientific studies; second, it deals with the implications of neuroscience for ethics (Levy, 2007). The latter is the focus of this chapter, as it addresses the question: What are the implications of neuroscience for the legal and psychiatric evaluation of a defendant's responsibility for his actions? This topic is an important theme in neuroethics, but also in the field of neurolaw, an area of research that examines the implications of neuroscience for the law (Spranger, 2012). In this chapter the approach is neuroethical as well as neurolegal, because the meaning of neuroscience for *criminal* responsibility is the topic of interest.

The chapter starts with an introductory section on the history of criminal responsibility or legal insanity and its place in criminal law. It will show that whether defendants are held legally accountable depends on two issues: first, their state of mind at the time of committing the criminal act; second, the legal criteria for insanity. Next, I discuss a central conceptual problem concerning criminal responsibility: free will and the brain. Some scholars argue that neuroscience threatens or even undermines the very concept of criminal responsibility, because it compromises free will. The next section considers some significant caveats as far as the application of neuroscience to assessments of criminal responsibility is concerned, and discusses an extraordinary neurolaw case of a 40-year-old schoolteacher with changed sexual interests. Finally, I explore how neuroscience could be helpful in the evaluations of defendants who raise an insanity defense.

The insanity defense: past and present

The insanity defense dates back to ancient times, when the Romans would exculpate defendants who were insane (Robinson, 1998; Simon and Ahn-Redding, 2006). This implies that the defense pre-dates the birth of psychiatry as a medical specialty. But how insanity is defined and how it is determined in practice has varied considerably over time.

For centuries, in the United Kingdom, the so-called "wild beast" test was being used (Robinson, 1998). In *Rex vs. Arnold*, Justice Tracy instructed the jury as follows about the "madman," a person "totally deprived of his understanding and memory so as not to know what he is doing, no more than an infant, brute or a wild beast" (Robinson, 1998; Simon and Ahn-Redding, 2006). In 1800, James Hadfield was tried after shooting at King George III. His lawyer, Thomas Erskine, argued that Hadfield should be considered insane (Robinson, 1998). He did so, introducing a new way of looking at insanity, in fact proposing a new legal criterion or standard for the insanity defense. According to Erskine, delusion – rather than the mental makeup of a wild beast – is the "true character" of insanity. Interestingly, the notion of delusion directly refers to psychopathology, and, unsurprisingly, *doctors* testified about Hadfield's state of mind (Robinson, 1998). What is interesting here, among other things, is that being insane or not depends on two crucial issues: (1) a defendant's state of mind at the time of the crime; and (2) the legal standard used to determine whether the state of mind *qualifies* as legal insanity. In other words, insanity is just as much about establishing psychological or psychiatric facts about the mental condition as it is about a legal norm. This is especially noteworthy when we realize that the legal norms may differ considerably between various legal systems (Meynen, 2013b).

In 1843, Daniel M'Naghten was found to be insane after, allegedly, attempting to kill the Prime Minister, Sir Robert Peel. However, due to what appears to have been a case of misidentification, he killed Edward Drummond, secretary to the prime minister, instead (Robinson, 1998). In the aftermath of this high-profile case, judges formulated the perhaps most influential legal insanity standard, the *M'Naghten Rule* (1843):

> At the time of committing the act, the party accused was laboring under such a defect of reason, from disease of the mind, as not to know the nature and quality of the act he was doing; or if he did know it, that he did not know what he was doing was wrong.
>
> *(M'Naghten's Case, 10 Cl. & Fin. 200, 8 Eng. Rep. 718, H.L. 1843)*

Looking more carefully at the M'Naghten Rule, it becomes evident that what counts – apart from a mental disorder resulting in a defect of reason – is a lack of *knowledge*: knowledge about the *nature and quality* or *wrongfulness* of the act. Many will agree that such knowledge *is* relevant to the question of criminal responsibility, and that a person who lacks such knowledge due to a mental disorder should be considered insane. For instance, a person may be convinced, because of a paranoid delusion, that his neighbor is on the verge of attacking him (while the neighbor is totally innocent), and at some point the person – terrified – may see no other option than to attack his neighbor in the stairway. According to the knowledge of this deluded person, his attack was an instance of self-defense. Because of such distorted knowledge, the person may eventually be considered insane.

Still, we may ask: are there other types of influence of mental disorders that may (partially) exculpate a defendant as well? In moral philosophy, it is usually recognized that, apart from the epistemic factor (having to do with knowledge), another factor is also relevant: control. If a person could not control their action or behavior, they may be exculpated, just as when their knowledge would have been affected.

Consider the case of commanding voices, as they may occur in schizophrenia. Some patients experience a particular type of verbal hallucinations: commanding voices (Braham *et al.*, 2004; Meynen, 2013a). The content of such commands may be either harmless or harmful. In some cases, obeying the voice may even constitute a crime, for instance, if the voice orders a patient to attack or kill another human being. What makes such voices especially salient from the perspective of the insanity defense is that, apparently, in some cases patients *cannot but obey* the command. So, irrespective of the patient's own wishes, desires, and motives, the patient has to obey the voice. For example, a woman hears such a commanding voice (the occurrence of which is beyond her control) ordering her to harm another person, which she does, even though this act goes against her own views and wishes (she doesn't hate the person, she isn't angry, in fact, she doesn't even know her victim). In this case, it is not the patient's distorted knowledge that exculpates her, but the fact that, due to a mental disorder, she could not control her behavior: after she heard the command, she could not but perform the attack (Meynen, 2013b).

So, while some disorders may affect human behavior by distorting a person's *knowledge* about her actions, other disorders may – not less profoundly – influence behavior by undermining a person's *control* of her actions. Many legal insanity standards contain a control prong in addition to a cognitive or epistemic element. For instance, the Model Penal Code:

> ...a person is not responsible for criminal conduct if at the time of such conduct as a result of mental disease or defect he lacks substantial capacity either to appreciate the criminality of his conduct or to conform his conduct to the requirements of the law.
>
> *(American Law Institute, 1962)*

Although the element of control is philosophically considered relevant to moral responsibility, and is reflected in various insanity standards, the influential legal scholar Stephen Morse criticizes it:

> Some criminal mental health law tests adopt such a standard, either alone or in addition to a rational capacity test. Lack of control is not well understood conceptually or scientifically in any of the relevant disciplines such as philosophy, psychology, and psychiatry, however, and we lack operationalized tests to accurately identify this type of lack of capacity. I have long been a critic of such standards for just these reasons.
>
> *(Morse, 2011b)*

Notably, Morse does not argue that the issue of control *per se* is irrelevant to a person's responsibility. He argues that the notion "lack of control" is insufficiently understood in the relevant disciplines (philosophy, psychiatry, and psychology), and that tests to accurately determine such a lack of control are not available.

In fact, concerning insanity standards, considerable differences between legal systems exist: some include a control prong, others don't (Simon and Ahn-Redding, 2006). And if we compare M'Naghten to the Model Penal Code, there are also differences with respect to knowledge (or "appreciation"). Furthermore, in Sweden there is no exculpation due to mental disorder in the strict sense, while in The Netherlands, legal insanity does exist, but there is no legal standard defining the insanity criteria (Meynen, 2013b). So, apparently, legally, there are different views on insanity and its criteria. In addition, in many legal systems, defendants are either sane or insane, but in other legal systems, there is a graded approach to insanity, for instance in The Netherlands. Morse has argued for diminished responsibility within the context of the US legal system (Morse, 2003). Meanwhile, the profound differences between legal systems make it hard, if not impossible, to define – in a way relevant for all these systems – what legal insanity is about.

A central conceptual issue: free will and the brain

What is so special about mental disorders that they can (partially) exculpate people? We may respond that this has to do with the fact that these disorders may *influence* our behavior. But this cannot be the (full) explanation, because many mental phenomena influence behavior, such as hate, hope, and greed. But if a crime is committed because of hate or greed, this does not exculpate a defendant at all. So, we should ask the question: what is so special about the influence of a mental disorder on our behavior that it may diminish guilt?

A considerable number of authors answer this question referring to free will (Meynen, 2010). For example, Reich writes, "the law recognizes that insanity compromises free will, and classifies someone without free will as legally not responsible for his or her actions" (Reich, 2005: 206). If this is indeed the rationale for the insanity defense, free will is crucial.

However, a problem may arise. In recent years, neuroscience has produced an incessant stream of findings that, allegedly, challenge free will (Pereboom, 2001; Levy, 2007; Kane, 2011). Perhaps the most famous of these findings has been produced by the Libet experiment. Libet observed that about half a second before a person becomes aware of an urge to act, a characteristic electrophysiological signal – the readiness potential – is detectable by EEG (Libet *et al.*, 1983; Libet, 1999). This finding, according to some, clearly indicates that free will is nothing but an illusory experience: the brain determines our actions, and our consciousness just follows brain activity (Spence, 1996; Wegner, 2002; Spence, 2009). Others doubt the far-reaching consequences people draw based on this experiment (Mele, 2009; Radder and Meynen, 2012). Still, Libet's results are no isolated findings; in other experimental designs, similar results have been obtained (Soon *et al.*, 2008; Fried *et al.*, 2011). Greene and Cohen have argued that neuro-findings such as these *will* have profound impact on (criminal) law, because the results undermine its basic assumption, namely that we are free and responsible agents (Greene and Cohen, 2004).

Philosophers, legal scholars, and neuroscientists have responded in different ways to this challenge (Meynen, 2009; Meynen and Oei, 2010). For instance, in 'The non-problem of free will in forensic psychiatry and psychology' Morse (2007) argues, in short, that free will is a troublesome concept, which, fortunately, on close examination, does not play a role in any doctrine relevant to legal responsibility (in the USA). Therefore Morse urges forensic psychiatrists and psychologists "to avoid all usage of free will in their forensic thinking and work product because it is irrelevant and spawns confusion" (Morse, 2007: 203). Furthermore, he proposes to use, instead of free will, the concept of "capacity for rationality." In fact, Morse is a compatibilist, in the sense that in his view (neuro)determinism does not threaten responsibility.

Renowned neuroscientist Gazzaniga holds that neuroscience-determinism does not threaten responsibility because the brain and responsibility are completely different matters (Gazzaniga, 2005). We will not find responsibility in the brain, because responsibility is a *social* concept and construct. It is therefore neither affected nor undermined by neuroscience findings; it is immune to such results.

In fact, the problem of free will and determinism is classic and complicated. From very different backgrounds and perspectives, the problem has been addressed by philosophers (Kane, 2002). For centuries, it has been a theological problem: given divine foreknowledge, can free will exist? And it has been a problem of physics: given Newton's laws, can free will exist? Vincent rightfully points to the many ways in which the term "determinism" can be understood in the free will debate, as "causation, necessitation, sufficiency, entailment, explanation, prediction, inevitability, fate, and predestination (this list is intended to be demonstrative not exhaustive)" (Vincent, 2013: 5). So, although neuroscience is clearly a new element in the "free will versus determinism" debate, it is not the case that neuroscience has evoked this problem by itself: it has a long history.

Philosophers have developed various solutions to at least some of the threats posed by determinism. Currently, many believe that, even if our actions are determined (e.g. by brain processes), we can still be free, or at least morally responsible. Examples are Harry Frankfurt (1971), Daniel Dennett (2003), and Peter Strawson (2003). Interestingly, some scholars even doubt whether we should think of our brains as fully determined. They point to the fact that, according to quantum mechanics, indeterministic events are part of nature. Couldn't such events take place in our brains as well, breaking the spell of determinism (Searle, 2007; Kane, 2011)? Ted Honderich, however, responds that quantum indeterminism cannot save free will (Honderich, 2002). First, according to Honderich, the exact implications of quantum mechanics remain fuzzy. Second, quantum events are relevant at the micro level of elementary particles, but at the level of the macro world – brains are part of that macro world – processes are deterministic, he argues. So, quantum indeterminism would be irrelevant to the problem of brain and free will.

Furthermore, even if brain quantum indeterminism would exist, does this provide room for free will? According to the philosopher Derk Pereboom, both determinism and quantum indeterminism are incompatible with free will (Pereboom, 2001). If my behavior is random, indeterministic, it cannot be considered free. It is clear that the free will debate is further complicated by such quantum arguments, because this opens up a new scientific domain, with which only few authors are really familiar.

The debate on the neuro-threat to free will is just one topic in neuroethics, but it is a central issue with respect to neuroscience and criminal responsibility. In the next sections we consider another area of neuroethical research and debate: the potential *positive* contribution of neuroscience to (assessments of) criminal responsibility. In other words, neuroscience not only threatens moral and legal responsibility, it may also support evaluations of a defendant's responsibility. This is because the law is interested in the defendant's mental functioning. Brain sciences may inform us about features of a defendant's brain that can tell us more about his or her mental condition and capacities. An example is a brain scan that is supportive of the diagnosis of Alzheimer's disease in an elderly defendant who killed his wife after a happy marriage of 40 years.

Applying neuroscience to criminal responsibility: caveats

Stephen Morse is a skeptic regarding the usefulness of neuroscience in criminal law procedures, coining the "brain overclaim syndrome" (Morse, 2005). In his paper on this "syndrome," he begins with a conceptual note:

> Brains do not commit crimes; people commit crimes. This conclusion should be self-evident, but, infected and inflamed by stunning advances in our understanding of the brain, advocates all too often make moral and legal claims that the new neuroscience does not entail and cannot sustain.
>
> *(Morse, 2005)*

So, the objects of the law (persons) and of neuroscience (brains) are very different, which fundamentally restricts the applicability of neuroscience in the courtroom. Meanwhile, Morse (2005) does not deny that neuroscience can be of help, although he makes clear that, at this point in time, neuroscience lacks sufficient explanatory power. He mentions several reasons why neuroscience studies do not deliver what some people claim. First, the small number of people included in neuroscience studies makes it hard to draw strong conclusions from this type of research. Second, statistically significant differences between groups do not at all exclude the possibility of considerable overlap between the individuals in these groups. For instance, although shoe sizes between men and women are statistically different, some women have larger feet than some men. Therefore, in general, it is hard to conclude whether a person is a man or a woman just based on shoe size. So, although differences may be statistically relevant at group level (and certainly worth publishing), they may not be helpful when looking at individual cases, as criminal law does. Furthermore, Morse argues that observing brain correlates of mental phenomena, such as consciousness, is not the same as providing a neuroscientific *explanation* of these phenomena. Certain EEG patterns associated with different levels of consciousness have been known for a long time, but these different levels themselves have not yet been sufficiently explained. In addition, Morse argues that

> Certain lesions can of course disable various human capacities, but few criminal responsibility cases in which the result is not already obvious based on behavioral evidence will involve a precise, identifiable neurological mechanism that will demonstrate that criminal responsibility was not present.
>
> *(Morse, 2005)*

So, even if neuroscience shows legally relevant brain abnormalities, this is likely to happen in cases in which behavioral evidence is already so clear that neuroscience will not really add anything.

In my view, this point may suggest a little too much confidence in behavioral evidence, though. In fact, although Morse's caveats are justified, they could give the impression that while behavioral evidence about a defendant's cognitive capacities is reliable, valid, and clear, in contrast neuroscience is unreliable and fishy. But many professionals in the forensic domain would like to have some neuro-tools or techniques to corroborate (or falsify) behavioral observations, and conclusions drawn from them. Not because current neuro-tools are 100 percent reliable, but because behavioral data are often less clear, less conclusive, and less reliable than they would wish. In contrast to Morse, the neurophilosopher Paul Churchland (1995) emphasized the fuzziness of courts in these respects: "But few will deny that courts are deeply unreliable at determining the many dimensions of cognitive, emotional and social competence in any defendant" (Churchland, 1995).

Although Morse is right about relevant weaknesses in neuroscience, there may be weaknesses in the current evaluation of defendants' mental capacities as well. For instance, the evaluations by psychiatrists can be considered subjective, prone to malingering or lying, and sometimes hard to replicate. And when we realize that our current behavioral and observational tools have certain flaws, we may be willing to consider neuroscientific tools, even if they still suffer from certain weaknesses.

Morse emphasizes the relevance of behavioral criteria for responsibility:

> If the person meets the behavioral criteria for responsibility, the person should be held responsible, whatever the brain evidence may indicate, such as the presence of an abnormality. If the person does not meet the behavioral criteria, the person should be held not responsible, however normal the brain may look.
>
> *(Morse, 2005)*

This is a strong statement, and in a way it is certainly true. Still, we should take into account that, in real life, conclusions about behavioral evidence may be less certain than Morse's view may suggest. For instance, a psychiatrist who has assessed a defendant may conclude that, taking into account the many findings of his or her assessment, this defendant should be considered legally sane. In this case, the conclusion is based on talking with the – maybe lying – defendant, reading contradictory eye-witness accounts, and reading conflicting information from previous treating physicians. But would this psychiatrist have reached the same conclusion if brain scans would have shown significant abnormalities? Suppose that brain scans would have drawn the psychiatrist's attention to specific details in the reports that he or she, without the scans, didn't pay much attention to. Morse is right that, in some cases, a brain scan abnormality may not make any difference. Still, in cases that are less clear, evidence of brain abnormalities could tip the balance. In other words, depending on how exactly a specific judgment about a particular defendant's insanity comes about, neuro-information may be relevant.

It may even be hard to determine exactly the impact of specific neurobiological aspects of a legal case on the final judgment. Consider *Roper v. Simmons* (2005), in which the US Supreme Court decided that the death penalty for juveniles was unconstitutional. To the court, via the *amicus briefs*, many neuroscience results were presented showing that brains continue to develop ("mature") after the age of 17. In Morse's interpretation, these neuroscience data did not have much impact on the Supreme Court's decision. He states:

> Now, we already knew from common-sense observation and rigorous behavioral studies that juveniles are on average less rational than adults. What did the neuroscientific evidence about the juvenile brain add? It was consistent with the undeniable behavioural data, and perhaps provided a partial causal explanation of the behavioural differences. The neuroscience data was therefore merely additive and only indirectly relevant.
>
> *(Morse, 2011a: 540)*

Let's assume that, indeed, the neuroscience evidence was "merely additive and only indirectly relevant." Still, the Supreme Court's decision was a *five-to-four* decision. In situations like these, "merely additive and only indirectly relevant" information might just turn the scale. Of course, we do not know for sure to which extent exactly neuro-evidence in this case influenced the judges' decisions. But for the sake of the argument, let us suppose that the *amicus briefs* would have provided the opposite neuroscience evidence to the Supreme Court, stating that *no*, the brain does not develop further after the age of 17, and that a 17-year-old brain is therefore *mature*. Perhaps, then, the Supreme Court's decision would have been different.

So, in spite of the justified theoretical and practical problems that are brought forward by Morse, in some cases neuroscience evidence could tip the balance by providing additive and indirect support for a particular legal outcome. Such potential neuroscientific impact calls for serious consideration of shortcomings and strengths of particular neuroscience evidence. In practice, it may even be hard to determine exactly at what point evidence is no longer "indirect and additive" but is direct and crucial. Let us explore this issue a little further by considering an exceptional case, which has been discussed frequently in neuroethics.

Case: a 40-year-old schoolteacher's changed sexual behavior

Although it is true that, in many cases, neuroscience may offer little more than statistically significant correlations, sometimes a causal relationship is likely based on neuroscientific findings.

> A 40-year-old, right-handed man in an otherwise normal state of health developed an increasing interest in pornography, including child pornography. He had a preexisting strong interest in pornography dating back to adolescence, although he denied a previous

> attraction to children and had never experienced related social or marital problems as a consequence. Throughout the year 2000, he acquired an expanding collection of pornographic magazines and increasingly frequented Internet pornography sites. Much of this prurient material emphasized children and adolescents and was specifically targeted to purveyors of child pornography. He also solicited prostitution at "massage parlors," which he had not previously done.
>
> *(Burns and Swerdlow, 2003)*

Notably, the man "went to great lengths to conceal his activities because he felt that they were unacceptable." Still, he continued his activities, "stating that 'the pleasure principle overrode' his urge restraint." At some point he started to make "subtle sexual advances toward his prepubescent stepdaughter, which he was able to conceal from his wife for several weeks." His wife did not notice any of this, until his stepdaughter informed her and she discovered his activities. He was diagnosed as having pedophilia, prescribed medication, and found guilty of child molestation. At a rehabilitation center for sexual addiction, as Burns and Swerdlow state, he "could not restrain himself from soliciting sexual favors from staff and other clients at the rehabilitation center and was expelled." So far the case merely describes a sudden and profound change in the sexual interests and (criminal) behavior of a 40-year-old man.

But now there is a dramatic turn in the story. At some point, he visits the emergency department of a hospital, because of a headache. "A nonphysiologic cause was suspected, and the psychiatry service admitted him with a diagnosis of pedophilia, not otherwise specified, after he expressed suicidal ideation and a fear that he would rape his landlady." Because the man complained about balance problems, he was examined neurologically and, during this examination, "he solicited female team members for sexual favors. He was unconcerned that he had urinated on himself" (Burns and Swerdlow, 2003). On examination, several neurological abnormalities were found concerning eye movements, reflexes, and the way he walked, but his memory was intact. A brain scan (MRI) was made, showing an impressive tumor displacing the right orbitofrontal lobe.

The tumor was resected, and his walking and bladder control improved. He then successfully followed a Sexaholics Anonymous program. After some time, he even returned home, because he was no longer considered dangerous for his stepdaughter. What makes this case extraordinary is that, after some time, a headache reoccurred, as did his secretly collecting pornography. An MRI revealed regrowth of the tumor, and again the tumor was resected. In their discussion of the case, Burns and Swerdlow state:

> Orbitofrontal lesion research suggests that sociopathic behavior results from a loss of impulse control rather than a loss of moral knowledge. [. . .] Our patient could not refrain from acting on his pedophilia despite the awareness that this behavior was inappropriate.
>
> *(Burns and Swerdlow, 2003)*

The obvious question, at least within the context of the present chapter, is twofold: should this schoolteacher be considered criminally responsible for his actions? What role does neuroscience play in this judgment? Morse discusses this case as well and, interestingly, he does not dismiss the relevance of the brain tumor found on MRI or the other neurological findings. In fact, he does not give a final conclusion about the patient's criminal responsibility, leaving it to the reader. Still, he provides some guidance how to evaluate the schoolteacher's criminal responsibility: we have to ask ourselves whether the teacher was able to control himself.

Humans are often confronted with urges, and many mental disorders may involve (extreme) urges, but if the teacher had the ability to control his behavior, he should have done so (Meynen, 2013c). Morse notes that we do not know the *strength* of his pedophilic urges, nor his *capacities* to control them, but the fact that, in the beginning, he was able to conceal the activities is indicative of some form of control, he argues. The case further shows that his ability to control himself "deteriorated quite rapidly" (Morse, 2011a). Indeed, apparently he was not even able to control his bladder.

Let us consider this case in a bit more detail, looking at different standards for legal insanity, firstly M'Naghten. The initial successful concealment of his activities at least suggests that he was aware of the inappropriateness – perhaps wrongfulness – of his behavior. Especially his own comment about the pleasure principle overriding urge restraint is an indication that that he was aware that these urges actually *required* urge restraint. Basically, in the beginning of the case there is little direct evidence that the man's knowledge or awareness that what he was doing was wrong was absent. But perhaps we should be more careful here. Maybe the patient's own statement under these extreme neuropathological conditions in terms of "a pleasure principle overriding the urge restraint" should be interpreted with caution. Given the tumor growth at that time and the behavior changes, the patient may have had problems interpreting his own behavior. Did he know what was actually happening to him? Could he explain his behavior, or was he actually at a loss as far as understanding what he was doing is concerned? The fact that he, for example, did not urinate on himself in the early phase does not necessarily imply that there were no grave problems interpreting and understanding what was happening to him at that time. Given the profound cognitive and neurological disturbances that eventually developed, for instance severe agraphia (inability to write), it may be wise to be careful using the patient's words "against him." Who knows which *relevant cognitive* capacities were already profoundly affected at an early stage?

Let us look at another element that is often present in insanity standards – a control prong, as in the Model Penal Code. First, as said, the fact that in the first stages the patients did not urinate on himself does not mean that other complex cognitive functions involved in action control were not decisively affected. So control issues may have played a (decisive) role from the beginning of the case. Later on, such issues appear to become more prominent, and eventually he apparently could not even control his bladder. Still, we should note that many people may at some point suffer from a loss of bladder control, and some may even give the impression of being unconcerned about it (just like this schoolteacher), but that is never *direct* evidence for legal insanity. The point I intend to make here is not that the fact that the patient is urinating on himself would be irrelevant; on the contrary. The point is that such elements of the case derive their relevance from the *specific constellation* of signs and symptoms in a particular patient. In fact there are many aspects of the case, many bits of information, that we will weigh evaluating this case (e.g. the schoolteacher's sudden and profound behavioral change, increasing neurological and neurocognitive problems, and, on top of that, an MRI showing a brain tumor). To this constellation we may add the fact that the legally relevant behavior stopped after the brain tumor had been removed, and that when such behavior restarted, regrowth was visible on MRI as well. The MRI is likely to play a role within the *constellation* of the other neurological and behavioral signs and symptoms, and in that way it may contribute to our judgment that the teacher should be considered not criminally responsible.

Is the MRI in this case direct or indirect evidence? Is it additive or crucial? Is it even possible to isolate neuroscientific elements – such as the MRI showing grave abnormalities – from the "rest" of the case? (Suppose, for the sake of the argument, that no abnormalities would have been found on MRI.) I will leave these questions to the reader. The point is that it may be hard to determine which facts are "merely additive" and which are "crucial" when forming a judgment about criminal responsibility in a case like this. In fact, even if neuroscience would be "additive" in nature, it can add something important to our judgment of a case.

Neuroscience and psychiatric evaluations of defendants

One might argue that, although in the teacher's case a psychiatric diagnosis was made, pedophilia, it is basically a neurological case, involving the effects of a brain tumor, rather than the effects of a strictly psychiatric disorder. How could neuroscience contribute to the evaluations of defendants suffering from a *psychiatric* disorder such as schizophrenia or bipolar disorder, rather than from neurological diseases like dementia or brain tumors? This question is be considered in this section. First, we look at the extent to which neuroscience is, at present, integrated in psychiatric diagnosis. Then we consider how neuroscience could contribute in the future to psychiatric assessments of defendants.

The contribution of neuroscience to diagnosis in psychiatry

Over the past decades, scientific psychiatric journals have published a steady stream of results on mental disorders and brain anatomy or functioning (Andreasen, 1984). But apart from neuropsychiatric diseases – such as Alzheimer's dementia, frontotemporal dementia, and vascular dementia – what has been the contribution of neuroscience to psychiatric diagnosis? This question can be addressed in different ways, but in the context of this chapter we look at the impact of neuroscience on the classification of mental disorders in the *Diagnostic and Statistical Manual of Mental Disorders* (DSM).

Since 1952, five editions of the DSM for psychiatric classification have been published, including some (text) revisions (Sadler, 2005; Bolton, 2008). The current edition is DSM-5, which appeared in 2013. In fact, already in 1999 (the DSM-IV had been introduced five years earlier), the initiative was formed to start working on a new edition, which would eventually become the DSM-5. The DSM-5 task force specifically aimed to incorporate scientific data, in particular those in neuroscience (Kupfer *et al.*, 2002). As Regier and colleagues put it:

> As we began the DSM-V developmental process in 1999, a major concern was to address a range of issues that had emerged over the previous 30 years [. . .] The opportunity to evaluate the readiness of neuroscientific advances in pathophysiology, genetics, pharmacogenomics, structural and functional imaging, and neuropsychology was also a priority.
>
> *(Regier et al., 2009)*

In spite of this priority to evaluate the readiness of neuroscientific advances in various domains, *none* of the criteria for any of the major psychiatric diagnoses in the DSM-5 (excluding dementias) has become neuroscientific in nature: diagnosing depression, schizophrenia, bipolar disorder, ADHD, borderline personality disorder, autism, panic disorder, continues to happen without any reference to neurobiological parameters – despite the task force's priority. For instance, in order to be diagnosed as suffering from a depression, a patient has to fulfill at least five out of nine criteria for a depressive episode (e.g. depressed mood, weight loss, sleep disturbance, loss of energy, difficulty concentrating).

So, although numerous neuroscientific results have been generated over the past decade(s), they are apparently not of a nature that justified incorporation in the DSM-5. As Urbaniok *et al.*

(2012: 174) state: "Although with the help of functional imaging techniques much progress has been made over the past 10 years in uncovering the causes of psychiatric diseases, conclusive explanatory models, even for disorders like schizophrenia, are still far off." In other words, at present and in general, neuroscience is not ready to produce the kind of data that could be helpful in an insanity defense as far as a psychiatric diagnosis in DSM terms is concerned.

Furthermore, the aspirations of the DSM-5 task force and – in contrast – the eventual results make clear that it might take considerably longer than anticipated before cognitive neuroscience is capable of adding relevant data within the context of diagnosing mental disorder. As Kupfer and Regier (2011) indicate ". . . we anticipated that these emerging diagnostic and treatment advances would impact the diagnosis and classification of mental disorders faster than what has actually occurred." Still, it seems not unreasonable to suppose that within a certain time frame neuroscience will begin to generate the kind of results that will directly support the diagnostic process in psychiatry. And if these become available, then they are likely to play a natural role in the psychiatrist's evaluation of the defendant as well.

Psychiatric evaluations of defendants: the future

In the future, neuroscience may help psychiatrists to diagnose defendants. But a psychiatric diagnosis is only one element of forensic psychiatric evaluations. Compared with general psychiatry, forensic psychiatry has a specific interest (Meynen, 2013a). In general psychiatry, patients often see a doctor because of what we may call subjective complaints, in the sense that it has to do with what the patient *experiences*. For instance, feeling depressed, experiencing panic attacks, fear of heights, feelings of emptiness and despair. Consequently, an important goal, if not a central goal, of general mental health care is to make the patient feel better, less depressed, less anxious, less empty.

Forensic psychiatry, in contrast, is often less interested in how a person feels. Crucial in forensic psychiatry is the person's *behavior*, especially legally relevant behavior. Breaking the law has to do with *actions*, rather than *feelings*. Forensic psychiatry, therefore, is especially interested in the impact of a disorder on law-breaking action. Of course, disorders, feelings, and action may be intertwined. For instance, in agoraphobia, a person's fearful avoidance behavior is part of the diagnostic criteria of the DSM, and the urge and act of stealing is an (essential) element of kleptomania. So there is no strict line between disorders, feelings, and actions. Nevertheless, forensic psychiatry has this specific interest in the disorder's impact on action. Therefore in the forensic psychiatric context it is helpful to distinguish between the disorder or psychopathological phenomenon on one hand, and its potential impact on a person's behavior on the other. For instance, the question may have to be answered whether psychotic symptoms made it impossible for a defendant to conform his or her behavior to the requirements of the law. How could neuroscience help to clarify the impact of the disorder on the behavior? This would be a particularly significant contribution of neuroscience to forensic psychiatry.

Let us consider the psychopathological phenomenon we already discussed, commanding voices. Why is it impossible for patients to disobey some of these voices? According to Braham *et al.* (2004), factors that may be associated with obeying the commands are severity of the command or familiarity of the voice. Shawyer *et al.* (2008) suppose that whether or not commanding voices are obeyed depends on a complex interaction between various factors. But much about these commanding voices and their impact on behavior is still unclear. Neuroscience would contribute substantially to forensic assessments if it could elucidate the neurobiological mechanisms through which commanding voices influence decision-making. Then it might become possible to assess, for example by fMRI, whether or not, in a particular defendant, this type of commanding voices is likely to exist or to have existed at the time of the crime. Such findings may have legal consequences: we may feel that if a defendant was suffering from voices they could not disobey – voices that made them commit a crime – they cannot be held legally responsible for their behavior.

In addition, if the insanity defense is successful, and the patient is admitted to a mental hospital, there may, in the future, be further opportunities provided by neuroscientific technique that may detect potential impact of voices on behavior. Neuroscience may reveal a way to sever the connection between the commanding voice and behavior. In other words, based on neuroscience, an *intervention* could become available that is specifically targeting the impact of the voices on the patient's behavior, even if treating the psychosis (voices) as such would be impossible in this patient.

Furthermore, in this way neuroscience could be helpful in *predicting* future impact of such voices on behavior. Let's assume that a defendant's brain could be scanned while he or she is in the mental hospital. Even without an intervention, it might be that, over time, the nature of the voices changes in such a way that the patient is no longer considered a threat to society because the dangerous aspect of his or her psychopathology is no longer present. Although the patient is still hearing voices, neuroimaging reveals that their nature is such that they are very unlikely to influence behavior. Hearing voices is just an example. In the future, neuroscience may contribute to psychiatric evaluations of defendants by elucidating the disorder-specific influence on a defendant's behavior in many psychopathological conditions, such as depression and delusion (Meynen, 2013a).

There could be further advantages of such future neuro-techniques. They are, as far as we know, less dependent on the defendant's willingness to share information, and less prone to manipulation. Because of the defendant's legal interests, and potential consequences of the trial, the possibility of a defendant's concealing or distorting information should always be taken into account. Of course, there are other sources of information as well, such as observation of a defendant's behavior, but also this behavior can, with different degrees of success, be "manipulated" by a defendant; he or she could, for instance, "play crazy." It would be valuable if forensic experts could make use of additional assessment tools (Aharoni *et al.*, 2008). This is what happens in other areas of medicine, where the patient interview is followed by physical examination, blood is taken, X-rays and MRI scans are made. Together these approaches yield the best information for diagnosis and treatment strategies. So brain imaging could be a valuable *extra* source of information for forensic psychiatrists, which would be hard to manipulate. Even if neuroscience would not provide 100 percent reliable results, it can still be valuable because it could strengthen other evidence, or draw attention to specific aspects of mental functioning, which can then be further behaviorally tested. Neuroscience is likely to be of increasing value for evaluations of criminal responsibility not because it offers 100 percent reliable results, but because it opens up a new field that can yield additional, relevant information about the defendant's likely state of mind at the time of committing the crime. Furthermore, neuroscience findings may support predictions about future criminal behavior and be helpful choosing intervention strategies.

Conclusion

The neuroethics of criminal responsibility is related to abstract matters – the freedom of the will, as well as practical issues – the use of neuroscience techniques, such as fMRI, in the court of law. In this chapter we focus on these practical issues. Although, in many legal cases, neurobiological findings may constitute no more than indirect evidence, this does not mean they are irrelevant. It may even be hard to determine the point where "indirect" evidence becomes "direct" evidence, and where "additive" evidence is eventually tipping the balance. Therefore prudent application of neuroscience in the courtroom is called for. At the same time, in the near future, neuroscience could increase the quality of psychiatric evaluations of defendants with respect to not only the question of criminal responsibility, but also prediction of criminal behavior and interventions that prevent future harm.

References

Aharoni, E., Funk, C., Sinnott-Armstrong, W., and Gazzaniga, M. (2008) 'Can neurological evidence help courts assess criminal responsibility? Lessons from law and neuroscience', *Annals of the New York Academy of Sciences*, 1124: 145–160.

American Law Institute (1962) *Model Penal Code*, Philadelphia, PA: American Law Institute.

Andreasen, N.C. (1984) *The Broken Brain: The Biological Revolution in Psychiatry*, New York: Harper & Row.

Bolton, D. (2008) *What is Mental Disorder? An Essay in Philosophy, Science, and Values*, Oxford: Oxford University Press.

Braham, L.G., Trower, P., and Birchwood, M. (2004) 'Acting on command hallucinations and dangerous behavior: a critique of the major findings in the last decade', *Clinical Psychology Review*, 24: 513–528.

Burns, J.M. and Swerdlow, R.H. (2003) 'Right orbitofrontal tumor with pedophilia symptom and constructional apraxia sign', *Archives of Neurology*, 60: 437–440.

Churchland, P.M. (1995) *The Engine of Reason, the Seat of the Soul: A Philosophical Journey into the Brain*, Cambridge, MA: MIT Press.

Dennett, D.C. (2003) *Freedom Evolves*, London: Penguin.

Frankfurt, H. (1971) 'Freedom of the will and the concept of a person', *Journal of Philosophy*, 68: 5–20.

Fried, I., Mukamel, R., and Kreiman, G. (2011) 'Internally generated preactivation of single neurons in human medial frontal cortex predicts volition', *Neuron*, 69: 548–562.

Gazzaniga, M.S. (2005) *The Ethical Brain*, New York: Dana Press.

Greene, J. and Cohen, J. (2004) 'For the law, neuroscience changes nothing and everything', *Philosophical Transactions of the Royal Society of London B, Biological Science*, 359: 1775–1785.

Honderich, T. (2002) 'Determinism as true, compatibilism and incompatibilism as false, and the real problem will', in R. Kane (ed.), *The Oxford Handbook of Free Will*, Oxford: Oxford University Press.

Kane, R. (2002) *The Oxford Handbook of Free Will*, Oxford: Oxford University Press.

Kane, R. (2011) *The Oxford Handbook of Free Will*, 2nd edn, Oxford: Oxford University Press.

Kupfer, D.J. and Regier, D.A. (2011) 'Neuroscience, clinical evidence, and the future of psychiatric classification in DSM-5', *American Journal of Psychiatry*, 168: 672–674.

Kupfer, D.J., First, M.B., and Regier, D.A. (eds) (2002) *A Research Agenda for DSM-V*, Arlington: American Psychiatric Association.

Levy, N. (2007) *Neuroethics*, Cambridge: Cambridge University Press.

Libet, B. (1999) 'Do we have free will?', *Journal of Consciousness Studies*, 6: 47–57.

Libet, B., Gleason, C.A., Wright, E.W., and Pearl, D.K. (1983) 'Time of conscious intention to act in relation to onset of cerebral activity (readiness-potential): the unconscious initiation of a freely voluntary act', *Brain*, 106: 623–642.

Mele, A.R. (2009) *Effective Intentions: The Power of Conscious Will*, New York: Oxford University Press.

Meynen, G. (2009) 'Should or should not forensic psychiatrists think about free will?', *Medicine, Health Care and Philosophy*, 12: 203–212.

Meynen, G. (2010) 'Free will and mental disorder: exploring the relationship', *Theoretical Medicine and Bioethics*, 31: 429–443.

Meynen, G. (2013a) 'A neurolaw perspective on psychiatric assessments of criminal responsibility: decision-making, mental disorder, and the brain', *International Journal of Law and Psychiatry*, 36: 93–99.

Meynen, G. (2013b) *Mad or bad? Over de grenzen van de psychiatrie* (inaugural lecture), Tilburg: Tilburg University.

Meynen, G. (2013c) 'Why mental disorders can diminish responsibility: proposing a theoretical framework', in A.W. Musschenga and A. van Harskamp (eds), *What Makes us Moral: On the Capacities and Conditions for Being Moral*, Berlin: Springer.

Meynen, G. and Oei, T.I. (2010) 'Free will and criminal responsibility', in M. Herzog-Evans (ed.), *Transnational Criminology Manual*, Vol. 1, Nijmegen: Wolf Legal Publishers, 193–207.

M'Naghten's Case, 10 Cl. & Fin. 200, 8 Eng. Rep. 718 (H.L. 1843).

Morse, S. (2003) 'Diminished rationality, diminished responsibility', *Ohio State Journal of Criminal Law*, 1: 289–308.

Morse, S. (2005) 'Brain overclaim syndrome and criminal responsibility: a diagnostic note', *Ohio State Journal of Criminal Law*, 3: 397–312.

Morse, S.J. (2007) 'The non-problem of free will in forensic psychiatry and psychology', *Behavioral Sciences & the Law*, 25: 203–20.

Morse, S.J. (2011a) 'Lost in translation? An essay on law and neuroscience', in M. Freeman (ed.), *Law and Neuroscience: Current Legal Issues*, Oxford: Oxford University Press.

Morse, S.J. (2011b) 'Mental disorder and criminal law', *Journal of Criminal Law and Criminology*, 101: 885–968.

Pereboom, D. (2001) *Living Without Free Will*, New York: Cambridge University Press.

Radder, H. and Meynen, G. (2012) 'Does the brain "initiate" freely willed processes? A philosophy of science critique of Libet-type experiments and their interpretation', *Theory & Psychology*, 23: 3–21.

Regier, D.A., Narrow, W.E., Kuhl, E.A., and Kupfer, D.J. (2009) 'The conceptual development of DSM-V', *American Journal of Psychiatry*, 166: 645–650.

Reich, W. (2005) 'Psychiatric diagnosis as an ethical problem', in S. Bloch, P. Chodoff, and S. Green (eds), *Psychiatric Ethics*, 3rd edn, Oxford: Oxford University Press.

Robinson, D.N. (1998) *Wild Beasts and Idle Humours: The Insanity Defense from Antiquity to the Present*, Cambridge, MA: Harvard University Press.

Roper v. Simmons, 125 S. Ct. 1183 (2005).

Sadler, J.D. (2005) *Values and Psychiatric Diagnosis*, Oxford: Oxford University Press.

Searle, J.R. (2007) *Freedom and Neurobiology Reflections on Free Will, Language, and Political Power*, New York: Columbia University Press.

Shawyer, F., Mackinnon, A., Farhall, J., Sims, E., Blaney, S., Yardley, P., *et al.* (2008) 'Acting on harmful command hallucinations in psychotic disorders: an integrative approach', *Journal of Nervous and Mental Disease*, 196(5): 390–398.

Simon, R.J. and Ahn-Redding, H. (2006) *The Insanity Defense, the World Over*, Lanham, MD: Lexington Books.

Soon, C.S., Brass, M., Heinze, H.J., and Haynes, J.D. (2008) 'Unconscious determinants of free decisions in the human brain', *Nature Neuroscience*, 11: 543–545.

Spence, S. (1996) 'Free will in the light of neuropsychiatry', *Philosophy, Psychiatry, & Psychology*, 3: 75–90.

Spence, S. (2009) *The Actor's Brain: Exploring the Cognitive Neuroscience of Free Will*, New York: Oxford University Press.

Spranger, T.M. (ed.) (2012) *International Neurolaw: A Comparative Analysis*, Heidelberg: Springer.

Strawson, P.F. (2003) 'Freedom and resentment', in G. Watson (ed.), *Free Will*, Oxford: Oxford University Press.

Urbaniok, F., Laubacher, A., Hardegger, J., Rossegger, A., Endrass, J., and Moskvitin, K., (2012) 'Neurobiological determinism: human freedom of choice and criminal responsibility', *International Journal of Offender Therapy and Comparative Criminology*, 56: 174–190.

Vincent, N.A. (2013) 'Law and neuroscience: historical context', in N.A. Vincent (ed.), *Neuroscience and Legal Responsibility*, New York: Oxford University Press, 1–24.

Wegner, D.M. (2002) *The Illusion of Conscious Will*, Cambridge, MA: MIT Press.

38

Alcohol and drug misuse as a biosocial source of crime

Christopher P. Salas-Wright and Jelena Todic

Introduction

The use and abuse of licit and illicit substances have far-reaching implications for the health and well-being of individuals and their communities. Worldwide, substantial proportions of the global burden of disease—that is, the years of life lost due to poor health and premature mortality—are associated with the use of alcohol, tobacco, and illicit drugs, with particularly large effects observed among socioeconomically marginalized populations (Rehm *et al.*, 2006, 2009). In the United States, the estimated economic costs related to substance-use disorders total more than half a trillion dollars per year due to losses in productivity and social expenditures related to healthcare, drug enforcement, incarceration, and crime (National Drug Intelligence Center, 2011; National Institute on Drug Abuse, 2012). An extensive body of literature has documented the links between substance misuse and the involvement of individuals in various manifestations of violent, antisocial, and criminal behavior (Krug *et al.*, 2002; Bennett *et al.*, 2008). Building upon this research, the purpose of this chapter is to highlight the manifold ways in which biosocial factors, substance misuse, and criminal behavior interrelate.

The epidemiology of substance misuse and crime

Prevalence of substance misuse

Substance misuse constitutes an important social and public health problem in the United States and across the globe (Danaei *et al.*, 2009; Degenhardt *et al.*, 2013). A recent Substance Abuse and Mental Health Service Administration (SAMHSA, 2013) report suggests that more than half of all Americans, or approximately 135 million people, regularly drink alcohol. Among these alcohol users, close to half (44 percent) report binge drinking (i.e. five or more drinks on the same occasion) and 6 percent report heavy drinking (i.e. five or more drinks on the same occasion on more than five days in the past month). Nearly one in ten Americans—approximately 24 million people—use illicit drugs, with more than one-third of these users reporting use of illicit drugs other than marijuana (37 percent). Additionally, a steadily increasing number of Americans

(3 percent) report regular nonmedical use of prescription-type psychotherapeutic drugs, with pain relievers identified as the most commonly used class of prescription-type drugs (SAMHSA, 2013).

Evidence suggests that substance misuse is particularly prevalent among young people in the United States. One in five teens (20.5 percent) report using alcohol before the age of 13, nearly three in four (70.1 percent) report trying alcohol at some point during adolescence, and one quarter (25.0 percent) of all late adolescents report having been drunk during the previous month (CDC, 2012; Johnston *et al.*, 2012). Illicit drug use is also widespread during adolescence. Indeed, one in five (20.1 percent) American adolescents report having tried illicit drugs by the 8th grade, a figure that steadily increases to approximately two in five (37.7 percent) by the 10th grade, and roughly half of all adolescents (49.9 percent) by the 12th grade (Johnston *et al.*, 2012). Adolescents are most likely to use marijuana (39.9 percent), followed by inhalants (11.4 percent), cocaine (6.8 percent), methamphetamine (3.8 percent), and heroin (2.9 percent) (CDC, 2012). The age of first use of illicit drugs and alcohol is important, as those who use alcohol and marijuana during early adolescence have been found to be more likely to experience drug and alcohol dependence later in life (SAMSHA, 2013).

Substance misuse and crime

A large body of evidence points to a connection between substance misuse and criminal behavior (Bennett *et al.*, 2008; Lundholm *et al.*, 2013). Indeed, compared with a rate of 8–10 percent in the general population (Compton *et al.*, 2007; SAMHSA, 2012), it is estimated that roughly half of all local (55 percent), state (53 percent), and federal (46 percent) inmates meet criteria for drug abuse or dependence (Chandler *et al.*, 2009). Criminal and antisocial behaviors associated with substance use/abuse include: drug selling (White *et al.*, 2008); violent and nonviolent crime (Krug *et al.*, 2002; White *et al.*, 2002); intimate partner violence (Moore *et al.*, 2008); prostitution (Yacoubian *et al.*, 2001); gang involvement (Hill *et al.*, 1999); and delinquent behaviors such as status offenses (Maynard *et al.*, 2012), and property crimes, drug crime, and interpersonal violence (Salas-Wright *et al.*, 2014). Notably, however, despite the strong association between substance use and crime, person-centered analyses have highlighted the fact that many individuals nevertheless use illicit substances without becoming involved in criminal behavior, and that some serious offenders report little or no substance use (Vaughn *et al.*, 2011, 2014c).

Biosocial models of substance misuse and crime

The interrelatedness of various biosocial factors, substance misuse, and crime is complex. A wide variety of theoretical frameworks can be depicted based on the various combinations of biological, environmental, and behavioral variables in question. Here, based on cutting-edge research, we put forth three overlapping models designed to highlight causal pathways related to the manifold relationships between biological and environmental factors, the use of licit and illicit substances, and the involvement of individuals and violent, antisocial, and criminal behavior.

Biosocial risk factors for substance misuse and crime

Research on the biosocial factors associated with substance misuse and crime suggests an important degree of overlap. That is, as illustrated in Figure 38.1, many of the biological factors found to be associated with substance misuse have also been found to have implications for various manifestations of violent, antisocial, and criminal behavior. For instance, we know that

Biological risk	×	**Socioenvironmental risk** *(interaction)*	→	**Substance misuse** and **crime**
Elevated prenatal testosterone	×	Family risk (e.g. physical abuse, neglect)		
Low resting heart rate	×	Social risk (e.g. substance misuse and crime opportunity)		

Figure 38.1 Biosocial risk factors related to both substance misuse and crime

temperamental factors such as negative emotionality have been shown to be related to both substance misuse and crime (DeLisi and Vaughn, 2014). We see similar links between various stress and sex hormones as well as genetic and neurobiological factors, particularly when in combination with psychosocial disadvantage and maltreatment (Vaughn *et al.*, 2014b). In this subsection, we examine the ways in which two increasingly important biosocial factors—digit ratio and low resting heart rate—are linked with both crime and substance misuse.

Segmental 2 : 4 digit ratio

An emerging body of literature has explored the relationship between segmental 2 : 4 digit ratio and antisocial behavior. Put simply, a number of studies have examined the question of whether or not the ratio between the lengths of the second digit (index finger) and fourth digit (ring finger) on an individual's hand can tell us anything about the likelihood of their involvement in criminal behavior. The rationale for such research is based in evidence suggesting that smaller digit ratios may function as a biomarker for prenatal testosterone levels (Galis *et al.*, 2010). Given the well established link between testosterone and externalizing, particularly with respect to aggression (Book *et al.*, 2001), it is thought that small digit ratios would be linked with the increased likelihood of aggressive behavior (Manning *et al.*, 1998).

Overall, the evidence seems to support such theorizing. A recent meta-analysis that examined 18 studies conducted between 2004 and 2009 found that digit ratio was significantly associated with indirect, direct, verbal, and physical aggressive antisocial behavior (Hönekopp and Watson, 2011). Moreover, beyond the direct associations with aggressive antisocial behavior, studies have also found an association between digit ratio and other factors known to be linked with crime, such as stimulation seeking (Wilson and Scarpa, 2011), attention deficit hyperactivity disorder (ADHD), and lower cognitive empathy (Romero-Martínez *et al.*, 2013). While a number of caveats can be highlighted in terms of the universality of findings (Voracek and Stieger, 2009), differential gender effects (Bailey and Hurd, 2005), and the effect size for meta-analytic data (Hönekopp and Watson, 2011), digit ratio remains an interesting, practical, and highly promising biomarker for criminal behavior.

And yet, for the purposes of this chapter, an important question remains: Is there a relationship between digit ratio and substance use? Noting the links between digit ratio and a variety of correlates of substance use—such as stimulation seeking, ADHD, and, of course, aggression—several recent studies have set out to explore this question. As was the case with crime, research seems to suggest that there is an important connection between digit ratio and substance use. For instance, Kornhuber and colleagues (2011) contrasted the digit ratio between individuals receiving inpatient treatment for alcohol dependency with those of healthy controls, and identified a clear relationship between lower digit ratio and alcohol dependency. Borkowska and Pawlowski (2013) examined the relationship between digit ratio and the use of alcohol and nicotine among a sample of college students in Poland. While no significant relationships were identified for alcohol use, digit ratio was positively associated with the use of nicotine among women. This is in contrast to the pattern observed for alcohol use, but nevertheless reflects a link between digit ratio—and presumably prenatal testosterone—and substance use. Notably, recent cross-national research also suggests that national averages in digit ratio are similarly related to alcohol use and smoking, such that nations with lower mean digit ratios report greater per capita alcohol consumption and lower per capita cigarette use (Manning and Fink, 2011).

Interestingly, it appears that the biosocial links between digit ratio, crime, and substance use may also be observed inter-generationally. Evidence suggests that the children, particularly boys, of mothers who smoke during pregnancy tend to have a smaller digit ratio compared with children of mothers who did not smoke during pregnancy (Rizwan *et al.*, 2007). This is consistent with studies showing that prenatal nicotine exposure can increase the testosterone levels of rat fetuses (Smith *et al.*, 2003), as well as studies highlighting a positive correlation between nicotine consumption and the blood testosterone levels of pregnant mothers (Kandel and Udry, 1999). Perhaps surprisingly, a rather different relationship has been observed between uterine exposure to alcohol and the digit ratio of offspring. That is, uterine alcohol exposure seems to be linked with higher, rather than lower, digit ratios. Although research on the relationship between maternal alcohol use and digit ratio among humans is scant, evidence in animal models has shown rather clearly that prenatal alcohol exposure disrupts (i.e. reduces) testosterone function (McGivern *et al.*, 1988). Indeed, uterine alcohol exposure has been found to be linked with greater forepaw digit ratios in rats (McMechan *et al.*, 2004). In all, research seems to suggest that the complex relationships between digit ratio, crime, and substance use can be observed not only between individuals, but also across generations.

Low resting heart rate

Low resting heart rate has consistently been identified as a risk factor for violent, antisocial, and criminal behavior (Lorber, 2004; Armstrong *et al.*, 2009). Ortiz and Raine (2004), in a meta-analysis of studies examining heart rate levels and antisocial behavior between 1971 and 2002, concluded that low resting heart rate may be the "best-replicated biological correlate of antisocial behavior in child and adolescent populations" (p. 159). While studies among adults are relatively limited in number, recent research suggests a link between heart rate and crime among persons during early and middle adulthood as well (Jennings *et al.*, 2013; Wilson and Scarpa, 2013). The conceptual link between heart rate and antisocial behavior is typically understood by means of stimulation seeking and fearlessness theories (Raine, 2002a,b). That is, individuals with lower resting heart rates are believed to be biologically predisposed to seek out highly stimulating behavior (e.g. high-risk crimes) and be less likely to experience fear in the face of potentially dangerous experiences (e.g. physical fights). It should be noted that recent evidence suggests that the links between low resting heart rate and antisocial behavior may be clearer among men than among women (Wilson and Scarpa, 2013). Whatever the case, a great deal of evidence suggests that persons toward the extreme low end of heart beats per minute tend to be more likely to possess dispositional factors linked with crime and violence.

Both stimulation seeking and fearlessness theories would suggest that low resting heart rate might also be associated with the initiation and abuse of licit and illicit substances. For instance, it might be surmised that individuals with low resting heart rate—and thus, presumably, a greater propensity for stimulation seeking and greater fearlessness—would be more likely to be drawn to the novelty of experimenting with alcohol and drugs. Or perhaps such individuals would be less fearful about the health risks or potentially legal consequences of such use. As has been suggested above, many biosocial factors linked with stimulation seeking have important implications for both substance misuse and crime. And yet, somewhat paradoxically, there is relatively little evidence to suggest an association between low resting heart rate and the use and abuse of licit and illicit substances. It is unclear, however, whether this lack of evidence is a function of a genuine gap in the research in this area or if, in fact, there is little to no relationship to be found between low resting heart rate and substance misuse. Raine (2013) has noted that one of the unique characteristics of the low resting heart rate–crime link is that it is diagnostically specific, inasmuch as most psychiatric conditions, including alcohol dependence, tend to be linked with higher rather than lower resting heart rate levels.

That said, a handful of recent studies have identified an inverse relationship between resting heart rate and alcohol consumption (Hsia *et al.*, 2009; Fernandes *et al.*, 2013). These studies, however, are relatively few in number, and have not systematically examined the mechanisms that might explain this association. Can we unequivocally state that low resting heart rate is linked with substance misuse as well? The current state of the science suggests that this very well may be the case—and certainly the links between low resting heart rate and stimulation seeking, as well as the comorbidity of substance misuse and crime suggest that it is likely the case—but further research is necessary before we can confidently state that such a link exists.

Socioenvironmental factors linked with biosocial risk for substance misuse and crime

Above we examined the ways in which particular biosocial factors are related to both substance misuse and crime. In this subsection, as illustrated in Figure 38.2, we build upon this concept by highlighting the ways in which exposure to socioenvironmental risk factors can have implications for human biology and, in turn, impact the likelihood of individuals misusing substances and becoming involved in criminal behavior. For example, recent studies suggest that chronic exposure to stressful experiences such as discrimination may lead to differences in diurnal cortisol rhythms among youth (DeSantis *et al.*, 2007). Variation in cortisol levels, in turn, has important implications for both substance misuse (Lovallo, 2006) and criminal behavior (van

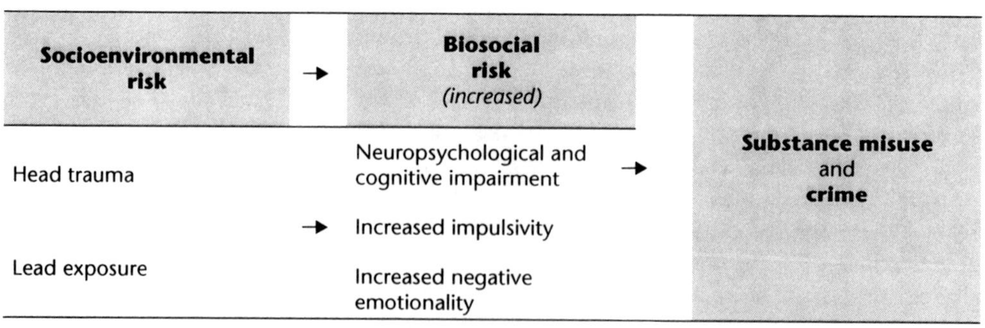

Figure 38.2 Socioenvironmental factors influence biological risk to impact substance misuse and crime

Goozen *et al.*, 2007). Similar examples can be found in reference to particular dietary factors (e.g. polyunsaturated fats) and—as evidenced above with respect to uterine nicotine exposure and digit ratio—prenatal exposure to licit and illicit substances (Hibbeln *et al.*, 2004; Borsonelo and Galduróz, 2008). In this subsection, we examine the ways in which socioenvironmental factors influence human biology and, in turn, increase the likelihood of both substance misuse and crime, by focusing on traumatic brain injury (TBI) and environmental exposure to lead.

Traumatic brain injury

Traumatic brain injury refers to head injuries that lead to either a loss of consciousness or amnesia, and are associated with subsequent impairment with respect to a cognitive and/or social functioning (Craswell *et al.*, 2004). Traumatic brain injury is remarkably widespread in the general population—it is estimated that more than one in ten (12 percent) individuals has experienced such an injury in their lifetime (Frost *et al.*, 2013). Notably, meta-analytic studies have estimated the prevalence of TBI to be between three and six times greater among individuals involved in the criminal justice system (Shiroma *et al.*, 2010; Farrer *et al.*, 2013). As is evident, these types of head injury are not randomly distributed in the population, but rather are disproportionately experienced by individuals at elevated risk for victimization through child abuse, violent altercations, intimate partner violence, and community violence exposure. Traumatic brain injuries are, by definition, the result of socioenvironmental experiences (e.g. violent victimization, traumatic accidents) that can alter our biology, most notably with respect to impaired neuropsychological and cognitive functioning (Schretlen and Shapiro, 2003), including declines related to self-control and self-regulation (Ganesalingam *et al.*, 2006).

A number of recent studies have identified elevated levels of both substance misuse and criminal behavior among individuals who have experienced a TBI. In a study of adolescent juvenile offenders in Phoenix, Arizona and Philadelphia, Pennsylvania, youth reporting having ever lost consciousness due to a head injury were found to be significantly more likely to meet criteria for alcohol and illicit drug use disorders, and to report involvement in delinquent and antisocial behavior (Vaughn *et al.*, 2014a). Findings from this study are consistent with those of other samples of juvenile offenders in the Midwestern United States and in the United Kingdom (Perron and Howard, 2008; Williams *et al.*, 2010). Several biosocial mechanisms might help to explain the relationship between TBI, substance misuse, and crime. For instance, Vaughn and colleagues (2014a) found that juvenile offenders reporting a TBI scored significantly higher in terms of impulsivity and negative emotionality, both of which are related to the neuropsychological consequences of head trauma and function as risk factors for misuse and crime. Cognitive impairment, which is linked with both substance misuse and crime, may also constitute an important biosocial mechanism in explaining the TBI–externalizing link (Kinnunen *et al.*, 2010). In all, while additional research could help to further elucidate the links between head trauma and substance misuse and crime, evidence seems to point to a dynamic in which socioenvironmental factors (e.g. trauma) impacts our biology (e.g. neuropsychological functioning) and, in turn, impacts the likelihood of involvement in substance misuse and crime.

Environmental exposure to lead

It is well established that exposure to lead has important implications for the neuropsychological and social development of children (Needleman, 2004). Recent research suggests that exposure—typically by means of the ingestion of lead-based paint and lead-contaminated dust—to even very low levels of lead can have important implications for key outcomes such as impulsivity, negative emotionality, intelligence, cognitive functioning, academic achievement, and neuropsychiatric disorders such as ADHD (Chiodo *et al.*, 2004; Bellinger, 2008a). Given that all the aforementioned outcomes are important correlates of criminal behavior, it is perhaps unsurprising that a rather robust body of evidence has accrued that underscores the relationship between lead exposure and problem behavior (Bellinger, 2008b). At the individual level, recent meta-analytic evidence examining the relationship between conduct problems and lead exposure identified a medium effect size ($r = 0.19$) across 19 studies of children and adolescents (Marcus *et al.*, 2010). Additional studies suggest that early lead exposure has important implications for delinquency and involvement in the criminal justice system during adolescence and early adulthood (Dietrich *et al.*, 2001; Wright *et al.*, 2008). At the societal level, a recent study by Mielke and Zahran (2012) comparing air lead emissions and aggravated assault across various cities in the United States suggests a powerful link between aggregate lead exposure and violent crime.

Despite the abundance of research on lead exposure and crime, relatively few studies have systematically examined the relationship between lead exposure and substance misuse. Dietrich and colleagues (2001), drawing from a prospective longitudinal birth cohort of roughly 200 inner-city youth, observed that prenatal/postnatal lead exposure was modestly correlated with adolescent alcohol and marijuana consumption; however, their analysis did not examine the longitudinal relationships between lead exposure and substance use initiation or misuse in any detail. Along the same lines, Fishbein and colleagues (2008) identified significant differences in tibial lead concentration between female injection drug users and a healthy community sample. However, several important methodological shortcomings—including a small sample size and the limitations of matching two independent samples—suggest that caution should be exercised in interpreting these results.

Although evidence of a lead–substance misuse link is relatively limited among humans, several studies have been carried out using animal models. For instance, Rocha and colleagues (2005) found that rats exposed to lead self-administered cocaine at significantly greater levels than nonexposed rats, suggesting that lead exposure may have implications for drug sensitivity. Similarly, exposure to lead has been found to be linked with differential reactivity for additional substances, including methamphetamine and opiates (Rocha *et al.*, 2005; Clifford *et al.*, 2009; Listos *et al.*, 2013). Given the links between lead exposure and neuropsychological and behavioral outcomes in humans, it is perhaps unsurprising that animal models reveal the relationship between exposure to lead and the use of illicit substances. While further research is needed, the current evidence on the neuropsychological and behavioral implications of lead exposure suggests, as was the case with head trauma, that lead exposure can impact human biology in ways that place individuals at a disproportionate risk for involvement in both substance misuse and crime.

Substance misuse as a source of biosocial liability for crime

Thus far, we have examined the ways in which particular biosocial factors relate to both substance misuse and crime, and how socioenvironmental risk factors impact our biology which, in turn, increases the likelihood of involvement in substance misuse and crime. In this final section, as illustrated in Figure 38.3, we explore the ways in which substance misuse impacts human biology such that individuals who use and abuse substances may be at increased risk for involvement in criminal behavior. In particular, we focus our attention on the distal and

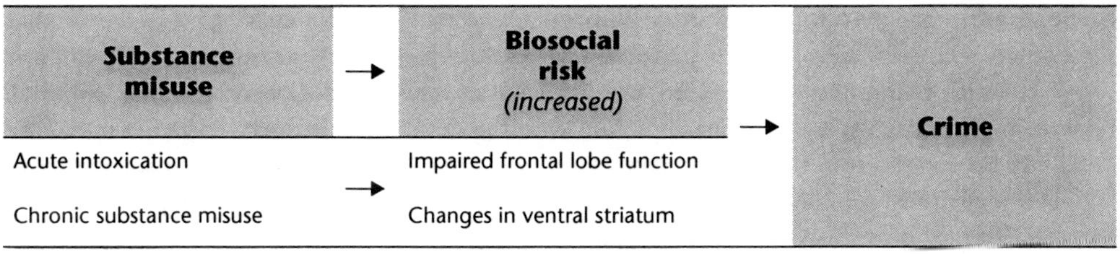

Figure 38.3 Substance abuse as source of biosocial liability for crime

proximal implications of substance misuse in terms of the capacity of individuals to exercise judgment and self-control which, in turn, are associated with the likelihood of involvement in high-risk and criminal behavior.

A substantial body of research suggests a link between acute alcohol and drug intoxication and crime. At a basic level, we know that, particularly with respect to impulsive or violent crimes, a substantial proportion of offenders are intoxicated while carrying out violent or criminal behaviors (Boles and Miotto, 2003). A recent study by Felson and Staff (2010) found that, controlling for chronic use and prior offending, alcohol intoxication was significantly associated with an extensive array of criminal behaviors, including homicide, physical and sexual assault, robbery, and burglary. Study findings suggest that, while even moderate alcohol consumption is associated with criminal behavior, a dose-dependent effect can also be identified as higher levels of intoxication were linked with increases in the likelihood of offending. Consistent with observational studies, experimental studies in controlled laboratory settings have demonstrated that the consumption of substances such as alcohol has been found to increase aggressive behavior, particularly among physically large men (Lipsey *et al.*, 1997; DeWall *et al.*, 2010). Several biosocial mechanisms have been identified that may explain the relationship between acute intoxication and criminal behavior. For instance, acute alcohol intoxication has been shown to impair frontal lobe functioning (e.g. planning, information processing, and self-control) and to alter neurochemical systems (e.g. stress-related activation, reward expectation) believed to be linked with aggression (Heinz *et al.*, 2011). Simply, evidence suggests that intoxication temporarily alters our capacity to think clearly, control our impulses, and respond to distress in ways that increase the likelihood of criminal and violent offending.

In addition to the link between acute intoxication and crime, chronic alcohol and drug use has been found to be associated with the increased likelihood of involvement in criminal and violent behavior (French *et al.*, 2000). Research indicates that chronic substance misuse has the capacity to have a longer-term impact on our biology in ways that can predispose individuals for involvement in criminal behavior. Several neurological mechanisms have been identified as potential mediating factors between chronic substance misuse and crime. For instance, recent meta-analytic evidence suggests that heavy social drinking may have longer-term implications for executive functioning (Montgomery *et al.*, 2012). Impaired executive functioning, which is known to be robustly associated with increased impulsivity (Morgan and Lilienfeld, 2000), is associated with an increased likelihood of individuals taking part in violent and criminal behavior. Similarly, Heinz and colleagues (2011) highlight the ways in which chronic alcohol use may be related to dysfunction in the amygdala and ventral stratum which, consequently, can predispose individuals to aggressive behavior.

Beyond substance misuse, some evidence suggests that prenatal exposure to substances such as cocaine can have important implications for self-control more than a decade later. For instance,

Coyle (2013), using magnetic resonance imaging with adolescents aged 13–15, found that adolescents who had been prenatally exposed to cocaine had less cortical thickness—which is correlated with impulsivity—compared with adolescents who had not been prenatally exposed. In other words, drug exposure *in utero* is linked with the maturation of areas of the brain that are known to be important in terms of impulsivity and self-control. This finding provides clear biological evidence that is in keeping with a robust body of research highlighting the links between prenatal drug exposure and impulsivity (Volkow, 2013), as well the relationship between impulsivity, self-control, and criminal behavior (de Ridder *et al.*, 2012). In all, evidence strongly suggests that exposure to a variety of licit and illicit substances—be it prenatally or via substance misuse during subsequent developmental periods—has the capacity to change our neurobiology in ways that can have implications for criminal behavior.

Conclusion

While each of these models shares a degree of conceptual overlap, they are intended to organize the complex and multifaceted biosocial processes linked with substance misuse and crime. Here we examine the ways in which particular biosocial factors, such as prenatal testosterone and low resting heart rate, are linked with both crime and substance misuse. We also look at how exposure to particular environmental risks, such as head trauma and the consumption of lead, can impact our biology and, in turn, impact the likelihood of involvement in crime and substance misuse. And we consider the ways in which the use and misuse of substances can impact our biology such that the likelihood of criminal involvement is increased. These overlapping frameworks point to a profound interconnectedness between our biology, the social environment, and neurological and psychosocial constructs related to substance misuse and crime. Alcohol and drug misuse certainly function as a biosocial source of crime, just as important biosocial and environmental risk factors are related to both the misuse of substances and involvement in criminal behavior.

Acknowledgement

This research was supported in part by grant number R25 DA026401 (PI: Valdez) from the National Institute on Drug Abuse at the National Institutes of Health.

References

Armstrong, T.A., Keller, S., Franklin, T.W., and MacMillan, S.N. (2009) 'Low resting heart rate and anti-social behavior: a brief review of evidence and preliminary results from a new test', *Criminal Justice and Behavior*, 36(11): 1125–1140.

Bailey, A.A. and Hurd, P.L. (2005) 'Finger length ratio (2D : 4D) correlates with physical aggression in men but not in women', *Biological Psychology*, 68(3): 215–222.

Bellinger, D.C. (2008a) 'Lead neurotoxicity and socioeconomic status: conceptual and analytical issues', *Neurotoxicology*, 29(5): 828–832.

Bellinger, D.C. (2008b) 'Neurological and behavioral consequences of childhood lead exposure', *PLoS Medicine*, 5(5): e115.

Bennett, T., Holloway, K., and Farrington, D. (2008) 'The statistical association between drug misuse and crime: a meta-analysis', *Aggression and Violent Behavior*, 13(2): 107–118.

Boles, S.M. and Miotto, K. (2003) 'Substance abuse and violence: a review of the literature', *Aggression and Violent Behavior*, 8(2): 155–174.

Book, A.S., Starzyk, K.B., and Quinsey, V.L. (2001) 'The relationship between testosterone and aggression: a meta-analysis', *Aggression and Violent Behavior*, 6(6): 579–599.

Borsonelo, E.C. and Galduróz, J.C.F. (2008) 'The role of polyunsaturated fatty acids (PUFAs) in development, aging and substance abuse disorders: review and propositions', *Prostaglandins, Leukotrienes and Essential Fatty Acids*, 78(4): 237–245.

Borkowska, B. and Pawlowski, B. (2013) 'Alcohol and nicotine intake and prenatal level of androgens measured by digit ratio', *Personality and Individual Differences*, 55: 685–687.

CDC (2012) 'Youth Risk Behavior Surveillance System: selected 2011 national health risk behaviors and health outcomes by race/ethnicity', Atlanta, GA: Centers for Disease Control and Prevention, www.cdc.gov/healthyyouth/yrbs/factsheets/index.htm

Chandler, R.K., Fletcher, B.W., and Volkow, N.D. (2009) 'Treating drug abuse and addiction in the criminal justice system: improving public health and safety', *Journal of the American Medical Association*, 301(2): 183–190.

Chiodo, L.M., Jacobson, S.W., and Jacobson, J.L. (2004) 'Neurodevelopmental effects of postnatal lead exposure at very low levels', *Neurotoxicology and Teratology*, 26(3): 359–371.

Clifford, P.S., Hart, N., Thompson, J., Buckman, S., Wellman, P.J., Bratton, G.R., and Nation, J. (2009) 'Prenatal lead exposure enhances methamphetamine sensitization in rats', *Pharmacology Biochemistry and Behavior*, 93(2): 165–169.

Compton, W.M., Thomas, Y.F., Stinson, F.S., and Grant, B.F. (2007) 'Prevalence, correlates, disability, and comorbidity of DSM-IV drug abuse and dependence in the United States', *Archives of General Psychiatry*, 64: 566–578.

Coyle, J.T. (2013) 'Brain structural alterations induced by fetal exposure to cocaine persist into adolescence and affect behavior', *JAMA Psychiatry*, 70(10): 1113–1114.

Craswell, K., Maughan, B., Davis, H., Davenport, F., and Goddard, N. (2004) 'The psychosocial needs of young offenders and adolescents from an inner city area', *Journal of Adolescence*, 27: 415–428.

Danaei, G., Ding, E.L., Mozaffarian, D., Taylor, B., Rehm, J., Murray, C.J.L., and Ezzati, M. (2009) 'The preventable causes of death in the United States: comparative risk assessment of dietary, lifestyle, and metabolic risk factors', *PLoS Medicine*, 6(4): e1000058.

Degenhardt, L., Whiteford, H.A., Ferrari, A.J., Baxter, A.J., Charlson, F.J., Hall, W.D., *et al.* (2013) 'Global burden of disease attributable to illicit drug use and dependence: findings from the Global Burden of Disease Study 2010', *The Lancet*, 382(9904): 1564–1574.

DeLisi, M. and Vaughn, M.G. (2014) 'Foundation for a temperament-based theory of antisocial behavior and criminal justice system involvement', *Journal of Criminal Justice*, 42(1): 10–25.

DeSantis, A.S., Adam, E.K., Doane, L.D., Mineka, S., Zinbarg, R.E., and Craske, M.G. (2007) 'Racial/ethnic differences in cortisol diurnal rhythms in a community sample of adolescents', *Journal of Adolescent Health*, 41(1): 3–13.

DeWall, C.N., Bushman, B.J., Giancola, P.R., and Webster, G.D. (2010) 'The big, the bad, and the boozed-up: weight moderates the effect of alcohol on aggression', *Journal of Experimental Social Psychology*, 46(4): 619–623.

Dietrich, K.N., Douglas, R.M., Succop, P.A., Berger, O.G., and Bornschein, R.L. (2001) 'Early exposure to lead and juvenile delinquency', *Neurotoxicology and Teratology*, 23(6): 511–518.

Farrer, T.J., Frost, R.B., and Hedges, D.W. (2013) 'Prevalence of traumatic brain injury in juvenile offenders: a meta-analysis', *Child Neuropsychology*, 19(3): 225–234.

Felson, R.B., and Staff, J. (2010) 'The effects of alcohol intoxication on violent versus other offending', *Criminal Justice and Behavior*, 37(12): 1343–1360.

Fernandes, R.A., Ronque, E.R.V., Venturini, D., Barbosa, D.S., Silva, D.P., Cogo, C.T., *et al.* (2013) 'Resting heart rate: its correlations and potential for screening metabolic dysfunctions in adolescents', *BMC Pediatrics*, 13(1): 48.

Fishbein, D.H., Todd, A.C., Ricketts, E.P., and Semba, R.D. (2008) 'Relationship between lead exposure, cognitive function, and drug addiction: pilot study and research agenda', *Environmental Research*, 108(3): 315–319.

French, M.T., McGeary, K.A., Chitwood, D.D., McCoy, C.B., Inciardi, J.A., and McBride, D. (2000) 'Chronic drug use and crime', *Substance Abuse*, 21(2): 95–109.

Frost, R.B., Farrer, T.J., Primosch, M., and Hedges, D.W. (2013) 'Prevalence of traumatic brain injury in the general adult population: a meta-analysis', *Neuroepidemiology*, 40(3): 154–159.

Galis, F., Ten Broek, C.M., Van Dongen, S., and Wijnaendts, L.C. (2010) 'Sexual dimorphism in the prenatal digit ratio (2D : 4D)', *Archives of Sexual Behavior*, 39(1): 57–62.

Ganesalingam, K., Sanson, A., Anderson, V., and Yeates, K.O. (2006) 'Self-regulation and social and behavioral functioning following childhood traumatic brain injury', *Journal of the International Neuropsychological Society*, 12(5): 609–621.

van Goozen, S.M., Snoek, H., Fairchild, G., and Harold, G.T. (2007) 'The evidence for a neurobiological model of childhood antisocial behavior', *Psychological Bulletin*, 133(1): 149–182.

Heinz, A.J., Beck, A., Meyer-Lindenberg, A., Sterzer, P., and Heinz, A. (2011) 'Cognitive and neurobiological mechanisms of alcohol-related aggression', *Nature Reviews Neuroscience*, 12(7): 400–413.

Hibbeln, J.R., Bissette, G., Umhau, J.C., and George, D.T. (2004) 'Omega-3 status and cerebrospinal fluid corticotrophin releasing hormone in perpetrators of domestic violence', *Biological Psychiatry*, 56(11): 895–897.

Hill, K.G., Howell, J.C., Hawkins, J.D., and Battin-Pearson, S.R. (1999) 'Childhood risk factors for adolescent gang membership: results from the Seattle Social Development Project', *Journal of Research in Crime and Delinquency*, 36: 300–322.

Hönekopp, J. and Watson, S. (2011) 'Meta-analysis of the relationship between digit-ratio 2D : 4D and aggression', *Personality and Individual Differences*, 51(4): 381–386.

Hsia, J., Larson, J.C., Ockene, J.K., Sarto, G.E., Allison, M.A., Hendrix, S.L., *et al.* (2009) 'Resting heart rate as a low tech predictor of coronary events in women: prospective cohort study', *British Medical Journal*, 338: b219.

Jennings, W.G., Piquero, A.R., and Farrington, D.P. (2013) 'Does resting heart rate at age 18 distinguish general and violent offending up to age 50? Findings from the Cambridge Study in Delinquent Development', *Journal of Criminal Justice*, 41(4): 213–219.

Johnston, L.D., O'Malley, P.M., Bachman, J.G., and Schulenberg, J.E. (2012) *Monitoring the Future. National Results on Adolescent Drug Use: Overview of Key Findings, 2011*, Ann Arbor, MI: Institute for Social Research, University of Michigan.

Kandel, D.B. and Udry, J.R. (1999) 'Prenatal effects of maternal smoking on daughters' smoking: nicotine or testosterone exposure?', *American Journal of Public Health*, 89(9): 1377–1383.

Kinnunen, K.M., Greenwood, R., Powell, J.H., Leech, R., Hawkins, P.C., Bonnelle, V., *et al.* (2010) 'White matter damage and cognitive impairment after traumatic brain injury', *Brain*, 134: 449–463.

Kornhuber, J., Erhard, G., Lenz, B., Kraus, T., Sperling, W., Bayerlein, K., *et al.* (2011) 'Low digit ratio 2D: 4D in alcohol dependent patients', *PloS One*, 6(4): e19332.

Krug, E.G., Mercy, J.A., Dahlberg, L.L., and Zwi, A.B. (2002) 'The world report on violence and health', *The Lancet*, 360(9339): 1083–1088.

Lipsey, M.W., Wilson, D.B., Cohen, M.A., and Derzon, J.H. (1997) 'Is there a causal relationship between alcohol use and violence? A synthesis of evidence', *Recent Developments in Alcoholism*, 13: 245–282.

Listos, J., Baranowska-Bosiacka, I., Talarek, S., Listos, P., Orzelska, J., Fidecka, S., *et al.* (2013) 'The effect of perinatal lead exposure on dopamine receptor d2 expression in morphine dependent rats', *Toxicology*, 310(9): 73–83.

Lorber, M.F. (2004) 'Psychophysiology of aggression, psychopathy, and conduct problems: a meta-analysis', *Psychological Bulletin*, 130(4): 531–552.

Lovallo, W.R. (2006) 'Cortisol secretion patterns in addiction and addiction risk', *International Journal of Psychophysiology*, 59(3): 195–202.

Lundholm, L., Haggard, U., Moller, J., Hallqvist, J., and Thiblin, I. (2013) 'The triggering effect of alcohol and illicit drugs on violent crime in a remand prison population: a case crossover study', *Drug and Alcohol Dependence*, 129(1/2): 110–115.

McGivern, R.F., Raum, W.J., Salido, E., and Redei, E. (1988) 'Lack of prenatal testosterone surge in fetal rats exposed to alcohol: Alterations in testicular morphology and physiology', *Alcoholism: Clinical and Experimental Research*, 12(2): 243–247.

McMechan, A.P., O'Leary-Moore, S.K., Morrison, S.D., and Hannigan, J.H. (2004) 'Effects of prenatal alcohol exposure on forepaw digit length and digit ratios in rats', *Developmental Psychobiology*, 45(4): 251–258.

Manning, J.T. and Fink, B. (2011) 'Digit ratio, nicotine and alcohol intake and national rates of smoking and alcohol consumption', *Personality and Individual Differences*, 50(3): 344–348.

Manning, J.T., Scutt, D., Wilson, J., and Lewis-Jones, D.I. (1998) 'The ratio of 2nd to 4th digit length: a predictor of sperm numbers and concentrations of testosterone, luteinizing hormone and oestrogen', *Human Reproduction*, 13(11): 3000–3004.

Marcus, D.K., Fulton, J.J., and Clarke, E.J. (2010) 'Lead and conduct problems: a meta-analysis', *Journal of Clinical Child and Adolescent Psychology*, 39(2): 234–241.

Maynard, B.R., Salas-Wright, C.P., Vaughn, M.G., and Peters, K.E. (2012) 'Who are truant youth? Examining distinctive profiles of truant youth using latent profile analysis', *Journal of Youth and Adolescence*, 41(2): 1671–1684.

Mielke, H.W. and Zahran, S. (2012) 'The urban rise and fall of air lead (Pb) and the latent surge and retreat of societal violence', *Environment International*, 43: 48–55.

Montgomery, C., Fisk, J.E., Murphy, P.N., Ryland, I., and Hilton, J. (2012) 'The effects of heavy social drinking on executive function: a systematic review and meta-analytic study of existing literature and new empirical findings', *Human Psychopharmacology: Clinical and Experimental*, 27(2): 187–199.

Moore, T.M., Sturart, G.L., Meehan, J.C., Rhatigan, D.L., Helmuth, J.C., and Keen, S.M. (2008) 'Drug abuse and aggression between intimate partners: a meta-analytic review', *Clinical Psychology Review*, 28: 247–274.

Morgan, A.B. and Lilienfeld, S.O. (2000) 'A meta-analytic review of the relation between antisocial behavior and neuropsychological measures of executive function', *Clinical Psychology Review*, 20(1): 113–136.

National Drug Intelligence Center (2011) *The Economic Impact of Illicit Drug Use on American Society*, Washington, DC: US Department of Justice.

National Institute on Drug Abuse (2012) *Drug Facts: Understanding Drug Abuse and Addiction*, Washington, DC: US Department of Justice.

Needleman, H. (2004) 'Lead poisoning', *Annual Reviews of Medicine*, 55: 209–222.

Ortiz, J. and Raine, A. (2004) 'Heart rate level and antisocial behavior in children and adolescents: a meta-analysis', *Journal of the American Academy of Child and Adolescent Psychiatry*, 43(2): 154–162.

Perron, B.E. and Howard, M.O. (2008) 'Prevalence and correlates of traumatic brain injury among delinquent youths', *Criminal Behavior and Mental Health*, 18(4): 243–255.

Raine, A. (2002a) 'Biosocial studies of antisocial and violent behavior in children and adults: a review', *Journal of Abnormal Child Psychology*, 30(4): 311–326.

Raine, A. (2002b) 'Annotation: the role of prefrontal deficits, low autonomic arousal, and early health factors in the development of antisocial and aggressive behavior in children', *Journal of Child Psychology and Psychiatry*, 43(4): 417–434.

Raine, A. (2013) *The Anatomy of Violence: The Biological Roots of Crime*, New York: Random House.

Rehm, J., Taylor, B., and Room, R. (2006) 'Global burden of disease from alcohol, illicit drugs and tobacco', *Drug and Alcohol Review*, 25: 503–513.

Rehm, J., Mathers, C., Popova, S., Thavorncharoensap, M., Teerawattananon, Y., and Patra, J. (2009) 'Global burden of disease and injury and economic cost attributable to alcohol use and alcohol-use disorders', *The Lancet*, 373(9682): 2223–2233.

de Ridder, D.T., Lensvelt-Mulders, G., Finkenauer, C., Stok, F.M., and Baumeister, R.F. (2012) 'Taking stock of self-control: a meta-analysis of how trait self-control relates to a wide range of behaviors', *Personality and Social Psychology Review*, 16(1): 76–99.

Rizwan, S., Manning, J.T., and Brabin, B.J. (2007) 'Maternal smoking during pregnancy and possible effects of in utero testosterone: evidence from the 2D : 4D finger length ratio', *Early Human Development*, 83(2): 87–90.

Rocha, A., Valles, R., Cardon, A.L., Bratton, G.R., and Nation, J.R. (2005) 'Enhanced acquisition of cocaine self-administration in rats developmentally exposed to lead', *Neuropsychopharmacology*, 30(11): 2058–2064.

Romero-Martínez, Á., Lila, M., Catalá-Miñana, A., Williams, R.K., and Moya-Albiol, L. (2013) 'The contribution of childhood parental rejection and early androgen exposure to impairments in socio-cognitive skills in intimate partner violence perpetrators with high alcohol consumption', *International Journal of Environmental Research and Public Health*, 10(8): 3753–3770.

Salas-Wright, C.P., Hernandez, L., Maynard, B.R., Saltzman, L., and Vaughn, M.G. (2014) 'Early alcohol use among Hispanic adolescents in the United States: an examination of behavioral risk and protective profiles', *Substance Use and Misuse*, 49: 864–877.

SAMHSA (2012) *Results from the 2011 National Survey on Drug Use and Health: Summary of National Findings*, Rockville, MD: Substance Abuse and Mental Health Services Administration.

SAMHSA (2013) *Results from the 2012 National Survey on Drug Use and Health: Summary of National Findings*, Rockville, MD: Substance Abuse and Mental Health Services Administration.

Schretlen, D.J. and Shapiro, A.M. (2003) 'A quantitative review of the effects of traumatic brain injury on cognitive functioning', *International Review of Psychiatry*, 15(4): 341–349.

Shiroma, E.J., Ferguson, P.L., and Pickelsimer, E.E. (2010) 'Prevalence of traumatic brain injury in an offender population: a meta-analysis', *Journal of Correctional Health Care*, 16(2): 147–159.

Smith, L.M., Cloak, C.C., Poland, R.E., Torday, J., and Ross, M.G. (2003) 'Prenatal nicotine increases testosterone levels in the fetus and female offspring', *Nicotine and Tobacco Research*, 5(3): 369–374.

Vaughn, M.G., DeLisi, M., Gunter, T., Fu, Q., Beaver, K.M., Perron, B.E., and Howard, M.O. (2011) 'The severe 5%: a latent class analysis of the externalizing spectrum in the United States', *Journal of Criminal Justice*, 39: 75–80.

Vaughn, M.G, Salas-Wright, C.P., DeLisi, M., and Perron, B.E. (2014a) 'Correlates of traumatic brain injury among juvenile offenders: a multi-site study', *Criminal Behaviour and Mental Health*, 24: 188–203.

Vaughn, M.G., Salas-Wright, C. P., and Maynard, B.R. (2014b) 'Drug abuse and addiction careers: a cell to society perspective', in B. Boutwell, K. Beaver, and J.C. Barnes (eds), *The Nurture versus Biosocial Debate in Criminology*, Thousand Oaks, CA: Sage.

Vaughn, M.G., Salas-Wright, C.P., DeLisi, M., and Maynard, B.R. (2014c) 'Violence and externalizing behavior among youth in the United States: is there a severe 5%?', *Youth Violence and Juvenile Justice*, 12(1): 3–21.

Volkow, N.D. (2013) 'Impact of drug exposure on the adolescent brain', *JAMA Pediatrics*, 167(4): 390–391.

Voracek, M. and Stieger, S. (2009) 'Replicated nil associations of digit ratio (2D : 4D) and absolute finger lengths with implicit and explicit measures of aggression', *Psicothema*, 21(3): 382–389.

White, H.R., Tice, P., Loeber, R., and Stouthamer-Loeber, M. (2002) 'Illegal acts committed by adolescents under the influence of alcohol and drugs', *Journal of Research in Crime and Delinquency*, 39: 131–152.

White, H.R., Loeber, R., and Farrington, D.P. (2008) 'Substance use, drug dealing, gang membership, and gun carrying and their predictive associations with serious violence and serious theft', in R. Loeber, D.P. Farrington, M. Stouthamer-Loeber, and H.R. White (eds), *Violence and Serious Theft: Development and Prediction from Childhood to Adulthood*, New York: Routledge, pp. 137–166.

Williams, W.H., Cordan, G., Mewse, A.J., Tonks, J., and Burgess, C.N. (2010) 'Self-reported traumatic brain injury in male young offenders: a risk factor for re-offending, poor mental health and violence?', *Neuropsychological Rehabilitation*, 20(6): 801–812.

Wilson, L.C. and Scarpa, A. (2011) 'The link between sensation seeking and aggression: a meta-analytic review', *Aggressive Behavior*, 37(1): 81–90.

Wilson, L.C. and Scarpa, A. (2013) 'Baseline heart rate, sensation seeking, and aggression in young adult women: a two-sample examination', *Aggressive Behavior*, 39(4): 280–289.

Wright, J.P., Dietrich, K.N., Ris, M.D., Hornung, R.W., Wessel, S.D., Lanphear, B.P., *et al.* (2008) 'Association of prenatal and childhood blood lead concentrations with criminal arrests in early adulthood', *PLoS Medicine*, 5(5): e101.

Yacoubian, G.S., Urbach, B.J., Larsen, K.L., Johnson, R.J., and Peters, R.J. (2001) 'A comparison of drug use between prostitutes and other female arrestees', *Journal of Alcohol and Drug Education*, 46(2): 12–26.

39

Empathy, crime, and justice

Chad Posick, Michael Rocque, and Matt DeLisi

Introduction

Justice is the cornerstone of any fair and democratic society. It should therefore come as little surprise that issues concerning the fairness of punishment, mechanisms of social control, and treatment of those who break societal rules are so important to people in society. Questions abound as to the origins of our sense of justice and how we can build a just society. One of the essential elements of justice includes the concept of empathy. Empathy is a focus in research efforts as an integral component of making laws, responding to lawbreakers, and preventing crime (Posick *et al.*, 2014). This chapter discusses how empathy is an essential component of each of these areas, demonstrating how it contributes to a holistic understanding of key concepts in the field of criminology. This foray into the importance of empathy in the study of criminology can help us further our understanding of why people break laws and, consequently, how to prevent those behaviors in the first place.

Decety (2010: 258) has used a simple definition of empathy as "the ability to recognize the feelings and emotions of others," while de Vignemont and Singer (2006: 435) provide a much more fine-tuned definition in which they state that empathy occurs when ". . . we have (a) an affective state, (b) which is isomorphic to another person's affective state, (c) which was elicited by observing or imagining another person's affective state, and (d) when we know that the other person's affective state is the source of our own affective state." This distinction between the cognitive dimension (recognition of others' emotions) and affective dimension (personal feelings in response to the recognition of others' emotions) is important in relating empathy to antisocial and delinquent behavior. It suggests that there are separate components of empathy which likely have different etiologies and biological/neurological correlates.

Empathy is often considered analogous to other emotions and traits, such as sympathy and compassion. While an extensive discussion on these distinctions would be distracting here, and debate continues whether such perceived hairsplitting between the terms is productive in the first place, we should be clear that our attention is on empathy as the ability to understand and relate to the feelings of another, which does not require a specific judgment or reaction. For example, Singer and Lamm (2009) point out that while empathy often does encompass compassion and altruism, it is distinct. What separates empathy from these other constructs is the notion of "shared feelings." These feelings are isomorphic to those of others, resulting in affective sharing ("feelings with") as opposed to merely "feeling for" another person. The point of bringing up this distinction here is that it may be important for the field of criminal justice to take note of because, for example, police officers should not necessarily be sympathetic to those who they deal with on a daily basis, whether they be the perpetrators or victims of crimes. However, they should be empathetic—that is, have the ability to understand and relate to others while remaining neutral in their actions. As these authors argue (along with many others who we discuss in this chapter), empathy is influenced by both contextual (sociological) and biological (psychological and neurological) mechanisms.

A more nuanced understanding of genetics, neuroscience, and other related fields has led social science researchers to believe there is utility in theoretical explanations of behavior that incorporate biology and cognitive research into sociological and psychological theories. However, these efforts, including emerging fields such as epigenetics and social neuroscience, are nascent. While we find much to celebrate in a multidisciplinary focus to the study of empathy in criminology, we must also acknowledge that there is much we do not yet know.

In the first of the following sections, we discuss empathy and justice from evolutionary, genetic, neuroscientific, and biosocial perspectives. We follow this review with a discussion of how empathy can be integrated into explanations of delinquent behavior, as well as the creation and enforcement of law and punishment. We conclude by offering suggestions on how delinquency prevention efforts and criminal justice responses can profit from an in-depth understanding and incorporation of empathy.

The origins of empathy and justice

It would be a mistake to assume that the concept of justice is unique to human culture. It would also be an oversight to conclude that the related concept of empathy is uniquely human—it isn't. For instance, primatologist Frans de Waal has studied various species of primates such as apes and chimpanzees, and observed elaborate reconciliation practices after an argument (de Waal, 1989). Often a female is the responsible one who reconciles two opposing males (de Waal, 2009). He has even observed communities of primates that develop a whole system of public safety, complete with "police" and other peacekeepers. These individual primates are responsible for keeping the peace and if they are absent, chaos ensues (de Waal has tested this by physically removing the peacekeepers).

De Waal believes that primates have a keen sense of justice as well. He describes several instances in *The Age of Empathy* (2009) where capuchin monkeys illustrate the importance of fairness and justice in their community. In one study, the monkeys were physically isolated from one of their peers so that they could not interact but could still see each other. One capuchin was offered two tokens: one that feeds only the individual, and a second token that feeds the individual and the peer. Capuchins almost always choose the token that will feed both. If the capuchin is paired with a complete stranger, or if they cannot see their peer, the chance that they choose the prosocial token decreases but still remains the more popular choice of the two.

In other studies, de Waal describes how he and his colleagues more specifically study justice and fairness. In one study, two capuchin monkeys were offered food in exchange for a pebble. Both monkeys would receive a cucumber (a delectable food for these monkeys) in exchange for the pebble. However, after a couple of exchanges, one monkey would begin to receive grapes (an even tastier treat). The monkey still receiving cucumbers would become outraged, even throwing the cucumber, at this injustice. Despite receiving a good piece of food, the relative injustice of the situation made it so that accepting what was given was not palatable. Not only do monkeys and other primates act out when they are the victims of injustice, but other examples show that they also have a general dislike for unfair treatment, exemplified by the primate that will not eat even when provided food because they observe others who are deprived food. Therefore it may be quite outrageous, even equally outrageous, to these primates to see others deprived of equal treatment

Empathy is an integral tool in the pursuit of justice. If individuals are unable to comprehend the feelings of others, then it is impossible to identify injustice when it occurs. And even if one does identify injustice when it occurs, if the observant does not care or cannot fathom experiencing the injustice (i.e. is not capable of placing themselves in someone else's shoes), they are very unlikely to be moved by the situation. It is not surprising then that de Waal (2009) finds empathy to be present in many primates such as apes, monkeys, and chimpanzees. Justice appears to be an integral piece to keeping individuals in society glued together. If injustice is rampant, and no-one is doing anything to improve fairness in society, society cannot flourish. Simplistically, society operates more smoothly with empathy.

What do these stories tell us about the origins of justice and empathy? They suggest that these concepts are part of our history and have evolutionary bases. Outrage against injustice is a way to protect the future of a species. An insightful discourse on the evolution of justice is offered by Walsh (2000). According to Walsh, our sense of justice has developed along the lines of the laws of natural selection. Concepts of justice are rooted in evolution as solutions to problems faced by early ancestors. Restorative justice approaches that hinge on reconciliation and empathy are favorable in an evolutionary sense for the human species in that they facilitate harmony among members and an expectation of "altruistic reciprocity" (Walsh, 2000: 847)—the idea that good-will, at some point, will be returned by the recipient (see also Trivers, 1971). Thus, in an evolutionary sense, it pays in the end to be empathetic and cooperative.

Reciprocal altruism is essential for the study of empathy and justice. Trivers (1971) acknowledges that humans have both altruistic and cheating tendencies which are, at times, equally important for survival. However, according to Trivers' model, in the long run altruism is the more beneficial strategy. Altruism was essential for the survival of the species in pre-modern life in which threatening situations were numerous. Those (e.g. cheaters) who fail to provide assistance eventually find themselves in a position in which they cannot find those willing to help them out of risky situations. Cheaters are thus selected out over time. In terms of justice, when we see an injustice in society, most of us feel compelled to act because we know—at least unconsciously—that down the line we may be the ones in need of assistance.

Hoffman (2000) argues that when a good person is wronged, observers feel that reciprocity has been violated, generating empathetic feelings of injustice. When there is a failure to repair the injustice, observers have feelings of guilt over inaction. Therefore injustice invokes actions that seek to remedy the wrong. Readers may be wondering how there is ever injustice over failed reciprocity if the outcome is always feelings of guilt for bystanders. Hoffman states that "blaming the victim" serves to reduce feelings of guilt. To put this into context, a law or policy that unfairly targets minorities (say stop-and-frisk policies) can be justified when individuals believe that minorities are responsible for the bulk of all crime, and thus deserve to be profiled and stopped.

While evolutionary psychology and related fields have enabled a sound understanding of the origins of empathy, genetics and biological sciences are also increasingly adding to our understanding of the heredity and development of empathy. Studies completed with replication suggest that traits such as empathy and altruism are at least partially inherited, showing that about 50 percent of variation in these traits are due to heredity (Rushton *et al.*, 1984, 1986). Related studies reveal that, while empathy is heritable, affective empathy is more genetically heritable than cognitive components of empathy (Zahn-Waxler *et al.*, 1992; Davis *et al.*, 1994). This is likely due to the different neurological underpinnings to these different empathetic traits (Hein and Singer, 2008; Knafo *et al.*, 2008, 2009; Vaske *et al.*, 2012). This tells us two important things: (1) empathy is partially heritable, and (2) there is a lot of variation in empathy that is due to environmental factors. This suggests that individuals low in empathy can improve their ability to understand others' emotions and share in their feelings.

Related to the second issue above, the social perspective argues that empathy, along with other features of personality, develops through socialization (Gottfredson and Hirschi, 1990; Unnever *et al.*, 2005; also Cialdini *et al.*, 1997). Parenting in particular has been implicated in the development of empathy. Schaffer *et al.* (2009) find that parenting is related to antisocial behavior through its impact on empathy. Permissive parenting, as opposed to authoritarian parenting, is associated with low levels of empathy. The authors argue that lax parenting (lack of discipline for poor behavior) hinders the development of empathetic qualities in the child. Similarly, Soenens *et al.* (2007) find that maternal support is important in the development of empathy in children. In fact, socialization in the early years of life is among the most important factors in developing the response to the suffering of others. Thus there is evidence that empathy may, in part, be a learned mechanism, in addition to having biological underpinnings. Research is continuing to uncover the origins of empathy, which will permit a deeper understanding of how this phenomenon is related to behavior.

Advances in neuroscience also allow us to understand how the inner workings of our brain determine our emotions. One of the influential insights from this field of study is the importance of mirror neurons, which enable individuals to understand what others are experiencing (Schulte-Rüther *et al.*, 2007; Frith and Singer, 2008; Iacoboni, 2008). Neurons are the "messengers" of our brain, sending signals to various parts of the brain. Mirror neurons are hypothesized to allow us to interpret the actions of others. These neurons develop at a very early age, suggesting that empathy and related traits are somewhat innate and important early on in life (Meltzoff and Moore, 1977). Some researchers (but certainly not all, or the majority) believe that mirror neurons are *the* most important part of our communication systems and are essential in understanding emotions. Marco Iacoboni (2008), in his book *Mirroring People*, explains that mirror neurons are the foundation for empathy in that they provide a mechanism for "inner imitation" in which an individual can simulate others' facial expressions. These signals are sent through the insula, which carries the signals (e.g. for feelings) to the brain's limbic system.

Mirroring can be both unconscious and conscious. Unconscious mirroring and contagion are present early in life and are thought to be more primitive constructs than empathy (or "precursors" to empathy, for some). Hoffman (2000) refers to empathy in the first months of life as "global empathy," evident when infants automatically cry in response to others' cries. This affective responsiveness is involuntary and is based on somato-sensory-motor resonance between the infants (Decety and Meyer, 2008). Essentially, affect is a rather ancient subcortical construct providing humans with a rudimentary way of matching the states of others. Later in life, humans attain the ability to consciously recognize and form an understanding of other peoples' emotions. This is an emotional understanding and requires knowledge about emotional processes (Decety and Michalska, 2012).

While mirror neurons help to partially explain empathy, an empathetic response requires reflection. This reflection is accomplished when the prefrontal regions of the brain become engaged and emotional signals are generated by the limbic system (Siegel, 2007). Underdevelopment or damage to these areas in the brain may impact the ability to recognize emotions in others and to reflect and display affective or behavioral empathy (see also Baron-Cohen, 2011a).

Several brain regions are relevant to empathy. This is often referred to as the "empathy circuit." Researchers rarely agree fully on the most important brain regions for empathy, and most if not all agree that there is no "empathy region" that totally controls the abstract concept of empathy. However, the empathy circuit can be a visual aid that illustrates regions that are responsible for everything from processing social information in the medial prefrontal cortex, to interpreting emotions in the inferior frontal gyrus, to mirroring activity such as emotions and action in the inferior parietal lobule. The latter part of the mirror neuron system is activated when performing and watching actions (for a full review of the empathy circuit, see Baron-Cohen, 2011b).

Brothers (1989) argues that empathy has developed in the primate central nervous system as a mechanism for identifying and responding to social cues in more adaptive ways to ensure species development and propagation. In this way of thinking, empathy is a biological mechanism that is central to species survival. Because empathy-related behaviors have been found in many mammalian species, it therefore follows that there is a biological element to its origin.

Empathy and criminology

Empathy is the ability to understand and feel others' emotions and interpret the intentions of others. Typically, empathy is viewed as an important trait that leads to more altruistic, cooperative, and prosocial behavior. Yet the distinction between cognitive (cold) and affective (hot) empathy is crucial here. Dutton (2012) argues that for those scoring high on psychopathy measures, it is not that empathy is entirely lacking—rather, it is the lack of affective empathy that allows psychopaths to feel no remorse for their actions. The ability to understand others' intentions (cognitive empathy), though, equips psychopaths to be able to take advantage of others. In addition, while altruism is generally thought to be a positive, Barbara Oakley and colleagues (Oakley, 2011; Oakley *et al.*, 2011) have shown how it can actually lead to devastation on the part of the individual and those around him/her. Further, the link between empathy and morality may not be altogether direct, since the expression of morality is situationally based—what is moral in one situation may not be so in another (Kitzinger, 1997; Tsang, 2002; Iacoboni, 2008). Nonetheless, research does tend to support the notion that empathy is related to prosocial behavior in general, which links the concept to crime and delinquency.

Empathy and delinquency

We now turn to the major focus of this chapter. Empathy, we argue here, can help us better understand all facets of criminology. This section focuses on the second of those components: the breaking of laws, better known as crime and delinquency. Perhaps the most dominant focus of criminology is to explain criminal behavior. Despite a relative neglect of empathy in explanations of crime and delinquency, several theories explicitly or implicitly stress the importance of emotions such as empathy in their perspectives. Two prime examples include self-control theory and situational action theory (SAT).

In Gottfredson and Hirschi's (1990) major work on self-control theory, they state that crimes are most likely to be committed by people with low self-control and that, because crime often results in pain or discomfort for victims, ". . . people with low self-control tend to be self-centered, indifferent, or insensitive to the suffering and needs of others" (p. 89). Gottfredson and Hirschi are clear that they believe the psychological approach to explaining crime is fatally flawed because criminals are different from non-criminals only to the extent that their levels of self-control differ—which are developed through socialization. Their concept of self-control clearly incorporates elements of empathy, particularly in describing people with low self-control as unconcerned about the feelings of others, as well as being self-centered without much ability to relate to others' emotions and feelings. What is ignored by neglecting other factors beyond socialization in the development and utilization of self-control is that self-control is influenced by a person's executive functioning and endocrine system. In order for a person to use self-control effectively, they must utilize glucose, and as blood glucose levels are depleted (through stress, regulating emotions, or extensive use of prior self-control) it becomes harder and harder to use self-control effectively (Muraven and Baumeister, 2000). It might therefore be quite an oversight to neglect the psychological and biological (e.g. the endocrine system) basis of self-control, as well as a trait such as empathy, when explaining criminal behavior.

Other hormones have also been found to be important in criminal and violent behavior. Shirtcliff and colleagues (2009) argue that cortisol and other neurohormones, such as oxytocin and vasopressin, are important in the empathy-to-behavior link. First, low cortisol levels correspond with low emotional arousal to the distress of others and the self. Second, neurohormones moderate the effect of the limbic system, particularly during stressful situations. Similarly, Bartz *et al.* (2011) suggest that oxytocin affects social cognition and behavior, but that these effects are influenced by the environment, indicating the need for a multilevel analysis of such relationships. This has important implications for how we integrate empathy into the study of crime.

In response to what he saw as a neglect of criminological theory to integrate the individual and the setting in explaining behavior, criminologist Per-Olof Wikström (2006) offers SAT as a theory of crime, which states that crime is the result of both the individual and the particular setting that the individual occupies. Central to the understanding of crime is how the individual interprets the setting, and how the individual decides to act according to the demands of the time and place. Importantly, as pointed out above, Wikström recognizes that what is appropriate behavior in one context might not be so in another. Wikström further states that "perception is thus an outcome of the *interaction* between the individual and the setting" (p. 82). Along with perception, Wikström places importance in morality and moral emotions. He views moral emotions as feelings of guilt or shame when an individual violates a rule of the setting. Thus it is integral to understand how personal morals interact with the setting to produce criminal behavior.

The role of empathy could provide a point of departure in understanding crimes related to an individual in a specific setting by revealing how an individual interprets the setting (cognitive empathy) and how the individual responds (affective and behavioral empathy). It has been established that these are two different processes with separate neurological pathways, and to understand the setting it appears more nuanced than just understanding the social cues and frames present at any given time. Further, moral emotions are inextricably linked to brain functioning. In particular, moral emotions are related to activation in the superior frontal gyrus, posterior cingulate gyrus, temporal poles, and amygdala (Ruby and Decety, 2004). Research shows that underdevelopment and damage to specific areas of the brain linked to moral judgment, such as the orbito-frontal cortex and ventromedial prefrontal cortex, leads to selfish behavior due to the failure of the individual to comprehend the intensions of others (Koenigs *et al.*, 2007; Young *et al.*, 2010; Baron-Cohen, 2011a). Approaching SAT only through the lens of sociology potentially limits true understanding of how, and through what processes, the individual translates cognition into action, which is the foundation of a theory of situated action.

Decision-making is also clearly established in brain functioning and the choice to commit a crime. With any theory that proposes the importance of decision-making, as both self-control theory and SAT do, it is imperative to account for the psychology of decision-making. Frith and Singer (2008) outline the three-pronged process of decision-making in social settings by linking it to mirror neurons. Mirror neurons allow us to understand the motor functions (e.g. gestures) of others, while empathy allows us to understand the emotions and feelings of others (Iacoboni, 2008). A higher-order concept, theory of mind, allows us to understand more abstract concepts such as the intentions of others (e.g. beliefs/wishes). While theory of mind is a complex human process, only a few brain regions are responsible for these functions, including the medial prefrontal cortex/paracingulate cortex, posterior superior temporal cortex, and the temporal poles. Understanding the involvement of these brain regions in the commission of violence and aggression, as well as how underdevelopment in or injury to these areas affects decisions, may assist in crime prevention and offender rehabilitation.

Several researchers have established an empirical link between lack of empathy and criminal behavior (see Jolliffe and Farrington, 2004, 2006), and conclude that empathy is an important trait that is responsible for explaining a portion of poor behavior by adolescents. Studies with diverse populations in different parts of the world have also associated cognitive and affective empathy to offending behavior, suggesting that low empathy may be related to offending regardless of social context (Luengo *et al.*, 1994; Kaukiainen *et al.*, 1999). In a recent study, we find that empathy is one of the strongest correlates of crime, even after accounting for the effects of other individual-level covariates (Posick *et al.*, 2014). However, some research finds that cognitive and affective empathy are differentially related to offending. Jolliffe and Farrington (2004) conclude that low cognitive empathy is strongly related to offending, while low affective empathy is related very weakly, and Mak (1991) reports that low affective empathy is a characteristic of delinquent females, but not males.

Quite interesting is the relationship between gender and violence, with males committing the bulk of all violent criminal activity. Criminology has struggled with explaining this phenomenon, but part of the answer may lie in the development of empathy. Empathy is consistently shown to be more pronounced in females (Rueckert and Naybar, 2008), and women have an easier time than men empathizing with individuals whom they dislike (Singer and Lamm, 2009). Perhaps it is the ability to empathize with others, as well as the ability to control the desire for revenge, that is responsible for a portion of the gap in violence between men and women. As the study by Mak (1991) discussed previously shows, a breakdown in this empathetic trait in females may result in delinquent behavior while males, who are not restrained in the first place, may not be affected much by such a breakdown.

Empathy is usually thought of as a positive trait, leading to strong social relationships, altruistic behavior, and good mental health. However, some research also shows that "too much" empathy might not be a positive trait. This line of work suggests that high levels of empathy and other altruistic behavior might have negative consequences that actually damage social ties and increase the likelihood of other poor health outcomes. Most relevant to this chapter is the empirical work that indicates that unusually high levels of empathy are related to increased risk of victimization. Individuals who exhibit high levels of empathy tend to be too trusting of others, and often expose themselves to opportunities to be victimized by neglecting to address the risks of their surroundings. Homant and Kennedy (2011), using a community sample of 100 individuals, found that extremely empathetic people are more likely to be victims of property crime and violent crime. Other researchers reach similar conclusions, using other outcomes such as guilt, anxiety, and depression (Zahn-Waxler and Van Hulle, 2011), that very high levels of empathy lead to poor mental health outcomes. They conclude that people who over-empathize and "feel" others' pain end up with emotions of distress over these shared feelings. Others who empathize, but who cannot help others, have feelings of guilt and stress over their inaction (see also Hoffman, 2000). What this literature highlights is that, while healthy levels of empathy are related to a host of positive outcomes, extreme levels of empathy (either too low or too high) are related to negative outcomes (for a different view on this conceptualization of empathy, see Baron-Cohen, 2011a). Therefore it may be that programs should aim to instill healthy levels of empathy, and ensure that people with unusually high levels of empathy learn to manage their emotions and control their behavioral responses.

Delivering justice: empathy, crime prevention, and the criminal justice system

Criminology is also focused on explaining phenomena related to the criminal justice system. Here, we believe empathy can be helpful as well. For example, empirical research has related empathy to punishment for offenders. While research has yielded much insight into the relationship between empathy and punishment, interpretation of the results from these studies is not straightforward. Support for the death penalty and harsh punishment for offenders are two areas that have been associated with empathy (Unnever *et al.*, 2005; Unnever and Cullen, 2009). They suggest that empathetic people view the death penalty as unduly harsh and unjust. Interestingly, in our own research, we find that empathetic people, while opposed to the death penalty, also report that they think existing laws too lenient (Posick *et al.*, 2014). This suggests a more complex relationship between empathy and justice than might appear at first glance. For example, in a situation involving a common crime, there are at least two groups to whom the public can relate. They might associate with the pain that a punishment may cause the accused; or they may associate with the pain that the offender caused the victim. Empathetic people, while viewing the death penalty as too harsh, may additionally empathize with victims, and support stiff penalties as a means to achieve justice.

One additional reason that it is so complex to uncover the exact relationship between empathy and support for punishment is the duality that Hoffman (2000) describes in combination with feelings of injustice about the death penalty. He suggests that individuals often have empathy for the victim by taking their perspective. If the victim is physically hurt or appears mentally anguished, the observer can match this state and become angry at the perpetrator. However, Hoffman also states that empathy for the victim does not require that the victim be angry or hurt. He calls this "anger-at-aggressor," while calling the former "empathetic-distress-for-victim." This framework allows for observers (i.e. society) to be outraged at the perpetrator of a murder even though the victim is no longer alive to feel distress or anger. Because it is possible to be empathetic toward an offender (for being the victim of an unjust or unduly harsh criminal justice system), it is equally plausible to empathize with the victim regardless of their actual feelings.

Neurological studies may lend additional insight into these findings. A common finding in fMRI studies is that empathy toward another is reduced (particularly in brain regions of men that "light up" when empathy is activated) when the "other" acts unfairly, supporting the idea of "altruistic punishment" or the punishment of someone who has broken societal rules (Singer *et al.*, 2006). Altruistic punishment is a mechanism that establishes and maintains cooperation in society by promoting long-term group interests over the short-term interests of individuals (Frith and Singer, 2008). The ability to empathize with others' pain is itself neurological—depending on activation of the anterior insula/fronto-insula cortex and the anterior cingulate cortex, which are the same regions of the brain that are activated when experiencing pain first-hand (Singer *et al.*, 2006). The activation of these regions also increases as the perceived level of pain increases, which, again, might suggest that extreme punishment such as the death penalty is seen to be "too harsh" (Hein and Singer, 2008). More research is needed to adequately investigate the relationship between empathy and punishment.

A relatively new area of study is the role of empathy in policing and criminal justice. While a long-standing literature on procedural justice and police legitimacy has found that police who act professionally and respectfully, and are knowledgeable during encounters with the community, are seen as more effective and trustworthy, less research has been done on the relationship between officer empathy and these perceptions. However, in our work, we show that officers who display empathetic understanding and concern toward citizens are viewed as more effective and trustworthy than other officers (Posick *et al.*, 2014). This suggests that being able to identify with the community and communicate understanding of serious issues will increase not only cooperation, but also perceptions of how well police do their job.

Because lack of empathy is implicated in poor behavior and other outcomes, prevention, intervention, and rehabilitation programs have been concerned with establishing and promoting empathy in individuals. An emotional focus to prevention and intervention has succeeded in producing favorable results, particularly when these efforts recognize the biological and psychological roots of cognition and emotionality. For example, mindfulness meditation has been shown to reduce anxiety as well as promote empathy by influencing regions in the brain (Holzel *et al.*, 2011). Epigenetics suggests that the brain is malleable, and that environmental influences can affect brain functioning by taking advantage of the plasticity of the brain. Mindfulness meditation has been linked with an increase in gray matter and subsequently has an effect on behavior. Empathy, a trait associated with gray matter in the brain, can be strengthened through programs that are able to influence the way the brain functions.

Cognitive behavioral therapy (CBT) is one of the most successful and most cited "best practices" in offender rehabilitation (Farrington and Welsh, 2007; Rocque *et al.*, 2012; Vaske *et al.*, 2012). In fact, increasing empathy has been a focus of cognitive behavioral programs (Farrington and Welsh, 2007; Vaske *et al.*, 2012). Like mindfulness meditation, one of the likely reasons that CBT works is that it "... corresponds to changes in brain functioning" (Vaske *et al.*, 2012: 91). In other words, CBT takes advantage of brain plasticity, the malleability of the brain to respond to environmental stimuli, by focusing on cognition and emotions. Here, CBT addresses both deficits in understanding (interpreting the actions of others) as well as broadening the individual's repertoire of acceptable solutions to a situation.

Other social interventions and practices have also been able to increase the empathetic response of individuals. Restorative justice, an approach to justice which incorporates dialogue between an offender, victim, and affected parties, has been linked to increased empathy by participating parties. Through story-telling, participants of restorative justice are able to better understand why a crime occurred, the thought process of the offender, the victims' feelings, and how the incident affected others. This type of dialogue increases the ability to understand and share in others' emotions and feelings, which increases empathy (Jackson, 2009). This type of finding is not new, and is similar to the conclusions of Batson and colleagues (1997, 2002) that story-telling can increase empathy toward offenders who are able to tell their story regarding why they committed a crime.

Finally, the family is an important social context that assists in positive development of empathy, and is a key element of social control theories of criminal behavior. Using longitudinal data from the Flourishing Families Project, Yoo *et al.* (2013) found that the parent–child relationship influences the development of empathy across time. In particular, they unraveled the impact of parental solicitation (knowing about children's friends and activities) and psychological control (using the Psychological Control Scale Youth Self Report [Barber, 1996]—e.g. "my parent tries to change how I feel or think about things") by uncovering their role through the adolescent's perceived balance connectedness. They found that a balance between controlling the child and warmly connecting with the child promotes empathy and other prosocial behaviors in youth. Similarly, Batanova and Loukas (2012) found that parent–child conflict reduced empathetic concern for others among females, and that connectedness to school promoted empathetic concern for both males and females in a school-based sample.

It is not only relationships with parents that matter, but also those with siblings. Lam *et al.* (2012) found that sibling warmth, measured using the Sibling Relationship Inventory (Stocker and McHale, 1992), for example "How often do you do nice things like helping out or doing favors for your sibling?"), was related to higher levels of empathy, while sibling conflict (e.g. "How often do you feel mad or angry at your sibling?") was related to lower levels of empathy. Practices and policies that promote warm and balanced bonding among family members—and other social institutions—would likely improve the development of empathy in adolescents.

While it is true that empathy appears to have direct neurological underpinnings, research has long shown that the expression of morality, and other emotions, is contingent on the social situation (Kitzinger, 1997; Mackenzie, 2006; Turner, 2009). Kitzinger (1997) reviews the famous social psychological studies of the 1950s–1970s which demonstrated that even ostensibly moral people can be persuaded by social context or social cues to be immoral. For example, in Stanley Milgrim's famous experiment, healthy subjects were convinced to give what they were led to believe were extremely dangerous electric shocks to other people simply for incorrectly answering queries. These studies indicate that our "ability to act on altruistic impulses and moral principles is sometimes inhibited by precisely these social pressures. At the very least it is obvious that this is sometimes true, and that we need to develop ways of recognizing and challenging those social pressures which result in apathetic or cruel behaviour in our everyday lives" (Kitzinger, 1997: 17). Similarly, Turner (2009), while recognizing that neurology is the primary system that underlies the expression of emotions, argues that the *way* in which emotions are expressed (or repressed) is related to culture. An emotion such as anger may be appropriately dealt with by "keeping it inside" in one culture, and by externalizing through shouting in another. This suggests that efforts to improve behavior should rely on not simply changing biological or neurological characteristics, but also addressing the way that we make decisions in particular social contexts.

Perhaps the most important policy implication of all this, and other developmental research, is to intervene as early as possible to encourage healthy development (Rocque *et al.*, 2012). Programs that begin early in life and that focus on parenting and social relationships are vital to establishing empathy. In fact, Gottfredson and Hirschi's (1990) self-control theory argues that this trait is established early in life, through effective parenting. Volbrecht *et al.* (2007) found that environmental influences were strongest (and genetic influences the weakest) predictors of empathy at the toddler stage, when parent socialization is most intense. At its core, socialization is about teaching the expectations of society and helping children think about others rather than acting on impulses. Further, pre-natal parenting programs have demonstrated positive impacts on infant biological development. For example, teaching mothers of the dangers of smoking or drinking alcohol during pregnancy can reduce the risk of neurological deficits at birth (Huizink and Mulder, 2006; Rocque *et al.*, 2012). Thus programs that help improve parenting practices are likely to be integral in the establishment and appropriate expression of empathy in children (Schaffer *et al.*, 2009).

Conclusion

In 2010, Francis Cullen addressed the American Society of Criminology, presenting his vision for the future of the field of criminology. He specifically envisioned, and encouraged, a more in-depth consideration of the role of emotions in behavior, using a multidisciplinary and developmental focus. He encouraged criminologists to investigate the role of both negative emotions, which is the common approach in contemporary criminology, and positive emotions in behavior (Cullen, 2011). The purpose of this chapter is to review some of the progress that various fields have made in understanding behavior and outcomes through the lens of empathy, and how the field of criminology can capitalize on this knowledge. As Cullen concedes, while there has been some attention to these issues, we still have a long way to go in identifying and adequately studying emotions.

Jolliffe and Murray (2012) view empathy as one of the emotions that has the potential to guide criminological research in the new millennia, and we agree. Along with these researchers, we see the field of criminology expanding its conceptualization of empathy and its components, as well as its effect on a multitude of outcomes and behaviors. New work is continually pushing the emotions and crime agenda (Karstedt *et al.*, 2011). However, there is still quite a bit of work to do in conceptualizing empathy in all of its components, fine tuning the measurement of empathy, and adequately studying the origins of empathy from a biosocial standpoint. Further, we have only scraped the tip of the iceberg in terms of using empathy as a key correlate of the many types of behavior that the field of criminology seeks to understand. Empathy may be related in various ways to internal harm, externalizing and aggressive behavior, victimization, and relationships with the criminal justice system. And as this chapter illuminates, the relationship of empathy to these outcomes might not be linear, with extremely high or low levels of empathy related differently from "healthy" levels of empathy. Further, deficits in certain components of empathy are related to different outcomes. This clearly has implications for prevention and intervention programs that seek to address the psychological and social influences on empathy and related traits. By continuing to refine measurement and broaden the use of empathy in criminology using a multidisciplinary research program, we can further the field's ability to explain behavior and design effective efforts to reduce harmful outcomes.

References

Barber, B.K. (1996) 'Parental psychological control: revisiting a neglected construct', *Child Development*, 67: 3296–3319.

Baron-Cohen, S. (2011a) *The Science of Evil: On Empathy and the Origins of Cruelty*, New York: Basic Books.

Baron-Cohen, S. (2011b) 'The empathy bell curve', Phi Kappa Phi Forum, Spring 2011, www.questia.com/magazine/1G1-267422895/the-empathy-bell-curve

Bartz, J.A., Zaki, J., Bolger, N., and Ochsner, K.N. (2011) 'Social effects of oxytocin in humans: context and person matter', *Trends in Cognitive Sciences*, 15: 301–309.

Batanova, M.D. and Loukas, A. (2012) 'What are the unique and interacting contributions of school and family factors to early adolescents' empathic concern and perspective taking', *Journal of Youth and Adolescence*, 41: 1382–1391.

Batson, C.D., Polycarpou, M.P., Harmon-Jones, E., Imhoff, H.J., Mitchener, E.C., Bednar, L.R., and Highberger, L. (1997) 'Empathy and attitudes: can feeling for a member of a stigmatized group improve feelings toward the group?', *Journal of Personality and Social Psychology*, 72: 105–118.

Batson, C.D., Chang, J., Orr, R., and Rowland, J. (2002) 'Empathy, attitudes, and action: can feeling for a member of a stigmatized group motivate one to help the group?', *Personality and Social Psychology Bulletin*, 28: 1656–1666.

Brothers, L. (1989) 'A biological perspective on empathy', *American Journal of Psychiatry*, 146: 10.

Cialdini, R.B., Brown, S.L., Lewis, B.P., Luce, C., and Neuberg, S.L. (1997) 'Reinterpreting the empathy–altruism relationship: when one into one equals oneness', *Journal of Personality and Social Psychology*, 73: 481–494.

Cullen, F.T. (2011) 'Beyond adolescence-limited criminology: choosing our future—The American Society of Criminology 2010 Sutherland Address', *Criminology*, 49: 287–330.

Davis, M.H., Luce, C., and Kraus, S.J. (1994) 'The heritability of characteristics associated with dispositional empathy', *Journal of Personality*, 62: 369–391.

Decety, J. (2010) 'The neurodevelopment of empathy in humans', *Developmental Neuroscience*, 32(4): 257–267.

Decety, J. and Meyer, M. (2008) 'From emotion resonance to empathic understanding: a social developmental neuroscience account', *Development and Psychopathology*, 20: 1053–1080.

Decety, J. and Michalska, K.J. (2012) 'How children develop empathy: the contribution of developmental affective neuroscience', in Decety, J. (ed.), *Empathy: From Bench to Bedside*, Cambridge, MA: MIT Press.

Dutton, K. (2012) *The Wisdom of Psychopaths: What Saints, Spies, and Serial Killers Can Teach Us about Success*, New York: Scientific American/Farrar, Straus and Giroux.

Farrington, D.P. and Welsh, B.C. (2007) *Saving Children from a Life of Crime: Early Risk Factors and Effective Interventions*, New York: Oxford University Press.

Frith, C.D. and Singer, T. (2008) 'The role of social cognition in decision making', *Philosophical Transactions of the Royal Society*, 363: 3875–3886.

Gottfredson, M.R. and Hirschi, T. (1990) *A General Theory of Crime*, Stanford, CA: Stanford University Press.

Hein, G. and Singer, T. (2008) 'I feel how you feel but not always: the empathic brain and its modulation', *Current Opinion in Neurobiology*, 18: 153–158.

Hoffman, M.L. (2000) *Empathy and Moral Development: Implications for Caring and Justice*, Cambridge, MA: Cambridge University Press.

Holzel, B.K., Carmody, J., Vangel, M., Congleton, M., Yerramsetti, S.M., Gard, T., and Lazar, S.W. (2011) 'Mindfulness practice leads to increases in regional brain gray matter density', *Psychiatry Research: Neuroimaging*, 191: 36–43.

Homant, R.J. and Kennedy, D. (2011) 'Does no good deed go unpunished? The victimology of altruism', in Oakley, B., Knafo, A., Madhavan, G., and Wilson, D.S. (eds), *Pathological Altruism*, New York: Oxford University Press.

Huizink, A.C. and Mulder, E.J. (2006) 'Maternal smoking, drinking or cannabis use during pregnancy and neurobehavioral and cognitive functioning in human offspring', *Neuroscience and Biobehavioral Reviews*, 30: 24–41.

Iacoboni, M. (2008) *Mirroring People: The New Science of How We Connect with Others*, New York: Farrar, Straus and Giroux.

Jackson, A.L. (2009) 'The impact of restorative justice on the development of guilt, shame, and empathy among offenders', *Southwest Journal of Criminal Justice*, 5: 184–206.

Jolliffe, D. and Farrington, D.P. (2004) 'Empathy and offending: a systematic review and meta-analysis', *Aggression and Violent Behavior*, 9: 441–476.

Jolliffe, D. and Farrington, D.P. (2006) 'Examining the relationship between low empathy and bullying', *Aggressive Behavior*, 32: 540–550.

Jolliffe, D. and Murray, J. (2012) 'Empathy and offending: implications for tomorrow's research and practice', in Loeber, R. and Welsh, B.C. (eds), *The Future of Criminology*, New York: Oxford University Press, pp. 62–69.

Karstedt, S., Loader, I., and Strang, H. (2011) *Emotions, Crime and Justice*, Oxford: Hart.

Kaukiainen, A., Bjorkvist, K., Lagerspetz, K., Osterman, K., Salmivalli, C., Rothberg, S., and Ahlbom, A. (1999) 'The relationship between social intelligence, empathy, and three types of aggression', *Aggressive Behavior*, 25: 81–89.

Kitzinger, C. (1997) 'Born to be good? What motivates us to be good, bad or indifferent towards others?', *New Internationalist*, 289: 15–17.

Knafo A., Zahn-Waxler, C., Van Hulle, C., Robinson, J.L., and Rhee, S.H. (2008) 'The developmental origins of a disposition toward empathy: genetic and environmental contributions', *Emotion*, 8: 737–752.

Knafo, A., Zahn-Waxler, C., Davidov, M., Van Hulle, C., Robinson, J.L., and Rhee, S.H. (2009) 'Empathy in early childhood', *Annals of the New York Academy of Sciences*, 1167: 103–114.

Koenigs, M., Young, L., Adolphs, R., Transel, D., Cushman, F., Hauser, M., and Damasio, A. (2007) 'A damage to the prefrontal cortex increases utilitarian moral judgments', *Nature*, 446: 908–911.

Lam, C.B., Solmeyer, A.R., and McHale, S.M. (2012) 'Sibling relationships and empathy across the transition to adolescence', *Journal of Youth and Adolescence*, 41: 1657–1670.

Luengo, M.A., Otero, J.M., Carrillo-de-la-Pena, M.T., and Miron, L. (1994) 'Dimensions of antisocial behavior in juvenile delinquency: a study of personality variables', *Psychology, Crime and Law*, 1: 27–37.

Mackenzie, S. (2006) 'Situationally edited empathy: an effect of socioeconomic structure on individual choice', *Critical Criminology*, 14: 365–385.

Mak, A. (1991) 'Psychosocial control characteristics of delinquents and non-delinquents', *Criminal Justice and Behavior*, 18: 287–303.

Meltzoff, A.N. and Moore, M.K. (1977) 'Imitation of facial and manual gestures by human neonates', *Science*, 198: 75–78.

Muraven, M. and Baumeister, R.F. (2000) 'Self-regulation and depletion of limited resources: does self-control resemble a muscle?', *Psychological Bulletin*, 126: 247–259.

Oakley, B. (2011) *Cold Blooded Kindness. Neuroquirks of a Codependent Killer, or Just Give Me a Shot at Loving You, Dear, and Other Reflections on Helping that Hurts*, Amherst, NY: Prometheus.

Oakley, B., Knafo, A., Madhavan, G., and Wilson, D.S. (eds) (2011) *Pathological Altruism*, New York: Oxford.

Posick, C., Rocque, M., and Rafter, N. (2014) 'More than a feeling: integrating empathy into the study of lawmaking, lawbreaking, and reactions to lawbreaking', *International Journal of Offender Therapy and Comparative Criminology*, 58: 5–26.

Rocque, M., Welsh, B.C., and Raine, A. (2012) 'Biosocial criminology and modern crime prevention', *Journal of Criminal Justice*, 40: 306–312.

Ruby, P. and Decety, J. (2004) 'How would you feel versus how do you think she should feel: a neuroimaging study of perspective-taking with moral emotions', *Journal of Cognitive Neuroscience*, 26: 988–999.

Rueckert, L. and Naybar, N. (2008) 'Gender differences in empathy: the role of the right hemisphere', *Brain and Cognition*, 67: 162–167.

Rushton, J.P., Fulker, D.W., Neale, M.C., Blizard, R.A., and Eysenck, H.J. (1984) 'Altruism and genetics', *Twin Research*, 33: 265–271.

Rushton, J.P., Fulker, D.W., Neale, M.C., Nias, D.K.B., and Eysenck, H. (1986) 'Altruism and aggression: the heritability of individual differences', *Journal of Personality and Social Psychology*, 50: 1192–1198.

Schaffer, M., Clark, S., and Jeglic, E.L. (2009) 'The role of empathy and parenting style in the development of antisocial behaviors', *Crime and Delinquency*, 55: 586–599.

Schulte-Rüther, M., Markowitsch, H.J., Fink, G.R., and Piefke, M. (2007) 'Mirror neuron and theory of mind mechanisms involved in face-to-face interactions: a functional magnetic resonance imaging approach to empathy', *Journal of Cognitive Neuroscience*, 19: 1354–1372.

Shirtcliff, E.A., Vitacco, M.J., Graf, A.R., Gostisha, A.J., Merz, J.L., and Zahn-Waxler, C. (2009) 'Neurobiology of empathy and callousness: implications for the development of antisocial behavior', *Behavioral Sciences and the Law*, 27: 137–171.

Siegel, D.J. (2007) *The Mindful Brain: Reflection and Attunement in the Cultivation of Wellbeing*, New York: Norton.

Singer, R. and Lamm, C. (2009) 'The social neuroscience of empathy', *Annals of the New York Academy of Science*, 1156: 81–96.

Singer, T., Seymour, B., O'Doherty, J.P., Stephan, K.E., Dolan, R.J., and Frith, C.D. (2006) 'Empathic neural responses are modulated by the perceived fairness of others', *Nature*, 439: 466–469.

Soenens, B., Duriez, B., Vansteenkiste, M., and Goossens, L. (2007) 'The intergenerational transmission of empathy-related responding in adolescence: the role of maternal support', *Personality and Social Psychology Bulletin*, 33: 299–308.

Stocker, C.M. and McHale, S.M. (1992) 'The nature and family correlates of preadolescents' perceptions of their sibling relationships', *Journal of Social and Personal Relationships*, 9: 179–195.

Trivers, R.L. (1971) 'The evolution of reciprocal altruism', *Quarterly Review of Biology*, 46: 35–57.

Tsang, J.A. (2002) 'Moral rationalization and the integration of situational factors and psychological processes in immoral behavior', *Review of General Psychology*, 6: 25–50.

Turner, J.H. (2009) 'The sociology of emotions: basic theoretical arguments', *Emotion Review*, 1: 340–354.

Unnever, J.D. and Cullen, F.T. (2009) 'Empathetic identification and punitiveness', *Theoretical Criminology*, 13: 283–312.

Unnever, J.D., Cullen, F.T., and Fisher, B.S. (2005) 'Empathy and public support for capital punishment', *Journal of Crime and Justice*, 28: 1–34.

Vaske, J., Gaylean, K., and Cullen, F.T. (2012) 'Toward a biosocial theory of offender rehabilitation: why does cognitive-behavioral therapy work?', *Journal of Criminal Justice*, 39: 90–102.

de Vignemont, F. and Singer, T. (2006) 'The empathic brain: how, when and why?', *Trends in Cognitive Science*, 10: 435–441.

Volbrecht, M.M., Lemery-Chalfant, K., Aksan, N., Zahn-Waxler, C., and Goldsmith, H.H. (2007) 'Examining the familial link between positive affect and empathy development in the second year', *Journal of Genetic Psychology*, 168: 105–130.

de Waal, F. (1989) *Peacemaking among Primates*, Cambridge, MA: Harvard University Press.

de Waal, F. (2009) *The Age of Empathy*, New York: Three Rivers Press.

Walsh, A. (2000) 'Evolutionary psychology and the origins of justice', *Justice Quarterly*, 17: 841–864.

Wikström, P.O.H. (2006) 'Individuals, settings, and acts of crime: situational mechanisms and the explanation of crime', in Wikström, P.O.H. and Sampson, R.J. (eds), *The Explanation of Crime: Context, Mechanisms, and Development*, Cambridge: Cambridge University Press, pp. 61–107.

Yoo, H., Feng, X., and Day, R.D. (2013) 'Adolescents' empathy and prosocial behavior in the family context: a longitudinal study', *Journal of Youth and Adolescence*, 42: 1858–1872.

Young, L., Bechara, A., Tranel, D., Damasio, H., Hauser, M., and Damasio, A. (2010) 'Damage to ventromedial prefrontal cortex impairs judgment of harmful intent', *Neuron*, 65: 845–851.

Zahn-Waxler, C. and Van Hulle, C. (2011) 'Empathy, guilt, and depression: when caring for others becomes costly to children', in Oakley, B., Knafo, A., Madhavan, G., and Wilson, D.S. (eds), *Pathological Altruism*, New York: Oxford University Press.

Zahn-Waxler C., Robinson J.L., and Emde R.N. (1992) 'The development of empathy in twins', *Developmental Psychology*, 28: 1038–1047.

40

Family interventions and their biosocial bases

Brandy R. Maynard and Matthew Larson

Introduction

The family has played a central role in theories of delinquency and crime since criminology emerged in the nineteenth century (Wells and Rankin, 1991), and research examining the etiology of antisocial behaviors historically has emphasized family factors (Heath *et al.*, 1997; Beaver *et al.*, 2007). Family-related factors have been identified as salient predictors of violent and criminal behavior (Farrington, 2002) and thus are considered important targets for delinquency and crime prevention and intervention. Much of the extant research, however, has focused on the interface between the individual and the environment, particularly the family, and has been informed by theories that focus on those interfaces (e.g. social learning theory, social control theory, life-course theory, strain theory), neglecting genetic and biological factors. While the family has historically received a great deal of attention, acceptance of biosocial research—research that examines and links psychosocial factors to genetics, neurobiology, and physiology, together with technological, methodological, and theoretical advances—has led to a burgeoning of biosocial research as it pertains to the study of human development, criminology, and family processes. As such, biosocial research has begun to advance our knowledge of the salience of genetic and other biological variables and interplay between biosocial factors that contribute, directly and indirectly, to delinquency and crime (Raine, 2002; Moffitt, 2005; Baker *et al.*, 2006; Raine and Yang, 2006; Vaughn *et al.*, 2009; Derzon, 2010) and family processes (see D'Onofrio and Lahey, 2010).

It is now well accepted that delinquent, criminal, and related behaviors (e.g. conduct disorder, oppositional defiant disorder, externalizing behavior problems, violence, antisocial behavior) result from numerous factors, including cognitive, psychological, familial, environmental, and biological (Brennan and Raine, 1997). Moreover, evidence suggests that dispositional and contextual contributory risk factors interact and/or cumulate in unique ways to form multiple pathways to a life of crime (Raine, 2002; Frick, 2004). Despite these advances in knowledge, inclusion of biological variables in prevention and intervention theory and science is lagging (Beauchaine *et al.*, 2008). Family-based interventions were developed largely prior to our advances in research and knowledge of biosocial factors, and although exceptions do exist, biosocial principles are largely absent from family-based prevention and intervention programs. Given more recent developments and advances in our understanding of the contribution and interplay of biological and social factors, we argue that moving toward a biosocial theory of family-based prevention and intervention will enrich our understanding of how and why family interventions work, and improve the effectiveness of interventions. Moreover, by embracing a biosocial framework, we can enhance family-based programs by more effectively and efficiently targeting biosocial variables and interactions that predict and moderate treatment response and advance identification, and matching of interventions to those most at risk and most likely to benefit (Beauchaine *et al.*, 2008).

While providing a thorough review of the biosocial determinants of crime and family processes is beyond the scope of this chapter, and has been covered elsewhere (Moffitt, 2005; Burt, 2009; D'Onofrio and Lahey, 2010, Raine, 2013; also see Chapter 31 in this volume), we begin this chapter with a broad overview of the literature related to the biosocial determinants of delinquency and crime, particularly as it relates to known family factors and processes. We then situate current family-based prevention and intervention programs within a biosocial framework by examining the biosocial bases of extant family-based prevention and intervention programs targeting delinquency and crime. We conclude this chapter with recommendations for incorporating biosocial factors in the development and testing of family-based prevention and intervention programs to advance the next generation of biosocially informed family-based interventions.

Family factors in the etiology of antisocial behavior

While the theoretical role of the family has developed over time as a result of research as well as shifts in ideology and family structures and functions in society, research on families and crime has focused largely on structural factors (i.e. physical arrangements and/or roles of family members), functional dynamics (i.e. relational dynamics or interactional processes amongst family members), or the interaction between them (Wells and Rankin, 1991). In terms of structural family factors, several have been examined and implicated in contributing to delinquency and crime. Family composition and intactness, family size, socioeconomic and employment status of the parents, deviance of parents or other family members, family size, and sibling birth order are some of the structural factors that have been identified as associated with delinquency and crime, either directly or indirectly (Loeber and Stouthamer-Loeber, 1986; Wells and Rankin, 1991; Lipsey and Derzon, 1998; Farrington, 2002; Ellis *et al.*, 2009; Derzon, 2010). The effects of structural factors, however, have been found to vary within and across studies due to methodological factors, methods of measurement, types of outcomes being measured, and participant characteristics (e.g. race, sex, gender, age; Derzon, 2010). Moreover, evidence suggests that while family structure may be associated with crime and delinquency, these effects are mediated by family processes and may be confounded or interact with genetic factors (DiLalla, 2002; Demuth and Brown, 2004; DeLisi *et al.*, 2009).

A host of family processes are also known to be associated with current or subsequent delinquency and crime (Loeber and Stouthamer-Loeber, 1986; Lipsey and Derzon, 1998; Farrington, 2002; Ireland and Smith, 2009; Derzon, 2010). For example, poor parental supervision and monitoring, and harsh or erratic discipline, have been found to be moderate-to-strong predictors of delinquency, violence, and crime (Haapasalo and Pokela, 1999; Griffin *et al.*, 2000; Herrenkohl *et al.*, 2000; Leschied *et al.*, 2008; Derzon, 2010). Relatedly, a protective effect of parental monitoring has predicted lower aggression levels (Griffin *et al.*, 2000). Lack of parental involvement, particularly lack of involvement by the father, has been associated with delinquency and aggression across several studies (Loeber and Stouthamer-Loeber, 1986). These effects, however, have been found to differ by family structure and gender (Griffin *et al.*, 2000) and may also be confounded, moderated, or mediated by genetic and other biological risk (Baker *et al.*, 2006; Button *et al.*, 2008).

While research on family risk and protective factors has been important to understanding the etiology and course of delinquency and crime, we need to transcend the risk factor paradigm and move to better understanding of causal processes (Moffitt, 2005). Biosocial research has provided indisputable evidence that all behavior is heritable, and delinquent and criminal behavior results from the interaction of biological and environmental factors (Raine, 2002). Biosocial research can advance our knowledge of the causal processes at play and thus provide more salient targets for intervention (Moffitt, 2005). Although a comprehensive review of biological factors is beyond the scope of this chapter, we provide a brief overview of biological correlates and predictors of crime and delinquency that will serve to provide context for examining the biosocial bases of family-based interventions.

Biological factors

Behavioral and molecular genetics research has provided substantial evidence of the salience of genetics on all behavioral outcomes (Turkheimer and Waldron, 2000; Plomin and Asbury, 2005), and delinquent and criminal behaviors are no exception (Rowe, 1983; Mednick *et al.*, 1984; Rowe and Osgood, 1984; Beaver *et al.*, 2007). The genetic basis of antisocial behavior stems from research examining concordance of delinquent and criminal behavior in twin, adoption, and sibling studies, and findings from research examining the genetic influence on behaviors and personality traits that are highly correlated with delinquency and crime. For example, genetic influence has been found for a wide range of traits and behaviors that are highly correlated to delinquency and crime, such as impulsivity, callous-unemotional traits, aggression and negative emotionality, and psychiatric outcomes such as antisocial personality disorder and substance abuse (Taylor *et al.*, 2000; DiLalla, 2002; Rhee and Waldman, 2002; Plomin and Asbury, 2005; Baker *et al.*, 2006; Leschied *et al.*, 2008; Beaver *et al.*, 2010; DeLisi and Vaughn, 2011). Genes, however, only account for approximately 40 to 50 percent of the variation in antisocial behavior (Rhee and Waldman, 2002). While behavioral genetics research has indeed provided evidence of the importance of genes in the etiology of delinquency and crime, this line of research has also effectively highlighted the importance of non-genetic factors.

Neuropsychological and cognitive deficits such as frontal lobe impairment and deficits in executive functioning, critical for self-control and response inhibition, have also been found to be related to aggression, delinquency, and criminal behavior (Morgan and Lilienfeld, 2000; Raine, 2002). Attention deficit hyperactivity disorder (ADHD) and impairments in impulsivity, self-control, and emotional regulation are strongly related to delinquency and criminal behavior, and linked to neurological impairment (Raine *et al.*, 2005; Ratchford and Beaver, 2009; Beaver *et al.*, 2010; DeLisi and Vaughn, 2011). Impairments in other areas of the brain have been found in delinquent youth and criminal adults, including the cortico-limbic circuitry, amygdala, hippocampus, temporal cortex, cingulate, and angular gyrus (Raine and Yang, 2006; Raine, 2008; Whitten, 2013). While neuropsychological and cognitive impairments have a significant genetic basis, some neurological impairment can be accounted for by environmental factors such as physical trauma, poverty, and stress.

Gene–family correlation and interaction

Research examining the relative contribution of genetic and environmental factors has provided indisputable evidence that both genotype and environment play important roles in influencing behavior; however, the respective roles of environmental factors and genotype in delinquency and crime are difficult to disentangle. Better understanding of the relationship between family and biological factors has important implications for prevention and intervention research and practice. Thus researchers have increasingly turned to studying the correlation and interplay of family and biological factors, moving from partitioning variance between genetic and environmental factors to a more nuanced and complex examination of interaction effects. We now turn our attention to examining the interplay of biological and family variables on delinquent and criminal behavior.

Research suggests that biological and social (e.g. family) risk and protective factors may work in additive ways (i.e. cumulative effect related to the number of risk factors present—the more risk factors, the greater likelihood of developing problematic behavior); through an interaction effect (i.e. certain risk factors interact with other risk factors to determine behavioral outcomes); or via transaction between multiple factors (i.e. the reciprocal exchange between individual and environmental risk factors produces certain outcomes; Kimonis and Frick, 2010). In the behavior genetics literature, the interplay between genetic and environmental variables has been examined in terms of genotype–environment correlation (rGE), i.e. the covariation of gene and environmental factors, and gene–environment interactions (G × E), i.e. the moderation of genetic effects as a function of environmental factors and *vice versa*.

Biosocial research has provided a corpus of studies pointing to relationships between genes and familial characteristics. However, these relationships are not as simple as might be suggested, as there are several ways in which genes and environments may covary: when genes and environment come from the same source (passive rGE); when the child's genotype influences the environment (reactive rGE); and when the environment is actively selected as a result of the child's genotype (active rGE). For example, studies have shown that bad parenting is associated with delinquent and criminal behavior (Derzon, 2010; Kimonis and Frick, 2010); however, this relationship could be explained in several ways. Parenting behavior could be associated with delinquent outcomes as a result of a child's inherited predisposition to antisocial behavior and the hostile environment created by the parents' abusive parenting practices (passive rGE). It is also possible that the child's difficult temperament (largely a function of genetics) evokes aggressive and hostile responses from parents, which may lead to delinquent behavior (i.e. reactive rGE; see Huh *et al.*, 2006; Beaver *et al.*, 2007). The third form of gene–environment correlation, active rGE, refers to when children select certain environments at least partly due to genotype. This type of rGE is less relevant to the discussion of the interplay between genetic and family factors, as children usually cannot actively select their parents. As we can see from these examples, genetic and environmental causes are often difficult to disentangle—is the child's genotype being expressed because of the environment, or is the child's environment a function of her genotype (DiLalla, 2002)?

The extent to which genetic and environmental effects vary as a function of the other (G × E) is also important to consider. While both family and biological variables have been found to account for delinquent and criminal behaviors, research has increasingly begun examining the interaction of genetic liability and family and environmental conditions on child outcomes. Research has found evidence of moderating effects of the family on genetic and environmental influences, as well as of genetic liability on family influences on delinquency and criminal outcomes (see Raine, 2002; Moffitt, 2005; Baker *et al.*, 2006; Horwitz and Neiderhiser, 2011). For example, children's genetic liability for antisocial behavior was found to provoke more harsh discipline from adoptive parents in families suffering adversity than families not suffering adversity (Riggins-Caspers *et al.*, 2003). In short, mounting evidence of G × E is beginning to elucidate complex and robust explanations of the relationship of biosocial factors that can provide both better causal explanations for delinquent and criminal behaviors, and potentially more specific and robust models for family-based prevention and intervention.

Biosocial bases of family-based prevention and intervention

The burgeoning corpus of biosocial research has vastly expanded our understanding of the etiology and development of delinquent and criminal behavior, while also highlighting the complex interplay of family influences with biological factors. From a biosocial perspective of prevention and intervention, "some of the best candidate environmental risk factors are those that are most likely to have an effect on the biological systems" (van Goozen and Fairchild, 2008: 959) involved in the development of antisocial behavior. However, although family-based interventions target seemingly salient environmental risk factors, they are primarily understood and predicated under what are now outdated theoretical and empirical models of family and crime. Indeed, biological and neurological principles are largely absent from family-based intervention theory and science. By explicitly examining and identifying the biosocial bases of family-based interventions, we can begin to more fully understand the mechanisms of family-based interventions, begin to elucidate which family-based interventions may be more effective for whom and under what conditions, and more fully integrate biosocial mechanisms into family-based interventions to improve the effectiveness and efficiency of prevention and intervention efforts.

For the remainder of this chapter, we examine the biosocial bases of family-based interventions that have some empirical support for their effectiveness in preventing or reducing delinquency, crime, or closely related outcomes. Some prevention and intervention models are explicitly informed, at least in part, by research on the biological as well as social and psychological bases of human behavior and risk. Most, however, were developed from social and psychological research and theory, but nevertheless likely do impact biomarkers. We focus first on early prevention programs during pregnancy, infancy, and the preschool years; then examine indicated prevention and interventions during childhood and adolescence; and finally review family-based interventions targeting adults. For each of the aforementioned developmental periods, we provide an overview of family-based interventions and discuss the empirical support and biosocial bases of the interventions.

Family-based prevention during pregnancy, infancy, and the preschool years

Several studies provide compelling evidence of the salience of biosocial risk factors and interactions during pregnancy, birth, infancy, and early childhood on later delinquent and criminal behavior (Raine, 2002). A substantial body of evidence from longitudinal studies also provides strong evidence that the earlier behavioral problems emerge in the life course, the more likely those behavioral problems will lead to more chronic and serious antisocial behaviors (White *et al.*, 1990; Moffitt, 1993; Piquero *et al.*, 2003; Piquero, 2008). Moreover, evidence suggests that children with earlier onset of problem behaviors become less amenable to change over time (Tremblay, 2000). All of this evidence underscores the need and opportunity for early family-based prevention efforts. Indeed, while early prevention programs often target proximal social factors, they do so in a way that, intentionally or unintentionally, has the potential to impact biological systems that can have far-reaching implications for behavior and outcomes well into adulthood.

Two of the most well known and supported family-based programs to prevent delinquency and crime are home visiting and parent-training programs. Numerous studies and several narrative and quantitative reviews have examined the effects of these programs on a myriad of behavioral outcomes. While findings are equivocal, outcome research on home visiting and parent-training programs has produced evidence of short- and long-term positive effects on disruptive and antisocial behaviors, delinquency, and criminal outcomes (Olds *et al.*, 1998; Farrington and Welsh, 1999; Tremblay and Japel, 2003; Piquero *et al.*, 2009).

Home visiting programs

Home visiting programs generally target at-risk mothers prior to or immediately following the birth of their baby. As the name implies, a trained interventionist, nurse, doctor, or other professional or paraprofessional visits the mother in her home. Depending on the specific program, the intervention involves providing parent education and advice about pre- and postnatal care, including proper nutrition and health behaviors (i.e. avoid smoking, alcohol, drugs). While home visiting interventions have the potential to target biological and social risk factors and causal mechanisms quite early, and thus prevent the emergence of antisocial behavior later in the life course, findings related to the effects of home visiting programs have, overall, been inconclusive (see Gomby, *et al.*, 1999; Bilukha *et al.*, 2005). Not all home visiting programs are the same, however; home visiting programs vary on their theoretical underpinnings, length and intensity of treatment, and targeted populations, among other variables. One of the most well known home visiting programs identified as an "effective" (www.ojjdp.gov/mpg) or "model" (www.blueprintsprograms.com) program is the Nurse–Family Partnership (NFP; Olds *et al.*, 1998; Olds, 2006).

The NFP provides low-income first-time mothers with prenatal and postnatal home visitation services by public health nurses until the child reaches the age of two. Visits last approximately 70–90 minutes, with approximately seven to nine visits during pregnancy (range 0–18) and an average of 23–26 visits from birth to two years of age (range 0–71; Olds, 2006). The program was originally designed to address parental health-related behaviors to improve birth outcomes and reduce child neurological impairment; to promote sensitive, competent care by the mother to reduce child abuse and neglect injuries; and to improve parental life-course outcomes (Olds, 2006). Although NFP was not explicitly developed to prevent delinquency and crime, studies have demonstrated positive effects on early antisocial behavior, and one study found positive effects on antisocial behavior at 15 years of age among children of the most at-risk mothers—those who were poor and unmarried (Olds *et al.*, 1998). Additional studies have found positive effects on other outcomes, such as prenatal health behavior, pregnancy and birth outcomes, and sensitive, competent care of the child; however, positive effects have not been consistent across all studies and outcomes (e.g. Kitzman *et al.*, 1997; Olds *et al.*, 1994, 1998, 2014; also see Olds, 2006).

The NFP targets several biological risk factors explicitly, including low birth weight, preterm birth, and neurological impairment, by intervening to reduce known environmental risk factors during pregnancy, such as prenatal smoking, alcohol use, and drug use; poor nutrition; and pregnancy and birth complications. Also, by intervening to promote sensitive and competent care by the mothers, early prevention efforts not only can improve the parent–child interaction and foster more adaptive behavior through psychosocial processes, but also has the potential to positively impact neurological substrates and functioning. Moreover, early intervention can "disrupt epigenetic transmission of problem behavior to future generations" by impacting gene expression and behavioral traits, which can have long-lasting effects on the child and future generations as well (van Goozen and Fairchild, 2008: 962).

Early parent-training programs

Parent training has been employed as a preventive intervention with parents of children ages 0–3, and involves providing some form of training to parents to modify parenting behaviors to enhance parenting skills, improve emotional and behavioral adjustment of the child, and improve parent–child interaction. While many of the outcome studies and reviews focus on parent-training interventions with older children and adolescents, evidence suggests that early parent training with parents of infants and toddlers can be effective in preventing later problematic and antisocial behavior (Bernazzani *et al.*, 2001; Barlow *et al.*, 2010).

Early parent-training programs are heterogeneous and vary in terms of delivery method (e.g. individual, group), settings, theoretical underpinnings, length and frequency of intervention, and whether other interventions are delivered along with parent training. Some parent-training programs are based on behavioral principles of contingency management, and train parents to use positive reinforcement and other behavioral principles to modify their child's behavior (Lundahl *et al.*, 2006), while some are based on other theories, such as family systems, Adlerian, and psychodynamic (Barlow *et al.*, 2010). Despite differences in theoretical underpinnings, techniques, and skills taught across various parent-training models, early parent-training programs generally aim to enhance the caregiving a child receives to provide an optimal environment for healthy child development and prevent negative outcomes that result from inadequate and harmful parenting practices.

While parent-training programs are based on psychosocial theories, and most don't explicitly target biological risk factors like NFP, parent-training programs do have the potential to affect biological processes, which may prevent delinquent and criminal outcomes later in life. From a biosocial framework, parent-training programs could impact a number of biological mechanisms implicated in delinquency and crime, such as emotion regulation, stress responsiveness, and cognitive development, by targeting environmental risk factors such as parenting practices and the quality of parent–child relationship and attachment.

Family-based interventions during childhood and adolescence

Family-based interventions with school-age youth generally serve as a selective or indicated prevention intervention with youth exhibiting disruptive behaviors to prevent delinquency, or as an intervention program targeting families of youth already involved in the justice system to reduce delinquent behavior and disrupt the path to more serious criminal behavior in adulthood. The empirical evidence of family-based interventions for the treatment of disruptive and antisocial behavior has been growing, and specific family-based interventions have been identified as "evidence-based" (Farrington and Welsh, 1999; Lipsey and Cullen, 2007; see also Blueprints for Healthy Youth Development, www.blueprintsprograms.com; Office of Juvenile Justice and Delinquency Prevention, www.ojjdp.gov/mpg). While there is growing evidence of effectiveness, family-based interventions vary by theoretical foundation, delivery model, intervention components, targeted population, and setting. Despite a number of differences across models, family-based interventions primarily target family structure and/or processes for intervention; biological factors are rarely explicitly targeted for change or as a risk factor for identification, assessment, or mediation of treatment. While there are numerous family-based interventions, we focus here on two types of family-based intervention model that have some empirical support—parent training and family therapy—and discuss the biosocial bases of these interventions.

Parent-training programs

While parent-training programs are used as a selective prevention intervention with at-risk parents of infants and toddlers, as described above, they are more frequently employed as an indicated prevention intervention with older children and adolescents exhibiting disruptive and externalizing behavior problems, and as an intervention with youth in the juvenile justice system. Parent-training programs are delivered primarily to the parent as the sole intervention component or in combination with other interventions. Like early parent-training programs, parent training with older youth is heterogeneous and varies considerably with regard to the theoretical underpinnings, program goals and components, delivery method, types of skills that are taught, and duration and frequency of the intervention. Many parent-training programs are founded on behavioral principles, and teach parents to use positive reinforcement and contingency management techniques to modify their child's behavior (Hogue and Liddle, 1999). Other parent-training programs are founded on other theoretical models, but aim to change the parenting skills and parent–child interactions to improve child behavior and parent–child communication and relationships.

Parent training is one of the better researched interventions, and has been found to be effective with youth across a range of child behavior problems including antisocial and delinquent behaviors (Serketich and Dumas, 1996; Kazdin, 1997; Farrington and Welsh, 1999; Piquero *et al.*, 2009). Several specific parent-training programs have been identified as an effective, model, or promising program by Blueprints for Healthy Youth Development (www.blueprintsprograms.com) and the Office of Juvenile Justice and Delinquency Prevention (www.ojjdp.gov/mpg), such as Parent–Child Interaction Therapy (PCIT), Incredible Years, and Triple P – Positive Parenting Program. While parent-training programs directly target social factors (i.e. parental behavior and parent–child interactions), from a biosocial perspective, parent-training programs likely affect brain–behavior interactions by altering activity in brain regions thought to be involved in antisocial and delinquent behavior, such as executive functions, sympathetic nervous system, emotion and reward drives, and emotion regulation.

Family therapy

Family therapy is a broad term describing family-based counseling interventions that generally aim to provide intervention to the entire family, often in a clinic or home setting, by a counseling professional (i.e. social worker, family therapist, psychologist) to affect change in family interactions, structure, and/or function. Family therapy interventions have been found to be effective with juveniles (Lipsey, 2009), but are quite diverse, are underpinned by a number of theories, and vary in terms of service delivery model, length and intensity of the intervention, and targeted populations. While most family therapy models target family processes and structures linked to risk factors for delinquency and crime, they vary in terms of the determinants targeted for change and the techniques used by therapists to facilitate change in the family and youth. Although there are a number of diverse family therapy interventions for families of at-risk youth, we focus here on two that have some empirical support with antisocial or delinquent youth in the prevention or desistance of antisocial or delinquent behaviors, and have been widely disseminated: Multisystemic Therapy (MST; Henggeler *et al.*, 1998) and Functional Family Therapy (FFT; Sexton, 2010).

MST is a widely known family-based intervention for youth exhibiting serious antisocial behavior. MST is a home-based intervention model grounded in systems and social ecological theories and informed by pragmatic models of the family and family therapy (i.e. structural and strategic family therapy models) as well as behavioral and cognitive models (Henggeler *et al.*, 1998). Unlike other manualized interventions that provide step-by-step instructions for each session, MST uses nine treatment principles to guide therapist behavior. In collaboration with the family, the MST therapist aims to address the multiple determinants of delinquency across the youth's ecology and to increase prosocial behavior. The determinants of delinquency and crime targeted for intervention by the MST therapist and family are informed by research on causes and correlates of antisocial behavior, but are selected through an analytic process and functional assessment of the specific drivers within the youth's ecology. Thus, while the process is driven by the treatment principles and assessment and analytic process, the specific interventions are unique to each family and directed across the ecology, including the youth, family, peer group, school, and other social systems.

FFT, a relatively brief (8–12 or up to 30 one-hour sessions) manualized family-based intervention delivered in an office or home setting, is recognized as an evidence-based intervention for youth exhibiting antisocial behavior, and has been widely disseminated (see Sexton and Alexander, 2005; Sexton, 2010). FFT is an integrative model grounded in systems theory and behavioral technologies that has evolved to incorporate attribution and information-processing theories and social constructionist and social influence ideas (Sexton and Alexander, 2005). The FFT model is relationally focused while also structured and protocol driven, with clearly articulated phases of intervention (i.e. engagement and motivation, behavior change, generalization) and specified goals, mechanisms of change, and therapist skills at each phase. The emphasis on the behavior change phase is grounded in risk and protective factor research, with particular emphasis on familial factors. While interventions may be individualized for the family, behavior change targets are focused on common risk and protective factors with the goal to build on protective family skills and improve the risk/protective factors ratio within the family (Sexton and Alexander, 2005).

Several studies point to positive effects of MST (e.g. Henggeler *et al.*, 1992; Borduin *et al.*, 1995; Timmons-Mitchell *et al.*, 2006; Butler *et al.*, 2011) and FFT (Sexton and Turner, 2010; White *et al.*, 2013) on antisocial behavior outcomes. As with most psychosocial interventions, positive effects are not consistently held across outcomes or studies, and effects have been found to vary based on the level of therapist adherence and fidelity with which the intervention is implemented (Henggeler *et al.*, 1997; Littell *et al.*, 2005), particularly with more difficult families (Sexton and Turner, 2010). Despite the equivocal evidence of family therapy approaches in general, and of the specific family therapy interventions discussed here, family-based interventions are seen as an important treatment modality for delinquency and crime prevention (Gottfredson *et al.*, 1997). Family therapy interventions aim to affect change in family process and structure, factors that are significantly associated with delinquency and later crime. While family therapy models rarely specifically target biological variables for change, family therapy likely has myriad effects on biological risk factors. By improving family interactions and warmth, increasing parenting monitoring and control, decreasing hostility and improving communication between family members, and improving the overall functioning of the family, family therapy may affect brain–behavior interactions by altering activity of the brain thought to be involved in delinquent and criminal behavior (e.g. executive functions, emotion-regulation, and self-control). Also, because the social environment can influence whether or not genes are "turned on" or "turned off," family therapy can help families improve the family environment, which can influence levels of transcription factors (e.g. cortisol) responsible for influencing gene expression (D'Onofrio and Lahey, 2010). For example, genetic vulnerabilities increase susceptibility to family risk factors, thus family factors may trigger an underlying genetic risk for antisocial behavior (Caspi *et al.*, 2002; Button *et al.*, 2008). With family therapy, family risk factors could be decreased, which could then prevent the underlying gene from being triggered and thus preventing the emergence of antisocial behavior.

Family interventions with adults

Family interventions that aim to reduce antisocial behavior center predominantly on children and adolescents (see Farrington and Welsh, 2007) due largely to the prospect of "child saving" and the broad social support that follows such efforts (Moon *et al.*, 2000; Cullen *et al.*, 2007). When it comes to reducing crime in adulthood, family-based interventions are few and far between. Instead, the preponderance of adult interventions entail variants of punishment, such as boot camps, intensive supervision, and intermediate sanctions (e.g. Aos *et al.*, 2001; Lipsey and Cullen, 2007), educational or vocational training (e.g. Wilson *et al.*, 2000; Visher *et al.*, 2005), or therapeutic efforts that neither involve nor are tied to the family (e.g. Pearson *et al.*, 2002). In short, the lack of family-based interventions in adulthood is largely a product of two factors: the family often serves as a risk factor for adult offenders (e.g. van Schellen *et al.*, 2012) and adult offenders do not always have an intact family (Laub and Sampson, 2003). In light of these realities, it should serve as no surprise that the most common and promising family interventions for adults concentrate on stabilizing the family by eliminating partner violence (Feder and Wilson, 2005).

Partner violence

Partner violence impacts more than 16 percent of married couples in the United States each year (Straus and Gelles, 1990), with lifetime prevalence rates indicating that over 32 percent of men and women are victimized by their partner (Tjaden and Thoennes, 2000). Partner violence has grave consequences for its victims (Coker *et al.*, 2000; Campbell, 2002), but also profound implications for the children who are exposed to it. According to McDonald and colleagues (2006), roughly 29 percent of American children live in a household in which partner violence occurred in the past year. Children who experience partner violence have higher levels of negative affect, negative emotions, and emotional dysregulation (Kitzmann *et al.*, 2003; Lloyd, 2013) and, importantly, are likely to remain enmeshed in a cycle of partner violence as adults (Simons *et al.*, 1995; Magdol *et al.*, 1998; Stith *et al.*, 2000; Ehrensaft *et al.*, 2003; Kwong *et al.*, 2003). Therefore the promise of developing effective family interventions that reduce partner violence is unquestionable, as they offer great benefits to the adults involved in the violence and the antisocial trajectories of children exposed to it. Given our interest in the biosocial bases of family interventions, it is pertinent to first understand why partner violence is so strongly concentrated within families. A range of explanations have been examined (e.g. Simons *et al.*, 1995), but this review restricts its attention to the general process of mate selection.

Biosocial origins of partner violence

Human mate selection has received considerable attention (e.g. Buss and Barnes, 1986; Shackelford *et al.*, 2005), with a growing body of research indicating that antisocial individuals are likely to select into relationships with antisocial others (Krueger *et al.*, 1998; Du Fort *et al.*, 2002; Knight, 2011). This selection process, known as assortative mating, is largely a product of phenotypic assortment, or the selection of partners based on observable characteristics such as personality or behavior (Farrington, 2011). Specifically, assortative mating is "the nonrandom coupling of individuals based on their similarity to each other on one or more characteristics" (Watson *et al.*, 2004: 1030). Though assortative mating is partly influenced by structural factors, it "is undoubtedly driven by personal preferences, which are often based on phenotypes, or the observable, genetically influenced characteristics" (Knight, 2011: 200). The relevance of this selection process is that individuals who choose partners based on behavioral and personality similarities that are correlated with partner violence, aggression, or substance dependence are, of course, at increased odds of involvement in partner violence (Moffitt *et al.*, 2001; Kim and Capaldi, 2004; Sakai *et al.*, 2004; Dick *et al.*, 2006). Furthermore, "over the course of generations that mate assortatively, trait-relevant genes become concentrated within families" (Krueger *et al.*, 1998: 182), which makes children from such fractious family circumstances both genetically and environmentally susceptible to analogous adult behavior themselves (e.g. Ehrensaft *et al.*, 2004). Therefore, as noted by Krueger and colleagues (1998: 184), "it is apparent that the role of assortative mating must be taken into account when investigating the development and persistence of antisocial behavior across the life course and across generations."

Biosocial underpinning of partner violence

Both the etiology and incidence of partner violence have much to do with biological and environmental factors and their interaction with one another. However, while a bulk of attention has been directed specifically at the environmental causes of partner violence, "biological and genetic contributors to violence have generally been overlooked in the family literature" (Kwong *et al.*, 2003: 299), though a few recent efforts have attempted to carefully explicate the biosocial nature of partner violence and thus offer important, integrated insight into the nature of the phenomenon (e.g. McKenry *et al.*, 1995; Soler *et al.*, 2000; George *et al.*, 2001; Powers *et al.*, 2006). For example, Booth and Dabbs (1993) examined testosterone levels in men, and found that testosterone negatively impacts marital relationships. Soler and colleagues (2000) also determined that testosterone directly influences rates of partner violence, but that controlling for social factors reduces the size of testosterone's effect. Specifically, they concluded that "for both types of abuse . . . the biosocial model had substantially higher predictive power than the biological model," which highlights the need for biosocial theorizing (Soler *et al.*, 2000: 736). Other biological variables, such as serotonin and cortisol, have also been implicated in the incidence of partner violence (George *et al.*, 2001; Powers *et al.*, 2006; see also Hellhammer *et al.*, 2009), and helped identify the biosocial significance of alcohol use in the etiology of partner conflict and violence (Leonard and Blane, 1992; Dick *et al.*, 2006). Importantly, family interventions centered on partner violence must appreciate that the emotional dispositions that give rise to partner violence are moderately heritable (Bouchard and Loehlin, 2001; Ehrensaft *et al.*, 2003) and have much to do with alcohol use and family discord (Caughlin *et al.*, 2000; Foran and O'Leary, 2008). For instance, evidence indicates that negative emotionality and neuroticism are strongly correlated with marital dissatisfaction and conflict (Karney and Bradbury, 1995; Robins *et al.*, 2000). Robins and colleagues (2002: 954) determined that "aggressive, stress-reactive, and alienated people have maladaptive relationships, and their relationships get progressively worse during their early 20s." Recognizing the impact of personality on relationship functioning and partner violence is crucial because it illuminates the significance to the phenomenon of underlying features such as temperament (see DeLisi and Vaughn, 2014). Taken together, these studies show that partner violence has biosocial roots that need to be considered by family-based interventions.

Biosocial bases of partner violence interventions

Family-based interventions that focus on the reduction of partner violence come in a variety of forms (Ehrensaft *et al.*, 2003; Feder and Wilson, 2005; Stover *et al.*, 2009). Few, if any, of the perpetrator-focused interventions have explicitly detailed the relevance of biological factors to the processes underlying their methods, but almost all have biosocial bases to at least some degree. This section reviews three of the most common partner violence interventions that have been widely assessed in recent decades, and explicitly details how each is inherently biosocial. In the end, this section will serve as a guide for future family interventions that aspire to improve their overall impact on partner violence and its various consequences.

Psycho-educational programs

Arguably the most common treatment intervention for batterers is known as the Duluth model (Pence and Paymar, 1993), which embraces a feminist psycho-educational approach concentrated on the notion of partner violence being a result of men actively attempting to control their female partners. According to Babcock *et al.* (2004), the Duluth model helps men develop more egalitarian relationships through a range of educational and consciousness-raising exercises that attend to the problems tied to intimidation and other forms of emotional coercion. In general, this intervention aims to improve empathizing, problem-solving skills, and stress-reduction abilities (Kelly and Johnson, 2008). Although the Duluth model makes no mention of biosocial processes, they are undoubtedly part of its approach. From a biosocial perspective, the education and consciousness-raising approach the model takes is tied to certain psychological factors that have biological underpinnings. Because emotionally coercive behavior is driven by jealousy and anger, psycho-educational approaches that center on the neural correlates of emotional regulation, and are deliberate about their connections with and impact on brain activity in those regions, will be primed to be most effective.

Cognitive behavioral therapy

Interventions that include cognitive behavioral therapy (CBT) are also common treatments for perpetrators of partner violence (see Feder and Wilson, 2005). Most CBT programs are group-based and aim to modify the cognitions that are associated with anger regulation (Austin and Dankwort, 1999). Specifically, they view partner violence as a learned behavior that can be unlearned through a series of skills training (e.g. communication) and anger management exercises (e.g. relaxation training) (Babcock *et al.*, 2004). Importantly, some versions of CBT are conjoint because they recognize that partner violence is often situational and perpetrated by both romantic partners due to their inability to settle arguments effectively (Moffitt *et al.*, 2001; Kelly and Johnson, 2008). Specifically, "couples counseling sees the couple as the reasons for the problem and, as such, works with both members to improve communication and conflict resolution skills" (Feder and Wilson, 2005: 241). From a biosocial perspective, it is clear that CBT efforts are tied to a host of neuropsychological factors, even though the mechanisms are not explicitly noted by this body of interventions (Vaske *et al.*, 2011). Further, it is likely that partner violence is operating through various psychological characteristics, such as temperament, that have biological roots (Moffitt *et al.*, 2001). Indeed, if CBT efforts are to have a real impact on slowing partner violence, it is pertinent that they be appropriately modified so that they are in tune with the various brain regions associated with emotional arousal that they are tied to (Vaske *et al.*, 2011). Without attention to those factors and their connection with social learning processes, CBT efforts in adulthood are likely to continue having a seemingly negligible influence on partner violence.

Substance-use treatment

Partner violence interventions aimed at reducing alcohol are also quite common (O'Leary *et al.*, 1989; Pihl *et al.*, 1993; Dunford, 2000; Fals-Stewart *et al.*, 2002; Jewkes, 2002; Stover *et al.*, 2009). In general, this is because "among all psychoactive substances, alcohol is arguably the most potent agent for eliciting aggression and reducing behavioral control in a subset of individuals" (Heinz *et al.*, 2011: 400). These programs are delivered with attention to reducing alcohol use by giving perpetrators skills to cope more effectively with their stress and negative moods, which overlaps with the approach taken by many CBT programs. From a biosocial perspective, there is little question surrounding the biological mechanisms that guide substance-use programs. Alcohol use is followed by lower inhibition and has a notable impact on higher-order cognition (Pihl *et al.*, 1993). Substance-use-focused treatments have shown promise (e.g. Stuart *et al.*, 2003; O'Farrell *et al.*, 2004; Klostermann *et al.*, 2010), but some evidence indicates that alcohol is unlikely to increase the likelihood of partner violence among individuals with a healthy temperament (Giancola, 2002). Thus the most effective alcohol-focused partner violence interventions will need to recognize that alcohol use is not only a product of environmental factors, but is also tied to personality dispositions with biological roots.

Conclusion

Findings from behavioral genetic and neurological research, in conjunction with research on family and environmental forces, effectively elucidate the salience of biosocial processes underlying delinquent and criminal behavior. As highlighted throughout this chapter, it is now well accepted that neither biological nor social factors alone are sufficient to explain the etiology of antisocial behavior, and thus both biological and environmental factors need to be considered in understanding, preventing, and desisting delinquent and criminal behaviors. Currently, however, family-based prevention and intervention programs are largely based on an incomplete understanding of biosocial underpinnings and causal processes, thus limiting their effectiveness and potentially wasting valuable resources (Moffitt, 2005). Therefore it seems prudent that we move toward a biosocial theory of family-based prevention and intervention, and more readily measure biological variables in intervention outcome research. Attending to familial and biological factors and tailoring preventions and interventions to youth and adults based on the unique genetic, neurological, and environmental risk is likely to be the most effective course of action. Moreover, considering biological and familial factors in intervention development and research can also lead to more effective and efficient targeting of interventions to those most at risk, and better matching of interventions with those most likely to respond favorably based on biological vulnerability and familial risk (Beauchaine *et al.*, 2008). In short, adopting a biosocial framework for prevention and intervention practice and research offers an incalculable degree of long-term and future potential. Without question, the future of delinquency and crime prevention and intervention involves greater attention to and understanding of the interplay of biological and social forces.

References

Aos, S., Phipps, P., Barnoski, R., and Lieb, R. (2001) *The Comparative Costs and Benefits of Programs to Reduce Crime*, Olympia, WA: Washington State Institute for Public Policy.

Austin, J.B. and Dankwort, J. (1999) 'Standards for batterer programs: a review and analysis', *Journal of Interpersonal Violence*, 14: 152–168.

Babcock, J.C., Green, C.E., and Robie, C. (2004) 'Does batterers' treatment work? A meta-analytic review of domestic violence treatment', *Clinical Psychology Review*, 23: 1023–1053.

Baker, L.A., Bezdjiam, S., and Raine, A. (2006) 'Behavioral genetics: the science of antisocial behavior, *Law and Contemporary Problems*, 69: 7–46.

Barlow, J., Smailagic, N., Ferriter, M., Bennett, C., and Jones, H. (2010) 'Group-based parent-training programmes for improving emotional and behavioural adjustment in children from birth to three years old', *Cochrane Database of Systematic Review*, 3: 1–95.

Beauchaine, T.P., Neuhaus, E., Brenner, S.L., and Gatzke-Kopp, L. (2008) 'Ten good reasons to consider biological process in prevention and intervention research', *Development and Psychopathology*, 20: 745–774.

Beaver, K.M., Wright, J.P., DeLisi, M., Walsh, A., Vaughn, M.G., Boisvert, D., and Vaske, J. (2007) 'A gene × gene interaction between DRD2 and DRD4 is associated with conduct disorder and antisocial behavior in males', *Behavioral and Brain Functions*, 3: 1–8.

Beaver, K.M., DeLisi, M., Vaughn, M.G., and Wright, J.P. (2010) 'The intersection of genes and neuropsychological deficits in the prediction of adolescent delinquency and low self-control', *International Journal of Offender Therapy and Comparative Criminology*, 54: 22–42.

Bernazzani, O., Cote, C., and Tremblay, R.E. (2001) 'Early parent training to prevent disruptive behavior problems and delinquency in children', *Annals of the American Academy of Political Social Science*, 578: 90–103.

Bilukha, O., Hahn, R.A., Crosby, A., Fullilove, M.T., Liberman, A., Moscicki, E., Snyder, S., Tuma, F., Corson, P., Schofield, A., and Briss, P.A. (2005) 'Effectiveness of early childhood home visitation in preventing violence: a systematic review', *American Journal of Preventive Medicine*, 28: 11–39.

Booth, A. and Dabbs, J.M. (1993) 'Testosterone and men's marriages', *Social Forces*, 72: 463–477.

Borduin, C.M., Mann, B.J., Cone, L.T., Henggeler, S.W., Fucci, B.R., Blaske, D.M., and Williams, R.A. (1995) 'Multisystemic treatment of serious juvenile offenders: long-term prevention of criminality and violence', *Journal of Consulting and Clinical Psychology*, 63: 569–578.

Bouchard Jr, T.J. and Loehlin, J.C. (2001) 'Genes, evolution, and personality', *Behavior Genetics*, 31: 243–273.

Brennan, P.A. and Raine, A. (1997) 'Biosocial bases of antisocial behavior: psychophysiological, neurological, and cognitive factors', *Clinical Psychology Review*, 17: 589–604.

Burt, S.A. (2009) 'A mechanistic explanation of popularity: genes, rule-breaking, and evocative gene–environment correlations', *Journal of Personality and Social Psychology*, 96: 783–794.

Buss, D.M. and Barnes, M. (1986) 'Preferences in human mate selection', *Journal of Personality and Social Psychology*, 50: 559–570.

Butler, S., Baruch, G., Hickey, N., and Fonagy, P. (2011) 'A randomized controlled trial of multisystemic therapy and statutory therapeutic intervention for young offenders', *Journal of the American Academy of Child and Adolescent Psychiatry*, 50: 1220–1235.

Button, T.M.M., Lau, J.Y.F., Maughan, B., and Eley, T. (2008) 'Parental punitive discipline, negative life events and gene–environment interplay in the development of externalizing behavior', *Psychological Medicine*, 38: 29–39.

Campbell, J.C. (2002) 'Health consequences of intimate partner violence', *The Lancet*, 359: 1331–1336.

Caspi, A., McClay, J., Moffitt, T.E., Martine, J., Craig, I.W., Taylor, A., and Poulton, R. (2002) 'Role of genotype in the cycle of violence in maltreated children', *Science*, 297: 851–854.

Caughlin, J.P., Huston, T.L., and Houts, R.M. (2000) 'How does personality matter in marriage? An examination of trait anxiety, interpersonal negativity, and marital satisfaction', *Journal of Personality and Social Psychology*, 78: 326–336.

Coker, A.L., Smith, P.H., Bethea, L., King, M.R., and McKeown, R.E. (2000) 'Physical health consequences of physical and psychological intimate partner violence', *Archives of Family Medicine*, 9: 451–457.

Cullen, F.T., Vose, B.A., Lero Jonson, C.N., and Unnever, J.D. (2007) 'Public support for early intervention: is child saving a "habit of the heart"?', *Victims and Offenders*, 2: 109–124.

DeLisi, M. and Vaughn, M.G. (2011) 'The importance of neuropsychological deficits relating to self-control and temperament to the prevention of serious antisocial beahvior', *International Journal of Child, Youth, and Family Studies*, 1/2: 12–35.

DeLisi, M., and Vaughn, M.G. (2014) 'Foundation for a temperament-based theory of antisocial behavior and criminal justice system involvement', *Journal of Criminal Justice*, 42: 10–25.

DeLisi, M., Beaver, K.M., Vaughn, M.G., and Wright, J.P. (2009) 'All in the family: gene × environment interaction between DRD2 and criminal father is associated with five antisocial phenotypes', *Criminal Justice and Behavior*, 36: 1187–1197.

Demuth, S. and Brown, S.L. (2004) 'Family structure, family processes, and adolescent delinquency: the significance of parental absence versus parental gender', *Journal of Research in Crime and Delinquency*, 41: 51–81.

Derzon, J.H. (2010) 'The correspondence of family features with problem, aggressive, criminal, and violent behavior: a meta-analysis', *Journal of Experimental Criminology*, 6: 263–292.

Dick, D.M., Agrawal, A., Schuckit, M.A., Bierut, L., Hinrichs, A., Fox, L., and Begleiter, H. (2006) 'Marital status, alcohol dependence, and GABRA2: evidence for gene–environment correlation and interaction', *Journal of Studies on Alcohol and Drugs*, 67: 185–194.

DiLalla, L.F. (2002) 'Behavior genetics of aggression in children: review and future directions', *Developmental Review*, 22: 593–622.

D'Onofrio, B.M. and Lahey, B.B. (2010) 'Biosocial influences on the family: a decade review', *Journal of Marriage and Family*, 72: 762–782.

Du Fort, G.G., Boothroyd, L.J., Bland, R.C., Newman, S.C., and Kakuma, R. (2002) 'Spouse similarity for antisocial behaviour in the general population', *Psychological Medicine*, 32: 1407–1416.

Dunford, F. (2000) 'The San Diego Navy experiment: an assessment of interventions for men who assault their wives', *Journal of Consulting and Clinical Psychology*, 68: 468–476.

Ehrensaft, M.K., Cohen, P., Brown, J., Smailes, E., Chen, H., and Johnson, J.G. (2003) 'Intergenerational transmission of partner violence: a 20-year prospective study', *Journal of Consulting and Clinical Psychology*, 71: 741–753.

Ehrensaft, M.K., Moffitt, T.E., and Caspi, A. (2004) 'Clinically abusive relationships in an unselected birth cohort: men's and women's participation and developmental antecedents', *Journal of Abnormal Psychology*, 113: 258–271.

Ellis, L., Beaver, K., and Wright, J. (2009) *Handbook of Crime Correlates*, Oxford: Academic Press.

Fals-Stewart, W., Kashdan, T.B., O'Farrell, T.J., and Birchler, G.R. (2002) 'Behavioral couples therapy for drug-abusing patients: effects on partner violence', *Journal of Substance Abuse Treatment*, 22: 87–96.

Farrington, D.P. (2002) 'Multiple risk factors for multiple problem violent boys', in R.R. Corrado, R. Roesch, S.D. Hart, and J.K. Gierowski (eds), *Multi-problem Violent Youth: A Foundation for Comparative Research on Needs, Interventions and Outcomes*, Amsterdam: IOS Press, pp. 23–34.

Farrington, D.P. (2011) 'Families and crime', in J.Q. Wilson and J. Petersilia (eds), *Crime and Public Policy*, Oxford: Oxford University Press, pp. 130–157.

Farrington, D.P. and Welsh, B.C. (1999) 'Delinquency prevention using family-based interventions', *Children and Society*, 13: 287–303.

Farrington, D.P. and Welsh, B.C. (2007) *Saving Children from a Life of Crime: Early Risk Factors and Effective Interventions*, New York: Oxford University Press.

Feder, L. and Wilson, D.B. (2005) 'A meta-analytic review of court-mandated batterer intervention programs: can courts affect abusers' behavior?', *Journal of Experimental Criminology*, 1: 239–262.

Foran, H.M. and O'Leary, K.D. (2008) 'Alcohol and intimate partner violence: a meta-analytic review', *Clinical Psychology Review*, 28: 1222–1234.

Frick, P.J. (2004) 'Developmental pathways to conduct disorder: implications for serving youth who show severe aggressive and antisocial behavior, *Psychology in the Schools*, 41: 823–833.

George, D.T., Umhau, J.C., Phillips, M.J., Emmela, D., Ragan, P.W., Shoaf, S.E., and Rawlings, R.R. (2001) 'Serotonin, testosterone and alcohol in the etiology of domestic violence', *Psychiatry Research*, 104: 27–37.

Giancola, P.R. (2002) 'Irritability, acute alcohol consumption and aggressive behavior in men and women', *Drug and Alcohol Dependence*, 68: 263–274.

Gomby, D.S., Culross, P.L., and Behrman, R.E. (1999) 'Home visiting: recent program evaluations – analysis and recommendations', *Future of Children*, 9: 4–26.

van Goozen, S.H.M. and Fairchild, G. (2008) 'How can the study of biological processes help design new interventions for children with severe antisocial behavior?', *Development and Psychopathology*, 20: 941–973.

Gottfredson, D., MacKenzie, D., Eck, J., Reuter, P., and Bushway, S. (1997) *Preventing Crime: What Works, What Doesn't, What's Promising: A Report to the United States Congress*, Washington, DC: US Department of Justice, Office of Justice Programs.

Griffin, K.W., Botvin, G.J., Scheier, L.M., Diaz, T., and Miller, N.L. (2000) 'Parenting practices as predictors of substance use, delinquency, and aggression among urban minority youth: moderating effects of family structure and gender', *Psychology of Addictive Behaviors*, 14: 174–184.

Haapasalo, J. and Pokela, E. (1999) 'Child-rearing and child abuse antecedents of criminality', *Aggression and Violent Behavior*, 4: 107–127.

Heath, A.C., Bucholz, K.K., Madden, P.A.F., Dinwiddie, S.H., Slutske, W.S., Bierut, L.J., Statham, D.J., Dunne, M.P., Whitfield, J.B., and Martin, N.G. (1997) 'Genetic and environmental contributions to alcohol dependence risk in a national twin sample: consistency of findings in women and men', *Psychological Medicine*, 27: 1381–1396.

Heinz, A.J., Beck, A., Meyer-Lindenberg, A., Sterzer, P., and Heinz, A. (2011) 'Cognitive and neurobiological mechanisms of alcohol-related aggression', *Nature Reviews Neuroscience*, 12: 400–413.

Hellhammer, D.H., Wüst, S., and Kudielka, B.M. (2009) 'Salivary cortisol as a biomarker in stress research', *Psychoneuroendocrinology*, 34: 163–171.

Henggeler, S.W., Melton, G.B., and Smith, L.A. (1992) 'Family preservation using multisystemic therapy: an effective alternative to incarcerating serious juvenile offenders', *Journal of Consulting and Clinical Psychology*, 60: 953–961.

Henggeler, S.W., Melton, G.B., Brondino, M.J., Scherer, D.G., and Hanley, J.H. (1997) 'Multisytemic therapy with violent and chronic juvenile offenders and their families: the role of treatment fidelity in successful dissemination', *Journal of Consulting and Clinical Psychology*, 65: 821–833.

Henggeler, S.W., Schoenwald, S.K., Bordun, C.M., Rowland, M.D., and Cunningham, P.B. (1998) *Multisystemic Treatment of Antisocial Behavior in Children and Adolescents*, New York: Guilford Press.

Herrenkohl, T.I., Maguin, E., Hill, K.G., Hawkins, J.D., Abbott, R.D., and Catalano, R.F. (2000) 'Developmental risk factors for youth violence', *Journal of Adolescent Health*, 26: 176–186.

Hogue, A. and Liddle, H.A. (1999) 'Family-based preventive intervention: an approach to preventing substance use and antisocial behavior', *American Journal of Orthopsychiatry*, 69: 278–293.

Horwitz, B.N. and Neiderhiser, J.M. (2011) 'Gene–environment interplay, family relationships, and child adjustment', *Journal of Marriage and Family*, 73: 804–816.

Huh, D., Tristan, J., Wade, E., and Stice, E. (2006) 'Does problem behavior elicit poor parenting? A prospective study of adolescent girls', *Journal of Adolescent Research*, 21: 185–204.

Ireland, T.O. and Smith, C.A. (2009) 'Living in partner-violent families: developmental links to antisocial behavior and relationship violence', *Journal of Youth and Adolescence*, 38: 323–339.

Jewkes, R. (2002) 'Intimate partner violence: causes and prevention', *The Lancet*, 359: 1423–1429.

Karney, B.R. and Bradbury, T.N. (1995) 'The longitudinal course of marital quality and stability: a review of theory, methods, and research', *Psychological Bulletin*, 118: 3–34.

Kazdin, A.E. (1997) 'Parent management training: evidence, outcomes, and issues', *Journal of the American Academy of Child and Adolescent Psychiatry*, 36: 1349–1356.

Kelly, J.B. and Johnson, M.P. (2008) 'Differentiation among types of intimate partner violence: research update and implications for interventions', *Family Court Review*, 46: 476–499.

Kim, H.K. and Capaldi, D.M. (2004) 'The association of antisocial behavior and depressive symptoms between partners and risk for aggression in romantic relationships', *Journal of Family Psychology*, 18: 82–95.

Kimonis, E.R. and Frick, P.J. (2010) 'Etiology of oppositional defiant disorder and conduct disorder: biological, familial and environmental factors identified in the development of disruptive behavior disorders', in R.C. Murrihy, A.D. Kidman, and T.H. Ollendick (eds), *Clinical Handbook of Assessing and Treating Conduct Problems in Youth*, New York: Springer, pp. 49–76.

Kitzman, H., Olds, D.L., Henderson, C.R., Jr., Hanks, C., Cole, R., Tatelbaum, R., and Barnard, K. (1997) 'Effect of prenatal and infancy home visitation by nurses on pregnancy outcomes, childhood injuries, and repeated childbearing: a randomized controlled trial', *Journal of the American Medical Association*, 278: 644–652.

Kitzmann, K.M., Gaylord, N.K., Holt, A.R., and Kenny, E.D. (2003) 'Child witnesses to domestic violence: a meta-analytic review', *Journal of Consulting and Clinical Psychology*, 71: 339–352.

Klostermann, K., Kelley, M.L., Mignone, T., Pusateri, L., and Fals-Stewart, W. (2010) 'Partner violence and substance abuse: treatment interventions', *Aggression and Violent Behavior*, 15: 162–166.

Knight, K.E. (2011) 'Assortative mating and partner influence on antisocial behavior across the life course', *Journal of Family Theory and Review*, 3: 198–219.

Krueger, R.F., Moffitt, T.E., Caspi, A., Bleske, A., and Silva, P.A. (1998) 'Assortative mating for antisocial behavior: developmental and methodological implications', *Behavior Genetics*, 28: 173–186.

Kwong, M.J., Bartholomew, K., Henderson, A.J.Z., and Trinke, S.J. (2003) 'The intergenerational transmission of relationship violence', *Journal of Family Psychology*, 17: 288–301.

Laub, J.H. and Sampson, R.J. (2003) *Shared Beginnings, Divergent Lives: Delinquent Boys to Age 70*, Cambridge, MA: Harvard University Press.

Leonard, K.E. and Blane, H.T. (1992) 'Alcohol and marital aggression in a national sample of young men', *Journal of Interpersonal Violence*, 7: 19–30.

Leschied, A., Chiodo, D., Nowicki, E., and Rodger, S. (2008) 'Childhood predictors of adult criminality: a meta-analysis drawn from the prospective longitudinal literature', *Canadian Journal of Criminology and Criminal Justice*, 50: 435–467.

Lipsey, M.W. (2009) 'The primary factors that characterize effective interventions with juvenile offenders: a meta-analytic overview', *Victims and Offenders*, 4: 124–147.

Lipsey, M.W. and Cullen, F.T. (2007) 'The effectiveness of correctional rehabilitation: a review of systematic reviews', *Annual Review of Law and Social Science*, 3: 297–320.

Lipsey, M.W. and Derzon, J.H. (1998) 'Predictors of violent or serious delinquency in adolescence and early adulthood: a synthesis of longitudinal research', in R. Loeber and D.P. Farrington (eds), *Serious and Violent Juvenile Offenders: Risk Factors and Successful Interventions*, Thousand Oaks, CA: Sage, pp. 86–105.

Littell, J., Forsythe, B., Popa, M., and Austin, S. (2005) 'Multisystemic therapy for social, emotional, and behavioral problems in youth aged 10–17: a systematic review', *The Campbell Collaboration Library*, 1.

Lloyd, S.A. (2013) 'Family violence', in G.W. Peterson and K.R. Bush (eds), *Handbook of Marriage and the Family*, New York: Springer, pp. 449–486.

Loeber, R. and Stouthamer-Loeber, M. (1986) 'Family factors as correlates and predictors of juvenile conduct problems and delinquency', *Crime and Justice*, 7: 29–149.

Lundahl, B., Risser, H.J., and Lovejoy, M.C. (2006) 'A meta-analysis of parent training: moderators and follow-up effects', *Clinical Psychology Review*, 26: 86–104.

McDonald, R., Jouriles, E.N., Ramisetty-Mikler, S., Caetano, R., and Green, C.E. (2006) 'Estimating the number of American children living in partner-violent families' *Journal of Family Psychology*, 20: 137–142.

McKenry, P.C., Julian, T.W., and Gavazzi, S.M. (1995) 'Toward a biopsychosocial model of domestic violence', *Journal of Marriage and the Family*, 57: 307–320.

Magdol, L., Moffitt, T.E., Caspi, A., and Silva, P.A. (1998) 'Developmental antecedents of partner abuse: a prospective-longitudinal study', *Journal of Abnormal Psychology*, 107: 375–389.

Mednick, S.A., Gabrielli, W.F., Jr, and Hutchings, B. (1984) 'Genetic influences in criminal convictions: evidence from an adoption cohort', *Science*, 224: 891–894.

Moffitt, T.E. (1993) 'Adolescence-limited and life-course-persistent antisocial behavior: a developmental taxonomy', *Psychological Review*, 100(4): 674–701.

Moffitt, T.E. (2005) 'The new look of behavioral genetics in developmental psychopathology: gene–environment interplay in antisocial behaviors', *Psychological Bulletin*, 131: 533–554.

Moffitt, T.E., Caspi, A., Rutter, M., and Silva, P.A. (2001) *Sex Differences in Antisocial Behaviour: Conduct Disorder, Delinquency and Violence in the Dunedin Longitudinal Study*, Cambridge, MA: Cambridge University Press.

Moon, M.M., Sundt, J.L., Cullen, F.T., and Wright, J.P. (2000) 'Is child saving dead? Public support for juvenile rehabilitation', *Crime and Delinquency*, 46: 38–60.

Morgan, A.B. and Lilienfeld, S.O. (2000) 'A meta-analytic review of the relation between antisocial behavior and neuropsychological measures of executive function', *Clinical Psychology Review*, 20: 113–136.

Olds, D.L. (2006) 'The nurse–family partnership: an evidence-based preventive intervention', *Infant Mental Health Journal*, 27: 5–25.

O'Farrell, T.J., Murphy, C.M., Stephan, S.H., Fals-Stewart, W., and Murphy, M. (2004) 'Partner violence before and after couples-based alcoholism treatment for male alcoholic patients: the role of treatment involvement and abstinence', *Journal of Consulting and Clinical Psychology*, 72: 202–217.

Olds, D.L., Henderson, C.R., Jr, and Kitzman, H. (1994) 'Does prenatal and infancy nurse home visitation have enduring effects on qualities of parental caregiving and child health at 25 to 50 months of life?', *Pediatrics*, 93: 89–98.

Olds, D., Henderson, C.R., Jr., Cole, R., Eckenrode, J., Kitzman, H., Luckey, D., and Powers, J. (1998) 'Long-term effects of nurse home visitation on children's criminal and antisocial behavior: 15-year follow-up of a randomized controlled trial', *Journal of the American Medical Association* 280: 1238–1244.

Olds, D.L., Holmberg, J.R., Donelan-McCall, N., Luckey, D.W., Knudtson, M.D., and Robinson, J. (2014) 'Effects of home visits by paraprofessionals and by nurses on children: follow-up of a randomized trial at ages 6 and 9 years', *JAMA Pediatrics*, 168: 114–121.

O'Leary, K.D., Barling, J., Arias, I., Rosenbaum, A., Malone, J., and Tyree, A. (1989) 'Prevalence and stability of physical aggression between spouses: a longitudinal analysis', *Journal of Consulting and Clinical Psychology*, 57: 263–268.

Pearson, F.S., Lipton, D.S., Cleland, C.M., and Yee, D.S. (2002) 'The effects of behavioral/cognitive-behavioral programs on recidivism', *Crime and Delinquency*, 48: 476–496.

Pence, E. and Paymar, M. (1993) *Education Groups for Men who Batter: The Duluth Model*, New York: Springer.

Pihl, R.O., Peterson, J.B., and Lau, M.A. (1993) 'A biosocial model of the alcohol–aggression relationship', *Journal of Studies on Alcohol and Drugs*, 11: 128–139.

Piquero, A.R. (2008) 'Taking stock of developmental trajectories of criminal activity over the life course', in A.M. Liberman (ed.), *The Long View of Crime: A Synthesis of Longitudinal Research*, New York: Springer, pp. 23–78.

Piquero, A.R., Farrington, D.P., and Blumstein, A. (2003) 'The criminal career paradigm', in M. Tonry (ed.), *Crime and Justice: A Review of Research*, Vol. 30, Chicago, IL: University of Chicago Press, pp. 359–506.

Piquero, A.R., Farrington, D.P., Welsh, B.C., Tremblay, R., and Jennings, W.G. (2009) 'Effects of early family/parent training programs on antisocial behavior and delinquency', *Journal of Experimental Criminology*, 5: 83–120.

Plomin, R. and Asbury, K. (2005) 'Nature and nurture: genetic and environmental influences on behavior', *Annals of the American Academy of Political and Social Science*, 600: 86–98.

Powers, S.I., Pietromonaco, P.R., Gunlicks, M., and Sayer, A. (2006) 'Dating couples' attachment styles and patterns of cortisol reactivity and recovery in response to a relationship conflict', *Journal of Personality and Social Psychology*, 90: 613–628.

Raine, A. (2002) 'Annotation: the role of prefrontal deficits, low autonomic arousal, and early health factors in the development of antisocial and aggressive behavior in children', *Journal of Child Psychology and Psychiatry*, 43: 417–434.

Raine, A. (2008) 'From genes to brain to antisocial behavior', *Current Directions in Psychological Science*, 17: 323–328.

Raine, A. (2013) *The Anatomy of Violence: The Biological Roots of Crime*, New York: Pantheon Books.

Raine, A. and Yang, Y. (2006) 'Neural foundations to moral reasoning and antisocial behavior', *Social Cognitive and Affective Neuroscience*, 1: 203–213.

Raine, A., Moffitt, T.E., Caspi, A., Loeber, R., Stouthamer-Loeber, M., and Lynam, D. (2005) 'Neurocognitive impairments in boys on the life-course persistent antisocial path', *Journal of Abnormal Psychology*, 114: 38–49.

Ratchford, M. and Beaver, K.M. (2009) 'Neuropsychological deficits, low self-control, and delinquent involvement: toward a biosocial explanation of delinquency', *Criminal Justice and Behavior*, 36: 147–162.

Rhee, S.H. and Waldman, I.D. (2002) 'Genetic and environmental influences on antisocial behavior: a meta-analysis of twin and adoption studies', *Psychological Bulletin*, 128: 490–529.

Riggins-Caspers, K.M., Cadoret, R.J., Knutson, J.F., and Langbehn, D. (2003) 'Biology–environment interaction and evocative biology–environment correlation: contributions of harsh discipline and parental psychopathology to problem adolescent behaviors', *Behavior Genetics*, 33: 205–220.

Robins, R.W., Caspi, A., and Moffitt, T.E. (2000) 'Two personalities, one relationship: both partners' personality traits shape the quality of their relationship', *Journal of Personality and Social Psychology*, 79: 251–259.

Robins, R.W., Caspi, A., and Moffitt, T.E. (2002) 'It's not just who you're with, it's who you are: personality and relationship experiences across multiple relationships', *Journal of Personality*, 70: 925–964.

Rowe, D.C. (1983) 'Biometrical genetic models of self-reported delinquent behavior: a twin study', *Behavior Genetics*, 13: 473–489.
Rowe, D.C. and Osgood, D.W. (1984) 'Heredity and sociological theories of delinquency: a reconsideration', *American Sociological Review*, 49: 526–540.
Sakai, J.T., Stallings, M.C., Mikulich-Gilbertson, S.K., Corley, R.P., Young, S.E., Hopfer, C.J., and Crowley, T.J. (2004) 'Mate similarity for substance dependence and antisocial personality disorder symptoms among parents of patients and controls', *Drug and Alcohol Dependence*, 75: 165–175.
van Schellen, M., Poortman, A.R., and Nieuwbeerta, P. (2012) 'Partners in crime? Criminal offending, marriage formation, and partner selection', *Journal of Research in Crime and Delinquency*, 49: 545–571.
Serketich, W.J. and Dumas, J.E. (1996) 'The effectiveness of behavioral parent training to modify antisocial behavior in children: a meta-analysis', *Behavior Therapy*, 27: 171–186.
Sexton, T.L. (2010) *Functional Family Therapy in Clinical Practice: An Evidence-based Treatment Model for Working with Troubled Adolescents*, New York: Routledge.
Sexton, T.L. and Alexander, J.F. (2005) 'Functional family therapy for externalizing disorders in adolescents', in J.L. Lebow (ed.), *Handbook of Clinical Family Therapy*, New Jersey: Wiley, pp. 164–191.
Sexton, T.L. and Turner, C.W. (2010) 'The effectiveness of functional family therapy for youth with behavioral problems in a community practice setting', *Journal of Family Psychology*, 24: 339–34.
Shackelford, T.K., Schmitt, D.P., and Buss, D.M. (2005) 'Universal dimensions of human mate preferences', *Personality and Individual Differences*, 39: 447–458.
Simons, R.L., Wu, C.I., Johnson, C., and Conger, R.D. (1995) 'A test of various perspectives on the intergenerational transmission of domestic violence', *Criminology*, 33: 141–172.
Soler, H., Vinayak, P., and Quadagno, D. (2000) 'Biosocial aspects of domestic violence', *Psychoneuroendocrinology*, 25: 721–739.
Stith, S.M., Rosen, K.H., Middleton, K.A., Busch, A.L., Lundeberg, K., and Carlton, R.P. (2000) 'The intergenerational transmission of spouse abuse: a meta-analysis', *Journal of Marriage and Family*, 62: 640–654.
Stover, C.S., Meadows, A.L., and Kaufman, J. (2009) 'Interventions for intimate partner violence: review and implications for evidence-based practice', *Professional Psychology: Research and Practice*, 40: 223–233.
Straus, M.A. and Gelles, R.J. (1990) 'How violent are American families? Estimates from the National Family Violence Resurvey and other studies', in M.A. Straus and R.J. Gelles (eds), *Physical Violence in American Families: Risk Factors and Adaptations to Violence in 8,145 Families*, New Brunswick, NJ: Transactional Publishing, pp. 95–112.
Stuart, G.L., Moore, T.M., Kahler, C.W., and Ramsey, S.E. (2003) 'Substance abuse and relationship violence among men court-referred to batterers' intervention programs', *Substance Abuse*, 4: 159–177.
Taylor, J., Iacono, W.G., and McGue, M. (2000) 'Evidence for a genetic etiology of early-onset delinquency', *Journal of Abnormal Psychology*, 109: 634–643.
Timmons-Mitchell, J., Bender, M.B., Kishna, M.A., and Mitchell, C.C. (2006) 'An independent effectiveness trial of multisystemic therapy with juvenile justice youth', *Journal of Clinical Child and Adolescent Psychology*, 35: 227–236.
Tjaden, P. and Thoennes, N. (2000) 'Prevalence and consequences of male-to-female and female-to-male intimate partner violence as measured by the National Violence Against Women Survey', *Violence Against Women*, 6: 142–161.
Tremblay, R.E. (2000) 'The development of aggressive behavior during childhood: what have we learned in the past century?', *International Journal of Behavioral Development*, 24: 129–141.
Tremblay, R.E. and Japel, C. (2003) 'Prevention during pregnancy, infancy, and the preschool years', in D.P. Farrington and J.W. Coid (eds), *Early Prevention of Adult Antisocial Behavior*, New York: Cambridge University Press, pp. 205–242.
Turkheimer, E. and Waldron, M. (2000) 'Nonshared environment: a theoretical, methodological, and quantitative review', *Psychological Bulletin*, 126: 78–108.
Vaske, J., Galyean, K., and Cullen, F.T. (2011) 'Toward a biosocial theory of offender rehabiltiation: why does cognitive-behavioral therapy work?', *Journal of Criminal Justice*, 39: 90–102.
Vaughn, M.G., Beaver, K.M., and DeLisi, M. (2009) 'A general biosocial paradigm of antisocial behavior', *Youth Violence and Juvenile Justice*, 7: 279–298.
Visher, C.A., Winterfield, L., and Coggeshall, M.B. (2005) 'Ex-offender employment programs and recidivism: a meta-analysis', *Journal of Experimental Criminology*, 1: 295–315.
Watson, D., Klohnen, E.C., Casillas, A., Nus Simms, E., Haig, J., and Berry, D.S. (2004) 'Match makers and deal breakers: analyses of assortative mating in newlywed couples', *Journal of Personality*, 72: 1029–1068.

Wells, E. and Rankin, J.H. (1991) 'Families and delinquency: a meta-analysis of the impact of broken homes', *Social Problems*, 38: 71–93.

White, J.L., Moffitt, T.E., Earls, F., Robins, L., and Silva, P.A. (1990) 'How early can we tell? Predictors of childhood conduct disorder and adolescent delinquency', *Criminology*, 28: 507–535.

White, S.F., Frick, P.J., Lawing, K., and Bauer, D. (2013) 'Callous-unemotional traits and response to functional family therapy in adolescent offenders', *Behavioral Sciences & the Law*, 31(2): 271–285.

Whitten, L.A. (2013) 'Translational neuroscience and potential contributions of functional magnetic resonance imaging (fMRI) to the prevention of substance misuse and antisocial behavior', *Prevention Science*, 14: 238–246.

Wilson, D.B., Gallagher, C.A., and MacKenzie, D.L. (2000) 'A meta-analysis of corrections-based education, vocation, and work programs for adult offenders', *Journal of Research in Crime and Delinquency*, 37: 347–368.

41

Action–reaction? Forensic evidence analysis and forensic awareness in sexual homicide

Eric Beauregard and Melissa Martineau

Introduction

DETECTIVE JENSEN: "*Did you know that by depositing those bones in Oregon you were gonna involve the FBI in this case?*"

GARY RIDGEWAY: "*Y . . . uh, yes.*"

DETECTIVE JENSEN: "*And you knew that that would screw us up really big time, didn't ya?*"

GARY RIDGEWAY: "*Yeah, yes. I would because a body's bein', goin' across the border, yeah.*"

(King County Police Confession Tape 40, 2003, p. 16, as cited in Guillén, 2007)

Gary Ridgeway, one of the most prolific serial killers of the USA, was able to avoid police detection between 1982 and 2001 whilst killing over 60 women. The "Green River" killer was able to elude detection even after a major task force was formed in 1984 consisting of over 50 investigators. But how did he do it? Part of the answer lies in what Ridgeway described above: forensic awareness. Ridgeway mentioned during his confession, which was provided over a six-month period, that he knew what evidence was present at the crime scene and what to do to remove it. Some of his strategies involved using plastic gloves to avoid leaving fingerprints, clipping victims' nails to avoid having some of his skin found there, stripping victims of their clothing and personal items to delay identification, and getting rid of any fiber evidence linking him with the victim (Guillén, 2007: 66–67). However, it was the relocation of the skeletal remains of several victims that was perhaps most significant in thwarting investigative efforts. This is illustrated by the quote taken from Ridgeway's confession at the beginning of this chapter. Essentially, Ridgeway revisited some of the body-dump sites of victims in King County to collect some of their remains. He then drove to Oregon and dropped the remains in different semi-rural areas, hoping to throw off the investigation by directing the focus of the search to Oregon. And it worked. Similar to Newton's law of motion – action–reaction – this chapter focuses on the role of forensic evidence in criminal investigations, and on how offenders reacted and adapted their forensic awareness in cases of sexual homicide.

The role of forensic evidence

Forensic evidence in the criminal investigation

One of the most important principles at the foundation of forensic sciences as we know it today – the Locard exchange principle (Edmond Locard, 1877–1966) – states that if an offender did not bring something to the crime scene, he certainly took something away, and proper forensic examination can reveal this (Walton, 2006: 8). According to Strom and Hickman (2010), the rise in the use of forensic evidence is mainly attributable to technological development and the diffusion of knowledge about the value of forensic evidence both in the criminal justice community and in popular culture (i.e. CSI-related crime dramas). In fact, forensic sciences and forensic evidence now form an integral part of criminal investigations.

In their extensive analytical review of approximately 2,700 cases distributed across four different jurisdictions in the USA, Peterson *et al.* (1984) identified six main reasons why forensic evidence was submitted for evaluation: (1) to establish an element of the crime (e.g. searching for the presence of semen from a rape victim to prove penetration and sexual intercourse); (2) to identify a suspect or victim; (3) to establish an association (e.g. use of evidence to link the offender to the murder weapon); (4) to test statements and alibis (e.g. analysis of a fingerprint left on a glass in the rape victim's apartment may be used to contradict the offender's statement); (5) to reconstruct the scene (e.g. use of blood-splatter analysis to indicate where the offender and victim were standing at the time of the first blow to the head with a blunt object); and (6) to corroborate (e.g. fingerprints and semen are found at a rape crime scene and analyzed in order to associate the suspect with the location and corroborate the victim's statement). In two out of the four jurisdictions analyzed by Peterson *et al.* (1984), the category of evidence most frequently collected in homicide cases was blood, whereas in the other two jurisdictions it was fingerprints. In relation to rape cases, three out of the four jurisdictions reported that semen was most frequently collected, whereas the remaining jurisdiction reported that hair was most common. In addition, their analyses revealed that incident variables (e.g. time lapse between discovery of the crime and police response; extent of physical interaction between the offender and the scene and/or victim; type of location where the crime occurred; and presence of witnesses) may affect the collection of forensic evidence. Arguably the most interesting aspect of this study is the finding that only a fraction of the forensic evidence collected was actually analyzed. For instance, in Kansas City, only 59 percent of aggravated assaults, 50 percent of rapes, 47 percent of robberies, and 25 percent of burglary cases submitted were analyzed. They concluded that the analysis of forensic evidence is of little value in cases where a sample from a suspect is not available for comparison; if the suspect has not been identified, the evidence is useless to identify an offender who is unknown. The previous findings suggest that, despite its proven usefulness, forensic evidence plays a role in only approximately half of criminal investigations.

Horvath and Meesig (1996) published a very interesting review on the role of forensic evidence in criminal investigations. The main findings can be summarized in the following way. First, police collect far more physical evidence than will be submitted for analysis, and a significant amount of collected scientific evidence is of little use in many cases (Greenwood and Petersilia, 1975). Second, forensic evidence is collected in only 5–14 percent of cases (Parker and Peterson, 1972; Ericson, 1981; Eck, 1983). Third, most forensic evidence does not lead to the identification of suspect(s) – without witness testimony – and is used primarily in the interrogation of the suspect in order to increase the likelihood of obtaining a confession (Wilson, 1976; Sanders, 1977; Simon, 1991). Fourth, the main reasons why evidence is not submitted for analysis are financial considerations, a suspect has been arrested, and perceived usefulness of the evidence type (Tilley and Ford, 1996).

These studies consistently show that in the majority of cases no forensic evidence is collected. In those cases where it is collected, much of it is not analyzed. If forensic evidence is analyzed, it rarely contributes to the investigation by identifying a suspect, but rather is used during the interview/interrogation of a suspect who has been identified through other investigative avenues. Although interesting and very informative, one major limitation of the studies reviewed above is that they are all dated. Considerable progress has been made in the area of forensic science, resulting in enhanced analytical capabilities and a reduction in the amount of evidence required for analysis. The National Institute of Justice (NIJ) has recently funded several studies to examine the analysis of forensic evidence in criminal investigation. Why is there an interest in the use of forensic evidence in criminal investigation at this time? Because, as surprising as it may seem, the best evidence currently available suggests that the collection and use of forensic evidence during the course of criminal investigations has not kept pace with advances in forensic science. Although we have the means and the expertise to collect more evidence, the reality is that very little collected evidence is actually being analyzed.

This is particularly true in cases of sexual assault. Untested sexual assault kits (SAK)[1] have been discovered in police evidence room across the USA: 10,000 in Los Angeles, 12,000 in Dallas, 10,500 in Detroit, to name a few (Ritter, 2011). In LA, the county and city crime labs did not have the capacity to test all SAKs that were stored, let alone new ones. When kits were analyzed, it was taking up to a year from the request for DNA testing to receive the final lab report (Human Rights Watch, 2009). A survey conducted by the NIJ in 2007 revealed that 18 percent of unsolved sexual assaults and 14 percent of unsolved homicides that occurred between 2002 and 2007 contained forensic evidence that remained in police custody and was not submitted for analysis to a crime lab (mainly consisting of serological/biological evidence and DNA; Strom *et al.*, 2007). Of course, not all forensic evidence in cases of sexual assault will be probative. For instance, in cases where consent is the issue (i.e. the suspect is not denying sexual contact with the victim but maintains it was consensual), or cases where charges have been dropped or the offender has pleaded guilty, the analysis of forensic evidence becomes less of a priority. However, the survey revealed that no less than 44 percent of law enforcement agencies (2,000 law enforcement agencies responded to the survey) mentioned that one of the reasons why they did not send evidence for analysis to the crime lab was that a suspect had not been identified, and an additional 15 percent explained that it had not been requested by a prosecutor (Strom and Hickman, 2010). Interestingly, 30 percent indicated they did not send evidence for analysis simply because they were uncertain of its usefulness (Strom *et al.*, 2007; Strom and Hickman, 2010).

Part of the problem stems from backlogs – which is different from untested evidence awaiting submission to a crime lab. NIJ defines a case as backlogged when it "has not been tested 30 days after submission to the crime laboratory" (Nelson, 2011: iii). Many suggest that there are two types of backlog in forensic science: "First, evidence submitted to crime laboratories might not be processed promptly, which creates a risk of justice delayed. Second, law enforcement agencies do not always submit forensic evidence from unsolved cases for testing which creates a risk of justice denied" (Cowan and Koppl, 2010: 411). Although laboratories are able to analyze more evidence than ever, more cases are submitted than can be analyzed. There is an increased demand for DNA testing, due mainly to the increased awareness of the potential for DNA evidence to help solve cases. This increased interest has translated into an increased amount of DNA evidence being collected in the course of criminal investigations (including property crimes); an increased capacity to test smaller samples, such as "touch DNA" samples, which occur when DNA is transferred by the simple touching of an object; an increased demand for testing in cold cases; and an increased effort to collect DNA samples from convicted felons and arrested persons (Roth, 2010).

After LA officials made the decision to perform DNA testing on all SAKs stored in the LAPD and LASD evidence rooms (nearly 11,000 SAKs), the NIJ was presented with a unique opportunity to examine backlog cases and better understand the backlog issue. What Peterson and colleagues (2012) found is that of the 1,948 untested SAKs, "DNA testing led to a suspect being identified in 90 cases; 70 in which there was a previously unidentified offender hit (4 percent of the total kits tested), and 20 in which there was a case-to-case known suspect hit (1 percent of the total kits tested)" (Ritter, 2012: 9). Although it can be concluded that testing all untested SAKs is unnecessary based on the study findings, what is unknown is whether future crimes could be solved by the uploading of all these additional DNA profiles into CODIS[2] (Peterson *et al.*, 2012). Based on some estimates from the United Kingdom's national DNA database, the hit rate for the year 2003–04 was 45 percent (Beaver, 2010). Beaver (2010) applied this estimate to the numbers found by Strom and Hickman (2010): of the 3,975 homicides and 27,595 rapes containing forensic evidence, 40 percent contained unanalyzed DNA evidence, which means that a total of 1,590 homicides and 11,038 rapes contained unanalyzed DNA evidence. Multiplied by the 45 percent hit rate found in the UK, the number of homicides and rapes with an actual match in the DNA database would be 715 and 4,967 respectively, meaning that a significant number of suspects could be identified (Beaver, 2010).

The findings from these studies are informative as they show that, although our abilities to find and analyze less substantial forms of forensic evidence have increased, unfortunately crime labs have not been able to meet the increased demand for analysis. Despite increased education in relation to the utility of forensic sciences, it seems that some investigators still do not fully understand how forensic evidence can contribute to their investigation – beyond use as an interrogation aid once the suspect has been identified. However, it is important to note that the studies discussed previously did not indicate whether the presence of forensic evidence was actually of investigative value.

A study funded by the NIJ and carried out in 2008 helps to shed some light on this important question. The study compared burglary investigations that used only traditional police practices (control group) with burglary investigations in which DNA evidence was also collected and analyzed (treatment group) (Roman *et al.*, 2008). Five hundred samples – evidence thought to contain human cells in the form of hair, tissue, bones, teeth, blood, or other body fluids – were collected from property crime scenes at five different locations (Denver, Topeka, Phoenix, Los Angeles, and Orange County; Ritter, 2008). All burglaries (57 percent residential, 29 percent commercial, and 13 percent theft from automobile) were investigated using traditional investigative methods, the only difference between the control and treatment groups being that cases from the treatment group were analyzed for DNA, whereas the cases from the control group were not analyzed for DNA for the first two months. The findings showed that when DNA was analyzed, more than twice as many suspects were identified, twice as many suspects were arrested, and more than twice as many cases were accepted for prosecution. Offenders identified using DNA evidence had significantly more serious criminal histories than those identified through traditional investigation (average of 5.6 prior felonies vs. 1.7). In addition, the study revealed that suspects were five times more likely to be identified through DNA evidence than fingerprints, and that blood and saliva evidence was more effective in solving burglary than other biological evidence (e.g. evidence from items touched by the suspect – "touch samples").

The use of forensic evidence in property crimes, more specifically burglaries, has a direct impact on the investigation of these crimes. Although the research available to date on personal crime (e.g. sexual assaults, homicides) does not allow one to reach similar conclusions to those for property crimes, it is clear that forensic evidence collected in cases of sexual assaults and homicides may be useful for present or future investigations, especially if the profile is uploaded in the CODIS database.

In summary, research is suggesting that forensic evidence cannot carry the weight it should in the criminal investigation, mainly due to problems related to the analysis of evidence (e.g. backlogs). While efforts are being made to solve these issues, it seems that criminals are also adapting their behavior in order to limit the amount of forensic evidence left at crime scenes. For example, it has been noted that offenders are increasingly wearing gloves to avoid leaving fingerprints, using tape to seal envelopes to avoid leaving DNA, and even using bleach (which destroys DNA) to clean up blood (Durnal, 2010). It has been suggested that the proliferation of crime dramas such as *CSI* has educated criminals, providing them with new strategies to learn how to avoid police detection – also known as the "CSI effect" (Cole and Dioso-Villa, 2007). It is also possible that, through experience, offenders recognize a need to increase their level of sophistication in the commission of their crime, and adapt their *modus operandi* (MO) accordingly. Such behaviors and strategies used by criminals to avoid detection have been associated to a relatively new concept in criminology: forensic awareness.

Forensic awareness

Forensic awareness is a term that has been used to describe the greater knowledge of law enforcement personnel as well as the public concerning forensic science and forensic evidence. However, this expression is now also being used to describe the greater knowledge and awareness of forensic science among offenders. Some offenders will adapt their MO, or take precautions before, during, or post-crime commission, to decrease their risk of apprehension. Arguably, this adaptation of crime strategy may be deemed an indication of evolving criminal sophistication on the part of the offender. Offenders who adapt their MO to thwart police investigative efforts may be said to be exhibiting "investigative awareness" (i.e. a knowledge or understanding of police investigative practice; Beauregard and Martineau, 2013). A related concept, coined in criminological research as "forensic awareness" (Davies, 1992), refers to an offender's knowledge or understanding of the importance of forensic evidence (e.g. DNA, fingerprints, dental impressions) to police investigation. In this context, forensic awareness could be defined as taking additional steps and adapting one's MO to hide evidence in order to ultimately avoid apprehension (Davies, 1992).

Davies (1992) identified a category of behavior – steps taken to ensure personal safety – that forms part of an offender's MO. The following elements were identified:

- avoiding interruption: gagging the victim (although relatively uncommon), targeting women who are alone, disconnecting the phone;
- protecting identity: preventing sighting of face (e.g. blindfolding the victim, attacking after nightfall, disabling the lights, wearing a mask, using verbal instructions about where the victim should look), trying to avoid leaving fingerprints or semen (e.g. using a condom, wiping off semen or objects that have been used during the sexual assault), lying to mislead investigators (e.g. lying about their name, concealing where the offender lives);

- any reference to the victim reporting: threatening to return and harm the victim, requesting to see pieces of identification (e.g. driver's license) so that victim could be located at a later date;
- steps taken to ensure safe departure: telling the victim not to move, telling the victim to count up to a certain number, threatening the victim not to look or follow, tying up the victim.

There is a dearth of research on the extent of forensic awareness among offenders and the impact that this awareness has on the crime process and investigation. Davies *et al.* (1997) found in their sample of 210 stranger rapists that 15 percent of offenders wore gloves or attempted to avoid leaving fingerprints, 5 percent attempted to clean up/remove semen, 28 percent took steps to hide their face, and 20 percent lied about their name and/or address. Moreover, Park *et al.* (2008) found that serial sex offenders were more likely to exhibit forensic awareness than single-offense sex offenders, but removing semen was not significantly related to offense history. Although 52.9 percent of serial sex offenders removed semen from the scene of the crime, fewer than 40 percent showed forensic awareness. What this seems to suggest is that some sex offenders are aware of forensic science and the potential of some forensic evidence to link them to their crime. However, surprisingly, forensic awareness strategies are used in a very limited way. Similar findings are evident in research related to homicide offenders.

Brown and Keppel's (2012) study of child abduction murderers reports interesting findings relating to the physical evidence destroyed by the offender at a crime scene location. First, fewer than half of the murderers (347 cases) destroyed evidence at the crime scene. In cases where evidence was destroyed, the offender destroyed clothing (6.9 percent), blood (4.7 percent), actual crime scene (4.4 percent), body or body parts (4.3 percent), other physical evidence (4.2 percent), other personal items (1.8 percent), murder weapon (1.0 percent), and bindings (0.5 percent). It was unknown whether or not the offender destroyed any physical evidence in 21.1 percent of the cases. Perhaps more interesting are the methods used by the offenders to destroy physical evidence. Although the method used is unknown in 21.6 percent of cases, the most common strategies used to destroy physical evidence at the scene include cleaning or washing (7.8 percent), burning (3.4 percent), hiding or burying (3.8 percent), cutting up (0.9 percent), throwing away (2.7 percent), removing from scene (2.5 percent), and other disposal (2.1 percent). Other studies conducted by Keppel and colleagues (Hanfland *et al.*, 1997; Brown *et al.*, 2006; Brown and Keppel, 2007; Brown, 2008) found that 24.2 percent of child abduction murderers discarded evidence after the murder. This discarded evidence was generally found along the roadway the offender had travelled to the murder and/or body disposal location, or to facilitate escape (36 percent), and was within one mile of where the body was recovered (between 56.5 and 59 percent of cases).

In their study of stranger rapists, Davies and Dale (1995) suggest that travelling longer distances to commit a crime may be an indicator of forensic awareness. An examination of sexual murderers led Beauregard and Field (2008) to suggest that offenders who moved the victim's body after the murder to delay detection and therefore make apprehension difficult showed a higher degree of forensic awareness. Moving the body complicates the investigation by decontextualizing the crime, potentially decreasing the likelihood that the victim will be found or, if found, identified. This action may also indicate forensic awareness as the offender is removing a significant source of forensic evidence from the scene of the homicide.

Two main hypotheses have been suggested to explain the formation of forensic awareness by offenders. First, the amount of publicity given to the importance of forensic evidence in the apprehension of criminals – the so-called "CSI effect" – may be increasing offenders' recognition of the need to become forensically aware. Second, forensic awareness may be an indication that the offender has had previous encounters with the criminal justice system, and may be educated in evidentiary procedure, and the necessity to remove evidence from a crime scene (Davies, 1992). The latter hypothesis seems to find support in two studies in particular. In a study comparing the behaviors of single and serial rapists, it was found that offenders with a more extensive criminal career were more likely to exhibit signs of forensic awareness (Park *et al.*, 2008). Similarly, Davies *et al.* (1997), in their examination of criminal antecedents of 210 cases of solved stranger sexual assaults, found that in cases of rape, offenders who removed semen and/or took fingerprint and sighting precautions were approximately 2 to 4 times more likely to have prior convictions for sexual offenses. On the other hand, if an offender did not take precautions, he was more likely to be a first-time sexual offender and may have been under the influence of drugs at the time of the crime, thus explaining why their mental capacity to be cautious and use precautions upon departure was diminished (Davies *et al.*, 1997).

In a more recent study on serial sex offenders, forensic awareness was considered to be evident during a rape if the offender did at least one of the following: concealed his identity, wore gloves, cleaned up semen, prevented his face from being seen, lied about his name, wore a condom, did not ejaculate in or on the victim, made the victim comb her pubic hair after the assault, and/or made the victim shower after the assault (Beauregard and Bouchard, 2010). Among the 222 rape events included in this study, offenders displayed some form of forensic awareness in 55.9 percent of cases. Offenders were not necessarily consistent in their use of forensic awareness from one event to the other. Preliminary analyses showed that for serial offenders, forensic awareness in a prior event was a mild predictor of forensic awareness in a subsequent event. More than half of the rape events included in the sample showed some signs of forensic awareness. The most common specific strategies used by offenders had to do with protecting their identity (i.e. concealing his identity, preventing his face from being seen, wearing gloves). DNA removal behaviors were scarcely found in this sample. For example, only 6.8 percent of offenders took precautions to remove DNA at the crime scene by wiping semen, and condoms were worn in only four cases (1.8 percent).

This finding alone raises some questions as to the sophistication of these offenders. On one hand, this seems to suggest that offenders are forensically aware and are trying to protect their identity during the crime-commission process. On the other hand, in only a minority of rape events do offenders take precautions to destroy or clean up the one specific type of evidence – DNA – that could link them to the victim or to previous crimes committed. This is also surprising considering the fact that offenders ejaculated at the scene in 63.5 percent of the rape events investigated. One potential explanation could be that offenders are capable of forensic awareness in the early stages of the crime-commission process (e.g. planning not to be recognized by the victim, wearing gloves not to leave fingerprints) but become less careful after commission of the rape.

In order to understand better why some offenders do not more effectively employ forensic awareness strategies while committing their crimes, Beauregard and Bouchard (2010) further analyzed the rationale provided by offenders during semi-structured interviews. Each of the 72 serial sex offenders included in the sample was asked an open-ended question as to why they did not take any specific action to avoid detection. The responses were content-analyzed and four distinct categories were identified. First, some offenders indicated that they did not even think about forensic awareness. The majority of serial sex offenders mentioned that they did not think about this as the crime was not planned; the acts were spontaneous, out of control. Also, some offenders mentioned that they were not forensically aware during their first offense, as they did not yet know what to expect in regard to the potential risks associated to forensic evidence. A second rationale provided by offenders was related to their mental state at the time of the crime commission. Offenders explained that in addition to being intoxicated, they were too confused, or even too angry, to think about avoiding detection. The third category appears as "irrational thinking," serial offenders reporting that they thought the victims consented; or that as there was no penetration, no semen could be found. For other offenders, it was important for their own vengeance that the victim saw their face. Finally, the fourth type of rationale provided by offenders was that an alternative to forensic awareness was used. According to these offenders, being forensically aware was not important or necessary; instead they used manipulation or bribes to make sure the victims did not report the crimes. These various findings seem to suggest that forensic awareness is situational and may vary according to the context of the crime-commission process. They also suggest that the use of forensic awareness strategies could be related to the type of crime. The next section examines specifically forensic awareness in the case of sexual homicide.

Forensic awareness in sexual homicide

Beauregard and colleagues (Beauregard and Martineau, 2013, 2014; Balemba *et al.* 2014) have conducted a series of studies investigating the use of forensic awareness strategies in cases of sexual homicide. Using data from a national database managed by the Royal Canadian Mounted Police (RCMP), we have identified a total of 350 sexual homicide cases. In more than half of the sexual murders, police could not identify the use of any forensic awareness strategies during the commission of the crime. However, 24.9 percent of offenders were found to use one strategy, 11.7 percent used two strategies, and a small group of offenders (9.1 percent) used three or more strategies during their crime. The most common strategy used was to destroy/remove evidence (30.6 percent); this includes wearing gloves, using a condom, setting fire to the scene, and cleaning the scene. This is followed by acting upon the victim and/or the environment (11.4 percent). Such precautions include threatening the victim not to report; disabling the lighting, the telephones, security system, and/or the victim's vehicle; administering a drug to the victim; tying up the victim; blocking access in and out of doors or windows; and killing the victim for the purpose of eliminating the witness. Other offenders simply disposed of the body (11.1 percent), while a minority attempted to protect their identity (4.3 percent) through wearing a mask, giving a false name, attempting to disguise or alter their appearance, and changing residence after the crime. Some offenders used "other" precautions (10.3 percent), which included using a scanner to ascertain police activities, using a device to alert them to anyone approaching, using a look-out or establishing a look-out location where the offender could observe the scene without being noticed, covering the victim's eyes, gagging the victim, and arranging an alibi. Staging of the crime scene was observed in only 0.9 percent of the cases. Interestingly, semen was found at the scene in slightly more than a quarter of the cases, presenting the possibility for DNA analysis. It is uncertain whether a lack of DNA evidence at the crime scene is the result of a conscious effort on the part of the offender to ensure no forensic evidence is left behind, or is simply the result of the murderer experiencing a sexual dysfunction during the crime (Ressler *et al.*, 1986a,b). However, the absence of DNA, at least in some cases, could be indicative of a level of forensic awareness on the part of the offender. The offender moved the victim's body from the offense scene to a body disposal site in 34 percent of the cases.

Taking additional steps and adapting the MO used in their crime specifically to hide evidence in order to ultimately avoid apprehension is not as common as we would have expected.

Only one-third of the offenders included in our sample of sexual murderers moved the victim's body after the murder. Moving the victim's body after the murder has been linked to an expressive crime scene, indicating an impulsive and personal aggression (Salfati, 2000; Santtila *et al.*, 2001). Moreover, Beauregard and Field (2008) found that sexual murderers presenting "organized" psychological characteristics are more likely to move the victim's body after the crime, compared with offenders presenting disorganized characteristics, which is congruent with previous literature in which offenders who are impulsive, unstable, and excessively violent, and who lack organization in their crime scenes, are more apt to leave the body at the crime scene (Beauregard and Proulx, 2002; Ressler *et al.*, 1986a,b, 1988). The finding is also consistent with the organized offender in Ressler *et al.*'s (1988) model; according to Ressler *et al.*, organized sexual murderers are in control during the crime, making sure not to leave evidence at the crime scene and transporting the victim's body to a different location to hide it. Therefore transporting the victim's body to a different location may serve the purpose of both obscuring any potential connections between the offender and the body dump site, and delaying discovery of the body.

Surprisingly, more than half of the sample used no specific precaution to avoid detection during the crime. This information is useful for the investigators of these crimes as it suggests that offenders were not particularly concerned about the evidence they may have left behind or were not criminally sophisticated enough to take precautionary measures. However, for those offenders whose actions exhibited concern about forensic evidence, it is noteworthy that most used two or more precautions. The most common strategy is destroying or removing evidence. Sexual murderers will wear gloves, a condom, or will set fire to or clean the scene following the murder.

One interesting question that has been overlooked in relation to forensic awareness strategies is their use over time. In other words, are more offenders more forensically aware today than back in the 1960s or 1970s? This question is particularly relevant, especially with the CSI effect hypothesis – which suggests that "watching television shows, such as *CSI*, has influenced the general public's attitudes, expectations, and decision making related to the use of scientific evidence in jury trials" (Baskin and Sommers, 2010b, p. 97). The "CSI effect" takes its name from *CSI: Crime Scene Investigation*, a police television show that focuses on the use of forensic sciences to solve crimes. For the rare individuals not familiar with this television drama, it first aired in 2000 and has been one of the top three television programs watched across America, with a reported 25 million viewers in the 2005/06 season (Cole and Dioso-Villa, 2007). Due to its great popularity, spin-offs were subsequently aired (*CSI Miami*, *CSI New York*), allowing crime drama of this type to be watched several times a week, even several times a day. Although empirical studies assessing whether or not the CSI effect really exists have come to opposing conclusions (see Maricopa County Attorney's Office, 2005; Podlas, 2005; Tyler, 2006; Schweitzer and Saks, 2007; Smith *et al.*, 2007; Stinson *et al.*, 2007; Kim *et al.*, 2009; Baskin and Sommers, 2010a; Holmgren and Fordham, 2011), one specific type of CSI effect suggested by Cole and Dioso-Villa (2007) – the *police chief's effect* – states that crime dramas such as *CSI* are in fact educational for criminals, providing them with new strategies to learn how to avoid police detection as well as an increased sophistication in the commission of their crime. According to this hypothesis, we should observe an increase in the use of forensic awareness strategies with the appearance of all these crime dramas. In order to examine this hypothesis, we have calculated the mean use of forensic awareness strategies for five-year periods from prior to 1960 to 2010 in cases of sexual homicide in Canada. Figure 41.1 shows that, prior to the 1960s, these strategies were totally absent from the offenders' repertoire (or were not detected by investigators). The use of detection-avoidance strategies began to increase significantly during the 1970s, to reach

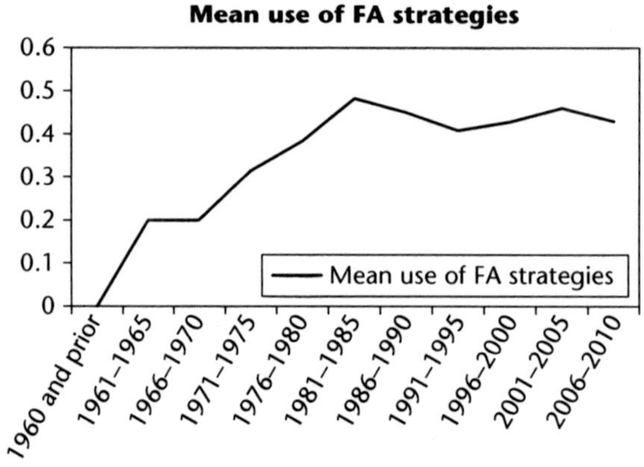

Figure 41.1 Mean use of forensic awareness strategies over time

a peak in the mid-1980s. Since then, the mean use of these strategies has remained relatively stable. Based on the current data, it is not possible to correlate an increase in the level of forensic awareness of sexual homicide offenders to the proliferation of televised crime dramas. As mentioned, most of these TV shows appeared after 2000. Another hypothesis that has been proposed to explain the increased use of forensic awareness strategies is that offenders who have been in contact with the criminal justice system develop forensic awareness through their experience in the system (Davies *et al.*, 1997). Thus, as offenders are convicted on the basis of forensic evidence, they may pass this information to fellow inmates as a form of expert/specialist knowledge (Coupe and Blake, 2006). Although it is not possible to test these two hypotheses empirically based on our actual data, it is nonetheless interesting to see that there has been an increase in the use of forensic awareness strategies over the years.

Another aspect of interest with regard to the use of forensic awareness strategies is whether the use of such strategies leads to the capacity for the offender to evade police detection. One of the problems related to the study of detection-avoidance strategies is that most research has been conducted on samples of convicted offenders, meaning that these offenders have been detected and apprehended by the police. In order to address this limitation found in previous studies, we analysed our sample of sexual homicide cases in a subsequent study by conducting separate analyses for the full sample (i.e. the 350 cases of sexual homicide) and for the unsolved cases only (i.e. 100 cases within the 350 that were still unsolved at the time to enter the information in the national database) (Balemba *et al.*, 2014). We examined various crime-scene factors directly observable in order to identify latent classes of sexual homicide type (i.e. case status, vaginal intercourse, offender took items, anal intercourse, semen located, mutilation of the victim, inanimate object inserted, victim's body moved, beating, restraints, blindfolds or gags, and strangulation). We argued that a theme approach is most compatible with this research due to the interrelatedness of many crime scene factors.

The first latent class analysis was performed on the full sample of sexual homicide cases and produced a three-cluster solution. The first cluster within the sample is the *sloppy/reckless* offender. The most distinctive features about this type of offender are that this type of homicide is most likely to be solved by law enforcement, and the crime scenes are most likely to contain semen evidence. Relatedly, these sexual homicides are also most likely to involve vaginal intercourse. Furthermore, the majority of offenders who commit such crimes do not take items from the victim or scene, they are least likely to inflict mutilation, and they do not beat nor strangle their victim. Overall, the *sloppy/reckless* sexual homicides appear to be the type of offense most easily solved by investigators due to the behavioural choices of the offender during crime commission. There appears to be either a lack of sufficient planning or a lack of concern for detection avoidance. The priority within this type of offense seems to be sex – or, perhaps more precisely, the sexual component of the rape.

The second cluster that emerged from the full sample is the *violent/sadistic* class. Although almost as likely to be solved as those crimes in the *sloppy/reckless* category, this is not what is most distinctive about this group. These are overall the most violent crimes that also display the highest frequency of torture tactics. Offenders who commit these types of crime are most likely to physically beat their victims, strangle them, and although not the majority, these crimes have the highest likelihood of involving mutilation of the victim. These sexual homicides are somewhat likely to involve vaginal intercourse and are the most likely to include the offender taking an item from the victim. Furthermore, semen is typically not left at the scene for investigators to find. Although offender behavior is not as reckless as in the first cluster, the focus does not appear to be on detection avoidance. Rather, the priority within this type of crime seems to be violence and torture. There seems to be a use of expressive, or simply excessive, amounts of violence within these offenses that is clearly not necessary to commit a sexual homicide. While committing these crimes, it appears that increasing the level of pain, suffering, and injury to the victim prior to the culmination of murder is the goal, regardless of the motivation for doing so. Moreover, it is evident that these offenders were stealing personal tokens from their victim, a type of behavior associated with the most violent sexual sadists (Gratzer and Bradford, 1995). However, it may be such lapses in judgment as stealing victim trophies that lead these offenders to be apprehended by police. The majority of the behavior during these crimes is dominated by a need to inflict pain, which could cloud the logical mind of these offenders. There exists the potential to leave a great deal more trace and biological evidence on or near the victim when there is more contact between the victim and offender.

The third and final cluster identified within the sample is the *forensically aware* offender group. Half of the crimes within this category remain unsolved, making the perpetrators of these offenses the least likely to be apprehended by law enforcement. Vaginal intercourse almost never transpires within such an offense and, no doubt due to this particular behavioral pattern, semen is rarely found at the scene. These crimes involve offenders who typically do not steal items and are the least likely to resort to physically beating, strangling, or mutilating their victims. Thus these types of crime involve offender behaviour that is more focused on avoiding apprehension (whether intentionally for that purpose or not) and do not involve behaviors that typically supply investigators with DNA or trace evidence. The victim is murdered, but with what is best interpreted as a predominance of instrumental violence, and a corresponding lack of expressive hostility. The offender completes the crime with as little extraneous activity as possible.

Although this three-cluster offense model is informative about sexual homicides in general, there is potentially valuable information to be derived from a closer look at the cases that were not solved by police. The first, significantly most prevalent cluster in the unsolved-only sample is the *forensically aware* class. The crimes in this category are most distinguished by the exceptionally low likelihood of vaginal intercourse and even less likelihood of semen being present at the scene. The offenders do not take items from the crime scene, do not mutilate their victims, and these sexual homicides do not involve physical beating nor do they involve strangulation of the victim. Thus these crimes are overwhelmingly behaviorally controlled and do not involve offender behaviors that are likely to leave physical evidence to tie the offender to the crime.

The second cluster within the unsolved-only sample is a very different type of crime. This group, the *lucky* class, involves behaviors that, realistically, should result in the offender's apprehension. These offenses are likely to involve vaginal penetration and semen is likely to be left at the scene. The offenders often take items from their crime, and are more likely than the *forensically aware* offender to mutilate their victims. These offenses are also less restrained with respect to violence; the victim is likely to have been physically beaten as well as strangled. Overall, these offenses resemble those of either the *sloppy/reckless* or the *violent/sadistic* clusters from the full sample analyses, and represent cases that should have been solved by police but, for whatever reason, were not solved early on. The question, then, is why are these particular cases so difficult for police to clear? These crimes are sloppy, unplanned, and unsophisticated, yet the offenders still manage to elude law enforcement, which is bizarre, given the police resources that are dedicated to murders (Klinger, 1997), particularly sexual murders. The soundest explanation is to simply acknowledge that sometimes criminals have luck on their side. As mentioned by Rossmo (2009: 3): "Not all crimes are solvable. In some circumstances, even solvable crimes are not cleared because of incompetence, misfeasance, nonfeasance, resource problems, or simple bad luck." It is possible that, in a case-by-case analysis, there would emerge some factor within each case that was especially detrimental to the investigation, such as the crime location presenting unforeseen difficulties, witnesses becoming uncooperative or lacking useful information, or key biological evidence being too degraded for DNA extraction. In such cases, perhaps the best one can hope is that the same offender will not be so lucky twice.

This study suggests that, through examination of observable factors regarding the state of the victim and the incident location, police and forensic analysts could determine the type of offense, whether *sloppy/reckless*, *violent/sadistic*, or *forensically aware*. Once this information is known, more can be extrapolated about the case, as well as potentially what other evidence may be found at the crime scene. Although crime scene investigators will analyze any crime scene with the same focus and energy, the typology identified can be used to predict *what* type of evidence, if any, could be expected to be found at the crime scene, thus increasing efficiency and allowing prioritization of evidence collection and analysis. However, despite its usefulness, this study does not really inform on the efficacy of using forensic awareness strategies for avoiding police detection. Therefore we conducted another study specifically addressing this gap in the literature.

Beauregard and Martineau (2014) examined whether the use of forensic awareness strategies increases the chance of avoiding police detection in cases of sexual homicide. According to Bouchard *et al.*,

> examining the outcome of criminal decision-making under a rational choice model is important because what makes a decision "rational" is also tied to results. Offenders may show much deliberation and thinking but still come up short—at least shorter than if they had made a "better decision". In other words, a rational choice approach should recognize that while most to all offenders show thinking and deliberation prior to offending, not all of them are as successful in their endeavors.
>
> *(Bouchard et al., 2013: 33)*

In our study, we examined the issue of avoiding police detection in sexual homicide from two different perspectives. In line with prior studies, the first dependent variable that we considered is the status of the case (0 = *unsolved*; 1 = *solved*) at the time the case was entered into the database, which is approximately 45 days into the investigation. Remember that our sample includes 250 cases that are solved and 100 cases that are unsolved. Moreover, the study looked at a complementary measure that has not been considered before in the research on avoiding detection, that is, the number of days before body recovery (i.e. by subtracting the date the victim was last seen alive from the date the body was recovered). It took an average of 71 days to recover a body in cases of sexual homicide (SD = 384.2; range = 6,491). As to the forensic awareness strategies, we divided the variables into two groupings. The first group of variables constitute the *specific precautions taken by the offender to avoid detection*. Those are specific forensic awareness strategies that can be identified in cases of sexual homicide. The second grouping of variables includes *MO behavior indicative of forensic awareness*. Two variables look specifically at evidence left at the crime scene: semen was found at the scene (27.7 percent) and availability of other biological evidence (24 percent), both presenting the possibility of DNA analysis. Three additional variables look specifically at actions performed upon the body, the offender moved the victim's body from the offense scene to a body disposal site (34 percent), the body was concealed (36 percent), and the body was dismembered (6.3 percent). Finally, in order to examine the level of risk taken by the sexual murderers while committing their crimes, we created risk scales (ranging from 0 to 2) for the contact, the offense, and the body recovery scenes (for more details, see Beauregard and Martineau, 2013). Although the risk taken by sexual murderers is on average close to 1 at the contact scene (SD = 0.9), the risk drops to 0.5 on average at the offense and body recovery scenes.

First, a look at the case status shows that when offenders leave no semen at the crime scene and their risk of detection is calculated as low, they are more likely to avoid police detection. However, the findings also indicate that taking any precautions to avoid detection is negatively related to the chance of avoiding detection, and that leaving biological evidence (blood, hair, etc.) at the crime scene is related to the case remaining unsolved. This is in the opposite direction of the working hypothesis. Two possible explanations may be offered to make sense of these contradictory findings. First, these counter-intuitive findings may be due to the nature of police data. Without access to the statements of victims, witnesses, or even the offender himself, information regarding the use of the forensic awareness strategies examined in this study may be limited, not reflecting the true prevalence of use of these strategies. Some forensic awareness strategies may be quite difficult to identify based upon careful study of the crime scene. Therefore police may be more aware of the use of forensic awareness strategies in solved cases. Another possibility is that the forensic awareness strategies that have been observed by police are not the most effective. It is possible that offenders who successfully avoid detection use alternative strategies not yet identified by law enforcement through interviews with offenders or through the study of crime scene evidence. The adoption of these more effective strategies and MOs (Beauregard *et al.*, 2007) may be enabling offenders to continue evading police detection.

Second, when investigating the number of days until body recovery, we notice that in cases where offenders leave semen at the crime scene or commit their crime when the risk of detection is higher, the police are likely to recover the victim's body in less time. However, offenders who conceal the body after the murder are likely to delay the discovery of the victim's body. The latter findings indicate that the use of precautions to avoid detection is related to case status, but not to the number of days until body recovery. The specific precautions taken by the offender to avoid detection investigated in the current study (e.g. destroying/removing evidence, protecting identity, acting upon victim/environment) have no direct impact on finding the body (unless an offender uses arson to destroy evidence that could be found on the body). Our findings show instead that in order to delay body recovery in cases of sexual homicide, the best strategy is to conceal the body after the murder – an MO indicative of forensic awareness (Beauregard and Field, 2008). Sexual murderers who choose to conceal the victim's body post-murder are likely to see the number of days until body recovery increase. This does not necessarily mean that the police will never apprehend the offender. However, delaying body recovery increases the chance of forensic evidence being lost and decreases the likelihood of the offender being connected to the crime. For this reason, we purport that the *number of days until body recovery* is a more appropriate measure of detection avoidance than case status. The number of days until body recovery is not biased by administrative rules or timing of data entry (e.g. information about the case should be entered after a specific number of days into the investigation). Moreover, while the two measures are significantly correlated, they yield very different findings.

Conclusion

In conclusion, it seems that the relationship between forensic evidence and the criminal justice system is a complex one. Technological advances have permitted the detection and analysis of smaller amounts of evidence collected on the victim's body or at the crime scene. This appears to have resulted in greater attention by various criminal justice professionals (investigators, prosecutors) on the potential probative value of forensic evidence. However, it also seems that this greater attention to forensic evidence and its analysis has not led to expected results. For one, despite greater evidence collection, a great deal of this evidence is either not submitted for analysis or not analyzed by crime labs. Many crime/forensic labs appear to lack the appropriate funding or resources to analyze all the evidence submitted for analysis, creating extremely long backlogs and sometimes causing delays. The inability to meet demands for analysis and subsequent need to prioritize submitted evidence has led in many jurisdictions to evidence remaining unanalyzed.

Despite an increase in the use of forensic awareness strategies over the years, some sexual murderers are still not using such strategies in order to avoid police detection. Those who are using such strategies have not necessarily been more successful at avoiding police detection. Instead, sexual murderers using specific MO strategies (i.e. concealing the victim's body) were able to delay body discovery. Despite the great publicity surrounding forensic science methods and capacities, it seems that not all offenders are capable of making use of forensic awareness strategies to their benefit. Crime context and criminal sophistication may both determine whether or not an offender can effectively utilize forensic awareness strategies to avoid detection.

The implications of these different findings are straightforward. First and foremost, in order to reap the benefits of advancements in forensic science, it is important to support the funding of forensic labs. Forensic evidence analysis can contribute meaningfully to criminal investigation and prosecution only if it is accessible to investigators within a reasonable period of time. Moreover, it is possible that there is a need to prioritize better the evidence sent to crime/forensic labs, based on the crime reconstitution. Therefore crime labs would have only a few pieces of evidence to analyze at a time, instead of being overwhelmed with all the evidence collected at the crime scene. Finally, the findings reported in this chapter have shown that, despite a better understanding of forensic sciences and analyses, there is still a need to continue the education of all the professionals within the criminal justice system on the full potential of such evidence to impact case outcome.

Research efforts should continue to investigate the different forensic awareness strategies used by offenders in order to understand better what offenders do in order to avoid detection.

Although we know many of the strategies available to offenders, it is possible that offenders have, over the years, developed other methods that have escaped us. Also, the research needs to take a look at the different factors related to the use – or not – of forensic awareness strategies by offenders. Prior studies have shown that in some specific contexts, offenders are not likely to use such strategies. This is important, but even more important is being able to identify particular contexts where offenders are more likely to use forensic awareness strategies, and whether these offenders present personal characteristics that differ from offenders who do not resort to using these strategies to avoid police detection.

Finally, it seems that, contrary to Newton's law of motion that for every action there is a reaction, sexual murderers have not adequately "reacted" to the improvement and technological advancements in forensic sciences. However, despite not making use of identified forensic awareness strategies in certain cases, it appears that some offenders have adopted other strategies in order to avoid detection.

Notes

1 Introduced in the 1960s, a SAK "is a box or envelope used to collect and store biological evidence from the victim of an alleged sexual assault. Evidence in a SAK can include vaginal, oral or anal swabs that may yield the perpetrator's DNA" (Ritter, 2011: 1).

2 CODIS (Combined DNA Index System) is a database operated by the FBI in which DNA profiles from crime scenes, convicted offenders, and (in some states) arrestees are stored and can be searched. As of 2010, CODIS contained more than 8.7 million offender profiles and approximately 330,000 profiles from crime-scene evidence, which led to more than 124,800 hits (Nelson, 2011).

References

Balemba, S., Beauregard, E., and Martineau, M. (2014) 'Getting away with murder: a thematic approach to solved and unsolved sexual homicides using crime scene factors', *Police Practice and Research*, 15(3): 221–233.

Baskin, D. and Sommers, I. (2010a) 'The influence of forensic evidence on the case outcomes of homicide incidents', *Journal of Criminal Justice*, 38: 1141–1149.

Baskin, D.R. and Sommers, I.B. (2010b) 'Crime-show-viewing habits and public attitudes toward forensic evidence: the "CSI effect" revisited', *Justice System Journal*, 31(1): 97–113.

Beauregard, E. and Bouchard, M. (2010) 'Cleaning up your act: forensic awareness as a detection avoidance strategy', *Journal of Criminal Justice*, 38: 1160–1166.

Beauregard, E. and Field, J. (2008) 'Body disposal patterns of sexual murderers: implications for offender profiling', *Journal of Police and Criminal Psychology*, 23: 81–89.

Beauregard, E. and Martineau, M. (2013) 'A descriptive study of sexual homicide in Canada: implications for police investigation', *International Journal of Offender Therapy and Comparative Criminology*, 57: 1454–1476.

Beauregard, E. and Martineau, M. (2014) 'No body, no crime? The role of forensic awareness in avoiding police detection in cases of sexual homicide', *Journal of Criminal Justice*, 42(2): 213–220.

Beauregard, E. and Proulx, J. (2002) 'Profiles in the offending process of nonserial sexual murderers', *International Journal of Offender Therapy and Comparative Criminology*, 46: 386–399.

Beauregard, E., Proulx, J., Rossmo, K., Leclerc, B., and Allaire, J.-F. (2007) 'Script analysis of hunting process in serial sex offenders', *Criminal Justice and Behavior*, 34: 1069–1084.

Beaver, K.M. (2010) 'The promises and pitfalls of forensic evidence in unsolved crimes', *Criminology and Public Policy*, 9: 405–410.

Bouchard, M., Beauregard, E., and Kalacska, M. (2013) 'Journey to grow: linking process to outcome in target site selection for cannabis cultivation', *Journal of Research in Crime and Delinquency*, 50: 33–52.

Brown, K.M. (2008) 'Child abduction murder: an analysis of the effect of victim age, victim race, victim gender, victim–offender relationship, forensic evidence, and time and distance separation on case solvability', PhD dissertation, Sam Houston State University, Huntsville, TX.

Brown, K.M. and Keppel, R.D. (2007) 'Child abduction murder: an analysis of the effect of time and distance separation between murder incident sites on solvability', *Journal of Forensic Sciences*, 52: 137–145.
Brown, K.M. and Keppel, R.D. (2012) 'Child abduction murder: the impact of forensic evidence on solvability', *Journal of Forensic Sciences*, 57: 353–363.
Brown, K.M., Keppel, R.D., Weis, J.G., and Skeen, M. (2006) *Investigative Case Management for Missing Children Homicides: Report II*, Seattle, WA: Attorney General (WA) and US Department of Justice Office of Juvenile Justice and Delinquency Prevention.
Cole, S.A. and Dioso-Villa, R. (2007) 'CSI and its effects: media, juries, and the burden of proof', *New England Law Review*, 41: 435–470.
Coupe, T. and Blake, L. (2006) 'Daylight and darkness: targeting strategies and the risks of being seen at residential burglaries', *Criminology*, 44: 431–464.
Cowan, E.J. and Koppl, R. (2010) 'An economic perspective on "unanalyzed evidence in law-enforcement agencies"', *Criminology and Public Policy*, 9: 411–419.
Davies, A. (1992) 'Rapists' behaviour: a three aspect model as a basis for analysis and the identification of serial crime', *Forensic Science International*, 55: 173–194.
Davies, A. and Dale, A. (1995) *Locating the Stranger Rapist* (Special Interest Series, Paper 3), London: Police Research Group, Home Office Police Department.
Davies, A., Wittebrood, K., and Jackson, J.L. (1997) 'Predicting the criminal antecedents of a stranger rapist from his offence behaviour', *Science and Justice*, 37: 161–170.
Durnal, E.W. (2010) 'Crime scene investigation (as seen on TV)', *Forensic Science International*, 199: 1–5.
Eck, J.E. (1983) *Solving Crimes: The Investigation of Burglary and Robbery*, Washington, DC: Police Executive Research Forum.
Ericson, R.V. (1981) *Making Crime: A Study of Detective Work*, Boston: Butterworth.
Gratzer, T. and Bradford, J.M. (1995) 'Offender and offense characteristics of sexual sadists: a comparative study', *Journal of Forensic Sciences*, 40: 450–455.
Greenwood, P. and Petersilia, J. (1975) *The Criminal Investigation Process: Vol. 1. Summary and Policy Implications*, Santa Monica, CA: Rand Corporation.
Guillén, T. (2007) *Serial Killers: Issues Explored through the Green River Murders*, Upper Saddle River, NJ: Pearson/Prentice Hall.
Hanfland, K.M., Keppel, R.D., and Weis, J.G. (1997) *Investigative Case Management for Missing Children Homicides*, Seattle, WA: Attorney General (WA) and US Department of Justice Office of Juvenile Justice and Delinquency Prevention.
Holmgren, J.A. and Fordham, J. (2011) 'The CSI effect and the Canadian and the Australian jury', *Journal of Forensic Sciences*, 56: 63–71.
Horvath, F. and Meesig, R. (1996) 'The criminal investigation process and the role of forensic evidence: a review of empirical findings', *Journal of Forensic Sciences*, 41: 963–969.
Human Rights Watch (2009) *Testing Justice: The Rape Kit Backlog in Los Angeles City and County*, New York: Human Rights Watch.
Kim, Y.S., Barak, G., and Shelton, D.E. (2009) 'Examining the "CSI-effect" in the case of circumstantial and eyewitness testimony: multivariate and path analyses', *Journal of Criminal Justice*, 37: 452–460.
King County Police Confession Tape 40, page 4, *State v. Ridgeway Case no. 01–1–10270–9*, Seattle, WA, June 19, 2003. Detective Jon Mattsen, Detective Ton Jensen.
Klinger, D.A. (1997) 'Negotiating order in patrol work: an ecological theory of police response to deviance', *Criminology*, 35: 277–306.
Maricopa County Attorney's Office (2005) *CSI Maricopa County: The CSI Effect and its Real-life Impact on Justice*, Phoenix, AZ: Maricopa County Attorney's Office.
Nelson, M. (2011) *Making Sense of DNA Backlogs, 2010 – Myths vs. Reality*, Washington, DC: US Department of Justice, Office of Justice Programs, National Institute of Justice.
Park, J., Schlesinger, L.B., Pinizzotto, A.J., and Davis, E.F. (2008) 'Serial and single-victim rapists: differences in crime-scene violence, interpersonal involvement, and criminal sophistication', *Behavioral Sciences and the Law*, 26: 227–237.
Parker, B. and Peterson, J.L. (1972) *Physical Evidence Utilization in the Administration of Criminal Justice*, Washington, DC: US Department of Justice, National Institute of Law Enforcement and Criminal Justice.
Peterson, J.L., Mihaljovic, S., and Gilliland, M. (1984) *Forensic Evidence and the Police: The Effects of Scientific Evidence on Criminal Investigations*, Washington, DC: US GPO.

Peterson, J., Johnson, D., Herz, D., Graziano, L., and Oehler, T. (2012) *Sexual Assault Kit Backlog Study*, Washington, DC: US Department of Justice, Office of Justice Programs, National Institute of Justice.

Podlas, K. (2005) 'The CSI effect: exposing the media myth', *Fordham Intellectual Property, Media and Entertainment Law Journal*, 16: 429–465.

Ressler, R.K., Burgess, A.W., Douglas, J.E., Hartman, C.R., and D'Agostino, R.B. (1986a) 'Sexual killers and their victims: identifying patterns through crime scene analysis', *Journal of Interpersonal Violence*, 1: 288–308.

Ressler, R.K., Burgess, A.W., Hartman, C.R., Douglas, J.E., and McCormack, A. (1986b) 'Murderers who rape and mutilate', *Journal of Interpersonal Violence*, 1: 273–287.

Ressler, R.K., Burgess, A.W., and Douglas, J.E. (1988) *Sexual Homicide: Patterns and Motives*, New York: Free Press.

Ritter, N. (2008) *DNA Solves Property Crimes (But Are We Ready for That?)*, Washington, DC: US Department of Justice, Office of Justice Programs, National Institute of Justice.

Ritter, N. (2011) *The Road Ahead: Unanalyzed Evidence in Sexual Assault Cases*, Washington, DC: US Department of Justice, Office of Justice Programs, National Institute of Justice.

Ritter, N. (2012) *Solving Sexual Assaults: Finding Answers through Research*, Washington, DC: US Department of Justice, Office of Justice Programs, National Institute of Justice.

Roman, J.K., Reid, S., Reid, J., Chalfin, A., Adams, W., and Knight, C. (2008) *The DNA Field Experiment: Cost-effectiveness Analysis of the Use of DNA in the Investigation of High-volume Crimes*, Washington, DC: US Department of Justice, Office of Justice Programs, National Institute of Justice.

Rossmo, D.K. (2009) *Criminal Investigative Failures*, Boca Raton, FL: CRC Press.

Roth, A. (2010) 'Database-driven investigations: the promise – and peril – of using forensics to solve "no suspect" cases', *Criminology and Public Policy*, 9: 421–428.

Salfati, C.G. (2000) 'The nature of expressiveness and instrumentality in homicide: implications for offender profiling', *Homicide Studies*, 4: 265–293.

Sanders, W. (1977) *Detective Work*, New York: Free Press.

Santtila, P., Canter, D.V., Elfgren, T., and Häkkänen, H. (2001) 'The structure of crime-scene actions in Finnish homicides', *Homicide Studies*, 5: 363–387.

Schweitzer, N. and Saks, M. (2007) 'The CSI effect: popular fiction about forensic science affects public expectations about real forensic science', *Jurimetrics*, 47: 357–364.

Simon, D. (1991) *Homicide: A Year on the Killing Streets*, Boston: Houghton Mifflin.

Smith, S.M., Patry, M.W., and Stinson, V. (2007) 'But what is the CSI effect? How crime dramas influence people's beliefs about forensic evidence', *Canadian Journal of Police and Security Services*, 5: 187–195.

Stinson, V., Patry, M.W., and Smith, S.M. (2007) 'The CSI effect: reflections from police and forensic investigators', *Journal of Police and Security Services*, 5: 125–133.

Strom, K.J. and Hickman, M.J. (2010) 'Unanalyzed evidence in law-enforcement agencies: a national examination of forensic processing in police departments', *Criminology and Public Policy*, 9: 381–404.

Strom, K.J., Ropero-Miller, J., Jones, S., Sikes, N., Pope, M., and Horstmann, N. (2007) *The 2007 Survey of Law Enforcement Forensic Evidence Processing*, Washington, DC: US Department of Justice, Office of Justice Programs, National Institute of Justice.

Tilley, N. and Ford, A. (1996) *Forensic Science and Crime Investigation*, Crime Detection and Prevention Series Paper 73, London: Home Office, Police Research Group.

Tyler, T. (2006) 'Viewing *CSI* and the threshold of guilt: managing truth and justice in reality and fiction', *Yale Law Journal*, 115: 1050–1085.

Walton, R.H. (2006) 'The cold case problem', in R.H. Walton (ed.), *Cold Case Homicides: Practical Investigative Techniques*, Boca Raton, FL: CRC Press, pp. 1–26.

Wilson, J.Q. (1976) *The Investigators: Managing FBI and Narcotics Agents*, New York: Basic Books.

42
Prison as an environmental pathogen

Sytske Besemer and Joseph Murray

Introduction

Considering the unprecedented number of incarcerated people in the United States, prison is a social pathogen that cannot be ignored when discussing biosocial influences on crime across the life-course. Mass incarceration is an embedded part of American society. One in every 99 Americans is currently in prison and, excluding children and the elderly, nearly one in 50 people is behind bars on any one day (Tonry, 2001; Johnson, 2005; Gottschalk, 2006; Warren, 2008; Walmsley, 2011). In this chapter we discuss how prison could act as an environmental pathogen on incarcerated individuals as well as on their children. We discuss theories and the latest empirical research on how prison might impact on individuals' reoffending and their children's criminal behavior. Does prison reduce or increase reoffending? Does parental incarceration increase the risks for children's own offending?

Pathogenic effects of prisons are not limited to reoffending outcomes of ex-prisoners, which has been the focus of debate about prison effects. Effects of imprisonment on the physical and mental health of prisoners are deeply concerning (Fazel and Baillargeon, 2011; Fazel and Seewald, 2012). Raphael (2014) describes how former inmates face considerable employment challenges on release because of limited work histories and stigma. Incarceration also affects the prisoner's family and the wider community. Incarceration is concentrated in disadvantaged and black communities in the United States, and contributes to poverty (Western, 2006; Clear, 2007; DeFina and Hannon, 2013) and a "host of destabilizing consequences – housing changes, school maladaptations, welfare problems, and strains on relationships" (Clear, 2008: 105). Already disadvantaged communities are negatively affected by incarceration and the consequent removal of a large proportion of adult males, serving to increase crime and decrease public safety (Clear, 2007, 2008). We recognize these collateral effects on innocent "quasi-inmates" (Comfort, 2003: 103), but we focus here on the effects of imprisonment on crime on the formerly incarcerated and their children. Also, because of space constraints, we do not discuss here macro-level differences in the use of imprisonment impact crime rates, or how incapacitation might prevent crime during the period of incarceration (for an overview of such studies see Cullen *et al.*, 2011; Durlauf and Nagin, 2011a,b; Nagin, 2013).

We first discuss the development of prison as a social institution and describe key elements of prison life. We then present a theoretical framework about how imprisonment affects re-offending or offending by prisoners' children. After that we discuss the current state of empirical science on these relationships. In the conclusion and summary we recommend actions for future research and policy.

Development of the prison

Prisons have not always been so prominent in the United States. Until the nineteenth century, several forms of corporal punishment and public humiliation were commonly used, such as the pillory or the stocks. At that time, prisons were used to hold offenders before their trial or awaiting their punishment, rather than as a form of punishment itself. In the early nineteenth century, so-called penitentiaries developed where offenders were held to separate them from criminal influences in their environment. Prison developed into a key criminal justice institution in American society (Rothman, 1971).

Although prison was a common institution in society, the imprisonment rates stayed very stable over a long period. Around 110 per 100,000 per population were incarcerated from the 1930s to the 1970s in the United States (Blumstein and Cohen, 1973). Interestingly, in the 1970s some scholars expected a decline in the use of prison as a criminal punishment (Sommer, 1976; Scull, 1977). However, from 1971 the United States increased its reliance on incarceration dramatically. Currently, around 2.3 million adults are being held in federal or state prisons or in jail on a daily basis, equivalent to an incarceration rate of 1 in every 99.1 adults, a tenfold increase since 1970 (Warren, 2008). Although not quite as dramatic as in the United States, many other high-income countries have also seen great increases in the use of incarceration, such as the United Kingdom and the Netherlands (von Hofer, 2003; Walmsley, 2011) as well as some low- and middle-income countries such as Brazil (Murray *et al.*, 2013). Our discussion in this chapter focuses on the United States because this country has the highest prison population rate in the world, as well as the highest absolute number of prisoners.

Several explanations have been suggested for this increase in the use of incarceration. One hypothesis is that a general shift in focus from rehabilitative to more punitive justice policies followed an increase in crime and dissatisfaction with contemporary crime policies. States started to adopt punitive offense-based policies instead of policies with an offender orientation based on rehabilitation and individualized sentencing (Tonry, 2009; Warren, 2009). The media and politicians fed the idea that crime was out of control, and politicians started to propose increasingly harsher sentences, a concept called penal populism. At the same time, the US embarked on four domestic wars: the war on drugs, the war on crime, the war on gangs, and the war on juveniles. Mandatory sentencing policies such as the California Three Strikes and You're Out Laws also tremendously increased the prison population. Furthermore, resources spent on social welfare were being allocated to the criminal justice system; and mentally ill patients, who used to be in the mental health system, were being fed into prisons and onto the streets (Cullen and Gilbert, 1982; Cullen and Jonson, 2014). An accumulation and interaction of these phenomena created a massive increase in the United States prison population (Beckett, 1997; Tonry, 2004; Gottschalk, 2006; Currie, 2013).

Incarceration experience

To understand how incarceration may influence the development of crime across the life-course, it is important to highlight what experiences incarceration involves for the person incarcerated and for their children left behind. Sykes (1958) identified five pains of imprisonment. Next to the deprivation of liberty, prisoners are also deprived of goods and services, heterosexual relationships, autonomy, and security. These latter four losses are, contrary to the deprivation of liberty, not the main purposes of imprisonment, but they are part of prison life and exert considerable stress on the prisoner. Goffman (1961) described how prisoners experience a "mortification" process by which they lose their sense of self due to characteristics of prison life. Irwin and Owen (2005) highlight several harms of imprisonment in terms of psychological damage due to the loss of agency and assaults on the self: anger, humiliation and feelings of injustice resulting from rules perceived as unfair, low priority given to healthcare, and economic exploitation such as disproportionately expensive telephone calls. Prison can literally be a pathogen to the incarcerated individuals in terms of its effects on health (Fazel and Baillargeon, 2011; Fazel and Seewald, 2012). Johnson (2005: 256) has described time in prison as "dead time spent in empty activities carried out in isolation from the larger world."

This environment also affects children with incarcerated parents, although the extent of the impact is likely to vary both with criminal justice and prison policies, and with children's prior relationships with their parents and alternative caregiving arrangements. If prisons organize special family and children's days, visits to prison are much less intimidating and frightening and can be very positive. Letters, telephone calls, and visits to prison can be vital to maintain relationships, but they are highly restricted (Poehlmann *et al.*, 2010). Children whose parents are removed from home by incarceration almost all experience grief, many having nightmares, worrying about their parent, and feeling lonely (Hissel *et al.*, 2011). While a large proportion – over 50 per cent – of children with incarcerated parents were not actually living with their parent before the incarceration (Glaze and Maruschak, 2008), many such children benefited from financial support and regular contact with their non-resident parent (Geller *et al.*, 2011; Geller, 2013).

It is important to realize that many children have been exposed to situations that pose a risk for their development before their parent's incarceration. Many have experienced parental substance abuse, mental illness, child abuse and neglect, poverty, or frequent house moves (Phillips *et al.*, 2002). Some children are present at the arrest of their parent, which can be frightening and bewildering. During and after the incarceration, children often experience changes in caregiver and house moves (Hissel *et al.*, 2011). It is difficult to remain in contact with the incarcerated parent because phone calls are expensive and parents are often incarcerated many miles away from home. Sometimes, however, parental incarceration can improve stability in children's lives, if children come from particularly chaotic, disadvantaged backgrounds influenced by parental antisocial behavior (Giordano, 2010; Hissel *et al.*, 2011).

Theoretical framework

Theories about incarceration and reoffending

The description above suggests that prison can act as an environmental pathogen causing harm to the persons involved. There are five theoretical positions that need to be reviewed to consider the possible effects of imprisonment on offending by ex-prisoners: specific deterrence, rehabilitation, prison as a learning environment, labeling theory, and defiance theory. We outline each in turn, before turning to theories about the effects of imprisonment on prisoners' children.

Specific deterrence

Specific deterrence refers to the possibility that people who have been through the criminal justice system will try to avoid the experience again, and so be less inclined to commit further crime. The idea of prison as a specific deterrent comes from the economic model of crime, in which individuals weigh the costs and benefits of offending in a rational choice process (Becker, 1968). It is hypothesized that a prison sentence would weigh heavily on the cost side of the equation, and thus reduce the probability of reoffending.

Although this rational choice model sounds straightforward, in reality the costs of offending are not always clear. Offenders often do not know what sentences they might face for different crimes, and deterrent effects depend on the perceived unpleasantness of incarceration, which may vary with individual characteristics. Furthermore, people who have recently been punished might perceive their chances of being caught again soon as lower than other people's chances due to the "gambler's fallacy" (Pogarsky and Piquero, 2003) – having a recent prison sentence might decrease instead of increase the offender's estimation of the risk of being caught again. Thus, although a rational choice model generally seems to predict a deterrent effect of incarceration, there are contingencies to this prediction depending on the perceived certainty and unpleasantness of punishment for a crime.

Rehabilitation and learning environment

To enable a reduction in reoffending, prisons should be designed as places to learn new skills or change problematic behavior, thereby fostering a non-criminal way of life. A prison sentence should include opportunities for incarcerated individuals to acquire personal as well as practical skills. Theoretically, in a rehabilitative prison environment, education and vocational training might increase the possibility that released offenders earn income through legitimate means, and reduce the probability of future offending (MacKenzie, 2002). Personal skills training could be targeted at drug addiction, emotion regulation, and violence reduction, and could also reduce the risk of re-offending. Unfortunately, with the increased use of imprisonment, money spent on education, training, and other rehabilitation programs has decreased and prisoners nowadays often do not have the opportunity to foster skills necessary to prevent reoffending (MacKenzie and Armstrong, 2012).

The opposite learning environment is also possible within prisons, which are often viewed as universities of crime. Prisons are communities in which individuals live together, sometimes for extended periods, and together they create a prison culture with associated values. These values are likely to be supportive of crime and are transmitted through daily interactions between incarcerated individuals (Irwin and Cressey, 1962; Irwin, 1980; Crewe, 2005, 2009). Especially in American prisons, gang membership and a violent street culture are imported into the prison (Wacquant, 2001; Irwin, 2005). Not only can people acquire or strengthen criminal values while incarcerated, they might develop or learn practical skills for committing certain crimes. Prison might thus represent a social learning environment in which people develop values as well as skills that are supportive of crime, and a prison sentence could thus nourish a "criminal career."

Labeling theory

Labeling theory also predicts an increase in offending as a consequence of incarceration. Classic labeling theory suggests that being publicly labeled a criminal causes increases in an individual's offending behavior, because people will act according to the label attached to them by society (Becker, 1963; Lemert, 1967). Revised versions of labeling theory divide this core idea into two key causes of offending following a labeling event. First, the experience of being labeled or stigmatized might influence someone's self-perception, attitudes, and beliefs, and increase their association with delinquent peers (Paternoster and Iovanni, 1989; Bernburg *et al.*, 2006). As such, incarceration might cause an increase in criminal behavior as a result of conforming to the stereotype of a "con."

Second, the stigma of a criminal label might push people into a criminal lifestyle by blocking conventional and non-criminal social and economic opportunities. For example, people might have trouble finding a stable job when they have a criminal record (Raphael, 2014). In the Rochester Youth Survey, Bernburg and Krohn (2003) found that being convicted had a negative impact on educational attainment, which in turn increased offending behavior.

Interestingly, undesirable labeling effects might depend on offender characteristics and on the type of criminal justice sentence received (Murray *et al.*, 2014). For example, labeling might be stronger with younger offenders, while personality and behavior is more malleable. Cullen and Jonson (2014) hypothesize that labeling will be stronger when sanctions are more punitively oriented. They also describe how inappropriate punishments that do not target dynamic risk factors – factors that are changeable – will have stronger labeling effects. Most prison sentences do not target dynamic risk factors and would be predicted to be ineffectual in that sense. Furthermore, Cullen and Jonson (2014) suggest that negative labeling effects are more likely to occur with low-risk or first-time offenders, who are at lower probability of continued criminality, in the absence of criminal justice interference. Thus, based on labeling theory, one would expect an increase in offending after incarceration, especially for young offenders.

Defiance theory

Defiance theory, proposed by Sherman (1993, 2010, 2014), is related to labeling theory, but focuses on different mechanisms. Sherman stresses the importance of emotions and legitimacy for the effectiveness of a criminal justice sentence, and the role that defiance plays in causing future criminal behavior if sentences are not perceived as fair. He describes that defiance will occur under four necessary conditions: the offender perceives a punishment as unfair; the offender feels alienated or is poorly bonded to the person or sanctioning agency; the offender perceives the sanction as stigmatizing and targeted at his person instead of at his law-breaking act; and the offender does not acknowledge the shame that the punishment caused him to suffer.

According to this theory, punishment should "shame the act, but not the actor" (Sherman, 2014: 156). When offenders define a prison sentence as unfair and feel excluded from the society that punished them, they will develop a certain pride that results in an increase and/or persistence of their criminal behavior. Sherman (2014: 155) states that "it may be sufficient to say that in the absence of fair treatment, offenders did not feel obligated to obey the law. This is not about anger; it is about a failure of the state to persuade its citizens to comply with a rule of law for the state's sake, as well as the citizens'."

To summarize the theories on the impact of incarceration on reoffending, deterrence theory and rehabilitation theory predict a decrease in reoffending after incarceration, whereas labeling theory, defiance theory, and the perspective of prison as a criminal learning environment predict an increase in reoffending as a result of incarceration. It is likely that the mechanisms described will not apply equally to all offenders; some might be deterred by the incarceration experience, while others may experience rehabilitation, labeling, or defiance effects.

Theories about effects of parental incarceration on children

Prisons were not designed with the aim of influencing the behavior of prisoners' children. Although the various strains associated with parental incarceration make it seem likely that incarceration might cause increases in children's own antisocial behavior, we first discuss how parental incarceration might impact positively on children's development.

Sometimes, parental incarceration could have a positive impact on prisoners' children. This could be the case if abusive, negligent, or violent parents are removed from the home by incarceration. For children in such environments, parental incarceration might increase their social capital and reduce social strain (Hagan and Dinovitzer, 1999; Hissel *et al.*, 2011). Families experiencing parental incarceration often have to cope with various problems such as unemployment, and financial and housing problems (Murray, 2005). This could go unnoticed for a long time, but parental incarceration could trigger social service intervention and support, which also could reduce social strain and improve children's social and emotional development. Although such considerations highlight that, at times, parental incarceration may benefit some children, Hagan and Dinovitzer (1999: 125) state that "it is more likely imprisonment is harmful to children even in dysfunctional families, because imprisonment will more often compound than mitigate pre-existing family problems."

Several theoretical perspectives suggest that parental incarceration is likely to have harmful effects on offspring and increase their own offending behavior. Strain theory holds that economic and psychological stresses motivate criminal behavior as a means to alleviate strain. Strain theories are relevant for considering the impact of parental incarceration because parental incarceration will almost inevitably produce economic and psychological strains for the children involved. Incarceration often causes a loss of family income and thus financial hardship. Sometimes families will receive financial support from the government, but this varies by country. Moreover, visiting incarcerated relatives, making phone calls, and writing letters can be costly and lead to additional economic strain. Moreover, social strain arises when children lose a parent to prison, because they lose "human and social capital" (Hagan and Dinovitzer, 1999: 124).

A remaining single parent may be unable to devote sufficient time and energy to their children, especially as they suffer from the loss of their partner and are forced to increase working hours to manage the loss in family income. If children have lived in single-parent families, children necessarily experience a change in caregiver and sometimes enter foster care. Through reductions in parental involvement during incarceration, children might become more likely to use peers as role models, which might also lead to more delinquent behavior. In her extensive research on the Ohio Life Course Study, Giordano illustrates how children might feel attracted to friends in a similar situation and find mutual solace:

> My friend Patrick, his mom had a drug problem. That's why he lives with his uncle and aunt . . . We understand each other. Like, I've met his mom before and then like . . . my mom and her friend dropped me off at Patrick's house before. And I felt weird 'cause she was all messed up on drugs . . . and she like, she like about fell out of the door. He was like . . . he was like, 'don't feel bad 'cause my mum does it too sometimes' . . . like I feel bad for him, but I know he feels bad for me too.
>
> *(Giordano, 2010: 154)*

Parental incarceration can also involve traumatic or stressful experiences leading to psychological strain. Some children are present when their parent is arrested. Prison visits can be frightening, because of the security precautions and restricted possibilities for contact during the visit

(Nesmith and Ruhland, 2008). Children whose parents are incarcerated experience grief, but cannot always express this openly, and often children are not told the true whereabouts of their incarcerated parent (Hissel *et al.*, 2011). From the perspective of attachment theory, these experiences can threaten a child's security about the reliable availability of their parent, leading to attachment problems which, in turn, could cause problem behavior (Murray and Murray, 2010).

Moreover, children of incarcerated parents might experience labeling and stigmatization. Prisoners' children can experience stigma and related bullying and teasing, which might increase problem behavior. Official bodies such as the police might be biased to pay more attention to prisoners' families (Farrington, 2011; Besemer *et al.*, 2013), increasing the chances that children of prisoners will be apprehended or convicted for illegal acts. This connects to the earlier mentioned labeling theory, which posits that people will behave according to the label society attaches to them. Stigmatization can also lead to emotional distress or strain. Hissel *et al.* describe that children often lie about the mother's whereabouts:

> "Children at school ask 'where is your mother?' Sometimes I say I don't know and sometimes I don't answer." Q: "How do you feel about them asking that?" CHILD: "I don't like it. Sometimes I'm a bit ashamed."
>
> *(Hissel et al., 2011: 355)*

An alternative explanation for any association between parental imprisonment and children's own offending is intergenerational transmission by other genetic or environmental mechanisms. Intergenerational transmission refers to the idea that some characteristic or behavior is seen in both the parent and the child. It is important to consider intergenerational transmission processes, because one explanation for the correlation between parental imprisonment and offspring offending is that other factors associated with parental criminality cause children themselves to offend (Murray and Farrington, 2008; Murray and Murray, 2010). Based on theories and studies of intergenerational transmission, we expect stronger transmission of criminal behavior from parent to child when parents have committed more serious criminal acts (Farrington, 2011; Besemer, 2012). This suggests that parental imprisonment may have no causal impact on children's offending, and instead the association between parental imprisonment and offspring offending is explained by the fact that parents who are imprisoned are more likely to be serious criminals with other background disadvantages that impact on the child. Environmental risk factors such as parental crime, parental mental illness, or drug abuse, or growing up in a disadvantaged neighborhood, might cause offspring offending, rather than the parental imprisonment itself. It is a major challenge for empirical research to test whether parental imprisonment has true causal effects on children over and above background risk factors.

To summarize the various points of view above, there might be undesirable effects of parental incarceration, causing increased likelihood of offending among prisoners' children, or there might be no causal impact of parental incarceration over and above background risks, or there may even be some deterrent or positive effects of parental incarceration for some children. In the next section we summarize the most recent empirical research on these issues.

Empirical findings

Incarceration and reoffending

Despite the abundant use of incarceration in the United States, rigorous scientific research demonstrating its effectiveness is scarce (for overviews see Gendreau *et al.*, 1999; Villettaz *et al.*, 2006;

Nagin *et al.*, 2009). Given background differences between incarcerated and non-incarcerated individuals, investigating the causal impact of imprisonment on reoffending is challenging if not impossible without an experimental design or a randomized controlled trial.

Nagin *et al.* (2009) provide an extremely useful review of the evidence on the possible effects of incarceration on reoffending. So far there have been just five studies randomly assigning custodial versus non-custodial sentences, of which one was a natural experiment in the Netherlands (van der Werff, 1979, 1981). One experiment was carried out in Switzerland in the Canton of Vaud (Killias *et al.*, 2000, 2010). The Directors of Corrections were interested in testing the effects of community service versus short prison sentences (up to 14 days) in a rigorous way. From July 1993 until the end of 1995, 123 offenders volunteering for the study were randomly assigned to one of these punishment options. Community service consisted of unpaid work such as working in nursing homes for the elderly, hospitals, schools, but also working in forests and cleaning natural resource areas. After two years, ex-prisoners had a higher arrest rate and had developed more unfavorable attitudes towards their sentence and the criminal justice system (Killias *et al.*, 2000).

In a longer-term follow-up after 11 years, reoffending was slightly more common among ex-prisoners, but this difference was not statistically significant. Interestingly, it appeared that ex-prisoners were better off after 11 years, they complied better with tax regulations, and were comparable in terms of employment history and marital status (Killias *et al.*, 2010). The authors concluded that short prison sentences did not appear to be damaging in the long term. It is important to emphasize, however, that offenders were incarcerated only for up to 14 days in this experiment. With these results, it is not possible to say anything about the impact of much longer sentences. Taken together, it appears that incarceration generally increases recidivism, although the evidence is weak (Nagin *et al.*, 2009).

Other studies point toward a criminogenic effect of incarceration, but the non-experimental studies with other designs provide two additional insights. As we discussed earlier, the problem with deterrence is that the costs of offending are generally not obvious to offenders, because they do not know what sentences they would face for different crimes. Helland and Tabarrok (2007) demonstrate that an imminent and extremely tangible threat in the form of life imprisonment appears to reduce reoffending. They investigated the impact of the California Three Strikes and You're Out Law to further consider the importance of imminence. They compared two groups of offenders: (a) individuals convicted of two "strikable" offenses; and (b) individuals who had been tried for a second "strikable" offense but were convicted of a non-strike-eligible offense. The imminent threat for the first group was the prospect of imprisonment for life if they committed a third offense. The second group did not face this threat – because they were convicted for one strikable offense – and had a 20 percent higher arrest rate compared with the first group.

In another study relevant to the issue of clarity and tangibility of punishment, Drago *et al.* (2009) investigated the impact of Italy's Collective Clemency Bill releasing more than 20,000 prisoners in 2006. Prisoners were released on condition that, if they were convicted for another crime within five years of release, they had to serve the remainder of the original sentence in addition to any new sentence. Prisoners with higher residual sentences had a lower recidivism rate. In this and the Californian study, individuals knew exactly what was going to happen if they committed another crime. Apparently, this clear information after incarceration did work to reduce recidivism, lending support for the idea of specific deterrence in the context of a clear and tangible threat.

On balance, the limited empirical evidence points toward a criminogenic effect of incarceration, supporting labeling theory, although high-quality research on specific labeling mechanisms is lacking. One illustration of labeling effects can be seen in the study by Laub and Sampson (1993) following Glueck's original study of 1,000 men. They show that there was no direct effect of the time incarcerated on later self-reported crime, but that longer incarceration periods predicted more job instability, which in turn predicted more criminal behavior in adulthood.

Despite the decreased spending on rehabilitation programs, criminologists continue to evaluate such efforts in the hope of finding humane and effective solutions to reducing reoffending. Cullen and Jonson (2014) reviewed several meta-analyses of intervention research and found that across all studies (including studies of ineffective programs), on average recidivism decreased by 10 percent among people involved in rehabilitation programs. Furthermore, they describe how treatment evaluation research shows heterogeneity in effects: some interventions are ineffective, while others reduce offending strongly and consistently.

Rehabilitation appears most effective when programs follow the three principles of effective correctional treatment (Andrews *et al.*, 1990; Smith *et al.*, 2009; Andrews and Bonta, 2010). Risk factors for crime can be static or dynamic. Static factors such as race and gender cannot be changed, but dynamic factors such as a drug addiction or antisocial attitudes are open to change. According to the *need principle*, interventions should focus on addressing dynamic criminogenic needs. Treatment will work only when it is responsive to these criminogenic needs, and this is called the *responsivity principle*. The *risk principle* states that interventions such as prison sentences should be given only to high-risk offenders, because they have many criminogenic needs that can be easily targeted. Low-risk offenders, in contrast, are more likely to stop offending if they do not become involved in the justice system through prison sentences. Prison does not always involve rehabilitation programs, and not every offender treatment is effective in reducing reoffending, but programs following these three principles are likely to reduce reoffending.

Parental incarceration

Murray *et al.* (2012a) conducted a systematic review and meta-analysis of 40 studies of parental incarceration and concluded that, similarly to the field of incarceration and reoffending, high-quality research using experimental methods on the impact of parental incarceration on offspring offending is lacking. The most rigorous studies demonstrated that parental incarceration was related to increased antisocial behavior among children. When they pooled the results from studies controlling for parental criminality or offspring's antisocial behavior before the parental incarceration, the results demonstrated that children who had experienced parental incarceration had about 10 percent increased risk for antisocial behavior compared with their peers.

Although parental incarceration clearly is a risk factor for delinquency among offspring, it is unclear whether or not the prison sentence is actually *causing* offspring delinquency (Murray and Farrington, 2008; Besemer *et al.*, 2011). It is also difficult to say what the intervening mechanisms are that explain why children of incarcerated parents have an increased risk of offending. A major push forward in elucidating these effects can be achieved using repeated observations of children and their family and peer environments both before and after parental incarceration, as was recently done in the Pittsburgh Youth Study (Murray *et al.*, 2012b). In this study, boys experiencing parental incarceration were matched to other boys on a range of background variables, including parental criminality, using propensity scores. Then changes in boys' behaviors were compared between the parental incarceration group and the propensity-matched control group. In the years after parental incarceration, boys who had experienced this event showed significantly increased levels of youth theft compared with the matched controls whose parents had not been incarcerated. Furthermore, Murray *et al.* (2012b) investigated mediating mechanisms linking parental incarceration and increases in youth theft. When parents went to prison, about half of the effect of parental incarceration on youth theft was explained by caretaker stress, less involvement of sons in the family, and involvement with delinquent peers. What was not measured in this study was stigmatization processes that might explain much of the remaining variance. In the absence of randomized experiments, investigations of how behaviors change through time, combined with matched controls, are vital to increase knowledge about mechanisms explaining the link between parental incarceration and offspring offending (Murray *et al.*, 2009; Murray, 2010).

In summary, the available evidence does not provide for firm causal conclusions about the effects of incarceration on reoffending, or on offending by prisoners' children. Incarceration seems to have a criminogenic or null effect on reoffending and is associated with an increased risk of offending among offspring.

Implications for research and policy

The most obvious implication from these empirical findings is that more rigorous research on the impact of incarceration on individuals' reoffending, as well as the offending of their children, is vital to answer the question whether incarceration should be regarded as an environmental pathogen. We recommend experimental as well as longitudinal studies of within-individual change to investigate these issues. In light of the possible criminogenic effects of incarceration, we suggest focusing resources on prevention efforts. From a human, moral, as well as economic perspective, it seems more defensible to spend money on preventing criminal behavior rather than responding with incarceration, which may cause considerable harm for individuals, their families, and their communities.

Interestingly, recently policy makers seem to recognize that prison sentences are often counterproductive, as stated by Attorney General Eric Holder in his speech in August 2013:

> We will start by fundamentally rethinking the notion of mandatory minimum sentences for drug-related crimes. Some statutes that mandate inflexible sentences – regardless of the individual conduct at issue in a particular case – reduce the discretion available to prosecutors, judges, and juries. Because they oftentimes generate unfairly long sentences, they breed disrespect for the system. When applied indiscriminately, they do not serve public safety. They – and some of the enforcement priorities we have set – have had a destabilizing effect on particular communities, largely poor and of color. And, applied inappropriately, they are ultimately counterproductive.
>
> *(Holder, 2013)*

It would be interesting to evaluate the impact of the changes Holder proposes. For example, one could compare people who currently receive a prison sentence with individuals who will not receive a prison sentence under the amended law. This is an important opportunity for a new natural experiment on the impact of incarceration. If families and children were also included in the study, the effects on children of the incarcerated could also be investigated.

Conclusion

Here we provide an overview of theories and empirical results on whether prison can be regarded as an environmental pathogen for the development of criminal behavior of prisoners, as well as of their children. Although rigorous investigations are scarce, the results seem to point to a criminogenic or null effect of incarceration on reoffending. Empirical evidence for special deterrent effects is weak, although an imminent threat, such as having to serve a suspended sentence, appears to work as a deterrent. Research suggests that labeling as well as defiance mechanisms may be important for understanding how incarceration impacts on individuals' future behavior. Specific rehabilitation programs might be able to counteract such harms and reduce reoffending. How are children with incarcerated parents affected by their parent's prison sentence? Although several theoretical mechanisms have been voiced, such as social, economic, and psychological strain, and stigmatization, research into these processes is extremely scarce. The available research points to parental incarceration being a risk factor with effects partly caused by caretaker stress and weakened family bonds.

References

Andrews, D.A. and Bonta, J. (2010) *The Psychology of Criminal Conduct*, Cincinnati, OH: Anderson.

Andrews, D.A., Zinger, I., Hoge, R.D., Bonta, J., Gendreau, P., and Cullen, F.T. (1990) 'Does correctional treatment work? A clinically relevant and psychologically informed meta-analysis', *Criminology*, 28: 369–404.

Becker, G.S. (1968) 'Crime and punishment: an economic approach', *Journal of Political Economy*, 76: 169–217.

Becker, H.S. (1963) *Outsiders: Studies in the Sociology of Deviance*, London: Free Press.

Beckett, K. (1997) *Making Crime Pay: Law and Order in Contemporary American Politics*, Oxford: Oxford University Press.

Bernburg, J.G. and Krohn, M.D. (2003) 'Labeling, life chances, and adult crime: the direct and indirect effects of official intervention in adolescence on crime in early adulthood', *Criminology*, 41: 1287–1318.

Bernburg, J.G., Krohn, M.D., and Rivera, C.J. (2006) 'Official labeling, criminal embeddedness, and subsequent delinquency', *Journal of Research in Crime and Delinquency*, 43: 67–88.

Besemer, S. (2012) *Intergenerational Transmission of Criminal and Violent Behaviour*, PhD dissertation, University of Cambridge.

Besemer, S., van der Geest, V.R., Murray, J., Bijleveld, C.C.J.H., and Farrington, D.P. (2011) 'The relationship between parental imprisonment and offspring offending in England and the Netherlands', *British Journal of Criminology*, 51: 413–437.

Besemer, S., Farrington, D.P., and Bijleveld, C.C.J.H. (2013) 'Official bias in intergenerational transmission of criminal behaviour', *British Journal of Criminology*, 53: 438–455.

Blumstein, A. and Cohen, J. (1973) 'A theory of the stability of punishment', *Journal of Criminal Law and Criminology*, 64: 198–206.

Clear, T.R. (2007) *Imprisoning Communities*, Oxford: Oxford University Press.

Clear, T.R. (2008) 'The effects of high imprisonment rates on communities', *Crime and Justice*, 37: 97–132.

Comfort, M.L. (2003) 'In the tube at San Quentin: the "secondary prisonization" of women visiting inmates', *Journal of Contemporary Ethnography*, 23: 77–107.

Crewe, B. (2005) 'Codes and conventions: the terms and conditions of contemporary inmate values', in A. Liebling and S. Maruna (eds), *The Effects of Imprisonment*, Cullompton, UK: Willan Publishing, pp. 177–208.

Crewe, B. (2009) *The Prisoner Society*, Oxford: Oxford University Press.

Cullen, F.T. and Gilbert, K.E. (1982) *Reaffirming Rehabilitation*, Cincinnati, OH: Anderson.

Cullen, F.T. and Jonson, C.L. (2014) 'Labeling theory and correctional rehabilitation: beyond unanticipated consequences', in D.P. Farrington and J. Murray (eds), *Empirical Tests of Labeling Theory: Advances in Criminological Theory*, Vol. 18, New Brunswick, NJ: Transaction Publishers, pp. 63–85.

Cullen, F.T., Jonson, C.L., and Nagin, D.S. (2011) 'Prisons do not reduce recidivism: the high cost of ignoring science', *Prison Journal*, 91: 48–65.

Currie, E. (2013) *Crime and Punishment in America*, New York: Picador.

DeFina, R. and Hannon, L. (2013) 'The impact of mass incarceration on poverty', *Crime and Delinquency*, 59: 562–586.

Drago, F., Galbiati, R., and Pietro, V. (2009) 'The deterrent effects of prison: evidence from a natural experiment', *Journal of Political Economy*, 117: 257–280.

Durlauf, S.N. and Nagin, D.S. (2011a) 'Imprisonment and crime', *Criminology & Public Policy*, 10: 13–54.

Durlauf, S.N. and Nagin, D.S. (2011b) 'The deterrent effect of imprisonment', in P.J. Cook, J. Ludwig, and J. McCrary (eds), *Controlling Crime: Strategies and Tradeoffs*, Chicago, IL: University of Chicago Press, pp. 43–94.

Farrington, D.P. (2011) 'Families and crime', in J.Q. Wilson and J. Petersilia (eds), *Crime and Public Policy*, New York: Oxford University Press, pp. 130–157.

Fazel, S. and Baillargeon, J. (2011) 'The health of prisoners', *The Lancet*, 377: 956–965.

Fazel, S. and Seewald, K. (2012) 'Severe mental illness in 33,588 prisoners worldwide: systematic review and meta-regression analysis', *British Journal of Psychiatry*, 200: 364–373.

Geller, A. (2013) 'Paternal incarceration and father–child contact in fragile families', *Journal of Marriage and Family*, 75: 1288–1303.

Geller, A., Garfinkel, I., and Western, B. (2011) 'Paternal incarceration and support for children in fragile families', *Demography*, 48: 25–47.

Gendreau, P., Coggin, C., and Cullen, F.T. (1999) *The Effects of Prison Sentences on Recidivism*, Ottawa: Department of the Solicitor General Canada, Public Works and Government Services.

Giordano, P.C. (2010) *Legacies of Crime: A Follow-up of the Children of Highly Delinquent Girls and Boys*, Cambridge: Cambridge University Press.

Glaze, L.E. and Maruschak, L.M. (2008) *Parents in Prison and their Minor Children*, Washington, DC: US Department of Justice.

Goffman, E. (1961) *Asylums: Essays on the Social Situation of Mental Patients and other Inmates*, Harmondsworth, UK: Penguin.

Gottschalk, M. (2006) *The Prison and the Gallows: The Politics of Mass Incarceration in America*, New York: Cambridge University Press.

Hagan, J. and Dinovitzer, R. (1999) 'Collateral consequences of imprisonment for children, communities and prisoners', in M. Tonry and J. Petersilia (eds), *Prisons*, Chicago, IL: University of Chicago Press, pp. 121–162.

Helland, E. and Tabarrok, A. (2007) 'Does three strikes deter? A nonparametric estimation', *Journal of Human Resources*, 42: 309–330.

Hissel, S.C.E.M., Bijleveld, C.C.J.H., and Kruttschnitt, C. (2011) 'The well-being of children of incarcerated mothers: an exploratory study for the Netherlands', *European Journal of Criminology*, 8: 346–360.

von Hofer, H. (2003) 'Prison populations as political constructs: the case of Finland, Holland and Sweden', *Journal of Scandinavian Studies in Criminology and Crime Prevention*, 4: 21–38.

Holder, E. (2013) 'Attorney General Eric Holder Delivers Remarks at the Annual Meeting of the American Bar Association's House of Delegates', www.justice.gov/iso/opa/ag/speeches/2013/ag-speech-130812.html

Irwin, J. (1980) *Prisons in Turmoil*, Boston. MA: Little, Brown.

Irwin, J. (2005) *The Warehouse Prison: Disposal of the New Dangerous Class*, Los Angeles, CA: Roxbury.

Irwin, J. and Cressey, D.R. (1962) 'Thieves, convicts and the inmate culture', *Social Problems*, 10: 142–155.

Irwin, J. and Owen, B. (2005) 'Harm and the contemporary prison', in A. Liebling and S. Maruna (eds), *The Effects of Imprisonment*, Cullompton, UK: Willan Publishing, pp. 94–117.

Johnson, R. (2005) 'Brave new prisons: the growing social isolation of modern penal institutions', in A. Liebling and S. Maruna (eds), *The Effects of Imprisonment*, Cullompton, UK: Willan Publishing, pp. 255–284.

Killias, M., Aebi, M., and Ribeaud, D. (2000) 'Does community service rehabilitate better than short-term imprisonment? Results of a controlled experiment', *Howard Journal of Criminal Justice*, 39: 40–57.

Killias, M., Gilliéron, G., Villard, F., and Poglia, C. (2010) 'How damaging is imprisonment in the long-term? A controlled experiment comparing long-term effects of community service and short custodial sentences on re-offending and social integration', *Journal of Experimental Criminology*, 6: 115–130.

Laub, J.H. and Sampson, R.J. (1993) 'Turning points in the life course: why change matters to the study of crime', *Criminology*, 31(3): 301–325.

Lemert, E.M. (1967) *Human Deviance, Social Problems, and Social Control*, Englewood Cliffs, NJ: Prentice Hall.

MacKenzie, D.L. (2002) 'Reducing the criminal activities of known offenders and delinquents: crime prevention in the courts and corrections', in L.W. Sherman, D.P. Farrington, B.C. Welsh, and D.L. MacKenzie (eds), *Evidence-based Crime Prevention*, London: Routledge, pp. 330–404.

MacKenzie, D.L. and Armstrong, G.S. (2012) 'The human experiment in treatment: a means to the end of offender recidivism', in R. Loeber and B.C. Welsh (eds), *The Future of Criminology*, Oxford: Oxford University Press, pp. 189–195.

Murray, J. (2005) 'The effects of imprisonment on families and children of prisoners', in A. Liebling and S. Maruna (eds), *The Effects of Imprisonment*, Cullompton, UK: Willan Publishing, pp. 442–462.

Murray, J. (2010) 'Longitudinal research on the effects of parental incarceration on children', in J.M. Eddy and J. Poehlmann (eds), *Children of Incarcerated Parents: A Handbook for Researchers and Practitioners*, Washington, DC: Urban Institute, pp. 55–73.

Murray, J. and Farrington, D.P. (2008) 'The effects of parental imprisonment on children', in M. Tonry (ed.), *Crime and Justice: A Review of Research*, Chicago, IL: University of Chicago Press, pp. 133–206.

Murray, J. and Murray, L. (2010) 'Parental incarceration, attachment, and child psychopathology', *Attachment & Human Development*, 12: 289–309.

Murray, J., Farrington, D.P., and Eisner, M.P. (2009) 'Drawing conclusions about causes from systematic reviews of risk factors: the Cambridge Quality Checklists', *Journal of Experimental Criminology*, 5: 1–23.

Murray, J., Farrington, D.P., and Sekol, I. (2012a) 'Children's antisocial behavior, mental health, drug use, and educational performance after parental incarceration: a systematic review and meta-analysis', *Psychological Bulletin*, 138: 175–210.

Murray, J., Loeber, R., and Pardini, D.A. (2012b) 'Parental involvement in the criminal justice system and the development of youth theft, marijuana use, depression, and poor academic performance', *Criminology*, 50: 255–302.

Murray, J., Cerqueira, D.R. de C., and Kahn, T. (2013) 'Crime and violence in Brazil: systematic review of time trends, prevalence rates and risk factors', *Aggression and Violent Behavior*, 18: 471–483.

Murray, J., Blokland, A., Farrington, D.P., and Theobald, D. (2014) 'Long term effects of conviction and incarceration on men in the Cambridge Study in Delinquent Development', in D.P. Farrington and J. Murray (eds), *Empirical Tests of Labeling Theory: Advances in Criminological Theory*, Vol. 18, New Brunswick, NJ: Transaction Publishers.

Nagin, D.S. (2013) 'Deterrence in the twenty-first century', *Crime and Justice*, 42(1): 199–263.

Nagin, D.S., Cullen, F.T., and Jonson, C.L. (2009) 'Imprisonment and reoffending', *Crime and Justice*, 38: 115–200.

Nesmith, A. and Ruhland, E. (2008) 'Children of incarcerated parents: challenges and resiliency, in their own words', *Children and Youth Services Review*, 30: 1119–1130.

Paternoster, R. and Iovanni, L. (1989) 'The labeling perspective and delinquency: an elaboration of the theory and an assessment of the evidence', *Justice Quarterly*, 6: 359–394.

Phillips, S.D., Burns, B.J., Wagner, H.R., Kramer, T.L., and Robbins, J.M. (2002) 'Parental incarceration among adolescents receiving mental health services', *Journal of Child and Family Studies*, 11: 385–399.

Poehlmann, J., Dallaire, D., Loper, A.B., and Shear, L.D. (2010) 'Children's contact with their incarcerated parents: research findings and recommendations', *American Psychologist*, 65: 575–598.

Pogarsky, G. and Piquero, A.R. (2003) 'Can punishment encourage offending? Investigating the "resetting" effect', *Journal of Research in Crime and Delinquency*, 40: 95–120.

Raphael, S. (2014) 'The effects of conviction and incarceration on future employment outcomes', in D.P. Farrington and J. Murray (eds), *Empirical Tests of Labeling Theory: Advances in Criminological Theory*, Vol. 18, New Brunswick, NJ: Transaction Publishers, pp. 237–262.

Rothman, D.J. (1971) *The Discovery of the Asylum: Social Order and Disorder in the New Republic*, Boston, MA: Little, Brown.

Scull, A.T. (1977) *Decarceration: Community Treatment and the Deviant: A Radical View*, Englewood Cliffs, NJ: Prentice Hall.

Sherman, L.W. (1993) 'Defiance, deterrence, and irrelevance: a theory of the criminal sanction', *Journal of Research in Crime and Delinquency*, 30: 445–473.

Sherman, L.W. (2010) 'Defiance, compliance and consilience: a general theory of criminology', in E. McLaughlin and T. Newburn (eds), *Handbook of Criminological Theory*, London: Sage, pp. 360–390.

Sherman, L.W. (2014) 'Experiments in criminal sanctions: labeling, defiance, and restorative justice', in D.P. Farrington and J. Murray (eds), *Empirical Tests of Labeling Theory: Advances in Criminological Theory*, Vol. 18, New Brunswick, NJ: Transaction Publishers, pp. 149–176.

Smith, P., Gendreau, P., and Swartz, K. (2009) 'Validating the principles of effective intervention: a systematic review of the contributions of meta-analysis in the field of corrections', *Victims & Offenders*, 4: 148–169.

Sommer, R. (1976) *End of Imprisonment*, New York: Oxford University Press.

Sykes, G.M. (1958) *The Society of Captives: A Study of a Maximum Security Prison*, Princeton, NJ: Princeton University Press.

Tonry, M. (2001) 'Symbol, substance, and severity in Western penal policies', *Punishment & Society*, 3: 517–536.

Tonry, M. (2004) *Thinking about Crime: Sense and Sensibility in American Penal Culture*, New York: Oxford University Press.

Tonry, M. (2009) 'Explanations of American punishment policies: a national history', *Punishment & Society*, 11: 377–394.

Villettaz, P., Killias, M., and Zoder, I. (2006) *The Effects of Custodial vs. Non-Custodial Sentences on Re-Offending: A Systematic Review of the State of Knowledge*, Campbell Systematic Review No. 13, Oslo: Campbell Collaboration.

Wacquant, L. (2001) 'Deadly symbiosis: when ghetto and prison meet and mesh', *Punishment & Society*, 3: 95–133.

Walmsley, R. (2011) *World Prison Population List*, 9th edn, London: International Centre for Prison Studies.

Warren, J. (2008) *One in 100: Behind Bars in America 2008*, Washington, DC: Pew Charitable Trusts.

Warren, R.K. (2009) 'Probation Reform in California: Senate Bill 678', *Federal Sentencing Reporter*, 22: 186–193.

van der Werff, C. (1979) *Speciale preventie, een penologisch onderzoek [Special prevention, a penological study]*, University of Amsterdam.

van der Werff, C. (1981) 'Recidivism and special deterrence', *British Journal of Criminology*, 21: 136–147.

Western, B. (2006) *Punishment and Inequality in America*, New York: Russell Sage Foundation.

43

Biosocial criminology

The future is here

Michael G. Vaughn and Matt DeLisi

Introduction

At the close of 2013, a cursory Google Scholar search yields over 4300 hits for the keywords "biosocial criminology." If the search parameters are restricted to the past five years, the yield is approximately 45 percent (N = 1880) of that total. This means that in the past five years, the scientific literature discussing biosocial criminology has nearly doubled, a result of increasing numbers—and types—of journals publishing papers on this subject. In addition to leading criminology and criminal justice journals, a wide swath of prestigious scientific journals has published or now regularly publishes results from studies on the biosocial underpinnings of crime and antisocial behavior. A partial rendering of these outlets includes *Proceedings of the National Academy of Sciences*, *International Journal of Epidemiology*, *American Journal of Psychiatry*, *JAMA Psychiatry* (formerly *Archives of General Psychiatry*), and *Nature Neuroscience*. The biosocial approach has even penetrated the galleys of the *American Sociological Review* and *American Journal of Sociology*.

There was a time in the not-so-distant past when it seemed obligatory to defend one's decision to take a biosocial approach to criminology. However, times are changing, and in our view—as many others agree—criminologists need not be so defensive. Today, biosocial criminology is thriving. The range of topics in the present volume alone points toward the vibrancy of the paradigm.

In this chapter, we outline the reasons why biosocial criminology has been successful. Specifically, we attempt to answer to the following question: What are the factors that explain the success of biosocial criminology? We believe there are four major factors. First, that biosocial criminology's roots stem from evolutionary theory. Second, that it possesses the qualities of holism and transdisciplinarity. Third, that major criminological constructs are fundamentally biosocial in nature. And fourth, that it offers value for prevention and policy. We term this the four-factor model of biosocial criminology paradigm success (see Figure 43.1).

Roots in evolutionary theory

Most people, when they hear the term "evolution," think of biology. Indeed, mammalian biology plays a significant role in evolution, for both humans and animals. However, the word itself refers

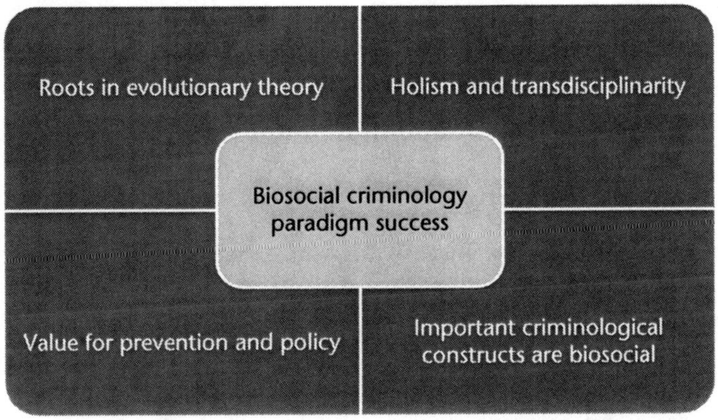

Figure 43.1 The four-factor model of biosocial criminology paradigm success

more broadly to change and adaptation over time. From this vantage point, major social transformations, such as the industrial revolution or the rise of the computer age, can be conceptualized as cultural evolution (Harris, 1979; Sanderson, 1990). Obviously, these social transformations have significantly altered the way humans live; however, our unique mammalian biology is always working in tandem, alongside any social or cultural changes. Studies from the field of primatology illustrate this notion of biological and social changes occurring in tandem. While chimpanzees are no doubt influenced by their genetic makeup, they also learn from other chimps when innovations occur, such as the ability to use tree leaves as water containers, or to use sticks or rocks as tools to crack open nuts. An investigation of the bonobo (an ape smaller than, though morphologically similar to, the chimpanzee) indicates that although these two species have evolved anatomically similar bodies, their evolved social constructs are markedly different. While chimpanzees are a male-dominant species, the bonobo are a profoundly female-centered species that readily employ sexual behavior to defuse interpersonal conflicts (de Waal, 1995). From this example—and many others—we can infer that biosocial processes are fundamental and have deep roots in evolution.

Both biological and cultural evolution are affected by the forces of selection processes. Many selective forces exist in our environments, but these forces are also susceptible to the forces of individual selection. Neighborhoods, for example, act to constrain behavioral options; however, people can select and change their neighborhood as well. Criminology needs to be wary of these "bottom-up" selection processes when formulating theory, and not merely assume "top-down" context-to-person effects.

In addition to selection, biological and cultural evolution is fundamentally shaped by basic biological and societal needs. For example, modern societies rely on technology to capture energy from a finite environment. We need energy both as individuals and collectives, whether in the form of electricity or food and water. There is no escaping these basic needs. Research has also shown that we have other basic biopsychological needs, such as sex and affection, sleep, language, and avoidance of stress from darkness, cold, heat, altitude, moisture, and lack of oxygen (Harris, 1979, 1994). Theories or paradigms that are built strictly upon social factors and do not consider these basic biological needs are limited and may ultimately prove futile.

The major barrier to applying evolutionary principles to human behavior and social problems appears to be biological determinism. Biosocial criminology acknowledges the major role of biological processes, but not in a simple, reductionist way. Further, determinism can run both ways (i.e. environmental determinism). The evolutionary paradigm is arguably one of the most successful in all of science. Incorporating evolutionary principles when developing biosocial theories of criminology can lead to a rich blueprint for understanding criminological science.

Holism and transdisciplinarity

Crime and antisocial behaviors are multifactorial in nature. Description and explanation of, and intervention in, these behaviors will necessarily involve a host of traditional disciplines (e.g. genetics, endocrinology, neuroscience, psychology, sociology, economics). The union of a wide swath of disciplines for the benefit of understanding a complex problem can be termed transdisciplinarity. Transdisciplinary approaches are gaining popularity because of the increasing recognition that understanding a given topic cannot be limited to a specific discipline—it is necessary to cut across multiple fields. Singular disciplinary contributions to understanding crime and antisocial behavior are valuable, but ultimately incomplete.

Looking at information and theories from a variety of disciplines is important to biosocial criminology, but integrating them into a complete picture, or holism, is equally important. The part-to-whole concept, or holism, is exemplified by the cell-to-society approach to human behavior developed by Vaughn *et al.* (2013). These scholars use the cell-to-society terminology to highlight that human behavior is impacted by processes occurring at all levels, from the smallest levels of biological organization (e.g. neurotransmitters, hormones) to the largest levels of social and physical environmental systems (e.g. institutions, social networks, technology, built environment). This kind of biosocial approach is broad and expansive, beginning with genes and moving to larger levels of physiological organization that impact human behavior, such as stress, emotion, and executive functions in the brain. These areas are ordered to reflect their evolutionary development. For instance, areas of the human brain that handle emotion evolved prior to those designed for rational thought and planning (MacLean, 1990; Massey, 2002). These emotional or "downstream" areas of the brain (and ultimately their DNA coding) influence the development of temperament and how individuals tend to respond to a given milieu. The influences are, of course, bidirectional and plastic, meaning that input from the environment molds and shapes us, but we in turn select for and shape our environments.

Although the need for transdisciplinary approaches in fully comprehending and subsequently intervening in criminal behavior is often acknowledged (Caspi and Moffitt, 2006; Laub, 2006), the effectiveness of transdisciplinary approaches is stymied by the devotion scholars have to their disciplinary training, and by lack of communication across seemingly disparate fields. As Vaughn *et al.* (2013: 7) state, "Such communication breakdowns bring to mind the image of two ships passing in the night, both laboring strenuously to advance in their desired direction but entirely unaware of one another's presence and the benefits their communication—and collaboration—might yield." Well articulated (and empirically supported) theories that take into account the interplay between biological factors and both proximal and distal environmental factors are likely to withstand the test of time. By taking these interactions into account, such theories in turn facilitate linkages with specific findings and theories across disciplines. There exists an enormous body of research published in the fields of psychology, psychiatry, pediatrics, public health, child development, developmental psychology, behavioral genetics, molecular genetics, neuroscience, and other fields of knowledge outside of traditional social science that are germane to criminology. In its holism and transdisciplinarity, biosocial criminology is taking advantage of these findings.

Many criminological constructs are biosocial

Many of the leading criminological constructs are explicitly biosocial. They involve combinations of social input and biological processes. It would be inefficient to discuss all such constructs here, so we have chosen three to highlight: low self-control, social learning, and callous-unemotionality.

The first construct, low self-control, is considered biosocial because self-control involves executive functions in the human brain that are shaped by social interactions and environmental opportunities and constraints. The modern concept of self-control in criminology was launched by Gottfredson and Hirschi in 1990, and a steady stream of corroborating empirical studies have followed. The success of the construct has even been described in monstrous terms (see DeLisi, 2011). Many criminologists, however, are seemingly unaware of analogous constructs, such as effortful control, impulsivity, and behavioral disinhibition among others, that have been studied widely in the psychological sciences for many decades. Although Gottfredson and Hirschi attributed the origins of low self-control to parenting practices, the direct links to executive governance are undeniable.

The celebrated case of Phineas Gage is considered the starting point for neuroscientific understanding of self-control. Gage was a railroad worker whose head was impaled by an iron rod during an unfortunate accident. Gage survived, but the iron rod destroyed his frontal lobe. Prior to the accident, Gage was described as cooperative, dutiful, and conscientious. Yet after the accident, Gage's behavior changed. He became aggressive, unreliable, profane, and lacking in restraint. Scientists surmised that his change in behavior resulted from the damage that had occurred to his frontal lobe, particularly his prefrontal cortex. Some executive functions of the prefrontal cortex include (but are not limited to) planning, delaying gratification, decision-making, judgment, and monitoring. Thus the executive functions of the prefrontal cortex help us to carry out goal-directed behavior and purposeful action. Deficits in these functions result in lower levels of self-control and a higher probability of antisocial behavior.

Social learning, our second construct, is biosocial because learning is processed in the human brain, is partly genetic, and is shaped by social interactions across the life-course. Social learning theories of crime have been employed in criminological research for over 70 years. In short, social learning theories of crime, or "deviance," state that individuals learn to commit deviant acts through their associations with others. Importantly, these interactions also include deviant messages and subsequent reinforcements. Although it seems obvious that social learning theories are a product of social interaction and socialization experiences, it is less obvious that the process of social learning is biosocial. Therefore it is important to point out that specific areas of the brain are involved in learning—including social learning. Furthermore, cognitive and emotional centers influence behavioral judgment and decision-making, both of which are involved in social learning. Finally, the highly plastic nature of the developing brain depends upon environmental input for its development. This includes the social environment as well as the physical, meaning that experiences of all kinds can affect the brain, particular early in life.

The third construct, callous-unemotionality (CU), has been shown to be highly heritable (Viding *et al.*, 2005) and linked to the size and function of the amygdala (Blair, 2010). This construct includes hallmark offender traits such as guiltlessness, lack of empathy, callous manipulation of others, and a tendency towards moral disengagement. These traits appear to be relatively stable over the course of a person's life and are a major part of psychopathy—an important aspect of career criminality (Vaughn *et al.*, 2008). A rising tide of research is showing the impact of CU traits on the development and stability of antisocial behavior (Frick and White, 2008). Although the origins of these traits are as yet unidentified, they are thought to derive from the emotional processing centers of the brain, specifically the amygdala and limbic system (DeLisi *et al.*, 2009). In particular, a study by Blair (2010) offered supporting evidence for this theory by demonstrating that amygdala dysfunction (e.g. hypoarousal) forms the neural basis for psychopathy and related "coldhearted" offenses. Moreover, brain imaging studies are increasingly being used to identify specific regions of the brain that are tied to antisocial behavior, and these studies have generally found that deficits in executive functioning and the amygdala are associated with antisocial behavior. Importantly, however, additional research is needed on CU traits in relation to situational environmental influences. In particular, pairing research by psychopathy and social science researchers would be especially valuable.

Prevention, treatment, and policy

In our view, almost all prevention and policy efforts that result in behavioral change are biosocial in nature. Beauchaine *et al.* (2008) points out ten reasons why biological processes matter in prevention. These include the use of biomarkers helping us to identify those most at risk, individual variability in neurobiology affecting response to treatment, and biological vulnerabilities moderating the effects of the environment and *vice versa*. Understanding these reasons and incorporating them into our studies help us to better identify, target, specify, and refine our prevention programs. In addition to identifying these biological processes, however, successful prevention of criminal and antisocial behavior will require greater progress in understanding the early developmental biosocial roots of effecting change (Rocque *et al.*, 2012). A simple but overarching question is: How do preventive interventions, which are enacted in a person's environment, work "under the skin"? Grasping the biological and environmental interplay is critical for at least two reasons. First, tests that improve our understanding of the underlying biosocial causality of certain behaviors can refine and thus increase the efficiency of prevention schemes that we develop, resulting in reduced social and economic costs down the road. Second, knowledge gained from these tests can be used to direct etiologic research conducted in prospective longitudinal or experimental designs.

The risk factors that are addressed in preventative measures involve both nature and nurture. Nurturing factors, such as low parental involvement during infancy and childhood, contribute to the rise and persistence of conduct problems in adulthood (Hiatt and Dishion, 2008). Research on children who were abused or neglected has found that a substantial proportion, perhaps as many as 20 percent, became antisocial in their adult years (Hiatt and Dishion, 2008). Nurturing factors also involve nature, in that parental behavior such as affection causes changes in both the infant's and the parent's brains. Prevention designs, now and in the future, will have the complex task of incorporating biological and environmental factors such as antisocial parents/siblings, harsh or affectionate child-rearing practices (as well as the consistency of these practices), low intelligence, effortful control, negative emotionality, nutrition, environmental toxins, and delinquent peers. This is especially challenging as these factors can all vary at different stages of a person's life, and will therefore be difficult to monitor and to predict how they affect a person's antisocial nature.

One interesting biosocial approach to prevention can be gleaned from genetic principles. Genes exert their influence on human behavior through complex interplay with the individual's environments (Lahey, 2008). When developing prevention efforts, two types of gene–environment relationship are useful: evocative gene–environment correlation; and active gene–environment correlation. Evocative gene–environment correlation occurs when a child's behavior changes, influences, or evokes behaviors from others that further their conduct problems (e.g. a child's difficult temperament evoking harsh parental retributions). Active gene–environment correlation occurs when a child's selection of environments is influenced by their genetic makeup. This can occur, for instance, when children form social bonds with other children who are similar to themselves (e.g. delinquent peers) (Lahey, 2008). Taking these interactions into account when designing prevention protocols may help enhance the protocols' effectiveness.

Many developmentally oriented prevention protocols already exist that have demonstrated effectiveness. Perhaps the best known of these is the nurse home visiting program developed by Olds (e.g. Olds *et al.*, 1998). Trials showed that when first-time low-income mothers were visited by nurses (who provided pre- and postnatal and child care information), their children demonstrated less abuse, neglect, crime, and violence at age 15 compared with the control group (Olds *et al.*, 1998). Consequently, this program is widely—and effectively—used today.

Most developmentally oriented prevention and treatment packages rely on cognitive and behavioral principles that engage neuro-circuitry. Such interventions are rooted in biosocial principles involving the pairing of a physiologic response (biological) to a stimulus (environmental). Various combinations of cognitive and behavioral treatments, such as teacher behavior therapy, token economies, group contingency therapy, peer management therapy, and self-management therapy, have all demonstrated large effects on disruptive behaviors that are often precursors to crime (Litschge *et al.*, 2010).

It might be possible that violent and aggressive behavior can be prevented via secondary treatments such as psychopharmacological agents or nutritional protocols. Some medications can reduce aggression directly; these include neuroleptics, lithium, and anti-convulsants for acute violence (Kruesi, 2007). Prevention scientists working in criminology may learn much from the field of addiction, which has long regarded abuse and dependence on psychoactive substances as a highly contextualized brain disease (Leshner, 1998). Many of the interventions in the addictions are explicitly biosocial, involving a combination of motivational, behavioral, and pharmacotherapy interventions. Although there is not likely to be a medication that will directly treat criminal behavior *per se*, there are a host of agents that treat co-occurring problems such as substance abuse. It is possible that these treatments designed for substance-use disorders may also simultaneously target many of the same mechanisms involved in criminal offending. Three medications that are commonly used for treating alcohol dependence, for instance, are naltrexone, acamprosate, and topiramate. These agents block or reduce the rewarding effects of alcohol or, in the case of acamprosate, reduce the excitation of neurons and resulting stress caused by alcohol withdrawal. Notably, these medications are intended to be used as part of an integrated treatment that also involves a psychosocial component. In addition to medication, there is some evidence that nutrition can be used to treat behavioral issues. For example, promising nutrition-based findings suggest that omega-3 fatty acids demonstrate cognitive effects that may impact mood stabilization and in turn reduce antisocial behavior (Gesch *et al.*, 2002). Furthermore, nutritional therapy may not only improve mental health and, consequently, behavioral problems, but it improves physical health as well.

With respect to policy, it is sometimes assumed that biosocial—especially biological—findings readily translate to harmful or *status quo* initiatives. Current evidence actually suggests that biosocial implications for policy are progressive and are actually politically liberal. One of the clearest recent examples involves the 2005 United States Supreme Court case of *Roper v. Simmons*, which ruled that the decision to employ the death penalty to juveniles was cruel and unusual punishment no matter how heinous the crime. Findings from research on neurological development and brain maturation during adolescence were a driving force behind the court's decision. As biosocial criminological findings continue to emerge, it seems likely that resulting policy implications will side in favor of early prevention and not in punishing the individual for biological vulnerability to environmental adversity.

Ongoing challenges

The biggest challenge to the continued advancement of biosocial criminology is adequate data. Most criminologists lack sufficient sources to conduct biosocial research. There has been an overreliance on extant data sources, which are few in number, creating some limitations to the biosocial criminology corpus. There are few secondary sources of data that have included requisite biological variables in conjunction with social factors and criminological outcomes. As a result, biosocial criminologists will need to partner with researchers in other disciplines—especially those associated with the medical sciences—to increase the data from which they can work. For example, including measures of brief violence or crime in neuropsychiatric studies can facilitate assessments of regions of the brain associated not only with compulsive drug seeking, but also with violence/crime. Although this requires relationship building and may slow productivity in the short-term, the long-term gain for biosocial criminology is evident. The creation of a multidisciplinary network to generate biosocial criminology data that can be shared via a singular web-based platform may be one avenue worth pursuing.

Conclusion

Biosocial criminology is alive and well, and won't go away any time soon. The factors that have contributed to its success are backed by research from a variety of disciplines, thereby providing sustainability to biosocial criminology. Researchers across the Academy, for instance, are becoming increasingly more interdisciplinary. There is some thought that the traditional "social-only" branch of criminology has exhausted itself. The success of the medical paradigm for prevention and treatment will likely have a corollary impact for biosocial criminologists who are primarily concerned with successful intervention. Twenty years ago—or perhaps even ten—texts such as this would have seemed inconceivable. Now, it is expected. The future is here.

References

Beauchaine, T.P., Neuhaus, E., Brenner, S., and Gatzke-Kopp, L. (2008) 'Ten good reasons to consider biological variables in prevention and intervention research', *Development and Psychopathology*, 20: 745–774.

Blair, R.J. (2010) 'Neuroimaging of psychopathy and antisocial behavior: a targeted review', *Current Psychiatry Reports*, 12: 76–82.

Caspi, A. and Moffitt, T.E. (2006) 'Gene–environment interactions in psychiatry: joining forces with neuroscience', *Nature Review: Neuroscience*, 7: 583–590.

DeLisi, M. (2011) 'Self-control theory: the *Tyrannosaurus rex* of criminology is poised to devour criminal justice', *Journal of Criminal Justice*, 39: 103–105.

DeLisi, M., Umphress, Z.R., and Vaughn, M.G. (2009) 'The criminology of the amygdala', *Criminal Justice and Behavior*, 36: 1231–1242.

Frick, P.J. and White, S.F. (2008) 'Research review: the importance of callous-unemotional traits for developmental models of aggressive and antisocial behavior', *Journal of Child Psychology and Psychiatry*, 49: 359–375.

Gesch, C.B., Hammond, S.M., Hampson, S.E., Eves, A., and Crowder, M.J. (2002) 'Influence of supplementary vitamins, minerals and essential fatty acids on the antisocial behaviour of young adult prisoners: randomised, placebo-controlled trial', *British Journal of Psychiatry*, 181: 22–28.

Gottfredson, M.R. and Hirschi, T. (1990) *A General Theory of Crime*, Stanford, CA: Stanford University Press.

Harris, M. (1979) *Cultural Materialism: The Struggle for a Science of Culture*, New York: Random House.

Harris, M. (1994) 'Cultural materialism is alive and well and won't go away until something better comes along', *Assessing Cultural Anthropology*, 12: 62–75.

Hiatt, K.D. and Dishion, T.J. (2008) 'Antisocial personality development', in T.P. Beauchaine and S. Hinshaw (eds), *Child and Adolescent Psychopathology*, Hoboken, NJ: John Wiley and Sons, pp. 370–404.

Kruesi, M.J. (2007) 'Psychopharmacology of violence', in D.J. Flannery, A.T. Vazsonyi, and I.D. Waldman (eds), *The Cambridge Handbook of Violent Behavior and Aggression*, New York: Cambridge University Press, pp. 618–635.

Lahey, B.B. (2008) 'Oppositional defiant disorder, conduct disorder and juvenile delinquency', in T.P. Beauchaine and S. Hinshaw (eds), *Child and Adolescent Psychopathology*, Hoboken, NJ: John Wiley and Sons, pp. 335–369.

Laub, J.H. (2006) 'Edwin H. Sutherland and the Michael Adler report: searching for the soul of criminology seventy years later', *Criminology*, 44: 235–258.

Leshner, A.I. (1998) 'Addiction is a brain disease, and it matters', *Science*, 278: 45–47.

Litschge, C.M., Vaughn, M.G., and McCrea, C. (2010) 'The empirical status of treatments for children and youth with conduct problems: an overview of meta-analytic studies', *Research on Social Work Practice*, 20: 21–35.

MacLean, P.D. (1990) *The Triune Brain in Evolution: Role in Paleocerebral Functions*, New York: Plenum Press.

Massey, D.S. (2002) 'A brief history of human society: the origin and role of emotion in social life', *American Sociological Review*, 67: 1–29.

Olds, D., Henderson, C.R., Cole, R., Eckenrode, J., Kitzman, H., Luckey, D., Pettit, L., Sidora, K., Morris, P., and Powers, J. (1998) 'Long-term effects of nurse home visitation on children's criminal and antisocial behavior: 15-year follow-up of a randomized controlled trial', *Journal of the American Medical Association*, 280: 1238–1244.

Rocque, M., Welsh, B.C., and Raine, A. (2012) 'Biosocial criminology and modern crime prevention', *Journal of Criminal Justice*, 40: 306–312.

Roper v. Simmons, 543 U.S. 551 (2005).

Sanderson, S. (1990) *Social Evolutionism: A Critical History*, Oxford: Basil Blackwell.

Vaughn, M.G., Howard, M.O., and DeLisi, M. (2008) 'Psychopathic personality traits and delinquent careers: an empirical examination', *International Journal of Law and Psychiatry*, 31: 407–416.

Vaughn, M.G., DeLisi, M., and Matto, H. (2013) *Human Behavior: A Cell to Society Approach*, Hoboken, NJ: John Wiley and Sons.

Viding, E., Blair, R.J.R., Moffitt, T.E., and Plomin, R. (2005) 'Evidence for substantial genetic risk for psychopathy in 7-year-olds', *Journal of Child Psychology and Psychiatry*, 46: 1–6.

de Waal, F.B.M. (1995) 'Bonobo sex and society: the behavior of a close relative challenges assumptions about male supremacy in human evolution', *Scientific American*, 272: 82–88.

Index

Page numbers in **bold** refer to figures, page numbers in *italic* refer to tables.